S0-BAL-171

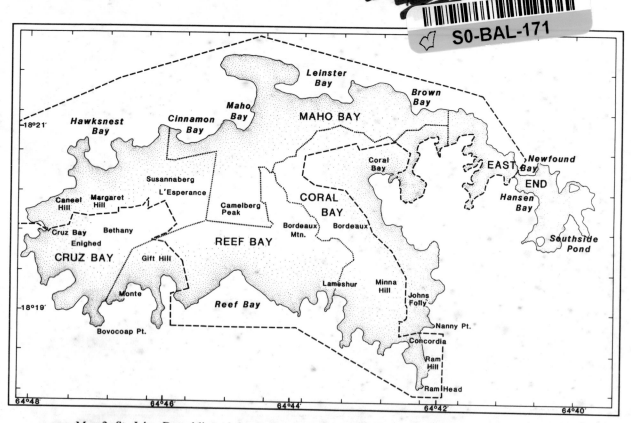

Map 2. St. John. Dotted lines define limits of quarters; dashed line defines National Park boundary.

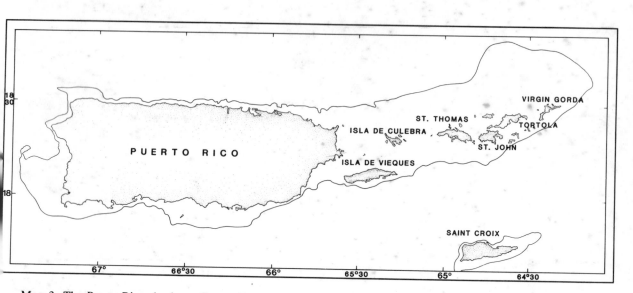

Map 3. The Puerto Rican bank, or Greater Puerto Rico. Contour line at the 100-fathom depth (within a few hundred feet of this line is the depth of 67–100 m).

Flora of St. John

Memoirs of The New York Botanical Garden

ADVISORY BOARD

PATRICIA K. HOLMGREN, *Director of the Herbarium*
The New York Botanical Garden

JAMES L. LUTEYN, *Senior Curator*
The New York Botanical Garden

SCOTT A. MORI, *Senior Curator*
The New York Botanical Garden

EDITORIAL BOARD

WILLIAM R. BUCK, *Editor*
The New York Botanical Garden
Bronx, New York 10458–5126

WM. WAYT THOMAS, *Associate Editor*
The New York Botanical Garden
Bronx, New York 10458–5126

THOMAS F. DANIEL (1991–1996)
Department of Botany
California Academy of Sciences
Golden Gate Park
San Francisco, California 94118

The MEMOIRS are published at irregular intervals
in volumes of various sizes and are designed to include
results of original botanical research by members of
The New York Botanical Garden's staff,
or by botanists who have collaborated in one or more of
The New York Botanical Garden's research programs.
Ordinarily only manuscripts of 100 or more typewritten pages
will be considered for publication.

Manuscripts should be submitted to the Editor.
For further information regarding editorial policy and
instructions for the preparation of manuscripts, contact the Editor.

Orders for published and forthcoming issues and volumes should be placed with:

Scientific Publications Department
THE NEW YORK BOTANICAL GARDEN
Bronx, New York 10458–5126 U.S.A.
(718) 817-8721
FAX (718) 817-8842
e-mail scipubs@nybg.org

Flora of St. John
U.S. Virgin Islands

Pedro Acevedo-Rodríguez
and Collaborators

Illustrated by Bobbi Angell

Pedro Acevedo-Rodríguez
Smithsonian Institution
Department of Botany, MRC-166
P.O. Box 37012
Washington, DC 20013-7012
acevedop@si.edu

17 December 1996
Memoirs of The New York Botanical Garden
Volume 78

THE NEW YORK BOTANICAL GARDEN
Bronx, New York 10458-5126, U.S.A.

Copyright © 1996, The New York Botanical Garden

Second printing 2005

Illustrations by Bobbi Angell: Copyright © 1996, Bobbi Angell
℅ The New York Botanical Garden

Published by
The New York Botanical Garden Press
Bronx, New York 10458

Manufacturing by The Maple-Vail Book Manufacturing Group

International Standard Serial Number 0071-5794

∞

The paper used in this publication meets the requirements of the
American National Standard for Information Sciences—Permanence
of Paper for Publications and Documents in Libraries
and Archives, ANSI/NISO (Z39.48—1992).

Printed in the United States of America on recycled paper, using soy-based ink

MetLife Foundation is a Leadership Funder of The New York Botanical Garden Press.

Library of Congress Cataloging-in-Publication Data

Acevedo-Rodríguez, Pedro.
    Flora of St. John, U.S. Virgin Islands / Pedro Acevedo-Rodríguez
and collaborators ; illustrated by Bobbi Angell.
       p. cm. — (Memoirs of The New York Botanical Garden, ISSN
0071-5794 ; v. 78)
    Includes bibliographical references and index.
    ISBN 0-89327-402-X
    1. Botany—Virgin Islands of the United States—Saint John—
Classification.   2. Botany—Virgin Islands of the United States—
Saint John—Identification. I. Title. II. Series.
QK1.N525   vol. 75
[QK231.V5]
581 s—dc20
[581.97297′22]   96-34052
                                               CIP

Flora of St. John

Contents

Introduction

The island of St. John, formerly known as St. Jan (Map 1), belongs to the Virgin Islands group, a natural appendage of the Puerto Rican bank. The islands making up the Virgin Islands group (St. Thomas, St. John, Tortola, Virgin Gorda, Anegada, and St. Croix) and Puerto Rico (Vieques, Culebra, and Puerto Rico) form a geographical, geological, and biological province with many shared natural features.

St. John, like most of the other Virgin Islands has a mountainous topography with very small intermountain valleys and coastal plains. The island has an approximate area of 31 km^2, with an east-west axis of 11 km, and a maximum north-south axis of approximately 5 km. The highest point on the island is Bordeaux Mountain, which reaches 387 m in elevation. The island has no permanent rivers and possesses only a few intermittent streams, which flow toward either the north or the south coast. For the most part, the soil is volcanic in origin and well drained, with depths to bedrock ranging from 25 to 50 cm.

Columbus discovered and named the Virgin Islands on his second trip to the New World in 1493. On November 14, Columbus and his crew anchored on an island that he named Santa Cruz (St. Croix later by the French). They found the island to be inhabited by Carib Indians, but they did not record how densely populated the island was. They left St. Croix for a group of islands already within view to the north. Columbus named the largest of these islands Santa Ursula (probably the island known today as Virgin Gorda) and the remaining ones the Eleven Thousand Virgins, which later became the Virgin Islands (Knox, 1852). Columbus reported that they were uninhabited, but they may have been visited frequently by the Carib Indians, who traveled widely within the Caribbean. Columbus never set foot on any of these small islands but nevertheless claimed them for the Spanish Empire. The Spaniards neglected the Virgin Islands and the Lesser Antilles and never set permanent colonies on any of them, possibly because of their small size or because of their lack of adequate water supply and of precious metals.

After it was discovered, St. John was occupied for short periods by small bands of Indians who engaged in warfare with the Spanish and by other Europeans engaged in modest agriculture or lumbering. During this period of low-level occupation, the island's natural resources were not severely affected.

When the Danes established the first European colony on St. John in 1718, the island had a dense forest cover (Tyson, 1984). The colonization caused changes in the physiognomy of the vegetation and in the utilization of natural resources on St. John. Relatively large populations of European planters and African slaves were involved in the production of agricultural staples for export. A large percentage of the land was converted into lucrative plantations, especially for sugar cane and cotton or later into range lands. During the first 10 years following the establishment of the colony, every large tree was harvested for building materials, and only small trees remained on the island (Tyson, 1984). By 1760, all arable land was under cultivation, and the settlements, woodcutters, and grazing livestock were encroaching upon the remaining natural vegetation. After 1765, sugar cane became the most important crop, and by the end of the century, 60% of the entire island was under sugar cultivation, with most of the remaining percentage devoted to secondary crops (cotton, tobacco, coffee). Such cultivation lasted until the 1830s, when the regional sugar economy collapsed because of rising production costs and falling market prices.

After 1848, the downfall of the plantation system continued with the abolition of slavery in the Danish Virgin Islands. Most of the land devoted to sugar cane plantations was abandoned, and agriculture shifted to secondary crops and livestock production. New small-scale forest industries were established, based on bay leaf harvesting and wood cutting for the production of charcoal. By 1879, St. John had regenerated an evident forest cover and was one of the most forested islands of the Virgin Islands group (Eggers, 1879). By the turn of the century, agriculture diminished further, allowing a sizable section of the landscape to revert to scrubland and secondary forests.

In 1917, the Danish Virgin Islands (St. John, St. Thomas, and St. Croix) were purchased by the government of the United States. The territory was acquired to consolidate U.S. Navy control of the Caribbean, but it soon became important for tourism. Much land was purchased by private parties from the U.S. mainland as real estate or for vacation homes. In 1956, the Virgin Islands National Park was created from land donated by Laurence Rockefeller. Today, with subsequent purchases, the park occupies about 65% of St. John's territory (Map 2), and its vegetation shows differing degrees of recovery. The parkland is protected from the most immediate impacts of development, but it is still vulnerable to some of the direct and indirect changes brought about by tourism and by grazing of feral animals. It is especially vulnerable along the park boundaries and in the privately owned inholdings within the limits of the park. The remaining 35% of the island is privately owned and has undergone a more marked deterioration of once-recuperating natural resources owing to an influx of tourists and year-round residents during the last several decades.

From the beginning of colonization, St. John was divided into quarters and these into numerous plantations. When agriculture reached its peak, the island was divided into eight quarters. These were later reduced to only five as agriculture deteriorated. Today, the quarter system is still in use, with Cruz Bay, Maho Bay, East End, Coral Bay, and Reef Bay quarters representing the official political divisions on St. John (Map 2). Cruz Bay Quarter is most developed; it has the only town, port, and hotels on the island, as well as numerous shops. The other quarters are relatively undeveloped, but new housing is flourishing at Fish Bay (Reef Bay Quarter), Coral Bay (Coral Bay Quarter), and East End (East End Quarter), the last being the most susceptible to disturbance because of the erosion associated with the steep hills. Most of the park land is contained within Maho Bay, Reef Bay, and Coral Bay quarters.

Geology

St. John belongs to the Puerto Rican bank, which was created by volcanism during the Cretaceous and became emergent largely through orogenic movements in the lower Eocene (35–40 Ma) (Meyerhoff, 1933). Subsequent sea flooding occurred in the Oligocene followed by uplift during a middle Miocene orogenesis (Meyerhoff, 1933). The continuity of the original mountain axis of the Puerto Rican bank is interrupted on the eastern end of Puerto Rico and reaches lower elevations through the islands of Culebra, Vieques, St. Thomas, St. John, and Tortola to Virgin Gorda. According to Mitchell (1954), during the Pliocene (2–3 Ma), this Cretaceous mountain axis underwent tilting to the northeast, allowing the Atlantic and Caribbean to flood part of the platform, thereby isolating the Virgin Islands from Puerto Rico. This notion has been challenged, however, by Weaver (1961), who maintains that the Caribbean area has been rather stable geologically since the Miocene and that the emergence of Puerto Rico and further isolation of the Virgin Islands resulted from eustatic sea-level changes.

Studies of changes in the Tertiary have shown sea level to be variable, with fluctuations above and below present day sea level (Vail & Hardenbol, 1979). During the late Oligocene (29 Ma), sea level was 150 m lower than it is at the present. More recently, during the many glaciations of the Pleistocene (the most recent one being 14,000 years ago), sea level dropped as much as 100 m below present-day sea level (Vail & Hardenbol, 1979). Such a lowering would have been sufficient to enable Puerto Rico and the Virgin Islands (except for St. Croix, which is separated by a deep sea trench) to exist as a continuous landmass, with approximately twice the area of present-day Puerto Rico (Map 3). After the last glaciation, the sea level rose, submerging once again a large portion of the Puerto Rican bank. As a result, many offshore islands disappeared, and this once-continuous landmass became dissected into smaller islands. In the last 6000 years, sea level fluctuations have changed the size of each island and the distances between the islands. Fragmentation of the ancestral biota may have fostered important changes in their populations by promoting local extinctions or speciation. But, because the isolation of the Puerto Rican bank fragments is relatively recent, very few new species are expected to have evolved from this event.

Climate

The climate of St. John is typical of many windward Caribbean islands, where most precipitation is due to convection caused by physical obstruction of mountains to trade winds. In addition, there is a considerable amount of rainfall that results from infrequent cold fronts and hurricanes. The precipitation in St. John ranges from 890 to 1400 mm per year, depending on the aspect and elevation of the site (Woodbury & Weaver, 1987). The eastern extreme of the island is the driest area, with an annual average rainfall of only 89–100 cm. This area is continuously swept by trade winds, which at low elevations have a drying effect. On the other hand, higher elevations are moister since most humidity is intercepted by the cooler mountains.

Mean annual rainfall for the weather station at Cruz Bay is 1130 mm, and the mean annual temperature is 26.9°C. These numbers are very similar for the station at Lameshur, with a mean annual precipitation of 1190 mm and a mean annual temperature of 26.3°C (Woodbury & Weaver, 1987). Precipitation is most predominant from May through November. The driest months are February and March, and the wettest, September.

Vegetation

The destruction of the natural vegetation on St. John has been extensive, spreading over nearly 90% of the island. Evidence of former cultivation is found in the half-buried stone terracing in much of the present-day mountain forest. The first 130 years of colonization were particularly harsh on St. John's natural resources. As a result, some of the native and endemic plant species have become extinct, or nearly extinct, with their populations reduced to a few individuals. Examples of these are *Solanum conocarpum*, *S. mucronatum*, *Malpighia infestissima*, *M. woodburyana*, and *Mammillaria nivosa*. In addition, the invasion by aggressive exotic plants may also have contributed to the demise of some of St. John's native plants. Today the most immediate threats to the regeneration of natural vegetation are development and the growing population of feral pigs, goats, and donkeys. Goats and donkeys are imposing selective changes on regeneration by grazing on palatable species. Pigs, on the other hand, are responsible for destroying shrubs and trees through rooting.

The present vegetation of St. John shows differing degrees of regeneration, ranging from recently disturbed to late-secondary successional forests, which may be as old as 100 years (Reilly, 1992). The new vegetation cover contains numerous introduced plants that have become established in dense stands or, more commonly, are intermixed with native species. Many of the weedy introduced species are particularly common in recently disturbed, open areas such as roadsides or waste grounds. There also are a few woody, adventive plants that are locally common, mostly because they have either a high endurance of harsh environments or a high degree of dispersibility. Examples of these are *Leucaena leucocephala*, *Melicoccus bijugatus*, *Calotropis procera*, and *Cryptostegia grandiflora*.

Eggers was probably the first to study the vegetation of the Virgin Islands. In 1876 he described the plant communities of St. Croix, and in 1879 he expanded his work to include St. Thomas, St. John, Tortola, Vieques, and Culebra. In the latter work, he considered all the Virgin Islands to have the same types of vegetation formations—i.e., littoral, scrub-land, forest, and cultivated areas—that he observed on St. Croix. For each formation, he listed the predominant species, and in some cases he named the formation after the most common species [e.g., *Croton* vegetation and *Eriodendron* (= *Ceiba*) vegetation]. His work of 1879 provides important information on land-use history as well.

In 1923, Børgesen refined Eggers's vegetation descriptions and recognized six sections (formations). His classification is similar to that of Eggers but elevates some of Eggers's hierarchical zones within the formations into sections. In summary, he recognized the following nine vegetation types: marine algal, sea-grass (in sea), coastal, sandy shore, rocky coast, thicket, forest, fresh water (in temporary pools), and cultivated land.

Woodbury and Weaver's (1987) study characterizes the vegetation of St. John under various formations that are highly influenced by humidity. These formations range from xerophytic cactus scrub to moist forests in the protected valleys and mountains. Woodbury and Weaver's study gives qualitative data with relative abundance of typical species, but it does not present data regarding density or the importance of species. They recognized 10 formations within two life zones (moist and dry) that are strongly characterized by the physiognomy of the vegetation but lack a distinctive floristic identity. Their formations seem artificial and at most, transitory. Recognition of two major vegetation types (one in the moist zone and one in the dry zone) seems preferable, because they show some degree of floristic distinctiveness. The moist forest is characterized by *Andira inermis*, *Buchenavia tetraphylla*, *Byrsonima spicata*, *Ceiba pentandra*, *Cestrum laurifolium*, *Cordia laevigata*, *Eugenia confusa*, *E. pseudopsidium*, *Gonzalagunia hirsuta*, *Hymenaea courbaril*, *Ilex urbaniana*, *Inga laurina*, *Manilkara bidentata*, *Miconia laevigata*, *Myrcia citrifolia*, *Myrciaria floribunda*, *Palicourea croceoides*, *Psychotria* spp., and *Ternstroemia peduncularis*. The dry vegetation is characterized by *Amyris elemifera*, *Cassine xylocarpa*, *Coccoloba uvifera*, *Crossopetalum rhacoma*, *Erithalis fruticosa*, *Jacquinia arborea*, *J. berterii*, *Plumeria alba*, *Sideroxylon obovatum*, and *S. salicifolium*.

The following classification has been modified from Woodbury and Weaver (1987) to reflect more natural vegetation types.

1. Dry formations. These were defined by Woodbury and Weaver as dry evergreen and were characterized under four types. They occupy 63% of the island territory and are combined here into three types.

A. *Dry evergreen woodland.* This is a widespread formation that occupies 33% of the island. It is characterized by dense growth, with a layer of trees that reach to 10 m tall and bear thick, sclerophyllous, small leaves. A layer of shrubs and herbs is also present, with a few species of vines. This formation contains numerous coastal species that occur along the lower portion of the hillsides. When this formation reaches the upper portions of the mountains, it grades into a forest with two strata that may contain additional species more characteristic of the uplands.

The only quantitative data for this type of formation are provided by the studies of Reilly et al. (1990) and Dallmeier et al. (1993) at two different sites. Differences in floristic composition and relative density between the two sites is probably due to their individual land-use histories. The site studied by Reilly, Earhart, and Prance has a total basal area of 13.26 m²/ha and was dominated by the introduced *Melicoccus bijugatus,* followed by (in order of importance) *Guapira fragrans, Ocotea coriacea, Bursera simaruba,* and *Eugenia monticola.* This forested area was described as having two strata, one that is 5–10 m in height and another that is 10–12 m in height, with occasional emergents from 15 to 20 m tall. The site studies of Dallmeier, Comiskey, and Ray show a higher total basal area of 25.15 m²/ha, with dominance shared by *Coccoloba microstachya, Maytenus laevigata, Guapira fragrans, Bourreria succulenta,* and *Tabebuia heterophylla.* Their data show trees that vary from 4 to 11 m without a clear distinction of strata.

B. *Dry evergreen thicket or scrub.* This is the second largest formation on St. John, occupying about 24% of the total land surface. This formation occupies similar, but moister, habitats and contains similar species to the previous formation, but the trees do not reach more than 5 m in height. In extremely severe environments subjected to the continuous action of winds, a stunted vegetation may develop. This usually has thicker or succulent leaves and may contain a few species of cacti (this area was defined by Woodbury and Weaver as rock pavement and coastal hedge). This formation has not been quantified yet, and such studies are needed.

C. *Thorn and cactus scrub.* This is a formation occupying about 6% of St. John's land area. It contains a few thorny, woody species and cacti with a maximum height of 5 m. This formation occurs in the driest areas of the island, which have poor, shallow soils. In many areas, this formation has an open aspect. There are no quantitative studies describing this type of formation.

2. Moist forest. The moist forest formations occupy only 16.5% of St. John and were classified by Woodbury and Weaver (1987) as upland, gallery, and basin. These divisions seem unnecessary because the formations so defined share a great number of species. The area characterized by Woodbury and Weaver as upland moist forest contains many more species than either the gallery or basin forest. This difference may be explained by the presence of more microhabitats and a larger elevational range. Most of the species found in gallery and basin forests can also be found in upland moist forests. This type of forest with local variations occurs throughout moist areas, in protected uplands, drainage areas (locally known as guts), and coastal protected valleys.

There are two quantitative studies of this formation: one by Reilly et al. (1990) and another by Forman and Hahn (1980). Reilly et al. studied two sites. The first was characterized as upland moist forest, and the second as gallery moist forest. The first site was located on the northwestern side of the Bordeaux Mountains and had a total basal area of 30.90 m²/ha. The 10 most important species (in order of importance) were *Guapira fragrans, Pimenta racemosa, Inga laurina, Byrsonima spicata, Acacia muricata, Ocotea coriacea, Tabebuia heterophylla, Faramea occidentalis, Chionanthus compacta,* and *Guazuma ulmifolia.* This forested area was described as having four strata: the lowest one less than 10 m in height; a midheight layer 10–15 m tall; a continuous canopy from about 15 to 20 m tall; and a top layer of a few emergent trees, reaching about 28 m in height. The second site, at L'Esperance, had a total basal area of 15.6 m²/ha. The 10 most important species (in order of importance) were *Ardisia obovata, Guapira fragrans, Andira inermis, Inga laurina, Ocotea coriacea, Chrysophyllum pauciflorum, Guettarda scabra, G. odorata, Tabebuia heterophylla,* and *Hymenaea courbaril.* According to Reilly et al. (1990), there were three distinct strata in this site; the lower stratum was less than 10 m tall, the middle one was made of a continuous canopy, and the highest consisted of emergent trees. They gave no measurements for the two latter strata, but they stated that trees were shorter than those of the first site.

The study by Forman and Hahn (1980) described a site located in the upper part of the Reef Bay valley. The forest was described as having a canopy from 25 to 30 m in height and a continuous understory layer ranging from 2 to 25 m tall. The most important species (in order of importance) were *Andira inermis, Amyris elemifera, Swietenia mahagoni, Melicoccus bijugatus, Casearia guianensis, Eugenia monticola, E. rhombea, Zanthoxylum monophyllum, Adenanthera pavonina,* and *Acacia muricata.*

The floristic composition and relative density in the three sites show differences that are probably the result of the particular land-use history of the site and do not necessarily indicate two different types of formations. For instance, the abundance of the introduced bay rum trees (*Pimenta racemosa*) in the Bordeaux study area may represent the remains of previous plantations. Selective logging for charcoal production and lumbering undoubtedly has played an important role in the composition of today's forests. In addition, these forests are too young to be considered a climax community.

3. Early successional vegetation. This successional stage was classified by Woodbury and Weaver (1987) as secondary vegetation. This term, however, is inappropriate since all vegetation formations on St. John are secondary in nature. Early successional vegetation covers about 15.4% of St. John and is found in areas subject to recent disturbance by humans. These include areas recently used for agriculture, pastures, areas along roadsides, trails, and dump sites.

4. Coastal wetlands. This zone occupies approximately 2.3% of St. John and was classified by Woodbury and Weaver (1987) into mangroves, salt flats, and lagoons. These formations are subject to prolonged, seasonal or tidally flooding by saline waters or by fresh waters after prolonged rains. Mangrove swamps on St. John, although containing the four typical Caribbean species of mangroves (*Rhizophora mangle, Avicennia germinans, Conocarpus erectus,* and *Laguncularia racemosa*) and associated halophytes, are poorly represented by a narrow strip of vegetation occurring in protected, shallow waters. Salt flats (seasonally or tidally flooded areas in the surroundings of the mangroves) and sandy coastal areas contain some of annual salt-loving species and a few other succulent annuals. The lagoon areas are seasonally flooded by heavy downpours during the wet months. These lagoons, usually lasting for a few months before drying up, commonly contain the aquatic species *Lemna aequinoctialis* and *Ruppia maritima*.

Flora and Floristic Affinities

The flora of St. John consists of 747 species of vascular plants (native and naturalized), of which 642 (86% of the total flora) are native to the island. There are 117 families of vascular plants represented on St. John (12 of which are introduced), with a total of 469 genera (55 of which are introduced). The pteridophytes (ferns and fern allies) are represented by 5 families and 16 genera, the dicots by 93 families and 372 genera, and the monocots by 19 families and 81 genera.

About 50% of the families are represented by 2–13 species; 39% of the families (46 families) are represented by a single species; and the remaining families (about 11%) have more than 15 species. The largest families are as follows:

Fabaceae	74 species
Poaceae	52 species
Euphorbiaceae	40 species
Asteraceae	37 species
Convolvulaceae	27 species
Rubiaceae	26 species
Polypodiaceae	25 species
Malvaceae	24 species
Cyperaceae	20 species
Boraginaceae	19 species
Myrtaceae	19 species
Solanaceae	16 species
Acanthaceae	16 species

Practically every species (99.7%) on St. John is also found on other islands of the Virgin Islands, with the exception of two endemic flowering plants (see Endemism, below). St. John's flora may be viewed as a subset of the Greater Puerto Rican flora, since 97% of its species also occur on Puerto Rico.

In spite of St. John's (and the Virgin Islands) being a natural appendage of the Greater Antilles, it shares roughly the same percentage of its flora with the Lesser Antilles (86.7%) as it does with the Greater Antilles (87.3%). These numbers are not totally surprising since St. John and the Virgin Islands are located at the boundary between the Greater and the Lesser Antilles, two groups of islands often considered to have different floristic affinities. A slightly smaller percentage (80%) of St. John's flora is found throughout the Greater Antilles and the Lesser Antilles. The percentage is even smaller when St. John's flora is compared with the pan-Caribbean region, with only 70% of its species found throughout the region. The percentage of St. John's flora shared with other areas beyond the Caribbean region drops sharply, with 29.5% found throughout the neotropics and 15.6% with pantropical distribution.

ENDEMISM. St. John has two endemic species, representing only 0.3% of its flora. They are *Eugenia earhartii* in the myrtle family and *Machaonia woodburyana* in the coffee family. In addition, St. John contains six species that are endemic to it and the other Virgin Islands, accounting for 0.9% of its native flora. These are *Anthurium* × *selloum, Croton fishlockii, Galactia eggersii, Malpighia infestissima, Solanum conocarpum,* and *Tillandsia lineatispica*.

The percentage of endemism increases when the whole phytogeographical province of Greater Puerto Rico (i.e., Puerto Rico and the Virgin Islands) is considered. There are 25 species of flowering plants that are present only on St. John, the Virgin Islands, and Puerto Rico, accounting for 3.8% of St. John's flora. These species are *Agave missionum, Argythamnia stahlii, Calyptranthes thomasiana, Chrysophyllum bicolor, C. pauciflorum, Coccothrinax alta, Crescentia linearifolia, Erythrina eggersii, Eugenia sessiliflora, E. xerophytica, Forestiera eggersiana, Malpighia woodburyana, Neea buxifolia, Opuntia repens, Ouratea litoralis, Pilea sanctae-crucis, Poitea florida, Psidium amplexicaule, Reynosia guama, Rondeletia pilosa, Roystonea borinquena, Scolosanthus versicolor, Solanum mucronatum, Stigmaphyllon flori-bundum,* and *Zanthoxylum thomasianum.*

It is not surprising for St. John and the Virgin Islands to have a very low percentage of endemism because they have been isolated from Puerto Rico for only short and intermittent periods followed by dry land connections. To start with, Puerto Rico has a very low percentage (12%) of endemic plants compared with other islands of the Greater Antilles (20–50%; fide Howard, 1973). It is not clear whether this low percentage of endemism is due to recent land-use history (the last 300–400 years of intensive deforestation) or to Puerto Rico's smaller size, fewer habitats, and greater distance from continental areas. If the latter factors are responsible for the less diverse nature of the Puerto Rican flora, they certainly should have an effect on speciation events in the Virgin Islands.

RARE AND ENDANGERED SPECIES.   The only species on St. John listed as endangered by the Federal Register of the U.S. Fish and Wildlife Service (August 1993) is *Zanthoxylum thomasianum.* The remaining known populations of this species on St. John have less than 20 individuals, all of which are mature. This species does not seem to propagate: no juveniles have been seen. In addition, a parasitic insect that deposits its eggs in the seeds of *Z. thomasianum* destroys most of its seed crop. The main culprit for the decline of this species may be habitat destruction. This has certainly reduced genetic variability, and with it, the ability of the species to cope with new challenges caused by the changing environment.

There are other species that, although not listed in the Federal Register, deserve attention and the prompt assessment of their populations. Some of the endemic species seem to be the most vulnerable because of their scarcity and restricted ranges. The destruction of their habitat would certainly mean their extinction, because they occur nowhere else. The Virgin Island endemics *Galactia eggersii, Tillandsia lineatispica,* and *Calyptranthes thomasiana* are examples. None of these species is common on St. John or elsewhere in the Virgin Islands. In addition, the Greater Puerto Rican endemics *Erythrina eggersii* and *Malpighia woodburyana* have the smallest populations, with very few individuals remaining on St. John.

The narrowly endemic *Machaonia woodburyana* and *Eugenia earhartii,* confined to St. John, could be characterized as threatened, especially the former, since its largest populations occur outside of park boundaries, where no protection exists. *Eugenia earhartii* is known from only two populations of very few individuals, which fortunately occur within the park limits.

The most endangered species on St. John seems to be *Solanum conocarpum,* which at present is known from only two individuals. This species was previously known only from the type specimen collected at Coral Bay in April of 1787 until recently when it was recollected. This species certainly needs to be studied and protected.

The flora of St. John has faced probable extinctions of some of its species. For example, *Solanum mucronatum* and *Malpighia infestissima* have not been located on St. John in recent times. Their populations outside St. John are rather small and restricted to some of the Virgin Islands and Puerto Rico.

Botanical History

PREVIOUS WORK.   Botanical exploration (Millspaugh, 1902; Urban, 1902) of St. John started later than that of many other West Indian islands. The oldest record is dated from 1767 by Gesch Oldendorp, who lived in the Danish West Indies until 1769. During this period, he studied different aspects of the colonies, including the natural history of these islands. His studies were published in 1777 in his book *History of the Mission of the Evangelical Brethren on the Caribbean Islands of St. Thomas, St. Croix, and St. John* (in German, translated into English in 1987). In his botanical section, Oldendorp mentioned numerous edible and useful plants present in the Danish West Indies. From that date to the beginning of the nineteenth century, St. John was visited by only a handful of plant collectors. One of the earliest collectors to visit the Danish West Indies was von Rohr, who lived on the islands in 1757–1791. However, it is not clear whether he collected on St. John. Hans West may have visited and collected plants on St. John before 1793, the year in which he published (in Danish) *An Enumeration of Plants from the Islands of St. Croix, St. Thomas, St. John, Tortola and Vieques.* This publication contains partial descriptions for some of the many new species described therein by him and Vahl.

During the nineteenth century, increasing numbers of botanists visited St. John and the Virgin Islands. The plants they collected were the basis for many new species described for the Caribbean region. Schlechtendal (1828–1831) published a series of papers on the flora of St. Thomas. In this work, he provided descriptions for 369 species of plants occurring on St. Thomas and the other Virgin Islands. In 1847, Krebs published a general description of the vegetation of St. Thomas. The most notable botanical work produced during this time was Eggers's (1879) *The Flora of St. Croix and the Virgin Islands.* Although it primarily concentrates on the then Danish Virgin Islands, it contains general information as well as descriptions of vegetation formations common to the islands. Eggers's flora contains brief diagnoses, phenological data, and common names for the vascular plants.

The twentieth century brought more interest in the floristics of the region. In 1902, Millspaugh published a new flora for St. Croix, including not only vascular plants but also marine algae. This work provided information on the abundance and distribution of species and contained new records of vascular plants based on more recent collections. In 1918, Britton published *The Flora of the American Virgin Islands* (previously the Danish West Indies, but the islands had been purchased by the United States in 1917). This work constituted a checklist for all vascular plants, cryptogams, and fungi for the newly acquired territory, with new records and descriptions of a few new species. A few years later, Britton and Wilson (1923–1926) published a complete flora for Puerto Rico and the Virgin Islands, under the title *Botany of Porto Rico and the Virgin Islands*. This publication, although outdated, is the most comprehensive floristic work published for the Greater Puerto Rican area, and it is a basic reference still widely used.

During the second half of this century, there has been a series of publications dealing with the Virgin Islands flora. These include works by Little and Wadsworth (1964) and Little et al. (1974) on the trees of Puerto Rico and the Virgin Islands. Furthermore, Liogier (1965) and Liogier and Martorell (1982) produced a series of publications updating the names of plants found in Puerto Rico and the Virgin Islands. Liogier also is engaged in the production of a new flora of Puerto Rico and adjacent islands, having completed three of the projected five volumes (Liogier 1985, 1988, & 1994). In 1985, Acevedo-Rodríguez and Woodbury published a field guide to the common vines of Puerto Rico that includes numerous species also occurring in the Virgin Islands. An enlarged and comprehensive revision of this work is under way and should be ready in a few years. The most recent publication dealing with the flora of the Virgin Islands is *Ferns of Puerto Rico and the Virgin Islands* (Proctor, 1989). Currently, Proctor is working on a revision of the monocots of Puerto Rico and the Virgin Islands, a work that should be ready within the next few years.

Numerous species have been described from specimens collected from the Virgin Islands over many decades. Although the scope of this book is St. John's flora, a listing of these names reveals the high level of scientific activity in the Virgin Islands during the last two centuries. Table 1 lists only species currently considered valid that are typified by Virgin Islands material. This listing represents only a fraction of the names typified by Virgin Islands material, since there are numerous names that have fallen under synonymy. Eighty-one species and one genus (*Fishlockia*) described from Virgin Islands material are currently accepted.

PLANT COLLECTORS. The following section lists the most prominent plant collectors, both past and present, active on St. John. This list is based on literature records and on searches of material in the U.S. National Herbarium and the herbarium of The New York Botanical Garden. Undoubtedly there are additional collectors who worked in St. John, but their specimens would be deposited in smaller herbaria.

John Ryan (?–1800). A physician and plantation owner on Monserrat during the second half of the eighteenth century. He was probably the first to collect and preserve plants from the island of St. John. In 1780, he made many collections on St. John; these are deposited at the Botanical Museum of Copenhagen (C).

Louis Claude M. Richard (1754–1821). French botanist, went to some of the French colonies in the New World to study useful plants with potential for introduction into the Old World. He made numerous collections in the New World (mostly Caribbean). During most of 1787, he visited St. John and some of the other Virgin Islands. His collections are deposited at the Museum of Natural History in Paris (P), with duplicates at C.

Paul E. Isert (1756–1789). Danish physician, collected plants on St. John, other Virgin Islands, and in the Lesser Antilles in 1787. His collections are deposited at C.

Peder Eggert Benzon (1788–1848). Danish pharmacist, lived on St. Croix from 1817 to 1848. He made numerous collections from the island of St. Croix but also collected on St. John and St. Thomas during his stay in that region. His collections are deposited at C, with duplicates at the Museum of Natural History at Stockholm (S) and at The New York Botanical Garden (NY). The duplicates at NY were examined by Britton during his preparatory studies for the *Flora of the American Virgin Islands*.

Peter Ravn (1783–1839). Norwegian surgeon, lived on St. Thomas from 1819 to 1839. During this time, he collected plants on St. John, St. Croix, St. Thomas, and Vieques. His collections are deposited at C, with duplicates at S, and at the herbarium of the Botanical Garden of Geneva (G). Some of his collections were studied by de Candolle and by Krebs.

Hans B. Hornbeck (1800–1870). Danish physician, lived on St. John from 1825 to 1844. During his stay on St. John, he collected numerous plants and other natural history specimens from St. John, St. Croix, St. Thomas, and Puerto Rico. His specimens are deposited at C, with duplicates at S and NY. The NY collections were studied by Britton for *The Flora of the American Virgin Islands*.

Johann C. Breutel (1788–1875). Born in Germany, became a steward of the Moravian church, later became a member of its board of directors. From December 1840 to July 1841 he visited the islands of St. John, St. Croix, St. Thomas, St. Kitts, and Antigua and made numerous collections (mostly cryptogams). His fern collections are deposited at the Leipzig herbarium (LZ) and the liverworts and mosses at Berlin (B, probably destroyed), with duplicates (mosses) at the British Museum (BM).

Henrik J. Krebs (1821–1907). Danish pharmacist, later became Swedish-Norwegian consul, and finally president of the Colonial Assembly in St. Thomas. He lived on St. Thomas from 1843 to 1870. From there he made numerous collecting trips to St. John, St. Croix, and throughout the Caribbean and northern South America. His collections are

Table I. Validly published species originally described from the Virgin Islands

Family	Scientific Name[a]	Year of publication	Island[b]
Acanthaceae	*Justicia microphylla* Lam.≡ **Oplonia microphylla** (Lam.) Stearn	1791	St. Croix
Agavaceae	**Agave eggersiana** Trel.	1913	St. Croix
	Agave missionum Trel.	1913	St.Thomas, St. John, Tortola
Amaranthaceae	**Amaranthus crassipes** Schltdl.	1831	St. Thomas
	Telanthera crucis Moq.≡ **Alternanthera crucis** (Moq.) Bold.	1849	St. Croix
Araceae	**Anthurium × selloum** K. Koch	1855	St. John
Asclepiadaceae	**Metastelma anegadensis** Britton	1925	Anegada
	Metastelma parviflorum R. Br. *ex* Schult.	1820	St. Croix, St. Kitts
Asteraceae	*Eupatorium triplinerve* Vahl≡ **Ayapana triplinervis** (Vahl) King & H. Rob.	1790	St. Croix
	Pectis humifusa Sw.	1788	St. Croix, Guadalupe
	Wedelia cruciana Rich.[1]	1807	St. Croix
Bignoniaceae	*Bignonia lactiflora* Vahl ≡ **Distictis lactiflora** (Vahl) DC.	1794	St. Croix
	Crescentia linearifolia Miers	1868	St. Croix
Boraginaceae	**Cordia ricksekeri** Millsp.	1902	St. Croix
Bromeliaceae	**Tillandsia lineatispica** Mez	1896	St. John
Buxaceae	*Crantzia laevigata* Vahl ≡ **Buxus vahlii** Baill.	1791	St. Croix?
Cactaceae	*Cereus trigonus* Haw. ≡ **Hylocereus trigonus** (Haw.) Saff.	1812	St. Croix
	Mammillaria nivosa Link *ex* Pfeiff.	1837	Tortola
	Opuntia antillana Britton & Rose	1918	St.Thomas, St.Croix, Tortola, Puerto Rico, Hispaniola, St. Kitts
Celastraceae	**Cassine xylocarpa** Vent.	1803	St. Thomas
	Rhamnus laevigatus Vahl ≡ **Maytenus laevigata** (Vahl) Griseb. *ex* Eggers	1794	St. Croix
Convolvulaceae	*Convolvulus acuminatus* Vahl ≡ **Ipomoea indica** var. **acuminata** (Vahl) Fosberg	1795	St. Croix
	Convolvulus quinquepartitus Vahl ≡ **Ipomoea quinquepartita** (Vahl) Roem. & Schult.	1798	St. Croix
	Exogonium eggersii House ≡ **Ipomoea eggersii** (House) D.F. Austin	1908	St. Thomas
Euphorbiaceae	**Croton betulinus** Vahl	1791	St. Thomas
	Croton fishlockii Britton	1920	Virgin Gorda
	Croton ovalifolius Vahl	1793	St. Thomas
	Ditaxis fasciculata Vahl *ex* A. Juss. ≡ **Argythamnia fasciculata** (Vahl *ex* A. Juss.) Müll. Arg.	1824	St. Thomas
	Euphorbia petiolaris Sims	1805	St. Thomas
	Poinsettia oerstediana Klotzsch & Garcke ≡ **Euphorbia oerstediana** (Klotzsch & Garcke) Boiss.	1860	St. Thomas
Fabaceae	*Acacia anegadensis* Britton ≡ **Fishlockia anegadensis** (Britton) Britton & Rose	1916	Anegada
	Erythrina eggersii Krukoff & Moldenke	1938	St. Croix
	Galactia eggersii Urb.	1900	St. John, St. Thomas
	Robinia aculeata Vahl ≡ **Pictetia aculeata** (Vahl) Urb.	1793	St. Croix
	Senna polyphylla var. **neglecta** H.S. Irwin & Barneby	1982	Anegada
Lamiaceae	**Salvia thomasiana** Urb.	1912	St. Thomas
Loranthaceae	**Dendropemon caribaeus** Krug & Urb.	1897	St. Thomas, St. John, St. Croix, Puerto Rico
Malpighiaceae	**Malpighia infestissima** Rich. *ex* Nied.	1899	St. Croix
	Malpighia woodburyana Vivaldi	1993	St. Thomas
Malvaceae	**Bastardia viscosa** (L.) Kunth var. **sanctae-crucis** R.E. Fries	1947	St. Croix
	Sida eggersii E. G. Baker ≡ **Bastardiopsis eggersii** (E. G. Baker) Fuertes & Fryxell	1892	Tortola
Melastomataceae	**Miconia thomasiana** O. Berg	1825	St. Thomas
Menispermaceae	*Cissampelos laurifolia* Poir. ≡ **Hyperbaena laurifolia** (Poir.) Urb.	1804	St. Thomas
Myrtaceae	**Calyptranthes thomasiana** O. Berg	1855	St. Thomas
	Eugenia earhartii Acev.-Rodr.	1993	St. John
	Eugenia floribunda H. West *ex* Willd. ≡ **Myrciaria floribunda** (H. West *ex* Willd.) O. Berg	1799	St. Croix
	Eugenia foetida Pers.	1806	St. Croix
	Eugenia sessiliflora Vahl	1794	St. Croix
	Psidium amplexicaule Pers.	1806	St. John
Nyctaginaceae	*Eggersia buxifolia* Hook.f. ≡ **Neea buxifolia** (Hook.f.) Heimerl	1883	St. Thomas
Ochnaceae	**Ouratea litoralis** Urb.	1899	St. Thomas, Puerto Rico

Table I. Validly published species originally described from the Virgin Islands *(continued)*

Family	Scientific Name[a]	Year of publication	Island[b]
Oleaceae	**Chionanthus compacta** Sw.	1788	St. Croix, Nevis
	Forestiera eggersiana Krug & Urb.	1893	St. Thomas
Piperaceae	**Peperomia humilis** A. Dietr.	1831	St. Croix
	Piper dilatatum Rich.	1792	St. Croix?
	Piper myrtifolia Vahl ≡ **Peperomia myrtifolia** (Vahl) A. Dietr.	1804	St. Croix
Poaceae	**Aristida cognata** Trin. & Rupr.	1842	St. Thomas
	Chloris foliosa Willd. ≡ **Gymnopogon foliosus** (Willd.) Nees	1806	St. Thomas
	Eragrostis ciliaris var. **laxa** Kuntze	1891	St. Thomas
	Panicum eggersii Hack. ≡ **Trichachne eggersii** (Hack.) Henr.	1901	St. Thomas
	Paspalum laxum Lam.	1791	St. Croix[2]
	Paspalum molle Poir.	1804	St. Thomas
	Paspalum notatum Flüggé	1810	St. Thomas
Polypodiaceae	*Adiantum rigidulum* Mett. *ex* Kuhn ≡ **Adiantum fragile** Sw. var. **rigidulum** (Mett. *ex* Kuhn) Proctor	1869	St. Croix, St. Thomas
Rhamnaceae	*Rhamnus ferreus* Vahl ≡ **Krugiodendron ferreum** (Vahl) Urb.	1793	St. Croix
Rubiaceae	**Catesbaea melanocarpa** Urb.	1899	St. Croix
	Diodia verticillata Vahl	1791	St. Croix
	Ixora coccinea L. var. **intermedia** Fosberg & Sachet[3]	1989	St. Croix
	Machaonia woodburyana Acev.-Rodr.	1993	St. John
	Scolosanthus versicolor Vahl	1797	St. Croix
Rutaceae	*Fagara thomasiana* Krug & Urb. ≡ **Zanthoxylum thomasianum** Krug & Urb.) P. Wilson	1896	St.Thomas
	Zanthoxylum punctatum Vahl	1793	St. Croix
Sapindaceae	**Cupania triquetra** A. Rich.	1845	St. John
Scrophulariaceae	*Vandellia diffusa* L. ≡ **Lindernia diffusa** (L.) Wettst.	1767	St. Thomas
Solanaceae	**Solanum conocarpum** Dunal	1913	St. John
	Solanum mucronatum O.E. Schulz	1909	St. John, St. Thomas, Puerto Rico
	Solanum polygamum Vahl	1894	St. Croix
	Solanum polygamum Vahl var. **thomae** Kuntze	1891	St. Thomas
Urticaceae	**Pilea sanctae-crucis** Liebm.	1851	St. Croix
Verbenaceae	**Vitex divaricata** Sw.	1788	St. Croix, St. Kitts
Violaceae	*Viola linearifolia* Vahl ≡ **Hybanthus linearifolius** (Vahl) Urb.	1793	St. Croix
Vitaceae	**Cissus obovata** Vahl	1794	St. Croix

Note: 1 = possibly a good species; 2 = possibly St. Croix; 3 = although the variety is named from St. Croix material, it is not native to the region.

[a]Plant names in bold letters are the currently accepted combinations.

[b]The islands mentioned are the type locality. When two or more islands are cited, the material represents syntypes.

deposited at C, with duplicates at NY and the Field Museum of Natural History (F). He published some contributions (1847) toward a flora of St. Thomas.

Henrik F.A. Baron von Eggers (1844–1903). Danish soldier and botanist, lived in the Virgin Islands (St. Croix and St. Thomas) from 1869 to 1887, from where he collected plants extensively throughout the Caribbean. His collections from 1870 to 1874 are from the Virgin Islands; those dating from 1880 to 1899 are mostly from the Greater Antilles, Bahamas, Dominica, Grenada, Tobago, Trinidad but also include collections from the Virgin Islands. Eggers collected on St. John in 1873 and in 1887. His first set of specimens is deposited at C, with duplicates distributed to many herbaria. His collections, the basis for his flora of St. Croix and the Virgin Islands, have been studied by numerous West Indian botanists, including Urban and Britton.

Johannes E. B. Warming (1841–1924). Danish botanist and naturalist, lecturer at the Copenhagen University and later director of the Botanical Garden of Copenhagen. He visited the Caribbean from October 1891 until March 1892. He made a few collections on St. John, St. Croix, and St. Thomas. His specimens are deposited at C, with duplicates in numerous herbaria.

Frederick C. E. Børgesen (1866–1956). Danish botanist, collected algae specimens in the Virgin Islands in January to April 1892, from 1895 to 1896, and 1905 to 1906. He published an article (1923) about the vegetation of the Danish West Indies and an account of their marine algae. His collections are deposited at C, with duplicates at F and at Munich (M).

Ove W. Paulsen (1874–1947). Danish botanist, curator at the Botanical Museum in Copenhagen, later professor of botany at the Danish pharmaceutical college. He visited the Danish Virgin Islands from December 1895 to February

1896, where he collected numerous specimens. His collections are deposited at C, with duplicates in numerous herbaria. The duplicates at NY were studied by Britton for *The Flora of the American Virgin Islands*.

Nathaniel L. Britton (1859–1934). American botanist, founder and first director of The New York Botanical Garden. He made about 40 expeditions throughout the West Indies, collected in the Virgin Islands starting in 1900 in St. Croix, and later in 1913 on St. Thomas, St. John, Tortola, Anegada, and nearby islands. His collections, deposited at NY, with duplicates at the U.S. National Herbarium (US), form the basis for his *Flora of the American Virgin Islands* (1918), *Botany of Porto Rico and the Virgin Islands* (Britton & Wilson, 1923–1926), and *The Vegetation of Anegada* (1916).

Christen C. Raunkiær (1860–1938). Danish botanist, professor of botany and director of the Botanical Garden of Copenhagen. Plants he collected in the former Danish West Indies (St. Croix, St. Thomas, and St. John) during 1905 to 1906 are deposited at C, with duplicates at NY and US.

C. F. Morrow (fl. 1920–1921). American ?, director of public schools in St. Thomas and hobbyist plant collector, one of the few women to collect plants in the U.S. Virgin Islands (St. John and St. Thomas). She collected on St. John from December 1920 to August 1921. Her flowering plant collections are deposited at US, and the fern collections are at C.

Ismael Vélez (1908–1970). Puerto Rican botanist, professor of botany at the Inter American University of Puerto Rico in San Germán. He collected plants on St. John and most of the other Virgin Islands from July to September of 1949 for his vegetation studies of the Lesser Antilles and Puerto Rico (Vélez, 1950, 1957). His collections are deposited at Louisiana State University herbarium (LSU), with duplicates at GH, K, NY, and US. Duplicates of his collections at the Inter American University in San Germán, Puerto Rico, were mostly destroyed by pests; the surviving specimens were recently transferred to the San Juan campus of the Inter American University.

Elbert L. Little Jr. (1907–   ). American dendrologist, while employed by the U.S. Department of Agriculture, collected numerous woody specimens from the Virgin Islands and Puerto Rico for his preparatory studies for his monumental two volumes on trees of Puerto Rico and the Virgin Islands (Little & Wadsworth, 1964; Little et al., 1974). He collected a few specimens from St. John during July 1954 and April 1967.

Roy O. Woodbury (1913–   ) American botanist, professor of botany and plant ecology at the University of Puerto Rico, Río Piedras. He made numerous collections of St. John plants from May 1982 through July 1983 for his preparatory studies for a checklist published by him and Weaver (1987). His collections are very important because, in many instances, they constitute records of the first specimens collected in this century for many earlier records. His collections also contain new records and new species. His work represents the starting point for the present flora of St. John. His specimens are deposited at the Biosphere Research Center at the Virgin Islands National Park Service (VINPS) on St. John, with a few duplicates at the herbarium of the Extension Service at the University of the Virgin Islands, St. Thomas Campus.

Scott A. Mori (1941–   ). American botanist, curator at The New York Botanical Garden. He visited St. John as an advisor for the new flora of St. John project. He collected many plants in June and November 1984. His specimens are deposited at NY and contain fine examples of the flora of St. John. They have been essential in the preparation of this book.

Ghillean T. Prance (1937–   ). British botanist, formerly curator and vice president for Science at The New York Botanical Garden, and currently director of The Royal Botanic Gardens at Kew. He visited St. John as an adviser for the projected new flora of St. John and collected many specimens in June 1984 and March 1985. His collections are deposited at NY and were examined during the preparatory studies for this book.

George R. Proctor (1920–   ). American botanist, a long-time employee of the Institute of Jamaica, presently director of the herbarium at the Department of Natural Resources in Puerto Rico. He visited St. John in 1983 while working on *Ferns of Puerto Rico and the Virgin Islands* (1989). His collections are deposited at the herbarium of the Department of Natural Resources in Puerto Rico (SJ), with duplicates at the Institute of Jamaica (IJ) and US.

History of the Project

I first visited the Virgin Islands in March 1985 when I joined a team from The New York Botanical Garden (NY), the University of Puerto Rico (UPR), and Yale University to do field work on the island of St. John. The main purpose of the trip was to establish permanent vegetation study plots in order to monitor vegetation changes in three different watersheds. The plots were marked and every tree or woody plant over 10 cm dbh, was identified and vouchered with a specimen deposited at NY. We also had the opportunity to explore most of the island and to collect numerous specimens, familiarizing ourselves with the flora of the island. Identifications of plants, both at the study sites and the material collected, were provided by Roy O. Woodbury, who at the time was working on a checklist of plants of St. John that had been commissioned by the National Park Service.

I and the other members of the team were extremely enthusiastic about being able to work on this important monitoring project. Early on, the importance of producing a full-fledged flora for the island became evident. Such a flora would be a cornerstone for further biological studies on this and nearby islands. We committed ourselves to this project and planned on pursuing the much needed field work. Numerous trips followed, and many people helped in this nascent project. Among them, John and Anne Earhart and Dr. Ghillean T. Prance have been very important in getting this new project off the ground through the financial support of the Homeland Foundation and The New York Botanical Garden. Initially, the project was to be a collaborative effort with Ghillean Prance, Roy Woodbury, John Earhart, Anne Reilly, and me, but eventually I took on full responsibility. After nine years of field work and research, the manuscript for the flora of St. John has been completed with the continuous support from the Homeland Foundation. This work also has been possible through the support of the Smithsonian Institution and the collaborative efforts of many botanists who provided treatments for some of the families.

Scope of the Flora

The flora treats all native vascular plant species from St. John, as well as naturalized, persistent, and common exotics found on the island of St. John. The distinction between naturalized and persistent is not readily apparent and in many cases may not be clear. For the purpose of the flora, only plants growing in the wild are considered naturalized (if spreading) or persistent (not spreading). Common ornamentals growing along roads are treated in full in this flora. Ornamentals with restricted distributions, however, are only mentioned after the last species of their genera in order to document their presence and their current status on the island. Species reported by previous workers but not documented by a voucher specimen are listed as doubtful records, and those based on misidentifications are excluded from the flora.

The flora is intended to be useful for both amateurs and students of the West Indian flora. In order to facilitate usage of the flora, especially by less experienced people, I use, when possible, nontechnical terms and include numerous illustrations and a glossary of technical terms. The complete descriptions and references to the species on St. John, and detailed distribution information, should render the work useful for the Virgin Islands as a whole and for the identification of plants in the adjacent dry districts of Puerto Rico as well.

Materials for the Flora

VOUCHER SPECIMENS.   This flora treats only those species (native or naturalized) for which we have seen at least one specimen. The principal collections used for this flora are those of N. L. Britton, H. F. A. Eggers, C. C. Raunkiær, C. F. Morrow, R. O. Woodbury, S. A. Mori, G. T. Prance, and myself. In the species descriptions under distribution, a few representative specimens from St. John are cited, except those for a few species of ferns, in which specimens could not be relocated. When species are rare, vouchers may be cited for other islands as well. The collections are cited either by a single-letter abbreviation of the collector's last name or by his or her initials, followed by the collection number. The collectors names are abbreviated as follows: P. Acevedo-Rodríguez (A), N. L. Britton (B), S. A. Mori (M), C. F. Morrow (CFM), G. T. Prance (GTP), G. R. Proctor (P), C. C. Raunkiær (R), I. Vélez (V), and R. O. Woodbury (W). These collections are deposited in different herbaria as follows. The collections of N. L. Britton, S. A. Mori, G. T. Prance, and myself (1985–1993) and duplicates of those of Raunkiær are deposited at NY. My collections from 1989 to 1993, and duplicates of Britton, Raunkiær, and Vélez collections are deposited at US. Finally, the main set of Woodbury specimens are deposited at the herbarium of the Virgin Islands National Park Service.

DESCRIPTIONS.   Every taxon from family to subspecies or variety is described. The family description is based on pertinent literature, with most information extracted from Cronquist (1981) and from Dahlgren et al. (1985). The descriptions embrace the variation present in the families as a whole but concentrate on features of the Caribbean taxa. References are often given for recent family treatments.

The generic concepts and descriptions are based on various sources; these include recent Caribbean floristic (see literature cited) and monographic treatments, descriptions from Engler's *Pflanzenreich,* and from Bentham and Hooker (1862–1883). The generic name is followed by the author's name and sometimes by the words "nom. cons." to indicate its conserved status. In many cases, the generic descriptions obtained from the literature are complemented by descriptions of specimens from the U.S. National Herbarium. Recent and pertinent references are often given.

The species concepts follow, when possible, modern floristic or monographic treatments. The name that appears in boldface type after a number represents the accepted name, followed by the author's name(s) and the bibliographical reference to the accepted publication date. Abbreviations of author's names follow Brummitt and Powell (1992), and bibliographical references follow Stafleu and Cowan (1976–1988) for books and Botanico-Periodicum-Huntianum (Lawrence et al., 1968) for serial publications. Names in italics in the lines that follow the boldfaced name represent the basionym and synonyms. Species descriptions are based almost exclusively on material from St. John, but specimens from nearby islands were used when inadequate material from St. John was available. The distribution of species is given for various localities on St. John. Reference to specimens collected there are cited in parentheses; the letter indicating the collector (see Voucher Specimens, above) is followed by the collection number. Distribution within the

Virgin Islands follows in the next sentence and is based on specimens from US, NY, or recent floras. The distribution given for species outside the Virgin Island group is based on modern floras for the region (Ackerman, 1992; Howard, 1974–1989; Nicolson et al., 1991; Proctor, 1989).

COMMON NAMES.   Only common names used on St. John or on other Virgin Islands are given. The main sources for these names are Eggers (1879) and Valls (1981, 1990). In addition, I have gathered some information directly from Lito Valls and from other residents of St. John. It is interesting to note that, despite the small size of these islands, usually more than one common name is applied to the same plant even on the same island.

ILLUSTRATIONS.   The excellent scientific illustrations that accompany the text were done by Bobbi Angell (NYBG). The quality of her work speaks for itself and nothing further needs to be said about it. Bobbi Angell is not only an excellent and sensible artist but also a fine botanist and keen observer who can perceive numerous details and particular features of individual plants. Every illustration is based on a voucher specimen, supplemented by photographs, pickled material, and/or sketches done by her from fresh material in the field. The specimens are deposited either at NY or at US (see Appendix 1). The vast majority of the illustrations were prepared for this project and paid for by a grant from the Homeland Foundation to The New York Botanical Garden. Some illustrations, also done by Bobbi Angell, were borrowed from Dr. Vicki A. Funk from her forthcoming treatment of Asteraceae for *The Flora of the Guianas,* from Dr. Scott A. Mori from his *Guide to the Vascular Plants of Central French Guiana,* from Dr. James Luteyn from his treatment of Plumbaginaceae for the *Flora of Ecuador,* from Dr. Noel Holmgren from his Scrophulariaceae for the *Flora of Ecuador,* from Dr. James W. Grimes from his treatment of the Leguminosae for the *Flora of the Guianas,* and from Dr. Arthur Cronquist's (1981) *An Integrated System of Classification of Flowering Plants.*

GLOSSARY.   A glossary is provided to help in the understanding of many technical botanical terms. The sources for these definitions are Jackson, 1928, and Little and Jones, 1980.

Contributors

James D. Ackerman, Departamento de Biología, Universidad de Puerto Rico, Río Piedras, Puerto Rico 00931. **Orchidaceae.**

Laurence J. Dorr, U.S. National Herbarium, Smithsonian Institution, Washington, D.C. 20560. **Sterculiaceae** and **Tiliaceae.**

Robert B. Faden, U.S. National Herbarium, Smithsonian Institution, Washington, D.C. 20560. **Commelinaceae** (reviewer).

Vicki A. Funk, U.S. National Herbarium, Smithsonian Institution, Washington, D.C. 20560. **Asteraceae.**

Paul A. Fryxell, U.S. Department of Agriculture in cooperation with Texas A&M University, College Station, Texas 77843. **Malvaceae.**

Lynn J. Gillespie, Canadian Museum of Nature, Ottawa, Ontario, Canada. **Euphorbiaceae** (reviewer).

James W. Grimes, The New York Botanical Garden, Bronx, New York 10458. **Fabaceae** (reviewer).

Alan Herndon, Department of Biological Sciences, Florida International University, Miami, Florida 33199. **Chamaesyce** (Euphorbiaceae, reviewer).

Scott LaGreca, Department of Botany, Duke University, Durham, North Carolina 27708. **Oleaceae.**

David B. Lellinger, U.S. National Herbarium, Smithsonian Institution, Washington, D.C. 20560. **Pteridophytes** (reviewer).

Gilberto N. Morillo, Herbario, Universidad de Los Andes, Mérida, Venezuela. **Asclepiadaceae** (reviewer).

Dan H. Nicolson, U.S. National Herbarium, Smithsonian Institution, Washington, D.C. 20560. **Araceae** (reviewer).

Paul M. Peterson, U.S. National Herbarium, Smithsonian Institution, Washington, D.C. 20560. **Poaceae.**

George R. Proctor, Departamento de Recursos Naturales, Area de Investigaciones Científicas, Puerta de Tierra, San Juan, Puerto Rico 00906. **Pteridophytes.**

John F. Pruski, U.S. National Herbarium, Smithsonian Institution, Washington, D.C. 20560. **Asteraceae.**

Rodolfo Quiros, 2254 Friley Hall, Iowa State University, Ames, IA 50012. **Piperaceae.**

Mark T. Strong, U.S. National Herbarium, Smithsonian Institution, Washington, D.C. 20560. **Agavaceae, Amaryllidaceae, Asphodelaceae, Cymodoceaceae, Cyperaceae, Dracaenaceae, Hydrocharitaceae, Hypoxidaceae,** and **Potamogetonaceae.**

Charlotte M. Taylor, Missouri Botanical Garden, P.O. Box 299, St. Louis, Missouri 63166. **Rubiaceae** (reviewer).

Dieter C. Wasshausen, U.S. National Herbarium, Smithsonian Institution, Washington, D.C. 20560. **Acanthaceae** (reviewer).

Robert D. Webster, Systematic Botany and Mycology Laboratory, Agriculture Research Service, BARC-West, Beltsville, Maryland 20705. **Poaceae.**

Acknowledgments

The preparation of this book has been a long process that has been possible only with the support of numerous people. Different aspects of this work, from fund raising, field work, literature searches, processing of collections, editing, and finally to publishing, have been aided by many colleagues and friends. I express my deepest appreciation to John and Anne Earhart for their continuous support toward the completion of this project, especially by making their St. John home available, assisting me in the field, and by providing financial support through the Homeland Foundation. My appreciation is also extended to Dr. Ghillean T. Prance, who got me started on this project and who laid the foundations for its completion. I am indebted to Dr. Roy O. Woodbury, who shared his knowledge of St. John flora and led me to some of the rarest plants on the island.

I am also indebted to my good friends Cristina López, Livia Colón, Olga Iris Alemán, Danilo Chinea, Anne Reilly, and Bobbi Angell, who assisted me with field work and made my stays on St. John more enjoyable. Also, Eleanor Gibney and Gary Ray accompanied me in the field and guided me to plants that were later found to be new records for the island.

I thank Virgin Islands National Park Service personnel, in particular, Dr. Caroline Rogers and Jennifer Bjork in helping with accommodations at the park dormitories, for providing access to their equipment and collections, issuing collection permits, and storing some of the field equipment at the park facilities and in their homes. Lito Valls has been a wonderful and enjoyable source of knowledge, particularly with common names of plants and some of their economic uses.

Back in Washington, I have been aided by some of the finest and dedicated colleagues and technicians at the Smithsonian Institution. I am indebted to Mark T. Strong and Marilyn Hansel for their performance of numerous tasks that included searching for specimens or literature, processing collections and photographs, editing, and offering their valuable comments. Dr. Dan H. Nicolson helped relieve the burden imposed by nomenclatural rules and bibliographic citations. John Pruski has been of valuable assistance in editing the introduction and part of the text. Dr. Laurence J. Dorr and Bobbi Angell also provided valuable comments on the introduction. I thank all the colleagues who wrote family treatments or reviewed my own treatments for the present flora.

My appreciation is extended to The New York Botanical Garden, where the project originated, for lending many specimens from the Virgin Islands, for providing logistical support for my artist, and for administrating a grant from the Homeland Foundation to pay for most of the scientific illustrations done for this project. My thanks are extended to Drs. Scott A. Mori, James Luteyn, Noel Holmgren, and James W. Grimes for lending illustrations. I also appreciate the support from the Biodiversity of the Guianas Program (Smithsonian Institution) through Dr. Vicki A. Funk, who paid for the illustrations of species in the Asteraceae. I am also indebted to the following specialists who provided determinations for many of the specimens collected on St. John: Rupert C. Barneby (NY), Robert B. Faden (US), Alan Herndon (FIU), Leslie R. Landrum (ASU), David B. Lellinger (US), Michael Nee (NY), Dan H. Nicolson (US), Terence D. Pennington (K), Charlotte M. Taylor (MO), Dieter Wasshausen (US), and Hendrik H. van der Werff (MO). I am further indebted to Dr. George R. Proctor for his editorial comments and to Dr. Thomas A. Zanoni for his intensive review of the whole manuscript.

Finally, I am much indebted to my wife, R. Amneris Siaca, who has assisted me on numerous occasions in the field and in my laboratory at the Smithsonian Institution.

Thanks are extended to all the people involved in this project and to those who always showed enthusiasm and interest, thereby helping to keep the flame lit over the years.

Selected References

Acevedo-Rodríguez, P. & R. O. Woodbury. 1985. Los bejucos de Puerto Rico. Vol. 1. U.S. Forest Serv. Gen. Techn. Rep. **SO-58**: 1–331.

Ackerman, J. D. 1992. The orchids of Puerto Rico and the Virgin Islands/ Las orquideas de Puerto Rico y las Islas Vírgenes. Editorial de la Universidad de Puerto Rico, Río Piedras, Puerto Rico. 168 pp.

Bentham, G. & J. D. Hooker. 1862–1883. Genera plantarum. A. Black, Hookerian Herbarium, Kew, England.

Børgesen, F. 1923. On the vegetation of the Virgin Islands of the United States formerly the Danish West Indies. Translated from the Danish by F. MacFarlane. Government Printing Office, St. Thomas.

Britton, N. L. 1916. The vegetation of Anegada. Mem. New York Bot. Gard. **6**: 565–580.

———. 1918. The flora of the American Virgin Islands. Brooklyn Bot. Gard. Mem. **1**: 19–118.

———. **& P. Wilson.** 1923–1926. Botany of Porto Rico and the Virgin Islands. Scientific Survey of Porto Rico and the Virgin Islands. New York Academy of Sciences, New York.

Brummitt, R. K. & C. E. Powell. 1992. Authors of plant names. Royal Botanic Garden, Kew.

Cronquist, A. 1981. An integrated system of classification of flowering plants. Columbia University Press, New York.

Dahlgren, R. M. T., H. T. Clifford & P. F. Yeo. 1985. The families of monocotyledons. Springer-Verlag, New York.

Dallmeier, F., J. Comiskey & G. Ray. 1993. User's guide to the Virgin Islands Biosphere Reserve Biodiversity Plot 01, U.S. Virgin Islands. The Smithsonian Institution/Man and Biosphere Biological Diversity Program, Washington, D.C.

D'Arcy, W. G. 1967. Annotated checklist of the dicotyledons of Tortola, Virgin Islands. Rhodora **69**: 385–450.

———. 1971. The island of Anegada and its flora. Atoll Res. Bull. **139**: 1–21.

Eggers, H. F. A. 1876. St. Croix's flora. Vidensk. Meddel. Dansk Naturhist. Foren. Kjøbenhavn **28**: 33–158.

———. 1879. The flora of St. Croix and the Virgin Islands. Government Printing Office, Washington, D.C.

Fishlock, W. C. 1912. The Virgin Islands B.W.I.: a handbook of general information. The West India Committee, London.

Forman, R.T.T. and D.C. Hahn. 1980. Spatial patterns of trees in a Caribbean semi-evergreen forest. Ecology **61**: 1267–1274.

Fosberg, F. R. 1976. Revisions in the flora of St. Croix, U.S. Virgin Islands. Rhodora **78**: 79–119.

Grisebach, A. H. R. 1859–1864. Flora of the British West Indian Islands. Lovell Reeve & Co., London.

Heatwole, H. & R. Levins. 1972. Biogeography of the Puerto Rican Bank: flotsam transport of terrestrial animals. Ecology **53**: 112–117.

—— & ——. 1973. Biogeography of the Puerto Rican Bank: species-turnover on a small cay, Cayo Ahogado. Ecology **54**: 1042–1055.

——, —— & M. D. Byer. 1981. Biogeography of the Puerto Rican Bank. Atoll Res. Bull. **251**: 1–62.

—— & F. MacKenzie. 1967. Herpetogeography of Puerto Rico. IV. Paleogeography, faunal similarity and endemism. Evolution **21**: 429–438.

Holmgren, P. K., N. H. Holmgren, & L. C. Barnett. 1990. Index herbariorum, part 1: the herbaria of the world. New York Botanical Garden, Bronx, New York.

Howard, R. A. 1973. The vegetation of the Antilles. Pages 1–38 in: A. Graham (ed.), Vegetation and vegetational history of northern Latin America. Elsevier, New York.

——. 1974–1989. Flora of the Lesser Antilles. Arnold Arboretum of Harvard University, Jamaica Plain, Massachusetts.

Jackson, B. D. 1928. A glossary of botanic terms with their derivation and accent. Reprinted in 1960. Hafner Publishing, New York.

Jadan, D. 1985. A guide to the natural history of St. John. Environmental Studies Program, St. John, St. Thomas Graphics, U.S. Virgin Islands.

Knox, J. P. 1852. A historical account of St. Thomas, W. I. Charles Scribner, New York.

Krebs, H. 1847. Et Bidrag til St. Thomas' Flora. Naturhist. Tidsskr. **1847**: 291–302.

Lawrence, G. H. M., A. F. G. Buchheim, G. S. Daniels & H. Dolezal. 1968. Botanico-Periodicum-Huntianum. Hunt Botanical Library, Pennsylvania.

Liogier, A. H. 1965. Nomenclatural changes and additions to Britton and Wilson's *Flora of Porto Rico and the Virgin Islands*. Rhodora **67**: 315–361.

——. 1985, 1988, 1994. Descriptive flora of Puerto Rico and adjacent islands. Spermatophyta. Vol. 1, Casuarinaceae to Connaraceae. Vol. 2, Leguminosae to Anacardiaceae. Vol. 3, Cyrillaceae to Myrtaceae. Editorial de la Universidad de Puerto Rico, Río Piedras, Puerto Rico.

—— & L. F. Martorell. 1982. Flora of Puerto Rico and adjacent islands: a systematic synopsis. Editorial de la Universidad de Puerto Rico, Río Piedras, Puerto Rico.

Little, E. L., Jr. 1969. Trees of Jost van Dyke (British Virgin Islands). U.S. Forest Serv. Res. Pap. ITF-9: 1–12.

——. 1976. Flora of Virgin Gorda (British Virgin Islands). U.S. Dept. Agric. Forest Serv. Res. Paper **ITF-21**: 1–36.

—— and F. H. Wadsworth. 1964. Common trees of Puerto Rico and the Virgin Islands. Agriculture Handbook No. 249. U.S. Forest Serv., Washington, D.C.

—— & R. O. Woodbury. 1980. Rare and endemic trees of Puerto Rico and the Virgin Islands. U.S.D.A. Conserv. Res. Rep. **27**: 1–26.

——, —— & F. H. Wadsworth. 1974. Trees of Puerto Rico and the Virgin Islands. Vol. 2. Agriculture Handbook No. 449. U.S. Forest Serv., Washington, D.C.

Little, R. J. & C. E. Jones. 1980. A dictionary of botany. Van Nostrand Reinhold, New York.

Low, R. H. & R. Valls. 1985. St. John backtime. Eyewitness accounts from 1718 to 1956. Eden Hill Press, U.S. Virgin Islands.

Meyerhoff, H. A. 1933. Geology of Puerto Rico. Univ. Puerto Rico Monogr. Ser. B, **1**: 1–306.

Millspaugh, C. F. 1902. Flora of the island of St. Croix. Publ. Field Columbian Mus., Bot. Ser. **68**: 441–546.

Mitchell, R. C. 1954. A survey of the geology of Puerto Rico. Univ. Puerto Rico Agric. Exp. Sta. Techn. Paper **13**: 1–167.

Nicolson, D. H. with R. A. DeFilipps, A. C. Nicolson, & others. 1991. Flora of Dominica, part 2: Dicotyledoneae. Smithsonian Contrib. Bot. **77**: 1–274.

Oakes, A. J. & J. O. Butcher. 1962. Poisonous and injurious plants of the U.S. Virgin Islands. U.S.D.A. Res. Serv. Misc. Publ. **882**: 1–97.

Oldendorp, C. G. A. 1987. History of the Mission of the Evangelical Brethren on the Caribbean Islands of St. Thomas, St. Croix, and St. John. Translated from the German edition 1777, by A. R. Highfield and V. Barac and edited by J. J. Bossard. Karoma Publishers, Ann Arbor, Michigan.

Proctor, G. R. 1989. Ferns of Puerto Rico and the Virgin Islands. Mem. New York Bot. Gard. **53**: 1–389.

Reilly, A. E. 1992. Impacts of natural and human disturbance on forests of St. John, U.S. Virgin Islands. Park Sci. **12**(2): 3, 4.

——, J. E. Earhart & G. T. Prance. 1990. Three sub-tropical secondary forests in the U.S. Virgin Islands: a comparative quantitative ecological inventory. Advances Econ. Bot. **8**: 189–198.

Schlechtendal, D. F. L. 1828–1831. Florula insulae Sti. Thomae Indiae Occidentalis. Linnaea **3**: 78–93, 251–276; **5**: 177–200, **5**: 682–688; **6**: 722–772.

Stafleu, F. A. & R. S. Cowan. 1976–1988. Taxonomic literature. Ed. 2. Bohn, Scheltema and Holkema, Utrecht.

Tyson, G. F. 1984. A history of land use on St. John, 1718–1950. Unpublished preliminary report prepared for the Virgin Islands National Park Service.

Urban, I. 1898–1904. Bibliographia indiae occidentalis botanica. Symb. Ant. **1**: 1–192 (1898); **2**: 1–8(1900); **3**: 1–13(1902); **5**: 1–16 (1904).

——. 1902. Notae bibiographicae peregrinatorum indiae occidentalis botanicorum. Symb. Ant. **3**: 14–158.

U.S. Fish and Wildlife Service. 1993. Endangered and threatened wildlife and plants. 50 CFR 17.11 & 17.12. August 23, 1993. U.S. Government Printing Office.

Vail, P.R. & J. Hardenbol. 1979. Sea-level changes during the tertiary. Oceanus **22**: 71–79.

Valls, I. 1981. What a pistarckle! A dictionary of Virgin Islands English Creole. Published by author, St. John, U.S. Virgin Islands.

——. 1990. What a pistarckle! A dictionary of Virgin Islands English Creole, new supplement. Published by author, St. John, U.S. Virgin Islands.

Vélez, I. 1950. Plantas indeseables en los cultivos tropicales. Editorial Universitaria, Río Piedras, Puerto Rico.

——. 1957. Herbaceous angiosperms of the Lesser Antilles. Inter American University of Puerto Rico, San Germán.

Weaver, J. D. 1961. Erosion surfaces in the Caribbean and their significance. Nature **190**: 1186–1187.

West, H. 1793. Bidrag til Beskrivelse over Ste. Croix med en kort Udsigt over St. Thomas, St. Jean, Tortola, Spanishtown, og Crabeneiland. Friderik Wilhelm Thiele, Copenhagen.

Woodbury, R. O. & E. L. Little. 1976. Flora of Buck Island Reef National Monument (U.S. Virgin Islands). U.S. Forest Serv. Res. Paper **ITF-19**: 1–27.

—— & P. L. Weaver. 1987. The vegetation of St. John and Hassel Island, U.S. Virgin Islands. U.S. Dept. Interior, Natl. Park Serv. Southeast Regional Office, Research/Resources Managem. Rep. **SER-83**: 1–103.

Woodworth, R. H. 1943. Economic plants of St. John, U.S. Virgin Islands. Bot. Mus. Leafl. **2**: 29–54.

Pteridophytes (Fern and Fern Ally Families)

By G. R. Proctor

Key to the Families of Pteridophytes

1. Leaves minute, scale-like and veinless or narrowly linear, with a single vein.
  2. Plant dichotomously branched, appearing leafless (the leaves minute scales); sporangia (synangia) trilobed, trilocular, scattered on branches.. 5. Psilotaceae

2. Plants with a main stem bearing lateral branches, with numerous needle-like leaves; sporangia subglobose, unilocular, congested into compact spikes at the end of lateral branches.................... 2. Lycopodiaceae
1. Leaves foliaceous, with branched veins.
   3. Sporangia produced on elongated spikelike structures .... 3. Ophioglossaceae
   3. Sporangia produced on lower side or margins of the leaf blade.
      4. Large palmlike ferns; caudex (stem) 3–10 m tall, 10–15 cm diam., with numerous large leaf scars;
         fronds (leaves) 2–3.5 m long .... 1. *Cyatheaceae*
      4. Small to medium-sized, < 2 m long or tall, herblike or vinelike ferns; stems usually inconspicuous
         (covered by leaf bases), shortened, or sometimes elongated, creeping or climbing .... 4. Polypodiaceae

1. **Cyatheaceae** (Tree-Fern Family)

Medium-sized to large ferns, mostly with a stout, erect, trunk like stem ("caudex"), or in some species the stem (rhizome) creeping or decumbent, clothed with hairs or scales and often spiny. Fronds small to very large, usually 1- to 4-pinnate (rarely simple), smooth or bearing spines on the stipe and sometimes also on rachis and midveins. Sori marginal or on lower surface, round or ellipsoid; indusium subtending and partially or entirely enclosing the sporangia, wanting or reduced to minute scales. Sporangia short-stalked, the stalk consisting of more than 3 rows of cells, borne on an elevated receptacle, and often accompanied by characteristic paraphyses; annulus oblique, not interrupted by the stalk; spores more or less globose, trilete, the surface smooth, ridged, faintly tuberculate or spinulose, or with pores or cavities. Gametophytes green, cordate, flat, monoecious, bearing multicellular, elongate, scale-like appendages.

In the broad sense as defined here, the family consists of about 12 genera and more than 700 species, of pantropical distribution, especially of cool, moist, tropical mountain climates.

REFERENCES: Maxon, W. R. 1912. The tree ferns of North America. Annual Rep. Board Regents Smithsonian Inst. 1911: 463–491, t. 1–12; Tryon, R. 1970. The classification of the Cyatheaceae. Contr. Gray Herb. 200: 3–53; Tryon, R. 1976. A revision of the genus *Cyathea*. Contr. Gray Herb. 206: 19–98.

1. CYATHEA Sm.

More or less palmlike plants, with an erect or less often decumbent or horizontal caudex, often with a dense growth of blackish adventitious roots toward the base, the apex clothed with scales characteristic for each species, and often spiny. Fronds arching-pendent to ascending, moderate to very large in size, 1- or 2-pinnate-pinnatifid, often pubescent and/or scaly; stipes thick, curved, and densely clothed with scales toward base, the scales more or less deciduous; veins free or (rarely) the basal ones joined to form costal areoles. Sori dorsal on veins, or else arising from vein axils; indusium present or absent, if present, subtending the sorus and various in size and shape, sometimes partially or wholly enclosing the sporangia, or else sometimes reduced to a mere scale; filamentous paraphyses often present; sporangia borne on an elevated receptacle; pedicel of sporangium short, consisting of four rows of cells; spores variously echinate, rugose, granulate, ridged, crested, or porous, always trilete.

A genus of perhaps 600 species, about evenly divided between the New World and Old World tropics.

1. Cyathea arborea (L.) Sm., Mém. Acad. Roy. Sci. Turin **5**: 417. 1793. *Polypodium arboreum* L., Sp. Pl. 1092. 1753. *Disphenia arborea* (L.) C. Presl, Tent. Pterid. 56. 1836. *Hemitelia arborea* (L.) Fée, Mém. Foug. **5**: 350. 1852. *Cormophyllum arboreum* (L.) Newman, Phytologist **5**: 238. 1854.

Cyathea serra Willd., Sp. Pl. **5**: 491. 1810. *Hemitelia serra* (Willd.) Desv., Mém. Soc. Linn. Paris **6**: 221. 1827.

Palmlike fern to 12 m tall. Trunk 10–15 m diam., with persistent oval frond scars, usually without spines, densely clothed toward apex with whitish or creamy, papery scales to 4 cm long. Fronds monomorphic; stipes 25–60 cm long, yellowish, minutely scurfy when young, usually more or less roughened-tuberculate with minute blunt projections but not armed with spines, densely clothed toward base with deciduous, whitish or creamy (rarely brownish or bicolorous), lanceolate to narrowly ovate-attenuate scales. Blades ovate, 2–3.5 m long, 2-pinnate-pinnatifid, acuminate at apex, slightly reduced at base, essentially glabrous, the tissue light green and of rather heavy texture; rachis slightly muriculate, glabrescent; pinnae alternate, 40–80 × 15–35 cm, stalked; pinnules lance-oblong to elliptic-lanceolate, mostly 1.5–2 cm broad, attenuate at apex, all but the lowest sessile; segments usually 25–32 pairs, narrowly oblong to nearly falcate, 2–5 mm broad, dilated at base, the margin serrate; midveins with 1 or 2 deciduous, white, bullate scales at base on lower side; veins 10–14 pairs, 1- to 3-forked. Sori usually 6–11 pairs per fertile segment, inframedial, their position marked by pits on adaxial surface; indusium saucer-shaped, firm, persistent; receptacles wedge-shaped to capitate; paraphyses a mixture of acicular hairs and shorter, capitate-glandular ones.

DISTRIBUTION: In moist open areas. Bordeaux Mountain (P40438). Also on St. Thomas and Tortola; Greater and Lesser Antilles and northernmost Colombia.

2. **Lycopodiaceae** (Clubmoss Family)

Terrestrial, epilithic, or epiphytic plants of erect, trailing, or pendent habit. Stems more or less elongate, simple or branched, densely clothed nearly throughout with numerous small, simple, 1-veined leaves. Sporophylls either like sterile leaves or modified to form compact spikes (strobili). Sporangia axillary, 1-loculed; spores all alike, minute, very numerous, more or less spheroid and trilete, the surface variously sculptured. Gametophytes fleshy or tuberous, with or without chlorophyll, monoecious.

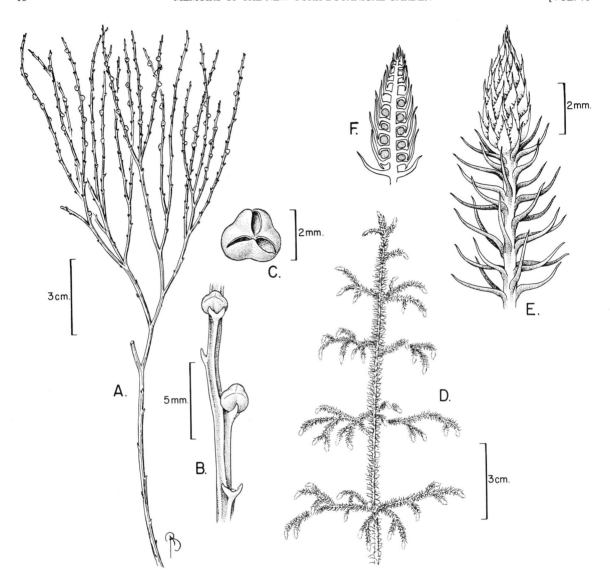

FIG. 1. **A–C.** *Psilotum nudum.* **A.** Habit. **B.** Detail of stem with sporangia subtended by bifid sporophylls. **C.** Sporangium. **D–F.** *Lycopodium cernuum.* **D.** Habit. **E.** Branch with strobilus. **F.** L.s. strobilus showing sporophylls and sporangia.

The family is here considered to consist of 2 living genera; of these, *Phylloglossum* (1 species) occurs only in Australia. The other is *Lycopodium,* a large and diverse group of worldwide distribution, divided by some authors into 3 or more genera (for instance see Øllgaard, 1989 and Wagner & Beitel, 1992).

REFERENCES: Øllgaard, B. 1987. A revised classification of the Lycopodiaceae s. lat. Opera Bot. 92: 153–178; Øllgaard, B. 1989. Index of the Lycopodiaceae. Biol. Skr. 34: 1–135. Øllgaard, B. 1992. Neotropical Lycopodiaceae—an overview. Ann. Missouri Bot. Gard. 79: 687–717; Underwood, L. M. & F. E. Lloyd. 1906. The species of *Lycopodium* of the American tropics. Bull. Torrey Bot. Club 33: 101–124. Wagner, W. H., Jr. & J. M. Beitel. 1992. Generic classification of modern North American Lycopodiaceae. Ann. Missouri Bot. Gard. 79: 676–686.

1. LYCOPODIUM L.

Characters as given for the family; plants perennial, with elongate stems; leaves spirally arranged in 4–24 ranks or more, rarely nearly verticillate or apparently opposite. Sporangia sessile or short-stalked; spores various, ranging from nearly smooth to foveolate, more or less ridged or reticulate, or rarely spinulose.

A genus of more than 250 species, the majority on tropical mountains.

1. **Lycopodium cernuum** L., Sp. Pl. 1103. 1753. *Lepidotis cernua* (L.) P. Beauv., Prodr. Aethéogam. 108. 1805. *Palhinhaea cernua* (L.) Franco & Vasc. in Vasc. & Franco, Bol. Soc. Brot. **41:** 25. 1967. *Lycopodiella cernua* (L.) Pic.-Serm., Webbia **23:** 165. 1968.   Fig. 1D–F.

Plants terrestrial, with horizontal-arching or assurgent main stems bearing at intervals several erect, feathery aerial branches, or sometimes the main stem itself erect, the lowermost branches decurved-spreading, elongate, and stoloniferous. Erect stems 30–100 cm long, giving rise to numerous spreading, freely branched, densely leafy lateral divisions, the ultimate branchlets (when fertile) ending in sessile cylindric-conic strobili. Leaves of main stem spirally arranged in 16–24 ranks, often appearing nearly whorled, subulate-attenuate, mostly entire, spreading or reflexed, keeled and decurrent adaxially at base. Lateral branches to 15 cm long, bearing leaves similar to those of the main stem but slightly smaller and strongly upcurved. Strobili often numerous, downwardly directed; sporophylls appressed-imbricate, in about 10 ranks, triangular-acuminate. Sporangia subglobose with unequal valves, concealed.

DISTRIBUTION: Moist, exposed, sunny banks in humid areas. Bordeaux Mountain (A2603). Also on Tortola and St. Thomas; pantropical.

3. Ophioglossaceae (Ophioglossum Family)

Terrestrial or epiphytic plants consisting of a short, fleshy rhizome bearing numerous more or less fleshy roots. Fronds 1 to several, erect or pendent, not articulate, consisting of a basal stalk bearing at the apex a simple or variously subdivided sterile blade and (when fertile) an erect or divergent spore-bearing spike or panicle (sporophyll), or rarely the sporophylls several. Sporangia sessile or subsessile, in 2 rows, naked, separate or laterally joined in a synangium, each opening by a transverse slit and lacking an annulus; spores equal (homosporous), without chlorophyll. Gametophytes subterranean, very small, tuberlike, mycorrhizic and usually without chlorophyll, bearing immersed gametangia.

A worldwide family of 3–5 genera and 60 or more species.

REFERENCE: Clausen, R. T. 1938. A monograph of the Ophioglossaceae. Mem. Torrey Bot. Club 19(2): 1–177, figs. 1–33.

1. OPHIOGLOSSUM L.

Perennial, and usually evanescent herbs, terrestrial or epiphytic. Rhizome short, usually erect, terminating in the erect exposed bud of the following season. Fronds erect or pendent, glabrous, somewhat fleshy, consisting of a common stalk, and separate sterile and fertile portions; sterile blade simple or palmately lobed, sessile or short-stalked, with reticulate venation, the primary areoles enclosing free veinlets and also sometimes secondary areoles. Sporophylls 1 or several simple, stalked spikes. Sporangia subglobose, coalescent in 2 marginal rows; spores globose, trilete, yellowish, thick-walled, the surface densely reticulate with fused minute cones and pits.

A genus of more than 30 species scattered almost throughout the world.

1. Ophioglossum reticulatum L., Sp. Pl. 1063. 1753.

Terrestrial fern 10–25 cm tall. Rhizome cylindric, erect, 0.5–1(–2) cm long, bearing persistent thick roots, these not stoloniferous. Fronds solitary or occasionally 2; stipes 5–10(–16) cm long. Sterile blade triangular-ovate to kidney-shaped, 3–6(–10) cm long, nearly sessile or with stalk to 10 mm long, usually broadly and shallowly cordate at base, the apex round, obtuse, or nearly acute; areoles with few to many free or partly joined included veinlets; stalk of fertile segment 5–18 cm long; fertile segment 1–5.5 cm long, usually bearing 20–60 pairs of fused sporangia. Spores 30–60 μ diam.

DISTRIBUTION: Occasional on shaded banks in moist forests, appearing after seasonal rains. Shaded slopes of Bordeaux Mountain (B536). Also on St. Thomas; pantropical.

4. Polypodiaceae (Fern Family)

Small to large, leafy plants of various habit, terrestrial, epilithic, or epiphytic, rarely aquatic; stem (rhizome) creeping to erect, with few to many roots, usually bearing hairs or scales. Fronds (leaves) stipitate or rarely sessile, uniform or dimorphic, the blades simple to variously subdivided (to 5-pinnate), rarely dichotomously fan-shaped, with or without indument, or else bearing scales, these appressed or spreading, glabrous or pubescent, sometimes glandular, and either basally or peltately attached; tissue sometimes provided with sessile or stalked glands, or in a few cases the lower surface covered with waxlike powder. Sporangia long-stalked (1–4 rows of cells), arranged in lines or clusters (sori), or completely covering the lower side of leafy or contracted fronds, or rarely scattered at random along veins or on nonvascular tissue; sori naked or protected by a scale-like indusium or an infolding or inrolling of the whole margin; annulus of the sporangium vertical, interrupted by a stomium of several thin-walled cells, rupturing transversely to release the spores explosively; spores bilateral and monolete, or globose to tetrahedral and trilete, smooth or wrinkled, ridged, rugose, papillose, or spinulose. Gametophytes green, flat, often more or less cordate, or ribbonlike to filamentous, glabrous or hairy, sometimes glandular, dioecious or monoecious, usually of short life-span, rarely perennial.

The family treated in a broad sense consists of more than 170 genera and perhaps as many as 10,000 species.

REFERENCES: Maxon, W. R. 1908. Asplenium salicifolium and confused species. Contr. U.S. Natl. Herb. 10: 475–481; Maxon W. R. 1913. Asplenium trichomanes and its American allies. Contr. U.S. Natl. Herb. 17: 133–153; Mickel, J. T. 1974. A redefinition of the genus Hemionitis. Amer. Fern J. 64: 3–12; Morton, C. V. & D. B. Lellinger. 1966. The Polypodiaceae subfamily Asplenioideae in Venezuela. Mem. New York Bot. Gard. 15: 1–49; Tryon, R. M. 1962. Taxonomic fern notes. II. Pityrogramma (including Trismeria) and Anogramma. Contr. Gray Herb. 189: 52–76; Tryon, R. M. & A. F. Tryon. 1982. Ferns and allied plants. Springer Verlag, New York.

Key to the Genera of Polypodiaceae

1. Fronds not divided (simple) .... 10. *Polypodium* (part)
1. Fronds divided.
    2. Sporangia in marginal sori (short or long), terminal on veins.

> 3. Blades pedate-pentagonal .... 5. *Doryopteris*
> 3. Blades pinnate or diffusely divided.
> > 4. Fronds with spiny axes, apical growth indeterminate (plants vinelike) .... 8. *Odontosoria*
> > 4. Fronds lacking spines; growth determinate (plants herb- or shrublike).
> > > 5. Fronds 1-pinnate, or 1-pinnate-pinnatifid with forked basal pinnae .... 11. *Pteris*
> > > 5. Fronds 3–5-pinnate .... 2. *Adiantum*
> 2. Sporangia dorsal (rarely submarginal).
> > 6. Sporangia completely covering abaxial side of fertile pinnae .... 1. *Acrostichum*
> > 6. Sporangia confined to veins.
> > > 7. Sporangia scattered along veins abaxially, not clumped in discrete sori.
> > > > 8. Blades essentially glabrous, but usually bearing a white or yellow waxlike powder abaxially;
> > > > veins all free .... 9. *Pityrogramma*
> > > > 8. Blades pubescent and without a waxlike powder; veins reticulate .... 6. *Hemionitis*
> > > 7. Sporangia densely clumped in discrete sori, with or without an indusium.
> > > > 9. Sori elongate.
> > > > > 10. Blades pinnate with several to many (to 20) pairs of pinnae; sori forming a continuous
> > > > > line on either side of and parallel to the costae .... 4. *Blechnum*
> > > > > 10. Blades tripartite to pentagonal; sori variable, in short lines oblique to the costae ...... 3. *Asplenium*
> > > > 9. Sori round or kidney-shaped.
> > > > > 11. Pinnae articulate to the rachis; rhizome producing long slender scaly stolons; veins always free .... 7. *Nephrolepis*
> > > > > 11. Pinnae continuous with the rachis, not articulate; rhizome without stolons; veins free or reticulate.
> > > > > > 12. Fronds more or less pubescent (hairs may be minute); plants terrestrial or occasionally epilithic .... 12. *Thelypteris*
> > > > > > 12. Fronds glabrous; plants epiphytic or epilithic .... 10. *Polypodium* (part)

1. ACROSTICHUM L.

Large, coarse ferns, chiefly of brackish or saline swamps at sea level, rarely in fresh-water swamps at low to middle elevations. Rhizomes woody, stout, erect or decumbent-ascending, bearing thick, spongy roots and with broad scales at the apex. Fronds clustered, erect, monomorphic or dimorphic; stipe stout; blades 1-pinnate with large simple pinnae, the veins closely reticulate without included free veinlets. Sporangia and paraphyses densely covering lower surface of fertile pinnae; indusium lacking; sporangia with annulus of 20–22 cells; spores tetrahedral-globose, trilete, the surface papillate with numerous minute diffuse rods or strands.

A small pantropical genus of several species.

1. Acrostichum danaeifolium Langsd. & Fisch., Pl. Voy. Russes Monde **1:** 5, t. 1. 1810.

Chrysodium lomarioides Jenman, Timehri **4:** 314. 1885. *Acrostichum lomarioides* (Jenman) Jenman, Bull. Bot. Dept., n.s. **5:** 154. 1898, nom. illegit., non *A. lomarioides* Bory, 1833. *Acrostichum excelsum* Maxon, Proc. Biol. Soc. Wash. **18:** 224. 1905 (based on *Chrysodium lomarioides* Jenman).

Terrestrial fern 0.5–1.5(–3.5) m tall. Rhizomes massive, clothed at the apex with rigid, spreading, linear or narrowly lanceolate, light- to dark-brown scales to 2 cm long. Fronds erect, dimorphic, the sterile ones 1.5–3.5 m long, the fertile ones taller and more rigidly erect; stipes very stout and subwoody, grooved along upper surface, usually dark brown, clothed at base with broadly linear, bicolorous scales, 2.5 × 0.3 cm. Pinnae very numerous, usually 20–40 pairs or more and a similar terminal one, both the sterile and fertile ones often crowded and subimbricate, often finely pubescent beneath, the margins lightly repand-cartilaginous; sterile pinnae chartaceous, the fertile pinnae rather fleshy in texture.

DISTRIBUTION: Occasional in intermittent ravines. Battery Gut (A4163). Also on St. Croix, St. Thomas, and Tortola; widespread in tropical and subtropical America.

DOUBTFUL RECORD: *Acrostichum aureum* L. was reported by Woodbury and Weaver from St. John (1987); however, no specimen was collected or cited by them, and attempts to locate the species have failed. *Acrostichum aureum* is normally found along the inland margins of mangrove swamps and in open brackish or salt marshes throughout the tropics. Although such habitats are limited on St. John, there is still a slight possibility that *A. aureum* exists on St. John. In any case, *A. aureum* differs from *A. danaefolium* by the fertile blades bearing sporangia only on the upper pinnae (vs. sporangia on all pinnae).

2. ADIANTUM L.

Terrestrial ferns of forest slopes, shady ravines, and rocky banks. Rhizomes short and suberect to slender and wide-creeping, bearing numerous narrow scales especially near the apex. Fronds densely clustered to widely spaced, monomorphic; stipes wiry, dark, and usually highly lustrous; blades suberect to deflexed or drooping, 1- to 5-pinnate (rarely simple), of various patterns of dissection, glabrous to minutely pubescent or distantly stellate-scaly, sometimes more or less glaucous beneath; pinnules sessile or stalked, often articulate, deciduous in some species, membranous to coriaceous; veins free, often fan-shaped branched, or less often reticulate. Sori appearing marginal, but the sporangia borne along (and sometimes between) the distal ends of ultimate veins, on the underside of the sharply reflexed, membranous margin of the pinnules or segments; paraphyses absent; annulus of about 18 cells; spores tetrahedral to globose, trilete, smooth to slightly granulate or papillate.

A pantropical genus of about 200 species, most numerous in South America, a few occurring or extending into temperate regions.

Key to the Species of *Adiantum*

1. Rhizome scales uniformly bright yellowish brown, margins denticulate-ciliolate; leaf blades lanceolate-ovate; ultimate pinnules quickly deciduous .... 1. *A. fragile*
1. Rhizome scales bicolorous, dark brown with pale margins, these fimbriate-ciliate; leaf blades triangular-ovate, subpentagonal, long-stalked; ultimate pinnules subpersistent .... 2. *A. tenerum*

1. Adiantum fragile Sw., Prodr. 135. 1788.

Adiantum parvifolium Fée, Mém. Foug. **7**: 27, t. 23, fig. 2. 1857.

Terrestrial, delicate fern 15–30(–50) cm tall. Rhizomes short, thick, often several-branched, densely clothed with bright yellowish brown, triangular-attenuate to lance-linear scales, 1–3.5 mm long, these with denticulate-ciliolate margins, and those of the extreme apex with contorted-filiform tips. Fronds closely clustered, ascending or spreading, 15–60 cm long; stipes lustrous purple-brown, much shorter than the blades, cylindrical, wiry, and glabrous. Blades lance-ovate, 10–45 cm long, 4–18 cm broad, 3- to 4-pinnate at base; pinnae delicately to firmly herbaceous, alternate, stalked, with distant oblique pinnules; ultimate pinnules obovate-cuneate, 4–20 × 3–13 mm, on stalks 1–3 mm long, very easily breaking off, especially when dry; sterile pinnules denticulate, the ultimate veinlets ending in the teeth. Sori nearly kidney-shaped to oblong, borne singly or in pairs on very shallow lobes of the round distal margin; indusioid flap pale brown, delicately translucent toward margin.

Key to the Varieties of *A. fragile*

1. Ultimate pinnules mostly 10–20 mm long; tissue light green, thin-membranous, glabrous, and nonglandular on both sides .... 1a. *A. fragile* var. *fragile*
1. Ultimate pinnules mostly 4–10 mm long; tissue green or slightly glaucous, firm-membranous, usually very minutely puberulous on abaxial side and with numerous minute resinous glands embedded in the abaxial epidermis .... 1b. *A. fragile* var. *rigidulum*

1a. Adiantum fragile var. fragile.   Fig. 2A, B.

As described in the key.

DISTRIBUTION: Common on shaded banks, ledges, and cliffs. Along road to Susannaberg (A696). Also on St. Croix, St. Thomas, and Tortola; Greater Antilles and Lesser Antilles.

1b. Adiantum fragile var. rigidulum (Mett. *ex* Kuhn) Proctor, Mem. New York Bot. Gard. **53**: 143. 1989.

Adiantum rigidulum Mett. *ex* Kuhn, Linnaea **36**: 76. 1869.

As described in the key.

DISTRIBUTION: Occasional in habitat similar to that of *A. fragile* var. *fragile*. Surroundings of Bethany (B263). Also on St. Croix and St. Thomas; Puerto Rico (including Vieques), Jamaica, and expected in Hispaniola.

2. Adiantum tenerum Sw., Prodr. 135. 1788.

Adiantum littorale Jenman, Ferns Brit. W. Ind. 96. 1899. *Adiantum tenerum* var. *littorale* (Jenman) Domin, Rozpr. Král. Ceské Spolecn. Nauk, Tř. Mat.-Přír. n.s. **2**: 141. 1929.

Terrestrial fern 18–50(82) cm tall. Rhizome short-creeping or ascending, 3–8 mm thick, densely clothed toward apex with triangular-acuminate or narrowly lance-acuminate scales, 1.5–3.0 mm long, these lustrous light to blackish brown with pale, fimbriate-ciliate margins. Fronds few, erect-spreading or sometimes pendent, 18–82 cm long; stipes lustrous dark purplish brown to black, shorter than or nearly equaling the blades, 7–35 cm long, cylindrical, glabrous. Blades triangular-ovate, subpentagonal, 12–46 × 9–35 cm, 3- to 5-pinnate at base; pinnae alternate, stalked; ultimate pinnules trapezoid or rhombic-oblong, subpersistent, terminal ones fan- to wedge-shaped, the fertile ones mostly 1–2 cm long and broad, on slender stalks 2–4 mm long; sterile pinnules often larger than the fertile ones, more or less deeply lobed or cleft, with margins lightly to sharply dentate, the ultimate veinlets ending in the teeth; tissue light to dull green, firmly membranous, glabrous or rarely minutely puberulous abaxially. Sori retuse-oblong, light brown, turgid, borne in pairs on each of the lightly bifid lobes of the fertile margins.

DISTRIBUTION: A common fern of shaded rocky hillsides, banks, and cliffs. Fish Bay Gut (A2494). Also on St. Croix and St. Thomas; southern United States, Bahamas, Greater Antilles, Cayman Islands, Lesser Antilles, Central America, and northern South America.

3. ASPLENIUM L.

Small to medium-sized, terrestrial, epilithic, or epiphytic ferns. Rhizomes creeping to erect, bearing clathrate scales, these often iridescent. Fronds clustered (rarely spaced), not articulate to the rhizome, monomorphic or rarely more or less dimorphic; blades simple, 1- to 4-pinnate, pinnatifid or decompound, anadromous, glabrous or rarely pubescent; pinnae often inequilateral, sometimes auriculate at base on the acroscopic side; veins usually forked, free, rarely a few joined; lower margins of pinnae or pinnules usually decurrent. Sori oval to narrowly linear, borne along the ultimate veins, occasionally doubled; indusium more or less elongate, membra-

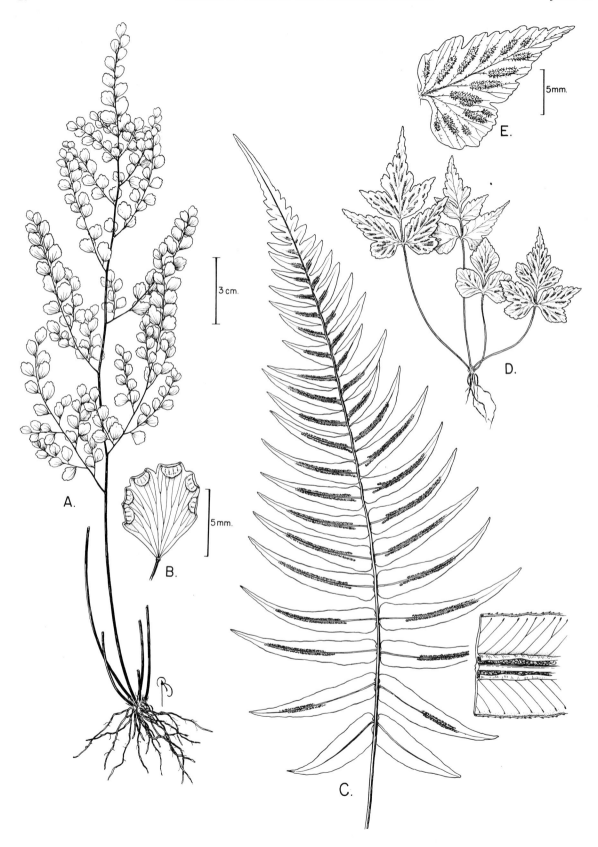

Fig. 2. **A, B.** *Adiantum fragile* var. *fragile*. **A.** Habit. **B.** Pinna with marginal indusia. **C.** *Blechnum occidentale:* frond and detail of pinna showing indusia borne along the costa. **D, E.** *Asplenium pumilum*. **D.** Habit. **E.** Fertile pinna showing venation and sori.

nous, laterally attached to vein, opening toward the apex; sporangia with annulus of 20 cells or more; spores more or less ellipsoid to nearly spheroid, monolete.

A cosmopolitan genus of more than 700 species.

1. Asplenium pumilum Sw., Prodr. 129. 1788.

Fig. 2D, E.

Asplenium anthriscifolium Jacq., Collectanea **2**: 103, t. 2, fig. 3, 4. 1788 [1789]. *Asplenium pumilum* var. *anthriscifolium* (Jacq.) Wherry, South. Fern Guide 346. 1964.

Terrestrial fern to 20 cm tall. Rhizome short, erect, 2–5 mm thick, with apical tuft of iridescent, gray-brown to blackish, linear-attenuate scales to 3 mm long with hairlike tips. Fronds seasonal, deciduous, mostly 5–20 cm long; stipes very slender, 1.5–12 cm long, often longer than the blades, sparsely clothed throughout with scattered, minute, whitish septate hairs. Blades variable, the smaller ones 3-lobed or 3-parted, the largest ones broadly pentagonal (to 9 cm broad), with 2–4 pairs of lateral pinnae and a triangular-acuminate, lobed apical one; lowest pinnae unequally triangular, coarsely lobed, the other lateral pinnae more or less lanceolate with crenate-dentate margins; tissue delicately herbaceous, with veins very oblique, subpinnately forked. Sori numerous, variable in length to 10 mm long; indusium whitish-translucent with minutely ciliate margin.

DISTRIBUTION: Common on shaded earthy or rocky banks. Maho Bay (A2526). Also on St. Thomas; Florida, Greater Antilles, Lesser Antilles, Tobago, and continental tropical America from Mexico to Peru and Brazil, reported in Africa.

4. BLECHNUM L.

Terrestrial or occasionally epiphytic, small to medium-sized ferns. Rhizomes ascending to erect, sometimes trunklike, or more or less elongate and scandent, usually densely clothed with scales, sometimes stoloniferous. Fronds mostly 1-pinnate or rarely simple, usually glabrous, monomorphic or dimorphic with the divisions of the fertile fronds strongly contracted; veins usually free except for the fertile veinlets. Sori elongate-linear, usually continuous, borne near or against the midvein or on an elongate transverse veinlet parallel to the midvein; indusium narrowly linear, continuous or nearly so, firm, opening toward the midvein; paraphyses absent; sporangia with annulus of 14–28 cells; spores ellipsoid, monolete, the surface nearly smooth to papillate, occasionally with minute spherical granules.

A worldwide genus of perhaps 180 species, the majority occurring in the Southern Hemisphere.

1. Blechnum occidentale L., Sp. Pl. 1077. 1753.

Fig. 2C.

Terrestrial fern 15–45(–75) cm tall. Rhizome erect or ascending, slender to stout, usually stoloniferous, densely clothed at the apex with brown, triangular-lanceolate to ovate scales mostly 2–4 mm long, these often with a blackish central stripe. Fronds ascending or spreading, mostly 15–75 cm long, the fertile ones tending to be slightly longer than the sterile though of similar form; stipes of fertile fronds often as long as the blades, strongly grooved along upper surface, bearing toward base few to numerous pale orange-brown, lanceolate scales, 3–5 mm long. Blades pinnate at base, pinnatisect distally, linear-oblong or lanceolate to narrowly triangular, widest at or near base, the apex acuminate or attenuate-subcaudate; pinnae linear-oblong or very narrowly triangular-oblong, scythe-shaped, 5–15 mm broad, the lowest pair usually short-stalked and deflexed, obtuse to acuminate at apex, subcordate and sometimes dilated at base; veins 1- or 2-forked, indistinct, terminating near the margins with enlarged tips. Indusium attached 0.2–1 mm from the midvein, sometimes decurved onto the rachis, the margin minutely denticulate-glandular.

DISTRIBUTION: Common on shaded banks or roadsides. Bordeaux Mountain (A2596). Also on St. Croix, St. Thomas, and Tortola; widespread from Florida to Argentina.

5. DORYOPTERIS J. Sm.

Small terrestrial ferns. Rhizomes decumbent to erect (rarely long-creeping and slender), scaly at the apex. Fronds monomorphic or dimorphic; stipes usually cylindric and lustrous black or purple-black. Blades usually pedately divided, rarely simple or 3-lobed; venation free or reticulate without free included veinlets. Sori marginal, usually continuous on a vascular commissure connecting the vein-tips; indusium abruptly differentiated from the recurved margin; paraphyses few or absent; sporangia with annulus of 14–22 cells; spores tetrahedral-globose, trilete, the surface nearly smooth to variously cristate.

A pantropical genus of about 25 species, most numerous in Brazil.

1. Doryopteris pedata (L.) Fée, Mém. Foug. 5: 133. 1852. *Pteris pedata* L., Sp. Pl. 1075. 1753. Fig. 3A–C.

Terrestrial fern 8–25(–50) cm tall. Rhizome short, decumbent to erect, at apex bearing small lanceolate-attenuate scales, these bicolorous with black median band and pale margins. Fronds moderately dimorphic, coriaceous, 8–50 cm long; stipes narrowly wing-angled or grooved adaxially, finely puberulous in the groove. Sterile blades with distinct hydathodes on upper surface and with crenate or crenulate margins; fertile blades 5–21 cm long, usually 2- to 3-pinnatifid, pentagonal in outline, with few to rather numerous triangular to lanceolate ultimate lobes or segments, the proximal primary basiscopic divisions often much larger than the second pair. Soral lines often broken by the sinuses.

DISTRIBUTION: Shaded rocky or earthy banks in humid areas. Coral Bay Quarter along Center Line Road (A2416), Susannaberg (A714). Also on St. Croix and St. Thomas; Greater and Lesser Antilles, Central and South America, and the Galapagos Islands.

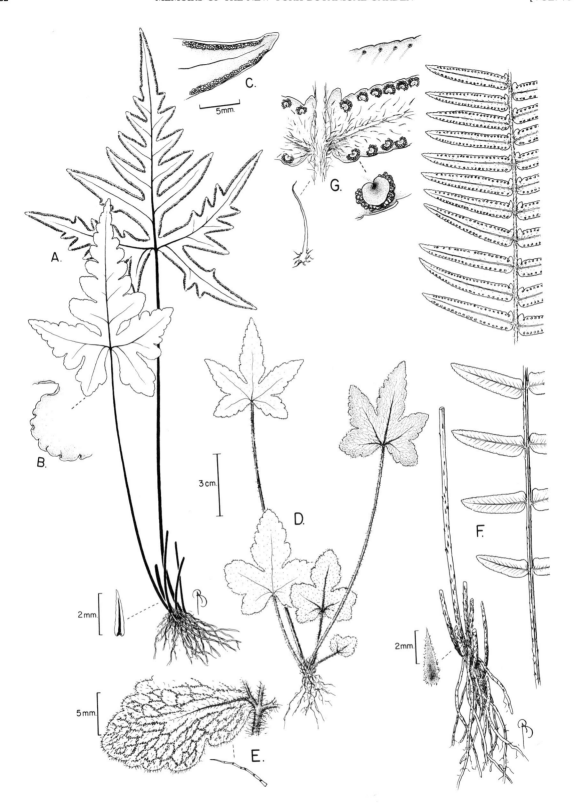

Fɪɢ. 3. A–C. *Doryopteris pedata*. **A.** Habit and detail of stipe scale. **B.** Detail of margin of sterile frond. **C.** Detail of fertile frond with marginal sori. **D, E.** *Hemionitis palmata*. **D.** Habit. **E.** Detail of abaxial side of fertile frond showing sori and trichome. **F, G.** *Nephrolepis multiflora*. **F.** Portion of rhizome, stipe bases, and detail of stipe scale, and portions of frond (above and below). **G.** Detail of frond, showing abaxial side with sori, marginal hydathodes on adaxial surface (above), and sorus and trichome (below).

6. HEMIONITIS L.

Small terrestrial, soft-herbaceous ferns. Rhizomes short-creeping to erect, bearing numerous lax, thin scales intermixed with pluricellular hairs. Fronds somewhat dimorphic, the sterile ones tending to be short-stalked and spreading, the fertile ones long-stalked and rigidly erect; stipes dark, lustrous, purple-brown or black. Blades palmately or pedately lobed or else pinnately compound; divisions broad and nearly or quite entire; veins free or anastomosing to form elongate-polygonal areoles lacking included free veinlets. Sporangia in many superficial lines, following the course of veins throughout the lower side of fertile blades; indusium and paraphyses absent; annulus of 14–20 cells; spores globose or tetrahedral-globose, trilete, the surface cristate, echinate, or tuberculate.

A genus of about 7 species from Mexico, Central America, West Indies, and one species from Southeast Asia.

REFERENCE: Mickel, J. T. 1974. A redefinition of the genus *Hemionitis*. Amer. Fern J. 64: 3–12.

1. Hemionitis palmata L., Sp. Pl. 1077. 1753. *Gymnogramma palmata* (L.) Link, Hort. Berol. **2**: 48. 1833. Fig. 3D, E.

Terrestrial fern 10–20(–35) cm tall. Rhizome short, suberect, 0.5–1 cm thick, loosely clothed with tawny, linear-attenuate scales. Fronds somewhat dimorphic; sterile fronds short-stipitate, spreading or nearly prostrate, with blades 3- or 5-lobed, the lobes short, round or acutish; fertile fronds rigidly erect, 15–35 cm long; stipes much longer than the blades, dark purple-brown, deciduously villous with lax or retrorse pluricellular hairs. Blades pentagonal in outline, 5–15 cm long and broad, 3-partite, the distal portion ovate to oblong-lanceolate and coarsely lobed below the acute apex; basal segments inequilateral, forked near the base (rarely tripartite), the upper division larger, both similar in outline to the apical division; veins usually dark brown on lower surface; tissue sparsely hirsute on upper surface, densely hirsute beneath.

DISTRIBUTION: Common on shaded earth banks or humus in secondary humid forests. Bordeaux Mountain (A2595), Cinnamon Bay along North Shore Road (A3518). Also on St. Croix, St. Thomas, and Tortola; Greater and Lesser Antilles, Trinidad, and continental tropical America from Mexico to Bolivia.

7. NEPHROLEPIS Schott

Medium-sized terrestrial, epiphytic or occasionally epilithic ferns. Rhizomes erect and short, sometimes decumbent, densely clothed with scales, and usually proliferating by means of elongate slender stolons, rarely producing underground tubers. Fronds clustered, crowded, not articulate to the rhizome, stipitate, the stipes usually short. Blades linear to linear-oblong, normally 1-pinnate, the apex often appearing to be of indeterminate growth or nearly so; pinnae numerous, articulate to the rachis and eventually deciduous; veins free, 1- to 4-branched, all but the fertile ones reaching nearly to the margins, with ends thickened (hydathodes) and often secreting a small, white dry exudate of calcium carbonate on the adaxial side. Sori usually terminal on the first distal vein branches, medial to submarginal, in a single row on either side of the midvein; indusium lunate to orbicular, attached at the sinus; paraphyses absent; spores ellipsoid, monolete, the surface irregularly tuberculate to rugose.

A pantropical genus of more than 20 species.

REFERENCE: Morton, C. V. 1958. Observation on cultivated ferns, V. The species and forms of *Nephrolepis*. Amer. Fern J. 48: 18–27.

Key to the Species of *Nephrolepis*

1. Costae glabrous on adaxial side or essentially so (rarely bearing a few deciduous, scattered fibrillose scales); rhizome and stipe clothed with spreading, narrow, yellowish brown, fiberlike scales; indusium round-cordate to kidney-shaped with open, U-shaped sinus .... 1. *N. exaltata*
1. Costae clothed on adaxial side with few to numerous very short (0.2–0.3 mm) brownish septate hairs (longer hairs present or absent); rhizome and base of stipe bearing scattered blackish, closely appressed scales with whitish minutely fimbriate margins; indusium round with narrow sinus .... 2. *N. multiflora*

1. Nephrolepis exaltata (L.) Schott, Gen. Fil., t. 3. 1834. *Polypodium exaltatum* L., Syst. Nat., ed. 10, **2**: 1326. 1759. *Aspidium exaltatum* (L.) Sw., J. Bot. (Schrader) **1800(2)**: 32. 1801. *Nephrodium exaltatum* (L.) R. Br., Prodr. 148. 1810.

Plants terrestrial or occasionally epiphytic. Rhizome short, suberect, concealed by the stout stipe bases, the apex clothed with a dense tuft of narrowly lanceolate-attenuate, glabrous, light orange-brown scales, ca. 8 × 0.8 mm, terminating in a long hairlike apex. Fronds closely clustered, suberect or spreading, 1–2.2 m long; stipes mostly 6–20 cm long, deciduously fibrillose-scaly, the scales spreading, linear-filiform, concolorous, pale orangebrown, basally attached. Blades linear, 50–100(–200) × 6–14 cm, slightly narrowed toward the base, the apex apparently of indeterminate growth; rachis light brown, deciduously fibrillosescaly, the scales glabrous; pinnae numerous, narrowly oblong or narrowly triangular-oblong and nearly scythe-shaped, 3–7 × 0.8–1.3 cm, acutish to subacuminate at apex, subcordate and obtusely auriculate at base, the auricle often overlapping the rachis, the margins bluntly serrulate to lightly crenate; tissue deciduously fibrillose to glabrate; veins 1- or 2-forked, rather close; indusium round-cordate to nearly kidney-shaped, the sinus usually open and U-shaped. Sori supramedial, rather close; indusium round-cordate to nearly kidney-shaped, the sinus usually open and U-shaped.

DISTRIBUTION: Uncommon on shaded banks, borders of thickets or in secondary forests. Upper slopes and summit of Bordeaux Mountain (P40433). Also on St. Croix, St. Thomas, and Tortola; cosmopolitan but perhaps naturally distributed in Florida, the Bahamas, Greater Antilles, and Mexico.

2. Nephrolepis multiflora (Roxb.) F.M. Jarrett *ex* C.V. Morton, Contr. U.S. Natl. Herb. **38**: 309. 1974. *Davallia multiflora* Roxb., Calcutta J. Nat. Hist. **4**: 515, t. 31. 1844. Fig. 3F, G.

Plants terrestrial or rarely epiphytic. Rhizome short, woody, erect, concealed by persistent stipe bases, clothed near apex with appressed, convex, lanceolate to ovate, dark orange-brown to black scales, mostly 2–3 × 0.5–1 mm, these peltately attached above the base, acuminate at apex, the narrow whitish margins finely fimbriate-ciliate; stolons numerous, wiry. Fronds erect, usually 30–120 cm long; stipes stout (to 4 mm diam.), 5–30 cm long with similar scales as the rhizome. Blades linear or lance-linear, to 1 m long or more and appearing to be of indeterminate growth, mostly 7–20 cm broad, somewhat narrowed toward base; rachis yellowish or grayish brown, rather densely clothed with pale brown, tortuous, filiform scales, these loosely long-ciliate especially around the peltate, dark-spotted base. Pinnae linear-oblong, usually 3.5–10 × 0.5–1.2 cm, acute to long-acuminate at apex, slightly inequilateral at base, the margins crenate to sharply and deeply serrate; tissue and veins beneath bearing numerous pale brown, loosely appressed, attenuate scales, these fimbriate-ciliate at the peltately attached base; veins usually 2-forked. Sori submarginal; indusium round, glabrous, with a narrow sinus.

DISTRIBUTION: A pioneer plant, occasional in open disturbed areas such as clearings, open banks and roadsides. Susannaberg (A2085). Also on Anegada, Guana Island, St. Croix, St. Thomas, Tortola, and Virgin Gorda; native to India and tropical Asia but widely naturalized in the neotropics.

8. ODONTOSORIA (C. Presl) Fée

Terrestrial ferns with scandent and rather rampant fronds of indefinite growth. Rhizomes slender, wide-creeping, densely clothed with narrow, almost hairlike scales. Fronds 2- to 4-pinnate; pinnae opposite; ultimate pinnules small and variously lobed or cleft; stipe and rachis subwoody, slender and flexuous, and usually spiny. Sori terminal on single veins, immersed in tissue; indusium joined to the opposed frond lobule to form an urn-shaped involucre, opening outward at margin; spores usually spherical, trilete, the surface smooth or slightly granulate.

A genus of about a dozen species with neotropical distribution.

1. **Odontosoria aculeata** (L.) J. Sm., Cult. Ferns 67. 1857. *Adiantum aculeatum* L., Sp. Pl. 1096. 1753, in greater part. *Davallia aculeata* (L.) Sm., Mém. Acad. Roy. Sci. Turin **5**: 415. 1793, in part. *Lindsayopsis aculeata* (L.) Kuhn, Festschr. 50 Jähr. Jubil. K. Realsch. Berlin 27. 1882. Fig. 4A–C.

Davallia dumosa Sw., Syn. Fil. 135, 353. 1806.

Terrestrial fern. Rhizome short-creeping, woody, 3–4.5 mm diam., densely clothed toward apex with lustrous reddish brown, curved, hairlike scales 3.5 × 0.3 mm. Fronds scandent, to 3 m long; stipes much shorter than the blades, light brownish or reddish brown, with many scattered, straight, narrow spines, 2–4 mm long. Pinnae numerous, in adjacent pairs, rigidly divaricate, 20–45 cm long, mostly ovate-triangular, abruptly reduced at base, evenly narrowed from above the base to the acute apex, the midveins flexuous; secondary midveins, and veinlets with scattered sharp, spreading, acicular spines; secondary pinnae numerous, close, spreading, alternate or subopposite, very unequal, those just above the basal pair the largest, triangular-oblong to ovate-triangular, 5–15 × 2–12 cm, the tertiary pinnae spreading or retrorse; ultimate pinnules irregularly rhombic, of 2–4 nearly cuneate segments, these cleft at the middle, each lobe emarginate and 2-soriate; tissue herbaceous, with evident veins.

DISTRIBUTION: Occasional in shaded, moist, disturbed areas. Along road to Bordeaux (A2095). Also on St. Thomas and Tortola; Puerto Rico, Cuba, and Hispaniola.

9. PITYROGRAMMA Link

Small to medium-sized terrestrial or epilithic ferns. Rhizomes short-creeping to erect, clothed with narrow, more or less attenuate, nonclathrate scales. Fronds fasciculate, all similar or somewhat dimorphic, not articulate to the rhizome; stipes usually dark, hard, and lustrous; blades 1- to 3-pinnate, narrowly oblong or lanceolate to ovate or triangular, usually covered on lower surface with white or yellow to nearly orange waxlike powder, otherwise naked or rarely with scales; veins free, pinnately forked. Sori indefinite; indusium absent; sporangia produced along the course of the veins and often at maturity forming confluent lines or masses; annulus of 20–24 cells; spores tetrahedral-globose, trilete, the surface coarsely tuberculate, reticulate, or ridged, rarely smooth and minutely granulate.

A genus of less than 20 species, most of them occurring in the neotropics.

Key to the Species of *Pityrogramma*

1. Blades lanceolate to oblong-triangular (the lowest pinnae usually not the longest); rhizome-scales long-filamentous at apex .... 1. *P. calomelanos*
1. Blades triangular (the lowest pinnae the longest); rhizome-scales short-acuminate . 2. *P. chrysophylla* var. *gabrielae*

1. **Pityrogramma calomelanos** (L.) Link, Handbuch **3**: 20. 1833. *Acrostichum calomelanos* L., Sp. Pl. 1072. 1753. *Gymnogramma calomelanos* (L.) Kaulf., Enum. Filic. 76. 1824. *Ceropteris calomelanos* (L.) Link, Fil. Spec. 141. 1841. *Neurogramme calomelanos* (L.) Diels *in* Engl. & Prantl, Nat. Pflanzenfam. **1**(4): 264. 1899. Fig. 4D, E.

Terrestrial fern 0.3–0.7(–1) m tall. Rhizome ascending to erect, clothed at apex with dark golden-brown, narrowly lanceolate scales mostly 3–3.5 mm long, these long-filamentous at apex. Fronds closely tufted, to 1 m long; stipes dark lustrous reddish brown, about as long as the blades or shorter. Blades lanceolate to oblong-triangular, 20–95 × 10–30 cm, 2(–3)-pinnate to 2-pinnate-pinnatifid, the lower surface covered with a whitish pow-

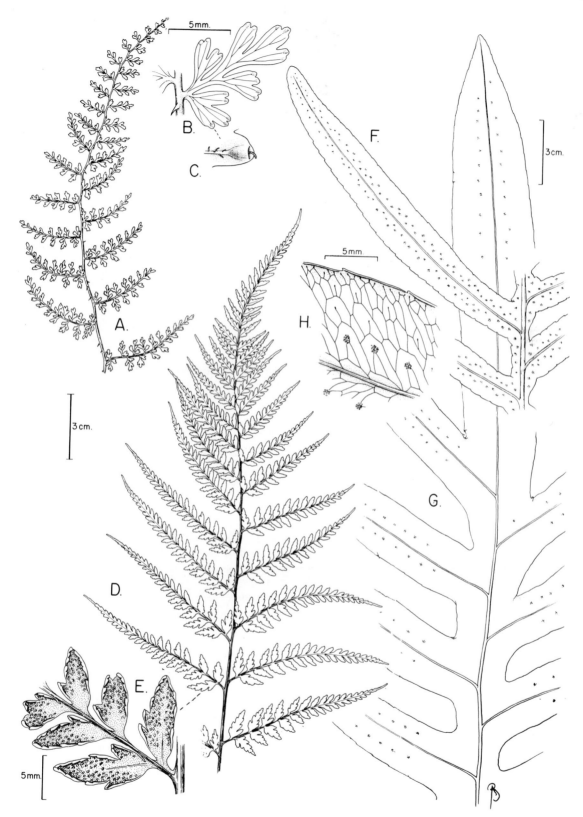

Fig. 4. A–C. *Odontosoria aculeata*. A. Portion of frond. B. Secondary pinna. C. Detail of pinnule showing vein with terminal sorus. D, E. *Pityrogramma calomelanos*. D. Portion of frond. E. Detail of fertile secondary pinnae. F–H. *Polypodium aureum*. F. Frond showing sori along costa and rachis. G. Frond showing sori along costa only. H. Detail of pinna showing sori and venation.

der; segments lanceolate to elliptic, usually at least sharp-pointed and often more or less sharply serrate or deeply incised; tissue firmly herbaceous. Sporangia numerous, often nearly concealing the undersurface of the pinnules; spores prominently ridged.

DISTRIBUTION: Occasional in rocky banks and in crevices of old buildings and ruins. Along road to Bordeaux (A3132), upper slopes and summit of Bordeaux Mountain (P40437), southeast of Peter Peak (P40430). Also on St. Croix, St. Thomas, and Tortola; tropical and subtropical America and Africa, naturalized elsewhere in warm regions.

2. Pityrogramma chrysophylla (Sw.) Link var. gabrielae

Domin, Rozpr. Král. České Spolecn. Nauk, Tř. Mat.-Přír. **2:** 151. 1929.

Terrestrial fern 15–30 cm tall. Rhizome ascending to erect, clothed at the apex with yellowish brown, narrowly lanceolate, subclathrate scales to 3 mm long. Fronds loosely tufted, 18–30 cm long, with lustrous, red-brown stipes shorter than to longer than the blades. Blades triangular-acuminate, or rarely oblong-acuminate, 9–20 × 6–12 cm, the lower surface whitish powdered; ultimate divisions lance-oblong to obliquely triangular, usually blunt, entire or little incised to more or less deeply lobed.

DISTRIBUTION: Occasional on rocky slopes and crevices of old masonry walls at low elevations near the sea. Annaberg Ruins (P40452). Also on St. Croix, St. Thomas, and Tortola; Puerto Rico and the Lesser Antilles.

10. POLYPODIUM L.

Small to large, epiphytic or epilithic ferns of varied habit. Rhizomes short- to long-creeping, often slender, more or less clothed with scales or sometimes nearly naked. Fronds articulate to the rhizome, monomorphic or somewhat dimorphic, mostly stipitate, glabrous or hairy, and with or without scales; blades simple to pinnatisect, sometimes pectinate, usually with entire margins; veins branched, free or variously reticulate, the areoles with or without included free veinlets. Sori round or oval (rarely somewhat linear), mostly terminal on free veinlets, not marginal, usually in one or more regular rows; indusium lacking; paraphyses present or absent; sporangia with annulus of 12–18 cells; spores ellipsoid, monolete, the surface usually verrucose to rugose or tuberculate, or in a few cases smooth with a minute granular deposit.

A genus of about 225 species with cosmopolitan distribution. The genus is broadly circumscribed here; other recent floras have recognized the segregate genera *Pleopeltis, Microgramma, Campyloneurum* for some of our taxa.

Key to the Species of *Polypodium*

1. Fronds pinnatisect .... 2. *P. aureum*
1. Fronds simple.
   2. Sori uniseriate on either side of the midrib (rachis); rhizomes slender and elongate; fronds scattered, not fasciculate.
      3. Sori round, bearing inconspicuous hairlike paraphyses, these simple or branched; fronds often somewhat dimorphic.
         4. Rhizome filiform, 0.5–1 mm thick; frond tissue of thin texture .... 3. *P. heterophyllum*
         4. Rhizome cordlike, 2–2.5 mm thick; frond tissue of firm or leathery texture .... 5. *P. lycopodioides*
      3. Sori oblong to linear, bearing scale-like peltate paraphyses (these soon deciduous); fronds not dimorphic .... 1. *P. astrolepis*
   2. Sori multiseriate on either side of the midrib (rachis); rhizomes short and subwoody; fronds fasciculate.
      5. Stipe virtually absent or usually <5 cm long, <1/10 the length of the blade; blades gradually attenuate at base, the margins plane, the tissue light green; primary vein-areoles usually divided into 2 secondary areoles by a single excurrent veinlet .... 6. *P. phyllitidis*
      5. Stipe usually 5–18 cm long, to 1/3 the length of the blade; blades cuneate or acuminate at base, the margins wavy, the tissue dark green; primary vein-areoles more or less irregularly divided into 3 unequal secondary areoles by 2 excurrent veinlets .... 4. *P. latum*

1. **Polypodium astrolepis** Liebm., Kongel. Danske Vidensk. Selsk. Naturvidensk. Math. Afh., ser. 2, **1:** 185. 1849. *Pleopeltis astrolepis* (Liebm.) E. Fourn. in Nyl. & Besch., Mexic. Pl. **1:** 87. 1872.

Grammitis lanceolata Schkuhr, Krypt. Gew. **1:** 9, t. 7. 1804, non *Polypodium lanceolatum* L., 1753. *Grammitis elongata* Sw., Syn. Fil. 22, 213. 1806, non *Polypodium elongatum* Aiton, 1789 (based on *Grammitis lanceolata* Schkuhr). *Polypodium elongatum* (Sw.) Mett., Abh. Senckenberg. Naturf. Ges. **2:** 88, t. 2, figs. 8, 9. 1857, non Aiton, 1789, nec Schrad., 1818. *Gymnogramma elongata* (Sw.) Hook., Sp. Fil. **5:** 157. 1864. *Polypodium lanceolatum* var. *elongatum* (Sw.) Krug, Bot. Jahrb. Syst.

24: 131. 1897. *Taenitis swartzii* Jenman, J. Bot. **17:** 263. 1879 (based on *Grammitis elongata* Sw.).

Grammitis revoluta Spreng. *ex* Willd., Sp. Pl. **5:** 139. 1810. *Pleopeltis revoluta* (Spreng. *ex* Willd.) A.R. Sm., Proc. Calif. Acad. Sci. **40:** 230. 1975.

Epiphytic fern. Rhizome slender, elongate, wide-creeping, 0.5–1.8 mm thick, bearing numerous somewhat matted roots, and covered with appressed, round, peltate, blackish scales, these nearly concealed by stiff, brown clustered hairs ca. 0.5 mm long. Fronds close to distant; stipes nearly lacking to 2 cm long, slightly flattened, narrowly green-marginate. Blades linear, lance-linear, or oblanceolate, 6–24 × 0.7–2 cm, simple, narrowed or attenuate at both ends, the apex obtuse to acute; rachis minutely scaly, greenish along upper surface, purple-black beneath, the scales widely scattered, lacerate-stellate with dark centers; veins reticulate; areoles sometimes containing free veinlets; tissue opaque and hygroscopic. Sori uniseriate, narrowly oblong to linear, 2–10 × 1–2 mm, parallel to the rachis, approximately medial, occupying the distal third or half of the blade; paraphyses few, scale-like, soon deciduous.

DISTRIBUTION: Occasional in moist forest. Based on a visual record by G. R. Proctor. Also in the Greater and Lesser Antilles, Tobago, Trinidad, and continental tropical America from Mexico to Bolivia and Brazil.

2. Polypodium aureum L., Sp. Pl. 1087. 1753. *Phlebodium aureum* (L.) J. Sm., J. Bot. (Hooker) **4**: 59. 1841. *Chrysopteris aurea* (L.) Link, Fil. Spec. 121. 1841. Fig. 4F–H.

Epiphytic fern. Rhizome creeping, 9–15 mm thick, densely clothed with numerous, pale orange or tawny, denticulate-ciliate scales, mostly 10–15 mm long, these linear-attenuate or filiform from a slightly enlarged and darker base, attached peltately by a minute brown stalk. Fronds arching or pendent, 0.5–1.7 m long, seasonally deciduous; stipes 20–70 cm long, lustrous dark brown to light reddish brown, glabrous. Blades ovate-oblong or broadly oblong, 35–100 × 22–45 cm, deeply and coarsely pinnatisect, often with a much larger terminal segment; segments 6–22 pairs, ligulate or lance-ligulate, 2–3.8 cm broad, obtuse to acuminate at apex, joined at base by a rachis-wing, 2–9 mm wide; margins subentire to minutely and distantly crenulate; veins mostly reticulate, forming a row of oblong to narrowly ob-triangular areoles without included veinlets, then 2 uneven and irregular series of larger areoles with included veinlets, mixed with smaller areoles lacking included veinlets; tissue usually green, rarely somewhat glaucous. Sori uniseriate, round or oval, located at the junction of 2 intra-areolate veinlets or sometimes at tip of a single free veinlet; very rarely a partial third row of sori present.

DISTRIBUTION: Common in moist forests. Bordeaux Mountain (A5110), Trail to Sieben (A2070). Also on St. Croix, St. Thomas, Tortola, and Virgin Gorda; Florida, Bahamas, Greater and Lesser Antilles, Trinidad, and continental tropical America.

3. Polypodium heterophyllum L., Sp. Pl. 1083. 1753. *Phymatodes heterophylla* (L.) Small, Ferns Florida 81. 1932. *Microsorium heterophyllum* (L.) Hawkes, Amer. Fern J. **41**: 52. 1951. *Craspedaria heterophylla* (L.) Diddell, Amer. Fern J. **43**: 114. 1953. *Microgramma heterophylla* (L.) Wherry, South. Fern Guide 346. 1964.

Polypodium serpens Sw., Prodr. 131. 1788, non G. Forst., 1786. *Anapeltis serpens* (Sw.) J. Sm., Cat. Cult. Ferns 5. 1857. *Polypodium swartzii* Baker *in* Hook. & Baker, Syn. Fil. 357. 1868 (based on *Polypodium serpens* Sw., 1788, non G. Forst., 1786).
Polypodium exiguum Heward, Mag. Nat. Hist., n.s., **2**: 458. 1838, non Griseb., 1864, nec Fée, 1869. *Phymatodes exiguum* (Heward) Underw., Torreya **3**: 18. 1903.

Epiphytic or epilithic fern. Rhizome filiform, wide-creeping, 0.5 mm wide, branched, densely clothed with tawny to reddish brown, denticulate, linear-attenuate scales, these 5–7 mm long, peltately attached far above the narrowed base. Fronds distant, extremely variable in form, glabrous; stipes 0.1–1.5 cm long,

marginate, naked. Blades dimorphic, simple, membranous; sterile blades oval or elliptic to narrowly elliptic, 1–3 cm long, or lanceolate to linear, 3–12 × 0.8–1.5 cm, narrowed at both ends; margins subentire, sinuate, or less often irregularly crenate to deeply incised; fertile blades mostly narrower than the sterile ones and more uniform, sometimes to 16 cm long; veins partly reticulate, forming 2 rows of areoles on either side of the rachis, the outer ones each with a single, included, free veinlet, the outermost veins partly reticulate, with many free, slightly enlarged veinlet-tips not reaching the margins. Sori uniseriate, round, medial, each located at the tip of a free areolar veinlet; paraphyses brown, few, hairlike, more than 1 cell wide at base.

DISTRIBUTION: Occasional, creeping on rocks or epiphytic in moist forests or in moist pockets in dry forests. Bordeaux Mountain (A1907), White Cliffs (A2037). Also on St. Croix and Virgin Gorda; Florida, Bahamas, Greater Antilles, Cayman Islands, and the northern Lesser Antilles.

4. Polypodium latum (T. Moore) T. Moore *ex* Sodiro, Crypt. Vasc. Quit. 371. 1894. *Campyloneurum latum* T. Moore, Index Fil. 225. 1861. *Polypodium phyllitidis* var. *β latum* (T. Moore) Hook., Sp. Fil. **5**: 38. 1863. *Polypodium phyllitidis* f. *latum* (T. Moore) Proctor, Bull. Inst. Jamaica, Sci. Ser. **5**: 49. 1953.

Epiphytic or epilithic fern. Rhizome short-creeping, to 10 mm thick, strongly nodose and often branched, enveloped in a dense mass of brown-tomentose rootlets; scales few, appressed, light brown, finely subclathrate, roundish-ovate, 3–5 mm long, the apex obtuse. Fronds closely fasciculate, 25–130 cm long; stipes stout, 5–18 cm long, subquadrangular. Blades simple, broadly ligulate, elliptic, or oblanceolate, to 115 cm long, mostly 5–13 cm broad near or above the middle, the apex abruptly acute, the base cuneate to acuminate, the margins wavy, narrowly cartilaginous; primary veins prominent, oblique-spreading, connected by numerous irregularly arching cross-veins forming areoles with several included free veinlets, these variously combined to form secondary areoles in 3 or 4 rows, an intermediate costular vein sometimes developed. Sori 2- to 4-seriate between the primary veins, located on free areolar veinlets below the tips.

DISTRIBUTION: Occasional on rocks and shaded tree trunks in most forests. Bordeaux Mountain (A2612). Also on St. Croix and St. Thomas; Florida, Greater and Lesser Antilles, Trinidad, and continental tropical America from Mexico to Colombia and Venezuela.

5. Polypodium lycopodioides L., Sp. Pl. 1082. 1753. *Pleopeltis lycopodioides* (L.) C. Presl, Tent. Pterid. 193. 1836. *Phlebodium lycopodioides* (L.) J. Sm. *ex* Hook., Companion Bot. Mag. **72**: 12. 1846. *Drynaria lycopodioides* (L.) Fée, Mém. Foug. **5**: 270. 1852. *Anapeltis lycopodioides* (L.) J. Sm., Cat. Cult. Ferns 6. 1857. *Niphobolus lycopodioides* (L.) Keyserl., Polyp. Herb. Bunge. 38. 1873. *Phymatodes lycopodioides* (L.) Millsp., Publ. Field Columbian Mus., Bot. Ser. **3**: 12. 1903. *Microgramma lycopodioides* (L.) Copel., Gen. Fil. 185. 1947.

Epiphytic fern. Rhizome slender, cordlike, wide-creeping, densely clothed with lance- or linear-attenuate, hair-pointed scales, these 6–10 mm long, reddish brown at first, becoming bicolorous with age, peltately attached and with loosely ciliate margins. Fronds simple, distant, somewhat dimorphic; stipes very short or lacking. Sterile blades linear-lanceolate to ligulate, 5–20 × 1–

2.5 cm, narrowed at both ends, the margins entire or slightly sinuate, narrowly cartilaginous and flat or minutely revolute; veins reticulate, forming a row of polygonal areoles mostly without included free veinlets along either side of rachis, then a series of much larger areoles of irregular shape, these mostly with to 6 included free veinlets, then a third uneven series of smaller polygonal areoles; tissue firm to leathery and semi-opaque; fertile blades often narrower than the sterile ones, elliptic to linear, 0.5–1 cm broad. Sori uniseriate, round, medial or supramedial, terminal on several inwardly pointing tips of free veinlets in the largest series of areoles; paraphyses few, concealed, hairlike.

DISTRIBUTION: Epiphytic on trees in moist forests. Upper slopes and summit of Bordeaux Mountain (P40434). Also on Tortola; pantropical.

6. Polypodium phyllitidis L., Sp. Pl. 1083. 1753. *Campyloneurum phyllitidis* (L.) C. Presl, Tent. Pterid. 190, t. 7, figs. 18–20. 1836. *Cyrtophlebium phyllitidis* (L.) J. Sm., J. Bot. (Hooker) **4**: 58. 1841.

Fig. 5A, B.

Epiphytic or less often epilithic fern. Rhizome short-creeping, 5–8 mm thick, enveloped in a mass of brown-tomentose rootlets, toward apex bearing appressed-imbricate, gray-brown, clathrate, ovate-oblong or broadly ovate scales, these 2–6 mm long, mostly acute to acuminate at apex. Fronds densely clustered; stipe nearly lacking or to 5 cm long, stout and subquadrangular, with green margins passing gradually upward into the long-decurrent blade. Blades simple, ligulate, narrowly oblong-linear, or lance-linear, chartaceous, 30–120 × 4–8 cm, the apex sharply acute or rarely obtuse, the base gradually attenuate, the margins entire, plane and narrowly cartilaginous; primary veins oblique, prominent on lower surface, connected by arching cross-veins forming areoles with usually 3 included free veinlets, the middle one usually prolonged and joined to the next higher cross-vein, thus forming 2 secondary areoles; the tissue light green. Sori 2-seriate between the primary veins, located on free areolar veinlets below the tips.

DISTRIBUTION: Occasional epiphyte in moist forests. Coral Bay (A2113), Susannaberg (A694). Also on St. Croix, St. Thomas, and Tortola; Florida, Bahamas, Greater Antilles, Cayman Islands, Lesser Antilles, Tobago, Trinidad, and continental tropical America from Mexico to Uruguay.

11. PTERIS L.

Small to large, coarse terrestrial ferns. Rhizomes creeping to erect, often stout and woody, clothed with scales at the apex. Fronds usually clustered, erect or arching, monomorphic or sometimes dimorphic, often long-stipitate; stipes narrowly grooved adaxially; blades 1- to 4-pinnate, elongate or pentagonal in outline, sometimes each of the basal pinnae as large as rest of the blade; midveins of penultimate divisions deeply but narrowly grooved adaxially, the parallel ridges often bearing minute spinelike awns, one each at the base of the ultimate midveinlets; venation free or reticulate; areoles without free included veinlets. Sori linear, marginal, the sporangia borne in a continuous line on a delicate transverse commissure connecting the ultimate vein-tips; paraphyses usually present and often numerous; indusium linear, continuous, formed by the modified reflexed margin, opening inwardly; sporangia with annulus of 16–34 cells; spores tetrahedral or globose, trilete, usually with a prominent equatorial flange, the surfaces tuberculate or reticulate.

A genus of perhaps 280 species, most of which are tropical.

Key to the Species of *Pteris*

1. Plant 15–30 cm tall; blades 1-pinnate, none of the pinnae lobed or forked; veins all free .... 2. *P. vittata*
1. Plant 1–1.5 m tall; blades 1-pinnate-pinnatisect, the basal pinnae forked; basal veins joined in a narrow
   costal arc, the other veins free .... 1. *P. biaurita*

1. Pteris biaurita L., Sp. Pl. 1076. 1753. *Campteria biaurita* (L.) Hook., Gen. Fil. t. 65-A. 1830. *Litobrochia biaurita* (L.) J. Sm., Cat. Cult. Ferns 37. 1857.

Fig. 5C–E.

Pteris biaurita var. *subpinnatifida* Jenman, Bull. Bot. Dept. **41**: 7. 1893.

Terrestrial fern, 1–1.5 m tall. Rhizome erect, woody, closely invested with stipe bases and densely clothed at apex with small triangular-ovate scales, these 1.5–3 mm long, bicolorous with glossy black, triangular-attenuate midrib and broad, pale, delicately clathrate, lacerate-ciliate margins. Fronds densely clustered, suberect, 0.6–1.5 m long; stipes stout, to 5 mm diam., straw-colored, grooved adaxially, naked except for a few bicolorous scales at base. Blades oblong or triangular-oblong, to 40 cm broad, 1-pinnate-pinnatisect, the basal pinnae forked; rachis glabrate and naked except for minute glandular papillae in the upper surface groove at axils of pinnae; pinnae 5–15 pairs, opposite, sessile or nearly so, all but the basal 2-partite ones, narrowly oblong-lanceolate to lanceolate-attenuate, broader on the basiscopic side; segments mostly 15–22 pairs, narrowly oblong to linear, 1–3 × 0.3–0.5 cm, joined by a costal wing 2–3 mm broad on either side, the fertile ones separated by round sinuses; veins prominulous, the

basal ones joined by a narrow costal arc with several excurrent simple branches, the others free, usually once-forked; tissue thin-herbaceous, light green, somewhat translucent. Indusium pale, 0.3–0.5 mm wide, with entire margin.

DISTRIBUTION: Occasional on ground of moist forest. Cinnamon Bay Trail (A4230). Also on St. Croix and St. Thomas; pantropical.

2. Pteris vittata L., Sp. Pl. 1074. 1753.

Terrestrial or epilithic fern, 15–30 cm tall. Rhizome decumbent to erect, densely clothed with pale greenish to pale brownish, linear, hair-pointed scales to 5 mm long. Fronds prostrate to ascending or arching, mostly 20–100 cm long; stipes pale brown, grooved along upper surface, much shorter than the blades and densely clothed with pale, hairlike scales. Blades usually 6–25 cm broad or more, oblanceolate to narrowly obovate, attenuate at base, at apex terminating abruptly in a long, linear pinna, usually larger than the lateral ones; lateral pinnae narrowly lance-linear, the bases often slightly dilated, the margins (without sori) finely and sharply serrulate; tissue thin-herbaceous, bright green, glabrous, the veins free, raised, mostly 1-forked near base. Indusium firm, greenish brown, entire or slightly wavy.

DISTRIBUTION: Occasional in crevices of old walls. Annaberg Ruins (A2924). Also on Guana Island, St. Croix, St. Thomas, and Tortola; native to the Old World tropics but naturalized in many parts of the Caribbean.

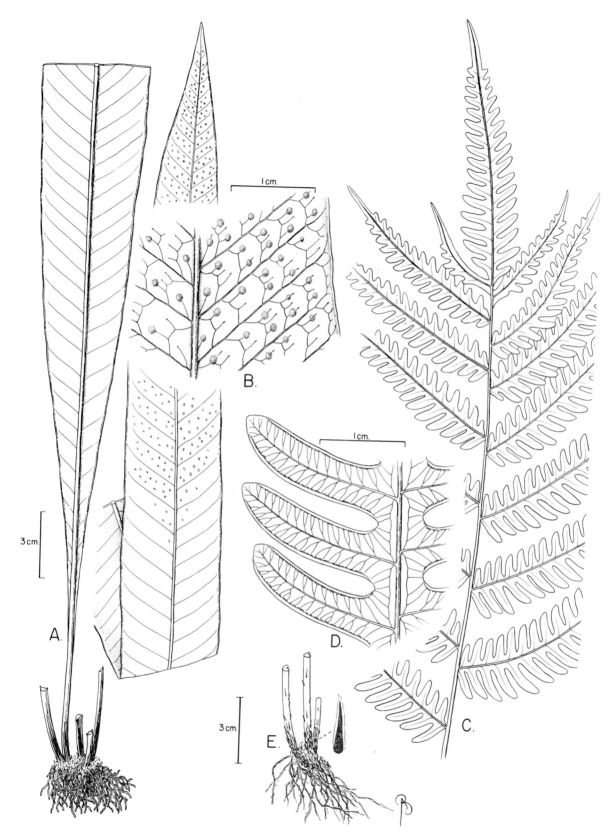

FIG. 5. A, B. *Polypodium phyllitidis*. A. Habit. B. Detail of distal portion of fertile frond showing sori. C–E. *Pteris biaurita*. C. Portion of frond. D. Detail of pinna with marginal sori. E. Rhizome and stipe bases and detail of scale.

12. THELYPTERIS Schmidel

Small to moderately large terrestrial or epilithic ferns. Rhizomes slender and wide-creeping to thick and erect, usually clothed at the apex with glabrous, hairy, or glandular scales. Fronds scattered to fasciculate; stipes not articulate; blades usually 1-pinnate or 1-pinnate-pinnatifid, not or slightly dimorphic, glabrous or commonly pubescent, the hairs simple, forked, or stellate, the veins free, more or less connivent at sinuses, or variously reticulate. Sori roundish or elliptic (rarely slightly elongate or diffuse), on lower surface of veins; indusium wanting or round to kidney-shaped, attached at the sinus; paraphyses usually absent; spores ellipsoidal, monolete, the perispore variously reticulate, crested, or winged, rarely papillate or spiny.

Treated in a broad sense, the genus contains nearly 1000 species distributed over much of the world.

Key to the Species of *Thelypteris*

1. Sori indusiate (at least when young); all hairs simple.
   2. Basal veins of some or all adjacent segments united below the sinus, producing an excurrent veinlet leading toward the sinus.
      3. Rhizome creeping; underside of costae bearing hairs of uniform length, to 0.2 mm long .... 1. *T. dentata*
      3. Rhizome erect or ascending; underside of costae bearing hairs of variable length, from 0.3 to 5 mm long .... 2. *T. hispidula* var. *inconstans*
   2. Basal veins of adjacent segments all free or connivent at the sinus; rhizome creeping, clothed at apex with narrow ciliate scales .... 3. *T. kunthii*
1. Sori lacking indusia; hairs very minute, forked or stellate, present at least on rhizome scales and toward base of stipe.
   4. Pinnae 6–18 pairs, 1–3 cm wide, the margins deeply lobed or pinnatifid; only lowermost pairs of adjacent veins joined.. 5. *T. tetragona*
   4. Pinnae 2–5(–6) pairs, 3–6 cm wide, the margins subentire to coarsely crenate; lower 3–5 pairs of adjacent veins joined, each pair giving rise to a free excurrent veinlet..................... 4. *T. poiteana*

1. Thelypteris dentata (Forssk.) E.P. St. John, Amer. Fern J. **26**: 44. 1936. *Polypodium dentatum* Forssk., Fl. Aegypt.-Arab. 185. 1775. *Dryopteris dentata* (Forssk.) C. Chr., Kongel. Danske Vidensk. Selsk. Naturvidensk. Math. Afh., ser. 5, **6**: 24. 1920. *Nephrodium dentatum* (Forssk.) G. Kumm., Magyar Bot. Lapok **32**: 60. 1933. *Cyclosorus dentatus* (Forssk.) Ching, Bull. Fan Mem. Inst. Biol. **8**: 206. 1938.   Fig. 6A–C.

Polypodium molle Jacq., Collectanea **3**: 188. 1789, non Schreb., 1771, nec All., 1785.
Aspidium molle Sw., J. Bot. (Schrader) **1800(2)**: 34. 1801. *Nephrodium molle* (Sw.) R. Br., Prodr. 149. 1810. *Polystichum molle* (Sw.) Gaudich., Voy. Uranie 326. 1828. *Hemesthenum molle* (Sw.) Gand. Fl. Eur. **27**: 178. 1891. *Dryopteris mollis* (Sw.) Hieron., Hedwigia **46**: 348. 1907.

Terrestrial fern 0.5–1.2 m tall. Rhizome short-creeping, ca. 6 mm thick, closely covered with persistent stipe bases, at apex bearing a tuft of dark brown, distantly ciliate, linear-attenuate scales, 4–10 mm long. Fronds slightly dimorphic, the fertile ones a little taller and with narrower pinnae than the sterile ones, to 1.2 m long; stipes 15–45 cm long, minutely puberulous throughout, and toward base bearing dark brown, linear-lanceolate scales, these ciliate and often also with surface hairs. Blades 40–90 × 14–34 cm, tapering evenly to the pinnatifid apex, at base the proximal 2–6 pairs of pinnae reduced and somewhat reflexed; rachis densely pubescent with curved hairs but without scales; largest pinnae 7–19 × 1.1–2.7 cm, attenuate at apex, sessile, pinnatifid; segments oblong or oblong-subfalcate, 2–4 mm broad, blunt to acute at apex, entire; veins 6–10 pairs, all simple; mid-

veins, midveinlets, and veins curved-pubescent on upper side, finely puberulous beneath, with hairs to 0.2 mm long; tissue light green, soft-herbaceous, not glandular. Sori nearly medial; indusium round to kidney-shaped, pubescent, nonglandular; sporangia glabrous but their stalks often bearing orange, stipitate glands.

DISTRIBUTION: Occasional in moist, disturbed habitats. Visually recorded by G. R. Proctor. Also on St. Croix, St. Thomas, and Tortola; apparently indigenous to the Old World but now pantropical.

2. Thelypteris hispidula (Decne.) C.F. Reed var. **inconstans** (C. Chr.) Proctor, Amer. Fern J. **70**: 89. 1980. *Dryopteris dentata* var. *inconstans* C. Chr., Kongl. Svenska Vetenskapsakad. Handl., ser. 3, **16(2)**: 27. 1936. *Thelypteris quandrangularis* var. *inconstans* (C. Chr.) A.R. Sm., Univ. Calif. Publ. Bot. **59**: 66. 1971.

Terrestrial fern 0.2–0.6(–1.2) m tall. Rhizome short, erect, 3–5 mm thick, bearing at apex a tuft of linear-lanceolate, lustrous dark brown scales 3–6(–8) mm long and <0.7 mm wide, minutely ciliate and often pubescent. Fronds spreading, not dimorphic; stipes mostly 10–40 cm long, straw-colored, sometimes darkened at base, pubescent and with a few scales near base. Blades oblong-elliptic or oblanceolate, 8–55 × 7–32 cm, tapering evenly to the short, pinnatifid apex; costae with hairs of different length (0.3–5 mm long); largest pinnae 3.5–16 × 0.8–2 cm, sessile, pinnatifid; segments oblong, slightly oblique or sometimes nearly scythe-shaped, mostly ca. 3 mm wide, entire or minutely crenulate toward apex; veins usually 5–9 pairs, simple except in enlarged basal segments. Sori nearly medial; indusium round to kidney-shaped, densely pubescent; sporangia glabrous but their stalks often bearing minute stipitate glands.

DISTRIBUTION: Occasional on shaded banks. Rosenberg (B299, B536). Also on St. Thomas and Tortola; Greater and Lesser Antilles.

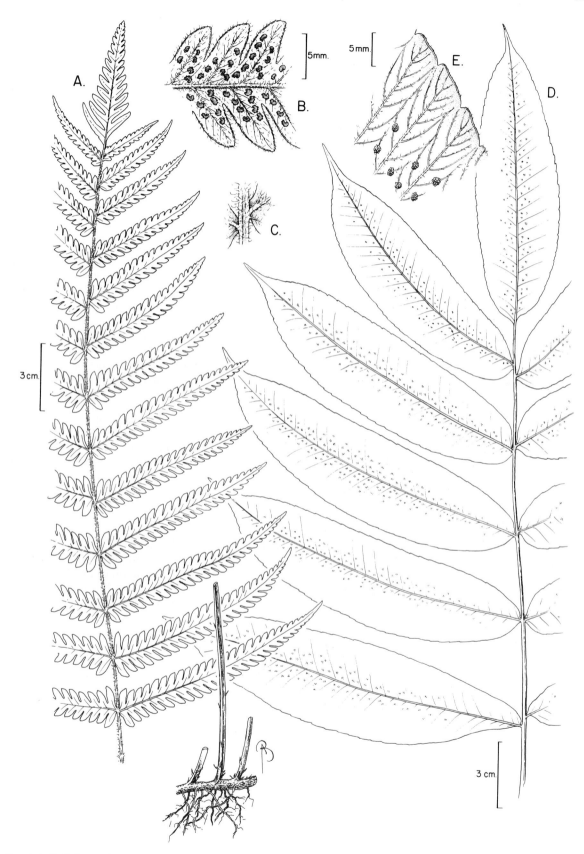

Fig. 6. A–C. *Thelypteris dentata*. A. Portion of frond and portion of rhizome and stipe bases. B. Detail of pinna showing sori. C. Detail of costa showing attachment of pinnae. D, E. *Thelypteris poiteana*. D. Portion of fertile frond. E. Detail of pinna showing sori, margin, and venation.

3. Thelypteris kunthii (Desv.) Morton, Contr. U.S. Natl. Herb. **38**: 53. 1967. *Nephrodium kunthii* Desv., Mém. Soc. Linn. Paris **6**: 258. 1827.

Nephrodium patens Jenman, Bull. Bot. Dept., n.s., **3**: 165. 1896; Ferns Brit. W. Ind. 240. 1908, non Desv. 1827, nec *Polypodium patens* Sw., 1788.
Dryopteris normalis C. Chr., Ark. Bot. **9(11)**: 31. 1910. *Thelypteris normalis* (C. Chr.) Moxley, Bull. S. Calif. Acad. Sci. **19**: 57. 1920.
Filix-mas augescens var. *normalis* (C. Chr.) Farw., Amer. Midl. Naturalist **12**: 253. 1931.

Terrestrial fern, 35–90 cm tall. Rhizome creeping, 4–8 mm thick (excluding stipe bases), clothed at apex with light brown, lance-linear or narrowly lanceolate-attenuate scales 4–7 mm long, these ciliate and with scattered minute hairs on surfaces; similar scales on stipe bases. Fronds closely distichous, mostly 35–90 cm long; stipes shorter than the blades, 11–40(–45) cm long, glabrate to sparsely pubescent. Blades lance-oblong to broadly ovate-oblong, to 60 cm long, 10–25 cm broad, acuminate at apex, very slightly narrowed at base, the lowest pinnae often deflexed at an angle to the plane of the blade; rachis and other vascular parts hirtellous on adaxial side, the whole underside whitish-pilose and minutely capitate-glandular; pinnae linear-attenuate, 0.8–2 cm broad, sessile, pinnatifid; segments more or less oblong-falcate, acute, the basal ones not or only slightly enlarged with slightly crenate margins, but never auriculate; veins 8–10 pairs, simple. Sori medial; indusium light brown, persistent, densely hairy; sporangial stalks often with minute stipitate glands.

DISTRIBUTION: Locally common on roadside banks and ditches, pastures, forest clearings, and among stones beside streams and rivers. Battery Gut (A5280). Also on St. Thomas and Tortola; Puerto Rico, Bahamas, Greater and Lesser Antilles (rare), Cayman Islands, southeastern United States, and southeastern Mexico to Central America.

4. Thelypteris poiteana (Bory) Proctor, Bull. Inst. Jamaica, Sci. Ser. **5**: 63. 1953. *Lastrea poiteana* Bory, Dict. Class. Hist. Nat. **9**: 233. 1825. *Dryopteris poiteana* (Bory) Urb., Symb. Ant. **4**: 20. 1903. *Goniopteris poiteana* (Bory) Ching, Sunyatsenia **5**: 239. 1940. Fig. 6D, E.

Polypodium crenatum Sw., Prodr. 132. 1788, non Forssk., 1775. *Goniopteris crenata* C. Presl, Tent. Pterid. 183. 1836. *Phegopteris crenata* (C. Presl) Mett., Fil. Hort. Bot. Lips. 84. 1856.

Terrestrial fern 0.5–1.2 m tall. Rhizome woody, creeping, 5–9 mm thick, bearing at apex a few, thin, light brown, subclathrate, minutely furcate-puberulous, lanceolate scales, 1.5–2 mm long.

Fronds few, slightly dimorphic, mostly 50–100 cm long; stipes 15–45 cm long (longer in fertile fronds), light brown, minutely furcate-puberulous chiefly toward base. Blades oval to broadly triangular-oblong, 20–35 × 18–30 cm; pinnae 2–5(–6) pairs, narrowly oblong to elliptic-linear or oblanceolate, sessile or nearly so, abruptly acuminate, 3–6 cm broad, the margins subentire to coarsely crenate; veins 6–9 pairs, the lower 3–5 adjacent ones united and giving rise to free excurrent veinlets, the rest connivent below or at the sinus. Sori in double rows between costules; indusium lacking; sporangia more or less setulose.

DISTRIBUTION: Occasional in moist forest understory. Visually recorded by G. R. Proctor. Also on St. Croix, St. Thomas, and Tortola; Greater and Lesser Antilles, Tobago, Trinidad, and continental tropical America from Guatemala to Peru and Brazil.

5. Thelypteris tetragona (Sw.) Small, Ferns S.E. States 256. 1938. *Polypodium tetragonum* Sw., Prodr. 132. 1788. *Aspidium tetragonum* (Sw.) Sw., J. Bot. (Schrader) **1800(2)**: 33. 1801. *Goniopteris tetragona* (Sw.) C. Presl, Tent. Pterid. 183. 1836. *Dryopteris tetragona* (Sw.) Urb., Symb. Ant. **4**: 20. 1903, non Kuntze, 1891.

Polypodium subtetragonum Link, Hort. Berol. **2**: 105. 1833. *Dryopteris subtetragona* (Link) Maxon *in* Britton & P. Wilson, Bot. Porto Rico **6**: 473. 1926. *Thelypteris subtetragona* (Link) E.P. St. John, Amer. Fern J. **26**: 44. 1936.
Polypodium smithianum Heward, Mag. Nat. Hist., n.s., **2**: 459. 1838.

Terrestrial fern 40–90 cm tall. Rhizome short-creeping, woody, 7–9 mm thick, at apex bearing a small tuft of brown, lance-attenuate scales, these 2–3 mm long, minutely puberulous with forked hairs. Fronds somewhat dimorphic, the fertile ones longer, narrower, and more stiffly erect; stipes of fertile fronds to 50 cm long, minutely puberulous with erect, forked hairs. Blades ovate-oblong to triangular-oblong, 30–45 × 10–25 cm; rachis with minute forked and stellate hairs, the other vascular parts on both sides with simple hairs only; pinnae 6–12 pairs, linear-oblong, sessile, acuminate to attenuate at apex, 1–3 cm broad, deeply lobed or pinnatifid, the fertile ones usually narrower than the sterile ones; segments oblong or triangular-oblong, mostly 4–6 mm broad at base, the apex obtuse or obliquely acutish; veins 6–10 pairs per segment, the lowermost adjacent ones joined and sending an excurrent veinlet to the sinus; tissue thinly herbaceous, glabrous. Sori inframedial; indusium absent; sporangia deciduously setulose, the setae simple.

DISTRIBUTION: Occasional in moist forests. Bordeaux Mountain (A3191), Maho Bay Gut (A2104). Also on St. Croix, St. Thomas, and Tortola; Florida, Greater and Lesser Antilles, Trinidad, and continental tropical America from Mexico to Ecuador and Brazil.

5. Psilotaceae

Plants epiphytic, epilithic, or sometimes terrestrial, the stems compactly branched in the substrate, the green aerial stems either simple and provided with small, 2-ranked leaves, or several-to many-times dichotomous and appearing leafless. Leaves minute, narrowly scale-like. Homosporous, the spores produced in synangia, these 2- or 3-loculed, dehiscing vertically, attached on adaxial base of minute bifid sporangiophores. Spores elongate-ellipsoid, monolete. Gametophytes subterranean or embedded in humus, tuberous, often elongate or branched, without chlorophyll.

A family of 2 genera and 3 species, with tropical and subtropical distribution.

1. PSILOTUM Sw.

Plants epiphytic or sometimes terrestrial. Subterranean stems short-creeping, dichotomously branched, beset with small, brownish, hairlike rhizoids; aerial stems loosely clustered, the lower unbranched part more or less elongate, dichotomous above into rather numerous narrow divisions. Leaves alternate, distichous or 3-ranked, minute, awl-shaped. Synangia depressed-globose, sessile, 3-lobed, 3-loculed; spores hyaline, the surface coarsely rugose to shallowly and compactly verrucate.

A genus of 2 species with pantropical distribution.

1. **Psilotum nudum** (L.) P. Beauv., Prodr. Aethéogam. 112. 1805. *Lycopodium nudum* L., Sp. Pl. 1100. 1753. Fig. 1A–C.

Hoffmannia aphylla Willd., Bot. Mag. (Römer & Usteri) **2(6):** 17. 1789.

Psilotum triquetrum Sw., J. Bot. (Schrader) **1800(2):** 109. 1801.

Bernhardia dichotoma Willd., J. Bot. (Schrader) **1800(2):** 132. 1801.

Psilotum floridanum Michx., Fl. Bor.-Amer. **2:** 281. 1803.

Bernhardia antillarum K. Müll., Bot. Zeitung (Berlin) **14:** 234. 1856.

Plants erect or ascending (rarely pendent), (5–)20–60 cm long, the main stalk 2–4 mm thick toward base; divisions dichotomous, numerous, the whole plant having a loosely brushlike appearance; branches distinctly 3-angled. Leaves remote, triangular-subulate, 1–2(–3) mm long, simple, the fertile ones (sporangiophores) smaller than the sterile or rudimentary, forked from base, 1 mm long. Synangia ca. 2 mm long and broad, yellow or brownish.

DISTRIBUTION: In understory of moist forests. Bordeaux Mountain (A1908). Also on St. Thomas, Tortola, and Virgin Gorda; pantropical.

Dicotyledons

Keys to Families of Dicotyledons

Key 1: Leaves Reduced or Wanting

1. Leaves wanting; plants green, not parasitic, succulent, globose, tubular, or flattened with joints, with numerous groups of spines.. 20. Cactaceae
1. Leaves reduced to scales (<2 mm long).
    2. Pinelike tree; twigs pine needle–like, green, cylindrical, ca. 2 mm wide, drooping; leaves whorled, forming a sheath at the node .... 25. Casuarinaceae
    2. Slender, vinelike parasites, anchoring to host plant by means of haustoria; stems yellowish, 3 mm in diam.; leaves alternate.
        3. Inflorescence a spike; anthers opening by 2 flaps; perianth of 3 tepals; style 1 .... 42. Lauraceae (*Cassytha*)
        3. Inflorescence a cyme; anthers opening by longitudinal slits; perianth of a tubular corolla; styles 2 .... 35. Cuscutaceae

Key 2: Leaves Compound

1. Leaves opposite.
    2. Inflorescence of long peduncled head; flowers minute.. 12. Asteraceae (*Bidens*)
    2. Inflorescence of panicles or cymes; flowers small to large.
        3. Petals 5, free; stamens 10.
            4. Leaves trifoliolate; leaflets aromatic, with numerous translucent oil glands; stipules wanting.. 75. Rutaceae (*Amyris*)
            4. Leaves pinnately compound; leaflets not aromatic, nor with translucent oil glands; stipules present... 93. Zygophyllaceae
        3. Petals 4–9, connate into a tubular corolla; stamens 2–5.
            5. Tendriled or twining liana.
                6. Corolla 4–9-lobed; ovary of 4 carpels; stamens 2; fruit a berry .. 60. Oleaceae (*Jasminum multiflorum*)
                6. Corolla 5-lobed; ovary of 2 carpels; stamens typically 4; fruit a 2-valvate capsule .... 15. Bignoniaceae
            5. Shrubs or small trees.
                7. Corolla 8–10 mm long, violet; ovary 4-locular; fruit a fleshy ellipsoid drupe 8–10 mm long, 1–4-seeded... 91. Verbenaceae (*Vitex*)
                7. Corolla at least 1.5 cm long, variously colored; ovary 2-locular; fruit a 2-valvate capsule; many-seeded.. 15. Bignoniaceae (*Tabebuia, Tecoma*)

1. Leaves alternate.
  8. Flowers zygomorphic.
    9. Tendriled vines; petals with a petaloid appendage at base within .... 76. Sapindaceae (part)
    9. Trees, shrubs, herbs or nontendriled vines; petals without appendages.
      10. Leaves palmately compound; petals 4; ovary borne on a gynophore .... 23. Capparaceae (*Cleome*)
      10. Leaves pinnately or ternately compound; petals 5; ovary not borne on a gynophore.
        11. Carpels 3; fruit a 3-valved capsule .... 53. Moringaceae
        11. Carpels 1; fruit a legume opening by 2 valves .... 38. Fabaceae (Caesalpinioideae and Faboideae)
  8. Flowers actinomorphic.
    12. Vines.
      13. Tendriled vines; leaves trifoliolate... 92. Vitaceae
      13. Twining vines; leaves palmately compound .... 32. Convolvulaceae (*Merremia,* part)
    12. Trees, shrubs or herbs.
      14. Herbs.
        15. Plant prostrate; leaves trifoliolate; corolla yellow; capsule cylindrical, loculicidal,
          5-locular .... 62. Oxalidaceae
        15. Plant erect; leaves pinnately dissected; corolla white; capsule circular to ovoid, schizo-
          carpic (separating into 2 mericarps), 2-locular.. 6. Apiaceae
      14. Trees or shrubs.
        16. Leaves palmately compound.
          17. Plant unarmed; stamens nearly sessile; flowers 4–5 mm long; ovary locules uniovu-
            lar; styles 5, distinct .... 9. Araliaceae
          17. Plant armed; stamens with long filaments; flowers 2.5–3 cm long; ovary locules
            multiovular; style 1... 16. Bombacaceae (*Ceiba*)
        16. Leaves pinnately compound or trifoliolate.
          18. Stamens connate into a column .... 49. Meliaceae
          18. Stamens free.
            19. Leaflets aromatic, with translucent oil glands...................................... 75. Rutaceae
            19. Leaflets not aromatic, without translucent oil glands.
              20. Flowers congested in globose, conical or cylindrical inflorescences; stamens
                numerous; fruit a flattened legume .... 38. Fabaceae (Mimosoideae)
              20. Flowers in panicles or racemes; stamens 3–10; fruit otherwise.
                21. Tree with aromatic resin; bark smooth, reddish, peeling off in paperlike
                  sheets... 19. Burseraceae
                21. Trees or shrubs, without aromatic resin; bark rough, brownish or gray-
                  ish, not peeling off as above.
                  22. Fruit a dehiscent capsule; seeds without a fleshy coat, or with
                    fleshy coat only at base.
                    23. Seeds not winged, arillate at base.............. 76. Sapindaceae (*Cupania*)
                    23. Seeds with a terminal wing, not arillate .... 49. Meliaceae (*Cedrela*)
                  22. Fruit an indehiscent drupe or 1-seeded berry; seeds or stone com-
                    pletely surrounded by a fleshy coat.
                    24. Fruits of 2–5 apically free, leathery, monocarps of equal size,
                      or with only 1 well-developed coccus and an abortive, smaller
                      one.
                      25. Monocarps 2–5, well developed, subtended by an en-
                        larged receptacle; leaves pinnately
                        compound .... 79. Simaroubaceae
                      25. Monocarps 2, usually one of them abortive; leaves trifoli-
                        olate....................................... 76. Sapindaceae (*Allophylus*)
                    24. Fruits ellipsoid to subglobose, leathery or fleshy, not separable
                      into monocarps.
                      26. Floral disk extrastaminal; fruit a leathery berry, green at
                        maturity .... 76. Sapindaceae (*Exothea, Melicoccus*)
                      26. Floral disk intrastaminal; fruit a fleshy or woody drupe,
                        yellow, red, or purple at
                        maturity .... 4. Anacardiaceae

Key 3: Leaves Simple

1. Leaves opposite or whorled.
  2. Plants producing a milky or yellowish latex.
    3. Leaves whorled.
      4. Leaves 6–11 per node, ovate, deltoid or rounded, long-petiolate; bark blackish brown, papery .... 37. Euphorbiaceae (*Euphorbia petiolaris*)
      4. Leaves 4 per node, oblong or elliptic, short-petiolate; bark grayish to light brown, thickened .... 7. Apocynaceae (*Rauvolfia*)
    3. Leaves opposite.
      5. Flowers unisexual, naked, minute, enclosed in a cuplike involucre (cyathium) .... 37. Euphorbiaceae (*Chamaesyce*)
      5. Flowers bisexual or unisexual, with calyx and a showy corolla, small to large, produced on cymose inflorescences, not subtended by an involucre.
        6. Flowers bisexual or unisexual; latex yellowish; corolla of free petals; stamens numerous; fruit a multilocular capsule; seeds covered by a fleshy coat .... 30. Clusiaceae
        6. Flowers bisexual; latex milky; corolla of connate petals, short to long tubular; stamens 5; fruit of 2 follicles; seeds usually with a tuft of long hairs at apex.
          7. Corona present; stamens fused to form a sheath around the ovary and united to the stigma head to form a gynostegium .... 11. Asclepiadaceae
          7. Corona wanting; the anthers distinct or more or less connivent around the style, not forming a gynostegium .... 7. Apocynaceae
  2. Plants not producing latex.
    8. Plants parasitic on shrub or tree branches .... 44. Loranthaceae
    8. Plants autotrophic (not parasitic).
      9. Flowers naked or perianth present, with calyx and corolla similar in size, texture, and color, sometimes petaloid, or the petals wanting.
        10. Flowers unisexual.
          11. Perianth of sepals connate into a funnel-shaped petaloid calyx; stems internally with islands of included phloem .... 57. Nyctaginaceae
          11. Perianth of free or basally connate tepals; stems internally without islands of included phloem.
            12. Woody shrubs, with coriaceous, ovate leaves; tepals 4–8, connate at base into a short cup-shaped perianth .... 60. Oleaceae (*Forestiera*)
            12. Herbs or weak-stemmed shrubs, with fleshy leaves; tepals 4, free to base.
              13. Weak-stemmed shrubs; leaves linear, cylindrical in cross section, 1–2 cm long; plants of salt flats and other coastal saline areas .... 14. Bataceae
              13. Fleshy herbs; leaves ovate, round, or obovate, fleshy or chartaceous; of diverse habitats but not as above .... 90. Urticaceae (*Pilea*)
        10. Flowers bisexual.
          14. Flowers without perianth .... 66. Piperaceae (*Peperomia* in part)
          14. Flowers with perianth.
            15. Tepals free to base; stamens with filaments basally connate into a short to elongate tube .... 3. Amaranthaceae
            15. Sepals basally or wholly connate; stamens with filaments free or adnate to the calyx.
              16. Plants fleshy, prostrate; sepals basally connate; fruits many-seeded; stems internally without islands of included phloem .... 2. Aizoaceae
              16. Plants herbaceous to woody (not fleshy), erect or scandent; sepals connate into a funnel-shaped structure; fruit a 1-seeded anthocarp; stems (internally) usually with islands of included phloem .... 57. Nyctaginaceae
      9. Flowers with distinct calyx and corolla.
        17. Stamens 2.
          18. Fruit fleshy, indehiscent .... 60. Oleaceae
          18. Fruit capsular, opening along longitudinal valves .... 1. Acanthaceae (part)
        17. Stamens 4 to numerous.
          19. Petals distinct.
            20. Ovary inferior.
              21. Leaf blades with scattered translucent dots; plants usually aromatic; stamens numerous .... 56. Myrtaceae
              21. Leaf blades without translucent dots; plants not aromatic; stamens 8–10.

22. Petals white, elongated, much longer than the calyx lobes; leaf blades
with 3–7 main arcuate veins; anthers lanceolate, opening by a terminal
pore or longitudinal slits .... 48. Melastomataceae
22. Petals greenish or pink (turning bright red with age), minute, as long as or
shorter than the calyx; leaf blades pinnately veined; anthers ovoid or ellipsoid opening by longitudinal slits .... 31. Combretaceae
20. Ovary superior.
23. Plants herbaceous, with succulent leaves and stems .... 70. Portulacaceae
23. Plants herbaceous or woody, not fleshy.
24. Stamens twice as many (sometimes as many) as the petals or more numerous; nectary disk annular or wanting.
25. Stipules foliaceous, covering the terminal buds, early deciduous, leaving a scar; trees of mangrove swamps .... 72. Rhizophoraceae
25. Stipules wanting; herbs, shrubs, or trees of terrestrial habitats.
26. Calyx with 4–5 pairs of glands at the outer base of the sepals;
plants usually with t-shaped hairs or bristles; petals long
clawed .... 46. Malpighiaceae
26. Calyx not glandular, forming a hypanthium; plant glabrous
or with simple hairs; petals not clawed, adnate to the
hypanthium .... 45. Lythraceae
24. Stamens as many as the petals; nectary disk annular, basal to the ovary.
27. Stamens opposite to petals; petals concave, surrounding the stamens;
leaves with entire margins .... 71. Rhamnaceae (*Krugiodendron, Reynosia*)
27. Stamens alternate to petals; petals plane, spreading; leaves with
crenate margins .... 27. Celastraceae (*Cassine, Crossopetalum*)
19. Petals connate into a tubular or funnel-shaped corolla.
28. Stipules present (may be early deciduous, check for stipule scars when in doubt).
29. Ovary superior .... 43. Loganiaceae
29. Ovary inferior .... 74. Rubiaceae
28. Stipules wanting.
30. Leaves with linear cystoliths (prominent whitish, short lines, mostly on upper
surface) .... 1. Acanthaceae
30. Leaves without cystoliths.
31. Leaves fleshy, usually producing plantlets (asexual reproduction) along the
margins .... 33. Crassulaceae
31. Leaves chartaceous to coriaceous, not fleshy, not producing plantlets at
margins.
32. Ovary inferior; flowers produced in dense heads .... 12. Asteraceae
32. Ovary superior; flowers in racemes, spikes, cymes or rarely in heads.
33. Ovary 2-locular.
34. Stamens 5; fruit an elongated follicle, or a drupe.
35. Erect herbs or subshrubs; leaves subopposite or whorled;
fruit a globose or 4-lobed drupe; inflorescence long,
scorpioid .... 17. Boraginaceae (*Heliotropium*)
35. Twining vines; leaves opposite; fruit of a pair of
elongated follicles; inflorescence a thyrselike
cyme .... 7. Apocynaceae (*Prestonia*)
34. Stamens 4; fruit capsular.
36. Capsule without hoodlike placental tissue .... 78. Scrophulariaceae
36. Capsule with persistent hoodlike placental
tissue .... 1. Acanthaceae (*Stenandrium, Thunbergia*)
33. Ovary 4-locular; corolla irregular, bilabiate; stamens 4.
37. Style borne from the base of carpels or between
locules .... 41. Lamiaceae
37. Style terminal on ovary .... 91. Verbenaceae
1. Leaves alternate.
38. Plants producing milky or yellowish latex or exudate.

39. Stipules present, persistent or deciduous leaving a conspicuous scar; flowers unisexual, usually minute or small.
    40. Stipules large, deciduous, contorted into a cone-shaped hood that protects the apical meristem... 52. Moraceae
    40. Stipules minute to large, persistent or deciduous, paired, not forming a hood that protects the apical meristem.. 37. Euphorbiaceae
39. Stipules wanting.
    41. Plant producing yellowish latex; corolla of free petals; stamens numerous.............. 63. Papaveraceae
    41. Plant producing milky latex; corolla tubular; stamens as many as the corolla lobes.
        42. Ovary superior.
            43. Flowers unisexual (plants usually dioecious); plants usually palmlike, with a weak trunk (usually unbranched and with numerous leaf scars) and a crown of large, long-petioled leaves.. 24. Caricaceae
            43. Flowers bisexual; plant not as above.
                44. Shrubs or trees; corolla cup-shaped or tubular <1.5 cm long; fruit fleshy, indehiscent; seeds with a large ventral scar... 77. Sapotaceae
                44. Herbs, twining vines, or shrubs; corolla funnel-shaped, trumpet-shaped, or saucer-shaped, >3 cm long (except for the saucer-shaped ones, which could be as small as 1 cm long); fruit dry, dehiscent; seeds without a scar.
                    45. Fruit a pair of spreading follicles, 10–14 cm long; seeds numerous, with a tuft of long hairs at apex .... 7. Apocynaceae (*Plumeria*)
                    45. Fruit capsular, 2-locular, <1.5 cm long; seeds 2–4, usually without hairs .... 32. Convolvulaceae
        42. Ovary inferior.
            46. Flowers solitary in leaf axils; corolla white, 8–18 cm long; leaves attenuate at base, the petiole not winged; fruit a capsule .... 21. Campanulaceae
            46. Flowers congested in dense heads; corolla yellow, 8–11 mm long; leaves clasping at base, or sometimes attenuate with a winged petiole; fruit an achene.. 12. Asteraceae (*Launaea, Sonchus*)
38. Plants not producing a colored latex or exudate.
    47. Vines or climbing shrubs.
        48. Plants climbing by tendrils.
            49. Tendrils opposite to leaves.
                50. Flowers unisexual; corolla tubular; ovary inferior, with parietal placentation; inflorescences of axillary, simple or compound racemes........................... 34. Cucurbitaceae
                50. Flowers bisexual; corolla of free petals; ovary superior, with basal placentation; inflorescences of cymes or panicles opposite the leaves............................... 92. Vitaceae
            49. Tendrils axillary, basal or distal on inflorescence.
                51. Tendrils distal on inflorescence; perianth of 5 similar tepals.... 69. Polygonaceae (*Antigonon*)
                51. Tendrils axillary or basal on inflorescence; perianth of distinct calyx and corolla or distinct calyx, corolla (sometimes wanting) and corona.
                    52. Corona wanting; ovary sessile; stamens borne on receptacle, opposite to petals; petals concave, surrounding the stamens; fruit dry, schizocarpic.. 71. Rhamnaceae (Gouania)
                    52. Corona present; ovary and stamens subtended by a stipe (androgynophore); stamens not opposite to the petals; petals not as above, or wanting; fruit fleshy, indehiscent or tardily dehiscent... 64. Passifloraceae
        48. Twining vines or scrambling shrubs, without tendrils.
            53. Plants bearing spines or thorns on stems and branches.
                54. Flowers unisexual, small, (<0.5 cm), lacking corolla; fruit a 1-seeded drupe .... 89. Ulmaceae (*Celtis iguanea*)
                54. Flowers bisexual, large (>1.5 cm); fruits with numerous seeds.
                  55. Perianth whitish, not differentiated into calyx and corolla, 2.5–5 cm wide; ovary inferior; berry yellow .... 20. Cactaceae (*Pereskia*)
                  55. Calyx green; corolla white, ca. 1 cm long; ovary superior; berry red-orange... 80. Solanaceae
            53. Plants not armed.

56. Twining vines.
   57. Flowers >4 cm long, very irregular, solitary at leaf axils; calyx corolla-like, inflated, constricted below the tube; corolla wanting .... 10. Aristolochiaceae
   57. Flowers <1 cm long, regular, in simple or branched inflorescences; calyx reduced; corolla present.
      58. Flowers unisexual, <3 mm long.
         59. Herbaceous vine; inflorescence racemose; leaves fleshy .... 13. Basellaceae
         59. Woody vine; inflorescence of cymes or of simple or branched spikes; leaves coriaceous to chartaceous .... 50. Menispermaceae
      58. Flowers bisexual, >5 mm long.
         60. Leaves rounded, cuneate or obtuse at base; flowers in scorpioid cymes .... 17. Boraginaceae (*Tournefortia*, part)
         60. Leaves cordate at base; flowers in short, axillary cymes .... 32. Convolvulaceae (*Convolvulus, Jacquemontia*)
56. Scrambling to climbing shrubs or herbs, not twining.
   61. Perianth of 4 tepals; cross section of stem showing xylem with discrete strands of phloem .... 65. Phytolaccaceae (*Trichostigma*)
   61. Perianth of calyx and corolla; cross section of stems showing xylem without included phloem.
      62. Inflorescence a scorpioid cyme; fruit fleshy, indehiscent, white .... 17. Boraginaceae (*Tournefortia hirsutissima*)
      62. Inflorescences not scorpioid; fruit leathery or dry, dehiscent, green, yellowish, or brown.
         63. Calyx without stipitate glands; corolla of free petals; ovary subtended by a stipe (gynophore); fruit elongate, many-seeded, opening along 1 suture .... 23. Capparaceae (*Capparis flexuosa*)
         63. Calyx with stipitate glands externally; corolla trumpet-shaped; ovary sessile; fruit a 1-seeded, valvate capsule .... 68. Plumbaginaceae
47. Herbs, shrubs or trees, not climbing.
   64. Ovary inferior.
      65. Corolla tubular.
         66. Flowers in heads; fruit dry, 1-seeded (achene), usually crowned by bristles, scales, or awns .... 12. Asteraceae
         66. Flowers in axillary cymes, racemes or solitary; fruit a drupe.
            67. Shrubs of sandy sea coasts; leaves rigid-coriaceous, entire; corolla strongly asymmetrical, the tube 1.5–2 cm long; drupe nearly globose, black .... 40. Goodeniaceae
            67. Trees of forested areas; leaves coriaceous, serrate; corolla symmetrical, the tube 5–7 mm long; drupe ellipsoid, dark blue .... 83. Symplocaceae
      65. Corolla of free petals or wanting.
         68. Trees or shrubs; flowers <5 mm long, corolla wanting .... 31. Combretaceae
         68. Herbs; flowers ca. 2 cm long; corolla of free yellow petals...................... 61. Onagraceae
   64. Ovary superior or partly inferior.
      69. Flowers unisexual.
         70. Ovary unilocular.
            71. Herbs or subshrubs.
               72. Subshrub, usually with stinging hairs; fruit an oblique achene .... 90. Urticaceae (*Laportea*)
               72. Herb, without stinging hairs; fruit an utricle.
                   73. Plants canescent; flowers solitary or in glomerules; utricles indehiscent .... 28. Chenopodiaceae
                   73. Plants not canescent; flowers in spicate inflorescences; utricles dehiscent .... 3. Amaranthaceae (*Amaranthus*)
            71. Trees, small trees, or shrubs.
               74. Fruit a crescent-shaped nut, subtended by a fleshy hypocarp (receptacle)... 4. Anacardiaceae (*Anacardium*)
               74. Fruit a drupe, a berry, or of multiple achenes.
                   75. Stipules wanting.

76. Trunk with branched spines; leaves serrate; fruit a red
      berry.. 39. Flacourtiaceae (*Xylosma*)
76. Trunk smooth, without spines; leaves entire; fruit a green or white
      drupe.
    77. Drupe subtended by a cupule; tepals 6, spreading; stamens 9,
       the anthers opening by 4 minute flaps ... 42. Lauraceae (*Ocotea* in part)
    77. Drupe not subtended by a cupule; calyx conical, of 4 connate
       sepals; stamens 8, the anthers opening by longitudinal
       slits... 86. Thymelaeaceae
  75. Stipules present.
    78. Rapid-growing, palmlike trees, with 1–3 crowns of long-petioled
       leaves; stipules large, deciduous, contorted into a cone-shaped
       hood that protects the apical meristem; flowers naked, in aments;
       fruits multiple (of fused achenes) .... 26. Cecropiaceae
    78. Shrubs or small, branched trees, with short-petioled leaves scat-
       tered along stems; stipules minute; flowers apetalous, in cymes or
       fascicles; fruit a drupe.
       79. Perianth of 5 sepals; stamens 5; styles 2; flowers in axillary
          cymes... 89. Ulmaceae
       79. Perianth of 4 sepals; stamens 3–6; style 1; flowers in axillary
          fascicles .... 37. Euphorbiaceae (*Drypetes*)
70. Ovary (2–)3(–5)-locular.
  80. Fruit dry, capsular.. 37. Euphorbiaceae
  80. Fruit fleshy, a drupe or a berry.
    81. Branches with axillary spines; leaves congested on short axillary branches;
       flowers paired or solitary; petals connate into a funnel-shaped
       corolla... 17. Boraginaceae (*Rochefortia*)
    81. Branches without spines; leaves not congested; flowers in cymes, fascicles
       or racemes; petals free or wanting.
       82. Corolla wanting; inflorescence racemose, cauliflorous; drupe
          6–8-lobed, yellow .... 37. Euphorbiaceae (*Phyllanthus acidus*)
       82. Corolla of free petals; inflorescence of axillary fascicles or cymes;
          drupe globose, ellipsoid or ovoid; red or yellow to orange.
          83. Ovary 4–5-locular; sepals imbricate; stigma sessile; drupe glo-
             bose, red, 4–8-seeded... 8. Aquifoliaceae
          83. Ovary 2-locular; stigma with 2 spreading lobes; drupe ellipsoid to
             ovoid, yellow to red, 2-seeded .... 27. Celastraceae (*Schaefferia*)
69. Flowers bisexual.
  84. Perianth reduced, the flower either naked or represented by a whorl of tepals or se-
    pals (the flower then apetalous).
    85. Flowers naked.. 66. Piperaceae
    85. Flowers with either tepals or sepals.
      86. Herbs.
        87. Perianth of 2 series of tepals .... 3. Amaranthaceae (*Amaranthus*)
        87. Perianth of one series of 4–5 sepals.
          88. Leaves in a rosette at base; inflorescences scapose; fruit a
            dehiscent capsule.. 51. Molluginaceae
          88. Leaves cauline, alternate along stems; inflorescences axillary; fruit
            a drupe, achene or an utricle.
            89. Filaments united at base into a short cup; fruit a circumscis-
              sile utricle...................................... 3. Amaranthaceae (*Celosia*)
            89. Filaments free to base; fruits indehiscent.
              90. Stamens as many as the sepals and opposite them; styles
                 3, free to base; fruit an utricle.................... 28. Chenopodiaceae
              90. Stamens 4 and alternate the sepals, or up to 8 and then in
                 2 series; style simple or wanting; fruit a drupe or a
                 hooked achene...................................... 65. Phytolaccaceae

86. Shrubs or small trees.
    91. Leaves whorled; calyx funnel-shaped; cross section of stem with included islands of phloem.. 57. Nyctaginaceae
    91. Leaves alternate; perianth cup-shaped; cross section of stem without included islands of phloem.
        92. Stipules wanting; anthers opening by 4 minute flaps; fruit usually subtended by a cupular structure .... 42. Lauraceae
        92. Stipules present; anthers opening by longitudinal slits; fruit not subtended by a cupular structure.
            93. Stipules minute; leaves serrate, with translucent lineations; fruit dehiscent or indehiscent; seeds few to many .... 39. Flacourtiaceae
            93. Stipules connate around the stem into an ocrea; leaves entire without lineations; fruit indehiscent; seed 1 per fruit .... 69. Polygonaceae (*Coccoloba*)
84. Perianth of calyx and corolla.
    94. Corolla gamopetalous.
        95. Corolla zygomorphic.
            96. Corolla trumpet-shaped .... 80. Solanaceae (*Brunfelsia*)
            96. Corolla funnel- or bell-shaped.
                97. Corolla 3.5–6 cm long; fruit >6 cm long................................. 15. Bignoniaceae (*Amphitecna, Crescentia*)
                97. Corolla to 2 cm long; fruit to 1.5 cm long.
                    98. Corolla ca. 2 cm long, yellow with purple blotches; fruit a yellowish, leathery drupe .... 54. Myoporaceae
                    98. Corolla <1 cm long, white; fruit a septicidal capsule.. 78. Scrophulariaceae
        95. Corolla actinomorphic.
            99. Leaves in a rosette at base; plant stemless; inflorescence a scapose spike .... 67. Plantaginaceae
            99. Leaves cauline, alternate or whorled.
                100. Leaves in whorls of 4.. 43. Loganiaceae
                100. Leaves alternate.
                    101. Flowers and or leaves with dark punctations or lineations.
                        102. Flowers punctate or lineate, in axillary racemes; petaloid staminodes present; filaments basally connate; leaves not punctate or lineate............... 85. Theophrastaceae
                        102. Flowers with orange punctations, in terminal panicles; staminodes wanting; filaments adnate to corolla tube; leaves dark punctate............................. 55. Myrsinaceae
                    101. Flowers and leaves not punctate or lineate.
                        103. Ovary 4-locular; flowers in scorpioid cymes... 17. Boraginaceae
                        103. Ovary 2- or 3-locular; flowers in cymes, racemes or solitary.
                            104. Bracteoles connate into an epicalyx; ovary 3-locular, partly inferior; stamens opposite the corolla lobes .... 59. Olacaceae (*Schoepfia*)
                            104. Bracteoles not forming an epicalyx; ovary 2-locular, superior; stamens alternate to corolla lobes.
                                105. Fruit a berry .... 80. Solanaceae
                                105. Fruit a capsule.
                                    106. Capsule many-seeded .... 78. Scrophulariaceae (*Capraria*)
                                    106. Capsule 4-seeded .... 32. Convolvulaceae (*Evolvulus*)
    94. Corolla of distinct petals.
        107. Stipules present.

108. Fleshy herbs... 70. Portulacaceae
108. Subshrubs, shrubs, or trees.
    109. Style gynobasic.
        110. Corolla white to cream; carpels connate into a fleshy
            drupe .... 29. Chrysobalanaceae
        110. Corolla yellow; carpels separating at maturity.
            111. Fruits of distinct monocarps subtended by a fleshy
                receptacle.. 58. Ochnaceae
            111. Fruits of distinct nutlets, not subtended by a
                fleshy receptacle............................... 82. Surianaceae
    109. Style distal on ovary.
        112. Ovary subtended by a gynophore; fruits narrow, elon-
            gated and opening along 1 suture or irregularly split-
            ting, or indehiscent and globose with numerous seeds
            inside.. 23. Capparaceae
        112. Ovary sessile; fruit otherwise.
            113. Petals with a petaloid appendage at base
                within .... 36. Erythroxylaceae
            113. Petals without appendages.
                114. Filaments connate into a tube around the
                    ovary.
                    115. Fruit indehiscent, slightly fleshy, fi-
                        brous, nearly globose to ellipsoid; tree
                        with many whorled horizontal
                        branches at different heights
                        .... 16. Bombacaceae (*Quararibea*)
                    115. Fruit dehiscent, dry, a capsule or a le-
                        gume, or if indehiscent then not fleshy
                        or fibrous; herbs, shrubs, or trees, not
                        branched as above.
                        116. Flower zygomorphic, with keel,
                          wing, and standard petals
                          .... 38. Fabaceae (Caesalpinioideae and
                                      Faboideae)
                      116. Flower actinomorphic, petals all
                        alike.
                        117. Anthers with 4 pollen sacs
                          (locules)............. 81. Sterculiaceae
                        117. Anthers with 2 pollen sacs
                          (locules)............... 47. Malvaceae
             114. Filaments free.
                118. Stamens numerous.
                    119. Shrubs, usually with star-shaped
                      trichomes; leaf blades not glandu-
                      lar; fruit dry, capsular......... 87. Tiliaceae
                    119. Trees; glabrous or nearly so; leaf
                      blades with a pair of glands near
                      the base; fruits fleshy.
                      120. Leaf blades with impressed
                        glands near the base; mar-
                        gins entire; buds with over-
                        lapping scales.......... 73. Rosaceae
                      120. Leaf blades with prominent
                        glands at the junction with
                        the petiole; margins serrate;
                        buds naked
                        .... 39. Flacourtiaceae (*Prockia*)
                118. Stamens 4 or 5.

      121. Petals concave, surrounding the
        stamens .... 71. Rhamnaceae
      121. Petals flat, alternating with the
        stamens .... 27. Celastraceae (*Maytenus*)
107. Stipules wanting.
  122. Stems with thorns or spines.
    123. Blades aromatic, due to the presence of translucent
     oil glands; fruit a globose berry or a capsule
     .... 75. Rutaceae (*Citrus, Zanthoxylum* in part)
    123. Blades not aromatic, without translucent oil glands; fruit a
     globose, leathery drupe .... 59. Olacaceae (*Ximenia*)
  122. Plants not spiny.
    124. Flowers fleshy, 3-merous.................................... 5. Annonaceae
    124. Flowers not fleshy, 4–5-merous.
      125. Ovary partly inferior, embedded in an elongated hypan-
        thium... 61. Onagraceae
      125. Ovary superior, not embedded in a hypanthium.
        126. Corolla of 4 petals .... 18. Brassicaceae
        126. Corolla of 5 petals.
          127. Plant spicy-aromatic, especially when
            bruised; calyx of 3 sepals; corolla red; sta-
            mens 10 with filaments connate into a col-
            umn; fruit a small (<1 cm in diam.), red
            berry....................................... 22. Canellaceae
          127. Plant not spicy nor aromatic; calyx of 2 or 5
            sepals; corolla yellow, yellow-green, reddish
            green or pink; stamens 5 or numerous; fruit
            a capsule or a nut.
            128. Fruit a crescent-shaped nut, subtended
              by a fleshy hypocarp (receptacle)
              ..... 4. Anacardiaceae (*Anacardium, Mangifera*)
            128. Fruit a capsule, not subtended by a
              fleshy hypocarp.
              129. Trees; corolla pink; stamens nu-
                merous, in 2–4 series; style sim-
                ple; seeds surrounded by a fleshy
                red coat .... 84. Theaceae
              129. Shrubs or subshrubs; corolla yel-
                low, pink, or magenta; stamens
                as many as or twice as many as
                the petals; styles 3–5, distinct or
                united at base; seeds without a
                fleshy red coat.
                130. Leaves nearly fleshy; seeds
                  tuberculate, without an aril-
                  lode..... 70. Portulacaceae (*Talinum*)
                130. Leaves chartaceous; seeds
                  reticulate, pitted, with a
                  unilateral dry aril.... 88. Turneraceae

1. Acanthaceae (Acanthus Family)

Reviewed by D. C. Wasshausen

Herbs, small shrubs, or climbing plants. Leaves opposite, simple, without stipules; blades usually with numerous cystoliths. Flowers bisexual, zygomorphic, solitary, in cymes, racemes, or panicles; calyx with 4 or 5 equal or unequal sepals; corolla funnel-shaped or salverform, 5-lobed or bilabiate; stamens 4 (equal, or 2 of them shorter) or 2, adnate to the corolla tube, the anthers lanceolate or ellipsoid, opening by longitudinal slits; ovary superior, 2-locular, of 2 connate carpels, usually subtended by a more or less cupular nectary disk, the style long, filiform, the stigma simple or lobed, ovules numerous, with axile placentation. Fruit capsular, opening along longitudinal valves; placental tissue usually hooklike and persistent after capsule dehisces.

About 250 genera and 2600 species with pantropical distribution.

REFERENCES: Long, R. W. 1970. The genera of Acanthaceae in the southeastern United States. J. Arnold Arbor. 51: 257–309; Stearn, W. T. 1971. A survey of the tropical genera *Oplonia* and *Psilanthele*. Bull. Brit. Mus. (Nat. Hist.) Bot. 4: 261–323; Stearn, W. T. 1971. Taxonomic and nomenclatural notes on Jamaican gamopetalous plants. J. Arnold Arbor. 52: 614–647.

Key to the Genera of Acanthaceae

1. Plants 2–5 m long, with climbing or trailing stems.
    2. Plants woody, with axillary spines; corolla 1–2 cm long .... 6. *Oplonia*
    2. Plants herbaceous without spines; corolla 2–5 cm long.
        3. Corolla funnel-shaped, lilac or light yellow; capsules ellipsoid .... 1. *Asystasia*
        3. Corolla salverform, white or orange; capsules subglobose, abruptly beaked at apex .... 10. *Thunbergia*
1. Plants usually <1 m tall, erect or decumbent.
    4. Plants woody, with paired axillary spines.
        5. Plant erect; corolla yellow .... 2. *Barleria*
        5. Plant erect with arching branches; corolla lavender .... 6. *Oplonia*
    4. Plants herbaceous, without spines.
        6. Plants rosettelike, with short internodes; petioles longer than the blades.......................... 9. *Stenandrium*
        6. Plants not rosettelike, with long internodes; petioles shorter than the blades.
            7. Flowers with 2 stamens.
                8. Flower solitary or few at leaf axil; corolla salverform; capsule club-shaped .... 8. *Siphonoglossa*
                8. Flowers in short or elongate, terminal or axillary inflorescences; corolla funnel-shaped; capsules ellipsoid or flattened ellipsoid.
                    9. Stems 6-angular; calyx subtended by 2 spatulate bracteoles united at base .... 4. *Dicliptera*
                    9. Stems cylindrical or 4-angular; calyx subtended by 2 distinct bracteoles .... 5. *Justicia*
            7. Flowers with 4 stamens.
                10. Flowers borne on dense, terminal, quadrangular spikes, with large foliaceous overlapping bracts; corolla (0.7–)1–1.5 cm long.. 3. *Blechum*
                10. Flowers solitary or in loose terminal or axillary cymes, bracts small, not overlapping; corolla 2.2–4.5 cm long.. 7. *Ruellia*

1. ASYSTASIA Blume

Perennial herbs. Leaves with abundant cystoliths at maturity. Flowers in terminal, unilateral spikes or racemes, each flower subtended by 2 small bracts and 2 bracteoles; calyx with 5 linear or lanceolate, equal sepals; corolla lilac, rose, white, or light yellow, funnel-shaped, zygomorphic, with 5 rounded lobes; stamens 4, didynamous; stigma 2-lobed or capitate. Capsule ellipsoid or club-shaped; seeds 2–4 per capsule, lenticular.

An Old World genus with approximately 40 species.

1. Asystasia gangetica (L.) T. Anderson *in* Thwaites, Enum. Pl. Zeyl. 235. 1860. *Justicia gangetica* L., Cent. Pl. II. 3. 1756.						Fig. 7A–E.

Decumbent or trailing herb, to 2 m long; stems 4-angular, rooting at nodes. Leaf blades 2.5–11 × 2.5–6.5(–8) cm, ovate or lanceolate, chartaceous, sparsely covered with short whitish hairs, especially on veins, the apex acuminate, acute or obtuse, the base obtuse, truncate, rounded or cordate, sometimes oblique, the margins crenulate; petioles 0.5–3(–4.5) cm long. Flowers covered with glandular hairs without and borne on spikes; bracts and bracteoles ovate, ciliate at margins. Calyx 5–7 mm long, with lanceolate sepals; corolla light yellow or lilac, 2.3–5 cm long; stamens and pistil not projecting beyond the tube. Capsule 2–2.5 cm long, club-shaped, turning from green to light brown, densely covered with glandular hairs. Seeds light brown, with irregular margins.

DISTRIBUTION: Introduced as an ornamental, found in gardens or escaped in disturbed areas. Cruz Bay (A3083). Also on St. Thomas and Tortola; cultivated throughout the tropics, often escaped.

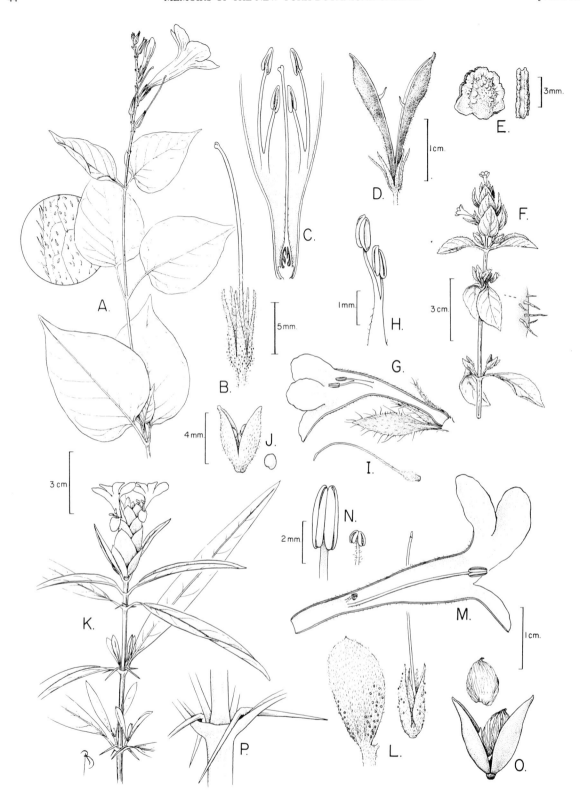

Fɪɢ. 7. **A–E.** *Asystasia gangetica.* **A.** Flowering branch and detail of leaf showing indument and margin. **B.** Calyx with pistil. **C.** L.s. corolla showing stamens and pistil. **D.** Dehisced capsule with hooklike placental tissue and persistent calyx. **E.** Seed, frontal and lateral views. **F–J.** *Blechum pyramidatum.* **F.** Flowering branch and detail of leaf margin showing trichomes. **G.** L.s. corolla with didynamous stamens. **H.** Stamens. **I.** Pistil. **J.** Dehisced capsule, and seed. **K–P.** *Barleria lupulina.* **K.** Flowering branch. **L.** Inflorescence bract and calyx with pistil. **M.** L.s. corolla with fertile and sterile stamens. **N.** Detail of fertile and sterile stamens. **O.** Dehisced capsule with seeds. **P.** Node with paired axillary spines.

2. BARLERIA L.

Armed or unarmed subshrubs, or herbs. Leaves short-petioled or sessile; blade with numerous minute, transversely oriented cystoliths. Flowers in axillary or terminal inflorescences; calyx with 4 unequal sepals; corolla yellow or blue, zygomorphic, with 5 lobes; stamens 4, didynamous; stigma simple. Capsule ovoid or oblong.

Primarily a paleotropical genus with approximately 100 species.

1. Barleria lupulina Lindl., Edward's Bot. Reg. **18**: pl. 1483. 1832.   Fig. 7K–P.

Spiny, erect subshrub, to 50 cm tall, many-branched from near the base; stems cylindrical, blackish, glabrous, except for younger parts, which are covered with minute whitish hairs; pith hollow. Leaves with a pair of spines at their axils, and usually short branches with clustered leaves. Leaf blades 4.5–13 × 0.7–2.2 cm, narrowly elliptic or linear, chartaceous, glabrous except for a few hairs on veins, lower surface lighter than the upper, the apex acute and mucronate, the base acute, the margins entire and revolute; petioles 0.5–1 cm long. Flowers clustered in dense terminal spikes with large, overlapping bracts covering the bases of the flowers; bracts obovate, to 1.7 cm long, glaucous, with minute glandular dots on both surfaces, the apex mucronate. Calyx with 4 unequal lanceolate sepals; corolla yellow, 3–5 cm long, projecting beyond the bracts with 5 lobes, one separating from the remaining 4 lobes; stamens 4, the 2 fertile exceeding the tube, the 2 sterile reduced; style exceeding the tube. Capsule flattened, ovate-lanceolate, glabrous, to 1.5 cm long. Seeds 5–5.5 mm long, ovate or lenticular, apiculate, with fibrous, golden covering.

DISTRIBUTION: Introduced from the Old World tropics as an ornamental, now escaped in open areas. Trail to Brown Bay (A1869, A4128). Also on St. Thomas; Bahamas, Jamaica, Hispaniola, Puerto Rico, and Lesser Antilles.

3. BLECHUM P. Browne

Perennial herbs. Leaves short-petiolate; blade with numerous minute cystoliths. Inflorescence terminal or axillary, quadrangular spikes or racemes, with large, foliaceous, overlapping bracts covering most of the flower; calyx with 5 equal or unequal, awl-shaped sepals; corolla actinomorphic, funnel-shaped, white or purplish; stamens 4, of equal length; style filiform, the stigma simple. Capsule wide ellipsoid or nearly globose; seeds rounded and flattened.

A small genus with 10 species, native to the American tropics.

1. Blechum pyramidatum (Lam.) Urb., Repert. Spec. Nov. Regni Veg. **15**: 323. 1918. *Barleria pyramidata* Lam., Encycl. **1**: 380. 1785.   Fig. 7F–J.

Blechum brownei Juss., Ann. Mus. Natl. Hist. Nat. **9**: 270. 1807.
Ruellia blechum L., Syst. Nat., ed. 10, **2**: 1120. 1759. *Blechum blechum* (L.) Millsp., Publ. Field Columbian Mus., Bot. Ser., **2**: 100. 1900.

Erect or decumbent herb, 20 to 50 cm tall, usually branching from near the base; stems slender, quadrangular, usually rooting at nodes, with lines of hairs along two of the sides. Leaf blades 1–3.5 × 0.7–1.8 cm, elliptic, ovate or lanceolate, chartaceous, covered with white, appressed hairs on both surfaces, the apex acute or obtuse, the base attenuate, obtuse or rounded, the margins entire or sinuate, and ciliate; petioles slender, 0.4–0.7 cm long. Flowers borne on dense, terminal spikes; bracts leaflike, ovate, chartaceous, green, pubescent, the margins with long white hairs. Calyx of 5 awl-shaped sepals, to 4 mm long; corolla pink, light mauve or lavender, 1–1.5 cm long, slightly projecting beyond the subtending bract, sparsely covered with minute, whitish hairs externally. Capsule 5–7 mm long, ellipsoid, slightly flattened, densely covered with short, whitish hairs, acuminate at apex, 12–16 seeded. Seeds ca. 1.5 mm long, lenticular, light brown and smooth.

DISTRIBUTION: Common throughout the island, in open areas. Ajax Peak (A2656), Francis Bay (W172-a), Johnson Bay (A2912). Also on St. Croix, St. Thomas, Tortola, and Virgin Gorda; from Mexico to South America, including the West Indies, introduced to the Old World tropics.

COMMON NAME: penguin balsam.

4. DICLIPTERA Juss., nom. cons.

Perennial herbs or shrubs; stems often hexagonal in cross section. Leaves petiolate; blades with numerous minute cystoliths. Flowers sessile, solitary or in axillary or terminal condensed cymes, forming a one-sided panicle or spike. Bracts minute and awl-shaped; bracteoles enlarged, covering most of the flower; calyx 5-lobed, translucent, cup-shaped with short or long lobes; corolla funnel-shaped, bilabiate; stamens 2; style filiform, the stigma simple or slightly lobed. Capsule ovoid to lenticular; seeds 2 or 4, lenticular.

A large genus with approximately 80 to 100 species with cosmopolitan distribution.

1. Dicliptera sexangularis (L.) Juss., Ann. Mus. Natl. Hist. Nat. **9**: 267. 1807. *Justicia sexangularis* L., Sp. Pl. 16. 1753.   Fig. 8A–E.

Justicia assurgens L., Sp. Pl. 23. 1753. *Dicliptera assurgens* (L.) Juss., Ann. Mus. Natl. Hist. Nat. **9**: 267. 1807. *Diapedium assurgens* (L.) Kuntze, Revis. Gen. Pl. **2**: 485. 1891.
Dicliptera portoricensis Nees *in* A. DC., Prodr. **11**: 498. 1847.

Dicliptera vahliana Nees *in* A. DC., Prodr. **11**: 498. 1847.

Erect subshrub, to 1.2 m tall, main stem with many lateral branches; stems hexagonal, glabrous except for a few hairs at nodes; young parts covered with glandular hairs. Leaf blades 1.5–9 × 0.5–5 cm, ovate or lanceolate, chartaceous, glabrous except for a few hairs along veins, both surfaces presenting numerous cystoliths, the apex acute or long acuminate, the base obtuse or acute, the margins entire and ciliate; petioles slender, 0.2–2 cm long. Flowers dispersed along terminal spiciform panicles; bracts

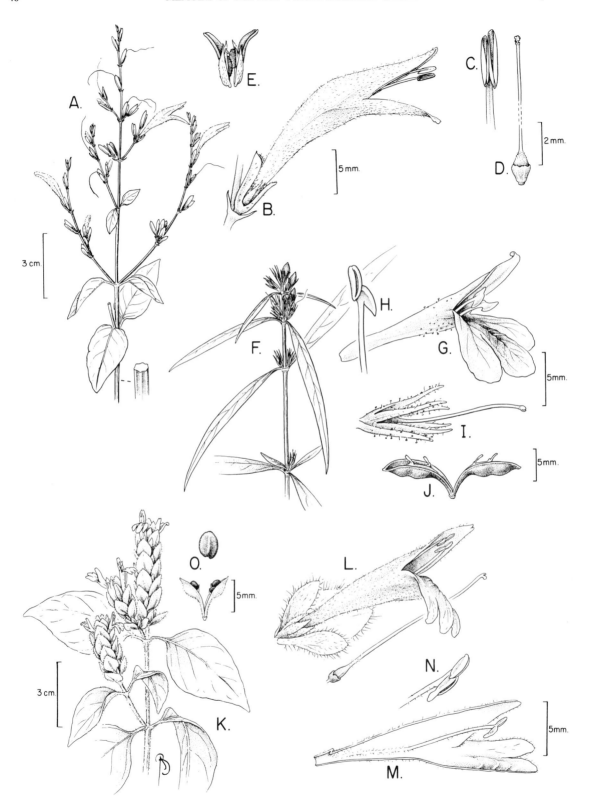

FIG. 8. A–E. *Dicliptera sexangularis*. **A.** Fertile branch and detail of hexagonal stem. **B.** Flower. **C.** Detail of stamen. **D.** Pistil with subtending nectary disk. **E.** Dehisced capsule with seeds. **F–J.** *Justicia periplocifolia*. **F.** Flowering branch. **G.** Corolla. **H.** Detail of stamen. **I.** Calyx and pistil. **J.** Dehisced capsule with hooklike placental tissue. **K–O.** *Justicia mirabiloides*. **K.** Flowering branch. **L.** Flower and pistil. **M.** L.s. corolla showing stamen. **N.** Detail of stamen. **O.** Dehisced capsule with seeds.

spatulate or awl-shaped, minute; bracteoles aggregate into an involucel, the outer connate at base, larger (0.5–0.7 cm), spatulate, with a prominent main vein, densely covered with glandular hairs. Calyx funnel-shaped, ca. 3 mm long, membranous, straw-colored, densely covered with appressed hairs; corolla scarlet or red-orange, 2–3 cm long. Capsule ca. 2 cm long, flattened, obovate, opening along 2 lateral valves. Seeds ca. 2 mm long, lenticular, brown, minutely spinulose, covered with glandular hairs.

DISTRIBUTION: A roadside weed found in open or disturbed areas. Cruz Bay (A3203), Lameshur (A3150). Also on St. Croix, St. Thomas, and Tortola; Bahamas, Greater Antilles, and from Mexico to northern South America to Brazil.

5. JUSTICIA L.

Erect or decumbent herbs. Leaves with numerous cystoliths; petioles slender and shorter than blade. Flowers in dense or sparse spikes, racemes, or panicles. Bracts and bracteoles usually awl-shaped; calyx funnel-shaped, usually with 5, awl-shaped sepals; corolla funnel-shaped, zygomorphic and bilabiate; stamens 2, slightly projecting beyond the tube; stigma capitate. Capsule ellipsoid or club-shaped, 2- or 4-seeded.

A tropical and subtropical genus with about 300 species.

Key to the Species of *Justicia*

1. Leaves narrowly lanceolate; bracts awl-shaped.
   2. Flowers spaced along elongate panicles; bracts 1–2 mm long; corolla 8–9 mm long .... 3. *J. pectoralis*
   2. Flowers clustered in short spikes; bracts 8–10 mm long; corolla 1.6–2.5 mm long .... 4. *J. periplocifolia*
1. Leaves ovate or elliptic; bracts spatulate, ovate, or deltate.
   3. Leaves ovate, pubescent; stems with lines of hairs along two sides; bracts foliaceous, ovate to deltate .... 2. *J. mirabiloides*
   3. Leaves elliptic, glabrescent; stems pubescent only at nodes; bracts spatulate .... 1. *J. carthaginensis*

1. Justicia carthaginensis Jacq., Enum. Syst. Pl. 11. 1760. *Adhatoda carthaginensis* (Jacq.) Nees *in* A. DC., Prodr. **11**: 403. 1847.

Justicia retusa Vahl, Symb. Bot. **2**: 8. 1791; Enum. Pl. **1**: 136. 1805.

Erect herb or subshrub to 1.5 m tall; stems cylindrical, with swollen, hairy collapsing nodes (when dried). Leaf blades 3–11 × 1.5–4.2 cm, elliptic, chartaceous, glabrous except for a few hairs along veins, with numerous cystoliths on both surfaces, the apex short or long acuminate, the base tapering, the margins revolute, entire or wavy and ciliate; petioles slender, 0.3–1 cm long. Flowers borne on dense axillary and terminal spikes; bracts spatulate. Calyx bell-shaped, with 5 deeply cleft, lanceolate-lineate sepals ca. 8 mm long; corolla reddish purple or violet, 3–4 cm long, with glandular hairs externally and often with whitish stripes within. Capsule 1.8 cm long, club-shaped or ellipsoid, straw-colored, minutely pubescent. Seeds 4, ca. 3 mm long, nearly globose, smooth, black or dark brown, shiny.

DISTRIBUTION: Uncommon, found in open disturbed areas. Enighed (A3927). Also on St. Croix, St. Thomas, and Tortola; widespread throughout the West Indies, Central America, Colombia, and Venezuela.

2. Justicia mirabiloides Lam., Tabl. Encycl. **1**: 39. 1791. *Drejerella mirabiloides* (Lam.) Lindau *in* Urb., Symb. Ant. **2**: 222. 1900. Fig. 8K–O.

Beloperone portoricensis Nees *in* A. DC., Prodr. **11**: 414. 1847.
Beloperone nemorosa sensu Eggers, Fl. St. Croix 80. 1879, non Nees, 1825.

Erect herb 0.5–1 m tall; stems cylindrical, drying blackish, with lines of hairs along two sides. Leaf blades 5–12 × 2.5–5.5 cm, ovate or widely lanceolate, chartaceous, with numerous cystoliths, pubescent, especially along the midvein, the apex acute or acuminate, the base deltoid or semirounded, the margins sinuate, revolute, and ciliate; petioles 1–3 cm long, slender. Flowers in dense terminal spikes; bracts foliaceous, ovate to deltoid, overlapping, with ciliate margins. Calyx funnel-shaped, with 5 lanceolate sepals, 4–5 mm long; corolla pink, 1.5–2 cm long. Capsule ca. 1 cm long, ellipsoid and pubescent. Seeds ca. 2.5 mm long, blackish, with glandular hairs.

DISTRIBUTION: In moist forest understory. Bordeaux area (A2597, A3171, A3175). Also on St. Croix and St. Thomas; the Greater Antilles.

3. Justicia pectoralis Jacq., Enum. Syst. Pl. 11. 1760. *Stethoma pectoralis* (Jacq.) Raf., Fl. Tellur. **4**: 61. 1838. *Dianthera pectoralis* (Jacq.) Murray, Syst. Veg., ed. 14, 64. 1784.

Decumbent herb, 30–40(–60) cm long, profusely branching from near the base; stems cylindrical, glabrous, except for a line of hairs along one side. Leaf blades 5–11 × 0.5–1.8 cm, lanceolate or narrowly lanceolate, membranous, glabrous, with numerous cystoliths, the apex long-acuminate, the base acute or obtuse and slightly unequal, the margins entire or minutely serrate; petioles slender, 5–8 mm long. Flowers borne on long terminal spiciform, unilateral panicles; bracts and bracteoles minute and awl-shaped, with minute glandular hairs. Calyx ca. 2.2 mm long, with 5 deeply cleft, lanceolate sepals; corolla lilac, purple, or pink, with white markings, 8–9 mm long, funnel-shaped. Capsule ca. 5 mm long, club-shaped, covered with glandular hairs. Seeds 4, 1.5 mm diam., reddish brown, lenticular, papillose.

DISTRIBUTION: Not common, mostly cultivated in moist grounds. Cinnamon Bay (A4097). Also on St. Croix and St. Thomas; widespread in tropical America.

COMMON NAMES: garden balsam, sweet mint.
USES: Locally used in tea for relief of cough.

4. Justicia periplocifolia Jacq., Collectanea **5**: 7. 1796 [1797]. Fig. 8F–J.

Justicia reflexiflora Vahl, Enum. Pl. **1**: 157. 1804.
Justicia reflexiflora var. *glandulosa* Eggers, Fl. St. Croix 80. 1879.

Erect herb to 50 cm tall, many-branched from a woody base; stems slender, cylindrical, glabrous and slightly swollen at nodes. Leaf blades 3–12 × 0.4–1.8 cm, lanceolate to linear, membranous, sparsely pubescent, with cystoliths, the apex long acuminate or acute, the base attenuate or obtuse, the margins entire and ciliate; petioles slender, 0.5–1.5 cm long. Flowers borne on dense axillary or terminal spikes; bracts and bracteoles linear, covered with glandular hairs. Calyx funnel-shaped, 8 mm long, with 5 deeply cleft, linear sepals; corolla lavender, 2.5 cm long, funnel-shaped, sparsely covered with glandular hairs externally, the lobes 1.2 cm long. Capsule obovoid, turning from green to straw-colored, densely pubescent. Seeds ca. 3 mm long, nearly globose, black, densely covered with minute, woolly, brown hairs.

DISTRIBUTION: A common weed from waste grounds and thickets. Bethany (B339), Bordeaux Mountain (A1894), Lameshur (A2747). Also on St. Croix, St. Thomas, and Tortola; Cuba, Hispaniola, Puerto Rico, and Lesser Antilles.

6. OPLONIA Raf.

Erect or scrambling shrubs, mostly with opposite axillary spines. Leaves small, short petioled or sessile; blades with numerous, transversely oriented cystoliths, secondary veins inconspicuous. Flowers heterostylous, axillary, solitary or in short condensed racemes; calyx with 5 awl-shaped, deeply parted lobes; corolla funnel-shaped, zygomorphic, with 5 lobes, one wider; stamens 2, included or exserted beyond the tube; staminodes 2; style filiform, the stigma simple. Capsule club-shaped, containing 2 or 4 lenticular seeds.

A genus with 19 species, 14 of which are native to tropical America and 5 to Madagascar.

Key to the Species of *Oplonia*

1. Axillary spines straight and ascending.. 1. *O. microphylla*
1. Axillary spines, at least some, curved, reflexed, or perpendicular to stem .... 2. *O. spinosa*

1. Oplonia microphylla (Lam.) Stearn, Bull. Brit. Mus. (Nat. Hist.), Bot. **4(7)**: 307. 1971. *Justicia microphylla* Lam., Tabl. Encycl. **1**: 37. 1791. *Anthacanthus microphyllus* (Lam.) Nees *in* A. DC., Prodr. **11**: 461. 1847.   Fig. 9E–J.

Erect or prostrate shrub with arching branches, 1–2 m long; stems obtusely quadrangular or cylindrical, densely covered with minute whitish hairs when young. Leaves usually clustered at nodes; blades 0.7–1.6 × 0.4–1.2 cm, ovate, obovate, elliptic, or narrowly elliptic, coriaceous, glabrous, with numerous cystoliths, the apex acute or obtuse, with a tuft of minute hairs, the base cuneate or attenuate, the margins revolute; petioles pubescent, ca. 1 mm long. Spines 0.4–1.2 cm long, straight, ascending, rarely perpendicular, pubescent. Flowers heterostylous, solitary or in condensed racemes, often with minute glandular hairs; bracts minute and awl-shaped. Calyx funnel-shaped, ca. 5 mm long, with 5 lanceolate, deeply cleft sepals; corolla lavender or pink, 1.8–2.0 cm long, the tube white within. Capsule 2–2.2 cm long, club-shaped and glabrous. Seeds ca. 4 mm long, lenticular, 4 per capsule.

DISTRIBUTION: Frequent in coastal thickets. Salt Pond (A760); Southside Pond (A1813, W600). Also occurring on St. Croix and St. Thomas; widely distributed in the West Indies, and occurring in Jamaica, Hispaniola, Puerto Rico, and Lesser Antilles south to the Grenadines.

2. Oplonia spinosa (Jacq.) Raf., Fl. Tellur. **4**: 65. 1838. *Justicia spinosa* Jacq., Enum. Syst. Pl. 11. 1760. *Anthacanthus spinosus* (Jacq.) Nees *in* A. DC., Prodr. **11**: 460. 1847.   Fig. 9K.

Scandent shrub to 2 m long; stems slender, obtusely quadrangular, densely covered with minute, whitish hairs when young. Leaves usually clustered at nodes; blades 0.5–2.2 × 0.4–2 cm, elliptic, obovate, or narrow elliptic, coriaceous, glabrous, with numerous cystoliths, the apex obtuse or rounded, slightly notched, with a tuft of minute hairs, the base cuneate or attenuate, the margins strongly revolute; petioles minute. Spines 0.4–1.2 cm long, recurved or rarely straight, perpendicular. Flowers solitary or in condensed racemes; bracts minute and awl-shaped. Calyx funnel-shaped, 3 mm long, with 5 deeply cleft, lanceolate sepals; corolla lavender or pink, 1.2–1.5 cm long. Capsule 1.2 cm long, club-shaped and glabrous. Seeds 3–4 mm long, lenticular, brown, 4 per capsule.

DISTRIBUTION: Common on open hillsides or in forests along ravines. Dittlif Point (A3975), Maria Bluff (A594). Also on St. Croix, Tortola, and Virgin Gorda; Bahamas, Cuba, Hispaniola, Puerto Rico, and the Lesser Antilles.

7. RUELLIA L.

Erect herbs or subshrubs. Leaves with numerous, transversely oriented cystoliths; petioles shorter than the blades. Flowers solitary or in axillary or terminal spikes, racemes, or panicles; calyx with 5 linear sepals, deeply cleft to the base; corolla funnel-shaped, regular; stamens 4, didynamous; stigma with 2 unequal lobes. Capsule oblong or club-shaped; seeds 8–32 (or more numerous) per capsule, lenticular.

Approximately 250 species with tropical and subtropical distribution, with a few species extending to the temperate zone.

Key to the Species of *Ruellia*

1. Flowers lilac, light blue, pink, or white; anthers not projecting beyond the tube; leaves with entire or crenate margins; petioles short (less than one-fourth of blade length).
    2. Roots tuberous; leaves elliptic, oblong, ovate, or obovate... 3. *R. tuberosa*
    2. Roots fibrous, not tuberous; leaves linear or narrowly elliptic .... 1. *R. brittoniana*

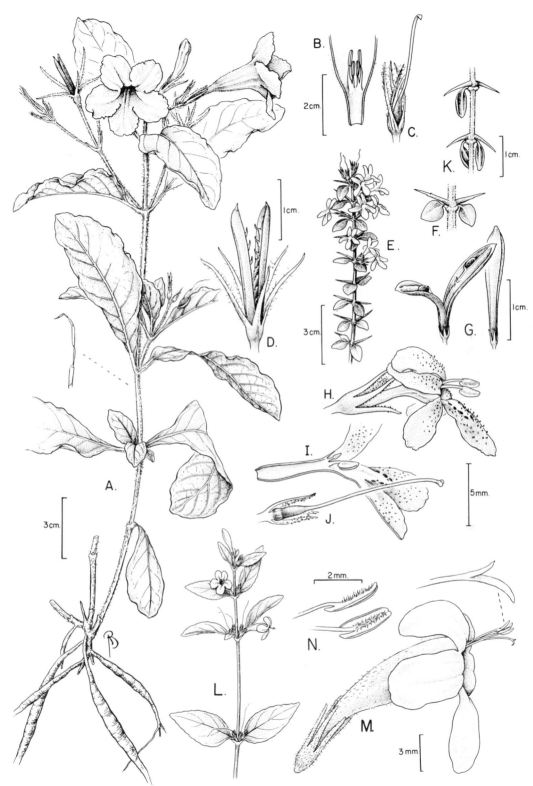

Fig. 9. A–D. *Ruellia tuberosa*. A. Habit and detail of trichome. B. L.s. corolla showing didynamous stamens. C. Calyx and pistil. D. Dehisced capsule with hooklike placental tissue and persistent calyx. E–J. *Oplonia microphylla*. E. Flowering branch. F. Node with axillary spines. G. Dehisced and closed capsule. H. Flower. I. L.s. corolla with stamen and staminode. J. Calyx with 2 sepals removed to show pistil. K. *Oplonia spinosa*; node with axillary spines. L–N. *Ruellia coccinea*. L. Fertile branch. M. Flower and detail of 2-lobed stigma. N. Anther, lateral and dorsal views.

1. Flowers red; anthers projecting beyond the tube; leaves crenate-dentate; petioles long (to open-third of blade length)... 2. *R. coccinea*

1. Ruellia brittoniana Leonard, J. Wash. Acad. Sci. **31:** 96. 1941.

Cryphiacanthus angustifolius Nees *in* A. DC., Prodr. **11:** 199. 1847, in part, non *Ruellia angustifolia* Sw.
Ruellia spectabilis Britton, Ann. New York Acad. Sci. **7:** 192. 1893, non Nicholson, 1886.
Ruellia tweedeana sensu I. Vélez, 1950 & H. Alain Liogier, 1982.

Subshrub to 1 m tall; stems obtusely quadrangular, glabrous except for a few hairs at nodes; nodes collapsing in dry specimens. Leaves ascending; blades 6–20 × 0.5–2 cm, linear or narrowly elliptic, chartaceous, glabrous, with prominent veins running along margins, upper surface with minute linear cystoliths transversely oriented, the apex and base long acuminate, the margins entire or wavy and revolute; petioles 0.5–2.5 cm long, slender. Flowers produced in short, axillary cymes; bracts linear, 1.5–2 cm long. Calyx with 5 equal, lanceolate sepals 1–1.3 cm long; corolla lilac or light blue, 3–4.5 cm long, with 5, rounded lobes. Capsule 2.3–3 cm long, club-shaped, apiculate at apex. Seeds ca. 2.5 mm wide, lenticular, light brown.

DISTRIBUTION: Uncommon, growing as an escape around settlements. Bordeaux Mountain (A3244), Lameshur (W774). Native to Mexico, cultivated throughout the West Indies, and often escaped.

2. Ruellia coccinea (L.) Vahl, Symb. Bot. **3:** 83. 1794.
Barleria coccinea L., Sp. Pl. 637. 1753. *Stemonacanthus coccineus* (L.) Griseb., Fl. Brit. W. I. 452. 1862. Fig. 9L–N.

Erect herb or subshrub to 1.5 m tall, usually branching at upper part; stems 4-angular, sulcate, glabrous, except for the node area. Leaf blades 3–15 × 1–5.5 cm, lanceolate or seldom ovate, chartaceous, upper surface glabrous or puberulent with many cystoliths, lower surface sparsely pubescent, especially on veins, the apex acuminate or acute, the base rounded, truncate or attenuate, the margins crenate-dentate and ciliate; petioles slender, 0.5–5 cm long. Flowers produced on short axillary cymes along main stem and short axillary branches; bracts and bracteoles minute and awl-shaped. Calyx 3 mm long with awl-shaped, deeply cleft sepals; corolla red, pink, or crimson, 2.2–3 cm long, puberulent externally; anthers projecting beyond the tube. Capsule 1.5 cm long, club-shaped, minutely pubescent. Seeds 1.5 mm long, lenticular, light brown, with whitish papillate margins, 8(–10) per capsule.

DISTRIBUTION: Uncommon in moist forest understory. Bordeaux area (A3174, W728). Also on St. Croix, St. Thomas, and Tortola; Hispaniola and Puerto Rico.

3. Ruellia tuberosa L., Sp. Pl. 635. 1753. Fig. 9A–D.

Herb to 50 cm tall, many-branched from a short woody base; roots thick, elongate, fusiform; stems quadrangular, sulcate, and covered with white, curly hairs when young, cylindrical and glabrescent at base. Leaves usually clustered at nodes; blades 4–9 × 2–3.7 cm, elliptic, oblong, ovate, or obovate, chartaceous, pubescent on both surfaces, especially on veins, the apex obtuse or rounded, usually retuse or shortly apiculate, the base long attenuate, the margins crenate and ciliate. Flowers produced on long or short axillary simple dichasia; bracts and bracteoles strap-shaped and ciliate at margins. Calyx of 5 lanceolate-linear, green, ciliate, deeply cleft sepals 0.7–1.5 cm long; corolla lilac, mauve to purplish blue (rarely white), 3–4.5 cm long, minutely pubescent externally, the lobes rounded and spreading. Capsule 1.2–2.2 cm long, nearly cylindrical or narrowly ellipsoid, glabrous, turning from green to brown. Seeds ca. 2 mm long, light brown, 20–32(–52) per capsule.

DISTRIBUTION: In open disturbed areas such as roadsides. Annaberg (A2923), Denis Bay (M16590), Lameshur (A2756). Also on St. Croix, St. Thomas, Tortola, and Virgin Gorda; widely distributed throughout tropical and subtropical America.

COMMON NAMES: Christmas pride, many roots, minie root.

8. SIPHONOGLOSSA Oerst.

Herbs or shrubs. Leaves petiolate; blades with numerous minute cystoliths. Bracts lacking. Flowers solitary or few at leaf axil; calyx with 4 sepals; corolla salverform, zygomorphic, with long, narrow tube and 2-lipped limbs; stamens 2, adnate at corolla mouth, included or projecting beyond the tube; stigma slightly 2-lobed. Capsule club-shaped, usually 4-seeded.

About 25 species, ranging from southern North America to northern South America, the West Indies, and South Africa.

1. Siphonoglossa sessilis (Jacq.) D.N. Gibson, Fieldiana, Bot. **34:** 82. 1972. *Justicia sessilis* Jacq., Enum. Syst. Pl. 11. 1760. *Rhytiglossa sessilis* (Jacq.) Nees *in* A. DC., Prodr. **11:** 345. 1847. *Dianthera sessilis* (Jacq.) Griseb., Fl. Brit. W. I. 455. 1862. Fig. 10K–P.

Justicia pauciflora Vahl, Eclog. Amer. **1:** 2. 1797.
Justicia borinquensis Britton *in* Britton & P. Wilson, Bot. Porto Rico **6:** 217. 1925.

Erect herb, 10–20 cm tall, usually many-branched from a semiwoody base; stems slender, cylindrical, with a line of hairs along one side. Leaf blades 2–5 × 0.8–2.6 cm, lanceolate, ovate or elliptic, chartaceous, pubescent or glabrescent, the apex acute, obtuse or rounded, sometimes with a short apiculum, the base obtuse or cordate, the margins ciliate, entire, less often crenate or lobed; petioles slender and short. Flowers solitary in leaf axil, but sometimes appearing congested because of short, lateral branches bearing flowers; bracts awl-shaped and minute. Calyx funnel-shaped, with 4 deeply cleft, lanceolate sepals, 4–5 mm long; corolla lavender, 1.2–2.5 cm long. Capsule ca. 1 cm long, club-shaped, with scattered glandular hairs. Seeds ca. 2 mm long, lenticular, yellowish brown and verrucose, 4 per capsule.

DISTRIBUTION: In open disturbed areas. Adrian Ruins (A2648), Lameshur (A2746, B641). Also on St. Croix, St. Thomas, Tortola, and Virgin Gorda; Puerto Rico, St. Martin to Guadeloupe, Trinidad, Margarita Island, and Colombia.

COMMON NAME: blossom.

USES: The plant is used locally for making a tea for cold relief.

FIG. 10. A–F. *Thunbergia fragrans*. **A.** Fertile branch. **B.** Floral tube and subtending bract. **C.** L.s. corolla with didynamous stamens. **D.** Pistil with style and detail of 2-lobed stigma. **E.** Capsule, and l.s. capsule. **F.** Seed. G–J. *Stenandrium tuberosum*. **G.** Fertile plant with detail of trichome. **H.** Flower with subtending bract. **I.** L.s. corolla, with stamens and pistil. **J.** Seed and dehisced capsule. K–P. *Siphonoglossa sessilis*. **K.** Habit. **L.** Node showing indument. **M.** Flower. **N.** Stamen adnate to corolla mouth and detail of anther. **O.** Calyx with 2 sepals removed to show pistil with cupular disk. **P.** Seed and capsule.

9. STENANDRIUM Nees, nom. cons.

Acaulescent low herbs. Leaves congested in a rosettelike configuration; blades with minute cystoliths; petioles usually elongate. Flowers borne on scapose spikes with imbricate bracts; calyx of 5, subequal sepals; corolla funnel-shaped; stamens 4, didynamous, adnate to the throat, not projecting beyond the lobes; stigma club-shaped or slightly lobed. Capsule ellipsoid to nearly cylindrical; seeds 2 or 4 per capsule.

A tropical and subtropical genus with approximately 50 species.

1. **Stenandrium tuberosum** (L.) Urb., Symb. Ant. **4:** 576. 1911. *Gerardia tuberosa* L., Sp. Pl. 610. 1753. Fig. 10G–J.

Ruellia rupestris Sw., Prodr. 93. 1788. *Stenandrium rupestre* (Sw.) Nees *in* A. DC., Prodr. **11:** 283. 1847.

Gerardia portoricensis Britton & P. Wilson, Bot. Porto Rico **6:** 214. 1925.

Herb 3–10 cm tall; roots usually tuberous; stems subterranean and short. Leaves congested, in a rosette at base; blades 1.5–4 × 0.8–1 cm, oblanceolate, obovate, spatulate, or ovate, membranous, densely or sparsely covered with white hairs, the cystoliths inconspicuous, the apex obtuse or rounded, the base tapering or nearly cuneate, the margins crenate and ciliate; petioles 1–5 cm long, slender, densely covered with white, long hairs. Flowers in scapose inflorescences; bracts lanceolate or awl-shaped, ca. 6 mm long, with long hairs. Calyx with 5, awl-shaped, greenish, 3 mm long sepals; corolla pinkish lavender, to 1.2 cm long. Capsule ca. 6 mm long, nearly cylindrical, puberulent to glabrous. Seeds ca. 1.5 mm long, lenticular, brown, with a few glandular hairs.

DISTRIBUTION: Rare, gregarious in forest understory. Bethany (B352), Fish Bay (W761). Also on St. Thomas; Puerto Rico, Hispaniola, St. Martin, and Guadeloupe.

10. THUNBERGIA Retz., nom. cons.

Herbaceous or woody climbers, often twining, or less often shrubs. Leaves without cystoliths. Flowers usually solitary at leaf axils, with long pedicels and a pair of foliaceous bracts covering the basal portion of the corolla; calyx short, cupuliform, truncate or many-toothed; corolla funnel-shaped or salverform, large, showy, regular or irregular; stamens 4, didynamous. Capsule globose at base, with a beaked apex; seeds nearly globose.

Approximately 200 species in the Old World tropics, with about a dozen cultivated or naturalized in the American tropics.

1. **Thunbergia fragrans** Roxb., Pl. Coromandel **1:** 47. 1796. Fig. 10A–F.

Thunbergia volubilis Pers., Syn. Pl. **2:** 179. 1806.

Herbaceous twiner, 2–3 m long; stems slender, cylindrical, striate, covered with minute appressed hairs. Leaf blades 6–11 × 1.8–6 cm, ovate or lanceolate, chartaceous, sparsely covered with appressed hairs, without cystoliths, upper surface slightly scabrous, the apex acute, the base truncate or nearly rounded, the margins entire and ciliate. Flowers solitary at leaf axil, not fragrant. Bracts green, ovate, 1.6–2 cm long. Calyx with numerous lanceolate sepals, 5 mm long; corolla white, 2.5–4 cm long, the tube narrowing at the base, yellow within. Capsule 1–2.5 cm long. Seeds ca. 5 mm diam., globose, brown, 2–4 per capsule.

DISTRIBUTION: Naturalized, found along roadsides. Annaberg (A1928), Bethany (B253), Susannaberg (A4665). Also on St. Croix, St. Thomas, and Tortola; native to India and Sri Lanka but naturalized throughout the tropics.

COMMON NAME: white Susan vine.

DOUBTFUL RECORD: *Thunbergia alata* Bojer *ex* Sims was reported from St. John by Britton and P. Wilson (1925) and by Woodbury and Weaver (1987), but no specimen has been located to confirm this report. In any case, this species with orange flowers does not seem to be persistent, as it has not been collected recently on the island.

2. Aizoaceae (Fig-Marigold Family)

Low succulent herbs, mostly prostrate or spreading. Leaves simple, opposite or alternate; stipules lacking or minute. Flowers bisexual, actinomorphic, axillary, solitary, clustered or in cymes; calyx with 4 or 5 fleshy sepals; corolla lacking in our taxa, but present in other members of the family; stamens 5 to numerous; ovary superior, with 2–5 or more united carpels; ovules 1 or more per carpel. Fruit a circumscissile capsule; seeds small, numerous.

Tropical or subtropical, with the majority of species in South Africa and Australia. The number of taxa recognized in this family varies from 10 to 130 genera and from 1200 to 2500 species according to different workers.

REFERENCE: Bogle, A. L. 1970. The genera of Molluginaceae and Aizoaceae in the southeastern United States. J. Arnold Arbor. 51: 431–462.

Key to the Genera of Aizoaceae

1. Primary branches short (usually <10 cm long); leaves minute (1–6 mm long); stamens 1–5; sepals without a dorsal keel .... 1. *Cypselea*
1. Primary branches long (>15 cm long); leaves 1–3 cm long; stamens 5 or more; sepals with a dorsal keel.
    2. Leaves oblanceolate, thickened (3–4 mm thick when fresh), acute at apex; stamens numerous .... 2. *Sesuvium*
    2. Leaves obovate or rounded, oblique at base, thin (<2 mm thick when fresh), obtuse, rounded and apiculate at apex; stamens 5–10 .... 3. *Trianthema*

1. CYPSELEA Turpin

A monospecific genus, characterized by the following species.

1. Cypselea humifusa Turpin, Ann. Mus. Natl. Hist. Nat. 7: 219. 1806. Fig. 11K–N.

Prostrate, mat-forming, perennial herb, many-branched from a taprooted base; main branches fleshy and almost cylindrical, to 10 cm long. Leaves opposite; blades 1–6 × 0.7–5 mm (one of them smaller), elliptic, oval, or obovate, fleshy, glabrous, the apex rounded, the base obtuse or rounded, the margins papillose; petioles 0.5–2 mm long. Stipules with fimbriate margins, decurrent from the petiole, forming a sheath around the stem. Secondary branches short, with clustered leaves, alternate along the main branch and opposite to the larger leaf of the opposite pair. Flowers solitary in leaf axil. Pedicels ca. 2 mm long; calyx funnel-shaped, whitish, ca. 2 mm long, not keeled, papillose externally; the sepals 5, 1 mm long; corolla lacking; stamens 5, white, adnate to the sepals; ovary rounded, 2-celled; stigmas short and papillose. Capsule thin, circumscissile at base, with axile placentation. Seeds numerous, lenticular, light brown, ca. 0.3 mm long, with thread-like placental tissue.

DISTRIBUTION: Uncommon, in marshy areas. Chocolate Hole (W709), Hurricane Hole (A2832). Also on Anegada and St. Thomas; southern United States, Greater Antilles, and Guadeloupe.

2. SESUVIUM L.

Succulent prostrate herbs. Leaves opposite, succulent and stipulate. Flowers solitary and axillary; calyx with 5 sepals, often colored, with a central keel on dorsal side; corolla lacking; stamens 5 to numerous; ovary 3–5 celled, with numerous ovules. Fruit a circumscissile capsule; seeds numerous, stalked, usually smooth.

A small genus with 8–10 species from tropical and subtropical, coastal, salty areas.

1. Sesuvium portulacastrum (L.) L., Syst. Nat., ed. 10, 2: 1058. 1759. *Portulaca portulacastrum* L., Sp. Pl. 446. 1753. Fig. 11A–D.

Decumbent, succulent, perennial herb, 10–30 cm long, many-branched from a taprooted base; stems cylindrical, succulent, yellowish green, reddish and rooting at nodes. Leaf blades 2–4 × 0.5–0.8 cm, oblanceolate, elliptic or oblong, fleshy (2–3 mm thick), glabrous, the apex acute or obtuse, the base acute, the margins entire; petioles 3–5 mm long, yellowish. Flowers solitary in leaf axils. Calyx showy, funnel-shaped, ca. 1 cm long, the sepals spreading, lavender or white within, green on lower surface, with a green hornlike projection; stamens numerous, in a columnar structure, the anthers deep pink or white; ovary ovoid with 3–5 distinct stigmas. Capsule conical, whitish, 9–11 mm long. Seeds lenticular, black, ca. 1 mm diam.

DISTRIBUTION: Common along saline beach dunes. Fish Bay (A2815), Ram Head Point (A2915). Expected to be on all islands of the Virgin Islands; a polymorphic species with pantropical distribution.

COMMON NAMES: bay flower, sea purslane, sea pusley.

3. TRIANTHEMA L.

Succulent prostrate herbs; branches appearing dichotomous. Leaves opposite (one of them smaller); stipules decurrent from base of petiole, forming a sheath around stem. Flowers solitary or in cymes; calyx funnel-shaped, sepals with a dorsal keel; stamens 5–10, usually of unequal length; ovary 1- or 2-celled. Fruit a membranous, circumscissile capsule; seeds lenticular-reniform and minute.

A genus of 20 species, with only 1 occurring in the neotropics.

1. Trianthema portulacastrum L., Sp. Pl. 223. 1753. Fig. 11E–J.

Prostrate, succulent, mat-forming perennial herb, many-branched from a swollen taprooted base; stems cylindrical, glabrous, yellowish, usually reddish at nodes, to 90 cm long. Leaf blades 0.5–3.5 × 0.5–3.2 cm, obovate, spatulate, rounded, elliptic, or ovate, fleshy, glabrous, the apex acute, apiculate, obtuse, or retuse, the base acute, obtuse, or rounded, the margins entire, with reddish hue; petioles 0.2–2 cm long. Secondary branches short, with clustered leaves, opposite to larger leaf of node; stipules lanceolate, ca. 3 mm long. Flowers solitary in leaf axils; calyx pink within, green externally with pink hornlike keel, the sepals ovate or oblong, ca. 5 mm long; stamens 10, pink; stigmas 2, white. Capsule, 4–5 mm long, truncate at apex with a 2-sided crest. Seeds kidney-shaped, blackish, with radiating ridges.

DISTRIBUTION: Along waste ground or open areas. Annaberg (W512), Cruz Bay (A2875), Johnson Bay (A3160). Also on St. Croix, St. Thomas, and Tortola; common throughout tropical and subtropical America.

3. Amaranthaceae (Amaranth Family)

Herbs, shrubs or woody vines. Leaves opposite or alternate, simple, without stipules. Flowers unisexual or bisexual, actinomorphic or nearly so, produced in cymules, spikes, heads, or panicles, subtended by a bract and 2 bracteoles. Tepals 3–5, subequal or innermost shorter; stamens as many as tepals with free or united filaments; pseudostaminodes frequently present; ovary superior, unilocular, of 2 or 3 united carpels, ovoid or nearly globose; style 1 or 2; stigma capitate, bifid or with 3(6) branches; ovules basal, mostly solitary. Fruit a circumscissile capsule (utricle), dehiscent or indehiscent; seeds kidney-shaped, lenticular, or nearly globose.

About 65 genera and between 800 and 900 species, nearly cosmopolitan.

REFERENCES: Eliasson, H. U. 1987. Amaranthaceae. In G. Harling & L. Anderson (eds.), Flora of Ecuador 28: 1–137; Eliasson, H. U. 1988. Floral morphology and taxonomic relations among the genera of Amaranthaceae in the New World and the Hawaiian Islands. J. Linn. Soc., Bot. 96: 235–283.

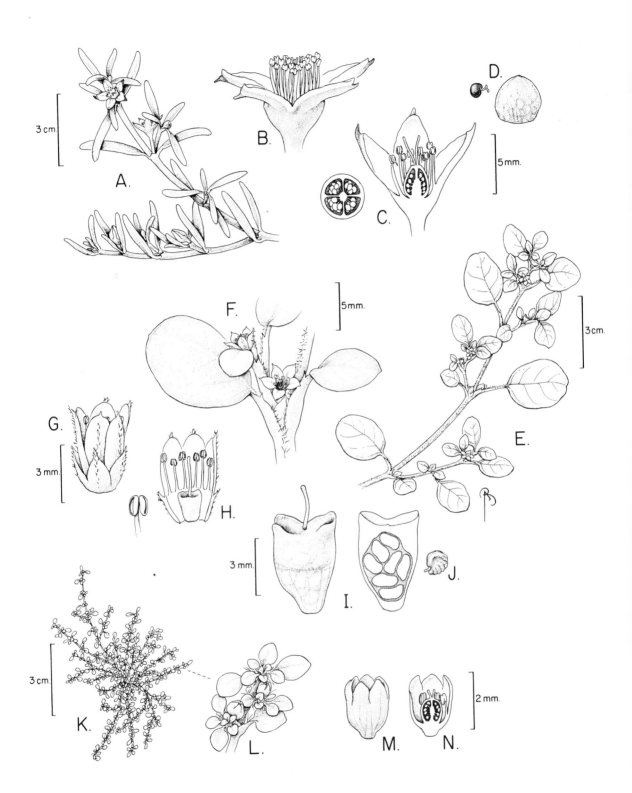

Fig. 11. A–D. *Sesuvium portulacastrum*. A. Flowering branch. B. Flower. C. C.s. ovary and l.s. flower. D. Stalked seed and capsule. E–J. *Trianthema portulacastrum*. E. Habit. F. Flowering branch; G. Flower. H. L.s. flower and detail of anther. I. Fruit and l.s. fruit. J. Seed. K–N. *Cypselea humifusa*. K. Habit. L. Fertile branch. M. Flower. N. L.s. flower.

Key to the Genera of Amaranthaceae

1. Leaves alternate.
    2. Flowers solitary on inflorescence axis; ovary with many ovules .... 5. *Celosia*
    2. Flowers congested along inflorescence axis; ovary with 1 ovule .... 3. *Amaranthus*
1. Leaves opposite.
    3. Plants prostrate or decumbent, usually mat-forming.
        4. Stigma capitate... 2. *Alternanthera*
        4. Stigma bifid.
            5. Plants fleshy; leaves oblong, linear, or lanceolate; inflorescence cylindrical, >1 cm long ... 4. *Blutaparon*
            5. Plants not fleshy; leaves elliptic; inflorescence globose, usually <1 cm long .... 6. *Gomphrena*
    3. Plants erect, 1 main stem or many-branched from base.
        6. Inflorescence terminal, paniculate, showing transition between leaves and bracts .... 7. *Iresine*
        6. Inflorescence terminal or lateral, of spikes or nearly globose heads.
            7. Inflorescence a spike; flowers early reflexed; bracteoles with spinelike apex .... 1. *Achyranthes*
            7. Inflorescence a nearly globose head; flowers ascending, bracteoles laterally compressed with a dorsal serrate crest.. 6. *Gomphrena*

1. ACHYRANTHES L.

Herbs or rarely shrubs. Leaves opposite and entire. Inflorescences of terminal and axillary spikes or spikelike. Bracts acuminate; bracteoles tapering to a spine. Flowers becoming reflexed; tepals (4–)5, persistent, connate at base; stamens 5, or rarely 2, alternating with pseudostaminodes and united at base into a cup; ovary with a single ovule per carpel, the style filiform and the stigma truncate. Fruit thin-walled, indehiscent, enclosed by a hardened perianth and persistent bracteoles, dispersing as a unit; seeds 1, cylindrical and brown.

A genus with less than 10 species, predominantly from the Old World.

1. Achyranthes aspera L., Sp. Pl. 204. 1753. *Centrostachys aspera* (L.) Standl., J. Wash. Acad. Sci. **5**: 75. 1915. Fig. 12E–J.

Achyranthes aspera var. *indica* L., Sp. Pl. 204. 1753. *Achyranthes indica* (L.) Mill., Gard. Dict., ed. 8. 1768. *Centrostachys indica* (L.) Standl., J. Wash. Acad. Sci. **5**: 75. 1915.

Erect annual herb or subshrub to 1 m tall; stems obtusely 4–angled, glabrous to densely pubescent. Leaf blades 3–15 × 1.5–6 cm, obovate, elliptic, or lanceolate, chartaceous, sparsely to densely covered with whitish hairs, especially along veins on lower surface, the apex acuminate or apiculate, the base tapering, sometimes oblique, the margins undulate; petioles 0.5–1.5 cm long. Flowers early reflexed, produced on long spikes with densely pilose axis. Bracts membranous, hyaline, ovate, long acuminate, with ciliate margins; bracteoles translucent, straw-colored, ciliate at margins, with main vein projecting as a spine. Tepals lanceolate, straw-colored or yellowish, stiff, glabrous and shiny, the outer ones longer (5–8 mm) than the innermost ones; stamens 2–4 mm long; pseudostaminodes rectangular; style 1–2 mm long, the stigma minute. Fruit 1–2 mm long, cylindrical, with truncate apex, enclosed by a hardened perianth that sticks to clothing, hair, and fur.

DISTRIBUTION: Common in open or disturbed areas, especially along the coast. Cruz Bay (A3086). Also on Tortola and Virgin Gorda; probably native to tropical Asia, but now cosmopolitan.

COMMON NAME: better man better.

2. ALTERNANTHERA Forssk.

Prostrate or scandent annual or perennial herbs, rarely shrubs. Leaves opposite and entire. Inflorescence of axillary or terminal, globose to cylindrical heads or short spikes. Bracts usually scarious; bracteoles boat-shaped. Tepals (4–)5, free, subequal, or outer longer; stamens 5, rarely 3 or 4, alternating with triangular interstaminal appendages, free or united into a short tube; ovary nearly globose or obovoid, of a single ovule carpel, the style of variable length, the stigma capitate. Fruit a thin-walled, flattened utricle, with persistent tepals, usually dispersed as a unit; seeds solitary, lenticular, smooth.

A predominantly neotropical genus, with about 150 species.

Key to the Species of *Alternanthera*

1. Most of inflorescences on long peduncles; largest leaf blades >5 cm long; petioles 0.5–5 cm long . 1. *A. brasiliana*
1. All inflorescences congested in leaf axils; leaves <5 cm long; petioles <2 cm long.
    2. Tepals lanceolate, with 3 main veins, entire margins and with simple or dendroid hairs.
        3. Spreading or ascending, robust herbs, branches usually reaching 1 m long; leaves elliptic to obovate; tepals with simple hairs .... 4. *A. tenella*
        3. Trailing, slender herbs rooting at nodes, branches usually to 70 cm long; leaves obovate or circular; tepals with dendroid hairs.. 3. *A. crucis*

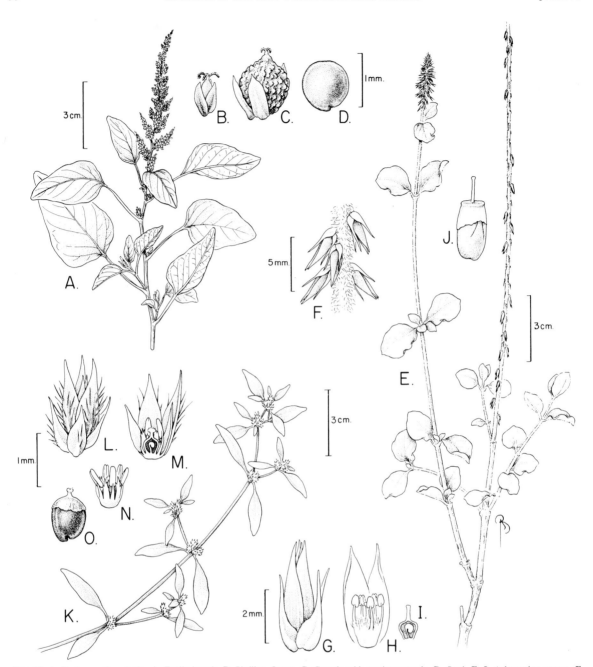

Fɪɢ. 12. A–D. *Amaranthus viridis.* **A.** Fertile branch. **B.** Pistillate flower. **C.** Capsule with persistent tepals. **D.** Seed. **E–J.** *Achyranthes aspera.* **E.** Fertile branch. **F.** Detail of inflorescence. **G.** Flower with subtending bracteoles. **H.** Flower with part of perianth removed. **I.** L.s. pistil. **J.** Seed with persistent upper portion of fruit wall and style. **K–O.** *Alternanthera tenella.* **K.** Fertile branch. **L.** Flower with subtending bracteoles. **M.** L.s. flower. **N.** Staminal ring, with stamens and interstaminal appendages. **O.** Seed with upper part of utricle present.

    2. Tepals ovate-lanceolate, mostly with 1(3) main vein, fimbriate margins and apex covered with hairs
       bearing retrorse barbs .... 2. *A. caracasana*

1. Alternanthera brasiliana (L.) Kuntze, Revis. Gen. Pl.
    2: 537. 1891. *Gomphrena brasiliana* L., Cent. Pl.
    II: 13. 1756.

    Decumbent subshrub, to 70 cm long; stems reddish, nearly
cylindrical, striate and rooting at lower nodes; young parts with

appressed white hairs. Leaf blades 3–11.5 × 1–6.5 cm, ovate,
elliptic, or lanceolate, chartaceous, upper surface glabrous, lower
surface pubescent or glabrescent, with reddish veins, the apex
acute or shortly acuminate, the base tapering, the margins entire
and ciliate; petioles 0.5–5 cm long. Flowers cylindrical to nearly
globose, axillary or terminal, long peduncled inflorescences; bracts

2 mm long; bracteoles 4 mm long, pubescent, with fimbriate upper margin. Tepals 4–5 mm long, oblong, straw-colored, pubescent, with 3 conspicuous veins, the midvein continuous to the tip; filaments united into a tube 1.5–2 mm long, the anthers lanceolate; staminodes exceeding the stamens, dentate at apex; ovary nearly conical, the style short, the stigma nearly capitate, white. Fruit flattened ellipsoid, with 2 protuberances at apex and a persistent style. Seeds lenticular, ca. 1 mm diam., light brown.

DISTRIBUTION: An escape, found only around settlements. Lameshur at VIERS (A3201). Also from Mexico to northern South America, Hispaniola, Puerto Rico, Lesser Antilles, Peru, Ecuador, and Brazil.

2. Alternanthera caracasana Kunth *in* Humb., Bonpl. & Kunth, Nov. Gen. Sp. **2**: 205. 1818.

Illecebrum peploides Willd. *ex* Roem. & Schult. , Syst. Veg. **5**: 517. 1819. *Alternanthera peploides* (Willd. *ex* Roem. & Schult.) Urb., Repert. Spec. Nov. Regni Veg. **15**: 168. 1918. *Achyranthes peploides* (Willd. *ex* Roem. & Schult.) Britton *in* Britton & P. Wilson, Bot. Porto Rico **5**: 279. 1924.

Alternanthera achyrantha R. Br. var. *parvifolia* Moq. *in* A. DC., Prodr. **13**(2): 359. 1849. *Alternanthera parvifolia* (Moq.) Fawc. & Rendle, Fl. Jamaica **3**: 139. 1914.

Alternanthera achyrantha sensu Griseb., Fl. Brit. W. I. 67. 1859, non R. Br., 1910.

Prostrate or spreading herb, 15–30 cm long, many-branched from a taprooted woody base; stems 4-angular, sparsely to densely covered with long white hairs, especially at the rooting nodes. Leaves congested at nodes, one of the pair smaller; blades 1–2.5 × 0.4–1.2 cm, elliptic, oblanceolate, or ovate-elliptic, chartaceous, puberulent, the apex acute or obtuse and mucronate, the base tapering, sometimes oblique, the margins entire or undulate; petioles 0.3–0.8 cm long, sericeous. Flowers produced in sessile, axillary, nearly globose heads; bracts and bracteoles 2–3.2 mm long, lanceolate or ovate. Tepals 3.5–4.5 mm long, ovate to lanceolate, whitish, with retrorsely barbed hairs, the margins hyaline and fimbriate, the single central vein projecting as a spine; filaments united at base, the anthers ovate; staminodes awl-shaped. Fruit compressed, circular. Seeds lenticular, 1 mm wide, yellowish brown.

DISTRIBUTION: In waste or open ground, especially along the coast. Calabash Boom (A2873, A3109). Also on St. Croix and St. Thomas; from eastern United States to Argentina, including the West Indies.

3. Alternanthera crucis (Moq.) Bold., Fl. Dutch W. Ind. Is. **1**: 58. 1909. *Telanthera crucis* Moq. *in* A. DC., Prodr. **13**(2): 362. 1849.

Alternanthera portoricensis Kuntze, Revis. Gen. Pl. **2**: 540. 1891.

Alternanthera culebrasensis Uline, Publ. Field Colombian Mus., Bot. Ser. **1**: 420. 1899.

Telanthera sintenisii Urb., Symb. Ant. **1**: 301. 1899. *Alternanthera sintenisii* (Urb.) Uline, Publ. Field. Colombian Mus., Bot. Ser. **1**: 421. 1899.

Telanthera dolichocephala Urb., Symb. Ant. **1**: 302. 1899. *Alternanthera dolichocephala* (Urb.) Urb., Symb. Ant. **4**: 221. 1905.

Prostrate herb, to 70 cm long, many-branched from base; branches trailing, rooting at nodes, slender, nearly cylindrical with long, woolly hairs. Leaf blades 1–2(–5) × 0.5–1.5(–4.5) cm, orbicular, or obovate, chartaceous, densely to sparsely sericeous, the apex rounded or obtuse and mucronate, the base tapering, rounded, nearly cordate or truncate; petioles 0.5–1(–2) cm long. Flowers in axillary, nearly globose or cylindrical, pedunculate inflorescences; bracts and bracteoles 1.5–2.5 mm long, ovate or lanceolate. Tepals 2.5–3.5 mm long, with 3 main veins, acute-apiculate, the outer ones ovate or lanceolate, covered with dendroid hairs, the inner ones smaller, conduplicate, glabrous or pubescent at base; anthers oblong; staminodes fimbriate at apex. Fruit ovoid, laterally compressed. Seeds ca. 1 mm diam., light brown.

DISTRIBUTION: On open ground and rocky hillsides, Little St. James Island (B1409). Also on Anegada, St. Croix, St. Thomas, and Tortola; Puerto Rico and in the Lesser Antilles.

4. Alternanthera tenella Colla, Mem. Reale Accad. Sci. Torino **33**: 131. 1829.   Fig. 12K–O.

Alternanthera polygonoides R. Br. var. *glabrescens* Griseb., Fl. Brit. W. I. 67. 1859.

Alternanthera ficoidea sensu authors in the West Indies, non R. Br., 1810.

Decumbent or scrambling, robust herb, to 1 m long, many-branched from a woody base; stems cylindrical or obtusely angular, rooting at nodes, usually with reddish hue, pubescent but becoming glabrous. Leaf blades 1.5–4 × 0.4–2 cm, one of the pair smaller, elliptic or spatulate, chartaceous, strigillose to sparsely strigillose, especially on lower surface, the apex acute or obtuse and mucronate, the base tapering and sometimes oblique, the margins entire and ciliate; petioles 0.4–0.8 cm long. Flowers in nearly globose sessile inflorescences; bracts and bracteoles 2–3.3 mm long, ovate. Tepals 2.5–4 mm long, with 3 main veins, the middle vein projecting as a spine at apex, densely or sparsely covered with ascending minutely spinulose hairs, the outer ovate or lanceolate, the inner smaller and conduplicate; anthers oblong; staminodes fimbriate at apex. Fruit ovoid. Seeds lenticular, ca. 1 mm wide, brown.

DISTRIBUTION: On open ground at low elevations, mostly along the coast. Coral Bay (A2830), Reef Bay (A2728), Waterlemon Bay (A1938). Also on St. Thomas; a polymorphic species distributed from Mexico to Argentina including the West Indies.

3. AMARANTHUS L.

Erect or prostrate annual herbs. Leaves alternate and petiolate, with entire or wavy margins. Flowers produced in dichasial cymes in spikelike, terminal or axillary inflorescences. Flowers unisexual, subtended by 1 bract and 2 bracteoles; plants dioecious or monoecious. Tepals greenish, 3–5; stamens (2–)3–5, the anthers with filaments free to base, interstaminal appendages absent; ovary ovoid, rounded or compressed, with a single ovule, the style short, the stigma slender, with 2 or 3 branches. Fruit a thin-walled, circumscissile capsule (utricle), with persistent style at apex; seeds vertically oriented, lenticular or nearly globose, smooth and shiny.

About 50–60 species, mostly native in the New World.

Key to the Species of *Amaranthus*

1. Plant prostrate; inflorescences nearly globose, aggregated in leaf axis .... 1. *A. crassipes*
1. Plant erect; inflorescence spikelike, elongate, axillary or terminal.
    2. Plant to 0.5 m tall; bracts acute or shortly acuminate at apex; stamens (2–)3; tepals of pistillate flowers 3, shorter than fruit; stigma as long as the verrucose fruit.. 3. *A. viridis*
    2. Plant to 1 m tall; bracts long acuminate at apex; stamens 5; tepals of pistillate flowers 5, longer than fruit; stigma much shorter than fruit .... 2. *A. dubius*

1. Amaranthus crassipes Schltdl., Linnaea 6: 757. 1831.

Prostrate herb, to 35 cm long, many-branched from a taprooted base; branches fleshy, light green. Leaf blades 0.9–1.9 × 0.3–1.2 cm, ovate, elliptic, or obovate, chartaceous, glabrous, the apex rounded and notched, the base tapering into a more or less elongate petiole, the margins entire. Flowers produced in congested, nearly globose axillary cymes; bracts and bracteoles minute, triangular. Tepals 1.5 mm long, oblong to spatulate with expanded base, greenish, involute, with hyaline margins; stamens 3–5, the anthers oblong, filaments glabrous; ovary ellipsoid, the style short, the stigmas 2, recurved, as long as the ovary. Fruit indehiscent, slightly flattened, ellipsoid. Seeds 1.2 mm long, flattened-ovate, reddish brown and shiny.

DISTRIBUTION: In open moist to dry areas. Coral Bay (A3108), Lameshur (A2573). Also on St. Croix, St. Thomas, Tortola, and Virgin Gorda; from southern United States to Peru, including the West Indies.

2. Amaranthus dubius Mart. *ex* Thell., Fl. Ad. Montpellier 203. 1912.

Erect herb to 1 m tall; stems nearly cylindrical, reddish with a few hairs. Leaf blades 2–12 × 0.7–6.8 cm, ovate, nearly deltate or lanceolate, chartaceous, glabrous except for a few scattered hairs on veins, the apex acute or obtuse, mucronate, usually notched, the base obtuse or tapering into a long (1.5–10 cm) petiole, the margins wavy. Flowers aggregated on short cymes along simple or branched spikelike inflorescences; bracts and bracteoles ovate, with long, sharp apex. Tepals lanceolate, ca. 3 mm long, with sharp apex, green midvein and hyaline margins; stamens 5, the anthers ellipsoid; ovary ovate, with 3 long papillose stigmas. Fruit 1.5 mm long, ovoid, warty. Seeds 1 mm long, lenticular, dark reddish brown, shiny.

DISTRIBUTION: A common weed of open, disturbed areas. Cruz Bay (A3077), Great Cruz Bay (A2366). Also on Tortola and Virgin Gorda; in continental tropical America, West Indies, and introduced to tropical Africa.

3. Amaranthus viridis L., Sp. Pl., ed. 2, 2: 1405. 1763. Fig. 12A–D.

Amaranthus gracilis Desf., Tabl. École Bot. 43. 1804.

Erect herb 30–50(–100) cm tall, with long taproot; stems cylindrical to obtusely 4-angular, reddish, with minute white hairs on young parts, becoming glabrous and furrowed when dried. Leaf blades 2–5.5 × 1–5 cm, lanceolate to rhombic, chartaceous, glabrous except for a few hairs on veins and margins, the apex obtuse, rounded or acute, mucronate, usually notched, the base obtuse to nearly truncate, slightly oblique, the margins whitish and slightly wavy; petioles 1–5 cm long. Flowers aggregated on slightly elongate cymes along panicles; bracts and bracteoles ovate with acuminate sharp apex and hyaline margins. Tepals greenish, ca. 2 mm long, elliptic, with acuminate apex and hyaline margins; stamens 2 or 3, the anthers oblong; ovary flattened, lenticular, the stigma 3, slightly curved. Fruit ca. 1.5 mm long, nearly globose, warty. Seeds 1.2 mm long, lenticular, dark brown and shiny.

DISTRIBUTION: A weed, mostly found in open disturbed areas. Cruz Bay (A2905), Johnson Bay (A3154), East End (B1206). Also on St. Croix, St. Thomas, and Virgin Gorda; widely distributed throughout the tropics.

COMMON NAMES: lumboo, whitey mary.

4. BLUTAPARON Raf.

Prostrate herbs, usually succulent. Leaves opposite and entire. Flowers aggregated in axillary or terminal heads. Tepals 5, thickened at base, subequal; stamens 5, the filaments united into a short cup at base; interstaminal appendages absent; ovary nearly spherical or flattened, with short style and 2 filiform stigmas. Fruit a thin-walled, dehiscent or indehiscent utricle; seeds solitary, lenticular to nearly circular.

A genus with 10 species from tropical and subtropical coastal areas.

1. Blutaparon vermiculare (L.) Mears, Taxon 31: 113. 1982. *Gomphrena vermicularis* L., Sp. Pl. 224. 1753. *Philoxerus vermicularis* (L.) Sm. *in* Rees, Cycl. 27. 1814. *Caraxeron vermicularis* (L.) Raf., Fl. Tellur. 3: 38. 1836 [1837]. Fig. 13J–M.

Prostrate or decumbent perennial herb, to 25 cm long; branches fleshy and cylindrical, reddish brown, yellowish and rooting at nodes, glabrous except for tufts of hairs at nodes. Leaf blades 1–6 × 0.2–0.8 cm, oblanceolate, fleshy, glabrous, the upper surface shiny, the lower surface dull, the apex acute or obtuse and mucronate, the base tapering into a clasping base, the margins strongly turned under. Flowers produced in dense cylindrical or globose heads, with tuft of hairs at base of pedicel; bracts and bracteoles ca. 3 mm long, cream, lanceolate. Tepals whitish, turning straw-colored with age, 3.5 mm long, oblong, glabrous; anthers lanceolate; ovary lenticular, the stigmas 2, short and straight. Seeds 1 mm long, lenticular, reddish brown.

DISTRIBUTION: Along sandy shores. Annaberg (A2921), Fish Bay (A3903). Also on Anegada, St. Croix, St. Thomas, Tortola, and Virgin Gorda; widespread throughout coastal areas of tropical America.

COMMON NAME: bay flower.

5. CELOSIA L.

Herbs or shrubs. Leaves alternate. Flowers bisexual, short pedicellate or sessile, produced in lateral cymes along axillary or terminal spikes, panicles, or thyrses; tepals 5, subequal, free; stamens 5, united at base into a cup; interstaminal appendages dentate or absent;

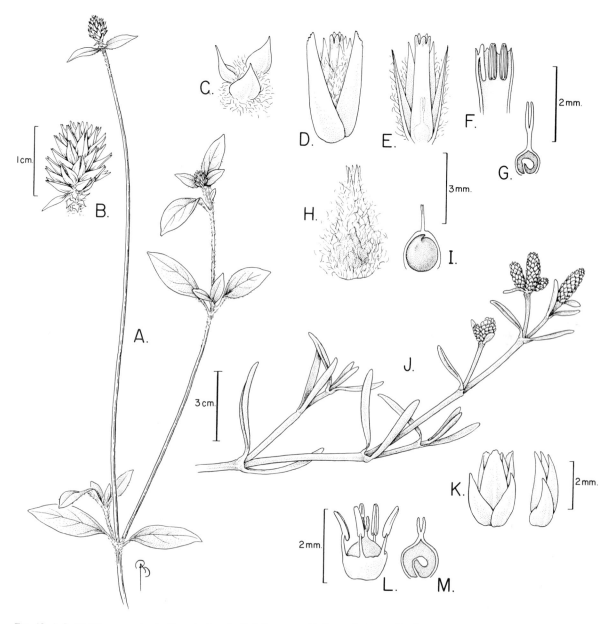

Fig. 13. A–I. *Gomphrena serrata*. **A.** Flowering branch. **B.** Inflorescence. **C.** Bracts. **D.** Flower. **E.** Flower with part of perianth removed to show staminal tube and pistil. **F.** L.s. summit of staminal tube showing anthers. **G.** L.s. pistil. **H.** Mature tepals enclosing fruit. **I.** Utricle and enclosed seed. **J–M.** *Blutaparon vermiculare*. **J.** Fertile branch. **K.** Flower, frontal and lateral views. **L.** Staminal cup and pistil. **M.** L.s. pistil.

ovary nearly globose, ovoid or cylindrical, the style well developed or short, the stigmas awl-shaped or capitate. Fruit a circumscissile utricle with 2 or more ellipsoid seeds.

About 50 species in tropical and subtropical areas.

1. Celosia nitida Vahl, Symb. Bot. **2:** 44. 1791.

Fig. 14G–K.

Erect perennial herb or subshrub, 30–50(–100) cm long, many-branched from a woody base, usually drying blackish; branches cylindrical, glabrous, striate, usually furrowed when dried. Leaf blades 2–5.5 × 1–3 cm, deltoid-lanceolate or lanceolate, chartaceous, glabrous, with scattered minute scales on upper surface, the apex acute, the base oblique, truncate, abruptly tapering into a more or less elongate petiole. Flowers alternate along inflores-

cence axis; bracts and bracteoles triangular, with central dark vein and hyaline margins. Tepals ca. 4 mm long, oblong-lanceolate, glabrous, whitish with reddish apex; filaments flattened, united into a 0.5 mm long, staminal cup; ovary ovoid and smooth, the stigmas 0.2 mm long. Fruit ca. 3 mm long, flattened-subglobose, smooth, opening irregularly. Seeds 9–19(–24) per fruit, 1.2 mm long, lenticular, dark brown, smooth and shiny.

DISTRIBUTION: In understory of coastal scrubs or thickets, on sandy grounds. Hansen Bay (A1807), Lameshur (B610). Also on St. Croix, St. Thomas, Tortola, and Virgin Gorda; in southern United States, Mexico, West Indies and northern South America.

F<small>IG</small>. **14. A–F.** *Iresine angustifolia.* **A.** Fertile branch. **B.** Detail of inflorescence. **C.** Flower with subtending bracts. **D.** Flower with removed perianth parts. **E.** Utricle with persistent staminal cup. **F.** Top of utricle and seed. **G–K.** *Celosia nitida.* **G.** Fertile branch. **H.** Detail of inflorescence. **I.** Flower with removed perianth parts. **J.** Utricle with persistent staminal cup. **K.** Seed.

6. GOMPHRENA L.

Prostrate or erect herbs, usually many-branched from base. Leaves opposite, short-petiolate or sessile, usually clasping at base. Flowers bisexual, nearly sessile, produced in terminal or axillary, globose or cylindrical heads; bracts and bracteoles membranous or nearly coriaceous, whitish or various-colored; tepals 5, free or united at base, usually pubescent; stamens 5, the filaments fused into a long tube, each filament with 2 apical lobes; interstaminal appendages absent; ovary nearly globose, with 1 ovule, the style of variable length, the stigmas 2, erect or divergent. Fruit an irregularly dehiscent, 1-seeded utricle; seeds lenticular.

About 100 species, mostly native to tropical America.

1. Gomphrena serrata L., Sp. Pl. 224. 1753.

Fig. 13A–I.

Gomphrena decumbens Jacq., Pl. Hort. Schoenbr. 4: 41. 1804.

Erect or procumbent perennial herb, to 30 cm tall, many-branched from a woody base, plant usually reddish tinged; stems obtusely 4-angular or cylindrical, collapsing when dried, pinkish at nodes, covered with ascending long hairs, especially on young parts and nodes. Leaf blades 2–4.5 × 0.8–1.2 cm, elliptic or oblanceolate, chartaceous, densely covered with appressed long hairs on both surfaces, the apex acute, mucronate, the base tapering into a clasping petiole <1 cm long, the margins entire and ciliate. Heads globose or nearly cylindrical, subtended by a pair of reduced leaves at base, with densely strigose axis; bracts triangular, membranous; bracteoles 5–7 mm long, lanceolate, involute, white, covering the flower. Pedicels with woolly hairs at base; tepals 4.5 mm long, lanceolate, greenish, with hyaline margins; staminal tube yellow; ovary white, smooth, lenticular. Fruit ca. 2 mm long, flattened to nearly globose, smooth, whitish. Seeds ca. 1 mm long, lenticular, smooth, yellowish and shiny.

DISTRIBUTION: In open or disturbed areas. Lind Point (A2935). Native to tropical America, perhaps introduced as ornamental on St. John; becoming a weed throughout the tropics.

7. IRESINE P. Browne, nom. cons.

Erect or scrambling herbs or shrubs; stems flattened at nodes. Leaves opposite or nearly so, petiolate. Flowers unisexual or bisexual, minute, with tuft of long hairs at base, produced in spikes along terminal or axillary panicles; bracts and bracteoles persistent; tepals 5; stamens 5, the filaments united at base into a cup; interstaminal appendages present or not; ovary with 1 ovule per carpel, the stigmas 2, sessile. Fruit thin-walled, circumscissile and nearly globose; seeds nearly globose to lenticular.

A genus with 40–70 species, probably confined to the New World.

1. Iresine angustifolia Euphrasén, Beskr. Ste. Barthél. 165. 1795.

Fig. 14A–F.

Iresine elatior Rich. *ex* Willd., Sp. Pl. **4**: 766. 1806.
Iresine celosioides Sw., Observ. Bot. 376. 1791, non L., 1763.

Erect or scrambling perennial herb or subshrub to 1.5 m long, with woody base and long taproot; stems cylindrical and striate, with swollen nodes. Leaf blades 2–13 × 0.5–4.5 cm, lanceolate to chartaceous, glabrous or with scattered hairs on veins, the apex long or shortly acuminate, the base obtuse or tapering, usually oblique, the margins entire and slightly revolute; petioles 0.5–2 cm long, with minute, curly hairs. Panicles to 30 cm long, with tomentose axes; bracts and bracteoles lanceolate, membranous, the bracteoles longer. Flowers with tuft of long (3 mm), woolly, white hairs, giving the inflorescence a fuzzy appearance; tepals ca. 1.5 mm long, greenish, elliptic; anthers 0.2 mm long, ellipsoid; ovary lenticular, yellowish. Fruit ca. 1 mm long, nearly globose and smooth. Seeds solitary, 0.8–1 mm long, nearly globose, smooth, blackish and shiny.

DISTRIBUTION: In open or disturbed areas from sea level to 250 m. Chocolate Hole (A779), Fish Bay (A2469), Reef Bay (A2726). Also on St. Croix, St. Thomas, and Tortola; from Mexico to Ecuador and Brazil, including the West Indies.

4. Anacardiaceae (Mango Family)

Trees or shrubs, usually resinous. Leaves alternate, or rarely opposite or whorled, compound or simple, without stipules. Flowers bisexual or unisexual, actinomorphic or zygomorphic, produced in panicles; calyx with 3–5 free or basally fused sepals; corolla with 3–5 free petals; stamens usually twice as many as petals (less often with less than to more than the number of petals), unequal, often some of them not functional, the filaments free or basally fused; nectary disk intrastaminal, annular; ovary superior, unilocular or rarely 4–5-locular, the placenta apical or basal, rarely lateral, the style 1(–4–5), the stigmas 3. Fruit a drupe or a nut.

About 70 genera and 600 species, with pantropical distribution but also extending into temperate regions.

REFERENCES: Barkley, F. A. 1957. Generic key to the sumac family (Anacardiaceae). Lloydia 20: 255–265; Mitchell, J. D. & S. A. Mori. 1987. The cashew and its relatives (*Anacardium*: Anacardiaceae). Mem. New York Bot. Gard. 42: 1–76.

Key to the Genera of Anacardiaceae

1. Leaves compound.
   2. Margins of leaflets spiny; flowers 3-merous .... 2. *Comocladia*
   2. Margins of leaflets without spines; flowers 5-merous.
      3. Shrub or small tree; fruits ellipsoid, red, 4–7 mm long .... 4. *Schinus*

3. Large or small trees; fruits nearly globose or ellipsoid, yellow, light orange or purple, >2 cm long .... 5. *Spondias*
1. Leaves simple.
   4. Leaves oblong, long-pointed at apex; petals with glandular ridges; fruits ovate, fleshy, >8 cm long, without fleshy receptacle (peduncle) .... 3. *Mangifera*
   4. Leaves obovate, rounded at apex; petals without glandular ridges; fruits ("seed") kidney-shaped, woody, <3 cm long, with a fleshy pedicel at base .... 1. *Anacardium*

1. ANACARDIUM L.

Shrubs to large trees. Leaves entire, alternate, aggregate toward branch tips, sessile or petiolate. Flowers bisexual or staminate; calyx 5-lobed, with lanceolate sepals; corolla bell-shaped, with 5 free reflexed petals; stamens 6–12, 1(–2) or 4 larger, the filaments basally fused into a staminal tube; ovary slightly obovoid, 1-locular, with 1 basal ovule, the style central or lateral, usually longer than the largest stamen(s), the stigmas punctiform or capitate; nectary disk absent. Fruit a kidney-shaped nut subtended by a fleshy and juicy peduncle.

A genus with 10 species, naturally distributed from Honduras to eastern Paraguay.

1. Anacardium occidentale L., Sp. Pl. 383. 1753.

Fig. 15A–E.

Small tree to 10 m tall. Leaf blades 10–18 × 5–12 cm, obovate to wide elliptic, coriaceous, glabrous, primary and secondary veins prominent on lower surface, vein angles with pit or depression, the apex rounded or obtuse and slightly notched, the base cuneate, the margins entire or slightly wavy; petioles 0.5–2 cm long. Inflorescence covered with ash-colored hairs; bracts and bracteoles ovate-deltoid. Calyx with 5 unequal, lanceolate or elliptic, 4–4.5 mm long sepals; petals 5, ca. 1 cm long, reflexed, turning from light yellow to pink; stamens 10, one thicker and projecting beyond the petals; ovary top-shaped, the style subapical, projecting beyond the petals, the stigma nearly capitate. Nut 2–3 cm long, woody, grayish, smooth, producing caustic oils, the pedicel fleshy (hypocarp or "cashew apple"), yellow or reddish and edible. Seeds ("cashew nut") solitary, kidney-shaped, to 1.5 cm long.

DISTRIBUTION: This plant, presumably native to northern South America, was introduced into the West Indies many centuries ago for its edible fruits. A few individuals were found along the roadside near Maria Bluff (A605).

COMMON NAMES: akaju, cashew, cashubaby, kushu.

NOTE: The latex produced by the plant and its fruit is caustic, causing severe dermatitis on contact. Cashew nuts must be roasted in order to make them edible. The fleshy pedicel ("cashew apple") generally is not toxic and can be eaten raw or preserved in syrup.

2. COMOCLADIA P. Browne

Shrubs or small trees, with poisonous resin. Leaves alternate, compound, odd-pinnate, with 9 to many sinuate or spinose leaflets. Flowers bisexual or unisexual, 3–4-merous, with free sepals and petals; petals red, twice as long as the sepals; stamens shorter than the petals, the anthers basifixed; ovary unilocular, with a cup-shaped disk at base, the stigmas 3, very short. Fruit an ellipsoid red or orange, fleshy drupe.

About 25 species distributed principally in the West Indies but also from Mexico to Central America.

1. Comocladia dodonaea (L.) Urb., Symb. Ant. **4:** 360. 1910. *Ilex dodonaea* L., Sp. Pl. 125. 1753.

Fig. 15F–I.

Comocladia ilicifolia Sw., Prodr. 17. 1788, nom. illegit.

Shrub to 5 m tall, with clear resinous sap; stems cylindrical and puberulent. Leaves clustered at branch tips, pinnately compound, with 9–21 leaflets; leaflets 1.5–3.5 × 1.3–2.7 cm, ovate or suborbicular, involute, usually wrinkled and reddish, chartaceous to coriaceous, glabrous except for a few hairs on veins and rachis, with prominent veins on lower surface, the apex acute or obtuse with long sharp spine, the base cuneate on terminal leaflet, nearly cordate on lateral ones, the margins spinulose, terminal leaflet with 2 spines on each side, lateral ones with 1; rachis nearly cylindrical, puberulent, the petiole 1–3 cm long. Inflorescence spikelike, much shorter than the subtending leaf; flowers aggregate in cymes along inflorescence rachis. Calyx ca. 0.5 mm long; petals 1 mm long, erect, triangular, with revolute margins; stamens <1 mm long, the anthers ellipsoid; nectary disk cup-shaped; ovary reduced in staminate flowers. Drupe ellipsoid, 1 cm long, red-orange and fleshy.

DISTRIBUTION: Very common in dry coastal scrubs. Lameshur (A745). Widespread on Jost van Dyke, St. Croix, St. Thomas, Tortola, and Virgin Gorda; Hispaniola, Puerto Rico, and in the Lesser Antilles.

COMMON NAMES: Christmas bush, Christmas tree, five finger, poison ash, pra pra.

NOTE: A poisonous shrub causing severe dermatitis on contact or inhalation of fumes from burning stems and leaves, similar to that produced by poison ivy.

3. MANGIFERA L.

Large trees. Leaves simple, alternate, aggregate at branch tips. Inflorescence terminal or in upper leaf axils. Flowers unisexual or bisexual, 4–5-merous; stamens 4–8, unequal, one larger, the remainder reduced and sterile; ovary unilocular, with a 5-lobed nectary disk at base, the carpel with 1 ovule, style eccentric or lateral, the stigma minute. Fruit an ovoid fleshy drupe; stones (seeds) large and compressed.

About 40 species native to Indo-Malesia, with 1 species cultivated throughout the tropics.

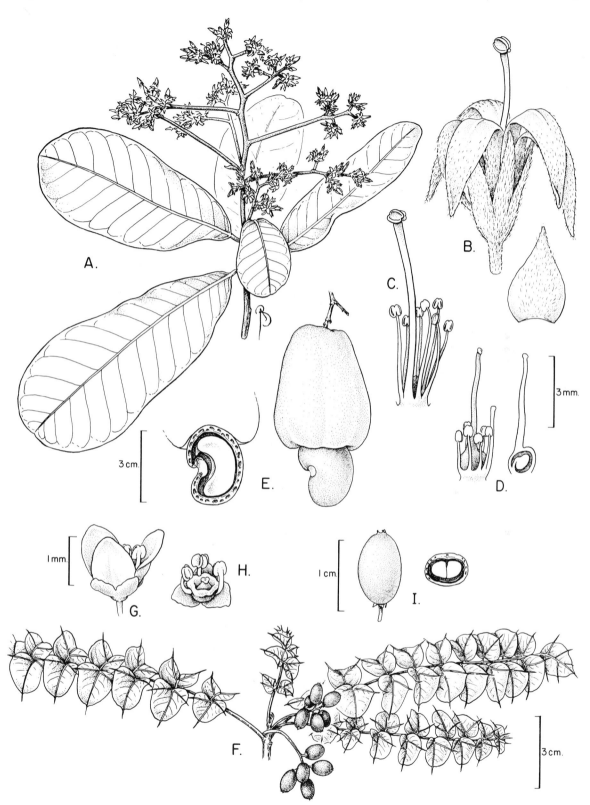

FIG. 15. A–E. *Anacardium occidentale.* **A.** Flowering branch. **B.** Flower and bract. **C.** Dimorphic stamens. **D.** Stamens and pistil and l.s. pistil. **E.** L.s. nut and nut with subtending fleshy pedicel. **F–I.** *Comocladia dodonaea.* **F.** Fruiting branch. **G.** Flower. **H.** Flower, petals removed to show stamens, nectary disk, and reduced pistil. **I.** Drupe and c.s. drupe.

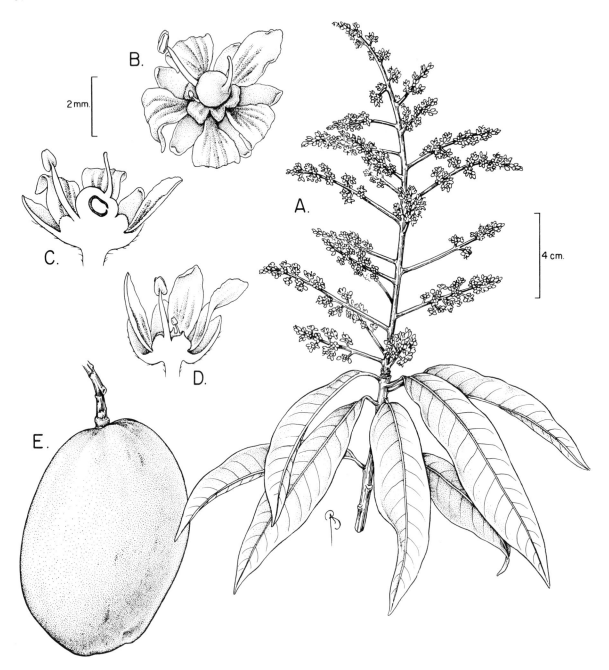

FIG. 16. *Mangifera indica*. **A.** Flowering branch. **B.** Bisexual flower. **C.** L.s. bisexual flower, showing single fertile stamen and pistil. **D.** L.s. staminate flower, showing stamens and reduced pistil. **E.** Drupe. (From Mori et al., in press, Guide to the Vascular Plants of Central French Guiana, Mem. New York Bot. Gard. 76.)

1. Mangifera indica L., Sp. Pl. 200. 1753.   Fig. 16.

Large tree, usually with clear resinous sap. Leaves simple, aggregate at branch tips; blades 10–25 × 2.5–6 cm, often falcate, oblong-elliptic, coriaceous, glabrous, with prominent primary and secondary veins, the apex acute or acuminate, the base oblique and tapering, the margins entire or slightly undulate; petioles 1–5 cm long. Panicles 20–40 cm long, the rachis with minute yellowish hairs; bracts and bracteoles 1–1.5 mm long, lanceolate, yellowish. Calyx with same indument as rachis, the sepals 5, lanceolate, 2 mm long; petals 5, spatulate or oblong, 3–3.5 mm long, turning from yellowish green to reddish green; 1 fertile stamen projecting beyond the petals; ovary nearly globose, sessile, the style sublateral and curved, the stigma punctiform. Fruit a drupe, 8–15 cm long, ovoid, fleshy, turning from green to yellow or orange. Stones (seeds) 5–8 cm long, flattened ovoid-reniform.

DISTRIBUTION: Commonly planted throughout the island, persistent in secondary forests in humid areas. Bordeaux (A934). Cultivated throughout the tropics for its edible fruits.

COMMON NAMES: mango, mango tree.

NOTE: The resin of the plant, including the fruit, may cause contact dermatitis in some people.

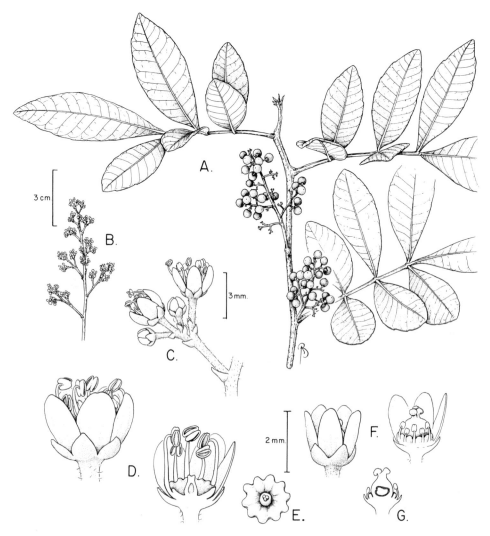

FIG. 17. *Schinus terebinthifolius*. **A.** Fruiting branch. **B.** Inflorescence. **C.** Detail of inflorescence. **D.** Staminate flower and l.s. staminate flower. **E.** Intrastaminal disk and pistillode. **F.** Pistillate flower and l.s. pistillate flower with reduced stamens; **G.** L.s. pistillate flower.

4. SCHINUS L.

Trees or shrubs. Leaves alternate and pinnately compound, the rachis often winged. Flowers bisexual or functionally unisexual, 4–5-merous, produced in short inflorescences; bract and bracteoles minute; calyx with short, triangular lobes; petals free, oblong; staminate flowers with 8–10 unequal stamens and a pistillode; pistillate flowers with reduced, sterile stamens and sessile ovary with cup-shaped disk at base, 1-locular, the locule with 1 ovule, the style terminal, trifid or simple, with 3 minute stigmas. Fruit a globose, red drupe.

About 27 species, native to South America.

1. Schinus terebinthifolius Raddi, Mem. Mat. Fis. Soc. Ital. Sci. Modena, Pt. Mem. Fis. **18:** 399. 1820.
Fig. 17.

Shrub or small tree to 5 m tall; stems cylindrical, minutely pubescent. Leaves with 5–7 leaflets and winged rachis; leaflets 3–6 × 1.3–2.5 cm, elliptic to oblanceolate, chartaceous, glabrous, the apex rounded, slightly notched, the base tapering, the margins entire or crenate and revolute, the petiolules very short. Inflorescence shorter than subtending leaf, the axis minutely pubescent; bracts and bracteoles triangular, 0.5–0.7 mm long. Calyx with 5 triangular lobes ca. 0.5 mm long; petals 5, oblong, 1.5 mm long, yellowish green, reddish at base within; stamens 10, 4 of which are larger; ovary nearly globose, the style single, with capitate stigma. Fruit a small drupe, 4–7 mm diam., globose, reddish.

DISTRIBUTION: Planted as an ornamental, a few individuals naturalized in Hawksnest area. Cruz Bay (A5055). Native of southern Brazil, Paraguay, and northern Argentina, becoming a weed in many tropical and subtropical areas.

COMMON NAMES: Brazilian pepper, Christmas berry, wild pepper.

NOTE: Fruits are known to cause allergic reaction in some people.

5. SPONDIAS L.

Small to large trees. Leaves compound, alternate, odd-pinnate; leaflets entire or crenate. Flowers bisexual or unisexual, 4–5-merous; calyx with short lobes; corolla with free petals, much longer than the sepals; stamens 8–10; ovary 4–5 locular, with short cup–shaped or annular nectary disk at base, each locule with a single ovule, the styles 4–5, the stigmas spatulate. Fruit a fleshy, ellipsoid to nearly globose drupe.

About 12 species native to southeastern Asia and tropical America. A few species cultivated throughout the tropics.

Key to the Species of *Spondias*

1. Leaf rachis winged; inflorescence <10 cm long; flowers red or purple; fruits purple .... *S. purpurea* (cultivated)
1. Leaf rachis not winged; inflorescence >25 cm long; flowers white; fruits yellow to light orange.
   2. Fruits ellipsoid to nearly globose, 4–8 cm long; bark smooth .... *S. dulcis* (cultivated)
   2. Fruits ellipsoid to obovoid, 2.5–4 cm long; bark warty.. *S. mombin*

1. Spondias mombin L., Sp. Pl. 371. 1753.   Fig. 18.

Spondias pseudomyrobalanus Tussac, Fl. Antill. **4:** 97. 1827.

Tree 10–15 m tall; bark gray to dark brown and warty; stems cylindrical with numerous whitish lenticels. Leaflets 15–19, opposite or alternate, 5–10 × 3–4 cm, falcate, elliptic-lanceolate, chartaceous, glabrous, primary veins prominent, the apex shortly acuminate, the base oblique, acute-rounded, the margins revolute, entire or crenulate, with marginal vein; petiolules ca. 5 mm long, the rachis cylindrical, not winged, minutely pubescent or glabrous, the petioles 5–10 cm long. Panicles terminal or axillary, 30–50 cm long, distinctly drooping when in fruit; bracts and bracteoles ovate to lanceolate. Flowers white or cream, fragrant; calyx lobes triangular, ca. 0.5 mm long; petals oblong to lanceolate, 2–3.5 mm long; stamens 9–10, ca. 2–3 mm long, the anthers lanceolate; ovary 5-locular, with cup-shaped, 10-lobed disk at base, the styles 5, short, erect. Fruit ellipsoid to obovoid, 3–4 cm long, yellow or light orange; stone developing only 1 seed.

DISTRIBUTION: Perhaps a native species, common along roadsides or in secondary moist forests. Cruz Bay (A1963). Native to tropical America, but widely cultivated in the tropics for its edible, tasty fruits.

COMMON NAME: hog plum.

CULTIVATED SPECIES: *Spondias dulcis* Sol. *ex* Parkinson, known as the pommecythere, and *Spondias purpurea* L., the purple plum, are both cultivated on St. John for their edible fruits; however, neither of them is common on the island.

5. Annonaceae (Soursop Family)

Trees, shrubs, or lianas. Leaves simple, alternate (2-ranked), without stipules. Flowers usually fleshy, bisexual or rarely unisexual, actinomorphic, produced in terminal axillary or cauliflorous inflorescences; calyx of 3 free or united sepals; corolla of 6 petals arranged in 2 series, the inner ones sometimes smaller, reduced to scales or wanting; stamens numerous, with short filaments or sessile anthers; ovary superior, of numerous carpels, united or free, with 1 to many ovules, the stigma capitate or club-shaped, sessile or on short style. Fruit of free or fused fleshy carpels; seeds 1 to many.

About 130 genera and 2300 species, mostly tropical.

1. ANNONA L.

Trees or shrubs. Leaves usually chartaceous to coriaceous. Flowers solitary or in clusters, terminal, but sometimes appearing as opposite the leaves; sepals valvate, much smaller than petals; petals fleshy; stamens numerous, with the fleshy, connective truncate, projecting beyond the anthers; carpels numerous, usually free, with a single ovule. Fruit fleshy, formed by fused carpels, with numerous seeds.

About 100 species, from tropical America and Africa; many species cultivated throughout the tropics for their edible fruits.

Key to the Species of *Annona*

1. Fruit smooth, ripening yellow.
   2. Leaves pubescent; mature flower buds linear-lanceolate, pubescent, greenish .... 3. *A. reticulata*
   2. Leaves glabrous; mature flower buds ovoid, glabrous, yellowish .... 1. *A. glabra*
1. Fruit with projections, ripening green or light brown.
   3. Mature flower buds 2.5 cm or longer, yellowish, and obtusely 3-angular-ovoid; fruit to 25 cm long, elongate-cordate, with spinelike projections .... 2. *A. muricata*
   3. Mature flower buds 1–1.5 cm long, green, obtusely 3-angular, linear-lanceolate; fruits to 12 cm long, heart-shaped, with rounded projections.. 4. *A. squamosa*

1. Annona glabra L., Sp. Pl. 537. 1753.   Fig. 19A–F.

Annona palustris L., Sp. Pl., ed. 2, **1:** 757. 1762.
Annona laurifolia Dunal, Monogr. Anonac. 65. 1817.

Tree 3–12 m tall; bark light gray and smooth; stems cylindrical, shiny, green, dark gray and striate when dried. Leaf blades 6–15 × 3.5–7 cm, elliptic, or ovate, chartaceous to coriaceous, involute, glabrous, the apex abruptly acuminate or acute, the base obtuse

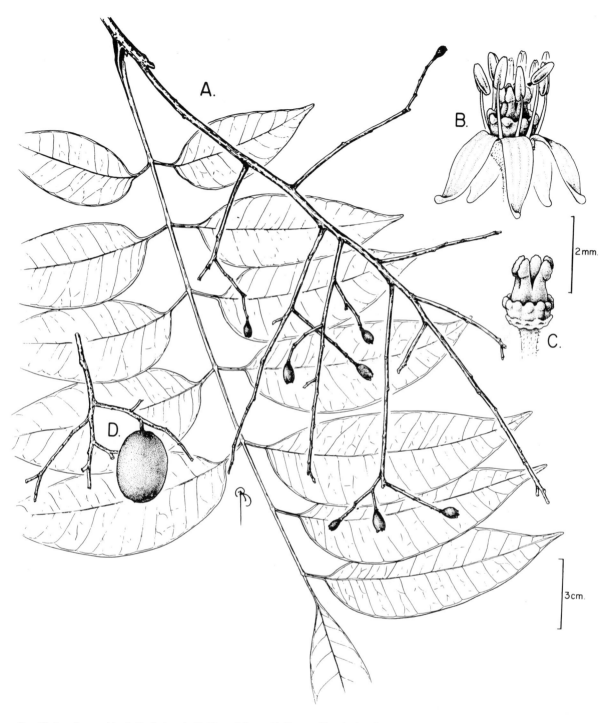

FIG. 18. *Spondias mombin.* **A.** Fertile branch. **B.** Bisexual flower. **C.** Flower with perianth and stamens removed to show cup-shaped disk and pistil. **D.** Drupe. (From Mori et al., in press, Guide to the Vascular Plants of Central French Guiana, Mem. New York Bot. Gard. 76.)

or nearly rounded, sometimes oblique, the margins entire and revolute; petioles 1–1.5 cm long. Flowers terminal but appearing lateral, pendent. Sepals ovate, 3.5–4.5 mm long; petals fleshy, ovate, concave, bright yellow on outer surface, light yellow with reddish base within, acute at apex, the outer petals 2–3 cm long, the inner 1–1.5 cm long; stamens club-shaped. Fruit fleshy, ovoid,

6–12 cm long, smooth, turning from green to light yellow, the pulp yellowish cream with sweet scent. Seeds numerous, ellipsoid, 1–1.5 cm long, light brown and smooth.

DISTRIBUTION: Near coastal mangrove swamps and inland along watercourses. Coral Bay (A2796). Also on Jost van Dyke, St. Croix, St. Thomas, Tortola, and Virgin Gorda; from Florida to Brazil, the West Indies and Africa.

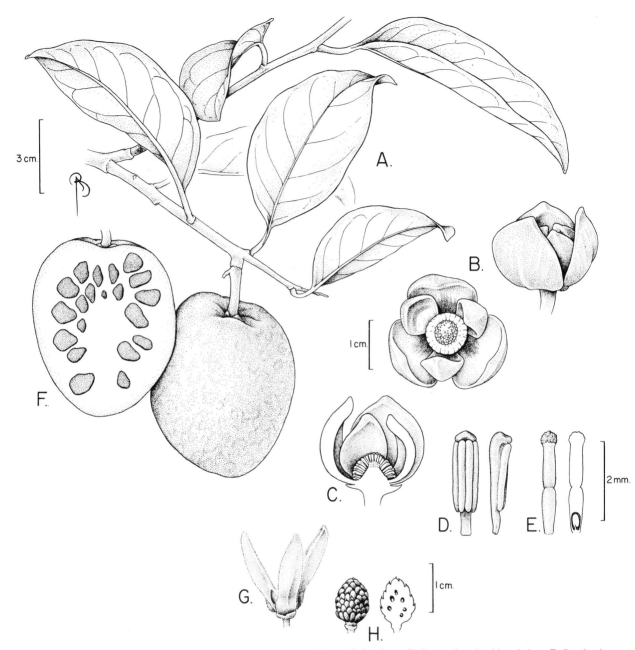

FIG. 19. A–F. *Annona glabra.* **A.** Fruiting branch. **B.** Flower, top and lateral views. **C.** L.s. flower. **D.** Stamen, frontal and lateral views. **E.** Carpel and l.s. carpel. **F.** L.s. fruit. **G, H.** *Annona squamosa.* **G.** Flower. **H.** Immature fruit and l.s. immature fruit.

COMMON NAMES: bonya, cork wood, dog apple, monkey apple, pond apple, wild soursop.

NOTE: Fruits usually not eaten by humans.

2. Annona muricata L., Sp. Pl. 536. 1753.

Tree 3–8 m tall; stems cylindrical, with numerous whitish, lineolate lenticels; young parts densely covered with rusty-brown hairs. Leaf blades 6–17 × 2.5–7 cm, elliptic, oblong or oblanceolate, chartaceous to coriaceous, glabrous except for a few hairs on primary veins and at vein angles, the apex abruptly acuminate, acute or obtuse, the base obtuse or tapering, the margins entire and slightly revolute; petioles 0.5–1.5 cm long, slightly swollen. Flowers terminal but appearing lateral, pendent. Sepals triangular, 3.5–4.5 mm long; petals 2.5–3.5 cm long, outer and inner nearly equal, cordate, greenish yellow, with acuminate and reflexed apex; stamens club-shaped. Fruit edible, fleshy, elongate-cordate, 15–30 cm long, the skin green, with spinelike projections, the pulp white and tart. Seeds numerous, elliptic, 1–1.7 cm long, dark brown, smooth.

DISTRIBUTION: Mostly found along roadsides and near settlements. Susannaberg (A2086). Also on Jost van Dyke, St. Croix, St. Thomas, Tortola, and Virgin Gorda; native to tropical America but perhaps introduced in the West Indies in pre-Columbian times.

COMMON NAMES: soursop, susaka, sweetsop.

3. Annona reticulata L., Sp. Pl. 537. 1753.

Tree 3–8 m tall; stems cylindrical, with numerous whitish, punctate lenticels; young leaves and stems densely covered with appressed yellowish hairs. Leaf blades 6–20 × 2.5–5 cm, elliptic, oblong or lanceolate, chartaceous, sparsely covered with minute yellowish hairs, especially on veins, the apex acute, less often obtuse, the base obtuse or rounded, the margins entire and slightly revolute; petioles 0.5–1.5 cm long, pubescent. Flowers pubescent, pendent, 2 or 3 aggregate on terminal, short, lateral branches. Sepals triangular, ca. 2 mm long; petals 1.3–1.5 cm long, lanceolate, greenish; stamens club-shaped. Fruit edible, fleshy, ovoid to globose, 6–12 cm long, turning from green to cream-yellow, smooth, the pulp cream and tart. Seeds elliptic, 1–1.3 cm long, light brown, smooth.

DISTRIBUTION: Introduced, mostly found along roadsides and near settlements. Susannaberg (A838). Also on St. Croix, St. Thomas, and Tortola; in tropical America, including the West Indies.

COMMON NAME: custard apple.

4. Annona squamosa L., Sp. Pl. 537. 1753.

Fig. 19G–H.

Tree 3–8 m tall; stems cylindrical, with numerous whitish, punctate lenticels; young parts pubescent; branches flexuous. Leaf blades 6–15 × 3–5 cm, elliptic, oblong-elliptic or lanceolate, chartaceous, sparsely covered with minute yellowish hairs, especially on veins, the apex acute or obtuse, the base obtuse or rounded, the margins entire and slightly revolute; petioles 0.5–1.5 cm long, pubescent. Flowers pubescent, pendent, 2 or 3 aggregate on terminal, short, lateral branches. Sepals triangular, ca. 2 mm long; petals 1.5–2 cm long, oblong-lanceolate, greenish; stamens club-shaped. Fruit edible, fleshy, cordate, 6–12 cm long, green, with rounded projections (overall aspect like a pine cone), the pulp cream and sweet. Seeds elliptic, 1–1.3 cm long, light brown, smooth.

DISTRIBUTION: Mostly found along roadsides and in secondary forests in semidry areas. Cruz Bay (A1962), Lameshur (A2692). Also on Jost van Dyke, St. Croix, St. Thomas, Tortola, and Virgin Gorda; in tropical America, including the West Indies.

COMMON NAME: sugar apple.

DOUBTFUL RECORD: *Guatteria caribaea* Urb. was reported by Woodbury and Weaver (1987) as occurring on St. John, but no specimen was collected at that time. No plant has been located to confirm their report in spite of many attempts.

6. Apiaceae (Carrot Family)

Aromatic, perennial or annual herbs, or rarely subshrubs. Leaves alternate or rarely opposite, pinnately compound, dissected or rarely simple, the petiole forming a sheathing base; without stipules. Flowers bisexual, rarely unisexual, actinomorphic, aggregated in simple or compound umbels, usually 5-merous; calyx of minute sepals; corolla of free minute petals; stamens alternating with the petals; ovary inferior, of 2 carpels, with annular disk at base, each carpel with 1 ovule pendulous from the apex. Fruit a dry schizocarp, separating into 2 units (mericarps).

About 300 genera and 3000 species, nearly cosmopolitan.

1. CYCLOSPERMUM Lag.

Annual herbs. Leaves pinnately dissected; petioles slender. Flowers white or greenish white, bisexual, produced in compound or simple umbels; bracts and bracteoles wanting; calyx with minute or obsolete sepals; corolla with 5 minute petals; ovary partially inferior, the style very short. Fruit with ovoid or circular outline, laterally compressed; each mericarp with 5 vertical ridges.

A genus of 2 or more species of South America, but 1 species with cosmopolitan distribution. Originally spelled as *Ciclospermum* but corrected to *Cyclospermum* to conform with classical Greek.

1. Cyclospermum leptophyllum (Pers.) Britton & P. Wilson, Bot. Porto Rico 6: 52. 1925. *Pimpinella leptophylla* Pers., Syn. Pl. 1: 324. 1805. *Apium leptophyllum* (Pers.) Benth., Fl. Austral. 3: 372. 1867.

Fig. 20.

Erect slender annual, glabrous herb 15–60 cm tall. Leaves 3–4-pinnately compound, 10–12 × 6–8 cm, the leaflets linear, membranous; petiole longer than the blade. Flowers produced in simple or compound umbels, the peduncles nearly sessile to 2 cm long. Calyx 0.4 mm long, bell-shaped, with inconspicuous sepals; petals white, 0.2–0.4 mm long, inflexed over the stamens; stamens short, the anthers ellipsoid. Fruit circular to ovoid, 1–3 mm long; mericarps readily separating.

DISTRIBUTION: Uncommon in moist, shaded areas. Bordeaux Mountain (W478). A pantropical weed, probably native to Brazil.

NOTE: See Kartesz and Gandhi (Phytologia 69: 129–130. 1990) for a discussion of authorship for the combination in *Cyclospermum*.

7. Apocynaceae (Dogbane Family)

Trees, shrubs or lianas, producing clear or usually milky latex. Leaves simple, opposite, whorled or rarely alternate, stipules lacking, or minute and interpetiolar. Flowers bisexual, actinomorphic, produced in cymose or racemose inflorescences, or solitary; calyx funnel-shaped with (4–)5 lobes; corolla usually showy, funnel-shaped or salverform, with (4–)5 lobes, the tube usually with a whorl of appendages within; stamens as many as the corolla lobes, attached to the corolla tube, the anthers distinct or more or less connivent around the style; ovary superior, with an annular or lobular disk at base, 2(–8)-carpellate, the carpels connate to various degrees, sometimes united only by their common styles and stigma, the placentation ventral, axile in syncarpous, 2-locular ovaries, or parietal in apocarpous, 1-locular ovaries, each locule with 2 to many ovules. Fruit diverse, usually a follicle or a berry; seeds often with a tuft of long hairs at apex.

About 200 genera and 2000 species, from tropical and subtropical areas.

REFERENCES: Leeuwenberg, A. J. M. 1983. Notes on *Nerium* L. and *Tabernaemontana* L. Meded. Rijks Landbouwhoogeschool 83: 57–60, Pagen, F. J. J. 1987. Oleander. *Nerium* L. and the oleander cultivars. Agric. Univ. Wageningen Pap. 87-2: 1–113; Rao, A. S. 1956. A revision of *Rauvolfia* with particular reference to the American species. Ann. Missouri Bot. Gard. 43: 253–354.

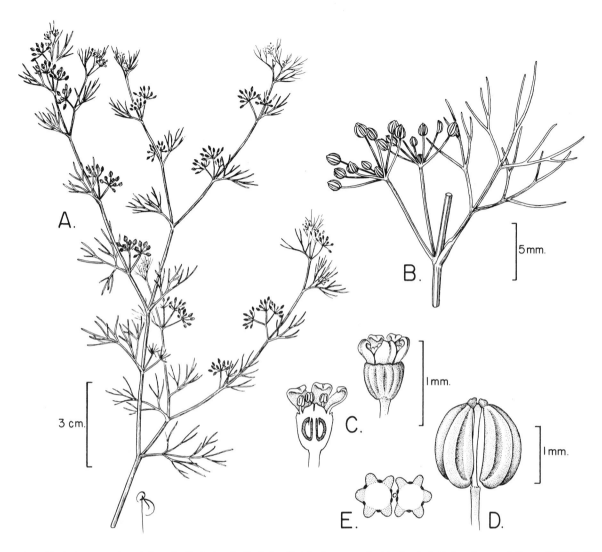

FIG. 20. *Cyclospermum leptophyllum.* **A.** Fertile branch. **B.** Node with infructescence and subtending leaf. **C.** L.s. flower and flower. **D.** Fruit. **E.** C.s. mericarps.

Key to the Genera of Apocynaceae

1. Twining, climbing plants.
   2. Plants woody, producing abundant milky latex; stems copper-colored; corolla bright yellow, 3–6 cm long .... 3. *Pentalinon*
   2. Plants herbaceous to subwoody, producing watery latex; stems green; corolla cream-colored, 4–6 mm long .... 5. *Prestonia*
1. Herbs, shrubs, or trees.
   3. Corolla pink or lilac.
      4. Plants herbaceous, to 1 m tall; leaves chartaceous or membranous, elliptic or oblanceolate, to 7 cm long .... 1. *Catharanthus*
      4. Plants woody, to 3 m tall; leaves coriaceous, oblong, elliptic, or oblanceolate, 10–30 cm long ..... 2. *Nerium*
   3. Corolla white, cream, or yellow.
      5. Corolla white or cream.
         6. Plants herbaceous .... 1. *Catharanthus*
         6. Plants woody (shrubs or trees).
            7. Leaves whorled (3 or 4 per node); corolla 3.5–8.5 mm long; fruit a fleshy berry .... 6. *Rauvolfia*
            7. Leaves alternate; corolla 3.3–6 cm long; fruit a woody follicle .... 4. *Plumeria*
      5. Corolla yellow .... *Thevetia* (cultivated)

1. CATHARANTHUS G. Don

Erect herbs or subshrubs. Leaves opposite, petiolate, with interpetiolar stipules. Flowers solitary or aggregate (to 4) in leaf axils; calyx of 5 elongate sepals; corolla salverform, the tube narrow, cylindrical, the lobes 5, rounded; stamens inserted in upper portion of tube, the filaments very short, the anthers lanceolate, free, without appendages; ovary with 2 nectary glands at base, the carpels united only by the common style. Fruit of 2 narrow cylindrical follicles; seeds numerous, without hairs.

A genus of 8 species, 1 restricted to India, the rest native to Madagascar, the following species widely cultivated throughout the tropics.

1. Catharanthus roseus (L.) G. Don, Gen. Hist. **4:** 95. 1837. *Vinca rosea* L., Syst. Nat., ed. 10, **2:** 944. 1759. *Lochnera rosea* (L.) Rchb., Consp. Regn. Veg. 134. 1828. *Ammocallis rosea* (L.) Small, Fl. S.E. U.S. 936. 1903. **Fig. 21E–H.**

Erect herb to 50 cm tall, producing a clear latex when cut; stems cylindrical and pubescent. Leaf blades 2–7 × 0.8–2.2 cm, elliptic or oblanceolate, membranous to chartaceous, sparsely pubescent, the apex rounded, obtuse and apiculate; the base tapering, the margins entire or crenulate; petiole 4–7 mm long, pubescent. Flowers nearly sessile, paired at leaf axils. Calyx bell-shaped, 5–6 mm long, lobes 3–4 mm long, awl-shaped; corolla pink or white, 2.5–3.5 cm long, the lobes 1.5–3 cm long, rounded; anthers 5–6 mm long; stigma capitate. Follicle 1–3 cm long, pubescent, with many ribs lengthwise. Seeds ca. 2 mm long, oblong-ellipsoid, black, sculptured.

DISTRIBUTION: Introduced as an ornamental, now escaped; widespread along the sandy coasts. East End (A4224). Also on St. Croix, St. Thomas, and Tortola; native to Madagascar, cultivated in tropical and subtemperate zones, becoming naturalized throughout the tropics.

COMMON NAMES: church flower, periwinkle.

2. NERIUM L.

Modern revisions consider the genus to be monospecific (Leeuwenberg, 1983; Pagen, 1987). The following description characterizes the genus.

1. Nerium oleander L., Sp. Pl. 209. 1753. **Fig. 21A–D.**

Shrub or small tree to 5 m tall, producing abundant, poisonous, clear, sticky latex. Leaves opposite or in whorls of 3(–4); blades 10–30 × 2–3.5 cm, oblong, elliptic or oblanceolate, coriaceous, glabrous, the apex acuminate or acute, the base tapering, the margins entire and revolute; petioles 2–4 mm long. Flowers produced along branched, terminal, thyrsoid inflorescences; peduncle usually longer than the flower-bearing portion, obtusely 3-angular; bracts and bracteoles ca. 5 mm long with colleters at base within. Calyx 4–6 mm long, with 5 deeply parted, lanceolate lobes; corolla 3.5–5 cm long, pink (in ours), white, or red, the narrowest part of the tube 0.5–1 cm long, the lobes 5, rounded, 2–2.5 cm long (corolla double in some cultivars), spreading; corona with filiform segments, projecting beyond the corolla; stamens inserted where the tube widens, the anthers with an apical feathery appendage, connivent around the stigma; ovary ovoid, of 2 connate carpels, with a lobular nectary disk at base, the carpel with many ovules. Fruit elongate, 10–20 cm long, of 2 follicles, splitting at maturity. Seeds oblong, 4–7 mm long, with a long tuft of hairs at apex.

DISTRIBUTION: Although not naturalized on St. John, it is included in this treatment because of its wide cultivation throughout the island. Coral Bay Quarter along Center Line Road (A4682). Also on Anegada, St. Croix, St. Thomas, and Tortola; native to the Mediterranean area, now widely cultivated throughout the tropics.

COMMON NAMES: leandra, oleander.

3. PENTALINON Müll. Arg.

Twining vines or scandent shrubs, producing milky latex. Leaves opposite, without glands, colleters present at axils. Flowers produced in dichasial axillary cymes; calyx bell-shaped, with 5, elongate sepals with colleters at base within; corolla funnel-shaped; anthers connivent around the stigma, with long filiform appendage; ovary with 5 glandular nectaries at base, the carpels 2, free, united by the common style. Fruit of 2 slender, cylindrical, apically connate follicles. Seeds numerous, narrowly elliptic, with tuft of long hairs at apex.

A neotropical genus of 2 species.

1. Pentalinon luteum (L.) Hansen & Wunderlin, Taxon **35:** 167. 1986. *Vinca lutea* L., Cent. Pl. 2: 12. 1756. *Urechites lutea* (L.) Britton, Bull. New York Bot. Gard. **5:** 316. 1907. **Fig. 22A–D.**

Echites suberecta Jacq., Enum. Syst. Pl. 13. 1760. *Urechites suberecta* (Jacq.) Müll. Arg., Linnaea **30:** 444. 1860.

Echites neriandra Griseb., Fl. Brit. W. I. 415. 1862.

Vine or climbing shrub, to 5 m long, many-branched from near the base; branches twining, cylindrical, pubescent. Leaf blades 3– 7 × 1.5–3.7 cm, elliptic or obovate, chartaceous to coriaceous, glabrous, or less often pubescent, the apex obtuse, rounded and apiculate, the base obtuse or rounded; petioles 0.3–0.7 cm long, pubescent. Sepals lanceolate, 10–12 mm long; corolla bright yellow, 3–6 cm long, the tube with reddish stripes within, the lobes rounded, 1.5–2 cm long. Follicle dark brown, smooth, 15–19 cm long.

DISTRIBUTION: Common in coastal thickets with sandy or rocky soil. Princess Bay (A2833). Also on Anegada, St. Croix, St. Thomas, Tortola, and Virgin Gorda; in Florida and throughout the West Indies.

COMMON NAME: wild allamanda.

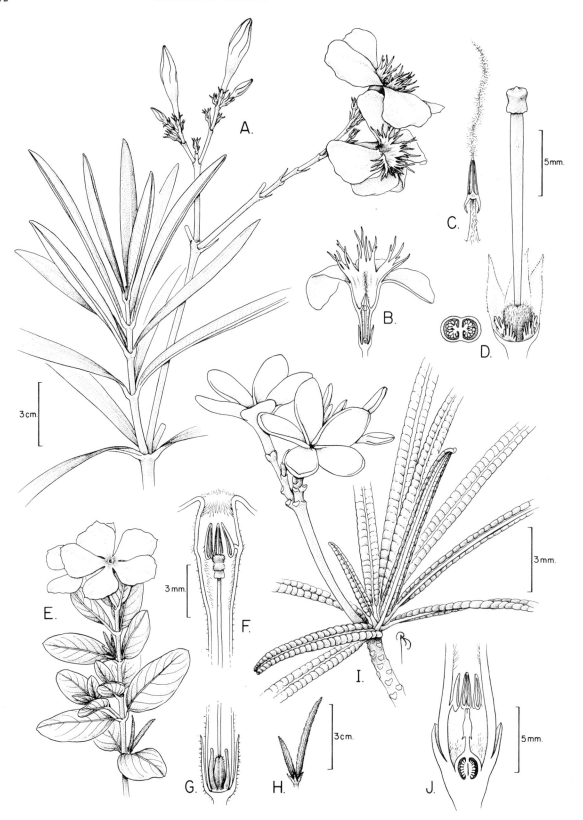

FIG. 21. A–D. *Nerium oleander*. **A.** Flowering branch. **B.** L.s. flower. **C.** Anther with apical appendage. **D.** C.s. ovary, and pistil with nectaries at base. **E–H.** *Catharanthus roseus*. **E.** Fertile branch. **F.** L.s. corolla, showing anthers and capitate stigma. **G.** L.s. calyx, showing nectaries and pistil. **H.** Follicles. **I, J.** *Plumeria alba*. **I.** Flowering branch. **J.** L.s. lower part of flower.

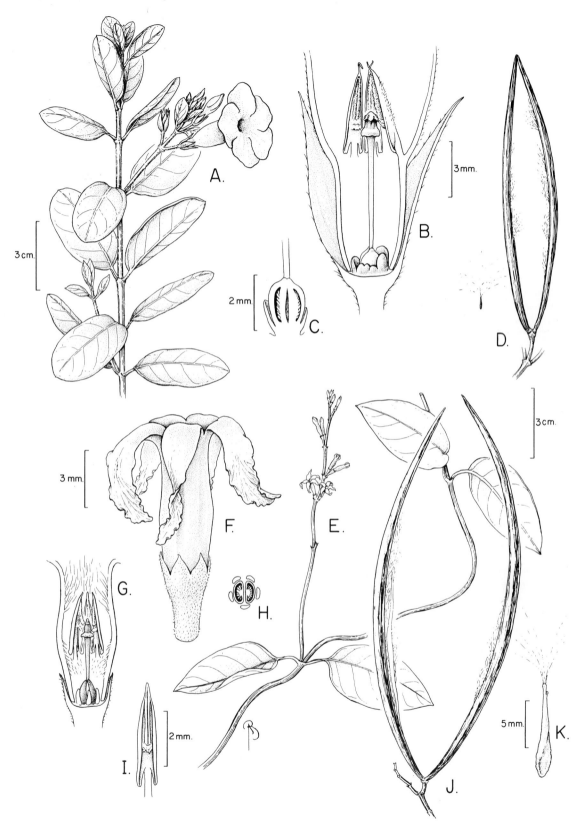

Fig. 22. A–D. *Pentalinon luteum*. A. Flowering branch. B. L.s. flower. C. L.s. ovary, with subtending nectaries. D. Dehiscing follicles and seed. E–
K. *Prestonia agglutinata*. E. Flowering branch. F. Flower. G. L.s. flower. H. C.s. carpels and nectary glands. I. Anther. J. Dehiscing follicles.
K. Seed.

4. PLUMERIA L.

Small trees, producing abundant milky latex. Leaves alternate, aggregate at branch tips. Flowers fragrant, produced in axillary, branched, thyrsiform inflorescences; peduncle longer than the flower-bearing portion; bracts minute, deciduous; calyx with 5 minute lobes; corolla funnel-shaped, with 5 equal, rounded, spreading lobes; stamens inserted at base of tube, the anthers free; ovary of 2 separate carpels, the carpels with many ovules. Fruit of 2 stout, spreading follicles; seeds numerous, compressed, with a tuft of long hairs at apex.

About 45 species native to tropical America. Spelled by Britton and P. Wilson (Bot. Porto Rico 6: 87–88. 1925) as *Plumiera*.

1. Plumeria alba L., Sp. Pl. 210. 1753.   Fig. 21I–J.

Small tree 3–7 m tall; bark light-colored and corky. Leaf blades 15–30 × 1.2–4 cm, linear, oblong or oblong-lanceolate, coriaceous, lower surface with prominent reticulate venation, sericeous, the apex acute or acuminate, the base obtuse or tapering, the margins strongly revolute; petioles 1–4 cm long, glabrous, furrowed. Calyx lobes ca. 3 mm long; corolla 3.3–6 cm long, white, yellow at center within, the tube 0.7–1.5 cm long, the lobes oblong, 2.5–4 cm long. Follicle 10–14 cm long, woody, brown to grayish, smooth.

DISTRIBUTION: Common in coastal thickets, sometimes cultivated for its fragrant flowers. Lind Point (A4240). Also on Anegada, Jost van Dyke, St. Croix, St. Thomas, Tortola, and Virgin Gorda; Puerto Rico, and in the Lesser Antilles (Anguilla to Grenada).

COMMON NAMES: frangipani, klang hout, milk tree, milky bush, paucipan, snake root, wild frangipani, wormy tree.

5. PRESTONIA R. Br.

Twining vines or lianas, producing clear or milky latex. Leaves opposite, petiolate, without glands; stipules interpetiolar and glandular (colleters). Flowers produced in congested or elongate (thyrselike) cymes; calyx of 5 sepals, each sepal with a glandular appendage at base within; corolla salverform, narrowed at base, distal portion of tube thickened; stamens 5, the filaments adnate to basal portion of corolla tube, the anthers sagittate; ovary of 2 carpels, with a 5-glandular nectary disk at base. Fruit of 2 follicles, with numerous seeds bearing a tuft of hairs at apex.

A neotropical genus with about 35 species.

1. Prestonia agglutinata (Jacq.) Woodson, Ann. Missouri Bot. Gard. **18**: 552. 1931. *Echites agglutinata* Jacq., Enum. Syst. Pl. 13. 1760.   Fig. 22E–K.

Echites circinalis Sw., Prodr. 52. 1788.

Herbaceous or subwoody, twining vine to 5 m long, producing a clear latex when cut; stems cylindrical, green. Leaves opposite; blades 4–19 × 3–9 cm, ovate, wide-elliptic, seldom orbicular or lanceolate, chartaceous, glabrous, the apex obtuse, acute, shortly acuminate and apiculate, the base obtuse, rounded or truncate, usually oblique, the margins entire or wavy and revolute; petioles 0.5–4 cm long, slender. Flowers paired along a thyrselike inflorescence; bracts and bracteoles ca. 1.5 mm long, lanceolate. Calyx 2–3 mm long with 5 ovate, deeply parted sepals; corolla cream, the tube 4–6 mm long, the lobes 2–4 mm long, long-pointed and reflexed. Follicle 10–25 cm long, linear, curved, hanging. Seeds numerous, 5–7 mm long, tear-shaped, light brown.

DISTRIBUTION: In open or disturbed areas along coastal thickets and in secondary forests. Bordeaux Mountain (A3142), Lameshur (A2045), Reef Bay (A2665). Also on St. Croix, St. Thomas, and Tortola; Hispaniola and Puerto Rico.

6. RAUVOLFIA L.

Shrubs or trees; aerial parts producing abundant milky latex. Leaves whorled (or opposite), bearing glands at axil and lower part of petiole; stipules interpetiolar and deciduous. Flowers produced in simple or compound, terminal dichasial cymes; bracts minute, deltoid; calyx with 5 deeply parted sepals, without glands; corolla salverform, the tube cylindrical; stamens inserted near middle of tube, the anthers free, not projecting beyond the tube; ovary of 2 separate or fused carpels, with an annular nectary disk at base. Fruit fleshy, of 2 fused or free drupes; seeds 1 per drupe.

A genus with pantropical distribution, with about 110 species.

Key to the Species of *Rauvolfia*

1. Trees; leaves with 14–20, nearly perpendicular, secondary veins, upper surface shiny; fruits 8–12 mm broad .... 1. *R. nitida*
1. Shrubs; leaves with 10–12, ascending, secondary veins, upper surface dull; fruits 5–7 mm broad .... 2. *R. viridis*

1. Rauvolfia nitida Jacq., Enum. Syst. Pl. 14. 1760.   Fig. 23.

Rauvolfia tetraphylla sensu Britton & P. Wilson, Bot. Porto Rico 6: 90. 1925, non L., 1753.

Tree to 20 m tall. Leaves whorled (4 per node); blades 6–15 × 1.8–4.5 cm, elliptic, oblong or rarely lanceolate, chartaceous to nearly coriaceous, upper surface shiny, secondary veins 14–20 on each side, almost perpendicular to central one, the apex acute or acuminate, the base obtuse or tapering, usually oblique, the margins entire or wavy and revolute; petioles 0.3–1 cm long. Calyx cup-shaped, ca. 2 mm long, the sepals ca. 1 mm. long, rounded; corolla cream or greenish white, the tube 4–5.5 mm long, pubescent within, the lobes 2–3 mm long, rounded, spreading. Fruit 8–12 mm broad, purple, shiny, smooth, with 2 fused drupes or one of them partially developed.

DISTRIBUTION: Common in dry coastal thickets to moist forests. Bor-

FIG. 23. *Rauvolfia nitida*. **A.** Flowering branch. **B.** Detail of inflorescence. **C.** L.s. corolla showing stamens. **D.** Calyx with removed corolla showing pistil with nectary disk at base. **E.** Drupe. **F.** C.s. drupe and seed.

deaux Mountain (A2878), Cinnamon Bay (A2092). Also on St. Croix, St. Thomas, and Tortola; Greater Antilles, St. Kitts, Guadeloupe, Martinique, and Trinidad.

COMMON NAMES: bitter ash, milk bush, milk tree.

2. Rauvolfia viridis Willd. *ex* Roem. & Schult., Syst. Veg. **4**: 805. 1819.

Rauvolfia lamarckii A. DC., Prodr. **8**: 337. 1844.

Shrub 1–2 m tall. Leaves whorled (3 or 4 per node); blades 4–15 × 2–5 cm, elliptic, chartaceous, upper surface dull, secondary veins 10–12 on each side, ascending, long-pointed at both ends, the base usually oblique, the margins entire or wavy and revolute; petioles 0.3–0.5 cm long. Calyx cup-shaped, ca. 1.5 mm long, the sepals ca. 0.5 mm long, rounded; corolla cream or white, the tube 2–3.5 mm long, pubescent within, the lobes 1.5–2 mm long, rounded, spreading. Fruit 5–7 mm broad, purple, shiny, smooth, with 2 fused drupes or one of them partially developed.

DISTRIBUTION: Common in dry coastal thickets or hillsides. Bethany (B193), Hurricane Hole (A2771). Also on St. Croix, St. Thomas, Tortola, and Virgin Gorda; Puerto Rico, Lesser Antilles, and northern South America.

COMMON NAMES: bitter bush, man better man.

CULTIVATED SPECIES: The small tree *Thevetia peruviana* (Pers.) Schum. with yellow flowers, locally known as lucky-nut, is found only in a few gardens.

8. Aquifoliaceae (Holly Family)

Shrubs or trees. Leaves alternate and simple; stipules usually present but deciduous. Flowers actinomorphic, 4–9-parted, unisexual or bisexual, produced in axillary cymes, or solitary; calyx minute; corolla with free or basally connate petals, usually deciduous; stamens alternating with the petals, inserted at base of corolla, the filament shorter than the petals; staminodes, when present, similar to stamens; ovary superior with 2–8 carpels, the stigma sessile and capitate, each carpel with a single ovule pendulous from the apex. Fruit a drupe, usually globose or ellipsoid, 2–8-seeded.

A family with cosmopolitan distribution, with 4 genera and 300–400 species.

1. ILEX L.

Shrubs or trees. Leaves petiolate, with minute stipules. Flowers staminate or pistillate on different plants (dioecious), solitary or in cymes; calyx minute, persistent; corolla with petals connate at base; stamens as long as the petals in staminate flowers, shorter and sterile in pistillate flowers; ovary sessile with 4–8 carpels; reduced to a pistillode in staminate flowers. Drupe fleshy, nearly globose or ellipsoid with 4–8 seeds.

A genus with cosmopolitan distribution, with ca. 300–400 species.

Key to the Species of *Ilex*

1. Leaves obovate to oblong, 6–12 cm long, chartaceous-coriaceous, the apex acute usually slightly notched, the margins often crenate-serrulate; inflorescence axes glabrous or puberulent; flowers 4(–5)-merous .... 1. *I. nitida*
1. Leaves elliptic, rarely obovate, 2.5–8 cm long, coriaceous, the apex rounded or deeply notched, the margins usually entire; inflorescence axes pubescent; flowers 4–5-merous .... 2. *I. urbaniana*

1. Ilex nitida (Vahl) Maxim., Mém. Acad. Imp. Sci. Saint Pétersbourg, ser. 7, **29(3):** 27. 1881. *Prinos nitidus* Vahl, Eclog. Amer. **2:** 26. 1798.

Tree 10–15 m tall; stems nearly cylindrical, glabrous, smooth. Leaf blades 6–12 × 4–6.5 cm, obovate to oblong, chartaceous-coriaceous, glabrous, midvein sunken on upper surface and prominent on lower surface, the apex acute, usually slightly notched, the base acute to nearly rounded, the margins crenate-serrate and revolute; petioles 6–10 mm long, furrowed, glabrous. Flowers solitary or fasciculate at leaf axils; axes of inflorescence glabrous or puberulent; bracts ca. 0.5 mm long, lanceolate. Calyx nearly disk-shaped, the lobes 4(–5), 1–1.5 mm long, rounded, with slightly ciliate margins; petals white to greenish, 3–4 mm long, free to base, obovate to oblong; stamens 2–3 mm long; ovary ovoid to nearly globose, 4-locular, the stigmas discoid. Fruit nearly globose, 5–8 mm wide, red at maturity.

DISTRIBUTION: Occasional in moist forest. Bordeaux Mountain (A4701, A5081). Also on Cuba, Jamaica, Hispaniola, Puerto Rico, Lesser Antilles, and Mexico.

2. Ilex urbaniana Loes., Bot. Jahrb. Syst. **15:** 316. 1893. Fig. 24.

Small tree 3–8 m tall; stems nearly cylindrical, glabrous, smooth, furrowed when young. Leaf blades 2.5–8 × 1.2–3.6 cm, elliptic, rarely obovate, coriaceous, glabrous, midvein sunken on upper surface and prominent on lower surface, the apex rounded or deeply notched, the base obtuse or rounded, the margins entire and strongly revolute; petioles 4–8 mm long, furrowed, pubescent when young. Flowers solitary or in short cymes at leaf axils; inflorescence axes pubescent; bracts ca. 0.5 mm long, lanceolate. Calyx cup-shaped, 1–1.5 mm long, 4–5-lobed, the lobes oblong, ciliate, obtuse at apex; petals white, 2–2.5 mm long, free to base, oblong; stamens 1.2–1.5 mm long, shorter and sterile in pistillate flowers; ovary conical to ovoid, reduced in staminate flowers, the stigmas discoid and lobed. Fruit ellipsoid to nearly globose, 4–5 mm wide, red at maturity.

DISTRIBUTION: An uncommon tree found in moist forests. Bordeaux Mountain (A4692, M17078). Also on St. Thomas and Tortola; Puerto Rico and Hispaniola.

DOUBTFUL RECORD: *Ilex sideroxyloides* (Sw.) Griseb. was reported by Woodbury and Weaver (1987) as occurring on St. John, but no specimen was collected by them. It seems that their record was based on an incorrect identification of *Ilex nitida*.

9. Araliaceae (Ginseng Family)

Trees, shrubs, lianas, or less often perennial herbs. Leaves alternate or rarely opposite or whorled, simple, pinnately or palmately compound or dissected, sometimes with stipular appendages. Flowers 5-merous, bisexual or unisexual, actinomorphic or zygomorphic, produced in umbels, arranged into racemes, spikes, or heads; calyx represented by small teeth or obsolete; corolla of (3–)5(–12) distinct petals; stamens free, as many as the petals or numerous, the anthers opening by longitudinal slits; ovary inferior, 2–5(–15)-carpellate, the carpels with 1 pendulous, axile ovule, the styles as many as the carpels, free or united. Fruit a drupe or a berry.

A family with about 70 genera and 700 species, principally in tropical and subtropical regions.

1. SCHEFFLERA J.R. Forst. & G. Forst.

Shrubs or trees. Leaves alternate, palmately compound, with stout, long petioles and stipular appendage at axil; leaflets radiating from petiole. Flowers pedicellate, unisexual or bisexual, regular, arranged in umbels along large, axillary panicles; calyx obconic or cup-shaped, sometimes with 5 apical teeth; petals valvate, sometimes connate forming a calyptra; nectary disk annular; stamens 5; styles 2–7, free. Fruit with fleshy exocarp, 5–10-ribbed, with persistent disk and styles at apex; seeds laterally compressed.

A genus of 150–200 species, with tropical distribution.

1. Schefflera morototoni (Aubl.) Maguire, Steyerm. & Frodin, Mem. New York Bot. Gard. **38:** 51. 1984. *Panax morototoni* Aubl., Hist. Pl. Guiane **2:** 949. 1775. *Didymopanax morototoni* (Aubl.) Decne. & Planch., Rev. Hort., ser. 4, **3:** 109. 1854. Fig. 25.

Fast-growing, andromonoecious tree, to 15 m tall, trunk straight, unbranched or branched only at top; bark grayish, smooth, with many large leaf scars. Leaves spirally arranged along stems; leaflets 10–12, 8–45 × 3–21 cm, elliptic, oblong, oblanceolate, chartaceous to coriaceous, glabrous or sparsely pubescent, lower

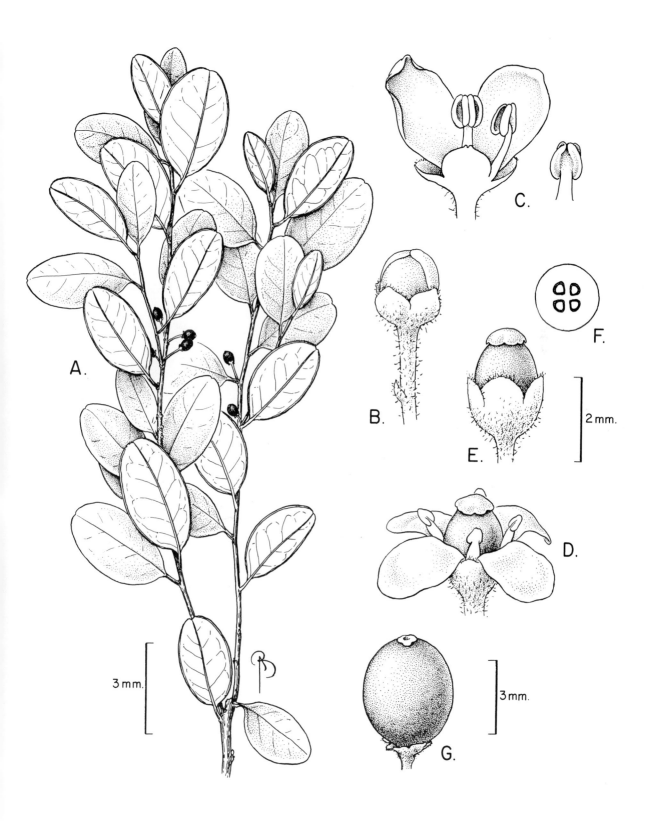

Fig. 24. *Ilex urbaniana*. **A.** Fruiting branch. **B.** Flowerbud. **C.** L.s. staminate flower and stamen. **D.** Pistillate flower with sterile stamens. **E.** Pistillate flowers with petals and stamens removed. **F.** C.s. ovary. **G.** Drupe.

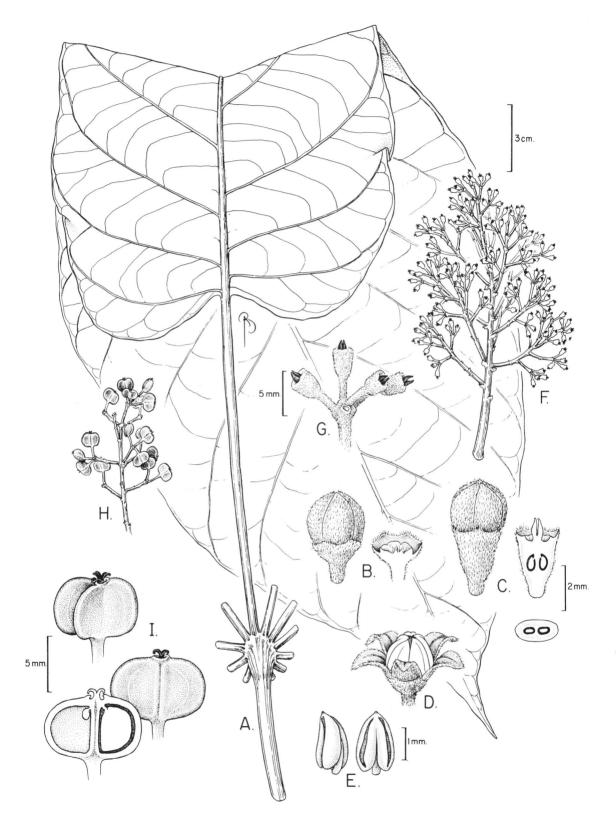

FIG. 25. *Schefflera morototoni*. **A.** Detail of compound leaf (all but one leaflet has been removed). **B.** Staminate flowerbud and l.s. nectary disk. **C.** Bisexual flowerbud and l.s. and c.s. hypanthium. **D.** Staminate flower. **E.** Stamen, lateral and frontal views. **F.** Inflorescence branch. **G.** Detail of inflorescence. **H.** Inflorescence with developing fruits. **I.** Fruits and l.s. fruit showing locules and developing seeds.

surface usually yellowish, the apex long- to short-acuminate, the base rounded or subcordate, unequal, the margins entire or finely serrate on young leaflets, the petiolules cylindrical, to 16 cm long; petioles to 1 m long, cylindrical and striate. Flowers staminate and bisexual, 4–7 in lateral umbels along many-branched inflorescences. Calyx cup-shaped, ca. 2 mm long in staminate flowers, forming a hypanthium 4–4.5 mm long in bisexual flowers, not toothed; petals ca. 4 mm long, reflexed at anthesis; stamens nearly sessile, the anthers lanceolate; styles 2, awl-shaped. Fruit dry, flattened kidney-shaped, 0.8–1.2 cm wide.

DISTRIBUTION: In moist secondary forests and disturbed areas such as roadsides. Bordeaux Mountain (A2887). Also on St. Thomas and Tortola; Cuba, Hispaniola, Puerto Rico, Mexico, to South America.

COMMON NAME: match wood.

10. Aristolochiaceae (Dutchman's Pipe Family)

The genus *Aristolochia* characterizes the family in the Caribbean region. A family with 8–10 genera and approximately 600 species, with tropical or subtropical distribution.

REFERENCES: González, F. 1990. Aristolochiaceae. Flora de Colombia. Vol. 12. Universidad Nacional de Colombia. Pfeifer, H. W. 1966. Revision of the North and Central American hexandrous species of *Aristolochia*. (Aristolochiaceae). Ann. Missouri Bot. Gard. 53: 115–196.

1. ARISTOLOCHIA L.

Herbaceous or woody twining vines or less often herbs or shrubs. Leaves alternate, simple, entire or trilobate, long-petioled, often with prominent pseudostipules. Flowers bisexual, zygomorphic or rarely actinomorphic, solitary in leaf axils or clustered on old stems; calyx corolla-like, inflated, constricted into a tube below apex, expanding above into 1 or 2 lobes; corolla wanting; stamens 5–6, sessile; ovary inferior or partially superior, 5–6-carpellate, each carpel with numerous ovules, with axile placentation, the styles 6, connate, the stigmas capitate. Fruit a capsule, usually opening along valves; seeds numerous, light and compressed.

A genus with approximately 450–550 species, chiefly with tropical or subtropical distribution.

Key to the Species of *Aristolochia*

1. Pseudostipules absent or inconspicuous; leaves ovate, lanceolate or hastate.............................. 2. *A. odoratissima*
1. Pseudostipules large, clasping the stem; leaves trilobate or kidney-shaped.
  2. Leaves trilobate (at least some); flowers with long filiform, hanging appendage .... 3. *A. trilobata*
  2. Leaves kidney-shaped; flowers without appendage .... 1. *A. elegans*

1. Aristolochia elegans Mast., Gard. Chron., n.s., **24**: 301. 1885.　　　　　　　　　　Fig.26 A–F.

Woody, twining vine to 7 m long; stems glabrous, cylindrical, smooth. Leaf blades 7–9 × 6–10 cm, reniform or ovate, with palmate venation, chartaceous, upper surface shiny, lower surface glaucous, with prominent veins, the apex obtuse, the base cordiform, the margins entire; petioles slender, 1–3 cm long, slightly swollen at base; pseudostipules pale green, clasping the stem, 1–1.5 cm long. Flowers hanging, solitary at leaf axils. Calyx forming a sigmoid tube, spreading at upper portion, forming a saucer-shaped limb, the tube 2.5–3.5 cm long, light green, the limb 5–7 cm wide, with irregular, maroon spots, the throat dark maroon, the anthers connivent. Fruit 5–6 cm long, turning from green to brown, nearly cylindrical, 5-ribbed, long-pointed at both ends, the apex with persistent, stigmatic surface; pedicels 3–4 cm long. Seeds numerous, 6–7 mm long, thin, flattened, triangular to ovate.

DISTRIBUTION: In secondary growth, as an escape. Apparently a recent introduction since it is restricted to Carolina (A4136). Also on Tortola; native to South America, but cultivated throughout the West Indies.

NOTE: The name *Aristolochia littoralis* Parodi has been used for this taxon in the Caribbean following Pfeifer (1966), who placed *A. elegans* into synonymy under *A. littoralis*. This notion has been challenged by F. González (1990), who considered *A. littoralis* a distinct species, restricted to Argentina.

2. Aristolochia odoratissima L., Sp. Pl., ed. 2, **2**: 1362. 1763.

Aristolochia pandurata Jacq., Pl. Hort. Schoenbr. **4**: 49. 1804.

Woody twining vine to 5 m long; stems cylindrical, glabrous, smooth. Leaf blades 6.5–11 × 3.5–6 cm, lanceolate, hastate or rarely ovate, with palmate venation, chartaceous to coriaceous, lower surface sparsely covered with minute whitish hairs, the apex acuminate, the base deeply cordiform, the margins wavy, revolute; petioles 1–3.5 cm long, furrowed, usually twisted; pseudostipules absent or inconspicuous. Calyx yellowish with purple network, the tube to 7 cm long, the limb spreading, 4–6 cm wide, without a hanging tail. Fruit 7–10 cm long, nearly cylindrical, 5-ribbed, arcuate, long-pointed at both ends. Seeds numerous, ca. 3 mm long, flattened, triangular.

DISTRIBUTION: In secondary growth, in semidry areas. Bethany (B338), Caneel Bay (A4098). In Mexico to Central America and the West Indies.

3. Aristolochia trilobata L., Sp. Pl. 960. 1753.

Fig. 26G–H.

Woody twining vine to 5 m long; stems cylindrical, glabrous, smooth, pinkish-tinged at nodes. Leaf blades 8–14 × 10–16 cm, trilobate, with palmate venation, coriaceous, lower surface densely covered with minute whitish hairs, the lobes deep or shallow, with obtuse apex, the base subcordiform or truncate, the margins entire or wavy; petioles 2–4.5 cm long, cylindrical, glabrous; pseudostipules to 1.5 cm long, rounded or reniform, clasping the stem. Calyx yellowish with purple network, the tube to 7 cm long, the limb with a long hanging lobe or tail. Fruit 6–8.5 cm long, nearly cylindrical, 6-ribbed. Seeds numerous, ca. 7 mm long, flattened, triangular.

DISTRIBUTION: In moist forest. Bordeaux Mountain (B583, A5080). Also on St. Croix, St. Thomas, and Tortola; from Central America to northern South America and the West Indies.

COMMON NAMES: pelican plant, tobacco pipe.

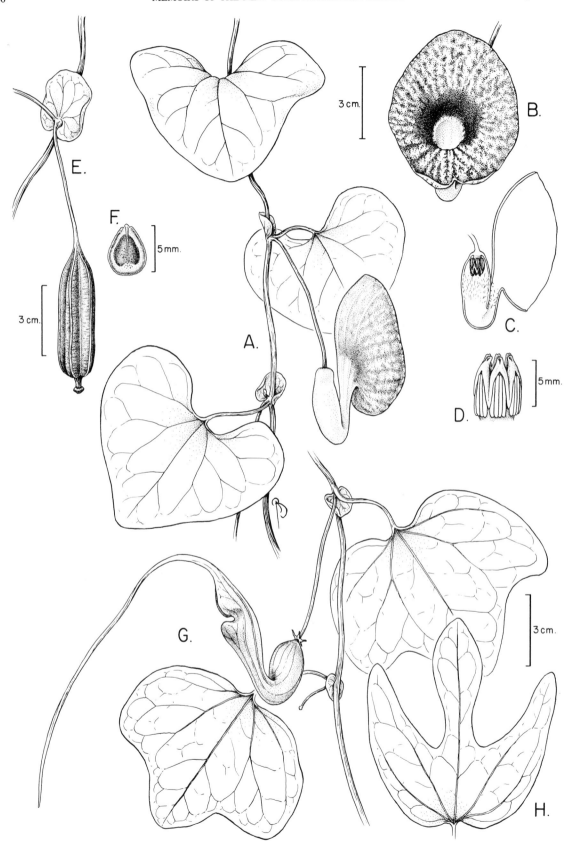

FIG. 26. A–F. *Aristolochia elegans*. **A.** Flowering branch. **B.** Flower, frontal view. **C.** L.s. flower. **D.** Connivent anthers. **E.** Fruit and pseudostipule. **F.** Seed. **G, H.** *Aristolochia trilobata*. **G.** Flowering branch. **H.** Deeply lobed leaf.

11. Asclepiadaceae (Milkweed Family)

Reviewed by G. N. Morillo

Herbs, woody or herbaceous twining vines, shrubs, or rarely trees, producing abundant milky latex. Leaves simple, opposite or whorled, usually with 2 or more small, digitate glands at base of blade; stipules lacking or minute. Flowers 5-merous, bisexual or less often functionally unisexual, actinomorphic, produced in umbelliform or racemiform cymes, sometimes solitary; calyx of connate sepals, usually reflexed; corolla with basally united petals, forming a short or elongate tube, the lobes spreading or reflexed; corona usually present; stamens inserted at the corolla tube, fused to form a sheath around the ovary and united to the stigma head to form a gynostegium, the anthers usually with apical appendages; ovary superior, of 2 distinct carpels, united only by the stigmatic surface, the carpels with many ovules, the ovules pendulous and imbricate upon the lateral placenta, the styles distinct. Fruit of 2 follicles, sometimes reduced to 1 by abortion, with many flattened, light seeds, these usually with a long tuft of hairs at apex.

A family with 250 genera and ca. 2000 species, predominantly tropical and subtropical.

REFERENCE: Woodson, R. E. 1941. The North American Asclepiadaceae. Ann. Missouri Bot. Gard. 28: 193–248.

Key to the Genera of Asclepiadaceae

1. Twining lianas, vines, or scandent shrubs.
  2. Robust liana; leaf blade without glands; corolla lavender, funnel-shaped, 5–7 cm long; stamens with distinct filaments; fruits winged or keeled .... 3. *Cryptostegia*
  2. Herbaceous to scarcely woody vines; leaf blade with a pair of minute glandlike projections at base on upper surface; corolla greenish or cream, nearly bell-shaped to rotate, <1 cm long; stamens connate to form a gynostegium.
    3. Stems slender (2–3 mm diam.); leaves lanceolate or ovate, to 2 cm long, the basal glands filiform; follicles linear-ellipsoid, smooth; corolla cream, bell-shaped.. 5. *Metastelma*
    3. Stems 5–7 mm diam.; leaves ovate, cordate at base, to 8 cm long, the basal glands triangular; follicles nearly ovoid, with blunt, spinelike projections; corolla greenish, rotate........................... 4. *Matelea*
1. Erect herbs or shrubs.
  4. Herbs; base of blade without glands; corolla orange, with reflexed, oblong petals; corona yellow; follicles spindle-shaped, 0.5 cm wide .... 1. *Asclepias*
  4. Shrubs; base of blade with numerous spinelike glands; corolla whitish, violet-tinged within, rotate, with spreading, obtuse petals; corona violet; follicles inflated, nearly ovoid, membranous, 6–7 cm wide... 2. *Calotropis*

1. ASCLEPIAS L.

Erect herbs. Leaves opposite or whorled, petiolate; stipules minute, interpetiolar. Flowers in axillary or terminal, long-peduncled umbels; calyx with 5 deeply parted sepals; corolla of 5 early reflexed petals, united at base; corona of parts, each composed of hood, horn, and appendage; stamens united with the fleshy stigma head; ovary of 2 free carpels, united at the stigma head, each carpel with many ovules. Fruit of 2 follicles with numerous seeds with a long tuft of hairs at apex.

A predominantly North American genus of 120 species.

1. Asclepias curassavica L., Sp. Pl. 215. 1753.

Fig. 27A–B.

Erect herb to 60 cm tall. Leaves opposite or less often in whorls of 3; blades glandless, 7–13 × 1.5–3.5 cm, elliptic, chartaceous, glabrous except for a few hairs on veins, the apex acuminate, the base obtuse or attenuate, the margins entire; petioles 0.5–1 cm long. Calyx 5–6 mm long, bell-shaped, pubescent, the sepals lanceolate; corolla orange, the petals oblong-elliptic, 4–6 mm long, reflexed; corona yellow; anthers lanceolate; stigma lobed. Follicle 6–9 × 0.5 cm, spindle-shaped, smooth, straw-colored at maturity. Seeds lenticular, brown, the hairs silky white.

DISTRIBUTION: A common weed along roadsides and in open or waste grounds. Bordeaux area (A1898), Susannaberg (A4680). Also on St. Croix, St. Thomas, Tortola, and Virgin Gorda; in Florida, continental tropical America, and the West Indies.

COMMON NAMES: kittie mcwanie, pretie guana, wild ipecacuana.

2. CALOTROPIS R. Br.

Shrubs or small trees. Leaves large, opposite, sessile or nearly so; stipules wanting. Flowers in axillary or terminal, long-peduncled umbels; calyx spreading, 5-parted, the sepals with many glands within at base; corolla nearly rotate, the lobes broad; corona fleshy, erect, with 5 scales; stamens attached to corolla base, with filaments fused into a short tube, the pollinia pendulous; ovary of 2 free carpels, united only at the capitate, 5-angled stigma. Fruit of 1 inflated follicle with fibrous mesocarp; seeds numerous, light, with tuft of hairs at apex.

A genus of 2 species, naturally distributed in warm and tropical areas of the Old World.

1. Calotropis procera (Aiton) W.T. Aiton, Hortus Kew.
2: 78. 1811. *Asclepias procera* Aiton, Hort. Kew.
1: 305. 1789.   Fig. 27C–G.

Shrub 1–4 m tall, many-branched from base; young parts and inflorescences densely covered with white woolly hairs; bark rough, grayish. Leaves sessile, 12–20 × 9.5–13.5 cm, wide elliptic, nearly rounded to pandurate, coriaceous, glabrous at maturity with shortly acuminate apex, the base cordate, clasping the stem, with many spinelike glands, the margins entire and ciliate. Calyx 5–6 mm long, bell-shaped, pubescent, the sepals ovate; corolla white, fleshy, the lobes ovate, light violet within, 1.5–2 cm long; corona deep violet, much smaller than the corolla; stigma light green. Follicle 8–13 × 6–7 cm, ovoid, inflated, membranous, slightly wrinkled, turning from green to straw-colored at maturity. Seeds lenticular, light brown, the hairs silky white.

DISTRIBUTION: Common along roadsides and open or waste grounds in dry areas. Coral Bay (A3770), Cruz Bay (A1949). Also on Jost van Dyke, St. Croix, St. Thomas, Tortola, and Virgin Gorda. Native to tropical Africa, now widespread throughout the tropics.

COMMON NAMES: cowheel, giant milkbush, milkweed, mountain cabbage.

NOTE: The abundant milky latex produced by the plant is said to be poisonous.

3. CRYPTOSTEGIA R. Br.

An Old World genus of 2 species, naturally distributed from Africa to India. The following species description characterizes it.

1. Cryptostegia grandiflora R. Br., Bot. Reg. 435. 1820.
Fig. 28A–D.

Twining liana or scandent shrub, to 10 m long; stems cylindrical, glabrous, with scattered lenticels. Leaves opposite; blades 4–10 × 2–4.7 cm, elliptic, oblong or rarely ovate, coriaceous, glabrous, with marked reticulate venation, the apex shortly acuminate, obtuse or rounded, the base obtuse or rounded, without glands, abruptly attenuate into the petiole, the margins entire; petioles 0.6–1.5 cm long; stipules minute and interpetiolar. Flowers produced in terminal long-peduncled cymes. Calyx bell-shaped, with 5 deeply parted, ovate sepals, 0.5–1.5 cm long; corolla funnel-shaped, light violet externally, 4–6 cm long, the lobes 1–2 cm long, spreading, the tube whitish within; corona of 5 filiform appendages; stamens inserted near the corolla base, the filaments short and free, the anthers connivent around the stigma. Fruit with 2 divergent, narrowly winged or keeled boat-shaped, woody follicles, 10–13 cm long, slightly wrinkled and straw-colored at maturity. Seeds numerous, brown, tear-shaped, with a long tuft of hairs at apex.

DISTRIBUTION: This aggressive exotic species was introduced on St. John in the 1950s as an ornamental. The plant has naturalized and is now well established on the island, and its range continues to expand. It is particularly abundant at Coral Bay (A4681) and at Lameshur. Also on Tortola; native to Africa, introduced into the West Indies.

COMMON NAMES: purple allamanda, rubber vine.

4. MATELEA Aubl.

Erect subshrubs or twining vines, with mixed indument, consisting of long nonglandular hairs and short glandular hairs. Leaves opposite, petiolate; without stipules. Flowers in axillary cymes; calyx 5-parted, the sepals often with glands at base within; corolla nearly rotate, deeply to shallowly parted; corona annular, short, with 10 appendages; stamens inserted near the corolla base, the filaments connate into a short tube, the anthers connivent around the style; ovary of 2, free carpels, the stigma capitate, projecting beyond the anthers. Fruit of a fusiform to ovoid follicle; seeds numerous, with tuft of hairs at apex.

A genus of approximately 100 species from southern United States to South America.

1. Matelea maritima (Jacq.) Woodson, Ann. Missouri
Bot. Gard. **28**: 222. 1941. *Asclepias maritima* Jacq.,
Enum. Syst. Pl. 17. 1760. *Cynanchum maritimum*
(Jacq.) Jacq., Select. Stirp. Amer. Hist. 83. 1763.
Gonolobus maritimus (Jacq.) R. Br., Mem. Wern.
Nat. Hist. Soc. **1**: 35. 1809. *Ibatia maritima* (Jacq.)
Decne. *in* A. DC., Prodr. **8**: 599. 1844.
Fig. 28E–H.

Ibatia muricata Griseb., Fl. Brit. W. I. 421. 1862.

Herbaceous to woody, twining vine, to 7 m long; stems cylindrical, densely pubescent when young, with grayish, corky bark when old. Leaf blades 4.5–10.5 × 3–7.5 cm, ovate, chartaceous, pubescent on both surfaces, the apex acute or acuminate, the base cordate, with a pair (rarely 3–7) of minute spinelike glands on the upper surface, at petiole junction, the margins entire; petioles 2–7 cm long, densely pubescent; stipules wanting. Calyx ca. 2 mm long, the sepals lanceolate or ovate, pubescent; corolla rotate, greenish, 4–6 mm wide, the petals ovate; corona yellowish green with rounded, short appendages; stamens nearly sessile, inserted near the corolla base; stigma projecting beyond the anthers. Follicle ovoid-fusiform, 5–8 × 3.5–4.5 cm, covered with many-tuberculate projections with white tips. Seeds numerous, flattened, tear-shaped, dark brown, with a tuft of long, white silky hairs at apex.

DISTRIBUTION: Common in coastal forests or scrub. East side of Bordeaux (A4674), Lameshur (A2737). Also on St. Croix, St. Thomas, Tortola, and Virgin Gorda; Cuba, Hispaniola, Puerto Rico, Lesser Antilles, and Panama to northern South America.

COMMON NAME: beach milk vine.

5. METASTELMA R. Br.

Herbaceous twining vines. Leaves opposite, small and petioled; stipules minute, interpetiolar. Flowers minute, produced in short-peduncled or sessile axillary cymes; calyx 5-parted, the sepals usually with glands at base within; corolla bell-shaped, the petals usually hairy within; corona with 5 appendages; stamens inserted at base of corolla, connivent around the conical stigma. Follicle linear-fusiform; seeds numerous, with a tuft of hairs at apex.

A neotropical genus with ca. 150 species.

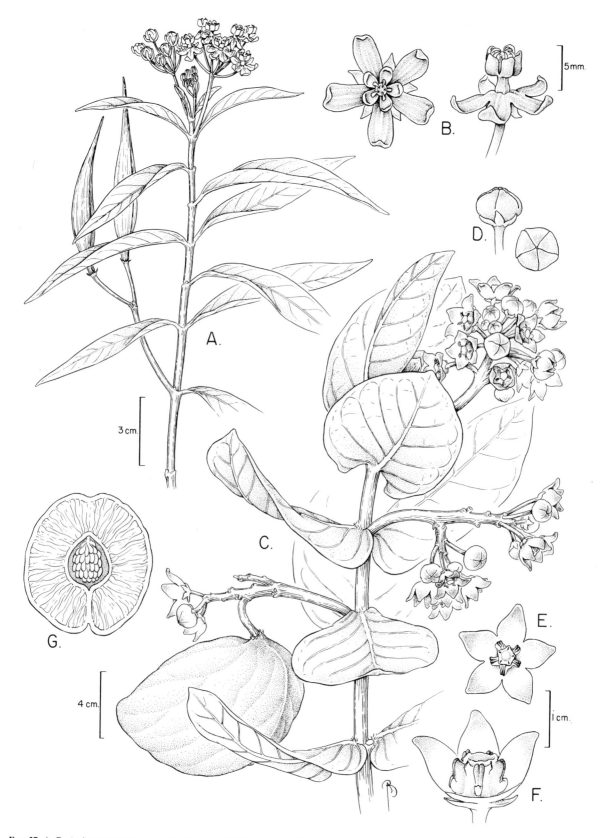

FIG. 27. A, B. *Asclepias curassavica*. A. Fertile branch. B. Flower, top and lateral views. C–G, *Calotropis procera*. C. Fertile branch. D. Flowerbud, lateral and top views. E. Flower. F. L.s. flower. G. C.s. fruit.

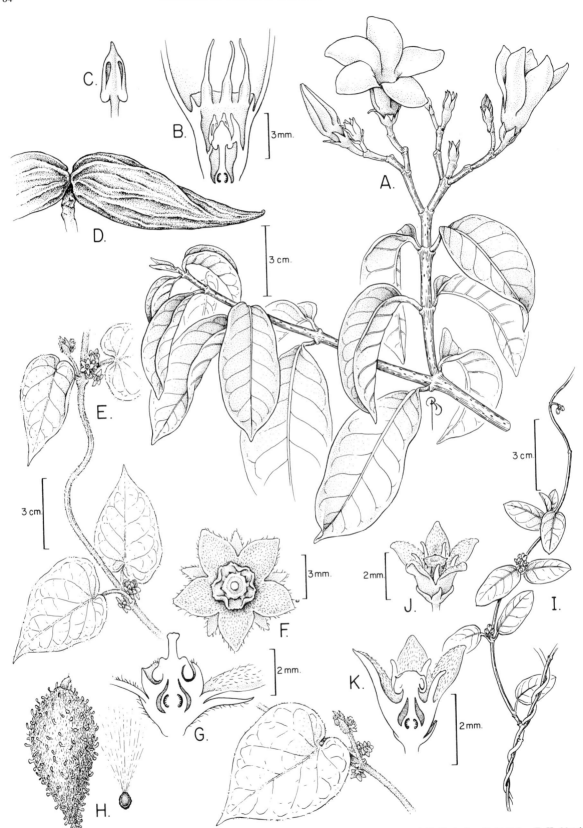

Fig. 28. A–D. *Cryptostegia grandiflora.* **A.** Flowering branch. **B.** l.s. flower with corona, stamens, and pistil. **C.** Anther. **D.** Follicles. **E–H.** *Matelea maritima.* **E.** Flowering branch. **F.** Flower. **G.** L.s. flower. **H.** Follicle and seed. **I–K,** *Metastelma grisebachianum.* **I.** Flowering branch. **J.** Flower. **K.** L.s. flower.

1. Metastelma grisebachianum Schltr. *in* Urb., Symb. Ant. **5**: 469. 1908. *Cynanchum grisebachianum* (Schltr.) Alain, Rhodora 67: 346. 1965.

Fig. 28I–K.

Metastelma decaisneanum Schltr. *in* Urb., Symb. Ant. **1**: 250. 1899, non Benth. & Hook., 1876. *Cynanchum decaisneanum* (Schltr.) Alain, Mem. Soc. Cub. Hist. Nat. "Pelipe Poey" **22**: 119. 1955, non R. W. Holm, 1953.

Metastelma albiflorum sensu Schltr. *in* Urb., Symb. Ant. **1**: 250. 1899, non Griseb., 1862.

Herbaceous twining vine to 6 m long; stems cylindrical, green, slender. Leaf blades 1–2 × 0.6–1.2 cm, ovate to lanceolate, chartaceous, glabrous, except for a few hairs on margins and petioles, the apex acute to obtuse and apiculate, the base subcordate or truncate, with a pair of minute filiform glands on upper surface, at petiole junction, the margins entire; petioles 0.3–0.5 cm long; stipules minute and interpetiolar. Calyx ca. 1 mm long; corolla bell-shaped, light green externally, whitish within, ca. 2.5 mm long, the lobes acute, spreading; corona of 5 filiform segments; stamens nearly sessile, inserted near the base of corolla. Fruit with 2 linear follicles, 3.5–5 cm long, smooth, turning from green to straw-colored at maturity. Seeds numerous, with a tuft of hairs at apex.

DISTRIBUTION: Common in most parts of the island. Southside Pond (A1815), Turner Bay (A3106). Also on St. Croix, St. Thomas, Tortola, and Virgin Gorda; Puerto Rico and the Lesser Antilles.

12. Asteraceae (Sunflower Family)
By V. A. Funk & J. F. Pruski

Annual or perennial herbs, shrubs, trees, or vines, frequently with hairs or glands, rarely with milky latex. Leaves alternate, opposite, or whorled, simple or less commonly deeply lobed or variously compound; stipules absent. Inflorescence generally determinate, of variously arranged compact indeterminate heads usually free from one another, but occasionally fused. Flowers sessile on a common, naked or paleate receptacle that is surrounded by an involucre of 2 to many bracts in 1 to several series, these commonly persistent. Heads 1- to many-flowered, homogamous (disciform, bilabiate, or ligulate) or heterogamous (radiate, disciform, or ligulate-pseudobilabiate). Individual flowers usually 5-merous, unisexual, bisexual, or neuter; calyx when present, represented on the top of the achenes by a persistent (or less commonly deciduous) pappus of scales, awns, setae, bristles, or by a crown; corollas sympetalous, actinomorphic or zygomorphic; stamens epipetalous; ovary inferior, bicarpellate, unilocular, with a single basal ovule, the style filiform, 2-branched. Fruit an achene.

The largest family of flowering plants, with approximately 1500 genera and 23,000 species occurring on all continents except Antarctica.

REFERENCES: Aristeguieta, L. 1964. Compositae. Flora de Venezuela. 10: 1–941; Cronquist, A. 1980. Vascular flora of the southeastern United States. Vol. 1. Asteraceae. The University of North Carolina Press, Chapel Hill; Cuatrecasas, J. 1969. Prima Flora Colombiana. 3. Compositae-Astereae. Webbia 24: 1–335; D'Arcy, W. G. (coordinator). 1975. Family 184. Compositae. In R. E. Woodson Jr., R. W. Schery, & collaborators, Flora of Panama, Part 9. Ann. Missouri Bot. Gard. 62: 835–1321; Keeley, S. C. 1978. A revision of the West Indian vernonias (Compositae). J. Arnold Arbor. 59: 360–413; King, R. M. & H. Robinson. 1987. The genera of the Eupatorieae (Asteraceae). Monogr. Syst. Bot. Missouri Bot. Gard. 22: 1–581; Solbrig, O. T. 1963. The tribes of Compositae in the southeastern United States. J. Arnold Arbor. 44: 436–461.

Key to the Genera of Asteraceae

1. Plants with milky latex; heads homogamous, ligulate.
    2. Heads narrowly cylindrical; achenes weakly beaked (in ours) .... 14. *Launaea*
    2. Heads bell-shaped; achenes not beaked .... 24. *Sonchus*
1. Plants without milky latex; heads heterogamous (radiate, disciform, or ligulate-pseudobilabiate) or homogamous (discoid).
    3. Leaves generally pinnately dissected to compound (in ours).
        4. Involucral bracts uniseriate; leaves and involucre with oil glands .... 27. *Tagetes*
        4. Involucral bracts bi- or few-seriate; leaves and involucre without oil glands.
            5. Heads inconspicuously radiate; involucral bracts monomorphic; achenes falling in an achene complex of 1 fertile ray achene (and the persistent corolla) and its subtending involucral bract attached to 2 disk flowers and their palea; corollas white; disk flowers functionally male or sterile; pappus of reduced awns (in ours) .... 18. *Parthenium*
            5. Heads obviously radiate (in ours); involucral bracts dimorphic; achenes falling individually; corollas white or yellow; disk flowers bisexual and fertile; pappus of 2–6 (in ours) retrorsely barbed awns .... 3. *Bidens*
    3. All leaves simple (in ours).
        6. All leaves opposite (in ours), not basal.
            7. Leaves and involucre dotted with oil glands; leaf base with stiff marginal cilia .... 19. *Pectis*
            7. Leaves and involucre not dotted with oil glands; leaf base without marginal cilia.
                8. Heads discoid; receptacle naked or with an involucral sheath totally surrounding each flower.

9. Leaves densely short-pilose on both surfaces; inflorescence glomerate; heads 1-flowered; each flower within a sheath; style branches recurved, fertile throughout their length; pappus a low lacerate crown.. 13. *Lagascea*

9. Leaves not densely short-pilose on both surfaces; inflorescence branching; heads 4- to many-flowered; flowers not enveloped within a sheath; style branches ascending, with a large sterile terminal appendage; pappus usually elongate.

    10. Twining vines (in ours); heads 4-flowered .... 16. *Mikania*

    10. Herbs to shrubs; heads many-flowered.

        11. Involucral bracts persistent, subequal, apically acuminate; pappus of 5 or 6 scales or crown-shaped .... 2. *Ageratum*

        11. Involucral bracts deciduous, graduated, apically broadly acute to rounded; pappus of about 40 bristles.. 5. *Chromolaena*

8. Heads radiate (in ours); receptacle paleate.

    12. Pappus of bristles or awns.

        13. Heads terminal, on long peduncles; ray corollas white; achenes monomorphic, obconic; pappus of many featherlike bristles .... 28. *Tridax*

        13. Heads axillary, subsessile or nearly so; ray corollas yellow; achenes dimorphic, outer ones flattened, winged; pappus of 2 or 3 awns .... 26. *Synedrella*

    12. Pappus absent or greatly reduced.

        14. Leaves lanceolate or less commonly elliptic, eglandular; corollas white; palea filiform .... 8. *Eclipta*

        14. Leaves elliptic to ovate, glandular; corollas pale yellow to orange; palea linear-lanceolate to lanceolate.

            15. Corollas pale yellow, those of the rays inconspicuous; disk flowers 5–9, functionally staminate; involucral fruit with uncinate prickles and topped by 2 prickles to 5 mm long.. 1. *Acanthospermum*

            15. Corollas yellow to orange, those of the rays conspicuous; disk flowers many, bisexual; achenes not covered by prickles.

                16. Procumbent herbs rooting at the nodes; leaves commonly 3-lobed; ray corolla limbs glandular on lower surface, shallowly 3-lobed at apex; achenes tuberculate.. 25. *Sphagneticola*

                16. Erect herbs or shrubs; leaves subentire to coarsely dentate; ray corolla limbs hispidulous on lower surface, deeply bilobed at apex; achenes pubescent, smooth .... 30. *Wedelia*

6. Leaves alternate or basal.

  17. Leaves basal (in ours).

    18. Inflorescence unbranched, without bracts; leaves white on lower surface, lyrate-pinnatifid at base; heads to 2(–3.5) cm tall; inner flowers slightly zygomorphic, pseudobilabiate.. 4. *Chaptalia*

    18. Inflorescence commonly branched, bracteolate; leaves green on both surfaces, attenuate at base; heads <1 cm tall; inner flowers actinomorphic, discoid .... 11. *Erigeron*

  17. Leaves cauline (in ours), alternate.

    19. Involucre uniseriate .... 10. *Emilia*

    19. Involucre 2- to several-seriate.

      20. Receptacle paleate.

        21. Leaves often trilobed, glandular below; inflorescence of many (>5) discoid heads (in ours); corollas yellow or greenish white; achenes cylindrical or nearly so; pappus of numerous erect bristles.. 17. *Neurolaena*

        21. Leaves crenate to dentate, eglandular; inflorescence of 1–5 radiate heads (in ours); corollas orange (in ours); achenes strongly winged; pappus of 2 awns, apically uncinate.. 29. *Verbesina*

      20. Receptacle naked.

        22. Pappus biseriate, the outer series of reduced squamellae, the inner series of elongate, deciduous or persistent bristles or scales.

            23. Herbs; inner pappus series of many fragile bristles; achenes nonribbed, appressed-pilose.. 7. *Cyanthillium*

            23. Subshrubs to shrubs (in ours); inner pappus series of several deciduous scales or numerous persistent bristles; achenes ribbed, glabrous, pilose, or sericeous.

1. ACANTHOSPERMUM Schrank

Annual herbs, prostrate to erect; stems appearing dichotomously branched, pubescent. Leaves simple, opposite; blades chartaceous, commonly pinnately 3-veined from near base, the margins subentire to pinnatifid. Inflorescence of solitary, sessile or short-pedunculate heads in the axils and forks of the stem. Heads radiate, several- to many-flowered; involucre hemispheric, of 2 dissimilar series of bracts, the outer involucral bracts green and foliaceous, equal in length and not overlapping, the inner ones light brown, equal in number to ray flowers and tightly enveloping ray achenes; receptacle slightly convex, paleate. Ray flowers in a single series, pistillate; corollas showy or inconspicuous, pale yellow, with a limb poorly or well developed. Disk flowers functionally male; corollas pale yellow; anthers black. Achenes inseparably enveloped in their involucral bracts forming an involucral false-fruit; fruit weakly or strongly laterally compressed, bearing straight or uncinate prickles; pappus lacking.

A primarily tropical American genus of 6 species, with 1 species (below) introduced to many ports of entry throughout the world.

1. Acanthospermum hispidum DC., Prodr. 5: 522. 1836.
Fig. 29.

Erect, annual herb to 60 cm tall, moderately branched above, covered with multicellular strigose hairs; stems with longitudinal lines. Leaves sessile, 1.5–9 × 0.6–4 cm, ovate to broadly elliptic or deltoid, 3-nerved from base, the apex acute to obtuse, the base acuminate to cuneate, the margins serrate. Inflorescence solitary, sessile in the leaf axils. Heads radiate; involucre bell-shaped at maturity, broadly expanding to 2 cm wide in fruit, the outer involucral bracts 5, herbaceous, 3.5–4.5 mm long, ovate to oblong with acute apices; receptacle with narrowly lanceolate palea. Ray flowers 6–8; corollas inconspicuous, cream-colored to pale yellow, 1.5–2 mm long. Disk flowers 5–9; corollas pale yellow, 1.5–2.5 mm long; anthers black. Achenes in an involucral fruit to 8 mm long, bearing scattered stiff, uncinate prickles with 2 conspicuous apical prickles to 5 mm long.

DISTRIBUTION: Occasional at sea level, on open roadsides and in dry scrublands. Coral Bay Quarter along Center Line Road (A3998), trail to Fortsberg (A5271), between Bethany and Rosenberg (B252). Also on St. Croix, St. Thomas, Tortola, and Virgin Gorda; throughout the Antilles, southeastern United States, Central America, South America, and naturalized throughout many areas of the world.

2. AGERATUM L.

Annual or perennial herbs (in ours) to shrubs; stems glabrous to pilose. Leaves simple, generally opposite, petiolate; blades commonly 3-nerved from base. Inflorescence generally terminal, in compact or open cymose or corymbiform clusters. Heads discoid, many-flowered, pedunculate; involucre bell-shaped, 2–3-seriate; involucral bracts weakly overlapping, subequal; receptacle conical, generally naked. Flowers bisexual; corollas actinomorphic, tubular, shortly 5-lobed, the triangular lobes papillose within; anthers in-

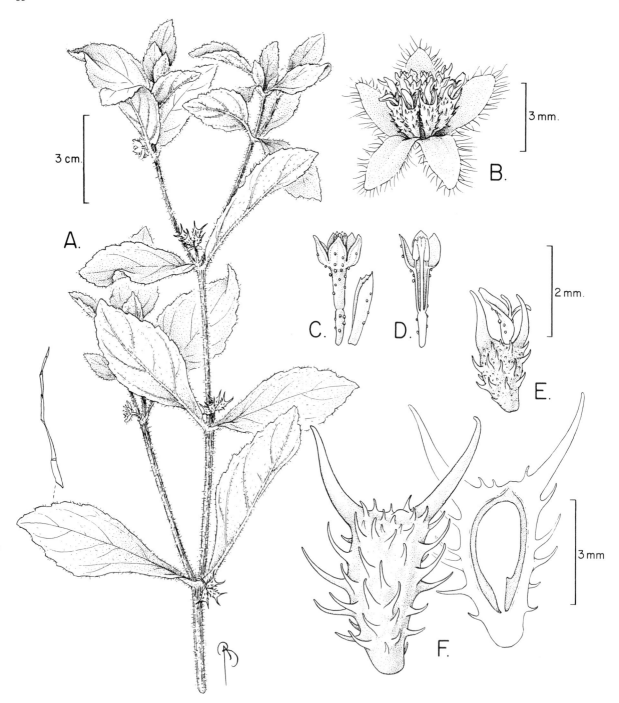

FIG. 29. *Acanthospermum hispidum.* **A.** Fertile branch and detail of trichome. **B.** Head with subtending bracts. **C.** Functional staminate disk flower and palea. **D.** L.s. disk flower. **E.** Pistillate ray flower. **F.** Involucral fruit and l.s. involucral fruit.

cluded; style branches long, greatly exserted from corollas, the upper half of style branches with a large sterile appendage. Achenes angled, glabrous or with bristles on the angles; pappus of 5(–6) apically tapering scales, crown-shaped, or absent.

A genus of about 40 species native to the neotropics, but introduced into the New World subtropics and the Paleotropics.

1. Ageratum conyzoides L., Sp. Pl. 839. 1753. Fig. 30.

Erect herb to 1 m tall, few-branched, pubescent; stems pilose. Leaf blades 2–8 × 1–5 cm, ovate, chartaceous, pilose, also glandular on lower surface, the apex mostly obtuse, the base acute to rounded, the margins crenate; petioles to 3 cm long, pilose. Inflorescence an irregularly branched panicle in the upper third of the plant, of several cymose clusters to 3 × 4 cm with to about

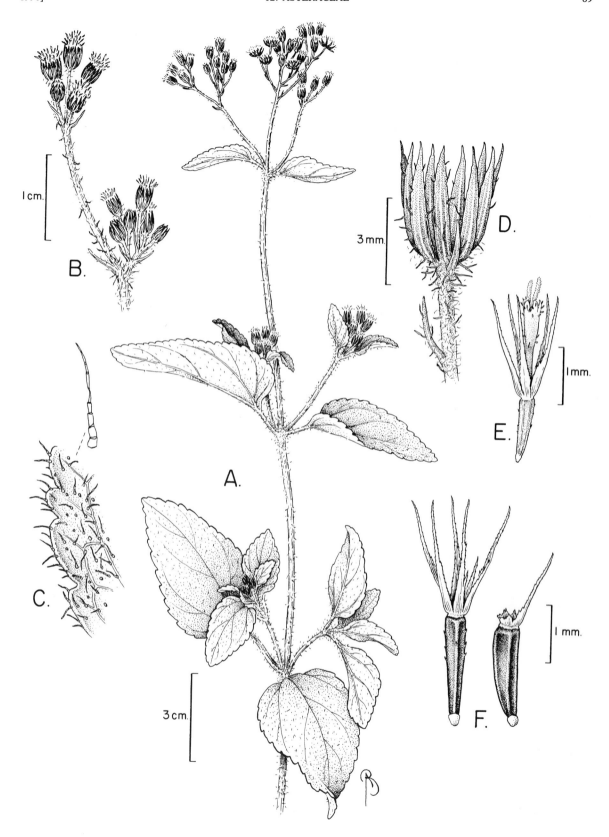

FIG. 30. *Ageratum conyzoides*. **A.** Fertile branch. **B.** Inflorescence branch. **C.** Leaf margin and trichome. **D.** Involucre and peduncle. **E.** Disk flower. **F.** Achenes.

25 heads. Heads ca. 40–70-flowered, to 5.5 mm tall and broad; involucre bell-shaped to hemispheric, to ca. 4 mm tall, 2-seriate; involucral bracts lanceolate, 2- or 3-striate, apically acuminate, sparsely pilose to glabrous; receptacle ca. 0.8 mm broad; peduncles 3–10 mm long, pilose, bracteolate. Corollas 1.5–2.0 mm long, glabrous or sparsely pilose, cream-colored to more commonly pale blue. Achenes 1.2–1.7 mm long, black, the angles with bristles; pappus awns 1.5–2.0 mm long, slightly longer than corollas, the pappus rarely reduced to a crown.

DISTRIBUTION: Occasional in open or disturbed areas. Along road to Bordeaux Mountain (A2858), Emmaus (A3149), Rosenberg (B291). Also on St. Croix, St. Thomas, and Tortola; native to the neotropics but adventive in the subtropics and in the paleotropics.

3. BIDENS L.

Annual or perennial herbs, less commonly shrubs or vines; stems usually conspicuously striate. Leaves simple to deeply lobed or variously compound, opposite or becoming alternate above; leaflets often ovate or lanceolate and serrate or more deeply lobed; petioles and rachis (when present) frequently narrowly winged. Inflorescence terminal, solitary or of a few long-peduncled heads. Heads with many flowers, radiate or less commonly discoid; involucre of 2 dissimilar series of bracts, the outer bracts green, narrow, the inner bracts brown with broadly hyaline margins, equal in length, broader than the outer bracts; receptacle convex or flat, paleate, the palea flat, usually resembling the inner involucral bracts. Ray flowers mostly sterile, in a single series, usually 5–12; corollas yellow or white. Disk flowers few to many, bisexual; corollas usually yellow; anthers usually black. Achenes weakly compressed tangentially, variously angled or ribbed, glabrous or pubescent; pappus of (0–)1–6, usually retrorsely barbed, erect or spreading awns.

A worldwide genus of about 75 species, mainly from Mexico, but also in North and South America and the Caribbean.

REFERENCE: Ballard, R. 1986. *Bidens pilosa* complex (Asteraceae) in North and Central America. Amer. J. Bot. 73: 1452–1465.

Key to the Species of *Bidens*

1. Pappus 2–awned; outer achenes straight; involucral bracts slightly spreading in fruit; ray corollas white .... 1. *B. alba* var. *radiata*
1. Pappus (3–)4(–6)-awned; outer achenes curved; involucral bracts greatly spreading in fruit; ray corollas yellow or white .... 2. *B. cynapiifolia*

1. Bidens alba (L.) DC. var. **radiata** (Sch.Bip.) Ballard *in* Melchert, Phytologia **32**: 295. 1975. *Bidens pilosa* L. var. *radiata* Sch.Bip. *in* Webb & Berthel., Hist. Nat. Iles Canaries III, **2**: 242. 1844.

Bidens pilosa of many authors, non L., 1753.
Coreopsis leucantha L. , Sp. Pl., ed. 2, **2**: 1282. 1763.
Bidens leucantha (L.) Willd., Sp. Pl. **3**: 1719. 1804.

Erect annual herb, 0.5–1.5 m tall, moderately branched; stems glabrous to sparsely villous, especially villous at the nodes. Leaves simple or trifoliate (rarely pinnately or bipinnately compound), opposite, to 10 cm long; blades (when leaf simple) or leaflets 3–6 × 1–3.7 cm, ovate, the apex acute to acuminate, the base attenuate to cordate or truncate, the margins serrate, the terminal leaflet essentially sessile; rachis (when present) to 1 cm long, commonly winged; petioles to 3 cm long. Inflorescence of few to many heads in terminal cymose clusters; peduncles 1–7 mm long, glabrous to sparsely villous. Heads radiate; involucre bell-shaped, generally not broadening greatly in fruit; outer involucral bracts 8–13, commonly spatulate, 3–4 mm long, the inner ones 8–10, ovate, 3–7 mm long. Ray flowers conspicuous, sterile, 5–8; corollas white. Disk flowers 20–65; corollas yellow, 3–5 mm long. Achenes usually straight, somewhat flattened to variously angled or ribbed, 4–13 mm long, glabrous to tuberculate-strigose; pappus of 2 awns, 1–2 mm long.

DISTRIBUTION: Common in open areas with secondary growth. Coral Bay Quarter along Center Line Road (A4012), Cruz Bay (A1953). Also on St. Croix and St. Thomas; widespread in the Caribbean, eastern Mexico, Central America, and northern South America.

COMMON NAMES: shepard's needle, Spanish needle.

2. Bidens cynapiifolia Kunth *in* Humb., Bonpl. & Kunth, Nov. Gen. Sp., folio ed. **4**: 185. 1820 [1818]. *Bidens bipinnata* L. var. *cynapiifolia* (Kunth) M. Gómez, Anales Soc. Esp. Hist. Nat. **19**: 275. 1890. Fig. 31.

Erect annual herb 0.3–1.5 m tall, sparsely to moderately branched; stems glabrous to puberulent, villous at the nodes. Leaves once to bipinnately divided or compound (rarely more finely dissected), opposite, ovate to elliptic in outline, 4–15 cm long; leaflets 1–6 × 0.4–2.5 cm, elliptic, the apex acuminate to acute, the base acuminate to cuneate, the margins serrate to deeply lobed; rachis to 5 cm long, often winged toward terminal leaflet; petioles to 3.5 cm long. Inflorescence of few to many terminally clustered heads; peduncles (3–)8–15 cm long. Heads radiate; involucre bell-shaped, becoming broadly expanded in fruit; involucral bracts in 2 dissimilar series, the outer ones 8–12, linear to oblanceolate, 3.6–6 mm long, the inner ones 8 or 9, lanceolate to ovate, 4–5.4 mm long. Ray flowers inconspicuous, 3–6; corollas yellow or white. Disk flowers 10–40; corollas yellow to yellowish orange, 3–6 mm long. Achenes usually dimorphic, mostly black, somewhat flattened to variously angled or ribbed, 4–15 mm long, glabrous; outer achenes 4–8 mm long, curved, densely tomentose; pappus of (3–)4(–6) awns, 1.5–3.5 mm long.

DISTRIBUTION: Common on open disturbed roadsides. Along road to Bordeaux Mountain (A2890, A3193). Also on St. Croix, St. Thomas, Tortola, and Virgin Gorda; throughout the Antilles, Central America, and South America, naturalized in Hawaii and in the Old World.

COMMON NAME: beggar ticks.

4. CHAPTALIA Vent., nom. cons.

Stemless perennial herbs, with many elongate fibrous roots. Leaves in basal rosettes, alternate (appearing verticillate); blades lanceolate to obovate, pinnately veined, commonly with a petiole-like base. Inflorescence monocephalous, scapose. Heads heterogamous, inconspicuously ligulate-pseudobilabiate, many-flowered; involucre mostly hemispheric; involucral bracts graduated, linear, api-

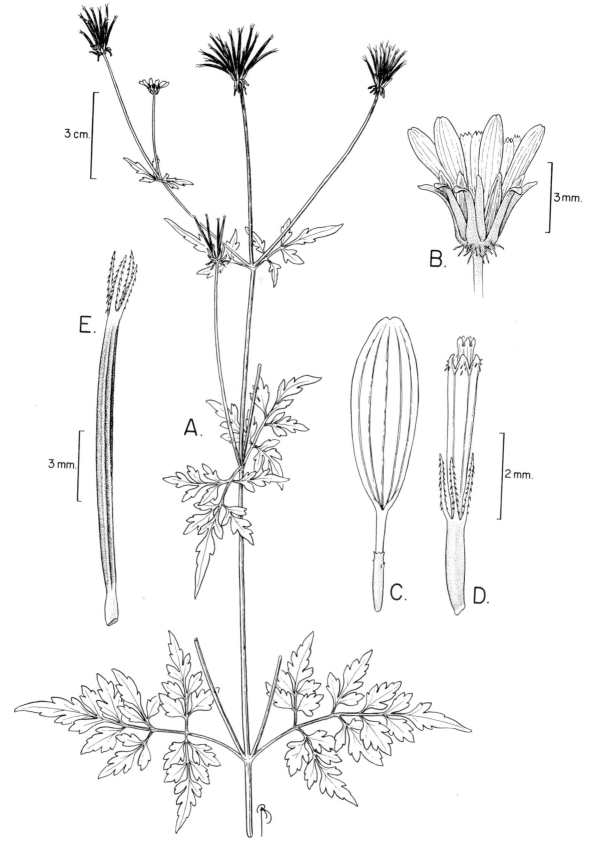

FIG. 31. *Bidens cynapiifolia*. **A.** Fertile branch. **B.** Head with subtending bracts. **C.** Ray flower. **D.** Disk flower. **E.** Disk achene.

cally acuminate; receptacle naked, foveolate. Flowers with heteromorphic, filiform, tubular corollas, commonly varying within a single head from outermost obscurely ligulate and pistillate to innermost somewhat pseudobilabiate, functionally staminate; anthers with basal, distinct, sterile tails; style branches short and obtuse. Achenes fusiform, commonly long-beaked or sometimes without a prolonged beak; pappus of many thin bristles about as long as the achenes and corollas.

A genus of about 55 species widely distributed in tropical and warmer regions of the New World.

1. Chaptalia nutans (L.) Pol., Linnaea **41:** 582. 1877.
Tussilago nutans L., Syst. Nat., ed. 10, **2:** 1214. 1759.   Fig. 32.

Chaptalia subcordata Greene, Leafl. Bot. Observ. Crit. **1:** 195. 1906.

Erect herb 10–45(–60) cm tall. Leaves all basal, sessile but narrowly attenuate into a petiole-like base, 5–30(–34) × 2.5–6(–10) cm, oblanceolate-spatulate and lyrate-pinnatifid, especially so near base, chartaceous, the upper surface green, glabrous or lingering cobwebby-woolly, the lower surface white-tomentose, the apex obtuse. Inflorescence of 1 to several peduncles, these cobwebby-woolly or becoming glabrous, naked, and without bracts or sometimes with a bracteole or two. Heads nodding when young, later more or less erect, ca. 150–320-flowered, to 2(–3.5) cm tall; involucre 12–25 mm tall, ca. 5-seriate; involucral bracts tomentose, reflexed when past fruit; receptacle flat, ca. 4 mm broad. Flowers trimorphic; outer flowers ligulate, pistillate; submarginal ones reduced, nearly actinomorphic, pistillate; inner ones somewhat irregularly 5-lobed, pseudobilabiate, appearing bisexual, but functionally staminate; all corollas white or nearly so, to 12 mm long. Fertile achenes long-beaked, 5–6-ribbed, lightly glandular, to 4 mm long, beak 10–15 mm long; pappus 10–12(–15) mm long, white or pink.

DISTRIBUTION: Occasional in open to shaded hills, roadsides, or moist banks. Along road to Bordeaux (A2859, A3181), between Bethany and Rosenberg (B240). Also on St. Croix, St. Thomas, and Tortola; throughout much of the West Indies, the neotropics, and parts of the New World subtropics.

COMMON NAME: white back.

5. CHROMOLAENA DC.

Erect or sprawling perennial herbs or shrubs; stems commonly pubescent. Leaves simple, generally opposite; blades commonly 3-nerved from base. Inflorescence mostly corymbiform. Heads discoid, 10–40-flowered; involucre constricted-cylindrical, ca. 4–7-seriate; involucral bracts overlapping, 3–5-veined, deciduous, the apex broadly acute to rounded; receptacle elongate, flat or conical on very top, generally naked, glabrous. Flowers bisexual; corollas actinomorphic, tubular, shortly 5-lobed, the lobes generally papillose or glandular; anthers largely included within corollas; style branches linear, elongate, greatly exserted from corollas, the upper half of style branches with a papillose sterile appendage. Achenes narrowly obconic, mostly 5-ribbed, sometimes glandular, the ribs pubescent or only vaguely so; pappus of ca. 40 bristles, about as long as the corollas and the achenes.

A genus of 166 species native from the southern United States to southern South America; 1 species introduced into the paleotropics.

Key to the Species of *Chromolaena*

1. Stems and leaves commonly canescent throughout; heads 8–10-flowered; corolla tubes and achenes glandular; pappus slightly enlarged apically; involucral bracts spreading in flower, outer ones persistent past fruit, the flat receptacle thereby visible only from above .... 3. *C. sinuata*
1. Stems and leaves densely pilose to puberulent; heads with 15 or more flowers; corolla tubes and achenes not glandular; pappus not apically enlarged; involucral bracts tightly overlapping in flower, all late deciduous, the barrel- or club-shaped receptacle thereby wholly exposed.
    2. Leaf blades deltoid to cordate, the apex acute to obtuse, the upper surface sparsely glandular; inflorescence compactly corymbiform; involucral bracts to 2 mm broad, rounded and glandular at apex; achenes 4–5 mm long, pappus 3.5–4.5 mm long .... 1. *C. corymbosa*
    2. Leaf blades deltoid to rhombic-ovate, the apex acuminate, the upper surface not glandular; inflorescence corymbiform to paniculate; involucral bracts <1.5 mm broad, not apically glandular, the inner ones acute at tips; achenes 3.5–4 mm long, pappus 5–6 mm long .... 2. *C. odorata*

1. Chromolaena corymbosa (Aubl.) R.M. King & H. Rob., Phytologia **20:** 200. 1970. *Eupatorium corymbosum* Aubl., Hist. Pl. Guiane **2:** 799. 1775. *Osmia corymbosa* (Aubl.) Britton & P. Wilson, Bot. Porto Rico **6:** 288. 1925.

Eupatorium atriplicifolium Lam., Encycl. **2:** 407. 1788.
Eupatorium atriplicifolium Vahl *in* H. West, Bidr. Beskr. Ste. Croix 302. 1793.

Shrub to 1.5 m tall; stems slightly succulent and branched above, pubescent, glandular. Leaf blades 1.3–6 × 0.8–5 cm, deltoid to cordate, succulent to chartaceous, 3-veined from near base, the upper surface puberulent, sparsely glandular to weakly so, the lower surface pubescent to puberulent, red-glandular, the apex acute to obtuse, the base truncate to nearly cordate, the margins subentire or more commonly crenate; petioles 0.3–1(–1.5) cm long, pubescent. Inflorescence terminal, only partly exserted above the subtending leaves, compactly corymbiform. Heads 16–21-flowered, 9.5–12 × 3.3–4 mm; involucre 7–9 mm long, 6–7-seriate; involucral bracts scarious toward base, the tips herbaceous, lightly puberulent, glandular, all with rounded apices, the inner involucral bracts broadly elliptic, 2–3 × 1.2–2 mm, grading to lanceolate inner ones, 6–7.5 × 1–1.5 mm; receptacle barrel-shaped, 1.5–2 × 0.8 mm; peduncles 0–8 mm long, puberulent. Corollas 3.5–4.8 mm long, cream-colored to pale lavender; style often pale lavender. Achenes 4–5 mm long; pappus bristles 3.5–4.5 mm long.

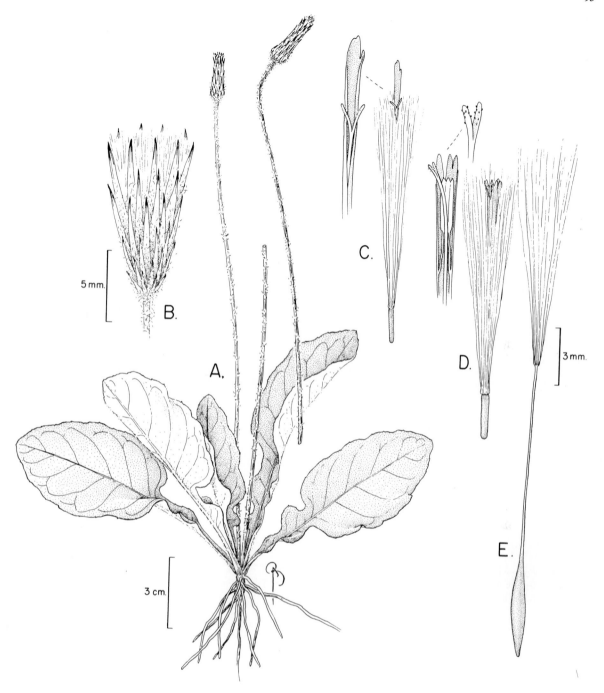

FIG. 32. *Chaptalia nutans.* **A.** Habit. **B.** Head. **C.** Pistillate outer flower and detail of apex of style. **D.** Bisexual inner flower, l.s. corolla showing stamens and style, and detail of style apex. **E.** Achene.

DISTRIBUTION: Common in rocky coastal areas and disturbed hillsides. Solomon Bay (A2280), Cob Gut (A3225), Dittlif Point (A3972). Also on Buck Island, Little St. James Island, St. Croix, St. Thomas, Tortola, and Virgin Gorda (fide Britton & P. Wilson, 1925); Bahamas, Hispaniola, Puerto Rico, and the Lesser Antilles.

2. Chromolaena odorata (L.) R.M. King & H. Rob., Phytologia **20:** 204. 1970. *Eupatorium odoratum* L., Syst. Nat., ed. 10, **2:** 1205. 1759. *Osmia odorata*

(L.) Sch.Bip., Jahresber. Pollichia **22–24:** 250, 252. 1866.

Fig. 33.

Eupatorium conyzoides Vahl, Symb. Bot. **3:** 96. 1794, non Mill., 1768.

Erect herb to shrub 1–3 m tall; stems branched above, densely pilose to puberulent. Leaf blades 2.5–12 × 1.2–7.5 cm, deltoid to rhombic-ovate, chartaceous, 3-veined from near base, the upper

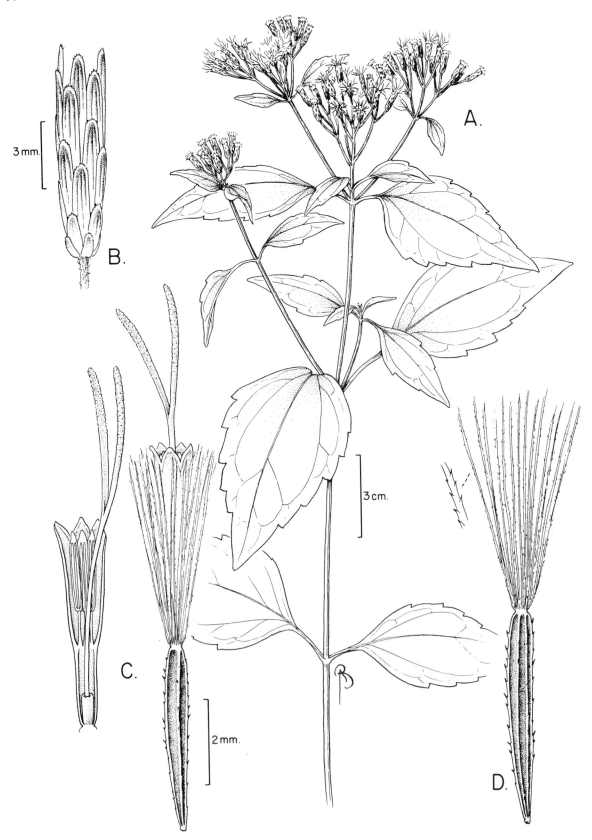

FIG. 33. *Chromolaena odorata*. **A.** Flowering branch. **B.** Involucre. **C.** L.s. disk corolla and disk flower. **D.** Achene and detail of pappus bristle.

surface puberulent to glabrous, the lower surface pubescent, densely red-glandular, the apex acuminate, the base cuneate-attenuate to subtruncate, the margins entire to serrate; petioles 0.4–2.5 cm long, pilose. Inflorescence terminal, corymbiform-paniculate, to 8 × 4 cm. Heads ca. 15–25-flowered, 9–12 × 3–4.5 mm; involucre 8–9.5 mm long, 5–6-seriate; involucral bracts ca. 15, glabrous to sparingly puberulent, scarious or with subherbaceous tips, the outer ones oblong-elliptic, ca. 2 × 1 mm, with rounded apex grading to the lanceolate inner ones, ca. 9 × 1.3 mm, with acute apex; receptacle shortly club-shaped, ca. 1.3 × 1 mm; peduncle 1–15 mm long, weakly pilose, sparingly glandular. Corollas 4.3–6 mm long, white to pale lavender. Achenes 3.5–4 mm long; pappus bristles 5–6 mm long.

DISTRIBUTION: An occasional roadside weed. Coral Bay Quarter at km 3.5 along Road 107 (A3767), Cruz Bay (A2360), Susannaberg (A3092). Also on St. Croix, St. Thomas, and Tortola; widespread in tropical and subtropical America from the southern United States to southern South America and introduced into the paleotropics.

COMMON NAMES: bitter bush, christmas bush, geritoo.

3. Chromolaena sinuata (Lam.) R.M. King & H. Rob., Phytologia 32: 283. 1975. *Eupatorium sinuatum* Lam., Encycl. 2: 407. 1788. *Osmia sinuata* (Lam.) Britton & P. Wilson, Bot. Porto Rico 6: 288. 1925.

Eupatorium canescens Vahl *in* H. West, Bidr. Beskr. Ste. Croix 302. 1793.

Erect shrub, 1–2 m tall, much-branched from base; stems peduncles, and leaves canescent. Leaf blades 0.6–3.5 × 0.4–2.5 cm, ovate, chartaceous, 3-veined from near base or palmately so from base, both surfaces canescent and glandular or infrequently solely glandular, the apex broadly acute to rounded, the base cuneate to truncate, the margins subentire to 2–5 sinuately lobed; petioles 2–9 mm long. Inflorescence terminal, several-headed, corymbiform, often appearing paniculate by dense stem branching. Heads 8–10-flowered, 7–10 × 2.5–3 mm; involucre 5–7 mm long, 3–5-seriate; outer involucral bracts elliptic, ca. 1.5 × 1 mm, canescent and glandular, persistent, greatly spreading with age, the inner involucral bracts lanceolate, to 6 × 1.3 mm, canescent and glandular or only so at tips, commonly deciduous; receptacle flat to dome-shaped on top, sides of receptacle obscured by persistent outer involucral bracts, 0.4 mm broad; peduncles 2–11 mm long. Corollas 3.5–4 mm long, glandular, cream-colored to lavender. Achenes 3–4 mm long, glandular and weakly puberulent; pappus bristles 3.5–4 mm long, slightly expanded at the tips.

DISTRIBUTION: Occasional in disturbed areas to rocky hillsides. Nanny Point (A2444), Lameshur (B635). Also on Guana Island, St. Croix, St. Thomas (fide de Candolle, Prodr. 5: 155. 1836), Tortola (fide D'Arcy, 1967), and Virgin Gorda; Cuba, Hispaniola, Puerto Rico, and the Lesser Antilles.

6. CONYZA Less., nom. cons.

Annual or biennial (rarely perennial) herbs, often simple-stemmed and weedy; stems and leaves more or less pubescent. Leaves simple, alternate, mostly sessile, entire to pinnatisect, pinnately veined. Inflorescence of several to many heads arranged in racemes or panicles, less commonly nearly spicate, heads rarely solitary. Heads inconspicuously heterogamous, small, disciform, subsessile or slender-pedunculate; involucre somewhat bell-shaped to hemispherical, 2- to several-seriate, soon spreading and reflexed at maturity; involucral bracts graduated, linear to lanceolate, with a herbaceous midvein and hyaline margins; receptacle mostly flat, naked, sometimes weakly foveolate-fimbriate. Outer flowers numerous and in several series, pistillate; corollas filiform, the apex denticulate or zygomorphic with a minute limb; style branches linear. Central flowers fewer, bisexual; corollas narrowly funnel-shaped or with a broadened (4–)5-lobed limb; style branches somewhat flattened, papillose. Achenes of pistillate and bisexual flowers uniform, compressed, the sides commonly 1-veined; pappus of 10–30 capillary bristles.

A genus of about 50 species mostly found in tropical or subtropical regions of both the New World and the Old World.

1. Conyza bonariensis (L.) Cronquist, Bull. Torrey Bot. Club 70: 632. 1943. *Erigeron bonariensis* L., Sp. Pl. 863. 1753. *Leptilon bonariense* (L.) Small, Fl. S.E. U.S. 1231, 1340. 1903.   Fig. 34.

Erigeron linifolius Willd., Sp. Pl. 3: 1955. 1804. *Leptilon linifolium* (Willd.) Small, Fl. S.E. U.S. 1231, 1340. 1903.

Annual pubescent herb 1–2 m tall; stems simple or branching in the inflorescence. Leaves cauline, numerous, tapering to a petiole-like base; blades 4–15 × 0.2–2 cm, oblanceolate to linear-lanceolate, gradually reduced above, chartaceous, indistinctly 3-nerved from well above base, pubescent on both surfaces, eglandular, the margins entire or nearly so to less commonly deeply serrate. Inflorescence an open bracteate pyramidal panicle in the upper half of the plant. Heads to 200- or more flowered, to 6–7 mm tall and broad; involucre hemispheric, 4–6 mm tall, 2–3-seriate; involucral bracts narrowly lanceolate, 2.2–6 × 0.5–0.7 mm, hispid, eventually completely reflexed; receptacle flat to subconvex, to ca. 2 mm broad; peduncles 3–20 mm long, hispid, bracteolate. Outer flowers with corollas weakly zygomorphic, 3.5–

4.5 mm long, tubular, white or purplish at top, topped by a 0.3–0.6 mm long inconspicuous limb. Inner flowers ca. 7–20(–30), filiform-cylindrical, ca. 3.5–4.5 mm long, shortly 5-lobed, yellowish. Achenes elliptic-oblong, 1.5–1.8 mm long, weakly puberulent; pappus of many bristles, 3–4 mm long.

DISTRIBUTION: An occasional weed of open disturbed areas such as roadsides. Along road to Bordeaux (A2882, A3135), Susannaberg (A852). Also on St. Thomas and Tortola; throughout much of the West Indies, neotropics, and warmer parts of nontropical America. Introduced into the paleotropics.

COMMON NAME: hairy horseweed.

DOUBTFUL RECORDS: *Conyza apurensis* Kunth [reported by Britton (1918) as *Erigeron spathulatus* Vahl and later by Britton and Wilson (1925) as *Leptilon chinense* (Jacq.) Britton] and *Conyza canadensis* (L.) Cronquist have been reported from St. John by Britton (1918), Britton and Wilson (1925), and Woodbury and Weaver (1987), but we found no vouchers at NY or US. *Conyza canadensis* (including *Erigeron pusillus* Nutt.) also has been reported from St. Thomas (fide Lessing *in* Schlechtendal, 1831; and Britton, 1918) and is known from St. Croix and Tortola. It may be distinguished from *C. bonariensis* by smaller heads, nearly glabrous involucral bracts, narrower leaves, and fewer pistillate flowers, these denticulate at apex and without a limb. *Conyza apurensis*, known in the Virgin Islands from the St. John Estate on St. Croix (possibly the source of the report from St. John Island), differs from both by basally branched stems, petiolate oblong leaves, fewer heads, and pistillate flowers with a limb to 1 mm long.

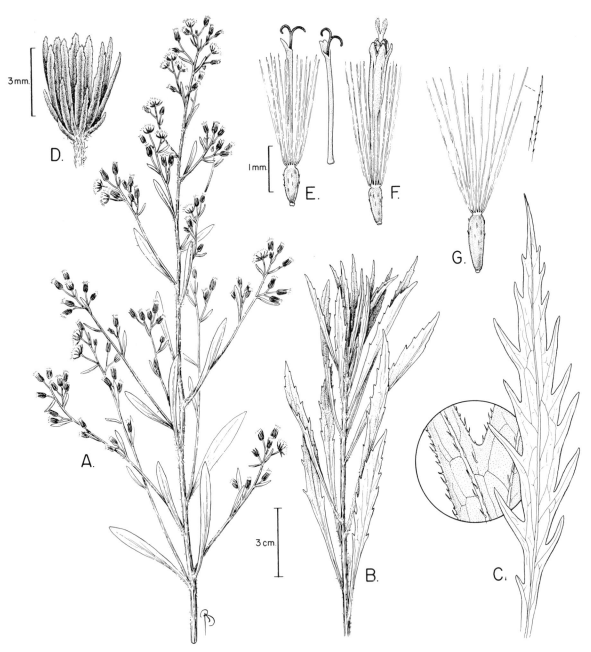

Fig. 34. *Conyza bonariensis*. **A.** Fertile branch. **B.** Cauline leaves. **C.** Basal leaf and detail of lower surface. **D.** Involucre. **E.** Pistillate outer flower and corolla. **F.** Disk flower. **G.** Achene and detail of pappus bristle.

7. CYANTHILLIUM Blume

Annual herbs; stems erect, puberulent. Leaves simple, alternate; blades pinnately veined. Inflorescence terminal, corymbiform or corymbiform-paniculate. Heads discoid; involucre bell-shaped, ca. 3-seriate, the involucral bracts moderately overlapping, narrowly lanceolate, sharply attenuate at tips, pilose, glandular, reflexed in fruit; receptacle naked. Flowers bisexual; corollas tubular to bell-shaped, shortly 5-lobed; anthers included, basally spurred; styles hispidulous in upper half, the branches lavender. Achenes nearly cylindrical, nonribbed, appressed-pilose, pustulate; pappus biseriate, white, the outer series of persistent, reduced squamellae, the inner series of many fragile bristles longer than the achenes.

A genus of 5 species, these distributed in Florida, Mexico, Central America, northern South America, West Indies, Africa, Asia, and the Pacific.

1. Cyanthillium cinereum (L.) H. Rob., Proc. Biol. Soc. Wash. **103**: 252. 1990. *Conyza cinerea* L., Sp. Pl. 862. 1753. *Vernonia cinerea* (L.) Less., Linnaea **4**: 291. 1829.

Erect herb to 80 cm tall; stems unbranched or more commonly few-branched. Leaf blades 1.5–3 × 1–1.5 cm, obovate, chartaceous, puberulent, also usually glandular below, the margins subentire or crenulate, tapering into a winged petiole to 1.5 cm long. Inflorescence of 5 to many heads on peduncles 0.5–1.5 cm long.

Heads ca. 15-flowered; involucre ca. 3.5 mm long; outer involucral bracts much reduced, the inner involucral bracts subequal. Corollas exserted from involucre, 3–4 mm long, the tube long narrow brown or cream-colored, the limb short, lavender, pilose. Achenes ca. 1.5 mm long; inner pappus bristles to 3 mm long, exserted from involucre and nearly as long as the corollas.

DISTRIBUTION: Common weed of disturbed areas, fields, and moist upland forests. Bordeaux Mountain (M17043), Lind Point (A3145), Lameshur (A2740). Also on St. Croix, St. Thomas, and Tortola; throughout the West Indies, Florida, Mexico, Central America, northern South America, Africa, Asia, and the Pacific.

8. ECLIPTA L., nom. cons.

Annual or perennial herbs, often prostrate or procumbent; stems usually much-branched, rough pubescent. Leaves simple, opposite, subsessile, entire to dentate. Inflorescence terminal and axillary, solitary or of a few short-pedunculate heads. Heads small, radiate; involucre of several foliaceous involucral bracts cupping the head until fruit abscission and then spreading to expose the receptacle; receptacle convex or flat, paleate, the palea filiform. Ray flowers many, pistillate; corollas white, early deciduous. Disk flowers many, bisexual; corollas usually yellow; anthers usually black. Achenes obconic, 4-angled, somewhat flattened, truncate at the apex, tuberculate; pappus absent or of a few short teeth.

A genus of 2 or 3 species native to the New World, with the following species semicosmopolitan, more common in warmer climates.

1. Eclipta prostrata (L.) L., Mant. Pl. **2**: 286. 1771. *Verbesina prostrata* L., Sp. Pl. 902. 1753.

Fig. 35.

Verbesina alba L., Sp. Pl. 902. 1753. *Eclipta alba* (L.) Hassk., Pl. Jav. Rar. 528. 1848.

Decumbent annual or short-lived perennial herb, to 1 m long, strigose. Leaves 2–11 × 0.4–3 cm, narrowly lanceolate or less commonly elliptic, 3-veined from near base, both surfaces strigose, eglandular, the apex acute to acuminate, the base narrowly cuneate and clasping the stem, the margins serrate to crenate. Inflorescence terminal, of solitary pedunculate heads; peduncles 1–5 cm long. Heads radiate; involucral bracts biseriate, slightly overlapping, 2.5–7 mm long, elliptic to lanceolate, the apex acute to caudate. Ray flowers many; corollas white, 1–2.5 mm long, the limb linear. Disk flowers many; corollas white, 1–2 mm long. Achenes 2–2.5 mm long, essentially glabrous with a few short hairs at apex; pappus of a few inconspicuous teeth or absent.

DISTRIBUTION: Occasional on open sandy coasts, roadsides, and in waste areas, often where soil is damp or wet. Carolina (A4141), Turner Bay (A3107), Enighed (A3926). Also on St. Croix, St. Thomas, and Tortola; Florida, pantropics.

9. ELEPHANTOPUS L.

Perennial rhizomatous herbs; stems erect, few-branched, pilose. Leaves cauline (in ours) or basal, alternate, sessile, clasping the stem, pinnately veined. Inflorescence of terminal corymbiform panicles, the heads glomerate, the glomerules dense, cupular, subtended by 2 or 3 leafy bracts. Heads discoid, 4-flowered; involucre cylindrical; involucral bracts 8, in 4 decussate pairs in 2 series, glandular, the outer four shorter than the inner four, the apices attenuate; receptacle naked. Flowers bisexual; corollas somewhat zygomorphic, irregularly 5-lobed, with a much deeper cut on the inner side; anthers white; styles hispidulous in upper half. Achenes obconic, 10-ribbed, hispidulous on veins; pappus of 5 straight subequal bristles that are commonly broadened at the base.

A genus of 25–29 species, primarily New World (principally tropical), also Africa, Asia, Fiji, and Australia.

1. Elephantopus mollis Kunth *in* Humb., Bonpl. & Kunth, Nov. Gen. Sp., folio ed. **4**: 20. 1820 [1818].

Fig. 36.

Erect herb 1–2 m tall, few-branched; stems pilose. Leaves cauline, 5–27 × 1.5–8(–10) cm, oblanceolate to obovate, chartaceous, pilose on upper surface, pilose and finely glandular on lower surface, the apex acute, the base long-attenuate, the margins crenulate. Inflorescence diffuse, peduncle 1–6 cm long, pilose. Glomerules 10–40-headed, the leafy bracts 3, cordiform to deltoid, to 1.5 × 1 cm, the lower surface pilose and glandular. Individual heads ca. 6–7 × 1–2 mm; outer 4 involucral bracts ca. 2–3 mm long, the inner 4 involucral bracts ca. 5–7 mm long. Corollas long- and narrowly tubular with a short bell-shaped limb, 5–6.5 mm long, white to cream-colored, sometimes with lavender lobes, glabrous or lobes occasionally glandular. Achenes 2–2.5 mm long; pappus bristles ca. 4 mm long, reaching to the base of the corolla limb.

DISTRIBUTION: Occasional along roadsides and in wet shaded areas. Along road to Bordeaux (A3141), between Bethany and Rosenberg (B259), Lameshur Trail (A3185). Also on St. Croix, St. Thomas, and Tortola; throughout the West Indies and much of the neotropics; introduced into tropical Africa, the Pacific (Fiji), and southeastern Asia.

COMMON NAME: elephant's foot.

10. EMILIA Cass.

Annual herbs with simple or few-branched stems. Leaves simple, alternate, variously petiolate or sessile to clasping the stem, pinnately veined. Inflorescence mostly terminal, loosely corymbiform, 1- to few-headed. Heads discoid, 15–50-flowered, short- to

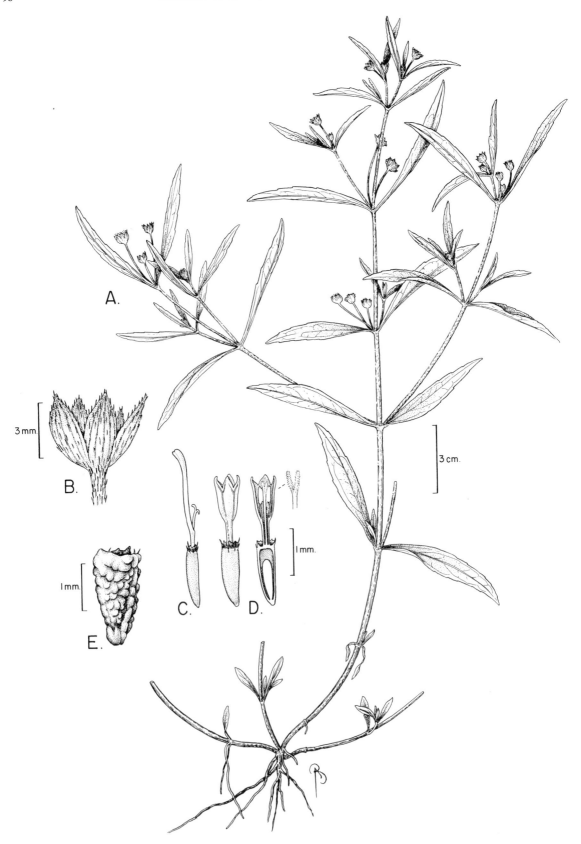

FIG. 35. *Eclipta prostrata.* **A.** Habit. **B.** Involucre. **C.** Ray flower. **D.** Disk flower, l.s. disk flower, and detail of style branches. **E.** Achene.

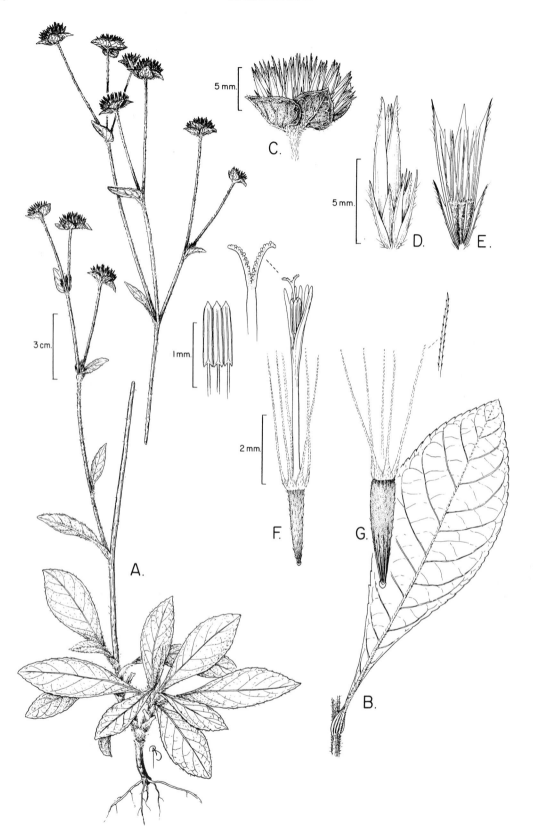

FIG. 36. *Elephantopus mollis*. **A.** Habit. **B.** Lower cauline leaf. **C.** Glomerate inflorescence. **D.** Head. **E.** L.s. head showing achenes. **F.** Zygomorphic disk flower and detail of anthers and style branches. **G.** Achene and detail of pappus bristle.

long-pedunculate; involucre cylindrical; involucral bracts uniseriate, connate but pulling apart with age; receptacle naked, more or less flat. Flowers bisexual; corollas tubular, lavender to purple, white or red; style branches weakly exserted, the apex papillose. Achenes cylindrical, 5-angled, minutely puberulent; pappus of many deciduous capillary bristles about as long as the corollas and much longer than the achenes.

A predominantly Old World genus of 30–45 species, also now weedy throughout the warmer parts of the New World.

REFERENCE: Nicolson, D. H. 1980. Summary of cytological information on *Emilia* and taxonomy of four Pacific taxa of *Emilia*. Syst. Bot. 5: 391–407.

Key to the Species of *Emilia*

1. Corollas red, extended above the tips of the involucral bracts; involucre 1–2 times longer than wide . 1. *E. fosbergii*
1. Corollas pink to lavender, not exserted from the involucral bracts; involucre 3–4 times longer than wide.. 2. *E. sonchifolia*

1. Emilia fosbergii Nicolson, Phytologia **32**: 34. 1975.

Fig. 37A–E.

Erect or ascending herb to 1 m tall; stems sometimes pubescent below, otherwise glabrous. Leaves sessile or on winged petioles, upper leaves sometimes reduced to bracts; blades greatly varying in size, 3–15 × 1–6.5 cm, but usually twice or more than twice as long as wide, membranous, glabrous, the apex acute, the base either auriculate and expanded or tapering, the margins serrate. Heads with ca. 20–50 flowers; involucre cylindrical, ca. 1 cm tall, about 1–2 times longer than wide; involucral bracts linear, green. Corollas red, ca. 1 cm long, extended above tips of the involucral bracts. Achenes slender, ca. 4 mm long, ribbed, puberulent on ribs; pappus of many silky white bristles, 6–7 mm long.

DISTRIBUTION: An occasional weed in disturbed sites. Bordeaux Mountain (M17024), Maria Bluff (A612), Maho Bay (A2408). Also on St. Croix and St. Thomas; tropical America, Florida, and some Pacific islands.

2. Emilia sonchifolia (L.) DC. *in* Wight, Contr. Bot. India 24. 1834. *Cacalia sonchifolia* L., Sp. Pl. 835. 1753.

Erect or ascending herb to 0.5 m tall; stems glabrous or somewhat pubescent. Leaves variable, sessile or with long-attenuate winged petioles; blades 4–11.5 × 1–5 cm, often elliptic-lanceolate or oblanceolate in outline, chartaceous, glabrous to thinly puberulent on both surfaces, the apex acute, the base either auriculate and expanded or tapering, the margins serrate to toothed or lobed with terminal lobe larger. Heads 15–30-flowered; involucre slender, ca. 1 cm tall, about 3–4 times longer than wide; involucral bracts 6–10, loosely pilose. Corollas pink to lavender, 7–8.5 mm long, not exserted from the involucral bracts. Achenes 3–4 mm long, ribbed, puberulent on ribs; pappus of many silky white bristles 5.5–7 mm long.

DISTRIBUTION: An occasional weed of disturbed sites. Carolina (A4140), Reef Bay (A2640). Also on St. Croix, St. Thomas, and Tortola; a pantropical weed of Old World origin.

COMMON NAME: scarlet paint bush.

11. ERIGERON L.

Perennial herbs, less commonly annuals, biennials, or shrubs; stems glabrous to pilose. Leaves simple, alternate, or rarely all basal (in ours), sessile or short-petiolate; blades entire to variously toothed or divided, pinnately veined. Inflorescence monocephalous, or few-headed and cymose, corymbiform, or paniculate. Heads radiate or rarely disciform, generally borne on leafless peduncles; involucre bell-shaped to hemispheric; involucral bracts subequal or 2(–5)-seriate, linear to lanceolate; receptacle flat or slightly convex, naked. Ray flowers nearly always present, pistillate, many, usually in 2 series, the limbs long, narrow. Disk flowers bisexual, numerous; corollas shortly 5-lobed, tubular to bell-shaped; style branches dorsally papillose, short. Achenes mostly compressed, the sides commonly strongly and thickly 1-nerved, the faces also sometimes variously ribbed; pappus of several to many fragile capillary bristles about as long as the disk corollas, sometimes an outer series of shorter bristles or scales or an outer crown-shaped pappus also present.

A genus of about 200 species of worldwide distribution, but more commonly found in temperate zones.

1. Erigeron cuneifolius DC., Prodr. **5**: 288. 1836.

Fig. 37F–J.

Annual or short-lived herb to 35 cm tall, acaulescent or nearly so. Leaves basal, subsessile, 1.5–4.5 × 0.5–1.4 cm, oblanceolate to obovate, chartaceous, pubescent on both surfaces, the apex obtuse, the base attenuate, the margins subentire or with 2 or 3 rounded teeth per side; peduncular bracts much reduced. Inflorescence loose and open, monocephalous or much more commonly of few to several cymosely arranged heads; peduncles 1 to several, simple or occasionally sparingly branched in upper half, to 35 cm long, slender, bracteate, the bracts ca. 3–10, well-spaced, sessile, to 15 × 4 mm, narrowly oblanceolate, entire. Heads radiate, 60–75-flowered, 4.5–5.5 mm tall and broad; involucre hemispheric,

broadly bell-shaped, to ca. 4–4.5 mm tall, 2–3-seriate; involucral bracts narrowly lanceolate, 1.5–4.5 × 0.2–0.5 mm, weakly and loosely pubescent, eventually completely reflexed; receptacle flat to subconvex, 1.9–2.3 mm broad. Ray flowers ca. 40–50 in several series, inconspicuous and not much exserted from involucre; corollas 3–4 mm long, the tube white, the limb violet, somewhat to tightly coiled; style branches linear. Disk flowers ca. 20–25; corollas tubular, 2.5–3 mm long, pale yellow; anthers included; style branches oblong. Achenes elliptic-oblong, 1.1–1.4 mm long, pilose at base, less so above; pappus simple, of ca. 20 bristles, 2.5–3 mm long.

DISTRIBUTION: Uncommon weedy plant of wet shaded banks. Cruz Bay (W680), Rosenberg (B293). Also on St. Thomas and Tortola; throughout the Greater Antilles; Philippines (Guy Nesom, pers. comm.).

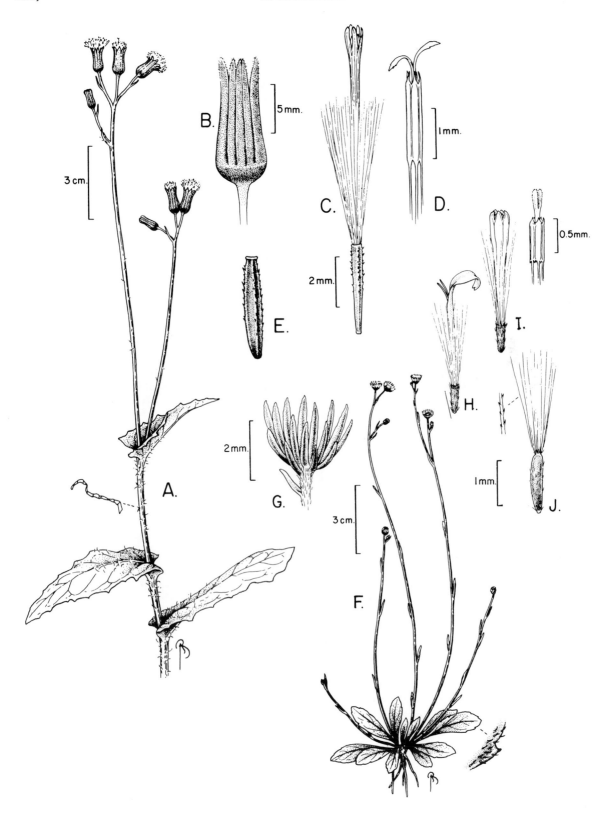

Fig. 37. A–E. *Emilia fosbergii.* A. Flowering branch and trichome. B. Involucre. C. Disk flower. D. Detail of anthers and style branches. E. Achene with pappus removed. F–J. *Erigeron cuneifolius.* F. Habit and detail of leaf margin with trichomes. G. Involucre. H. Ray flower. I. Disk flower and detail of anthers and style branches. J. Achene and detail of pappus bristle.

12. GNAPHALIUM L.

Annual or perennial herbs, usually woolly or tomentose; stems commonly simple and erect. Leaves simple, alternate, linear to narrowly elliptic. Inflorescence mostly terminal, cymose-corymbiform or paniculate. Heads disciform; involucre cylindrical or bell-shaped; involucral bracts graduated, overlapping, membranous, translucent; receptacle generally naked. Flowers heterogamous; corollas actinomorphic; styles bifid, branches linear, apex truncate. Outer flowers many, pistillate, the corollas filiform. Central flowers 5–10 (–25), presumably bisexual, the corollas tubular, shortly 3–5-lobed, lobes often glandular-pubescent, the anthers caudate. Achenes obovoid, glabrous or papillose; pappus of several to many free, subequal, barbellate capillary bristles, these usually deciduous.

A cosmopolitan genus of 80–150 species with centers of diversity in the highlands of Mexico, South America, and Africa.

1. Gnaphalium domingense Lam., Encycl. 2: 743. 1788.

Gnaphalium portoricense Urb., Symb. Ant. 3: 409. 1903.

Annual or short-lived perennial herb to 1 m tall; stems cobwebby-woolly, simple but branching above in the inflorescence. Leaves sessile, somewhat clasping the stem, 3–9 × 0.3–1.3 cm, lanceolate to oblanceolate, chartaceous, with obscure venation, the upper surface cobwebby to puberulent, sometimes glandular, the lower surface white-tomentose, the margins entire to crenulate. Inflorescence corymbiform, of several more or less compact clusters of 3–10 sessile or nearly sessile heads. Heads ca. 6 × 3–5 cm; involucre cylindrical to bell-shaped, 3–4-seriate, ca. 6 mm tall; involucral bracts several, cream-colored or pale yellow, glabrous or the outermost cobwebby at base, elliptic or lanceolate; receptacle flat, 2 mm broad; peduncles lacking or to 1 mm long. Outer flowers ca. 30–40; corollas 3–3.5 mm long, cream-colored. Inner flowers ca. 10; corollas 3–3.5 mm long, pale yellow, the lobes lightly glandular. Achenes 0.75 mm long, glabrous, brown; pappus bristles ca. 12, 3–3.5 mm long, white, readily deciduous.

DISTRIBUTION: Rare coastal weed. Caneel Bay (W673). Throughout the Greater Antilles.

13. LAGASCEA Cav., nom. cons.

Annual herbs or shrubs; stems gray, yellow-green, or purple, glabrous to pilose and often stipitate-glandular. Leaves simple, opposite, petiolate to sessile; blades ovate to oblanceolate, the apex acute to acuminate, the base obtuse to nearly auriculate, the margins more or less serrate. Inflorescence of many 1-flowered (rarely more) heads aggregated secondarily into terminal, bell-shaped, solitary, cymose, or racemose clusters; inflorescence bracts 4–6, separate, herbaceous; involucral bracts connate laterally into an involucral tube (sheath), this with free lobes on top, surrounding and enclosing the achene. Flowers discoid, bisexual; corollas yellow, white, pink, or red; anthers yellow. Achenes obconic, hidden within the lobed involucral tubes; pappus lacking or crown-shaped.

Principally a Mexican genus of 8 species distributed from Arizona south to Nicaragua. One species has been introduced to the West Indies, South America, and the Old World.

1. Lagascea mollis Cav., Anales Ci. Nat. 6: 331. 1803.

Nocca mollis (Cav.) Jacq., Fragm. Bot. 58. 1805.

Annual herb to 1 m tall, moderately branched above, pubescent to densely so. Leaf blades 1.2–7 × 0.5–4.5 cm, ovate to narrowly ovate, chartaceous, 3-nerved from near base, the apex acute to acuminate, the base obtuse to slightly attenuate, the margins entire to serrate; petioles to 2.7 cm long. Inflorescence solitary, terminal, bell-shaped, ca. 1 cm tall, made of 8–25 1-flowered heads; peduncle ca. 3–5 mm long; bracts 5, 5–15 mm long, lanceolate to ovate; involucre of the individual heads tubular, 4–5 mm long, of 4 or 5 involucral bracts each with 2 or 3 rows of 5–8 small resinous glands, densely pilose, the apical lobes >1 mm long, acuminate. Corollas inconspicuous, white to purple-pink, 3 mm long. Achenes obovoid, 3 mm long; pappus a lacerate crown.

DISTRIBUTION: In open disturbed areas. Coral Bay Quarter along Center Line Road (A3996). Also on St. Croix, St. Thomas, and Tortola; throughout much of the West Indies, Florida, Arizona, coastal Mexico, northern Central America, South America, and in the Old World.

COMMON NAME: silkleaf.

14. LAUNAEA Cass.

Annual or perennial herbs or spiny dwarf shrubs, with milky latex; stems solitary or few, freely and dichotomously branched. Leaves simple, alternate, cauline or basal, usually with a long-attenuate winged petiole, pinnately veined, the margins toothed or lobed. Heads few to numerous; involucral bracts in several rows, overlapping, with scarious margins; receptacle naked. Flowers ligulate, bisexual; corollas yellow. Achenes cylindrical or flattened, margin without ascending hairs, ribbed but not beaked or sometimes weakly beaked (in ours); pappus of several rows of simple hairs.

A genus of 30 species found from the Mediterranean to eastern Africa and the Canary Islands; 1 species (below) a pantropical weed.

REFERENCE: Jeffrey, C. 1966. Notes on Compositae: I. The Cichorieae in east tropical Africa. Kew Bull. 18: 427–486.

1. Launaea intybacea (Jacq.) Beauverd, Bull. Soc. Bot. Genève 2: 114. 1910. *Lactuca intybacea* Jacq., Icon. Pl. Rar. 1: tab. 162. 1784. *Brachyrampus intybaceus* (Jacq.) DC., Prodr. 7: 177. 1838.　　　Fig. 38E–H.

Erect annual or biennial herb to 1 m tall; stems simple or branched, glabrous. Leaves 5–28 × 1.5–11 cm, membranous, the apex acuminate, the margins deeply or shallowly toothed, often purple on lower surface; lower ones ovate to oblanceolate, clustered, with a long-attenuate winged petiole; upper leaves lanceolate, spread along stem, sessile and clasping. Inflorescence with heads solitary or few in a cluster, these short-pedunculate. Heads 15–35-flowered; involucre pyriform to narrowly cylindrical, 10–12 mm high; involucral bracts scariously marginate, the outer ones small and ovate, the inner ones linear-lanceolate, 2–3 times larger than the outer ones; corollas bright to pale yellow, 8–9 mm long. Achenes dimorphic, inner ones more or less cylindrical, outer ones flattened, linear, 4–5 mm long, longitudinally grooved, warty, somewhat constricted at the apex into a weak beak; pappus of numerous white bristles ca. 7–9 mm long, some bristles hairlike, thereby resulting in an obscurely dimorphic pappus.

DISTRIBUTION: Occasional in open dry areas, usually along sandy coasts. Cruz Bay (A2346), Chocolate Hole (A4067), East End (A684). Also on St. Croix, St. Thomas, and Tortola; a pantropical and subtropical weed of lowland areas.

COMMON NAME: wild lettuce.

NOTE: This species is difficult to place generically because it has characteristics of both *Lactuca* and *Launaea* and because both of these genera partly overlap with *Sonchus*. All the specimens of this species that we examined lacked ascending hairs on the achene margins. Therefore we have followed Jeffrey (1966) and provisionally retained this species in *Launaea*, although some achenes are compressed and the pappus is not strongly dimorphic. This species is often treated in New World floras as *Lactuca*.

15. LEPIDAPLOA (Cass.) Cass.

Annual or perennial, pubescent, often glandular herbs to subshrubs (in ours); stems erect, often much-branched. Leaves simple, alternate, sessile or petiolate, pinnately veined. Inflorescence terminal or axillary, cymose, heads single or clustered, often subtended by bracteate leaves. Heads discoid, 8–35-flowered, sessile or nearly so; involucre bell-shaped, 3–6-seriate; involucral bracts persistent, graduated, lanceolate, the outer involucral bracts often spreading, slender with aristate apices, the inner involucral bracts commonly erect with acute apices; receptacle naked. Flowers bisexual; corollas actinomorphic, narrowly bell-shaped, commonly violet, often pubescent or glandular, especially so on the 5 elongate lobes; anthers spurred; styles hispidulous in upper half. Achenes angled, commonly 8–10-ribbed, generally pilose, often also glandular; pappus biseriate, persistent, the outer series of several free short squamellae, the inner series of numerous long capillary bristles.

A genus of about 120 species found in the neotropics.

Key to the Species of *Lepidaploa*

1. Leaves rounded at apex, glandular on upper surface; outer involucral bracts deltoid to triangular, with narrowly acute apices; corolla lavender or violet; pappus cream-colored .... 1. *L. glabra*
1. Leaves acute to acuminate at apex, nonglandular on upper surface; outer involucral bracts linear-lanceolate, with attenuate apices; corolla white; pappus brownish .... 2. *L. sericea*

1. Lepidaploa glabra (Willd.) H. Rob., Proc. Biol. Soc. Wash. **103:** 487. 1990. *Conyza glabra* Willd., Sp. Pl. **3:** 1940. 1804.   Fig. 38A–D.

Vernonia albicaulis Vahl *ex* Pers., Syn. Pl. **2:** 404. 1807.
Vernonia longifolia Pers., Syn. Pl. 2: 404. 1807.
Vernonia punctata Sw. *ex* Wikstr., Kongl. Vetensk. Acad. Handl. 1827: 72. 1828.
Vernonia thomae Benth. *ex* Oerst., Vidensk. Meddel. Dansk Naturhist. Foren. Kjøbenhavn **5–7:** 66. 1852.

Erect subshrub to shrub 1–2 m tall; stems densely pubescent. Leaf blades 1.5–7 × 0.7–2.5 cm, elliptic to ovate, chartaceous, the upper surface puberulent to weakly so, glandular, the lower surface pubescent to puberulent, glandular, the apex rounded, obtuse, or less commonly acute, the margins entire; petioles 2–7 mm long, pubescent. Inflorescence often condensed, 4–10-headed, main axis <4 cm long. Heads 10–20-flowered, 6–8 mm tall; involucre 4–5.5 × 3.5–5 mm, ca. 4-seriate; involucral bracts densely pubescent to weakly so, sometimes glandular, the outer ones shortly deltoid to triangular, with narrowly acute apices, the inner ones 4–5 × 1 mm, lanceolate; receptacle dome-shaped, 1.5–2 mm broad; corollas 5–6 mm long, lavender or violet, the lobes nearly half the length of entire corolla, commonly glandular at tips when young, glabrous when mature. Achenes 2–2.5 mm long; pappus cream-colored, the outer series ca. 0.7 mm long, the inner series 4–4.5 mm long.

DISTRIBUTION: Common in disturbed areas and moist upland forests. Maria Bluff (A3148), Annaberg Ruins (A1930), Center Line Road, km 4.5 (A2910). Also on St. Croix, St. Thomas, Tortola, and Virgin Gorda; Puerto Rico, Lesser Antilles.

2. Lepidaploa sericea (Rich.) H. Rob., Proc. Biol. Soc. Wash. **103:** 492. 1990. *Vernonia sericea* Rich., Actes Soc. Hist. Nat. Paris **1:** 112. 1792.

Erect subshrub to shrub 1–1.3 m tall; stems densely pubescent. Leaf blades 1–7.5 × 0.6–2.8 cm, lanceolate, chartaceous, the upper surface pilose to strigose, eglandular, the lower surface densely sericeous, the apex acute to acuminate, the margins entire to crenulate; petioles sericeous, (0–)1–5 mm long. Inflorescence diffuse, a leafy nearly scorpioid cyme, main branches commonly 10–20 cm long, with heads sessile, single or paired in the axils of to 20 leaves spaced about 1 cm apart. Heads 11–15-flowered, 7–8 mm tall; involucre 5–6 × 3–4 mm, ca. 3-seriate; involucral bracts strigose to weakly so with age, loosely appressed, greatly spreading in fruit, the outer involucral bracts 2–3 × 0.4 mm, linear-lanceolate, the apices attenuate, the inner involucral bracts 5–6 × 1 mm, lanceolate; receptacle weakly dome-shaped, 1–1.5 mm broad; corollas 4–5 mm long, white, glabrous. Achenes 2 mm long, sericeous, also commonly glandular; pappus brownish, the outer series ca. 0.7 mm long, the inner series 4–5.5 mm long.

DISTRIBUTION: A common weed of roadsides to wet shaded banks. Coral Bay (A3139), Rosenberg (B303), Bordeaux (A3855). Also on St. Croix, St. Thomas, Tortola, and Virgin Gorda; Cuba, Hispaniola, and Puerto Rico.

16. MIKANIA Willd., nom. cons.

Twining herbaceous to woody vines; stems cylindrical or hexagonal. Leaves simple, opposite, generally petiolate. Inflorescence terminal or axillary, ultimate branches spiciform, racemiform, glomerate, or paniculate-corymbiform. Heads discoid, 4-flowered; involucre cylindrical, of 4 subequal, weakly overlapping involucral bracts, often subtended by a smaller bract; receptacle flat, naked,

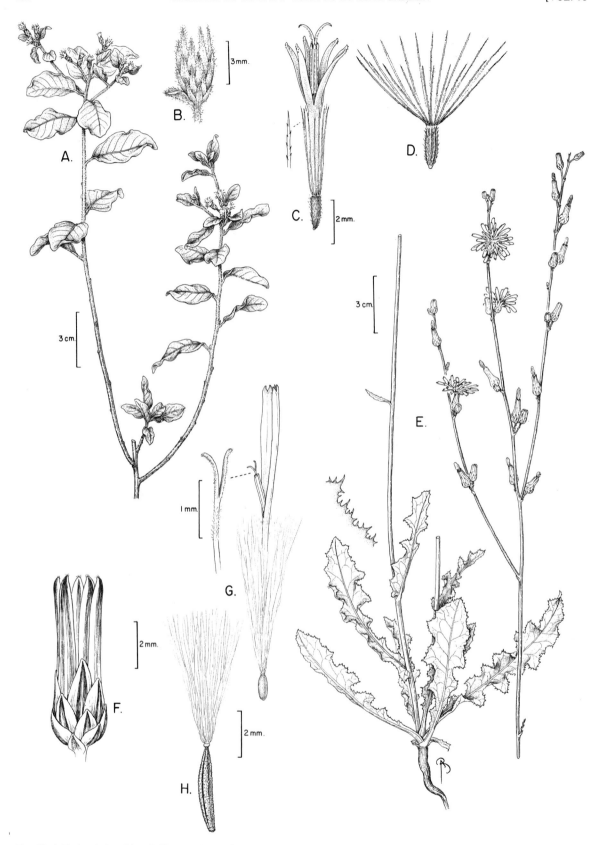

FIG. 38. A–D. *Lepidaploa glabra*. A. Flowering branch. B. Involucre. C. Disk flower and detail of pappus bristle. D. Achene. E–H. *Launaea intybacea*. E. Habit. F. Involucre. G. Ligulate flower and detail of upper portion of style. H. Achene.

glabrous. Flowers bisexual; corollas actinomorphic, tubular, cream-colored, throat urn-shaped to broadly bell-shaped; style branches ascending and elongate, cream-colored, the upper half with a large sterile appendage. Achenes commonly pentagonal, brownish to black; pappus of numerous bristles, these about as long as the corollas and longer than the achenes.

A genus of about 400 species primarily in the New World tropics and subtropics, but also 11 species in the New World temperate zone or warmer regions of the Old World.

REFERENCE: Holmes, W. C. 1993. The genus *Mikania* (Compositae: Eupatorieae) in the Greater Antilles. Sida, Bot. Misc. 9: 1–69.

1. Mikania cordifolia (L.f.) Willd., Sp. Pl. **3**: 1746. 1804. *Cacalia cordifolia* L.f., Suppl. Pl. 351. 1781.

Fig. 39.

Mikania gonoclada DC., Prodr. **5**: 199. 1836.

Herbaceous to semiwoody vine, to 7 m long; stems twining, hexagonal, pubescent. Leaf blades 2–11 × 1.4–11 cm, cordiform, chartaceous, palmately 3(or 5)-veined from base, the upper surface puberulent, the lower surface pubescent to tomentose, glandular, the margins subentire to dentate; petioles to 7 cm long, those of same node commonly connected by a stipule-like appendage. Inflorescence a corymbiform panicle of many to numerous heads.

Heads 7–9.3 mm long, often subtended by a broadly elliptic subinvolucral bract 3–4 mm long; involucre 5.5–8 × 1–2 mm; involucral bracts oblanceolate, the outer pair distinctly pubescent, the inner pair weakly puberulent; peduncles 2–7 mm long, pubescent. Corollas ca. 4–4.5 mm long, sparingly glandular, the throat bell-shaped, cleft to about middle into narrow lobes, these ca. 1.5 mm long. Achenes ca. 3.5 mm long, commonly puberulent and sparingly glandular; pappus to ca. 4.5 mm long, exserted from the involucre.

DISTRIBUTION: Occasional in disturbed areas. Along road to Bordeaux (A4109). Also on St. Croix, St. Thomas, and Tortola; throughout much of the West Indies and the lowland neotropics, also in Gulf coastal United States.

17. NEUROLAENA R. Br.

Erect herbs or shrubs; stems rough pubescent. Leaves simple, alternate, short-petiolate or sessile; blades subentire to dentate and sometimes deeply lobed. Inflorescence terminal and axillary, in large, corymbiform panicles. Heads discoid (in ours) or radiate; involucre bell-shaped, 3–4-seriate; involucral bracts graduated, scarious, the apex obtuse or acute; receptacle paleate. Ray flowers (when present) pistillate. Disk flowers bisexual; corollas yellow; anthers black. Achenes oblong; pappus of numerous persistent bristles.

A genus of 11 species from the American tropics, principally in northern Central America.

1. Neurolaena lobata (L.) Cass. *in* F. Cuvier, Dict. Sci. Nat. **34**: 502. 1825. *Conyza lobata* L., Sp. Pl. 862. 1753. *Calea lobata* (L.) Sw. , Prodr. 113. 1788.

Fig. 40.

Erect coarse herb 1–4 m tall; stems soft and pithy, sparsely branched, striate, densely pubescent when young. Leaves tapering to a short petiole; blades varying in length with the upper stem leaves 8–25 × 1.5–10 cm, the lower ones much larger, oblanceolate to narrowly elliptic, chartaceous, pinnately veined, pubescent, also glandular on lower surface, the margins commonly serrate-

denticulate to entire or lobed, the lower leaves usually with 2 distinct lobes toward base. Inflorescence a series of loose to dense pedunculate clusters of corymbiform cymes. Heads discoid, ca. 30-flowered, ca. 1 cm or less tall; involucre 4–5-seriate; involucral bracts oblong; receptacle convex, paleate, the palea thin. Corollas pale yellow or greenish white, puberulent, ca. 5 mm long. Achenes cylindrical or obscurely 5-ribbed, lightly pubescent, ca. 2–4 mm long; pappus of numerous white bristles, ca. 5 mm long.

DISTRIBUTION: Rare shrub in open wet areas. Top of Bordeaux Mountain (W114). Also on St. Thomas and Tortola; southern Mexico, Central America, West Indies, and northern South America.

18. PARTHENIUM L.

Annual or perennial erect herbs or shrubs; stems usually much branched, tomentose or scabrous. Leaves alternate, simple, and entire to dentate or pinnately dissected. Inflorescence solitary in terminal corymbs or panicles. Heads small, radiate; involucre biseriate or few-seriate, broad, appressed, dry, subequal or the outer ones gradually shorter; receptacle small, convex, paleate, the palea membranous, subtending the outer disk flowers. Ray flowers 5, pistillate; corollas usually white, ca. 1 mm long. Disk flowers many, all except those of the outer row sterile and falling as a unit, at least some flowers of outer row functionally male and attached to the ray flowers; anthers white or yellow. Achenes dorsally compressed, those of the rays fertile and each ray achene forming a unit by fusing with the subtending involucral bract and 2 outer disk flowers and their palea; pappus of 2 or 3 short or elongate awns or lacking.

A genus of 16 or more species, all but 1 species restricted to North America and Mexico.

1. Parthenium hysterophorus L., Sp. Pl. 988. 1753.

Fig. 41.

Erect annual herb 30–75 cm tall; stems much-branched, strigose and often scabrous throughout, conspicuously striate at least on younger parts. Leaves sessile above or lower ones with a winged petiole; upper stem leaves much reduced, lanceolate in outline, lower stem leaves with blades to 20 × 12 cm, ovate to oblong in outline, pinnatifid or bipinnatifid, the segments 2–6, paired, to 7 × 3 cm, linear or lanceolate, dentate or lobate. Inflorescence paniculate, with many small heads. Heads radiate;

involucral bracts biseriate, the outer involucral bracts 5, separate, narrowly ovate, 1.5–2 mm long, acute, palea elliptic, ca. 1 mm long. Ray flowers inconspicuous; corollas cream-colored, persistent on achenes, the limb ca. 0.5 mm long, the tube obscure. Disk flowers many; corollas light yellow, ca. 1–2 mm long; ovary sterile; some outer flowers functionally male. Achenes flattened, 2 mm long; pappus of 2 broad awns 0.5 mm long; achenes falling in a complex of 1 ray achene (and the persistent corolla) and its subtending involucral bract and 2 disk flowers with their palea.

DISTRIBUTION: Common in waste or open disturbed areas. Enighed (A3930). Also on St. Croix, St. Thomas, and Tortola; throughout much of

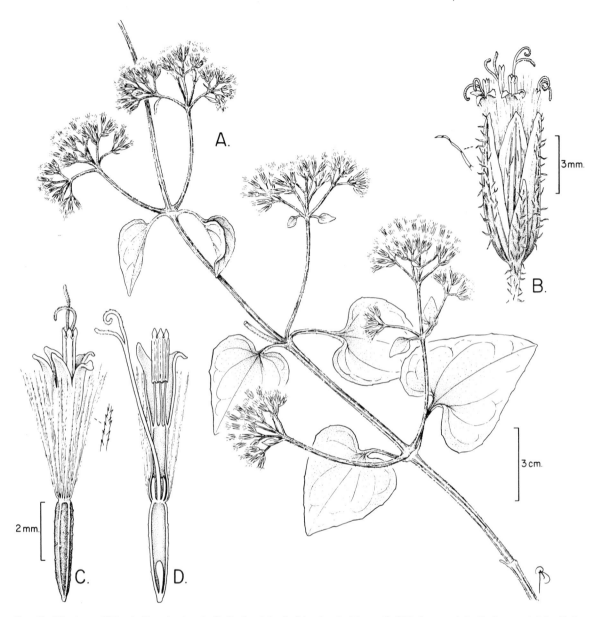

Fig. 39. *Mikania cordifolia*. **A.** Flowering branch. **B.** Head and detail of involucral trichome. **C.** Disk flower and detail of pappus bristle. **D.** L.s. disk flower.

the West Indies, primarily Mexican, but ranging from the southern United States into Central America, South America, and the paleotropics.

COMMON NAMES: false ragweed, feverfew, mule weed, quinine weed, whitehead broom.

19. PECTIS L.

Annual or perennial herbs or subshrubs, generally strong-scented; stems profusely branched, often purplish. Leaves simple, opposite, sessile, narrow, uninervate, surface variously punctate with conspicuous oil glands, the base connected by a narrow connate rim, the margins usually entire with basal pairs of bristly cilia. Inflorescence with heads solitary to densely cymose. Heads radiate, small; involucre uniseriate; involucral bracts 3–12 (usually 5), equal in length, distinct, variously punctate with usually elongate oil glands; receptacle naked and flat or convex. Ray flowers pistillate, usually equal in number to and inserted at base of involucral bracts; corollas yellow to purplish. Disk flowers few to many, bisexual; corollas yellow to purplish; style branches often reduced in size. Achenes cylindrical, surface roughened with many inconspicuous ribs, glabrous to variously pubescent; pappus various.

About 85 species distributed throughout most of the warmer parts of the New World.

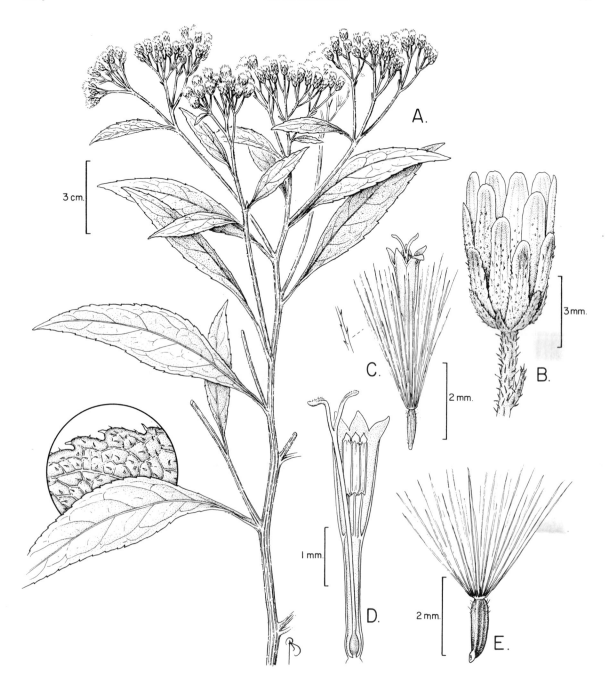

FIG. 40. *Neurolaena lobata.* **A.** Flowering branch and detail of leaf surface. **B.** Involucre. **C.** Disk flower and detail of pappus bristle. **D.** L.s. disk flower. **E.** Achene.

Key to the Species of *Pectis*

1. Prostrate small annual herb or more often perennial with more or less woody base; leaves oblanceolate to obovate; pappus a combination of scabrid bristles and shorter scales or slender bristles .... 1. *P. humifusa*
1. Erect, finely branching annual herb; leaves narrowly linear to lanceolate; pappus of 1–4 spreading awns flattened at the base .... 2. *P. linifolia*

1. Pectis humifusa Sw., Prodr. 114. 1788. Fig. 42A–F.

Pectis serpyllifolia Less., Linnaea **6:** 715. 1831.

Prostrate, spreading, mat-forming herb, more or less woody and much-branched at the base; stems to 25 cm long, often branched toward apex as well, densely leafy, puberulent. Leaves

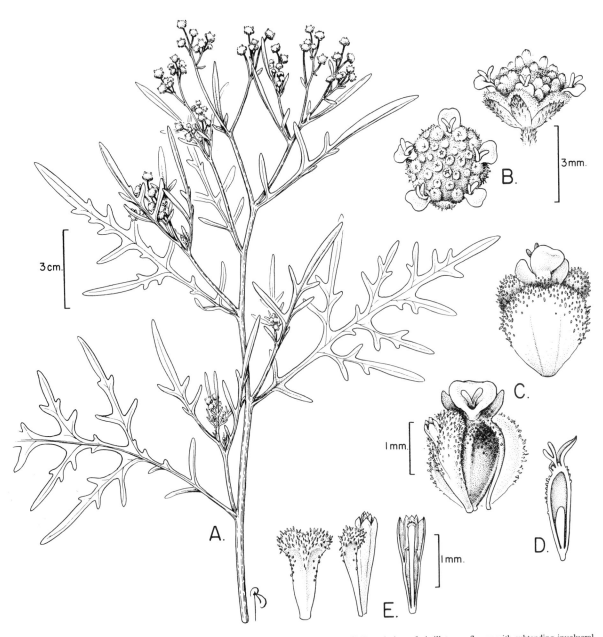

FIG. 41. *Parthenium hysterophorus.* **A.** Fertile branch. **B.** Head, top and lateral views. **C.** Dorsal view of pistillate ray flower with subtending involucral bract and frontal view of pistillate ray flower with subtending staminate disk flower and palea. **D.** L.s. ray flower. **E.** Palea, staminate disk flower and palea, and l.s. disk flower.

4–13 × 1.5–3.5 mm, oblanceolate to obovate, chartaceous, glabrous with sunken oil glands scattered on lower surface, the apex rounded or obtuse, the margins with 2–8 pairs of bristles toward base. Inflorescence with heads terminal and solitary or in few-headed cymes. Heads radiate; involucre bell-shaped to cylindrical, ca. 6 mm tall; involucral bracts 5, obovate, with oil glands scattered on lower surface; peduncles 0–12 mm long, bracteolate. Ray flowers 5, very reduced; corollas yellow to purple, with oil glands. Disk flowers (6–)10–20; corollas yellow to purple, ca. 3 mm long. Achenes all fertile, cylindrical, slightly constricted at base, 2.5–3.5 mm long, strigose to minutely villous; pappus of ray achenes with 2–3 scabrid, bristle-tipped scales and to 10 shorter scales or slender bristles; pappus of disk achenes with 4–

15 scabrid bristles or slender scales and up to 12 shorter bristles or scales; both pappus types whitish to brown or yellow.

DISTRIBUTION: Rare in coastal or open areas. Dittlif Point (A3968). Also on St. Croix and St. Thomas (fide Lessing *in* Schlechtendal, 1831; Britton, 1918; and Britton & P. Wilson, 1925); from Puerto Rico eastward and southward throughout the Lesser Antilles to the coast of Guyana; also collected once in Florida.

2. Pectis linifolia L., Syst. Nat., ed. 10, **2:** 1221. 1759.

Pectis punctata Jacq., Enum. Syst. Pl. 28. 1760.

Erect annual herb 5–85 cm tall, profusely branched above; stems often purple, glabrous to short pubescent near nodes. Leaves

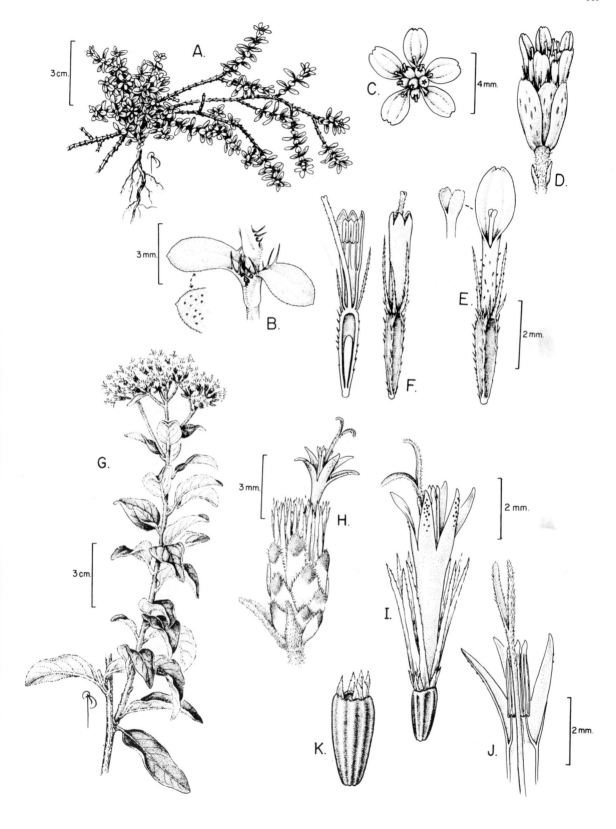

FIG. 42. A–F. *Pectis humifusa*. **A.** Habit. **B.** Leaves with basal cilia and detail of gland-dots on lower surface. **C.** Head, top view. **D.** Head, lateral view.
E. Ray flower and detail of style branches. **F.** L.s. disk flower and disk flower. **G–K.** *Piptocoma antillana*. **G.** Flowering branch. **H.** Head. **I.** Disk
flower. **J.** L.s. upper part of disk flower showing style branches and anthers. **K.** Achene with inner series of pappus scales removed.

1–6.5 cm × 1–2 mm, linear to lanceolate, chartaceous, glabrous with sunken oil glands scattered on lower surface, the apex acute, the base sometimes narrowed and appearing petiole-like, the margins revolute with 1 or 2 pairs of bristles toward base. Inflorescence cymose. Heads radiate; involucre cylindrical, ca. 5 mm tall; involucral bracts 5, linear, with oil glands on outer surface; peduncles 10–30 mm long, slightly expanded at the apex, sometimes purple. Ray flowers 5, very reduced; corollas yellow to purple, with oil glands. Disk flowers 2–5; corollas yellow to purple, ca. 2.5–4

mm long. Achenes all fertile, cylindrical, slightly constricted at base, 3–5.5 mm long, puberulent; pappus of (1–)3(–4) spreading, brown to purple awns flattened at the base.

DISTRIBUTION: Rare along hillsides and openings in forests. Southside Pond (A4072), Lameshur (B629). Also on Buck Island, St. Croix, St. Thomas, Tortola (fide D'Arcy, 1967), and Virgin Gorda; this species is native to three distinct areas: the Sonoran Desert of the southwestern United States and adjacent Mexico, southern Mexico and the Caribbean, and the west coast of Ecuador and Peru and the Galapagos Islands. It has recently become naturalized in Hawaii. Variety *hirtella* Blake is endemic to Mexico.

20. PIPTOCOMA Cass.

Erect to scandent shrubs or subshrubs; stems tomentose when young, irregularly angled. Leaves simple, alternate, petiolate; blades narrowly lanceolate to ovate or cordate, pinnately veined, the upper surface becoming glabrous, sometimes glandular, the lower surface stellate-tomentose. Inflorescence terminal, corymbiform or glomerate, of several to many heads, these short-pedunculate. Heads discoid, 4–12-flowered; involucre cylindrical; involucral bracts overlapping, graduated; receptacle subconvex to flat, naked, often ridged, very rarely 1- or 2-awned. Flowers bisexual; corollas actinomorphic, funnelform; anthers spurred; styles hispidulous in upper half, branches slender. Achenes obconic, 5-angled, glabrous; pappus biseriate, the outer series persistent, of ca. 10 free to seemingly connate short squamellae, the inner series deciduous, of ca. 7–15 elongate scales.

A genus of 3 species distributed in Hispaniola, Puerto Rico, and the Virgin Islands.

REFERENCE: Stutts, J. G. & M. A. Muir. 1981. Taxonomic revision of *Piptocoma* Cass. (Compositae: Vernonieae). Rhodora 83: 77–86.

NOTE: Current studies may show *Piptocoma* and the closely related Central and South American genus *Pollalesta* to be congeneric. Because *Pollalesta* is a later name, it may require the transfer of its species into *Piptocoma* increasing the size of *Piptocoma* to 18 species.

1. **Piptocoma antillana** Urb., Ark. Bot. 23A, **11:** 50. 1931. Fig. 42G–K.

Erect shrub 0.4–4 m tall, much-branched. Leaf blades 2–7 × 0.6–2.2 cm, lanceolate to narrowly elliptic, chartaceous, the upper surface glandular and puberulent, the lower surface often gray or rust-colored, the base attenuate and tapering onto a petiole, the margins entire; petioles 2–8 mm long. Inflorescence subcompact, to 4 × 6 cm, with to ca. 35 heads. Heads (4–)5–6(–7)-flowered, to 7.5 mm tall; involucre 4.5–6 × 2–3 mm, 4–5-seriate; involu-

cral bracts elliptic to elliptic-lanceolate, apically rounded, outer 2 or 3 herbaceous, pubescent, inner ones scarious and glabrous, except for the pubescent and glandular tips; peduncles 1–11 mm long, occasionally absent. Corollas 5–6 mm long, lavender, the lobes and sometimes the tube glandular. Achenes 1.5–2.5 mm long; pappus cream-colored, outer series ca. 0.5–1 mm long, inner series ca. 4–5 mm long.

DISTRIBUTION: Common in sandy or rocky areas behind shore vegetation. Coral Bay (R1916), East End (A666), Southside Pond (A1816). Also on Peter Island (fide D'Arcy, 1967), St. Thomas, Tortola, and Virgin Gorda; Puerto Rico.

21. PLUCHEA Cass.

Annual or perennial herbs or shrubs, commonly aromatic; stems erect, sometimes winged by decurrent leaf bases. Leaves simple, alternate; blades pinnately veined, both surfaces commonly glandular and pubescent. Inflorescence terminal, often paniculate; peduncles commonly pubescent. Heads disciform; flowers very numerous; involucral bracts graduated, overlapping, reflexed after fruit; receptacle flat, naked. Flowers heterogamous, unisexual; corollas actinomorphic, commonly purplish. Outer flowers very numerous, pistillate; corollas filiform, apically denticulate; styles shortly bifid. Central flowers few to several, functionally staminate, but appearing bisexual; corollas tubular, bell-shaped, shortly 5-lobed; anthers caudate; styles weakly bifid or entire, the ovary sterile. Achenes very small, cylindrical, often 3–6-ribbed or ribbing obscure, brown; pappus of several barbellate capillary bristles, about as long as the corollas.

A genus of 40–80 species widely distributed from United States to Mexico, Central America, tropical and subtropical South America, West Indies, Africa, southeastern Asia, Indo-Malay area, Australia, and the Pacific.

REFERENCES: Gillis, W. T. 1977. *Pluchea* revisited. Taxon 26: 587–591; Khan, R. & C. E. Jarvis. 1989. The correct name for the plant known as *Pluchea symphytifolia* (Miller) Gillis (Asteraceae). Taxon 38: 659–662.

Key to the Species of *Pluchea*

1. Shrubs; stems pubescent to sometimes glabrate; leaves chartaceous to firmly so, commonly moderately pubescent on lower surface, the margins entire or nearly so; inner flowers 15 or more .... 1. *P. carolinensis*
1. Herbs; stems puberulent to nearly glabrous; leaves chartaceous to thinly so, commonly puberulent on lower surface, the margins subentire or serrate; inner flowers 12 or fewer ....:....... 2. *P. odorata*

1. **Pluchea carolinensis** (Jacq.) G. Don *in* Sweet, Hort. Brit., ed. 3, 350. 1839. *Conyza carolinensis* Jacq., Collectanea. **2:** 271. 1788 [1789].

Pluchea odorata sensu Godfrey, J. Elisha Mitchell Sci. Soc. **68:** 247. 1952, non (L.) Cass., 1826.

Pluchea symphytifolia of many authors, non (Mill.) Gillis, 1977.

Erect shrub 1–2 m tall, with many lateral branches; stems pubescent to sometimes glabrate, irregularly angled. Leaf blades 4–15 × 1.5–6 cm, broadly elliptic to lanceolate, chartaceous to nearly coriaceous, the upper surface puberulent to nearly glabrous, the lower surface pubescent to puberulent, weakly glandular, the margins entire to subentire; petioles 0.5–1.5(–2.5) cm long, pubescent to densely so. Inflorescence a broadly rounded corymbiform panicle to 15 × 20 cm, with numerous heads. Heads 6–8 × 8–10 mm; involucre hemispheric to bell-shaped, 3–4-seriate, ca. 5 mm tall; involucral bracts many, pubescent, occasionally weakly glandular, becoming glabrous, especially toward tips of inner involucral bracts, the outer ones ovate, apically rounded, the inner ones linear-lanceolate, apically attenuate to acuminate, very innermost late-deciduous; receptacle glabrous, ca. 2 mm broad; peduncles 4–10 mm long, densely pubescent. Outer flowers with corollas 3.5–4 mm long. Inner flowers 15 or more; corollas 4–5 mm long, lightly glandular at tips. Achenes filiform, 0.5–1 mm long, weakly puberulent, ribs obscure or 1 or 2 apparent; pappus of ca. 10 bristles, ca. 4 mm long.

DISTRIBUTION: Common in coastal scrubs and disturbed areas. Emmaus (A2440, A3239), Maria Bluff (A609). Also on St. Croix, St. Thomas, Tortola, and Virgin Gorda; throughout much of the West Indies, Florida, Mexico, Central America, northern South America, and the Pacific region.

COMMON NAMES: cattle tongue, ovra bla, ramgoat bush, sourbush, sweet scent.

2. Pluchea odorata (L.) Cass., Dict. *in* F. Cuvier, Sci. Nat. **42**: 3. 1826. *Conyza odorata* L., Syst. Nat., ed. 10, **2**: 1213. 1759. Fig. 43.

Conyza purpurascens Sw., Prodr. 112. 1788. *Pluchea purpurascens* (Sw.) DC., Prodr. **5**: 452. 1836.

Erect herb, 0.5–1.5 m tall, much-branched from base; stems puberulent to more commonly glabrous. Leaf blades 3.5–12 × 1–6 cm, lanceolate to ovate, chartaceous, puberulent, sometimes weakly glandular on both surfaces, the margins subentire to serrate; petioles 0.3–2.5 cm, pubescent to puberulent, sometimes winged. Inflorescence a diffuse panicle of many flat-topped corymbiform clusters 3–4 × 2–3 cm, with 10 or more heads. Heads 5–7 × 4–6 mm; involucre bell-shaped, ca. 3-seriate, ca. 4 mm tall; involucral bracts many, puberulent, the outer ones long-triangular, apically acute, the inner ones lanceolate, apex attenuate; receptacle weakly puberulent, 1.5–2 mm broad; peduncles 1–6(-8) mm long, pubescent. Outer flowers with corollas 3–3.5 mm long. Inner flowers ca. 5–9(–12); corollas 3.5–4 mm long, glabrous or very weakly puberulent at tips. Achenes ca. 1 mm long, weakly puberulent and weakly glandular, ca. 5-ribbed, the ribs cream-colored; pappus of ca. 15 bristles, ca. 3.5 mm long.

DISTRIBUTION: Common in disturbed open areas. Coral Bay (A733), Emmaus (A1996, A3240). Also on Anegada, St. Croix, St. Thomas, and Tortola; throughout much of the West Indies, United States, Mexico, Central America, and northern South America.

COMMON NAME: marsh fleabane.

22. PSEUDELEPHANTOPUS Rohr

Perennial herbs; stems few-branched, pilose. Leaves simple, alternate, sessile, commonly clasping the stem, pinnately veined, much reduced in the inflorescence. Inflorescence terminal, strongly bracteate, spicate-racemose, heads in axillary glomerules. Heads discoid, 4-flowered; involucre cylindrical, involucral bracts 8, in 4 decussate pairs; receptacle convex, naked. Flowers bisexual; corollas tubular, irregularly and deeply 5-lobed; anthers basally sagittate; styles hispidulous in upper half, branches linear. Achenes obconic, ca. 10-ribbed; pappus of ca. 4–10 setae, subequal and spirally twisted or unequal with 6 shorter and 4 longer, 2 of the longer apically twice-folded.

A genus of 2 species, native to and widespread in the neotropics, occasionally introduced into the subtropics and the paleotropics.

1. Pseudelephantopus spicatus (Juss. *ex* Aubl.) C.F. Baker, Trans. Acad. Sci. St. Louis **12**: 45, 54, 56. 1902. *Elephantopus spicatus* Juss. *ex* Aubl., Hist. Pl. Guiane **2**: 808. 1775. Fig. 44.

Erect herb to ca. 1 m tall; stems striate, angled toward apex, but cylindrical toward base. Leaves cauline, 4–17 × 0.7–4.2 cm, oblanceolate (on vegetative parts of stems) or lanceolate (in flowering branches), chartaceous, sparsely puberulent to glabrous on both surfaces, also glandular on lower surface, the apex acute, the base attenuate, often narrowly so and resembling a winged petiole, the margins sinuate to more commonly serrulate. Inflorescence of several spicate-glomerate leafy branches, 1–3 heads sessile or nearly so in each axil of the several to many lanceolate

leafy bracts. Heads to 12 × 2–3.5 mm; involucre to 11 × 2–3.5 mm, 2–3-seriate; involucral bracts lanceolate, green-tipped, straw-colored at base, inner ones lightly glandular at apex, involucral bracts otherwise glabrous, the outer 4 in 2 unequal pairs about half the length of the subequal inner 4 involucral bracts. Corollas 6–7 mm long, cream-colored, glabrous. Achenes 4–7 mm long, pubescent on the ribs, glandular in the furrows; pappus of 8–10 subequal bristles, the 4 longer ca. 6 mm long, 2 of these twice-folded near tips.

DISTRIBUTION: Occasional along roadsides and in other disturbed areas. Cruz Bay Quarter along Center Line Road (A4160). Also on St. Croix, St. Thomas, and Tortola; throughout much of the West Indies and the neotropics, introduced into Florida, Africa, Asia, the Pacific region, and Australia.

COMMON NAME: bulltongue bush.

23. PTEROCAULON Elliott

Perennial herbs to subshrubs; stems winged by decurrent leaf bases. Leaves simple, alternate, sessile. Inflorescence terminal, spicate or glomerate. Heads disciform, sessile; involucre bell-shaped, involucral bracts tomentose at least at base; receptacle small, flat, naked. Flowers heterogamous, unisexual; corollas actinomorphic, cream-colored to yellow. Outer flowers many, pistillate; style bifid, branches glabrous. Inner flowers few, functionally staminate by sterility of the ovary; anther base with sterile tails; style weakly bifid, pubescent distally, the ovary sterile. Achenes cylindrical, fusiform, or plump; pappus of many capillary bristles about as long as the corollas.

A genus of 18 species distributed in the eastern United States, Mexico, Central America, West Indies, tropical South America, southeastern Asia, Indo-Malay area, and Australia.

REFERENCE: Cabrera, A. L. & A. M. Ragonese. 1978. Revisión del género *Pterocaulon* (Compositae). Darwiniana 21; 185-257

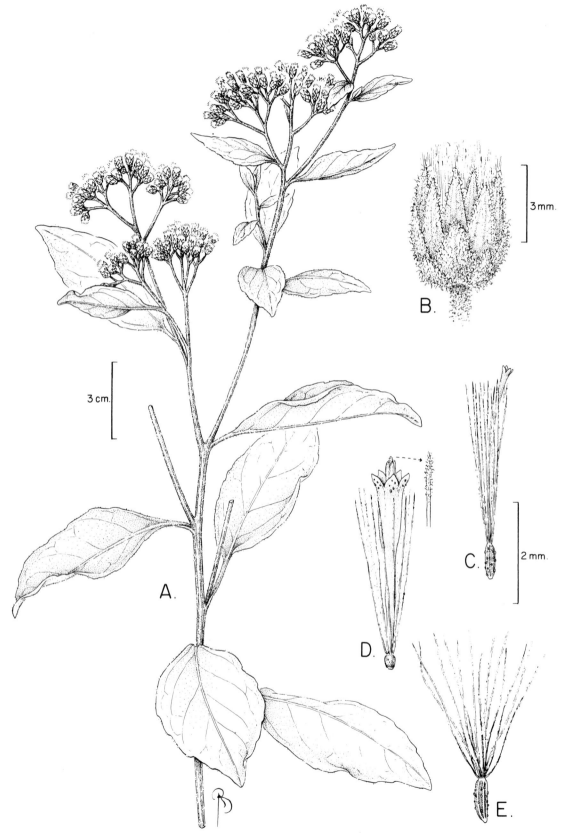

FIG. 43. *Pluchea odorata.* **A.** Flowering branch. **B.** Head. **C.** Outer pistillate flower. **D.** Functionally inner staminate flower and detail of unbranched style. **E.** Achene with innermost elongate scales removed.

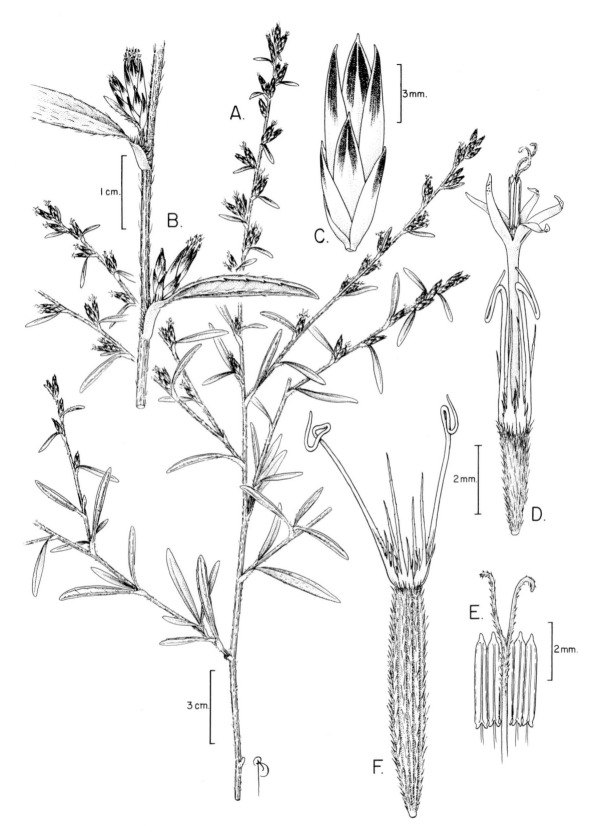

Fɪɢ. 44. *Pseudelephantopus spicatus*. **A.** Flowering branch. **B.** Axillary glomerate inflorescences. **C.** Involucre. **D.** Disk flower. **E.** Upper portion of style and anthers. **F.** Achene.

1. Pterocaulon virgatum (L.) DC., Prodr. **5:** 454. 1836. *Gnaphalium virgatum* L., Syst. Nat., ed. 10, **2:** 1211. 1759.   Fig. 45.

Erect herb, 50–90 cm tall, few-branched; stems white-tomentose. Leaves 3–9 × 0.2–1(–2) cm, linear-lanceolate distally, narrowly elliptic proximally, chartaceous, pinnately veined, the upper surface weakly cobwebby to nearly glabrate, the lower surface white-tomentose, the margins subentire or weakly toothed. Inflorescence an interrupted spike to 30 cm long, simple or less commonly branched, heads single or more commonly clustered in groups of 3–6. Heads many-flowered, 6–8(–9) × 4–7 mm; involucre 5–7 mm tall, 4–5-seriate; involucral bracts many, graduated, overlapping, spreading with age, 1-veined, outer ones triangular, tomentose, inner ones linear-lanceolate, often deciduous, tomentose proximally, becoming glabrous at the attenuate apex. Outer flowers many; corollas ca. 5 mm long, filiform, very shortly lobed, cream-colored. Inner flowers 2–3(–5); corollas 4–5 mm long, tubular, 5-lobed, yellowish, often papillose above. Achenes 0.5–1.5 mm long, pubescent, ribbed; pappus white, 5–6 mm long.

DISTRIBUTION: Occasional in open or disturbed areas. Rosenberg (B314). Also on St. Croix, St. Thomas, and Tortola; Greater Antilles, Texas, Mexico, Honduras, and South America.

24. SONCHUS L.

Annual or perennial herbs to small trees, with milky latex. Stems hollow, few-branched, often with gland-tipped bristles. Leaves simple, alternate and cauline or basal, sessile and sheathing or clasping the stem, spinose-dentate to deeply toothed; lower and midstem leaves generally pinnatifid; upper stem leaves ovate-lanceolate. Inflorescence terminal mostly paniculate. Heads ligulate, 1 to many; involucre bell-shaped, 2- to several-seriate; involucral bracts graduated; receptacle naked. Flowers ca. 100, bisexual; corollas yellow. Achenes flattened, narrowed at both ends with 1–4 ribs on each face, not beaked; pappus of white bristles or of numerous, persistent capillary hairs and fewer, thicker, deciduous bristles.

A genus of 30–70 species native to Eurasia, tropical Africa, and some Atlantic islands and introduced into the New World.

1. Sonchus oleraceus L., Sp. Pl. 794. 1753.   Fig. 46.

Erect herb to 1 m tall, simple or sometimes branched; stems hollow, striate. Leaves 5–30 × 1.3–12 cm, obovate to oblanceolate, membranous, glabrous, mostly incised, denticulate, apical segment more or less truncate, the apex acute to obtuse, basal segment often narrowed into a distinct, winged, mostly entire-margined petiolar base, expanded into auricles clasping the stem, the teeth callose-mucronate. Inflorescence an open panicle, peduncles with gland-tipped bristles. Heads showy, 20 or more flowered; involucre bell-shaped, 10–20 mm tall; involucral bracts many in several, graduated, overlapping series, the outermost series small and obovate with acuminate apex, the innermost series nearly the length of the head and more linear. The outermost series of flowers slightly larger than the inner ones. Corollas yellow, ca. 11 mm long, the tube pubescent. Achenes flattened, ca. 3.5 mm long, rugose; pappus dimorphic, ca. 5–11 mm long, of numerous white, persistent hairs and fewer, thicker, strigillose, deciduous bristles.

DISTRIBUTION: Uncommon weed of moist soils. Cinnamon Bay (W681). Also found on St. Croix, St. Thomas (fide Lessing *in* Schlechtendal, 1831), and Tortola; native to temperate regions of the Old World, but widely distributed as a weed in temperate and tropical regions.

COMMON NAMES: sow thistle, wild salad.

25. SPHAGNETICOLA Hoffm.

Procumbent perennial, puberulent to pubescent, sometimes succulent herbs; stems rooting at the nodes, elongating sympodially. Leaves simple, opposite, sessile or shortly petiolate. Inflorescence terminal (usually appearing axillary when heads laterally displaced), 1- to few-headed. Heads radiate, many-flowered, long-pedunculate; involucre weakly 2(–3)-seriate; involucral bracts foliaceous or inner ones only apically so; receptacle paleate. Ray flowers pistillate; corollas yellow or orange. Disk flowers bisexual; corollas yellow or orange; anthers black; style branches nearly erect or slightly reflexed. Achenes black and tuberculate at maturity, the tubercules tan; pappus fimbriate, fimbriae obscured by a corky crown-shaped collar, this much shorter than the achene body.

A genus of 4 species, 3 of these native to the New World, lowland tropics, and subtropics; a fourth species native to southeastern Asia and adjacent Pacific islands.

1. Sphagneticola trilobata (L.) Pruski, comb. nov. *Silphium trilobatum* L., Syst. Nat., ed. 10, **2:** 1233. 1759. *Wedelia carnosa* Rich. *in* Pers., Syn. Pl. **2:** 490. 1807, nom. illegit. (based on *Silphium trilobatum*). *Wedelia trilobata* (L.) Hitchc., Annual Rep. Missouri Bot. Gard. **4:** 99. 1893.   Fig. 47.

Procumbent herb, 2 m or longer. Leaves sessile or subsessile, those of a node commonly inconspicuously basally connate; blades 3–10 × 2.5–6 cm, oblanceolate to rhombic, often 3-lobed, chartaceous to fleshy, pinnately 3-veined from above base, pubescent to puberulent, also glandular on lower surface, gradually tapering into a subpetiolar base, the margins subentire or toothed, each margin often with a prominent medial lobe; petioles 0–5 mm long. Heads solitary, bell-shaped, on peduncles 3.5–14 cm long; involucre obconic, ca. 10–14 mm tall and broad; involucral bracts 12–15, subequal, green, ca. 10–14 × 2.5–4.5 mm, oblanceolate to oblong, strigose and weakly glandular or inner ones merely puberulent; palea oblanceolate. Ray flowers 4–10; limbs to 15 mm long, shallowly 3-lobed at apex, the lower surface glandular, exserted from involucre. Disk flowers many; corollas 4.5–5.5 mm long, shortly 5-lobed, the lobes strongly pubescent-papillose within or marginally, occasionally glandular on outer surface. Achenes pyriform, ca. 3 mm long; the collar to ca. 1.1 mm long.

DISTRIBUTION: Occasional in strand vegetation, roadsides, to dry evergreen hillsides, but more common in wet areas. Along road to Bordeaux Mountain (A2894), Hillside above Cinnamon Bay (M17091). Also on St. Croix, St. Thomas, and Tortola; throughout much of the West Indies, southeastern United States, Mexico, Central America, and South America. Widely cultivated, escaping and becoming naturalized throughout tropical regions of both Eastern and Western hemispheres (except Africa).

COMMON NAMES: creeping ox eye, wild marigold, wedelia.

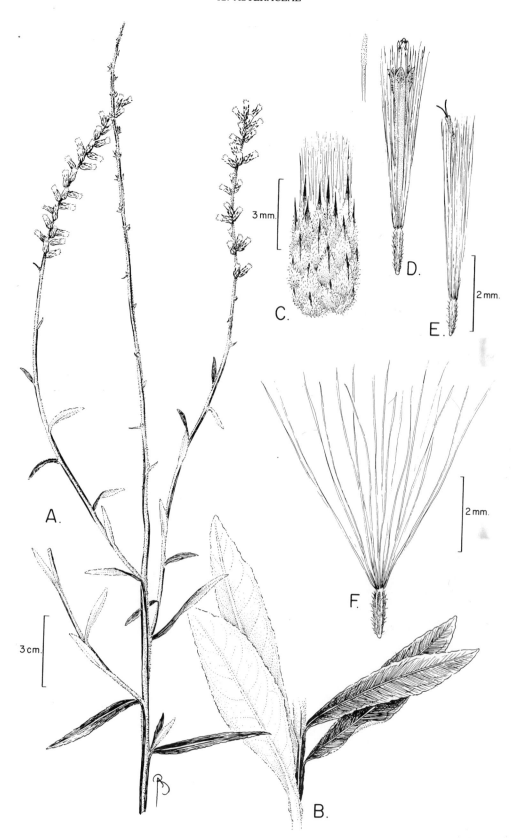

FIG. 45. *Pterocaulon virgatum.* **A.** Flowering branch. **B.** Cauline leaves. **C.** Flowering head. **D.** Functionally inner staminate flower and detail of unbranched style. **E.** Outer pistillate flower. **F.** Achene.

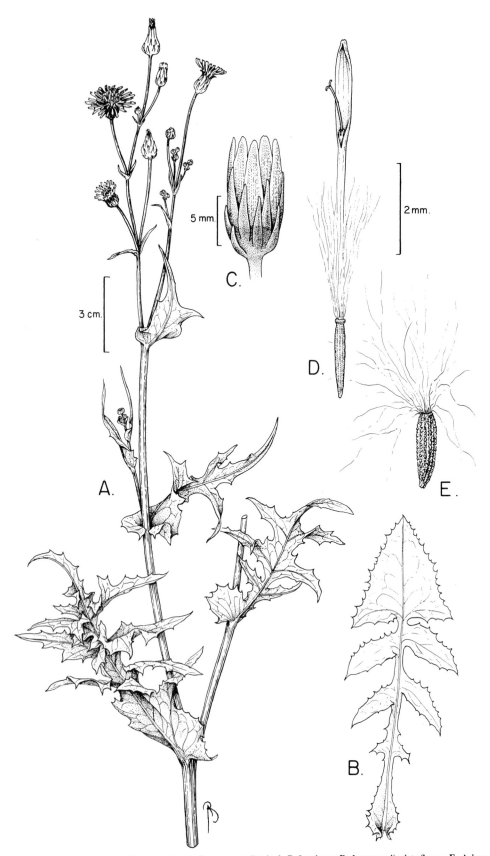

Fig. 46. *Sonchus oleraceus*. **A.** Flowering branch. **B.** Lower cauline leaf. **C.** Involucre. **D.** Immature ligulate flower. **E.** Achene.

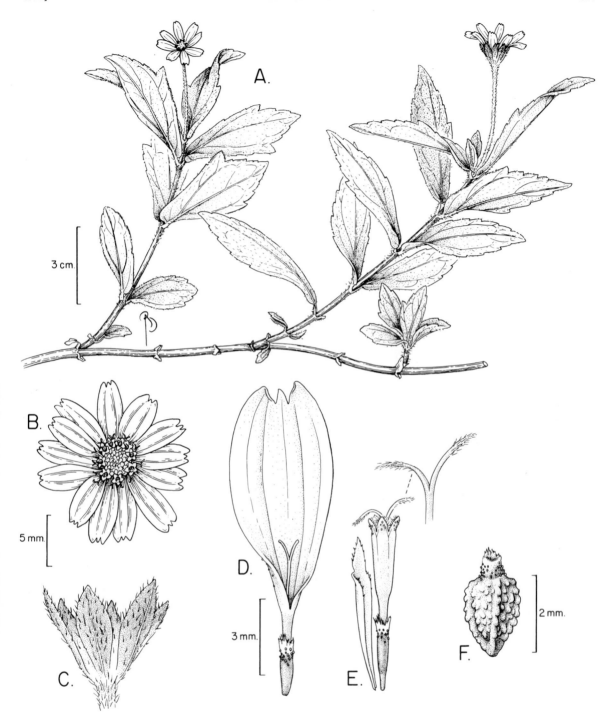

FIG. 47. *Sphagneticola trilobata.* **A.** Flowering branch. **B.** Head. **C.** Involucre. **D.** Ray flower. **E.** Disk flower with subtending palea and detail of style branches. **F.** Achene.

26. SYNEDRELLA Gaertn., nom. cons.

Monospecific; the characters are those of the species, given below.

1. Synedrella nodiflora (L.) Gaertn., Fruct. Sem. Pl. **2:** 456. 1791. *Verbesina nodiflora* L., Cent. Pl. I. 28. 1755.   Fig. 48.

Erect or procumbent, annual or short-lived perennial herb, to 1(–1.5) m tall; stems weakly branched, pubescent, eglandular. Leaves simple, opposite; blades 3–11 × 1.5–6 cm, elliptic to

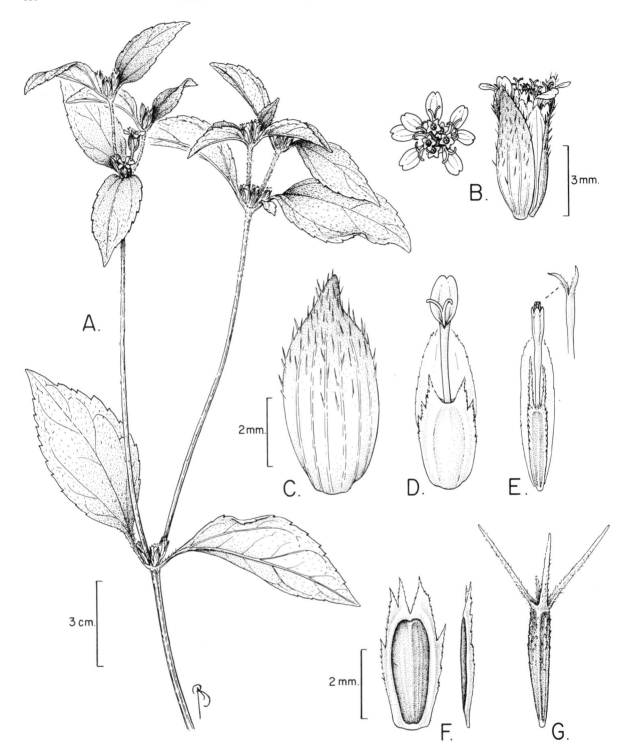

FIG. 48. *Synedrella nodiflora.* **A.** Flowering branch. **B.** Head, top and lateral views. **C.** Outer involucral bract. **D.** Ray flower with subtending involucral bract. **E.** Disk flower with subtending palea and detail of style branches. **F.** Ray achene, frontal and lateral views. **G.** Disk achene.

ovate, chartaceous, eglandular, strigose, the upper surface weakly so, pinnately 3-veined from near base, the apex nearly acuminate to obtuse, the base abruptly contracted and attenuate, the margins subentire to somewhat serrate; petioles obscure or to 2.5 cm long, commonly winged. Inflorescence of 1 to several heads clustered in leaf axils, subsessile or nearly so, rarely long-pedunculate. Heads radiate, 9–18-flowered, ca. 9 × 4 mm; involucre cylindrical; involucral bracts few, subequal, striate, elliptic-lanceolate, strigose, the outer 1–3 foliaceous; receptacle convex, minute, paleate, the palea ca. 6 × 1 mm, narrowly elliptic, concave;

peduncle 0.5(–4) cm long. Ray flowers 3–6, pistillate; corollas yellow, 3–4 mm long, glabrous, the limb shortly exserted. Disk flowers ca. 8, bisexual; corollas yellow, to 3.5 mm long, shortly 4- or 5-lobed, the lobes pubescent; anthers black. Achenes dimorphic; ray achenes 3–5 mm long, oblong-ovoid, flattened, winged, the wings deeply cut and grading into 2 stout pappus awns, the awns and the marginal teeth ca. 1 mm long; disk achenes ca. 3 mm long, obconic, slightly compressed, with 2(–3) stout divaricate awns, the awns to ca. 3 mm long.

DISTRIBUTION: Common in disturbed open areas. Coral Bay Quarter along Road 107 (A4041), near mouth of Fish Bay Gut (A3888). Also on St. Croix, St. Thomas, and Tortola; Florida, widespread throughout the West Indies, Mexico, Central America, tropical South America, and parts of the Old World tropics and subtropics.

COMMON NAMES: fatten barrow, node weed.

27. TAGETES L.

Strongly scented annual or perennial herbs or shrubs; stems often much branched, often straw-colored, usually glabrous. Leaves simple to pinnately compound (in ours), opposite or alternate above, sessile or petiolate; blades with scattered elongate or round oil glands, the margins entire to serrate. Inflorescence with heads solitary or in leafy cymes; peduncles short to elongate and often inflated above. Heads varying in size, radiate (in ours) or discoid; involucre uniseriate; involucral bracts 3–10, equal in length, fused, oil glands usually in 1–3 rows; receptacle flat or convex, naked. Ray flowers in 1 to several series, often showy; corollas white, yellow, orange, or red. Disk flowers few to many, bisexual; corollas yellow to orange. Achenes elongate, cylindrical, the surface roughened with many inconspicuous ribs, glabrous to variously pubescent, those of rays often smaller than those of the disks; pappus of awns or scales.

A genus of about 40 species with a natural range from the southwestern United States to northern Argentina. Several species are cultivated throughout the world.

1. Tagetes erecta L., Sp. Pl. 887. 1753.

Erect annual herb, 30–70(180) cm tall; stems profusely branched above, straw-colored, ribbed, glabrous to puberulent or villous in the axils. Leaves pinnately compound, opposite below and sometimes alternate above, 3.5–15 cm long, broadly elliptic in outline; leaflets 9–17, opposite or alternate, sessile, 1–4.5 cm × 2–6 mm, gradually reduced toward petioles, lanceolate, the surfaces glabrous with sunken oil glands scattered or in 2 or more submarginal rows, the apex and the base usually acute, the margins serrate; rachis 3–9(–10) cm long, winged; petioles to 2.5 cm long, winged. Inflorescence solitary, erect; peduncles 4–12 cm long, strongly inflated apically. Heads radiate; involucre bell-shaped, 17–21 × 12–17 mm; involucral bracts uniseriate, 6–10, oil glands elongate in 2 submarginal rows (sometimes scattered). Ray flowers 5–10 (or often in multiple series in cultivated forms); corollas well exserted, yellow to orange or red, with oil glands. Disk flowers many to numerous; corollas yellow to red or purple, 10–15 mm long. Achenes all fertile, black, 6–10 mm long, cylindrical, slightly constricted at base, glabrous to puberulent on angles; pappus of 1 or 2 lanceolate scales ca. 10 mm long and a crown of squamellae 3–5 mm long.

DISTRIBUTION: Escaped from cultivation in open sunny areas. Along road to Bordeaux (A2860, A2888). Commonly cultivated throughout the world and naturalized in many areas; native to Mexico, but all or most extant populations are probably derived from cultivated individuals.

COMMON NAME: African marigold.

28. TRIDAX L.

Annual or perennial herbs; stems erect to decumbent, occasionally rooting at the nodes, weakly to densely pubescent. Leaves simple (in ours) to pinnatifid, opposite or sometimes alternate above, petiolate or sessile. Inflorescence terminal, of 1 to several cymosely arranged heads on long naked peduncles. Heads radiate (in ours) or discoid; involucre hemispheric to bell-shaped, 2–5-seriate; involucral bracts subequal to unequal and overlapping, outer ones herbaceous or apically so, to scarious and purple, inner ones with purple scarious margins; receptacle weakly convex or more commonly conical, paleate, the palea lanceolate, boat-shaped. Ray flowers (when present) few, pistillate; corolla limbs inconspicuous to conspicuous, often apically lobed, white, yellow, red, or purplish. Disk flowers many, bisexual; corollas actinomorphic or occasionally pseudobilabiate, commonly yellow, shortly to deeply 5-lobed; anthers brown; style branches recurved. Achenes obconic, glabrous to densely pubescent; pappus rarely wanting, commonly of about 20 featherlike bristles or scales about as long as the disk corollas.

A genus of 29 species native to the neotropics and concentrated in Mexico but with 1 species introduced and weedy in the tropics and subtropics of the Old World.

1. Tridax procumbens L., Sp. Pl. 900. 1753.   Fig. 49.

Decumbent perennial herb, to 1 m long; stems pilose, rooting at the nodes. Leaves simple; blades 3–6.5 × 1–5 cm, lanceolate to ovate, chartaceous, eglandular, hirsute on both surfaces, 3-nerved from very near the base, the margins serrate to deeply incised; petioles 4–20 mm long. Heads commonly solitary on pilose peduncles to 25 cm long, radiate, 10–12 × 7–11 mm; involucre bell-shaped, 7–8 mm tall, 2(–3)-seriate; involucral bracts few, weakly overlapping, nearly equal in length, narrowly ovate, the outer ones herbaceous, pilose, the inner ones scarious and puberulent; receptacle convex to short-conic; palea ca. 6–8 mm long, hyaline, but with a thicker medial vein. Ray flowers 3–6; corollas white to cream, 7–8 mm long, the limb apically 3-lobed, sometimes nearly to base, the lower surface of the limb pubescent on the 2 main veins. Disk flowers 25 or more; corollas tubular, yellow, ca. 6 mm long, the lobes pubescent. Achenes 2–2.5 mm long, black, pubescent; pappus of 20 unequal featherlike bristles about as long as the corolla tubes, those of the rays thereby shorter.

DISTRIBUTION: Common weed of open disturbed areas. Calabash Boom (A2870, A4025). Also on St. Croix, Tortola, and Water Island; widely distributed throughout the neotropics and introduced into the warmer regions of the Old World.

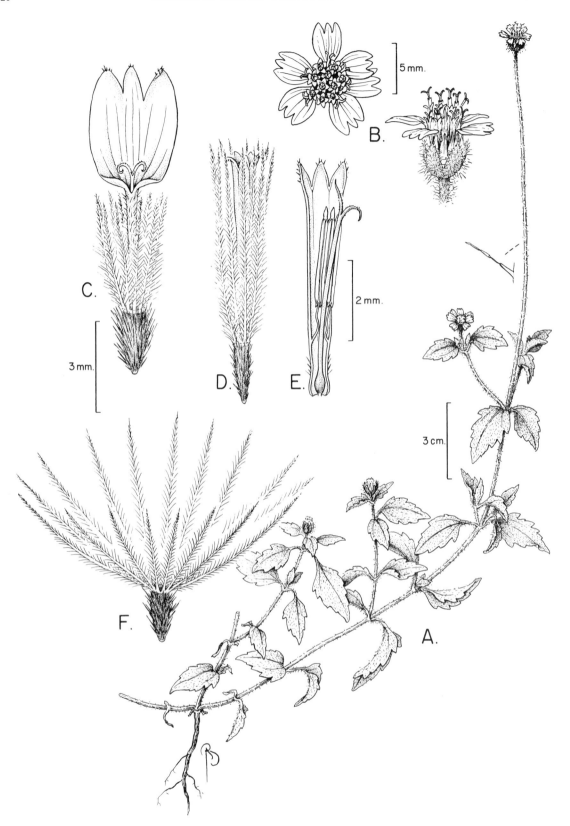

FIG. 49. *Tridax procumbens.* **A.** Habit and detail of stem trichome. **B.** Head, top and lateral views. **C.** Ray flower. **D.** Disk flower. **E.** L.s. disk corolla. **F.** Achene.

29. VERBESINA L.

Herbs, shrubs, or small trees; stems sometimes winged. Leaves simple, alternate or sometimes opposite; blades chartaceous to nearly coriaceous, entire, lobed, or pinnatifid, frequently decurrent forming wings on the petioles and stems, pinnately veined, more or less scabrous, seldom glabrous, the lower surface occasionally tomentose or discolorous; petioles sometimes winged. Inflorescence mostly of corymbiform panicles of several to many heads, but heads occasionally solitary on long peduncles. Heads radiate (in ours) or occasionally discoid; involucre hemispheric to bell-shaped, sometimes subtended by bracteoles; involucral bracts subequal to 2–6-seriate, ovate to oblanceolate or linear, often herbaceous; receptacle commonly conical, paleate, the palea concave, folded around the outer edge of the laterally compressed disk achenes. Flowers with corollas orange-yellow to white, these glabrous to more commonly pilose, especially on the tubes. Ray flowers (when present) pistillate or less commonly sterile; corolla tubes commonly shorter than limbs, the limb apically 3-dentate or entire. Disk flowers bisexual; corollas tubular, shortly 5-lobed; anthers black, the apical appendage cream-colored; style branch tips commonly pilose. Achenes strongly flattened, winged or less commonly wingless, glabrous or more commonly upwardly pubescent, sometimes tuberculate, oblong or obovoid; pappus mostly of 2 deciduous or persistent well-developed awns, 1 or both rarely obsolete.

A genus of about 300 species restricted to the New World, with most of the species in the tropics.

1. **Verbesina alata** L., Sp. Pl. 901. 1753. *Tepion alatum* (L.) Britton, Bull. New York Bot. Gard. **8**: 408. 1917.

Erect or weakly ascending perennial herb to 1 m tall, sparingly branched; stems winged, puberulent. Leaves sessile or more commonly abruptly narrowed into a subpetiolar base to 3 cm long; blades 3–14 × 1.5–7 cm, lanceolate to ovate, chartaceous, eglandular, the upper surface subglabrous to weakly strigose, the lower surface strigose to densely so, the margins crenate to irregularly dentate. Inflorescence loose and open, monocephalous or with 2–5 cymosely arranged heads; peduncles to 15 cm long, hispidulous. Heads radiate, ca. 60 or more flowered, globose, to 1.5 × 2

cm; involucre hemispheric, to 5 × 10 mm, weakly 2–3-seriate; involucral bracts subequal, ca. 4–5 × 1 mm, lanceolate to oblanceolate, strigose; receptacle conical, the palea ca. 4.5 × 1 mm, oblanceolate, herbaceous, puberulent at apex. Ray flowers pistillate, ca. 35; corollas orange, to 3 mm long. Disk flowers ca. 60; corollas funnel-shaped, orange, to 2 mm long. Achenes strongly winged, ovate in outline, 3.7–5.5 × 1.2–2 mm, sparsely pubescent over entire surface, the wings 1.2–1.6 mm wide; pappus of 2 unequal, persistent awns, the tips deciduous or uncinate, the smaller awn to 1.8 mm long, the longer to 3.7 mm long.

DISTRIBUTION: Occasional on open hillsides. American Hill [*Raunkiær s.n.*, 15 Mar 1906 (US)]. Also on St. Croix, St. Thomas, and Tortola; Greater Antilles and Lesser Antilles.

30. WEDELIA Jacq., nom. cons.

Rough-pubescent perennial herbs or shrubs; stems erect, less commonly weakly ascending, rarely prostrate. Leaves simple, opposite; blades ovate to less commonly linear, pubescent and also often glandular, 3-veined from base or from above base; petioles usually present, seldom winged. Inflorescence open, of a few to several terminal or axillary pedunculate heads, less commonly monocephalous; peduncles naked, pubescent. Heads radiate, rarely discoid; involucre 2–4-seriate; involucral bracts several, overlapping, graduated or subequal, the outer ones often foliar and longer than the inner ones; receptacle convex, paleate, the palea folded lengthwise. Ray flowers fertile, rarely sterile or absent; corollas commonly yellow, the limb apically 2- or 3-toothed. Disk flowers bisexual; corollas commonly yellow, shortly 5-lobed, the lobes often puberulent on both surfaces; anthers commonly black; style branches usually papillose. Achenes often oblong, but narrowed above into a neck, usually unwinged, pubescent, eglandular, the base with a large foot and often with 1 or 2 oil glands; pappus a fimbriate crown, commonly without awns.

A genus of about 60 species found from the southwestern United States, through Central America to tropical and subtropical South America and the West Indies, also Africa.

1. **Wedelia fruticosa** Jacq., Enum. Syst. Pl. 28. 1760. *Wedelia frutescens* Jacq., Select. Stirp. Amer. Hist. 217, t. 130. 1763, orth. var. of *W. fruticosa*.

Wedelia calycina Rich. *in* Pers., Syn. Pl. **2**: 490. 1807.
Wedelia jacquini Rich. *in* Pers., Syn. Pl. **2**: 490. 1807.
Wedelia parvifolia Rich. *in* Pers., Syn. Pl. **2**: 490. 1807.
Wedelia affinis DC., Prodr. **5**: 541. 1836.
Anomostephium buphthalmoides DC., Prodr. **5**: 560. 1836.
Wedelia buphthalmoides (DC.) Griseb., Fl. Brit. W. I. 372. 1861.

Perennial herb or shrub, 0.4–3 m tall, erect; stems few- to much-branched, hispid. Leaf blades 1.5–15 × 0.6–6.5 cm, elliptic to ovate, chartaceous, hirsute and stipitate-glandular on both surfaces, pinnately 3-veined from above base, the apex acute to attenuate, the base acute to rounded, attenuate onto petioles, the

margins subentire to coarsely dentate; petioles 2–20 mm long, those of a node inconspicuously connate. Inflorescence of 1–3 terminal heads on peduncles 0.5–4 cm long. Heads radiate, ca. 28–40-flowered; involucre bell-shaped, 7–15 × 6–12 mm, 2-seriate; involucral bracts green, lanceolate, the outer series wholly or partly herbaceous, hispid, the inner series scarious or herbaceously tipped, hispidulous to nearly glabrous; receptacle weakly convex, ca. 2 mm broad; palea lanceolate, 6 mm long. Ray flowers 8–12; corolla tubes 2 mm long, the limb 7–13 × 3–4 mm, the lower surface of the limb hispidulous on the 2 larger veins, apically bilobed to deeply so, the lobes 1–3 mm deep. Disk flowers 20–30; corollas 4.5–5.5 mm long. Achenes obconic to obovoid at maturity, 3–4 mm long; pappus an irregularly cut crown 0.5–1 mm tall.

DISTRIBUTION: Occasional in rocky hills, thickets, and disturbed areas. Hart Bay (A801), Dittlif Point (A3978). Also on Anegada, Little St. James Island, St. Croix, St. Thomas, Tortola, and Virgin Gorda; throughout much of the West Indies (but not Jamaica), southern Central America to northern South America.

NOTE: This species as interpreted here is polymorphic and includes both ovate large-leaved herbaceous and elliptic small-leaved shrubby plants, but it does not include *Wedelia cruciana* L. C. Rich. of St. Croix or the lanceolate-leaved species common in the West Indies. *Wedelia fruticosa* is most readily recognized by its apically bilobed ray corolla limbs and its elliptic to ovate leaves.

DOUBTFUL RECORDS: *Cosmos caudatus* Kunth was reported on St. John by Britton (1918), Britton and Wilson (1925), and Woodbury and Weaver (1987), but no voucher was located at NY, US, or VINPS. It is possible that this species occurs on St. John only in cultivation.

Eleutheranthera ruderalis (Sw.) Sch.Bip. was reported by Britton (1918) as occurring on St. John but is not described above because we have seen no voucher at NY or US from St. John. However, a sheet from the St. John River of Cuba (*Rugel 285,* NY), determined by Britton, possibly is the source of his report of the species on St. John Island. *Wedelia*

discoidea Less., typified by material from St. Thomas, is a synonym of *E. ruderalis.*

Xanthium strumarium L. was cited by Eggers (1879), using the name *Xanthium macrocarpum* DC., as occurring on all the Virgin Islands and also by Britton and Wilson (1925) as *Xanthium chinense* Mill. for St. John, but no voucher of this weedy species from St. John has been located.

Zinnia peruviana (L.) L. was reported by Britton (1918) as *Crassina multiflora* (L.) Kuntze and by Woodbury and Weaver (1987) as *Zinnia multiflora* L. for St. John and indeed may escape from cultivation. However, we saw no vouchers from the flora area and thus the naturalized occurrence on St. John of this species is questionable.

EXCLUDED SPECIES: *Galinsoga quadrangularis* Ruíz & Pav. [reported as *Galinsaga ciliata* (Raf.) Blake] cited from St. John by Woodbury and Weaver (1987), was based on a misidentification of a specimen of *Ageratum conyzoides* L.

13. Basellaceae (Basella Family)

Twining or scrambling plants with fleshy underground rhizomes. Leaves alternate, simple, entire, somewhat succulent; stipules wanting. Flowers bisexual or unisexual, actinomorphic, in axillary or terminal panicles, racemes, or spikes; calyx of 2 distinct sepals; corolla of 5 distinct or basally united petals; stamens 5, the filaments adnate to the base of petals, the anthers opening by longitudinal slits or terminal pores; ovary superior, of 3 united carpels, 1-locular, with solitary basal ovule, the styles 3 or rarely 1. Fruit a utricle, usually fleshy, and covered by a persistent corolla or accrescent calyx; seeds 1 per fruit.

A family with 4 genera and ca. 20 species, with tropical or subtropical distribution.

REFERENCE: Sperling, C. R. 1987. Systematics of the Basellaceae. Unpublished Ph.D. dissertation. Harvard University, Cambridge, Massachusetts.

1. ANREDERA Juss.

Twining or scandent herbaceous vines. Leaves alternate, slightly succulent and petiolate. Flowers bisexual or functionally unisexual, produced in axillary or terminal hanging racemes. Calyx much shorter than the corolla, the sepals free nearly to base, adnate to petals at base; petals connate at base; stamens with filaments nearly free to base or connate and adnate to the petals, the anthers oblong, opening by longitudinal slits; ovary 3-carpellate, the styles 3, free, as long as the ovary. Fruit a utricle.

A New World genus with ca. 12 species.

1. **Anredera vesicaria** (Lam.) C.F. Gaertn., Suppl. Carp. 176, t. 213. 1807. *Basella vesicaria* Lam., Encycl. 1: 382. 1785.   Fig. 50.

Boussingaultia leptostachys Moq. *in* A. DC., Prodr. 13(2): 229. 1849. *Anredera leptostachys* (Moq.) Steenis, Fl. Males. 5: 303. 1957.

Succulent, slender, glabrous, twining vine to 10 m long; young stems and petioles pinkish. Leaf blades 3–11 × 2–6.5 cm, ovate to lanceolate, succulent, chartaceous when dried, venation inconspicuous, the apex acute or acuminate, the base obtuse, rounded or cordate, attenuate, seldom oblique, the margins entire,

sometimes wavy; petioles 0.4–2 cm long. Racemes usually terminal; bracteoles boat-shaped, longer than the corolla. Flowers functionally unisexual; calyx 0.4 mm long, whitish, the sepals elliptic, boat-shaped; petals white, 1.5–2.2 mm long; stamens 3–4 mm long, lanceolate; ovary glabrous, shorter than the petals. Fruit obovoid, slightly triangular, 1.1–1.3 mm long, crowned by the enlarged fleshy beaked style (not seen in St. John populations).

DISTRIBUTION: In open, disturbed areas, such as roadsides or along secondary coastal scrubs. Cruz Bay (A2512), trail to Brown Bay (A1870), Mary Point (A4228). Also on St. Croix, St. Thomas, and Tortola; southern United States (Texas and Florida) to northern South America, including the West Indies.

NOTE: This treatment follows C. R. Sperling (1987), who considered *A. leptostachys* to be a synonym of *A. vesicaria.*

14. Bataceae (Saltwort Family)

A family with a single genus and 2 species; the following genus characterizes the family.

1. BATIS P. Browne

Monoecious or dioecious subshrubs or shrubs. Leaves opposite, simple, succulent, linear; stipules minute and deciduous. Flowers unisexual, solitary or clustered in axillary or terminal spikes; staminate flowers actinomorphic, covered with a saccate organ splitting into 2 lobes or along one side; tepals 4; stamens 4, alternating with the tepals, the anthers opening by longitudinal slits; pistillate flower without perianth; ovary superior, 2-carpellate, each carpel with 2 locules, the locule with a single ovule, the placentation basal, the stigmas 2. Fruit fleshy, with 4 pyrenes that are embedded in and dispersed with entire pistillate inflorescence.

Distributed along coastal areas; one species in tropical and subtropical America, the Galapagos Archipelago and Hawaii, the other in New Guinea and Australia.

1. **Batis maritima** L., Syst. Nat., ed. 10, 2: 1289. 1759.   Fig. 51.

Dioecious prostrate subshrub, to 30 cm tall; stems weak, 4-angular, rooting at nodes; bark peeling off in flakes. Leaf

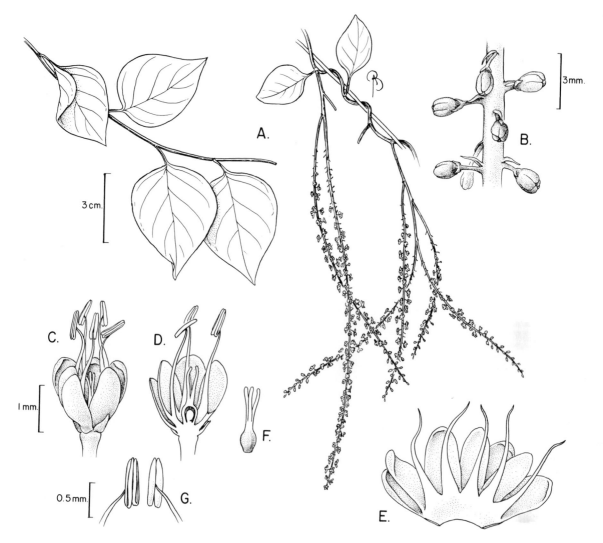

FIG. 50. *Anredera vesicaria.* **A.** Branch and inflorescence branches. **B.** Detail of inflorescence. **C.** Flower. **D.** L.s. flower. **E.** Corolla spread open showing adnate filaments. **F.** Pistil. **G.** Anthers.

1–2 × 0.2–0.6 cm, linear, oblanceolate, cylindrical or 3–4-angled in cross section, succulent, the apex acute and mucronate, the base clasping, with a basal, reflexed appendage, the margins entire. Flowers in axillary, short inflorescences; tepals spatulate, minute; anthers ovoid; ovary with 4 unequal locules. Fruit fleshy, with 1 seed per locule. Seeds 1.8–2.0 mm long, club-shaped to ellipsoid, minutely papillose with a costa along one edge.

DISTRIBUTION: Common along coastal areas in salt marshes. Newfound Bay (A4150, A4257). Also on Anegada, St. Croix, St. Thomas, and Tortola; throughout tropical America, the Galapagos Archipelago and Hawaii.

COMMON NAMES: salt plant, saltwort.

15. Bignoniaceae (Calabash Tree Family)

Trees, shrubs, or tendriled lianas. Leaves opposite, whorled or rarely alternate, compound, less often simple, without stipules. Flowers bisexual, zygomorphic, produced in terminal, axillary ramiflorous or cauliflorous panicles or racemes; calyx cup-shaped, usually truncate, 5-denticulate, bilabiate or spathelike; corolla of fused petals, of variable sizes, narrowly tubular, funnel-shaped, salverform, or bell-shaped, usually with 5 lobes, slightly bilabiate; stamens typically 4 (didynamous) but sometimes 2 or 5, a staminode sometimes present; ovary superior, bilocular or rarely unilocular, with axile placentation, usually with an annular disk at base, the style 1, the stigma bifid. Fruit usually a 2-valved capsule or indehiscent; seeds numerous, flattened, usually winged.

A family with ca. 100 genera and 800 species, predominantly neotropical in distribution.

REFERENCES: Gentry, A. H. 1980. Bignoniaceae. Part 1. Fl. Neotrop. Monogr. 25(1): 1–131. Gentry, A. H. 1982. Bignoniaceae. Flora de Venezuela 8(4): 7–433. Gentry, A. H. 1992. Bignoniaceae, Part 2. Fl. Neotrop. Monogr. 25(2): 1–369.

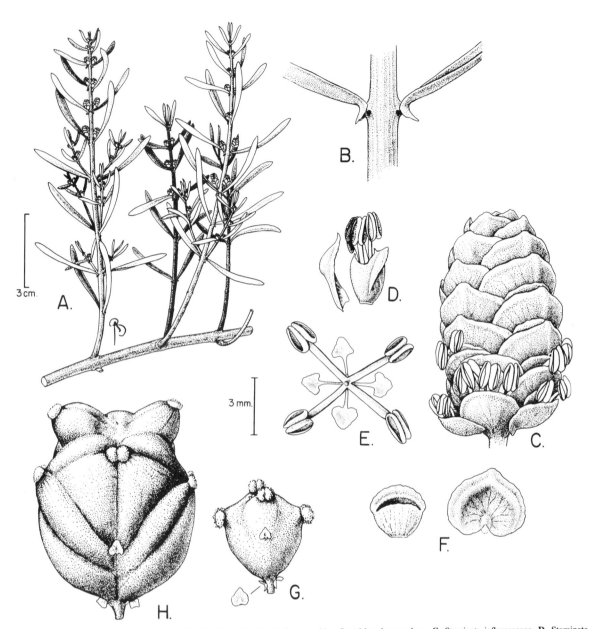

Fig. 51. *Batis maritima.* **A.** Flowering branches. **B.** Node showing leaf bases with reflexed basal appendage. **C.** Staminate inflorescence. **D.** Staminate flower with its enclosing saccate organ and subtending bract, lateral view. **E.** Staminate flower, top view. **F.** Dorsal view of saccate organ that encloses the staminate flower and dorsal view of bract of staminate inflorescence. **G.** Pistillate inflorescence at anthesis. **H.** Multiple fruit. (From Cronquist, 1981.)

Key to the Genera of Bignoniaceae

1. Tendriled lianas.
  2. Tendrils trifid, clawlike; corolla yellow; stems cylindrical when young .... 5. *Macfadyena*
  2. Tendrils simple, filiform, coiling; corolla pinkish to lavender or violet; stems 4-angled when young.
    3. Corolla funnel-shaped, 1.5–3 cm long, violet; mature stems 4-angled, sulcate; cross section of stem with 4 marginal, dark phloem arms; foliage drying reddish .... 2. *Arrabidaea*
    3. Corolla tubular to bell-shaped, 2.5–7.5 cm long, pinkish to lavender with yellowish throat; mature stems cylindrical; cross section of stem with 8 marginal, dark phloem arms; foliage drying green or brown .... 4. *Cydista*
1. Trees or shrubs.
  4. Leaves simple, alternate or clustered; fruits indehiscent, broadly ellipsoid to globose.

5. Leaves alternate to nearly opposite; fruits relatively fragile, easily crushed .... 1. *Amphitecna*
5. Leaves clustered on short lateral branches; fruits woody, hard-shelled .... 3. *Crescentia*
   4. Leaves compound (or unifoliolate), opposite; fruits elongate, dehiscent.
      6. Capsules erect angular, boat-shaped; corolla orange .... 6. *Spathodea*
      6. Capsules hanging, linear, flattened; corolla white, pink, or yellow.
         7. Leaves pinnately compound; leaflets serrate; corolla yellow, funnel-shaped........................ 8. *Tecoma*
         7. Leaves palmately compound or simple; leaflets entire; corolla white, pink, or yellow, salverform
      .... 7. *Tabebuia*

1. AMPHITECNA Miers

Small to medium-sized trees. Leaves simple and alternate. Flowers solitary or a few congested, terminal on branches or on short lateral twigs; calyx large, rupturing irregularly, usually deeply cleft in 2 elongate sepals; corolla zygomorphic, tubular, funnel-shaped, or bell-shaped, thick, inconspicuously transversely folded at throat; stamens 4, nearly exserted; staminode shorter than the stamens; ovary ovoid-ellipsoid, incompletely 2-locular. Fruit large, soft to hard shelled, with fleshy pulp inside; seeds wingless, numerous, flattened, >1.3 cm long.

A genus with 18 species, distributed from Mexico to coastal Venezuela, Ecuador, and the West Indies.

1. Amphitecna latifolia (Mill.) A. H. Gentry, Taxon **25**: 108. 1976. *Crescentia latifolia* Mill., Gard. Dict., ed. 8. 1768. *Enallagma latifolia* (Mill.) Small, Fl. Miami 171. 1913.   Fig. 52 E–G.

Crescentia cucurbitina L., Mant. Pl. **2**: 250. 1771.
Crescentia lethifera Tussac, Fl. Antill. **4**: 50. 1827.
Crescentia toxicaria Tussac, Fl. Antill. **4**: 50. 1827.
Crescentia cucurbitina var. *heterophylla* Kuntze, Revis. Gen. Pl. **2**: 479. 1891.

Tree to 10 m tall; bark light brown; young branches 3–4-angular, becoming cylindrical at age. Leaves alternate to nearly opposite, simple; blades 7–19 × 3–10 cm, obovate, oval, or elliptic, chartaceous to coriaceous, glabrous, the apex shortly acuminate or rarely obtuse or rounded, the base tapering into a short swollen petiole, the margins entire and revolute. Flowers in clusters of 2 or 3 or solitary at end of lateral branches. Calyx green, 2.5–3.5 cm long, glabrous, split halfway into 2 elongate, concave lobes; corolla pale yellow to greenish white, 3.5–6 cm long, funnel-shaped, the lobes deltoid, slightly reflexed; stamens 4, nearly exserted, of equal length, the staminode very short; ovary covered with minute scales. Fruit broadly ellipsoid to nearly globose, 6–9 cm long, relatively fragile, easily crushed, turning from green to brown. Seeds ca. 1.3 cm long, lenticular.

DISTRIBUTION: Along ravines and in coastal forests. Battery Gut (A4161), Fish Bay Gut (A2491). Also on St. Croix and St. Thomas; Florida, Mexico, Central America, West Indies, and northern South America.

COMMON NAME: jumbie calabash

2. ARRABIDAEA DC.

Lianas or rarely small trees or shrubs; stem 4-angular when young to cylindrical when mature, in cross section with 4 phloem arms. Leaves opposite, trifoliolate, more often with terminal leaflet replaced by a long filiform coiling tendril. Flowers produced in large axillary or terminal panicles; calyx cup-shaped, truncate, bilabiate or minutely 5-denticulate at apex; corolla reddish, pink, or violet, salverform, funnel- or bell-shaped, pubescent without; stamens 4, didynamous; ovary 2-locular, with a cup-shaped disk at base. Fruit a linear, flattened capsule, the valves parallel to the partitioning wall, midvein prominent; seeds 2-winged, the wing hyaline.

A neotropical genus with ca. 70 species.

1. Arrabidaea chica (Humb. & Bonpl.) Verl., Rev. Hort. **40**: 154. 1868. *Bignonia chica* Humb. & Bonpl., Pl. Aequinoct. **1**: 107. 1807.   Fig. 53A–E.

Adenocalymna portoricense A. Stahl, Estud. Fl. Puerto Rico **6**: 186. 1888.

Liana to 15 m long; stem 4-angular, 4-sulcate, reaching 6 cm diam., cross section with 4 marginal dark phloem arms. Leaves drying reddish, 2–3-foliolate with terminal leaflet replaced by a long, unbranched, filiform coiling tendril, the tendril early deciduous; lateral leaflets 3.4–11 × 1.5–5 cm, ovate, chartaceous, glabrous or with scattered hairs, midvein prominent on lower surface, the apex acuminate or acute, the base truncate, rounded to nearly cordate, the margins entire; petioles and petiolules cylindrical, drying blackish, the petiolules 0.3–4.5 cm long, the petiole 1.5–7 cm long. Flowers fragrant, produced in terminal panicles. Calyx 3–5 mm long, puberulous, truncate, or minutely denticulate at apex; corolla zygomorphic, violet, 1.5–3 cm long, funnel-shaped, the lobes rounded; stamens and staminodes included, inserted near the base of corolla; ovary cylindrical, with minute scales. Fruit 12–233 cm long, linear to narrowly elliptic, smooth, glabrous, semiwoody, acute or obtuse at both ends. Seeds numerous, 7–9 mm long.

DISTRIBUTION: Along ravines in gallery forests. Maho Bay Gut (A2105). Throughout the neotropics.

3. CRESCENTIA L.

Small to medium-sized trees. Leaves simple or trifoliolate, congested on short lateral twigs; petiolate. Flowers solitary or in pairs, produced at the trunk or on short lateral branches; calyx large, split into 2, elongate, concave sepals; corolla zygomorphic and bell-shaped, thick, transversely folded near the bottom at one side; stamens 4, nearly exserted; ovary ovoid, 1-locular, with numerous multiseriate ovules. Fruit large, hard-shelled, smooth, with fleshy pulp inside; seeds numerous, flattened, nearly cordate, to 1 cm.

A genus with 6 species, distributed from Mexico to Amazonian Brazil and the West Indies.

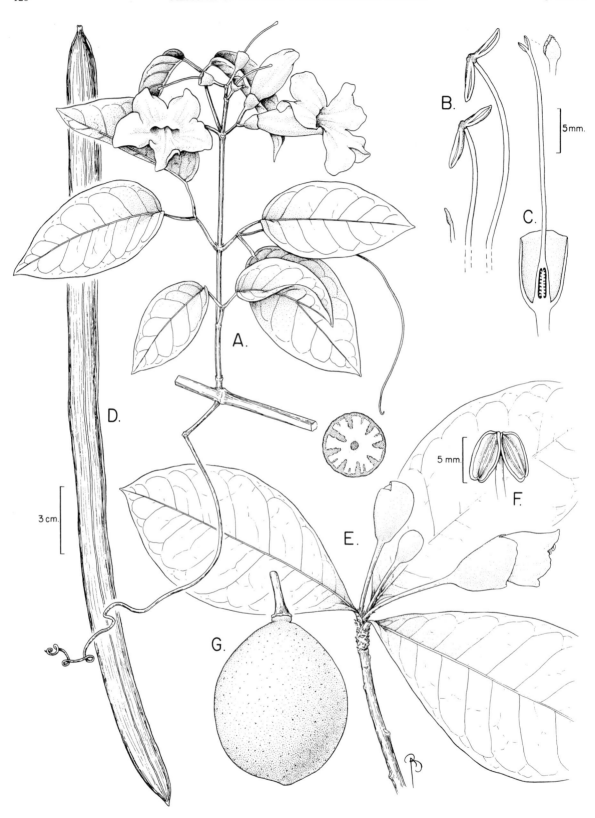

FIG. 52. A–D. *Cydista aequinoctialis.* **A.** Flowering branch and c.s. mature stem. **B.** Staminode and upper portion of stamens showing anthers. **C.** L.s. pistil and detail of style-branch. **D.** Fruit. **E–G.** *Amphitecna latifolia.* **E.** Fertile branch. **F.** Anthers. **G.** Fruit.

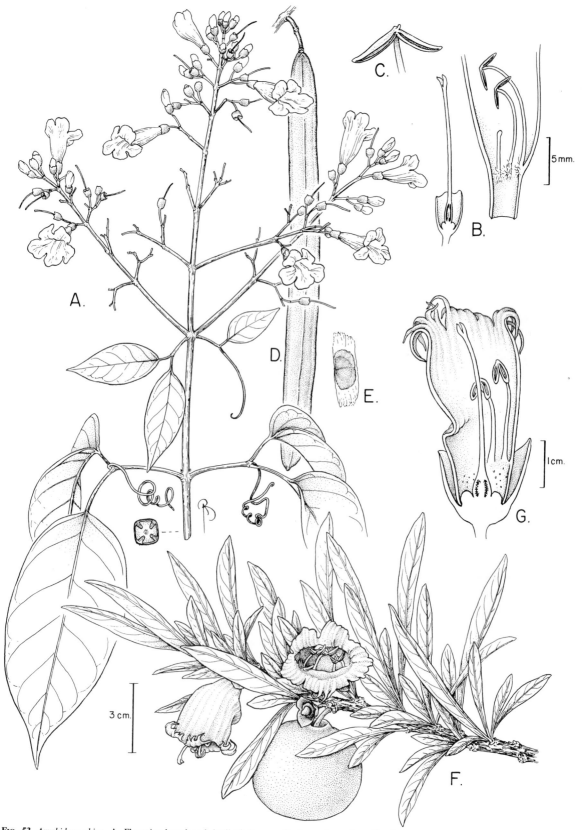

Fig. 53. *Arrabidaea chica.* **A.** Flowering branch and detail of stem c.s. **B.** L.s. calyx and corolla. **C.** Anther. **D.** Fruit. **E.** Seed. **F, G.** *Crescentia linearifolia.* **F.** Fertile branch. **G.** L.s. flower.

Key to the Species of *Crescentia*

1. Leaves oblanceolate or obovate, >2.5 cm wide; tertiary venation coarsely reticulate .... 1. *C. cujete*
1. Leaves linear or very narrowly oblanceolate, 1–1.8 cm wide; tertiary venation finely reticulate ..... 2. *C. linearifolia*

1. Crescentia cujete L., Sp. Pl. 626. 1753.

Crescentia angustifolia Willd. *ex* Seem., Trans. Linn. Soc. London **23**: 20. 1862.

Small tree to 10 m tall, with hanging branches; lateral branches short, verrucose. Leaves fascicled on lateral branches, simple; blades 4–20 × 2.5–7.5 cm, oblanceolate or obovate, chartaceous, glabrous except for a few hairs along midvein on lower surface, both surfaces with many dotlike scales, tertiary venation coarsely reticulate, the apex obtuse, acute, or shortly acuminate, the base tapering into a more or less elongate petiole, the margins entire or slightly wavy and revolute; petioles swollen at base. Flowers solitary or paired at end of the lateral branches. Calyx green, 2.5–3 cm long, glabrous, tubular, thick, deeply parted into 2 obovate, concave lobes; corolla yellowish to greenish, reddish-tinged along nerves, 4–7.5 cm long, bell-shaped with a transverse folding near the base, the lobes shortly deltoid, with caudate apex; stamens nearly exserted; ovary conical, 5–7 mm long, with an annular disk at base. Fruit globose, depressed-globose or ellipsoid, 10–20 cm long (often much larger elsewhere), turning from green to brown, usually hanging from trunks or branches. Seeds ca. 1 cm long, whitish.

DISTRIBUTION: Common throughout the island. Susannaberg (A2087). Also on Jost van Dyke, St. Croix, St. Thomas, Tortola, and Virgin Gorda; widely cultivated in tropical America, probably native to Belize (Gentry, 1982).

COMMON NAMES: calabash tree, gobi, gobi tree.

2. Crescentia linearifolia Miers, Trans. Linn. Soc. London **26**: 172. 1868.　　Fig. 53F–G.

Crescentia microcarpa Bello, Anales Soc. Esp. Hist. Nat. **10**: 294. 1881.

Small tree to 7 m tall, with hanging branches; lateral branches short and verrucose. Leaves fascicled on lateral branches, simple; blades 0.5–8.5 × 0.1–1.8 cm, linear, or narrowly oblanceolate, chartaceous, glabrous except for dotlike scales on lower surface, tertiary venation finely reticulate, the apex obtuse, acute, or acuminate, the base tapering into a swollen end, without a discrete petiole, the margins entire or slightly wavy and revolute. Flowers solitary or paired at end of the lateral branches. Calyx green, 2.5–3 cm long, glabrous, tubular, thick, deeply parted into 2, obovate, concave lobes; corolla greenish, 4–7.5 cm long, bell-shaped, with a transverse folding near the base, the lobes shortly deltoid, with caudate apex; stamens nearly exserted; ovary conical, 5–7 mm long, with an annular disk at base. Fruit globose, depressed-globose or ellipsoid, 5–10 cm long, turning from green to brown, usually hanging from trunks or branches. Seeds <1 cm long, whitish.

DISTRIBUTION: Common throughout the island, especially in dry areas. Holter Road (A2342) Lameshur (B601). Also on St. Thomas; Puerto Rico.

COMMON NAMES: black calabash, calabash tree, jumbie calabash, minie gobi.

4. CYDISTA Miers

Tendriled lianas; branches cylindrical when mature or 4-angular when young; cross section of stem with 8 dark phloem arms. Leaves opposite, simple or 2-foliolate, with a filiform coiling tendril replacing a third leaflet. Flowers in axillary or terminal racemes or panicles; calyx cup-shaped, truncate or slightly bilobed at apex; corolla white, pink, or purple, tubular to funnel-shaped; stamens 4, didynamous; ovary cylindrical, 2-locular, the ovules in 2 rows. Fruit a linear flattened capsule, with valves parallel to the partitioning wall, the midvein inconspicuous; seeds flattened, 2-winged, brown.

A neotropical genus with 6 species.

1. Cydista aequinoctialis (L.) Miers, Proc. Roy. Hort. Soc. London **3**: 191. 1863. *Bignonia aequinoctialis* L., Sp. Pl. 623. 1753.　　Fig. 52A–D.

Bignonia hostmannii E. Mey., Nova Acta Phys.-Med. Acad. Caes. Leop.-Carol. Nat. Cur. **12**: 779. 1825.
Cydista amoena Miers, Proc. Roy. Hort. Soc. London **3**: 191. 1863.

Liana to 20 m long, many-branched from base; bark grayish and fissured; young stems 4-angled, mature stems cylindrical. Leaves 2-foliolate, the unbranched tendril deciduous, 4–20 cm long; leaflets drying brown or greenish, 5.5–16 × 2.5–9 cm, ovate, oblong or elliptic, chartaceous, glabrous, lower surface sparsely covered with dotlike scales, the apex obtuse, acute, or acuminate, the base obtuse to nearly cordate, the margins entire; petioles and petiolules 4-angular, the petiolules 0.8–4 cm long, pubescent or glabrous, the petiole 0.9–4.5 cm long. Flowers fragrant, in axillary or terminal panicles. Calyx cup-shaped with truncate or denticulate apex, 4–10 mm long, pubescent or covered with dotlike scales, yellowish; corolla zygomorphic, tubular to bell-shaped, 2.5–7.5 cm long, the tube slightly flattened, narrower at base, white without, yellowish within, the lobes rounded, lavender, the upper lobes nearly reflexed, the lower expanded, the throat yellowish, with vertical, dark violet lines; stamens included; ovary cylindrical, 2–3 mm long, without a disk at base, the stigma bilobed. Capsule 21–43 × 1.5–2.4 cm, with slightly prominent margins. Seeds 1.5–2 cm long, brown, with a membranous wing.

DISTRIBUTION: Along coastal plains, usually near mangrove swamps. Fish Bay (A2810). Also on St. Croix and St. Thomas; throughout tropical America.

COMMON NAME: guard wiss.

5. MACFADYENA A. DC.

Tendriled lianas, often with aerial roots at nodes; stem cylindrical, in cross section with many phloem arms, radially distributed, the arms deeply (reaching the pith) to shallowly embedded. Leaves opposite, 2-foliolate, with a 3-fid, clawlike tendril on distal position.

Flowers in axillary cymes or panicles; calyx bell-shaped; corolla yellow, tubular to bell-shaped; stamens 4; ovary linear-cylindrical, 2-locular, with an annular disk at base, the locules with 2–4 ovules. Fruit a narrow, elongate, and flattened capsule, with valves parallel to the partitioning wall; seeds numerous, light, 2-winged.

A neotropical genus with about 4 species.

1. Macfadyena unguis-cati (L.) A. H. Gentry, Brittonia 25: 236. 1973. *Bignonia unguis-cati* L., Sp. Pl. 623. 1753. *Bignonia unguis* L. *emend.* DC. *in* A. DC., Prodr. 9: 146. 1845 (new name for *B. unguis-cati*). *Batocydia unguis* (L.) DC. *in* A. DC., Prodr. 9: 146. 1845. *Doxantha unguis* (L.) Miers, Proc. Roy. Hort. Soc. London 3: 190. 1863. *Doxantha unguis-cati* (L.) Miers *emend.* Rehder, Mitt. Deutsch. Dendrol. Ges. 1913: 262. 1913. Fig. 54A–C.

Liana 10–15 m long; stem cylindrical, with numerous lenticels, reaching 6 cm diam. Leaflets 6–16 × 1.2–7 cm, elliptic, oblong, or obovate, chartaceous, glabrous, or with a few dotlike scales, the apex acute or acuminate, the base obtuse, cuneate, or rounded, usually oblique, the margins wavy; tendrils deciduous, much shorter than the leaflets; petioles and petiolules glabrous, the petiolule 0.5–2.5 cm long, the petiole 1–4.5 cm long; pseudostipules in leaf axils. Flowers solitary or in pairs in leaf axils. Calyx bell-shaped, 12–16 mm long, green, with 5 unequal lobes; corolla yellow, zygomorphic, funnel-shaped, 4–8 cm long, the tube slightly flattened, the lobes rounded, 2–4 cm long, the two upper lobes slightly reflexed; stamens 4, didynamous, included, inserted at base of corolla, the staminode much shorter than the stamens; ovary 6–7 mm long, covered with dotlike scales. Capsule slightly woody, brown, 25–95 cm long, covered with dotlike scales, hanging from the branches. Seeds 1–1.8 cm long, with membranous wing.

DISTRIBUTION: Abundant on trees, in dry or moist forests or in open, disturbed areas such as roadsides. Cruz Bay Quarter along Center Line Road (A2110), Susannaberg (A3958). Also on St. Croix, St. Thomas, Tortola, and Virgin Gorda; throughout tropical and subtropical America.

COMMON NAMES: cat claw, cat paw, monkey earring, wist.

6. SPATHODEA P. Beauv.

A monospecific genus characterized by the following species.

1. Spathodea campanulata P. Beauv., Fl. Oware 1: 47. 1805. Fig. 54D–H.

Large tree to 30 m tall, sometimes reaching 1 m diam; bark smooth and grayish; wood white and soft; stem cylindrical, with numerous lenticels. Leaves opposite, pinnately compound, with 9–15 leaflets; leaflets 5–12 × 2.5–5 cm, elliptic or oblong, chartaceous, glabrous except for a few hairs along the prominent veins on lower surface, the apex acute or acuminate, the base obtuse, acute, or rounded and oblique, the margins slightly wavy and revolute; petiolules short, the petiole cylindrical, the rachis slightly flattened along upper surface. Flowers in terminal racemes. Calyx nearly fusiform, subwoody, curved toward one side, 5.5–8 cm long, densely covered with rusty-brown, soft hairs; corolla orange, yellow within toward base and along margins, 6–9 cm long, zygomorphic, bell-shaped, the lobes rounded or obtuse, 2.5–3.5 cm long; stamens nearly exserted, inserted at base of corolla; ovary oblong, 2-locular, with many ovules. Fruit an erect, oblong-ellipsoid, woody capsule, opening along one side, 17–25 cm long, the opened capsule boat-shaped. Seeds numerous, thin, 2-winged, the wings hyaline.

DISTRIBUTION: Introduced as an ornamental, only a few individuals known on St. John. Maho Bay Quarter along Center Line Road (A2511). Also on St. Thomas and Tortola; native to tropical Africa but widely cultivated and naturalized throughout the tropics.

COMMON NAME: African tulip tree.

7. TABEBUIA DC.

Shrubs or trees. Leaves opposite, simple or palmately compound, with (1–)3–7 leaflets. Flowers in terminal panicles or racemes; calyx cup-shaped, bell-shaped, or tubular, truncate, 2–5-lobed at apex, corolla of various colors, zygomorphic, salverform to funnel-shaped; stamens 4, didynamous; ovary 2-locular, oblong, the locule with many ovules. Fruit a nearly cylindrical capsule, smooth or slightly warty, the valves perpendicular to the partitioning wall; seeds numerous, light, 2-winged or rounded without wings.

A neotropical genus with ca. 100 species.

Key to the Species of *Tabebuia*

1. Corolla yellow; leaves with (3–)5–7 leaflets; petiolules 1–4.5 cm long .... 1. *T. aurea*
1. Corolla pink to white; leaves with 1–5 leaflets; petiolules 0.2–2 cm long .... 2. *T. heterophylla*

1. Tabebuia aurea (Silva Manso) S. Moore, Trans. Linn. Soc. London, Bot. 4: 423. 1895. *Bignonia aurea* Silva Manso, Enum. Subst. Braz. 40. 1836.

Tecoma caraiba Mart., Flora 24 (Beibl. 2): 14. 1841. *Tabebuia caraiba* (Mart.) Bureau, Vidensk. Meddel. Dansk Naturhist. Foren. Kjøbenhavn 1893: 113. 1893.

Small tree to 10 m tall; branches nearly cylindrical or angular, densely covered with dotlike scales. Leaves digitate-compound, with (3–)5–7, long-petioluled leaflets; leaflets 5.5–14 × 1.5–2.7 cm, oblong, narrowly elliptic or oblanceolate, chartaceous to coriaceous, densely covered with dotlike scales on both surfaces, midvein prominent on lower side, the apex obtuse or acute, the base obtuse or rounded, sometimes oblique, the margins entire and strongly revolute; petioles and petiolules furrowed on upper

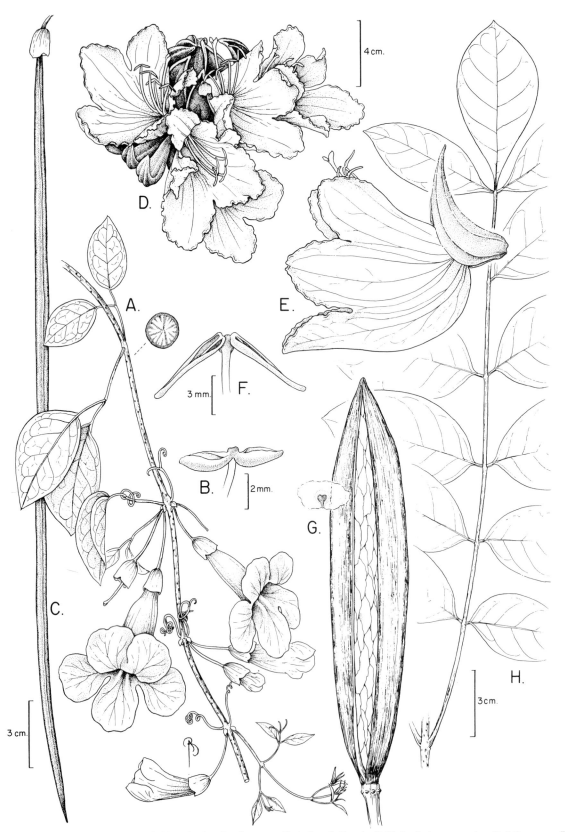

Fig. 54. **A–C.** *Macfadyena unguis-cati.* **A.** Flowering branch and c.s. stem. **B.** Anther. **C.** Capsule. **D–H.** *Spathodea campanulata.* **D.** Inflorescence. **E.** Flower. **F.** Anther. **G.** Opening capsule and seed. **H.** Leafy branch.

surface; the petiolules 1–4.5 cm long, swollen at apex, the petiole 4.5–9 cm long, swollen at base. Flowers on short terminal racemes. Calyx bell-shaped with 5 irregular lobes, ca. 1.5 cm long, densely covered with dotlike scales; corolla yellow, 5–6 cm long, funnel-shaped, the lobes ca. 1.5 cm long, rounded; stamens included; ovary oblong, lepidote, the style solitary, the stigma peltate. Fruit 10–12 × 1–1.2 cm, grayish, oblong, beaked at apex. Seeds oblong, 2 cm long, with membranous hyaline whitish wings.

DISTRIBUTION: Uncommon, introduced species, planted in gardens and along roadsides. Coral Bay (A2027, V3120). Native to Brazil but widely planted throughout tropical America.

COMMON NAME: yellow oak.

2. Tabebuia heterophylla (DC.) Britton, Ann. Missouri Bot. Gard. **2**: 48. 1915. *Raputia heterophylla* DC., Mém. Mus. Hist. Nat. **9**: 153. 1822. Fig. 55F–H.

Tabebuia lucida Britton, Ann. Missouri Bot. Gard. **2**: 48. 1915.

Tree 3–15 m tall. Leaves digitate-compound with 1–5 leaflets; leaflets 3.5–9 × 1.2–3.5 cm, elliptic or oblanceolate, coriaceous,

densely covered with dotlike scales on both surfaces, the apex shortly acuminate, obtuse or rounded, the base cuneate, obtuse or nearly rounded, the margins wavy; petioles and petiolules furrowed along upper surface, densely covered with dotlike scales, the petiolules 0.2–2 cm long, not swollen, the petioles 1.5–5 cm long, swollen at base. Flowers in terminal short racemes. Calyx bell-shaped, 0.8–1 cm long, with 5 obtuse lobes at apex; corolla pink or nearly white, funnel-shaped, 3.5–5 cm long, the tube paler than the lobes; stamens included; ovary oblong, slightly flattened, densely covered with scales, the style longer than the ovary. Capsules semiwoody, 6–11 × 0.6–1 cm long, brown, covered with scales, nearly cylindrical, with tapering apex, and persistent calyx. Seeds thin, flattened, oblong-elliptic, 7–9 × 4–5 mm wide, the wings 6–7 mm long, white and hyaline.

DISTRIBUTION: Common throughout the island, from coastal scrub to moist forests. East End (GTP29320), Lameshur (B638). Also on Anegada, Jost van Dyke, St. Croix, St. Thomas, Tortola, and Virgin Gorda; Hispaniola, Puerto Rico, and the Lesser Antilles, naturalized in Florida and the Bahamas.

COMMON NAMES: black cedar, pink cedar, pink manjack tooshee, white cedar.

NOTE: This is the official tree of the British Virgin Islands.

8. TECOMA Juss.

Shrubs or trees. Leaves opposite, pinnately compound, or rarely simple. Flowers in terminal racemes or panicles; calyx tubular or bell-shaped, 5-toothed at apex; corolla yellow or orange, tubular to funnel-shaped; stamens 4, exserted or included; ovary narrowly cylindrical, with a cup-shaped disk at base. Fruit a linear, flattened capsule with dehiscence perpendicular to the partitioning wall; seeds numerous, thin, 2-winged, the wings hyaline.

A genus of 14 species, 2 in Africa and the remaining 12 distributed throughout tropical America, especially in the Andean region.

1. Tecoma stans (L.) A. Juss. *ex* Kunth *in* Humb., Bonpl. & Kunth, Nov. Gen. Sp. **3**: 144. 1819. *Bignonia stans* L., Sp. Pl. ed. 2, **2**: 871. 1763. Fig. 55A–E.

Shrub or small tree to 7 m tall. Leaves 3–9-foliolate; leaflets 2.5–15 × 0.8–6 cm, gradually larger toward the distal end; lanceolate, membranous to chartaceous, glabrous except for a few hairs on lower surface along veins and vein angles, the apex long-acuminate, the base cuneate, the margins serrate; petioles 4.5–7 cm long, furrowed at upper surface. Flowers in terminal racemes; calyx bell-shaped, green, 3–7 mm long; corolla yellow, 4–5 cm long, zygomorphic, funnel-shaped, the tube very narrow at base,

gradually opening toward the throat, with 2 longitudinal folds along the lower side, the lobes rounded, one of them projecting to the front. Capsule pendent, linear, 7–21 × 5–8 cm long, flattened (nearly cylindrical when fresh), smooth, tapering toward both ends. Seeds thin, elliptic to heart-shaped 5–7 × 4–5 mm long, the wings whitish, oblong, 6–7 mm long.

DISTRIBUTION: Commonly cultivated in St. John, becoming naturalized. Johnson Bay (A2125). Also on Jost van Dyke, St. Croix, St. Thomas, Tortola, and Virgin Gorda; native to the neotropics but widespread throughout the tropics through cultivation.

COMMON NAMES: catapult tree, ginger thomas, yellow cedar.

NOTE: This is the official flower of the U.S. Virgin Islands.

16. Bombacaceae (Balsa Wood Family)

Shrubs to trees; bark sometimes with deciduous prickles. Leaves alternate, simple or palmately compound, with deciduous stipules. Flowers usually large, actinomorphic, bisexual, produced in axillary cymes, clustered or solitary; bracteoles usually forming an epicalyx; calyx, bell- or cup-shaped, 5-lobed or parted to base; corolla of 5 distinct petals; stamens 5 to numerous, the filaments fused into a short column, adnate to base of corolla, the anthers opening by longitudinal slits; ovary superior or partially inferior with 2–5(–8) carpels, each carpel with 2 to many ovules, the placentation axile, the style simple, the stigma capitate. Fruit indehiscent or a 5-valved capsule; seeds usually embedded in hairy tissue.

A family of 20–30 genera and ca. 200 species, with pantropical distribution.

Key to the Genera of Bombacaceae

1. Leaves palmately compound; fruit a fusiform capsule, 8–10 cm long; bark with deciduous thorns .... 1. *Ceiba*
1. Leaves simple; fruit nearly globose, indehiscent, <2 cm long; bark smooth .... 2. *Quararibea*

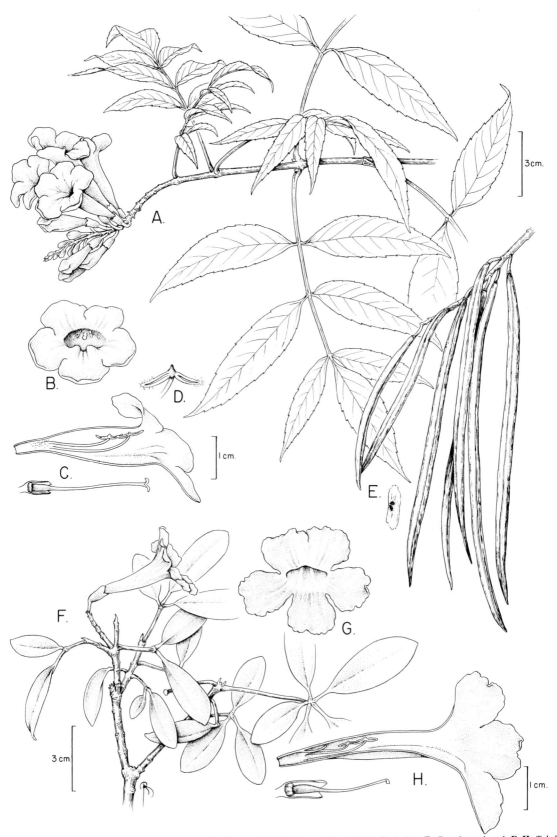

Fig. 55. A–E. *Tecoma stans*. A. Flowering branch. B. Corolla, top view. C. L.s. corolla and pistil. D. Anther. E. Capsules and seed. F–H. *Tabebuia heterophylla*. F. Flowering branch. G. Corolla, top view. H. L.s. corolla and pistil.

1. CEIBA Mill.

Large trees, with prominent buttress roots; bark with deciduous thorns. Leaves palmately compound with 5–9 leaflets, long petio-late. Flowers clustered at branch ends; bracteoles not forming an epicalyx; calyx bell-shaped, with short, valvate sepals; petals free to base, thickened; stamens 5; ovary nearly inferior, with 5 multiovular locules, the stigma capitate. Fruit a woody capsule, opening by 5 valves; seeds black, embedded in woolly fibers.

About 10 species, apparently native to the New World but widespread throughout the tropics.

1. **Ceiba pentandra** (L.) Gaertn., Fruct. Sem. Pl. **2**: 244. 1791. *Bombax pentandrum* L., Sp. Pl. 511. 1753. Fig. 56A–F.

Ceiba anfractuosa (DC.) M. Gómez *in* M. Gómez & Roíg, Fl. Cuba 66. 1914.

Eriodendron anfractuosum DC. var. *caribaeum* DC., Prodr. **1**: 479. 1824. *Ceiba pentandra* (L.) Gaertn. var. *caribaeum* (DC.) Bakh., Bull. Jard. Bot. Bui-tenzorg, ser. 3, **6**: 196. 1924.

Tree to 25 m tall, usually >1 m diam.; bark grayish or cream-colored. Leaflets 4–18 × 1.5–5 cm, oblong, oblanceolate, or elliptic, chartaceous to subcoriaceous, glabrous, the apex acute to long-acuminate, the base attenuate to cuneate, the margins entire and slightly wavy, the petiolules 0.4–1 cm long; petioles 7–19 cm long, cylindrical. Flowers in axillary clusters at end of branches. Calyx bell-shaped, 1.5–2 cm long, thickened, glabrous without, pubescent within, the lobes triangular; petals white within, 2.5–3 cm long, spatulate, densely covered with appressed, rusty-brown hairs without. Capsule pendulous, fusiform, 8–10 cm long, covered with rusty-brown scales. Seeds 3–5 mm long, black, smooth; woolly fibers light brown.

DISTRIBUTION: In moist forests or on coastal plains with deep soils. Cinnamon Bay Campground (A4231). Also on Jost van Dyke, St. Croix, St. Thomas, Tortola, and Virgin Gorda; native to tropical America, natural-ized in the Old World tropics.

COMMON NAMES: jumbie tree, kapok tree, silk cotton tree.

2. QUARARIBEA Aubl.

Shrubs to small trees with whorled branching. Leaves simple; stipules deciduous. Flowers solitary or a few clustered in leaf axils; calyx hardened, conical or top-shaped, irregularly lobed, accrescent; petals narrow, twice as long as the calyx; stamens numerous, the filaments fused into an elongate tube; ovary slightly inferior, 2–3-locular, each locule with 2 ovules, the style projecting beyond the stamens, the stigma disk-shaped. Fruit slightly fleshy, fibrous, indehiscent; seeds 1 or 2, slightly angular.

A genus of 30–50 species, distributed from southern Mexico to northern South America, including the West Indies.

1. **Quararibea turbinata** (Sw.) Poir. *in* Lam., Encycl. Suppl. **4**: 636. 1816. *Myrodia turbinata* Sw., Prodr. 102. 1788. Fig. 56G–K.

Shrub or tree 10–12 m tall, with 1 main trunk and many lateral, horizontal, whorled branches at different heights; twigs brown, cylindrical, usually with numerous lenticels, the young parts sparsely covered with stellate hairs. Leaf blades 8–26 × 4.5–12.5 cm, elliptic, oval, or obovate, chartaceous to subcoriaceous, glabrous, with prominent vein network on lower surface, often with small depressions at secondary vein angles, the apex short- to long-acuminate, the base obtuse, rounded or cuneate, often unequal, the margins entire and slightly wavy; petioles 1–1.5 cm long; stipules awl-shaped, ca. 5 mm long. Calyx conical (cup-shaped in fruits), 0.8–1.5 cm long, with stellate hairs; petals cream, 1.8–2 cm long, spatulate, densely covered with stellate hairs; staminal tube 1.3–1.5 cm long; style angular and densely stellate. Fruit globose, depressed-globose or ellipsoid, 1.5–2 cm long, apiculate at apex, green, with prominent, parallel fibers throughout its length. Seeds whitish, obtusely angular.

DISTRIBUTION: Common as an understory shrub or small tree in moist forests. East End (A2115); Cinnamon Bay (A2091). Also on St. Croix; from Hispaniola southeast to the Lesser Antilles.

COMMON NAMES: garrot, swizzle stick tree.

17. Boraginaceae (Borage Family)

Herbs, shrubs, trees, or vines. Leaves alternate or rarely whorled, simple, without stipules. Flowers bisexual or rarely unisexual, actinomorphic or slightly zygomorphic, produced in cymes, panicles, or spikes, often coiled at distal end (scorpioid); calyx of 5 distinct sepals; corolla tubular or trumpet- or funnel-shaped; stamens 5, inserted on the tube, the anthers opening by longitudinal slits; ovary superior, 2-carpellate, usually with an annular disk at base, each carpel divided into 2 chambers, each chamber with 1 ovule, the placentation axile, the style 1 or 2, usually branched into 2 stigmas, or sometimes arising from the base of the ovary. Fruit fleshy drupes or capsules; seeds (1–)4.

A family with approximately 100 genera and 2000 species with cosmopolitan distribution.

Key to the Genera of Boraginaceae

1. Flowers solitary or paired in leaf axils; branches spiny .... 5. *Rochefortia*
1. Flowers in inflorescences; branches without spines.
    2. Inflorescences an unbranched or 2-branched spikes, coiled (scorpioid) at distal end; herbs or low sub-shrubs.
        3. Leaves somewhat succulent; fruit fleshy, of 2 nutlets, initially hollowed at base .... 1. *Argusia*
        3. Leaves not succulent; fruit a dry capsule, not hollowed at base .... 4. *Heliotropium*

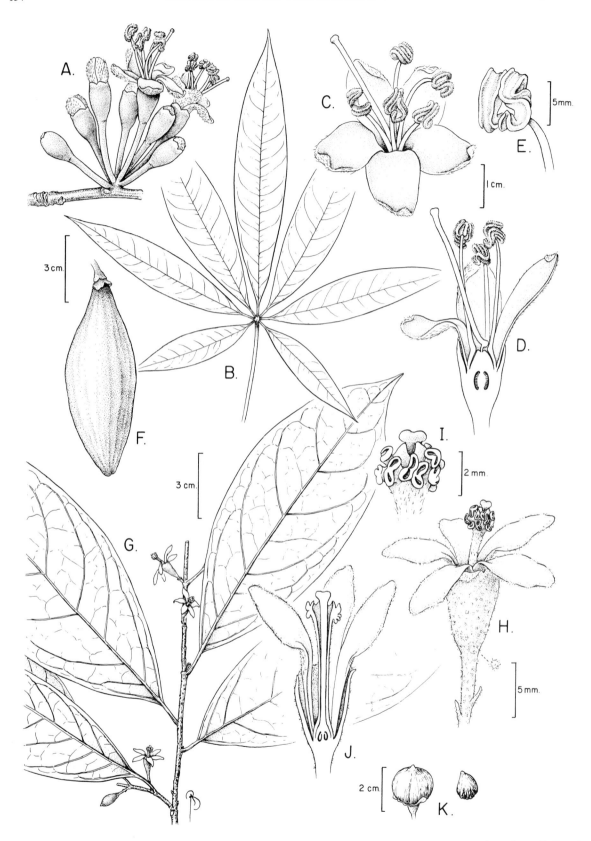

FIG. 56. A–F. *Ceiba pentandra.* **A.** Inflorescence. **B.** Leaf. **C.** Flower. **D.** L.s. flower. **E.** Anther. **F.** Capsule. **G–K.** *Quararibea turbinata.* **G.** Flowering branch. **H.** Flower and detail of scale. **I.** Summit of staminal tube surrounding the style. **J.** L.s. flower. **K.** Fruit and seed.

2. Inflorescences many-branched (panicles, cymes, or heads), coiled (scorpioid) or not at distal end; shrubs, trees, or vines.
    4. Style bifid, with 4 stigmatic branches .... 3. *Cordia*
    4. Style simple or crowned by 2 stigmatic branches.
       5. Twining vines, lianas, scandent or erect shrubs; style unbranched; stamens and style not projecting beyond the corolla; corolla lobes reflexed or spreading; fruit white or pale orange ... 6. *Tournefortia*
       5. Trees; style with 2-lobed stigma; stamens and style projecting beyond the corolla; corolla lobes spreading but not reflexed; fruit red or red-orange .... 2. *Bourreria*

1. ARGUSIA Boehm.

Shrubs or small trees, covered with appressed whitish hairs. Leaves alternate, somewhat fleshy and crowded. Flowers 5-parted, in scorpioid cymes or spikes; calyx with sepals free to base; corolla trumpet-shaped with short tube and reflexed lobes; anthers sessile or nearly so; ovary of 2 carpels, each carpel with 2 chambers and 2 ovules, the stigma sessile. Fruit fleshy, ovoid, hollowed at the base, separating into 2 nutlets.

A genus with 3 species, 2 from the Old World and 1 from the neotropics.

1. **Argusia gnaphalodes** (L.) Heine, Fl. Nouv. Caled. 7: 108. 1976. *Heliotropium gnaphalodes* L., Syst. Nat., ed. 10, 2: 913. 1759. *Tournefortia gnaphalodes* (L.) Roem. & Schult., Syst. Veg. 4: 538. 1819. *Messerschmidia gnaphalodes* (L.) I.M. Johnst., J. Arnold Arbor. 16: 165. 1935.
Fig. 57E–I.

Erect shrub to 1.5 m tall, many-branched; all parts densely covered with white, silky hairs; older stems glabrous, brown, with numerous leaf scars. Leaves ascending when young; blades 2–10 × 0.5–0.9 cm, linear-spatulate, densely covered with silky hairs, somewhat succulent, the apex obtuse or rounded, the base tapering, the margins entire. Sepals obovate, 3–4 mm long; corolla white above, 5–6 mm long, glabrous within; anthers inserted at upper portion of tube. Fruit fleshy becoming dry, ovoid, or conical, 5–7 mm long, splitting into 2 bony halves.

DISTRIBUTION: Common along sandy coasts. Johns Folly Bay (A4062), Europa Bay (A747), Lameshur (B652). Also on Anegada, St. Croix, St. Thomas, Tortola, and Virgin Gorda; throughout the West Indies.

COMMON NAMES: bay lavender, crab bush, sea lavender.

2. BOURRERIA P. Browne, nom. cons.

Shrubs or small trees. Leaves alternate and petiolate. Flowers in terminal corymbose or paniculate inflorescences; calyx bell-shaped, 5-lobed at apex; corolla trumpet-shaped with spreading lobes; stamens projecting beyond the corolla tube; ovary lobed, with 4 uniovular chambers, the style branched into 2 short or elongate stigmas, usually projecting beyond corolla. Fruit a globose, fleshy drupe; seeds 4 per fruit.

A genus of 15–20 species with neotropical distribution.

1. **Bourreria succulenta** Jacq., Enum. Syst. Pl. 14. 1760.
Fig. 57A–D.

Bourreria recurva Miers, Contr. Bot. 2: 234. 1869.

Shrub or tree to 8 m tall; bark smooth grayish or light brown. Leaf blades 5–15 × 4–8 cm, elliptic, oblong, or ovate, glabrous, chartaceous to coriaceous, the apex obtuse, acute, notched and mucronate, the base tapering, the margins entire and slightly revolute; petioles 0.5–1.5 cm long. Inflorescences with straight branches. Calyx green, bell-shaped, 4–5 mm long, with 5 ovate lobes, irregularly splitting and persistent in fruit; corolla white, the tube 5–6 mm long, greenish yellow within at base, the lobes 6–8 mm long, rounded; anthers and style exserted, the stigma 2-lobed, each lobe nearly globose. Fruit turning from green to red-orange, 5–12 mm diam., shiny, nearly globose or ovoid, apiculate at apex.

DISTRIBUTION: One of the commonest trees on the island, in dry to moist areas. Hawksnest (A2667). Also on Anegada, Jost van Dyke, St. Croix, St. Thomas, Tortola, and Virgin Gorda; in Florida, the West Indies, and Venezuela.

COMMON NAMES: chink, chinkwood, juniper, pigeon berry, pigeon wood, spoon tree.

NOTE: Sterile specimens of *Cordia laevigata* Lam. are often confused with *B. succulenta*. However, *B. succulenta* has leaves that are glabrous beneath, whereas *C. laevigata* has leaves that are canescent beneath.

3. CORDIA L.

Shrubs or trees. Leaves alternate and petiolate. Flowers usually heterostylous, in terminal or axillary, corymbose, paniculate, cymose, or spikelike inflorescences; calyx bell-shaped to tubular, 5-toothed or 5–10-lobed at apex, usually striate, and often enlarging in fruit; corolla funnel- or trumpet-shaped; stamens shorter or longer than the corolla tube; ovary with 4 uniovular chambers, the style branched into 4 stigmas. Fruit a globose, fleshy drupe with 1 to 4 stones.

A genus of approximately 300 species with pantropical distribution.

REFERENCES: Borhidi, A., E. Gondár & Zs. Orosz-Kovács. 1988. The reconsideration of the genus *Cordia* L. Acta Bot. Acad. Sci. Hung. 34: 375–423; Taroda, N. & P. Gibbs. 1986. Studies on the genus *Cordia* L. (Boraginaceae) in Brazil. 1. A new infrageneric classification and conspectus. Revista Bras. Bot. 9: 31–42.

NOTE: The genus *Cordia* has been treated in different ways, and to 10 segregate genera have been recognized. In this flora, however, I favor a more comprehensive point of view, where a single genus is recognized (but see references for other points of view).

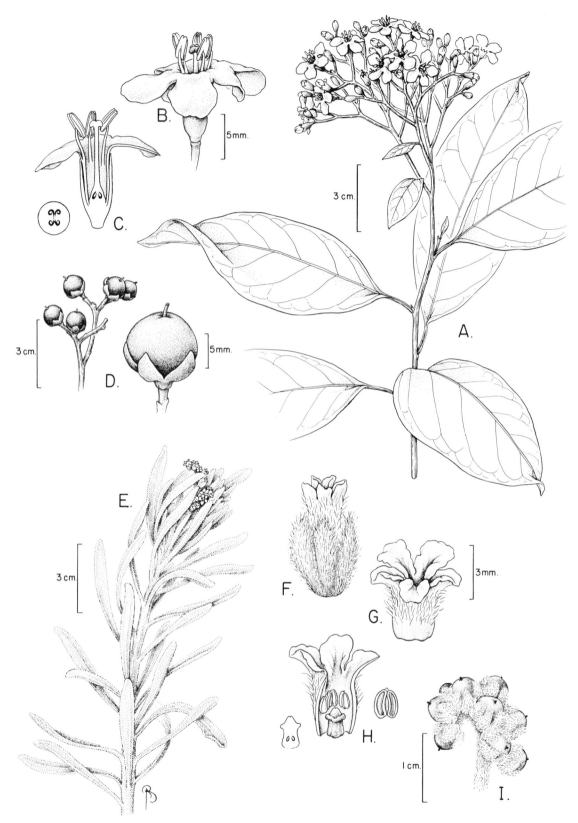

FIG. 57. A–D. *Bourreria succulenta*. **A.** Flowering branch. **B.** Flower. **C.** C.s. ovary and l.s. flower. **D.** Branch of infructescence and mature fruit. **E–I.** *Argusia gnaphalodes*. **E.** Flowering branch. **F.** Flower. **G.** Corolla. **H.** L.s. corolla showing stamens and pistil, and stamen. **I.** Detail of cyme with maturing fruits.

Key to the Species of *Cordia*

1. Leaf margins serrate or dentate.
   2. Inflorescence of globose cymes; plant a climbing or scandent shrub; fruits red, <4 mm long.. 5. *C. polycephala*
   2. Inflorescence a paniculate cyme; plant a small tree or shrub; fruits white, >6 mm long .... 3. *C. dentata*
1. Leaf margins entire or wavy.
   3. Plant densely covered with stellate hairs, especially on young parts; calyx tubular, striate.......... 1. *C. alliodora*
   3. Plant variously pubescent but hairs not stellate; calyx bell-shaped or tubular, not striate.
      4. Calyx tubular or nearly so, accrescent in fruit; corolla >1.5 cm long, bright orange, trumpet-shaped.
         5. Calyx long-tubular (1.3–2.3 cm long), pubescent or puberulent; fruit ovoid to globose .... 7. *C. sebestena*
         5. Calyx tubular (1–1.3 cm long), glabrous or nearly so; fruits obconical...................... 6. *C. rickseckeri*
      4. Calyx bell-shaped, not accrescent; corolla, <1 cm long, white, funnel-shaped.
         6. Fruit white or cream; leaves (9–)15–45 cm long, lanceolate, elliptic, or ovate .... 8. *C. sulcata*
         6. Fruit red or red-orange; leaves 8–20 cm long, elliptic or obovate.
            7. Corolla >6 mm long, with ascending lobes; leaves elliptic, with alternate secondary veins
            .... 4. *C. laevigata*
            7. Corolla <4 mm long, with reflexed lobes; leaves mostly obovate, with opposite (at least in some of the leaves), ascending, prominent secondary veins................................ 2. *C. collococca*

1. Cordia alliodora (Ruíz & Pav.) Oken, All. Naturgeschichte Bot. **3:** 1098. 1841. *Cerdana alliodora* Ruíz & Pav., Fl. Peruv. **2:** 47. 1799.

Cordia gerascanthus sensu Griseb., Fl. Brit. W. I. 478. 1862, non L., 1759.

Tree 4–15 m tall; bark dark gray and rough; branches covered with stellate hairs. Leaf blades 7–18.5 × 3–7 cm, elliptic, lanceolate, or oblanceolate, slightly scabrous above, densely covered with stellate hairs especially on lower surface, chartaceous to coriaceous, the apex long-acuminate, the base obtuse and unequal, the margins wavy; petioles 1–2.5 cm long, densely covered with stellate hairs. Inflorescences paniculate, with lateral cymes, the flowers closely packed. Calyx green, tubular to bell-shaped, 6–7 mm long, densely stellate and striate; corolla white, trumpet-shaped, and persistent, the tube 3–7 mm long, the lobes spatulate, ascending, as long as the tube; anthers and styles exserted beyond the tube, the styles 2, fused most of their length, the stigmas 4. Fruit dry, 5 mm long, cylindrical to fusiform, enclosed by dried flower parts.

DISTRIBUTION: Commonly found in moist areas but also planted along streets. Annaberg (A1925), Lameshur (B527). Also on Jost van Dyke, St. Thomas, Tortola, and Virgin Gorda; Cuba, Hispaniola, Puerto Rico, Lesser Antilles, Mexico, Central America, and northern South America.

COMMON NAME: copper.

2. Cordia collococca L., Amoen. Acad. **5:** 377. 1760.
Fig. 58J–M.

Tree 10–12 m tall. Leaf blades 6–20 × 3–9 cm, obovate, oblanceolate, or rarely elliptic, glabrous, or with scattered hairs on lower surface, chartaceous to coriaceous, secondary veins ascending and prominent on lower surface, the apex shortly acuminate, obtuse or rounded, the base tapering or obtuse, the margins slightly wavy; petioles 0.5–2 cm long. Inflorescences corymbose or paniculate with coiled lateral branches. Calyx bell-shaped, 1.7–2 mm long, densely covered with rusty-brown hairs; corolla 4–6 mm long, white, funnel-shaped, the tube 2–3 mm long, hairy at base within, the lobes oblong, reflexed, as long as the tube; anthers and styles exserted beyond the tube; ovary conical, the styles 2, fused most of their length, the stigmas 4. Fruit red or red-orange,

1 cm diam., globose or depressed-globose, fleshy, with 1 large irregular stone.

DISTRIBUTION: Common tree in moist areas and coastal plains with deep soils. Gift Hill (A830), Lameshur (A2745), Reef Bay Trail (P29264). Also on Jost van Dyke, St. Croix, St. Thomas, Tortola, and Virgin Gorda; in the Greater and Lesser Antilles, Mexico to northern South America.

COMMON NAMES: manjack, red manjack.

NOTE: This species was originally published as *Cordia callococca*, but the spelling was changed by subsequent workers to *C. collococca*.

3. Cordia dentata Poir. *in* Lam., Encycl. **7:** 48. 1806.

Small tree or shrub to 10 m tall; branches sympodial, ending in an inflorescence. Leaf blades 4–12 × 4–7.5 cm, ovate, oval, or rounded, scabrous above, pubescent below, with tufts of hairs at vein angles, chartaceous, the apex obtuse or shortly acuminate and mucronate, the base rounded or obtuse, the margins dentate-mucronate; petioles 1–2 cm long. Inflorescence a paniculate cyme with slightly curved branches. Calyx green, bell-shaped, 3–4 mm long, strigose, slightly striate; corolla white, funnel-shaped, 8–11 mm long; anthers and styles exserted, the styles 2, fused at base, the stigmas 4. Fruit 6–8 mm diam., ellipsoid, turning from green to white, fleshy, with sticky pulp.

DISTRIBUTION: Apparently introduced, found mostly along roadsides. Along road from Fish Bay to Coccoloba (A2816), Susannaberg (A3806). Also on St. Croix and St. Thomas; the West Indies, Central America, and northern South America.

COMMON NAMES: flute boom, white manjack.

NOTE: The name *Varronia alba* Jacq. (1760) [and its derivatives *Cordia alba* (Jacq.) Roem. & Schult. and *Calyptracordia alba* (Jacq.) Britton], once applied to this taxon, was based on discordant elements published by J. Commelijn (Horti. Med. Amstelod. 155, t. 80. 1697) and later typified by Johnston (J. Arnold Arbor. 21: 347. 1940) solely on Commelijn's plate. The plate has recently been identified as *Trema micranthum* (L.) Blume (Ulmaceae); as a result, *Varronia alba* becomes a synonym of *Trema micranthum*.

4. Cordia laevigata Lam., Tabl. Encycl. **1:** 422. 1792.

Cordia nitida Vahl *in* H. West, Bidr. Beskr. Ste. Croix 275. 1793.

Tree 10–15 m tall; young parts densely covered with appressed, rusty-brown hairs. Leaf blades 6–13 × 2.5–6.5 cm, elliptic, rarely oval or ovate, upper surface shiny and glabrous, lower surface with scattered hairs on veins, chartaceous to coriaceous,

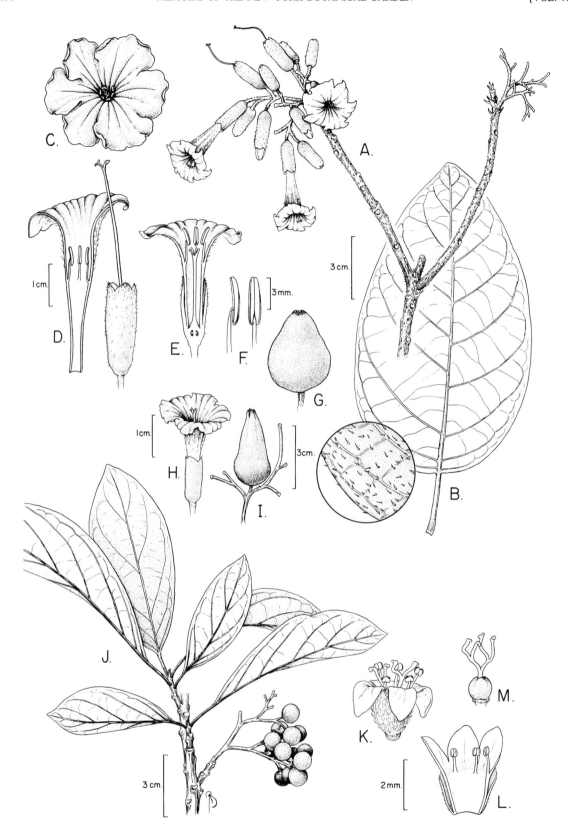

Fig. 58. A–G. *Cordia sebestena*. **A.** Inflorescence. **B.** Leaf with detail of adaxial surface. **C.** Flower. **D.** L.s. corolla of long-styled flower and calyx with corolla removed showing style. **E.** L.s. short-styled flower. **F.** Stamen, lateral and frontal views. **G.** Fruit. **H, I.** *Cordia rickseckeri*. **H.** Flower. **I.** Fruit. **J–M,** *Cordia collococca*. **J.** Fruiting branch. **K.** Flower. **L.** L.s. corolla. **M.** Pistil.

secondary veins curved and slightly prominent on lower surface, the apex acute, obtuse or acuminate, the base acute or obtuse, the margins slightly wavy; petioles 0.5–2 cm long. Inflorescences corymbose cymes with coiled lateral branches. Calyx bell-shaped, 4–5 mm long, sparsely covered with rusty-brown hairs; corolla white, funnel-shaped, 6–7 mm long, the lobes rounded, ascending, ca. 4 mm long; anthers and style exserted beyond the tube, the filaments hairy at base; ovary conical, the style 1, with 4–8 branches. Fruit bright red, 0.8–1 cm diam., globose, fleshy, with 1 large irregular stone.

DISTRIBUTION: Uncommon in dry to moist forests. America Hill (R1431), Bordeaux Mountain (A4261). Also on Anegada, St. Croix, St. Thomas, and Tortola; Greater Antilles and Central America.

COMMON NAMES: bastard capa, manjack, West Indian cherry, wild capa.

5. Cordia polycephala (Lam.) I.M. Johnst., J. Arnold Arbor. **16**: 33. 1935. *Varronia polycephala* Lam., Tab. Encycl. **1**: 418. 1792.

Lantana corymbosa L., Sp. Pl. 628. 1753, non *Cordia corymbosa* Willd. *ex* Roem. & Schult., 1819. *Varronia corymbosa* (L.) Desv., J. Bot. (Desvaux) **1**: 275. 1808.

Scandent shrub to 5 m long; young parts pubescent. Leaf blades 4–15 × 2–5.5 cm, lanceolate, ovate, or elliptic, upper surface more or less scabrous, lower surface lighter, densely to slightly pubescent, with slightly prominent veins, chartaceous to coriaceous, the apex acute or acuminate, the base obtuse or rounded, usually unequal, the margins serrate; petioles 0.5–1 cm long, pubescent. Flowers sessile, in axillary or terminal, simple or branched inflorescences, with lateral globose cymes. Calyx cup-shaped, 2.5–3 mm long, densely covered with rusty-brown hairs, persistent in fruit; corolla white, bell-shaped, 4–5 mm long, the lobes very short; anthers and style included; style 1, with 4 branches. Fruit bright red, 3–4 mm diam., globose, fleshy, covered by an enlarged calyx.

DISTRIBUTION: Common in open disturbed areas. Along road to Bordeaux (A3834), Bethany (B188, B248). Also on St. Croix, St. Thomas, Tortola, and Virgin Gorda; Hispaniola, Puerto Rico, Lesser Antilles, and northern South America.

6. Cordia rickseckeri Millsp., Publ. Field Columbian Mus., Bot. Ser. **l**: 522. 1902. *Sebesten rickseckeri* (Millsp.) Britton *in* Britton & P. Wilson, Bot. Porto Rico **6**: 124. 1925.   Fig. 58H, I.

Cordia sebestena L. var. *brachycalyx* Urb., Symb. Ant. **1**: 389. 1899. *Sebesten brachycalyx* (Urb.) Britton, Bull. Torrey Bot. Club **43**: 457. 1916.

Tree to 8 m tall; bark gray, rough; young parts pubescent. Leaf blades 10–18 × 4–11 cm, ovate, elliptic, oval, or obovate, upper surface more or less scabrous, lower surface densely to sparsely pubescent with prominent veins, coriaceous, the apex obtuse, rounded, acute, or notched, the base obtuse, rounded or nearly cordate, usually unequal, the margins wavy and revolute; petioles 1–4 cm long, pubescent. Flowers in axillary or terminal, corymbose cymes. Calyx orange, tubular or nearly so, 1–1.3 ×

0.4–0.6 cm, glabrous or nearly so, enlarging in fruit; corolla orange or scarlet-red, trumpet-shaped, the tube 2–3 cm long, striate and pubescent without, the lobes 5–6, rounded and wavy, 0.7–1 cm long; anthers and style slightly exserted; style 1, with 4 branches. Fruit obconical, 3–3.5 cm long, turning brown, covered with enlarged fleshy calyx, style persistent. Stones containing 1–4 white seeds.

DISTRIBUTION: A common tree, especially on the coastal plains with deep soil. Coral Bay (L21924), Lameshur (A3957), Pen Point (A2011). Native to the Virgin Islands (Jost van Dyke, St. Croix, St. Thomas, Tortola, and Virgin Gorda) and Puerto Rico

COMMON NAMES: black manjack, dog almond.

7. Cordia sebestena L., Sp. Pl. 190. 1753. Fig. 58A–G.

Tree to 10 m tall; young parts pubescent. Leaf blades 8–20 × 4–13.5 cm, ovate, elliptic, oval, or obovate, upper surface more or less scabrous, lower surface densely to sparsely pubescent with prominent veins, coriaceous, the apex obtuse, rounded, acute, or shortly acuminate, the base obtuse, rounded or nearly cordate, usually unequal, the margins wavy and revolute; petioles 1–3.2 cm long, pubescent. Flowers heterostylous, in axillary or terminal, corymbose cymes. Calyx orange, tubular or nearly so, 1.3–2.3 × 0.5–0.8 cm, pubescent or puberulent, enlarging in fruit; corolla orange or scarlet-red, trumpet-shaped, the tube 2.3–3.5 cm long, striate and pubescent without, the lobes 5–6 rounded and wavy, 1–2.5 cm long; anthers and style slightly exserted; style 1, with 4 branches. Fruit ovoid to nearly globose, turning white, 2.5–3 cm long, covered with enlarged fleshy calyx, persistent style. Stones containing 1–4 white seeds.

DISTRIBUTION: Commonly planted along roadsides and in gardens. Coral Bay (A4065), Lameshur (A738). Apparently native to the Virgin Islands (St. Croix, St. John, St. Thomas, Tortola, and Virgin Gorda); the Greater Antilles and the Bahamas, introduced elsewhere as an ornamental.

COMMON NAMES: bay almond, geiger, scarlet cordia, trumpet tree.

8. Cordia sulcata A. DC., Prodr. **9**: 488. 1845.

Cordia macrophylla sensu Griseb., Fl. Brit. W. I. 479. 1862, non L., 1762.

Small tree or shrub to 10 m tall; young branches and inflorescence axes covered with rusty-brown hairs. Leaf blades 9–45 × .5–19 cm, lanceolate, elliptic, or ovate, upper surface more or less scabrous, lower surface, densely to sparsely pubescent with prominent reticulate veins, chartaceous to coriaceous, the apex obtuse, rounded, or acuminate, the base obtuse, rounded, or nearly cordate, usually unequal, the margins crenate; petioles 1–4 cm long, densely pubescent. Flowers in axillary or terminal, corymbose or paniculate cymes. Calyx bell-shaped, 2–3 mm long, densely covered with rusty-brown hairs; corolla white or cream, bell-shaped with reflexed lobes, the tube 3–3.5 mm long, the lobes 1–1.5 mm long, oblong to triangular; anthers and style slightly exserted, the filaments hairy at base; style 1, with 4 branches. Fruit globose or depressed-globose, 5–7 mm diam., turning green to light yellow to white, fleshy.

DISTRIBUTION: A common tree of secondary forests and disturbed moist areas. Bordeaux Mountain (A1885, B586), Cinnamon Bay (A1090). Also on Jost van Dyke, St. Croix, St. Thomas, and Tortola; Cuba, Hispaniola, Puerto Rico, and the Lesser Antilles.

COMMON NAMES: mucilage manjack, white manjack.

4. HELIOTROPIUM L.

Erect or prostrate herbs or shrubs. Leaves alternate or rarely subopposite or whorled. Flowers in scorpioid (coiled) cymes, spikes or racemes; calyx with 5 deeply parted, lanceolate lobes; corolla trumpet- or funnel-shaped, with spreading lobes; stamens included in corolla tube, not exserted; ovary 4-locular, the style terminal, short or absent, the stigma conical. Fruit dry, splitting into 2–4 nuts.

A genus with approximately 250 species, with cosmopolitan distribution.

Key to the Species of *Heliotropium*

1. Leaves succulent, narrowly oblanceolate, without petioles; plant glabrous .... 2. *H. curassavicum*
1. Leaves not succulent, variously shaped but not oblanceolate, petiolate; plant pubescent.
    2. Plants densely sericeous (appressed, long, white hairs); leaves lanceolate, linear, or elliptic, mostly sub-opposite or in whorls of 3; fruits subglobose... 4. *H. ternatum*
    2. Plants pubescent or glabrescent; leaves elliptic or ovate, alternate or subopposite; fruits 2-lobed, emarginate or slightly compressed.
        3. Leaves 4–17 cm long; petioles 4–10 cm long; stems 4-angled; corolla 4–5 mm long; fruits emarginate at apex .... 3. *H. indicum*
        3. Leaves 2–5 cm long; petioles 1–1.5 cm long; stems terete; corolla 0.5–2.5 mm long; fruits slightly compressed .... 1. *H. angiospermum*

1. Heliotropium angiospermum Murray, Prodr. Stirp. Gott. 217. 1770. *Schobera angiosperma* (Murray) Britton *in* Britton & P. Wilson, Bot. Porto Rico **6:** 134. 1925.

Heliotropium parviflorum L., Mant. Pl. **2:** 201. 1771.

Erect herb, to 1 m tall, many-branched from a woody base; branches cylindrical; young parts sericeous. Leaves alternate; blades 2–5 × 1.2–2 cm, lanceolate, ovate, or elliptic, sparsely covered with white, appressed hairs, chartaceous, often rugose, the apex obtuse, acute, or acuminate, mucronate, the base obtuse or attenuate, usually unequal, the margins crenate and ciliate; petioles 1–1.5 cm long, pubescent. Flowers in axillary or terminal, scorpioid spikes (sometimes branched), 2–14 cm long. Sepals awl-shaped, 1–1.5 mm long; corolla white, funnel-shaped, the tube 1.5–2 mm long, pubescent within, the lobes 0.5–1 mm long; anthers and stigma sessile. Fruit conical, slightly compressed ca. 3 mm diam. at base, splitting into 4 nuts.

DISTRIBUTION: A common weed along roadsides and in disturbed areas. East End (A665), along trail to Fortsberg (A4095), Maria Bluff (A2336). Also on St. Croix, St. Thomas, Tortola, and Virgin Gorda; southeastern United States to South America, including the West Indies.

COMMON NAME: eyebright.

NOTE: A specimen of this species was erroneously reported as *Heliotropium procumbens* Mill. by Woodbury and Weaver (1987).

2. Heliotropium curassavicum L., Sp. Pl. 130. 1753.
Fig. 59A–G.

Erect or prostrate, glabrous herb or subshrub, to 30 cm tall, many-branched from a woody taprooted base. Leaves alternate or nearly opposite, sessile, 1.5–4.5 × 0.2–0.4 cm, narrowly oblanceolate, succulent (ca. 1 mm thick), usually with grayish or bluish hue (glaucous), only main vein conspicuous, the apex obtuse or rounded, the base cuneate, the margins entire. Flowers in terminal, coiled, branched or unbranched spikes, 1–8 cm long. Sepals green, awl-shaped, 1.5 mm long, erect; corolla white, urceolate, the tube 1.5–2 mm long, the lobes spreading, 0.5–1 mm long; anthers and stigma sessile. Fruit depressed globose, slightly lobed, ca. 2.5 mm diam., splitting into 4 nuts.

DISTRIBUTION: A common herb from open, disturbed areas, especially along sandy beaches. Great Cruz Bay (A795, A3104), East End (A2426). Also on St. Croix, St. Thomas, Tortola, and Virgin Gorda; southeastern United States to Central America, the West Indies and Asia.

3. Heliotropium indicum L., Sp. Pl. 130. 1753. *Tiaridium indicum* (L.) Lehm., Pl. Asperif. Nucif. 14. 1818.

Erect herb, to 1 m tall, many-branched at upper portion; stems 4-angled, young parts with long white hairs. Leaves alternate or nearly opposite; blades 4–17 × 2–9 cm, ovate or elliptic, chartaceous, sparsely covered with long white hairs, vein prominent on lower surface, the apex acute or short-acuminate, the base obtuse to nearly rounded, tapering into a 4–10 cm long petiole, the margins wavy. Flowers in terminal, coiled, unbranched spikes, 15–25 cm long. Sepals green, awl-shaped, 2–3 mm long with setulose hairs; corolla white, lilac, or light blue, urceolate, the tube 3–4 mm long, pubescent without, the lobes spreading, 0.5–1 mm long; anthers and stigma included in tube. Fruit pyriform, emarginate at apex, with 2 nutlets separating at apex, 2.5 mm long.

DISTRIBUTION: An uncommon herb in open and disturbed areas. Lameshur (A2572, A2736). Also on St. Croix, St. Thomas, Tortola, Virgin Gorda; throughout the Greater Antilles and Trinidad.

4. Heliotropium ternatum Vahl, Symb. Bot. **3:** 21. 1794.

Heliotropium fruticosum sensu Griseb., Fl. Brit. W. I. 486. 1862, non L., 1759.

Densely sericeous subshrub or herb to 50 cm tall, many-branched from a woody base. Leaves nearly opposite, in whorls of 3 or seldom alternate; blades 1–2.5 × 0.4–0.8 cm, linear elliptic or lanceolate, coriaceous, sericeous (especially on lower surface), main vein prominent on lower surface, the apex acute, the base obtuse, the margins strongly revolute; petioles 1–2 mm long. Flowers in terminal, coiled, unbranched spikes, 2–3 cm long. Sepals awl-shaped, 4–5 mm long, sericeous; corolla white, trumpet-shaped, the tube 4–5 mm long, yellowish inside, the lobes spreading, 1.5–2 mm long; anthers and stigma included in tube. Fruit nearly globose, slightly lobed, ca. 2 mm long, densely strigose, splitting into 2 nutlets.

DISTRIBUTION: In open, dry areas. Dittlif Point (A3977), Lameshur (B602), Maria Bluff (A2333). Also on St. Croix, St. Thomas, and Virgin Gorda; throughout the West Indies and northern South America.

5. ROCHEFORTIA Sw.

Dioecious shrubs or small trees with spiny branches. Leaves clustered on short, axillary branches. Flowers unisexual, in terminal, short cymes or solitary in leaf axils; calyx with 4–5, deeply parted sepals, corolla trumpet- or funnel-shaped, the lobes 4–5, much longer than the tube; anthers as many as corolla lobes, exserted; ovary conical, 4-chambered, the styles 2, free, the stigma discoid. Fruit globose, with 4 nutlets.

A genus of 6–12 species with Caribbean distribution.

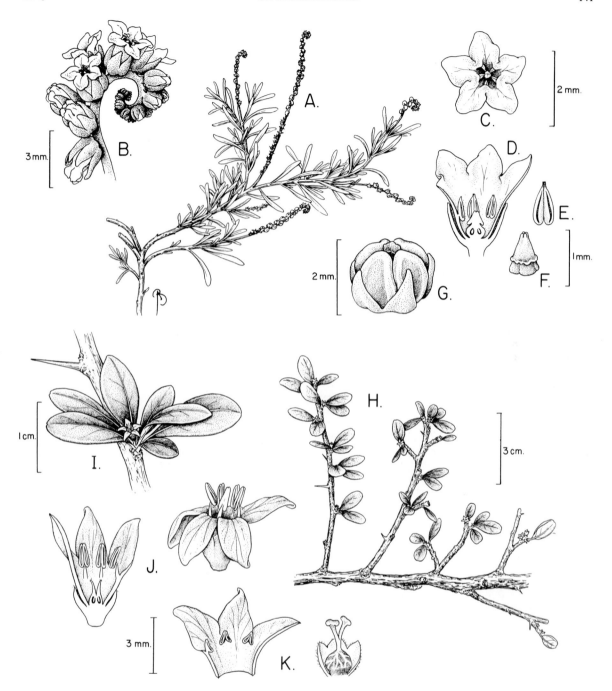

FIG. 59. A–G. *Heliotropium curassavicum*. **A.** Flowering branch. **B.** Detail of cyme. **C.** Flower. **D.** L.s. flower. **E.** Anther. **F.** Pistil. **G.** Fruit of 4 nuts. H–K. *Rochefortia acanthophora*. **H.** Flowering branch. **I.** Detail of flowering branch. **J.** L.s. staminate flower with pistillode and staminate flower. **K.** L.s. corolla of pistillate flower with sterile stamens and calyx with 2 sepals removed showing pistil.

1. Rochefortia acanthophora (A. DC.) Griseb., Fl. Brit. W. I. 482. 1862. *Ehretia acanthophora* A. DC., Prodr. **9:** 510. 1845. Fig. 59H–K.

Shrub or small tree to 7 m tall; branches usually flexuous, with axillary spines. Leaves congested on short axillary branches; blades 0.8–2.5 × 0.4–1 cm, obovate to oblanceolate, coriaceous, sparsely covered with minute hairs, especially on lower surface, slightly scabrous on upper surface, main vein prominent on lower surface, the apex rounded and notched, the base cuneate, the margins entire and revolute; petioles 1–3 mm long. Flowers sessile, paired or solitary in leaf axils. Sepals oblong–ovate, 4–5 mm long, pubescent; corolla white to yellowish, funnel-shaped, the tube 1.5–2 mm long, striate, the lobes spreading or ascending, oblong, 2.5–3.5 mm long; stamens unequal, exserted, the anthers ellipsoid, with short filaments; ovary conical, glabrous, the styles

2 mm long, the stigma dilated. Fruit nearly globose, 4–5 mm diam.

DISTRIBUTION: A common tree of dry coastal scrub. West slopes of

Fish Bay (A3909). Also on St. Croix, St. Thomas, and Virgin Gorda; the Greater and Lesser Antilles.

6. TOURNEFORTIA L.

Vines, lianas, trees, or shrubs. Leaves alternate. Flowers bisexual, produced on terminal, coiled, branched cymes; calyx with 5, deeply parted, awl-shaped lobes; corolla trumpet-shaped, the tube slightly swollen, the lobes reflexed or spreading; stamens inserted on tube, not exserted; ovary 2-carpellate, each carpel with 2 chambers, the style simple, 2-lobed, not exserted. Fruit fleshy, globose or 2–4-lobed, with 4 or fewer nutlets.

A genus with approximately 150 species with pantropical distribution.

Key to the Species of *Tournefortia*

1. Erect shrub or small tree .... 2. *T. filiflora*
1. Twining vines or lianas.
    2. Plants hirsute .... 3. *T. hirsutissima*
    2. Plants glabrous or pubescent.
       3. Plants slender (stems usually <8 mm diam.); leaves 1–3.5 cm long; fruits nearly globose, white, with 4 black, round spots .... 4. *T. microphylla*
       3. Plants robust (stems >8 mm diam.); leaves 3–9 cm long; fruits globose, white .... 1. *T. bicolor*

1. Tournefortia bicolor Sw., Prodr. 40. 1788.

Tournefortia laevigata Lam., Tabl. Encycl. **1:** 416. 1792.

Twining woody vine, to 10 m long; young branches and inflorescences strigose. Leaf blades 4–15 × 2.5–8 cm, lanceolate or elliptic, chartaceous, glabrous, except for a few hairs on veins and margins, veins slightly prominent on lower surface, the apex acute, shortly acuminate, or obtuse, the base cuneate or obtuse, the margins entire or slightly wavy; petioles 5–15 mm long. Flowers sessile, in branched cymes. Sepals awl-shaped, 1.5–2 mm long, pubescent; corolla white, trumpet-shaped, the tube 4–6 mm long, strigose without on upper half, the lobes spreading, lanceolate, ca. 1 mm long; stamens included; ovary ovoid, 4-celled, obscurely 4-lobed, the stigma sessile. Fruit globose, 8 mm diam., fleshy, white, separating into 4 nutlets.

DISTRIBUTION: Uncommon, mostly found in gallery forests. Caneel Bay Trail to Turtle Point (A4121). Also on St. Thomas and Tortola; throughout tropical America.

2. Tournefortia filiflora Griseb., Fl. Brit. W. I. 483. 1862.

Shrub or tree to 6 m tall; young parts strigillose. Leaf blades 15–35 × 8–13 cm, elliptic or oblong-elliptic, chartaceous, glabrous, except for a few hairs on veins, main veins slightly raised on lower surface, the apex acuminate or acute, the base cuneate or obtuse, the margins entire or slightly wavy; petioles 2–4 cm long. Flowers sessile, in cymes with spikelike branches. Sepals awl-shaped, ca. 1 mm long, pubescent; corolla white, trumpet-shaped, the tube 2–3 mm long, the lobes reflexed, triangular, 0.5–1 mm long; stamens included; ovary conical-bilobed, the style short. Fruit nearly globose, 4 mm diam., fleshy, yellowish to white.

DISTRIBUTION: A rare tree, found only in a few locations. Trunk Bay (W174). Also on St. Croix; Puerto Rico and the Lesser Antilles.

3. Tournefortia hirsutissima L., Sp. Pl. 140. 1753.

Fig. 60A–E.

Twining woody vine or climbing shrub to 10 m long; bark light brown, rough; stem densely covered with long stiff hairs. Leaf blades 10–22 × 2–11 cm, elliptic, ovate, or rarely obovate, chartaceous, hirsute, the apex acuminate or acute, the base obtuse, rounded or cuneate, often unequal, the margins entire; petioles 1–1.5 cm long. Flowers sessile, in branched cymes. Sepals awl-shaped, ca. 2–3 mm long, pubescent; corolla white, trumpet-shaped, the tube 4–5 mm long, strigose, the lobes expanded or reflexed, 2 mm long, depressed along central vein; anthers sessile, inserted on middle of tube; ovary conical, the style short, the stigma conical. Fruit white, nearly globose, 5–6 mm diam., fleshy, with persistent stigma at apex.

DISTRIBUTION: Common in open or disturbed areas. Coral Bay (A2458), Cruz Bay Quarter along Center Line Road (A2865). Also on St. Croix, St. Thomas, and Tortola; Greater and Lesser Antilles and Trinidad.

COMMON NAMES: chichery grape, chiggernit, giniper.

4. Tournefortia microphylla Bertero *ex* Spreng., Syst. Veg. **1:** 644. 1824. *Tournefortia volubilis* L. var. *microphylla* (Bertero *ex* Spreng.) A. DC., Prodr. **9:** 523. 1845.   Fig. 60F–J.

Twining vine to 7 m long; stem slender, <8 mm diam., with lenticels, young parts pubescent. Leaf blades 1.5–3.5 × 0.5–1.5 cm, elliptic, ovate, or lanceolate, chartaceous, upper surface with short hairs, the apex acute or obtuse and often mucronate, the base obtuse or rounded often unequal, the margins entire; petioles ca. 4 mm long. Flowers short-pedicellate, on cymes with coiled branches. Calyx ca. 1.5 mm long, the sepals awl-shaped, ca. 1 mm long, puberulent; corolla white, trumpet-shaped, the tube 2 mm long, strigose, swollen at middle, the lobes reflexed, linear, 1 mm long; ovary conical, the style longer than the ovary, the stigma elongate. Fruit white, ovoid to nearly globose, 4 mm long, fleshy, with 4, black round spots.

DISTRIBUTION: Common in open or disturbed areas, such as roadsides and secondary forests. Along trail to Brown Bay (A1881), Cruz Bay (A2326), East End (A3793). Also on Anegada, St. Croix, St. Thomas, Tortola, Virgin Gorda; Hispaniola, Puerto Rico, and the Lesser Antilles.

18. Brassicaceae (Mustard Family)

Herbs. Leaves alternate or rarely opposite, simple or less often pinnately dissected or lobed, rarely palmately lobed, without stipules. Flowers bisexual, actinomorphic, produced in racemes or rarely solitary; bracts usually absent; calyx of 4, deeply parted

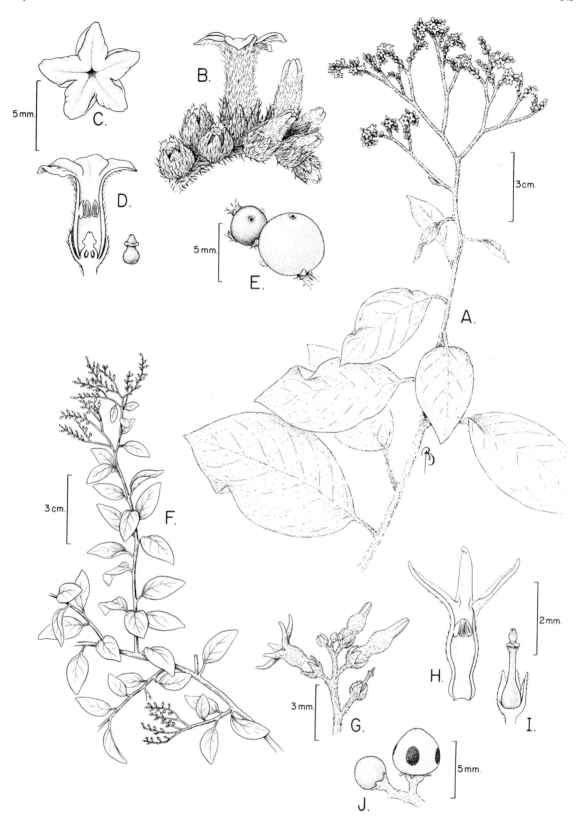

FIG. 60. A–E. *Tournefortia hirsutissima*. A. Flowering branch. B. Detail of cyme. C. Flower, top view. D. L.s. flower and pistil. E. Fruit. F–J. *Tournefortia microphylla*. F. Flowering branch. G. Detail of cyme. H. L.s. corolla with adnate stamens. I. Calyx with pistil. J. Fruits.

sepals; corolla of 4, free petals, alternating with the sepals, and usually very narrow at base; stamens 6 (the outer two shorter) or rarely 4, the anthers opening by longitudinal slits; ovary superior, usually with 4 nectary glands at base, with 2 united carpels, the carpel usually bilocular with 1 to many ovules, the placentation parietal, the style 1 or wanting, the stigmas capitate or bilobed. Fruit dry, indehiscent, or usually with deciduous valves that open, leaving a central placental partition (replum), or rarely transversely jointed; seeds attached to the replum.

A family of 350 genera and approximately 3000 species, mainly from temperate zones.

Key to the Genera of *Brassicaceae*

1. Fruit fusiform, 20–25 mm long; flowers 6–8 mm long .... 1. *Cakile*
1. Fruit lenticular, to 5 mm long; flowers 1–1.5 mm long .... 2. *Lepidium*

1. CAKILE Mill.

Annual or perennial, fleshy herbs. Leaves alternate. Flowers in terminal racemes; sepals hyaline margined; petals obovate or spatulate, white to purple; stamens 6; nectary glands present; ovary fusiform, the stigma capitate. Fruit fusiform, lomentum, transversely separating into 1 or 2 1-seeded segments.

A genus of 7 species primarily distributed along sandy beaches around the world.

1. Cakile lanceolata (Willd.) O. E. Schulz *in* Urb., Symb. Ant. **3**: 504. 1903. *Rhaphanus lanceolatus* Willd., Sp. Pl. **3**: 562. 1800.   Fig. 61A–F.

Cakile domingensis Tussac, Fl. Antill. **1**: 119. 1808.

Erect to decumbent herb, to 50 cm tall, many-branched from a woody taprooted base, glabrous. Leaf blades 3–10 × 1–3 cm, oblanceolate, fleshy, glabrous, the apex acute, the base tapering, the margins serrate to sinuate; petioles 0.4–1 cm long. Flowers in terminal racemes; sepals oblong, 2.8–3.3 mm long; petals 6–8 mm long, white, spatulate; stamens 6, two shorter, the anthers lanceolate; nectary glands rounded; ovary fusiform, the stigma capitate. Fruit turning from green to straw-colored when dry, fusiform, 2.5–3 cm long, the lower segment ca. 5 mm long.

DISTRIBUTION: Common along sandy beaches or in disturbed coastal areas. Chocolate Hole (A780), Dittlif Point (A3969), trail to Salt Pond (A2047). Also on St. Croix, St. Thomas, Tortola, and Virgin Gorda; widespread throughout the Caribbean basin.

2. LEPIDIUM L.

Annual to perennial herbs. Leaves alternate, rarely simple, pinnately divided to tripinnate. Flowers in terminal racemes; sepals hyaline; petals obovate or spatulate, white, rarely pink or yellow; stamens 6, 4, or 2; nectary glands present; ovary with 2 ovules, the style short or absent, the stigma capitate. Fruit a lenticular, ovate, elliptic, or oblong silique, the septum transverse; seeds pendulous, 1 per locule.

A genus of 175 species with cosmopolitan distribution, primarily in temperate regions.
REFERENCE: Hitchcock, C. L. 1945. The Mexican, Central American, and West Indian Lepidia. Madroño 8: 118–143.

1. Lepidium virginicum L., Sp. Pl. 645. 1753.   Fig. 61G–K.

Erect herb to 90 cm tall, glabrous or sparsely pubescent, branching in upper portion. Leaves alternate; blades 2.5–10 × 0.3–1.5 cm, oblanceolate, obovate, or spatulate, chartaceous, glabrous or sparsely pubescent, the apex acute or obtuse, the base tapering, the margins serrate or deeply incised in upper half; petioles not distinctive. Flowers in terminal racemes; sepals green, oblong, 0.7 mm long; petals white, 1–1.5 mm long, spatulate; stamens 2; nectary glands minute; ovary lenticular, the stigma sessile and capitate. Fruit straw-colored when dry, lenticular to circular, flattened, ca. 4 mm long, the valves falling away, leaving the replum. Seeds ellipsoid, light brown and pendulous.

DISTRIBUTION: A common weed of open areas, especially on sandy beaches. Bordeaux Road (A2891), along road to Herman Farm (A866). Also on St. Croix, St. Thomas, and Tortola; native to North America, introduced and naturalized elsewhere.

19. Burseraceae (Frankincense Family)

Trees or shrubs, often producing resin. Leaves alternate, pinnately compound or less often unifoliolate, stipules lacking. Flowers 3–5-merous, bisexual or unisexual, actinomorphic, in panicles; calyx with sepals connate to base; petals distinct or seldom wanting; stamens twice as many as petals, the anthers opening by longitudinal slits, the filaments free, borne outside an annular nectary disk; ovary superior, with 3–5 united carpels, the style 1 and terminal, the stigmas lobed or capitate, the ovules 2 or rarely 1 in each cell, pendulous, the placentation axile. Fruit a drupe with 1–5, 1-seeded pyrenes.

A family of approximately 20 genera and 600 species, with pantropical distribution.

1. BURSERA L., nom. cons.

Trees or shrubs to 15 m tall, producing clear fragrant resin; bark smooth, copper-colored, peeling off in thin flakes; branches usually contorted. Leaves imparipinnate. Flowers bisexual or unisexual, in axillary panicles; calyx 4–5-lobed; corolla of 4 or 5 free petals; stamens 8 or 10; ovary ovoid, 3-locular. Fruit a leathery drupe, with exocarp falling away, leaving a white to beige pyrene.

A genus of approximately 80 species with neotropical distribution.

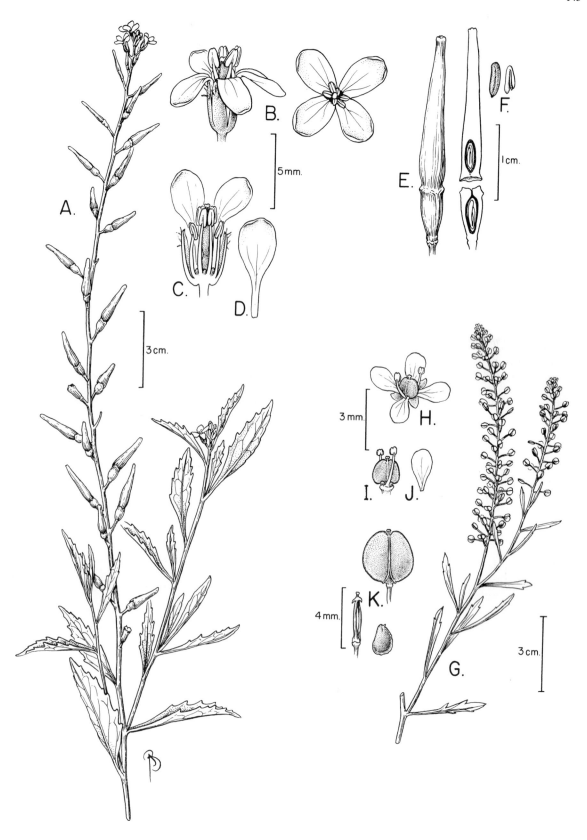

FIG. 61. **A–F.** *Cakile lanceolata.* **A.** Fertile branch. **B.** Flower, lateral and top views. **C.** L.s. flower. **D.** Petal. **E.** Fruit and l.s. fruit showing seeds. **F.** Seed and embryo. **G–K.** *Lepidium virginicum.* **G.** Fertile branch. **H.** Flower. **I.** Flower with perianth removed showing stamens and pistil. **J.** Petal. **K.** Silique, replum, and seed.

1. Bursera simaruba (L.) Sarg., Gard. & Forest **3**: 260. 1890. *Pistacia simaruba* L., Sp. Pl. 1026. 1753. *Elaphrium simaruba* (L.) Rose, N. Amer. Fl. **25**: 246. 1911.　　　　　　　　　　Fig. 62.

Deciduous tree or shrub to 15 m, and 30 cm diam.; bark reddish, smooth, peeling in large, thin flakes; resin sometimes accumulating at base of trunk. Leaves 5–9-foliolate; leaflets opposite, 3–7 × 1.5–4.5 cm, ovate, oblong, lanceolate, or elliptic, chartaceous, glabrous, the apex long-acuminate, the base rounded or obtuse, usually unequal, the margins entire; petiolules 5–10 mm long; petioles longer than the leaflets. Flowers unisexual (pistillate appearing as bisexual) in axillary racemes, at ends of branches; calyx ca. 2 mm long, sepals 5, minute, triangular; petals 5, white, 3.5–4 mm long, lanceolate and reflexed; nectary disk annular; stamens 10, the anthers lanceolate; ovary conical, of 3 carpels, each carpel with 2 ovules, the stigmas 3, sessile. Drupe ellipsoid-trigonous, 1–1.2 cm long, turning from green to reddish, with resinous aroma. Pyrenes 1-seeded, trigonous, beige.

DISTRIBUTION: A common tree of dry forested hills. Susannaberg (A840, A1861). Also on Jost van Dyke, St. Croix, St. Thomas, Tortola, and Virgin Gorda; throughout the Caribbean, also Florida, and Mexico to South America.

COMMON NAMES: cachibou, gommier, gumbo limbo, naked indian, red bellytree, takantin, tourist tree.

20. Cactaceae (Cactus Family)

Succulent shrubs, trees, or vines, terrestrial or epiphytic; stems green, of variable shapes, tissue accumulating water, usually covered with spines (reduced leaves). Normal foliage leaves persisting only in a few members (*Pereskia*), then alternate and simple. Spines restricted to well-defined areas, usually accompanied by a tuft of barbed hairs or bristles (glochids). Flowers bisexual (seldom unisexual), actinomorphic, usually large and fleshy, mostly solitary at areoles or terminal on branches; tepals united at base forming a perianth tube; stamens numerous, the anthers opening by longitudinal slits; ovary inferior, forming a hypanthium, with numerous, united carpels, carrying numerous ovules, the placentation parietal, the style solitary and the stigmas as many as the carpels. Fruit commonly fleshy, with numerous seeds.

A family of approximately 200 genera and 1000–2000 species, almost all native to the New World.

REFERENCES: Anderson, F., W. Bartholett, V. Eggli, D. R. Hunt, R. Kiesling, B. E. Levenberger, H. Sánchez-Mejorada & N. P. Taylor. 1986. The genera of Cactaceae: toward a new consensus. Bradleya **4**: 65–78; Britton, N. L. & J. N. Rose. 1919–1923. The Cactaceae. Carnegie Inst. Wash. Publ. 248(1–4).

Key to the Genera of Cacaceae

1. Stem terete, without ribs; normal foliage leaves present in addition to spines .... 5. *Pereskia*
1. Stem various, but not as above; leaves in the form of spines only.
　2. Plants spherical, not elongate, with longitudinal ribs.
　　3. Plants crowned by an elongate cephalium; the cephalium covered with numerous, slender red spines; plants to 60 cm tall; flowers reddish pink .... 3. *Melocactus*
　　3. Plants rounded, not crowned by a cephalium; plant to 15 cm tall; flowers yellow .... 2. *Mammillaria*
　2. Plants elongate; stems either flattened or variously ribbed.
　　4. Branches flattened, jointed and constricted at joints .... 4. *Opuntia*
　　4. Branches cylindrical, variously ribbed, but not flattened, jointed, or constricted at joints.
　　　5. Plants lianas, climbing by means of adventitious roots; stems deeply ribbed; flowers >12 cm long.
　　　　6. Stems with 3 or 4 prominent ribs; spines conical, <7 mm long, with swollen bases, 2–3 mm wide; hypanthium glabrous without .... 1. *Hylocereus*
　　　　6. Stems with 5–8 slightly prominent ribs; spines acicular, 5–8 mm long, >1 mm wide at base; often with long hairs; hypanthium densely woolly without .... 7. *Selenicereus*
　　　5. Plants shrubs or small trees, erect, columnar; stems nearly cylindrical with many shallow ribs.
　　　　7. Stems with 7–11 longitudinal ribs; areoles with tuft of whitish, long, woolly hairs, especially at apex of stem; spines yellowish, 1–3 cm long .... 6. *Pilosocereus*
　　　　7. Stems with 9–14 longitudinal ribs; areoles felted, but without long woolly hairs; spines yellow but turning grayish, 3–7 cm long .... 8. *Stenocereus*

1. HYLOCEREUS Britton & Rose

Lianas climbing by means of adventitious roots, with many elongate, lateral branches; stem angular, with 3 or 4 ribs. Spines clustered around areoles, along rib margins. Flowers solitary, sessile, usually at ends of branches; hypanthium elongate, funnel-shaped; outer perianth segments scale-like; inner segments petal-like and narrow; stamens numerous; style elongate, stout, not protruding, the stigmas numerous and filiform. Fruit with persistent fleshy scales and numerous minute seeds.

A genus with approximately 20 species with Caribbean distribution.

1. Hylocereus trigonus (Haw.) Saff., Annual Rep. Board Regents Smithsonian Inst. **1908**: 553. 1909. *Cereus trigonus* Haw., Syn. Pl. Succ. 181. 1812.　　　　　Fig. 63A–C.

Cereus triangularis sensu Eggers, Fl. St. Croix. 57. 1879, non L., 1753.

Liana to 10 m long, with many lateral branches; stem 1–3 cm wide, angled, with 3 or 4 ribs, these with wavy margins; surface

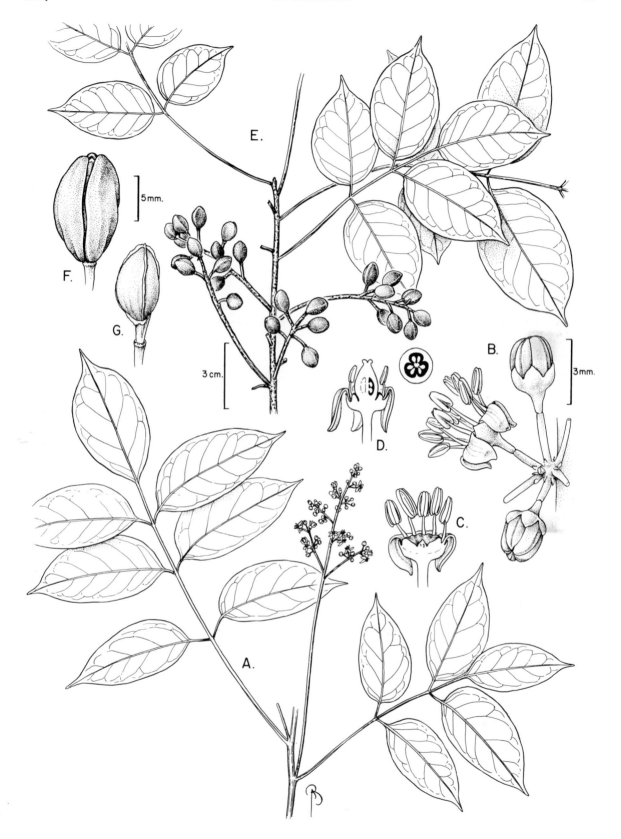

FIG. 62. *Bursera simaruba.* **A.** Flowering branch. **B.** Staminate flower and buds. **C.** L.s. staminate flower. **D.** L.s. pistillate flower with sterile stamens and c.s. ovary. **E.** Fruiting branch. **F.** Drupe. **G.** Pyrenes attached to axis.

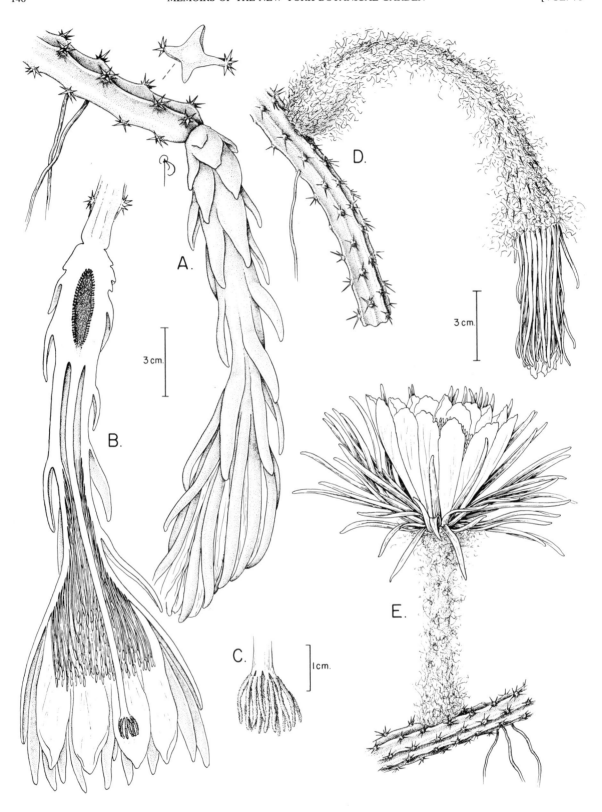

FIG. 63. A–C. *Hylocereus trigonus.* **A.** Flowering branch and c.s. stem. **B.** L.s. flower. **C.** Detail of style-branches. **D, E.** *Selenicereus grandiflorus.* **D.** Branch with senescent flower. **E.** Branch with freshly opened flower.

with many lenticular whitish dots; spines conical, 4–7 mm long, usually in groups of 8, light brown or grayish, slightly swollen at base. Flowers hanging, along the stem or terminal on branches. Perianth glabrous, funnel-shaped, with segments connate into an elongate tube projecting beyond the hypanthium; outer segments fleshy, green, scale-like; inner segments white, petal-like; stamens numerous, in 2 rows, the filaments slender and yellowish. Fruit fleshy, ovoid or ellipsoid, 12–14 × 5–7 cm, purplish pink with white pulp, with numerous minute, black seeds.

DISTRIBUTION: Common throughout dry areas. Calabash Boom (A2874). Also on St. Thomas, Tortola, and Virgin Gorda; Puerto Rico, Mona, and the Lesser Antilles, cultivated and escaped elsewhere.

COMMON NAMES: chickenet, chigger apple, nightblooming cereus.

CULTIVATED SPECIES: *Hylocereus undatus* (Haw.) Britton & Rose is found occasionally growing in gardens.

2. MAMMILLARIA Haw.

Plants spherical or nearly cylindrical. Spines on spirally arranged tubercles, the tubercles cylindrical and green, usually with white hairs or bristles in axils. Flowers small, solitary, borne axillary to the tubercles. Perianth yellow or white; stamens numerous, united into a tube; style not protruding, the stigmas filiform. Fruit club-shaped or pyriform, pink or red, smooth, with numerous minute brown or black seeds.

A genus of about 150 species, ranging from southwestern United States to northern South America, including the West Indies.

1. Mammillaria nivosa Link *ex* Pfeiff., Beschr. Synon. Cact. 12. 1837. *Coryphantha nivosa* (Link *ex* Pfeiff.) Britton & Rose, Ann. Missouri Bot. Gard. **2**: 45. 1915. *Neomammillaria nivosa* (Link *ex* Pfeiff.) Britton & Rose, Cact. 4: **71**. 1923.   Fig. 64I–L.

Plant globose, to 15 cm tall, with long taproot, forming colonies, and producing scanty milky sap. Tubercles conical, spirally arranged, 10–12 mm long, with a crown of radially arranged spines and tufts of woolly white hairs at axils (the area between the tubercles densely hairy), the spines 8–12 per tubercle, 4–22 mm long, turning from yellow to dark brown. Flowers funnel-shaped, 2 cm long, with very short tube; outer perianth segments scale-like; inner perianth segments petal-like, yellow; stigma lanceolate and papillose. Fruit 10–12 mm long, pyriform to club-shaped, smooth, pinkish red, with pink watery pulp. Seeds brown, minute.

DISTRIBUTION: A rare plant, only a few colonies known from the north shore. New Found Bay (A4149). Also on Tortola; the Puerto Rican islands of Mona and Desecheo and on Turk, Caicos, Bahamas, Antigua, Barbuda, St. Martin, and St. Barts.

3. MELOCACTUS Link & Otto

Plants depressed-globose, spherical to cylindrical, with 10–25 longitudinal ribs, crowned by a cylindrical head (cephalium) bearing spirally arranged, reddish spines and tufts of whitish hairs; ribs with areoles along equidistant clusters of radiating spines; glochids wanting. Flowers small, salverform, pink, spirally arranged at the cephalium. Perianth segments spreading; stamens numerous on upper portion of short hypanthium; ovary inferior, the style slender, with numerous linear stigmas. Fruit fleshy, club-shaped to linear, with numerous exarillate seeds.

A genus with about 20 species, in the Greater and Lesser Antilles and Mexico to northern South America.

1. Melocactus intortus (Mill.) Urb., Repert. Spec. Nov. Regni Veg. **16**: 35. 1919. *Cactus intortus* Mill., Gard. Dict., ed. 8. 1768.   Fig. 64A–H.

Melocactus atrosanguineus Pfeiff., Enum. Diagn. Cact. 44. 1837.

Melocactus communis sensu Eggers, Fl. St. Croix 57. 1879, non Link & Otto, 1827.

Melocactus portoricensis Suringar, Verslagen Meded. Afd. Natuurk. Kon. Akad. Wetensch. ser. 3, **9**: 408. 1891.

Plant usually solitary, globose, depressed-globose or ovoid, to 60 cm tall, crowned by a cylindrical cephalium (sometimes more than one); ribs 10–25; spines in clusters of 10–20, brown to gray, stout, unequal, 1–6 cm long; cephalium nearly globose to long-cylindrical, densely covered with whitish, cottony fibers and spirally arranged clusters of reddish spines. Flowers reddish pink, spirally arranged on cephalium, surrounded by a cluster of reddish spines. Floral tube to 2 cm long, hypanthium ca. 1 cm long; stamens of unequal length; style as long as the perianth tube, the stigmas 5, narrowly lanceolate. Fruit club-shaped, juicy, 1.5–2 cm long, reddish pink (usually fading white), with persistent style at apex. Seeds black, minute.

DISTRIBUTION: Common in dry coastal areas. Along trail to Drunk Bay (A4679). Also on Anegada, St. Croix, St. Thomas, Tortola, and Virgin Gorda; Puerto Rico, Bahamas and the Lesser Antilles.

COMMON NAMES: barrel cactus, pope's head, turk's cap.

4. OPUNTIA Mill.

Erect, creeping shrubs to small trees with jointed flattened stems or trees with an erect cylindrical trunk, usually covered with clusters of long spines. Joints succulent, variously shaped, usually longer than wide, with cluster of acicular spines and glochids surrounding the areoles. Leaves minute and early deciduous. Flowers solitary and sessile, usually along joint margins. Receptacle green, usually with clusters of spines and glochids; hypanthium forming a short tube; perianth segments petal-like and spreading; stamens numerous, usually shorter than the perianth, but seldom projecting beyond; style stout, the stigmas numerous, filiform. Fruit fleshy, with numerous seeds, each surrounded by an aril.

A genus of approximately 200 species, native to the New World.

REFERENCE: Howard, R. A. & M. Touw. 1982. *Opuntia* species in the Lesser Antilles. Cact. Succ. J. (Los Angeles) **54**: 170–179.

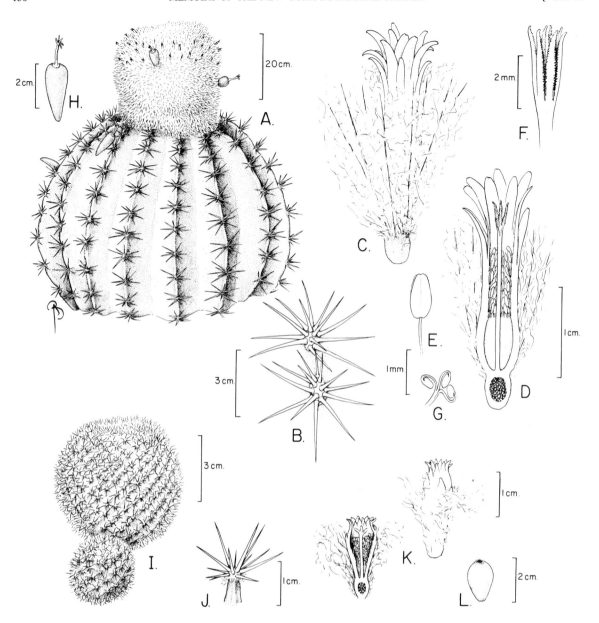

FIG. 64. A–H. *Melocactus intortus*. A. Habit. B. Spine clusters. C. Flower. D. L.s. flower. E. Stamen. F. Stigmas. G. Ovules on stalked placentas. H. Fruit. I–L. *Mammillaria nivosa*. I. Habit. J. Spine cluster. K. L.s. flower, and complete flower. L. Fruit.

Key to the Species of *Opuntia*

1. Plant treelike, 2–3 m tall, with a single main cylindrical trunk; joint branches on upper portion, spreading
   or descending.. 4. *O. rubescens*
1. Plant shrubby or creeping, with or without a short, main cylindrical stem.
   2. Joints oblong to linear .... 3. *O. repens*
   2. Joints obovate, oval, or nearly round.
      3. Perianth red or crimson and erect; stamens and style longer than petals; areoles lacking spines
         .... 1. *O. cochenillifera*
      3. Perianth yellow and spreading; stamens and style shorter than petals; areoles with or lacking spines.
         4. Areoles usually lacking spines or a few present .... *O. ficus-indica* (cultivated)
         4. Areoles with acicular clusters of long (1–3 cm) yellowish spines.................................. 2. *O. dillenii*

1. Opuntia cochenillifera (L.) Mill., Gard. Dict., ed. 8. 1768. *Cactus cochenillifer* L., Sp. Pl. 468. 1753. *Nopalea cochenillifera* (L.) Salm-Dyck, Cact. Hort. Dyck. **849:** 64. 1850.

Erect shrub to 4 m tall, with a short cylindrical trunk and many lateral, oblong, fattened branches, to 50 cm long, with lateral joints. Joints obovate, 10–30 cm long; areoles 1.5–3.5 cm apart, with whitish, minute woolly fibers and deciduous, yellow glochids; spines usually lacking. Flowers 5-7 cm long. Perianth red or crimson, erect, broad, obtuse; stamens exserted, the filaments bright pink, the anthers light yellow; style light yellow, longer than the stamens, the stigmas erect, filiform, yellow. Fruit red, to 5 cm long, pear-shaped, with persistent glochids. Seeds kidney-shaped, 5 mm wide.

DISTRIBUTION: Uncommon in dry coastal scrub, persistent after cultivation. Lameshur (A4145). Also on St. Croix, St. Thomas, Tortola, and Virgin Gorda; possibly native to Jamaica, now widespread in tropical areas through cultivation.

COMMON NAME: French prickly pear.

2. Opuntia dillenii (Ker Gawl.) Haw., Suppl. Pl. Succ. 79. 1819. *Cactus dillenii* Ker Gawl. *in* Edwards, Bot. Reg. **3: pl.** 255. 1818. *Opuntia stricta* Haw. var. *dillenii* (Ker Gawl.) L. D. Benson, Cact. Succ. J. (Los Angeles) **41:** 126. 1969.

Opuntia horrida Salm-Dyck *ex* DC., Prodr. **3:** 472. 1828.

Opuntia tuna sensu Eggers, Fl. St. Croix 58. 1879, non (L.) Mill., 1768.

Erect or spreading shrub, to 3 m tall, with numerous lateral branches; basal stem nearly cylindrical, lateral branches flattened, grayish. Joints obovate, rounded or oblanceolate, 7–36 cm long; areoles 4–5 cm apart, with grayish, woolly fibers, brown glochids and clusters of 3–7 spines, the spines awl-shaped, slightly flattened, straight or slightly curved, yellow to yellowish brown, 3–6 cm long. Flowers 5–7 cm long; receptacle conical. Outer perianth segments narrower, greenish; inner segments yellow, obovate, obtuse; stamens shorter than petals, light yellow; style stout, whitish, the stigmas fingerlike. Fruit reddish, 4–7 cm long, pear-shaped, with persistent glochids. Seeds kidney-shaped, yellowish brown, 4–5 mm wide.

DISTRIBUTION: One of the most commonest *Opuntia* species in dry coastal scrub vegetation. Hurricane Hole (A2831). Also on Anegada, St. Croix, St. Thomas, Tortola, and Virgin Gorda; native to southeastern United States to eastern Mexico, and the West Indies, introduced in northern South America.

COMMON NAMES: bull sucker, miss blyden, prickly pear.

3. Opuntia repens Bello, Anales Soc. Esp. Hist. Nat. **10:** 277. 1881.   Fig. 65A, B.

Opuntia curassavica sensu Eggers, Fl. St. Croix 58. 1879, non (L.) Mill., 1768.

Erect or decumbent shrub to 50 cm tall, forming dense clumps, to a few meters wide. Joints oblong or linear, 5–15 cm long, flattened; areoles ca. 5 mm apart, with whitish, minute, woolly fibers, and deciduous, light yellow glochids; spines 6–8 per areole, thin, acicular, pinkish when young, then turning brown and finally fading gray, 1.5–4.5 cm long. Flowers 4 cm wide; receptacle usually with spines. Perianth bright yellow, spreading, broad, obtuse; stamens shorter than petals. Fruit red, 2–3 cm long, pear-shaped, few-seeded.

DISTRIBUTION: Common in dry coastal scrub. Drunk Bay (A4135), Southeast Pond (A2907). Also on St. Croix, St. Thomas, Tortola, and Virgin Gorda; Puerto Rico.

COMMON NAME: suckers.

4. Opuntia rubescens Salm-Dyck *ex* DC., Prodr. **3:** 474. 1828. *Consolea rubescens* (Salm-Dyck *ex* DC.) Lem., Rev. Hort. **1862:** 172. 1862.   Fig. 65C–G.

Opuntia catacantha Link & Otto *ex* Pfeiff., Enum. Diagn. Cact. 166. 1837.

Opuntia spinosissima sensu Eggers, Fl. St. Croix 58. 1879, non Mill., 1768.

Treelet 6 m tall, with a single cylindrical trunk, ca. 20 cm diam., with persistent clusters of spines; spines grayish, 2–12.5 cm long; bark reddish or grayish brown, flaking off; branches flattened, spreading or descending, with lateral joints. Joints lanceolate, oblong or elliptic, with wavy margins, 20–50 cm long; areoles 1–2 cm apart, with yellowish, minute, woolly fibers and deciduous brown glochids; spines 6–12, thin, acicular, grayish, 1–3 cm long. Flowers 2–3 cm wide; receptacle flattened-obovoid to flattened-ellipsoid, slightly tuberculate, with numerous spines. Perianth yellow, turning orange with age, erect, broad, obtuse with apiculate apex; stamens shorter than petals; style stout, as long as the stamens. Fruit green, to 6 cm long, flattened-obovoid, with persistent spines. Seeds rounded, 6–8 mm wide.

DISTRIBUTION: Common in dry coastal scrub areas. Waterlemon Bay (A4132). Also on St. Croix, St. Thomas, Tortola, and Virgin Gorda; Puerto Rico and the Lesser Antilles.

COMMON NAMES: blyden bush, buds an' rice, prickly pear.

CULTIVATED SPECIES: *Opuntia ficus-indica* (L.) Mill. is found occasionally growing in gardens as an ornamental plant but has not naturalized.

DOUBTFUL RECORD: *Opuntia antillana* Britton & Rose was reported for St. John by Woodbury and Weaver (1987), but no collection was made then and further attempts to find the species have failed.

5. PERESKIA Mill.

Shrubs, trees, or scandent shrubs; stem cylindrical, with axial areoles surrounded by clusters of spines; glochids wanting. Leaves persistent, alternate, somewhat succulent and petiolate. Flowers solitary or in racemes, panicles, or cymes. Receptacle with lower and upper sets of bracteoles, and areoles; perianth with sepaloid and petaloid segments; stamens numerous, usually of unequal length and shorter than the perianth; ovary superior or medial, the style stout, the stigmas numerous, fingerlike, papillose. Fruit fleshy, with persistent bracteoles and areoles; seeds few, not arillate.

A neotropical genus of 17 species.

REFERENCE: Leuenberger, B. E. 1986. *Pereskia* (Cactaceae). Mem. New York Bot. Gard. **41:** 1–141.

1. Pereskia aculeata Mill., Gard. Dict., ed. 8. 1768.   Fig. 66A–E.

Cactus pereskia L., Sp. Pl. 469. 1753. *Pereskia pereskia* (L.) H. Karst., Deut. Fl. 888. 1882.

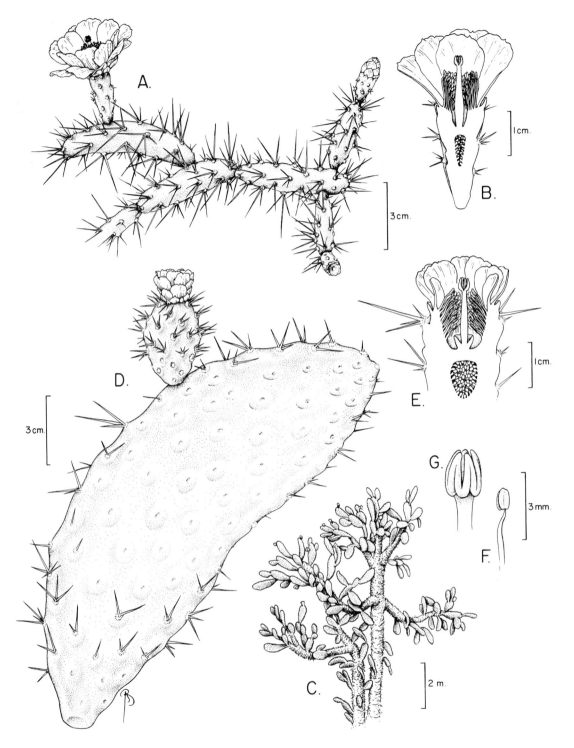

Fɪɢ. 65. A–B. *Opuntia repens.* A. Flowering branch. B. L.s. flower. C–G. *Opuntia rubescens.* C. Habit. D. Fertile joint. E. L.s. flower. F. Stamen. G. Stigmas.

Climbing shrub or liana, to 15 m long, many-branched from a woody base; stem 2–3 cm diam., with recurved paired spines, usually becoming spineless with age; bark light brown and smooth; young branches spiny, elongate, clambering or hanging; areoles with brown or gray woolly fibers; spines dimorphic, to 1.5 cm long. Leaf blades 4–7(–11) × 1.5-5 cm, lanceolate, ovate, or oblong, fleshy, glabrous, the apex shortly acuminate, the base cuneate to rounded, the margins entire; petioles 3–7 mm long.

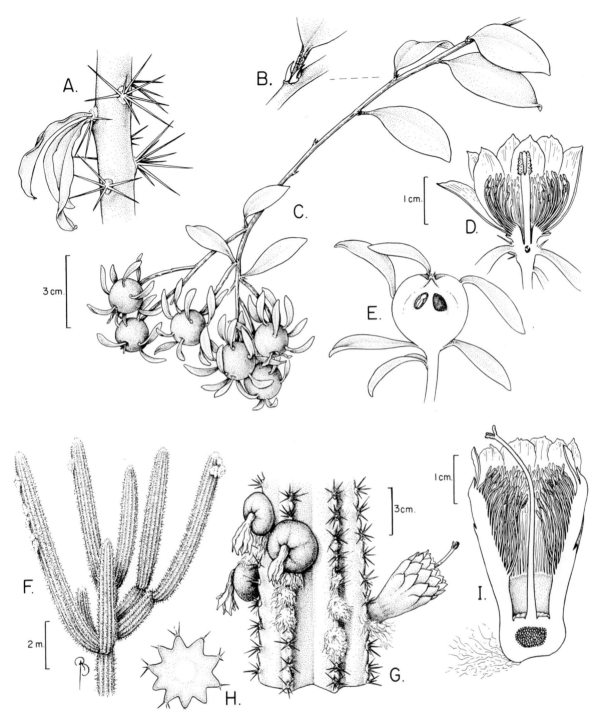

FIG. 66. A–E. *Pereskia aculeata.* **A.** Mature stem with leaves and spine clusters. **B.** Detail of branch showing paired spines at base of leaf petiole. **C.** Fruiting branch. **D.** L.s. flower. **E.** L.s. fruit. **F–I.** *Pilosocereus royenii.* **F.** Habit. **G.** Portion of fertile branch. **H.** C.s. stem. **I.** L.s. flower.

Flowers 2.5–5 cm wide; receptacle cup-shaped, with a few bracteoles and areoles with acicular spines. Perianth white, spreading, broad, with obtuse apex; stamens in 2 series, yellowish; ovary superior, cylindrical, not distinct from the style. Fruit smooth, yellow to orange, ca. 2 cm diam., globose, with persistent bracteoles and spines. Seeds 4–5 mm wide, lenticular to rounded, dark brown and smooth.

DISTRIBUTION: Apparently spontaneous after cultivation. Along road from Bordeaux to Coral Bay (A4267). Also on St. Croix and St. Thomas; now throughout tropical America, its place of origin unknown.

COMMON NAME: Barbados gooseberry.

6. PILOSOCEREUS Byles & G.D. Rowley

Shrubs or small trees, with erect, columnar branches; stems cylindrical, with 10–12 longitudinal ribs; areoles along margins of ribs, usually with long white woolly hairs, surrounded by clusters of spines and lacking glochids. Leaves wanting. Flowers solitary, sessile and bell- to funnel-shaped, with long hypanthium; outer perianth segments scale-like; inner segments petaloid; stamens numerous, inserted on hypanthium; ovary medial, the style stout. Fruit depressed-globose, with numerous exarillate, black seeds.

A genus of approximately 50 species, distributed from Mexico to northern Peru and central Brazil, including the West Indies.

1. Pilosocereus royenii (L.) Byles & G.D. Rowley, Cact. Succ. J. Gr. Brit. **19**: 67. 1957. *Cactus royenii* L., Sp. Pl. 467. 1753. *Cephalocereus royenii* (L.) Britton & Rose, Contr. U.S. Natl. Herb. **12**: 419. 1909. Fig. 66F–I.

Cereus armatus Otto *ex* Pfeiff., Enum. Diagn. Cact. 81. 1837.
Cereus floccosus Otto *ex* Pfeiff., Enum. Diagn. Cact. 81. 1837.
Cereus leiocarpus Bello, Anales Soc. Esp. Hist. Nat. **10**: 276. 1881.

Shrub or small tree, to 7 m tall; branches ascending, bluish green, arising from a short trunk to 30 cm diam.; ribs 7–11; areoles 1–1.5 cm apart, with long, white, woolly hairs, especially near apex of stems; spines 10–12, acicular, yellowish, 1–3 cm long. Flowers funnel-shaped, 5–6 cm long; perianth tube yellowish green at base, the outer segments yellowish to white with pinkish margins, the inner segments white; stamens shorter than the perianth; the stigma exserted, light yellow. Fruit 4–5.5 cm wide, depressed-globose, turning from green to purplish red when mature, the pulp crimson. Seeds nearly kidney-shaped, ca. 3 mm wide, black and shiny.

DISTRIBUTION: Common in dry, coastal scrub areas. Salt Pond (A4668). Also on Anegada, Jost van Dyke, St. Croix, St. Thomas, Tortola, and Virgin Gorda; Puerto Rico and the Lesser Antilles.

COMMON NAMES: dacta dul dul, didledoo, pipe organ.

7. SELENICEREUS Britton & Rose

Lianas or trailing shrubs, climbing by means of adventitious roots; stems elongate, 4–12 -angled or -ribbed, irregularly producing aerial roots; areoles small, usually elevated, densely covered with short woolly hairs; spines acicular and short. Flowers nocturnal, solitary, large, funnel-shaped with long tube; tube with scale-like segments, the areoles with long hairs; upper segments narrow and petal-like; stamens numerous, shorter than the perianth segment; style stout, the stigma numerous. Fruit large, reddish, covered with basal scales, spines, or bristles.

A genus of 16 species distributed from southern United States, through Mexico to northern South America and throughout the West Indies.

1. Selenicereus grandiflorus (L.) Britton & Rose, Contr. U.S. Natl. Herb. **12**: 430. 1909. *Cactus grandiflorus* L., Sp. Pl. 467. 1753. Fig. 63D, E.

Liana to 6 m long, with many lateral branches, climbing by means of adventitious roots; stem 2–2.5 cm wide, 5–8-ribbed; areoles 1–1.5 cm apart along stem ribs, prominent, covered with whitish minute woolly hairs; spines acicular, 5–8 mm long, usually in groups of 5, yellowish. Flowers 15–18 cm long, hanging, produced along the stem or terminal on branches. Perianth funnel-shaped, with an elongate, densely woolly tube; outer segments linear, salmon-colored, the basal ones scale-like and short; inner segments light yellow or whitish, petal-like, narrow; stamens numerous, in 2 rows; style longer than the inner segments. Fruit fleshy, ovoid, to 8 cm long, red or yellowish, with persistent scales, pulp white. Seeds numerous, minute and black.

DISTRIBUTION: Common in dry disturbed areas and along coastal trails (Cruz Bay to Lind Point). Hurricane Hole (A2835). Also on St. Croix, St. Thomas, and Tortola; native to Jamaica and Cuba, now widely planted and naturalized throughout tropical areas.

COMMON NAME: nightblooming cereus.

8. STENOCEREUS (Berger) Riccob.

Shrubs or small trees, with erect, columnar branches; stems cylindrical, with 9–14 longitudinal ribs; areoles along margins of ribs, usually felted, surrounded by clusters of spines and lacking glochids. Leaves wanting. Flowers solitary, sessile and bell- to funnel-shaped, with long hypanthium; outer perianth segments scale-like; inner segments petaloid; stamens numerous, inserted on hypanthium; ovary medial, the style stout. Fruit wide ellipsoid to globose, spiny at first but the spines deciduous, with numerous exarillate, black seeds.

A genus of 25 species, distributed from southwestern United States to Mexico, the Greater Antilles and northern South America.
REFERENCE: Kiesling, R. 1982. Problemas nomenclaturales en el género *Cereus* (Cactaceae). Darwiniana **24**: 443–453.

1. Stenocereus peruvianus (Mill.) R. Kiesling, Darwiniana **24**: 446. 1982. *Cereus peruvianus* Mill., Gard. Dict., ed. 8. 1768.

Cereus hystrix Salm-Dyck, Obs. Bot. Horto Dyck **3**: 7. 1822. *Lemaireocereus hystrix* (Salm-Dyck) Britton & Rose, Contr. U.S. Natl. Herb. **12**: 425. 1909.

Shrub or small tree, to 10 m tall; branches ascending, bluish green, arising near the base from a short trunk, this with deciduous spines; ribs 9–14; areoles 1–2 cm apart; spines usually 13, acicular, yellow turning gray with brown tips, 3–7 cm long. Flowers funnel-shaped, 8–9 cm long; perianth tube with fleshy salmon-colored segments, the inner segments white to greenish; stamens shorter than the perianth; the stigma club-shaped, white. Fruit 5–

6 cm long, ellipsoid, turning from green to scarlet, covered with clusters of deciduous spines, the pulp red.

DISTRIBUTION: Uncommon in dry, coastal scrub. Drunk Bay (A5061).

Also occurs on Cuba, Jamaica, Hispaniola, and the Puerto Rican islands of Desecheo, Muertos, and Mona.

COMMON NAME: pipe organ.

21. Campanulaceae (Bellflower Family)

Herbs or shrubs, with watery or milky exudate. Leaves alternate or rarely opposite, simple, entire or serrate; stipules wanting. Flowers bisexual or rarely unisexual, actinomorphic or zygomorphic, solitary or in racemose or cymose inflorescences; calyx of 5 distinct sepals; corolla bell-shaped or highly zygomorphic with a more or less prolonged tube, 5-lobed or bilabiate; stamens 5, the filaments attached to a nectary disk or to the corolla tube, the anthers connate or connivent, opening by longitudinal slits; ovary inferior to partially superior, commonly crowned by a nectary disk, 2–5-carpellate, each carpel commonly unilocular, with numerous, axile ovules, the style 1, with bilobed stigmas. Fruit a capsule, opening by pores or slits, or rarely a berry, with numerous seeds.

A family of about 70 genera and 2000 species with cosmopolitan distribution.

NOTE: The family consists of 2 well-defined subfamilies, the Lobelioideae and the Campanuloideae, sometimes considered as distinct families.

1. HIPPOBROMA G. Don

A monospecific genus, characterized by the following species.

1. **Hippobroma longiflora** (L.) G. Don, Gen. Hist. **3:** 717. 1834. *Lobelia longiflora* L., Sp. Pl. 930. 1753. *Isotoma longiflora* (L.) C. Presl, Prodr. Monogr. Lobel. 42. 1836.   Fig. 67.

Erect or decumbent herb, to 50 cm tall, producing abundant milky exudate. Leaves alternate; blades 3–12 × 2–6 cm, lanceolate, elliptic, or oblanceolate, chartaceous, glabrous, the apex obtuse to acute and mucronate, the base attenuate, the margins repand-dentate. Flowers solitary. Calyx green, bell-shaped, with 5 linear, denticulate and ciliate margined lobes; corolla white,

salverform, long-tubular (8–16 cm long), with 5 oblong spreading lobes; stamens 5, usually exserted, the anthers white; ovary of 5 locules with numerous ovules, the style long, projecting beyond the anthers. Fruit capsular, nearly bell-shaped, green, 6–9 mm diam. Seeds ellipsoid, ca. 1 mm long, brown.

DISTRIBUTION: Uncommon along moist disturbed areas. Along road to Bordeaux (A3130), along trail to Sieben (A2065a). Also on St. Croix, St. Thomas, and Tortola; throughout the West Indies, southern Florida and tropical continental America, naturalized in the Old World.

NOTE: The plant is said to be toxic to grazing animals and the milky exudate to be an irritant to skin and eyes.

22. Canellaceae (Wild Cinnamon Family)

Trees or shrubs with aromatic leaves and bark. Leaves alternate, simple, with translucent punctations, without stipules. Flowers bisexual, actinomorphic, produced in terminal or axillary corymbiform cymes; calyx of 3 distinct sepals; corolla of 5–12 distinct, imbricate petals; stamens 6 to numerous, united into a column around the ovary, the anthers toward outside of the column, opening by longitudinal slits; ovary superior, 2–6-carpellate, each carpel unilocular, with 2 to many ovules on each placenta, the placentation parietal, the style 1, with as many stigmas as carpels. Fruit a berry with 2 to many seeds. A family of 6 genera and about 17 species, with disjunct distribution, from Florida to Brazil, including the West Indies, Africa, and Madagascar.

REFERENCE: Wilson, T. K. 1960. The comparative morphology of the Canellaceae. Synopsis of the genera and wood anatomy. Trop. Woods **112:** 1–27.

1. CANELLA P. Browne, nom. cons.

A monospecific genus, characterized by the following species.

1. **Canella winterana** (L.) Gaertn., Fruct. Sem. Pl. 1: 373. 1788. *Laurus winterana* L., Sp. Pl. 371. 1753.   Fig. 68.

Canella alba Murray, Syst. Veg., ed. 14, 443. 1784, nom. illegit.

Small tree or shrub to 6 m tall; bark dark brown, with numerous lenticels, aromatic, spicy. Leaf blades 3–10.5 × 2–4 cm, oblanceolate, obovate, or oblong, coriaceous, glabrous, the apex obtuse or rounded, the base attenuate, the margins entire; petioles 0.4–1

cm long. Flowers fleshy; calyx green, bell-shaped, with 3 free, rounded, 3 mm long sepals; corolla reddish, with 5 ovate petals, 4.5–6 mm long, united at base; stamens 10, as long as the corolla, the staminal tube red, the anthers yellow; ovary of 2 locules with 2 ovules each, the style short, the stigmas truncate. Fruit globose, turning from green to red or purple, 6–10 mm diam. Seeds 4, nearly globose, 5–6 mm long, black.

DISTRIBUTION: Common along moist coastal forests. Cruz Bay Quarter along Center Line Road (A2936), Maho Bay Quarter along North Shore Road (A2841). Also on Anegada, St. Croix, and St. Thomas; southern Florida, the West Indies, and northern South America.

COMMON NAMES: caneel, pepper cinnamon, white bark, wild cinnamon.

23. Capparaceae (Caper Family)

Erect or scandent shrubs, trees, or herbs. Leaves alternate, simple, or palmately compound (3–7 leaflets); stipules minute. Flowers bisexual, actinomorphic, in racemes, corymbs, or solitary and axillary; calyx with 4 imbricate to valvate sepals; corolla with 4 white, pink, or yellow distinct petals; stamens 4 to many; ovary superior, of 2 carpels, borne on a stipe (gynophore); ovules few to many

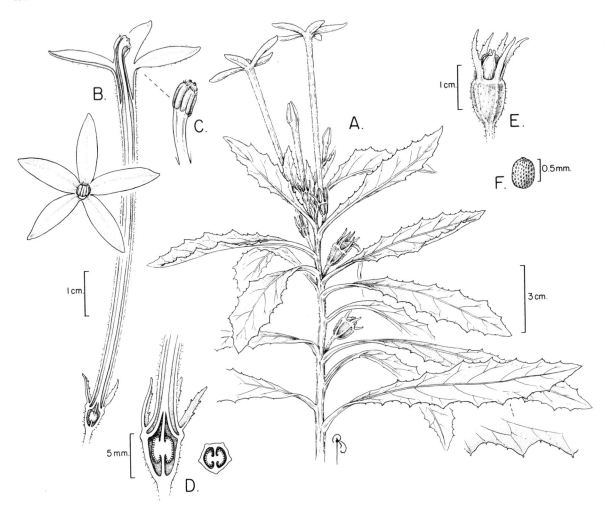

FIG. 67. *Hippobroma longiflora.* **A.** Fertile branch. **B.** Corolla from above and l.s. flower. **C.** Portion of stamens with connivent anthers. **D.** L.s. lower part of flower showing ovary and c.s. ovary. **E.** Fruit. **F.** Seed.

with parietal placentation; stigma capitate. Fruit fleshy, dry or woody, opening along longitudinal valves, irregularly, or not opening, more or less elongate, or globose in *Morisonia.*

About 45 genera and 800 species, from tropical or subtropical regions.
REFERENCE: Ernst, W. R. 1963. The genera of Capparaceae and Moringaceae in the southeastern United States. J. Arnold Arbor. 44: 81–95.

Key to the Genera of Capparaceae

1. Herbs or subshrubs; leaves palmately compound (with 3–7 leaflets), membranous to chartaceous; fruits
   opening by valves, leaving a border where the seeds are attached (replum)..................... 2. *Cleome*
1. Shrubs or small trees; leaves simple, leathery; fruits opening or not, not leaving a replum.
   2. Fruit globose .... 3. *Morisonia*
   2. Fruit nearly cylindrical, elongate .... 1. *Capparis*

1. CAPPARIS L.

Small trees, erect or rarely scandent shrubs. Leaves simple, usually with budlike extrafloral nectaries at axils. Flowers bisexual and regular, in short racemes or solitary. Stamens numerous, much longer than the petals (2–3 times as long); ovary oblong, on a more or less elongate stipe (gynophore). Fruit elongate and opening along one side or short cylindrical and indehiscent or irregularly rupturing; seeds numerous, in 2 rows along fruit wall, covered with a fleshy coat.

A genus with tropical or subtropical distribution and about 150–350 species.
REFERENCE: Prado, D. E. 1993. Lectotypification of *Capparis baducca* L. (Capparaceae). Taxon 42: 655–660.

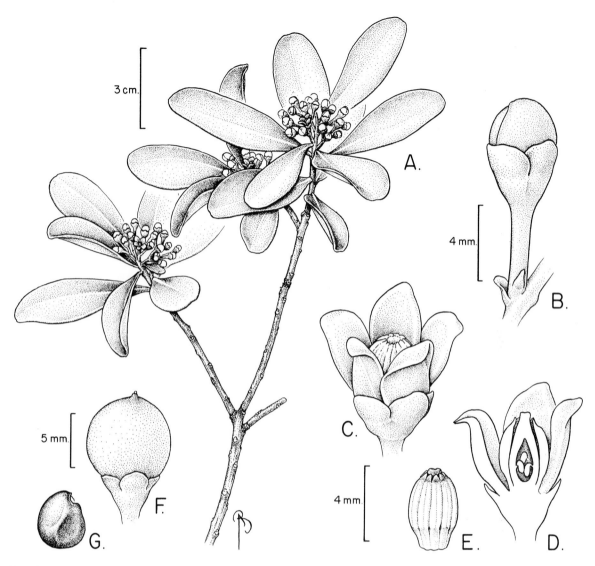

FIG. 68. *Canella winterana.* **A.** Flowering branch. **B.** Flowerbud. **C.** Flower. **D.** L.s. flower. **E.** Connate stamens surrounding pistil. **F.** Fruit. **G.** Seed.

Key to the Species of *Capparis*

1. Underside of leaves silvery to golden (covered with star-shaped scales).
   2. Leaves leathery, margins strongly revolute; flower buds 4-angular, with sepals enclosing the petals; petals with disk-shaped scales on lower surface .... 3. *C. cynophallophora*
   2. Leaves chartaceous, margins slightly or not revolute; flower buds ellipsoid, with short sepals not enclosing the petals; petals densely covered with white stellate hairs on both surfaces .... 6. *C. indica*
1. Underside of leaves green, without hairs or scales.
   3. Petioles to 5.5 cm long, swollen at both ends; leaves obovate to elliptic, tapering to an auriculate base
   .... 2. *C. baducca*
   3. Petioles <1 cm long, not swollen; leaves not as above.
      4. Leaves chartaceous, usually elliptic, with acute or acuminate apex; fruit woody, ellipsoid, to 6 cm long; stipe as long as or longer than the fruit .... 1. *C. amplissima*
      4. Leaves thick-leathery, with rounded, obtuse, or acute apex; fruit fleshy, oblong, >8 cm long; stipe shorter than the fruit.
         5. Leaves broadly oblong, oval, or obovate, with obtuse or subcordate base, with 5–7 pairs of secondary veins, lower surface with minute cavities (visible with a hand lens); fruit 1–1.3 cm wide
         .... 5. *C. hastata*

5. Leaves variable, linear, oblong, lanceolate or obovate, with rounded or obtuse base, with 7–11
pairs of secondary veins, lower surface smooth; fruits 6–8 mm wide .... 4. *C. flexuosa*

1. Capparis amplissima Lam., Encycl. 1: 607. 1785.

Capparis portoricensis Urb., Symb. Ant. 1: 309. 1899.

Small glabrous tree or shrub, to 5 m tall; bark gray, smooth or slightly fissured, forming rectangular plates, with strong fetid smell; wood whitish. Leaf blades 6–10.5 × 3.3–4.5 cm, chartaceous, elliptic, acute or acuminate at apex, obtuse or rounded at base, both surfaces dull green, the margins entire, slightly revolute and wavy; petioles furrowed, 6–7 mm long. Axillary glands cylindrical to trigonous and yellowish. Flowers glabrous; calyx with 4 rounded sepals with hyaline margins, the outer smaller than the inner ones; petals oblong to obovate, without claws, 2–3 cm long, white, coriaceous, glabrous, with hyaline margins; stamens numerous; gynophore brown. Fruit woody, to 6 cm long, ellipsoid, shiny green, turning grayish with age, irregularly rupturing; stipe as long as or longer than the fruit. Seeds ovate, 6–7 mm long.

DISTRIBUTION: In dry and moist forests from sea level to 300 m elevation. Coral Bay (A1848), Bordeaux Mountain (A1899). Also on St. Thomas and Tortola; Hispaniola, Puerto Rico, and Dominica.

2. Capparis baducca L., Sp. Pl. 504. 1753.

Capparis frondosa Jacq., Enum. Syst. Pl. 23. 1760.

Glabrous shrub or small tree to 6 m; bark brown and smooth. Leaf blades 10–25 × 3.5–11.5 cm, leathery, obovate or elliptic, both surfaces dull, with prominent reticulate venation, the apex acute or acuminate, the base auriculate, the margins entire or slightly revolute; petioles 1–5.5 cm long (longer in lower leaves), swollen at both ends. Inflorescences axillary or terminal, ascending corymbs with 5–10 flowers. Flowers glabrous; calyx cup-shaped, 2.5–3 mm long, with 4 rounded lobes, 1–1.5 mm long; petals obovate, greenish to white, notched at apex, glabrous, reflexed, 1.2 cm long; stamens unequal, ca. 1.5 cm long; ovary 0.5–0.7 cm long, oblong to club-shaped, nearly as long as the gynophore. Fruit fleshy, nearly cylindrical with constricted outline, reddish at maturity, 3.5–6 cm long, opening along one side or irregularly, few-seeded; stipe shorter than the fruit. Seeds bean-shaped, ca. 0.7 cm long, with white, fleshy pulp.

DISTRIBUTION: Common in the understory of moist to dry forests. Along road to Bordeaux (A4259), Susannaberg (A4663). Also on Jost van Dyke, St. Croix, St. Thomas, and Tortola; southern Mexico to northern South America, including the West Indies

COMMON NAME: rat bean.

3. Capparis cynophallophora L., Sp. Pl. 504. 1753.
Fig. 69A–E.

Capparis jamaicensis Jacq., Enum. Syst. Pl. 23. 1760.

Small tree or shrub, to 8 m tall and 10 cm diam.; bark light gray and smooth. Branches flattened, densely covered with rusty-brown or golden peltate scales. Leaf blades 5–8 × 1.3–3 cm, leathery, elliptic or oblanceolate, acute or acuminate at apex, attenuate at base, the upper surface glabrous and shiny, the lower surface densely covered with golden-silvery peltate scales, the margins revolute; petioles 0.8–1.5 cm long, densely covered with peltate scales. Flowers fragrant, borne on short axillary cymes, the peduncle and pedicels flattened and rusty-colored; flower buds white, 4-angled; sepals 4, ovate to lanceolate, 0.8–1 cm long, whitish without, light green within; petals obovate, white turning pink with age; filaments white or pink; anthers green or light

yellow; gynophore white, the ovary oblong, light green, the stigma bright orange. Fruit leathery, linear-cylindrical, constricted between seeds, 8–15 cm long, surface rusty-brown, orange within, opening along one side, stipe brown, ca. 5.5 cm long, glabrous. Seeds bean-shaped, shiny black, with orange fleshy aril at base.

DISTRIBUTION: A common tree found in moist or dry coastal forests. East End (A1798, A1821). Also on St. Croix, St. Thomas, Tortola, and Virgin Gorda; throughout the West Indies, Florida to Central America.

COMMON NAMES: black willie, black witty, Jamaican caper, linguam tree.

4. Capparis flexuosa (L.) L., Sp. Pl., ed. 2, 1: 722. 1762.
Morisonia flexuosa L., Pl. Jamaic. Pug. 14. 1759.
Fig. 69F–H.

Capparis saligna Vahl, Symb. Bot. 3: 66. 1794.

Erect or liana-like, glabrous shrub, to 10 m long, many-branched from base; bark dark gray, smooth to rough, inner bark beige. Branches usually zigzag and hanging. Leaf blades 5–10 × 1.5–2.5 cm, leathery, linear, oblong, lanceolate, or obovate, acute or obtuse at apex, obtuse or rounded at base, blade with 7–11 pairs of secondary veins, the margins entire, revolute; petioles 8 mm or shorter. Flowers opening at night or early morning, ephemeral after one night, produced in short corymbs at end of branches; calyx with yellowish, rounded sepals, 6–10 mm long; petals green or yellowish green, obovate, 1.5–3 cm long; stamens numerous, white, 4–6 cm long; gynophore turning from white to deep pink, the ovary light green, oblong. Fruit cylindrical, constricted between seeds, 15–25 cm long, surface turning from light, shiny green to light yellow, with a noticeable red line along two sides, inner walls bright red, irregularly splitting; fruit wall usually coiling away from the placental tissue still containing the seeds; stipe brown, shorter than fruit. Seeds bean-shaped, 1 cm long, with white, fleshy aril.

DISTRIBUTION: Common in dry and coastal forests. Lind Point (A2312). Also on Jost van Dyke, St. Croix, St. Thomas, Tortola and Virgin Gorda; throughout the West Indies and Florida to South America.

COMMON NAMES: bottle wiss, dog caper, limber caper.

5. Capparis hastata Jacq., Enum. Syst. Pl. 23. 1760.
Capparis flexuosa (L.) L. forma *hastata* (Jacq.) Dugand, Caldasia 1(2): 51. 1941.

Capparis cynophallophora var. *latifolia* Griseb., Fl. Brit. W. I. 18. 1859. *Capparis latifolia* (Griseb.) A. Stahl, Estud. Fl. Puerto-Rico 2: 186. 1884.
Capparis coccolobifolia Eichler *in* Mart., Fl. Bras. 13(1): 284. 1865.

Small tree to 6 m tall and 8 cm diam.; bark dark gray; young parts covered with whitish papillae and lenticels. Branches zigzag and spreading. Leaf blades 5–11.5 × 4–6.5 cm, leathery, oval, broadly oblong or obovate, both surfaces dull, with prominent reticulate venation, the lower surface with minute cavities (visible with a hand lens), the apex usually rounded or obtuse, less often acute, notched or mucronate, the base obtuse or subcordate, the margins entire and yellowish; petioles <1 cm long, verrucose. Gland in leaf axil globose or obovoid. Inflorescences terminal racemes, few- to 10-flowered. Flowers nocturnal, ephemeral, lasting one night, glabrous; calyx with 4 distinct, green, rounded sepals with hyaline margins, the outer sepals smaller (6–8 mm);

Fɪɢ. 69. A–E. *Capparis cynophallophora.* **A.** Branch of juvenile plant and flowering branch of mature plant with detail of peltate scales. **B.** Flowerbud and detail of peltate scales. **C.** L.s. flower. **D.** Petal with detail of peltate scale. **E.** Fruit. **F–H.** *Capparis flexuosa.* **F.** Flowering branch. **G.** Leaves. **H.** Fruit. **I–K.** *Morisonia americana.* **I.** Flowering branch. **J.** L.s. flower and details of stellate scales. **K.** Fruits.

petals greenish white, obovate, concave, ca. 2.3 cm long; stamens white, the filaments to 4.5 cm long; ovary cylindrical, yellowish, the gynophore longer than the filaments. Fruit 8–30 cm long, cylindrical with constricted outline, straight or curved, turning from green to pinkish yellow, bright red within, usually opening along one side. Seeds bean-shaped, ca. 1.5 cm long, with white, fleshy aril.

DISTRIBUTION: Common in dry seasonal and coastal forests, from sea level to 500 m elevation. Along trail to Brown Bay (A1879), Princess Bay (A4685). Also on Jost van Dyke, St. Croix, and St. Thomas; Hispaniola, Puerto Rico, Lesser Antilles, to northern South America.

COMMON NAME: caper tree.

6. Capparis indica (L.) Druce, Bot. Exch. Club Soc. Brit. Isles **3**: 415. 1914 [Feb]. *Breynia indica* L., Sp. Pl. 503. 1753. *Capparis indica* (L.) Fawc. & Rendle, J. Bot. **52**: 144. 1914 [Jun].

Shrub or small tree to 7 m tall and 8 cm diam.; bark gray to dark gray, smooth to slightly fissured, inner bark light yellow and bitter; young parts densely covered with golden or silvery scales; branches flattened and angular when young. Leaf blades 5–10 × 2.2–4.5 cm, chartaceous, elliptic (rarely obovate), linear and much longer in sprout shoots, glabrous except for a few stellate hairs on midrib and margins, lower surface densely covered with silvery or golden scales, secondary veins noticeable, the apex acute or obtuse and mucronate, rarely rounded or notched, the base obtuse or tapering, the margins slightly wavy and slightly or not revolute; petioles 1.5 cm or shorter, densely covered with golden scales. Inflorescences short, axillary corymbs with to 12 flowers. Flower buds ellipsoid, densely covered with golden scales, the sepals not covering the petals. Flowers fragrant; calyx of 4 equal, triangular or lanceolate sepals to 2 mm long; sepaloid appendages opposite to the sepals, triangular and slightly smaller than the sepals; petals white, obovate or spatulate, ca. 1.2 cm long, densely covered with whitish stellate hairs on both surfaces; filaments glabrous, except for few hairs at base; gynophore and ovary densely covered with silvery golden scales. Fruit linear-cylindrical, with outline constricted between seeds, 10–20 cm long and 0.8 cm wide, surface rusty-brown, orange within, opening along one side; stipe much shorter than the fruit, also rusty-brown. Seeds ellipsoid, ca. 7 mm long, with fleshy orange aril.

DISTRIBUTION: In dry forests or thickets, from sea level to 330 m elevation. Along road to Bordeaux (W117); Cruz Bay Quarter along Center Line Road (5036), Lameshur (A5444). Also on Jost van Dyke, St. Croix, St. Thomas, Tortola, and Virgin Gorda; southern Mexico to northern South America, including the West Indies.

COMMON NAME: linguam.

2. CLEOME L.

Annual or perennial herbs or subshrubs, seldom reaching more than 1 m. Leaves alternate, palmately compound, with 3–7 leaflets and petioles almost as long as the blades. Flowers produced in racemes at the end of stem, or solitary in leaf axils, the racemes showing transition between leaves and bracts. Flowers zygomorphic; calyx of 4 free sepals; petals 4, clawed, arranged toward one side of the flower; stamens 6 to many, sometimes on elongate stipe (androgynophore); ovary nearly cylindrical, usually borne on elongate stipe (gynophore) or less often sessile (as in *C. viscosa*), with 2 carpels, the stigma capitate. Fruit an elongate, cylindrical, dry capsule, opening along 2 valves, leaving a border where the seeds were attached (replum); seeds kidney-shaped, 1.5–3 mm long, numerous, along 2 lines on fruit wall.

A cosmopolitan genus of about 200 species.

REFERENCES: Iltis, H. H. 1959. Studies in the Capparidaceae VI. *Cleome* sect. *Physostemon*: taxonomy, geography and evolution. Brittonia 11: 123-162; Iltis, H. H. 1960. Studies in the Capparidaceae VII. Old World cleomes adventive in the New World. Brittonia 12: 279–294.

Key to the Species of *Cleome*

1. Petals yellow, stamens not projecting beyond the petals; plants viscous .... 3. *C. viscosa*
1. Petals white, stamens projecting beyond the petals; plants not viscous.
   2. Stipules spiny; floral bracts simple; fruit stipe without articulation.. 2. *C. spinosa*
   2. Stipules wanting; floral bracts trifoliolate; fruit stipe with articulation at its middle .... 1. *C. gynandra*

1. Cleome gynandra L., Sp. Pl. 671. 1753. *Gynandropsis gynandra* (L.) Briq., Annuaire Conserv. Jard. Bot. Genève **17**: 382. 1914.

Cleome pentaphylla L., Sp. Pl., ed. 2, **2**: 938. 1763.

Herb or subshrub ca. 0.5 m tall; young stems covered with minute glandular hairs, older parts becoming glabrous. Leaves membranous, with 5 or 7 leaflets; leaflets elliptic or oblanceolate, 2.5–5 cm long, the terminal longer, glabrous, except for few hairs on veins and margins, the margins finely serrate or entire; petioles almost as long as the blades, more or less covered with white, glandular hairs. Flowers in terminal racemes, showing a transformation of leaves to trifoliolate bracts; calyx green, covered with white glandular hairs externally, the sepals lanceolate, 5 mm long; petals white, 1.5–2 cm long, the base 1–1.5 cm long, often pink or purple; stamens 6, the filaments turning from purple to white, 5 cm long, the anthers purple; gynophore purple, as long as the filaments, the ovary green, cylindrical. Fruit glabrous, turning from green to straw-colored, linear-cylindrical, with slightly wavy outline and prominent reticulate venation, tapering toward both ends, rugulose, 3–6.5 cm long, borne on a stipe (gynophore + androgynophore) 3.5–5 cm long, with an articulation (scar from stamens) at its middle; peduncle 1.5–2.7 cm long. Seeds kidney-shaped, beige, 2 mm wide, verrucose-reticulate.

DISTRIBUTION: In open, disturbed areas. Emmaus (W326). Also on St. Croix, St. Thomas, and Tortola; introduced from Africa into the Caribbean at an early date, now from southeastern United States to Brazil.

COMMON NAMES: massambee, small spider flower.

2. Cleome spinosa Jacq., Enum. Syst. Pl. 26. 1760.

Fig. 70A–E.

Cleome pungens Willd., Hort. Berol. t. 18. 1804.

Herb or subshrub to 1 m, with many lateral branches; stem finely striate, sparsely to densely covered with slender, glandular hairs; stipules minute, spiny, often deciduous. Leaves palmately compound with 3 or 5 leaflets, membranous to chartaceous; leaflets

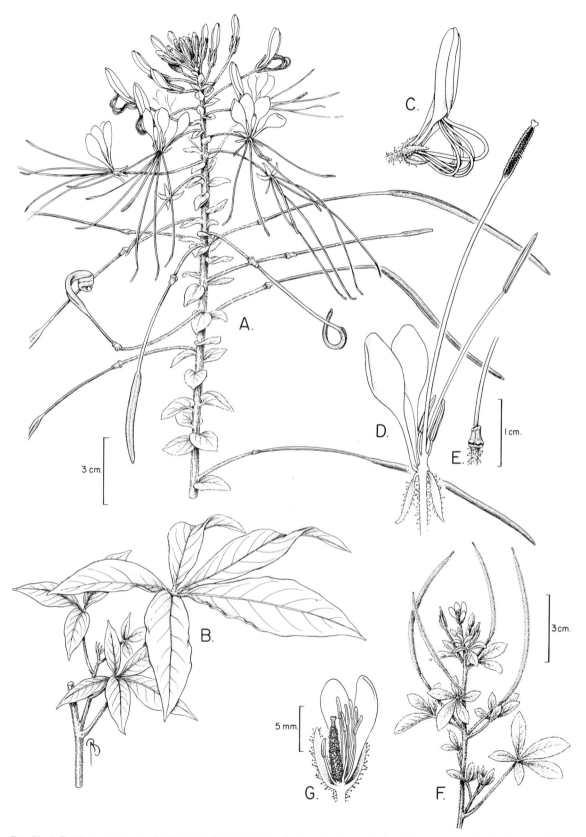

FIG. 70. A–E. *Cleome spinosa.* **A.** Fertile branch. **B.** Leafy branch. **C.** Expanding flower. **D.** L.s. fully open flower. **E.** Detail of receptacle with gynophore. **F, G.** *Cleome viscosa.* **F.** Fertile branch. **G.** L.s. flower.

elliptic or lanceolate, 3–8 1–1.8 cm, acuminate at apex, attenuate, obtuse, or rounded at base, both surfaces glabrous or more or less covered with white, glandular hairs; petiolules and petioles densely covered with glandular hairs, petioles as long or longer than the leaflets. Flowers borne in terminal racemes, showing a transition of leaves to simple bracts; pedicels ca. 2 cm long, densely covered with white, glandular, erect hairs. Calyx of 4, free, awl-shaped, reflexed sepals, ca. 8 mm long, densely covered with same type of hairs as pedicel; petals white, glabrous, ca. 2 cm long, oval to nearly rounded, tapering into elongate (1–1.2 cm long) narrow base; stamens 6, the filaments glabrous, purple, ca. 5 cm long, the anthers purple, ca. 3 mm long; ovary cylindrical, purple, 5–6 mm long, densely covered with minute papillae, not projecting beyond the petals (short gynophore). Fruit turning from green to straw-colored, cylindrical, with slightly wavy outline, to 10 cm long, with obtuse apex; stipe purple, shorter than the fruit. Seeds beige, kidney-shaped, smooth or warty, ca. 2 mm long, numerous.

DISTRIBUTION: In open disturbed areas and along roadsides. Emmaus (A1993, A3777). Also on St. Croix and St. Thomas; widely distributed in the New World.

COMMON NAMES: spider flower, wild massambee.

NOTE: This is not the cultivated *C. spinosa* of authors (*Cleome hassleriana* Chodat.) that has pink petals and glabrous ovaries.

3. Cleome viscosa L., Sp. Pl. 672. 1753.   Fig. 70F, G.

Herb or subshrub to 1 m tall; viscous, especially on young parts; stem green, striate; stipules minute, covered with glandular hairs, early deciduous. Leaves chartaceous, palmate, with 3 or 5 leaflets; leaflets elliptic, oblong or oblanceolate, 1–7 × 0.5–2.2 cm (the central one larger), both surfaces covered with minute, whitish, glandular hairs, the apex acute or acuminate, the base attenuate or cuneate, the margins wavy and ciliate; petioles shorter or as long as the blades. Flowers solitary in leaf axil; calyx light green, densely covered with glandular hairs, the sepals 4, oblong to lanceolate, 5 mm long; petals yellow, obovate, 1 cm long, tapering into a 3 mm long, narrow base, usually purplish; stamens (10–)16–26, the filaments yellow, 7 mm long, the anthers greenish; ovary light yellow, cylindrical, sessile, densely covered with glandular hairs, the style minute, the stigma yellow and capitate. Fruit cylindrical, 6–8 cm long, striate, turning from green to straw-colored, without stipe; peduncle 1.5–2.3 cm long. Seeds brown, kidney-shaped, 1.5 mm long, transversely ridged.

DISTRIBUTION: Common in open, disturbed areas. Cruz Bay (A1954, A2353), Emmaus (A1997). Also on St. Croix, St. Thomas, Tortola, and Virgin Gorda; a temperate and tropical weed native to the Old World tropics.

3. MORISONIA L.

Small trees or shrubs. Leaves alternate, simple. Flowers regular, produced in corymbs or racemes; calyx bell-shaped, splitting into 2–4 lobes, petals 4; stamens numerous; ovary on long gynophore; stigma capitate. Fruit indehiscent, globose, with numerous seeds attached to the fruit wall.

A neotropical genus of 4 species.

1. Morisonia americana L., Sp. Pl. 503. 1753.

Fig. 69I–K.

Small tree or shrub, to 10 m tall; bark gray, smooth or warty; young parts covered with rusty-brown, disk-shaped scales. Leaf blades 6–23 × 2.3–10 cm, leathery, oblong, ovate, lanceolate, or elliptic, involute, glabrous, except for a few scales over the reticulate, prominent veins, the apex acute, obtuse rounded or acuminate, the base obtuse, rounded, or auriculate, less often cuneate, the margins wavy; petioles 1–6.5 cm long, swollen at both ends, usually covered with rusty-brown scales. Flowers in short, axillary racemes, to 10-flowered; calyx ca. 1 cm long, densely covered with stellate scales, splitting into 2 or 4 rounded lobes; petals elliptic, ca. 1.5 cm long, densely covered with stellate scales on lower surface and with stellate hairs on inner surface; stamens numerous, on a short stipe (androgynophore), the filaments glabrous, ca. 1.5 cm long; gynophore much longer than the ovary, densely covered with rusty-brown scales, the ovary obconical, with same scales as gynophore. Fruit leathery, globose, light brown, indehiscent, 2.5–4 cm diam., usually eaten by rats. Seeds numerous, bean-shaped, ca. 1 cm long, embedded in fleshy pulp.

DISTRIBUTION: A common species in moist to dry forests. Along road to Bordeaux (A4678), Lameshur (B514, V3100). Also on Jost van Dyke, St. Croix, St. Thomas, and Tortola; from western Mexico to northern South America, including the West Indies.

COMMON NAME: rat apple.

24. Caricaceae (Papaya Family)

Generally unbranched, medium to small, soft-stemmed trees, with a terminal cluster of long petiolate leaves, producing a milky exudate. Leaves alternate, mostly palmately lobed or palmately compound or rarely entire or pinnatisect; stipules when present spinelike. Flowers actinomorphic, usually unisexual, produced in axillary cymes or rarely solitary; calyx small, 5-lobed; corolla long-tubular in male flowers or short-tubular in female flowers, with 5 spreading, twisted petals; stamens 10, in 2 whorls or rarely only 5 stamens in a single whorl, adnate to corolla tube; ovary superior, unilocular, of 5 united, multiovular carpels with parietal placentation, the style short or wanting, the stigmas 5. Fruit a fleshy, large berry with numerous seeds.

A neotropical family of 4 genera and 30 species.

1. CARICA L.

Small trees with palmate, entire, or palmately compound leaves, without stipules, polygamous, dioecious or rarely monoecious. Male flowers on long, panicle-like cymes; female flowers on short, few-flowered cymes. Other characters typical of the family.

A genus of 22 species, naturally occurring from southern Mexico to northern Argentina. One species (*C. papaya*) is widely cultivated throughout the tropics for its edible fruits.

1. Carica papaya L., Sp. Pl. 1036. 1753.   Fig. 71.

Tree to 6 m tall, usually dioecious, unbranched (or with few branches at upper portion), producing abundant milky exudate; stem with numerous, large leaf scars; bark grayish and smooth. Leaf blades 20–35 cm diam., chartaceous to leathery, palmately lobed (7–11 lobes), with rounded outline, glabrous, lower surface lighter, with prominent palmate venation, the base cordate, the

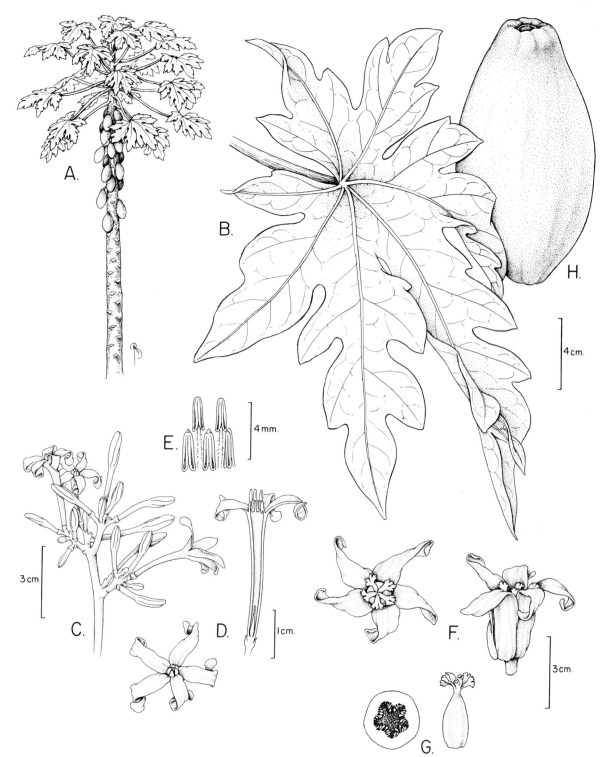

FIG. 71. *Carica papaya*. **A.** Habit. **B.** Leaf. **C.** Detail of staminate inflorescence. **D.** Staminate flower from above and l.s. staminate flower. **E.** Stamens of 2 whorls. **F.** Pistillate flower, top and lateral views. **G.** C.s. ovary and pistil. **H.** Fruit.

margins wavy; petioles 30–90 cm long, hollow. Flowers creamy-white to light yellow; calyx cup-shaped, 1–1.5 mm long; corolla long-tubular (2–4 cm), salverform, with reflexed, lanceolate lobes in male flowers, or nearly bell-shaped, with nearly free petals in

female flowers; stamens 10, pilose, the outer ones with longer filaments; ovary ovoid, the style very short, the stigmas ca. 1 cm long, flattened. Fruit variously shaped, usually ellipsoid to rounded, 5–45 × 5–15 cm, turning from green to yellow, with yellow to light orange pulp. Seeds black, ca. 4–6 mm long, ellipsoid, wrinkled, with fleshy, translucent aril.

DISTRIBUTION: Mainly cultivated, but also spontaneous in moist, dis-

turbed areas. Gift Hill (A820), Great Cruz Bay (A2365). Native to tropical America, now widely cultivated throughout the tropics.

COMMON NAME: papaya.

25. Casuarinaceae (Australian Pine Family)

Trees or shrubs, monoecious or dioecious. Leaves small, scale-like, whorled, more or less connate, forming a sheath at each node. Twigs drooping, green, resembling *Equisetum* (horse-tail). Inflorescence of two kinds, the staminate amentiferous, the pistillate capitate. Flowers small, wind-pollinated, without perianth, unisexual. Staminate flowers of 1 stamen, subtended by a single bract and 2 bracteoles. Pistillate flowers with a single pistil of 2 carpels, each containing 2 ovules (usually 1 carpel develops). Fruit 1-seeded, samaroid, initially enclosed by woody bracteoles.

Four genera with 90 species, mostly from Australia.

REFERENCE: Wilson, K. L. & L. A. S. Johnson. 1986. Casuarinaceae. *In* A. S. George (ed.), Flora of Australia 3: 100–174.

1. CASUARINA L.

Characters of the family. A genus with approximately 55 species.

REFERENCE: Friis, I. 1980. The authority and date of publication of the genus *Casuarina* and its type species. Taxon 29: 499–501.

1. Casuarina equisetifolia L., Amoen. Acad. **4:** 143. 1759.

Monoecious tree, reaching 30 m tall and 45 cm diam.; bark grayish brown and scaly; branches elongate, hanging. Leaves 6–8 per node, scale-like, lanceolate, with ciliate margins. Staminate inflorescences a spike, 2–3 cm long, with spirally imbricate flowers, the stamens brown; pistillate inflorescence a very short spike,

3–5 mm long, with congested flowers at its distal part, the style short, bearing 2 reddish stigmas. Fruit (infructescence) woody, conelike, ellipsoid, 1.5–1.8 cm long, turning from green to brown, with numerous dehiscent, 1-seeded locules. Seeds elliptic, winged, 6–8 mm long.

DISTRIBUTION: An uncommon ornamental tree. Turner Bay (A4661). Native to Australia, now widely planted throughout the tropics.

COMMON NAME: Australian pine, weeping willow.

26. Cecropiaceae (Trumpet Tree Family)

Dioecious trees, shrubs, or lianas, rarely producing milky sap. Leaves alternate, simple, entire to deeply lobed, with large stipules. Flowers unisexual, minute, with reduced perianth, produced in aments; calyx of 2–4 distinct or connate sepals; petals lacking; stamens 2–4; ovary superior, of a single carpel, with a basal ovule. Fruit fleshy, simple or multiple.

A family of 6 genera and ca. 275 species, with tropical distribution.

1. CECROPIA Loefl., nom. cons.

Trees, not producing milky sap; branches few, with septate diaphragmal pith; stipules large, early deciduous. Leaves palmately lobed to entire, peltate, lower surface usually whitish (tomentose); petioles elongate. Flowers in cylindrical aments, the aments clustered at distal portion of peduncles and subtended by a large bract. Fruit an achene, with persistent perianth parts.

A genus of 60–100 species from tropical America.

1. Cecropia schreberiana Miq. *in* Mart., Fl. Bras. **4:** 150. 1853. Fig. 72.

Cecropia antillarum Snethl., Notizbl. Bot. Gart. Berlin-Dahlem **8:** 364. 1923.

Cecropia urbaniana Snethl., Notizbl. Bot. Gart. Berlin-Dahlem **8:** 366. 1923.

Cecropia sericea Snethl., Notizbl. Bot. Gart. Berlin-Dahlem **8:** 368. 1923.

Tree to 20 m tall and 35 cm diam.; trunk usually branched only at upper portion; bark grayish and smooth. Leaf blades 25–40 cm across, leathery, with rounded outline, 7–9 palmately lobed, the upper surface scabrous, the lower surface whitish tomentose,

the venation prominent, the base cordate to peltate, the margins wavy; petioles 20–70 cm long, pubescent. Staminate aments numerous, 4–5 × 0.4 cm, with short individual peduncles; pistillate aments 2–6, 5–6 × 0.6–1 cm, sessile; bracts ovate, densely covered with long whitish hairs covering the immature inflorescences; peduncle 4–9 cm long, pubescent. Achene whitish, ca. 2 mm long.

DISTRIBUTION: A fast-growing tree of moist secondary forests and disturbed areas such as roadsides and landslides. Cruz Bay Quarter, along Center Line Road (A2845). Also on St. Croix, St. Thomas, and Tortola; throughout the West Indies.

COMMON NAMES: trumpet tree, trumpet wood.

NOTE: Many West Indian botanists have referred this species to *Cecropia peltata* L., a species that is restricted to northern South America, Central America, and Jamaica.

27. Celastraceae (Bittersweet Family)

Trees, shrubs, or woody vines. Leaves simple, alternate to opposite, rarely whorled; stipules minute, early deciduous or wanting. Flowers bisexual or unisexual, actinomorphic, produced on cymose or racemose, axillary or terminal inflorescences; calyx minute, with 4–5 distinct or connate sepals; corolla of 4–5 distinct, spreading petals; stamens as many as the petals and alternating with them; ovary

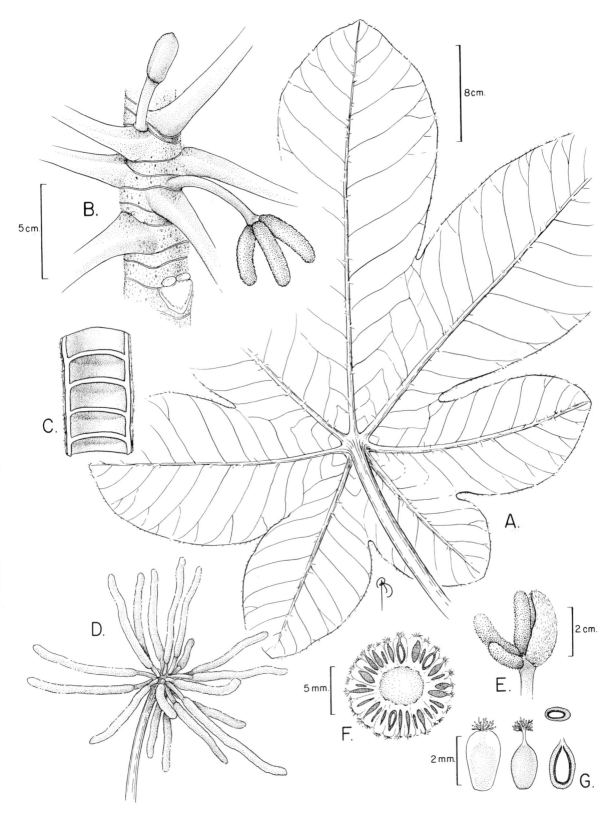

Fig. 72. *Cecropia schreberiana*. **A.** Leaf. **B.** Detail of branch with axillary pistillate inflorescences. **C.** L.s. stem showing diaphragmal pith. **D.** Staminate inflorescence. **E.** Pistillate inflorescence. **F.** C.s. pistillate ament. **G.** Pistils and c.s. and l.s. ovary.

superior, with an annular nectary disk at its base (seldom fused to ovary), the locules 2–5, with axile placentation, each with 2 ovules, the style short or lacking, stigmas 1–5. Fruit a drupe or capsule, usually with arillate seeds.

A primarily pantropical family of about 50 genera and 800 species, with members extending into temperate zones.

REFERENCE: Brizicky, G. K. 1964. The genera of Celastrales in the southeastern United States. J. Arnold Arbor. 45: 206–234.

Key to the Genera of Celastraceae

1. Leaves opposite to subopposite; flowers in axillary, dichasial cymes.
    2. Flower 5-merous; fruits >2 cm long, leathery .... 1. *Cassine*
    2. Flowers 4-merous; fruits <5 mm long, fleshy .... 2. *Crossopetalum*
1. Leaves alternate; flowers fasciculate in leaf axils.
    3. Leaf margins strongly revolute and white; flower 5- merous; fruits ellipsoid to nearly globose, green, 1–1.2 cm long, capsular; seed with white aril .... 3. *Maytenus*
    3. Leaf margins slightly revolute but not white; flowers 4-merous; fruits ellipsoid to nearly globose, with a median longitudinal groove, yellow to orange, 4–5 mm long, drupaceous; seed not arillate ...... 4. *Schaefferia*

1. CASSINE L.

Dioecious shrubs or small trees. Leaves opposite, subopposite, or alternate, coriaceous, entire or crenulate; stipules minute. Flowers unisexual, produced on axillary cymes; calyx of 5 distinct sepals; corolla of 5 distinct, greenish white petals; stamens 5, shorter than the petals, modified into petaloid staminodes in pistillate flowers; nectary disk flattened and crenulate; ovary partially immersed on disk, with 2–5 locules, the stigma 2–5-lobed. Fruit a leathery drupe, with exarillate seeds.

A genus of approximately 80 species, most of which are native to Africa, 6 species in the Caribbean.

1. **Cassine xylocarpa** Vent., Choix Pl. 23, t. 23. 1803.

Elaeodendron xylocarpum (Vent.) DC., Prodr. **2:** 11. 1825. Fig. 73E–J.

Elaeodendron dioicum Griseb., Fl. Brit. W. I. 709. 1864.

Shrub or rarely a small tree, 2–4 m tall, many-branched from base; bark dark gray and smooth. Leaves opposite or subopposite; blades 2.5–10 × 1.5–5 cm, leathery, elliptic, ovate, rounded, or oblong, glabrous, the venation reticulate, slightly prominent or inconspicuous, the apex obtuse or rounded, less often acute or retuse, the base rounded or obtuse, less often cuneate, the margins strongly revolute, entire or crenate-serrate; petioles 0.5–1 cm long. Flowers with short pedicels, produced in short, dichasial cymes. Calyx cup-shaped, with rounded, 1 mm long, greenish sepals; petals ovate, ca. 1.5 cm long; stamens 1 mm long; staminodes of pistillate flowers oblong; ovary depressed-ovoid, the stigmas 2- or 3-lobed. Fruit ellipsoid, ovoid, or globose, 2–3.2 cm long, obtuse or shortly apiculate at apex, turning from green to yellow. Stones with 2 or 3 ovate-elliptic brown seeds.

DISTRIBUTION: Common in dry forests along the coast. Cinnamon Bay (M17083), Salt Pond (A2951), Southside Pond (A1824). Also on Anegada, Jost van Dyke, St. Croix, St. Thomas, Tortola, and Virgin Gorda; in eastern Mexico, Panama, Venezuela, and the West Indies.

COMMON NAMES: nothing nut, nut muscat, spoon tree, ton ton.

2. CROSSOPETALUM P. Browne

Shrubs or small trees. Leaves opposite or whorled, chartaceous, crenulate to serrate; stipules minute. Flowers bisexual, produced on axillary cymes; calyx of 4 connate sepals; corolla of 4 distinct, greenish, reflexed petals; stamens 4, shorter than the petals; nectary disk cup-shaped, 4-lobed; ovary of 2–5 locules, the stigma 4-lobed. Fruit a fleshy drupe, with exarillate seeds.

A genus of 15 species from the American tropics.

1. **Crossopetalum rhacoma** Crantz, Inst. Rei Herb. **2:** 321. 1766. Fig. 73A–D.

Rhacoma crossopetalum L., Syst. Nat., ed. 10, **2:** 896, 1361. 1759.

Myginda pallens Sm. *in* Rees, Cycl. **24:** 1813.

Shrub or rarely a small tree, 2–4 m tall, many-branched from base; bark gray and smooth; young branches 4-angled. Leaves opposite or rarely subopposite; blades 1.5–3.7 × 1.2–2.3 cm, chartaceous, ovate, elliptic, lanceolate, or rounded, glabrous, the venation reticulate and inconspicuous, the apex obtuse or rounded, less often retuse, the base rounded or obtuse, the margins revolute, crenate-serrate; petioles 1–2 mm long. Flowers on short, simple, dichasial cymes; pedicels 3–4 mm long, with a pair of bracteoles at base. Calyx cup-shaped, ca. 1 mm long, sepals rounded; petals oblong, ca. 1.5 mm long; stamens <1 mm long; ovary ovoid, the stigmas 4-lobed. Fruit asymmetrically obovoid, slightly angular, 3–5 mm long, turning from green to red. Stones with 2 or 3 ellipsoid, light brown seeds.

DISTRIBUTION: Common in dry forests. Fish Bay (A2372), Salt Pond (M17063). Also on Anegada, St. Croix, St. Thomas, Tortola, and Virgin Gorda; Mexico to northern South America, including the West Indies.

3. MAYTENUS Molina

Shrubs or small trees. Leaves alternate, chartaceous to coriaceous, entire to crenulate; stipules minute, deciduous. Flowers functionally unisexual, pediceled, clustered or solitary in leaf axils; calyx of 5 distinct greenish sepals; corolla of 5 distinct greenish or white

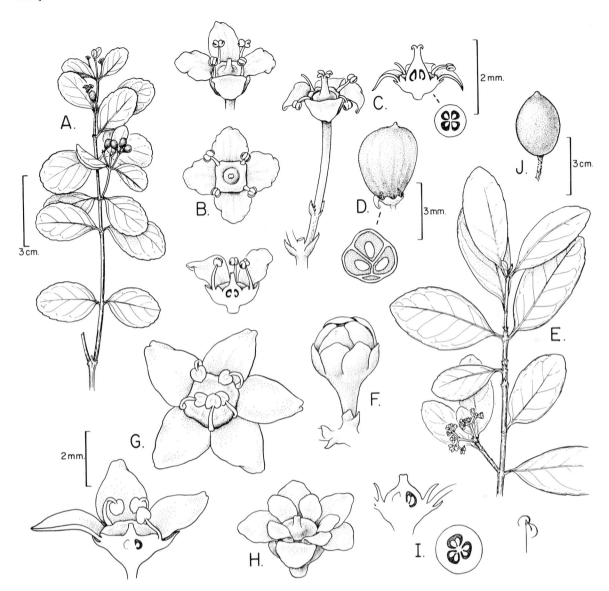

FIG. 73. A–D. *Crossopetalum rhacoma.* **A.** Fertile branch. **B.** Flower with immature pistil, lateral and top views, and l.s. flower. **C.** Flower with mature pistil, l.s. flower with mature pistil, and c.s. ovary. **D.** Fruit and c.s. fruit. E–J. *Cassine xylocarpa.* **E.** Flowering branch. **F.** Flowerbud. **G.** Staminate flower from above and l.s. staminate flower. **H.** Pistillate flower with petaloid staminodes. **I.** L.s. pistillate flower and c.s. ovary. **J.** Drupe.

petals; stamens shorter than petals; nectary disk flattened and 5-lobed; ovary partially immersed on disk, 2-locular, each locule with 1 or 2 ovules, the style short or wanting, the stigma 2-lobed. Fruit a dehiscent capsule, with 1 or 2 arillate seeds.

A genus of about 200 species, mostly in South America.

1. Maytenus laevigata (Vahl) Griseb. *ex* Eggers , Vidensk. Meddel. Dansk Naturhist. Foren. Kjøbenhavn **8:** 109. 1876. *Rhamnus laevigatus* Vahl, Symb. Bot. **3:** 41. 1794. Fig. 74A–E.

Senacia elliptica Lam., Tabl. Encycl. **2:** 96. 1797. *Maytenus elliptica* (Lam.) Krug & Urb. *ex* Duss, Fl. Phan. Antill. Franç. 145. 1897.

Small tree 3–9 m tall; bark gray and smooth. Leaf blades 3–11 × 1.5–7.5 cm, leathery, ovate, elliptic, or obovate, glabrous, the apex obtuse or rounded, usually notched, the base obtuse, rounded, or truncate, less often cuneate, the margins strongly revolute, entire and whitish; petioles 3–6 mm long. Flowers with short pedicels (2 mm), solitary or fascicled. Calyx cup-shaped, with rounded, 0.4 mm long, greenish sepals, with ciliate margins; petals ovate, ca. 2 mm long, whitish or cream; stamens 1 mm long; ovary ellipsoid, the stigmas short. Fruit ellipsoid to globose, 1–1.2 cm long, yellowish green, opening along 2 sides to expose a large, ellipsoid seed covered with a white aril.

DISTRIBUTION: Common in moist forests and in dry evergreen woodlands. Lameshur (A2716), along trail to Sieben (A2069). Also on Jost van Dyke, St. Croix, and St. Thomas; throughout the Greater and Lesser Antilles.

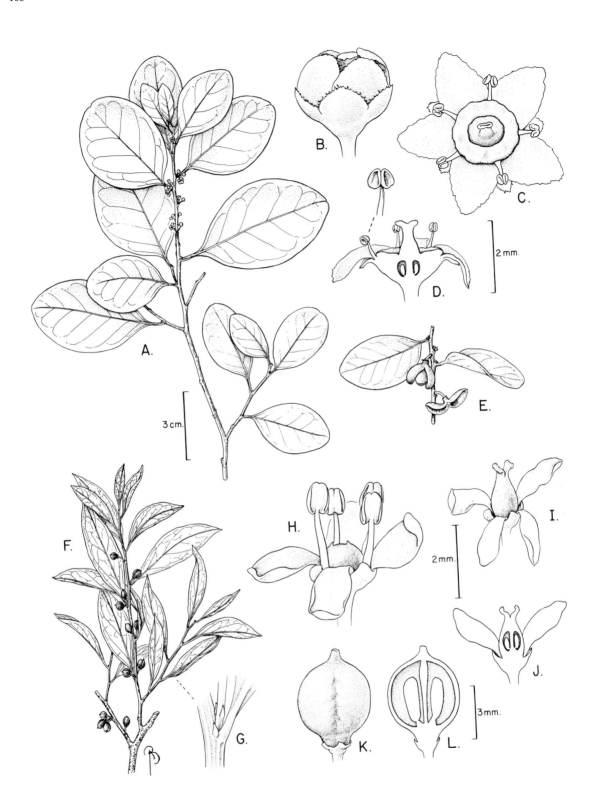

Fig. 74. A–E. *Maytenus laevigata*. **A.** Flowering branch. **B.** Flowerbud. **C.** Pistillate flower. **D.** L.s. pistillate flower and detail of sterile anther. **E.** Detail of branch with dehisced fruits. **F–L.** *Schaefferia frutescens*. **F.** Fruiting branch. **G.** Detail of stem showing stipules. **H.** Staminate flower. **I.** Pistillate flower. **J.** L.s. pistillate flower. **K.** Drupe. **L.** L.s. drupe.

4. SCHAEFFERIA Jacq.

Dioecious shrubs or small trees. Leaves alternate, chartaceous, entire; stipules minute. Flowers unisexual by abortion, pediceled, clustered in leaf axils, or less often solitary; calyx cup-shaped, greenish, with 4 lobes; corolla of 4 distinct petals; stamens as long as the petals; nectary disk wanting or reduced; ovary 2-locular, each locule with 1 ovule, the style short, the stigmas with 2 spreading or reflexed lobes. Fruit an indehiscent drupe, with 2 exarillate seeds.

A genus of about 16 species, with neotropical distribution.

1. Schaefferia frutescens Jacq., Enum. Syst. Pl. 33. 1760.
Fig. 74F–L.

Schaefferia completa Sw., Prodr. 38. 1788.

Shrub 2–4 m tall, many-branched from base; bark gray; young branches striate. Leaf blades 2.5–7 × 1–3.2 cm, chartaceous, elliptic, oblanceolate, or spatulate, glabrous, secondary veins ascending; the apex acute or acuminate or less often obtuse and notched, the base tapering into a short swollen petiole, the margins slightly revolute and entire. Flowers with short pedicels (1.5 mm), fascicled in leaf axils. Calyx cup-shaped, with oblong, ca. 0.5 mm long, greenish sepals; petals oblong, ca. 1.5 mm long, whitish or cream; ovary ellipsoid in pistillate flowers, rudimentary in staminate flowers, the stigmas whitish, spreading. Drupe flattened, ellipsoid to ovoid, 4–5 mm long, turning from green to yellow or orange, with persistent reflexed stigmas at apex. Seeds 3–4 mm long, nearly ellipsoid.

DISTRIBUTION: Common in dry forests and dry evergreen woodlands. Monte (A2375), Southside Pond (A1859). Also on Jost van Dyke, St. Croix, St. Thomas, Tortola, and Virgin Gorda; Greater and Lesser Antilles, southern Mexico to South America.

COMMON NAME: boxwood.

DOUBTFUL record: *Gyminda latifolia* (Sw.) Urb. was reported by Woodbury and Weaver (1987) for St. John, but no specimen was collected by them and recent attempts to collect the species have failed.

28. Chenopodiaceae (Goosefoot Family)

Herbs or less often shrubs or small trees. Leaves simple, alternate or rarely opposite; stipules wanting. Flowers small, bisexual or unisexual, mostly actinomorphic, solitary to many, in leaf axils or in spikes, panicles, or cymes; calyx of usually 5 distinct sepals; petals wanting; stamens as many as the sepals and opposite them, with distinct or connate filaments; the anthers bilocular, longitudinally dehiscent, incurved, dorsifixed; nectary disk annular or glandular, internal to the stamens only in bisexual flowers; ovary superior, of 2 or 3(–5) united carpels forming a unilocular ovary with a single basal ovule, the styles distinct or connate, the stigmas as many as the carpels. Fruit a dehiscent or indehiscent utricle, often with persistent calyx or bracteoles; seed usually lenticular.

A family of about 100 genera and 1500 species with cosmopolitan distribution, but frequently in saline habitats.

Key to the Genera of Chenopodiaceae

1. Leaves 1–3 cm long, farinaceous, densely covered with whitish hairs on lower surface; inflorescence a short cluster of flowers .... 1. *Atriplex*
1. Leaves 2–9 cm long, glabrous or nearly so; inflorescence elongate, with distinct clusters of flowers spaced along the rachis .... 2. *Chenopodium*

1. ATRIPLEX L.

Monoecious or dioecious shrubs or herbs, more or less covered with canescent hairs. Leaves alternate or opposite, petiolate to sessile. Flowers solitary or in glomerules, staminate and pistillate usually in separate glomerules; bracteoles 2, accrescent only in pistillate flowers, persistent on fruits; calyx with 3–5 sepals; stamens 3–5, with free or connate filaments. Utricle indehiscent; seeds lenticular.

A genus of about 100, salt–tolerant species with cosmopolitan distribution.

1. Atriplex cristata Humb. & Bonpl. *ex* Willd., Sp. Pl. 4: 959. 1806.
Fig. 75E–G.

Spinacia littoralis Jacq., Enum. Syst. Pl. 33. 1760. *Atriplex littoralis* (Jacq.) Fawc. & Rendle, J. Bot. 64: 15. 1926, non L., 1753.

Axyris pentandra Jacq., Select. Stirp. Amer. Hist. 244. 1763, nom. illegit. *Atriplex pentandra* (Jacq.) Standl., N. Amer. Fl. 21: 54. 1916, nom. illegit.

Monoecious, erect or decumbent subshrub to 1 m tall. Leaves alternate; blades 1–3 × 0.3–1.5 cm, chartaceous, lanceolate, oblong, or oblanceolate, lower surface densely covered with whitish, farinaceous hairs, the apex acute or obtuse, and mucronate, the base tapering into a short petiole, the margins repand-dentate. Staminate flowers in short terminal spikes; pistillate flowers fascicled in leaf axils; bracteoles rounded, thick, crested. Calyx 5-lobed, ca. 0.5 mm long; stamens 5, 0.5 mm long; ovary lenticular, of 2 locules, the style short and thick, the stigmas 2, filiform. Fruit 5 mm long, covered with accrescent bracts with hornlike projections. Seeds 3–4 mm long, lenticular, yellowish brown.

DISTRIBUTION: Rare, occasionally found along sandy beaches. Newfound Bay (W285F). Also on St. Croix, St. Thomas, and Tortola; Florida, Bahamas, Greater Antilles, Curaçao and northern South America.

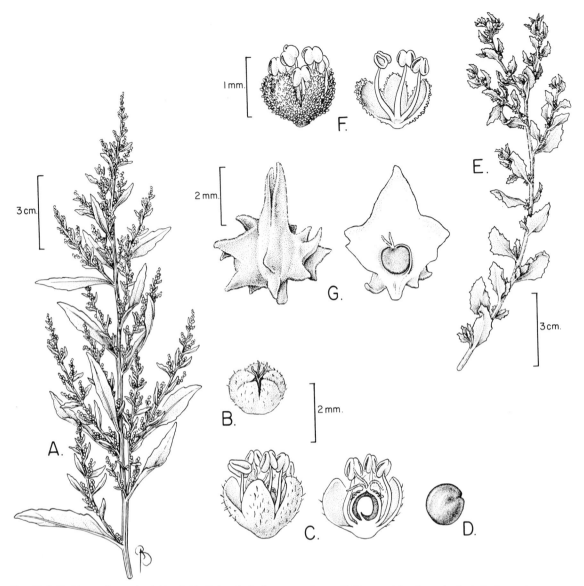

FIG. 75. A–D. *Chenopodium ambrosioides*. **A.** Fertile branch. **B.** Protogynous flower. **C.** Flower and l.s. flower. **D.** Seed. **E–G.** *Atriplex cristata*. **E.** Fertile branch. **F.** Staminate flower and l.s. staminate flower. **G.** Accrescent bracteoles with enclosed fruit and l.s. accrescent bracteoles showing fruit.

2. CHENOPODIUM L.

Herbs or rarely subshrubs, more or less farinaceous or glabrous. Leaves alternate, entire to sinuate, petiolate to sessile. Flowers bisexual, in glomerules, either solitary or aggregated on spikes; bracteoles lacking; calyx of 5 sepals; stamens 5, with flattened membranous filaments; styles 2 or 3. Utricle indehiscent, membranous; seeds nearly lenticular.

A genus of 100 to 150 species, predominantly temperate in distribution.

1. Chenopodium ambrosioides L., Sp. Pl. 219. 1753.
  Chenopodium anthelminthicum L., Sp. Pl. 220.
  1753.   Fig. 75A–D.

Erect subshrub to 1 m tall, with strong, fetid smell, many-branched from a woody base; stem ribbed to cylindrical, more or less pubescent. Leaf blades 2–9 × 0.6–3.8 cm, chartaceous, lanceolate or oblanceolate, glabrous or nearly so, lower surface with abundant yellowish gland dots, the apex obtuse or acute, the base tapering into a more or less elongate (to 2 cm), winged petiole, the margins deeply lobed or serrate to entire on upper leaves. Flowers minute, greenish, protogynous, in axillary glomerules or in spikes of glomerules, the spikes 1–2 cm long. Calyx greenish, ca. 1 mm long, the sepals oblong; stamens ca. 1 mm long; styles 3, whitish. Utricle whitish, ca. 1 mm long, covered with persistent sepals. Seeds 1 mm long, nearly lenticular, reddish brown.

DISTRIBUTION: In open waste grounds. Cinnamon Bay (W634). Also on St. Croix, St. Thomas, and Tortola; a weedy species native to Central America, now cultivated and naturalized worldwide in warm regions.

COMMON NAMES: wormseed, wormwood.

29. Chrysobalanaceae (Coco Plum Family)

Trees or shrubs. Leaves alternate, simple, coriaceous; petioles usually with 2 glands; stipules minute to large, deciduous or persistent. Flowers bisexual or less often unisexual, actinomorphic or zygomorphic, produced in racemose, paniculate, or cymose inflorescences; bracts usually 2; receptacle cup-shaped; calyx lobes 5; corolla of (4–)5 distinct petals, seldom lacking, inserted at margin of a nectary disk; stamens 2 to many, inserted at margin of disk; ovary superior, of 3 uniovular carpels (usually only 1 developing), the placentation axile, the style filiform, gynobasic, the stigmas 3. Fruit a dry or fleshy drupe; stone fibrous or bony with a single seed.

A family of about 17 genera and 420 species with pantropical distribution.

REFERENCE: Prance, G. T. 1972. Chrysobalanaceae. Fl. Neotrop. Monogr. 9: 1–410.

1. CHRYSOBALANUS L.

Shrubs or small trees. Leaves glabrous or with a few appressed hairs; stipules minute and deciduous. Flowers bisexual, actinomorphic, produced on terminal or axillary cymes; bracts and bracteoles small; calyx with 5 acute lobes; corolla of 5 distinct petals, longer than the calyx lobes; stamens 12 to 26, the filaments united at base; ovary of 2 carpels, inserted at base of receptacle, the style pubescent. Fruit a fleshy drupe; stone with 4–8 prominent, longitudinal ridges.

A genus of 2 species, distributed in tropical America and Africa, naturalized in Asia and Pacific Islands.

1. Chrysobalanus icaco L., Sp. Pl. 513. 1753.   Fig. 76.

Shrub to small tree to 5 m tall; branches lenticellate, glabrous, pubescent when young. Leaf blades 2–8 × 1.2–6 cm, coriaceous, rounded to ovate-elliptic, glabrous, veins inconspicuous, the apex rounded and notched, the base cuneate to rounded, the margins entire; petioles slightly swollen, 2–4 mm long. Flowers short-pediceled, the cymes pubescent, axillary on short lateral branches. Calyx cup-shaped, 4–5 cm long, densely rusty-brown pubescent; petals spatulate, ca. 4 mm long, white to cream; stamens 5 mm long, the filaments densely pubescent, united in small groups; ovary nearly globose, densely pubescent, the style 5 mm long, densely pubescent. Drupe ellipsoid, ovoid, or globose, 1.5–3 cm long, with edible, fleshy mesocarp, turning from green to yellow or reddish when mature. Seeds 1–2.8 cm long, nearly ellipsoid.

DISTRIBUTION: Uncommon in moist forests. Bordeaux (B564), Rosenberg (B286). Also on St. Croix, St. Thomas, Tortola, and Virgin Gorda; Florida to coastal southern Brazil, including the West Indies, the Pacific coast of Central America and the western coast of Africa.

COMMON NAMES: cacos, coco plum, fat pork, icaco, Spanish nectarine.

30. Clusiaceae (Mammee Apple Family)

Polygamous or dioecious trees, shrubs, or lianas, producing abundant, yellowish latex. Leaves opposite or whorled, simple and petiolate; stipules wanting. Flowers bisexual or less often unisexual, actinomorphic, solitary or in cymose inflorescences; bracteoles present; calyx of 4–5(–6) distinct or connate sepals; corolla of 4–6(–9) distinct, imbricate petals; stamens numerous, free to fascicled; ovary superior, of (1–)4–5(–9) carpels with usually 2 ovules per carpel, the placentation axile, rarely basal or parietal, the styles as many as the carpels or united into a single style with a peltate or lobed stigma. Fruit a berry, drupe, or capsule; seeds often arillate.

A family of about 50 genera and 1200 species with pantropical distribution.

Key to the Genera of Clusiaceae

1. Free-standing or strangling tree, usually with numerous stilt roots; flowers 8–10 cm diam.; fruit a green capsule, 3.5–5 cm diam.; seeds numerous, 5 mm long, with reddish aril .... 1. *Clusia*
1. Free-standing tree, without stilt roots; flowers 2.5–3 cm diam.; a tan, fleshy berry, 7–15 cm diam.; seeds 2–4, 5–8 cm long, without aril .... 2. *Mammea*

1. CLUSIA L.

Dioecious or polygamous trees or shrubs, usually stranglers or less often climbers. Leaves glabrous, thick, leathery. Flowers unisexual or bisexual, large, solitary or produced on terminal cymes; bracts 2–6, small; calyx of 4–6 rounded, overlapping sepals; corolla of 4–9 distinct, overlapping petals; stamens numerous, the filament usually united at base; staminodes present (in pistillate flowers) or wanting (in staminate flowers); ovary of 4–12 carpels, with numerous ovules, the style thick, the stigmatic surface peltate. Fruit a fleshy capsule, splitting along numerous valves; seeds minute, covered with bright, reddish aril.

A neotropical genus of about 150 species.

1. Clusia rosea Jacq., Enum. Syst. Pl. 34. 1760.
Fig. 77A–E.

Dioecious, free-standing or strangling tree, to 12 m tall, usually with many upright branches from base, and numerous stilt roots; bark gray and smooth. Leaves opposite; blades 5–20 × 4–11 cm, thick coriaceous, spatulate, glabrous, the apex rounded or truncate, the base tapering or rounded, often unequal, the margins entire and revolute; petioles stout, ca. 1 cm long. Flowers 8–10 cm diam., solitary at ends of branches or in cymes; bracts 2, rounded. Calyx reddish, with 2 pairs of dissimilar, rounded, concave sepals, to 2 cm long; petals 6, spatulate, 3–4 cm long, white, with pink band within at middle; staminodial ring brown; stigma light green, sticky. Fruit globose, leathery, capsule 3.5–5 cm diam., green

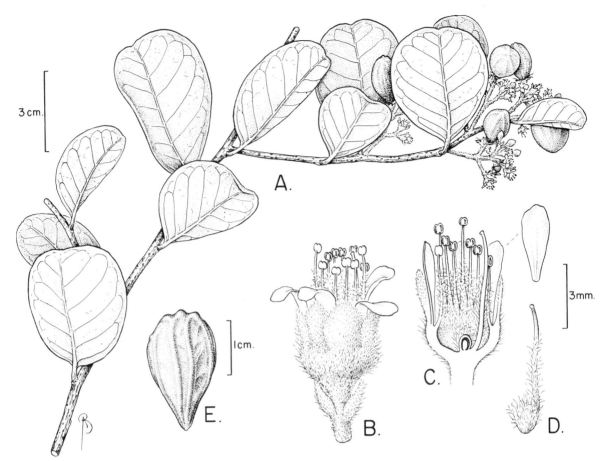

Fig. 76. *Chrysobalanus icaco.* **A.** Fertile branch. **B.** Flower. **C.** L.s. flower and detail of petal. **D.** Pistil. **E.** Immature drupe.

with abundant yellow latex. Seeds numerous, 5 mm long, ellipsoid, the aril reddish.

Distribution: Common in moist forests, but also present in dry coastal vegetation. Bordeaux (A1919), White Cliffs (A756). Also on St. Croix, St. Thomas, Tortola, and Virgin Gorda; the Bahamas, Greater Antilles, Anguilla, and St. Martin.

Common names: autograph tree, chigger, false mammee, pitch apple.

2. MAMMEA L.

Polygamous trees. Leaves glabrous, thick, leathery. Flowers unisexual or bisexual, large, solitary or clustered in leaf axils; bracteoles minute; calyx irregularly splitting into 2 concave sepals; corolla of 4–6 distinct, overlapping petals; stamens numerous, the filaments free or united at base; ovary of 2 carpels, each carpel with 2 ovules, the style short, the stigma peltate. Fruit a fleshy berry with leathery skin; seeds 2–4, large, with woody cover.

A genus of about 50 species mostly distributed from Madagascar to the Pacific tropics and the New World tropics.

1. Mammea americana L., Sp. Pl. 512. 1753.

Fig. 77F–K.

Tree to 20 m tall and 45 cm diam.; bark brown and rough. Leaves opposite; blades 9–20 × 5–11 cm, coriaceous, elliptic or obovate, glabrous, with numerous translucent dots, the apex rounded or nearly truncate, often retuse, the base cuneate, obtuse or rounded, the margins entire and revolute; petioles stout, ca. 1 cm long. Flowers 2.5–3 cm diam., bisexual and staminate, clustered in leaf axils. Calyx pale yellow, 0.8–1.2 cm long; petals 6, obovate, ca. 2 cm long, white; stamens free, the anthers bright yellow, linear; stigma peltate. Fruit globose, tan, 7–15 cm diam.; edible pulp light orange. Seeds 2–4, 5–8 × 3–5 cm.

Distribution: Uncommon in moist forests. Coral Bay Quarter along Center Line Road (A2842). Also on St. Croix, St. Thomas, and Tortola; throughout the West Indies, Mexico, Central America and Trinidad.

Common name: mammee apple.

31. **Combretaceae** (Gree Gree Tree Family)

Trees, shrubs, or lianas. Leaves opposite, alternate or whorled, simple and petiolate; stipules wanting. Flowers bisexual or less often unisexual, actinomorphic, or slightly zygomorphic, produced in axillary or terminal racemes, spikes, or panicles; calyx of 4–5

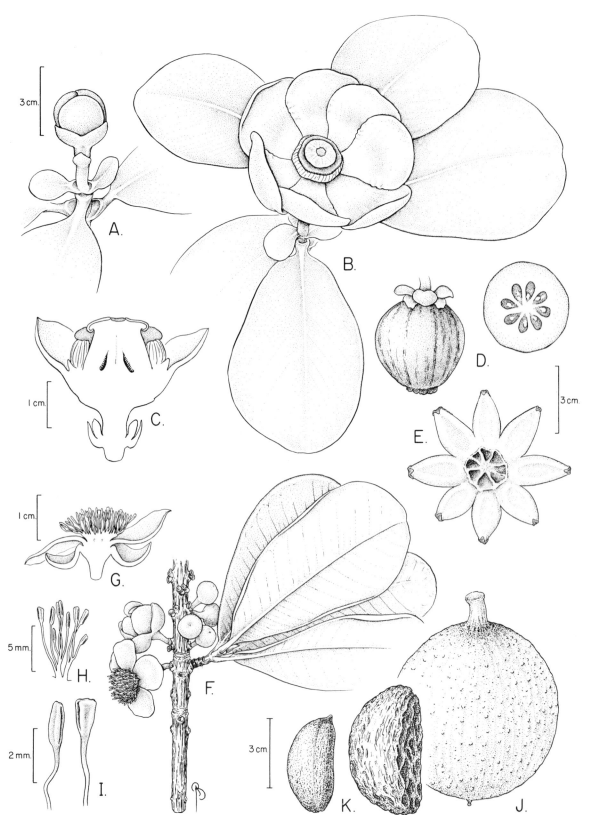

Fɪɢ. 77. A–E. *Clusia rosea.* **A.** Detail of flowering branch with flowerbud. **B.** Pistillate flower. **C.** L.s. pistillate flower. **D.** Fruit and c.s fruit. **E.** Dehisced fruit. **F–K.** *Mammea americana.* **F.** Fertile branch. **G.** L.s. staminate flower. **H.** Stamen group. **I.** Detail of dimorphic stamens. **J.** Fruit. **K.** Embryo and seed.

sepals, distally on a hypanthium; corolla of 4–5 distinct, small petals or wanting; stamens twice as many as sepals; intrastaminal disk present; ovary inferior, of 2–5 carpels united to form a unilocular ovary with 2(–6) pendulous ovules, the style 1, the stigma capitate. Fruit a drupe or a samara, with a single seed.

A family of about 20 genera and 400 species with pantropical distribution.

Key to the Genera of Combretaceae

1. Liana or climbing shrub; hypanthium bright red, long-tubular .... 5. *Quisqualis*
1. Trees; hypanthium greenish to whitish, cup-shaped.
    2. Fruit multiple, pine cone-like .... 3. *Conocarpus*
    2. Fruit solitary, smooth or ridged, not pine cone-like.
        3. Trees growing in permanent or seasonally flooded areas .... 4. *Laguncularia*
        3. Trees growing on nonflooded land.
            4. Fruit 4.5–6 cm long, hard; leaves obovate, 12–30 cm long .... 6. *Terminalia*
            4. Fruit <2.5 cm long, fleshy; leaves spatulate, <7 cm long.
                5. Leaf blade without glands; fruits ellipsoid, 2–2.5 cm long, smooth; inner bark yellowish
                .... 1. *Buchenavia*
                5. Leaf blade with a pair of impressed glands at base; fruits conical, 4–6 mm long, 5-angled;
                inner bark light brown .... 2. *Bucida*

1. BUCHENAVIA L., nom. cons.

Trees or shrubs. Leaves alternate, entire, crowded at ends of short sympodial branches. Flowers bisexual, regular, with long hypanthium, borne in axillary capitate spikes; calyx early deciduous, cup-shaped, with 5 minute lobes on distal portion; corolla wanting; stamens 10, exserted, the anthers adnate to the filament; nectary disk annular, surrounding the style; ovary inferior, at the base of the hypanthium, the style long exserted. Fruit a 1-seeded, fleshy drupe.

A neotropical genus of 21 species.

1. **Buchenavia tetraphylla** (Aubl.) R.A. Howard, J. Arnold Arbor. **64:** 266. 1983. *Cordia tetraphylla* Aubl., Hist. Pl. Guiane **1:** 224, t. 88. 1775.

Fig. 78H–M.

Bucida capitata Vahl, Eclog. Amer. **1:** 50, t. 8. 1797.
Buchenavia capitata (Vahl) Eichler, Flora **49:** 165. 1866.

Tree to 25 m tall, with spreading, nearly horizontal, sympodial branches; bark light brown and smooth, inner bark yellowish. Leaf blades 3–6 × 1.2–3.2 cm, coriaceous, obovate to spatulate, lower surface with appressed, rusty-brown hairs, especially along main vein, foveate at vein axils, the apex rounded, obtuse, or retuse, often mucronate, the base cuneate, the margins entire; petioles 0.4–1 cm long. Hypanthium densely covered with rusty-brown hairs; calyx greenish, early deciduous, ca. 2 mm long; filaments 4–5 mm long, adnate to base of calyx; nectary disk covered with rusty-brown hairs. Drupe ellipsoid to ovoid, 2–2.5 cm long, turning from green to yellowish orange.

DISTRIBUTION: Common in moist forests. Bordeaux (A4702). Also on Tortola; the West Indies and Panama to South America.

2. BUCIDA L.

Trees or shrubs. Leaves alternate, entire, crowded at ends of short sympodial branches. Flowers bisexual or rarely some staminate, regular, with long hypanthium, borne in axillary spikes; calyx persistent, cup-shaped, with 5 minute lobes on distal portion; corolla wanting; stamens 10, exserted, the anthers versatile; nectary disk annular, surrounding the style; ovary inferior, unilocular, basal at hypanthium, the style long-exserted. Fruit a 1-seeded drupe, with persistent calyx at apex.

A genus of 3 species, distributed in the West Indies, southern Florida, and from Mexico to Central America.

1. **Bucida buceras** L., Syst. Nat., ed. 10, **2:** 1025. 1759.

Fig. 78A–G.

Tree to 20 m tall, with sympodial branches, sometimes spiny; bark grayish to light brown, rough, with rectangular plates, inner bark light brown. Leaf blades 2.5–8 × 1.5–4 cm, coriaceous, obovate, elliptic, spatulate, or oblanceolate, lower surface with a few appressed, yellowish hairs, becoming glabrous, the apex rounded, obtuse and rarely retuse, the base cuneate, with a pair of glands, the margins entire; petioles 4–13 mm long. Hypanthium densely covered with rusty-brown hairs; calyx greenish, pubescent to glabrous, persistent, ca. 2 mm long; filaments 5–8 mm long, adnate to base of calyx; disk rufous tomentose. Drupe obliquely ovoid, 4–6 mm long, densely sericeous, usually damaged by insects and developing elongate hornlike galls.

DISTRIBUTION: Common in moist and coastal forests. Adrian Ruins (A722), Lameshur (A4047). Also on Jost van Dyke, St. Croix, St. Thomas, and Tortola; throughout the West Indies, Florida and Central America.

COMMON NAMES: gree gree tree, gregery, gri gri.

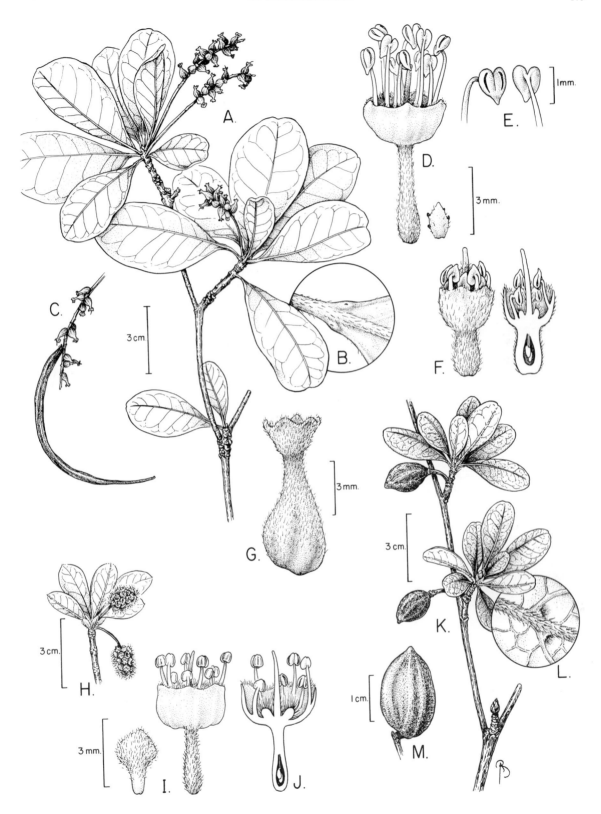

Fig. 78. A–G. *Bucida buceras*. A. Fertile branch. B. Detail of leaf base showing glands. C. Fruiting raceme with elongated, galled fruit. D. Staminate flower and subtending bract. E. Upper portion of stamens showing anthers. F. Bisexual flower and l.s. bisexual flower. G. Drupe. H–M. *Buchenavia tetraphylla*. H. Flowering branch. I. Flower bract and flower. J. L.s. flower. K. Fruiting branch. L. Detail of lower surface of leaf. M. Fruit.

3. CONOCARPUS L.

Trees or shrubs. Leaves alternate and entire; petioles short, with a pair of glands near the blade. Flowers bisexual, regular, with a short hypanthium, borne in short heads in axillary or terminal racemes; hypanthium laterally flattened and winglike at base; calyx of 5 ovate sepals; corolla wanting; stamens 5–10, exserted, the anthers versatile; nectary disk well developed, surrounding the style; ovary inferior, unilocular, surrounded by the hypanthium, the style long-exserted. Fruit drupaceous, leathery, laterally flattened, grouped into a conelike head.

A genus of 2 species, distributed in tropical America and Africa.

1. Conocarpus erectus L., Sp. Pl. 176. 1753.

Fig. 79A–G.

Small tree or shrub to 8 m tall; bark light brown, rough, with irregular plates. Leaf blades 2.5–9 × 1.5–3 cm, coriaceous, narrowly elliptic, elliptic, or oblanceolate, glabrous, or with a few (to many) appressed hairs, the blades with numerous dark punctations and foveolate at vein axils, the apex acute or acuminate or rarely obtuse, the base tapering into a more or less defined petiole, the margins entire; petioles 4–10 mm long, with 2 marginal glands. Flowers in small (5–10 mm diam.), dense heads. Hypanthium short, densely sericeous; calyx with 5 ovate lobes, ca. 1.5 mm long; stamens 10, the filaments 2 mm long; disk densely sericeous. Drupes dry, 2-winged, slightly concave, grouped into globose, conelike head, to 1 cm long. Seeds nearly lenticular, ca. 1.5 mm long.

DISTRIBUTION: Common along the coast, sometimes in swampy areas but also in dry soils or even on rocky shores. Fortsberg (A4082). Also on St. Croix, St. Thomas, Tortola, and Virgin Gorda; throughout tropical America and the western coast of Africa.

COMMON NAMES: button mangrove, button wood.

4. LAGUNCULARIA C.F. Gaertn.

A monospecific genus, characterized by the following species.

1. Laguncularia racemosa (L.) C.F. Gaertn., Suppl. Carp. 209, t. 217, fig. 3. 1807. *Conocarpus racemosus* L., Syst. Nat., ed. 10, **2**: 930. 1759.

Fig. 79H–K.

Small tree or shrub, to 15 m tall; root system with branched pneumatophores, sticking out of the ground. Leaves opposite; blades 3–8.5 × 2–4.6 cm, coriaceous, elliptic or seldom ovate, glabrous, the blade with submarginal, glandular pits, the apex rounded, rarely slightly notched, the base obtuse, the margins entire; petioles 1.5–2 cm long, with 2 marginal, slightly prominent glands. Flowers bisexual, regular, with a short hypanthium, produced in axillary spikes; hypanthium densely sericeous, 5 mm long; calyx of 5 lobes, ca. 2 mm long; corolla of 5 minute (1.5 mm long), deciduous, obovate petals; stamens 10, exserted, the anthers versatile; nectary disk flattened, surrounding the style; ovary inferior, unilocular, surrounded by the hypanthium, the style long-exserted. Fruit a leathery, densely sericeous drupe, slightly flattened, ellipsoid to club-shaped and ridged, with persistent calyx at apex, 1.5–2 cm long.

DISTRIBUTION: Common along the coast in mangrove swamps and along sandy beaches. Great Cruz Bay (A791), Fish Bay (A3902). Also on Jost van Dyke, St. Croix, St. Thomas, Tortola, and Virgin Gorda; throughout tropical America and western Africa.

COMMON NAME: white mangrove.

5. QUISQUALIS L.

Twining lianas. Leaves opposite with persistent petioles forming spines on old stems. Flowers bisexual, regular, with a long, tubular hypanthium, crowned by the calyx, borne in terminal or axillary spikes; calyx of 5 valvate sepals; corolla of 5, imbricate petals; stamens 10, exserted, the filaments adnate to the upper portion of hypanthial tube, the anthers versatile; nectary disk present or absent, surrounding the style; ovary inferior, unilocular, basal at hypanthium tube, the style exserted. Fruit dry, winged.

A paleotropical genus of 17 species, with some species extensively cultivated throughout the tropics.

1. Quisqualis indica L., Sp. Pl., ed. 2, **1**: 556. 1762.

Fig. 80A-E.

Liana to 6 m long, many-branched from base; branches slightly flattened at nodes. Leaf blades 6–15 × 2.2–6 cm, chartaceous, elliptic, oblong, or lanceolate, lower surface with appressed, rusty-brown hairs, especially along veins, the apex acuminate, the base rounded, the margins entire; petioles 5–12 mm long. Hypanthium narrow, 4–6.5 cm long, pubescent, green to light green; calyx lobes green, triangular, ca. 2 mm long; petals 5, pink, turning bright red with age, oblong or oblanceolate, 1–1.5 cm long; stamens slightly exserted, the filaments unequal in length; disk wanting; style exserted. Fruit with elliptic outline, 5-winged, to 3 cm long.

DISTRIBUTION: Uncommon ornamental plant, in gardens, persistent after cultivation. Mary Pt. (W535). Commonly persisting after cultivation throughout the tropics.

COMMON NAMES: heart o'man, Rangoon creeper.

6. TERMINALIA L.

Trees or shrubs. Leaves alternate, entire, crowded at end of short sympodial branches. Flowers bisexual or bisexual and staminate, regular, with short hypanthium, borne in axillary spikes; calyx persistent, cup-shaped, with 5 lobes on distal portion; corolla wanting; stamens 10, the anthers versatile; nectary disk 5-lobed, densely pubescent; ovary inferior, at the base of the hypanthium, the style exserted. Fruit a 1-seeded dry or fleshy drupe.

A genus of about 200 species with pantropical distribution, but mainly represented in the paleotropics.

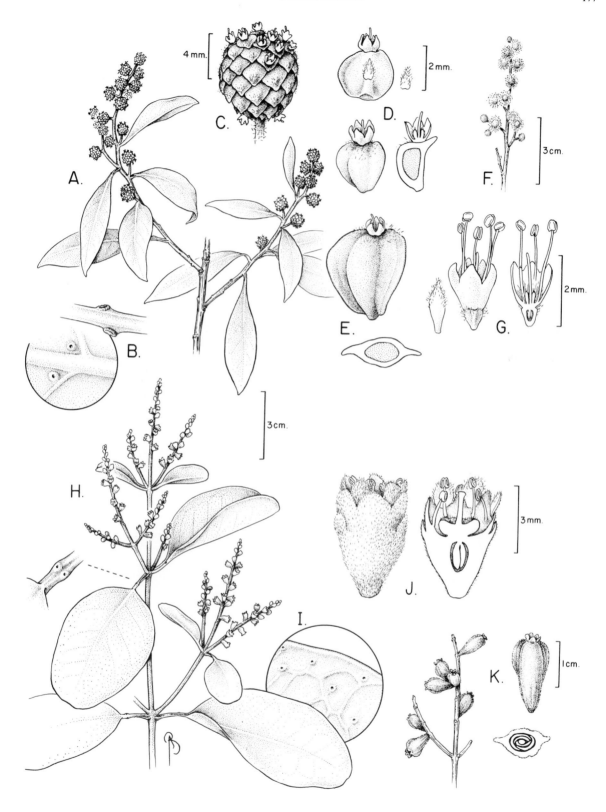

Fɪɢ. **79. A–G.** *Conocarpus erectus.* **A.** Fertile branch. **B.** Detail of glands on leaf petiole and detail of pits at vein axils on lower surface of leaf. **C.** Inflorescence. **D.** Developing fruit with subtending bract; bract, dorsal view; and l.s. of developing fruit. **E.** Fruit and c.s. fruit. **F.** Inflorescence. **G.** Flower and subtending bract and l.s. flower. **H–K.** *Laguncularia racemosa.* **H.** Flowering branch and detail of glands on leaf petiole. **I.** Detail of upper surface of leaf showing glandular pits. **J.** Flower and l.s. flower. **K.** Infructescence, drupe, and c.s. drupe.

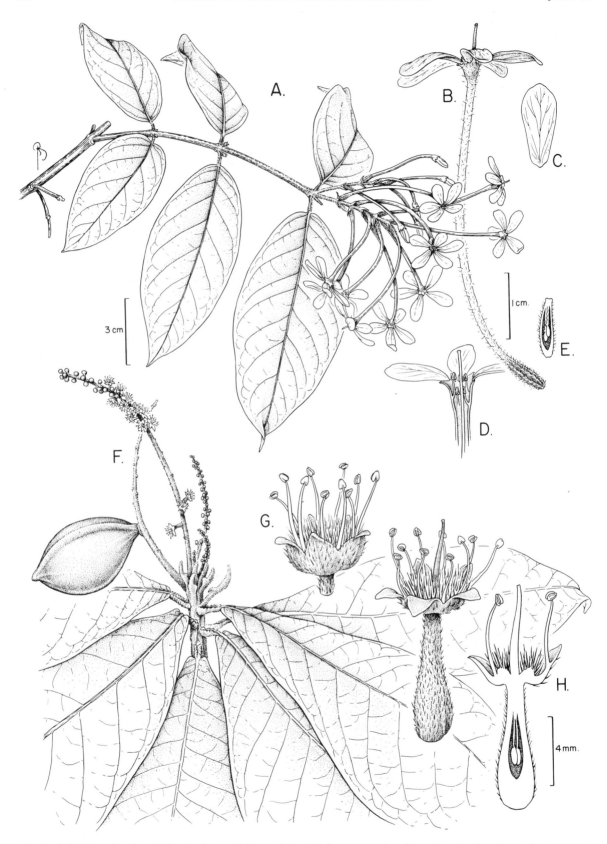

FIG. 80. A–E. *Quisqualis indica.* **A.** Flowering branch. **B.** Flower. **C.** Petal. **D.** L.s. upper portion of hypanthium showing calyx, petals, stamens, and style. **E.** L.s. ovary. **F–H.** *Terminalia catappa.* **F.** Fertile branch. **G.** Staminate flower. **H.** Bisexual flower and l.s. flower.

1. Terminalia catappa L., Syst. Nat., ed. 12, **2**: 674. 1767.   Fig. 80F–H.

Tree to 20 m tall, with spreading, nearly horizontal, sympodial branches; bark grayish to light brown. Leaf blades 12–30 × 6–18 cm, coriaceous, obovate, lower surface with appressed, rusty-brown hairs, especially along the main vein, the blade foveate at vein axils, the apex obtuse or rounded, usually ending in an apiculum, the base tapering, cuneate, or subcordate, the margins

entire; petioles stout, 0.5–2 cm long. Hypanthium densely covered with rusty-brown hairs; calyx yellowish, 2–2.5 mm long; filaments 3–4 mm long; disk densely rufous-tomentose. Drupe slightly fleshy, obliquely ellipsoid and somewhat flattened toward the margins, 4.5–6 cm long, turning from green to red then brown.

DISTRIBUTION: Commonly planted along roads. Cruz Bay Quarter in front of Park Service Headquarters (A2917). Also on St. Croix, St. Thomas, and Tortola; native to Malaysia, now naturalized mainly in coastal areas throughout the neotropics.

COMMON NAMES: almond tree, West Indian almond.

32. Convolvulaceae (Morning Glory Family)

Twining vines or prostrate herbs, seldom erect shrubs, trees, or lianas usually with milky sap. Leaves alternate, simple, entire or lobed; stipules wanting. Flowers large, showy, actinomorphic, bisexual or rarely unisexual, in axillary dichasia or seldom solitary; calyx usually of 5 distinct sepals; corolla gamopetalous, funnel-shaped, bell-shaped, or rotate, usually with 5 more or less prominent lobes; stamens usually 5, the filaments unequal, adnate to the base of corolla tube, the anthers sagittate, opening by longitudinal slits; ovary superior, with an annular nectary disk at base, of 2(–3–5) carpels, usually with 2 ovules per locule, the placentation axile, the styles 1 or 2. Fruit a loculicidal or indehiscent capsule, sometimes with accrescent sepals; seeds (1–)2–4, per fruit, large, naked or hairy.

A family of 50 genera and ca. 1500 species, with cosmopolitan distribution, but best developed in tropical and subtropical areas.

REFERENCES: Austin, D. F. 1982. Convolvulaceae. Flora de Venezuela 8(3): 15–226; Wilson, K. A. 1960. The genera of Convolvulaceae in the southeastern United States. J. Arnold Arbor. 41: 298–317.

Key to the Genera of Convolvulaceae

1. Styles 2; prostrate to erect herbs or subshrubs .... 2. *Evolvulus*
1. Style 1; twining vines (except for *Ipomoea carnea,* which is an erect shrub).
  2. Stigma globose or nearly so; fruits indehiscent or 4-valvate.
    3. Fruits indehiscent, seeds freed by dissolution of fruit wall .... 6. Stictocardia
    3. Fruit a dehiscent capsule, opening along 4 valves.
      4. Anthers twisted after dehiscence; leaves usually palmately compound (except *M. umbellata*); sepals elongate in fruit .... 5. *Merremia*
      4. Anthers not twisted; leaves entire or lobed; sepals not elongate in fruit .... 3. *Ipomoea*
  2. Stigma filiform, oblong, or ellipsoid; fruits 4-valvate but each valve splitting in two.
    5. Stigma filiform; corolla white .... 1. *Convolvulus*
    5. Stigma oblong; corolla blue, white, or red .... 4. *Jacquemontia*

1. CONVOLVULUS L.

Erect or prostrate herbs, shrubs, or twining vines. Leaves entire or lobed. Flowers bisexual, regular, produced in axillary cymes; calyx of 5 equal or unequal sepals; corolla bell-shaped; stamens 5, the anthers sagittate; style 1, the stigmas 2, filiform. Fruit a 4-seeded capsule, 4-valvate but each valve spitting in two; seeds with 2 flattened sides and a convex one.

A genus of about 250 species, mostly from warm, temperate regions of the Old World, also from western North America and South America.

1. Convolvulus nodiflorus Desr. *in* Lam., Encycl. **3**: 557. 1792. *Jacquemontia nodiflora* (Desr.) G. Don, Gen. Hist. **4**: 283. 1838.   Fig. 81A–E.

Convolvulus albiflorus Vahl *in* H. West, Bidr. Beskr. Ste. Croix 271. 1793, nomen nudum.

Subwoody, twining vine to 5 m long, not producing milky sap; stems cylindrical, slender, densely covered with whitish, 3-branched hairs. Leaf blades 2–4.5 × 1–3 cm, chartaceous, lanceolate, ovate, elliptic to rounded, covered with 3-branched hairs, especially on lower surface, the apex obtuse, acute, acumi-

nate or seldom retuse, usually mucronate, the base rounded, cordate, or truncate, the margins entire; petioles slender, 0.5–1.5 cm long. Calyx light green, cup-shaped, with equal, ovate sepals, 3–3.5 mm long; corolla funnel-shaped, white, 1.2–1.5 cm long, with 5 obtuse lobes; stamens white, the filaments 8–10 mm long; ovary white, the stigmas exserted. Capsule thin-walled, ovoid, straw-colored, ca. 5 mm long.

DISTRIBUTION: Common in dry open areas. Coral Bay (A4000), Fish Bay (A2378), Salt Pond (A3165). Also on St. Croix, St. Thomas, Tortola, and Virgin Gorda; throughout the West Indies, Central and South America.

COMMON NAME: clashi mulat.

2. EVOLVULUS L.

Prostrate or erect herbs or shrubs, not producing milky sap (at least in our species). Leaves small and entire. Flowers bisexual, regular, borne in axillary, dichasial cymes or solitary; calyx of 5 equal or subequal sepals; corolla rotate, shortly tubular at base or

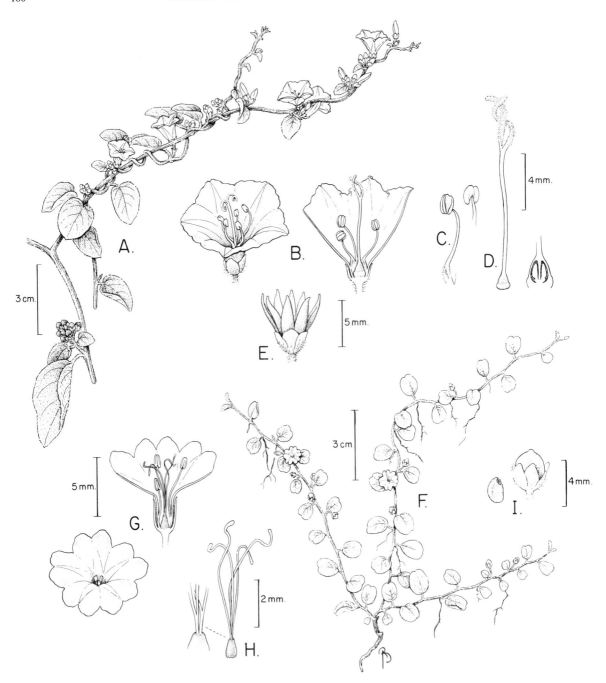

FIG. 81. **A–E.** *Convolvulus nodiflorus.* **A.** Fertile branch. **B.** Flower and l.s. flower. **C.** Stamen, lateral and dorsal views. **D.** Pistil and l.s. ovary. **E.** Dehisced capsule. **F–I.** *Evolvulus nummularius.* **F.** Habit. **G.** Corolla from above and l.s. flower. **H.** Pistil and detail of branched style bases. **I.** Seed and capsule.

funnel-shaped; stamens 5, the anthers ovate or oblong; ovary with 2 distinct or basally united styles, each style divided into 2 filiform stigmas. Fruit a 4-seeded (or less), ovoid or globose, 4-valvate capsule; seeds usually smooth, 3-angled.

A neotropical genus of about 100 species.

REFERENCE: Ooststroom, S. J. van. 1934. A monograph of the genus *Evolvulus.* Meded. Bot. Mus. Herb Rijks Univ. Utrecht 14: 1–267.

Key to the Species of *Evolvulus*

1. Plants creeping, rooting at nodes; leaf apex usually notched; flowers sessile or nearly so.......... 3. *E. nummularius*
1. Plants erect to decumbent; leaf apex obtuse, rounded, or seldom notched and mucronate; flowers or inflorescences on elongate peduncle.

2. Leaf oblong to obovate or nearly rounded, mucronate at apex; corolla 8–10 mm diam.; flowers solitary .... 1. *E. convolvuloides*

2. Leaves lanceolate or linear, acute at apex; corolla 3–5 mm diam.; flowers solitary or several in cymes .... 2. *E. filipes*

1. **Evolvulus convolvuloides** (Willd. *ex* Schult.) Stearn, Taxon **21**: 649. 1972. *Nama convolvuloides* Willd. *ex* Schult., Syst. Veg. **6**: 189. 1820.

Evolvulus glaber Spreng., Syst. Veg. **1**: 862. 1824.

Spreading herb to 30 cm long, many-branched from a woody taproot; branches cylindrical, slender, usually densely covered with appressed, whitish hairs, especially when young. Leaf blades 7–23 × 5–13 mm, chartaceous, obovate, oblong, elliptic to nearly rounded, lower surface punctate, densely covered with appressed, whitish hairs, especially when young, the apex obtuse, rounded, or seldom retuse, usually mucronate, the base obtuse to subcordate, usually oblique, the margins entire and ciliate; petioles slender, 0.5–2 mm long. Flowers solitary, on long, pubescent peduncle (1–1.7 cm long), the bracteoles lanceolate, 1.5–2 mm long. Calyx green, pubescent, the sepals lanceolate, 3–3.5 mm long; corolla saucer-shaped, white, lavender, or pale blue, 8–10 mm wide, with 10 rounded lobes, pilose externally; stamens white; ovary and styles white. Capsule thin-walled, ovoid or nearly globose, straw-colored, ca. 4 mm long. Seeds 4 or fewer, brown, 2 mm long, papillose.

DISTRIBUTION: Uncommon herb found in moist disturbed areas. East End Quarter along Center Line Road (A2780), Southside Pond (W604A). Also on Anegada, St. Croix, St. Thomas, Tortola, and Virgin Gorda; throughout tropical America.

NOTE: This species was reported erroneously as *Evolvulus linifolius* L. by Woodbury and Weaver (1987).

2. **Evolvulus filipes** Mart., Flora **24**(Beibl. 2): 100. 1841.

Evolvulus linifolius sensu Britton & P. Wilson, Bot. Porto Rico **6**: 104. 1925, non L., 1762.

Decumbent or prostrate herb, to 20 cm long, with many lateral branches; branches cylindrical, slender, densely covered with appressed, whitish hairs, especially when young. Leaf blades 10–25 × 4–5 mm, chartaceous, lanceolate to linear, densely covered with appressed, whitish hairs to glabrous, the apex acute, the base obtuse, sometimes oblique, the margins entire and ciliate; petioles slender, 0.5–2 mm long. Flowers solitary or several, 2–2.5 cm long, pubescent peduncle; bracteoles lanceolate, minute. Calyx green, pilose, the sepals ovate, ca. 2 mm long; corolla saucer-shaped, blue, 3–4.5 mm wide, stamens white; ovary and styles white. Capsule thin-walled, globose, straw-colored, ca. 2 mm long. Seeds 4, brown, 2 mm long, papillose.

DISTRIBUTION: An uncommon herb found in disturbed areas. Enighed (A4269). Also on St. Croix, St. Thomas, and Virgin Gorda; Mexico to South America, Jamaica, and the Lesser Antilles.

3. **Evolvulus nummularius** (L.) L., Sp. Pl., ed. 2, **1**: 391. 1762. *Convolvulus nummularius* L., Sp. Pl. 157. 1753.   Fig. 81F–I.

Creeping herb to 30 cm long, many-branched from a taproot; branches cylindrical, slender, rooting at nodes, densely covered with appressed, whitish hairs, especially when young. Leaf blades 7–16 × 5–14 mm, chartaceous, obovate or almost rounded, lower surface punctate, densely covered with appressed, whitish hairs, especially when young, the apex rounded and usually notched, the base rounded to subcordate, usually oblique, the margins entire and ciliate; petioles slender, 4–9 mm long. Flowers solitary, on very short peduncles, the bracteoles oblanceolate, ca. 2 mm long. Calyx green, pilose, the sepals lanceolate, ca. 2.5 mm long; corolla saucer-shaped, white or pale blue, 8–10 mm wide, with 5 rounded lobes, sparsely pilose without; stamens white; ovary and styles white. Capsule thin-walled, nearly globose, straw-colored, ca. 4 mm long. Seeds 2, brown, 3-angled, with two sides concave and one convex, 2 mm long, papillose.

DISTRIBUTION: Found along disturbed areas. Lameshur (A3151, A4045). Also on St. Croix and St. Thomas; throughout tropical America, Africa, Madagascar, and India.

3. IPOMOEA L.

Twining or creeping vines, seldom shrubs, usually producing abundant milky sap (at least in our species). Leaves simple, lobed, palmately or pinnately compound. Flowers bisexual, regular, produced in axillary, simple or compound dichasial cymes, or solitary; calyx of 5 equal or unequal sepals; corolla bell-shaped, funnel-shaped, or salverform, with entire or lobed margin; stamens 5, included or rarely exserted, the filaments usually unequal, the anthers lanceolate; ovary with 2–4 locules, the style single, the stigmas nearly globose. Fruit a 4-seeded, ovoid or globose, 4-valvate capsule; seeds 3-angled, glabrous or pubescent.

A pantropical genus of about 500 species.

Key to the Species of *Ipomoea*

1. Plant an erect shrub .... 1. *I. carnea* subsp. *fistulosa*
1. Plants creeping or twining vines.
   2. Plants creeping, rooting at nodes, seldom climbing; leaves deeply notched at apex; found along sandy beaches .... 7. *I. pes-caprae*
   2. Plants twining vines, usually not rooting at nodes; leaves otherwise; inland, rarely along the coast.
      3. Corolla yellow, with burgundy center .... 6. *I. ochracea*
      3. Corolla variously colored, but not yellow.
         4. Corolla trumpet-shaped (usually with exserted stamens and stigmas).
            5. Corolla scarlet to orange-red or bright pink, diurnal.
               6. Corolla bright pink, thick, with 5 reflexed, acute lobes; stamens pink; sepals pink, rounded; leaves usually ovate to lanceolate, with a nearly truncate base .... 8. *I. repanda*

6. Corolla scarlet to red-orange, thin, the lobes shallow and rounded, not reflexed; stamens white or pink; sepals green, lanceolate; leaves ovate, 3–5-lobed, cordate at base ... 3. *I. hederifolia*
    5. Corolla white, nocturnal... 13. *I. violacea*
  4. Corolla funnel- or bell-shaped; stamens and stigmas included.
    7. Leaves kidney-shaped to lyrate, <1.5 cm long, congested on short axillary branches; petioles often longer than the blades .... 2. *I. eggersii*
    7. Leaves variously shaped, but not kidney-shaped or lyrate, >2.5 cm long, seldom congested on lateral branches; petioles shorter than the blades.
      8. Corolla <1.5 cm long .... 12. *I. triloba*
      8. Corolla 4 cm or longer.
        9. Corolla 5–7 cm long, 5.5–6.6 cm wide.
          10. Sepals 12–18 mm long, without hyaline margins, the apex acuminate; corolla tube darker within.
            11. Sepals with 3 prominent main, parallel veins, projecting as keels; corolla pink .... 9. *I. setifera*
            11. Sepals without veins projecting as keels; corolla reddish violet .... 4. *I. indica* var. *acuminata*
          10. Sepals 6–8 mm long, without conspicuous veins, the margins hyaline; corolla blue, turning violet with age.. 11. *I. tricolor*
        9. Corolla 4–4.5 cm long, 3.5–5 cm wide.
          12. Sepals lanceolate, 2–3 cm long, hirsute at base; corolla pale blue, lilac, or fuchsia, the tube whitish within .... 5. *I. nil*
          12. Sepals ovate, 8–12 mm long, glabrous; corolla pink, the tube darker within .. 10. *I. tiliacea*

1. Ipomoea carnea Jacq. subsp. **fistulosa** (Mart. *ex* Choisy) D.F. Austin, Taxon **26**: 237. 1977. *Ipomoea fistulosa* Mart. *ex* Choisy in A. DC., Prodr. **9**: 349. 1845.

Erect subwoody shrub to 2 m tall, producing abundant milky sap; stems slightly angular, slender, puberulent. Leaf blades 8–21 × 3.5–8.5 cm, chartaceous, lanceolate, the lower surface densely covered with minute, velvety hairs, especially along the prominent veins, the apex long-acuminate, the base cordate to truncate, the margins entire; petioles slender, 2–7 cm long. Calyx light green, with 5 ovate sepals, the apex obtuse, 6–7 mm long; corolla funnel-shaped, pink or light purple, the tube darker within, 5–8 cm long, with rounded lobes; stamens and pistil pink, included. Capsule thin-walled, ovoid, straw-colored, 1.5 mm long. Seeds usually 4 per capsule, brown, covered with woolly hairs.

DISTRIBUTION: Uncommon, mostly found under cultivation as an ornamental, with a few naturalized populations. Maria Bluff (A1988). Cultivated throughout the tropics.

2. Ipomoea eggersii (House) D.F. Austin, Ann. Missouri Bot. Gard. **64**: 335. 1979. *Exogonium eggersii* House, Bull. Torrey Bot. Club **35**: 104. 1908.

Fig. 82E, F.

Ipomoea arenaria sensu Urb., Symb. Ant. **4**: 508. 1910, non Choisy, 1838.

Subwoody twining vine, to 5 m long, many-branched from base; branches producing abundant milky sap, cylindrical, slender, reddish brown or green, glabrous. Leaves congested on short, axillary branches; blades 6–15 × 4–13 mm, chartaceous, kidney-shaped, lyrate, or many-lobed, glabrous, the central vein prominent on lower surface, projecting as a mucron beyond the deeply notched apex, the base cordate, truncate, or rounded, the margins sinuate; petioles slender, 5–30 mm long. Calyx light green, 5–6 mm long, the sepals ovate to nearly rounded, unequal; corolla funnel-shaped, 4–5 cm long, the tube light green without, usually

whitish within, the limb pink or light purple, 3–4 cm wide, with 5 obtuse lobes; stamens and pistil white, included. Capsule ellipsoid, 1.5 cm long, light brown. Seeds 4, dark brown, ca. 6 mm long, with a tuft of long brownish hairs at apex.

DISTRIBUTION: Common in dry coastal scrub and along beach areas. Lind Point (A3095), Salt Pond (A762), White Cliffs (A2711). Also on St. Croix, St. Thomas, Tortola, and Virgin Gorda; the Lesser Antilles.

NOTE: This species was erroneously identified by Woodbury and Weaver (1987) as *Ipomoea steudelii* Millsp.

3. Ipomoea hederifolia L., Syst. Nat., ed. 10, **2**: 925. 1759.

Fig. 83J–M.

Quamoclit coccinea sensu Urb., Symb. Ant. **4**: 514. 1910, non (L.) Moench, 1794.

Ipomoea coccinea sensu authors, non L., 1753.

Subwoody twining vine to 5 m long, producing watery sap; stems cylindrical, slender, glabrous. Leaf blades 4–13 × 4.5–11 cm, chartaceous, ovate to deeply repand 3–5-lobed, glabrous, minutely dotted, the lobes acuminate at apex, the base cordate or sagittate; petioles slender, 4.5–9 cm long. Flowers in long, dichasial cymes with long, lateral 1-sided racemes. Calyx light green, 5–8 mm long, the sepals lanceolate, rounded at base, with a long, awl-shaped appendage; corolla salverform, scarlet to orange-red, 4–4.5 cm long, thin, with 5 obtuse lobes, the limb 2.5 cm diam.; stamens and pistil white or pinkish-tinged, exserted. Capsule thin-walled, globose, light brown, ca. 8 mm diam. Seeds light brown, 5 mm long, pubescent.

DISTRIBUTION: Uncommon in disturbed open areas. Cruz Bay (A2347, A3082), Enighed (A3101). Also on St. Croix and St. Thomas; throughout the West Indies, Central and South America, introduced into the Old World tropics.

COMMON NAME: sweet william.

4. Ipomoea indica (Burm.) Merr. var. **acuminata** (Vahl) Fosberg, Bot. Not. **129**: 38. 1976. *Convolvulus acuminatus* Vahl, Symb. Bot. **3**: 26. 1794. *Ipomoea*

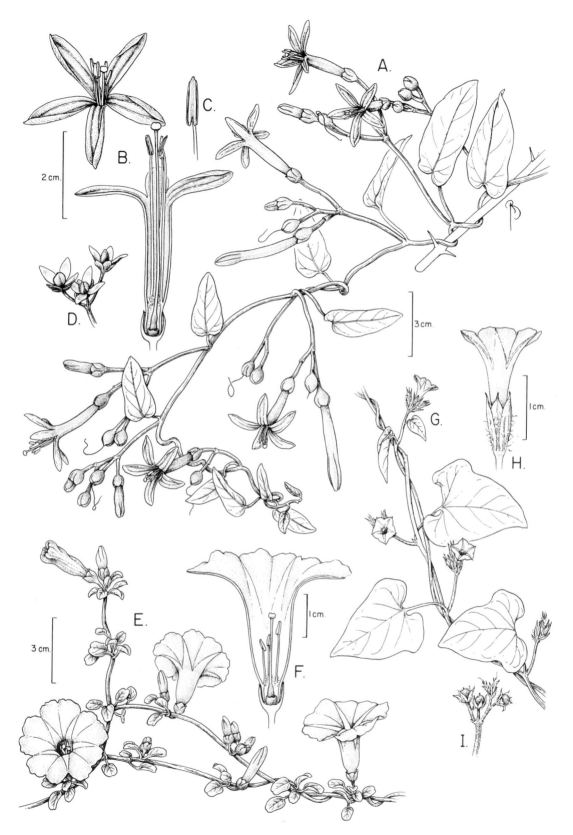

FIG. 82. A–D. *Ipomoea repanda.* **A.** Fertile branch. **B.** Corolla from above and l.s. flower. **C.** Anther. **D.** Dehisced capsules. **E, F.** *Ipomoea eggersii.* **E.** Flowering branch. **F.** L.s. flower. **G–I.** *Ipomoea triloba.* **G.** Flowering branch. **H.** Flower. **I.** Infructescence.

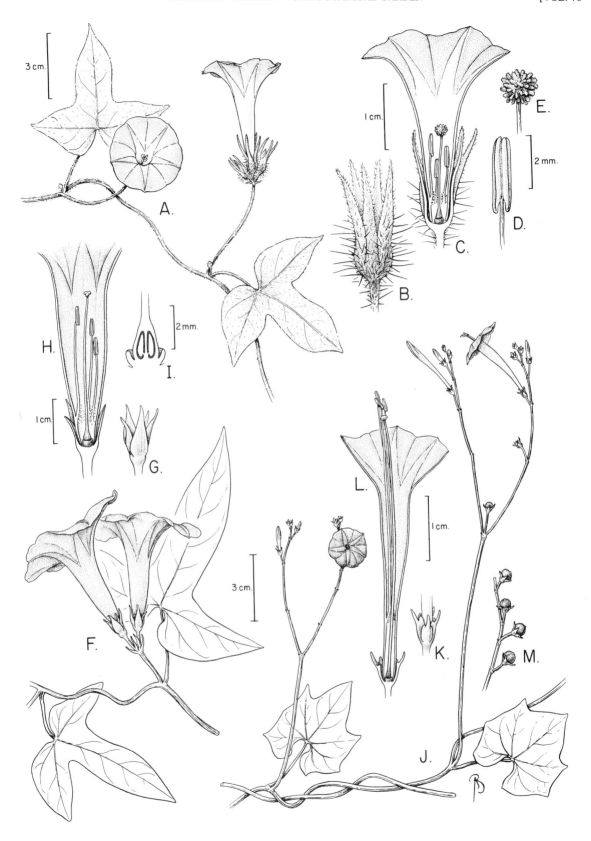

FIG. 83. A–E. *Ipomoea nil.* **A.** Flowering branch. **B.** Calyx. **C.** L.s. flower. **D.** Anther. **E.** Stigma. **F–I.** *Ipomoea indica* var. *acuminata.* **F.** Flowering branch. **G.** Calyx. **H.** L.s. flower. **I.** L.s. ovary and disk. **J–M.** *Ipomoea hederifolia.* **J.** Fertile branch. **K.** Detail of calyx. **L.** L.s. flower. **M.** Infructescence.

acuminata (Vahl) Roem. & Schult., Syst. Veg. **4:** 228. 1819.   Fig. 83F–I.

Herbaceous twining vine to 5 m long, producing abundant milky sap; stems cylindrical, slender, glabrous. Leaf blades 6–11 × 4–8 cm, chartaceous, ovate or 3–lobed, the apex acute or shortly acuminate, the base cordate, the margins entire; petioles slender, as long or longer than the blades. Flowers in simple dichasia or solitary on long, axillary peduncles; bracts lanceolate, 1.8–2 cm long. Calyx light green, 15–18 mm long, the sepals subequal, lanceolate, with long-acuminate apex; corolla funnel-shaped, reddish violet with darker tube within, 5–7 cm long, with 5 obtuse rounded lobes; stamens included; stigmas white, barely exposed. Capsule globose, 10–15 mm diam. Seeds black, smooth with tuft of short hairs at base.

DISTRIBUTION: Not common, found in open, disturbed areas near the coast. Along trail to Brown Bay (A4127); Johnson Bay (A4666), Maho Bay Camp (A4051). Also on St. Croix, St. Thomas, and Tortola; throughout tropical America.

5. Ipomoea nil (L.) Roth, Catal. Bot. **1:** 36. 1797. *Convolvulus nil* L., Sp. Pl., ed. 2, **1:** 219. 1762.

Fig. 83A–E.

Subwoody, twining, pilose vine to 5 m long, not producing milky sap; stems cylindrical, slender, densely covered with reflexed yellowish hairs. Leaf blades 4.5–13.3 × 5.7–15 cm, chartaceous, deeply 3-lobed, the lobes ovate to lanceolate, with acuminate apex, the base cordate or deeply cordate; petioles slender, 2.5–8 cm long. Flowers in simple dichasia or sometimes solitary on peduncle; peduncle as long as the petioles or shorter. Calyx light green, hirsute at base, with 5 equal, lanceolate sepals, 2–3 cm long; corolla funnel-shaped, 4–4.5 cm long, light blue, lavender or fuchsia, with white tube, the lobes rounded; stamens and pistil white and included. Capsule thin-walled, nearly globose, ca. 5 mm long, straw-colored, glabrous, with persistent sepals. Seeds ca. 4 mm long, blackish brown, with a few grayish hairs.

DISTRIBUTION: Uncommon in open disturbed areas. Enighed (A3100), Lind Point (A2315), Lameshur Bay (A3161). Also on St. Croix, St. Thomas, and Tortola; native to Mexico, now widespread throughout the tropics.

COMMON NAME: morning glory.

6. Ipomoea ochracea (Lindl.) G. Don, Gen. Hist. **4:** 270. 1838. *Convolvulus ochraceus* Lindl., Bot. Reg. **13:** 1060. 1827.

Subwoody twining vine to 5 m long, producing scanty watery sap; stems cylindrical, slender, densely covered with minute, whitish hairs. Leaf blades 3–10 × 2.5–7.5 cm, chartaceous, ovate, glabrous except for a few hairs along veins, the apex acuminate and usually mucronate, the base cordate, the margins entire to sinuate; petioles slender, pubescent, 2–6 cm long. Calyx light green, 5–7 mm long, with 5 ovate, membranous, glabrous sepals; corolla funnel-shaped, 3–4 cm long, light yellow, with base of tube burgundy within, the limb with 5 rounded lobes; stamens and stigmas white, included. Capsule thin-walled, ovoid, straw-colored, 1.3–1.6 cm long. Seeds black, dull, 4–5 mm long, glabrous.

DISTRIBUTION: Apparently a recent introduction, found only in Cruz Bay Quarter along North Shore Road close to vicinity of Caneel Bay (A2607, A3096, A4099). Also on St. Croix; possibly native to tropical Africa, now cultivated throughout the tropics.

7. Ipomoea pes-caprae (L.) R. Br. *in* Tuckey, Narr. Exped. Zaire 477. 1818. *Convolvulus pes-caprae* L., Sp. Pl. 159. 1753.

Subwoody creeping vine or seldom climbing, to 10 m long, producing abundant milky sap; stems cylindrical, glabrous, producing adventitious roots along the swollen nodes. Leaf blades 6.5–12 × 4–7.3 cm, thick coriaceous, oblong to nearly rounded, glabrous, involute, the apex deeply notched and mucronate, the base rounded, cordate, or truncate, the margins entire; petioles stout, 3–8 cm long, slightly flattened along upper surface, swollen near the blade. Flowers solitary or in simple dichasia; peduncle shorter than the petioles; bracts ca. 5 mm long, awl-shaped. Calyx yellowish green, 1.2–1.4 cm long, the sepals unequal, ovate, with rounded and mucronate apex; corolla funnel-shaped, pink or lavender, 4.5–5 cm long, the limb to 6 cm diam., with 5 rounded lobes; stamens and pistil included. Capsule thin-walled, globose, light brown, 0.8–1 cm diam., with persistent, reflexed sepals. Seeds 4, dark brown, pubescent.

DISTRIBUTION: Common along sandy beaches. Salt Pond Bay (A2052). Also on Anegada, St. Croix, St. Thomas, and Tortola; throughout the tropics.

COMMON NAMES: bay vine, beach morning glory, goat foot morning glory.

8. Ipomoea repanda Jacq., Enum. Syst. Pl. 13. 1760. *Exogonium repandum* (Jacq.) Choisy, Convolv. Diss. Sec. 128. 1838.   Fig. 82A–D.

Herbaceous to slightly woody twining vine to 8 m long, producing abundant milky sap; stems cylindrical, slender, glabrous, lineolate when young and lenticellate with age. Leaf blades 5–18 × 1.5–13 cm, chartaceous, ovate to lanceolate, glabrous, the apex acute or acuminate and mucronate, the base nearly cordate, truncate, or rounded, the margins entire; petioles slender, furrowed along upper surface, 1.5–6 cm long. Flowers in compound axillary dichasial cymes; bracts early deciduous. Calyx pink, cup-shaped, 8–10 mm long, the sepals unequal, ovate, with rounded apex; corolla salverform, bright pink to scarlet, 4–5.5 cm long, the limb with 5 elongate, reflexed lobes, with obtuse apices; stamens and stigmas pink or the stigmas white, exserted. Capsule thin-walled, ellipsoid, light brown, 12–15 mm long. Seeds 4, black, 5–6 mm long, with long, brownish hairs along two edges.

DISTRIBUTION: One of the commonest *Ipomoea* species on the island, mostly found in moist areas. Center Line Road by km 3.4 (A2400), along road to Bordeaux (A3123). Also on St. Thomas and Tortola; Puerto Rico and in the Lesser Antilles.

COMMON NAME: Mari de Lugo.

9. Ipomoea setifera Poir. *in* Lam., Encycl. **6:** 17. 1804.

Convolvulus ruber Vahl, Eclog. Amer. **2:** 12. 1798. *Ipomoea rubra* (Vahl) Millsp., Publ. Field Colombian Mus., Bot. Ser. **2:** 86. 1900, non Murray, 1784.

Herbaceous twining or creeping vine to 10 m long, producing abundant milky sap; stems cylindrical, slender, pilose, sometimes reddish-tinged. Leaf blades 4.5–16(–24) × 2.7–6.5 cm, chartaceous, oblong to ovate, glabrous, venation usually reddish-tinged, the margins revolute, the apex obtuse, notched and mucronate, the base cordate or sagittate, the margins entire to sinuate; petioles slender, 10–20 cm long. Flowers in axillary, compound dichasia; peduncle thick, much shorter than the petioles; bracts ovate, 1.5–2 cm long. Calyx light green, 12–17 mm long, the sepals ovate, unequal, with 3 prominent parallel veins, the apex long-acuminate; corolla funnel-shaped, pinkish violet, the tube darker within, 4–7 cm long, the limb 4–5 cm diam., with 5 rounded lobes; stamens pinkish, included; stigmas pinkish, slightly exserted. Capsule thin-walled, ovoid, ca. 1 cm diam., covered by persistent sepals. Seeds light brown, tomentose.

DISTRIBUTION: Uncommon in moist open areas. Center Line Road by km 1 (A3094). Also in tropical America and tropical Africa.

10. Ipomoea tiliacea (Willd.) Choisy *in* A. DC., Prodr.
9: 375. 1845. *Convolvulus tiliaceus* Willd., Enum.
Pl. 203. 1809.

Herbaceous twining vine to 10 m long, producing abundant
milky sap; stems cylindrical, slender, glabrous or puberulent. Leaf
blades 5–20 × 4–14 cm, chartaceous, ovate, or seldom 3–5-
lobed, the apex acuminate, the base cordate, the margins ciliate;
petioles slender, puberulent 4–20 cm long. Flowers in compound
dichasia; peduncle smooth, shorter than the petioles, the pedicels
smooth; bracts early deciduous. Calyx light green, 8–12 mm long,
the sepals unequal, oblanceolate, glabrous, with apiculate apex;
corolla funnel-shaped, pink, 5–6 cm long, the limb 4–5 cm diam.,
with 5 rounded lobes; stamens and pistil included. Capsule thin-
walled, depressed-globose, brown, 8–10 mm diam. Seeds 4 or
fewer, blackish brown, smooth, 3.5–4 mm long.

DISTRIBUTION: Common in moist open areas. Cruz Bay Quarter along
Road 104 by km 3.4 (A2650), Fish Bay (A2468). Also on St. Croix, St.
Thomas, Tortola, and Virgin Gorda; throughout tropical America, intro-
duced into the Old World tropics.

COMMON NAME: willy vine.

11. Ipomoea tricolor Cav., Icon. 3: 5, t. 208. 1795.

Ipomoea violacea sensu authors, non L., 1753.

Herbaceous twining vine to 5 m long, not producing milky
sap; stems cylindrical, slender and glabrous. Leaf blades 4–12 ×
4–14 cm, chartaceous, cordate, glabrous, the apex acuminate and
mucronate, the base deeply cordate, the margins entire; petioles
slender, 2.5–10 cm long. Flowers in axillary, simple or compound
dichasial cymes; peduncle as long as or shorter than the petioles;
bracts minute. Calyx light green, 6–8 mm long, the sepals equal,
lanceolate, glabrous, with hyaline margins; corolla funnel-shaped,
blue when fresh, turning purple with age, the throat white, the
tube white without, yellowish within, 6–7 cm long, the limb with
5 rounded lobes; stamens and pistil white, included. Capsule
conical, with persistent style, straw-colored, 1.3–1.6 cm long.
Seeds 4, black, dull, 9–10 mm long.

DISTRIBUTION: Introduced, probably as an ornamental, found in a few
open disturbed areas. Bethany (B210), Susannaberg (A3119). Also on St.
Croix and St. Thomas; native to Central America, now widely cultivated
and naturalized throughout the tropics.

12. Ipomoea triloba L., Sp. Pl. 161. 1753. Fig. 82 G–I.

Herbaceous twining or creeping vine to 3 m long, producing
scanty milky sap; stems cylindrical, slender and puberulent. Leaf
blades 3–6 × 3–5 cm, chartaceous, ovate to subrounded, usually

3(–5)-lobed, glabrous or puberulent, the lobes more or less deeply
cut, acute or acuminate at apex, usually mucronate, the base
cordate or sagittate; petioles slender, 2–7 cm long, with 2 glands
at base of blade. Flowers in compound or simple, axillary dichasial
cymes; peduncle tuberculate, usually longer than the subtending
petiole; bracts minute; pedicels tuberculate. Calyx light green, ca.
8 mm long, the sepals unequal, ovate to oblanceolate, acuminate
or acute at apex, pilose; corolla funnel- to bell-shaped, pink or
lavender, usually reddish within at base, 1.4–1.6 cm long, with 5
obtuse lobes; stamens and pistil white, included. Capsule depressed-globose with persistent style, brown, pilose, ca. 8 mm
diam. Seeds 5 mm long, dark brown to black, glabrous.

DISTRIBUTION: A common weed of open disturbed areas. Enighed
(A3099), Lind Point (A2314), trail to Sieben (A2056). Also on St. Croix,
St. Thomas, Tortola, and Virgin Gorda; native to tropical America, now
widespread as a weed throughout the tropics.

13. Ipomoea violacea L., Sp. Pl. 161. 1753.

Convolvulus tuba Schltdl., Linnaea 6: 735. 1831. *Calo-
nyction tuba* (Schltdl.) Colla, Mem. Nov. Sp. Calo-
nyction 15. 1840.

Ipomoea macrantha Roem. & Schult., Syst. Veg. 4:
251. 1819.

Subwoody twining vine to 15 m long, producing abundant
milky sap; stems cylindrical, reaching 2–3 cm diam., with numerous concentric rings of phloem and xylem; bark grayish to whitish
with numerous lenticels. Leaf blades 9–15 × 7–11 cm, thick
coriaceous, cordate, seldom 3-lobed, involute and glabrous, the
apex acute or acuminate and mucronate, the base cordate to deeply
cordate, the margins wavy; petioles stout, longer than the blade,
swollen at base of blade. Flowers nocturnal, solitary or in axillary,
simple dichasia; peduncle to 7 cm long. Calyx yellowish green,
2–2.5 cm long, the sepals thick, glabrous, unequal, with obtuse
to rounded apex; corolla thick, salverform, white, 5–7 cm long,
the tube yellowish without, the limb 6–7 cm diam., with 5 obtuse
lobes; stamens and pistil included. Capsule globose or depressed-
globose, straw-colored, 2.5–3 cm diam., with persistent reflexed
sepals. Seeds obtusely 3-angled, 1–1.2 cm long, velvety brown,
with a line of brownish hairs along two edges.

DISTRIBUTION: Common in coastal areas. Emmaus (A4007), Nanny
Point (A2451). Also on St. Croix, St. Thomas, and Tortola; throughout
the tropics.

NOTE: *Ipomoea batatas* (L.) Lam. has been collected from spontaneous
individuals (A3243); however, it does not seem to be persistent.

DOUBTFUL RECORD: *Ipomoea alba* L. was reported for St. John by
Woodbury and Weaver (1987), but no specimen has been located to confirm
this report.

4. JACQUEMONTIA Choisy

Twining or creeping vines, herbs, or shrubs, usually with branched hairs. Leaves simple, entire or lobed, usually with dark punctations. Flowers bisexual, regular, produced in axillary, usually congested, compound, dichasial cymes; calyx of 5 equal or unequal
sepals; corolla bell-shaped, funnel-shaped, rotate, or salverform, with entire or lobed margins; stamens 5, included or rarely exserted,
the filaments usually unequal, the anthers lanceolate; ovary with 2 locules and a single style, the stigmas 2, oblong to ellipsoid, slightly
flattened. Fruit a 4-seeded, ovoid to globose capsule, 4-valvate but each valve splitting in two; seeds glabrous or pubescent, 3-angled.

A predominantly neotropical genus with about 100 species.

Key to the Species of *Jacquemontia*

1. Corolla tubular, 2–2.5 cm long, scarlet . . . . 4. *J. solanifolia*
1. Corolla rotate or funnel-shaped, <1.5 cm long, blue or white.
　　2. Corolla funnel-shaped, white . . . . 2. *J. havanensis*
　　2. Corolla rotate, blue or violet-blue, with whitish star-shaped center.

3. Plant ferruginous, tomentose .... 1. *J. cumanensis*
3. Plant finely pubescent.. 3. *J. pentanthos*

1. Jacquemontia cumanensis (Kunth) Kuntze, Revis. Gen. Pl. **2**: 441. 1891. *Convolvulus cumanensis* Kunth *in* Humb., Bonpl. & Kunth, Nov. Gen. Sp. **3**: 99. 1819.   Fig. 84E.

Subwoody twining vine to 2 m long, many-branched from base, not producing sap; stems cylindrical, slender, densely covered with soft, ferruginous tomentose, with many short, lateral branches. Leaf blades 1.2–4 × 0.8–2.6 cm, chartaceous, cordate, involute, tomentose on both surfaces, the apex obtuse or acute, mucronate or seldom retuse, the base cordate to truncate, the margins wavy; petioles slender, 0.8–2 cm long, tomentose. Flowers in axillary, congested dichasia; peduncle 1.5–3 cm long; bracts elliptic, ca. 5 mm long. Calyx green, 8–10 mm long, the sepals tomentose, wide-ovate to deltoid, with acuminate apex; corolla thin, rotate, violet or blue, with whitish star-shaped center inside tube, 2.2–2.5 cm wide, the limb with a nearly pentagonal outline, the margins reflexed; stamens and pistil white, exserted. Capsule globose, ca. 5 mm diam., brown, covered by persistent sepals. Seeds obtusely 3-angled, 4 mm long, brown, glabrous.

DISTRIBUTION: Occasional along coastal areas. Dittlif Point (A3964, A3971), east side of Minna Hill (A4237). Also on St. Croix, St. Thomas, and Tortola; Puerto Rico, the Netherlands Antilles, and Venezuela.

2. Jacquemontia havanensis (Jacq.) Urb., Symb. Ant. **3**: 342. 1902. *Convolvulus havanensis* Jacq., Observ. Bot. **2**: 25, t. 45, fig. 3. 1767.   Fig. 84F–I.

Convolvulus jamaicensis Jacq., Observ. Bot. **3**: 6. 1768. *Jacquemontia jamaicensis* (Jacq.) Hallier f. *in* Soler., Syst. Anat. Dicot. 641. 1899.

Subwoody twining vine to 2 m long, many-branched from base, not producing sap; stems cylindrical, slender, densely to sparsely covered with whitish, stellate hairs. Leaf blades 2.4–5 cm long, nearly coriaceous, lanceolate, ovate to linear, sparsely covered with stellate hairs on both surfaces, the apex obtuse and mucronate, the base truncate to rounded, usually oblique, the margins revolute; petioles slender, 0.6–1.2 cm long, densely covered with stellate hairs. Flowers in axillary, compound cymes shorter than the subtending leaf; peduncle densely stellate. Calyx green, 4–5 mm long, the sepals unequal, puberulent, ovate to oblanceolate, with apiculate apex; corolla thin, funnel-shaped, 1–1.3 cm long, white, sometimes with pinkish hue along borders, the lobes obtuse, spreading; stamens and pistil white, slightly exserted. Capsule ovoid to ellipsoid, 5–6 mm long, brown, with persistent sepals at base. Seeds 3-angled, 2–2.3 mm long, brown, glabrous, with a marginal, membranous, short wing.

DISTRIBUTION: Uncommon along coastal areas. Chocolate Hole (A2348), Harbor Point (A4077). Also on St. Croix, St. Thomas, Tortola,

and Virgin Gorda; throughout the West Indies, Florida, and eastern Mexico to Belize.

3. Jacquemontia pentanthos (Jacq.) G. Don, Gen. Hist. **4**: 283. 1837. *Convolvulus pentanthos* Jacq., Collectanea **4**: 210. 1790 [1791].   Fig. 84A–D.

Subwoody twining vine to 2 m long, many-branched from base, not producing sap; stems cylindrical, slender, sparsely covered with 3-branched, whitish hairs, becoming glabrous. Leaf blades 1.8–6 × 1.2–3.7 cm, chartaceous, ovate to lanceolate, sparsely covered with 3-branched hairs on both surfaces, the apex obtuse, acute, or mucronate or seldom retuse, the base cordate to truncate, the margins entire or slightly wavy; petioles slender, 1–4 cm long. Flowers in axillary, congested dichasia; peduncle longer than the subtending leaf; bracts elliptic, ca. 5 mm long. Calyx green, 8–10 mm long, the sepals sparsely covered with brown, short hairs, wide-ovate to deltoid, with acuminate apex; corolla thin, rotate, violet or blue, with whitish, star-shaped center inside tube, the limb with nearly pentagonal outline, 2.2–2.5 cm wide, the margins reflexed; stamens and pistil white. Capsule globose ca. 5 mm diam., brown, covered by persistent sepals. Seeds obtusely 3-angled, 4 mm long, brown, glabrous.

DISTRIBUTION: Very common throughout the island. Bordeaux (B592), East End Point (A687), trail to Margaret Hill (A2306). Also on St. Croix, St. Thomas, Tortola, and Virgin Gorda; throughout the neotropics, introduced in Malaysia and Sri Lanka.

COMMON NAMES: clashie malashie, wild daisy.

4. Jacquemontia solanifolia (L.) Hallier f., Bot. Jahrb. Syst. **16**: 542. 1893. *Ipomoea solanifolia* L., Sp. Pl. 161. 1753.

Subwoody, twining vine to 2 m long, many-branched from base, not producing sap; stems cylindrical, slender, densely covered with 3-parted, whitish hairs when young, becoming glabrous and brown. Leaf blades 3–6 × 1.5–3.7 cm, chartaceous, lanceolate, sparsely covered with 3-parted hairs on both surfaces, the apex obtuse or acute and mucronate, the base cordate, rounded, obtuse, or truncate, the margins revolute; petioles slender, furrowed on upper surface, 1.5–4 cm long, pubescent. Flowers in axillary, lax, elongate cymes; peduncle 2–4 cm long; bracts awl-shaped, ca. 5 mm long. Calyx green, 4–5 mm long, the sepals glabrous, unequal, ovate to rounded; corolla tubular with spreading lobes, scarlet, 2–2.5 cm long, the lobes obtuse, with mucronate apex; stamens and stigma white, exserted. Capsule ovoid to conical 7–8 mm long, light brown, with persistent sepals at base. Seeds 3-angled, 4 mm long, brown, glabrous, with a marginal fringe around two of the corners.

DISTRIBUTION: Uncommon, found in a few locations in coastal forests. Eastern side of Reef Bay (A2663, A3989). Also on St. Croix and St. Thomas; Puerto Rico and the Lesser Antilles.

5. MERREMIA Endl.

Twining vines, sometimes with glandular hairs. Leaves palmately compound, palmately lobed or simple. Flowers bisexual, regular, produced in axillary, compound, dichasial cymes or solitary; calyx of 5 subequal sepals, elongate in fruit; corolla bell-shaped or funnel-shaped, with entire or slightly lobed margins; stamens 5, included, the filaments subequal, the anthers lanceolate, twisted after dehiscence; ovary with 2 locules and a single style, the stigmas 2, nearly globose. Fruit a 4-seeded, ovoid to globose capsule, 4-valvate, 4-valvate or opening irregularly; seeds obtusely 3-angled, glabrous or velvety pubescent.

A pantropical genus of about 80 species.

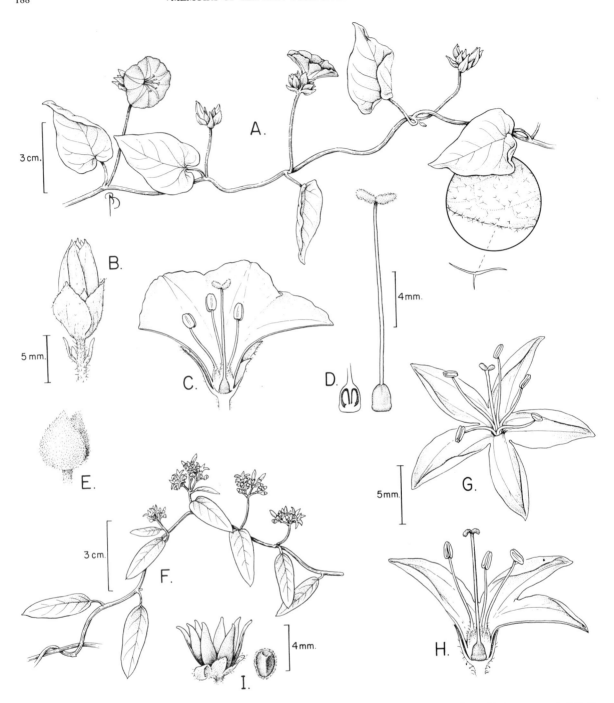

FIG. 84. A–D. *Jacquemontia pentanthos.* A. Flowering branch and detail of indument. B. Flowerbud. C. L.s. flower. D. L.s. ovary and pistil. E. *Jacquemontia cumanensis;* calyx. F–I. *Jacquemontia havanensis.* F. Flowering branch. G. Corolla, top view. H. L.s. flower. I. Dehisced capsule and seed.

Key to the Species of *Merremia*

1. Leaves simple; inflorescence umbel-like .... 5. *M. umbellata*
1. Leaves 5–7-palmately compound or -palmately lobed; flowers solitary or a few in lax dichasial cymes.
   2. Leaves 7-palmately-lobed.
      3. Plant glabrous; corolla bright yellow... 4. *M. tuberosa*
      3. Plant hispid; corolla white with reddish or pinkish center .... 2 *M. dissecta*

2. Leaves 5-palmately compound.
    4. Margins of leaflets entire; inflorescence axes and sepals hispid...................................... 1. *M. aegyptia*
    4. Margins of leaflets serrate or lobed; inflorescence axes and sepals glabrous .... 3. *M. quinquefolia*

1. Merremia aegyptia (L.) Urb., Symb. Ant. **4:** 505. 1910. *Ipomoea aegyptia* L., Sp. Pl. 162. 1753.

Fig. 85A–E.

Subwoody, twining or creeping vine to 3 m long, producing scanty milky sap; all parts except the corolla covered with hispid yellowish hairs; stems cylindrical, slender. Leaves 5-palmately compound, chartaceous; leaflets 4–14 × 2–6 cm, oblanceolate to elliptic, sparsely covered with long, hispid yellowish hairs on both surfaces, the apex and base acuminate, the margins ciliate; petioles slender, furrowed on upper surface, 6–8 cm long, hispid. Flowers in axillary, lax, dichasial cymes; peduncle shorter than petioles; bracts deciduous. Calyx green, 1.5–2 cm long, the sepals hispid at base, unequal, lanceolate to elliptic; corolla bell-shaped, white, 2.5–3 cm long; stamens and stigma white. Capsule nearly globose, ca. 1 cm diam., with accrescent, spreading sepals at base. Seeds light brown, glabrous.

DISTRIBUTION: Occasional in open disturbed areas. Cruz Bay (A3080). Also on St. Croix, St. Thomas, Tortola, and Virgin Gorda; throughout the tropics.

2. Merremia dissecta (Jacq.) Hallier f., Bot. Jahrb. Syst. 16: 552. 1893. *Convolvulus dissectus* Jacq., Observ. Bot. **2:** 4, t. 28. 1767.

Subwoody twining vine to 2 m long, many-branched from base, producing scanty milky sap; stems cylindrical, striate, covered with long, stiff, yellow hairs when young, becoming glabrous at age. Leaves 7-palmately-lobed, lower leaflets once divided, chartaceous; leaflets 4–7 × 1.2–3 cm, glabrous, with deeply lobed margins and acuminate and mucronate apex, the base tapering, unequal; petioles slender, 1.5–4 cm long, hispid. Flowers solitary or a few on axillary peduncles; bracts early deciduous. Calyx green, 2–2.2 cm long, the sepals glabrous, unequal, lanceolate to rounded; corolla bell-shaped, white, with reddish center within, 3–4 cm long, the limb slightly reflexed, with rounded lobes; stamens and stigma yellow, included. Capsule conical, to 1.5 cm long, light brown, with elongate, persistent sepals at base. Seeds obtusely 3-angled, ca. 7 mm long, dull black, glabrous.

DISTRIBUTION: Occasional in open, disturbed areas. Coral Bay (A1987). Also on St. Croix, St. Thomas, and Tortola; throughout the tropics.

COMMON NAME: noyan vine.

3. Merremia quinquefolia (L.) Hallier f., Bot. Jahrb. Syst. **16:** 552. 1893. *Ipomoea quinquefolia* L., Sp. Pl. 162. 1753.

Fig. 85F–K.

Herbaceous twining or creeping vine to 3 m long, many-branched from base, producing watery sap; stems cylindrical, slender, covered with long, stiff, yellowish hairs or glabrous. Leaves 5-palmately compound, chartaceous; leaflets 2–7 × 0.5–1.2 cm, elliptic or lanceolate, glabrous, with serrate margins and acuminate-mucronate apex, with tapering base; petioles slender, 1–3 cm long, hispid or glabrous. Flowers solitary or on simple, axillary dichasia; peduncle as long as the petioles, covered with

minute, glandular hairs; bracts minute. Calyx green, ca. 1 cm long, the sepals glabrous, unequal; corolla funnel-shaped, white to light yellow, ca. 2–2.2 cm long, the limb slightly reflexed, with obtuse lobes; stamens and stigma light yellow, included. Capsule nearly globose, 0.7–1 cm long, straw-colored, with persistent sepals covering most of the fruit. Seeds obtusely 3-angled, ca. 3 mm long, black, with woolly, whitish hairs.

DISTRIBUTION: A common weed, found in open areas. Cruz Bay (A3081), Fish Bay (A3895), Lind Point (A2296). Also on St. Croix, St. Thomas, Tortola, and Virgin Gorda; throughout tropical America.

4. Merremia tuberosa (L.) Rendle *in* Dyer, Fl. Trop. Afr. **4:** 104. 1905. *Ipomoea tuberosa* L., Sp. Pl. 160. 1753.

Twining liana 10–15 m long, producing milky sap; stems cylindrical, stout, glabrous. Leaf blades 7-palmately lobed, 7–12 × 6–11 cm, chartaceous, glabrous or puberulent, the lobes elliptic, long-acuminate at apex, the base cordate, the margins revolute; petioles stout, glabrous, as long as the blade. Flowers solitary or a few, in axillary dichasial cymes; peduncle as long as the petioles, becoming woody; bracts deciduous, minute. Calyx yellowish green, 2–3 cm long, the sepals glabrous, fleshy, unequal, accrescent and becoming woody in fruit; corolla funnel-shaped, yellow, to 5 cm long, the limb 4–5 cm diam.; stamens and stigma included. Capsule ovoid 1.5–2.5 cm long, light brown, with long, reflexed, persistent sepals. Seeds obtusely 3-angled, to 1.5 cm long, black, velvety pubescent.

DISTRIBUTION: Uncommon, cultivated or found as an escape. Lower Cruz Bay Gut (W787). Apparently native to tropical America, now widespread in cultivation and escaped throughout the tropics.

COMMON NAME: wood rose.

5. Merremia umbellata (L.) Hallier f., Bot. Jahrb. Syst. **16:** 552. 1893. *Convolvulus umbellatus* L., Sp. Pl. 155. 1753.

Fig. 86A–E.

Subwoody twining vine to 5 m long, producing scanty milky sap; stems cylindrical, slender, glabrous or pubescent, with a pair of spinelike projections at nodes. Leaf blades simple, 4–17 × 5–12 cm, chartaceous, ovate to lanceolate, glabrous, the apex obtuse, acute or short-acuminate and mucronate, the base cordate or sagittate, the margins wavy; petioles usually longer than blades, cylindrical. Flowers in axillary, umbel-like cymes, with short woody peduncles. Calyx green, 1–1.5 cm long, the sepals glabrous, unequal, overlapping, ovate to rounded; corolla funnel-shaped, bright yellow, 2.5–3 cm long, the limb ca. 3 cm wide, with obtuse lobes; anthers white and included; stigmas green and slightly exserted. Capsule globose, ca. 1 cm long, brown, covered by persistent, accrescent sepals. Seeds obtusely 3-angled, 8 mm long, brown, velvety pubescent, with a line of longer hairs around two of the corners.

DISTRIBUTION: Common in open, disturbed areas. Fish Bay (A2467, A2472), Peter Peak (A4103). Also on St. Croix, St. Thomas, and Tortola; throughout the tropics.

6. STICTOCARDIA Hallier f.

Twining vines, glabrous or pubescent. Leaves simple, ovate to rounded, gland dotted on lower surface. Flowers bisexual, regular, solitary or a few in axillary, dichasial cymes; calyx of 5 subequal sepals; corolla funnel–shaped, with star shape on limb; stamens and

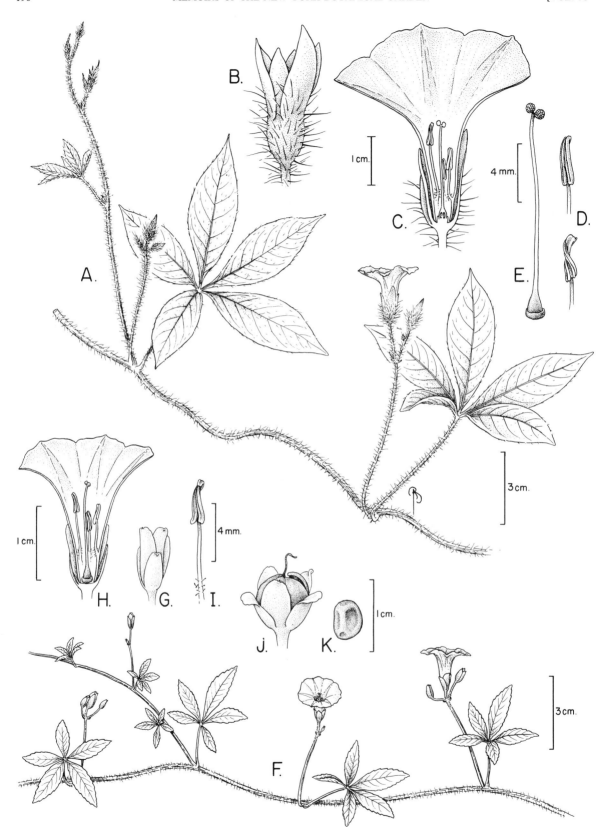

FIG. 85. A–E. *Merremia aegyptia.* **A.** Flowering branch. **B.** Calyx. **C.** L.s. flower. **D.** Anther before and after dehiscing. **E.** Pistil. **F–K.** *Merremia quinquefolia.* **F.** Flowering branch. **G.** Calyx. **H.** L.s. flower. **I.** Stamen. **J.** Capsule. **K.** Seed.

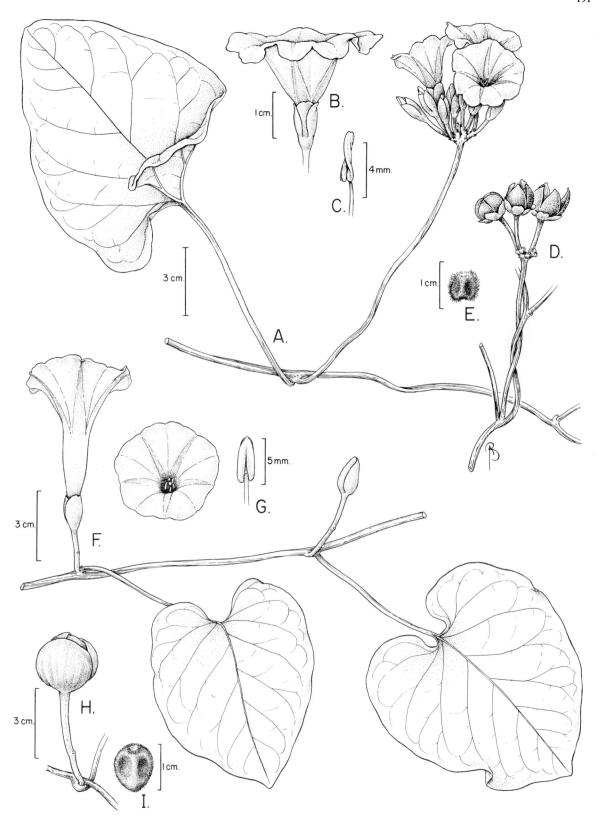

FIG. 86. A–E. *Merremia umbellata*. A. Flowering branch. B. Flower. C. Anther. D. Infructescence. E. Seed. F–I. *Stictocardia tiliifolia*. F. Flowering branch. G. Corolla from above and anther. H. Fruit with accrescent sepals. I. Seed.

pistil included, the anthers lanceolate; stigmas 2, nearly globose. Fruit 4-seeded, indehiscent, with fibrous pericarp; seeds obtusely 3-angled, velvety pubescent.

A paleotropical genus with about 12 species, one of which is widely cultivated throughout the tropics.

1. Stictocardia tiliifolia (Desr.) Hallier f., Bot. Jahrb. Syst. **18:** 159. 1894. *Convolvulus tiliaefolius* Desr. in Lam., Encycl. **3:** 544. 1792. *Rivea tiliaefolia* (Desr.) Choisy. in A. DC., Prodr. **9:** 325. 1845. Fig. 86F–I.

Rivea campanulata sensu House, Muhlenbergia **5:** 72. 1909, not as to type, non (L.) House, 1909.

Subwoody twining vine to 5 m long, usually with numerous short lateral branches, producing scanty milky sap; stems cylindrical, slender, glabrous or pubescent. Leaf blades simple, 5–17 × 5–12 cm, chartaceous, ovate to subrounded, glabrous, the apex obtuse, acute, or short-acuminate and mucronate, the base cordate or sagittate, the margins wavy; petioles usually longer than blades, slightly furrowed. Flowers solitary or few in axillary, short-peduncled inflorescence. Calyx green, 1.7–2 cm long, the sepals glabrous, unequal, overlapping, ovate to rounded; corolla funnel-shaped, pink or lavender, the tube violet within, to 8 cm long, the limb to 6 cm wide, with rounded lobes; anthers and stigmas pink, included. Fruit globose, 2.5–3 cm long, brown, covered by persistent accrescent sepals, late dehiscent by dissolution of fruit wall. Seeds obtusely 3-angled to rounded, 1 cm long, brown, velvety pubescent.

DISTRIBUTION: Uncommon, found in open disturbed areas. Adrian Ruins (A2898, A3120, A4008). Also on St. Croix, St. Thomas, and Tortola; native to the Asian tropics, now widespread through cultivation.

33. Crassulaceae (Life Plant Family)

Succulent herbs or subshrubs. Leaves alternate, opposite, or rarely whorled, simple and entire; stipules wanting. Flowers actinomorphic, bisexual or rarely unisexual, most often 5-merous (but also 3 to many), borne in terminal cymes or seldom solitary; calyx of distinct sepals or rarely connate at base; corolla of distinct petals, seldom forming a tube; stamens usually twice as many as the petals, in 2 cycles, the filaments of equal length, distinct or connate at base, the anthers opening by longitudinal slits; ovary superior; carpels as many as the petals, distinct or connate at base, with a nectariferous appendage, the placentation marginal, with 1 to numerous ovules, the styles distinct. Fruit mostly of separate follicles; seeds small, 1 to many.

A family of 25 genera and ca. 900 species, with cosmopolitan distribution, but best developed in southern Africa and Mexico.

1. BRYOPHYLLUM Salisb.

Fleshy erect herbs. Leaves opposite, simple, entire, lobed or pinnately compound. Flowers bisexual, regular, pendent in terminal panicles; calyx long-tubular, with 4 lobes; corolla tubular, constricted above the base; stamens 8, in 2 series, adnate to the corolla base, included, the anthers lanceolate; pistil of 4 distinct or partially united carpels, the stigma terminal. Fruit of 4 many-seeded follicles.

A genus native to Madagascar, with about 20 species, one of which is widely cultivated and naturalized throughout the tropics.

1. Bryophyllum pinnatum (Lam.) Oken, Allg. Naturgesch. **3:** 1966. 1841. *Cotyledon pinnata* Lam., Encycl. **2:** 141. 1786. *Kalanchoe pinnatum* (Lam.) Pers., Syn. Pl. **1:** 146. 1805. Fig. 87.

Erect, fleshy herb, to 1 m tall, mostly with 1 main, unbranched stem, glabrous. Leaf simple to deeply lobed (the segments often distinct and pinnately compound); blades (or segments) 6–13 × 2–8 cm, fleshy, lanceolate, elliptic, oblong, or oblanceolate, the apex obtuse, rounded and retuse, the base obtuse, rounded, or truncate, the margins deeply crenate, usually reddish, often producing plantlets; petioles usually longer than blades. Flowers pendulous on large, terminal panicles. Calyx slightly inflated, green to reddish green, 2.5–4 × 1–1.5 cm; corolla dull reddish to salmon-colored, 4–5 cm long, much narrower than the calyx, developing late, the buds accumulating water; stamens and pistil included, each carpel with a nectary gland at base. Follicles ellipsoid, 1–1.5 cm long, beaked, brown, covered by persistent, accrescent sepals. Seeds linear, ca. 1 mm long, dark brown.

DISTRIBUTION: A common weed found in open, disturbed areas. Susannaberg (A2384, A3959). Also on St. Croix, St. Thomas, and Tortola; native to Madagascar, now cultivated and naturalized throughout the neotropics.

COMMON NAMES: clapper bush, green love, leaf of life, life plant, wonderful leaf.

34. Cucurbitaceae (Cucumber Family)

Monoecious or dioecious, tendriled climbing or trailing herbs or soft woody shrubs, usually with abundant watery sap; tendrils opposite to leaves. Leaves alternate, simple, trifoliolate or palmately compound; petioles elongate and usually with nectariferous glands; stipules wanting. Flowers large or small, unisexual or rarely bisexual, actinomorphic, solitary or in axillary simple or compound racemes; calyx of (3–)5(–6) sepals or lobes at upper portion of hypanthium; corolla tubular, with 5 more or less prominent lobes or with (3–)5(–6) nearly distinct petals, the corollas usually dissimilar in staminate and pistillate flowers; stamens 5, the filaments adnate to the base of corolla tube, the anthers opening by longitudinal slits; ovary inferior, with usually (2–)3(–5) carpels, each with 1 to many ovules, the placentation parietal or axile, the styles simple, with bilobed stigmas; nectary disk central in staminate flowers, in pistillate flowers, distal on the ovary. Fruit a berry or a leathery capsule (a pepo); seeds numerous and usually flattened.

A family of 90 genera and ca. 700 species, predominantly in tropical and subtropical areas.

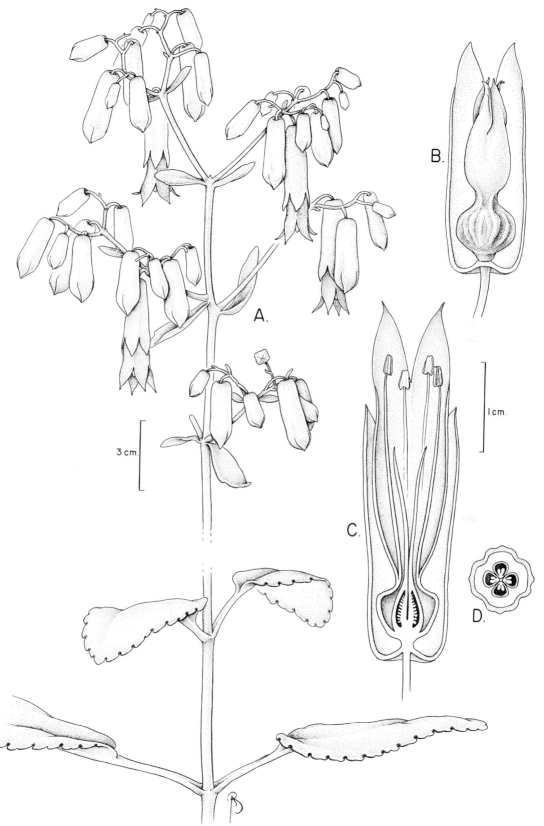

Fig. 87. *Bryophyllum pinnatum.* **A.** Flowering branch. **B.** L.s. calyx showing unopened immature corolla. **C.** L.s. mature flower. **D.** C.s. lower part of corolla showing ovary.

Key to the Genera Cucurbitaceae

1. Fruit ellipsoid-angular to fusiform-angular, orange-yellow, with thornlike projections (muricate) along 6 ridges, opening along 3 valves, exposing numerous seeds covered with red arils .... 6. *Momordica*
1. Fruit ellipsoid, obovoid to cylindrical, not angular, variously colored but not orange-yellow, smooth or with spinelike projections, indehiscent; seeds not covered with a red aril.
  2. Fruit with numerous spinelike projections (spinulose); vine mostly creeping.............................. 2. *Cucumis*
  2. Fruit smooth; vine climbing or creeping.
    3. Fruit 20-45 cm long; corolla bright yellow.
      4. Fruits nearly ellipsoid, with fibrous mesocarp.. 4. *Luffa*
      4. Fruits rounded, depressed-globose or obovoid, with fleshy mesocarp .... (cultivated) *Cucurbita*
    3. Fruit 1–2 cm long with fleshy mesocarp; corolla white or light yellow, to 1.5 cm wide.
      5. Fruit congested in leaf axils .... 3. *Doyerea*
      5. Fruit solitary or in racemes.
        6. Plant slender, to 2 m long; leaves membranous; corolla light yellow, 5 mm across .... 5. *Melothria*
        6. Plant robust, to 10 m long; leaves chartaceous; corolla whitish or yellowish green, 1.5 cm across .... 1. *Cayaponia*

1. CAYAPONIA Silva Manso, nom. cons.

Monoecious or dioecious, tendriled herbaceous vines. Leaves simple, entire, lobed, or palmately divided. Flowers unisexual, actinomorphic, in axillary racemes, panicles, fascicles, or solitary; calyx bell-shaped, with 5 minute lobes; corolla short-tubular at base, with spreading or reflexed lobes; staminate flowers with a pistillode and 3 exserted stamens, the filaments adnate to the corolla base, the anthers sigmoid and connivent; pistillate flowers with 3 staminodes, the ovary 3-lobed, with a long, terminal style, 3-branched at apex. Fruit a small berry; seeds 2 or 3, flattened and beaked.

A predominantly neotropical genus of about 60 species.

1. Cayaponia americana (Lam.) Cogn. *in* A. DC. & C. DC., Monogr. Phan. **3**: 785. 1881. *Bryonia americana* Lam., Encycl. **1**: 498. 1785.

Fig. 88E, F.

Monoecious vine to 10 m long, with many lateral branches; stems angular or nearly cylindrical, with numerous longitudinal ribs, glabrous or puberulent, swollen at nodes; tendrils axillary, simple or branched. Leaf blades 5–20 × 5–18 cm, ovate, chartaceous, usually with coarse hairs on both surfaces, palmately lobed, with 3–5 deep sinuses, the lobes oblong-lanceolate, oblanceolate to linear, with obtuse, acuminate, or rounded apices, the base lyrate, the margins entire or crenate; petioles much shorter than blades, furrowed. Flowers axillary, solitary or in few-flowered racemes. Calyx green, 5–8 mm long; corolla nearly white or yellowish green, the tube ca. 1 cm long, the lobes to 1.5 cm long, oblong, reflexed; stamens with green filaments and yellow anthers. Berry smooth, ellipsoid, 1.5–2 cm long, turning from dark green to yellow and finally red. Seeds flattened-ovoid, 0.8–1.2 cm long, without aril.

DISTRIBUTION: Common in open moist areas. Bethany (B205), Susannaberg (A702), trail to Sieben (A2065). Also on St. Croix, St. Thomas, Tortola, and Virgin Gorda; throughout the West Indies.

2. CUCUMIS L.

Monoecious, tendriled herbaceous vines. Leaves simple, dentate or palmately lobed. Flowers unisexual, regular, in axillary fascicles or solitary; calyx bell-shaped, with 5 minute lobes at apex; corolla bell-shaped, yellow; staminate flowers with 3 stamens and a pistillode, the filaments adnate to the corolla base, the anthers free or connivent; pistillate flowers with 3 staminodes, the ovary of 3 locules with numerous horizontal ovules; style 3–5-branched at apex. Fruit a small, ellipsoid to cylindrical berry; seeds numerous, elliptic, small.

A genus of about 30 species, native to the Old World; some are widely cultivated, such as the melon (*Cucumis melo* L.) and the cucumber (*C. sativus* L.).

REFERENCE: Kirkbride, J. H. 1993. Biosystematic monograph of the genus *Cucumis* (Cucurbitaceae). Parkway Publishers, Boone, North Carolina.

1. Cucumis anguria L., Sp. Pl. 1011. 1753.

Fig. 88G–J.

Creeping vine to 2 m long, many-branched from base; stems angular and hirsute; tendrils axillary and simple. Leaf blades 3–10 × 3.5–10 cm, chartaceous, usually with coarse hairs on both surfaces, 3–5-palmately lobed with deep sinuses, the lobes oblong to oblanceolate, with obtuse or rounded apices, the base lyrate, the margins crenate and denticulate; petioles shorter than blades.

Flowers solitary or in axillary fascicles. Calyx yellowish, 5–6 mm long; corolla light yellow, ca. 1 cm long, the lobes acute. Berry ellipsoid to obovoid, spinulose, 4–5 cm long, light yellowish green, edible. Seeds elliptic, beige, 1–1.3 cm long.

DISTRIBUTION: Not common, found in a few disturbed sites. Pen Point (A4023). Also Anegada, St. Croix, St. Thomas, and Tortola; native to Africa, now spontaneous throughout the West Indies, Central America, and South America.

COMMON NAME: wild cucumber.

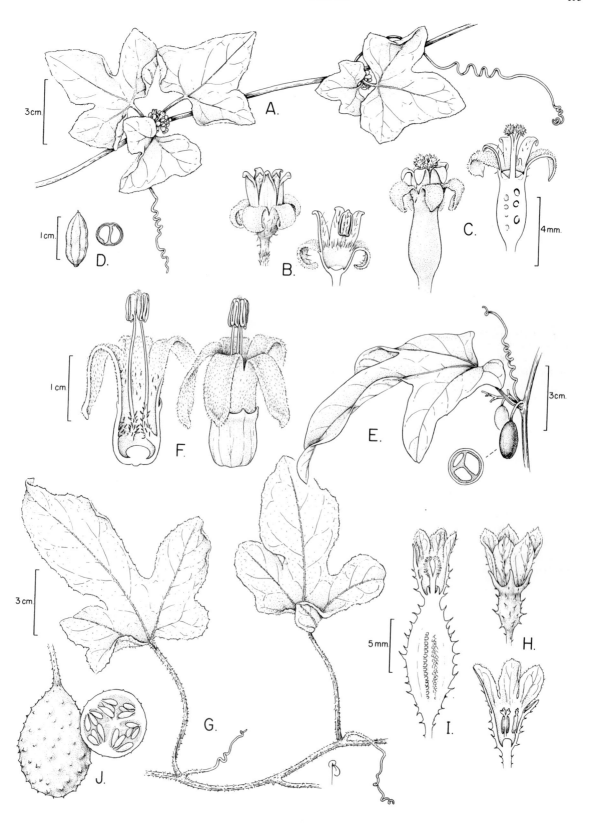

FIG. 88. A–D, *Doyerea emetocathartica*. **A.** Flowering branch. **B.** Staminate flower, and l.s. staminate flower. **C.** Pistillate flower, and l.s. pistillate flower. **D.** Berry, and c.s. berry. **E, F.** *Cayaponia americana*. **E.** Fruiting branch and c.s. berry. **F.** L.s. staminate flower and flower. **G–J.** *Cucumis anguria*. **G.** Leafy stem. **H.** Staminate flower and l.s. staminate flower. **I.** L.s. pistillate flower. **J.** Berry and c.s. berry.

3. DOYEREA Grosourdy

A monospecific genus (or perhaps of 2 species) characterized by the following species.

1. Doyerea emetocathartica Grosourdy, Med. Bot. Criollo **2**: 338. 1864. *Corallocarpus emetocatarticus* (Grosourdy) Cogn., Bull. Soc. Roy. Bot. Belgique **30**: 279. 1891.   Fig. 88A–D.

Anguria glomerata Eggers, Fl. St. Croix 55. 1879.

Dioecious, perennial tendriled vine to 10 m long, many-branched from a tuberous base; stems nearly cylindrical, fragile, green or grayish, glabrous, producing abundant watery sap; tendrils axillary and simple. Leaf blades 5–10 × 5.5–10 cm, with ovate to wide-ovate outline, usually 3-lobed, chartaceous, upper surface slightly rough, lower surface pubescent, the apex acute, obtuse, or acuminate and often mucronate, the base cordiform to lyrate, the margins minutely toothed; petioles to 4 cm long, furrowed and pubescent. Flowers unisexual, in short, axillary, cymose inflorescences. Calyx green, bell-shaped, 3.5–4 mm long, with 5 reflexed sepals; corolla nearly white or yellowish green, the tube ca. 1 cm long, with 5 erect to reflexed, oblong, pilose petals, to 1.5 mm long; stamens 2, opposite to 2 of the petals, the filaments short, the anthers connivent; ovary inferior and elongate, the style bifurcating near the apex, the stigmas bifid, nearly globose and slightly exserted. Berry asymmetrically ellipsoid, 1–1.2 cm long, smooth, light green with dark green spots, finally turning orange. Seeds ovoid, ca. 4 mm long, mottled brown.

DISTRIBUTION: Common in coastal scrubs. Haulover Pt. (A4226), Lameshur (A4044), Maho Bay (A1944). Also on St. Croix and St. Thomas; Mexico to northern South America including the West Indies.

4. LUFFA Mill.

Monoecious, tendriled, herbaceous vines. Leaves simple, 5–7-lobed. Flowers unisexual, large and actinomorphic, calyx bell-shaped, with 5 elongate lobes; corolla of 5 free petals; staminate flowers in axillary racemes, with 3–5 free stamens; pistillate flowers solitary, with 3 staminodes, the ovary elongate, with numerous horizontal ovules. Fruit rounded to obtusely trigonous in cross section, opening by apical pores, mesocarp fibrous; seeds many, flattened, winged or unwinged.

A genus of about 10 species, native to the Old World.

1. Luffa aegyptiaca Mill., Gard. Dict., ed. 8. 1768.
                                      Fig. 89A–C.

Momordica luffa L., Sp. Pl. 1009. 1753.
Momordica cylindrica L. , Sp. Pl. 1009. 1753. *Luffa cylindrica* (L.) M. Roem. *ex* T. Durand & H. Durand, Syll. Fl. Congol. 299. 1909, non M. Roem., 1846.
Luffa cylindrica M. Roem., Prospect Fam. Nat. Syn. Monogr. **2**: 63. 1846.

Monoecious herbaceous vine to 10 m long, with many lateral branches; stems nearly cylindrical or angular, with numerous longitudinal ribs, glabrous or puberulent; tendrils axillary, 3-branched. Leaf blades 11–25 × 7–25 cm, chartaceous, with coarse hairs on both surfaces, palmately 3–7-lobed, with deep sinuses, the lobes lanceolate to ovate, with acute or acuminate apices, the base cordate, the margins entire to serrate; petioles as long as or longer than the blade, furrowed. Flowers solitary or in axillary racemes, the pistillate ones on long receptacles. Calyx green, 1.2–1.5 cm long; corolla yellow, the petals obovate to cuneiform, rounded or notched at apex, 2–4.5 × 1–4 cm long; stamens 5, included. Fruit smooth, nearly cylindrical, 20–45 cm long, green, straw-colored when dried. Seeds elliptic, 1–1.3 cm long, nearly black, smooth.

DISTRIBUTION: Spontaneous after cultivation. Turner Bay (A4066). Also on St. Croix and St. Thomas; native to the Old World, but cultivated and now spontaneous throughout the tropics.

COMMON NAMES: sponge cucumber, strainer vine.

5. MELOTHRIA L.

Monoecious, tendriled, herbaceous vines. Leaves simple, entire or lobed. Flowers actinomorphic, unisexual; staminate flowers in axillary racemes; pistillate flowers solitary. Calyx cup-shaped, with 5 minute lobes at apex; corolla short-tubular at base, with spreading, deeply parted lobes. Staminate flowers with a pistillode; stamens 3, the filaments free, the anthers free or connivent. Pistillate flowers usually with 3 staminodes; ovary ovoid to fusiform, with numerous horizontal ovules, the style short, with 3 linear stigmas at apex. Fruit a small ovoid to ellipsoid berry; seeds numerous and minute.

A neotropical genus of about 10 species.

1. Melothria pendula L., Sp. Pl. 35. 1753. Fig. 89D–F.

Bryonia guadalupensis Spreng., Syst. Veg. **3**: 15. 1826. *Melothria guadalupensis* (Spreng.) Cogn. *in* A. DC. & C. DC., Monogr. Phan. **3**: 580. 1881.
Melothria pergava Griseb., Fl. Brit. W. I. 289. 1860.

Monoecious, herbaceous vine to 2 m long; stems slender, nearly cylindrical and striate, glabrous or puberulent; tendrils axillary and simple. Leaf blades 6–12 × 4–8.5 cm, ovate to triangular, entire to deeply 3–5-palmately lobed, chartaceous, usually with coarse hairs on both surfaces, the lobes obtuse to acuminate, the central one longer, the base lyrate or cordate, the margins crenate to repand and denticulate; petioles shorter than blades, furrowed. Calyx bell-shaped, green, to 5 mm long; corolla bell-shaped, light yellow, ca. 5 mm long, with 5 spreading petals from a short tube; stamens almost sessile, adnate to corolla tube, the anthers free; pistillate flowers with an annular, 2-lobed disk. Berry smooth, ovoid, 15–18 cm long, turning from light green to blackish green. Seeds elliptic, beige, ca. 2 mm long.

DISTRIBUTION: Occasional in open, moist areas. Maho Bay Quarter along Center Line Road (A2413). Also St. Croix, St. Thomas, Tortola, and Virgin Gorda; throughout tropical America.

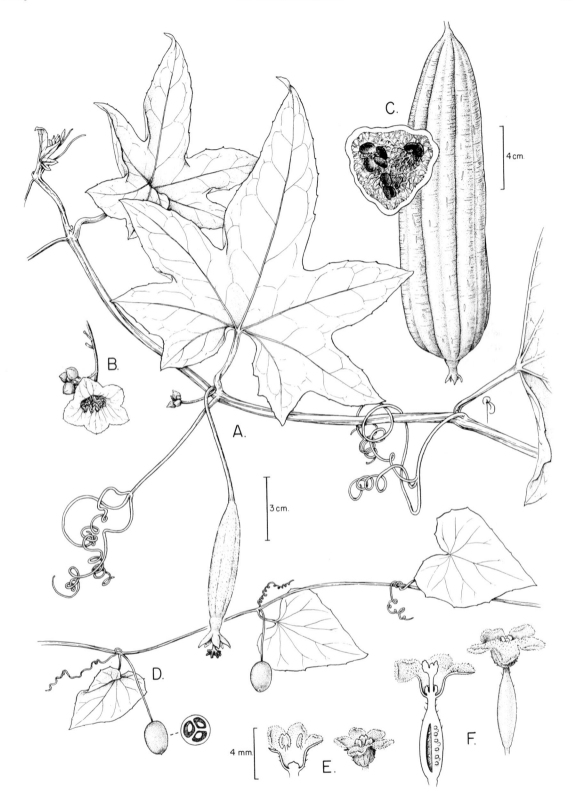

Fig. 89. A–C. *Luffa aegyptiaca*. A. Fertile branch. B. Staminate inflorescence. C. Fruit and c.s. fruit. D–F. *Melothria pendula*. D. Fruiting branch and c.s. berry. E. L.s. staminate flower and staminate flower. F. L.s. pistillate flower and pistillate flower.

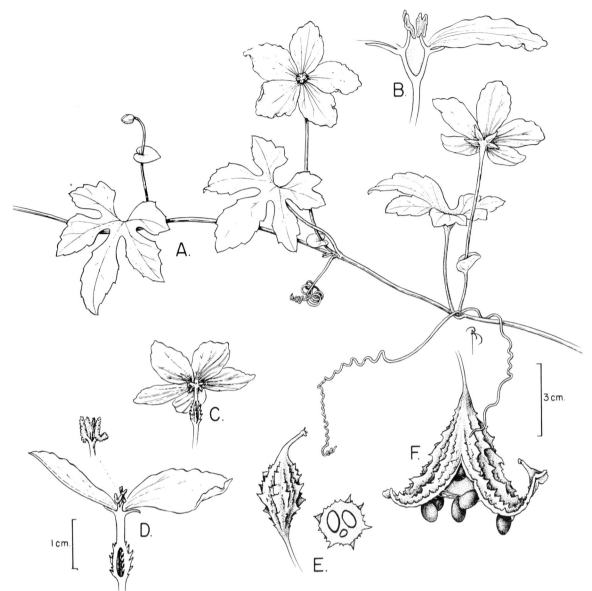

FIG. 90. *Momordica charantia.* **A.** Flowering branch. **B.** L.s. staminate flower. **C.** Pistillate flower. **D.** L.s. pistillate flower and detail of stigmas. **E.** Capsule and c.s. capsule. **F.** Dehiscing capsule showing seeds. (From Mori et al., in press, Guide to the Vascular Plants of Central French Guiana, Mem. New York Bot. Gard. 76).

6. MOMORDICA L.

Monoecious, tendriled herbaceous vines. Leaves simple, entire, or palmately lobed. Flowers unisexual and actinomorphic; calyx bell- or funnel-shaped, with 5 minute lobes at apex; corolla with spreading petals or bell-shaped. Staminate flowers in racemes or solitary; stamens 3; pistillode lacking or glandular. Pistillate flowers solitary in leaf axils, long-peduncled, subtended by a leaflike bract; staminodes wanting, ovary ellipsoid with numerous horizontal ovules, the style terminal, with 3 stigmas at apex. Fruit a fleshy capsule opening along 3 valves; seeds usually many, flattened, beaked and arillate.

A predominantly African genus of about 45 species.

1. Momordica charantia L., Sp. Pl. 1009. 1753.

Fig. 90.

Herbaceous vine to 8 m long; stems slender, nearly cylindrical, striate, glabrous to puberulent, many-branched; tendrils axillary and simple. Leaf blades 3–11 × 3–10 cm, membranous, puberu-lent on both surfaces, chartaceous, with coarse hairs on both surfaces, deeply 5-lobed, the lobes elliptic to obovate-ovate, with obtuse or acute apices, the base cordate, the margins crenate and denticulate; petioles slender, as long as the blade. Flowers solitary, long-pediceled; bract ovate to kidney-shaped, on lower portion of pedicel. Calyx bell-shaped, yellowish green, 10–12 mm long, with lanceolate lobes; corolla light yellow, rotate, with 5 spreading petals, 1–1.7 cm long; stamens free, with short filaments; ovary ellipsoid and muricate. Capsule ellipsoid-angular to fusiform-angular, with thornlike projections (muricate), 3–5 cm long, turning from green to orange-yellow, opening from the apex along 3 valves, exposing the seeds, which hang freely from the fruit wall. Seeds few to many, flattened and beaked, covered with a bright red, juicy aril.

DISTRIBUTION: Common in open, moist areas. Lameshur (A2734), Johnson Bay (A4671). Also on St. Croix, St. Thomas, Tortola, and Virgin Gorda; originally native to the Old World tropics, now pantropical in distribution.

COMMON NAMES: jumbie pumkin, maiden apple, old maid.

NOTE: The ground seeds are said to be poisonous.

CULTIVATED SPECIES: The common squash or pumpkin [*Cucurbita moschata* (Lam.) Poir.] has been collected from spontaneous individuals (A3118); however, it does not seem to be persistent.

35. Cuscutaceae (Love Plant Family)

A family with a single genus and about 150 species, with cosmopolitan distribution, but best developed in the neotropics. The following genus characterizes the family.

REFERENCE: Yuncker, T. G. 1920. Revision of the North American and West Indian species of *Cuscuta*. Univ. Illinois Biol. Monogr. 6(2–3): 1–142.

1. CUSCUTA L.

Twining parasitic herbs, attached to host by haustoria. Leaves alternate, much reduced, scale-like; stipules wanting. Flowers minute, bisexual, actinomorphic, produced in dense headlike or spikelike inflorescences; calyx cup-shaped, with 5 or 4 distinct or connate sepals; corolla white or pink, tubular, with 5 or 4 lobes; stamens 5 or 4, the filaments adnate to the corolla tube, bearing a scale at base within, the anthers opening by longitudinal slits; ovary superior, compound of 2 carpels, each with 2 ovules, the placentation axile, the styles distinct. Fruit a circumscissile, intrastylar or irregularly dehiscent capsule or indehiscent berry; seeds (1–)2–4, per fruit, minute.

A nearly cosmopolitan genus of 150 species.

Key to the Species of *Cuscuta*

1. Plant robust; stems 1.5–2.5 mm diam., forming a tangle a few meters long; sepals rounded; capsules mostly 1-seeded... 1. *C. americana*
1. Plant slender; stems to 0.5 mm diam., <30 cm long; sepals lanceolate; capsules 4-seeded............ 2. *C. umbellata*

1. Cuscuta americana L., Sp. Pl. 124. 1753.   Fig. 91.

Vine to 5 m long, with many entangled, lateral branches; stems golden-yellow, cylindrical, smooth and glabrous, 1.5–2.5 mm diam. Leaves reduced to 1–2 mm long, yellowish scales. Flowers bisexual, sessile, on short, axillary cymes. Calyx cup-shaped, whitish, membranous, 2.3–2.5 mm long, with 5 rounded, minute lobes; corolla whitish, cup-shaped to tubular, ca. 2.3 mm long, with 5 minute, rounded lobes; stamens 5, attached to upper portion of corolla tube, the filaments with a fringed scale at base; ovary depressed-globose. Fruit thin-walled, nearly globose, 1.5–2 mm long, with persistent styles, opening along the intrastylar region; seeds 1 or rarely 2 per fruit, nearly globose to lenticular, ca. 1.5 mm long, smooth, light brown.

DISTRIBUTION: Common in open, dry areas. Fish Bay (A2466), Lameshur (A2569), Nanny Point (A2453). Also on St. Croix, St. Thomas, Tortola, and Virgin Gorda; a common weed throughout the New World.

COMMON NAMES: love, love bush, love vine, yellow dodder, yellow love.

2. Cuscuta umbellata Kunth *in* Humb., Bonpl. & Kunth, Nov. Gen. Sp. **3**: 121. 1819.

Vine to 30 cm long; stems light yellow, cylindrical, smooth and glabrous, to 0.5 mm diam. Flowers bisexual and sessile, on short axillary cymes. Calyx cup-shaped, whitish, membranous, 1.2–1.5 mm long, with 5 lanceolate lobes; corolla whitish, cup-shaped, ca. 2 mm long, with 5 lanceolate lobes; stamens 5, attached to the upper portion of corolla tube, the filaments with a fringed scale at base; ovary depressed-globose. Fruit thin-walled, nearly globose, 1.5–2 mm long, the styles persistent, with intrastylar opening. Seeds 4 per capsule, nearly globose to angled, 1 mm long, smooth, light brown.

DISTRIBUTION: Rare, found only in a few localities. Annaberg (W784). Also throughout tropical America.

NOTE: A collection of this species was cited in error as *Cuscuta indecora* Choisy by Woodbury and Weaver (1987).

36. Erythroxylaceae (Coca Family)

A family with 4 genera and about 250 species, with pantropical distribution, but best developed in the neotropics. The single genus *Erythroxylum* in the New World characterizes the family in the region.

1. ERYTHROXYLUM P. Browne

Shrubs or small trees; young branches with overlapping leaf scales. Leaves simple and alternate, often with a light line along both sides of main vein; stipules intrapetiolar. Flowers small, bisexual, actinomorphic, usually heterostylous, in axillary fascicles or seldom solitary; pedicels with minute bracteoles; calyx of 5 sepals connate into a short tube; corolla of 5 distinct, deciduous petals, usually with a petaloid appendage within; stamens 10, the filaments connate into a more or less short tube, the anthers opening by longitudinal

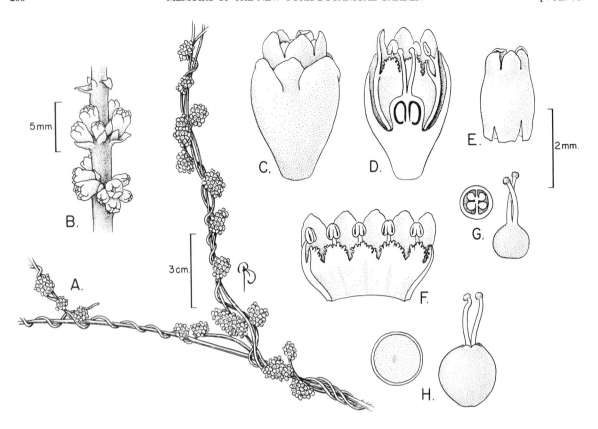

Fɪɢ. 91. *Cuscuta americana.* **A.** Fertile branch. **B.** Detail of fertile branch. **C.** Flower. **D.** L.s. flower. **E.** Corolla. **F.** Corolla spread apart to show stamens and fringed scales at base of filaments. **G.** Pistil and c.s. ovary. **H.** C.s. fruit and fruit.

slits; ovary superior, of 3(–2) carpels, each with a single ovule, the placentation subapical, the styles distinct or rarely connate into a single one. Fruit a single-seeded elongate drupe.

A predominantly neotropical genus with about 230 species.

1. Erythroxylum brevipes DC., Prodr. **1:** 573. 1824.

Fig. 92.

Tree or shrub to 8 m tall, with drooping foliage; bark light brown, much-fissured into small, rectangular plates; short lateral branches glabrous, with overlapping, brown leaf scales. Leaf blades 1.5–3.5 × 0.8–2 cm, obovate to oblanceolate, chartaceous, lines parallel to main vein inconspicuous, lower surface lighter, the apex rounded or slightly notched and mucronulate, the base cuneate to obtuse, the margins entire; petioles slender, 1–2 mm long. Flowers 1–3, axillary, usually on short, leafless branches. Calyx green, cup-shaped, 2.5 mm long, the sepals ovate; petals white, obovate, 2.5–3 mm long, crenate at margins, with a 1 mm long appendage within, at base; stamens white, unequal, the staminal tube 1 mm long; ovary ellipsoid to cylindrical, the styles white, filiform, the stigmas greenish and globose. Drupe ellipsoid to oblong-ellipsoid, 7–8 mm long, turning from green to bright red, becoming sulcate when dry. Seeds 1 per fruit, ellipsoid, ca. 7 mm long.

Dɪsᴛʀɪʙᴜᴛɪᴏɴ: A common shrub from moist forests to coastal scrub. Cinnamon Bay (M17010), Europa Bay (A742), Lind Point (A2696). Also on Jost van Dyke, St. Croix, St. Thomas, Tortola, and Virgin Gorda; Hispaniola, Puerto Rico, and St. Barts.

Cᴏᴍᴍᴏɴ ɴᴀᴍᴇs: brazilette, brizzlet, wild cherry.

37. **Euphorbiaceae** (Poinsettia Family)

Reviewed by L. J. Gillespie

Monoecious or dioecious herbs, shrubs, trees, vines, or lianas, usually with stellate hairs, producing abundant milky or clear latex. Leaves alternate, opposite, or whorled, simple, entire, or lobed; stipules usually present. Flowers unisexual, actinomorphic, produced in cymes, racemes, spikes, or pseudanthia; perianth 3–6-merous or lacking; calyx of valvate lobes; corolla of 2–6 distinct petals; stamens 2 to many or even reduced to 1, the filaments distinct or variously united, the anthers opening by longitudinal slits or seldom by apical pores; nectary disk intrastaminal, extrastaminal, or wanting; ovary superior, compound of (2–)3(–many) carpels, each with

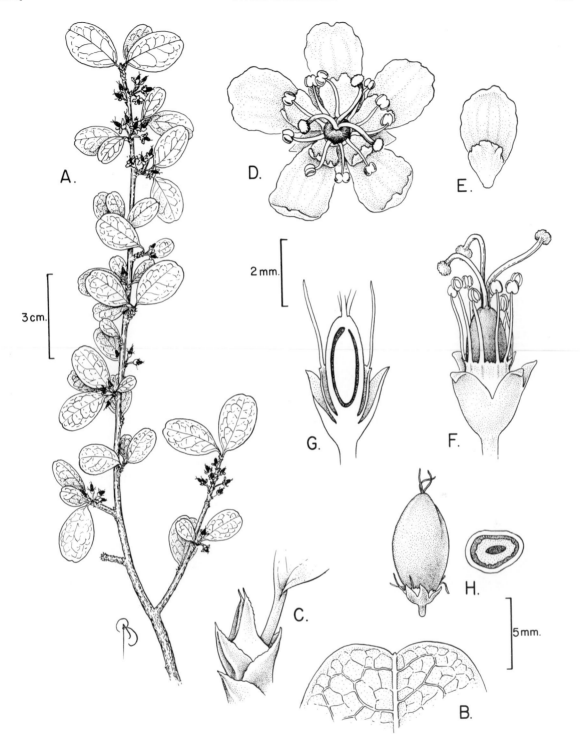

FIG. 92. *Erythroxylum brevipes*. **A.** Fertile branch. **B.** Detail of emarginate leaf apex. **C.** Detail of overlapping bud scales and intrapetiolar stipules at base of leaf. **D.** Flower. **E.** Petal with petaloid appendage. **F.** Flower with petals removed. **G.** L.s. flower showing ovary. **H.** Drupe and c.s. drupe.

1 or 2 ovules, the placentation axile, the styles distinct or united into a single one. Fruit typically a schizocarpous capsule or less often a drupe or a samara; seeds 1 or 2 per locule.

A family of 300 genera and about 7500 species, with cosmopolitan distribution, but predominantly in tropical and subtropical regions.

REFERENCE: Webster, G. L. 1967. The genera of Euphorbiaceae in the southeastern United States. J. Arnold Arbor. 48: 363–430.

Key to the Genera of *Euphorbiaceae*

1. Flowers in pseudanthia.
   2. Plants twining climbers, not producing milky latex; pseudanthial bracts foliaceous .... 6. *Dalechampia*
   2. Plants erect or prostrate, not twining or climbing, producing abundant milky latex; pseudanthia cup-shaped (cyathia) or elongate.
      3. Cyathia elongate, with a basal spur, reddish, showy, without apical glands .... 15. *Pedilanthus*
      3. Cyathia cup-shaped and not spurred, mostly green, usually with 4 or 5 apical glands bearing a petal-like appendage.
         4. Plants prostrate to erect; leaves opposite, oblique at base, usually with serrate margins; pistillate flowers lacking calyx .... 4. *Chamaesyce*
         4. Plants erect; leaves alternate, symmetrical at base, usually with entire margins; pistillate flowers with minute or rudimentary calyx .... 8. *Euphorbia*
1. Flowers in spikes, cymes, cymules, racemes, panicles, or fascicles.
   5. Ovary with 2 ovules per locule.
      6. Nectary disk intrastaminal; ovary of 1 carpel; fruit an ellipsoid drupe .... 7. *Drypetes*
      6. Nectary disk extrastaminal or wanting; ovary of 3–many carpels; fruit a capsule or a many-lobed drupe, flattened at poles.
         7. Corolla present .... 19. *Savia*
         7. Corolla absent.
            8. Lateral branch tips sharp, spinelike; staminate flower subsessile (pedicel to 1 mm long).... 9. *Flueggea*
            8. Plants without spines; staminate flowers distinctly pedicellate.
               9. Plants dioecious; seeds with fleshy bluish seed coat .... 14. *Margaritaria*
               9. Plants monoecious; seeds without a fleshy or bluish coat .... 16. *Phyllanthus*
   5. Ovary with 1 ovule per locule.
      10. Plants dioecious .... 2. *Adelia*
      10. Plants monoecious.
         11. Plant covered with stellate hairs or usually with lepidote scales .... 5. *Croton*
         11. Plant glabrous or pubescent but not as above.
            12. Corolla present.
               13. Corolla red, pink, or rarely white; nectary disk annular, cup-shaped, or lobed; stamens 8–15, in 2 whorls; style simple or bifid; plant usually with glandular viscidulous hairs or glabrous .... 13. *Jatropha*
               13. Corolla white; nectary disk divided into 4–5 glands; stamens 4–5, in 1 whorl; style twice bifid; plant glabrous or covered with T-shaped hairs .... 3. *Argythamnia*
            12. Corolla wanting.
               14. Plants climbing or twining, covered with stinging hairs................................. 20. *Tragia*
               14. Plants erect herbs, shrubs, or trees, without stinging hairs.
                  15. Bracts nonglandular.
                     16. Herbs; leaf margins serrate; calyx present, stamens 4–8 .... 1. *Acalypha*
                     16. Shrubs or small trees; leaf margins toothed to crenate; calyx wanting or rudimentary, stamens 2 or 3.. 10. *Gymnanthes*
                  15. Bracts of staminate flowers biglandular.
                     17. Plant with clear latex; petioles not glandular .... 17. *Ricinus*
                     17. Plant with copious milky latex; petioles with 1 or 2 apical glands.
                        18. Staminate flowers of numerous stamens in 2 or 3 whorls .... 12. *Hura*
                        18. Staminate flowers of 2 stamens in 1 whorl.
                           19. Fruit a 3-locular capsule, 5–6 cm long; petioles with 2 apical glands.. 18. *Sapium*
                           19. Fruit a 6–10 locular drupe, 1–1.5 cm long; petioles with 1 apical gland.. 11. *Hippomane*

1. ACALYPHA L.

Monoecious or dioecious herbs, shrubs, or small trees, not producing milky latex. Leaves alternate, often with serrate margins; stipules deciduous. Flowers borne in axillary or terminal, unisexual or bisexual spikes or panicles. Allomorphic flowers produced distally on inflorescence in a few species. Staminate flowers several at each node, subtended by a minute bract; calyx 4-lobed; corolla and nectary disk wanting; stamens 4–8 with free or connate filaments, the anthers elongate and twisted; pistillode present. Pistillate flowers 1(–3) per node, subtended by a foliaceous bract; calyx 3–5-lobed; ovary of 3 uniovular carpels, the styles simple or many-branched, filiform. Fruit a 3-lobed capsule with 1 seed per locule.

A genus of about 400 species, with tropical and subtropical distribution.

1. Acalypha poiretii Spreng., Syst. Veg. 3: 879. 1826.
Fig. 93H–O.

Acalypha macrostachya Poir. *in* Lam., Encycl. 6: 208. 1804, non Jacq., 1797.

Erect herb or subshrub to 70 cm tall, branching along main stem, pubescence of long and erect, short and curved hairs, some glandular. Leaf blades 2.5–6 × 1–3 cm, ovate, chartaceous, pubescent on both surfaces, the apex acute, the base rounded to cuneate, the margins serrate-crenate; petioles slender, 1.5–4 cm long; stipules nearly linear, deciduous. Flowers in axillary, bisex-

ual spikes; allomorphic pistillate flowers present at distal portion of inflorescence; staminate portion (distal) <1 cm long; pistillate flowers on lower portion of inflorescence, the bracts ca. 5 mm long, dentate. Calyx green, of 4 sepals, 0.2 mm long; ovary 3-lobed, pubescent, the styles 3, free to base, the stigmas threadlike. Capsule 3-lobed, ca. 1.5 mm long, puberulent. Seeds nearly ellipsoid 1–1.5 mm long, brown to gray.

DISTRIBUTION: A weed of open areas, introduced from the Old World tropics. Annaberg Ruins (A2925), Center Line Road by entrance to Bordeaux (A3846), Johnson Bay (A3156). Also on St. Thomas and St. Croix; Puerto Rico and the Lesser Antilles.

NOTE: This species was reported in error as *Acalypha chamaedrifolia* and *A. indica* by Woodbury and Weaver (1987).

2. ADELIA L., nom. cons.

Dioecious shrubs or small trees, not producing milky latex. Leaves simple, pellucid-punctate, clustered on short, lateral, woolly shoots; stipules minute. Flowers clustered in leaf axils; calyx deeply parted; corolla lacking; nectary disk annular. Staminate flowers on short pedicels; calyx 4- to 5-lobed; stamens 8–15, the filaments united into a short stipe, the anthers short, ellipsoid; pistillode absent. Pistillate flowers on long pedicels; calyx 5–6-lobed; ovary 3-lobed, of 3 uniovular carpels, the styles free, laciniate. Fruit a 3-lobed capsule with 1 seed per locule.

A small genus of 10 species from tropical America.

1. Adelia ricinella L., Syst. Nat., ed. 10, 2: 1298. 1759.
Ricinella ricinella (L.) Britton *ex* P. Wilson, Bull. New York Bot. Gard. 8: 395. 1917.   Fig. 93P–U.

Shrub or small tree to 10 m tall; bark light gray; branches usually hanging, short lateral branches ending in a spinose tip. Leaf blades 1.5–7 × 1–3.5 cm, obovate to oblanceolate, chartaceous, pellucid dotted, lower surface barbate at vein angles, the apex rounded or slightly notched, the base cuneate to obtuse, the margins entire; petioles slender, 1–5 mm long; stipules minute.

Flowers 1 to many in leaf axils. Calyx green, 3.5–4 mm long, the sepals lanceolate, reflexed; corolla wanting. Staminate flowers on short pedicels (<1 cm). Pistillate flowers on pedicels to 3(–4) cm long. Capsule green, 3-lobed, 6–8 mm long, on curved peduncle, septicidal, then loculicidal, leaving a central columella. Seeds nearly globose, smooth, ca. 3 mm long, gray.

DISTRIBUTION: Occasional in coastal thickets. Caneel Bay (A5261), Coral Bay (*Raunkiær s.n.*). Also on Tortola and Virgin Gorda; Greater Antilles, Lesser Antilles, and northern Venezuela.

3. ARGYTHAMNIA P. Browne

Monoecious or less often dioecious shrubs or subshrubs, glabrous or variously pubescent, usually with malpighiaceous (T-shaped) hairs, not producing milky latex. Leaves simple, alternate, or less often clustered on short, lateral shoots; stipules minute. Flowers clustered or in short, axillary bisexual racemes; calyx of 4 or 5 deeply parted sepals; corolla of 4 or 5 distinct petals or less often wanting; nectary disk divided into 4 or 5 glands. Staminate flowers distal on inflorescence; stamens 4 or 5, on a short stipe, the anthers short, ellipsoid; pistillode absent. Pistillate flowers proximal; ovary 3-lobed, of 3 uniovular carpels, the styles 3, twice bifid. Fruit a 3-lobed capsule with 1 seed per locule.

A genus of about 80 species of tropical America.

REFERENCE: Ingram, J. 1967. A revisionary study of *Argythamnia* subgenus *Argythamnia* (Euphorbiaceae). Gentes Herb. 10: 1–38.

Key to the Species of *Argythamnia*

1. Low erect or decumbent subshrub; pistillate flowers apetalous .... 3. *A. stahlii*
1. Erect shrubs or subshrubs; pistillate and staminate flowers with petals.
   2. Leaves alternate, not congested; stamens 4 .... 1. *A. candicans*
   2. Leaves congested on short lateral shoots; stamens 10 .... 2. *A. fasciculata*

1. Argythamnia candicans Sw., Prodr. 39. 1788.
Fig. 93A–G.

Erect shrub or subshrub to 1 m tall, branching along main stem, with appressed malpighiaceous (T-shaped) hairs. Leaf blades 1.3–8.5 × 0.9–3.6 cm, elliptic or seldom obovate, chartaceous, pubescent, becoming glabrous, the apex acute, acuminate or obtuse

and mucronate, the base cuneate to obtuse, with 3 main veins from base, the margins entire or finely denticulate; petioles slender, 1.5–2.5 mm long, pubescent; stipules minute, triangular. Racemes to 1 cm long. Calyx of elliptic sepals, 1.5–2 mm long, sericeous without; corolla of obovate, greenish petals, densely sericeous without. Staminate flowers 4-merous, distal on inflorescence; stamens 4. Pistillate flowers 5-merous, proximal on inflorescence.

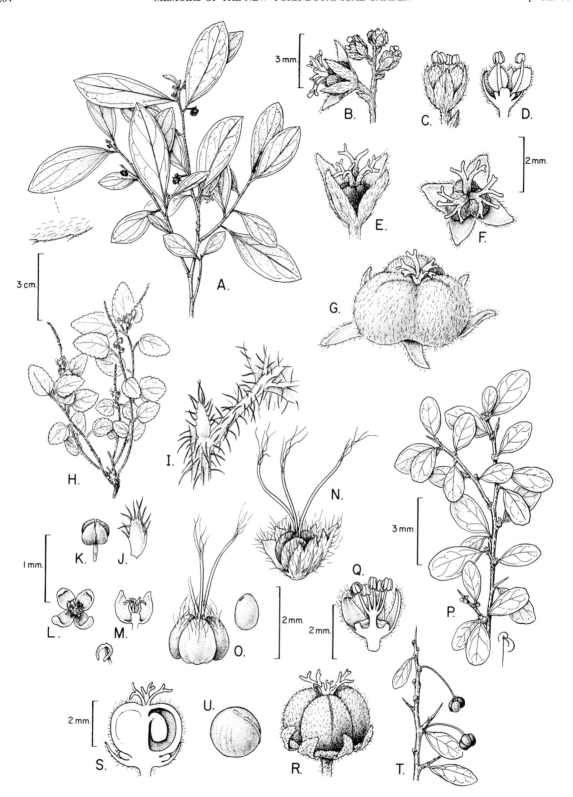

Fig. 93. A–G. *Argythamnia candicans*. A. Fertile branch and detail of leaf margin trichomes. B. Inflorescence. C. Staminate flower. D. L.s. staminate flower. E, F. Pistillate flower. G. Capsule. H–O, *Acalypha poiretii*. H. Habit. I. Stipule. J. Staminate flower bract. K. Staminate flowerbud. L. Open staminate flower. M. L.s. staminate flower and detail of stamen. N. Pistillate flower. O. Capsule and seed. P–U. *Adelia ricinella*. P. Flowering branch. Q. L.s. staminate flower. R. Capsule. S. L.s. capsule showing nectary disk and persistent sepals. T. Fruiting branch. U. Seed.

Capsule green, 4–6 mm high, villous. Seeds globose, reticulate, ca. 2 mm long, brown.

DISTRIBUTION: Common as an understory shrub in coastal woodlands. Bethany (B199), Europa Bay (A3195). Also on Anegada, St. Croix, St. Thomas, Tortola, and Virgin Gorda; Bahamas, Greater Antilles, and rare in the Lesser Antilles.

2. Argythamnia fasciculata (Vahl *ex* A. Juss.) Müll. Arg., Linnaea 34: 146. 1865. *Ditaxis fasciculata* Vahl *ex* A. Juss., Euphorb. Gen. 110, pl. 7, fig. 24. 1824.

Erect shrub 1–1.5 m tall, many-branched from base; young parts covered with appressed, T-shaped hairs. Leaves congested on short, lateral shoots, oblanceolate to spatulate; blades 1–3 × 0.5–1.3 cm, chartaceous to coriaceous, puberulent, the apex obtuse, rounded, or notched and seldom mucronate, the base cuneate to obtuse, the margins entire; petioles slender, to 2 mm long; stipules minute, triangular. Flowers solitary; calyx of 5 lanceolate sepals, 6–7 mm long, densely sericeous without; corolla white (turning reddish when dry), the petals 5, obovate, glabrous to 6 mm long, reflexed at upper portion; stamens 10, in 2 whorls, the filaments and staminal column white, the anthers bluish; ovary ellipsoid, sericeous, the styles bifid, sericeous, the stigmatic surface capitate. Capsule green, deeply 3-lobed, 4–6 mm broad, villous. Seeds globose, ca. 2 mm long, blackish.

DISTRIBUTION: A common shrub in coastal woodlands. Great Cruz Bay (A2374), Emmaus (A3117), southern slopes of Fish Bay (A3113). Also on St. Croix, St. Thomas, Tortola, and Virgin Gorda; Puerto Rico, Cuba, and St. Martin.

3. Argythamnia stahlii Urb., Symb. Ant. 1: 336. 1899.

Erect or decumbent subshrub to 30 cm tall, many-branched from a woody taprooted base; young parts covered with appressed, T-shaped hairs. Leaves alternate; blades 1–3.5 × 0.7–1.4 cm, ovate or elliptic, chartaceous to coriaceous, puberulent, the apex obtuse or rounded, the base cuneate to obtuse, with 3–5 ascending veins, the margins entire, with T-shaped hairs; petioles slender, to 4 mm long; stipules minute, narrowly triangular. Racemes to 1 cm long. Staminate flowers terminal on inflorescence; calyx of 4 elliptic sepals, 1–2 mm long, densely sericeous without; corolla of 4 greenish petals; nectary disk of 4 glands; stamens 4. Pistillate flowers on lower portion of racemes; calyx of 5 elliptic-oblong sepals, 2.5–3 mm long; corolla wanting; ovary 3-lobed, densely sericeous, the styles 3 times bifid, the stigmatic surface papillose. Capsule deeply 3-lobed, 4–6 mm broad, villous, turning from green to straw-colored. Seeds globose, ca. 2 mm long, blackish and rough.

DISTRIBUTION: An uncommon shrub of coastal open areas. East End Pond (A4070), eastern slopes of White Cliffs (A2033). Also on Anegada; Puerto Rico.

4. CHAMAESYCE Gray

Reviewed by A. Herndon

Monoecious herbs, subshrubs or less often shrubs or small trees, prostrate to erect, usually profusely branched from a woody base, glabrous or variously pubescent, producing abundant milky latex. Leaves simple, opposite, usually inequilateral and oblique at base, with entire or serrate margins; stipules minute. Flowers without perianth, produced within a cyathium, the cyathia solitary or clustered in cymules in leaf axils; cyathium cup-shaped, 5-toothed at apex, with 4 glands, each bearing a petal-like appendage. Staminate flowers few to many, consisting of 1 stamen. Pistillate flowers solitary, central in cyathium; ovary of 3 carpels, pedicel elongate, the carpels with a solitary ovule; styles 3, free or united at base, bifid. Fruit a 3-lobed capsule with 1 seed per locule, on a long exserted, reflexed pedicel.

A cosmopolitan genus of about 250 species, but principally in the American tropics.

Key to the Species of *Chamaesyce*

1. Shrub to small tree; leaves linear to narrowly oblong .... 1. *C. articulata*
1. Herbs or low shrubs; leaves variously shaped but not linear.
   2. Capsules glabrous.
      3. Leaves with entire margins.
         4. Plants erect, to 50 cm tall; leaves nearly coriaceous, ascending, nearly overlapping, 5–13 mm long, with truncate to cordate base .... 5. *C. mesembrianthemifolia*
         4. Plants usually prostrate (seldom erect), to 20 cm long; leaves membranous, spreading, not overlapping, 2–6 mm long, with rounded to subcordate base .... 8. *C. serpens*
      3. Leaves with serrate margins.
         5. Stipules lanceolate, 1–2.5 mm long, usually spreading at maturity .... 3. *C. hypericifolia*
         5. Stipules wide-triangular, 0.5 mm long, not reflexed .... 4. *C. hyssopifolia*
   2. Capsules pubescent.
      6. Plant usually prostrate and spreading, glabrous or tomentose (pubescence of minute curly hairs).
         7. Capsule pubescent only on angles; leaves glabrous .... 7. *C. prostrata*
         7. Capsule with scattered hairs; leaves pubescent .... 9. *C. thymifolia*
      6. Plant usually erect or decumbent; pubescence of two types of hairs (whitish, minute, curly or appressed; or long, erect with reddish or orange cross-walls).
         8. Plant many-branched from base; inflorescence on short axillary densely clustered glomerules; cyathia glands with rounded appendages .... 2. *C. hirta*

8. Plant pseudodichotomously branched; inflorescence terminal on leafy branches; cyathia glands without appendages .... 6. *C. opthalmica*

1. Chamaesyce articulata (Aubl.) Britton, Mem. New York Bot. Gard. **6**: 574. 1916. *Euphorbia articulata* Aubl., Hist. Pl. Guiane **1**: 480. 1775.

Euphorbia linearis Retz., Observ. Bot. **3**: 32. 1783. *Chamaesyce linearis* (Retz.) Millsp., Publ. Field Columbian Mus., Bot. Ser. **2**: 410. 1916.

Erect shrub to small tree 1–2.5 m tall, glabrous, with numerous branches from a woody base; branches reddish or yellowish with conspicuous swollen nodes. Leaf blades 2–6 × 0.2–1.2 cm, linear to elliptic, nearly coriaceous, the apex acute, obtuse or seldom rounded, the base rounded and unequal, the margins entire and revolute; petioles 1–1.5 mm long; stipules triangular and ciliate. Cyathia solitary, cone-shaped, on peduncles to 7 mm long; glands green, the appendages broadly elliptic, with a white band along the margins. Pistillate flower exserted; ovary glabrous, the styles elongate, jointed for 1–2 mm at their bases. Capsule glabrous, 3-lobed, ovoid, ca. 4 mm long. Seeds ovoid, ca. 2 mm long, light brown.

DISTRIBUTION: A common shrub of open sandy coasts and coastal thickets. Emmaus (A3116), Lameshur (A603). Also on Anegada, St. Croix, St. Thomas, Tortola, and Virgin Gorda; Puerto Rico, the Bahamas, and the Lesser Antilles.

2. Chamaesyce hirta (L.) Millsp., Publ. Field Columbian Mus., Bot. Ser. **2**: 303. 1909. *Euphorbia hirta* L., Sp. Pl. 454. 1753.

Euphorbia pilulifera L., Sp. Pl. 454. 1753. *Euphorbia globulifera* Kunth *in* Humb., Bonpl. & Kunth, Nov. Gen. Sp. **2**: 56. 1817.

Erect to decumbent (sometimes prostrate) herb, to 30 cm tall, many-branched from a woody taproot; stems reddish, covered with two types of trichomes, these either whitish, minute, and curly or appressed, or long and erect with reddish or orange cross-walls. Leaf blades 5–40 × 3–15 mm, ovate to lanceolate, nearly coriaceous, the apex acute, acuminate, or obtuse, the base rounded or cuneate and oblique, the margins sharply serrate; petioles 1–2 mm long; stipules lacerate, minute. Cyathia densely clustered in axillary glomerules, cone-shaped; peduncles to 15 mm long; gland appendages rounded, minute, whitish. Pistillate flower exserted; ovary strigose. Capsule strigose, 3-lobed, ovoid, ca. 1 mm long. Seeds cuneiform, 0.7–0.9 mm long, light brown.

DISTRIBUTION: A common weed of open and disturbed areas. Lind Point (A1973), Maria Bluff (A3110). Also on St. Croix, St. Thomas, Tortola, and Virgin Gorda; a pantropical weed.

COMMON NAME: milkweed.

3. Chamaesyce hypericifolia (L.) Millsp., Publ. Field Columbian Mus., Bot. Ser. **2**: 302. 1909. *Euphorbia hypericifolia* L., Sp. Pl. 454. 1753.

Chamaesyce glomerifera Millsp., Publ. Field Columbian Mus., Bot. Ser. **2**: 377. 1913.

Annual erect herb or subshrub, to 50 cm tall, many-branched from a woody taproot; stems reddish, glabrous. Leaf blades 5–27 × 2–11 mm (those of lateral shoots smaller), oblong, ovate, or lanceolate, membranous, the apex obtuse or rounded, the base rounded-cuneate and oblique, the margins minutely serrate; petioles 1–2 mm long; stipules to 2.5 mm long, lanceolate, often

spreading. Cyathia mostly in terminal, leafless dichasia, cone-shaped; peduncles to 10 mm long; glands rounded, the appendages ear-shaped, whitish, 5–7 mm long. Pistillate flowers exserted; ovary glabrous. Capsule glabrous, deeply 3-lobed, ovoid, 1.4–1.8 mm long. Seeds cuneiform, 0.8–0.9 mm long, wrinkled, light brown.

DISTRIBUTION: A common weed. Fish Bay Gut (A3872), Johnson Bay (A3159). Also on St. Croix, St. Thomas, and Tortola; throughout tropical America.

4. Chamaesyce hyssopifolia (L.) Small, Bull. New York Bot. Gard. **3**: 429. 1905. *Euphorbia hyssopifolia* L., Syst. Nat., ed. 10, **2**: 1048. 1759.   Fig. 94A–D.

Annual erect or decumbent herb, to 50 cm tall, many-branched from a woody taprooted base; stems yellowish or reddish, glabrous. Leaf blades 5–25 × 2–10 mm (those of lateral shoots smaller), oblong to lanceolate or seldom elliptic, membranous, the apex obtuse or rounded, the base rounded-cuneate and oblique, the margins minutely serrate; petioles 1–2 mm long; stipules triangular, ca. 0.5 mm long, minute, with glandular margins. Cyathia in terminal or axillary, leafy glomerules, cone-shaped; peduncles to 10 mm long; glands rounded, the appendages ear-shaped, pinkish or whitish, 2–3 mm long. Pistillate flower exserted; ovary glabrous. Capsule glabrous, deeply 3-lobed, ovoid, ca. 1.6 mm long. Seeds cuneiform, 1.0–1.1 mm long, blackish, with 2–3 transverse ridges and 1 longitudinal lighter ridge.

DISTRIBUTION: A common weed. Along road to Bordeaux (A2590), Cruz Bay (A3090); Fish Bay (A2499). Also on St. Thomas and Tortola; throughout tropical America.

5. Chamaesyce mesembrianthemifolia (Jacq.) Dugand, Phytologia **13**: 385. 1966. *Euphorbia mesembrianthemifolia* Jacq., Enum. Syst. Pl. 22. 1760.   Fig. 94E–J.

Euphorbia buxifolia Lam., Encycl. **2**: 421. 1788. *Chamaesyce buxifolia* (Lam.) Small, Fl. S.E. U.S. 711. 1903.

Perennial erect subshrub to 50 cm tall, glabrous, many-branched from base, the branches sometimes decumbent, reddish with persistent stipules. Leaves ascending, nearly overlapping; blades 5–13 × 2–8 mm, ovate to elliptic, nearly coriaceous, glaucous beneath, the apex obtuse, the base truncate to nearly cordate and slightly oblique, the margins entire; petioles 1 mm long; stipules triangular to ovate, whitish, persistent in leafless stems. Cyathia axillary, solitary, cone-shaped; peduncles to 3 mm long; glands rounded, the appendages white, broadly elliptic, to 0.5 mm long. Pistillate flower exserted; ovary glabrous. Capsule glabrous, 3-lobed, ovoid, ca. 2 mm long. Seeds ovoid, ca. 1.3 mm long, light brown, smooth.

DISTRIBUTION: Common along sandy coasts. Ram Head Point (A2916), Fortsberg (A4079), Brown Bay (A1871). Also on Anegada, St. Croix, St. Thomas, and Virgin Gorda; the West Indies, Venezuela, and Colombia.

6. Chamaesyce opthalmica (Pers.) D. G. Burch, Ann. Missouri Bot. Gard. **53**: 98. 1966. *Euphorbia opthalmica* Pers., Syn. Pl. **2**: 13. 1806.

Annual erect or decumbent herb, to 15 cm tall, tomentose, with multicellular hairs, many-branched from a nearly woody base;

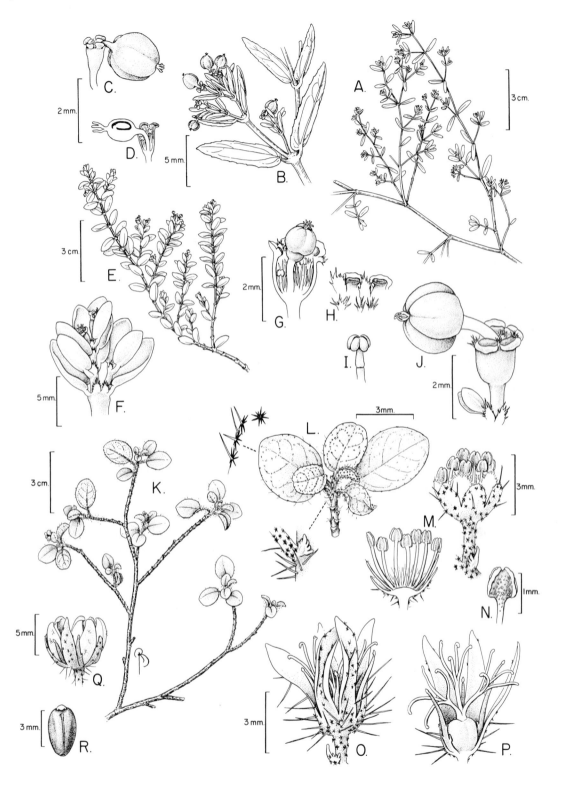

Fɪɢ. 94. A–D. *Chamaesyce hyssopifolia*. A. Flowering branch. B. Detail of fruiting branch. C. Cyathium with maturing capsule. D. L.s. cyathium showing staminate flowers and immature capsule. E–J. *Chamaesyce mesembrianthemifolia*. E. Flowering branch. F. Detail of branch with inflorescences. G. L.s. cyathium showing staminate flowers and immature capsule. H. Detail of summit of cyathium cup showing glands with appendages. I. Staminate flower. J. Cyathium with maturing capsule. K–R. *Croton fishlockii*. K. Flowering branch. L. Leafy branch with details of stellate hairs and stipule. M. L.s. flower showing stamens and staminate flower. N. Anther. O. Pistillate flower. P. Pistillate flower with 2 sepals and 2 petals removed to show pistil. Q. Dehisced capsule. R. Seed.

stems reddish, with pseudodichotomous branching. Leaf blades 5–18 × 3–9 mm, elliptic to rhombic, membranous, usually reddish-tinged, the apex obtuse or acute, the base rounded-cuneate, oblique, the margins finely serrate; petioles ca. 1 mm long; stipules attenuate, minute. Cyathia in terminal, leafless, densely flowered dichasia, cone-shaped; peduncles to 5 mm long; glands minute, without appendages. Pistillate flower exserted; ovary pubescent. Capsule pubescent, 3-lobed, ca. 1 mm long. Seeds cuneiform, ca. 0.8–0.9 mm long, light brown.

DISTRIBUTION: Common in disturbed, open areas. Johns Folly (A3941), Lind Point (A4238). Also throughout the West Indies and in Florida.

7. Chamaesyce prostrata (Aiton) Small, Fl. S.E. U.S. 713. 1903. *Euphorbia prostrata* Aiton, Hort. Kew. **2:** 139. 1789.

Annual prostrate to decumbent herb, to 20 cm long, many-branched from a taproot; stems reddish, glabrous. Leaf blades 4–8 × 2–3 mm, broadly elliptic or obovate, membranous, pellucid punctate, usually reddish-tinged at margins, the apex obtuse or rounded, the base rounded, nearly symmetrical, the margins obscurely serrate; petioles ca. 1 mm long; stipules minute. Cyathia solitary in axils of short lateral shoots; glands and appendages minute, white. Pistillate flower exserted; ovary pubescent at angles. Capsule pubescent at angles, 3-lobed, ca. 1–1.2 mm long. Seeds cuneiform, ca. 0.9–1.0 mm long, gray.

DISTRIBUTION: Uncommon in disturbed, open areas. Johns Folly (A3940, A3942). Also on Tortola, Virgin Gorda; throughout the West Indies, Florida, Mexico, and South America.

8. Chamaesyce serpens (Kunth) Small, Fl. S.E. U.S. 709. 1903. *Euphorbia serpens* Kunth *in* Humb., Bonpl. & Kunth, Nov. Gen. Sp. **2:** 52. 1817.

Euphorbia pileoides Millsp., Publ. Field Columbian Mus., Bot. Ser. **2:** 62. 1900.

Perennial prostrate, mat-forming herb, to 20 cm long, many-branched from a taproot; stems pinkish, glabrous, alternately branching their entire length, rooting at nodes. Leaf blades 2–6 × 1–4 mm, ovate to rounded, membranous, the apex rounded or retuse, the base rounded to subcordate, subsymmetrical, the margins entire or obscurely crenate, light-colored; petioles ca. 0.5 mm long; stipules minute, deltoid but conspicuous. Cyathia solitary in axils of short, lateral shoots; glands purplish, the appendages white, ear-shaped; ovary glabrous. Capsule glabrous, 3-lobed, 1.2–1.4 mm long. Seeds ovoid, ca. 1 mm long, smooth, gray or light brown.

DISTRIBUTION: A common roadside weed. Cruz Bay (A3089), Enighed (A3102), Johnson Bay (A4029). Also on Anegada, St. Croix, St. Thomas, and Tortola; Mexico, Florida, and the West Indies.

9. Chamaesyce thymifolia (L.) Millsp., Publ. Field Columbian Mus., Bot. Ser. **2:** 412. 1916. *Euphorbia thymifolia* L., Sp. Pl. 454. 1753.

Annual prostrate herb, to 20 cm long, many-branched from a taproot, pubescent to glabrescent; stems not rooting at nodes. Leaf blades 4–10 × 2–5 mm, oblong-elliptic, membranous, pubescent, the apex rounded or acute, the base rounded-cuneate, oblique, the margins serrate, especially toward apex; petioles 0.5–1 mm long; stipules minute, linear. Cyathia in pairs or clustered in leaf axils, cone-shaped, densely pubescent, on short peduncles; glands purplish, the appendages ovate, pinkish; ovary pubescent. Capsule pubescent, 3-lobed, ca. 1.1 mm long, not exserted from the involucre. Seeds ovoid to conical ca. 0.8 mm long, smooth, light brown.

DISTRIBUTION: A common roadside weed. Gift Hill (A2482). Also on St. Croix and St. Thomas; Florida, Mexico to South America including the West Indies.

5. CROTON L.

Monoecious herbs, shrubs, or small trees, with stellate hairs or lepidote scales, producing a clear to orange watery latex, sometimes with fetid smell. Leaves simple and alternate, often densely hairy on lower surface, the petioles often with a pair of glands; stipules minute to foliaceous. Flowers in axillary or terminal spikes or racemes; calyx of 4–6 deeply parted sepals; corolla of 4–6 distinct petals, or wanting; nectary disk annular or dissected into glands. Staminate flowers distal on inflorescence; calyx of 4–6 sepals; stamens 10 to many, free; pistillode absent. Pistillate flowers proximal on inflorescence; calyx of 5 sepals; petals usually wanting; ovary 3-lobed, of 3 uniovular carpels, the styles 3, bifid to many times divided. Fruit a 3-lobed capsule with seed per locule.

A genus of about 600 species with worldwide distribution, predominantly in the tropics.

Key to the Species of *Croton*

1. Underside of mature leaves densely covered with yellowish or whitish stellate hairs.
   2. Leaf margins entire, obscurely serrate or crenate; plant with strong or weak fetid smell.
      3. Petioles with a pair of glands near the blade; leaf margins entire or crenulate; plant producing copious orange watery latex .... 4. *C. flavens* var. *rigidus*
      3. Petioles without glands; leaf margins finely serrate to crenate; plant producing clear latex .... 1. *C. astroites*
   2. Leaf margins crenate; plant not fetid .... 2. *C. betulinus*
1. Mature leaves sparsely stellate pubescent to glabrescent.
   4. Leaves deeply 3(–5)-lobed .... 5. *C. lobatus*
   4. Leaves simple, entire.
      5. Leaves elliptic; stellate hairs white; sepals fringed by glandular margins .... 6. *C. ovalifolius*
      5. Leaves ovate to rounded; lateral arms of stellate hairs reddish brown; sepal with entire, nonglandular margins .... 3. *C. fishlockii*

1. Croton astroites Aiton, Hort. Kew. 3: 375. 1789.

Croton phlomoides Pers., Syn. Pl. 2: 585. 1806.
Croton venosus Spreng., Neue Entdeck. Pflanzenk. 3: 24. 1822.

Erect shrub to 5 m tall, densely covered with stellate hairs, producing watery latex and a weak fetid smell; bark gray and smooth. Leaf blades 2.5–7.2 × 1.3–5 cm, ovate to lanceolate, coriaceous, densely covered with yellowish stellate hairs, the apex acute to obtuse and mucronate, the base rounded to cordate, the margins finely serrate to crenate; petioles 1–4 cm long, usually with rusty-brown pubescence; stipules early deciduous, awl-shaped; senescent leaves orange. Racemes terminal, 2–6 cm long; flowers 5-merous, densely stellate. Staminate flowers distal on inflorescence; calyx 2.5–3 mm long; petals 2.5–3 mm long; stamens white, numerous, to 5 mm long. Pistillate flowers on proximal portion of inflorescence; calyx 5–7 mm long; corolla wanting; ovary 3-lobed, densely stellate, the styles 3 times bifid. Capsule stellate pubescent, 3-lobed, ca. 5 mm long. Seeds ellipsoid to bean-shaped, trigonous in cross section, 4.5–5 mm long, smooth, black mottled, carunculate.

DISTRIBUTION: A common shrub of coastal scrub and disturbed areas. Fish Bay (A2388), vicinity of White Cliff (A2713), Lameshur (A2759). Also on St. Croix, St. Thomas, Tortola, and Virgin Gorda; Puerto Rico and Lesser Antilles.

COMMON NAMES: marang, white marang.

2. Croton betulinus Vahl, Symb. Bot. 2: 98. 1791.

Erect shrub to 2 m tall, densely covered with stellate hairs when young; bark dark brown, peeling off. Leaf blades 1–2.5 × 0.7–1.5 cm, ovate to lanceolate, coriaceous, densely covered with white, stellate hairs on lower surface, the apex obtuse, seldom notched and mucronate, the base truncate, rounded, or nearly cordate, the margins crenate; petioles 0.3–1 cm long, densely stellate pubescent, with 2 elongate glands at apex; stipules early deciduous. Racemes simple or branched, terminal, 1–3 cm long; flowers 5-merous, densely covered with white and yellow stellate hairs. Staminate flowers distal on inflorescence; calyx 1.5–2 mm long; petals 1.5 mm long; stamens white, 10, to 3 mm long. Pistillate flowers on lower portion of inflorescence; calyx greenish, the sepals to 2.5 mm long, spreading, trilobed at apex; corolla wanting; ovary 3-lobed, densely covered with yellow, stellate hairs (appearing as hirsute), the styles bifid near the base, white, linear. Capsule with yellow, stellate pubescence, 3-lobed to ellipsoid, ca. 4 mm long. Seeds ellipsoid to bean-shaped, trigonous in cross section, 3 mm long, smooth, grayish, carunculate.

DISTRIBUTION: A common shrub of coastal scrubs and disturbed areas. Margaret Hill (A2309), Southside Pond (A1838), Susannaberg (A836). Also on Anegada, St. Croix, St. Thomas, Tortola, and Virgin Gorda; Puerto Rico, Cuba, Hispaniola, and Lesser Antilles.

COMMON NAMES: broom bush, pistarckle broom.

3. Croton fishlockii Britton, Torreya 20: 84. 1920.

Fig. 94K–R.

Erect subshrub to 1 m tall, with many lateral branches along main stem, densely covered with long, white, stellate hairs; twigs brown to dark gray. Leaf blades 1–1.8 × 0.7–1.7 cm, ovate, wide elliptic or rounded, membranous, sparsely covered with stellate hairs (becoming glabrous at age), the hairs with a central, long, white arm and numerous lateral, short, reddish brown arms, the apex obtuse to rounded, seldom notched, the base rounded to nearly cordate, the margins entire, ciliate and darker; petioles

0.7–1.4 cm long, with long stellate hairs; stipules minute, early deciduous. Racemes few-flowered, terminal; flowers 5-merous, with same pubescence as leaves. Staminate flowers with calyx 1.5–2 mm long, the sepals oblong, with ciliate margins; petals 1.5 mm long, ovate, with ciliate margins; stamens 10, short. Pistillate flowers with sepals to 7 mm long, oblong, spreading, white and glabrous within; corolla of awl-shaped petals; ovary 3-lobed, green, puberulent, the styles 3 times bifid, white, linear. Capsule glabrescent, trigonous-ellipsoid, ca. 6 mm long. Seeds ellipsoid to bean-shaped, trigonous in cross section, 4.5 mm long, smooth, light brown, carunculate.

DISTRIBUTION: A rare subshrub found in the southeastern scrub of St. John. Concordia (A4251), Minna Hill (A4236), Salt Pond (A757). Also on Anegada, Great Camanoe, Tortola, and Virgin Gorda; introduced on Guana Island.

4. Croton flavens L. var. rigidus Müll. Arg. *in* A. DC., Prodr. 15(2): 613. 1866. *Croton rigidus* (Müll. Arg.) Britton *in* Britton & P. Wilson, Bot. Porto Rico 5: 481. 1924.

Erect shrub to 3 m tall, densely covered with yellowish, stellate hairs, producing orange, watery latex, and a strong fetid smell. Leaf blades 2–10 × 1–4 cm, ovate to ovate-lanceolate or elliptic, coriaceous, densely covered with yellowish or white, stellate hairs, the apex acute to acuminate and mucronate, the base subcordate, rounded, or nearly tapering, the margins entire or crenulate; petioles 1–5 cm long, densely covered with yellowish, stellate hairs, bearing a pair of stipitate glands near the apex; stipules minute, early deciduous. Racemes terminal, 5–10 cm long; flowers 5-merous, densely stellate. Staminate flowers distal on inflorescence; calyx ca. 1.5 mm long; petals 1.5–1.7 mm long, spatulate, white; stamens white, 16(–19). Pistillate flowers proximal on inflorescence; calyx 2–2.5 mm long; corolla of reduced acicular petals or wanting; ovary nearly globose, densely stellate pubescent, the styles twice bifid, brown. Capsule stellate pubescent, nearly globose, ca. 5 mm long. Seeds ellipsoid to bean-shaped, trigonous in cross section, 3–3.5 mm long, smooth, grayish, carunculate.

DISTRIBUTION: A common shrub of dry coastal scrub and disturbed areas. Brown Bay (A1865), Dittlif Point (A3967), Salt Pond (A3772). This variety endemic to Puerto Rico and the Virgin Islands (St. Croix, St. Thomas, Tortola, Virgin Gorda).

COMMON NAMES: marang, yellow marang.

NOTE: The species occurs throughout the West Indies and in Colombia and Venezuela, presenting a great deal of variation in color and density of pubescence and in shape and texture of leaves. The populations in Puerto Rico and the Virgin Islands are distinctive in that their leaves are lanceolate to elliptic and taper into a short, nearly cordate base. On the other hand, populations on other islands have mostly lanceolate leaves, with cordate to truncate bases. Recognition of the Puerto Rican and Virgin Islands populations as a distinct species is not warranted because this difference is considered to be minor.

5. Croton lobatus L., Sp. Pl. 1005. 1753.

Annual erect herb to 60 cm tall, usually with pseudodichotomous branching, sparsely covered with long, stellate, white hairs, especially when young. Leaf blades 4–7 × 6–8 cm, deeply 3(–5)-lobed, membranous, with a few scattered stellate hairs, especially along the veins, the base cordate or truncate; lobes elliptic to oblanceolate, with acuminate to long-acuminate apex, the margins crenate-serrate; petioles 3–8 cm long, almost glabrous; stipules 5–6 mm long, awl-shaped, with laciniate, glandular margins. Racemes simple, terminal, 5–11 cm long; flowers 5-merous, sparsely pilose. Staminate flowers distal on inflorescence; calyx ca. 1 mm long, the sepals ovate; petals to 1.5 mm long; stamens

10, white. Pistillate flowers on lower portion of inflorescence; calyx greenish, the sepals to 7 mm long, oblong, with glands along margins; petals to 7 mm long white or pinkish; ovary 3-lobed, sparsely pilose, green, the styles 3–8-branched, linear. Capsule sparsely pilose, 3-lobed to ellipsoid, 5–6 mm long. Seeds oblong, trigonous in cross section, ca. 4 mm long, warty, brown to grayish, carunculate.

DISTRIBUTION: A common herb of disturbed areas. Cruz Bay Quarter along Center Line Road (A645), Enighed (A3103), Fish Bay (A3874). Also on St. Croix, St. Thomas, and Tortola; throughout the West Indies (except Jamaica), Mexico, and Central America.

6. Croton ovalifolius Vahl *in* H. West, Bidr. Beskr. Ste. Croix 307. 1793.

Erect or prostrate subshrub to 50 cm tall, many-branched from a woody taproot base, secondary branching pseudodichotomous, covered with long, white, stellate hairs, especially when young. Leaf blades 1.5–3.4 × 0.5–1.5 cm, narrowly elliptic to elliptic, membranous, becoming glabrous at age, the apex obtuse or acute, seldom notched, the base obtuse, the margins entire, ciliate; petioles 0.3–2.5 cm long, with long, stellate hairs; stipules minute, awl-shaped, with glandular margins, early deciduous. Racemes terminal, flat-topped (appearing as a cymose inflorescence), bracteoles with glandular margins; flowers 5-merous. Staminate flowers on distal portion of inflorescence; calyx ca. 1.5 mm long; petals 1.5 mm long, white; stamens white, 10. Pistillate flowers on proximal part of inflorescence; sepals to 5 mm long, awl-shaped, green, with stipitate glands along the margins; corolla wanting; ovary ovoid, hirsute, the styles 6-branched, linear. Capsule pilose, trigonous to nearly globose, ca. 3–5 mm long. Seeds ellipsoid to bean-shaped, trigonous in cross section, ca. 3 mm long, smooth, grayish or reddish brown, carunculate.

DISTRIBUTION: A common weed of open disturbed areas. Emmaus (A1998), Hurricane Hole (A2768); Lameshur (B640). Also on St. Croix, St. Thomas, Tortola, and Virgin Gorda; Greater Antilles, Lesser Antilles, and northern South America.

RECENTLY INTRODUCED SPECIES: Two individuals of *Croton discolor* Willd. were pointed out to me by Eleanor Gibney and Gary Ray. These are found along North Shore Road in the area around Hawksnest. They seem to be recent introductions from seeds carried in imported landfill (possibly from St. Thomas) for road improvement.

6. DALECHAMPIA L.

Monoecious, herbaceous, twining vines, or less often subshrubs, sometimes with stinging hairs, producing scanty watery latex. Leaves simple, alternate, 3–5-lobed; stipules minute. Flowers apetalous, clustered in short cymes, the cymes on a long peduncle, with 2 large foliaceous bracts forming an axillary pseudanthium. Staminate flowers on distal cymules; bractlets modified into resinous glands; calyx of 4–6 valvate lobes; stamens numerous, clustered on a short stipe into a globose head, the anthers short, opening along longitudinal slits; pistillode absent. Pistillate flowers on proximal cymules, subtended by a bract; calyx of 8–12 lobes with glandular margins; ovary 3-lobed, of 3 uniovular carpels, the styles united, the stigmas peltate. Fruit a 3-lobed capsule with 1 seed per locule, with persistent calyx lobes, often bearing stinging hairs.

A genus of about 100 species with predominantly neotropical distribution.

1. Dalechampia scandens L., Sp. Pl. 1054. 1753.

Fig. 95A–E.

Twining herbaceous vine to 5 m long, sparsely pilose. Leaf blade 5–7 × 6–12 cm, deeply 3-lobed, membranous, pilose, especially on lower surface, the base cordate; lateral lobes ovate and unequal, the terminal one oblong to obovate, with acute to acuminate apex, the margins minutely toothed; petioles 6–8 cm long, pilose; stipules ovate to lanceolate, to 1 cm long. Pseudanthia on peduncles to 4 cm long; bract 3(–5)-lobed, to 2.5 cm long, foliaceous, with serrate margins and stinging glandular hairs. Staminate flowers with calyx of 4 sepals to 1.5 mm long; stamens numerous, grouped into a nearly globose head. Pistillate flowers with calyx of 10–12 sepals, margins with glandular hairs; ovary 3-lobed, depressed-globose, the styles united. Capsule puberulent, deeply 3-lobed, depressed at poles, to 5 mm long. Seeds globose, 3.3 mm diam., black-mottled.

DISTRIBUTION: Not very common, mostly found in open disturbed areas. Brown Bay (A1882), East End Point (A660), Lameshur (B511). Also on St. Croix, St. Thomas, Tortola, and Virgin Gorda; throughout the West Indies (except Jamaica), and Mexico to South America.

7. DRYPETES Vahl

Dioecious shrubs or trees, not producing milky latex. Leaves alternate, simple, entire or denticulate; stipules minute to foliaceous. Flowers produced in axillary fascicles; calyx of 4–6 imbricate lobes; corolla wanting. Staminate flowers with 4–11 stamens with free filaments; nectary disk intrastaminal; pistillode tomentose. Pistillate flowers with an ovary of 1 or 2 carpels, the carpels with 2 ovules, stigmas sessile or nearly so. Fruit a small 1-seeded drupe.

A variable genus of about 200 species; ca. 10 species in the neotropics, the remaining species in the Old World tropics.

1. Drypetes alba Poit., Mém. Mus. Hist. Nat. **1:** 157. 1815.

Fig. 95F–I.

Tree to 8 m tall, glabrous; bark light gray, smooth. Leaf blades 5–12 × 1.5–3.5 cm, elliptic, coriaceous, reticulate-veined, the apex acute or shortly acuminate, the base obtuse, slightly asymmetrical, the margins minutely toothed; petioles 5–9 mm long; stipules 2, minute, 1 mm long. Flowers yellowish green, clustered in leaf axils; calyx of 4 oblong sepals to 1.5 mm long; stamens 3–6, ca. 3 mm long; ovary 1-locular, obliquely ellipsoid, pubescent, the style minute, the stigma discoid. Drupe obliquely obovoid, pubescent, 8–10 mm long.

DISTRIBUTION: Rare, found only in moist forests. Bordeaux Mountain (A4704, A4710). Also in Cuba, Hispaniola, Puerto Rico, and Jamaica.

8. EUPHORBIA L.

Monoecious erect herbs, shrubs, or small trees, glabrous or variously pubescent, producing abundant milky latex. Leaves alternate, opposite, or whorled, less often reduced to scales; stipules deciduous or persistent. Flowers produced within a cyathium, the cyathia

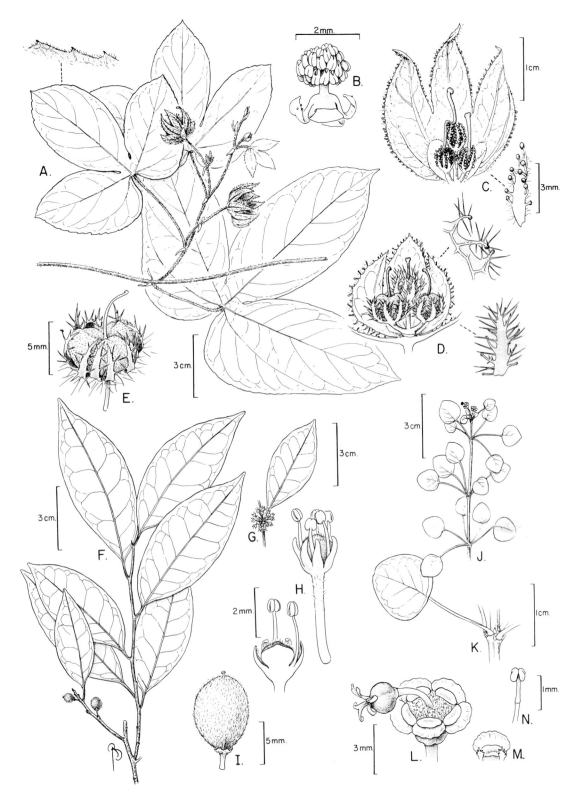

Fɪɢ. 95. A–E. *Dalechampia scandens*. **A.** Flowering branch and detail of leaf margin. **B.** Staminate flower. **C.** Pistillate cyme with one foliaceous bract removed to show pistillate flowers and their subtending bracts, and detail of sepal. **D.** Maturing pistillate cyme, with one foliaceous bract removed, and details of bract margin and sepal. **E.** Capsule. **F–I.** *Drypetes alba*. **F.** Fruiting branch. **G.** Inflorescence. **H.** L.s. staminate flower, and staminate flower. **I.** Drupe. **J–N.** *Euphorbia petiolaris*. **J.** Flowering branch. **K.** Detail of stipules. **L.** Cyathium and young fruit. **M.** Gland with appendage. **N.** Staminate flower.

solitary or clustered in cymules in leaf axils; cyathium cup- to bell-shaped, 2–5-toothed at apex, with (1–2–)4–5 glands, often bearing a petal-like appendage. Staminate flowers naked, few to many, consisting of 1 stamen. Pistillate flowers, solitary, central on cyathium; calyx minute or rudimentary; ovary of 3 carpels on elongate pedicel, the carpels with a solitary ovule; styles connate throughout most of their length, bifid at apex. Fruit a 3-lobed capsule with 1 seed per locule, on a long-exserted, reflexed pedicel.

A genus of about 1000 species with cosmopolitan distribution.

Key to the Species of *Euphorbia*

1. Trees; leaves congested at nodes or reduced to scales.
    2. Stems and branches succulent, green; leaves reduced to scales or lacking.
        3. Stems cylindrical, without spines; branches whorled; leaves obovate................................ 4. *E. tirucalli*
        3. Stems 4-ribbed, with spines along margins; branches alternate; leaves wanting .... *E. lactea* (cultivated)
    2. Stems not succulent, with blackish brown, papery bark; leaves long-petioled, congested at swollen nodes.. 3. *E. petiolaris*
1. Herbs; leaves alternate.
    4. Cyathia glands 2, with petaloid appendages .... 2. *E. oerstediana*
    4. Cyathia gland solitary, without petaloid appendages.. 1. *E. heterophylla*

1. Euphorbia heterophylla L., Sp. Pl. 453. 1753.

Erect herb to 50 cm, branching along main stem, glabrous or pubescent, with a thick rootstock. Leaves alternate (uppermost opposite); blades 4–10 × 2–5 cm, ovate to elliptic, seldom linear or rounded, chartaceous, glabrous or puberulent, lower surface usually glaucous, the apex obtuse or acute, the base tapering, obtuse, or rounded, the margins entire, dentate or sinuate and often reddish-tinged; petioles 5–22 mm long, puberulent, usually reddish-tinged; stipules minute, triangular, early deciduous. Cyathia funnel-shaped, on terminal cymes; gland solitary, conical, green or pinkish, with annular aperture. Staminate flowers many per cyathium, some exserted. Capsule 3-lobed, ovoid glabrous, chartaceous, reddish-tinged, to 6 mm long. Seeds ovoid-angular, 2 mm long, dark brown.

DISTRIBUTION: A common roadside weed. Fish Bay (A3875), Gift Hill (A821), Peter Peak (A4104). Also on St. Croix, St. Thomas, and Tortola; throughout the tropics.

2. Euphorbia oerstediana (Klotzsch & Garcke) Boiss. *in* A. DC., Prodr. **15(2):** 59. 1862. *Poinsettia oerstediana* Klotzsch & Garcke, Monatsber. Königl. Preuss. Akad. Wiss. Berlin **1860:** 103. 1860. *Dichylium oerstedianum* (Klotzsch & Garcke) Britton *in* Britton & P. Wilson, Bot. Porto Rico **5:** 499. 1924.

Erect herb 10–25(–50) cm tall, glabrous, with a swollen taproot; stems usually flexuous, green or reddish-tinged. Leaves alternate (uppermost opposite); blades 2.5–9 × 0.8–4 cm, ovate, elliptic, or lanceolate, chartaceous, glabrous or puberulent, lower surface usually glaucous, the apex acute or acuminate, the base rounded, obtuse, or truncate and seldom oblique, the margins entire or slightly wavy; petioles slender, reddish-tinged, 0.3–1.7 mm long; stipules early deciduous. Cyathia funnel-shaped, on terminal cymules, the glands 2, bilabiate, with a green, petaloid, erect appendage. Staminate flowers many per cyathium, some exserted. Pistillate flowers solitary; ovary pubescent, the stigma bifid, short, fingerlike. Capsule 3-lobed, ovoid, emarginate at apex, puberulent, chartaceous, turning from green to straw-colored, to 4 mm long. Seeds ellipsoidal, 2.5 mm long, warty, yellowish.

DISTRIBUTION: Not very common; found mostly in coastal forest understory. Fish Bay (A3883), Hawksnest Beach (A2902). Also on St. Croix and St. Thomas; Greater Antilles, Lesser Antilles, and northern South America.

3. Euphorbia petiolaris Sims, Bot. Mag. **23:** pl. 883. 1805. *Aklema petiolare* (Sims) Millsp., Ann. Missouri Bot. Gard. **2:** 43. 1915.　　Fig. 95J–N.

Euphorbia verticillata Poir. *in* Lam., Encycl. Suppl. **2:** 611. 1812.

Small tree 2–7 m tall, producing abundant milky, caustic, latex; bark blackish brown, shiny, peeling off in papery layers. Leaves in whorls of 6–11; blades 0.7–3 × 0.6–2.5 cm, ovate, deltoid to rounded, chartaceous, glabrous, the apex obtuse to rounded and slightly notched, the base rounded to truncate, the margins entire; petioles slender, 1–3.5 cm long, reddish brown; stipules minute, triangular, early deciduous. Cyathia cup-shaped, in terminal cymules; peduncles <1 cm long; glands 5, disk-shaped, yellowish, the appendages nearly rounded, light green with crenate apex. Staminate flowers numerous, exserted. Pistillate flowers with ovoid ovary, glabrous, the styles bifid, united at base. Capsule glabrous, 3-lobed, ovoid, emarginate at apex, to 5 mm long. Seeds globose to ovoid, 3 mm diam., grayish or brown, warty.

DISTRIBUTION: A common tree of dry coastal forests and scrub. Bethany (B234), Susannaberg (A701, A844). Also on Anegada, St. Croix, St. Thomas, Tortola, and Virgin Gorda; Jamaica, Hispaniola, Puerto Rico, and Lesser Antilles.

COMMON NAME: black manchineel.

NOTE: A highly poisonous tree with caustic milky sap.

4. Euphorbia tirucalli L., Sp. Pl. 452. 1753.

Succulent shrub or small tree to 5 m tall, producing abundant milky, caustic, latex; bark dark green, smooth, branches whorled, green, cylindrical, spineless, forking upward. Leaves alternate, early deciduous; blades 1.5–2.5 cm long, oblanceolate, acute or obtuse at apex. Cyathia reddish, cup-shaped, ca. 3 mm across, in terminal yellowish, stalkless heads, the glands 5. Staminate flowers numerous; stamens 1 per flower. Pistillate flowers with ellipsoid, pubescent ovary, the styles bifid, united at base. Capsule glabrous, 3-lobed, ovoid, puberulous to 8 mm diam. Seeds ellipsoid, 3 mm long, black, with whitish margins.

DISTRIBUTION: An occasional shrub of dry open areas, persistent after cultivation. Lind Point (A4241). Also St. Thomas and Tortola; native to tropical and southern Africa, now widespread through cultivation.

COMMON NAMES: milk bush, pencil-bush.

CULTIVATED SPECIES: *Euphorbia lactea* Roxb. and *Euphorbia leucocephala* Lotsy are occasionally found in gardens or are persistent after cultivation, but they are not naturalized.

DOUBTFUL RECORDS: *Euphorbia cyathophora* Murray was reported to occur on St. John by Britton and P. Wilson (1925) and by Woodbury and Weaver (1987). However, no specimen has been located to confirm these records. A collection of *E. heterophylla* (W682) was incorrectly identified as *E. cyathophora* by Woodbury.

9. FLUEGGEA Willd.

Dioecious (rarely monoecious) shrubs or small trees, not producing milky latex; branch tips usually spiny. Leaves simple, alternate, entire; stipules minute. Flowers solitary or fascicled in leaf axils; calyx of 4–7 imbricate, biseriate sepals; corolla lacking. Staminate flowers distinctively pediceled to nearly sessile; nectary disk extrastaminal, dissected into 4–7 lobes; stamens 4–7, the filaments free, the anthers short, ellipsoid; pistillode present. Pistillate flowers on short pedicels; nectary disk rounded; ovary of 3 (rarely 2 or 4) biovular carpels, the styles free, bifid. Fruit a 3-locular, dry or fleshy capsule, when dry the locules separating from the central columella.

A small genus of 12 species, 9 in the Old World.

REFERENCE: Webster, G. L. 1984. A revision of *Flueggea* (Euphorbiaceae). Allertonia 3: 259–312.

1. Flueggea acidoton (L.) G.L. Webster, Allertonia **3**: 299. 1984. *Adelia acidoton* L., Syst. Nat., ed. 10, **2**: 1298. 1759. *Securinega acidoton* (L.) Fawc. & Rendle, J. Bot. **57**: 68. 1919.   Fig. 96H–K.

Dioecious shrub to 3 m tall, glabrous; bark gray, more or less fissured; branches flexuous and spiny. Leaves deciduous, congested on short lateral twigs; blades 0.5–1.2 × 0.5–1 cm, obovate, coriaceous to rigid, glabrous, the apex rounded or notched, the base cuneate, the margins entire; petioles 1–2 mm long; stipules lanceolate, minute. Staminate flowers in dense axillary cymules; pedicel to 1 mm long; calyx 4–5-merous, the sepals elliptic, ca. 1 mm long; nectary disk of 4–5 segments; stamens 4–5, the filaments free; pistillode present. Pistillate flowers solitary; pedicel to 1 cm long in fruit; calyx 5-merous, the sepals elliptic, ca. 1.5 mm long; nectary disk annular-angular; ovary ribbed, of 3 carpels, the styles bifid. Capsule glabrous, 3-lobed, depressed at poles, to 4 mm long. Seeds trigonous, smooth, ca. 2 mm long, pale yellow, usually brown-mottled.

DISTRIBUTION: Rare, in dry forests and coastal scrubs. Nanny Point (A2448), Little St. James Island (B1403). Also on St. Croix and St. Thomas; the Greater Antilles and the Bahamas.

10. GYMNANTHES Sw.

Monoecious shrubs or small trees, glabrous, producing a clear or milky latex. Leaves simple and alternate, entire or less often serrate; stipules minute. Flowers in axillary, bisexual and or staminate racemes, the racemes conelike when young owing to overlapping bracts, the bisexual racemes bearing a pistillate flower at base; calyx wanting, corolla and nectary disk wanting. Staminate flowers in groups of 2 or 3, subtended by a bracteole; calyx rudimentary; stamens 2–3, with filaments united into a short stipe; pistillode absent. Pistillate flower solitary, subtended by 2 or 3 alternate bracteoles; sometimes with sterile anthers; ovary 3-lobed, of 3 uniovular carpels, the styles free or united at base, simple and recurved. Fruit a 3-lobed capsule with 1 seed per locule; columella persistent after dehiscence, the locule separating from the columella and then splitting.

A genus of 12 species with Caribbean and circum-Caribbean distribution.

1. Gymnanthes lucida Sw., Prodr. 96. 1788. *Ateramnus lucidus* (Sw.) Rothm., Feddes Repert. Spec. Nov. Regni Veg. **53**: 5. 1944.   Fig. 96A–G.

Shrub or small tree to 7 m tall, glabrous, producing a clear latex when cut; bark dark gray, smooth, forming rectangular plates. Leaf blades 5–11 × 3–5 cm, oblanceolate to elliptic, coriaceous, glabrous, with prominent reticulate venation when dried, the apex obtuse to acute, slightly notched, the base tapering, with 1 or a few dotlike glands underneath, the margins minutely toothed, crenate, or entire; petioles 3–10 mm long; stipules ovate-lanceolate, minute, early deciduous. Racemes axillary, bisexual, cone-shaped when immature. Staminate flowers distal on racemes; stamens 2 or 3. Pistillate flower on long curved pedicels at base of raceme; ovary nearly globose, glabrous, the styles united at base, recurved. Capsule glabrous, depressed-globose, ca. 1 cm diam., with 6 longitudinal furrows. Seeds nearly globose, 5 mm long, light brown.

DISTRIBUTION: A common tree of coastal dry forests and scrubs. Cob Gut (A3219), Gift Hill (A817), Maria Bluff (A2328). Also on St. Croix, St. Thomas, Tortola, and Virgin Gorda; Florida and throughout the West Indies.

COMMON NAMES: crab wood, goat wood.

11. HIPPOMANE L.

Monoecious, glabrous shrubs or trees, producing abundant, toxic, milky latex. Leaves simple and alternate, serrate; petioles long, with a single gland at apex; stipules minute, deciduous. Flowers apetalous, in terminal, bisexual spikes. Staminate flowers congested at distal nodes, subtended by biglandular bracts; calyx of 2–3 minute sepals; stamens 2, the filaments coherent; pistillode and disk absent. Pistillate flowers usually solitary or paired at lower node of spike; calyx of 3 sepals; ovary ovoid, of 6–10, uniovular carpels, the styles as many as the carpels, simple and recurved. Fruit a depressed-globose drupe, with 6–10 seeds inside the stone.

A genus of 3 or 4 species within the Caribbean.

1. Hippomane mancinella L., Sp. Pl. 1191. 1753.   Fig. 97A–G.

Tree 8–12 m tall, glabrous, producing abundant poisonous, milky latex; bark gray, smooth. Leaf blades 5–14 × 3–8 cm, ovate to elliptic, nearly coriaceous, glabrous, the apex acute, obtuse, or rounded, the base obtuse or rounded, the margins crenate-dentate; petioles 3–5 cm long, with an annular gland at apex; stipules ovate to lanceolate, ca. 2.5 mm long, early deciduous. Spikes 3–12 cm long. Staminate flowers numerous at inflo-

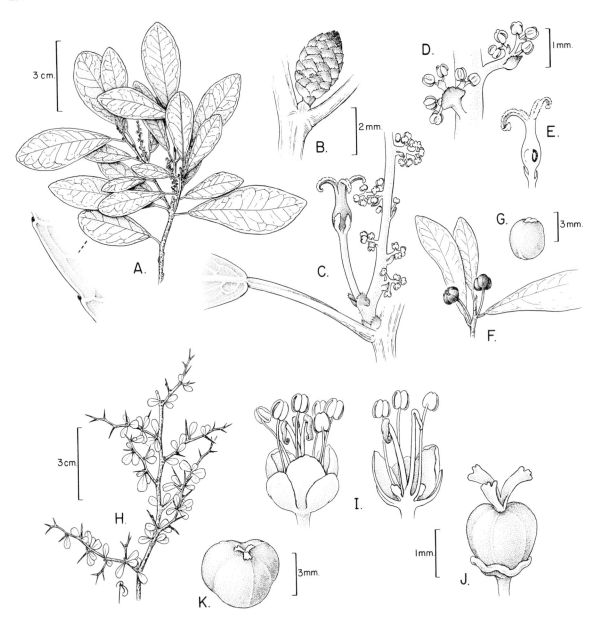

FIG. 96. A–G. *Gymnanthes lucida*. A. Flowering branch and detail of leaf margin. B. Immature inflorescence. C. Base of mature inflorescence showing solitary pistillate flower and upper staminate flowers. D. Detail of staminate flowers. E. L.s. pistillate flower. F. Fruiting branch. G. Seed. H–K. *Flueggea acidoton*. H. Branch. I. Staminate flower and l.s. staminate flower. J. Pistillate flower with calyx removed to show pistil and nectary disk. K. Capsule.

rescence nodes. Pistillate flowers solitary or paired at proximal nodes of spikes; ovary nearly ovoid, glabrous, the styles united at base, recurved. Capsule glabrous, depressed-globose, 2–2.5 cm diam., turning from green to yellow, with sweet smell. Seeds flattened, 5 mm long.

DISTRIBUTION: A common tree of coastal sandy beaches. Chocolate Hole (A781). Also on Anegada, St. Croix, St. Thomas, Tortola, and Virgin Gorda; a Caribbean and circum-Caribbean species.

COMMON NAMES: death apple, manchineel, mangineedle, poison apple.

NOTE: The latex and fruits from this tree are deadly poisonous if ingested and not treated in time. The latex causes severe dermatitis or temporary blindness when it comes in contact with the skin or eyes. The plant is dangerous because the fruits resemble a small apple and have a sweet smell, enticing people to eat them.

12. HURA L.

Monoecious, large trees, producing copious, poisonous, milky latex; bark with short conical thorns. Leaves simple, alternate, and serrate; petioles long, with 2 round glands at apex; stipules large, deciduous. Flowers unisexual; calyx cup-shaped; corolla wanting;

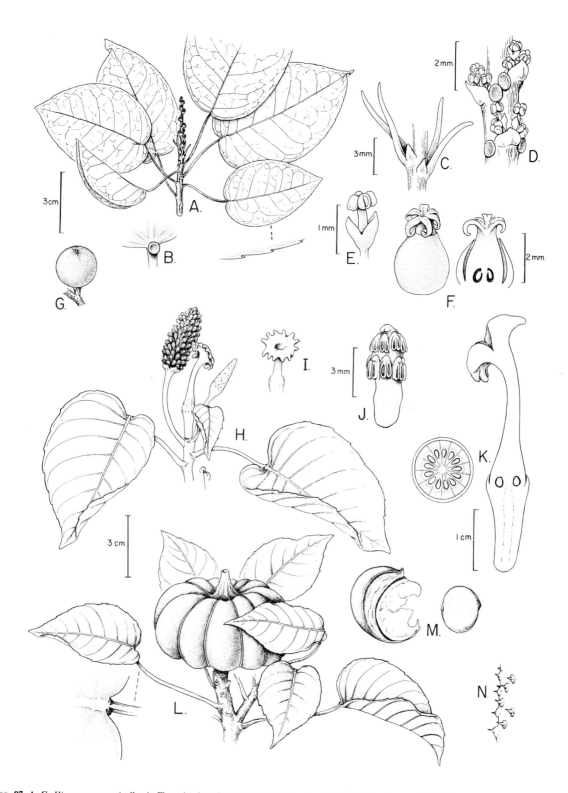

FIG. 97. A–G. *Hippomane mancinella*. **A.** Flowering branch and detail of leaf margin. **B.** Detail of gland at summit of leaf petiole. **C.** Detail of proximal portion of spike showing pistillate bracts. **D.** Detail of distal part of spike showing staminate flowers and their subtending biglandular bracts. **E.** Staminate flower. **F.** Pistillate flower and l.s. pistillate flower. **G.** Drupe. **H–N.** *Hura crepitans*. **H.** Flowering branch. **I.** Pistillate flower. **J.** Staminate inflorescence. **K.** C.s. ovary, and l.s. pistillate flower. **L.** Capsule and detail of leaf petiole showing glands at apex. **M.** Capsule valve and seed. **N.** Detail of bark showing thorns. (From Mori et al., in press, Guide to the Vascular Plants of Central French Guiana, Mem. New York Bot. Gard. 76.)

nectary disk wanting. Staminate flowers in long pedunculate, axillary spikes; stamens numerous, with connate filaments forming a tube with 2 or 3 whorls of anthers; pistillode absent. Pistillate flowers solitary in leaf axils, much larger than the staminate ones; ovary ovoid, with 5–20 carpels, these with a single ovule, the styles connate into an elongate tube, with spreading, radiating lobes at apex. Fruit a strongly depressed, woody capsule, with fluted, rounded equatorial outline, the locules with explosive dehiscence, the columella persistent; seeds nearly coin-shaped.

A genus of 2 or 3 species of tropical America.

1. Hura crepitans L., Sp. Pl. 1008. 1753. Fig. 97H–N.

Tree to 20 m tall, glabrous, producing abundant poisonous milky latex; bark gray, with many conical thorns. Leaf blades 12–25 × 6–15 cm, ovate to rounded, coriaceous, glabrous, the apex acuminate to long-acuminate, the base cordate, the margins crenate, with minute, glandular teeth; petioles 8–15 cm long, with a pair of rounded glands at apex; stipules oblong-lanceolate, densely pubescent, to 1 cm long, early deciduous. Staminate spikes to 10 cm long; bracts reddish; peduncle to 5 cm long, reflexed.

Pistillate flowers solitary or borne at base of staminate inflorescence; calyx green; stylar column green, stigmatic arms burgundy. Capsule explosively dehiscent, glabrous, woody, depressed at poles, 12–15-lobed, 7–8 cm diam., turning from green to brown. Seeds flattened, nearly circular, to 2 cm long, light brown, smooth.

DISTRIBUTION: Not very common, found mostly as a roadside tree in humid areas. Adrian Ruins (A730, A4010), Enighed (A2368). Also on St. Croix, St. Thomas, and Tortola; common throughout tropical America.

COMMON NAMES: monkey no climb, monkey pistol, sand box.

NOTE: The latex of this tree is poisonous and caustic.

13. JATROPHA L.

Monoecious shrubs, subshrubs, or trees, glabrous or with simple or branched, glandular hairs, producing copious clear or whitish to yellowish, poisonous latex. Leaves alternate, simple to palmately compound, with entire, serrate, or dissected margins; petioles long; stipules small to large, glandular. Flowers in axillary or terminal, bisexual, dichasial cymes; calyx of 5 imbricate lobes; corolla of 5 discrete petals; nectary disk extrastaminal, annular to cup-shaped or lobed. Staminate flowers lateral on individual dichasia; stamens 8–15, unequal, in 2 whorls, the filaments connate to various degree, the anthers often twisted, bearing a glandular appendage at apex; pistillode absent. Pistillate flowers terminal on individual dichasia; ovary ellipsoid to globose, of 3 carpels, these with a single ovule, the styles free or connate at base, simple or bifid, the stigmas thickened. Fruit a 3-lobed or globose, fleshy to dry capsule; seeds nearly bean-shaped.

A genus of about 175 species from tropical America, Africa, and India.

Key to the Species of *Jatropha*

1. Leaves palmately compound, with 10–12 lobes, the segments dissected (corolla rose to scarlet)
   .... *J. multifida* (cultivated)
1. Leaves 3–5-lobed or entire.
   2. Plants with glandular viscidulous hairs (corolla dark red)... 1. *J. gossypifolia*
   2. Plants not glandular.
      3. Corolla white.. *J. curcas* (cultivated)
      3. Corolla bright red... *J. integerrima* (cultivated)

1. Jatropha gossypifolia L., Sp. Pl. 1006. 1753. Fig. 98.

Erect subshrub to 1 m tall, many-branched from base, producing abundant clear or milky latex; foliage and young branches with viscidulous, glandular dendroid hairs. Leaf blades 3–15 × 5–17 cm, 3–5-palmately lobed, chartaceous, pubescent, the base cordate, the lobes ovate, oblong, or elliptic with glandular margins and acuminate apices; petioles 2.5–18 cm long, densely covered with glandular, branched hairs; stipules dissected, with glandular margins, to 1 cm long. Flowers in terminal, dichasial cymes, the bracts awl-shaped, with glandular margins; calyx pubescent, the sepals ovate, with glandular margins; corolla red, with white center, the petals obovate, 3.5–4 mm long. Staminate flowers

numerous. Pistillate flowers few, terminal on dichasial units or basal to the lateral branches; ovary hispid, the styles free, bifid, the stigmas thickened. Capsule globose to obovoid, hispidulous, 10–12 mm long, turning from green to dry, semiwoody. Seeds nearly bean-shaped, ca. 8 mm long, light brown, blackish mottled, with a large caruncle.

DISTRIBUTION: A common roadside weed. Emmaus (A3776), Salt Pond area (A3166), Susannaberg (A832). Also on Anegada, St. Croix, St. Thomas, Tortola, and Virgin Gorda; throughout tropical America.

COMMON NAMES: belly ache bush, body catta, wild physicnut.

CULTIVATED SPECIES: The following species are growing in gardens but are not known to be spontaneous: *Jatropha curcas* L., *Jatropha integerrima* Jacq., and *Jatropha multifida* L.

14. MARGARITARIA L.f.

Dioecious shrubs or small trees, not producing milky latex. Leaves simple, alternate, and entire; petioles short; stipules minute. Flowers in pairs, in leaf axils on short lateral branches; calyx of 4 sepals; corolla lacking, nectary disk annular. Staminate flowers long-pediceled; stamens 4, slightly unequal, the filaments free; pistillode absent. Pistillate flowers short-pediceled; ovary ovoid to cylindrical, with 2–6 carpels, these with 2 ovules each, the styles free or connate at base, bifid. Fruit a capsule, with irregular dehiscence; each locule 1-valved; seeds 2 per locule, oblong, with metallic blue outer coat (sarcotesta).

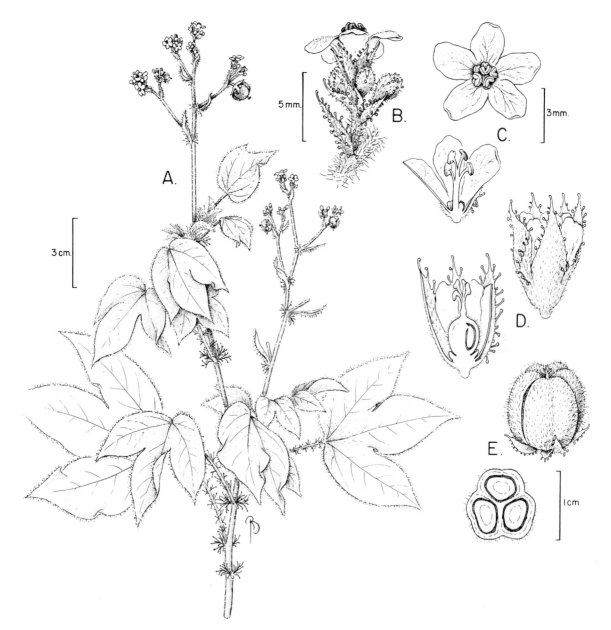

FIG. 98. *Jatropha gossypifolia*. **A.** Fertile branch. **B.** Detail of cyme with staminate flowers. **C.** Staminate flower and l.s. staminate flower. **D.** L.s. pistillate flower and pistillate flower. **E.** Capsule and c.s. capsule.

A tropical genus of about 14 species.
REFERENCE: Webster, G. L. 1979. A revision of *Margaritaria* (Euphorbiaceae). J. Arnold Arbor. 60: 403–444.

1. Margaritaria nobilis L.f., Suppl. Pl. 428. 1782. *Phyllanthus nobilis* (L.f.) Müll. Arg. *in* A. DC., Prodr. 15(2): 414. 1866.   Fig. 99A–G.

Shrub or small tree to 12 m tall, glabrous; bark light to dark brown, rough, peeling off in large rectangular plates; twigs dark brown with numerous grayish lenticels. Leaf blades 6–14 × 1.8–5.5 cm, elliptic or rarely oblanceolate, membranous, glaucous underneath, the apex acuminate to acute, the base obtuse and revolute, the margins entire; petioles 4–5 mm long; stipules min-ute, lanceolate, persistent. Staminate flowers few to many in short axillary fascicles. Pistillate flowers usually paired or a few on short axillary branches; pedicel 1–1.3 cm long; styles connate at base, bifid at apex. Capsule depressed-globose semiwoody, 0.7–1 cm diam., usually 5-locular. Seeds oblong, 3–4 mm long, with metallic blue fleshy outer coat (sarcotesta).

DISTRIBUTION: A common tree of moist to dry forests. Adrian Ruins (A2844), Enighed (A828). Also on St. Croix, St. Thomas, Tortola, and Virgin Gorda; throughout tropical America.

COMMON NAME: gonglehout.

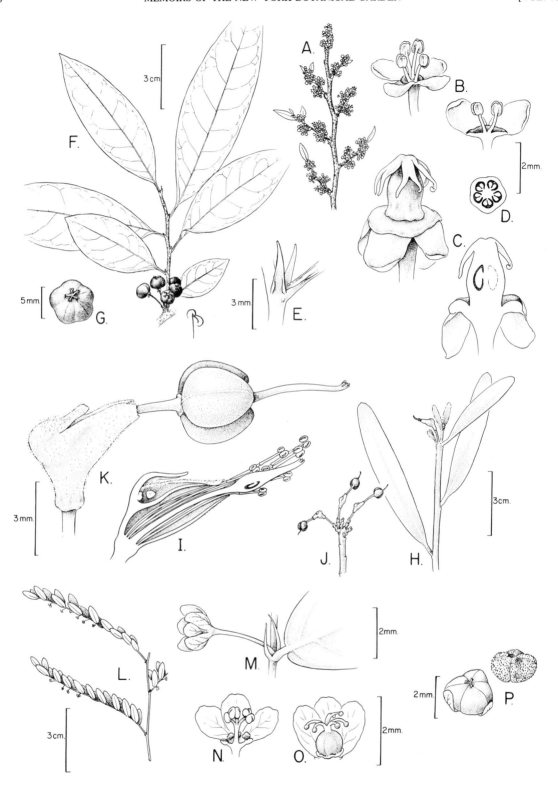

Fɪɢ. 99. A–G. *Margaritaria nobilis.* **A.** Flowering branch. **B.** Staminate flower and l.s. staminate flower. **C.** Pistillate flower and l.s. pistillate flower. **D.** C.s. ovary. **E.** Detail of stipules. **F.** Fruiting branch. **G.** Capsule. **H–K.** *Pedilanthus tithymaloides* subsp. *angustifolius.* **H.** Flowering branch. **I.** L.s. cyathium showing staminate flowers, pistillate flowers and spur. **J.** Detail of cymes. **K.** Cyathium and capsule. **L–P.** *Phyllanthus niruri.* **L.** Flowering branch. **M.** Detail of node with axillary flower and stipules. **N.** Staminate flower. **O.** Pistillate flower. **P.** Capsule and seeds.

15. PEDILANTHUS Poit.

Monoecious shrubs or less often small trees, sometimes with succulent stems, glabrous or with simple hairs, producing copious milky latex. Leaves simple, alternate, entire, sometimes succulent and deciduous; petioles short; stipules minute or lacking. Flowers borne within a cyathium, the cyathia subtended by 2 clasping bracts and borne in compound axillary or terminal dichasial cymes; cyathia bright red, elongate, bilaterally symmetric, with a spur at base and 5 elongate lobes toward the distal portion, the spur containing 4 glands within. Staminate flowers numerous, naked, of 1 stamen. Pistillate flowers, solitary, central on cyathium; ovary of 3 uniovulate carpels on an elongate pedicel; styles 3, free or united at base, usually bifid. Fruit a 3-lobed capsule with 1 seed per locule, on a long exserted, reflexed pedicel; dehiscence explosive, leaving a central columella.

A genus of 14 species native to the New World, with most species restricted to Mexico.

REFERENCE: Dressler, R. L. 1957. The genus *Pedilanthus* (Euphorbiaceae). Contr. Gray Herb. 182: 1–188.

1. Pedilanthus tithymaloides (L.) Poit., Ann. Mus. Natl.
   Hist. Nat. **19**: 393. 1812. *Euphorbia tithymaloides*
   L., Sp. Pl. 453. 1753.

Key to the Subspecies of *Pedilanthus tithymaloides*

1. Leaves linear or narrowly elliptic, underside without a distinctive keel on midvein; styles <5 mm long
   .... 1a. *P. tithymaloides* subsp. *angustifolius*
1. Leaves lanceolate to ovate, underside with keel along midvein; styles >7 mm long
   .... *P. tithymaloides* subsp. *tithymaloides* (cultivated)

1a. Pedilanthus tithymaloides (L.) Poit. subsp. **angusti-folius** (Poit.) Dressler, Contr. Gray Herb. **182**: 161. 1957. *Pedilanthus angustifolius* Poit., Ann. Mus. Natl. Hist. Nat. **19**: 393. 1812.   Fig. 99H–K.

Leaning shrub with numerous, hanging branches to 1 m long, nearly glabrous; stems cylindrical, green, slightly succulent. Leaf blades 2.5–10 × 0.4–1 cm, linear or narrowly elliptic, succulent and deciduous, lower surface puberulent, the apex obtuse or acute, the base obtuse, the margins entire; petioles 1–2 mm long; stipules inconspicuous. Cyathia green with reddish hue, in terminal, dichasial cymes. Staminate flowers numerous, some projecting beyond the cyathium. Pistillate flowers partially exserted; ovary nearly globose; styles <5 mm long, united for most of their length. Capsule trigonous, 5–6 mm long, green with reddish hue and persistent, reddish styles. Seeds bean-shaped, ca. 3 mm long, light brown, smooth, with a ventral, elongate suture.

DISTRIBUTION: Persistent after cultivation, in scrub areas. Minna Hill (A3224), Lameshur (B619). Also on St. Croix, St. Thomas, and Tortola; throughout the Greater Antilles.

COMMON NAME: Christmas candle.

NOTE: The sap produced by this species is said to be poisonous.

CULTIVATED SPECIES: *Pedilanthus tithymaloides* (L.) Poit. subsp. *tithymaloides* is occasionally found in gardens.

16. PHYLLANTHUS L.

Monoecious herbs, shrubs, or small trees, glabrous, or seldom with simple hairs, not producing milky latex; stems sometimes flattened into a photosynthetic phylloclade. Leaves simple, alternate and entire or reduced to scales; petioles short; stipules minute, persistent or deciduous; Flowers solitary in leaf axils, or in bisexual, axillary cymes or terminal thyrses, or cauliflorous; calyx of 4–6 sepals; corolla lacking, nectary disk extrastaminal, annular or lobed. Staminate flowers long-pediceled (usually longer than those of pistillate flowers); stamens (2–)6, the filaments free or connate to various degrees; pistillode absent. Pistillate flowers short-pediceled; ovary trilobed, with 3(–12) carpels, these with 2 ovules each, the styles free or connate, with simple or bifid branches. Fruit usually 3-lobed or depressed-globose, thin-walled capsule with explosive dehiscence, or a drupe; each locule 2-valved; seeds 2 per locule, trigonous.

A tropical genus of about 650 species.

REFERENCE: Webster, G. L. 1956–1958. A monographic study of the West Indian species of *Phyllanthus*. J. Arnold Arbor. 37: 92–122, 217–256, 340–357; 38: 51–79, 170–198, 295–373; 39: 49–100, 111–212.

Key to the Species of *Phyllanthus*

1. Small trees or shrubs; ovary of 3 or 4 carpels; flowers on leafless branches; fruit a fleshy drupe .... 1. *P. acidus*
1. Herbs; ovary of 3 carpels; flowers on leafy branches; fruit a dry capsule.
   2. Leaves unequal (obtuse-rounded) at base; seeds verrucose .... 3. *P. niruri*
   2. Leaves symmetrical at base; seeds 5–6-ribbed .... 2. *P. amarus*

1. Phyllanthus acidus (L.) Skeels, U.S.D.A. Bur. Pl. Industr. Bull. **148**: 17. 1909. *Averrhoa acida* L., Sp. Pl. 428. 1753.

Cicca disticha L., Mant. Pl. **1**: 124. 1767. *Phyllanthus distichus* (L.) Müll. Arg. *in* A. DC., Prodr. **15**(2): 413. 1866.

Small tree or shrub to 8 m tall, glabrous; bark grayish, smooth. Leaf blades 4–9 × 2.5–4 cm, alternate, on short lateral deciduous branches, ovate to lanceolate, chartaceous, the apex acute or shortly acuminate, the base rounded to nearly truncate, seldom unequal, the margins entire; petioles 2.5–5 mm long; stipules minute, triangular. Inflorescence cauliflorous, racemes with lateral bisexual cymules; calyx reddish, of 5 sepals. Staminate flowers numerous; stamens 5, the filament free and very short. Pistillate flowers 1 per cymule; ovary of 3 or 4 carpels, styles connate at base, bifid. Drupe oblate, 6–8-lobed, yellow, with crisp acid mesocarp, 1.5–2 cm diam.

DISTRIBUTION: Persistent after cultivation but apparently not spreading. Emmaus (A2010). Also on St. Croix, St. Thomas, and Tortola; native to South America but cultivated throughout the tropics.

COMMON NAME: gooseberry.

2. Phyllanthus amarus Schum., Beskr. Guin. Pl. 421. 1827.

Erect herb 0.3–1 m tall, glabrous. Leaf blades 5–11 × 3–6 mm, alternate, on short lateral deciduous branches, oblong to obovate, membranous, the apex rounded, the base obtuse or rounded, equal, the margins entire; petioles <0.5 mm long; stipules ca. 2 mm long, lanceolate. Cymules of 1 pistillate and 1 staminate flower; calyx of 5 sepals; stamens 3, the filament connate into a short column; ovary of 3 carpels, styles free at base, bifid. Capsule oblate, with rounded outline, ca. 2 mm diam., straw-colored. Seeds wedge-shaped, ca. 1 mm long, light brown, with 5–6 longitudinal ribs along dorsal side.

DISTRIBUTION: Common weed occurring along roadsides and in open disturbed areas. Bordeaux (A3138), Cruz Bay (A3091), Fish Bay (A3871). Also on Tortola and Virgin Gorda; throughout the tropics as a weed.

3. Phyllanthus niruri L., Sp. Pl. 981. 1753. Fig. 99L–P.

Erect herb to 50 cm tall, with numerous branches along main stem, glabrous. Leaf blades 8–17 × 3–9 mm, alternate, on short lateral deciduous branches, oblong to elliptic, membranous, glaucous on lower surface, the apex obtuse, the base obtuse-rounded, unequal, the margins entire; petioles <0.7 mm long; stipules 1.5–2 mm long, awl-shaped. Staminate flowers in cymules, proximal on lateral branches; calyx 1 mm long, of 5 sepals; stamens 3, the filament connate into a short column. Pistillate flowers solitary on distal portion of branches; calyx 2 mm long, the sepals spatulate; ovary of 3 carpels, styles free at base, bifid. Capsule oblate, with rounded outline, 3.5–4 mm diam., straw-colored. Seeds wedge-shaped, ca. 2 mm long, light brown, verrucose all over.

DISTRIBUTION: Common roadside weed. Coral Bay Quarter along Center Line Road (A4245), Lameshur (W585), along road to Peter Peak (A4107). Also on St. Croix, St. Thomas, Tortola, and Virgin Gorda; typical subspecies restricted to the West Indies (except Jamaica, Cayman Islands, and Bahamas).

COMMON NAMES: cane piece senna, creole chinine, creole senna, gale of wind.

NOTE: Woodbury and Weaver (1987) erroneously reported a specimen of this species as *Phyllanthus stipulatus* (Raf.) Webster.

17. RICINUS L.

A monospecific genus, characterized by the following species.

1. Ricinus communis L., Sp. Pl. 1007. 1753.

Fig. 100A–G.

Monoecious shrub or small tree 1–2.5 m tall, glabrous, producing a clear watery latex; stems soft-wooded, usually reddish when young. Leaves alternate; blades 7–9-palmately lobed, with rounded outline, to 20 cm wide, membranous, peltate, the venation prominent, glabrous, the base cordate, the lobes lanceolate, with acuminate apex and serrate-glandular margins; petioles as long as the blades, reddish; stipules to 1.5 cm long, ovate-lanceolate. Flowers in cymules disposed along a terminal, bisexual, erect panicle; calyx calyptrate, splitting into 3–5 sepals, to 9 mm long; corolla and nectary disk wanting. Staminate flowers in proximal cymules; stamens numerous, the filaments connate into fascicles; pistillode absent. Pistillate flowers in distal cymules on inflorescence; ovary 3-lobed, of 3 uniovular carpels, the styles connate at base, bifid, papillose. Fruit a 3-lobed ovoid capsule, 2 cm diam., covered with soft spinelike projections, with 1 seed per locule. Seeds ellipsoid, 1–1.8 cm long, brown, black mottled.

DISTRIBUTION: A weed of disturbed areas. Fish Bay (A2502). Also on St. Croix, St. Thomas, Tortola, and Virgin Gorda; native to Africa but cultivated and naturalized throughout tropical America.

COMMON NAMES: castorbean, castor oil plant, castornut, ricin.

18. SAPIUM P. Browne

Monoecious, glabrous shrubs or trees, producing abundant, toxic, milky latex. Leaves simple, alternate, entire; petioles short, with a pair of glands at apex; stipules minute, deciduous. Flowers apetalous, without nectary disk, in terminal, bisexual spikes. Staminate flowers congested at distal nodes, subtended by a biglandular, minute bract; calyx of 2 minute sepals; stamens 2, the filaments free; pistillode absent. Pistillate flowers solitary at nodes of lower half of spike; calyx of 2 or 3 sepals; ovary ovoid, of 3 uniovulate carpels, the styles connate at base. Fruit a nearly globose capsule, with 1-seeded locules, dehiscent from the apex; seeds with a fleshy, reddish coat, often only 1 seed fully developed.

A genus of about 80 species of tropical America.

1. Sapium caribaeum Urb., Symb. Ant. 3: 308. 1902.

Fig. 100H–M.

Tree 10–20 m tall, glabrous, producing abundant poisonous, milky latex; bark gray, smooth, with numerous rectangular plates. Leaf blades 6.5–22 × 3–5.6 cm, elliptic or seldom oblanceolate, membranous, glabrous, the apex long- or short-acuminate, the base obtuse, the margins entire or finely serrate; petioles 1–3 cm long, with a pair of minute, rounded glands at apex; stipules ovate ca. 2 mm long, early deciduous. Spikes solitary, to 14 cm long. Staminate flowers numerous at distal nodes of inflorescence. Pistillate flowers solitary or paired at proximal nodes of spikes; ovary nearly ovoid. Capsule glabrous, 2-lobed, ovoid, 5–6 mm long, turning from green to brown. Seeds nearly globose, ca. 3 mm long, with fleshy red coat.

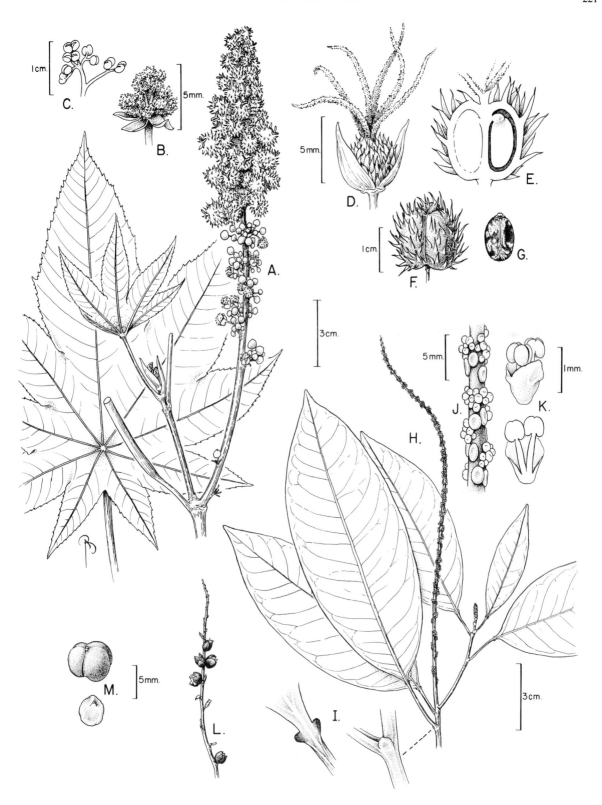

FIG. 100. A–G. *Ricinus communis.* **A.** Leaf and fertile branch. **B.** Staminate cymule. **C.** Stamens. **D.** Pistillate flower. **E.** L.s. pistillate flower. **F.** Capsule. **G.** Seed. **H–M.** *Sapium caribaeum.* **H.** Flowering branch and detail of branch showing stipule at base of leaf petiole. **I.** Detail of leaf petiole showing glands at summit. **J.** Detail of distal part of inflorescence showing staminate flowers and subtending biglandular bracts. **K.** Staminate flower and l.s. staminate flower. **L.** Infructescence showing dehisced capsules. **M.** Capsule and seed.

DISTRIBUTION: A common tree of moist areas. Maho Bay Gut (A2127). Also on Tortola; the Lesser Antilles.

COMMON NAME: milk tree.

NOTE: The latex from this tree is caustic.

DOUBTFUL RECORD: *Sapium laurocerasus* Desf. was reported for St. John by Little and Wadsworth (1964), but no specimen has been located to confirm this record.

19. SAVIA Willd.

Monoecious or dioecious shrubs or small trees, glabrous, or with simple hairs, not producing milky latex. Leaves simple, alternate, and entire; petioles short; stipules minute, deciduous. Flowers in axillary, unisexual cymes or racemes; calyx of 5 imbricate sepals; corolla of 5 petals, nectary disk annular, extrastaminal. Staminate flowers subsessile; stamens 5, the filaments free or connate to various degrees; pistillode absent. Pistillate flowers short-pediceled; ovary trilobed, with 3(–12) carpels, these with two ovules each, the styles free or connate, with simple or bifid branches. Fruit usually a 3-lobed or depressed-globose, thin-walled capsule with explosive dehiscence or a drupe; each locule 2-valved; seeds 2 per locule, trigonous.

A genus of about 25 species occurring in Greater Antilles, South America, Africa, and Madagascar.

1. Savia sessiliflora (Sw.) Willd., Sp. Pl. **4**: 771. 1806.

Croton sessiliflorum Sw., Prodr. 100. 1788.

Fig. 101A–D.

Dioecious shrub or small tree to 7 m tall, not producing milky latex; young twigs pubescent; bark grayish brown, fissured. Leaf blades 4–8 × 2–3.5 cm, ovate to ovate-lanceolate, chartaceous, glabrous, the apex acute to long-acuminate, the base rounded or obtuse, the margins entire; petioles 1–2 mm long; stipules lanceolate, ca. 2 mm long, early deciduous. Staminate flowers subsessile, congested in leaf axils; calyx ca. 1.2 mm long; filaments free. Pistillate flowers short-pediceled, solitary in leaf axils; calyx 1.5–2 mm long; ovary trilobed, of 3 carpels, the styles free, bifid at apex. Capsule 3-lobed, ovoid, 6–8 mm long, turning from green to dark brown. Seeds trigonous, 4–5 mm long, light brown, smooth.

DISTRIBUTION: A common tree of coastal dry forests and scrub. Battery Gut (A4168), Cob Gut (A3209), Lameshur (B521). Also on Jost van Dyke, St. Croix, and St. Thomas; the Greater Antilles.

20. TRAGIA L.

Perennial monoecious, erect or climbing (usually twining) herbs, covered with stinging hairs; latex clear. Leaves simple or trilobed, alternate, serrate, or entire; petioles short; stipules minute, deciduous. Flowers in terminal or axillary, bisexual racemes; corolla and nectary disk lacking. Staminate flowers on short pedicels, numerous, usually 1 per raceme node; calyx of 3 or 4 valvate sepals; stamens (2–)3–5, the filaments connate at base or frequently half of their length; pistillode small. Pistillate flowers long-pediceled; calyx of 2 whorls of 3 imbricate sepals; ovary of 3 uniovulate carpels, the styles connate, with simple branches, the stigma papillate. Fruit usually a 3-lobed thin-walled capsule with explosive dehiscence; seeds 1 per locule, nearly spherical, smooth or slightly rough.

A genus of about 150 species of tropical, subtropical, and warm temperate regions.

1. Tragia volubilis L., Sp. Pl. 980. 1753.   Fig. 101E–I.

Twining herbaceous vine, many-branched from a semiwoody base; branches to 3 m long, covered with appressed, minute hairs and also glandular, stinging hairs. Leaves alternate, often congested on short, lateral branches; blades 2–8 × 0.9–4 cm, oblong, ovate, or lanceolate, membranous, upper surface usually with glandular stinging hairs, the apex acuminate, the base nearly truncate or cordate, the margins serrate; petioles slender, 0.3–6 cm long; stipules awl-shaped, ca. 4 mm long. Racemes axillary, with greenish flowers. Staminate flowers numerous, 1 per node, distributed along entire inflorescence; calyx 0.8–1 mm long; stamens 2 or 3, short. Pistillate flowers on pedicel at base of racemes; calyx ca. 1 mm long, pilose; ovary hispid. Capsule of 3 rounded cocci, 7 mm diam., covered with stinging, hispid hairs. Seeds globose, ca. 2 mm diam., brown, smooth.

DISTRIBUTION: A common weed of disturbed dry areas, but also in humid forests. Bethany (B204), Cinnamon Bay (M17008), Susannaberg (A703). Also on St. Croix, St. Thomas, and Tortola; throughout tropical America.

COMMON NAMES: bran nettle, creeping cowitch, stinging nettle.

NOTE: Contact with this plant causes a severe rash and a burning sensation that can last for nearly half an hour.

CULTIVATED SPECIES: One individual of *Cnidoscolus aconitifolius* (Mill.) I.M. Johnst. was observed at Cruz Bay.

38. Fabaceae (Legume Family)
Reviewed by J. W. Grimes

Trees, shrubs, herbs, vines, or lianas. Leaves alternate or rarely opposite, pinnate or bipinnate, unifoliolate or rarely simple; stipules mostly and stipels sometimes present. Flowers large to minute, actinomorphic or zygomorphic, bisexual or rarely unisexual, produced in very diverse inflorescences; calyx of 5 distinct or connate sepals; corolla of 5 distinct unequal petals, or tubelike, with 5 more or less prominent lobes; stamens mostly 10 or more numerous, the filaments unequal or equal, distinct or connate into a tube, the anthers opening by longitudinal slits or terminal pores; ovary superior, of a single carpel, usually with an annular nectary disk at base, the carpel multiovular, the ovules 1 to many, attached in 2 alternating rows to the coalescent margins of the adaxial suture, the style terminal, simple. Fruit usually dry and dehiscent along both sutures; seeds variable in shape, size, and texture.

A family with about 640 genera and ca. 17,000 species, with cosmopolitan distribution. The family as treated here consists of 3 distinct subfamilies.

REFERENCES: Britton, N. L. & J. N. Rose. 1930. Caesalpiniaceae. N. Amer. Fl. 23(4): 201–349. Polhill, R. M. & P. H. Raven (eds.). 1981. Advances in Legume Systematics. Parts 1 & 2. Royal Botanic Gardens, Kew, England.

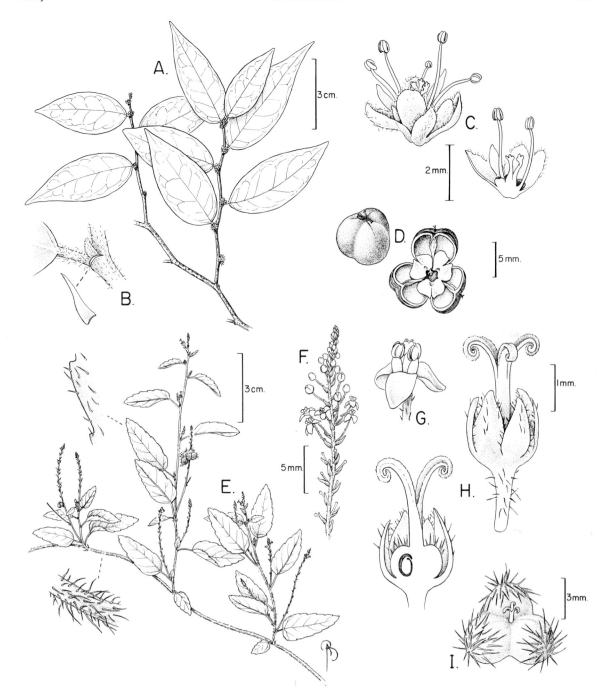

FIG. 101. **A–D.** *Savia sessiliflora*. **A.** Flowering branch. **B.** Detail of branch showing leaf petiole and stipule. **C.** Staminate flower and l.s. flower. **D.** Capsule and dehisced capsule. **E–I.** *Tragia volubilis*. **E.** Fertile branch, detail of leaf margin showing trichomes, and detail of stem showing trichomes. **F.** Inflorescence. **G.** Staminate flower. **H.** L.s. pistillate flower and pistillate flower. **I.** Capsule.

Key to the Subfamilies of Fabaceae

1. Corolla of connate petals, actinomorphic; flowers usually in heads or dense spikes .... III. *Mimosoideae*
1. Corolla of distinct petals, zygomorphic or only slightly so; flowers usually in racemose inflorescences.
    2. Corolla with an extended central petal (standard), 2 lateral ones connate into a keel, and 2 smaller
       lateral petals .... II. *Faboideae*
    2. Corolla with 5 equal or subequal petals .... I. *Caesalpinioideae*

I. Caesalpinioideae (Tamarind Subfamily)

Trees, shrubs, lianas, or less often herbs. Leaves alternate, once or twice pinnate or less often simple; stipules usually present, stipels usually lacking. Flowers small to large and showy, in racemes or cymes; calyx of 5 distinct sepals (sometimes the upper 2 connate) or cup-shaped with 5 lobes; corolla of 5 distinct petals, the innermost one usually smaller than the adjacent ones; stamens 10 or fewer, all alike or heteromorphic, the filaments distinct or connate, the anthers opening by longitudinal slits or by apical or basal pores; ovary usually surrounded by an annular nectary disk at base.

The subfamily consists of approximately 150 genera and 2200 species distributed throughout tropical and subtropical regions.

REFERENCES: Irwin, H. S. & R. C. Barneby. 1982. The American Cassiinae. A synoptical revision of Leguminosae tribe Cassieae subtribe Cassiinae in the New World. Mem. New York Bot. Gard. 35: 1–918. Robertson, K. R. & Y.-T. Lee. 1976. The genera of Caesalpinioideae (Leguminosae) in the southeastern United States. J. Arnold Arbor. 57: 1–90.

Key to the Genera of Caesalpinioideae

1. Leaves simple (2 fused leaflets) with deeply notched apex .... *Bauhinia* (cultivated)
1. Leaves compound.
   2. Leaves bipinnate.
      3. Rachis of pinna flattened and laminar; leaflets alternate, minute, <5 mm long; fruit torulose.. 5. *Parkinsonia*
      3. Rachis of pinna cylindrical; leaflets opposite, 6 mm long or longer; fruits flattened or slightly inflated.
         4. Trees, not spiny; legume flattened, woody or semiwoody.
            5. Stipules pinnatisect (leaflike); corolla 6–8 cm wide, orange or red; legume 30–60 cm long, woody .... 3. *Delonix*
            5. Stipules minute, early deciduous; corolla 4.5–5 cm wide, yellow; legume 7.5–10 cm long, slightly woody, winged at margins.. 6. *Peltophorum*
         4. Erect or scandent spiny shrubs; legume flattened or semi-inflated (then spiny), leathery or papery .... 1. *Caesalpinia*
   2. Leaves once-pinnate.
      6. Leaves of 2 opposite leaflets; corolla white, to 6 cm wide; fruit wall thick and woody........... 4. *Hymenaea*
      6. Leaves of numerous leaflets; corolla yellow, to 5 cm wide; fruit wall thin, papery, leathery, or woody.
         7. Erect or decumbent subshrubs; petioles with stipitate glands................................. 2. *Chamaecrista*
         7. Trees, erect shrubs, or scandent shrubs; petioles lacking stipitate glands.
            8. Fruits crustaceous, torulose; seeds covered by fleshy tart pulp............................... 8. *Tamarindus*
            8. Fruits cylindrical, compressed, 4-winged, leathery or fleshy, not torulose.
               9. Petals 2.2–2.5 cm long; anthers opening through longitudinal slits; legumes 20–35 cm long, woody, cylindrical .... *Cassia* (cultivated)
               9. Petals 0.8–2.2 cm long; anthers opening by terminal pores; legumes not as above........... 7. *Senna*

1. CAESALPINIA L.

Trees, erect shrubs, or climbing shrubs. Leaves bipinnate; pinnae opposite, the leaflets opposite or alternate; petioles and rachis without stipitate glands; stipules minute to foliaceous. Flowers unisexual or bisexual, in axillary or terminal racemes; pedicels jointed at apex; calyx nearly bell-shaped, 5-lobed at apex; corolla of various colors, of 5 free petals; stamens 10, the filaments flattened, free, of equal length, the anthers opening along longitudinal slits; ovary 1-locular, sessile or shortly stipitate, with many ovules. Legume variously shaped, dehiscent or indehiscent; seeds 1 to few, variously shaped.

A tropical genus with about 100 species.

Key to the Species of *Caesalpinia*

1. Scandent shrubs; corolla yellow, 1–2 cm wide; fruits slightly inflated, spiny.
   2. Stipules large, foliaceous; bracts persistent; fruits densely spiny; seeds gray............................ 1. *C. bonduc*
   2. Stipules minute; bracts deciduous; fruits sparsely spiny; seeds yellow to orange-brown .... 2. *C. ciliata*
1. Erect shrubs; corolla (yellow) orange or red, 3–5 cm wide; fruits flattened, not spiny .... 3. *C. pulcherrima*

1. Caesalpinia bonduc (L.) Roxb., Fl. Ind., ed. 1832, **2:** 362. 1832. *Guilandina bonduc* L., Sp. Pl. 381. 1753. Fig. 102A–C.

Caesalpinia crista sensu Urb., Symb. Ant. **2:** 269. 1900, and sensu Britton & P. Wilson, Bot. Porto Rico **5:** 378. 1924, non L., 1753.

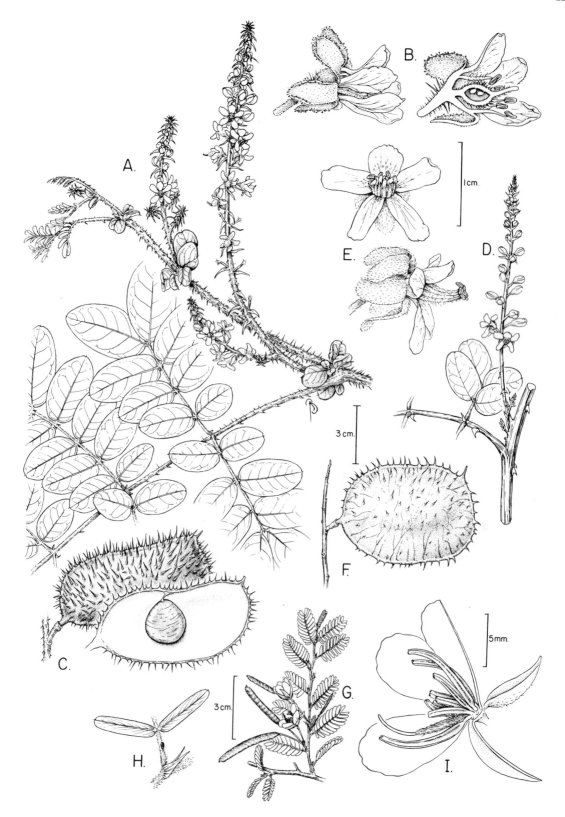

Fɪɢ. 102. A–C. *Caesalpinia bonduc.* **A.** Flowering branch. **B.** Pistillate flower and l.s. pistillate flower. **C.** Legume and legume with valve removed to show seed. **D–F.** *Caesalpinia ciliata.* **D.** Flowering branch. **E.** Staminate flower, top and lateral views. **F.** Legume. **G–I.** *Chamaecrista glandulosa* var. *swartzii.* **G.** Fertile branch. **H.** Detail of stem showing leaf petiole with discoid gland and stipules at base. **I.** L.s. flower.

Scandent shrub to 6 m long, profusely branching along main stem; stems cylindrical, to 2.5 cm diam., densely covered with prickles. Leaves 30–75 × 10–34 cm; pinnae 4–8 pairs, the leaflets 3–7 pairs per pinna, 2.5–5(–6.8) × 1.2–3 cm, ovate, lanceolate, oblong, or elliptic, chartaceous, glabrous or with scattered hairs along veins, midvein prominent on lower surface, the apex obtuse and mucronate, the base truncate, rounded to nearly cordate, the margins entire, revolute, ciliate; petiolules short, cylindrical; petioles and leaf rachises spiny, with a pair of recurved spines at the base of leaflets; stipules foliaceous, persistent, of 3–5 segments. Flowers functionally unisexual on long axillary and terminal racemes, the rachis rufous-tomentose and densely prickled; bracts lanceolate, 8–14 mm long, persistent. Calyx forming a nearly bell-shaped hypanthium, 4–6 mm long, rufous-tomentose, the sepals reflexed, oblong; petals yellow, 5.5–8 mm long, spatulate; stamens 10, only 4 fertile in staminate flowers, all sterile in pistillate flowers, the filaments tomentose at base; ovary sessile, rudimentary in staminate flowers, the stigma terminal. Legume oblong, 4–7.5 × 2–4 cm, semi-inflated, the valve densely spiny, tardily dehiscent through both sutures. Seeds few, 1.5–2 cm long, ovoid to rounded, smooth, gray, hard.

DISTRIBUTION: Along sandy beaches. Johnson Bay (A4021), Little Lameshur Bay (A3932). Also on St. Croix, St. Thomas, Tortola, and Virgin Gorda; throughout tropical coastal areas.

COMMON NAMES: gray nicker, neckar, nickel, nicker nut, scorcher.

2. Caesalpinia ciliata (Bergius *ex* Wikstr.) Urb., Symb. Ant. 2: 275. 1900. *Guilandina ciliata* Bergius *ex* Wikstr., Kongl. Vetensk. Acad. Handl. **1825**: 431. 1825.   Fig. 102D–F.

Caesalpinia bonduc sensu Griseb., Fl. Brit. W. I. 204. 1860, non L., 1753.

Guilandina melanosperma Eggers, Fl. St. Croix 46. 1879.

Caesalpinia divergens Urb., Symb. Ant. 2: 276. 1900.

Scandent shrub to 6 m long, profusely branching along main stem; stems cylindrical, to 2.5 cm diam., sparsely covered with recurved prickles. Leaves 20–35 × 10–28 cm; pinnae 3–10 pairs, the leaflets 4–8 pairs per pinna, 1.5–3 × 1–1.7 cm, elliptic, ovate, or rounded, chartaceous, glabrous or with scattered hairs,

midvein prominent on lower surface, the apex rounded and mucronate, the base obtuse or cuneate, slightly unequal, the margins entire, revolute, ciliate; petiolules short, cylindrical, tomentose; petioles and leaf rachis spiny, with a pair of recurved spines at the base of leaflets; stipules lanceolate-acuminate, 0.5–1 mm long. Flowers unisexual on axillary and terminal long racemes, the rachis rufous-tomentose, sparsely prickled; bracts lanceolate, 3.5–6 mm long, deciduous. Calyx forming a nearly bell-shaped hypanthium, 4–5 mm long, rufous-tomentose, the sepals reflexed, oblong; petals yellow, 4.5–6.5 mm long, spatulate; stamens 10, only 4 fertile in staminate flowers, all sterile in pistillate flowers, the filaments hairy; ovary sessile, rudimentary in staminate flowers, the stigma terminal. Legume oblong to nearly rounded, 5–7 × 3–5.5 cm, semi-inflated, the valves sparsely spiny, tardily dehiscent through both sutures. Seeds few, 1.5–2 cm long, ovoid to rounded, smooth, yellowish to orange-brown (less often black).

DISTRIBUTION: Along sandy beaches. Drunk Bay (A4134), Haulover Point (A4227). Also on St. Croix, Tortola, and Virgin Gorda; Hispaniola, Puerto Rico, and the Lesser Antilles.

COMMON NAME: yellow nicker.

3. Caesalpinia pulcherrima (L.) Sw., Observ. Bot. 166. 1791. *Poinciana pulcherrima* L., Sp. Pl. 380. 1753.

Erect shrub to 3 m tall; stems prickly. Leaves 18–35 × 14–22 cm; pinnae 5–8 pairs, the leaflets 4–9 pairs per pinna, 1–2.5 × 0.6–1.2 cm, oblong to oblanceolate, chartaceous, glabrous, the apex rounded or slightly notched, the base obtuse or cuneate, unequal, the margins entire; petiolules short, cylindrical, glabrous; petioles and leaf rachises usually without spines. Flowers bisexual on long, terminal racemes, the rachis glabrous; bracts minute, early deciduous. Calyx of 4 spreading, petal-like green to orange, glabrous sepals 1–2 cm long; petals yellow, red, or orange, 1.5–2.5 cm long, clawed; stamens 10, red, long-exserted, the filaments hairy below; ovary sessile, flattened, reddish, the stigma terminal. Legume oblong, 8–13 × 1.5–2 cm, flattened, glabrous, without spines, dehiscent through both sutures by twisting valves, the apex beaked. Seeds few, 0.8–1 cm long, flattened, cuneate, smooth, brown.

DISTRIBUTION: Cultivated throughout the island, sometimes escaping cultivation. Along road to Ajax Peak (A2655). Also on St. Croix, St. Thomas, and Tortola; native to South America but widely cultivated and sometimes naturalizing throughout the Caribbean.

COMMON NAMES: Barbados pride, dudeldu, dwarf poinciana.

2. CHAMAECRISTA Moench

Annual or perennial herbs or shrubs. Leaves paripinnate; leaflets opposite; petioles usually grooved along upper surface, pulvinulate at base and bearing 1 or 2 discoid or stipitate glands; rachis grooved, ending in a filiform segment, often with additional glands at the bases of the leaflets; stipules minute to foliaceous. Flowers bisexual, solitary or in a few species in supra-axillary racemes; pedicels bracteolate near or above middle; calyx of 5 unequal sepals; corolla yellow, of 5 free, heteromorphic petals; stamens 5–10, usually heteromorphic, the filaments shorter than the anthers; ovary 1-locular, sessile or shortly stipitate, with many ovules. Legume dehiscent through both sutures, the valves coiling; seeds many, flattened.

A tropical genus with 265 species.

Key to the Species of *Chamaecrista*

1. Perennial erect shrubs or subshrubs; corolla to 3 cm wide; pedicels 7–28 mm long... 1. *C. glandulosa* var. *swartzii*
1. Annual or biennial herbs, erect or decumbent, sometimes with semiwoody base; corolla to 1 cm wide; pedicels 0.5–4 mm long .... 2. *C. nictitans* var. *diffusa*

1. Chamaecrista glandulosa (L.) Greene var. **swartzii** (Wikstr.) H.S. Irwin & Barneby, Mem. New York Bot. Gard. **35**: 784. 1982. *Cassia swartzii*

Wikstr., Kongl. Vetensk. Acad. Handl. **1825**: 430. 1825.   Fig. 102G–I.

Erect subshrub to 1 m tall; stems solitary, branching at upper end. Leaves 2–6 cm long; leaflets 12–17 pairs, 6–12 × 1.5–2.5 mm, oblong to spatulate, chartaceous, glabrous or with a few scattered hairs on lower surface, the apex rounded with a minute mucro, the base unequal, rounded-cuneate, the margins entire; petioles with a stipitate gland at middle; rachis puberulent, without glands; stipules subulate 2–7 mm long. Flowers solitary or congested, to 3 cm wide; bracts awl-shaped, minute, persistent; pedicels 7–28 mm long. Calyx of linear-lanceolate sepals, 8–14 mm long; petals yellow, unequal, the lowest petal twice as large as the others, 1–1.3 cm long; stamens 10, unequal; ovary sessile, flattened, pubescent. Legume dark brown, 30–45 × 4–5 mm, linear-oblong, flattened, puberulent, with slightly wavy margins, beaked at apex. Seeds few, 2–3 mm long, flattened, rectangular, brown.

DISTRIBUTION: A common weed of open areas. Annaberg Ruins (A1926), Bordeaux (A3172), Maria Bluff (A598). Also on St. Croix, St. Thomas, Tortola, and Virgin Gorda; Puerto Rico and the Lesser Antilles.

2. Chamaecrista nictitans (L.) Moench subsp. nictitans var. diffusa (DC.) H.S. Irwin & Barneby, Mem. New York Bot. Gard. 35: 833. 1982. *Cassia diffusa* DC., Mém. Soc. Phys. Genève 2: 130. 1824. *Chamaecrista diffusa* (DC.) Britton, Ann. Missouri Bot. Gard. 2: 41. 1915.

Erect or decumbent herb, to 40 cm long, many-branched, sometimes from a semiwoody base. Leaves 1.5–6 cm long; leaflets 8–20 pairs, 6–16 × 1–2 mm, linear to oblong, chartaceous, glabrous or puberulent on lower surface, the apex rounded or truncate, with a minute mucro, the base unequal, rounded-cuneate, the margins entire, ciliate; petioles with a stipitate gland at middle; rachis puberulent, without glands; stipules subulate 3–11 mm long. Flowers solitary, to 1 cm wide; bracts awl-shaped, minute, persistent; pedicels 0.5–4 mm long. Calyx of linear-lanceolate sepals, 3.8–6 mm long; petals yellow, of equal length, 3.7–6 mm long; stamens 10, unequal; ovary sessile, flattened, pubescent. Legume 35–40 × 3–4 mm, dark brown, linear, flattened, puberulent, with slightly wavy margins, beaked at apex. Seeds few, ca. 2 mm long, flattened, nearly rectangular, brown.

DISTRIBUTION: A common weed of open areas. Fish Bay (A3893), Lameshur (B502). Also on St. Croix; throughout the West Indies, northern Venezuela, and Colombia.

3. DELONIX Raf.

Trees. Leaves twice pinnate; pinnae opposite, the leaflets opposite; petioles and rachis without nectariferous glands; stipules pinnatisect, deciduous. Flowers bisexual, large, showy, in axillary or terminal flat-topped racemes; pedicels elongate; calyx of 5 valvate sepals; corolla of various colors, of 5 free, long-clawed petals; stamens 10, the filaments free, of equal length, the anthers opening along longitudinal slits; ovary sessile with many ovules, the style filiform, the stigma truncate. Legume oblong, flattened, woody, the valves septate, elastically dehiscent through both sutures; seeds numerous, oblong, transversely oriented.

A genus of 3–8 species, from Madagascar, eastern Africa, and Asia.

1. Delonix regia (Bojer *ex* Hook.) Raf., Fl. Tellur. 2: 92. 1837. *Poinciana regia* Bojer *ex* Hook., Bot. Mag. 56: pl. 2884. 1829. Fig. 103.

Tree to 25 m tall; bark light brown, smooth. Leaves 22–50 cm long; pinnae 9–25 pairs; leaflets 10–35 pairs per pinna, 6–15 mm long, oblong or elliptic, chartaceous, glabrous, the apex and base acute or rounded, the margins entire; stipules deciduous. Flowers in axillary or terminal racemes; bracts ovate, 5–8 mm long, deciduous; pedicels 4–7 cm long. Calyx forming a nearly cup-shaped hypanthium, 4–12 mm long, the sepals oblong to elliptic, 2–3 cm long; petals orange, red, or rarely yellow, 4.5–6 cm long, long-clawed, obovate, the upper central petal slightly larger with whitish or yellowish spots; stamens long exserted, the filaments hairy on lower portion; ovary pubescent. Legume 30–60 × 4–5.5 cm, oblong, glabrous, dark brown, the valves tardily dehiscent. Seeds many, ca. 2 cm long, flattened-cylindrical, dark brown with light brown mottling.

DISTRIBUTION: A common tree, planted along roads and gardens. Along Center Line Road by km 7.5 (A2076). Native to Madagascar, now cultivated and naturalized throughout the tropics.

COMMON NAMES: flamboyant tree, poinciana.

4. HYMENAEA L.

Large, resinous trees. Leaves compound of 2 opposite leaflets; stipules early deciduous. Flowers bisexual, large, in terminal or axillary panicles; pedicels short, jointed at apex; calyx bell-shaped, with 4 or 5 imbricate sepals; corolla white, of 5 free, clawed or sessile, equal petals; stamens 10, the filaments free, of equal length, the anthers opening along longitudinal slits; ovary shortly stipitate, flattened, with few ovules, the style filiform, eccentric. Legume elongate, hard-shelled, woody, slightly flattened, indehiscent; seeds 1 to several, ellipsoid-flattened, hard, covered with a starchy pulp.

A genus of 14 species, most of which (13) are from tropical South America.

REFERENCE: Lee, Y.-T. & J. H. Langenheim. 1975. Systematics of the genus *Hymenaea* L. (Leguminosae, Caesalpinioideae, Detariae). Univ. Calif. Publ. Bot. 69: 1–109.

1. Hymenaea courbaril L., Sp. Pl. 1192. 1753. Fig. 104A–E.

Large tree 25 m tall; bark smooth, light brown to gray. Leaflets 2, 4–12 cm long, asymmetric, one margin ovate, the other elliptic, coriaceous, glabrous, the apex acute to acuminate, the base obtuse to rounded, asymmetric, the margins entire; petiole cylindrical, thickened, 0.7–1.5 cm long; stipules 2–2.5 cm long, covering the young leaf, early deciduous. Flowers in short axillary or terminal panicles, opening in the evening and dropping in the morning; pedicels 2–5 mm long; bracts minute. Calyx nearly bell-shaped, 1–2 cm long, the sepals 5, oblong, 1.2–1.8 cm long, velvety pubescent without, early deciduous, leaving a wide receptacle; petals white, 1.3–2 cm long, ovate; stamens exserted; ovary glabrous, oblong, the stigma capitate. Legume 5–15 × 3–6 cm, glabrous, oblong-ellipsoid, dark reddish brown, slightly flattened.

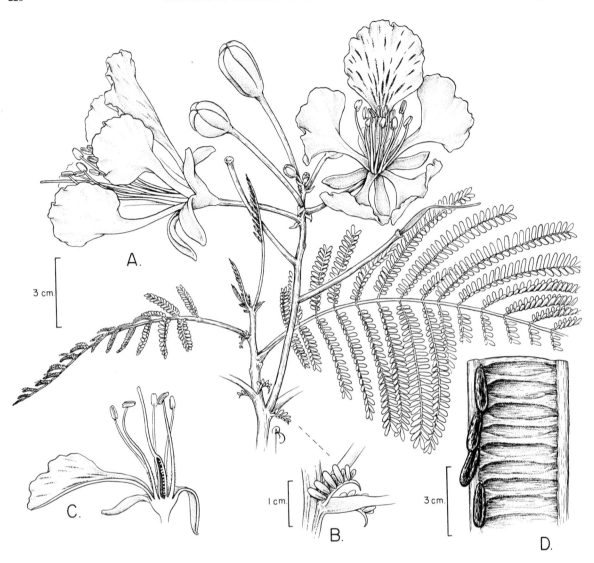

FIG. 103. A–D. *Delonix regia.* **A.** Flowering branch. **B.** Detail of stem showing pinnatisect stipules at base of leaf petiole. **C.** L.s. flower. **D.** Portion of dehisced legume showing seeds.

Seeds 2–6, 1.5–2.2 cm long, oblong-ellipsoid, dark brown, smooth, woody, surrounded by an edible, starchy, dry, odorous pulp.

DISTRIBUTION: Common tree of humid areas. Annaberg (A2868), Reef Bay Trail (P29270). Also on St. Croix, St. Thomas, Tortola, and Virgin Gorda; throughout tropical America.

COMMON NAMES: locust tree, stinking toe.

5. PARKINSONIA L.

Spiny shrubs or small trees. Leaves bipinnate; pinnae 1 or 2 pairs, opposite, the leaflets minute; stipules minute. Flowers bisexual, medium-sized, in axillary racemes; pedicels jointed near the apex; calyx of 5 nearly equal sepals; corolla yellow, of 5 free, spreading, equal petals; stamens 10, the filaments pubescent, free, of equal length, the anthers opening along longitudinal slits; ovary shortly stipitate, with few ovules, the style filiform. Legume nearly cylindrical, torulose, leathery, indehiscent; seeds many, nearly ellipsoid, hard.

A genus of 2 species from tropical America and Africa.

1. Parkinsonia aculeata L., Sp. Pl. 375. 1753.

Fig. 104F–I.

Shrub or small tree to 8 m tall; bark rough, fissured, light brown to gray; branches long-spiny. Leaves 15–30 cm long, main rachis 0.5–1 cm long, stiff, spin-like, persistent after pinnae have been shed; pinnae 1 or 2, with flattened, laminar, green rachis; leaflets numerous, alternate, 2–5 mm long, oblanceolate, chartaceous, glabrous, the apex obtuse to rounded, the base cuneate, the margins entire; stipules early deciduous, minute. Flowers in axil-

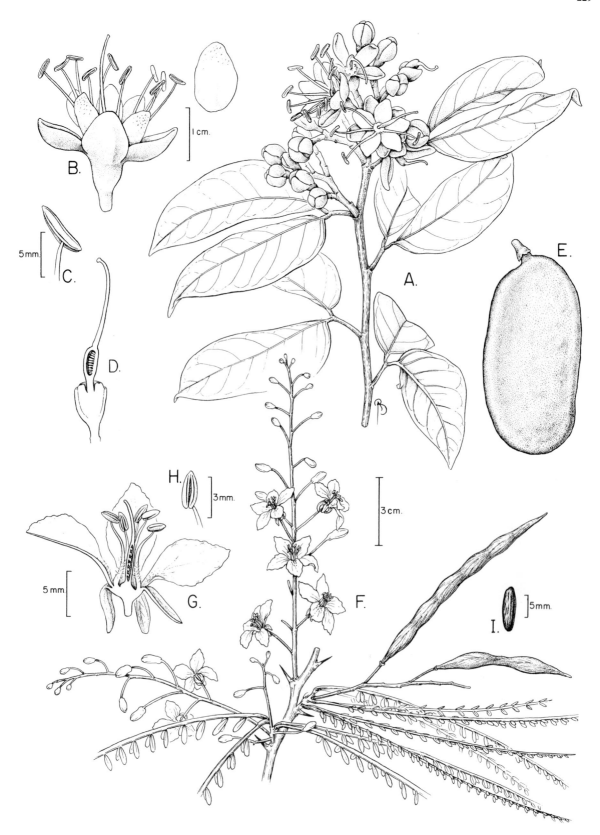

Fɪɢ. 104. A–E. *Hymenaea courbaril*. **A.** Fertile branch. **B.** Flower and petal. **C.** Anther. **D.** L.s. flower receptacle showing pistil. **E.** Legume. **F–I.** *Parkinsonia aculeata*. **F.** Fertile branch. **G.** L.s. flower. **H.** Anther. **I.** Seed.

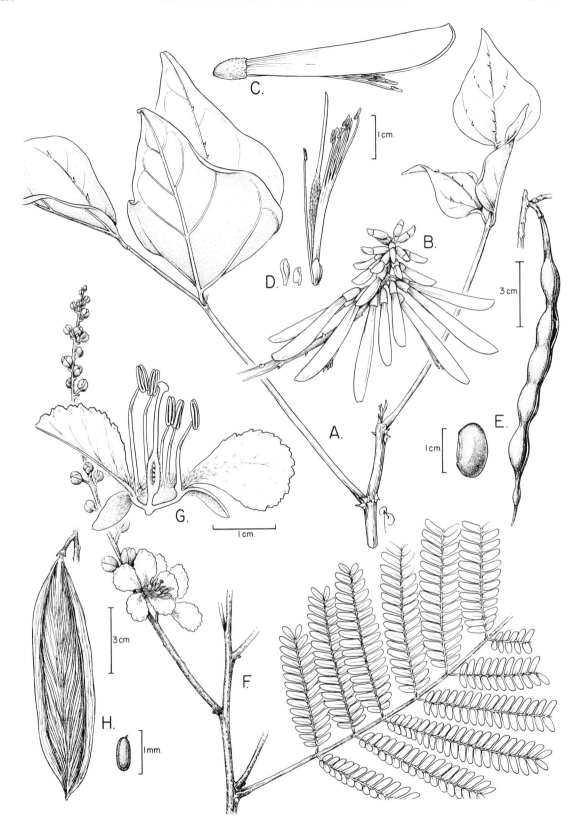

Fig. 105. A–E. *Erythrina eggersii*. A. Leafy branch. B. Inflorescence. C. Flower. D. Wing and keel petal and flower with standard removed showing pistil and diadelphous stamens. E. Seed and legume. F–H. *Peltophorum pterocarpum*. F. flowering branch. G. L.s. flower. H. Legume and seed.

lary racemes; pedicels 0.7–1.2 cm long; bracts minute, deciduous. Calyx forming a nearly cup-shaped hypanthium, 2–3 mm long, the sepals 5, oblong, 4–6 mm long, glabrous, reflexed, deciduous; petals yellow, 1.3–1.5 cm long, ovate, acute at apex, the lower central petal with red spots; stamens 10, the filaments shorter than the petals; ovary pubescent, oblong, the stigma punctate. Legume 4–14 cm long, linear, torulose, yellowish brown.

Seeds 1 to several, 8–9 mm long, ellipsoid, light brown, smooth, woody.

DISTRIBUTION: Common shrub of dry areas. Fortsberg (A4040), Lameshur (A736), Waterlemon Bay (A1932). Also on Anegada, Jost van Dyke, St. Croix, St. Thomas, Tortola, and Virgin Gorda; apparently native to southern United States and Mexico, cultivated throughout the Caribbean region.

COMMON NAMES: horse bean, Jerusalem thorn.

6. PELTOPHORUM Walp., nom. cons.

Trees. Leaves bipinnate; pinnae opposite, the leaflets opposite; stipules minute, deciduous. Flowers bisexual, small to large, in terminal racemes or panicles; bracts minute; pedicels short; calyx tubular, 5-lobed at apex; corolla yellow, irregular, of 5 unequal, free, clawed petals; stamens 10, the filaments free, pilose, the anthers dorsifixed, opening by longitudinal slits; ovary sessile, with 1 to few ovules. Legume oblong, elliptic to nearly lanceolate, winged along both margins, indehiscent; seeds 1–4, flattened.

A tropical and subtropical genus of 15 species.

1. Peltophorum pterocarpum (DC.) Baker *ex* K. Heyne, Nutt. Pl. Ned.-Ind., ed. 2, **2**: 755. 1927. *Inga pterocarpa* DC., Prodr. **2**: 441. 1825.

Fig. 105F–H.

Caesalpinia inermis Roxb., Fl. Ind., ed. 1832, **2**: 367. 1832. *Peltophorum inerme* (Roxb.) Llanos *in* Blanco, Fl. Filip., ed. 3, **4**: 69, t. 335. 1880.

Tree to 20 m tall; bark light brown with many lenticels; branches rusty-brown pubescent when young. Leaves 20–40 cm long; pinnae 11–30 pairs; leaflets 13–32 pairs, 1.4–2 × 0.5–0.8 cm, oblong, coriaceous, puberulent on lower surface, the apex rounded and notched, the base strongly asymmetric, rounded-cuneate, the margins entire; rachis cylindrical, furrowed, puberulent; stipules deciduous. Flowers in terminal panicles; pedicels 0.5–0.7 cm long; bracteoles 3–4 mm long, oblong. Calyx cup-shaped, 9–10 mm long, the sepals reflexed, rusty pubescent without, ovate, 5–7 mm long; petals 2–2.5 cm long, obovate, clawed, tomentose along the middle; stamens as long as the petals; ovary rusty tomentose, ovoid, the style elongate. Legume 7.5–10 cm long, oblong or slightly constricted between seeds, woody, reddish brown, with margins thinned into a wing, beaked at apex, indehiscent. Seeds 10–12 mm long, oblong, flattened, beige.

DISTRIBUTION: Mostly planted along roadsides but becoming naturalized. Cruz Bay (A4662). Also on St. Thomas and Tortola; native to tropical Asia and Australia but widely cultivated throughout the tropics.

7. SENNA Mill.

Trees, erect or climbing shrubs or herbs. Leaves pinnate; leaflets opposite; petioles and rachis grooved, usually with a stipitate gland; stipules persistent. Flowers bisexual, small to large, in axillary or terminal racemes or panicles or solitary; bracts minute to foliaceous; pedicels short to elongate; calyx of 5 sepals; corolla yellow, irregular, of 5 unequal, dimorphic, free, clawed petals; stamens 10, 3 usually smaller and not fertile, the filaments flattened, free, the anthers basifixed, opening by terminal pores; ovary shortly stipitate, with few ovules. Legume variously shaped, dehiscent or indehiscent; seeds many, variously shaped.

A genus of about 260 species, most of which are neotropical.

Key to the Species of *Senna*

1. Trees to 15 m tall; legumes 15–30 cm long .... 5. *S. siamea*
1. Erect or climbing shrubs, 0.5–3 m tall; legumes to 19 cm long.
  2. Legumes with 2 longitudinal wings .... 1. *S. alata*
  2. Legumes not winged.
    3. Scandent shrubs; legumes cylindrical, fleshy .... 2. *S. bicapsularis*
    3. Erect shrubs; legumes flattened, dry, semiwoody.
      4. Legumes turgid, nearly cylindrical (5–6 mm wide); leaf rachis with an elongate, stipitate gland between the lowermost pair of leaflets; petiole without glands .... 3. *S. obtusifolia*
      4. Legumes flattened (7–9 mm wide); leaf rachis without glands; petiole with a sessile, rounded gland at base .... 4. *S. occidentalis*

1. Senna alata (L.) Roxb., Fl. Ind., ed. 1832, **2**: 349. 1832. *Cassia alata* L., Sp. Pl. 378. 1753. *Herpetica alata* (L.) Raf., Sylva Tellur. 123. 1838.

Shrub to 3 m tall, many-branched from base. Leaves 20–35 cm long; leaflets opposite, 5–12 pairs, 6–16 × 3–8.5 cm, oblong to obovate, chartaceous, puberulent on lower surface, the apex obtuse to rounded, notched, the base strongly asymmetric, rounded-cuneate, the margins entire; rachis nearly quadrangular, without glands; stipules persistent, lanceolate to triangular, asymmetric at base, with one side nearly auriculate. Flowers congested along axillary racemes; pedicels 0.7–1.1 cm long; bracts 2.2–2.5 cm long, petal-like, yellow, early deciduous. Sepals yellow, petal-like, concave, oblong-obovate, 1–1.2 cm long, early deciduous; petals yellow, 1.5–2.2 cm long, concave, obovate, clawed, obtuse at apex, the lower central petal longer; stamens 10, dimorphic, 7 of them longer and fertile, all much shorter than the petals; ovary pubescent, the style incurved, the stigma green. Legume 10–17 cm long, oblong, with a longitudinal wing along dorsal side of each valve, septate, beaked at apex, dehiscent. Seeds wedge-shaped, ca. 6 mm long, brown.

DISTRIBUTION: Uncommon shrub of moist open areas, sometimes grown in gardens. Cruz Bay (A4063). Also on St. Croix, St. Thomas, and Tortola; native probably to the Guianas, Orinoco, and Amazon basins in Venezuela, Brazil, and Colombia but naturalized throughout the tropics.

COMMON NAMES: candle bush, Christmas candle, fleiti, golden candle-stick, ringworm bush, saparilla.

2. Senna bicapsularis (L.) Roxb., Fl. Ind., ed. 1832, 2: 342. 1832. *Cassia bicapsularis* L., Sp. Pl. 376. 1753.   Fig. 106A–F.

Scandent shrub to 3 m long, many-branched from base; branches horizontal or pendulous. Leaves 3–8 cm long; leaflets 2–4 pairs, 1.5–4 × 1–2.5 cm, obovate to oblanceolate, slightly succulent, glabrous, the apex truncate, notched and usually mucronate, the base asymmetric, cuneate to subcordate, the margins entire; rachis with a stipitate, ellipsoid, nectariferous gland between the lowest pair of leaflets; stipules lanceolate, minute, deciduous. Flowers in axillary racemes; pedicels 1–3.5 mm long, articulate; bracts lanceolate to awl-shaped, deciduous. Calyx forming am obconic hypanthium, 1.5–3.4 mm long, the sepals 5, oblong-elliptic, 5–8 mm long, glabrous, yellowish green; petals yellow, concave, 1–1.5 cm long, cuneate at base, the lower, central petal longer, obovate; the others oblong; stamens 10, dimorphic, 3 of them longer and fertile, the other 7 sterile, the filaments of all 10 shorter than the petals; ovary glabrous, the style incurved. Legume 8–19 × 1–2 cm, oblong, cylindrical, fleshy, septate, indehiscent. Seeds numerous, 4–6 mm long, ellipsoid, light brown, smooth.

DISTRIBUTION: Common roadside shrub. Adrian Ruins (A3121), Cruz Bay (B191), Fish Bay (A3917). Also on Anegada, St. Croix, St. Thomas, Tortola, and Virgin Gorda; probably native to the Greater Antilles, now widespread throughout the tropics.

COMMON NAMES: stiverbush, styver bla.

3. Senna obtusifolia (L.) H.S. Irwin & Barneby, Mem. New York Bot. Gard. 35: 252. 1982. *Cassia obtusi-folia* L., Sp. Pl. 377. 1753.

Emelista tora sensu Britton & P. Wilson, Bot. Porto Rico 5: 371. 1924, non L., 1753.

Erect subshrub to 50 cm tall. Leaves 4–8 cm long; leaflets 3 pairs, 3–5 cm long, oblanceolate, chartaceous, glabrous or pubescent, the apex obtuse, mucronulate, the base asymmetric, cuneate-rounded, the margins entire, ciliate; rachis with an elongate fusiform gland between the lowest pair of leaflets; stipules linear, lanceolate, 6–8 mm long, late deciduous. Flowers solitary or paired in axillary racemes; pedicels 1–2 cm long, pubescent; bracts lanceolate, deciduous. Calyx green, of 5 expanded, oblanceolate, unequal sepals, 5–9 mm long; petals yellow, 0.8–1.1 cm long, obovate, rounded at apex, the lower central petal larger, with notched apex; stamens 10, dimorphic, 7 of them fertile, the filaments shorter than the petals; ovary pubescent, oblong, the style slightly recurved, the stigma cup-shaped. Legume 7.5–15 ×

0.5–0.6 cm, linear, turgid, nearly cylindrical, straight, but flat-tened and strongly curved when young, semiwoody, dehiscent. Seeds 25–30, 3.8–4.5 mm long, rhombic, light brown, smooth.

DISTRIBUTION: Common roadside weed. Coral Bay Quarter along Center Line Road (A3995). Also on St. Croix, St. Thomas, and Tortola; weedy throughout the tropics.

4. Senna occidentalis (L.) Link, Handbuch 2: 140. 1831. *Cassia occidentalis* L., Sp. Pl. 377. 1753.

Erect herb to subshrub 0.5–1.5 m tall. Leaves 10–16 cm long, malodorous; leaflets 4–6 pairs, 4–9 cm long, lanceolate to lance-ovate, chartaceous, with a few glandular hairs on lower surface, the apex acuminate, the base asymmetric, rounded to cordate, the margins entire, ciliate; petiole with a sessile, rounded gland at base; stipules lanceolate, 4–6 mm long, deciduous. Flowers few in axillary racemes; pedicels 0.7–1 cm long, pubescent; bracts lanceolate, 1–1.5 cm long, deciduous. Calyx green, of 5 expanded, obovate to oblanceolate sepals, 5–9 mm long; petals yellow, 0.9–1.8 cm long, obovate, rounded at apex, the lower central petal larger, with notched apex; stamens 10, dimorphic, all of equal size, but only 6 of them fertile, the filaments of all shorter than the petals; ovary pubescent, oblong, the style recurved, the stigma pubescent. Legume ascending, 6–13 × 0.7–0.9 cm, linear, flat-tened, straight or slightly curved, coriaceous, beaked at apex, dark brown, with ribbed, light margins, semiwoody, dehiscent. Seeds 40–60, 4–5 mm long, ovate to rounded, compressed, light brown, smooth.

DISTRIBUTION: Common roadside weed. Along dirt road to Bordeaux (A1902), Fish Bay (A3905). Also on Anegada, St. Croix, St. Thomas, Tortola, and Virgin Gorda; weedy throughout the neotropics.

COMMON NAMES: stinking weed, wild coffee.

5. Senna siamea (Lam.) H.S. Irwin & Barneby, Mem. New York Bot. Gard. 35: 98. 1982. *Cassia siamea* Lam., Encycl. 1: 648. 1785. *Sciacassia siamea* (Lam.) Britton *in* Britton & Rose, N. Amer. Fl. 23(4): 252. 1930.

Tree to 15 m tall; bark smooth, grayish. Leaves 15–30 cm long; leaflets 3–10 pairs, 4.5–7.5 cm long, oblong-elliptic to oblong-lanceolate, chartaceous, glabrous or puberulent, the apex obtuse, truncate, notched or mucronate, the base rounded, the margins entire; rachis without glands; stipules awl-shaped, decidu-ous. Flowers in terminal panicles; pedicels 1.7–3.0 cm long; bracts lanceolate, 5–6 mm long, deciduous. Calyx of 5 expanded, unequal, concave, rounded, green sepals, 5–9 mm; petals yellow, 1–1.8 cm long, obovate to oblanceolate, clawed, rounded at apex, 2 of them smaller; stamens 10, of equal size, but only 7 fertile, the filaments of all shorter than the petals; ovary pubescent, oblong, the style curved. Legume 15–30 cm long, linear-oblong, flattened, woody, brown, the margins thickened, the valves with alternate mound and depressions over the seeds, dehiscent along both sutures. Seeds 6–8 mm long, lenticular, light brown, smooth.

DISTRIBUTION: Common roadside tree of humid areas. Cruz Bay (A1989). Also on St. Thomas; native to Burma and Thailand, cultivated or naturalized throughout the tropics.

COMMON NAME: yellow cassia.

8. TAMARINDUS L.

A monospecific genus characterized by the following species.

1. Tamarindus indica L., Sp. Pl. 34. 1753.
Fig. 106G–J.

Tree to 10 m tall. Leaves once pinnate; leaflets opposite, 1–2.3 cm long, chartaceous, oblong, the apex obtuse or truncate,

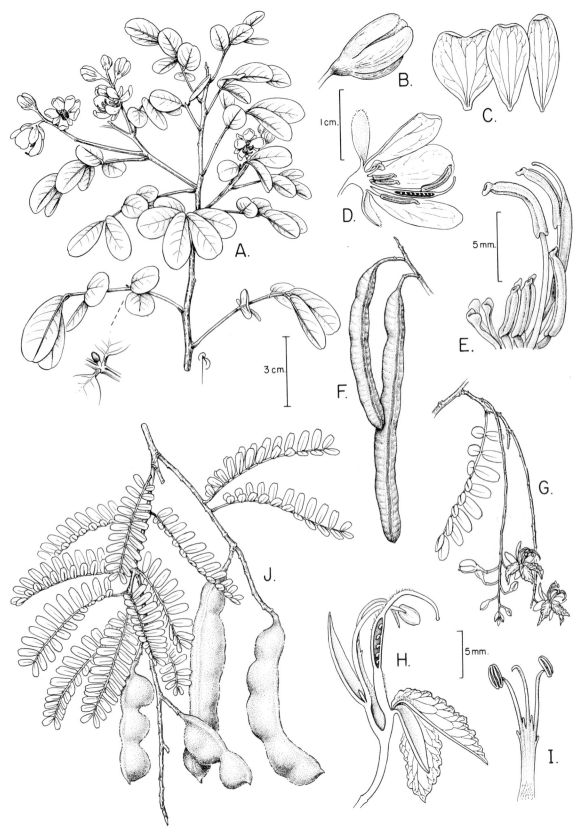

Fig. 106. A–F. *Senna bicapsularis*. A. Flowering branch and detail of leaf rachis showing nectariferous gland. B. Flowerbud. C. Petals. D. L.s. flower. E. Detail of stamens and pistil. F. Legumes. G–J. *Tamarindus indica*. G. Flowering branch. H. L.s. flower. I. Androecium. J. Fruiting branch.

the base asymmetric, rounded, the margins entire; stipules minute, early deciduous. Flowers bisexual, in axillary or terminal racemes; pedicels short; calyx forming a short funnel-shaped hypanthium at base, the sepals 4, expanded, petal-like, 0.8–1.1 cm long; corolla yellow, of 3 showy petals and 2 inconspicuous scale-like petals, the showy petals with reddish veins, dissimilar, the lateral ones larger, 1–1.3 cm long; stamens 7 or 8, 4–5 reduced to staminodes, the 3 fertile ones with filaments incurved, fused half of their length into a flattened structure, the anthers opening along longitudinal slits; ovary 1-locular, short–stipitate, with many ovules, the stipe adnate to the hypanthium, the stigma nearly capitate. Legume 10–15 cm long, oblong, straight or curved, with one torulose side, the walls crustaceous (hard and fragile), indehiscent, tan to light brown. Seeds several, 1–1.7 cm long, wedge-shaped, dark brown, shiny, embedded in brown, tart, edible pulp.

DISTRIBUTION: A common tree of open dry areas. Lameshur (A1810). Also on other islands of the Virgin Islands; apparently native to India, now cultivated and naturalized throughout the tropics.

COMMON NAME: tamarind.

CULTIVATED SPECIES: *Bauhinia monandra* Kurz (with pink flowers), *Bauhinia tomentosa* L. (with white or yellow flowers), and *Cassia fistula* L. (with golden-yellow flowers) are occasionally cultivated.

DOUBTFUL RECORDS: *Haematoxylum campechianum* L. was reported by Britton and Wilson (1924) and by Woodbury and Weaver (1987) as occurring on St. John, but I have not been able to locate any specimen to confirm these records.

II. Faboideae (Papilionoideae) (Pea Subfamily)

Trees, shrubs, herbs, vines, or lianas. Leaves alternate or seldom opposite, pinnate or trifoliolate; stipules and stipels usually present. Flowers bisexual, small to large, strongly irregular, produced in racemes, spikes, or seldom heads; calyx tubular, with 5 lobes or bilobed; corolla of 5 petals, the uppermost (standard) larger and usually spreading, the lateral ones (wings) usually smaller, reflexed or spreading, the central petals (keel) usually connate or imbricate enfolding the androecium and gynoecium; stamens 10, the filaments free or diadelphous, with 9 stamens connate into a laminar structure around the ovary and the tenth stamen free, or monadelphous with all filaments united into a tube, the anthers opening by longitudinal slits; ovary surrounded at base by an annular nectary disk.

This subfamily consists of approximately 440 genera and 12,000 species of cosmopolitan distribution.

Key to the Genera of Faboideae

1. Leaves trifoliolate.
    2. Twining vines or scandent herbs (*M. lathyroides* sometimes erect).
        3. Keel spirally twisted.
            4. Corolla burgundy; legume linear, cylindrical .... 27. *Macroptilium*
            4. Corolla pink to light purple; legume linear, flattened .... 28. *Phaseolus*
        3. Keel recurved, not spirally twisted.
            5. Corolla yellow.
                6. Corolla <1 cm long; underside of leaflet with numerous, yellow, resinous dots; legumes oblong, flattened .... 33. *Rhynchosia*
                6. Corolla >1.2 cm long; underside of leaflet without resinous dots; legumes linear, slightly flattened .... 39. *Vigna*
            5. Corolla pink, white, red, or lavender.
                7. Standard 3–4 cm wide; corolla lavender .... 15. *Centrosema*
                7. Standard <1.5 cm wide; corolla white, pink, red, or lavender.
                    8. Legumes to 5 cm long.
                        9. Corolla white .... 38. *Teramnus*
                        9. Corolla pink or red .... 22. *Galactia*
                    8. Legumes >6 cm long.
                        10. Legumes pubescent, 3.5–4 mm wide .... 32. *Pueraria*
                        10. Legumes glabrous, 1.5–4 cm wide.
                            11. Corolla white; terminal leaflet deltoid; legumes with verrucose margins .... 25. *Lablab*
                            11. Corolla pink; terminal leaflet elliptic; legumes with smooth margins .... 14. *Canavalia*
    2. Erect shrubs, trees, or prostrate herbs.
        12. Erect or prostrate herbs.
            13. Corolla pink, lavender, or violet, sometimes with light yellowish margins; legumes indehiscent, covered with hooked hairs (that stick to cloth), separating into 1-seeded segments, the apex apiculate .... 20. *Desmodium*
            13. Corolla yellow; legume dehiscent, glabrous, hooked at apex .... 36. *Stylosanthes*
        12. Shrubs or trees.
            14. Corolla bright red to red-orange, the standard tubular, 4–6 cm long; branches usually spiny .... 21. *Erythrina*
            14. Corolla yellow, the standard nearly rounded, reflexed, 2–3.5 cm long; branches not spiny.
                15. Legumes slightly inflated, laterally depressed between seeds; seeds few, rounded, slightly flattened .... 13. *Cajanus*

15. Legumes nearly cylindrical, inflated; seeds numerous, kidney-shaped .... 18. *Crotalaria*
1. Leaves pinnate or unifoliolate.
  16. Twining woody vines or scandent shrubs.
    17. Corolla blue-violet.. 16. *Clitoria*
    17. Corolla pink or pinkish purple.
      18. Twining woody vines to 3 m long; stems thin; leaflets opposite; seeds bright red with black
        spot... 9. *Abrus*
      18. Scandent shrubs, woody; leaflets alternate or only 1; seeds brown.
        19. Plants to 15 m long; stems to 12 cm diam., spiny; leaflets alternate; fruit circular-curled
          .... 26. *Machaerium*
        19. Plants 2–4 m long; stems to 3 cm diam., not spiny; leaflet only 1; fruit flattened, nearly
          circular .... 19. *Dalbergia*
  16. Shrubs, herbs, or trees, not climbing.
    20. Leaf with a single leaflet.
      21. Corolla <7 mm long, pink; legume cylindrical, splitting into cylindrical joints .... 11. *Alysicarpus*
      21. Corolla 1–1.7 cm long, yellow or lavender; legume slightly inflated, opening along one side
        .... 18. *Crotalaria*
    20. Leaf once-pinnate.
      22. Corolla yellow.
        23. Legume torulose .... 35. *Sophora*
        23. Legume not torulose.
          24. Leaflets ending in a sharp spiny tip... 29. *Pictetia*
          24. Leaflets not spiny tipped.
            25. Legume flattened, the lower margin notched between seeds .... 10. *Aeschynomene*
            25. Legume cylindrical .... 34. *Sesbania*
      22. Corolla pink.
        26. Trees to 10 m tall.
          27. Legumes indehiscent, ellipsoid to globose .... 12. *Andira*
          27. Legumes elongate, splitting along both sutures or breaking into segments.
            28. Legumes 4–winged, the wings papery, breaking into transverse segments .... 30. *Piscidia*
            28. Legumes flattened, not winged, opening along both sutures .... 23. *Gliricidia*
        26. Shrubs, small tree (to 6 m), or decumbent herbs.
          29. Legumes nearly cylindrical, linear, curved.. 24. *Indigofera*
          29. Legumes flattened, straight or curved at apex.
            30. Inflorescence terminal or opposite to leaves .... 37. *Tephrosia*
            30. Inflorescence axillary.
              31. Legumes transversely septate; plant herbaceous or scarcely woody (sub-
              shrub) .... 17. *Coursetia*
              31. Legumes not septate; plant a shrub or small tree.............................. 31. *Poitea*

9. ABRUS Adans.

Slender, twining, woody vines. Leaves once-pinnate; leaflets opposite; stipules minute to foliaceous. Flowers small, congested in swollen nodes along axillary or terminal pseudoracemes; bracts minute; pedicels short; calyx bell-shaped, with 5 teeth at apex; corolla pink, white, or reddish; stamens 9, the filaments united into a long tube; ovary nearly sessile, with many ovules, the style smooth, the stigma capitate. Legume flattened, dehiscent; seeds many, ellipsoid to nearly globose, bright red with a black spot.

A genus of 17 species native to the Old World tropics.

REFERENCE: Verdcourt, B. 1970. Studies in the Leguminosae-Papilionoideae for "Flora of tropical East Africa": II. Kew Bull. 24: 235–342.

1. Abrus precatorius L., Syst. Nat., ed. 12, **2:** 472. 1767.
Fig. 107A–E.

Twining slender, woody vine to 3 m long, many-branched from base. Leaves once-pinnate, 3–5 cm long; leaflets 8–15 pairs, 8–1.5 × 0.3–0.7 cm, oblong to oblanceolate, membranous, puberulent on lower surface, the apex rounded and mucronulate, the base rounded, the margins entire; rachis without glands, ending in a minute mucro; stipules linear, 3–5 mm long, persistent. Flowers nearly sessile, in axillary or terminal racemes; bracts minute, deciduous. Calyx bell-shaped, 3 mm long, puberulent;

corolla pink, the standard oval, to 1 cm long, wings and keel as long as the standard, clawed. Legume 2–3.5 cm long, oblong, slightly inflated, the margins slightly wavy, beaked at apex, dehiscent through both sutures. Seeds 3–6, ca. 6 mm long, subglobose, bright red with black spot at point of attachment.

DISTRIBUTION: Common roadside weed of disturbed areas. Great Cruz Bay (A2370). Also on St. Croix, St. Thomas, Tortola, and Virgin Gorda; native to the Old World, now naturalized throughout the tropics and subtropics.

COMMON NAMES: crab's eye, jumbie bead, rosary bead, scrubber, wild liquorice.

NOTE: Seeds of this vine are deadly poisonous if chewed and swallowed.

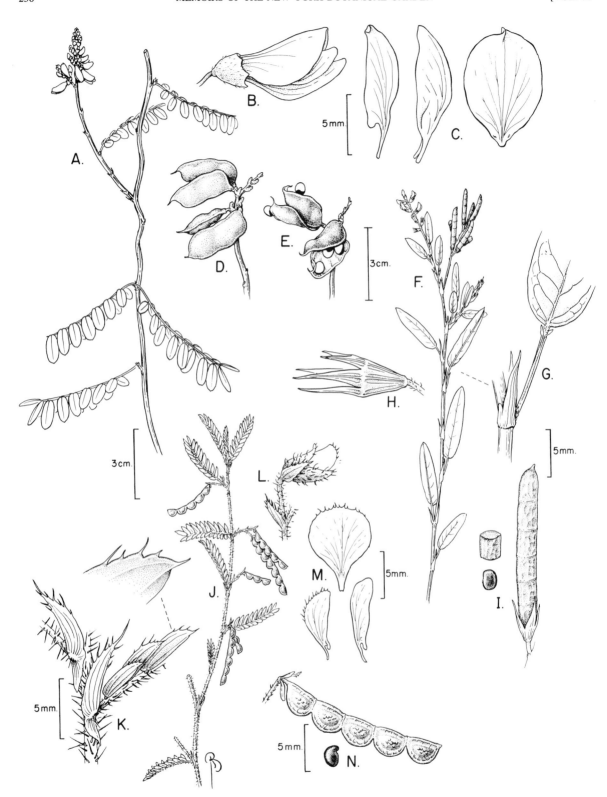

FIG. 107. A–E. *Abrus precatorius*. A. Flowering branch. B. Flower. C. Wing, keel, and standard petals. D. Legumes. E. Dehiscing legumes showing seeds. F–I. *Alysicarpus vaginalis*. F. Fertile branch. G. Detail of stem showing leaf petiole and stipules. H. Calyx. I. Legume section, seed, and legume. J–N. *Aeschynomene americana*. J. Fruiting branch. K. Detail of stem showing stipules at base of leaf petiole and detail of leaflet margin. L. Flower. M. Standard, wing, and keel petals. N. Legume and seed.

10. AESCHYNOMENE L.

Erect or trailing herbs or shrubs. Leaves pinnate; leaflets opposite, numerous; stipules peltate. Flowers small, in axillary or terminal racemes, bracteolate at base; bracts persistent; pedicels not jointed; calyx bell-shaped or bilabiate, with 5 unequal sepals; corolla yellow, red, or purple; stamens 10, diadelphous; ovary shortly stipitate, with few ovules, the style glabrous, the stigma capitate. Legume with 2 to many joints, flattened, indehiscent; seeds few, kidney-shaped, minute, dark brown or black.

A genus of 150 species, most of which are from the Americas and Africa.

REFERENCE: Rudd, V. E. 1955. The American species of *Aeschynomene*. Contr. U.S. Natl. Herb. 32: 1–172.

1. Aeschynomene americana L., Sp. Pl. 713. 1753.

Fig. 107J–N.

Aeschynomene glandulosa Poir. *in* Lam., Encycl. Suppl. 4: 76. 1816. *Aeschynomene americana* var. *glandulosa* (Poir.) Rudd, Contr. U.S. Natl. Herb. 32: 26. 1955.

Aeschynomene americana var. *depila* Millsp., Publ. Field Columbian Mus., Bot. Ser. 1: 494. 1902.

Erect subshrub to 75 cm tall, glabrous to hispid, many-branched from a woody base. Leaves 3–7 cm long, weakly sensitive (folding when touched); leaflets numerous, alternate, 5–15 × 1–2 mm, oblong, chartaceous, glabrous, the apex acute and mucronulate, the base strongly asymmetric, cuneate-cordate, the margins hispid; rachis without glands; stipules peltate, the upper portion lanceolate, 5–12 mm long, ciliate at margins, persistent. Flowers bisexual, in axillary or terminal racemes; bracts and bracteoles lanceolate, persistent. Calyx nearly bell-shaped, 3–5 mm long, hispid; corolla yellow, the standard widely ovate, with a reddish ring at base, to 1 cm long, the wings longer than the keel, clawed. Legume 10–30 × 3–4 mm, oblong, flattened, 3–9 articulate, glabrous. Seeds 2–3 mm long, kidney-shaped, brown.

DISTRIBUTION: Common roadside weed of humid areas. Fish Bay (A3892), Lameshur (B613), Rosenberg (B313). Also on St. Croix, St. Thomas, and Tortola; throughout warm areas of the New World.

11. ALYSICARPUS Desv.

Erect or decumbent herbs. Leaves unifoliolate; stipules free or connate; stipels usually present. Flowers small, in axillary or terminal pseudoracemes; bracts lanceolate, ca. 3 mm long, deciduous; pedicels not jointed at apex; calyx of 4 or 5 unequal, elongate sepals; corolla violet or pink, the standard nearly circular, larger than the keel and wings; stamens 10, monadelphous; ovary shortly stipitate, with few ovules, the stigma capitate. Legume nearly cylindrical, with numerous separating joints, indehiscent; seeds 1 per joint, nearly globose.

An Old World genus of 116 species.

1. Alysicarpus vaginalis (L.) DC., Prodr. 2: 353. 1825.

Hedysarum vaginale L., Sp. Pl. 746. 1753.

Fig. 107F–I.

Hedysarum nummularifolium L., Sp. Pl., ed. 2, 1051. 1763. *Alysicarpus nummularifolius* (L.) DC., Prodr. 2: 353. 1825.

Decumbent or trailing herb, many-branched from a taprooted base; branches to 50 cm long. Leaf blades 0.5–4 × 0.4–1.3 cm, rounded to lanceolate, chartaceous, glabrous or puberulent, the apex obtuse to rounded, seldom mucronulate, the base rounded to subcordate, the margins entire; stipules lanceolate, 4–6 mm long, straw-colored, persistent. Flowers bisexual; calyx obconical, 4–5 mm long, puberulent; corolla pink, the standard widely obovate, yellowish at base, to 1 cm long, the wings as long as the keel. Legume 1–2 cm long, linear, cylindrical, 3–8 articulate, puberulent, the articulations cylindrical, wrinkled, 1-seeded. Seeds 1 mm long, oblong to rounded.

DISTRIBUTION: Common roadside weed of humid areas. Susannaberg (A4157). Also on St. Croix, St. Thomas, Tortola, and Virgin Gorda; native to tropical Asia, now weedy throughout the tropics.

12. ANDIRA Juss.

Trees. Leaves pinnate; leaflets opposite or alternate; stipules minute, deciduous; stipels 1 per leaflet. Flowers small, in axillary or terminal panicles; bracts minute, deciduous; pedicels short; calyx bell-shaped, truncate, or with 5 minute teeth at apex; corolla pink to purple, the standard circular, wings and keel of similar size; stamens 10, diadelphous; ovary shortly stipitate, with few ovules, the stigma punctiform. Fruit globose to globose-ellipsoid, 1-seeded, indehiscent.

A genus of 20 species, with most species native to the neotropics.

1. Andira inermis (W. Wright) Kunth *ex* DC., Prodr. 2: 475. 1825. *Geoffraea inermis* W. Wright, Philos. Trans. 62: 513. 1777.

Fig. 108.

Geoffraea jamaicensis W. Wright, Philos. Trans. 62: 512. 1777. *Andira jamaicensis* (W. Wright) Urb., Symb. Ant. 4: 298. 1905.

Vouacapoua americana sensu Millsp., Publ. Field Columbian Mus., Bot. Ser. 1: 1902, non Aubl. 1775.

Tree to 15 m tall; bark moderately rough, grayish to brown. Leaves imparipinnate, 16–39 cm long; leaflets 9–15, one distal on rachis, the others opposite, 4–15 × 2–7.6 cm, oblong or elliptic, chartaceous, glabrous, the apex short- to long-acuminate, the base obtuse to rounded, the margins entire; petiolules swollen; stipels minute, filiform; stipules awl-shaped, 3 mm long, early deciduous. Panicles as long as the leaves; bracts minute, early deciduous; pedicels short. Calyx bell-shaped, 3–4 mm long, minutely toothed, ferruginous-tomentose, sometimes pinkish-tinged;

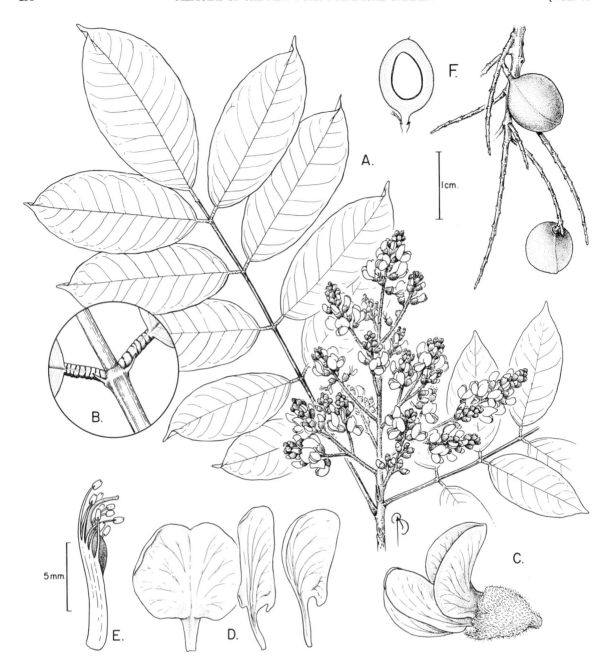

FIG. 108. *Andira inermis*. **A.** Flowering branch. **B.** Detail of leaf rachis node showing stipels and swollen petiolules. **C.** Flower. **D.** Standard, wing, and keel petals. **E.** Monadelphous stamens and pistil. **F.** L.s. fruit and fruiting branch.

corolla pink to purple, the standard nearly circular, to 1 cm long, involute, the wings as long as the keel; ovary elliptic, glabrous to puberulent, the style incurved, the stigma punctiform. Legume 3–3.5 × 2.5 cm, ellipsoid to subglobose, glabrous, the walls green and fleshy, usually eaten by bats, 1-seeded.

DISTRIBUTION: Common tree of moist areas, abundant in secondary forests or along roadsides. Emmaus (A3775), trail to Sieben (A2083). Also on St. Croix, St. Thomas, and Tortola; throughout warm areas of the New World.

COMMON NAMES: bastard mahogany, dog almond, hon kloot, pig turd.

13. CAJANUS DC., nom. cons.

Shrubs. Leaves trifoliolate; stipules persistent. Flowers medium-sized, in axillary or terminal pseudoracemes; bracts early deciduous; pedicels elongate; calyx bell-shaped, of 5 minute sepals; corolla yellow, the standard clawed, auriculate at base, wings and keel of similar size; stamens 10, monadelphous; ovary sessile, pubescent, the style curved, the stigma capitate. Legume elongate, flattened, dehiscent; seeds 3–7, circular.

A genus of 3 species, native to tropical Africa. The following species is widely cultivated throughout the tropics for its edible seeds.

1. Cajanus cajan (L.) Millsp., Publ. Field Columbian Mus., Bot. Ser. **2**: 53. 1900. *Cytisus cajan* L., Sp. Pl. 739. 1753.   Fig. 109A–E.

Cajan cajan (L.) Huth, Helios **11**: 133. 1893, nom. inadmiss.

Shrub to 3 m tall. Leaflets 2.5–10 × 1.5–3.5 cm, elliptic to lanceolate, chartaceous, lower surface densely covered with whitish, woolly hairs, and a few yellow, resinous dots, the apex acute, the base cuneate, the margins entire; petioles 1.5–5 cm long; stipules ovate, to 3 mm long, persistent. Flowers bisexual, in densely ferruginous-tomentose pseudoracemes; bracts densely ferruginous-tomentose; pedicels 1–1.5 cm long. Calyx nearly bell-shaped, 8–10 mm long, tomentose, glandular; corolla yellow, the standard widely ovate, reddish without, to 1.5 cm long, the wings as long as the keel, clawed. Legume 4.5–10 × 0.7–1.3 cm, oblong, glandular-pubescent, slightly inflated with constrictions between seeds, tardily dehiscent. Seeds 6–8 mm long, depressed-globose, turning from green to light brown.

DISTRIBUTION: Spontaneous after cultivation. Cruz Bay (A1952). Also on St. Croix, St. Thomas, and Tortola; presumably native to Africa, widely cultivated throughout the tropics.

COMMON NAMES: gungo pea, pigeon pea, vendu bountje.

14. CANAVALIA DC., nom. cons.

Woody to herbaceous twining or trailing vines. Leaves trifoliolate; stipules minute, deciduous. Flowers in axillary or terminal pseudoracemes; bracts 2 per flower, deciduous; pedicels borne on swollen nodes; calyx bell-shaped, with 5 minute, unequal sepals, 2 larger; corolla pink, purple, or violet, the standard obovate, clawed, auriculate at base, reflexed, thickened at base along midvein, the wings and keel of similar size; stamens 10, monadelphous; ovary sessile or stipitate, pubescent, the style filiform, the stigma capitate. Legume elongate and flattened or slightly inflated, longitudinally keeled along both sutures and on the ventral side, dehiscent or indehiscent; seeds 3 to many, usually bean-shaped.

A genus of 50 species, most of which are native to the New World.

REFERENCE: Sauer, J. D. 1964. Revision of *Canavalia*. Brittonia 16: 106–181.

1. Canavalia rosea (Sw.) DC., Prodr. **2**: 404. 1825. *Dolichos roseus* Sw., Prodr. 105. 1788.

Fig. 109L–P.

Dolichos maritimus Aubl., Hist. Pl. Guiane **2**: 765. 1775. *Canavalia maritima* (Aubl.) Urb., Repert. Spec. Nov. Regni Veg. **15**: 400. 1919. *Canavalia maritima* Thouars, J. Bot. Agric. **1**: 80. 1813.

Dolichos obtusifolius Lam., Encycl. **2**: 295. 1786, non Jacq., 1768. *Canavalia obtusifolia* (Lam.) DC., Prodr. **2**: 404. 1825.

Trailing or less often twining vine to 6 m long, many-branched from a woody base. Leaves trifoliolate; leaflets 4–10.4 × 3–10 cm, oblong, ovate, to orbicular, chartaceous, puberulent, the apex obtuse, rounded, notched, and mucronulate, the base broadly cuneate to rounded, the margins entire; rachis 3–4.5 cm long; petioles 4–6 cm long; stipules triangular, persistent. Flowers paired at swollen nodes along erect, elongate pseudoracemes; bracts minute, deciduous. Calyx 8–11 mm long, greenish, puberulent; corolla pink to purplish pink, the standard oval, 2–2.5 cm long, reflexed, white at center, wings and keel shorter than the standard, clawed. Legume 10–17 × 2.3–2.5 cm, oblong, nearly woody, ribbed along suture, beaked at apex, dehiscent along both sutures, the valves twisting. Seeds many, 1.5–1.8 cm long, ellipsoid, slightly compressed, brown, with white hilum.

DISTRIBUTION: A common coastal vine. Chocolate Hole (A776), Little Lameshur Bay (A2018). Also on Tortola and Virgin Gorda; throughout warm coastal areas of the world.

COMMON NAME: bay bean.

15. CENTROSEMA Benth., nom. cons.

Herbaceous twining vines. Leaves trifoliolate; stipules minute, persistent. Flowers large, axillary, solitary or a few on long, bracteate peduncles; bracts appressed to the calyx; calyx bell-shaped, with 5 elongate, unequal or nearly equal sepals; corolla pink, lavender, or white, the standard ovate to rounded, clawed, the wings and keel of similar length, much smaller than the standard; stamens 10, diadelphous or monadelphous; ovary nearly sessile, the style curved, pubescent, the stigma capitate or truncate. Legume linear, flattened, with thickened or ribbed margins, dehiscent along both sutures, the valves twisting; seeds many, small, bean-shaped.

A genus of about 45 species, with pantropical distribution.

1. Centrosema virginianum (L.) Benth., Comm. Legum. Gen. 56. 1837. *Clitoria virginiana* L., Sp. Pl. 753. 1753.   Fig. 109F–K.

Twining woody vine to 2 m long; stems slender, wiry. Leaves trifoliolate; leaflets 2.5–6.5 × 1–4 cm, very variable in shape, lanceolate, ovate, or oblong, chartaceous, glabrous except for a few hairs along veins, the apex acute or seldom obtuse, mucronulate, the base broadly cuneate to rounded, the margins entire; rachis 0.7–1.5 cm long; petioles 2.3–2.8 cm long; stipules linear, straw-colored, persistent. Flowers solitary or paired, on elongate peduncles; bracts ovate, persistent. Calyx 10–15 mm long, greenish, puberulent, the sepals nearly equal; corolla lavender, the standard widely ovate, 3–4 cm wide, white at center, with magenta stripes, the wings and keel clawed, much shorter than the standard. Legume 10–13 × 0.3–0.4 cm, linear-oblong, ribbed along both

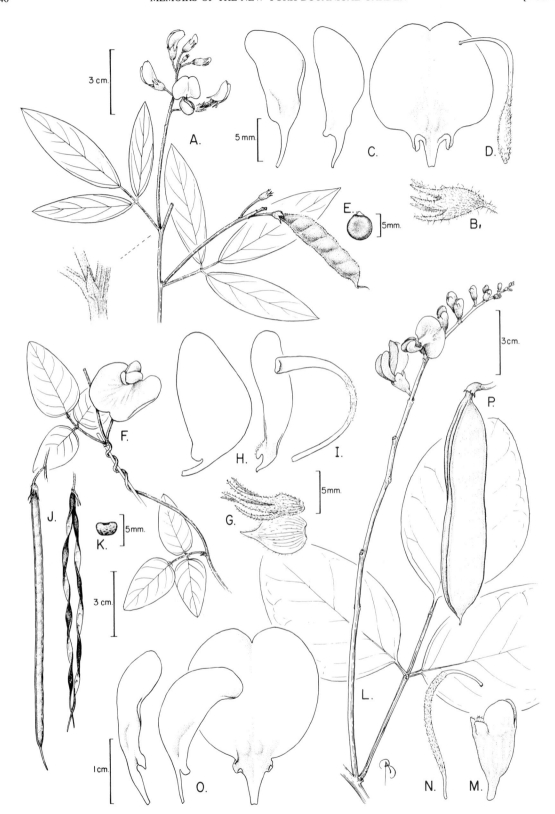

Fɪɢ. 109. A–E. *Cajanus cajan*. A. Fertile branch and detail of stem showing stipule at base of leaf petiole. B. Calyx. C. Wing, keel, and standard petals. D. Pistil. E. Seed. F–K. *Centrosema virginianum*. F. Flowering branch. G. Calyx and subtending bract. H. Wing and keel petals. I. Pistil. J. Legume and twisted valves of legume after dehiscence. K. Seed. L–P. *Canavalia rosea*. L. Flowering branch. M. Calyx. N. Pistil. O. Keel, wing, and standard petals. P. Legume.

margins, long-beaked at apex. Seeds numerous, ca. 2 mm long, bean-shaped, flattened, brown.

DISTRIBUTION: Common weed throughout the island. Fish Bay (A3896),

trail to Solomon Bay (A2318). Also on all islands of the Virgin Islands; native throughout tropical and subtropical America.

COMMON NAME: wist vine.

16. CLITORIA L.

Herbaceous to woody, twining vines or trees. Leaves imparipinnate; leaflets 3–9, opposite; stipules minute, persistent. Flowers large, solitary or a few in axillary racemes, bracteoles large, persistent, appressed to the base of the calyx; calyx nearly bell-shaped, fluted, with 5 unequal or nearly equal sepals; corolla blue-violet, white, or red, the standard rounded, wrinkled, longer than the other petals, the wings longer than the keel; stamens 10, diadelphous or monadelphous; ovary stipitate, the style curved, pubescent inside, the stigma truncate. Legume linear to oblong, flattened, dehiscent; seeds few, rounded to oblong.

A genus of about 30 species, most of which are from the New World.

REFERENCE: Fantz, P. R. 1990. *Clitoria* (Leguminosae) Antillarum. Moscosoa 6: 152–166.

1. Clitoria ternatea L., Sp. Pl. 753. 1753.

Fig. 110A–D.

Twining vine to 2 m long. Leaves imparipinnate; leaflets 5–7, opposite, 1.5–4.5 × 1–3.5 cm, elliptic or seldom ovate or oblong, chartaceous, puberulent, the apex mucronate, obtuse, the base obtuse, the margins entire; rachis 2–7 cm long; petioles 2–4 cm long; stipules linear, straw-colored, persistent. Flowers solitary on short peduncles; bracts ovate, persistent. Calyx 15–20 mm long,

greenish, puberulent, the sepals nearly equal; corolla bluish violet, the standard widely ovate, 3.5–5 cm long, notched at apex, the wings and keel clawed, much shorter than the standard. Legume 9–11 × 1 cm, oblong, ribbed along both margins, long-beaked at apex, dehiscent by twisting valves. Seeds numerous, 5–6 mm long, oblong, flattened, dark brown.

DISTRIBUTION: Spontaneous after cultivation. Johnson Bay (A2623). Also on St. Croix, St. Thomas, and Tortola; native to the Old World but cultivated and naturalized throughout the tropics.

COMMON NAMES: blue vine, butterfly pea.

17. COURSETIA L.

Erect or scandent shrubs, perennial herbs, or trees. Leaves impari- or paripinnate or unifoliolate; leaflets opposite, usually sericeous or woolly; stipules minute, persistent, or deciduous; stipels usually present. Flowers small, in axillary racemes or panicles, the bracts minute, deciduous or persistent, the bracteoles wanting; calyx bell-shaped, with 5 elongate sepals; corolla of various colors, the standard rounded to kidney-shaped, notched at apex, auriculate, the wings and keel of similar length, the keel acute at apex; stamens 10, diadelphous or monadelphous; ovary sessile, pubescent, with 12–30 ovules, the style curved, pubescent, the stigma capitate. Legume linear to oblong, flattened, with straight or sinuous margins and fine cross-septa between the seeds, dehiscent; seeds slightly compressed.

A genus of 38 species from the neotropics.

REFERENCE: Lavin, M. 1988. Systematics of *Coursetia* (Leguminosae-Papilionoideae). Syst. Bot. Monogr. 21: 1–167.

1. Coursetia caribaea (Jacq.) Lavin, Adv. Leg. Syst. **3**: 63. 1987. *Galega caribaea* Jacq., Select. Stirp. Amer. Hist. 212, t. 125. 1763. *Cracca caribaea* (Jacq.) Benth. *in* Benth. & Oerst., Vidensk. Meddel. Dansk Naturhist. Foren. Kjøbenhavn **1853**: 9. 1854.

Fig. 110K–M.

Erect subshrub to 1.5 m tall, silky pubescent, especially on young parts. Leaves imparipinnate; leaflets 9–17, opposite, 0.5–3 cm long, elliptic, membranous, densely covered with silky hairs, the apex rounded to acute, mucronate, the base rounded to cuneate,

the margins entire; stipules linear, 5–8 mm long, persistent; stipels minute. Racemes 1 to few-flowered; bracts deciduous. Calyx 5–5.5 mm long, silky, the sepals awl-shaped; corolla white to pale pink, the standard clawed, orbicular, 10–12 mm long, the wings and keel clawed, as long as the standard. Legume 4–6 × 0.25 cm, linear, impressed between the seeds, reddish brown, dehiscent through both sutures by twisting valves. Seeds numerous, ca. 2 mm, nearly quadrangular, slightly flattened, reddish brown.

DISTRIBUTION: Occasional in wooded understory. Caneel Bay (A4117), Frank Bay (A4219). Also on St. Croix, St. Thomas, and Tortola; Puerto Rico, Hispaniola, the Lesser Antilles, and from Mexico to northern South America.

18. CROTALARIA L.

Erect herbs or shrubs. Leaves imparipinnate or unifoliolate; leaflets 3–7, opposite; stipules minute to large, persistent; stipels wanting. Flowers small to medium-sized, in terminal or axillary racemes, the bracts minute, deciduous or persistent, paired bracteoles at the base of the calyx tube; calyx bell-shaped, with 5 elongate, unequal sepals; corolla yellow, blue, or pinkish, the standard rounded to kidney-shaped, notched at apex, sometimes with cordate base, the wings and keel of similar length, the keel acute at apex; stamens 10, monadelphous; ovary sessile, pubescent, with many ovules, the style curved, bearded, the stigma capitate. Legume cylindrical or nearly so, linear to oblong, dehiscent; seeds small, ellipsoid.

A genus of about 550 species, most of which are from Africa.

Key to the Species of *Crotalaria*

1. Leaves simple.
    2. Corolla light violet to lavender; leaves broadly ovate; stipules foliaceous; legume tan when dry
    .... 5. *C. verrucosa*

Fig. 110. **A–D.** *Clitoria ternatea*. **A.** Flowering branch. **B.** Flower. **C.** Keel and wing petals. **D.** Twisted valves of legume after dehiscence. **E–J.** *Crotalaria pallida* var. *ovata*. **E.** Flowering branch. **F.** Flower. **G.** Keel, wing, and standard petals. **H.** Monadelphous stamens. **I.** Pistil. **J.** Legume and twisted valve. **K–M.** *Coursetia caribaea*. **K.** Fertile branch. **L.** Flower. **M.** Wing, standard, and keel petals.

2. Corolla yellow; leaves oblanceolate to spatulate; stipules minute, early deciduous; legume blackish when dry .... 4. *C. retusa*
1. Leaves trifoliolate.
3. Inflorescence terminal (sometimes appearing opposite to leaves), longer than subtending leaf.
4. Inflorescence 15–25 cm long, densely flowered .... 3. *C. pallida* var. *ovata*
4. Inflorescence 5–15 cm long, sparsely flowered .... 1. *C. incana*
3. Inflorescence axillary, shorter than the subtending leaf.. 2. *C. lotifolia*

1. Crotalaria incana L., Sp. Pl. 716. 1753.

Erect subshrub, 30–70 cm tall, many-branched from base or branched only distally; branches pilose when young. Leaves trifoliolate; leaflets 1.5–5 × 0.9–4 cm, obovate to elliptic, chartaceous, lower surface with minute, appressed hairs, the apex mucronulate, rounded, the base cuneate to obtuse, the margins entire; petioles 2–5 cm long; stipules early deciduous. Pseudoracemes 5–15 cm long, terminal, sparsely flowered; bracts minute, early deciduous. Flowers hanging; calyx 7–9 mm long, pubescent; corolla yellow, the standard nearly obovate, 10–13 mm long, the keel incurved, as long as the standard, the wings shorter than the standard. Legume 2–3.5 × 0.7–1 cm, hanging, oblong, pubescent, slightly inflated, impressed along both sutures, dehiscent, tan-colored. Seeds numerous, ca. 2 mm wide, kidney-shaped to cordate, reddish brown, shiny.

DISTRIBUTION: A common weed of open disturbed areas. Coral Bay (A4244), Maho Bay Quarter along Center Line Road (A1853). Also on St. Croix, St. Thomas, Tortola, and Virgin Gorda; widespread throughout the tropics.

COMMON NAME: rattle bush.

2. Crotalaria lotifolia L., Sp. Pl. 715. 1753.

Erect subshrub to 1 m tall, branching along main stem; young parts densely sericeous. Leaves trifoliolate; leaflets 1–4 × 0.4–0.9 cm, oblanceolate, chartaceous, lower surface densely sericeous, the apex rounded or notched, the base attenuate, the margins entire; petioles 0.5–3.5 cm long, slender; stipules early deciduous. Pseudoracemes to 2 cm long, axillary, few-flowered; bracts minute, early deciduous. Calyx 5–7 mm long, sericeous; corolla yellow, the standard ovate, clawed, 9–11 mm long, with reddish veins, the keel and wings as long as the standard. Legume 2–3 × 0.4–0.7 cm, hanging, oblong, pubescent, slightly inflated, impressed along both sutures, dehiscent, tan. Seeds numerous, ca. 3 mm long, kidney-shaped, light brown, dull.

DISTRIBUTION: A common weed of open dry areas. White Cliffs (A2038), along trail from Lameshur to Reef Bay (A2719). Also on Anegada, St. Croix, St. Thomas, and Virgin Gorda; throughout the West Indies, Mexico, and Central America.

NOTE: The original spelling of this name was published as *latifolia* but was corrected to *lotifolia* by subsequent authors.

3. Crotalaria pallida Dryand. var. ovata (G. Don) Polhill, Kew Bull. 22: 365. 1968. *Crotalaria ovata* G. Don, Gen. Hist. 2: 138. 1832.   Fig. 110E–J.

Crotalaria falcata Vahl *ex* DC., Prodr. 2: 132. 1825.

Erect subshrub to 1.5 m tall, branching along main stem; branches pubescent. Leaves trifoliolate; leaflets 3–7 × 1.5–4 cm, elliptic to obovate, chartaceous, lower surface with minute, appressed hairs, the apex mucronulate, notched, the base cuneate, the margins entire; petioles 4–5.5 cm long; stipules early decidu-

ous. Pseudoracemes 15–25 cm long, terminal, densely flowered; bracts minute, early deciduous. Flowers hanging; calyx 5–6 mm long, greenish, densely covered with silky hairs; corolla yellow, the standard elliptic to ovate, 10–12 mm long, with reddish veins, the keel incurved, acuminate, as long as the standard, the wings shorter than the standard. Legume 3–4 × 0.5–0.6 cm, hanging, linear, slightly inflated, impressed along both sutures, tardily dehiscent, tan when dry. Seeds numerous, ca. 2 mm wide, kidney-shaped, light brown, shiny.

DISTRIBUTION: A common weed of open, disturbed areas. Emmaus (A3931), Lameshur (A2026). Also on St. Croix; Bahamas and Puerto Rico (including Vieques); native to the Old World tropics.

4. Crotalaria retusa L., Sp. Pl. 715. 1753.

Erect subshrub to 1 m tall, branching near the base or along main stem; young parts sericeous. Leaves simple; blades 3–10 × 1–3.5 cm, oblanceolate to spatulate, chartaceous, lower surface densely sericeous, the apex mucronulate, rounded or notched, the base attenuate, the margins entire; petioles 2–3 mm long, slender; stipules minute, persistent. Racemes to 25 cm long, terminal, few-flowered; bracts minute, early deciduous. Calyx broadly bell-shaped, 12–14 mm long, puberulent; corolla yellow, the standard ovate, clawed, to 18 mm long, with reddish veins within, the keel and wings as long as the standard. Legume 2.5–4.5 × 1.1–1.3 cm, oblong, glabrous, nearly cylindrical, impressed along both sutures, dehiscent, blackish at maturity. Seeds numerous, ca. 4 mm long, kidney-shaped to cordate, dark brown, shiny.

DISTRIBUTION: A common weed of open disturbed areas. Cruz Bay (A3085), Bovocop Point (A1975). Also on St. Croix, St. Thomas, Tortola, and Virgin Gorda; originally from Asia, now pantropical.

COMMON NAMES: rattle box, yellow lupine.

5. Crotalaria verrucosa L., Sp. Pl. 715. 1753.

Erect subshrub to 1 m tall, branching along main stem; branches 3- to 4-angled; young parts sericeous. Leaves simple; blades 3–12 × 1.7–7.5 cm, broad ovate, chartaceous, lower surface sericeous to puberulent, the apex mucronulate, obtuse and often notched, the base attenuate, cuneate to rounded, the margins entire or slightly wavy; petioles 2–4 mm long, slender; stipules sickle-shaped to ovate, foliaceous, 6–18 mm long, persistent. Racemes to 15 cm long, terminal, erect, few-flowered; bracts minute, early deciduous. Calyx bell-shaped, 0.8–10 mm long, puberulent; corolla blue-violet to lavender, the standard ovate, clawed, to 15 mm long, lavender, the keel and wings shorter and darker than the standard. Legume 3–4 × 0.9–1.1 cm, oblong, hanging, nearly cylindrical, impressed along both sutures, sparsely covered with appressed hairs, dehiscent, tan. Seeds numerous, ca. 5 mm long, kidney-shaped, greenish brown.

DISTRIBUTION: A common weed of open disturbed areas. Coral Bay (A2621), Lameshur (A2044). Also on St. Croix, St. Thomas, Tortola, and Virgin Gorda; originally from Asia, now pantropical.

19. DALBERGIA L.f., nom. cons.

Trees, scandent shrubs, or lianas. Leaves imparipinnate or unifoliolate; leaflets alternate; stipules minute, deciduous. Flowers small, in short, axillary or terminal racemes or panicles, the bracts and bracteoles minute, deciduous or persistent; calyx bell-shaped, with 5

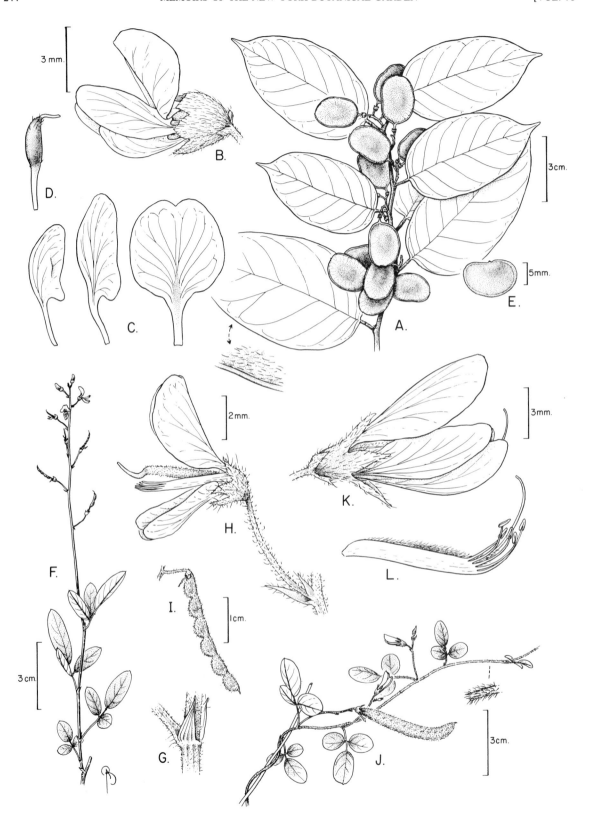

FIG. 111. A–E. *Dalbergia ecastaphyllum*. **A.** Fruiting branch and detail of leaf margin showing trichomes. **B.** Flower. **C.** Keel, wing, and standard petals. **D.** Pistil. **E.** Seed. **F–I.** *Desmodium incanum*. **F.** Fertile branch. **G.** Detail of stem showing stipules at base of leaf petiole. **H.** Flower. **I.** Legume. **J–L,** *Galactia dubia*. **J.** Habit and detail of stem showing indument. **K.** Flower. **L.** Monadelphous stamens.

short to elongate, unequal sepals; corolla white, yellow, or pink, the standard rounded to ovate, notched at apex, narrowed at base, the wings usually longer than the keel; stamens 9 or 10, all fused or one of them free; ovary stipitate, pubescent, with few ovules, the style usually curved, the stigma minute. Legume flattened, linear to rounded, usually with membranous margins, indehiscent; seeds small, lenticular.

A primarily tropical genus with about 100 species.

1. Dalbergia ecastaphyllum (L.) Taub. *in* Engl. & Prantl, Nat. Pflanzenfam. **3(3):** 335. 1894. *Hedysarum ecastaphyllum* L., Syst. Nat., ed. 10, **2:** 1169. 1759. Fig. 111A–E.

Hecastophyllum brownei sensu Duss, Fl. Phan. Antill. Franç. 22. 1897, non Pers., 1807.

Arching or scandent shrub, 1–3 m long, many-branched from base. Leaves unifoliolate; blades 6–12 × 4.5–8 cm, ovate or oblong, coriaceous, glabrous, the apex obtuse to shortly acuminate, the base rounded to nearly cordate, the margins entire; petioles 5–10 mm long; stipules awl-shaped to lanceolate, 0.5–1 cm long, late deciduous. Panicles 1–3 cm long, densely ferruginous-tomentose; bracts minute, early deciduous. Calyx bell-shaped, ca. 2.5 mm long, ferruginous-tomentose; corolla pinkish to white, the standard nearly rounded, clawed, notched at apex, to 7 mm long, the wings longer than the standard. Legume 2–2.3 × 1.5–2 cm, nearly rounded, flattened, turning from green to brown, indehiscent. Seed 1, to 1.7 cm long, oblong.

DISTRIBUTION: A common coastal species. White Cliffs (A2039). Also on St. Croix, St. Thomas, Tortola, and Virgin Gorda; throughout tropical America and tropical Africa.

20. DESMODIUM Desv., nom. cons.

Erect or prostrate perennial herbs. Leaves mostly trifoliolate; stipules minute, straw-colored, persistent or deciduous; stipels present. Flowers small, in axillary pseudoracemes or panicles, the bracts and bracteoles minute, deciduous or persistent; calyx bell-shaped, with 5 short sepals; corolla of various colors, but mostly pink, yellow, or bluish, the standard oblong to rounded, notched at apex, narrowed at base, the wings and keel of similar length, the wings appressed against the keel; stamens 10, diadelphous or all fused; ovary stipitate or sessile, pubescent, with few ovules, the style inflexed, the stigma minute. Legume a linear, flattened or twisted loment, the valves deeply constricted between the seeds along the margins, or only the ventral one, indehiscent, but the segments individually breaking away; the individual segments or the whole legume usually sticking to the cloth or fur of transient people or animals (because of hooked hairs); seeds small, bean-shaped.

A genus of 300 species distributed in temperate and tropical regions.

Key to the Species of *Desmondium*

1. Legumes notched between segments along one margin.
2. Plant creeping, rooting at the nodes; leaflets <1 cm long .... 4. *D. triflorum*
2. Plant decumbent or scandent, seldom rooting at the nodes; leaflets 3–9 cm long .... 2. *D. incanum*
1. Legumes notched between segments along both margins.
3. Legumes puberulent, with 1 fertile segment .... 1. *D. glabrum*
3. Legumes with hooked hairs and 4–8 fertile segments .... 3. *D. procumbens*

1. Desmodium glabrum (Mill.) DC., Prodr. **2:** 338. 1825. *Hedysarum glabrum* Mill., Gard. Dict., ed. 8. 1768.

Hedysarum molle Vahl, Symb. Bot. **2:** 83. 1791. *Desmodium molle* (Vahl) DC., Prodr. **2:** 232. 1825. *Meibomia mollis* (Vahl) Kuntze, Revis. Gen. Pl. **1:** 198. 1891.

Erect subshrub to 1 m tall, branching along main stem. Leaves trifoliolate; leaflets 3.5–7.5 × 1.5–4.5 cm, ovate to lanceolate, chartaceous, pubescent, the apex acute to obtuse, the base cuneate to rounded, the margins entire or crenulate; petioles 2–4 cm long; stipules ovate to lanceolate, 5–6.5 mm long, late deciduous; stipels awl-shaped, minute. Flowers fascicled along axillary or terminal pseudopanicles; peduncles to 8 cm long, densely glandular pubescent; bracts minute. Calyx nearly bell-shaped, ca. 1.5–2 mm long, pubescent; corolla pink. Legume notched along both sutures, puberulent; segments 3, the fertile one 6–8 mm long, elliptic, emarginate on one side. Seeds 2.5–3.5 mm long, oblong-elliptic.

DISTRIBUTION: A weedy species of open disturbed areas. Rendezvous Pt. (A3990). Also in St. Croix, St. Thomas, Tortola, and Virgin Gorda; a West Indian and circum-Caribbean species.

2. Desmodium incanum DC., Prodr. **2:** 332. 1825. Fig. 111F–I.

Hedysarum canum J.F. Gmel., Syst. Nat. **2:** 1124. 1791, nom. illegit. *Desmodium canum* (J.F. Gmel.) Schinz & Thell. *in* Schellenb., Schinz & Thell., Mém. Soc. Sci. Nat. Neuchâtel **5:** 371. 1913. *Hedysarum supinum* Sw., Prodr. 106. 1788, non Vill., 1779. *Meibomia supina* (Sw.) Britton, Ann. New York Acad. Sci. **7:** 83. 1892.

Decumbent or ascending subshrub to 1 m long, many-branched from a woody base. Leaves trifoliolate; leaflets 3–9 × 0.8–3.8 cm, elliptic, ovate, oblong, or lanceolate, chartaceous, puberulent, especially on lower surface, the apex mucronulate, acute to obtuse, the base obtuse to rounded, the margins entire; petioles 0.7–3 cm long; stipules ovate to lanceolate, 6–10 mm long, persistent; stipels awl-shaped, minute. Pseudoracemes terminal, few-flowered, 10–18 cm long, pubescent; bracts minute, persistent. Calyx nearly bell-shaped, 2.5–3 mm long, pubescent; corolla pink to pinkish purple, the standard ovate, 5–6 mm long, the wings and keel as long as the standard. Legume 2–4 cm long, of 5–8 fertile segments, puberulent, notched only along one suture. Seed 1 per segment, 2.5–3.5 mm long, oblong-elliptic, light brown.

DISTRIBUTION: Common in understory of woodlands and in open disturbed areas. Solomon Bay (A2291), East End (A2785). Also on St. Croix, St. Thomas, Tortola, and Virgin Gorda; a pantropical species.

3. Desmodium procumbens (Mill.) Hitchc., Annual Rep. Missouri Bot. Gard. **4**: 76. 1893. *Hedysarum procumbens* Mill., Gard. Dict., ed. 8. 1768.

Hedysarum spirale Sw., Prodr. 107. 1788, nom. illegit. *Desmodium spirale* (Sw.) DC., Prodr. **2**: 332. 1825. *Meibomia spiralis* (Sw.) Kuntze, Revis. Gen. Pl. **1**: 198. 1891.

Decumbent herb to 50 cm long, many-branched from a sub-woody base. Leaves trifoliolate; leaflets 1.5–9 × 0.4–1 cm, oblong-elliptic to lanceolate, chartaceous, puberulent, the apex mucronulate, obtuse, the base obtuse to rounded, the margins entire; petioles 1–5 cm long; stipels awl-shaped, minute; stipules ovate to awl-shaped, to 6 mm long, persistent. Pseudoracemes axillary, few-flowered, 10–15 cm long, pubescent; bracts minute, persistent. Calyx nearly bell-shaped, 1.5–2 mm long, puberulent; corolla greenish; ovary torulose, the style abruptly incurved. Legume 2–3 cm long, twisted, densely covered with hooked hairs, of 4–5 fertile segments, notched between seeds along both margins, the segments rhomboid. Seed 1 per segment, ca. 2 mm long, oblong-elliptic.

DISTRIBUTION: A weed of open disturbed areas. Coral Bay (A3999). Also on St. Croix, St. Thomas, and Tortola; a West Indian and circum-Caribbean species, introduced into the Philippines and tropical Africa.

4. Desmodium triflorum (L.) DC., Prodr. **2**: 334. 1825. *Hedysarum triflorum* L., Sp. Pl. 749. 1753. *Meibomia triflora* (L.) Kuntze, Revis. Gen. Pl. **1**: 197. 1891. *Sagotia triflora* (L.) Duch. & Walp., Linnaea **23**: 738. 1850.

Prostrate, spreading herb to 50 cm long, with numerous lateral branches, rooting at nodes; young parts pilose. Leaves trifoliolate; leaflets 5–10 × 5–11 mm, obovate to obcordate, coriaceous, pilose on lower surface, the apex notched or truncate, the base cuneate, the margins entire; petioles 3–7.5 mm long; stipules 2.5–3 mm long, asymmetrically ovate, with truncate base, persistent. Racemes axillary, 2–3 cm long, with 1–4 pairs of flowers. Calyx nearly bell-shaped, 2.5–3 mm long, pilose; corolla violet, the standard to 3 mm long. Legume 2–3 cm long, flattened, curved, densely covered with hooked hairs, of 3–6 fertile articulations, notched between seeds along one of the margins. Seed 1 per segment, ca. 2 mm long, lenticular, light brown.

DISTRIBUTION: Common roadside weed. Bordeaux area (A2616), Fish Bay Gut (A3898), Susannaberg (A4158). Also on St. Croix, St. Thomas, Tortola, and Virgin Gorda; a pantropical weed.

DOUBTFUL RECORD: *Desmodium scorpiurus* (Sw.) Desv. was reported by Woodbury and Weaver (1987) as occurring on St. John, but no specimen has been located to confirm this record.

21. ERYTHRINA L.

Trees or shrubs, usually armed with woody thorns. Leaves trifoliolate; stipules minute, deciduous; stipels minute, often glandular. Flowers large, showy, clustered in axillary or terminal pseudoracemes, the bracts and bracteoles minute, deciduous or persistent; calyx nearly bell-shaped or tubular, truncate at apex; corolla red, orange, or yellow, the standard elongate, oblong, involute, narrowed at base, much longer than the other petals, keel petal coherent; stamens 10, monadelphous or diadelphous; ovary stipitate, with 2 to several ovules, the style curved, the stigma capitate. Legume nearly cylindrical, constricted between seeds, dehiscent; seeds bean-shaped, often bright orange.

A genus of 100 species, with tropical distribution.

Key to the Species of *Erythrina*

1. Leaflets without spines (cultivated) .... *E. corallodendrum* (cultivated)
1. Leaflets with spines on both surfaces .... 1. *E. eggersii*

1. Erythrina eggersii Krukoff & Moldenke, Phytologia **1**: 289. 1938. Fig. 105A–E.

Erythrina horrida Eggers, Fl. St. Croix 45. 1879, non DC., 1825.

Small tree or shrub to 7 m tall; trunk, stems, branches and leaves with conical spines. Leaves trifoliolate; leaflets 8–10 × 6–7.5 mm, broadly ovate to triangular, chartaceous, with yellowish, conical spines on both surfaces, the apex shortly acuminate, the base obtuse to truncate, the margins entire; petiolules swollen, with a pair of elongate glands at base; petioles and rachis spiny; stipules triangular, straw-colored, persistent. Racemes terminal,

5–10 cm long; bracts minute, persistent. Calyx nearly bell-shaped, truncate, 6–8 mm long; corolla bright red, the standard to 4–5 cm long, the wings and keel very small; stamens diadelphous. Legume to 12 cm long, woody, slightly flattened, deeply constricted between seeds, beaked at apex. Seeds to 12 mm long, oblong-elliptic, bright red.

DISTRIBUTION: A rare or nearly endangered species, found in few localities. Bordeaux (A1923). Also on St. Croix and St. Thomas; Puerto Rico (including Vieques).

COMMON NAME: cockspur.

CULTIVATED SPECIES: *Erythrina corallodendrum* L. is known from a few individuals growing along roads.

22. GALACTIA P. Browne

Herbaceous to woody twining vines. Leaves trifoliolate or unifoliolate; petiolules and petioles swollen at base; stipules minute, deciduous; stipels minute or absent. Flowers small, clustered in axillary or terminal pseudoracemes, the bracts and bracteoles minute, deciduous or persistent; calyx bell-shaped, of 4 elongate sepals; corolla mostly pink to lavender or white, less often red, the standard elliptic to rounded, reflexed, narrowed at base, the wings coherent to the keel; stamens 10, diadelphous, unequal; ovary pubescent, sessile, with many ovules, the style curved and glabrous, the stigma capitate. Legume flattened, linear, often slightly curved, beaked at apex, dehiscent by twisting valves; seeds small, few, ovoid, brown.

A genus of 50 species, primarily distributed in tropical regions of the New World.

Key to the Species of *Galactia*

1. Corolla bright red.. 2. *G. eggersii*
1. Corolla pink to lavender.
   2. Legumes 5–5.5 mm wide; standard 12–15 mm long .... 1. *G. dubia*
   2. Legumes 6–9 mm wide; standard 8–10 mm long .... 3. *G. striata*

1. Galactia dubia DC., Prodr. **2**: 238. 1825.

Fig. 111J–L.

Herbaceous to woody twining vine to 2 m long, covered with appressed whitish hairs; stems slender, wiry, pubescent. Leaves trifoliolate; leaflets 1–4 × 0.8–2 cm, elliptic, oblong, or obovate, subcoriaceous, pubescent and paler on lower surface, the apex mucronate, notched or rounded, the base rounded, the margins entire; petioles 1–2 cm long; stipules lanceolate, 2–3 mm long, persistent; stipels wanting. Pseudoracemes axillary, few-flowered, 1–4 cm long, pubescent; bracts minute, persistent. Calyx nearly bell-shaped, 5–10 mm long, pilose; corolla iight pink to lavender, the standard oblong-elliptic, greenish at base within, 12–15 mm long, reflexed, the wings and keel slightly shorter than the standard. Legume 3–6 cm long, 5–5.5 mm wide, oblong-linear, flattened, slightly curved, pubescent. Seeds ca. 4 mm long, bean-shaped, dark brown.

DISTRIBUTION: Common in open disturbed areas and in coastal scrub or forests. Fish Bay (A3114), Maria Bluff (A2331), Salt Pond (M17062). Also on Anegada, St. Croix, St. Thomas, and Tortola; Puerto Rico (including Mona, Vieques, Culebra, and Icacos), the Lesser Antilles.

COMMON NAME: iron weed.

2. Galactia eggersii Urb., Symb. Ant. **2**: 311. 1900.

Herbaceous twining vine to 2 m long, covered with appressed whitish hairs when young; stems slender, wiry, pubescent, becoming glabrous. Leaves trifoliolate; leaflets 1–3 × 0.7–2.1 cm, elliptic to oblong, subcoriaceous, pubescent and paler on lower surface, the apex mucronate, notched, or rounded, the base rounded, the margins entire; petioles 0.5–2 cm long; stipules lanceolate, 2–3 mm long, persistent; stipels awl–shaped. Pseudoracemes axillary, 1- to few-flowered, 1–2.5 cm long, pubescent;

bracts minute, persistent. Calyx nearly bell-shaped, 8–12 mm long, pilose, the sepals unequal; corolla bright red, the standard elliptic, reflexed, 14–18 mm long, the wings and keel as long as the standard; style white. Legume 5–6 cm long, 5–6 mm wide, oblong-linear, flattened, slightly curved, pubescent, with wavy margins. Seeds 5 mm long, kidney-shaped, dark brown, shiny.

DISTRIBUTION: Occasional in dry woodlands and in open, disturbed areas. Carolina (A4138). Endemic to Guana Island, St. John, St. Thomas, and Tortola.

3. Galactia striata (Jacq.) Urb., Symb. Ant. **2**: 320. 1900. *Glycine striata* Jacq., Hort. Bot. Vindob. **1**: 32, t. 76. 1770.

Galactia berteriana DC., Prodr. **2**: 238. 1825.

Semiwoody twining vine to 5 m long; stems slender, wiry, and pubescent. Leaves trifoliolate; leaflets 3–6 × 2–3 cm, elliptic to ovate, chartaceous, pubescent on both surfaces, the apex mucronate and rounded or seldom obtuse, the base rounded, the margins entire; petioles to 4.5 cm long; stipules filiform, 2–3 mm long, persistent; stipels awl-shaped, minute. Pseudoracemes axillary, few- to several-flowered, 4–14 cm long, densely pubescent, the flowers 2 or 3 at swollen nodes; bracts minute, persistent. Calyx to 10 mm long, pubescent; corolla pink to lavender, the standard 8–10 mm long, elliptic, purple, white-striped, yellowish at base within. Legume 4–8 cm long, 6–9 mm wide, oblong, flattened, slightly curved at apex, pubescent. Seeds 4–5 mm long, flattened, kidney-shaped to oblong, dark brown, dull.

DISTRIBUTION: A common species from open disturbed areas or secondary moist forests. Lameshur (B607), Maho Bay Quarter along Center Line Road (A3816). Also on St. Croix, St. Thomas, Tortola, and Virgin Gorda; throughout the neotropics.

23. GLIRICIDIA Kunth

Trees. Leaves imparipinnate; leaflets opposite; stipules minute, persistent; stipels wanting. Flowers medium-sized, clustered in axillary racemes, the bracts and bracteoles minute; calyx bell-shaped, truncate or with minute teeth at apex; corolla pink, the standard elliptic to rounded, reflexed, notched at apex, narrowed at base, with 2 callosities by the midvein at base, the wings as long as the keel; stamens 10, diadelphous; ovary stipitate, with many ovules, the style inflexed and glabrous, the stigma capitate. Legume flattened, oblong, or widening near the beaked apex, dehiscent; seeds rounded, flattened.

A genus of 9 species native to Central and South America.

1. Gliricidia sepium (Jacq.) Kunth *ex* Walp., Repert. Bot. Syst. **1**: 679. 1842. *Robinia sepium* Jacq., Enum. Syst. Pl. 28. 1760.

Fig. 112F–I.

Tree to 10 m tall. Leaves imparipinnate; leaflets opposite, 7–15, 3–6 × 2–2.5 cm, elliptic, oblong, or lanceolate, chartaceous, pubescent on lower surfaces, the apex acute, the base acute to rounded, the margins entire; petiolules to 4.5 mm long; stipules ovate, 2–3 mm long, persistent. Racemes nearly terminal on lateral

branches, many-flowered, 12–15 cm long, puberulent; bracts minute, persistent. Calyx to 4–5 mm long, truncate, puberulent; corolla pink (to lavender), the standard to 16 mm long, obovate, notched at apex. Legume 7–15 × 1.2–1.7 cm, oblong, flattened, glabrous, beaked, dehiscent by twisting valves. Seeds to 1 cm long, circular, flattened, light brown.

DISTRIBUTION: Not very common, persistent after cultivation. Caneel Bay (A4119). Native to South America but widespread in cultivation (usually reproduced by cuttings) throughout the neotropics.

COMMON NAMES: pea tree, quick stick.

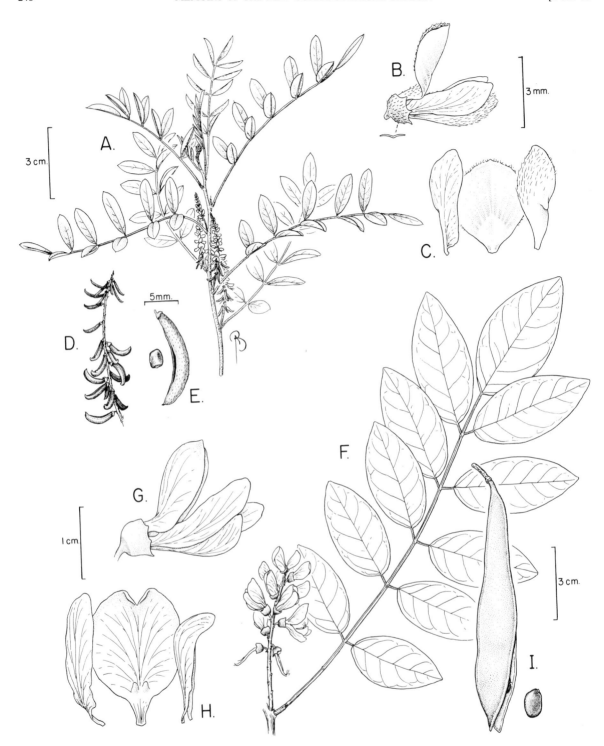

FIG. 112. A–E. *Indigofera suffruticosa*. A. Fertile branch. B. Flower and detail of trichome. C. Wing, standard, and keel petals. D. Infructescence. E. Seed and legume. F–I. *Gliricidia sepium*. F. Flowering branch. G. Flower. H. Wing, standard, and keel petals. I. Dehiscing legume and seed.

24. INDIGOFERA L.

Herbs or shrubs. Leaves imparipinnate; leaflets 3 to several, opposite or alternate; stipules minute, persistent; stipels minute. Flowers small, in axillary racemes or spikes, not clustered, the bracts minute; calyx cup-shaped with 5 minute, subequal teeth at apex; corolla pink or red, the standard rounded or obovate, narrowed at base, the wings as long as the keel, adherent to the keel petal; stamens 10, diadelphous, the anthers glandular at apex; ovary sessile, with 1 to many ovules, the style inflexed, the stigma capitate. Legume cylindrical, flattened, or 4-angular, linear, often curved, with septa between seeds, dehiscent; seeds rounded, ellipsoid, or angular.

A genus of 700 species, of warm temperate zones.

Key to the Species of *Indigofera*

1. Legumes 3–3.5 cm long, nearly straight...2. *I. tinctoria*
1. Legumes 1–1.5 cm long, strongly curved ....1. *I. suffruticosa*

1. Indigofera suffruticosa Mill., Gard. Dict., ed. 8. 1768.
Fig. 112A–E.

Erect shrub to 1.5 m tall, many-branched from base or along main stem. Leaves imparipinnate; leaflets 9–13, opposite, 1.5–3.7 × 0.5–1.7 cm, oblanceolate, elliptic, or oblong, membranous, pubescent especially on lower surface, the apex mucronate, acute to obtuse, the base tapering, the margins entire; petiolules 1–1.5 mm long; stipules awl-shaped, 2–3 mm long, persistent. Racemes axillary, many-flowered, 6–10 cm long, sericeous; bracts minute, persistent. Calyx bell-shaped, 1 mm long, puberulent; corolla pink, the standard to 5 mm long, wide-elliptic, the wings and keel as long as the standard; style curved, the stigma capitate. Legume 1–1.5 cm long, strongly curved, cylindrical, pubescent, beaked, dehiscent along one suture. Seeds ca. 2 mm long, square to oblong.

DISTRIBUTION: A common roadside weed. Bordeaux (A3183), Coral Bay Quarter and along Center Line Road (A4246). Also on St. Croix, St. Thomas, Tortola, and Virgin Gorda; native to the neotropics, introduced into the Old World.

2. Indigofera tinctoria L., Sp. Pl. 751. 1753.

Erect shrub to 1.5 m tall, many-branched from base. Leaves imparipinnate; leaflets opposite, 11–13, 1.5–2.5 × 0.7–1 cm, oblong, elliptic, or oblanceolate, membranous, puberulent on lower surface, the apex mucronulate, rounded or seldom obtuse, the base tapering to obtuse, the margins entire; petiolules 1–1.5 mm long; stipules awl-shaped, 2–3 mm long, persistent. Racemes axillary, many-flowered, 5–10 cm long, sericeous; bracts minute, persistent. Calyx bell-shaped, 1–1.5 mm long, pubescent; corolla pink, tomentose without, the standard to 5 mm long, broadly elliptic, the wings and keel as long as the standard. Legume 3–3.5 cm long, curved only at apex, cylindrical, becoming glabrous, tardily dehiscent; seeds ca. 2 mm long, square to oblong.

DISTRIBUTION: An occasional roadside weed. Round Bay (W667). Also on St. Croix and St. Thomas; native to the Old World tropics, now naturalized throughout the tropics.

COMMON NAME: indigo.

25. LABLAB Adans.

A monospecific genus, characterized by the following species.

1. Lablab purpureus (L.) Sweet, Hort. Brit., ed. 1, 481. 1826. *Dolichos purpureus* L., Sp. Pl., ed. 2, **2:** 1021. 1763.
Fig. 113A–F.

Dolichos lablab L., Sp. Pl. 725. 1753.

Herbaceous to subwoody twining vine to 5 m long. Leaves trifoliolate; leaflets 5–16 × 4.5–12 cm, broadly ovate to rhomboid, chartaceous, pubescent especially along veins, the apex acute to acuminate, the base cuneate or truncate, unequal on lateral leaflets, the margins entire, ciliate; petiolules swollen, pubescent; stipules lanceolate, to 5 mm long, persistent; stipels awl-shaped, minute. Pseudoracemes axillary, erect, few-flowered, to 25 cm long, the flowers 2 or 3 at swollen nodes; bracts minute, persistent. Calyx bell-shaped, 6–7 mm long, with 4–5 unequal sepals; corolla white or purple, the standard 1.4–2 cm long, rounded, the wings obovate, the keel incurved; stamens diadelphous; ovary flattened, the style curved, the stigma terminal. Legume 5–10 × 2–3 cm, nearly oblong, widened at apex, flattened, the upper suture warty, beaked at apex, tardily dehiscent through both sutures. Seeds 3–5, to 1 cm long, ovate to elliptic, slightly flattened, tan or light brown, with a linear white funicular remnant.

DISTRIBUTION: An occasional roadside weed. Cruz Bay Quarter along Center Line Road (A2899, A2908). Native to Africa, now naturalized throughout the tropics.

COMMON NAME: bona wiss.

26. MACHAERIUM Pers., nom. cons.

Shrubs or lianas. Leaves imparipinnate; leaflets alternate; stipules spinescent, persistent, even along main stems; stipels lacking. Flowers medium-sized, in axillary or terminal racemes or panicles, the bracts minute; calyx asymmetrically bell-shaped, with 5 minute, unequal teeth at apex; corolla purple to white, the standard rounded to kidney-shaped, shortly clawed, the wings and keel subequal, incurved; stamens 10, monadelphous; ovary shortly stipitate, with 1 or 2 ovules, the style inflexed, the stigma punctiform. Legume flattened, curled, with circular outline; seeds solitary, kidney-shaped.

A genus of 150 species primarily of tropical America.

1. Machaerium lunatum (L.f.) Ducke, Arch. Jard. Bot. Rio de Janeiro **4:** 310. 1925. *Pterocarpus lunatus* L.f., Suppl. Pl. 317. 1782. *Drepanocarpus lunatus* (L.f.) G. Mey., Prim. Fl. Esseq. 238. 1818.
Fig. 113G–L.

Scandent shrub to 15 m long; stems to 12 cm diam., with numerous recurved stipular spines. Leaves imparipinnate; leaflets 5–7, alternate or rarely opposite, 2–7 × 0.5–3 cm, oblong or oblanceolate, chartaceous, glabrous, the apex notched, the base obtuse to rounded, the margins entire; petiolules 2–3 mm long; stipules curved, spinelike, to 1 cm long, persistent on mature stems. Racemes axillary or terminal, few- to many-flowered, 5–15 cm long; bracts minute, persistent. Calyx 3.5–5 mm long, glabrous; corolla pink to pinkish purple, the standard to 8 mm long, notched at apex, the wings falcate, the keel incurved.

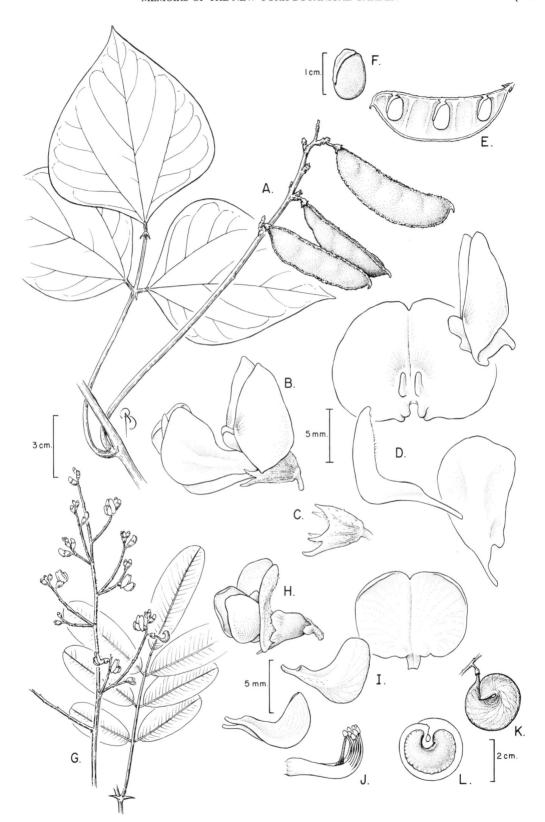

Fig. 113. A–F. *Lablab purpureus*. A. Fruiting branch. B. Flower. C. Calyx. D. Petals: standard, frontal and lateral views (above); keel (below left); and wing (below right). E. Legume with valve removed to show seeds. F. Seed. G–L. *Machaerium lunatum*. G. Fertile branch and leaf. H. Flower. I. Keel, wing, and standard petals. J. Monadelphous stamens and style. K. Legume. L. L.s. legume showing seed.

Legume 2–3 cm wide, flattened, curled with circular outline, indehiscent. Seed 1, nearly kidney-shaped.

DISTRIBUTION: Common in coastal swamps bordering mangroves. Coral Bay (A2829). Also on St. Croix and St. Thomas; native to tropical America.

27. MACROPTILIUM Urb.

Erect, scandent, or trailing herbs. Leaves trifoliolate; stipules and stipels minute. Flowers medium-sized, paired at nodes of axillary pseudoracemes, the bracts minute; calyx nearly bell-shaped to tubular, with 5 equal or unequal teeth at apex; corolla usually burgundy, the standard rounded, reflexed, clawed, the wings long-clawed, much longer than the other petals, the keel long-clawed, twisted on upper portion, fused to staminal tube; stamens 10, diadelphous; ovary subsessile, flattened, pubescent, with several ovules, the style thickened at base, twisted, the stigma punctiform. Legume linear, cylindrical, dehiscent by twisting valves; seeds numerous, small, oblong-cylindrical.

A genus of 20 species, native to the New World.

1. **Macroptilium lathyroides** (L.) Urb., Symb. Ant. **9**: 457. 1928. *Phaseolus lathyroides* L., Sp. Pl., ed. 2, **2**: 1018. 1763. Fig. 114A–E.

Erect or scandent herb to 2(–3) m long, many-branched from base or along main stem. Leaves trifoliolate; leaflets 3–5 × 2–3 cm, elliptic, ovate, or lanceolate, chartaceous, pubescent to puberulent on lower surface, the apex mucronulate, acute or obtuse, the base obtuse or cuneate, the margins entire; petiolules 2–3 mm long, pubescent; rachis 1–1.5 cm long; stipules lanceolate, straw-colored, to 1 cm long; stipels awl-shaped, minute, persistent. Pseudoracemes axillary, to 30 cm long, the flowers shortly pedicellate, borne on swollen nodes; bracts minute, persistent. Calyx 5–6 mm long, pubescent; corolla burgundy, the standard to 1.5 cm long, the keel spirally twisted. Legume 6–12 cm long, linear, cylindrical, dehiscent by twisting valves. Seeds numerous, 2–3 mm long, oblong.

DISTRIBUTION: A common roadside weed. Cruz Bay (A860), Fish Bay (A3879), Lameshur (A2757). Also on St. Croix, St. Thomas, and Tortola; the West Indies and Central America.

COMMON NAME: wild-bush-bean.

28. PHASEOLUS L.

Erect herbs or twining vines. Leaves trifoliolate; leaf rachis more or less elongate; stipules striate, persistent; stipels minute. Flowers medium-sized, clustered at swollen nodes in axillary pseudoracemes, the bracts minute, persistent; calyx nearly bell-shaped, 2-lipped, with 5 unequal, minute teeth at apex; corolla white, pink, red, purple, or yellow, the standard nearly rounded, short-clawed, reflexed, the keel spirally twisted, narrow; stamens 10, diadelphous, the anthers alternately basifixed and versatile; ovary subsessile, linear, with 1 to several ovules, the style curved, introrsely bearded, the stigma flattened. Legume linear to oblong, straight, dehiscent; seeds bean-shaped.

A genus of about 200 species with worldwide distribution.

1. **Phaseolus peduncularis** Kunth *in* Humb., Bonpl. & Kunth, Nov. Gen. Sp. **6**: 447. 1824. *Vigna peduncularis* (Kunth) Fawc. & Rendle, Fl. Jamaica **4(2)**: 68. 1920. Fig. 115A–E.

Twining herbaceous vine, 3–5 m long, many-branched along main stem; stems reddish-tinged when young. Leaves trifoliolate; leaflets 3–9 × 2–7 cm, broadly ovate to deltoid, the lateral ones asymmetric, chartaceous, glabrous to puberulent, the apex acuminate, the base cuneate to truncate, the margins entire; petiolules 4–5 mm long; rachis 2–2.5 cm long; stipules lanceolate, straw-colored, 6–8 mm long; stipels awl-shaped, minute, persistent. Pseudoracemes axillary, to 30 cm long, the flowers clustered at swollen nodes; bracts minute, persistent. Calyx 4–5 mm long, puberulent; corolla light purple, the standard to 1.5–1.8 cm long, the keel S-shaped. Legume 6–14 cm long, linear, flattened, beaked at apex, puberulent, dehiscent by twisting valves. Seeds several, 2–3 mm long, oblong, brown.

DISTRIBUTION: An occasional roadside weed. Along road from Bordeaux to Coral Bay (A3796, A3868). Also in the Greater Antilles and northern South America.

29. PICTETIA DC.

Shrubs or small trees. Leaves imparipinnate or rarely unifoliolate; leaflets 3 to many, opposite or alternate, with a spiny bristle at apex; stipules rigid, persistent; stipels minute. Flowers medium-sized, in axillary racemes, the bracts minute; calyx bell-shaped, with 5 unequal, minute teeth at apex; corolla yellow, the standard nearly rounded, short clawed, reflexed, the wings clawed, falcate, the keel slightly shorter than the wings; stamens 10, monadelphous; ovary stipitate, with several ovules, the style glabrous, incurved, the stigma punctiform. Legume oblong, flattened, straight or slightly curved, indehiscent; seeds lenticular.

A genus of 6–8 species, from the Greater Antilles.

1. **Pictetia aculeata** (Vahl) Urb., Symb. Ant. **2**: 294. 1900. *Robinia aculeata* Vahl *in* H. West, Bidr. Beskr. Ste. Croix 300. 1793. Fig. 114J–N.

Aeschynomene aristata Jacq., Pl. Hort. Schoenbr. **2**: 59, pl. 237. 1797. *Pictetia aristata* (Jacq.) DC., Ann. Sci. Nat. (Paris) **4**: 94. 1825.

Robinia squamata Vahl, Symb. Bot. **3**: 88, pl. 69. 1794. *Pictetia squamata* (Vahl) DC., Ann. Sci. Nat. (Paris) **4**: 94. 1825.

Shrub or small tree to 12 m tall; branches ascending; bark rough, peeling off in large flakes. Leaves imparipinnate; leaflets 9–25, 1–2 cm long, rounded, oblong to obovate, coriaceous, glabrous or puberulent, the apex notched or nearly truncate, ending

FIG. 114. A–E. *Macroptilium lathyroides*. **A.** Flowering branch. **B.** Detail of stem showing flower, subtending bract, and swollen inflorescence node. **C.** Standard, wing, and keel petals. **D.** Legume and twisted valves of legume after dehiscence. **E.** Seed. **F–I.** *Piscidia carthagenensis*. **F.** Inflorescence. **G.** Flower. **H.** Wing and keel petals. **I.** Legumes and branch with leaf (background). **J–N.** *Pictetia aculeata*. **J.** Flowering branch and detail of branch showing spinelike stipules at base of leaf petiole. **K.** Leaflet. **L.** Flower. **M.** Wing, standard, and keel petals. **N.** Legume.

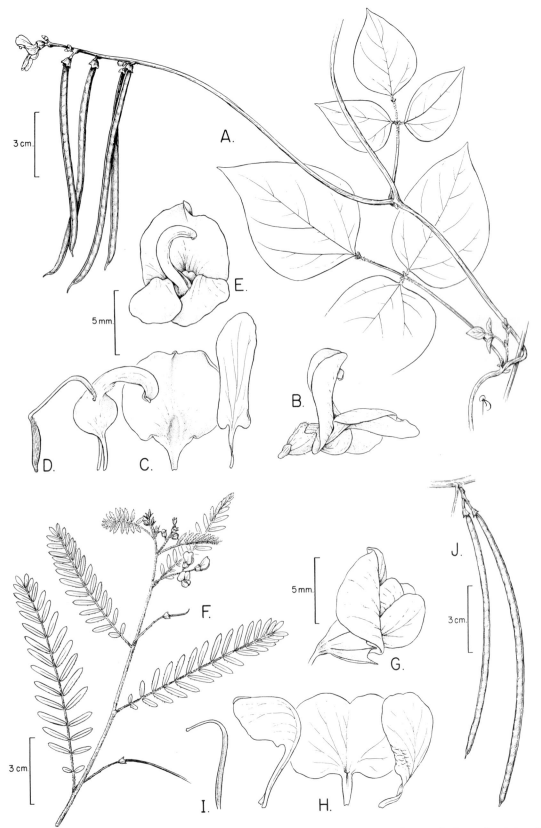

Fɪɢ. 115. A–E. *Phaseolus peduncularis*. A. Fertile branch. B. Flower. C. Keel, standard, and wing petals. D. Pistil. E. Flower, frontal view. F–J. *Sesbania sericea*. F. Fertile branch. G. Flower. H. Keel, standard, and wing petals. I. Pistil. J. Legumes.

in a sharp spiny tip, the base asymmetric, obtuse to nearly cordate, the margins entire; petiolules to 1 mm long; stipules rigid, spine-like, 5–10 mm long; stipels absent. Racemes axillary, shorter than subtending leaf, few-flowered; pedicels 0.7–2 cm long; bracts minute, persistent. Calyx 5–6 mm long, puberulent; corolla bright yellow, the standard nearly rounded, to 2 cm long, the wings and keel shorter than the standard. Legume 2–5 × 0.5–0.6 cm, linear, flattened, constricted between seeds, beaked at apex. Seeds numerous, 2–3 mm long, oblong, brown.

DISTRIBUTION: A common shrub of dry coastal scrub. Bethany (B219), East End (A668). Also on all islands of the Virgin Islands and Puerto Rico (including Vieques and Culebra).

COMMON NAME: fustic.

30. PISCIDIA L., nom. cons.

Small trees. Leaves imparipinnate; leaflets 5–25, opposite; stipules deciduous; stipels lacking. Flowers small, in axillary panicles or racemes, the bracts deciduous; calyx bell-shaped, with 5 equal, minute teeth at apex; corolla pink to white, the standard nearly rounded, the wings clawed, falcate, coherent to the keel, the keel long-clawed; stamens 10, monadelphous; ovary sessile, densely pubescent, with several ovules, the style glabrous, incurved, the stigma punctiform. Legume oblong, 4-winged, the wings with sinuate margins, indehiscent, breaking irregularly into a few transverse units that are wind-dispersed; seeds few, oblong to kidney-shaped.

A genus of 10 species from Florida, Central America, and the West Indies.

REFERENCE: Rudd, V. E. 1969. A synopsis of the genus *Piscidia* (Leguminosae). Phytologia 18: 473–499.

1. Piscidia carthagenensis Jacq., Enum. Syst. Pl. 27. 1760. Fig. 114 F–I.

Piscidia erythrina sensu Eggers, Fl. St. Croix. 45. 1879, non L., 1759.

Small tree to 15 m tall; bark dark gray, smooth or fissured. Leaves imparipinnate, deciduous during flowering time; leaflets 5–15, 6–17 × 3–11 cm, ovate, elliptic, obovate, or lanceolate, chartaceous, white-pubescent on lower surface, the apex obtuse to shortly acuminate, the base rounded, the margins entire; petiolules swollen, 4–5 mm long; rachis 2–2.5 cm long; stipules ovate, to 5 mm long. Flowers few to many, in axillary panicles to 30 cm long; bracts minute. Calyx 5–8 mm long, velvety-pubescent; corolla pink, the standard 10–15 mm long, pubescent without. Legume to 12 cm long, oblong, beaked at apex, the wings 1–2 cm wide. Seeds few, ca. 2 mm long, elliptic.

DISTRIBUTION: An occasional tree of coastal forests and scrub. Bethany (B336). Also on Jost van Dyke, St. Croix, St. Thomas, Tortola, and Virgin Gorda; occurring naturally from Mexico to Peru and from Puerto Rico to the Lesser Antilles, apparently introduced in Cuba.

COMMON NAMES: dogwood, fish poison tree, stink tree.

31. POITEA Vent.

Trees or erect or ascending shrubs. Leaves paripinnate or imparipinnate; leaflets numerous, opposite; stipules lanceolate, deciduous or persistent; stipels minute. Flowers medium-sized, in axillary fascicles or racemes, the bracts minute, persistent; calyx tubular to bell-shaped, truncate or bilobed; corolla pink, red, or purple, the standard rounded to kidney-shaped, slightly notched at apex, clawed and auriculate at base, shorter than the keel petals, the wings and keel clawed, of similar length; stamens 10, diadelphous, of unequal length; ovary stipitate, with several ovules, the style glabrous and incurved, the stigma punctiform. Legume flattened, linear, with thickened margins, beaked at apex, dehiscent; seeds several, flattened, ovate.

A genus of 12 species from Cuba, Hispaniola, Puerto Rico, the Virgin Islands, and Dominica.

REFERENCE: Lavin, M. 1993. Biogeography and systematics of *Poitea* (Leguminosae). Syst. Bot. Monogr. 37: 1–87.

1. Poitea florida (Vahl) Lavin, Syst. Bot. Monogr **37**: 61. 1993. *Robinia florida* Vahl *in* H. West, Bidr. Beskr. Ste. Croix 300. 1793. *Sabinea florida* (Vahl) DC., Ann. Sci. Nat. (Paris) **4**: 92. 1825. Fig. 116K–N.

Shrub or small tree to 6 m tall, usually leafless when flowering; trunk to 10 cm diam., the bark gray, slightly fissured. Leaves imparipinnate, 5–9 cm long; leaflets 5–15 pairs, 0.5–1.7 × 0.2–0.8 cm, oblong to obovate, chartaceous, puberulent on lower surface, the apex mucronulate, rounded, the base rounded to nearly truncate, the margins revolute; petiolules ca. 1 mm long; rachis grooved along upper surface; stipules narrowly lanceolate, 4–6 mm long. Flowers fascicled on very short axillary racemes. Calyx funnel-shaped, 4–5 mm long, with 5 minute teeth at apex; corolla pink or lavender, the standard to 18 mm long, widely ovate, the wings slightly longer than the standard. Legume 8–11 × 0.4–0.6 cm, linear, flattened, long-stipitate, long-tipped at apex, dehiscent by twisting valves. Seeds many, 4–5 mm long, lenticular, dark brown.

DISTRIBUTION: Common in moist to dry forest. Susannaberg (A2799, M16572). Also on St. Thomas, Tortola, Virgin Gorda; Puerto Rico (including Culebra and Vieques).

COMMON NAMES: soldier whip, wattapama.

32. PUERARIA DC.

Herbaceous to woody, twining, stout vines. Leaves trifoliolate; stipules ovate to linear, persistent; stipels minute. Flowers medium-sized, clustered along nodes of axillary or terminal pseudoracemes, the bracts minute; calyx bell-shaped, with 5 unequal teeth at apex; corolla blue to purple, the standard obovate, notched at apex, clawed and auriculate at base, the wings clawed, spurred, the keel slightly longer than the wings; stamens 10, monadelphous or diadelphous; ovary sessile, with several ovules, the style glabrous, incurved, the stigma capitate. Legume linear, compressed, dehiscent by twisting valves; seeds numerous, oblong.

A genus of 20 species native to southern Asia, 1 species introduced into the New World as a forage plant.

1. Pueraria phaseoloides (Roxb.) Benth., J. Linn. Soc., Bot. **9**: 125. 1865. *Dolichos phaseoloides* Roxb., Fl. Ind., ed. 1832, **3**: 316. 1832. Fig. 117A–E.

Herbaceous twining vine to 15 m long, profusely branched, forming dense tangles; branches cylindrical, pubescent. Leaves trifoliolate; leaflets 3–12 cm long, ovate to rhombic, the lateral

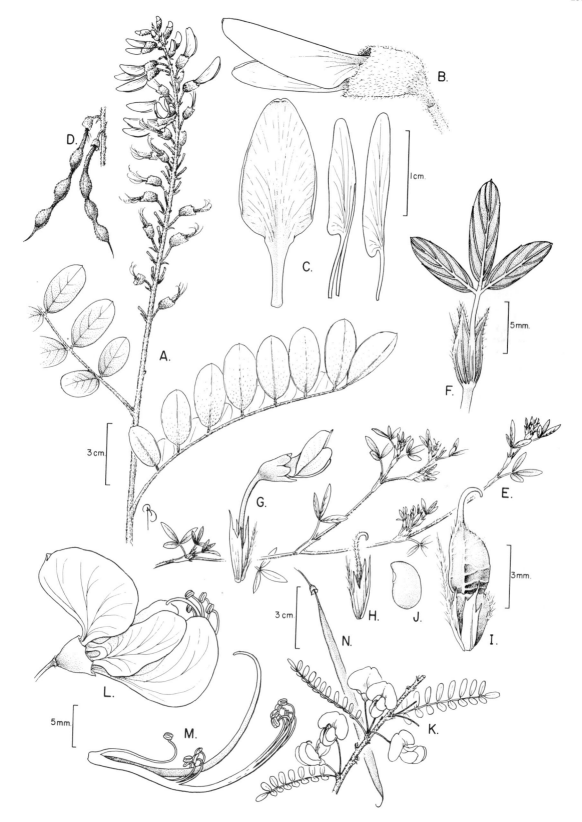

Fig. 116. A–D. *Sophora tomentosa*. **A.** Flowering branch. **B.** Flower. **C.** Standard, keel, and wing petals. **D.** Legumes. **E–J.** *Stylosanthes hamata*. **E.** Flowering branch. **F.** Stipules adnate to base of leaf petiole. **G.** Flower with subtending bracts. **H.** Calyx and pistil. **I.** Legume with subtending calyx and bracteole. **J.** Seed. **K–N.** *Poitea florida*. **K.** Flowering branch. **L.** Flower. **M.** Diadelphous stamens and pistil. **N.** Legume.

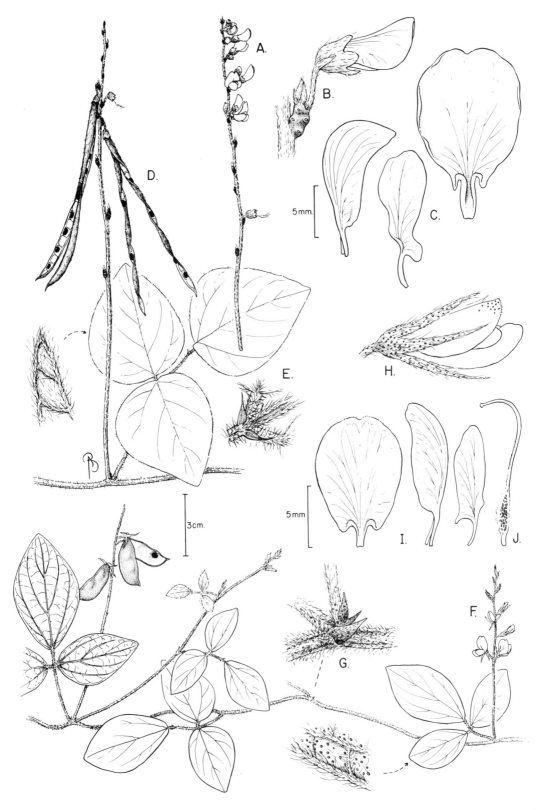

Fig. 117. A–E. *Pueraria phaseoloides*. **A.** Pseudoraceme. **B.** Detail of pseudoraceme with flower and swollen node. **C.** Keel, wing, and standard petals. **D.** Fruiting branch and detail of underside of leaflet showing indument. **E.** Detail of leaf rachis showing stipels. **F–J.** *Rhynchosia reticulata*. **F.** Fruiting branch and detail of underside of leaf margin showing glands and indument. **G.** Detail of stem showing stipules at base of leaf petiole. **H.** Flower. **I.** Standard, keel, and wing petals. **J.** Pistil.

ones asymmetric, chartaceous, silvery-pubescent on lower surface, the apex acute, the base cuneate in central leaflet, rounded-obtuse in laterals, the margins entire; petiolules swollen, 4–5 mm long, pubescent; rachis 2–2.5 cm long; petiole grooved, pubescent, to 12 cm long; stipules narrowly lanceolate, 3–5 mm long; stipels awl-shaped, minute, persistent. Pseudoracemes axillary, to 25 cm long, the flowers 2 or 3, borne on swollen nodes; bracts minute, persistent. Calyx ca. 5 mm long, pubescent; corolla lavender, the

standard to 1.5 cm long, the wings lavender, the keel whitish. Legume 6–9 cm long, linear, flattened. Seeds few, ca. 3 mm long, oblong.

DISTRIBUTION: An occasional roadside weed. Along Center Line Road by entrance to Ajax Peak (W630), Maho Bay Quarter along Center Line Road (A2411). This aggressive species, native to Southeast Asia, has been introduced into the New World as a forage and erosion-control plant; it has become naturalized in some areas of the neotropics.

33. RHYNCHOSIA Lour., nom. cons.

Twining or trailing herbaceous to woody vines. Leaves trifoliolate, lower surface with yellow, resinous dots; stipules deciduous; stipels minute. Flowers small, in axillary pseudoracemes, the bracts minute, persistent or deciduous; calyx bell-shaped, with 4–5 elongate sepals; corolla yellow, the standard obovate to rounded, slightly notched at apex, clawed and auriculate at base, the wings clawed, spurred, the keel slightly longer than the wings; stamens 10, monadelphous or diadelphous; ovary shortly stipitate, with few ovules, the style glabrous, incurved, the stigma capitate. Legume oblong, flattened, beaked at apex, usually dehiscent by twisting valves; seeds few to many, flattened, rounded to elliptic.

A genus of 200 species with pantropical distribution.

REFERENCE: Grear, J. W. 1978. A revision of the New World species of *Rhynchosia* (Leguminosae-Faboideae). Mem. New York Bot. Gard. 31: 1–168.

Key to the Species of *Rhynchosia*

1. Leaflets prominently reticulate below; calyx 6–10 mm long .... 2. *R. reticulata*
1. Leaflets not prominently reticulate below; calyx 2.5–3 mm long .... 1. *R. minima*

1. Rhynchosia minima (L.) DC., Prodr. **2**: 385. 1825. *Dolichos minimus* L., Sp. Pl. 726. 1753.

Herbaceous twining vine to 3 m long, many-branched from a woody base; branches slender, cylindrical, pubescent. Leaves trifoliolate; leaflets 1.7–3 × 1.7–3 cm, widely ovate to rhombic, the lateral ones asymmetric, chartaceous, glabrous or puberulent, densely gland-dotted on lower surface, the apex acute to obtuse, the base cuneate on central leaflet, rounded-obtuse on laterals, the margins entire; petiolules minute, pubescent; rachis 4–8 mm long; petiole grooved, pubescent, to 3 cm long; stipules narrowly lanceolate, 2–3 mm long; stipels minute. Pseudoracemes axillary, to 10 cm long, few-flowered. Calyx ca. 23.5–3 mm long; corolla yellow, the standard 3.5–4 mm long, the wings and keel as long as the standard. Legume 1.2–1.6 × 0.4–0.5 cm, oblong, flattened, pubescent, dehiscent by twisting valves, brown when mature. Seeds 1 or 2, 3–4 mm long, kidney-shaped, light brown.

DISTRIBUTION: Common along roadsides and open ground. Enighed (A3921), Maria Bluff (A2337). Also on St. Croix, St. Thomas, Tortola, and Virgin Gorda; common throughout the Caribbean and the tropics.

2. Rhynchosia reticulata (Sw.) DC., Prodr. **2**: 385. 1825. *Glycine reticulata* Sw., Prodr. 105. 1788.

Fig. 117F–J.

Subwoody, twining vine to 3 m long; branches slender, cylindrical, pubescent. Leaves trifoliolate; leaflets 2–8(–10) × 1–5 (–7) cm, ovate, lanceolate, or rhombic, the lateral ones asymmetric, coriaceous, pubescent, the lower surface densely gland-dotted, with prominently reticulate veins, the apex acute to obtuse, the base cuneate to obtuse in central leaflet, rounded-obtuse in laterals, the margins entire; petiolules 2–3 mm long, pubescent; rachis 4–8 mm long; petiole grooved, pubescent, to 6 cm long; stipules narrow lanceolate, to 10 mm long; stipels minute. Pseudoracemes axillary, to 10 cm long, few-flowered. Calyx 6–10 mm long; corolla yellow, the standard to 10 mm long, the wings and keel nearly as long as the standard. Legume 2–2.2 × 0.8–1 cm, oblong, flattened, pubescent, dehiscent. Seeds 1 or 2, 5 mm long, lenticular, light brown, mottled.

DISTRIBUTION: Common along roadsides and open ground. Cob Gut (A3227), Fish Bay Gut (A3889), Lind Point (A2304). Also on St. Croix, St. Thomas, Tortola, and Virgin Gorda; throughout the Caribbean, Mexico, Central America, and South America.

34. SESBANIA Scop.

Herbs, shrubs, or small trees. Leaves paripinnate; leaflets numerous, opposite; rachis ending in a filiform segment; stipules deciduous; stipels minute. Flowers small to large, in axillary racemes, the bracts minute; calyx broadly bell-shaped, 5-lobed; corolla of various colors, the standard ovate to rounded, clawed at base, the wings and keel clawed, of similar length; stamens 10, diadelphous, bent at base; ovary stipitate, with several ovules, the style glabrous and incurved, the stigma capitate. Legume linear, nearly cylindrical, transversely septate, beaked at apex, dehiscent; seeds numerous, oblong-cylindrical, brown.

A genus of 50 species with pantropical distribution.

1. Sesbania sericea (Willd.) Link, Enum. Hort. Berol. Alt. **2**: 244. 1822. *Coronilla sericea* Willd., Enum. Pl. 773. 1809.   Fig. 115F–J.

Shrub to 3 m tall, many-branched along main stem, the stems reddish-tinged, sericeous-villous. Leaves 5–12 cm long; leaflets 5–15 pairs, 1–2.5 × 0.3–0.8 cm, oblong to elliptic, chartaceous,

sericeous on lower surface, the apex mucronulate, rounded, the base rounded, slightly asymmetric, the margins entire; petiolules ca. 1 mm long; rachis grooved along upper surface; stipules awl-shaped, 4–5 mm long. Racemes axillary, to 6 cm long, few-flowered. Calyx broadly bell-shaped, 3–5 mm long; corolla yellow, the standard to 12 mm long, widely ovate, the wings and keel as long as the standard. Legume 12–16 cm long and 2.5–3 mm wide, linear, slightly curved, cylindrical, long-tipped at apex, tardily dehiscent. Seeds many, ca. 3 mm long, cylindrical, light brown.

DISTRIBUTION: An occasional weed of open disturbed areas. Enighed (A3925), Fish Bay (A2475). Also on St. Croix and St. Thomas; native to Sri Lanka, now a widespread weed throughout the tropics.

35. SOPHORA L.

Shrubs or small trees. Leaves imparipinnate; leaflets numerous, opposite; stipules minute, deciduous; stipels wanting. Flowers small to medium, in terminal racemes or panicles, the bracts minute; calyx bell-shaped, with 5 minute teeth at apex; corolla white or yellow, the standard ovate to rounded, clawed at base, the wings and keel clawed, of similar length; stamens 10, free; ovary stipitate, with several ovules, the style incurved, the stigma punctiform. Legume linear, nearly cylindrical with constrictions between seeds, beaked at apex, tardily dehiscent; seeds numerous, ovoid to globose.

A genus of 50 species with pantropical and subtropical distribution.

1. Sophora tomentosa L., Sp. Pl. 373. 1753.

Fig. 116A–D.

Shrub or small tree to 5 m tall, whitish, grayish, or yellowish tomentose, especially when young. Leaves 12–20 cm long; leaflets 9–19, opposite or subopposite, 2.3–5 × 1–2.7 cm, ovate to elliptic, chartaceous, densely tomentose on lower surface, the apex rounded or obtuse, sometimes apiculate, the base obtuse, asymmetric, the margins revolute; petiolules 2–3 mm long; rachis cylindrical, tomentose; stipules linear-triangular, ca. 3 mm long.

Racemes terminal, to 25 cm long, few- to many-flowered. Calyx asymmetrically bell-shaped, 7–8 mm long, tomentose; corolla yellow, the standard to 18 mm long, narrowly ovate, clawed, the wings and keel as long as the standard. Legume 8–12 cm long, torulose, long-stipitate, long-tipped at apex. Seeds 4–8, 5–6 mm long, nearly globose, yellowish brown.

DISTRIBUTION: Occasional along the sandy coast. Dittlif Point (W805), Haulover Bay (A4146). Also on Anegada, St. Croix, St. Thomas, Tortola, and Virgin Gorda; pantropical.

36. STYLOSANTHES Sw.

Erect or prostrate herbs. Leaves trifoliolate; stipules adnate to the petiole, conspicuously veined, persistent; stipels wanting. Flowers small, in axillary, short spikes, the bracts large, appressed against the calyx; calyx bell-shaped to tubular, 4 sepals united, 1 free; corolla yellow, the standard rounded, notched at apex, the wings oblong, as long as the keel, the keel incurved; stamens 10, monadelphous; ovary shortly stipitate, with 2 or 3 ovules, the style incurved, the stigma punctiform. Legume flattened, constricted between seeds, separating into 2 or 3 segments, hooked at apex, indehiscent; seeds nearly lenticular.

A genus of 50 species, with most species native to the New World and a few species from Africa and Asia.

REFERENCE: Mohlenbrock, R. H. 1958. A revision of the genus *Stylosanthes*. Ann. Missouri Bot. Gard. 44: 299–355.

1. Stylosanthes hamata (L.) Taub., Verh. Bot. Vereins Prov. Brandenburg **32:** 22. 1890. *Hedysarum hamatum* L., Syst. Nat., ed. 10, **2:** 1170. 1759.

Fig. 116E–J.

Erect to prostrate subshrub, many-branched from a woody base; branches to 30 cm long. Leaflets 0.7–2 × 0.2–0.5 cm, elliptic, chartaceous, glabrous or puberulent, the veins ascending, prominent and whitish on lower surface, the apex mucronulate, obtuse to acute, the base acute to obtuse, slightly asymmetric, the margins denticulate, ciliate; petioles 3–8 mm long; stipules to 9

mm long. Flowers axillary, solitary at ends of condensed short branches. Calyx bell-shaped, 3–4 mm long; corolla yellow, the standard 4–5 mm long, rounded, clawed, the wings and keel slightly shorter than the standard. Legume of 2 articles, 5–7 mm long, the apex hooked. Seeds ca. 2 mm long, flattened, oblong-elliptic.

DISTRIBUTION: A weed commonly found in open disturbed areas. East End (A2778), Lind Point (A2303). Also on Anegada, St. Croix, St. Thomas, Tortola, and Virgin Gorda; throughout the West Indies and from Mexico to northern South America.

COMMON NAMES: donkey weed, mother sea gel.

37. TEPHROSIA Pers., nom. cons.

Erect or prostrate herbs or subshrubs. Leaves imparipinnate; the leaflets numerous, mostly opposite; stipules subulate, persistent; stipels wanting. Flowers small, in terminal, axillary or opposite to leaf racemes, the bracts minute; calyx bell-shaped, with 5 elongate, subequal sepals; corolla yellow, red, or purple, the standard nearly rounded, notched at apex, pubescent on lower surface, the wings as long as the keel, clawed, the keel incurved; stamens 10, diadelphous, of unequal length; ovary sessile, with several ovules, the style incurved, the stigma capitate. Legume linear, flattened, dehiscent by twisting valves; seeds flattened, rounded.

A genus of 250–300 species with cosmopolitan distribution.

Key to the Species of *Tephrosia*

1. Leaflets 11–15, elliptic, oblanceolate to elliptic, obtuse, rounded, or acute at apex .... 1. *T. cinerea*
1. Leaflets 5–9, oblong to obovate, notched at apex .... 2. *T. senna*

1. Tephrosia cinerea (L.) Pers., Syn. Pl. **2**: 328. 1807. *Galega cinerea* L., Syst. Nat., ed. 10, **2**: 1172. 1759. Fig. 118G–K.

Prostrate or erect herb, many-branched from a woody base; branches to 25 cm long; young foliage silvery pubescent. Leaflets 11–15, opposite, 1.3–3.8 × 0.2–0.9 cm, oblanceolate to elliptic, chartaceous, silvery pubescent on lower surface, the veins ascending, the apex mucronulate, obtuse, rounded, or acute, the base obtuse, the margins ciliate; petiolules ca. 1 mm long; stipules awl-shaped, to 4 mm long. Racemes terminal or opposite to leaves, few-flowered, to 10 cm long. Calyx bell-shaped, 5–7 mm long; corolla pink to pinkish violet, the standard 7–8 mm long, oval, clawed, the wings and keel as long as the standard. Legume 4–6 cm long, linear, slightly flattened. Seeds few, 3–4 mm long, bean-shaped.

DISTRIBUTION: A common weed of open dry areas. Ram Head (A2049, W28). Also on St. Croix, St. Thomas, Tortola, and Virgin Gorda; throughout the neotropics.

2. Tephrosia senna Kunth *in* Humb., Bonpl. & Kunth, Nov. Gen. Sp. **6**: 458. 1824.

Galega cathartica Sessé & Moc., Fl. Mexic., ed. 2, 175. 1894. *Tephrosia cathartica* (Sessé & Moc.) Urb., Symb. Ant. **4**: 283. 1905.

Erect subshrub to 2 m tall; young foliage densely pubescent. Leaflets 5–9, opposite, 2–4 × 0.6–1.6 cm, oblong to obovate, chartaceous, silvery pubescent on lower surface, the veins ascending, the apex mucronulate, notched, the base obtuse, the margins ciliate; petiolules ca. 1 mm long; stipules awl-shaped, to 7 mm long. Racemes terminal or opposite to leaves, few-flowered, to 12 cm long. Calyx bell-shaped, ca. 4 mm long; corolla purple or reddish purple, the standard to 6 mm long, oval. Legume 3–5 cm long, linear, slightly flattened. Seeds few, 3 mm long, bean-shaped.

DISTRIBUTION: An occasional weed of the sandy coast. Cinnamon Bay (W635B). Also on Tortola and Virgin Gorda; throughout the West Indies and northern South America.

38. TERAMNUS P. Browne

Herbaceous twining or trailing vines. Leaves trifoliolate; stipules awl-shaped, persistent; stipels minute. Flowers small, in axillary pseudoracemes, the bracts small; calyx bell-shaped, with 4 or 5 elongate, nearly equal sepals; corolla white or yellow, the standard obovate, the wings longer than the keel, clawed, the keel incurved; stamens 10, monadelphous, the anthers of unequal size; ovary sessile, with several ovules, the style short, pubescent, the stigma capitate. Legume linear, flattened, curved at apex, dehiscent by twisting valves; seeds several, flattened, oblong.

A genus of 8 species with pantropical distribution.

1. Teramnus labialis (L.f.) Spreng., Syst. Veg. **3**: 235. 1826. *Glycine labialis* L.f., Suppl. Pl. 325. 1782. Fig. 118L–P.

Twining vine 3–5 m long; stems slender, sparsely pubescent; young parts hirsute. Leaflets 1.5–6.2 × 0.6–3.5 cm, elliptic to ovate, chartaceous, with appressed hairs, especially on lower surface, the apex mucronulate, obtuse to rounded, the base obtuse to rounded, the margins ciliate; petiolules ca. 2–3 mm long;

stipules lanceolate, to 3 mm long. Pseudoracemes axillary, few-flowered, to 11 cm long. Calyx bell-shaped, 3–5 mm long, hirsute; corolla white, the standard to 5 mm long. Legume 3–5 cm long, linear, slightly curved, flattened, pubescent. Seeds several, 2.5–3 mm long, oblong.

DISTRIBUTION: A common weed of open and disturbed areas. Cruz Bay (A1956), Lameshur (B645). Also on St. Croix, St. Thomas, Tortola, and Virgin Gorda; throughout the West Indies and Central America.

COMMON NAME: blue wiss.

39. VIGNA Savi

Herbaceous climbing or trailing, twining vines. Leaves trifoliolate; stipules variously shaped, persistent; stipels subulate. Flowers small to medium, in axillary or terminal pseudoracemes, the bracts small; calyx nearly bell-shaped, 5-lobed, 2-lipped, with 2 upper lobes almost completely fused; corolla yellow or blue, the standard rounded, the wings spurred, as long as the keel, clawed, the keel recurved, truncate; stamens 10, diadelphous, the anthers of equal size; ovary sessile, with several ovules, the style curved, pubescent on distal portion, the stigma lateral. Legume linear to oblong, compressed to almost cylindrical, dehiscent by twisting valves; seeds compressed, square to kidney-shaped.

A genus of 100 species, mostly from tropical Africa and Asia.

1. Vigna luteola (Jacq.) Benth. *in* Mart., Fl. Bras. **15(1)**: 194, t. 50, fig. 2. 1859. *Dolichos luteolus* Jacq., Hort. Bot. Vindob. **1**: 39, t. 90. 1770. Fig. 118A–F.

Dolichos repens L., Syst. Nat., ed. 10, **2**: 1163. 1759. *Vigna repens* (L.) Kuntze, Revis. Gen. Pl. **1**: 212. 1891, non Baker, 1876.

Trailing or climbing, twining vine, to 10 m long, many-branched from base; stems cylindrical, slender, sparsely pubescent. Leaflets 2.5–7.5 × 1–5 cm, ovate or lanceolate, chartaceous,

sparsely pubescent along veins, the apex acute, the base cuneate to rounded, the margins ciliate; petioles 3.5–9 cm long; petiolules ca. 3–2 mm long; stipules lanceolate, ca. 3 mm long. Racemes axillary, few-flowered, to 30 cm long. Calyx bell-shaped, 4–5 mm long; corolla yellow, the standard 1.5–2 cm long, nearly rounded, the wings and keel as long as the standard. Legume 5–7 × 0.5–0.7 cm, linear-oblong, slightly flattened, constricted between seeds. Seeds few, 5 mm long, oblong to square, reddish brown.

DISTRIBUTION: A common coastal vine, sometimes in open disturbed areas. Enighed (A4005). Also on St. Croix, St. Thomas, and Tortola; throughout tropical and subtropical America.

COMMON NAMES: goat wiss, wild pea.

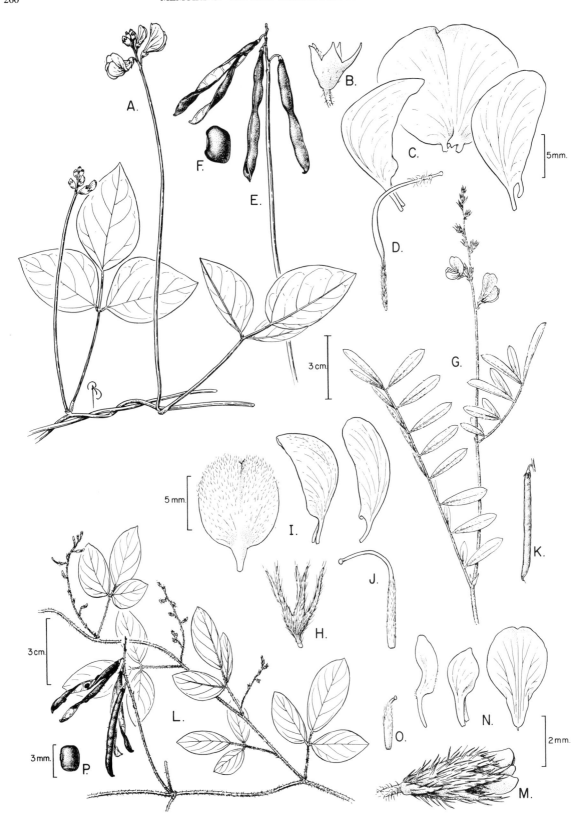

FIG. 118. A–F. *Vigna luteola*. A. Flowering branch. B. Calyx. C. Keel, standard, and wing petals. D. Pistil. E. Legumes. F. Seed. G–K. *Tephrosia cinerea*. G. Flowering branch. H. Calyx. I. Standard, keel, and wing petals. J. Pistil. K. Legume. L–P, *Teramnus labialis*. L. Fertile branch. M. Flower. N. Wing, keel, and standard petals. O. Pistil. P. Seed.

III. Mimosoideae (Sensitive Plant Subfamily)

Trees, shrubs, herbs, vines, or lianas. Leaves alternate, twice- or once-pinnate; stipules usually present, stipels wanting. Flowers usually small, regular, produced in heads, spikes, or rarely racemes; calyx tubular, with (3–)5 lobes; corolla usually tubular, 4–5-lobed; stamens 10, usually long exserted, the filaments free or connate, the anthers opening by longitudinal slits.

This subfamily consists of about 50 genera and 3000 species mostly from tropical and subtropical regions.

REFERENCE: Elias, T. S. 1974. The genera of Mimosoideae (Leguminosae) in the southeastern United States. J. Arnold. Arbor. 55: 67–118.

Key to the Genera of Mimosoideae

1. Filaments free or united only at the very base.
    2. Petals connate into a tubular or funnel-shaped corolla; plants spiny (if not spiny, then inflorescence an elongate spike or a panicle of heads).
        3. Fruit indehiscent, breaking away from ribbed, thickened margins (replum); petiole and rachis without nectary glands .... 46. Mimosa
        3. Fruit dehiscent lengthwise, not leaving a replum; petiole and rachis usually with nectary glands ... 40. *Acacia*
    2. Petals free to base; plant not spiny (inflorescence of a single globose head) .... 45. *Leucaena*
1. Filaments connate into a more or less elongate tube.
    4. Petiole and rachis without nectary glands.
        5. Seeds bright orange-red, angular, partially compressed; leaflets alternate .... 41. *Adenanthera*
        5. Seeds brown, lenticular or elliptic, flattened; leaflets opposite.
            6. Inflorescence globose, bright red .... *Calliandra* (cultivated)
            6. Inflorescence obconical, white .... 49. *Zapoteca*
    4. Petiole and rachis with nectary gland(s).
        7. Plant spiny; fruits turgid, twisting upon dehiscence and exposing several black, pendulous seeds with a red aril .... 47. *Pithecellobium*
        7. Plant not spiny; fruits and seeds otherwise.
            8. Leaves once-pinnate; seeds covered with a white fleshy pulp .... 44. *Inga*
            8. Leaves twice pinnate; seeds without a fleshy covering.
                9. Small to large trees; legumes >1 cm wide.
                    10. Legumes oblong, papery to fleshy, tardily dehiscent or indehiscent; seeds lighter along margins.
                        11. Corolla light yellow; legumes papery, flattened, ca. 2 mm thick, 3–4 cm wide, late dehiscent; rachis with a single nectariferous gland at base of petiole .... 42. *Albizia*
                        11. Corolla pink; legumes fleshy-leathery, compressed, ca. 4 mm thick, 1.4–1.5 cm wide, indehiscent; leaf rachis with a discoid nectariferous gland between each pair of pinna .... 48. *Samanea*
                    10. Legumes compressed, curled and nearly circular, indehiscent; seeds reddish brown .... *Enterolobium* (cultivated)
                9. Subshrubs to 1 m tall; legumes 0.3–0.4 cm wide .... 43. *Desmanthus*

40. ACACIA Mill.

Trees, shrubs, or lianas; stems smooth or with prickles. Leaves twice-pinnate; pinnae opposite, the leaflets usually small, numerous, opposite; petioles and rachis usually with nectariferous glands; stipules usually spinelike, persistent. Flowers bisexual, in globose heads, cylindrical spikes, or short racemes, the bracts usually small; calyx bell-shaped, of 5 united or distinct sepals; corolla yellow or white, tubular, 4–5-lobed; stamens numerous, exserted, the filaments free or united at base; ovary sessile or stipitate, with several ovules, the styles filiform, curved, pubescent on distal portion. Legume linear to linear-oblong, straight to curved, compressed to almost cylindrical, dehiscent or indehiscent; seeds variously shaped, most often compressed.

A genus of 500 species, mostly from tropical America, tropical Africa, and Australia.

Key to the Species of *Acacia*

1. Lianas.
    2. Plant not spiny; stems cylindrical or nearly so; legume 2.5–3 cm wide .... 5. *A. vogeliana*
    2. Stem and leaf rachis with prickles; stems obtusely 4-angled, usually splitting longitudinally into 4 segments when old; legume 1.5–2 cm wide .... 3. *A. retusa*
1. Small to medium-sized trees.
    3. Plant not spiny; inflorescence an elongate spike; corolla cream-colored .... 2. *A. muricata*

3. Plant spiny; inflorescence a globose head; corolla bright yellow.
    4. Pinnae 8–18 pairs; leaf rachis with 1–3 nectary glands toward distal end; legume 9–10 mm wide, flattened, with constricted margins between seeds.. 1. *A. macracantha*
    4. Pinnae 2–6 pairs; leaf rachis with 1 or 2 nectary glands at distal end; legume 5–6 mm wide, nearly cylindrical, usually with constricted margins between seeds .... 4. *A. tortuosa*

1. Acacia macracantha Humb. & Bonpl. *ex* Willd., Sp. Pl. **4**: 1080. 1806.

Shrub or small tree to 10 m tall, with many spreading branches from trunk; bark grayish brown, smooth to fissured; branches cylindrical, densely pubescent, becoming glabrous with age, lenticellate, zigzag-shaped. Leaves 2.5–16 cm long; pinnae 8–18 pairs; rachis furrowed, densely pubescent, with 1 basal gland and 1–3 distal glands; leaflets 18–32 pairs per pinna, 3–4 mm long, oblong, chartaceous, the apex obtuse, the base rounded, unequal, the margins ciliate; stipules straight, spinose, 1–4.5 cm long, persistent. Heads globose, 5–8 mm diam., congested in leaf axils; peduncles 1–2 cm long. Calyx bell-shaped, ca. 1 mm long, greenish; corolla bell-shaped, ca. 1.5 mm long, bright orange-yellow; stamens long-exserted, orange-yellow. Legume 9–11 × 0.9–1 cm, flattened, straight or slightly curved, puberulent, the margins constricted between the seeds, indehiscent. Seeds ca. 6 mm long, widely elliptic, slightly compressed.
DISTRIBUTION: A common tree of dry coastal scrub. Bordeaux (A2617), Fish Bay (A3915), Hansen Bay (A1805). Also on Jost van Dyke, St. Croix, St. Thomas, and Tortola; the Greater Antilles, Lesser Antilles, and Venezuela.
COMMON NAME: stink-casha.

2. Acacia muricata (L.) Willd., Sp. Pl. **4**: 1058. 1806. *Mimosa muricata* L., Syst. Nat., ed. 10, **2**: 1311, 1504. 1759. *Senegalia muricata* (L.) Britton & Rose, N. Amer. Fl. **23**: 113. 1928.

Acacia nudiflora Willd., Sp. Pl. **4**: 1058. 1806.

Tree 2–10(–15) m tall; bark grayish, forming rectangular plates; branches cylindrical, lenticellate, tomentose when young. Leaves 12–16 cm long; pinnae 3–6 pairs; rachis furrowed, puberulent, with a discoid gland between each pair of pinnae; leaflets 10–14 pairs per pinna, 1–1.7 cm long, oblong, chartaceous, the apex rounded, the base rounded-cordate, unequal, the margins entire; stipules minute, deciduous. Spikes axillary, cylindrical, 7–15 cm long. Calyx bell-shaped, 1–1.2 mm long, pubescent; corolla bell-shaped, cream-colored, 2–2.2 mm long; stamens long-exserted. Legume 9–15 × 1.7–2.2 cm, flattened, usually curved, glabrous, the margins constricted between the seeds or entire, elastically dehiscent from top to bottom. Seeds 1.5 cm long, oblong, flattened.
DISTRIBUTION: Common in moist to dry forests and coastal scrub. Along road to Bordeaux (M17056), Hansen Bay (A1802), Lameshur (B513). Also on Jost van Dyke, St. Croix, St. Thomas, Tortola, and Virgin Gorda; the Greater and Lesser Antilles.
COMMON NAME: amaret.

3. Acacia retusa (Jacq.) R.A. Howard, J. Arnold Arbor. **54**: 459. 1973. *Mimosa retusa* Jacq., Enum. Syst. Pl. 34. 1760; Select. Stirp. Amer. Hist. 267. 1763. Fig. 119A-D.

Mimosa paniculata H. West *ex* Vahl, Eclog. Amer. **3**: 39. 1809, non Willd., 1806. *Acacia westiana* DC., Prodr. **2**: 464. 1825. *Senegalia westiana* (DC.) Britton & Rose, N. Amer. Fl. **23**: 119. 1928.

Acacia riparia sensu Britton & P. Wilson, Bot. Porto Rico **5**: 353. 1924, non Kunth, 1823.

Liana to 15 m long; stems 4-dangled, deeply furrowed, sometimes splitting into 4 longitudinal sections when old; branches slightly angled, with numerous recurved spines. Leaves 6–12 cm long; pinnae 3–8 pairs; rachis flattened, pubescent, with basal and distal stipitate, annular glands, sometimes with recurved spines; leaflets 16–22 pairs per pinna, 3–5 mm long, oblong, chartaceous, the apex obtuse, the base cordate-obtuse, unequal, the margins entire; stipules minute, early deciduous. Heads globose, 1–1.5 cm diam., white, in terminal panicles. Calyx bell-shaped, 0.6–1 mm long, glabrous; corolla bell-shaped, 2–2.5 mm long; stamens white, exserted. Legume 6–15 × 1.5–2 cm, flattened, straight, glabrous or tomentose, the margins ribbed, chartaceous, dehiscent along both sutures. Seeds 4–6 mm long, lenticular, reddish brown.
DISTRIBUTION: A common weed of dry districts and open disturbed areas. East End (A3791), Bordeaux area (A4055). Also on St. Thomas, Tortola, and Virgin Gorda; Hispaniola (a doubtful record), Puerto Rico (including Vieques and Culebra), Lesser Antilles, and northern South America.
COMMON NAMES: catch and keep, white police.

4. Acacia tortuosa (L.) Willd., Sp. Pl. **4**: 1083. 1806. *Mimosa tortuosa* L., Syst. Nat., ed. 10, **2**: 1312. 1759.

Shrub or small tree to 7 m tall, many-branched from base; bark brown, smooth; branches cylindrical, finely lenticellate, glabrous with age, zigzag-shaped, spiny. Leaves 3–6 cm long; pinnae 4–6 pairs; rachis furrowed, pubescent, with 1 basal gland and 2 distal glands; leaflets 8–18 pairs per pinna, 3–5 mm long, oblong, coriaceous, with midvein prominent on lower surface, the apex obtuse, the base cordate-obtuse, unequal, the margins ciliate; stipules spinescent, 1–6 cm long, persistent, whitish with age. Heads globose, 6–8 mm diam., bright yellow, solitary or congested in leaf axils; peduncles 1–2 cm long. Calyx bell-shaped, 1.2–1.5 mm long, pubescent on upper portion; corolla bell-shaped, 2 mm long; stamens long-exserted, bright yellow. Legume 8–16 × 0.5–0.6 cm, linear-oblong, nearly cylindrical, constricted between the seeds, slightly curved, puberulent, dark brown, nearly woody, indehiscent. Seeds ca. 4 mm long, ovoid.
DISTRIBUTION: A common tree of dry coastal scrub. Hurricane Hole (A2765). Also on St. Croix, St. Thomas, and Tortola; the Greater Antilles, Lesser Antilles, and northern South America.
COMMON NAME: casha.

5. Acacia vogeliana Steud., Nomencl. Bot., ed. 2, **1**: 1840. *Lysiloma vogeliana* (Steud.) Stehlé, Bull. Mus. Hist. Nat. (Paris), ser. 2, **18**: 193. 1946.

Acacia ambigua Vogel, Linnaea **10**: 600. 1836, non Hoffmanns., 1826. *Lysiloma ambigua* (Vogel) Urb., Ark. Bot. **22(8)**: 28. 1929.

Liana to 10 m long, with numerous lateral, twining branches; stems cylindrical, striate and puberulent when young. Leaves 6–13 cm long, puberulent; pinnae 5 pairs, pulvinulate, with a pair of linear glands above the pulvinulus; rachis furrowed, puberulent,

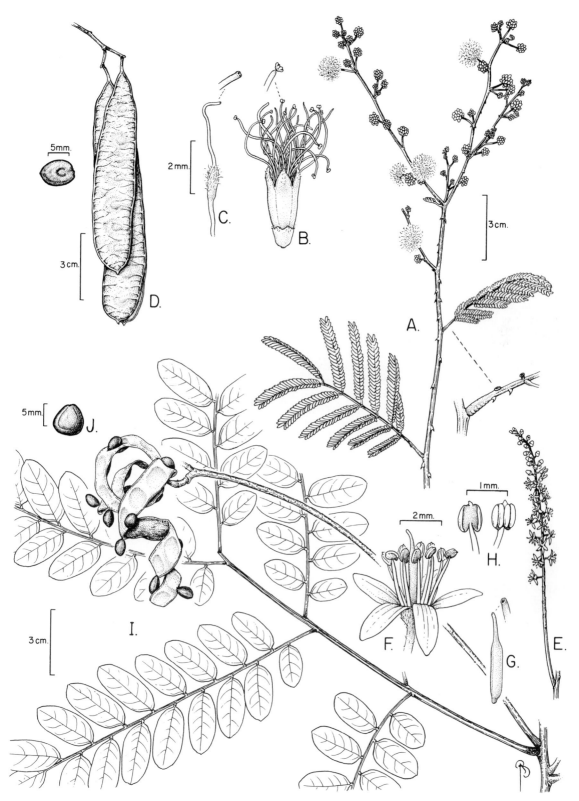

FIG. 119. A–D. *Acacia retusa*. **A.** Flowering branch and detail of stem showing leaf petiole with annular glands. **B.** Flower and detail of tip of stamen showing anther. **C.** Pistil and detail of tip of style showing stigma. **D.** Seed and legumes. **E–J.** *Adenanthera pavonina*. **E.** Inflorescence. **F.** Flower. **G.** Pistil and detail of stigma. **H.** Apex of stamens showing dorsal and frontal views of anthers. **I.** Fruiting branch **J.** Seed.

with annular glands between the two proximal and distal pairs of pinnae; leaflets 15–16 pairs per pinna, 5–10 mm long, oblong-lanceolate, chartaceous, the apex obtuse, mucronulate, the base rounded-obtuse, unequal, the margins entire; stipules minute, early deciduous. Heads globose, ca. 1 cm diam., white to cream-colored, in terminal panicles to 20 cm long. Calyx bell-shaped, 0.6–1 mm long, strigose; corolla white to cream-colored, bell-shaped, strigose, 2–2.5 mm long; stamens white, exserted. Legume 9–16 × 2.5–3 cm, oblong, flattened, chartaceous, glabrous, stipitate, the margins ribbed and slightly wavy. Seeds 4–6 mm long, elliptic, light brown, prominent on fruit walls.

DISTRIBUTION: Uncommon plant, known only from a single locality. Coral Bay Quarter along dirt road to Bordeaux, a few meters from Center Line Road (A3794). Also on Hispaniola, Puerto Rico, and probably Martinique.

NOTE: This species has been long considered to be a member of the genus *Lysiloma*. However, Rupert Barneby (NY) on studying the specimens belonging to this taxon has determined that it properly belongs to *Acacia* rather than *Lysiloma*.

DOUBTFUL RECORD: *Acacia farnesiana* (L.) Willd. was reported by Britton and Wilson (1924) to occur on St. John, but no specimen has been located and the species has not been recently collected on the island to confirm this record.

41. ADENANTHERA L.

Trees. Leaves twice pinnate; pinnae opposite, the leaflets usually medium-sized, numerous, alternate; petioles and rachis without nectariferous glands; stipules wanting or early deciduous. Flowers bisexual, in narrow racemes; calyx bell-shaped, with 5 teeth at apex; corolla yellow, short-tubular or with 5 free petals; stamens 10, exserted, the filaments united at base, the anthers with a deciduous apical gland; ovary sessile, with several ovules, the style filiform, the stigma punctiform. Legume linear, straight, compressed, distended over seeds, dehiscent by twisting valves; seeds rounded, slightly compressed, bright red.

A genus of 8 species native to tropical Asia, Australia, and the Pacific islands.

1. Adenanthera pavonina L., Sp. Pl. 384. 1753.

Fig. 119E–J.

Tree to 13 m tall; bark brown, smooth, slightly fissured; branches cylindrical, smooth, glabrous. Leaves 15–28 cm long; pinnae 3–5 pairs; rachis furrowed, glabrous, without glands; leaflets alternate, 11–21 per pinna, 1–4.3 cm long, oblong to ovate, chartaceous, the apex rounded and mucronulate, the base rounded-obtuse, asymmetrical, the margins entire; stipules lacking. Racemes axillary, to 25 cm long. Calyx 1–1.2 mm long, puberu-

lent; petals free to base, cream to white, 2.7–3 mm long; stamens as long as the corolla. Legume 15–20 × 1.3–1.5 cm, linear, flattened, glabrous, dehiscent from top to bottom by twisting valves. Seeds 8–9 mm long, ovoid-compressed, slightly angled, bright orange-red, shiny.

DISTRIBUTION: Occasional in moist forests. Along trail to Petroglyphs (A2930). Native to tropical Asia, persistent after cultivation; naturalized throughout the tropics.

COMMON NAMES: coquelicot, jumbie bead.

42. ALBIZIA Durazz.

Trees. Leaves twice-pinnate; pinnae opposite, the leaflets numerous, opposite, with palmate venation; petioles and rachis with nectariferous glands; stipules deciduous. Flowers in globose heads or narrow racemes, these axillary, solitary or in panicles; heads with bisexual flowers or with bisexual flowers and 1–2 central, staminate, larger flowers; calyx bell-shaped, with 5 teeth at apex; corolla yellow or pink, bell-shaped or funnel-shaped, 5-lobed; stamens numerous, exserted, the filaments united at base into a short tube, ovary nearly sessile, with several ovules, the style filiform, the stigma capitate, minute. Legume oblong, flattened, papery to leathery, tardily dehiscent; seeds elliptic to rounded, flattened.

A genus of 75–150 species, native to the tropics.

1. Albizia lebbeck (L.) Benth., London J. Bot. **3**: 87. 1844. *Mimosa lebbeck* L., Sp. Pl. 516. 1753. *Acacia lebbeck* (L.) Willd., Sp. Pl. **4**: 1066. 1806.

Fig. 120A–E.

Tree to 12 m tall; bark gray, smooth; branches cylindrical, lenticellate, glabrous. Leaves 10–20 cm long; pinnae 2–5 pairs; rachis nearly cylindrical, glabrous to puberulent, with a solitary elliptic gland near the base of the petiole; leaflets opposite, 4–16 per pinna, 2.5–5 cm long, oblong to ovate, chartaceous, the apex rounded and notched, the base rounded-obtuse, asymmetrical, the margins entire; stipules early deciduous. Heads axillary, hemi-

spherical, with a central larger, staminate flower, the peduncles to 10 cm long. Calyx 2.5–3 mm long, pubescent; corolla 8–10 mm long, light yellow; stamens yellow, the filaments 4 times longer than the corolla. Legume 13–24 × 3–4 cm, flattened, straw-colored, the margins thickened, straight or depressed between seeds, dehiscent along one suture. Seeds 9–10 mm long, elliptic, flattened, light brown.

DISTRIBUTION: A few individuals along roadsides. Lind Point (A2516, A3981). Also on St. Croix, St. Thomas, and Tortola; native to tropical Asia, cultivated and naturalized in tropical America.

COMMON NAMES: kitty katties, rattle bush, sheck sheck, Tibet tree, tibit, tipit, woman's tongue.

43. DESMANTHUS Willd., nom. cons.

Erect or scandent herbs or subshrubs. Leaves twice-pinnate; pinnae opposite, the leaflets opposite; rachis canaliculate, with nectariferous glands between lower pair of pinnae, distally ending in minute setae; stipules subulate, persistent. Flowers bisexual, unisexual (staminate), or sterile, in axillary, long-pedunculate globose heads; bracts subulate, strongly 1-nerved; calyx bell-shaped, with 5 teeth at apex; corolla white or greenish, of 5 petals, these free, united at base or lacking; stamens 10 or 5, exserted, the filaments free; ovary subsessile, with several ovules, the style filiform, dilated at apex, the stigma truncate, concave, or punctiform. Legume linear, flattened, straight or curved, dehiscent; seeds transverse, ovate.

A genus of 24 species native to the neotropics.

REFERENCE: Luckow, M. 1993. Monograph of *Desmanthus* (Leguminosae: Mimosoideae). Syst. Bot. Monogr. 38: 1–166.

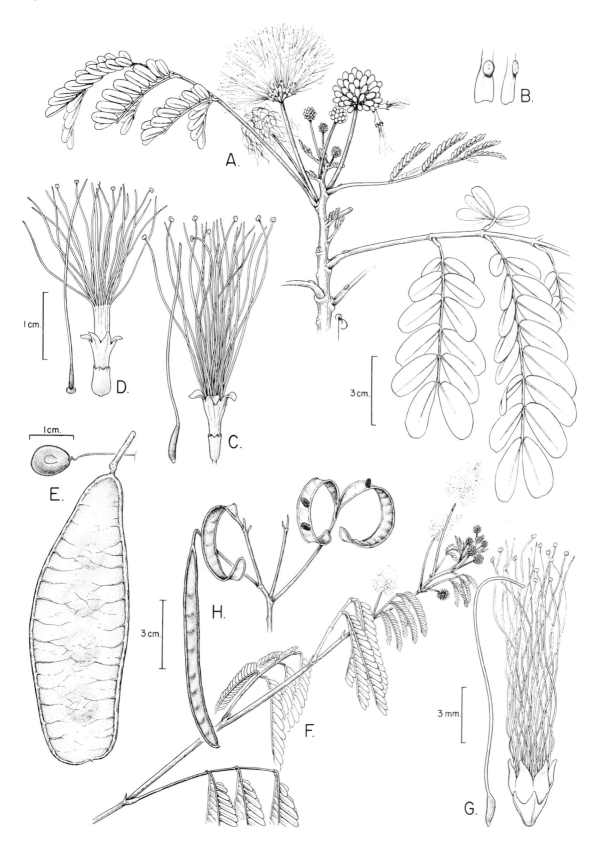

FIG. 120. A–E. *Albizia lebbeck*. **A.** Flowering branch. **B.** Detail of leaf petiole bases showing glands. **C.** Pistil and bisexual flower. **D.** Pistillode and terminal staminate flower. **E.** Seed and legume. F–H. *Zapoteca portoricensis*. **F.** Flowering branch. **G.** Pistil and flower. **H.** Dehiscing legumes.

1. Desmanthus virgatus (L.) Willd., Sp. Pl. **4:** 1047. 1806. *Mimosa virgata* L., Sp. Pl. 519. 1753. *Acuan virgatum* (L.) Medik., Theodora 62. 1786. *Acacia virgata* (L.) Gaertn., Fruct. Sem. Pl. **2:** 317. 1791. Fig. 121.

Acuan bahamense Britton & Rose, N. Amer. Fl. **23:** 132. 1928.

Desmanthus depressus Humb. & Bonpl. *ex* Willd., Sp. Pl. **4:** 1046. 1806.

Erect or decumbent herb or subshrub, 0.5–2 m tall, few-branched from a taprooted base; branches angled, with corky ridges, reddish or green, sparsely pubescent. Leaves 4.4–11 cm long; pinnae 2–4 pairs; rachis canaliculate, sparsely pubescent, with an annular gland at base of lowest pair of pinnae; leaflets opposite, 9–21 pairs, 5–10 × 1.2–3.3 mm, oblong to linear, chartaceous, the apex rounded and mucronate, the base rounded-acute, asymmetrical, the margins ciliate; stipules 3–6 mm long, bristle-like, auriculate at base, usually pubescent. Inflorescences (heads) axillary, on peduncles 1–4.5 cm long, with bisexual, male and sterile, sessile flowers; bracts triangular. Calyx 1–3 mm long, glabrous, with white margins; petals oblanceolate, 3–4 mm long, white; stamens 10, 5–7 mm long; ovary glabrous, linear. Legume 4.5–8.5 × 0.3–4 cm, linear, flattened, the valves brown, lighter along the margins, dehiscent with the sutures curling back over the valves. Seeds 13–22 per pod, transversally inserted, 2.4–3.2 mm long, flattened, black or reddish brown.

DISTRIBUTION: Common in open and disturbed areas. Cruz Bay (A2316), Fish Bay (A3904), Reef Bay Trail (A2722). Also on Anegada, St. Croix, St. Thomas, Tortola, and Virgin Gorda; throughout the West Indies, Guyana, and Surinam to the coast of northeastern Brazil.

NOTE: The material herein recognized under *D. virgatus* has recently been annotated by Dr. Denton as *D. leptophylus* Kunth, *D. pernambucanus* (L.) Thell., and *D. virgatus*. In reexamining this material, I could see no strong differences that would lead me to recognize three species rather than just *D. virgatus*. The only character that would separate the St. Johnian material into more than one species is the habit, a rather variable feature that seems to be related to habitat.

44. INGA Mill.

Trees. Leaves once-pinnate, large; leaflets opposite; rachis usually winged, with nectariferous glands between leaflets; stipules minute. Flowers bisexual, in axillary spikes, globose heads, racemes, or umbels; calyx bell-shaped or tubular, with 5 minute teeth at apex; corolla white, funnel-shaped, with 5 more or less elongate lobes or these united at base; stamens numerous, long-exserted, the filaments of unequal length united into a tube at base; ovary sessile, with several ovules, the style filiform, the stigma terminal, small, capitate or punctiform. Legume oblong, straight, flattened, swollen or angled, indehiscent; seeds transverse, ellipsoid, surrounded by fleshy, white pulp.

A genus of 200 species native to the Neotropics.

1. Inga laurina (Sw.) Willd., Sp. Pl. **4:** 1018. 1806. *Mimosa laurina* Sw., Prodr. 85. 1788. Fig. 122G–K.

Mimosa fagifolia L., Sp. Pl. 516. 1753. *Inga fagifolia* (L.) Willd. *ex* Benth., Trans. Linn. Soc. **30:** 607. 1875, non G. Don, 1832.

Shrub or small tree to 10 m tall; bark gray, smooth, with horizontal lines of lenticels; branches cylindrical, lenticellate, glabrous. Leaves 10–20 cm long; rachis furrowed and slightly flattened along upper surface, with a discoid gland between the pairs of leaflets; leaflets opposite, (1–)2(–3) pairs, 5–12 cm long, elliptic to ovate, coriaceous, the apex acuminate to obtuse, the base obtuse-attenuate, asymmetrical, the margins entire; stipules early deciduous. Spikes axillary, 4–12 cm long. Calyx 1–1.5 mm long, puberulent; corolla 3.5–4 mm long, green; filaments white, 3 times longer than the corolla. Legume 10–15 × 2.5–3 cm, coriaceous, oblong, flattened, green, the thickened margins slightly wavy. Seeds several ca. 2 cm long, oblong, flattened.

DISTRIBUTION: A common tree of moist forests. Bordeaux area (A2877, B561). Also on St. Croix, St. Thomas, Tortola, and Virgin Gorda; Greater and Lesser Antilles and Trinidad.

COMMON NAMES: lady finger tree, pomshock, Spanish oak, sweet pea.

45. LEUCAENA Benth.

Shrubs or small trees. Leaves twice-pinnate; pinnae opposite; rachis with a nectariferous gland between the two lower pinnae; leaflets numerous, opposite; stipules minute, deciduous. Flowers bisexual, in axillary, globose heads; peduncle with 2 fused bracts at or near the apex; calyx bell-shaped to funnel-shaped, with 5 minute teeth at apex; corolla white, of 5 free, oblanceolate petals; stamens 10, long-exserted, the filaments of unequal length, free, the anthers pilose; ovary stipitate, with several ovules, the style filiform, the stigma punctiform. Legume oblong, flattened, straight, with ribbed margins, dehiscent; seeds transverse, lenticular or ovate.

A genus of 10 species native to Central America.

REFERENCE: National Research Council. 1984. *Leucaena:* Promising forage and tree crop for the tropics. Ed. 2. National Academy Press, Washington, D.C.

1. Leucaena leucocephala (Lam.) de Wit, Taxon **10:** 54. 1961. *Mimosa leucocephala* Lam., Encycl. **1:** 12. 1785. Fig. 122A–F.

Acacia glauca Willd., Sp. Pl. **4:** 1075. 1806. *Leucaena glauca* (Willd.) Benth., J. Bot. (Hooker) **4:** 416. 1842.

Small tree to 5 m tall; bark light brown to grayish, smooth, slightly fissured, usually with lenticels; branches cylindrical, lenti-cellate, pubescent when young. Leaves 14–22 cm long; pinnae opposite, 5–7 pairs; rachis furrowed along upper surface, with a discoid gland between the lower pair of pinnae; leaflets opposite, 9–14 pairs per pinna, 7–14 mm long, oblong, chartaceous, the apex acute, the base rounded-obtuse, asymmetrical, the margins entire; stipules minute. Inflorescence axillary, the peduncles 3–4 cm long, the heads white, 1.5–2 cm diam. Calyx 1.2–1.5 mm long, pubescent; petals 2.5–2.9 mm long; filaments twice as long as the petals. Legume 15–20 × 1.5–2 cm, oblong, flattened, coriaceous, turning from green to straw-colored, the margins thick-

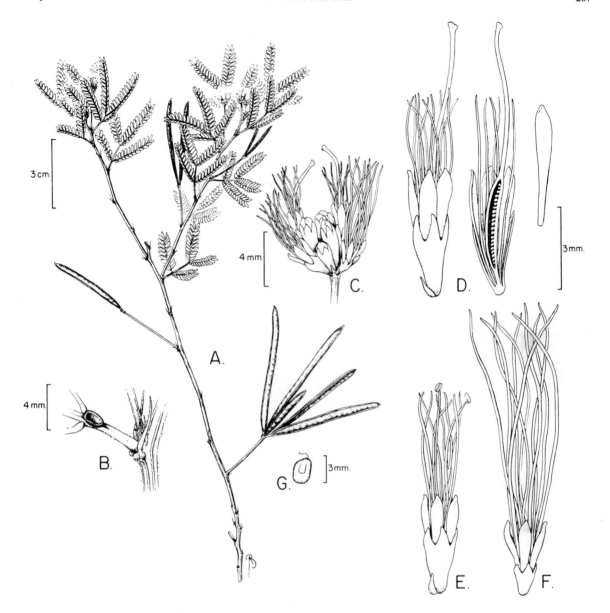

FIG. 121. *Desmanthus virgatus*. **A.** Fertile branch. **B.** Detail of stem showing leaf petiole with gland between lower pair of pinnae and stipules at base. **C.** Head. **D.** Bisexual flower, l.s. bisexual flower, and petal. **E.** Staminate flower. **F.** Sterile flower. **G.** Seed. (From Grimes, J. and R. Barneby, in prep., Mimosaceae. In A. R. A. Görts-van Rijn (ed.), Flora of the Guianas.)

ened, straight or slightly wavy, tardily dehiscent from top to bottom. Seeds 7–8 mm long, oblong-elliptic, flattened, light brown, shiny.

DISTRIBUTION: Common throughout the island, from moist forests to dry coastal scrubs. Susannaberg (M16576), Lameshur (A2753). Common on most of the Virgin Islands; native to Central America but widely cultivated throughout the tropics.

COMMON NAMES: tan tan, tanty, wild tamarind.

46. MIMOSA L.

Herbs or erect or scandent shrubs; stems with numerous prickles. Leaves twice-pinnate; pinnae opposite, the leaflets numerous and opposite; petiole and rachis without nectariferous glands; stipules minute, deciduous or persistent. Flowers bisexual or wholly staminate, in axillary or terminal globose heads; bracts small, usually shorter than the corolla, subpersistent; calyx minute nearly cup-shaped, with 5 minute teeth at apex; corolla yellow or pink, funnel-shaped with 3–6 petals; stamens as many or twice as many as the petals, long-exserted, the filaments free; ovary stipitate, with several ovules, the style filiform, the stigma punctiform. Legume oblong, flattened, papery, straight, breaking away from ribbed entire or spiny margins, beaked at apex, indehiscent; seeds transverse, lenticular to widely ovate.

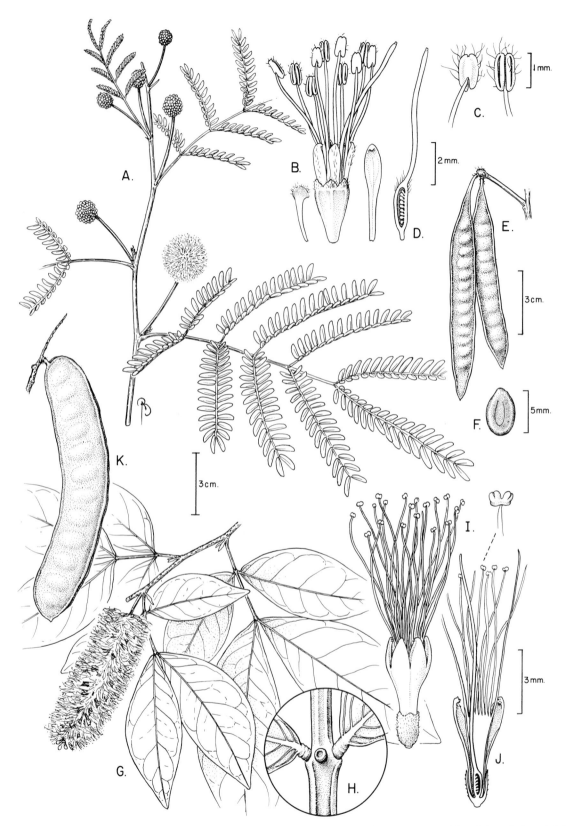

Fig. 122. A–F. *Leucaena leucocephala*. **A.** Flowering branch. **B.** Flower bract, flower, and petal. **C.** Anther, dorsal and frontal views. **D.** L.s. pistil. **E.** Legumes. **F.** Seed. **G–K.** *Inga laurina*. **G.** Flowering branch. **H.** Detail of leaf petiole showing gland between lower leaflets. **I.** Flower. **J.** L.s. flower and detail of stamen. **K.** Legume.

A genus of about 450 species, with pantropical distribution, predominantly from the neotropics.

REFERENCE: Barneby, R. C. 1991. Sensitivae Censitae. A description of the genus *Mimosa* L. (Mimosaceae) in the New World. Mem. New York Bot. Gard. 65: 1–835.

Key to the Species of *Mimosa*

1. Scandent shrub to 6 m long; stems glabrous, 4-angled, striate; leaves with spiny rachis, the pinnae 4 or 5 pairs; corolla 3-merous .... 1. *M. ceratonia*
1. Erect to decumbent herb to 50 cm tall; stems pilose, cylindrical; leaf rachis without spines, the pinnae 1 or 2 pairs; corolla 4-merous .... 2. *M. pudica*

1. Mimosa ceratonia L., Sp. Pl. 523. 1753. *Lomoplis ceratonia* (L.) Raf., Sylva Tellur. 118. 1838.
Fig. 123A–F.

Scandent shrub to 6 m long, many-branched from base; branches obtusely 4-angled, glabrous, striate, with numerous recurved prickles. Leaves 7–15 cm long; pinnae 4 or 5 opposite pairs; rachis furrowed along upper surface, with numerous prickles; leaflets opposite, 3–8 pairs per pinna, 1–1.5 cm long, obliquely obovate to nearly rounded, chartaceous, the apex rounded, the base rounded-obtuse, asymmetrical; stipules 8–10 mm long, lanceolate. Heads arranged on terminal racemes, the peduncles 1–2 cm long, with numerous prickles, the heads 1.3–1.7 cm diam. Calyx 0.7–1 mm long, glabrous; corolla 3-merous, ca. 2 mm long, pink (seldom white); filaments white, twice as long as the corolla. Legume 4–6 × 1.5–1.7 cm, coriaceous, oblong, flattened, turning from green to straw-colored, dehiscent by separation of the valves from a persistent, thickened, spiny margin. Seeds 7–8 mm long, oblong-elliptic, flattened, dark brown.

DISTRIBUTION: Common weed of dry coastal scrub. Bethany (B200). Also on St. Croix, St. Thomas, Tortola, and Virgin Gorda; Hispaniola, Puerto Rico, and the Lesser Antilles.

COMMON NAMES: amaretsteckel, black amaret.

2. Mimosa pudica L., Sp. Pl. 518. 1753.

Erect to decumbent herb to 50 m tall, with woody base; stems cylindrical, reddish when young, sparsely pilose, with a few recurved prickles, rooting at nodes on contact with the soil. Leaves 7–15 cm long, sensitive to touch; pinnae 1 or 2 opposite pairs on long pilose petiole; leaflets opposite, 14–25 pairs per pinna, 7–12 mm long, obliquely oblong-oblanceolate, chartaceous, the apex obtuse, the base rounded-obtuse, asymmetrical, the margins ciliate; stipules 10–12 mm long, awl-shaped, ciliate. Heads on long (1.5–2.5 cm), pilose peduncles, solitary or clustered in leaf axils, the heads 7–8 mm diam. Flowers 4-merous; calyx 0.1–0.2 mm long, membranous, reddish; corolla 2.2–2.5 mm long, pinkish; filaments pink, twice as long as the corolla. Legume 5–20 × 4 mm, oblong, flattened, constricted between seeds, dehiscent by the separation of the valves from a persistent, thickened margin with straight bristles. Seeds several, ca. 1.7 mm long, oblong-elliptic, flattened.

DISTRIBUTION: Common weed of open disturbed areas. Maho Bay Quarter along Center Line Road (A2410). Also on St. Croix, St. Thomas, Tortola, and Virgin Gorda; widespread throughout the tropics.

COMMON NAMES: grishi grishi, gritchee, sensitive plant.

47. PITHECELLOBIUM Mart., nom. cons.

Trees or shrubs. Leaves twice-pinnate; pinnae and leaflets opposite, 1 to several pairs; petiole and rachis with nectariferous glands; stipules usually spiny, persistent. Flowers bisexual, in globose heads, these long-pedunculate, solitary or in axillary or terminal racemes; calyx minute, cup-shaped to bell-shaped, with 5 minute teeth at apex; corolla yellow or pink, bell-shaped, 5- or 6-lobed; stamens numerous, long-exserted, the filaments basally fused into a tube, this free from the corolla; ovary stipitate, with several ovules, the style filiform, the stigma punctiform. Legume oblong, turgid, leathery, curved and twisted, dehiscent by twisting valves; seeds variously shaped, usually arillate and hanging from valves.

A genus of about 200 species with pantropical distribution.

1. Pithecellobium unguis-cati (L.) Benth., London J. Bot. 3: 200. 1844. *Mimosa unguis-cati* L., Sp. Pl. 517. 1753.
Fig. 123G–M.

Shrub or small tree to 6 m tall; bark light brown to grayish, smooth, slightly fissured and warty, sometimes with persistent spines; branches usually arched, cylindrical, striate, armed with persistent stipular spines. Leaves 3–10 cm long; pinnae 2, opposite; leaflets opposite, 2 per pinna, 2.5–6 × 1.5–5 cm long, obliquely obovate to nearly rounded, coriaceous, glabrous, the apex obtuse or rounded, the base rounded-obtuse, asymmetrical, the margins entire; stipules 5–10 mm long, spinelike. Heads subglobose, 1.5–2.5 cm wide, few-flowered, borne on axillary racemes. Flowers 5-merous; calyx 1.5–2 mm long, green, glabrous; corolla 4–5 mm long, yellow; the filaments yellow, twice as long as the petals. Legume 5–10 × 1 cm, twisted, torulose, turgid, turning from green to brown, dehiscent from top to bottom by twisting valves exposing several hanging seeds. Seeds ca. 8 mm long, ellipsoid, shiny black, covered with a red aril on lower half.

DISTRIBUTION: Common shrub of coastal scrublands and open disturbed areas. Great Cruz Bay (A2355), Enighed (A4006). Also on Anegada, Jost van Dyke, St. Croix, St. Thomas, and Virgin Gorda; Florida, throughout the West Indies, Venezuela.

COMMON NAMES: blackbead, bread and cheese, crabprickle, goatbush.

48. SAMANEA (Benth.) Merr.

Large trees. Leaves twice-pinnately compound; pinnae opposite, 3–4(–7) pairs, the leaflets numerous, opposite, with pinnate venation; petioles and rachis with nectariferous glands; stipules herbaceous, lanceolate, early deciduous. Flowers heteromorphic, produced in axillary or terminal, umbelliferous heads, the peripheral flowers at least shortly pedicellate, the distal one thicker, sessile; calyx of

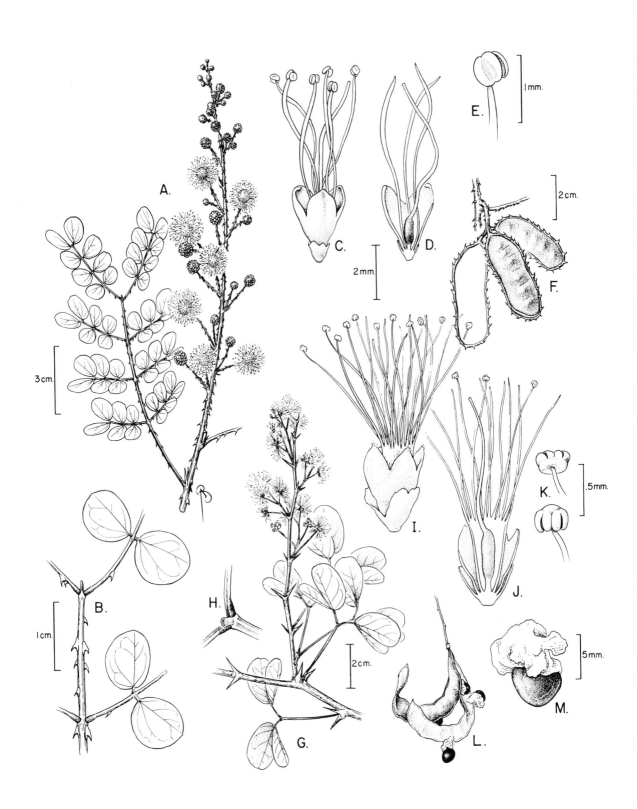

Fɪɢ. 123. A–F. *Mimosa ceratonia*. **A.** Flowering branch. **B.** Detail of leaf showing recurved prickles on rachis. **C.** Flower. **D.** L.s. flower. **E.** Anther. **F.** Legumes and persistent fruit margin. **G–M.** *Pithecellobium unguis-cati*. **G.** Flowering branch. **H.** Detail of leaf rachis showing gland. **I.** Flower. **J.** L.s. flower. **K.** Anthers. **L.** Legume dehiscing. **M.** Seed with aril.

peripheral flowers vase-shaped; corolla pink, bell-shaped or funnel-shaped, 5-lobed; stamens 20–36, exserted, the filaments united at base into a short tube; ovary (sub)sessile, narrowly ellipsoid, with several ovules, the style filiform, the stigma minute, capitate. Legume subsessile, broad-linear, straight or nearly so, valves thick pulpy when mature, either adherent or narrowly septiferous between seeds, indehiscent; seeds transverse, oblong-ellipsoid, hard opaque.

A genus of 3 species native to tropical continental America, but 1 species (*S. saman*) planted and naturalized in the West Indies and the Old World.

REFERENCE: Merrill, E. D. 1916. Systematic position of the rain tree. J. Wash. Acad. Sci. 6: 46–47.

1. Samanea saman (Jacq.) Merr., J. Wash. Acad. Sci. **6**: 47. 1916. *Mimosa saman* Jacq., Fragm. Bot. 15. 1800. *Pithecellobium saman* (Jacq.) Benth., London J. Bot. **3**: 216. 1844. *Albizia saman* (Jacq.) F. Muell., Select Pl. 12. 1876.

Tree 8–25 m tall; bark gray, rough; branches widely spreading to form a wide crown; stems cylindrical and pubescent when young. Leaves 25–35 cm long; pinnae 3–6 pairs; rachis furrowed along upper surface, pubescent, with a discoid gland between each pair of pinnae; leaflets opposite, 4–7 pairs per pinna, 2–4.5 cm long, obliquely obovate to oblong, coriaceous, pubescent on lower surface, the apex obtuse, the base rounded-obtuse, asymmetrical,

the margins entire; stipules minute, early deciduous. Inflorescence axillary or terminal, the peduncles 4–6 cm long. Calyx 6–8.5 mm long, tomentose; corolla 11–13 mm long, pink; filaments white at base but pink toward the tips, 3 times longer than the corolla. Legume 15–22 × 1.4–1.6 cm, straight or slightly curved, compressed, fleshy-leathery, the margins thickened, apiculate at apex. Seeds several, 5–8 mm long, oblong to nearly square, flattened, darker at the center.

DISTRIBUTION: Cultivated on St. John, with a few spontaneous individuals. Coral Bay (A2631, A2817), Lameshur Bay (A737). Apparently native to northern South America, planted and naturalized in the West Indies and throughout the tropics.

COMMON NAMES: licorice, rain tree.

49. ZAPOTECA H. M. Hern.

Erect or scandent shrubs. Leaves twice-pinnate; pinnae opposite, of several pairs; leaflets opposite, numerous, small; petioles seldom with a nectariferous glands; stipules leafy, persistent. Flowers bisexual, sessile, in globose heads, these long-pedunculate, axillary, solitary or congested; bracts small; calyx cup-shaped, with 5 or 6 minute teeth at apex; corolla white, bell-shaped to funnel-shaped, with 5 or 6 reflexed lobes; nectary disk present; stamens numerous, long-exserted, the filaments basally fused into a tube, this free from the corolla; ovary shortly stipitate, with several ovules, the style filiform, the stigma distal, cup-shaped. Legume oblong, straight, leathery, with ribbed margins, elastically dehiscent from apex to base; seeds nearly elliptic to ovate.

A genus of 8 species, distributed from southwestern United States to northern Argentina, including the Greater Antilles.

REFERENCE: Hernández, H.M. 1986. *Zapoteca:* a new genus of neotropical Mimosoideae. Ann. Missouri Bot. Gard. 73: 755–763.

1. Zapoteca portoricensis (Jacq.) H.M. Hern., Ann. Missouri Bot. Gard. **73**: 758. 1986. *Mimosa portoricensis* Jacq., Collectanea **4**: 143. 1790 [1791]. *Acacia portoricensis* (Jacq.) Willd., Sp. Pl. **4**: 1069. 1806. *Calliandra portoricensis* (Jacq.) Benth., London J. Bot. **3**: 99. 1844. *Anneslia portoricensis* (Jacq.) Britton, Brooklyn Bot. Gard. Mem. **1**: 50. 1918. Fig. 120F–H.

Erect shrub 1–2(–3) m tall, many-branched from near the base; bark light brown, smooth; branches cylindrical, smooth, glabrous or puberulous. Leaves 7–15 cm long; pinnae 2–4, opposite; leaflets opposite, 15–30 pairs per pinna, 8–15 × 1.5–2.5 mm, oblong, asymmetric, chartaceous, the apex obtuse or rounded, the base truncate-attenuate, very asymmetrical, the margins ciliate; stipules 5–10 mm long, lanceolate, persistent. Heads obconical,

3–3.5 cm wide, white, on long (3–6 cm) peduncles, solitary or clustered in leaf axils. Flowers 5-merous; calyx 2 mm long, glabrous; corolla 3–4 mm long, light green; the filaments white, to 2 cm long. Legume 7.5–10 × 0.8–1 cm, oblong, flattened, straight or curved, the margins entire, thickened. Seeds 5–6 mm long, nearly elliptic, mottled brown.

DISTRIBUTION: A common shrub of dry areas, also found in open disturbed areas. Along road to Ajax Peak (A2654), Bordeaux area (A3238). Also on St. Thomas and Tortola; Jamaica, Hispaniola, Puerto Rico, Grenada, continental tropical America.

CULTIVATED SPECIES: *Calliandra haematocephala* Hassk. with bright red flowers and *Enterolobium cyclocarpum* (Jacq.) Griseb. with pinkish flowers are occasionally used as ornamental plants.

DOUBTFUL RECORD: *Prosopis juliflora* (Sw.) DC. was reported by Woodbury and Weaver (1987) as occurring on St. John, but no specimen has been located nor has the species been recently collected on the island to confirm this record.

39. Flacourtiaceae (Flacourtia Family)

Trees or shrubs. Leaves alternate or rarely opposite, simple, entire or glandular serrate; blades usually glandular punctate or lineate; stipules early deciduous or lacking. Flowers small to large, actinomorphic, bisexual or rarely unisexual, usually in axillary cymes or racemes, seldom solitary; calyx of 3–9 distinct to connate sepals; corolla of 3–8 distinct petals, sometimes poorly differentiated from the sepals or wanting; staminodes usually present; stamens numerous or less often only 4, centrifugal, the filaments distinct or connate into clusters, the anthers opening by longitudinal slits or terminal pores; nectary disk usually annular, external or internal to the stamens; ovary superior (rarely inferior), compound of 2–10 multiovular parietal placentae, unilocular or seldom multilocular with axile placentation, the styles distinct, or united. Fruit a berry, or less often a loculicidal capsule or a drupe, sometimes with accrescent sepals.

A tropical family with about 85 genera and 800 species.

REFERENCE: Sleumer, H. O. 1980. Flacourtiaceae. Fl. Neotrop. Monogr. 22: 1–499.

Key to the Genera of Flacourtiaceae

1. Leaves with 3 main veins arising from a rounded to cordate base; blades usually with a pair of rounded glands close to the petiole; stipules 1–1.5 cm long, deciduous, leaving a conspicuous scar; inflorescence terminal, racemose .... 2. *Prockia*
1. Leaves with a single midvein, the base obtuse, attenuate, cuneate or rounded; blades without glands at base; stipules <5 mm long, usually early deciduous or wanting; inflorescence axillary, fasciculate.
    2. Trunk with branched spines; leaves rigid-coriaceous; stipules wanting; flowers unisexual; fruit a berry .... 4. *Xylosma*
    2. Trunk smooth, without spines; leaves coriaceous-chartaceous; stipules minute, usually early deciduous; flowers bisexual; fruit a fleshy capsule.
        3. Lower surface of leaf glabrous or puberulent, tertiary veins not prominent; sepals free or only connate at base, reflexed; staminodes alternating with filaments, filaments free .... 1. *Casearia*
        3. Lower surface of leaf pubescent, tertiary veins very prominent on lower surface; sepals connate into a tubular calyx; staminodes wanting; filaments connate into a tube adnate to the calyx tube .... 3. *Samyda*

1. CASEARIA Jacq.

Trees or shrubs. Leaves alternate, entire to serrate; blades glandular punctate or lineate; petioles present; stipules minute, deciduous. Flowers bisexual, sessile or pedunculate, in axillary fascicles; bracts small, persistent; calyx of 5–9 sepals, connate at base; corolla wanting; staminodes internal to or between the stamens; stamens 6–24, the filaments distinct, the anthers opening by longitudinal slits; ovary unilocular, with 3 multiovular wall placentae, the style simple or 3-branched at apex, the stigma capitate. Fruit a fleshy or dry, usually 3-valved capsule; seeds few to numerous, covered with a fleshy coat.

A pantropical genus of 180 species, with numerous species in the New World.

Key to the Species of *Casearia*

1. Style short, stigmas 3, glabrous, the ovary glabrous; sepals and pedicels glabrous .... 3. *C. sylvestris*
1. Style elongate and simple, the stigma capitate and pubescent, the ovary pubescent; sepals and pedicels pubescent to puberulent.
    2. Lower surface of leaves puberulent along veins, not barbate in axils of veins; stipules pubescent; sepals puberulent; stamens 8; staminodes lanceolate; fruits ellipsoid, ridged, brown, the pericarp thick .... 2. *C. guianensis*
    2. Lower surface of leaves barbate (tuft of hairs) in axils of veins; stipules glabrous; sepals pubescent; stamens 10; staminodes oblong–club-shaped; fruits globose, smooth, yellow, the pericarp thin .... 1. *C. decandra*

1. Casearia decandra Jacq., Enum. Syst. Pl. 21. 1760. Fig. 124A–D.

Casearia parvifolia Willd., Sp. Pl. **2**: 628. 1799.

Shrub or small tree 4–6 m tall; bark light brown to gray, smooth; branches cylindrical, pubescent when young. Leaf blades 2.5–9 × 1.5–3.5 cm, elliptic, ovate, or lanceolate-elliptic, chartaceous, pellucid-punctate, glabrous except for a tuft of hairs in the axils of veins along lower surface, the apex acute or acuminate, the base obtuse, asymmetrical, the margins finely serrate; petioles pubescent to glabrous, 3–7 mm long; stipules 2–4 mm long, oblong to linear, tardily deciduous. Flowers in axillary fascicles; pedicels puberulent, 5–8 mm long. Calyx whitish to greenish white, 3.5–4 mm long, pubescent, the sepals 5, oblong-lanceolate, reflexed at anthesis; staminodes pubescent-pilose, oblong-club-shaped; stamens 10, the filaments pilose, exceeding the sepals; ovary ovoid, tomentose, the style elongate, the stigma capitate, pubescent-papillose. Capsule globose, 1–1.4 cm diam., splitting from bottom to top, turning from green to light yellow, apiculate at apex, the pericarp thin. Seeds 5–6 mm long, nearly ellipsoid, covered with a tan aril.

DISTRIBUTION: A common shrub of moist areas, usually along roadsides and open disturbed areas. Emmaus (A735), Susannaberg (A2805). Also on St. Croix, St. Thomas, Tortola, and Virgin Gorda; common in the West Indies south from Hispaniola and from Honduras to Brazil and Argentina.

2. Casearia guianensis (Aubl.) Urb., Symb. Ant. **3**: 322. 1902. *Iroucana guianensis* Aubl., Hist. Pl. Guiane **1**: 329, t. 127. 1775.

Casearia ramiflora Vahl, Symb. Bot. **2**: 50. 1791.

Shrub or small tree 2–6 m tall; bark light brown to gray; branches cylindrical, pubescent when young. Leaf blades 5–15 × 2.5–7 cm, elliptic, obovate or oblong, chartaceous, pellucid punctate, puberulent along lower surface, especially along veins, the apex short, usually abruptly acuminate, the base obtuse, asymmetrical, the margins serrate; petioles tawny pubescent, 3–10(–13) mm long; stipules ca. 2 mm long, awl-shaped, pubescent, tardily deciduous. Flowers in axillary fascicles; pedicels 4–6 mm long, puberulent. Calyx white to greenish yellow, 4–5 mm long, appressed pubescent without, the sepals 5, oblong; staminodes densely pilose, lanceolate, reflexed in anthesis; stamens 8, the filaments glabrous, shorter than the sepals; ovary ovoid, tomentose, the style elongate, the stigma capitate, pubescent-papillose. Capsule ellipsoid, with 6 longitudinal ribs, 6–10 mm diam., splitting from top to bottom, turning from green to brown, the pericarp thick. Seeds 5–6 mm long, nearly ellipsoid, covered with a bright red aril.

DISTRIBUTION: A common tree of moist secondary forests and open disturbed areas. Adrian Ruins (A2700), trail to Genti Bay (P29266). Also

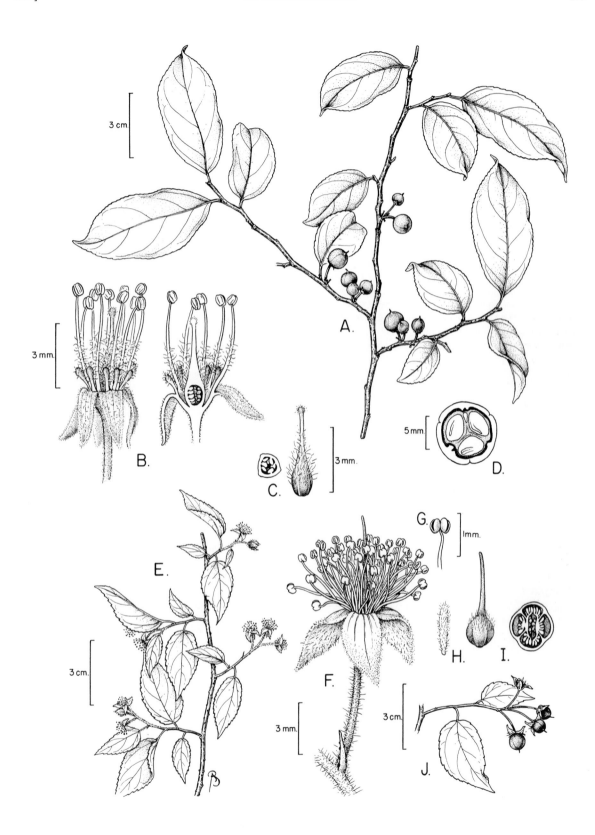

Fɪɢ. 124. A–D. *Casearia decandra.* A. Fruiting branch. B. Flower and l.s. flower. C. C.s. ovary and pistil. D. C.s. fruit. E–J. *Prockia crucis.* E. Flowering branch. F. Flower and subtending bract. G. Stamen. H. Petal. I. Pistil and c.s. ovary. J. Fruiting branch.

on St. Croix, St. Thomas, and Tortola; the Greater Antilles, Lesser Antilles, Panama to northern South America south to Brazil.

3. Casearia sylvestris Sw., Fl. Ind. Occid. 2: 752. 1798.

> Samyda parviflora L., Syst. Nat., ed. 10, 2: 1025. 1759, non Loefl. ex L., 1758, nec Sessé & Moc., 1893. Casearia parviflora (L.) J.F. Gmel., Syst. Nat. 2: 700. 1791, non Jacq., 1780.

Shrub or small tree 2–6 m tall; bark beige, smooth, lenticellate; branches cylindrical, glabrous. Leaf blades 5–14 × 2–5 cm, elliptic to lanceolate, chartaceous to coriaceous, pellucid punctate, glabrous, the apex long-acuminate, the base obtuse-rounded, asym-metrical, the margins finely serrate to crenulate; petioles glabrous, 3–8 mm long; stipules minute, triangular. Flowers in axillary fascicles; pedicels glabrous, 2–5 mm long. Calyx white, ca. 2 mm long, glabrous, the sepals 5, ovate, erect at anthesis; stami-nodes slightly pilose, oblong; stamens 10, the filaments puberulent, equaling the sepals; ovary ovoid, glabrous, the style short, 3-branched, the stigma glabrous, smooth. Capsule globose, smooth, 4–5 mm diam., splitting from top to bottom, turning from green to purple, the pericarp thin. Seeds ca. 2 mm long, angled-ellipsoid, covered with a bright red-orange aril.

DISTRIBUTION: A common tree of moist secondary forests and open disturbed areas. Adrian Ruins (A2109), Susannaberg (A713, A2801). Also on St. Croix, St. Thomas, and Tortola; the Greater Antilles, the Lesser Antilles and continental tropical America.

2. PROCKIA P. Browne ex L.

Shrubs or small trees. Leaves alternate; blades serrate, with 3 main veins from base, not punctate or lineate, the base usually with a pair of glands; petioles present; stipules foliaceous, deciduous or persistent. Flowers bisexual, pedunculate, in terminal racemes; bracts minute, persistent; calyx of 3 free sepals; petals 3, smaller and alternate with the sepals; staminodes wanting; stamens numerous, the filaments distinct, the anthers opening by longitudinal slits; ovary 3–5-locular, with multiovular wall placentae, the style simple, the stigma capitate. Fruit a drupe, with a persistent style; seeds numerous, embedded in white pulp, not arillate.

A genus of 2 species, 1 in Venezuela, the other widespread throughout tropical America.

1. Prockia crucis P. Browne ex L., Syst. Nat., ed. 10, 2: 1074. 1759.   Fig. 124E–J.

Shrub or small tree to 8 m tall; bark light gray, smooth; branches cylindrical, glabrous. Leaf blades 3.5–13 × 2–7 cm, ovate to broadly elliptic, chartaceous, not punctate, glabrous or with a few hairs, the apex long-acuminate, the base cordate, usually with a pair of glands at the junction with the petiole, the margins serrate; petioles 0.5–3 cm long; stipules 1–1.5 cm long, lanceolate-falcate, serrate. Calyx yellowish, the sepals 7–8 mm long, pubescent within, broadly ovate; petals yellowish, 3–3.5 mm long, lanceolate, pubescent, early deciduous; stamens numer-ous, the filaments glabrous, of unequal length; ovary ovoid to nearly globose, glabrous, the style elongate. Drupe globose, gla-brous, 5–7 mm diam., black. Seeds ca. 1.5 mm long, angled-ellipsoid, light brown, smooth.

DISTRIBUTION: An uncommon tree of moist forests. Bordeaux Mountain (W218), Fish Bay (A2671). Also on St. Thomas; tropical America.

3. SAMYDA Jacq.

Shrubs or small trees. Leaves alternate, entire to serrate, pellucid-punctate or lineate; petioles present; stipules deciduous or persis-tent. Flowers bisexual, pedunculate, solitary or 2–3 per leaf axil; bracts minute, persistent; calyx tubular or bell-shaped, 4–6-lobed; petals wanting; staminodes wanting; stamens 8–20, in 1 whorl, the filaments connate for most of their length, adnate to the calyx tube, the anthers opening by longitudinal slits; ovary unilocular, with 3–5 multiovular wall placentae, the style simple, the stigma lobed to capitate. Fruit a fleshy capsule, dehiscing by 3–5 valves; seeds numerous, covered with a fleshy aril.

A genus of 9 species, found in Mexico, Central America, and the West Indies.

1. Samyda dodecandra Jacq., Enum. Syst. Pl. 21. 1760.   Fig. 125A–D.

> Samyda decandra Jacq., Enum. Syst. Pl. 21. 1760.
> Samyda serrulata L., Sp. Pl., ed. 2, 1: 558. 1762.

Shrub 1–3 m tall; bark light brown, smooth; branches cylindri-cal, densely ferruginous pubescent. Leaf blades 3–11 × 1.5–6 cm, elliptic or ovate, coriaceous, pellucid punctate, pubescent, especially on the lower surface, the apex acute to rounded, often mucronate, seldom retuse, the base rounded or obtuse, the margins sharply serrate; stipules ca. 1 mm long, lanceolate, early decidu-ous; petioles to 5 mm long. Calyx bell-shaped, 1.5–1.8 cm long, pubescent, green without, white within, the sepals 5, spreading, slightly reflexed, unequal, 6–8 mm long, fleshy, oblong; stamens 10–12, the staminal tube 3–5 mm long, pubescent without; ovary ovoid, densely pubescent, the style equal to or exceeding the staminal tube, the stigma nearly cup-shaped, green. Capsule fleshy, ovoid to globose, pubescent, 1.2–1.5 cm diam., green with reddish tinge without, orange to red within, splitting from top to bottom, the pericarp thick. Seeds 2.5–3.3 mm long, nearly ovoid, with a fleshy yellowish aril.

DISTRIBUTION: Found in dry forests and in coastal scrub. Bethany (B340), trail to Brown Bay (A1875). Also on St. Croix, St. Thomas, and Tortola; the Greater Antilles (except Jamaica) and the Lesser Antilles.

4. XYLOSMA G. Forst., nom. cons.

Shrubs or small trees; stems often armed with branched, axillary spines. Leaves alternate, entire to serrate, not pellucid-punctate or lineate; petioles present; stipules wanting. Flowers unisexual, pedunculate, in axillary racemes or fascicles; bracts minute, persistent; calyx of 4 or 5 shortly connate sepals; petals wanting; staminodes wanting; nectary disk glandular, external to the stamens. Staminate flowers with 8 or more stamens, the filaments free, the anthers opening by longitudinal slits; pistillodes wanting. Pistillate flowers lacking stamens, the ovary unilocular, with 2–3, few-ovulate wall placentae; the style simple or 2–branched, the stigma flattened. Fruit an indehiscent, few-seeded berry.

A genus of about 100 species, most of which occur in the neotropics.

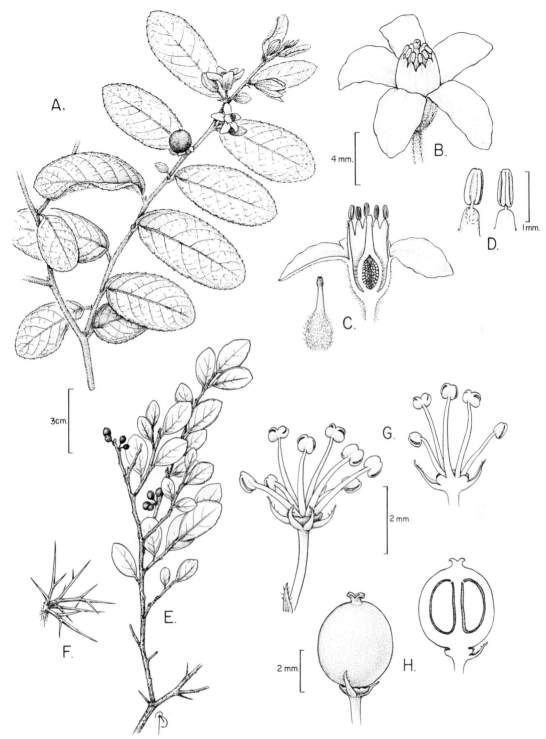

Fig. 125. A–D. *Samyda dodecandra*. A. Fertile branch. B. Flower. C. Pistil and l.s. flower. D. Detail of lobes of staminal tube showing anthers. E–H. *Xylosma buxifolia*. E. Fruiting branch. F. Detail of stem showing branched spines. G. Staminate flower and l.s. staminate flower. H. Fruit and l.s. fruit.

1. Xylosma buxifolia A. Gray *in* Griseb., Pl. Wright. 155. 1860. *Myroxylon buxifolium* (A. Gray) Krug & Urb., Bot. Jahrb. Syst. 15: 300. 1892.   Fig. 125E–H.

Dioecious shrub 2–8 m tall; trunk armed with branched spines; branches angled, glabrescent, usually with axillary spines, the spines 4–15 mm long. Leaf blades 1.5–5 × 0.8–3.2 cm, elliptic to obovate, rigid coriaceous, glabrous, the apex acute to rounded, often mucronate, seldom retuse, the base cuneate or obtuse, the margins revolute, entire or crenulate-serrate; petioles 1–3 mm long. Flowers in axillary fascicles; sepals lanceolate, 1–1.5 mm long, ciliate, reflexed; nectary disk lobed, glabrous; stamens 8–20, the filaments 2–3 mm long; ovary ovoid, glabrous, the styles

2, connate at base, the stigma 2-lobed. Berry ovoid to globose, glabrous, 4–5 mm diam., turning from red to black. Seeds 2–4 per fruit, ca. 3 mm long, nearly ovoid.

DISTRIBUTION: An uncommon shrub found in moist forests to coastal thickets. Bordeaux Mountain (W686). Also on St. Croix and Virgin Gorda; throughout the West Indies (except Jamaica).

40. Goodeniaceae (Goodenia Family)

Perennial herbs or scarcely woody shrubs. Leaves alternate, rarely opposite or whorled, simple and entire; stipules wanting. Flowers medium-sized, strongly zygomorphic, bisexual, in axillary cymes or racemes, seldom solitary; calyx tubular, 5-lobed; corolla tubular, bilabiate; stamens 5, the anthers connate into a tube where the pollen is collected; nectary disk of 1 or 2 glands internal to the stamens; ovary inferior or rarely superior, compound, of 2(–4) carpels, the carpels with 1 to many ovules, the placentation basal, the style single. Fruit a capsule or drupe.

A tropical and subtropical family with 14 genera and about 300 species, mainly in Australia, New Zealand, and the Pacific Islands.

1. SCAEVOLA L., nom. cons.

Shrubs or subshrubs. Leaves alternate, sessile or petiolate, fleshy, glabrous, or stellate pubescent. Flowers bisexual, solitary or in cymes in leaf axils; calyx tubular or annular, entire or 5-lobed at apex; corolla of various colors, unilabiate, the tube split to the base along one side, the lobes 5, opposite to the split; stamens with free filaments, anthers connate, opening along longitudinal slits; ovary inferior, of 2 uniovular locules, the stigma simple or 2-lobed, usually hairy. Fruit a 1- or 2-seeded, fleshy drupe.

A genus of about 100 species, primarily found in Western Australia, 2 species widely distributed throughout coastal areas.

1. Scaevola plumieri (L.) Vahl, Symb. Bot. **2**: 26. 1791. *Lobelia plumieri* L., Sp. Pl. 929. 1753. Fig. 126.

Weak shrub to 1 m tall, many-branched from base, the lower branches decumbent; branches cylindrical to slightly angular, glabrous, becoming square and hollow with age, with numerous leaf scars. Leaves usually clustered at ends of branches; blades 5–8 × 2–5 cm, obovate or oblanceolate, fleshy, glabrous, the apex rounded, the base tapering into a semiclasping base, the margins entire. Flowers in axillary cymes; hypanthium cup-shaped, calyx annular or 5-lobed, 1–1.5 mm long, glabrous; corolla 5-lobed, 1.5–2 cm long, the tube densely pubescent within, the lobes lanceolate, white; stamens much shorter than the tube; ovary ovoid, the style as long as the floral tube, slightly arched. Drupe bluish black, ovoid to globose, glabrous, 6–8 mm diam., 1-seeded.

DISTRIBUTION: An uncommon shrub found along sandy coasts. Hansen Bay (A4222). Also on Anegada, St. Croix, and Tortola; widespread on tropical sandy coasts.

41. Lamiaceae (Mint Family)

Herbs or shrubs. Leaves opposite or rarely whorled, simple, entire or serrate, usually aromatic; stipules wanting. Flowers small to large, strongly zygomorphic, bisexual or rarely unisexual, produced in terminal or axillary cymes, heads, racemes, or spikes; calyx more or less tubular, 5-lobed; corolla tubular, strongly zygomorphic, usually bilabiate or seldom with 1 or 4 lobes; stamens 4, 2 with longer filaments or rarely 2 stamens modified into staminodes, the filaments adnate to the corolla tube, the anthers opening by longitudinal slits; ovary superior, sometimes on a gynophore, with an annular or unilateral nectary disk at base, the ovary of 2 carpels, 4-locular, the style gynobasic or between the ovary lobes, ovules 1 per locule, the placentation basal, the stigmas 2, minute. Fruit of 4, 1-seeded nutlets, drupaceous or dry, calyx sometimes persistent in fruit.

A cosmopolitan family with about 200 genera and 3200 species.

Key to the Genera of Lamiaceae

1. Stamens declined or resting on lower lip of corolla.
   2. Calyx conspicuously 2-lipped; lower corolla lobe planar or spreading .... 4. *Ocimum*
   2. Calyx lobes equal; lower corolla lobe pouchlike... 1. *Hyptis*
1. Stamens erect or ascending, not declined or resting on lower lip of corolla.
   3. Stamens 4; upper corolla lobe concave.
      4. Leaves deeply 3–5-lobed; inflorescence a glomerule to 1.5 cm wide; corolla lavender .... 3. *Leonurus*
      4. Leaves not lobed, serrate; headlike inflorescence to 4.5 cm wide; corolla orange..................... 2. *Leonotis*
   3. Stamens 2; upper corolla lobe not concave .... 5. *Salvia*

1. HYPTIS Jacq., nom. cons.

Herbs or subshrubs. Leaves opposite, usually serrate. Flowers bisexual, produced in heads, spikes, or panicles; calyx tubular or bell-shaped, regular, usually enlarging in fruit, the lobes 5, equal or nearly so; corolla purple or white, bilabiate, the upper lip 2-lobed, the lower lip 3-lobed, with middle lobe pouchlike; stamens 4, declined onto lower lip; ovary on enlarged base, the style elongate. Fruit of 2 or 4, 1-seeded, ovoid, cylindrical or flattened nutlets.

A genus of 400 species, native to tropical America.

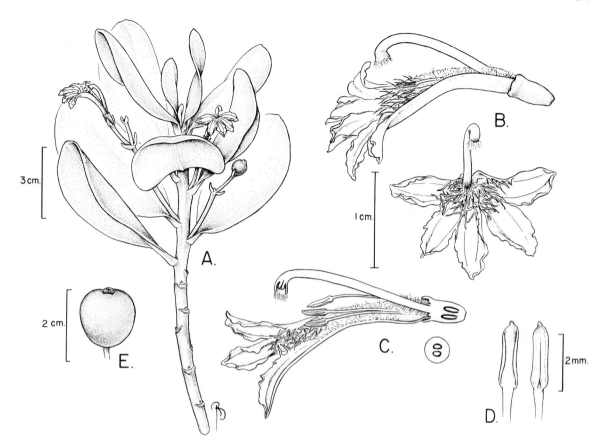

FIG. 126. *Scaevola plumieri.* **A.** Fertile branch. **B.** Flower, lateral and top views. **C.** L.s. flower and c.s. ovary. **D.** Anther, dorsal and frontal views. **E.** Fruit.

Key to the Species of *Hyptis*

1. Inflorescence of globose heads .... 1. *H. capitata*
1. Inflorescence racemose or cymose, not forming globose heads.
   2. Flowers congested in leaf axils or in terminal spikes; calyx puberulent, with subulate-lanceolate lobes .... 4. *H. verticillata*
   2. Flowers in axillary racemes or terminal panicles; calyx pubescent, with bristle-like lobes.
      3. Stems pilose; calyx 9–10 mm long .... 3. *H. suaveolens*
      3. Stems puberulent; calyx 5.5–6 mm long .... 2. *H. pectinata*

1. Hyptis capitata Jacq., Icon. Pl. Rar. **1:** 11, t. 114. 1781.

Erect herb or subshrub to 2 m tall; stem 4-angled, hollow, sparsely pubescent. Leaf blades 3–12 × 2–6 cm, ovate or lanceolate, chartaceous, sparsely pubescent, with numerous glandular dots, the apex acute, the base tapering or cuneate, the margins deeply and irregularly serrate; petioles 1–3 cm long. Flowers sessile, in axillary, peduncled heads, the heads 1.5–3 cm diam.; peduncle 1–3.5 cm long; bracts oblong-lanceolate, truncate at apex; calyx tubular, 6–8 mm long, sparsely pubescent, the lobes bristle-like, 1.5–2 mm long; corolla white, 3–5 mm long, pubescent at apex; stamens slightly exserted. Nutlets 4, ca. 1.5 mm long, angular-ellipsoid, light brown, smooth.

DISTRIBUTION: A common roadside weed. Bethany (B261). Also on St. Croix, St. Thomas, and Tortola; apparently native to tropical America, established in the Old World tropics.

COMMON NAME: wild hops.

2. Hyptis pectinata (L.) Poit., Ann. Mus. Natl. Hist. Nat. **7:** 474. 1806. *Nepeta pectinata* L., Syst. Nat., ed. 10, **2:** 1097. 1759.　　　　Fig. 127D–F.

Erect herb or subshrub to 2 m tall; branches lateral, ascending; stem obtusely 4-angled, hollow, sparsely puberulent. Leaf blades 3–8 × 2–4.5 cm, ovate, aromatic, chartaceous, pubescent, especially along lower surface, with numerous glandular dots, the apex acute to acuminate, the base obtuse to nearly truncate, the margins serrate to double serrate; petioles 1–3.5 cm long. Flowers short-pedicellate, in axillary racemes or terminal panicles; peduncle ca. 2 mm long; bracts filiform, ca. 5 mm long; calyx green, nearly bell-shaped, 5.5–6 mm long, whitish-pubescent, the lobes bristle-like, 1.5–2 mm long, with tufts of hairs between lobes; corolla lavender to pink, 1.5–2 mm long, pubescent without; stamens

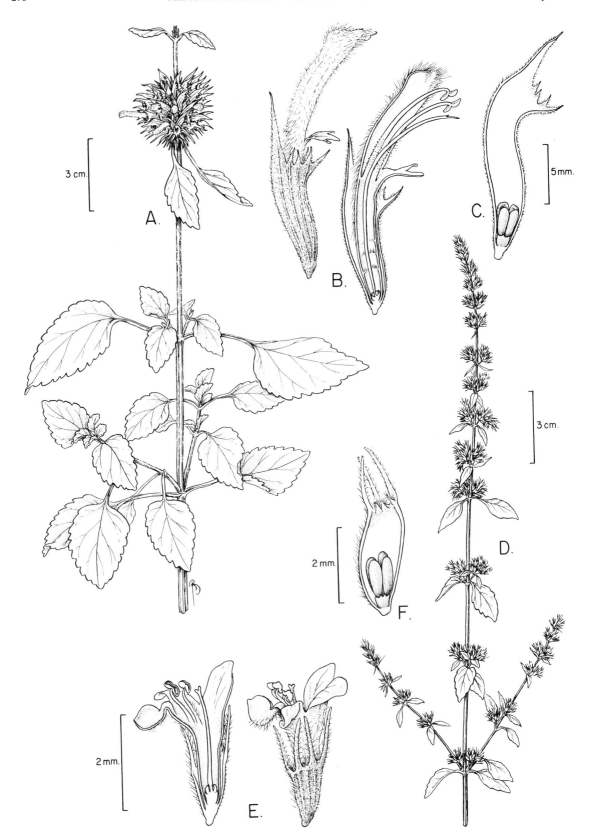

FIG. 127. A–C. *Leonotis nepetifolia*. **A.** Fertile branch. **B.** Flower and l.s. flower. **C.** L.s. calyx showing maturing nutlets. **D–F.** *Hyptis pectinata*. **D.** Fertile branch. **E.** L.s. flower and flower. **F.** L.s. calyx showing maturing nutlets.

slightly exserted. Nutlets 4, ca. 0.9 mm long, oblong, black, smooth.

DISTRIBUTION: A common weed of disturbed, open ground. Bordeaux area (A2587). Also on St. Croix, St. Thomas, Tortola, and Virgin Gorda; a pantropical weed.

COMMON NAME: French tea.

3. Hyptis suaveolens (L.) Poit., Ann. Mus. Natl. Hist. Nat. 7: 472. 1806. *Ballota suaveolens* L., Syst. Nat., ed. 10, 2: 1100. 1759.

Erect herb or subshrub to 1.5 m tall, usually many-branched from base; stem obtusely 4-angled, hollow, pilose, becoming glabrous. Leaf blades 1.3–8 × 1.2–5 cm, ovate, chartaceous, aromatic, pubescent, the apex obtuse or rounded, the base rounded to subcordate, the margins serrate; petioles 0.5–4 cm long. Flowers short-pedicellate, in axillary racemes or terminal panicles; peduncle 4.5–5.5 mm long; bracts filiform, ca. 2 mm long; calyx green, bell-shaped, with 10 prominent ribs, 9–10 × 4–4.5 mm long, whitish-pilose, with a few glandular hairs, the lobes bristle-like, 2–3 mm long, with numerous hairs between lobes; corolla light violet, 6.5–8 mm long, pubescent externally; stamens not exserted.

Nutlets 2, 4–4.5 mm long, oblong, beaked, flattened, blackish, rugulose.

DISTRIBUTION: A common weed of disturbed, open ground. Johnson Bay (A4027), Lameshur (B503). Also on St. Croix, St. Thomas, Tortola, and Virgin Gorda; throughout tropical America, naturalized in the Old World.

4. Hyptis verticillata Jacq., Icon. Pl. Rar. 1: 11. 1781.

Erect herb or subshrub to 2.5 m tall with numerous lateral branches; stem obtusely 4-angled, not hollow, puberulent, becoming glabrous. Leaf blades 2.5–9.5 × 0.5–3 cm, lanceolate, chartaceous, puberulent on lower surface, the apex acute or acuminate, the base acute, the margins irregularly serrate; petioles 0.5–2 cm long. Flowers congested in leaf axils or in long terminal spikelike racemes; pedicels 0.5–0.6 mm long; bracts filiform, ca. 0.5 mm long; calyx green, bell-shaped, 1.5–2 mm long, puberulent, the lobes subulate-lanceolate, 0.6–0.8 mm long; corolla funnel-shaped, pale blue or whitish, 1.5–2 mm long, glabrous; stamens not exserted. Seeds 4, ca. 1 mm long, ellipsoid-angled, light brown.

DISTRIBUTION: A common weed of disturbed, open ground. Cinnamon Bay (M17089). Also on St. Thomas; throughout tropical America.

2. LEONOTIS R. Br.

Herbs or subshrubs. Leaves opposite, usually serrate. Flowers bisexual, mostly orange, produced in globose glomerules, these solitary or 2 to several in terminal interrupted spikes; calyx tubular, recurved at apex, the lobes 8, unequal, spinelike; corolla tubular, bilabiate, the upper lip entire, concave, the lower lip 3-lobed; stamens 4, ascending, subequal, included. Fruit of 4, 1-seeded, oblong, or 3-angled nutlets.

A pantropical genus of 40 species.

1. Leonotis nepetifolia (L.) W.T. Aiton, Hortus. Kew. 3: 409. 1811. *Phlomis nepetaefolia* L., Sp. Pl. 586. 1753.   Fig. 127A–C.

Erect herb to 2 m tall, aromatic, with short, axillary branches, with congested leaves; stem obtusely 4-angled, furrowed, pubescent. Leaf blades 4–12 × 2–10 cm, broadly ovate, chartaceous, pubescent especially along veins, the apex acute or obtuse, the base subcordate, constricted into a winged petiole, the margins deeply crenate; petioles 2.5–6 cm long. Flower glomerules globose, 3–6 cm wide, solitary or 2 to several on terminal spikes; bracts 1.2–1.4 cm long, lanceolate, spine-tipped; calyx green, 10–12 mm long, pubescent, the lobes spine-tipped; corolla tubular, orange, 1.5–2 cm long, pubescent, the upper lip much longer than the lower one; stamens exserted. Seeds 4, 3 mm long, 3-angled, wedge-shaped, dark brown to black.

DISTRIBUTION: A common weed of disturbed, open ground. Fish Bay (A3878). Also on St. Croix, St. Thomas, Tortola, and Virgin Gorda; native to tropical Africa but widespread throughout the tropics.

COMMON NAMES: hollow stalk, hollow stock, rabbitfood.

3. LEONURUS L.

Herbs. Leaves opposite, deeply lobed, the lobes incised. Flowers bisexual, mostly lavender, produced in axillary glomerules or terminal spikes; calyx top-shaped, the lobes 5, subequal, awl-shaped with spiny apex; corolla tubular, bilabiate, the upper lip rounded or notched at apex, the lower lip 3-lobed; stamens 4, erect, didynamous, included. Fruit of 4, 1-seeded, oblong, or 3-angled nutlets.

A genus of 14 species, native to temperate Asia and Europe.

1. Leonurus sibiricus L., Sp. Pl. 584. 1753.
                                                    Fig. 128A–C.

Erect herb to 50 cm tall; main stem obtusely 4-angled, puberulent, with numerous axillary branches. Leaf blades 4–12 × 5–15 cm, deeply 3–5-lobed, chartaceous, becoming glabrous, the lobes deeply incised or lobed, puberulent along lower veins, the apex acute, the base narrowed, the margins ciliate; petioles 2.5–10 cm long. Flower glomerules 1.5–2 cm wide, in leaf axils; bracts 2–3 mm long, awl-shaped; calyx green, bell-shaped, 4.5–8.5 mm long, puberulent, the lobes spine-tipped; corolla deeply bilabiate, lavender, 1.1–1.2 cm long, pubescent, the upper lip slightly longer than the lower one; stamens slightly shorter than the upper lip, the anthers violet. Seeds 4, 2.5 mm long, 3-angled, wedge-shaped, light brown, smooth.

DISTRIBUTION: A fairly common weed of disturbed, open, moist ground. Carolina (A2584). Also on St. Croix, St. Thomas, and Tortola; native to Asia but widespread from tropical to subtemperate America.

4. OCIMUM L.

Aromatic herbs or subshrubs. Leaves opposite, entire to serrulate. Flowers bisexual, pedicellate, in verticils at the nodes of terminal racemes or panicles; calyx bell-shaped, enlarged and reflexed in fruit, bilabiate, the upper lip entire, rounded or concave, the lower lip

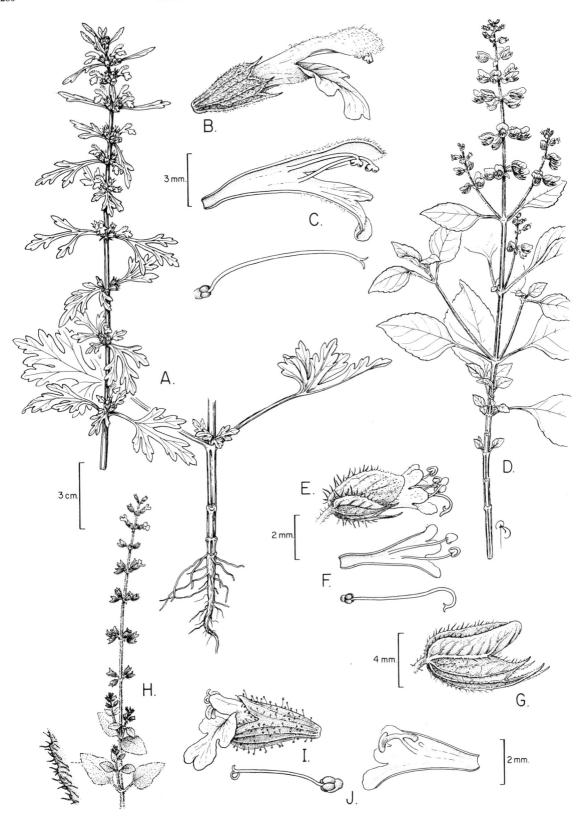

Fɪɢ. **128. A–C.** *Leonurus sibiricus*. **A.** Habit. **B.** Flower. **C.** L.s. corolla and pistil. **D–G.** *Ocimum campechianum*. **D.** Fertile branch. **E.** Flower. **F.** L.s. corolla and pistil. **G.** Calyx. **H–J.** *Salvia serotina*. **H.** Flowering branch and detail of leaf margin showing indument. **I.** Flower. **J.** Pistil and l.s. corolla.

with 4 unequal, awl-shaped teeth; corolla tubular, bilabiate, the upper lip 4-lobed, the lower lip entire, planar or spreading; stamens 4, didynamous, exserted, declinate. Fruit of 4, 1-seeded, globose to obovoid nutlets.

A tropical genus of about 100 species.

1. Ocimum campechianum Mill., Gard. Dict., ed. 8. 1768. Fig. 128D–G.

Ocimum micranthum Willd., Enum. Pl. 630. 1809.

Erect subshrub to 80 cm tall; branching pseudodichotomous or less often many-branched from base; stems obtusely 4-angled, cylindrical at base, pubescent at corners, glandular when young. Leaf blades 2.5–5 × 1.5–3.5 cm, ovate to diamond-shaped, chartaceous, glabrous except for hairs along veins, with numerous resin dots, the apex acute, the base narrowed, the margins serrate; petioles 1–3 cm long. Inflorescence to 15 cm long; bracts 3–3.5 mm long, ovate, leaflike; calyx green, bell-shaped, 3.5–8 mm long, pubescent, with numerous resin dots, persistent and covering the nutlets; corolla pink, lavender, or blue, 4–4.5 mm long, puberulent, the upper lip as long as the lower one; stamens slightly exserted. Seeds 4, ca. 2 mm long, ellipsoid, light brown.

DISTRIBUTION: A common weed of disturbed, open ground, also in understory of secondary vegetation. Coral Bay Quarter along Center Line Road (A3771). Also on St. Croix, St. Thomas, Tortola, and Virgin Gorda; widespread through tropical America.

COMMON NAMES: bellyache balsam, bitter bush plant, passia balsam.

5. SALVIA L.

Aromatic herbs or subshrubs. Leaves opposite, entire-crenate or serrate. Flowers bisexual, in verticils at the nodes of terminal racemes or spikes; calyx bell-shaped or tubular, bilabiate, the upper lip entire (less often 3-lobed), the lower lip 2-lobed; corolla tubular, bilabiate, the upper lip entire, the lower lip 3-lobed; stamens 2, erect, included or exserted. Fruit of 4, 1-seeded, oblong-ovoid nutlets.

A genus of about 500 species, with cosmopolitan distribution.

Key to the Species of *Salvia*

1. Plant puberulent to glabrescent, decumbent; leaf base narrowed; leaf margins serrate; bracts ovate .... 2. *S. occidentalis*
1. Plant pubescent or pilose, erect; leaf base cordate or truncate; leaf margins crenate-dentate; bracts elliptic to lanceolate.
    2. Plant pilose; lower leaf surface densely covered with soft, curled, white hairs; corolla white or lavender .... 3. *S. serotina*
    2. Plant pubescent (curved hairs); lower leaf surface glabrous except for hairs along veins; corolla blue or violet .... 1. *S. micrantha*

1. Salvia micrantha Vahl, Enum. Pl. 1: 235. 1804.

Erect herb to 40 cm tall, with many lateral branches along main stem; stems obtusely 4-angled, puberulent, with minute, curved hairs. Leaf blades 1.5–4.5 × 1–3.5 cm, ovate to triangular, chartaceous, glabrous except for hairs along veins on lower surface, the apex obtuse to acute, the base cordate to truncate, the margins crenate-dentate; petioles 0.5–2.5 cm long. Inflorescence to 15 cm long; bracts elliptic to lanceolate, ca. 2 mm long; calyx green, bell-shaped, 5–7 mm long, pubescent, with many capitate-glandular hairs and a few scattered resin dots; corolla with light violet or blue lobes, 7–8 mm long, glabrous, the upper lip shorter than the lower one; stamens shorter than upper lip, the anthers violet. Seeds 4, ellipsoid-flattened, 2–2.2 mm long, grayish, smooth.

DISTRIBUTION: Common herb, found in open disturbed areas. Mandal (A2575), Susannaberg (A2806). Also on St. Croix; the Greater Antilles, Lesser Antilles, Florida, and Panama.

2. Salvia occidentalis Sw., Prodr. 14. 1788.

Decumbent herb to 40 cm tall, many-branched from a thickened base; stems obtusely 4-angled, puberulent or glabrescent, reddish-tinged, swollen at nodes. Leaf blades 1.5–6 × 1–3 cm, ovate to triangular, chartaceous, sparsely pubescent on lower surface, the apex acute, the base narrowed, the margins serrate; petioles 0.5–1 cm long. Inflorescence to 20 cm long; bracts ovate, 2–2.5 mm long; calyx green, tubular, 3–3.5 mm long, densely covered with capitate-glandular hairs; corolla violet or light blue, 5–6.5 mm long, glabrous, the upper lip slightly shorter than the lower one; stamens included. Seeds 4, ellipsoid, 1.7–2 mm long, light brown, smooth.

DISTRIBUTION: Uncommon herb, found in open, moist areas. Bordeaux area (4059), along road to Peter Peak (A4108). Also on St. Croix, St. Thomas, and Tortola; widespread throughout the tropics.

3. Salvia serotina L., Mant. Pl. 1: 25. 1767.

Fig. 128H–J.

Erect subshrub to 25 cm tall, many-branched from a woody base; stems obtusely 4-angled, cylindrical at base, pilose, especially on young parts. Leaf blades 1–2.5 × 0.7–1.5 cm, ovate to triangular, chartaceous, densely pubescent on lower surface, the apex obtuse, the base rounded, truncate or nearly cordate, the margins entire to crenate-serrate; petioles 3–17 mm long. Inflorescence to 10 cm long; bracts elliptic-lanceolate, 3–4 mm long, ciliate; calyx green, bell-shaped, 4–5.5 mm long, densely covered with capitate-glandular hairs; corolla white or lavender, 6–7.5 mm long, glabrous, the upper lip shorter than the lower one; stamens included. Seeds 4, 3-angled, ellipsoid, 1.5–1.7 mm long, light brown, smooth.

DISTRIBUTION: Common herb, found in open fields and on sandy beaches. Long Bay (A671). Also on Anegada, St. Croix, St. Thomas, and Tortola; throughout the West Indies and Central America.

DOUBTFUL RECORD: The red-flowered *Salvia coccinea* Edlinger was listed by Britton (1918) as occurring on hillsides and in waste and cultivated ground. However, no specimen has been located or recollected to confirm this record.

42. Lauraceae (Avocado Family)

Trees, shrubs, or rarely twining herbs. Leaves alternate, rarely opposite, whorled or absent, simple, entire, usually aromatic; stipules wanting. Flowers small, actinomorphic, bisexual or rarely unisexual, produced in axillary or terminal cymes, racemes, or panicles, rarely solitary; tepals 6, in 1 or 2 whorls, or sometimes 9 in whorls of 3; stamens (3–)9(–12), the filaments often with a pair of nectariferous appendages, the anthers opening by 2–4 flaps or valves at the base; ovary superior or rarely inferior, unilocular, with a single pendulous ovule, the style terminal, the stigma capitate or discoid. Fruit a 1-seeded berry or drupe, sometimes with an enlarged receptacle.

A family with about 50 genera and 2000 species, with tropical and subtropical distribution.

REFERENCES: Howard, R. A. 1981. Nomenclatural notes on the Lauraceae of the Lesser Antilles. J. Arnold Arbor. 62: 45–62; Kostermans, A. J. G. H. 1957. Lauraceae. Reinwardtia 4: 193–256; Rohwer, J. G. 1993. Lauraceae: *Nectandra*. Fl. Neotrop. Monogr. 60: 1–333.

Key to the Genera of Lauraceae

1. Parasitic herbaceous twining vines; leaves scale-like, 0.1–0.2 cm long or wanting; inflorescence of short spikes .... 1. *Cassytha*
1. Nonparasitic trees or shrubs; leaves with typical regular blade, 8–25 cm long; inflorescence paniculate.
  2. Leaves with 3 main veins from near the base (lower pair of secondary veins stout and ascending).. 2. *Cinnamomum*
  2. Leaves with 1 main vein bearing weak secondary veins (pinnately veined).
    3. Fruit <7 cm long; pericarp fleshy .... *Persea* (cultivated)
    3. Fruit 0.8–2.5 cm long; pericarp leathery.
      4. Fertile stamens 3; fruiting receptacle double-rimmed; leaves strongly aromatic; bark peeling off in long thick flakes.. 3. *Licaria*
      4. Fertile stamens 9; fruiting receptacle not rimmed; leaves weakly aromatic; bark smooth .... 4. *Ocotea*

1. CASSYTHA L.

Parasitic, twining, herbaceous vines. Leaves reduced to scales. Flowers bisexual, sessile in axillary spikes; tepals 6, in 2 unequal series; fertile stamens 9, in 3 whorls; ovary globose, covered by the perianth parts. Fruit fleshy, berrylike.

A cosmopolitan genus of 20 species.

1. Cassytha filiformis L., Sp. Pl. 35. 1753.
                         Fig. 129A–F.

Herbaceous, twining, parasitic vine, 1–5 m long, densely branched, adhering to host by haustoria; stems cylindrical, to 3 mm diam., yellowish or yellow-green, glabrous. Leaves 1–2 mm long, scale-like, or wanting. Spike few-flowered, sometimes headlike; tepals white, ovate, ca. 2 mm wide. Fruit globose, light green or white, 5–7 mm diam.

DISTRIBUTION: Uncommon herb, found in open fields. Maria Bluff (W802). Also on St. Croix, St. Thomas, Tortola, and Virgin Gorda; a cosmopolitan species.

2. CINNAMOMUM Schaeff., nom. cons.

Trees or shrubs. Leaves usually aromatic, alternate or congested near the ends of branches, entire, with 3 main veins from near the base, coriaceous. Flowers bisexual, in axillary or terminal panicles; tepals 6, persistent in fruit; staminodes present; stamens 9, in 3 whorls, the inner whorl with stipitate glands; ovary sessile, the style slender. Fruit an ellipsoid berry subtended by a short cupule.

A genus of about 250 species, mostly from the Old World tropics.

REFERENCE: Kostermans, A. J. G. H. 1961. The New World species of *Cinnamomum* Trew (Lauraceae). Reinwardtia 6: 17–24.

1. Cinnamomum elongatum (Nees) Kosterm., Rein-
      wardtia **6**: 21. 1961. *Phoebe elongata* Nees, Syst.
      Laur. 116. 1836.   Fig. 129G–J.

Tree to 20 m tall; branches obtusely angled, puberulent. Leaves alternate, 4–12 × 2–6 cm, lanceolate or elliptic, chartaceous-coriaceous, with 3 main veins from near the base, puberulent, the apex acuminate or obtuse, the base acute or obtuse, the margins entire; petioles 1–2 cm long. Flowers glabrous, in short, axillary racemes or panicles; tepals ovate, ca. 3 mm long. Fruit cupule narrowly obconical, with persistent tepals, the fruit ellipsoid, black, glabrous, 1–1.5 cm long.

DISTRIBUTION: Common in moist forests. Bordeaux (A4264). Also on St. Thomas; Cuba, Hispaniola, Puerto Rico,and the Lesser Antilles.

3. LICARIA L.

Trees. Leaves alternate, rarely opposite, entire, pinnately veined, coriaceous. Flowers bisexual, in axillary or terminal racemes or panicles; tepals 6, nearly equal, early deciduous; staminodes 6, scale-like, or wanting; stamens 3, with basal glands; ovary included in perianth tube. Fruit included at base in cupular, enlarged, double-rimmed receptacle.

A neotropical genus of 45 species.

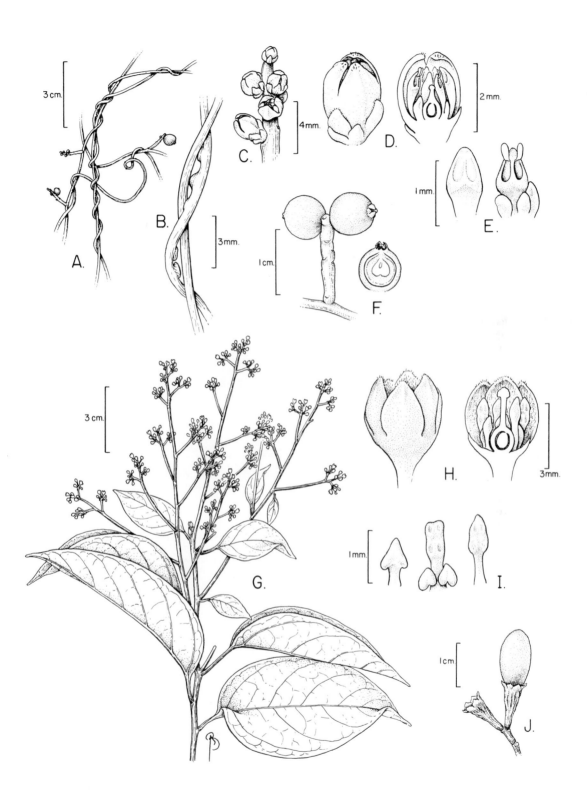

Fig. 129. A–F. *Cassytha filiformis*. **A.** Habit. **B.** Detail of stem showing haustoria. **C.** Inflorescence. **D.** Flower and l.s. flower. **E.** Anther and dehisced anther. **F.** Infructescence and l.s. fruit. **G–J.** *Cinnamomum elongatum*. **G.** Flowering branch. **H.** Flower and l.s. flower. **I.** Detail of staminode, stamen with glands, and stamen without glands. **J.** Fruiting branch.

Key to the Species of *Licaria*

1. Stamens free; staminodes present; fruit cupule to 8 mm diam.; leaves lanceolate to elliptic-lanceolate .... 1. *L. parvifolia*
1. Stamens connate into a short tube; staminodes wanting; fruit cupule 12–19 mm diam.; leaves ovate-elliptic .... 2. *L. triandra*

1. Licaria parvifolia (Lam.) Kosterm., J. Sci. Res. (Jakarta) **1**: 149. 1952. *Laurus parvifolia* Lam., Encycl. **3**: 451. 1792.

Laurus salicifolius Sw., Fl. Ind. Occid. **2**: 709. 1798. *Acrodiclidium salicifolium* (Sw.) Griseb., Fl. Brit. W. I. 280. 1860. *Licaria salicifolia* (Sw.) Kosterm., Recueil Trav. Bot. Néerl. **34**: 597. 1937.

Tree to 15 m tall; bark dark gray to brown, lenticellate, aromatic, peeling off in irregular plates; branches nearly cylindrical, ferruginous tomentose, lenticellate or smooth. Leaf blades aromatic, 5–14 × 1–3.7 cm, lanceolate, elliptic-lanceolate or elliptic when young, coriaceous, reticulate-veined, ferruginous-pubescent, but becoming glabrous with age, the apex obtuse to acuminate, the base obtuse, sometimes slightly unequal, the margins entire; petioles 6–10 mm long. Flowers ferruginous-tomentose, in short axillary racemes; hypanthium 1.5–1.7 mm long, tepals ovate, 0.5–0.7 mm long; stamens 3, the anthers 2-celled, each opening by an uplifted valve. Fruit cupule nearly woody, 5–7 mm wide, obconical, double-rimmed at apex, the fruit ellipsoid, glabrous, 10–15 mm long.

DISTRIBUTION: Occasional in moist forests. Bordeaux Mountain (A4696). Also on St. Croix, St. Thomas, and Tortola; probably in Hispan-iola (fide Lamarck, 1792), Puerto Rico (including Vieques), and the Lesser Antilles.

2. Licaria triandra (Sw.) Kosterm., Recueil Trav. Bot. Néerl. **34**: 588. 1937. *Laurus triandra* Sw., Prodr. 65. 1788. *Misanteca triandra* (Sw.) Mez, Jahrb. Königl. Bot. Gart. Berlin **5**: 103. 1889.

Fig. 130H–K.

Tree to 20 m tall; trunk to 25 cm diam.; bark dark gray or brown, lenticellate, aromatic, peeling off in irregular plates; branches angled, ferruginous-puberulent when young, lenticellate. Leaf blades aromatic, 4–12 × 1.5–5 cm, elliptic or oblong-elliptic, chartaceous–coriaceous, reticulate-veined, puberulent, glabrescent, the apex acuminate, the base obtuse, slightly unequal, the margins entire; petiole 1–1.5 cm long. Flowers ferruginous-tomentose, in terminal panicles; hypanthium 1.7–2 mm long, the tepals ovate, 0.7–1 mm long; stamens 3, the filaments connate into a short tube, the anthers 2-celled, each opening by an uplifted valve. Fruit cupule woody, green, 1.2–1.5 cm wide, cup-shaped, warty, double-rimmed at apex, the fruit ellipsoid, green, glabrous, 2–2.5 cm long.

DISTRIBUTION: Common in moist forests. Trail to Petroglyph (A4122). Also in the Greater Antilles, Lesser Antilles, and Florida.

4. OCOTEA Aubl.

Trees or shrubs. Leaves usually slightly aromatic, alternate or congested near the ends of branches, entire, pinnately veined, coriaceous. Flowers bisexual or unisexual, in axillary panicles; tepals 6, equal or unequal, early deciduous; staminodes present or wanting; stamens 9, the inner series with a pair of glands; ovary sessile, depressed-globose, the style slender. Fruit an ellipsoid to globose berry, seated on or enclosed by a cupule.

A genus of more than 500 species, mostly from tropical and subtropical America but also in tropical and southern Africa.

NOTE: The distinction between *Ocotea* and *Nectandra* has always been problematic. The characters used for distinguishing these two genera show numerous instances of intermediate stages, making the distinction of them rather unclear. A modern revision by Rohwer (1993) of some of the species in this complex recognized *Nectandra* as distinct. However, he recognized a large number of overlapping characters between the two putative genera as well. Since most species in the flora area have been treated as either genera, I opt for a simpler approach, recognizing only *Ocotea*.

Key to the Species of *Ocotea*

1. Leaf blades gall-infected (at least some of the leaves); fruit cupule warty .... 3. *O. leucoxylon*
1. Leaf blades not gall-infected; fruit cupule smooth.
  2. Leaves coriaceous, with upper shiny surface; tepals oblong or oblanceolate, densely pubescent within; flowers bisexual.
    3. Tepals oblong to oblanceolate, 4.5–5.5 mm long; stems glabrescent, shiny, grayish .... 1. *O. coriacea*
    3. Tepals oblong, 2–2.5 mm long; stems with appressed white or yellowish hairs .... 4. *O. patens*
  2. Leaves chartaceous-coriaceous, upper surface dull; tepals ovate, 2–2.5 mm long, sparsely pubescent; flowers functionally unisexual; stems with appressed white or yellowish hairs .... 2. *O. floribunda*

1. Ocotea coriacea (Sw.) Britton in Britton & Millsp., Bahama Fl. 143. 1920. *Laurus coriacea* Sw., Prodr. 65. 1788. *Nectandra coriacea* (Sw.) Griseb., Fl. Brit. W. I. 281. 1860. Fig. 130A–G.

Small tree to 10 m tall; bark dark gray or brown, smooth, spicy; branches slightly pubescent when young, becoming glabrous, grayish and slightly shiny. Leaf blades slightly aromatic, 5–15 × 1.7–5.5 cm, oblong-lanceolate, elliptic or oblong-elliptic, coriaceous, reticulate-veined, glabrous, the upper surface shiny, with primary and secondary veins drying yellowish, the apex acute or acuminate, the base obtuse, slightly unequal, the margins entire; petiole 0.5–1.5 cm long. Flowers white, fragrant, in axillary panicles, the axes reddish and sparsely pubescent; hypanthium puberulent without, very shallow, the tepals oblong to oblanceolate, 4.5–5.5 mm long, densely papillate, pubescent within; stami-

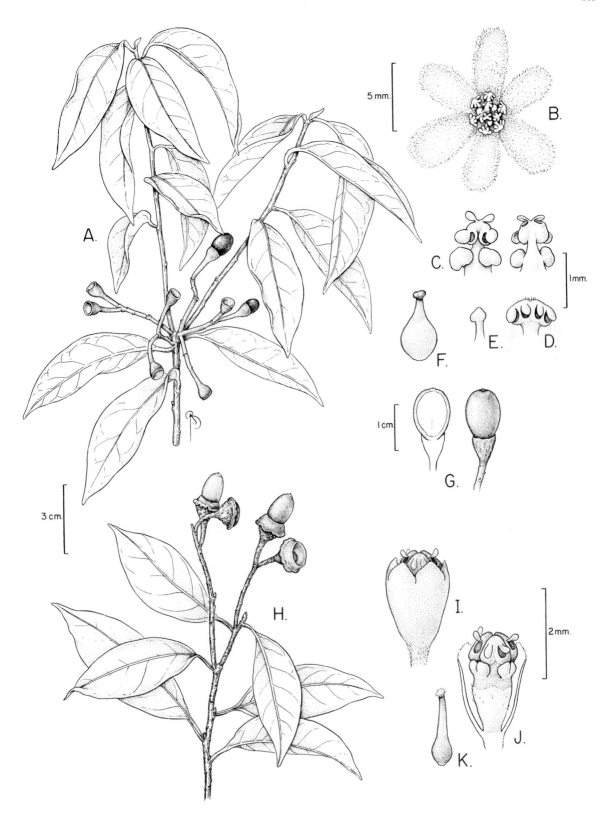

Fɪɢ. 130. A–G. *Ocotea coriacea*. A. Fruiting branch. B. Flower, top view. C. Frontal and dorsal views of inner stamens showing glands and anthers with flaps. D. Detail of anther of outer stamen showing flaps. E. Staminode. F. Pistil. G. L.s. fruit and fruit. H–K. *Licaria triandra*. H. Fruiting branch. I. Flower. J. Flower with portion of calyx removed showing stamens. K. Pistil.

nodes shorter than the stamens; stamens 9, the filaments of 6 outer ones pubescent. Fruit cupule 4–6 mm wide, funnel-shaped, bright red, smooth, with truncate apex, the fruit globose or globose-ellipsoid, 1–1.5 cm long, turning from green to black, glabrous.

DISTRIBUTION: A very common species throughout the island, occurring in dry to moist forest understory. Bordeaux area (A4260), Susannaberg (A847). Also on Jost van Dyke, St. Croix, St. Thomas, Tortola, and Virgin Gorda; Florida, from Mexico to northern South America, and the West Indies.

COMMON NAME: pepper cillament.

2. Ocotea floribunda (Sw.) Mez, Jahrb. Königl. Bot. Gart. Berlin **5**: 325. 1889. *Laurus floribunda* Sw., Prodr. 65. 1788. *Strychnodaphne floribunda* (Sw.) Griseb., Fl. Brit. W. I. 283. 1860.

Dioecious tree to 20 m tall; bark grayish or brown, smooth, becoming slightly fissured; branches sparsely pubescent when young, becoming glabrous, drying blackish. Leaf blades aromatic, 5–15 × 1.5–7 cm, elliptic, elliptic-lanceolate, or oblong-elliptic, chartaceous to nearly coriaceous, reticulate-veined, the primary and secondary veins drying dark brown, lower surface with a few appressed, rusty hairs, glabrescent, the lower surface dull, the apex obtuse to acuminate, the base acute to obtuse, the margins entire; petioles 1–1.3 cm long. Flowers functionally unisexual, in axillary panicles or racemes, the axes sparsely pubescent; masculine flowers with a pistillode; female flowers with indehiscent stamens; tepals ovate, 2–2.5 mm long, sparsely papillate-pubescent within; stamens 9; stigma truncate. Fruit cupule 5–7 mm wide, cup-shaped, brown, 2-rimmed, the fruit globose, green, 1–1.5 cm long, shiny, glabrous.

DISTRIBUTION: An uncommon species of moist forests. Bordeaux (M17052, W619, A4694). Also on Tortola; Central America to northern South America, including the Greater and Lesser Antilles.

3. Ocotea leucoxylon (Sw.) Laness., Pl. Util. Col. Franç. 158. 1886. *Laurus leucoxylon* Sw., Prodr. 65. 1788.

Dioecious tree to 20 m tall; trunk to 30 cm diam.; bark grayish or brown, smooth, becoming slightly fissured; branches ferruginous-tomentose when young, becoming glabrous. Leaf blades 8–25 × 3–8 cm, oblong-lanceolate, elliptic, or oblong-

elliptic, coriaceous, reticulate-veined, glabrous, usually with numerous insect galls, the apex obtuse or acute, the base obtuse to rounded, the margins revolute; petioles 1–1.5 cm long. Flowers unisexual with remnant parts of opposite sex, yellowish green in axillary panicles, the axes sparsely pubescent; tepals ovate 2–2.2 mm long, glabrate to sparsely papillate within; stamens 9. Fruit cupule 5–8 mm wide, funnel-shaped, brown, warty, with truncate apex, the fruit globose, 8–10 mm long, turning from green to black, glabrous.

DISTRIBUTION: A common species of moist secondary forests. Bordeaux area (A2099), Susannaberg (A847). Also on St. Croix, St. Thomas, Tortola, and Virgin Gorda; Greater Antilles, Lesser Antilles.

4. Ocotea patens (Sw.) Nees, Hufeland. Ill. 10. 1833. *Laurus patens* Sw., Prodr. 65. 1788. *Nectandra patens* (Sw.) Griseb., Fl. Brit. W. I. 281. 1860.

Tree to 15 m tall; bark grayish or brown, smooth, spicy; branches puberulent when young, becoming glabrous, drying blackish. Leaf blades 8–17 × 2.5–6 cm, elliptic, elliptic-lanceolate, or oblong-elliptic, coriaceous, reticulate-veined, glabrous, the upper surface shiny, the apex obtuse to acuminate, the base obtuse, the margins entire; petioles 0.5–2 cm long. Flowers bisexual, in axillary panicles, the axes sparsely pubescent; tepals oblong 2–2.5 mm long, papillose-pubescent within; stamens 9. Fruit cupule 5–6 mm wide, funnel-shaped, brown, smooth, with truncate apex, the fruit ellipsoid, 2–2.5 cm long, turning from green to black, glabrous.

DISTRIBUTION: An uncommon species of moist secondary forests. Bordeaux area (A5116). Also in Hispaniola, Puerto Rico, and the Lesser Antilles.

DOUBTFUL RECORD: *Ocotea turbacensis* Kunth (as *Nectandra sintenisii* Urb.) was reported by Woodbury and Weaver (1987) as occurring on St. John, but no specimen has been located to confirm this record. It is otherwise known in the Virgin Islands for St. Thomas and Tortola.

EXCLUDED SPECIES: *Nectandra hihua* (Ruíz & Pav.) Rohwer (as *Nectandra antillana* Meisn.), reported by Woodbury and Weaver (1987), was based on misidentifications of *Ocotea floribunda* and *Ocotea coriacea*.

CULTIVATED SPECIES: *Persea americana* Mill., the avocado, is cultivated on the island but not known to be spontaneous.

43. Loganiaceae (Logania Family)

Shrubs or vines, rarely trees or herbs. Leaves opposite or whorled, rarely alternate, simple, entire or toothed; stipules interpetiolar and leafy or a pair of lobes flanking each petiole. Flowers actinomorphic or zygomorphic, bisexual or rarely unisexual, 4–5-merous, in terminal or axillary cymes or rarely solitary; calyx lobes free or connate at base, unequal; corolla tubular, cylindrical, funnel-shaped or urn-shaped; stamens as many as the corolla lobes and alternating with them, the filaments inserted on the corolla tube, the anthers opening by longitudinal slits; ovary superior or partly inferior, bilocular, with numerous axial ovules, the style terminal, simple or 2–4-lobed. Fruit a septicidal or loculicidal capsule, rarely a berry; seeds small and numerous, variable, sometimes winged.

A family of about 29 genera and 450 species, with tropical and subtropical distribution.

REFERENCE: Rogers, G. K. 1986. The genera of Loganiaceae in the southeastern United States. J. Arnold Arbor. 67: 143–185.

1. SPIGELIA L.

Herbs or subshrubs; stems usually with longitudinal ridges; internal phloem present. Colleters present at base of leaves, bracts, and sepals. Leaves opposite or nearly whorled, sessile or nearly so, pinnately veined; stipules interpetiolar, membranaceous. Flowers bisexual, 5-merous, in terminal cymes; calyx lobes free or connate at base; corolla funnel-shaped; stamens included or exserted; ovary superior, 2-lobed, the style much longer than the ovary, the stigma simple or scarcely 2-lobed. Fruit a septicidal or loculicidal capsule, strongly bilobed at apex; seeds rugose or warty.

A genus of 50 species, from tropical and subtropical America.

1. Spigelia anthelmia L., Sp. Pl. 149. 1753. Fig. 131.

Erect, glabrous herb to 50 cm tall, branching near the bases. Leaves whorled; blades 4–15 × 2–5 cm, lanceolate or ovate-

lanceolate, chartaceous, the apex acute or acuminate, the base tapering to nearly rounded, the margins entire; petioles 0–2 mm long. Flowers in terminal scorpioid cymes; calyx lobes narrowly lanceolate, 1.5–2 mm long; corolla greenish white or pinkish, 5–

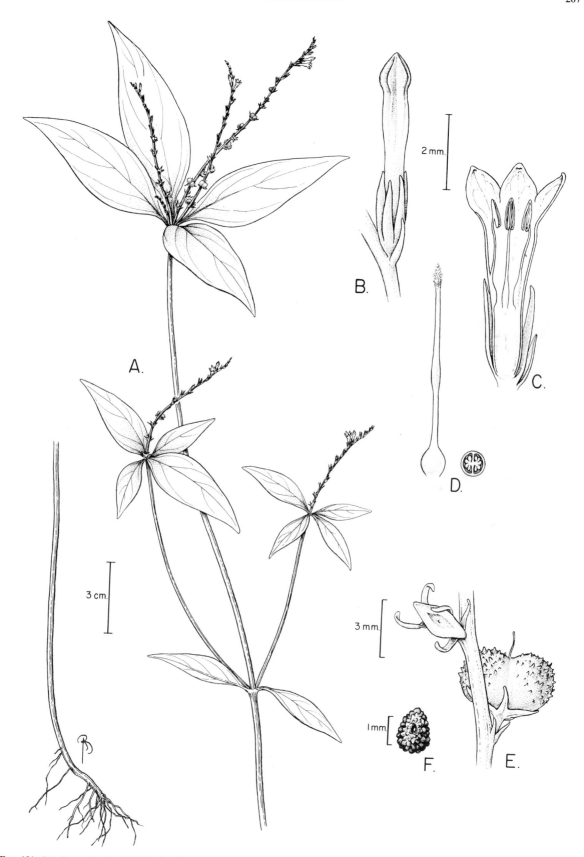

FIG. 131. *Spigelia anthelmia.* **A.** Habit. **B.** Flowerbud. **C.** L.s. flower. **D.** Pistil and c.s. ovary. **E.** Detail of fruiting branch showing capsule and persistent calyx. **F.** Seed.

9 mm long, stamens 5, included in corolla tube; ovary globose, green, papillose, the style simple. Capsule bilobed, 5–6 mm wide, tuberculate, turning from green to straw-colored. Seeds ellipsoid-ovoid, 1–2 mm long, tuberculate, brown.

DISTRIBUTION: An occasional roadside weed of moist areas. Carolina (A4139), Lind Point (A3087). Also on St. Croix, St. Thomas, and Virgin Gorda; a circum-Caribbean species, including the West Indies.

COMMON NAMES: water grass, worm grass.

44. Loranthaceae (Mistletoe Family)

Parasitic or semiparasitic shrubs or vines. Leaves opposite or alternate, simple, entire; stipules lacking. Flowers actinomorphic, bisexual or unisexual, produced in monads or triads arranged in axillary umbels, corymbs, spikes, or racemes; calyx adnate to the ovary, the sepals forming a ring at apex of ovary; corolla cup-shaped, petals 6(4–8), free or connate, valvate; stamens 6, 3 longer, filaments adnate to the corolla, the anthers opening by longitudinal slits; ovary inferior, unilocular, with 4–12 ovules, the placentation free, central, rarely basal or axile. Fruit a 1-seeded berry.

A family of 30–70 genera and about 940 species, with worldwide distribution.

1. DENDROPEMON Rchb.

Semiparasitic shrubs; stems green, cylindrical, flattened or quadrangular. Leaves opposite, coriaceous or fleshy, simple, entire. Flowers bisexual, pedicellate or nearly sessile, subtended by a bract and 2 bracteoles forming a calyxlike structure; corolla of 5 or 6 valvate petals; stamens adnate to the base of the petals; the style short, the stigma simple, capitate. Fruit a small, fleshy berry.

A West Indian genus of 15 species.

1. **Dendropemon caribaeus** Krug & Urb., Bot. Jahrb. Syst. **24**: 27. 1897. *Struthanthus caribaeus* (Krug & Urb.) Stehlé, Bull. Soc. Bot. France **34**: 32. 1954. *Phthirusa caribaeus* (Krug & Urb.) Engl. *in* Engl. & Prantl, Nat. Pflanzenfam., Nachtr. **2–4**: 135. 1897.   Fig. 132.

Loranthus emarginatus sensu Griseb., Fl. Brit. W. I. 312. 1860, non Sw., 1788.

Semiparasitic shrub, growing on branches of host shrubs or trees, many-branched from near the base, the branches hanging or erect, to 30 cm long. Leaf blades 2.5–5 × 1.5–3 cm, obovate, ovate to nearly rounded, nearly fleshy to fragile, the apex rounded, obtuse, truncate and usually mucronulate, the base tapering, the margins entire; petioles 2–5 mm long. Flowers bisexual, in short, axillary spikelike racemes; bracts green, covering the calyx; calyx cup-shaped, reddish; petals reflexed, nearly lanceolate, reddish, 2–2.5 mm long; stamens exserted, white; ovary globose, reddish-tinged, style white. Berry cylindrical, truncate at apex, purplish black, 6–8 mm long.

DISTRIBUTION: Common in moist and dry forests. Maria Bluff (A599), Fish Bay (A3907). Also on Anegada, St. Croix, St. Thomas, Tortola, and Virgin Gorda; Puerto Rico and the Lesser Antilles.

COMMON NAMES: bass an'boom, gadamighty.

45. Lythraceae (Loosestrife Family)

Trees, shrubs, or herbs; stems usually quadrangular. Leaves opposite, rarely subopposite or alternate, simple, entire; stipules axillary, minute. Flowers actinomorphic or zygomorphic, bisexual, 4–6-merous, in axillary racemes or cymes or terminal panicles; calyx bell-shaped or tubular, forming a hypanthium at base, the sepals valvate, usually alternating with accessory teeth; corolla of free petals, alternating with the sepals, adnate to the hypanthium; stamens as many as the petals or numerous, equal or unequal, the anthers opening by longitudinal slits; ovary superior, 2–4-locular, the ovules numerous with axile placentation, usually surrounded by a disk at base. Fruit a septicidal or loculicidal capsule or indehiscent, enclosed by a persistent hypanthium; seeds usually numerous, minute.

A tropical and subtropical family of about 27 genera and 450 species.

Key to the Genera of Lythraceae

1. Shrub over 2 m tall; stipules spiny; corolla white .... 2. *Ginoria*
1. Herb 0.35–1.0 m tall; stipules wanting; corolla white, pink, or purple-pink .... 1. *Ammannia*

1. AMMANNIA L.

Low glabrous annual herbs; stems 4-angled. Leaves opposite, sessile, lanceolate or oblanceolate, entire; stipules wanting. Flowers 4–5-merous, in axillary fascicles or cymes; hypanthium bell-shaped, becoming globose in fruit; calyx with accessory teeth or these lacking; petals 4 or wanting, early deciduous; stamens 4 or 8; style exserted or included. Capsule irregularly dehiscent, with thin, smooth wall.

A genus of about 25 species with worldwide distribution.

REFERENCE: Graham, S. A. 1985. A revision of *Ammannia* (Lythraceae) in the Western Hemisphere. J. Arnold Arbor. 66: 395–420.

Key to the Species of *Ammannia*

1. Style >1.5 mm long; corolla purple-pink; cymes 1–5-flowered .... 1. *A. coccinea*
1. Style to 0.5 mm long; corolla wanting or light pink to white; cymes 1–3-flowered .... 2. *A. latifolia*

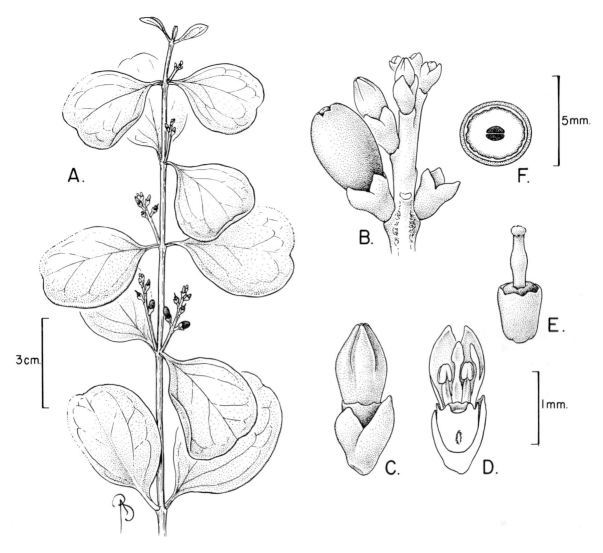

Fig. 132. *Dendropemon caribaeus*. **A.** Fertile branch. **B.** Detail of inflorescence with flowers and fruit. **C.** Flower with subtending bracteoles. **D.** L.s. flower. **E.** Calyx and pistil. **F.** C.s. berry.

1. Ammannia coccinea Rottb., Pl. Horti Univ. Rar. Progr. (Hafn.) 7. 1773.   Fig. 133G–K.

Erect, glabrous subshrub to 50 cm tall, many-branched from near the base; stems 4-angled, reddish, with 4 longitudinal ribs, one along each corner. Leaves sessile, ascending, 1.5–5 × 0.2–1 cm, linear, linear-lanceolate, oblong-elliptic, coriaceous, becoming reddish with age, the apex acute or obtuse, the base auriculate, clasping the stem, the margins strongly revolute. Flowers 1–5 in axillary cymes; bracts minute, oblong, persistent; peduncle 0–2 mm long, stout; hypanthium bell-shaped, green, 8-keeled; calyx teeth triangular, alternating with accessory teeth; petals 4, purple-pink, ca. 2 mm long, ascending, obovate, early deciduous; stamens 4–7, as long as the petals, the filaments purple-pink; ovary ovoid, reddish-tinged, the style filiform 1.5–2.5 mm long, the stigma capitate, yellow. Capsule nearly globose, 4–5 mm diam., the hypanthium persistent, with 4–8 faint, longitudinal lines. Seeds numerous, nearly ovoid, ca. 0.4 mm long, yellowish brown.

DISTRIBUTION: Occasional in moist open areas and in salt flats along the coast. Fortsberg (A4080), Lameshur (A3199). Also on St. Croix and

St. Thomas; eastern North America to northern South America, including the West Indies.

2. Ammannia latifolia L., Sp. Pl. 119. 1753.

Erect, glabrous subshrub to 1 m tall, unbranched or with lateral branches on distal portion of main stem; stems 4-angled, reddish, with 4 longitudinal ribs, one along each corner. Leaves sessile, ascending, 1.5–7.5 × 0.5–1.5 cm, lanceolate or elliptic-lanceolate, chartaceous, the apex acute or obtuse, the base of lower leaves cuneate, those of upper leaves auriculate, clasping the stem, the margins entire. Flowers 1–3 in axillary cymes; bracts minute, oblong, persistent; peduncles 0–3 mm long, stout; hypanthium urn-shaped, faintly 8-keeled; calyx teeth triangular, alternating with accessory teeth; petals 4, pale pink to white, ca. 1 mm long, early deciduous; stamens 4–8, included; style to 0.5 mm long, the stigma capitate. Capsule nearly globose, 4–6 mm diam., the hypanthium persistent, smooth. Seeds numerous, nearly ovoid, 0.4–0.5 mm long, light brown.

DISTRIBUTION: Occasional in salt flats along coastal areas. Annaberg (A2922). Also on St. Croix, St. Thomas, and Tortola; eastern North America, the West Indies, and scattered throughout South America.

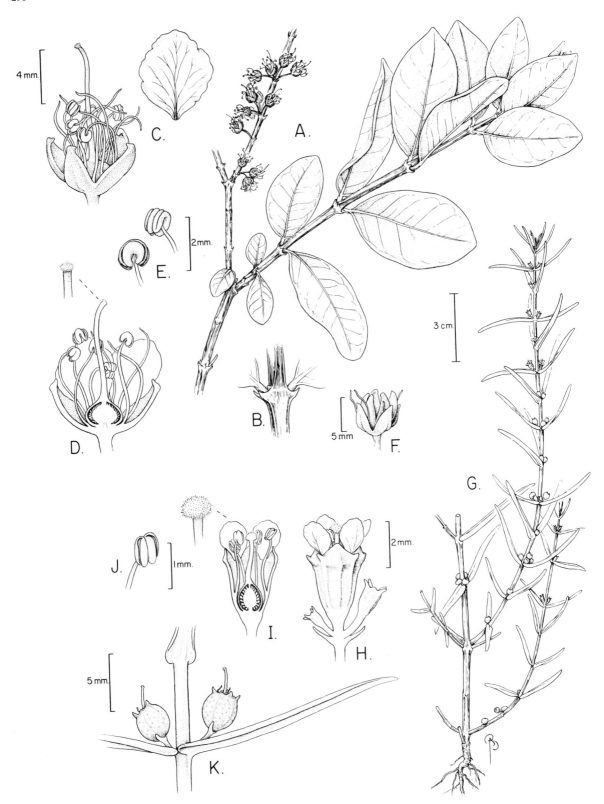

Fig. 133. A–F. *Ginoria rohrii*. A. Flowering branch. B. Detail of node showing stipular spines. C. Flower with petals removed and petal. D. L.s. flower and detail of stigma. E. Detail of upper portion of stamens showing anthers. F. Capsule after dehiscing. G–K. *Ammannia coccinea*. G. Habit. H. Detail of inflorescence. I. L.s. flower and detail of stigma. J. Detail of upper portion of stamen showing anther. K. Detail of branch showing capsules.

2. GINORIA Jacq.

Glabrous shrubs, usually armed with 4 stipular spines at nodes. Leaves opposite, petiolate, simple, entire, pinnately veined. Flowers 4–6-merous, pedicellate, solitary or in axillary racemes; calyx nearly globose, with ovate or triangular lobes; petals 4–6, pink to white, erose; stamens 10–23, inserted on the calyx tube; ovary nearly globose, 2–5-locular, the stigma capitate. Capsule septicidal, with 2–5 valves and smooth wall.

A genus of 10 species with the majority of species in Cuba and others in Hispaniola, Puerto Rico, and the Virgin Islands.

1. Ginoria rohrii (Vahl) Koehne, Bot. Jahrb. Syst. **3:** 351. 1882. *Antherylium rohrii* Vahl, Skr. Naturhist.-Selsk. **2:** 211. 1792. Fig. 133A–F.

Shrub or small tree 3–6 m tall, with ascending branches; bark light brown, rough, in longitudinal plates, the inner bark cream-colored; stems 4-angled when young, glabrous, bearing 2–4 prickles at nodes. Leaf blades 3–10 × 1.5–6 cm, elliptic or ovate, chartaceous, the apex obtuse to rounded, the base narrowed, the margins entire; petioles glabrous, 0–2 mm long. Flowers long-pedicellate, numerous, in axillary, short racemes; bracts minute, oblong, persistent; pedicels 8–10 mm long, slender; calyx lobes 4–5, ovate-lanceolate, 4–5 mm long, parted to the base; petals 4–5, white, 5.5–7 mm long, obovate, deciduous; stamens numerous, exserted; ovary nearly globose, the style much longer than the stamens. Capsule globose to ovoid, 4–5 mm long. Seeds numerous, wedge-shaped, minute.

DISTRIBUTION: A common shrub of coastal dry forests or thickets. Cob Gut (A3221), East End (A682). Also on St. Croix, St. Thomas, Tortola, and Virgin Gorda; Hispaniola and Puerto Rico.

COMMON NAMES: prickle wood, sugar ant.

46. Malpighiaceae (Barbados Cherry Family)

Trees, shrubs, or vines; pubescence simple or with T-shaped (malpighiaceous) hairs. Leaves opposite, simple and entire, often bearing glands on petioles or blades; stipules minute. Flowers subtly to strongly zygomorphic, bisexual, in axillary or terminal racemes, corymbs, cymes, or panicles; sepals 4–5, usually 4 of them with 1 or 2 extrafloral nectaries; corolla of 5 free petals, alternating with the sepals, clawed, the posterior petal usually smaller than the remaining four; stamens 10 or fewer by reduction, the anthers opening by longitudinal slits; ovary superior, of 2–3 uniovular carpels, the placentation axile, the styles free or less often connate into 1. Fruit a fleshy drupe or dry, samaroid schizocarp; seeds 1 per carpel.

A family of about 70 genera and 1250 species, pantropical but predominantly in the New World.

Key to the Genera of Malpighiaceae

1. Trees or shrubs; fruits wingless, fleshy or dry.
  2. Leaf blades conspicuously glandular; plant without bristles or stiff hairs.
    3. Leaf glands 2, cone-shaped, on lower portion of leaf margins; sepals without glands; stamens with free filaments; fruit capsular.. *Galphimia* (cultivated)
    3. Leaf glands 2, impressed at base of blade between midrib and margins; sepals with a pair of elongate glands on lower surface; fruit a fleshy drupe... 1. *Bunchosia*
  2. Leaf blades eglandular or if glandular then the plant with bristles or stiff hairs.
    4. Styles slender and subulate, the stigma minute; fruits turning from green to yellow; plant not covered with stiff hairs .... 2. *Byrsonima*
    4. Styles stout and cylindrical, the stigma elongate; fruits red when mature; plant with stiff hairs causing dermatitis on contact (not very obvious in the cultivated *M. emarginata*) .... 4. *Malpighia*
1. Twining vines or lianas; fruit winged, samaroid schizocarps.
  5. Corolla yellow; inflorescence as wide or wider than long; leaves (at least some) >5 cm long ... 5. *Stigmaphyllon*
  5. Corolla pink; inflorescence longer than wide; leaves <2 cm long....................................... 3. *Heteropteris*

1. BUNCHOSIA Kunth

Shrubs or trees, usually armed with simple or T-shaped hairs. Leaves entire, pinnately veined, with a pair of impressed glands on lower portion of blade; petioles glandless; stipules small, free, borne at base of petiole. Flowers pedicellate, in axillary or terminal racemelike inflorescences; bracteoles with a large, fleshy gland; calyx of 5 sepals, all or 4 of them bearing a pair of fleshy glands without; petals yellow or nearly white; stamens 10, the anthers uniform; ovary of 2–3 connate carpels, the style free to completely connate. Fruit a fleshy drupe containing 2 or 3 1-seeded pyrenes.

A genus of about 55 species throughout tropical and subtropical areas of the New World.

1. Bunchosia glandulosa (Cav.) DC., Prodr. **1:** 581. 1824. *Malpighia glandulosa* Cav., Diss. **8:** 411. 1789. Fig. 134L–Q.

Shrub or small tree 3–7 m tall; bark gray, fissured, rough, with strong fetid smell when cut; stems cylindrical, gray, with numerous whitish lenticels, pubescent when young. Leaf blades 3.5–10.5 × 1.4–5.6 cm, elliptic, chartaceous, glabrous, the apex obtuse acute or acuminate, the base narrowed, with a pair of impressed glands, the margins entire; petioles glabrescent, 3–10 mm long. Flowers long-pedicellate, in axillary, racemelike cymes; bracteoles minute, with a large nectary; pedicels 5–10 mm long,

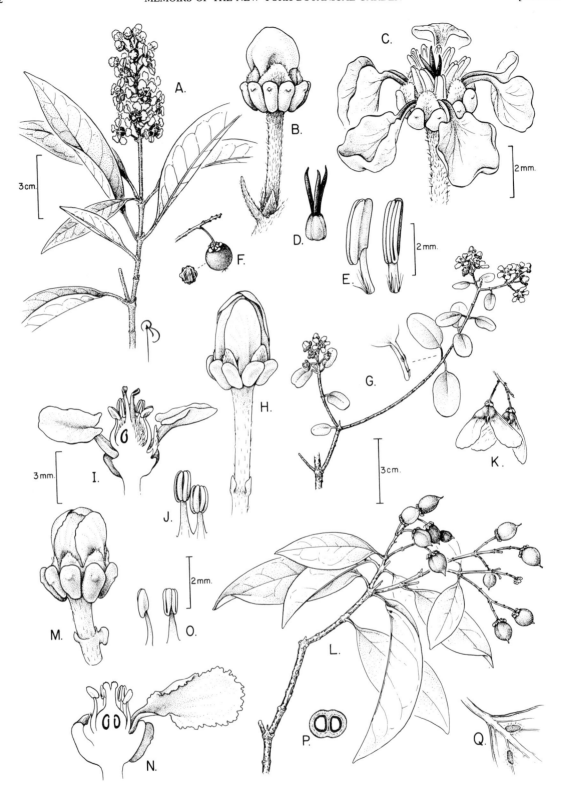

Fig. 134. A–F. *Byrsonima spicata*. A. Flowering branch. B. Flowerbud showing biglandular sepals and bracts at base of pedicel. C. Flower. D. Pistil. E. Stamen, lateral and frontal views. F. Drupe and stone. G–K. *Heteropterys purpurea*. G. Flowering branch and detail of leaf petiole. H. Flowerbud showing biglandular sepals and paired bracts at base of pedicel. I. L.s. flower. J. Stamens. K. Winged mericarps. L–Q. *Bunchosia glandulosa*. L. Fruiting branch. M. Flowerbud showing biglandular sepals and bracteoles with nectary gland. N. L.s. flower. O. Stamen, lateral and frontal views. P. C.s. fruit. Q. Detail of blade with basal glands.

slender; calyx bearing 8 fleshy glands; petals light yellow, obovate to ovate-lanceolate, 8–9 mm long, long-clawed; filaments connate at base; ovary of 2 carpels, glabrous, the style single. Drupe ovoid to globose, 8–10 mm long, dark orange.

DISTRIBUTION: A common shrub of coastal dry forests or thickets. Carolina (A2819), Lameshur (B646). Also on Anegada, Jost van Dyke, St. Croix, and St. Thomas; the Bahamas, Hispaniola, Puerto Rico, and Lesser Antilles.

2. BYRSONIMA Kunth

Shrubs or trees, usually covered with simple or T-shaped hairs. Leaves entire, pinnately veined, eglandular; petioles eglandular; stipules small, free or partially to completely connate, borne at base of petiole or intrapetiolar. Flowers pedicellate, in axillary or terminal racemelike inflorescences; bracteoles eglandular. Calyx of 5 sepals, all bearing a pair of fleshy glands or all eglandular; petals yellow, white, pink, or reddish, the posterior petal usually smaller; stamens 10, the anthers uniform; ovary of 3 connate carpels, the styles free to completely connate. Fruit a fleshy drupe containing a trilocular stone.

A genus of about 150 species throughout tropical and subtropical America, mostly South American.

Key to the Species of Byrsonima

1. Leaves elliptic to oblong-lanceolate, 5–15 cm long, secondary veins prominent on lower surface; petals yellow .... 2. *B. spicata*
1. Leaves obovate or oblanceolate, 1.5–4 cm long, secondary veins inconspicuous; petals turning from white to pink, becoming red at maturity .... 1. *B. lucida*

1. Byrsonima lucida (Mill.) DC., Prodr. **1**: 580. 1824. Malpighia lucida Mill., Gard. Dict., ed. 8. 1768.

Malpighia cuneata Turcz., Bull. Soc. Imp. Naturalistes Moscou **31**: 390. 1858. *Byrsonima cuneata* (Turcz.) P. Wilson, Bull. New York Bot. Gard. **8**: 394. 1917.

Shrub or small tree (1–)3–4(–7) m tall, usually many-branched from base; bark gray, smooth; stems cylindrical, gray, with numerous lenticels, densely sericeous when young, hairs fading to white. Leaf blades 1.5–4 × 0.8–1.5 cm, obovate or oblanceolate, coriaceous, sparsely covered with appressed, rusty hairs, the apex rounded or obtuse, the base attenuate, the margins entire; petioles 3–10 mm long. Flowers long-pedicellate, in terminal racemes; bracts minute, subulate; pedicels 8–12 mm long, slender, with reddish brown pubescence; calyx pinkish, of 5 biglandular sepals, the glands fleshy, elongate, whitish to yellowish; petals spreading, turning from white to pink, becoming red at maturity, 6–8 mm long, long-clawed, the limb widely ovate to reniform; anthers white at margins; ovary glabrous, tricarpellate, the styles 3, free to base. Drupe globose or ovoid, 8–12 mm diam., glabrous, apiculate at apex, turning from green to yellow and finally to yellowish brown.

DISTRIBUTION: A common shrub of dry to moist forests or coastal thickets. Coral Bay Quarter along Center Line Road (A2942), White Cliffs (A2035). Also on Anegada, St. Croix, and St. Thomas; Florida, Bahamas, Greater and Lesser Antilles.

COMMON NAME: gooseberry.

NOTE: *Byrsonima lucida* seems to hybridize with *B. spicata*. A few individuals with intermediate characters between these two species have been collected at a few localities on St. John. The plants have leaves resembling *B. lucida* in shape and *B. spicata* in having prominent secondary veins, but leaf size is intermediate between the two. Coloration of petals is yellow to orange as in *B. spicata*, and the posterior petal lacks the claw glands, a character typical of *B. spicata* but absent in *B. lucida*. None of the individuals collected on St. John is known to set fruit, although I sought

fruit repeatedly. This fact may indicate that the individuals belong to a sterile population. Both species seem to hybridize in the Lesser Antilles (Howard, 1988; Nicolson et al., 1991). The names *Byrsonima ophiticola* Britton and *Byrsonima horneana* Britton & Small have been applied to similar plants in western Puerto Rico.

2. Byrsonima spicata (Cav.) DC., Prodr. **1**: 580. 1824. *Malpighia spicata* Cav., Diss. **8**: 409, pl. 237. 1789. *Byrsonima coriacea* Sw. var. *spicata* (Cav.) Nied. *in* Engl., Pflanzenr. IV, **141**: 700. 1928.

Fig. 134A–F.

Byrsonima coriacea sensu Little & Wadsworth, 1964, non Sw., 1788.

Tree 3–15 m tall; bark gray, smooth or finely fissured; twigs cylindrical, densely sericeous when young. Leaf blades 5–15 × 1.8–5.5 cm, elliptic to oblong-lanceolate, chartaceous, sparsely covered with appressed, rusty hairs especially along midvein, secondary veins prominent on lower surface, the margins entire obtuse to acute, the base obtuse; petioles 5–15 mm long, sericeous. Flowers in terminal, ascending racemes, 4–10 cm long; bracts minute, subulate; pedicels 7–8 mm long, slender, sericeous; calyx yellow, of 5 biglandular sepals, the glands fleshy, elongate, whitish to yellowish; petals spreading, yellow, turning orange in age, 5–8 mm long, long-clawed, with a widely ovate to reniform limb, the posterior petal with 2 or more marginal glands at apex of the claw; anthers yellowish; ovary tricarpellate, sericeous, the styles 3, glabrous, free to base. Drupe nearly globose to obovoid, 10–12 mm diam., glabrous, apiculate at apex, turning from green to yellow, becoming yellowish brown at maturity.

DISTRIBUTION: A common tree of moist forests. Bordeaux area (A3820), Cruz Bay Quarter along Center Line Road (A2939), Rosenberg (B319). Also on St. Croix, St. Thomas, Tortola, and Virgin Gorda; Cuba, Hispaniola, Puerto Rico, Lesser Antilles, northern South America, south to Bolivia.

3. HETEROPTERIS Kunth

Twining woody vines, shrubs, or small trees, usually covered with simple or T-shaped hairs. Leaves entire, pinnately veined, usually with impressed glands on blade; petioles often bearing a pair of stipitate glands; stipules minute or wanting. Flowers pedicellate, in axillary or terminal umbels, corymbs, or pseudoracemes; calyx of 5 basally connate sepals, 4 of them bearing a pair of fleshy glands externally, or seldom glandless; petals yellow, pink, or white; stamens 10, somewhat unequal, all fertile, the filaments connate at base, the anthers ellipsoid to oblong; ovary of 3 connate carpels, the styles free, erect, thickened. Fruit a schizocarp containing 3 samaras, each with a well-developed dorsal wing.

A genus of about 125 species distributed in the neotropics, except for a single African species.

1. Heteropteris purpurea (L.) Kunth *in* Humb., Bonpl. & Kunth, Nov. Gen. Sp. **5:** 164. 1821 [1822]. *Banisteria purpurea* L., Sp. Pl. 427. 1753.

Fig. 134G–K.

Banisteria parvifolia Vent., Choix Pl. 51. 1808. *Heteropteris parvifolia* (Vent.) DC., Prodr. **1:** 591. 1824.

Twining, slender woody vine 3–5 m long, with numerous lateral branches; stems cylindrical, slender, wiry, with numerous lenticels, puberulent. Leaf blades 1.5–2 × 1–1.5 cm, oblong, elliptic, or ovate, chartaceous, puberulent, the apex obtuse or rounded, the base obtuse or rounded, the margins entire; petioles 3–10 mm long, with a pair of stipitate glands at middle. Flowers long-pedicellate, in terminal or axillary racemes; bracts minute, eglandular; pedicels 5–6 mm long, slender, reddish brown, pubescent; calyx bearing 8 fleshy, elongate, green glands; petals pink, widely ovate, 4.5–5 mm long, long-clawed; filaments glabrous, the anthers yellowish; stigma green. Samara winged on distal portion, 1.5–3 cm long, turning from green to brown when dried.

DISTRIBUTION: Common in dry open areas and coastal scrub. Coral Bay (A2658), Lameshur (B627). Also on St. Croix, St. Thomas, Tortola, and Virgin Gorda; Florida, Bahamas, Greater and Lesser Antilles.

DOUBTFUL RECORD: *Heteropteris laurifolia* (L.) A. Juss. Was listed by Woodbury and Weaver (1987) as occurring on St. John; however, no specimens have been located. Numerous attempts to find this species on St. John have failed.

4. MALPIGHIA L.

Small to medium-sized shrubs, usually covered with simple or T-shaped hairs. Leaves entire, wavy, or toothed, pinnately veined, usually with impressed glands on blade; petioles eglandular; stipules minute, interpetiolar. Flowers pedicellate, solitary or in axillary, elongate racemes or congested into a corymb-or umbel-like inflorescence; calyx of 5 free, ovate sepals, bearing 1 or 2 fleshy glands externally (for a total number of 6, 8, or 10); petals pink, violet, light yellow, or white, the posterior petals usually larger; stamens 10, in 2 whorls, the external whorl usually larger, all fertile, the filaments slightly connate at base; ovary globose, glabrous, 3-locular, of 3 connate carpels, the styles connate at base, erect, thickened. Fruit a fleshy drupe, containing 3 pyrenes, some of which may fail to develop.

A neotropical genus of about 40 species, most of which occur in the West Indies.

REFERENCE: Vivaldi, J. L. 1979. The systematics of *Malpighia* L. (Malpighiaceae), Dissertation, Cornell University, Ithaca, New York.

NOTE: Most species of *Malpighia* have bristles or sharp stiff hairs that may penetrate the skin on contact, causing irritation or dermatitis.

Key to the Species of *Malpighia*

1. Leaves <2.5 cm long, the margins sinuate-dentate, with marginal bristles; corolla pink .... 1. *M. coccigera*
1. Leaves >2.5 cm long, the margins entire; corolla white, light yellow, or pink.
  2. Corolla pink.
    3. Leaves linear or narrowly elliptic, 4–10 cm long, with acute apices; fruits 1–1.3 cm wide ..... 3. *M. linearis*
    3. Leaves elliptic, 2.5–7 cm long, with retuse apices; fruits 1.5–2.5 cm wide, edible
    .... *M. emarginata* (cultivated)
  2. Corolla whitish, sometimes yellowish- or pinkish-tinged.
    4. Leaf margins with perpendicular, T-shaped bristles, 2.5–5 mm long; leaf blade scabrous on upper surface (because of V-shaped hairs), lower surface strigose, with T-shaped hairs 2.5–5 mm long; corolla white to yellowish-tinged .... 4. *M. woodburyana*
    4. Leaf margins without bristles; leaf blades nearly smooth (hairs T-shaped, 1–1.5 mm long), lower surface strigillose, with T-shaped hairs 1–1.5 mm long; corolla white to pinkish-tinged .... 2. *M. infestissima*

1. Malpighia coccigera L. , Sp. Pl. 426. 1753.

Erect (seldom prostrate) shrub 1–5 m tall, many-branched from base; bark light brown, fissured; branches reddish-tinged when young, glabrous to strigillose. Leaf blades 1–2.5 × 0.8–2 cm, oblong, ovate to obovate, coriaceous, glabrous, or the lower surface sparsely covered with T-shaped hairs, the apex obtuse, truncate, or emarginate, the base obtuse, the margins sinuate-dentate, with T-shaped bristles; petioles 0.5–1.5 mm long; stipules minute, glabrous or strigillose. Flowers 2–5, congested in leaf axils; bracteoles minute, ciliate; pedicels 9–20 mm long, slender; calyx bearing (6–)8–10 fleshy glands; petals light pink, 11–12.5 mm long, long-clawed; filaments glabrous; anterior style short, slender; posterior styles longer, thicker. Drupe 5–15 mm diam., depressed-globose, 3-lobed, fleshy, bright red.

DISTRIBUTION: A rare species of coastal scrub and dry deciduous forests. Salt Pond Trail (M17064), trail to Drunk Bay (A4133). Also in Cuba, Hispaniola, and Puerto Rico, cultivated elsewhere.

NOTE: Populations of *M. coccigera* on St. John seem to belong to subspecies *coccigera*, but they differ from typical material by their robust habit. Subspecies *coccigera* is restricted to Hispaniola, Puerto Rico, and the Virgin Islands and is known only from cultivation in the Lesser Antilles. Two other subspecies (*horrida* and *apiculata*) are known from Cuba and Hispaniola. They differ from subspecies *coccigera* by their smaller leaves and thinner, shorter posterior styles.

2. Malpighia infestissima Rich. *ex* Nied., Malpighia 15. 1899.

Malpighia pallens Small, N. Amer. Fl., **25(2):** 157. 1910.

Malpighia thompsonii Britton & Small *in* Britton & P. Wilson, Bot. Porto Rico **5:** 443. 1924.

Shrub 2–2.5 m tall, many-branched from base; bark light brown to grayish, slightly fissured; stems nearly cylindrical, grayish, puberulent. Leaf blades 3.5–8(–11) × (1–)2–5 cm, elliptic to ovate, chartaceous to nearly coriaceous, sparsely covered with T-shaped hairs on both surfaces, the hairs 1–1.5 mm long, or glabrescent, lower surface strigillose, the apex obtuse to acute, emarginate and sometimes mucronate, the base rounded to subcordate, with a pair of impressed glands, the margins without bristles; petioles 1–3 mm long; stipules minute. Flowers in short, axillary racemes or panicles; bracteoles minute; pedicels 9–17 mm long, slender; calyx bearing 10 fleshy glands, unequal; petals white to pinkish-tinged, reflexed, 8–11 mm long, clawed; filaments glabrous; anterior style slightly shorter. Drupe ca. 2 cm diam., depressed-globose, obscurely 9-lobed, red when mature.

DISTRIBUTION: A rare species of coastal scrub and dry deciduous forests. Probably extirpated on St. John; known only from a single collection by Vest in 1906. Also on Buck Island and St. Croix.

COMMON NAMES: mad dog, stinging bush, touch me not.

3. Malpighia linearis Jacq., Enum. Syst. Pl. 21. 1760.
Fig. 135A–F.

Erect shrub 1–2 m tall, many-branched from base; bark light brown, fissured; branches slightly flattened, grayish, strigillose, soon glabrescent. Leaf blades 4–8(–10) × 0.4–1.2 cm, linear to narrowly elliptic, nearly coriaceous, sparsely covered with T-shaped hairs, especially on lower surface, upper surface nearly smooth, lower surface strigillose, with T-shaped hairs 1–1.5 mm long, the apex acute and mucronate, the base tapering, the margins entire, revolute; petioles 2–4 mm long; stipules minute, strigillose. Flowers few, in short axillary racemes; bracteoles minute; pedicels 11–18 mm long, slender; calyx green, bearing 6(–10) fleshy glands; petals bright pink, 5.5–9 mm long, the claw white; fila-

ments glabrous, white, the anthers yellow; styles green, the anterior shorter. Drupe 1–1.3 cm diam., depressed-globose, red when mature.

DISTRIBUTION: An uncommon species of coastal scrub and dry deciduous forests. Lameshur (A3229). Also on Anegada, St. Thomas, and Water Island; Culebra Island (Puerto Rico) and the Lesser Antilles.

4. Malpighia woodburyana Vivaldi in Acev.-Rodr., Brittonia 45: 130. 1993.

Erect shrub 1–2.5(–5) m tall, many-branched from base; bark light brown, fissured; branches green, strigillose, with V-shaped hairs. Leaf blades 6–8(–14) × 4–6(–9) cm, ovate or elliptic, chartaceous, upper surface scabrous from V-shaped, stiff hairs, lower surface sparsely covered with T-shaped hairs, 2.5–5 mm long, the apex obtuse or emarginate, the base nearly cordate or obtuse, the margins entire, revolute, armed with T-shaped bristles 2.5–5 mm long; petioles 1–3 mm long, strigillose; stipules minute. Flowers 2–4, in short, axillary, umbel-like racemes; bracteoles minute; pedicels 12–20 mm long, slender; calyx green, strigillose, bearing 6–8(–10) fleshy glands; petals white to yellow-tinged, clawed, 7–11 mm long; filaments glabrous; styles with uncinate apex, the anterior style slightly smaller. Drupe 1.5–2 cm diam., with ovoid outline and 6–9 longitudinal ridges, green.

DISTRIBUTION: An uncommon species of coastal scrub and dry deciduous forests. Bethany (B179), Hawksnest Bay (A2311). Also on Anegada, Buck Island, St. Thomas, and Water Island; Puerto Rico including Magueyes, Culebra, and Vieques.

CULTIVATED SPECIES: *Malpighia emarginata* DC., commonly known as the Barbados cherry, is cultivated for its red, edible, tasty fruits, exceptionally rich in vitamin C.

5. STIGMAPHYLLON Juss.

Twining woody vines, usually covered with simple or T-shaped hairs. Leaves entire, pinnately veined, usually with a pair of glands at base or below the base at petiole; stipules minute, eglandular, deciduous. Flowers pedicellate, in axillary umbels, corymbs, or pseudoracemes; calyx of 5 free, ovate sepals, 4 bearing 2 fleshy glands externally and the posterior sepal glandless; petals 5, yellow, sometimes reddish-tinged, clawed, the posterior petals with a longer claw; stamens 10, in 1 whorl, subequal or unequal, the filaments connate at base; ovary 3-locular, of 3 connate carpels, the styles free to base, erect or slightly recurved, the anterior and posterior styles heterogeneous. Fruit a samaroid schizocarp, of 3 winged mericarps, the wing dorsally thickened along frontal margin.

A genus of about 100 species, in the neotropics from eastern Mexico to northern Argentina, including the West Indies.

REFERENCE: Anderson, C. 1987. *Stigmaphyllon* (Malpighiaceae) in Mexico, Central America and the West Indies. Contr. Univ. Michigan Herb. 16: 1–48.

Key to the Species of *Stigmaphyllon*

1. Young branches tomentose; leaves densely covered with appressed hairs on lower surface (tomentose to sericeous); blades with conspicuous reticulate venation; samaras tomentose, greenish when young
.... 2. *S. floribundum*
1. Young branches strigillose; leaves glabrescent on both surfaces (strigillose when young); blades with pinnate venation (3° veins not conspicuous); samaras becoming glabrous with age, usually reddish when young .... 1. *S. emarginatum*

1. Stigmaphyllon emarginatum (Cav.) A. Juss., Ann. Sci. Nat. Bot., ser. 2, 13: 290. 1840. *Banisteria emarginata* Cav., Diss. 9: 425. 1790.
Fig. 135G–N.

Banisteria periplocifolia Desf. ex DC., Prodr. 1: 589. 1824. *Stigmaphyllon periplocifolium* (Desf. ex DC.) A. Juss., Ann. Sci. Nat. Bot., ser. 2, 13: 290. 1840. *Stigmaphyllon lingulatum* (Poir.) Small, N. Amer. Fl. 25(2): 140. 1910.

Twining liana to 15 m long; bark reddish brown, rough; stems usually numerous from base, twining together, forming a ropelike structure, nearly cylindrical, copper-colored, with numerous lenticels, strigillose when young. Leaf blades (1.1–)2.5–10(–13) × (0.5–)1–5.5(–10.5) cm, lanceolate, ovate, oblong, linear, or seldom rounded, coriaceous, with a few appressed hairs, especially on lower surface, becoming glabrous, the apex acute or obtuse, emarginate-mucronate, the base truncate to cordate, the margins entire; petioles 0.2–2 cm long, with a pair of disk-shaped glands below the blade. Flowers long-pedicellate, 15–25 in congested, axillary umbels or pseudoracemes; bracts minute, eglandular; pedi-

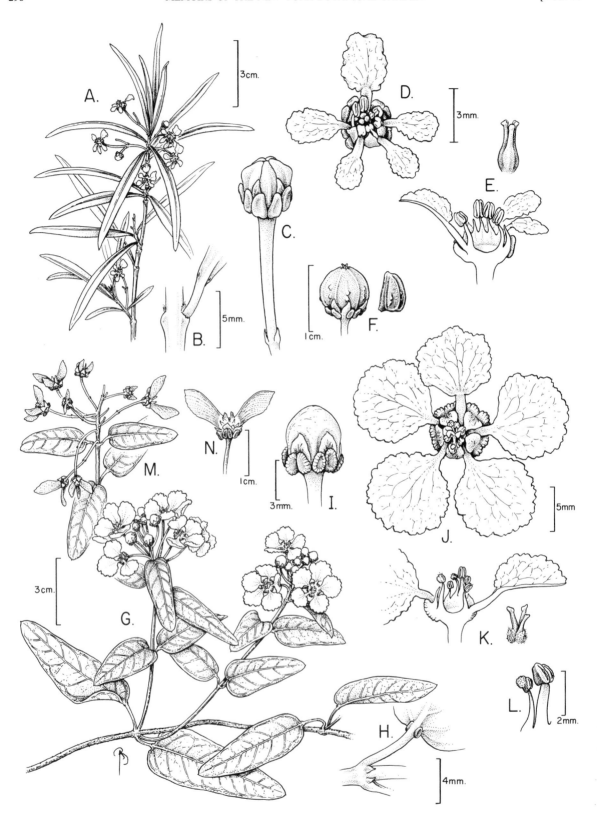

FIG. 135. A–F. *Malpighia linearis*. **A.** Flowering branch. **B.** Detail of stem node showing leaf petiole and glands on base of leaf blade. **C.** Flowerbud. **D.** Flower. **E.** Pistil and l.s. flower. **F.** Drupe and pyrene. **G–N.** *Stigmaphyllon emarginatum*. **G.** Flowering branch. **H.** Detail of stem showing leaf petiole with disk-shaped glands at base of leaf blade. **I.** Flowerbud showing calyx glands. **J.** Flower. **K.** L.s. flower and detail of pistil. **L.** Detail of sterile and fertile stamens. **M.** Fruiting branch. **N.** Winged mericarps.

cels 3–23 mm long, slender; calyx bearing 8–10 fleshy, elongate, green glands; petals bright yellow, rounded, 11–15 mm long, long-clawed; stamens unequal, the anthers glabrous or puberulent; anterior style shorter, the stigmas green. Samara wing 1.5–2.2 cm long, reddish-tinged when young, strigillose, becoming glabrous with age.

DISTRIBUTION: Common in dry open areas and coastal scrub. Dittlif Point (A3965), Salt Pond (A4042). Also on Anegada, St. Croix, St. Thomas, Tortola, and Virgin Gorda; Jamaica, Hispaniola, Puerto Rico (including Culebra Island), and the Lesser Antilles, south to St. Lucia.

2. Stigmaphyllon floribundum (DC.) C.E. Anderson, Syst. Bot. **11:** 128. 1986. *Banisteria floribunda* DC., Prodr. **1:** 589. 1824.

Banisteria tomentosa Desf. *ex* DC., Prodr. **1:** 589. 1824. *Stigmaphyllon tomentosum* (DC.) Nied., Stigmatoph. 5. 1899, non Juss., 1824.

Twining liana to 15 m long; bark reddish brown, rough; stems usually numerous from base, twining together, forming a ropelike structure, nearly cylindrical, with numerous lenticels, tomentose when young. Leaf blades 3.6–18 × (1–)2.5–15.5 cm, elliptic, broadly elliptic, oblong, or sometimes lanceolate or rounded, coriaceous, glabrous or sparsely pubescent above, tomentose to sericeous below, venation conspicuously reticulate, the apex obtuse, truncate, emarginate-mucronate, the base obtuse, truncate, or nearly cordate, the margins entire; petioles 1–2.5 cm long, pubescent, with a pair of disk-shaped glands below the blade. Flowers long-pedicellate, 20–25 in congested axillary umbels or pseudoracemes; bracts minute, eglandular; pedicels 10–22 mm long, slender; calyx bearing 8 fleshy, elongate, green glands; petals bright yellow, rounded, 11–15 mm long, long-clawed; stamens unequal, the anthers glabrous or puberulent; anterior style shorter, the stigmas green. Samara wing 1.8–3.2 cm long, turning from green to brown when dry, tomentose.

DISTRIBUTION: Common in moist to deciduous, coastal forests. Bordeaux area (A2854). Also on Virgin Gorda; Puerto Rico (including Mona island).

COMMON NAME: bull wiss.

CULTIVATED SPECIES: *Galphimia gracilis* Bartl., with yellow flowers, although not very common, is cultivated in gardens.

47. Malvaceae (Cotton Family)
By P. A. Fryxell

Herbs, shrubs, or trees, often stellate-pubescent; roots fibrous or woody, sometimes fleshy in perennial herbs; stems erect or procumbent. Leaves alternate, ovate to lanceolate, sometimes lobed or dissected, with hairs stellate, simple, sometimes prickly or glandular, or rarely lepidote; stipulate. Inflorescences racemes or panicles, sometimes spikes or scorpioid cymes, rarely umbels or heads or flowers solitary or fasciculate in the leaf axils. Flowers with involucel present or absent; calyx gamosepalous, truncate, 5-lobed or 5-parted; petals 5, distinct, usually clawed; androecium monadelphous, the anthers reniform, numerous (rarely reduced to 5), 2-locular; ovary superior, 3–40-carpellate, the styles 1–40, the stigmas truncate, capitate, or decurrent. Fruit schizocarpic or capsular, sometimes a berry; seeds reniform or turbinate, pubescent or glabrous.

A family of more than 100 genera and perhaps 2000 species, principally tropical and subtropical in distribution but with a few temperate-zone genera.

REFERENCES: Fryxell, P. A. 1979. The natural history of the cotton tribe. Texas A&M Press, College Station, Texas; Fryxell, P. A. 1988. Malvaceae of Mexico. Syst. Bot. Monogr. 25: 1–522.

Key to the Genera of Malvaceae

1. Flowers subtended by an involucel at base of calyx.
  2. Involucel 3-parted.
    3. Involucellar bracts broadly cordate-ovate, foliaceous .... 3. *Gossypium*
    3. Involucellar bracts linear, spatulate, or ligulate.
      4. Leaves cordate-ovate, entire, coriaceous; corolla 4–6 cm long; fruits capsular, indehiscent, 5-celled .... 10. *Thespesia*
      4. Leaves truncate, serrate, crenate, or dentate; corolla to 1 cm long; fruits schizocarpic, mericarps 10–14 .... 6. *Malvastrum*
  2. Involucel 5–9(or more)-parted.
    5. Styles and stigmas 5; fruit a 5-celled capsule .... *Hibiscus* (cultivated)
    5. Styles and stigmas 10; fruit a schizocarp of 5 mericarps.
      6. Leaves with nectary on underside at base of midrib; corolla lavender; mericarps with numerous glochids .... 11. *Urena*
      6. Leaves lacking nectaries; corolla yellow; mericarps apically 3-spined, the spines retrorsely barbed .... 7. *Pavonia*
1. Flowers lacking an involucel.
  7. Fruits globose, inflated, hispid, pendent on slender pedicels; plants decumbent to ascending, often scandent; corolla white .... 4. *Herissantia*
  7. Fruits neither inflated nor pendent; plants usually erect shrubs, rarely trees (sometimes prostrate herbs), seldom scandent; corolla of various colors, often yellow.
    8. Carpels 3-seeded, often acute or acuminate apically.

9. Mericarps 3–5, divided into 2 cells by a medial constriction, the lower cell indehiscent, 1-seeded, the upper cell dehiscent, 2-seeded, apically bulbous-apiculate .... 12. *Wissadula*

9. Mericarps 6–8, undivided, with apical spines 2–4 mm long .... 1. *Abutilon*

8. Carpels 1-seeded, blunt or spined apically.

    10. Flowers congested in few-flowered heads subtended by specialized flora bracts; styles and stigmas twice as many (10) as mericarps (5) .... 5. *Malachra*

    10. Flowers solitary, fasciculate, or disposed in inflorescences but not subtended by specialized floral bracts; styles and stigmas of same number as mericarps (5–14).

        11. Plants arborescent .... *Bastardiopsis* (cultivated)

        11. Plants subshrubs or herbs.

            12. Calyx 4–5 mm long, unribbed; mericarps essentially indehiscent, fragile-walled.... 9. *Sidastrum*

            12. Calyx usually >5 mm long, often ribbed or angular; mericarps usually more or less dehiscent, often indurate.

                13. Calyx basally 10-ribbed; mericarps 5–14, indurate, usually more or less reticulate laterally, often apically 2-spined; plants variously pubescent, viscid or not .... 8. *Sida*

                13. Calyx obscurely 5-nerved; mericarps 6–8, essentially unornamented, blunt at apex; plant notably viscid and malodorous .... 2. *Bastardia*

1. ABUTILON Mill.

Subshrubs or shrubs to small trees, glabrescent or pubescent, sometimes glandular-pubescent. Leaves elliptic, ovate, or cordate, sometimes lobed or parted, usually crenate or dentate. Flowers solitary in the leaf axils or aggregated into racemes or panicles, rarely in umbels; involucel absent; calyx gamosepalous, the lobes lanceolate, ovate, or cordate; petals often yellow or yellow-orange, less often white, lavender, or rose; staminal column filamentiferous at apex, the anthers numerous; styles 5 to many, the stigmas capitate. Fruit schizocarpic (sometimes pseudocapsular), the mericarps 5 to many, usually apically acute or acuminate to spinescent (rarely rounded); seeds usually 3–6 per mericarp, glabrous or slightly pubescent.

A genus of ca. 200 species from the Americas, Africa, Asia, and Australia.

1. Abutilon umbellatum (L.) Sweet, Hort. Brit., ed. 1, 53. 1826. *Sida umbellata* L., Syst. Nat., ed. 10, **2:** 1145. 1759.   Fig. 136A–D.

Shrub 1–2 m tall, densely stellate-pubescent. Leaf blades broadly ovate or weakly 3-lobulate, 6–12 × 4–10 cm, the apex acute or acuminate, the base cordate, the margins crenate or serrate, minutely stellate-pubescent, more densely so beneath; petioles 1.5–3 cm long; stipules 8–10 mm long. Inflorescences in few-flowered axillary umbels, these sometimes aggregated api-

cally. Calyx 6–8 mm long, the base rounded, half-divided, prominently hirsute, the hairs sometimes 2–3 mm long; petals yellowish, 5–8 mm long. Fruit 6–10 mm diam., mericarps 6–8, with divergent apical spines 2–4 mm long, hirsute. Seeds 2 mm long, papillate.

DISTRIBUTION: Common in open disturbed areas. Fish Bay near mouth of Fish Bay Gut (A3899), Johns Folly (A3933), Ram Head Trail (W194a). Also on St. Croix, St. Thomas, Tortola, and Virgin Gorda; the West Indies and the United States (Texas) to Central and South America (Venezuela to Peru).

2. BASTARDIA Kunth

Herbs or shrubs 0.5–3 m tall, stellate-pubescent and sometimes with long simple hairs, glandular hairs, or both. Leaves petiolate; blades cordate-ovate, the apex acuminate, the margin subentire or serrate, pubescent or viscid. Flowers solitary or paired in the leaf axils or aggregated into terminal leafy panicles; involucel absent; calyx 5-nerved, 5-lobed, divided almost to the base, the lobes acuminate; corolla yellowish. Fruit pseudocapsular (septicidal dehiscence suppressed), blunt or beaked at apex; mericarps 5–8, dorsally rounded; seeds solitary, minutely pubescent to subglabrous.

A tropical American genus of 3 or 4 species, extending from the southern United States (Texas) to Brazil and Peru, including the West Indies.

1. Bastardia viscosa (L.) Kunth *in* Humb., Bonpl. & Kunth, Nov. Gen. Sp. **5:** 256. 1822. *Sida viscosa* L., Syst. Nat., ed. 10, **2:** 1145. 1759.

Shrub 0.3–1.5 m tall; stems viscid (often malodorous) or not. Leaf blades cordate-ovate, stellate-tomentose, or softly tomentose

or tomentulose above and beneath. Flowers solitary or paired, sometimes forming terminal panicles; pedicels slender; calyx 4–5 mm long, deeply 5-lobed, each lobe apiculate with prominent midrib; corolla yellowish. Fruit mericarps 6–8, loculicidally dehiscent, tomentulose; seeds solitary per locule.

Key to the Varieties of *B. viscosa*

1. Plants stellate-tomentose (glandular hairs absent), the long simple hairs few or absent; corolla ca. 4 mm long .... 1a. *B. viscosa* var. *sanctae-crucis*

1. Plants with short glandular hairs and with long simple hairs, more or less throughout; corolla 5–7 mm long .... 1b. *B. viscosa* var. *viscosa*

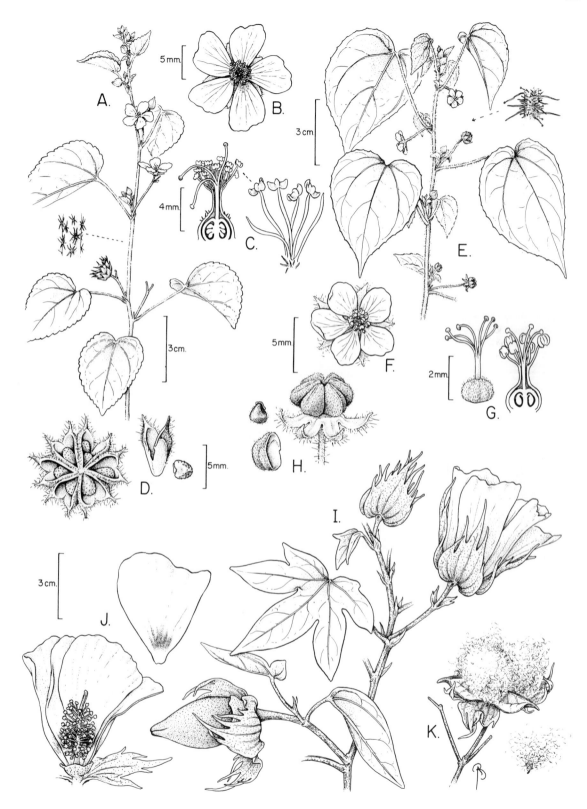

FIG. 136. A–D. *Abutilon umbellatum.* **A.** Fertile branch and detail of stem indument. **B.** Flower. **C.** L.s. staminal column showing pistil, and upper portion of staminal tube with free filaments. **D.** Schizocarp, mericarp, and seed. **E–H.** *Bastardia viscosa* var. *viscosa.* **E.** Fertile branch and detail of stem indument. **F.** flower. **G.** Pistil and l.s. flower. **H.** Seed, mericarp, and schizocarp. **I–K.** *Gossypium barbadense.* **I.** Fertile branch. **J.** L.s. flower, and petal. **K.** Opened capsule and seed covered with cotton fibers.

1a. Bastardia viscosa var. **sanctae-crucis** R.E. Fr., Kongl. Svenska Vetenskapsakad. Handl., ser. 3, **24:** 22, t. 5, fig. 1. 1947.

Shrub 30–75 cm tall; stems nonviscid, densely stellate-tomentose, with few or no long simple hairs. Leaf blades 1–4 × 0.8–2.5 cm, the apex acute or acuminate, the base cordate, the margins serrate, softly tomentose above and beneath; petioles 1–1.5 cm long. Calyx tomentose, without glandular hairs; corolla ca. 4 mm long. Fruit 4–5 mm diam.; mericarps 6 or 7, minutely apiculate. Seeds 1.8–2 mm long, glabrate or sparsely pubescent.

DISTRIBUTION: A common roadside weed. Annaberg (W780), Coral Bay (A2787), Hurricane Hole (A2436). Also on St. Croix; throughout the West Indies.

1b. Bastardia viscosa (L.) Kunth var. **viscosa.**

Fig. 136E–H.

Shrub 0.5–1.5 m tall; stems viscid, sometimes also with long simple hairs. Leaf blades 4–12 × 3–10 cm, softly tomentose above and beneath, often viscid above, the apex acuminate, the base cordate, the margins subentire to serrate; petioles 2–6 cm long. Calyx tomentose and also with long simple hairs; corolla 5–7 mm long. Fruit 5–7 mm diam.; mericarps 6–8, blunt. Seeds 2–2.2 mm long, minutely pubescent.

DISTRIBUTION: A common weed of open, disturbed sites. Coral Bay (A2707), Hawksnest Beach (A2901); Lameshur (W573). Also on St. Croix, St. Thomas, and Virgin Gorda; West Indies and the southern United States (Texas) to South America (Peru).

3. GOSSYPIUM L.

Shrubs or sometimes trees, stellate-pubescent or glabrate, more or less gland-dotted throughout. Leaves petiolate; blades ovate to deeply parted, the margin entire, usually with 1 or more nectaries on principal veins beneath. Flowers solitary in the leaf axils or in sympodial inflorescences; pedicels usually surmounted by trimerous nectaries; involucel 3-parted, the bracts small and scale-like or large and foliaceous, enclosing the bud, cordate, laciniate or entire, and persistent; calyx gamosepalous, truncate or 5-dentate (sometimes 5-lobed); corolla usually large and showy, white, cream, yellow, or rose, sometimes with dark center; androecium included, apically 5-dentate, the anthers numerous; style solitary with 3–5 decurrent stigmatic lobes, more or less exceeding the androecium. Capsule 3–5-celled, usually glabrous, often prominently gland-dotted; seeds usually pubescent or lanate.

A tropical and subtropical genus of 50 species from relatively arid areas of Africa, the Middle East, Australia, Mexico, and South America; some species are widely grown in cultivation as the source of commercial cotton.

REFERENCE: Fryxell, P. A. 1976. A nomenclator of *Gossypium:* the botanical names of cotton. Techn. Bull. U.S.D.A. 1491: 1–114.

1. Gossypium barbadense L., Sp. Pl. 693. 1753.

Fig. 136I–K.

Shrub 1–3 m tall, sometimes arborescent, sparsely stellate-pubescent or glabrate, gland-dotted throughout, the glands usually dark-pigmented. Leaf blades 3–7-lobed, 5–20 × 9–20 cm, the lobes ovate to lanceolate; petioles 3.5–15 cm long; stipules often prominent, subulate to falcate, 1–5 cm long. Flowers solitary or in sympodial inflorescences; bracts of involucel 4–6 cm long, broadly ovate, foliaceous, laciniate. Calyx 8–10 mm long, truncate; petals to 8 cm long, usually yellow with a dark red spot at base. Capsule usually narrowly elongate, 3.5–6 cm long, 3-celled, prominently pitted, glabrous. Seeds several per locule, free or fused together, lanate, the hairs (cotton) usually white.

DISTRIBUTION: Occasional to common in open dry areas. Maria Bluff (A613), North Shore Road (A3235), Center Line Road near turnoff to Bordeaux Mountain (*Mori & Woodbury 17079*). Also on Jost van Dyke, St. Croix, and St. Thomas; native to South America but cultivated now in many parts of the world as a commercial crop and as a garden plant throughout the tropics.

COMMON NAME: cotton.

4. HERISSANTIA Medik.

Herbs, subshrubs, or shrubs, erect or decumbent, pubescent or hirsute, sometimes viscid. Leaves ovate, the base cordate, the margin dentate. Flowers axillary, solitary, or borne on several-flowered peduncles; involucel absent; calyx 5-lobed, the lobes lanceolate or ovate; corolla white or yellow; androecium included, bearing anthers at apex. Fruit schizocarpic, oblate, inflated, pendent; mericarps 10 or more, dehiscent, with fragile walls, 1–3-seeded, pubescent or hispid; seeds glabrous or minutely scabridulous.

A neotropical genus of 6 or more species, one of which (the following) is widely distributed, the remainder of local distribution.

1. Herissantia crispa (L.) Brizicky, J. Arnold Arbor. **49:** 279. 1968. *Sida crispa* L., Sp. Pl. 685. 1753. *Gayoides crispum* (L.) Small, Fl. S.E. U.S. 764. 1903. *Bogenhardia crispa* (L.) Kearney, Leafl. W. Bot. **7:** 120. 1954.   Fig. 137A–D.

Perennial subshrub, erect to decumbent, sometimes scandent, to 1 m long, minutely stellate-puberulent. Leaves petiolate below to subsessile apically; blades cordate-ovate, 3.5–9 × 2–5 cm, minutely stellate-puberulent above and beneath, the apex acute or acuminate, the base cordate, the margins crenate; petioles 0.5–2.5 cm long. Flowers solitary in the leaf axils, the pedicels slender, sometimes exceeding the subtending leaves. Calyx 4–7 mm long; corolla white, 6–11 mm long; androecium yellow, included. Fruit globose, 13–20 mm diam.; mericarps 10–12, laterally compressed, dehiscent dorsally, papery, hispid. Seeds 3 per locule, minutely scabridulous, 1.7 mm long.

DISTRIBUTION: Infrequent in coastal scrubs. Dittlif Point (W759). Also on St. Thomas; the southern United States to South America (Argentina), adventive in the Old World.

5. MALACHRA L.

Herbs or subshrubs, sometimes puberulent, more commonly hispid, often with urticating hairs. Leaves petiolate; blades suborbicular to broadly ovate, angular, palmately lobed to palmately divided, usually pubescent, the apex acute to obtuse, the base truncate, the margins serrate; stipules filiform. Flowers in condensed, bracteate, headlike racemes, the "heads" axillary or terminal; specialized bracts

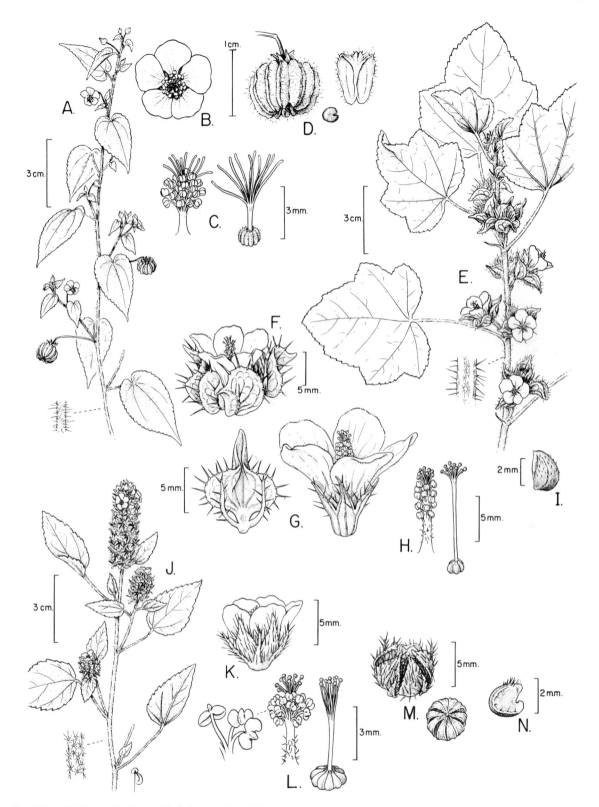

FIG. 137. A–D. *Herissantia crispa.* **A.** Fertile branch and detail of stem indument. **B.** Flower. **C.** Staminal column with protruding stigmas and pistil. **D.** Schizocarp, seed, and mericarp. **E–I.** *Malachra alceifolia.* **E.** Fertile branch and detail of stem indument. **F.** Flower and involucral bracts. **G.** Involucral bract and flower. **H.** Staminal column with protruding stigmas and pistil. **I.** Seed. **J–N.** *Malvastrum americanum.* **J.** Flowering branch and detail of stem indument. **K.** Flower. **L.** Staminal column with protruding stigmas, detail of stamen group, and pistil. **M.** Schizocarp with and without subtending calyx. **N.** Mericarp.

subtending the inflorescences broadly ovate-cordate, sessile, often with prominent nerves alternating with white intercostal tissue basally, green and foliaceous distally; involucel usually absent; calyx small, deeply 5-lobed; corolla often inconspicuous, white, yellow, or lavender; staminal column with 5 apical teeth; styles 10, the stigmas capitate. Fruit a schizocarp, glabrous or puberulent; mericarps 5, smooth, 1-seeded, blunt or spiny at apex; seeds glabrous.

A genus of 8–10 species originating in the New World but now also occurring in the Old World.

1. Malachra alceifolia Jacq., Collectanea 2: 350. 1789.
Fig. 137E–I.

Malachra rotundifolia Schrank, Pl. Rar. Hort. Monac. t. 56. 1820.

Herb or subshrub 0.5–2 m tall; stems usually hispid, sometimes very sparsely so. Leaf blades orbicular to ovate, 3–5-angled or slightly lobulate, 6–9 × 4–9 cm, sparsely hispid to glabrate, the apex acute to obtuse, the base truncate, the margins serrate; petioles 4–6 cm long. Inflorescence a series of compact, axillary, few-flowered glomerules, short-pedunculate to sessile; specialized floral bracts several, subsessile, the base cordate, the apex acuminate, often with white reticulate tissue at base, often hispid; involucel absent. Calyx deeply 5-lobed, 6–8 mm long, hispid; petals yellow, 1–1.5 cm long; staminal column subequal to corolla. Fruit minutely puberulent, the mericarps ca. 3 mm long, brownish with reticulate veins. Seeds solitary per locule, glabrous.

DISTRIBUTION: Often in disturbed ground. Lameshur (W515), Enighed (A3098). Also on St. Croix and St. Thomas; West Indies and southern Mexico, Central America, and northern South America.

6. MALVASTRUM A. Gray

Perennial shrubs or subshrubs (sometimes annual), erect, with patent or appressed stellate pubescence. Leaves petiolate, ovate or lanceolate, usually simple, serrate, crenate, or dentate; stipules lanceolate or falcate. Flowers solitary in the leaf axils or aggregated in apical spikes or racemes; involucel of 3 linear, spatulate, or ligulate bractlets; calyx 5-lobed; corolla yellowish or orangish, rarely with red center; androecium included, bearing anthers at apex; styles 7–18, the stigmas capitate. Fruit schizocarpic, oblate; mericarps 7–18, indehiscent, horseshoe–shaped, hispid, with prominent ventral notch, sometimes 2–3-cuspidate, 1-seeded, setose or pubescent; seeds glabrate.

A genus of 14 species, principally American but adventive in the Old World.

Key to the Species of *Malvastrum*

1. Hairs of upper leaf surface simple; flowers all axillary, scattered along branches, manifestly pedicellate, the pedicels 2–10 mm long; stem pubescence of appressed stellate hairs with 4 rays, 2 antrorse and 2 retrorse; mericarps with 3 prominent cusps; staminal column glabrous .... 3. *M. coromandelianum*
1. Hairs of upper leaf surface stellate; flowers sometimes solitary but usually in terminal spicate inflorescences, sessile or subsessile; stem pubescence of stellate hairs with 3–12 more or less spreading radii; mericarps with rounded dorsal wall lacking cusps (or these obscure); staminal column pubescent.
   2. Stem hairs mostly 3–5-rayed, appressed stellate; inflorescence a spike usually no more than 2 cm long; flowers with pedicels ca. 1 mm long; lateral walls of mericarp more or less pubescent ...... 2. *M. corchorifolium*
   2. Stem hairs mostly 5–12-rayed, tufted stellate; inflorescence a dense spike usually >3 cm long; flowers sessile; lateral walls of mericarp nearly glabrous .... 1. *M. americanum*

1. Malvastrum americanum (L.) Torr. *in* Emory, Rep. U.S. Mex. Bound. 2: 38. 1859. *Malva americana* L., Sp. Pl. 687. 1753.
Fig. 137J–N.

Herb, subshrub, or shrub 0.5–1.5 m tall, stellate-pubescent, the hairs 5–12-rayed. Leaf blades ovate or weakly lobulate, 2–9 × 1–8 cm, minutely stellate-pubescent above and beneath, the apex acute, the base truncate, the margins serrate; petioles 1–6 cm long. Inflorescences dense spikes, usually >3 cm long, terminal or terminating lateral axillary branches; flowers sessile, bracteate, subtended by an involucel of 3 linear bractlets. Calyx 5–6 mm long (accrescent to 6–10 mm in fruit), hirsute; petals 8–9 mm long, yellow-orange; androecium yellowish, the column stellate-pubescent. Fruit oblate, 5–6 mm diam.; mericarps 11–14, apically setose, dorsally smooth and glabrous, laterally ribbed, without cusps. Seeds 1.5 mm long.

DISTRIBUTION: Common along open and disturbed ground. East End (A5389), Johns Folly (A3943), Southside Pond (A4069). Also on St. Croix, St. Thomas, and Virgin Gorda (fide Britton & Wilson, 1924: 550); West Indies, southern United States (Florida, Texas) south to Mexico, Central America, and South America (Argentina), also adventive in the Old World.

NOTE: This species was treated incorrectly by some previous workers (e.g., Britton, 1918; Britton & Wilson, 1924; Eggers, 1897; Millspaugh, 1902) as *Malvastrum spicatum* (L.) A. Gray, a plant that is in fact a *Melochia*.

2. Malvastrum corchorifolium (Desr.) Britton *ex* Small, Fl. Miami 119. 1913. *Malva corchorifolia* Desr. *in* Lam., Encycl. 3: 755. 1792, based on *Malva scoparia* Jacq., Collectanea 1: 59. 1787, non L'Hér., 1786.

Erect subshrub 25–60 cm tall; stems with 3–5-rayed appressed stellate hairs. Leaf blades ovate, 2–6 × 1–3 cm, the margins dentate, the upper surface with 4–5-rayed stellate hairs; petioles 1–4 cm long. Flowers solitary in the axils of the upper leaves, subsessile, becoming congested and spicate on branch tips, the spikes usually no more than 2 cm long, the individual flowers bracteate, subtended by an involucel of 3 lanceolate bractlets. Calyx 5–6 mm long (accrescent to 7–11 mm in fruit), moderately pubescent; corolla exceeding calyx, yellow; androecium yellowish, the column stellate-pubescent. Fruit oblate, 5–6.5 mm diam.;

mericarps 11–13, apically setose, laterally pubescent, with cusps greatly suppressed.

DISTRIBUTION: Common in open disturbed and waste ground. In vicinity of Enighed (A3924), Coral Bay (A2578), Maho Bay Quarter along Center Line Road (A1852). Also on St. Croix and Virgin Gorda (fide Britton & Wilson, 1924: 550); West Indies, United States (Florida), Mexico, and Central America.

3. Malvastrum coromandelianum (L.) Garcke, Bonplandia 5: 295. 1857. *Malva coromandeliana* L., Sp. Pl. 687. 1753.

Malva tricuspidata W.T. Aiton, Hortus. Kew. 4: 210. 1812, nom. superfl. *Malvastrum tricuspidatum* (W.T. Aiton) A. Gray, Pl. Wright. 1: 16. 1852.

Erect subshrub or sometimes annual herb 0.3–1 m tall; stems with appressed, 4-rayed stellate hairs, the rays 2 antrorse, 2 retrorse. Leaf blades ovate or lanceolate, 2–8.5 × 1–4 cm, the margins dentate, the upper surface usually with simple hairs; petioles 1–3 cm long . Flowers axillary, usually solitary on short pedicels (2–10 mm long); bractlets of involucel 3, linear or narrowly spatulate, shorter than calyx. Calyx 5–7 mm long (accrescent to 8–11 mm in fruit), hirsute; corolla exceeding the calyx, yellow; androecium yellowish, the column glabrous. Fruit oblate, 6–7 mm diam.; mericarps 10–12, setose, 3-cusped (1 cusp apical, 2 on dorsal wall).

DISTRIBUTION: Occasional to common in open areas. Fish Bay (A3918), Lameshur (A3200). Also on St. Croix, St. Thomas (fide Britton, 1918), and Tortola; a pantropical weed, from the United States (Florida, Texas) to South America (northern Argentina).

7. PAVONIA Cav.

Prostrate perennial herbs, erect subshrubs, or shrubs (rarely arborescent), often stellate-pubescent, sometimes viscid, sometimes glabrate. Leaves petiolate; blades ovate, elliptic, lanceolate, oblanceolate, or deltoid, sometimes lobed or asymmetrical, the margin dentate or crenate (rarely entire). Flowers solitary, or paired in the leaf axils, or aggregated in racemes, panicles, or heads; involucel present, 4–24-parted, the bractlets distinct or basally connate; calyx 5-lobed; corolla white, lavender, purple, or yellow, sometimes with a dark center; staminal column included or exserted, sometimes declined with secund anthers, surmounted by 5 apical teeth; styles 10, exceeding the androecium, the stigmas capitate. Fruit schizocarpic; mericarps 5, usually indehiscent, sometimes winged, spined, rugose, or otherwise ornamented, minutely pubescent or glabrous; seeds solitary.

A tropical and subtropical genus of ca. 250 species, ca. 220 of which occur in the New World.

1. Pavonia spinifex (L.) Cav., Diss. 3: 133, t. 45, fig. 2. 1787. *Hibiscus spinifex* L., Syst. Nat., ed. 10, 2: 1149. 1759. Fig. 138A–E.

Subshrubs 1–2 m tall; stems minutely pubescent to glabrate, the hairs often in well-defined longitudinal rows. Leaf blades ovate, 4–12 × 2–8 cm, sparsely stellate-pubescent to glabrate beneath, with simple hairs above, the apex acute, the base truncate or somewhat cordate, the margins irregularly crenate-serrate; petioles 0.5–3(–5) cm long. Flowers solitary, axillary; pedicels 1–3(–4) cm long; bracts of involucel 5–8, lanceolate or slightly spatulate, subequal to calyx. Calyx 8–11 mm long, ca. half-divided, the lobes ciliate with hairs 1–2 mm long; petals 18–27 mm long, yellow; androecium slightly exserted from corolla; styles slender, exserted. Fruit schizocarpic, 8–10 mm diam.; mericarps 3-spined, the spines retrorsely barbed, 6 or more mm long, the central spine erect, the lateral spines divergent.

DISTRIBUTION: Occasional in open disturbed areas such as roadsides and waste grounds. Solomon Bay close to Research Center (A2293), road to Bordeaux (A3836). Also on St. Croix, St. Thomas, and Tortola; the southeastern United States, the Bahaman Archipelago, and the Greater and Lesser Antilles, but not in South America (where it has frequently been cited as a result of misidentifications of related species).

8. SIDA L.

Perennial herbs or subshrubs, erect or prostrate, glabrous or pubescent, sometimes viscid. Leaves petiolate to subsessile; blades ovate (sometimes lobed), elliptic, rhombic, or linear, the margin usually dentate. Flowers solitary in the leaf axils, in axillary glomerules, or in dense or open terminal inflorescences; involucel absent; calyx 5-lobed, often 10-ribbed at base and angularly plicate in bud; corolla white, yellow, orangish, rose, or purple, sometimes with a dark red center; androecium included, usually pallid, bearing anthers at apex; styles 5–14, the stigmas capitate. Fruit schizocarpic, glabrous or pubescent; mericarps 5–14, usually indurate, usually laterally reticulate, basally indehiscent with well-differentiated dorsal wall, apically more or less dehiscent with 2 more or less differentiated spines or muticous; seeds solitary, glabrous.

A genus of ca. 150 species from the Americas, Africa, Asia, and Australia.

Key to the Species of *Sida*

1. Plants prostrate herbs.
  2. Leaves ovate–cordate; calyx lobes trullate, the margins dark green; plants rooting at the nodes; corolla yellow .... 8. *S. repens*
  2. Leaves narrowly elliptic; calyx lobes lacking dark green margins; plants not rooting at the nodes; corolla pink .... 2. *S. ciliaris*
1. Plants erect or ascending (sometimes scandent) herbs or subshrubs.
  3. Leaf disposition (and branching pattern) distichous; stipules prominent, often falcate, several–veined.
    4. Styles and mericarps 8–10; calyx 6–8 mm long .... 1. *S. acuta*
    4. Styles and mericarps 5; calyx 4–7 mm long.
      5. Calyx 4–5 mm long; plants sparsely pubescent; leaves narrowly lanceolate, acute; mericarps minutely apiculate .... 5. *S. glomerata*

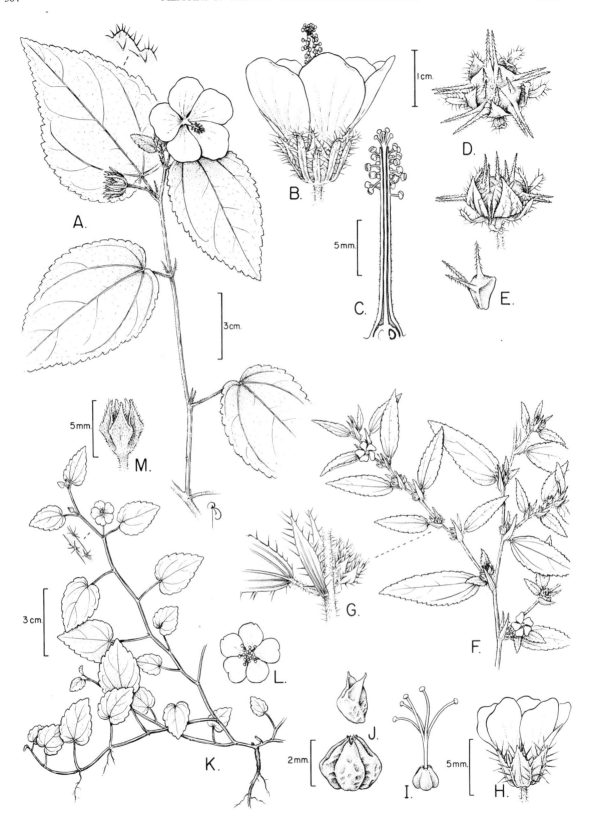

Fig. 138. A–E. *Pavonia spinifex*. A. Fertile branch and detail of leaf margin. B. Flower. C. L.s. staminal column and pistil. D. Schizocarp, top and lateral views. E. Mericarp. F–J. *Sida glomerata*. F. Fertile branch. G. Detail of stem showing stipules at base of leaf petiole and inflorescence. H. Flower. I. Pistil. J. Mericarp and schizocarp. K–M. *Sida repens*. K. Habit and detail of stem indument. L. Flower, top view. M. Fruiting calyx.

5. Calyx 6–7 mm long; plants densely and softly stellate–pubescent; leaves broadly ovate–elliptic, often obtuse; mericarps with beaks 1–1.5 mm long .... 7. *S. jamaicensis*

3. Leaves and branches spirally disposed; stipules inconspicuous, usually 1–veined.

 6. Flowers and fruits subsessile, aggregated into dense axillary glomerules (or in axillary peduncu- late "heads"); stems often hispid, the hairs 2–3 mm long; mericarps muticous, glabrous; stems often reclining or scandent .... 10. *S. urens*

 6. Flowers and fruits usually pedicellate, not glomerulate (usually solitary); stems not hispid (ex- cept sometimes in *S. aggregata*); mericarps usually apically spined, often pubescent; stems usu- ally erect (except often reclining in *S. glabra*).

  7. Styles and mericarps 8–14; calyx manifestly 10-ribbed .... 3. *S. cordifolia*

  7. Styles and mericarps 5–8; calyx 5- or 10-angled.

   9. Leaves ovate, lanceolate, or narrowly oblong; plant minutely stellate-pubescent; calyx 5–7 mm long, stellate-tomentose, the margins sometimes reddish .... 9. *S. spinosa*

   9. Leaves cordate-ovate; plant sometimes viscid; calyx 4–6 mm long, the margins dark green.

    10. Calyx lobes lanceolate, longer than wide, equaling or exceeding the calyx tube; spines of mericarps 1.5–2 mm long; stems stellate-pubescent; flowers axillary ...... 4. *S. glabra*

    10. Calyx lobes triangular, wider than long, shorter than the calyx tube; spines of meri- carps 0.5–1 mm long; stems viscid; flowers in panicles .... 6. *S. glutinosa*

1. Sida acuta Burm.f., Fl. Indica 147. 1768.

Sida carpinifolia L.f., Suppl. Pl. 307. 1782.

Erect shrub or subshrub ca. 1 m tall, with distichous branching pattern and leaf arrangement, more or less hirsute to glabrate. Leaf blades lanceolate to ovate, 3–9 × 0.5–4 cm, concolorous, hirsute to glabrate; petioles 2–5 mm long, the apex acute, the base truncate, the margins serrate (at least distally); stipules prominent, several-veined, broadly falcate and often exceeding the petiole. Flowers solitary or paired in the leaf axils; pedicels more or less equaling the petioles. Calyx 6–8 mm long, ca. half-divided, basally 10-costate, often ciliate; petals 7–10 mm long, white, yellow, or yellow-orange (often polymorphic for color in a single population); androecium included, pallid; styles 8–10. Fruit gla- brous; mericarps 8-10, laterally reticulate, the apical spines vari- ously developed.

DISTRIBUTION: A common weed of open areas such as roadsides and waste grounds. Annaberg (M17080), Cruz Bay (A2273), trail to Sieben (A2061). Also on St. Croix, and St. Thomas, and Tortola; pantropical.

2. Sida ciliaris L., Syst. Nat., ed. 10, **2:** 1145. 1759.

Sida bellidifolia Gand., Bull. Soc. Bot. France **71:** 629. 1924.

Sida microtricha Gand., Bull. Soc. Bot. France **71:** 629. 1924.

Sida ononidifolia Gand., Bull. Soc. Bot. France **71:** 629. 1924.

Prostrate perennial herb to 25 cm long, freely branching; stems with appressed stellate hairs, these usually 4-armed and aligned longitudinally with the stem axis. Leaves mostly crowded at apices of branches because of marked shortening of apical internodes; blades narrowly elliptic, 0.5–2 × 0.5–0.7 cm, glabrous above, stellate-pubescent beneath, the apex acute or obtuse, the margins few-toothed apically; petioles 5–6 mm long; stipules linear to oblanceolate, 4–12 mm long, crowded, partially adnate to petiole, ciliate. Flowers and fruits subsessile, crowded among leaves and stipules at branch apices. Calyx 4–6 mm long, half-divided, hirsute; corolla 5–11 mm long, pink (or less often red-orange or yellowish). Fruit conical, more or less muricate; mericarps 5–8, essentially indehiscent.

DISTRIBUTION: Recently disturbed sites. Bethany (B329), Frank Bay (A4217), Nanny Point (A2441). Also on St. Croix, St. Thomas, Tortola, Virgin Gorda, and Anegada; the southern United States to Argentina.

3. Sida cordifolia L., Sp. Pl. 684. 1753.

Erect shrub or subshrub to 1.5 m tall with ascending branches, stellate-tomentose throughout. Leaf blades broadly cordate to lance-ovate, 1.5–9 × 1–6.5 cm, reduced upward, soft-tomentose above and beneath, the apex acute, the base cordate, the margins serrate to near the base; petioles 2–3 cm long. Flowers and fruits axillary but usually aggregated terminally into a congested paniculate or corymbiform inflorescence. Calyx prominently 10- ribbed, 6–7 mm long, tomentose; corolla 8–11 mm long, yellow- orange; styles 8–14. Fruit 6–7 mm diam., apically pubescent; mericarps 8–14, each with 2 apical retrorsely pubescent spines (variably developed, rarely suppressed).

DISTRIBUTION: Occasional in moist or marshy areas. Lameshur (W677), without locality (CFM174). Also on St. Croix, St. Thomas, and Virgin Gorda; pantropical.

4. Sida glabra Mill., Gard. Dict., ed. 8. 1768, non Nutt., 1834.

Erect herb 0.5–1 m tall; stems minutely stellate-pubescent, the hairs 0.2–0.4 mm long. Leaf blades narrowly ovate, 4–6 × 2–3 cm, sparsely stellate-pubescent above and beneath (hairs <0.5 mm long), the apex acuminate, the base cordate, the margins serrate; petioles 1.5–2.5 cm long; stipules linear, 3–4 mm long. Flowers solitary in the leaf axils; pedicels slender, 10–16 mm long, pubescent (including some glandular hairs). Calyx 5–6 mm long, stellate-pubescent, 5-angled, half-divided, the lobes lanceo- late-acuminate, 3–3.5 mm long, marginally hispid-ciliate, the margins dark green; petals 5 mm long, yellow-orange; staminal column 2 mm long, pallid, pubescent, the filaments 1 mm long; styles 5, exceeding the androecium, the stigmas capitate. Fruit 4 mm diam., pubescent; mericarps 5, ca. 4 mm long (including spines), with 2 erect apical antrorsely pubescent spines, 1.5–2 mm long.

DISTRIBUTION: Occasional in open habitats. Lameshur (A3230), Ram Head (W197), between Salt Pond and Coral Bay (A2793). Also on St. Croix, St. Thomas, and Tortola; the West Indies and Central America to Colombia and Venezuela.

5. Sida glomerata Cav., Diss. **1:** 18, t. 2, fig. 6. 1785.
Fig. 138F–J.

Subshrub 0.5–1 m tall, with distichous branching and leaf arrangement; stems arching, sparsely pubescent with stellate or sometimes simple hairs ca. 1 mm long. Leaves short-petiolate to subsessile, sometimes appearing imbricate because of short internodes; blades lanceolate to elliptic, 1.5–6 × 1–2.5 cm, sparsely ciliate above and beneath, the apex acute, the base obtuse to subcordate, the margins serrate, ciliate; petioles to 4 mm long; stipules prominent, broadly falcate, several-nerved, usually exceeding the petioles, ciliate. Flowers short-pedicellate or subsessile, solitary or in axillary glomerules. Calyx 4–5 mm long, ciliate; corolla 6–8 mm long, white or yellowish; androecium shorter than the petals, glabrous, the anthers yellowish; styles 5. Fruit subglabrous; mericarps 5, with 2 minute apical spines.

DISTRIBUTION: Common widespread weed of open habitats. Bethany to Rosenberg (B213), Bordeaux (A3850), Coral Bay (A2447). Also on St. Croix, St. Thomas, Tortola, and Virgin Gorda; West Indies and Central America (as far north as Nicaragua) to much of South America.

6. Sida glutinosa Cav., Diss. 1: 16, t. 2, fig. 8. 1785.

Erect herb to 2 m tall; stems densely viscid with minute glandular hairs 0.2–0.4 mm long, also with scattered long simple hairs 1.5 mm long. Leaf blades ovate, 4–7(–9) × 2.5–5(–7) cm, stellate-pubescent above and beneath (hairs >0.5 mm long), more densely so beneath, the apex acuminate, the base deeply cordate, the margins crenate-serrate; petioles 1.5–4 cm long; stipules subulate, 1–2 mm long. Flowers in profuse open panicles, mostly terminal and more or less leafy; pedicels slender, 7–17 mm long, viscid. Calyx 4 mm long, viscid, 10-angled, less than half-divided, the lobes acuminate, 1–1.5 mm long, the margins dark green, hispid-ciliate; petals 5 mm long, yellow-orange; staminal column 2 mm long, pallid, pubescent, the filaments 1 mm long; styles 5, exceeding the androecium, purplish, the stigmas capitate. Fruit 4 mm diam., pubescent; mericarps 5, ca. 3 mm long (including spines), with 2 erect apical antrorsely pubescent spines, 0.5–1 mm long.

DISTRIBUTION: Occasional in open, disturbed habitats. Fish Bay (W738). Also on St. Croix; the West Indies, Mexico, and Central America to Colombia and Venezuela.

7. Sida jamaicensis L., Syst. Nat., ed. 10, **2:** 1145. 1759.

Subshrub 0.5–1 m tall, with distichous branching pattern and leaf arrangement, softly stellate-pubescent throughout. Leaf blades broadly ovate-elliptic, 2–5 × 1–3 cm, somewhat discolorous, soft-pubescent above and beneath, the apex acute or obtuse, the margins serrate-crenate; petioles 4–6 mm long. Flowers 1 or more per leaf axil; pedicels shorter than the petioles. Calyx 6–8 mm long, the lobes acuminate, ciliate; corolla slightly exceeding the calyx, white. Fruit apically pubescent; mericarps 5, 2-spined (the spines 1–1.5 mm long), blackish at maturity.

DISTRIBUTION: East End (A3214), Fortsberg (A4086), Lameshur (B632). Also on St. Croix, St. Thomas, and Virgin Gorda; the West Indies and Mexico to northern South America.

8. Sida repens Dombey *ex* Cav., Diss. **1:** 7. 1785.
Fig. 138K–M.

Sida pilosa Cav., Diss. **1:** 9, t. 1, fig. 8. 1785, non Mill., 1768. *Sida javensis* Cav. ssp. *expilosa* Borss. Waalk., Blumea **14:** 185. 1966 (based on *Sida pilosa* Cav.).

Sida humilis sensu Britton & P. Wilson, Bot. Porto Rico 5: 553. 1924, non Cav., 1788.

Prostrate herb to 30 cm long; stems slender, pubescent to glabrate, rooting at the nodes. Leaf blades ovate, 1.2–1.8 × 1.1–1.6 cm, with appressed simple hairs above and minute stellate hairs (or glabrate) beneath, the apex rounded-acute, the base cordate, the margins crenate to the base; petioles 1–1.8 cm long. Flowers solitary in the leaf axils; pedicels slender, exceeding the subtending petioles. Calyx 5–7 mm long, the lobes trullate, the margins dark green, ciliate; corolla 6–8 mm long, yellowish. Fruit ca. 3 mm diam., minutely pubescent; mericarps 5, short-spined.

DISTRIBUTION: Occasional to common, in open and disturbed habitats. Bethany (B331), Lameshur (A4046), Solomon Beach (A2290). Also on St. Croix, St. Thomas, Tortola, and Virgin Gorda; Central America, northeastern South America, and West Indies; possibly introduced from Malaysia (Java to the Philippines).

9. Sida spinosa L., Sp. Pl. 683. 1753.

Shrub to 1(–2) m tall, erect, with ascending branches; stems minutely stellate-pubescent. Leaf blades broadly ovate, lanceolate, or narrowly oblong, 2–4 × 0.7–2 cm, smaller upward, glabrate above, minutely stellate-tomentose beneath, the apex usually acute, the margins serrate to the base, sometimes marginally reddish, discolorous; petioles 0.5–1.5 cm long (sometimes with a spur ["spine"] at base of petiole). Flowers solitary or in small groups in the leaf axils, scattered along the stem and crowded at apices; pedicels to 1 cm long. Calyx 5–7 mm long, the lobes triangular, sometimes red-margined, minutely stellate-tomentose; corolla ca. 5 mm long, yellow or yellow-orange. Fruit 4–5 mm diam.; mericarps 5, apically 2-spined, the spines ca. 1 mm long, antrorsely pubescent.

DISTRIBUTION: Occasional in open disturbed areas such as roadsides and waste grounds. Coral Bay (A4243), Enighed (A3920). Also on St. Croix, St. Thomas, and Virgin Gorda; pantropical and extending into the temperate zones as an annual.

10. Sida urens L., Syst. Nat., ed. 10, **2:** 1145. 1759.

Erect or reclining subshrub, sometimes scandent, 1–1.5 m tall; stems usually hispid with simple hairs 2–3 mm long together with shorter stellate hairs, sometimes with only stellate hairs. Leaf blades ovate, 2–9 × 2–4 cm, more or less stellate-pubescent beneath or with simple hairs above that are more or less antrorsely oriented, the apex acuminate, the base cordate, the margins crenate-serrate; petioles 1–4 cm long. Flowers and fruits sessile or short-pedicellate, crowded in axillary glomerules or pedunculate "heads"; pedicels shorter than calyces. Calyx 5–8 mm long, the lobes trullate with dark green margins, setose; corolla ca. 5 mm long, orange (fading rose), often with a dark red center. Fruit 3.5–4 mm diam.; mericarps 5, glabrous, muticous, essentially indehiscent.

DISTRIBUTION: Uncommon, known from a single collection. Bordeaux Mountain (W695). Also on St. Thomas; Mexico, West Indies, Central and South America, and Africa.

DOUBTFUL RECORD: *Sida rhombifolia* L. was cited as occurring on St. John by Britton and Wilson (1924) and was reported by Woodbury and Weaver (1987), but no specimens have been seen nor has it been recently collected on the island.

EXCLUDED SPECIES: *Sida aggregata* C. Presl, reported by Woodbury and Weaver (1987), was based on an incorrect identification of *Bastardia viscosa* (L.) Kunth var. *sanctae-crucis* R.E. Fr.

9. SIDASTRUM Baker f.

Erect shrubs or subshrubs, stellate-pubescent to glabrate. Leaves petiolate, the blades lanceolate or elliptic to ovate, the base cuneate, truncate, or cordate, the apex acute, the margin crenate, serrate, or subentire, pubescent or glabrate. Flowers in axillary fascicles or spikes or in terminal racemes or panicles; involucel absent; calyx small, basally rounded, unribbed, ca. half-divided; petals barely exceeding calyx to twice as long, yellow, orangish, or purple; androecium subequal to petals, bearing anthers at apex, the anthers few (sometimes only 10); styles 5–10, the stigmas capitate. Fruit oblate or subconical, often stellate-pubescent; mericarps 5–10, essentially indehiscent, smooth or slightly reticulate laterally, 1-seeded; seeds sparsely pubescent to subglabrous.

A genus of 7 New World species, with additional species to be added from Australia and the Pacific.

1. **Sidastrum multiflorum** (Jacq.) Fryxell, Brittonia **31:** 298. 1979. *Sida multiflora* Jacq., Observ. Bot. **2:** 23, t. 45, fig. 1. 1767. Fig. 139E–H.

Sida acuminata DC., Prodr. **1:** 462. 1824.
Sida acuminata var. *macrophylla* Schltdl., Linnaea **3:** 269. 1828.

Subshrub or shrub to 1.5 m tall; stems with yellowish stellate pubescence. Leaf blades ovate–elliptic, 3–6 × 2–3 cm, discol-orous, minutely stellate-pubescent above and beneath, the apex acute, the base truncate, the margins finely crenate-serrate; petioles 0.4–1.5 cm long. Flowers 1–3 per axil; pedicels slender, to 1 cm long. Calyx 4–5 mm long, unribbed, yellowish-pubescent; corolla 5 mm long, yellow; androecium included. Fruit oblate, 4–5 mm diam., stellate-pubescent; mericarps 8–10, fragile-walled.

DISTRIBUTION: Common in open, disturbed habitats such as clearings and roadsides. Coral Bay (A2788), L'Esperance (A2558), trail to Seiben (A2646). Also on St. Croix, St. Thomas, Guana Island (Tortola), and Virgin Gorda.

10. THESPESIA Sol. *ex* Corrêa

Shrubs or more commonly trees, glabrous or pubescent, usually punctate. Leaves petiolate; blades ovate or trilobulate, the apex obtuse to acuminate, the margin entire, often with abaxial foliar nectaries. Flowers large and showy, borne singly on axillary pedicels (sometimes aggregated apically), sometimes with bracteate articulation; involucel of 3 to many elements, sometimes subtended by trimerous nectaries; calyx truncate to 5-lobed; corolla white, yellow, or rose, with or without dark center; staminal column apically 5-dentate, usually included, pallid; style single, the 3–5 stigmatic lobes decurrent. Fruit capsular, 3–5-celled, coriaceous or ligneous, dehiscent or indehiscent; seeds several per locule, glabrous or pubescent.

A genus of 17 species from New Guinea, Asia, Africa, and the West Indies.

1. **Thespesia populnea** (L.) Sol. *ex* Corrêa, Ann. Mus. Natl. Hist. Nat. **9:** 290, t. 8, fig. 2. 1807. *Hibiscus populneus* L., Sp. Pl. 694. 1753. Fig. 139A–D.

Small tree or shrub 3–12 m tall; stems minutely lepidote to glabrate; older trunks with fissured bark. Leaf blades ovate, 5–18 × 3–11 cm, coriaceous, glabrous above, glabrate beneath, with elongate nectariferous zone on midrib beneath, the apex acuminate, the base deeply cordate, the margins entire; petioles 3.5–6 cm long. Flowers axillary, usually solitary, with erect pedicels 1–5 cm long; involucel of 3 ligulate bracts, these often exceeding the calyx, deciduous, leaving scars. Calyx truncate, 6–10 mm long, lepidote; corolla 4–6 cm long, yellow, with or without purple center; androecium columnar, 1.5–2 cm long, bearing anthers throughout, yellowish; style clavate, exceeding androecium. Capsule 5-celled, oblate, 3.5 cm diam., indehiscent, minutely lepidote. Seeds 8–9 mm long, short-pubescent.

DISTRIBUTION: Common along sandy beaches. Coral Bay (A4028), Solomon Bay (A2286), Waterlemon Bay (A2522). Also on Jost van Dyke, St. Croix, and St. Thomas; pantropical in littoral habitats; sometimes cultivated as an ornamental shade tree.

11. URENA L.

Shrubs 0.5–2 m tall, more or less stellate-pubescent. Leaves petiolate, variable: blades often 3–5-angled, -lobed, or -parted, less often ovate, oblong, or lanceolate, the margin serrate, with 1 or more prominent foliar nectaries on principal veins of abaxial surface. Flowers solitary or glomerulate in the leaf axils or forming terminal racemes; pedicels usually short; involucel 5-lobed, the lobes alternate with the lobes of the calyx; calyx 5-lobed; petals rose or lavender; androecium included, apically 5-dentate, the filaments short; styles 10, slender, the stigmas capitate. Fruit schizocarpic, 5-lobed, the lobes convex; mericarps 5, prominently glochidiate (in ours) or smooth, essentially indehiscent; seeds solitary, glabrous.

A genus of about 6 species, principally South American, of which 2 species are nearly pantropical.

1. **Urena lobata** L., Sp. Pl. 692. 1753. Fig. 140A–F.

Shrub 1–2 m tall; stems minutely stellate-pubescent. Leaf blades ovate, angular, or shallowly lobed, 3–9 × 1–10 cm (smaller and narrower upward), discolorous, minutely stellate-pubescent above and beneath, with small but prominent nectary at base of midrib beneath (sometimes several nectaries on the princi-pal nerves), the apex acute, the base truncate to cordate, the sinuses (if lobed) narrow, the margins irregularly serrate; petioles 0.5–7.5 cm long (reduced upward), pubescent like stem; stipules subulate, 3–5 mm long. Flowers axillary, usually solitary; pedicels shorter than subtending petioles; involucellar bracts 5, 5–8 mm long, connate basally, the lobes linear-lanceolate to triangular. Calyx slightly shorter than to equaling the involucel; petals 18–20 mm long, lavender; androecium with anthers subsessile on column, purplish. Fruit oblate, 8–10 mm diam., with numerous glochidi-ate spines.

DISTRIBUTION: Occasional in disturbed sites along roadsides. Bordeaux Mountain (A3177, A3849). Also on St. Croix, St. Thomas, and Tortola; pantropical.

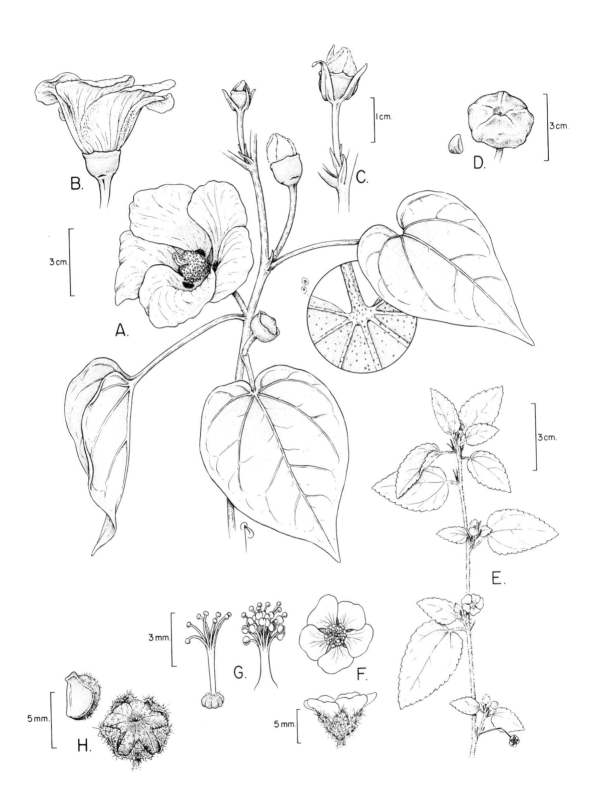

FIG. 139. A–D. *Thespesia populnea*. **A.** Fertile branch and details of abaxial base of leaf blade and leaf indument. **B.** Flower. **C.** Flowerbud with subtending epicalyx and bracteoles. **D.** Seed and fruit. E–H. *Sidastrum multiflorum*. **E.** Fertile branch. **F.** Flower, top and lateral views. **G.** Pistil and staminal column. **H.** Mericarp and schizocarp.

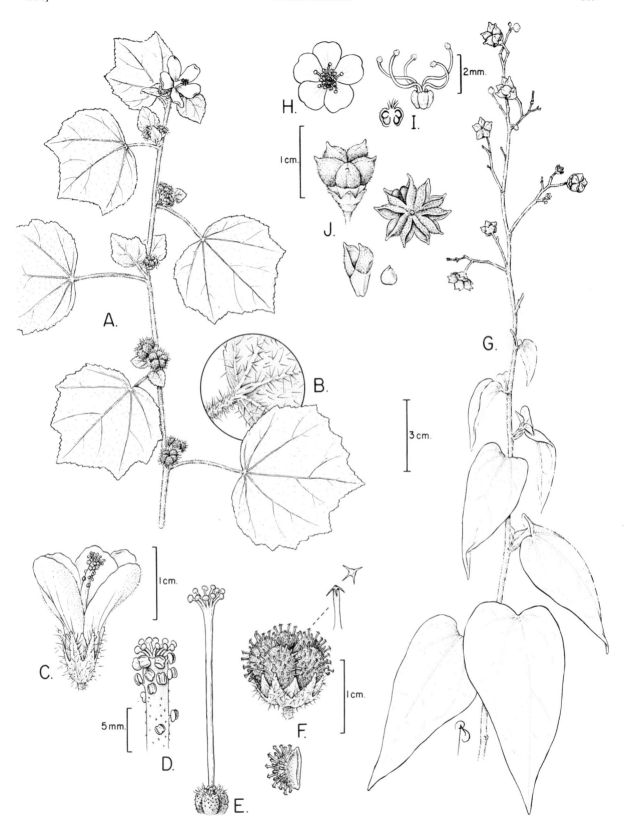

FIG. 140. A–F. *Urena lobata*. A. Fertile branch. B. Detail of abaxial base of leaf blade. C. Flower. D. Detail of staminal column apex. E. Pistil. F. Schizocarp, details of glochidiate spine and mericarp. G–J. *Wissadula periplocifolia*. G. Fertile branch. H. Flower, top view. I. L.s. ovary and pistil. J. Schizocarp, lateral and top views, and mericarp and seed.

12. WISSADULA Medik.

Herbs or subshrubs, usually erect, stellate-pubescent or sometimes glabrate. Leaves long- or short-petiolate; blades broadly ovate to narrowly triangular, the margin entire or crenate-dentate. Flowers sometimes solitary in the leaf axils, usually in condensed or open terminal panicles; involucel absent; calyx 5-lobed, usually small; petals usually small, yellowish; androecium bearing anthers at the apex, included; styles 3–6, usually slender, the stigmas capitate. Fruit schizocarpic but septicidal dehiscence sometimes incomplete (the fruit thus pseudocapsular); mericarps 3–6, divided into indehiscent lower cell and dehiscent upper cell by a constriction, apically bulbous-apiculate; seeds (1–)3 with 1 in lower cell and usually 2 in upper cell, the lower seed relatively densely pubescent, the upper seeds less so.

Principally a South American genus of about 25 species, extending northward to the West Indies and southern Texas, and also in Africa.

Key to the Species of *Wissadula*

1. Petioles to 1.5 cm long; leaf blades shallowly cordate, more than twice as long as wide, ovate-triangular, the margins relatively straight .... 2. *W. periplocifolia*
1. Petioles 4–7 cm long; leaf blades deeply cordate, less than twice as long as wide, orbiculate- or ovate-acuminate, the margins curved throughout .... 1. *W. amplissima*

1. Wissadula amplissima (L.) R.E. Fr., Kongl. Svenska Vetenskapsakad. Handl. **43(4)**: 48. 1908. *Sida amplissima* L., Sp. Pl. 685. 1753.

Shrub to 2 m tall, usually unbranched below the inflorescence, with stipitate stellate hairs (notably at apex of petioles) and also minutely puberulent. Leaf blades orbiculate or ovate, 4–12 × 5.5–8 cm, discolorous, the apex acuminate, the base deeply cordate, the margins entire; petioles 4–7 cm long (reduced upward). Inflorescence an open terminal panicle, more or less leafless, with long-pedicellate flowers. Calyx 3 mm long, ca. half-divided; petals 5 mm long, yellowish; androecium included, yellowish, the filaments longer than the column. Fruit minutely puberulent; mericarps 3–5, constricted below, bulbous-apiculate above, 6–7 mm long, minutely puberulent. Seeds 3, 2.5 mm long, patchily pubescent.

DISTRIBUTION: A common roadside weed. Between Bethany and Rosenberg (B216), Coral Bay (A4001), Fish Bay area (W676). Also on St. Croix and Tortola (fide Britton, 1918; Britton & Wilson, 1924: 548); Texas and the West Indies to northern South America and parts of Africa.

2. Wissadula periplocifolia (L.) C. Presl *ex* Thwaites, Enum. Pl. Zeyl. 27. 1858. *Sida periplocifolia* L., Sp. Pl. 684. 1753.   Fig. 140G–J.

Shrub 1–2 m tall, more or less branched, stellate-pubescent, the hairs frequently stipitate (especially on younger growth). Leaf blades ovate-triangular, 3–7 × 1–3 cm, glabrate above, the apex acute, the base shallowly cordate, the margins straight, entire, markedly discolorous; petioles to 1.5 cm long. Inflorescence an

open, more or less leafy panicle. Calyx 2.5–3 mm long; corolla 3–4 mm long, yellow or white, sometimes with red center; androecium included, the filaments longer than the column. Fruit minutely puberulent; mericarps 4–5, constricted below, bulbous-apiculate above, 5–6 mm long. Seeds 3, 2.5 mm long, patchily to densely pubescent.

DISTRIBUTION: Common in open disturbed habitats. Road to Bordeaux Mountain (A3127), trail to Brown Bay (A1866), trail to Fortsberg (A4093). Also on St. Croix; the West Indies and Mexico to South America and the paleotropics.

CULTIVATED SPECIES:

Bastardiopsis eggersii (Baker f.) Fuertes & Fryxell, comb. nov. *Sida eggersii* Baker f., J. Bot. **30**: 139. 1892 (based on *Eggers 3183*), non *Sida eggersii* Gand., Bull. Soc. Bot. France **71**: 630. 1924 (based on *Eggers 195*).

Abutilon virginianum Krapov., Bonplandia **3**: 44. 1969.
DISTRIBUTION: Caneel Bay Hotel ground (A4113). Endemic to Culebra and Tortola.

Hibiscus rosa-sinensis L. var. *rosa-sinensis*, the common hibiscus, is commonly cultivated but unknown in the wild.

DOUBTFUL RECORD: *Hibiscus tiliaceus* L. was reported on St. John by Little and Wadsworth (1964: 326). Previously, Eggers (1879) cited the species as *Paritium tiliaceum* A. St.-Hil. from Fish Bay on St. John. Neither Britton (1918) nor Woodbury and Weaver (1987) included it in the floristic lists for the island. The correct name for this species in our region is *Hibiscus pernambucensis* Arruda. No specimens or living plants have been found, so this species is considered to be a dubious report for our flora.

48. Melastomataceae (Melastome Family)

Herbs, shrubs, trees, or rarely plants climbing by roots; stems usually quadrangular. Leaves opposite, simple, with 3–9 subparallel veins; stipules mostly wanting or vestigial. Flowers actinomorphic, bisexual or rarely unisexual, large and showy, 4– to 5-merous, produced in panicles, thyrses, or corymbs; bracts usually showy; calyx valvate or calyptrate or seldom forming a hypanthium; corolla of distinct petals or these rarely connate at base; stamens in 2 whorls, twice as many as the petals, the anthers opening by a terminal pore or less often by 2 terminal pores or longitudinal slits; ovary inferior, of 3–5 connate carpels, with numerous ovules with axile placentation, the style long, filiform, the stigma capitate, punctate or lobed. Fruit a loculicidal capsule or berry, with numerous minute seeds.

About 200 genera and over 4000 species with tropical and subtropical distribution.

REFERENCES: Cogniaux, A. 1886. Melastomaceae et Cucurbitaceae Portoricenses. Jahrb. Königl. Bot. Gart. Berlin 4: 276–285; Judd, W. S. & J. D. Skean Jr. 1991. Taxonomic studies in the Miconieae (Melastomataceae) IV. Generic realignments among terminal-flowered taxa. Bull. Florida State Mus., Biol. Sci. 36(2): 25–84.

Key to the Genera of Melastomataceae

1. Lower surface of leaf densely covered with stellate hairs, surface rusty-brown or grayish white .... 2. *Tetrazygia*
1. Lower surface of leaf green, only sparsely covered with stellate hairs, surface greenish .... 1. *Miconia*

1. MICONIA Ruíz & Pav., nom. cons.

Shrubs or small trees, variously pubescent but usually with stellate hairs present. Leaves entire or toothed, 3–7-veined, petiolate or sessile. Flowers small to medium-sized, 4–5-merous, pedicellate, in axillary or terminal panicles or thyrses; calyx tubular or bell-shaped, the sepals minute; petals white to pink; stamens 8–10, variously shaped, the anthers opening by 1 or 2 terminal pores; ovary 2–8-locular. Fruit a small berry with numerous seeds.

A genus of about 1000 species in the neotropics.

1. **Miconia laevigata** (L.) D. Don *in* Sweet, Hort. Brit. 159. 1826. *Melastoma laevigatum* L., Syst. Nat., ed. 10, **2**: 1022. 1759. Fig. 141I–N.

Erect treelet 1–2 m tall, with many lateral branches along main stem; stems slightly flattened, densely covered with minute stellate hairs when young, becoming glabrous. Leaf blades 11–24 × 5–10 cm, elliptic to lanceolate, chartaceous, with 5 subparallel veins, sparsely covered with minute, stellate hairs, the apex long-acuminate, the base obtuse to nearly rounded, the margins toothed;

petioles 1.5–3.5 cm long, stellate-pubescent. Flowers short-pedicellate, in terminal panicles; bracts minute, subulate. Calyx bell-shaped, stellate-pubescent, yellowish green; petals white, oblanceolate, 4–4.5 mm long; stamens white, the anthers longer than the filaments, opening by a terminal pore; style white, glabrous, the stigmas capitate. Berry purple, ovoid, 4.5 mm long.

DISTRIBUTION: Common in moist disturbed areas. Road to Bordeaux (A2594), Maho Bay Quarter along Center Line Road (A2406). Also on St. Croix, St. Thomas, and Tortola; Central America to northern South America, including the West Indies.

2. TETRAZYGIA DC.

Shrubs or small trees, variously pubescent but usually with dendritic or lepidote hairs. Leaves of a pair unequal, entire or toothed, 3–5-veined, petiolate. Flowers, medium-sized, 4–5-merous, pedicellate, in terminal panicles or corymbs; calyx constricted just below the apex, urn- or bell-shaped, the sepals minute; petals white, obovate; stamens 8–10, isomorphic, the anthers opening by a terminal pore; ovary 4–5-locular. Fruit a small berry with numerous seeds.

A genus of about 25 species, in the West Indies, except for 1 species in Florida.

Key to the Species of *Tetrazygia*

1. Lower surface of leaves ash-colored (grayish white); petals 1.2–1.4 cm long; calyx urn-shaped, greenish, sparsely covered with minute stellate hairs .... 2. *T. eleagnoides*
1. Lower surface of leaves rusty-brown; petals ca. 3 mm long; calyx bell-shaped, densely covered with rusty-brown, stellate hairs.. 1. *T. angustifolia*

1. **Tetrazygia angustifolia** (Sw.) DC., Prodr. **3**: 172. 1828. *Melastoma angustifolia* Sw., Prodr. 71. 1788. *Miconia angustifolia* (Sw.) Griseb., Fl. Brit. W. I. 258. 1860.

Shrub or small tree, 1.5–6 m tall; bark light gray, rough; stems nearly 4-angled, densely covered with minute, rusty-brown, stellate hairs when young, becoming glabrous. Leaf blades 4–8 × 0.4–1.4 cm, narrowly elliptic to linear-lanceolate, coriaceous, with 3 subparallel veins, lower surface densely covered with rusty-brown, stellate hairs, the apex acuminate, the base obtuse to nearly rounded, the margins strongly revolute; petioles 4–10 mm long, stellate-pubescent. Flowers short-pedicellate, in lateral cymes along short, terminal, panicle-like inflorescences; bracts minute, ovate. Calyx bell-shaped, 1.5–2 mm long, densely covered by stellate-peltate hairs, yellowish green; petals white, rosy-tinged, asymmetrically obovate, clawed, 2.5–3 mm long; stamens 8, glabrous; style white, glabrous, the stigma punctiform. Berry turning from green to red, becoming purple at maturity, ovoid to nearly globose, 4–5-lobed, 4–5 mm long.

DISTRIBUTION: Common in moist forests. Road to Bordeaux (A2600, A3822). Also on St. Thomas, Tortola, and Virgin Gorda; Jamaica, Puerto Rico, and the Lesser Antilles.

COMMON NAMES: sprat wood, stinking fish.

2. **Tetrazygia eleagnoides** (Sw.) DC., Prodr. **3**: 172. 1828. *Melastoma eleagnoides* Sw., Prodr. 72. 1788. Fig. 141A–H.

Shrub or small tree, 3–10 m tall; bark light gray, thick, rough, separating into narrow flakes; stems nearly 4-angled, densely covered with rusty-brown, stellate-peltate scales when young, becoming glabrous. Leaf blades 6–14.2 × 1–3.5 cm, elliptic to lanceolate, subcoriaceous, with 3 subparallel veins from near the base, lower surface densely covered with grayish white, stellate-peltate hairs, the apex acute or obtuse, mucronulate, the base obtuse to nearly rounded, the margins entire; petioles 5–15 mm long, stellate-pubescent. Flowers in lateral cymes, along terminal, panicle-like inflorescences; bracts minute. Calyx urn-shaped, 5–6 mm long, sparsely covered by stellate-peltate hairs, greenish; petals white, obovate, clawed, 12–14 mm long; stamens 8, glabrous, yellow; style white, glabrous, curved, the stigma punctiform. Berry turning from green to purple, depressed-globose, 4-lobed, 8–10 mm diam.

DISTRIBUTION: Common in moist and dry evergreen forests. Susannaberg (A3809), trail to Genti Bay (GTP29289). Also on St. Croix, St. Thomas and Virgin Gorda; Hispaniola and Puerto Rico (including Vieques).

COMMON NAME: kre kre.

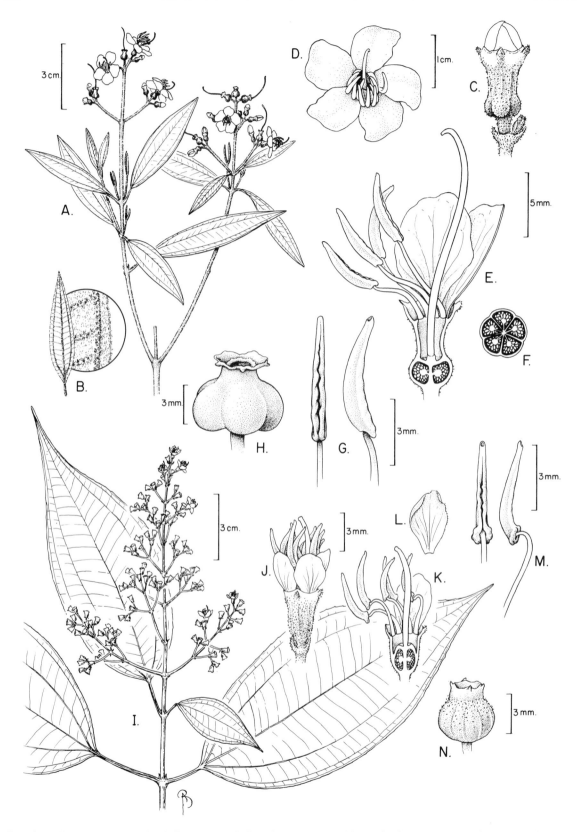

Fig. 141. A–H. *Tetrazygia eleagnoides*. **A.** Flowering branch. **B.** Leaf and detail of abaxial side of leaf showing indument. **C.** Flowerbud. **D.** Flower.
E. L.s. flower. **F.** C.s. ovary. **G.** Anther, frontal and lateral views. **H.** Berry. **I–N.** *Miconia laevigata*. **I.** Flowering branch. **J.** Flower. **K.** L.s.
flower. **L.** Petal. **M.** Anther, frontal and lateral views. **N.** Berry.

49. Meliaceae (Mahogany Family)

Trees or shrubs. Leaves alternate, pinnately compound; leaf rachis with a distal leaflet, or with an apical bud, or tardily developing into new leaves; leaflets opposite or alternate, entire or rarely serrate; stipules wanting. Flowers small to medium-sized, actinomorphic, bisexual or rarely unisexual, produced in various types of axillary or cauliflorous inflorescences; calyx of 3–5 sepals, usually connate at base to form a tube; corolla of free petals or less often connate at base, these as the sepals; stamens as many or twice as many as the petals, the filaments free or connate into a tube, sometimes adnate to the corolla, the anthers opening by longitudinal slits; ovary superior, sometimes on a gynophore, with an annular nectary disk at base, the ovary of 2–5 carpels, ovules 2 or more, rarely 1 in each locule, with axile placentation, the style terminal, the stigmas headlike. Fruit a septicidal or loculicidal capsule or less often a drupe or a berry.

A family with about 51 genera and 1400 species, chiefly with tropical and subtropical distribution.

REFERENCE: Pennington, T. D. 1981. Meliaceae. Fl. Neotrop. Monogr. 28: 1–472.

Key to the Genera of Meliaceae

1. Fruit a drupe; leaflets serrate.
  2. Leaves 2–3-pinnate; flowers 0.8–1.2 cm long; staminal column violet .... 2. *Melia*
  2. Leaves once-pinnate; flowers 4–5 mm long; staminal column white .... *Azadirachta* (cultivated)
1. Fruit a capsule; leaflets entire.
  3. Capsule ovoid, 6–10 cm long, with thick pericarp.. 3. *Swietenia*
  3. Capsule ellipsoid, 2.5–3 cm long, with thin pericarp .... 1. *Cedrela*

1. CEDRELA P. Browne

Deciduous, large trees, glabrous or with simple hairs. Leaves once-pinnate, paripinnate, seldom imparipinnate, the leaflets entire, petiolulate. Flowers medium-sized, 5-merous, unisexual, with vestigial organs of opposite sex, pedicellate, in terminal thyrses; calyx cup-shaped, shallow to deeply lobed; petals adnate one-half to one-third of their length to a long androgynophore; stamens 5, the filaments free, adnate to an androgynophore at base; ovary 5-locular, each locule with 8–14 ovules. Fruit a capsule opening from apex by 5 valves, leaving a central columella; seeds with a terminal wing.

A neotropical genus of about 8 species.

1. Cedrela odorata L., Syst. Nat., ed. 10, **2**: 940. 1759. Fig. 142H–N.

Cedrela sintenisii C. DC., Annuaire Conserv. Jard. Bot. Genève **10**: 169. 1907.

Tree to 15 m tall; bark light gray or light brown, fissured, bitter; stems cylindrical, glabrous or sparsely pubescent, usually with rounded lenticels. Leaves paripinnate, rarely with an abortive terminal leaflet; leaflets 6–12 pairs, opposite to alternate, sessile to petiolulate, 7–15 × 3–5 cm, ovate-lanceolate, oblong-lanceolate, or ovate, often falcate, chartaceous, with strong fetid smell, the apex acute or acuminate, the base asymmetrical, obtuse-rounded, the margins entire. Flowers in terminal thyrses, as long as or shorter than the subtending leaf; bracts minute, viscidulous. Calyx cup-shaped, 1–1.5 mm long, sparsely pubescent or glabrous, splitting along one or two sides; petals white, oblong, adnate to gynophore one-half of their length; stamens 5, free, 2–3 mm long, glabrous; ovary globose, 5-locular, each locule with 10–14 ovules, the style 1–1.5 mm long. Capsules oblong-ellipsoid, with thin woody pericarp, brown, with numerous light lenticels, 2.5–3 cm long. Seeds light brown, 2–3 cm long.

DISTRIBUTION: Uncommon in moist deep soil of forests. Trunk Bay (A2608, A2850). Common throughout the neotropics.

COMMON NAMES: acajou, red cedar.

2. MELIA L.

Shrubs or small trees, with simple and stellate hairs. Leaves 2–3-pinnate, imparipinnate; leaflets serrate, petiolulate. Flowers, medium-sized, 5–6-merous, bisexual or unisexual, pedicellate, in axillary, panicle-like inflorescences; calyx deeply lobed; petals free to base; stamens 10–12, the filaments connate into a staminal tube with appendages projecting beyond the anthers; nectary disk at base of ovary; ovary 3–6-locular, each locule with 2 ovules, the style columnar, with a 5–6-lobed stigma. Fruit a drupe, the stone 1–5-lobed.

An Old World genus of 2–5 species, planted and naturalized elsewhere.

1. Melia azedarach L., Sp. Pl. 384. 1753. Fig. 142A–G.

Melia sempervirens Sw., Prodr. 67. 1788.

Shrub or small tree to 10 m tall; bark reddish brown, fissured, bitter; stems cylindrical, sparsely stellate-pubescent, becoming glabrous. Leaves 2–3 times pinnate, less often once pinnate, imparipinnate; pinnae opposite, with 2–6 pairs of opposite leaflets; leaflets petiolulate, 3–8 × 0.8–3 cm, ovate to lanceolate, membranous, the apex acute to long-acuminate, the base acute to rounded, the margins serrate or lobed. Flowers in axillary panicles, as long or longer than the subtending leaf; bracts minute. Calyx 1–1.5 mm long, sparsely pubescent; petals white to light violet, oblanceolate, 8–12 mm long, reflexed; staminal tube nearly cylindrical, dark violet, 7–8 mm long, stamens 10; ovary glabrous, the

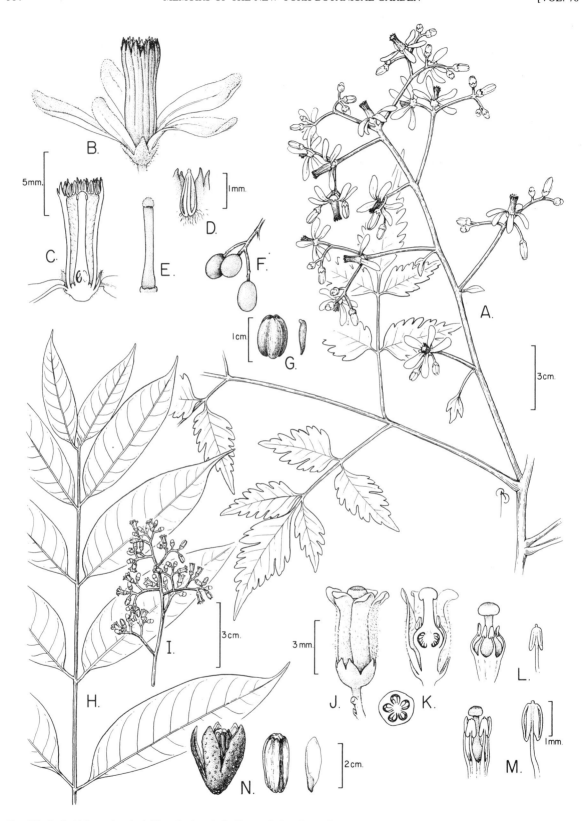

Fig. 142. A–G. *Melia azedarach*. **A.** Flowering branch. **B.** Flower. **C.** L.s. flower. **D.** Detail of summit of staminal tube showing anther. **E.** Pistil. **F.** Fruits. **G.** Stone and seed. **H–N.** *Cedrela odorata*. **H.** Leaf. **I.** Inflorescence. **J.** Flower. **K.** L.s. flower and c.s. ovary. **L.** Pistil with surrounding staminal tube and detail of vestigial stamen. **M.** Staminal tube surrounding vestigial pistil and detail of stamen. **N.** Dehiscing capsule, column with attached seeds, and seed.

style columnar. Drupe fleshy, ellipsoid, 1–1.5 cm long, yellow.

DISTRIBUTION: A common roadside shrub, naturalized on St. John. Carolina (A4137), Great Cruz Bay (A2371). Also on St. Croix, St. Thomas, and Tortola; cultivated and naturalized throughout the tropics and subtropics of the New World.

COMMON NAMES: chinaberry, hagbush, lilac.

3. SWIETENIA Jacq.

Large trees, deciduous, glabrous. Leaves once-pinnate, paripinnate, seldom imparipinnate; leaflets entire, petiolulate. Flowers unisexual with vestigial organs of opposite sex, pedicellate, in short, axillary thyrses; calyx 5-lobed to about middle; petals 4–5 free, spreading; stamens 8–10, with filaments connate into an urn-shaped staminal tube, the anthers slightly projecting; nectary disk annular, surrounding ovary at base; ovary (4–)5(–6)-locular, each locule with 9–16 ovules. Fruit an erect capsule, opening upward from the base by the valves, leaving a central columella; seeds flattened, with a terminal wing.

A genus of 3 species, native to Central America, South America, Cuba, Jamaica, and Hispaniola.

1. Swietenia mahagoni (L.) Jacq., Enum. Syst. Pl. 20. 1760. *Cedrela mahagoni* L., Syst. Nat., ed. 10, **2**: 940. 1759.   Fig. 143.

Tree to 15 m tall; bark light brown, fissured; stems cylindrical, glabrous. Leaves paripinnate, rarely with an abortive terminal leaflet, 12–25 cm long; leaflets 2–5 pairs, opposite or subopposite, petiolulate, 5–6 × 2.5–3 cm, ovate-lanceolate to elliptic-ovate, often falcate, chartaceous, the apex acute or acuminate, the base very asymmetrical, cuneate-rounded, the margins entire. Flowers unisexual, in terminal or axillary thyrses, as long as or shorter than the subtending leaf; bracts minute. Calyx cup-shaped, 0.8–1.5 mm long, glabrous; petals greenish, expanded, concave, oblong; staminal tube urn-shaped, 3–4 mm long, glabrous, the anthers included; nectary disk reddish; ovary globose, 4–5-locular, each locule with 12–14 ovules, the style 1–1.5 mm long, the stigma discoid. Capsule erect, woody, ovoid, dark brown, minutely lenticellate, 6–10 cm long, the pericarp thick. Seeds flattened, reddish brown, 4–5 cm long, including the wing.

DISTRIBUTION: Persistent after cultivation; Cruz Bay Town (A2514). Native to southern Florida, the Bahamas, Cuba, Cayman Islands, Jamaica, and Hispaniola; introduced into Puerto Rico, the Virgin Islands, and the Lesser Antilles.

COMMON NAME: mahogany.

CULTIVATED SPECIES: *Azadirachta indica* Juss. is known only in cultivation on St. John.

DOUBTFUL RECORD: *Trichilia hirta* L. was reported for St. John by Britton and P. Wilson (1924); however, no collection was made then, and further attempts to find the species have failed.

50. Menispermaceae (Moonseed Family)

Woody or herbaceous twining vines, or less often scandent to erect shrubs; cross section of stems usually with anomalous arrangements of vascular tissue. Leaves alternate, simple, palmately veined, long-petiolate; stipules wanting. Flowers minute, actinomorphic, 3–6-merous, usually unisexual, produced in axillary or cauliflorous cymes, panicles, or pseudoracemes; calyx of distinct sepals; corolla of 6 distinct petals, sometimes more numerous or wanting; stamens commonly 6 or from many to fewer, the filaments free or connate at base, the anthers opening by longitudinal or rarely transversal slits; carpels 3, or rarely 6, free, often on a gynophore, ovules 2, becoming reduced to 1 by abortion, attached to the ventral suture, the stigma sessile or on a short style. Fruit a drupe with a curved stone.

A family with 70 genera and 400 species, mainly with tropical and subtropical distribution.

Key to the Genera of Menispermaceae

1. Herbaceous, pubescent vine to 5 m long; leaves widely ovate to nearly rounded, cordate or truncate and peltate at base; fruits globose, bright red-orange .... 1. *Cissampelos*
1. Glabrous liana to 15 m long; leaves ovate to widely ovate; fruits obovoid, purple .... 2. *Hyperbaena*

1. CISSAMPELOS L.

Dioecious, herbaceous, twining vines, usually with simple hairs. Leaves rounded to ovate, peltate or cordate at base, palmately veined, long-petiolate. Flowers minute, unisexual, pedicellate, in axillary inflorescences. Staminate flowers in corymbs; calyx of 4 free sepals; corolla cup-shaped; stamens 4, connate into a short tube, with sessile anthers. Pistillate flowers zygomorphic, in elongate cymes with foliaceous bracts; calyx and corolla of a single sepal and a single petal, both toward same side; ovary sessile, unilocular, with a single basal ovule, the stigma lobed. Drupe globose, fleshy, with a verrucose stone; seeds horseshoe-shaped.

A genus of 19 species, with tropical distribution.

REFERENCE: Rhodes, D. G. 1975. A revision of the genus *Cissampelos*. Phytologia 30: 415–484.

1. Cissampelos pareira L., Sp. Pl. 1031. 1753.   Fig. 144.

Twining vine to 5 m long; stems cylindrical, glabrous to pilose. Leaf blades 4–12 × 4–10 cm, widely ovate to nearly rounded, chartaceous, lower surface sparsely to densely pubescent, the apex obtuse, rounded or emarginate, usually mucronulate, the base cordate or truncate, usually peltate, the margins wavy and revolute; petioles sparsely to densely pubescent, 2–7 cm long. Flowers 1–1.5 mm long, greenish, in axillary cymes. Calyx pilose; bracts of pistillate inflorescence foliaceous, ovate, pilose. Drupe globose,

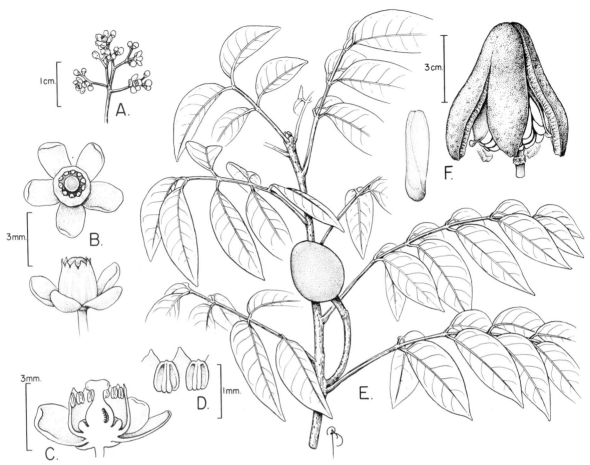

FIG. 143. *Swietenia mahagoni.* **A.** Portion of inflorescence. **B.** Flower, top and lateral views. **C.** L.s. flower. **D.** Detail of apex of staminal tube showing anthers. **E.** Fruiting branch. **F.** Seed and dehiscing fruit.

juicy, bright red-orange, pilose, 4–5 mm diam.; stone horseshoe-shaped, verrucose.

DISTRIBUTION: A common weedy vine of open disturbed, moist areas or secondary forests. Road to Bordeaux (A3126), trail to Brown Bay (A1868). Also on St. Croix, St. Thomas, and Tortola; throughout the tropics.

COMMON NAMES: pataka, velvet leaf.

2. HYPERBAENA Miers *ex* Benth., nom. cons.

Twining lianas, scandent shrubs or trees, glabrous or with simple hairs. Leaves coriaceous, rounded to ovate, entire, peltate or cordate at base, pinnately veined, long-petiolate. Flowers minute, greenish or white, unisexual, 6-merous, pedicellate, in axillary spicate to paniculate inflorescences; sepals unequal; petals 6, rarely fewer or absent, free to base; stamens longer than the petals, the anthers opening by longitudinal or transverse slits; staminodes 6; ovary of 2–5 free ovoid to globose carpels, the stigma sessile. Drupe of various colors; stone and seeds U-shaped.

A genus of about 20 species, with neotropical distribution.

REFERENCE: Mathias, M. E. and W. Theobald. 1981. A revision of the genus *Hyperbaena* (Menispermaceae). Brittonia 33: 81–104.

1. Hyperbaena domingensis (DC.) Benth., J. Proc. Linn. Soc., Bot. 5, Suppl. **2**: 50. 1861. *Cocculus domingensis* DC., Syst. Nat. **1**: 528. 1817.   Fig. 145.

Twining liana to 15 m long, or rarely a scandent shrub, with many lateral branches; stems cylindrical, glabrous, in cross section with concentric rings. Leaf blades 4–20 × 2.5–12 cm, ovate to widely ovate, coriaceous, glabrous, lower pair of veins arising from near the base; the apex acute to acuminate, rarely obtuse, the base cordate, the margins entire; petioles 1–6 cm long, swollen at apex. Flowers minute, whitish; staminate inflorescence clustered; pistillate inflorescence usually solitary. Calyx of staminate flowers 0.6–0.8 mm long, slightly larger in pistillate flowers; petals ca. 0.5 mm long; ovary of 3 ellipsoid carpels. Drupe asymmetrically obovoid, purple, glabrous, 10–23 mm long.

DISTRIBUTION: An uncommon vine of moist forests; only 1 male plant known at Bordeaux (A3173). Also on St. Thomas; the Greater and Lesser Antilles, northern South America to Bolivia.

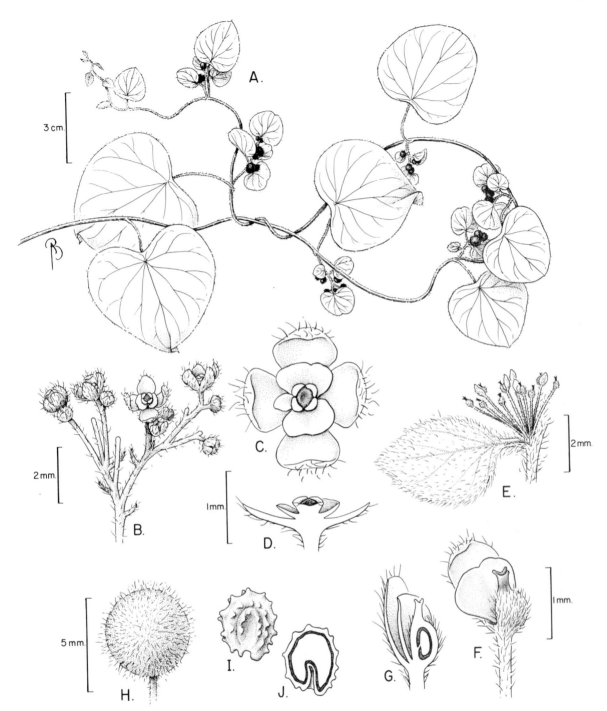

Fig. 144. *Cissampelos pareira*. **A.** Fertile branch. **B.** Staminate cyme. **C.** Staminate flower, top view. **D.** L.s. staminate flower. **E.** Axillary fascicle of pistillate flowers. **F.** Pistillate flower. **G.** L.s. pistillate flower. **H.** Drupe. **I.** Seed. **J.** L.s. seed.

51. **Molluginaceae** (Carpet-Weed Family)

Herbs or less often shrubs. Leaves opposite, alternate, or whorled, simple, sessile or short-petiolate; stipules wanting or deciduous. Flowers minute, actinomorphic, bisexual or rarely unisexual, in axillary cymes or solitary; calyx of 4–5 distinct sepals or these connate at base; corolla wanting or minute, often tubular at base; stamens usually twice as many as the sepals or the second whorl wanting, the filaments free or connate at base, the anthers opening by longitudinal slits; ovary superior, of 2–5 connate carpels, with an annular disk at base, the placentation axile, with 1 to many ovules, the styles distinct. Fruit a loculicidal capsule or indehiscent.

A family of 13 genera and about 100 species, with tropical and subtropical distribution, especially in Africa.

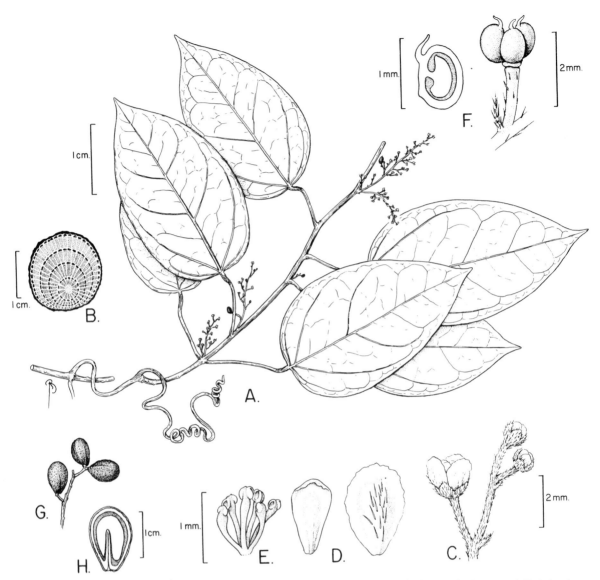

FIG. 145. *Hyperbaena domingensis*. **A.** Flowering branch. **B.** C.s. old stem. **C.** Detail of inflorescence. **D.** Inner petal and outer petal. **E.** Androecium. **F.** L.s. carpel and gynoecium. **G.** Fruits. **H.** L.s. fruit.

1. MOLLUGO L.

Annual herbs. Leaves basal and cauline; basal leaves forming a rosette; cauline leaves opposite or whorled; stipules deciduous. Flowers minute, greenish or white, bisexual, pedicellate, in axillary cymes or solitary; calyx of 5 sepals; petals wanting; stamens 3–10, the filaments connate at base; ovary of 3–5 locules, each locule with numerous ovules. Capsule thin, membranous, with numerous minute seeds.

A genus of 20 species, with tropical and subtropical distribution.

1. Mollugo nudicaulis Lam., Encycl. **4:** 234. 1797.

Fig. 146.

Herb with a long taproot and basal rosette of leaves; acaulescent. Leaf blades 4–6 × 1–1.5 cm, chartaceous, obovate to oblanceolate, glabrous, the apex rounded, the base attenuate, the margins entire; petioles glabrous, 5–30 mm long. Flowers minute, whitish, produced in pseudodichotomous, axillary cymes; inflo-

rescence glandular, to 30 cm long. Sepals ca. 2 mm long, oblong, membranous; stamens 4–5, glabrous, white. Capsule ellipsoid, 2–2.5 mm long, membranous. Seeds ca. 0.5 mm long, slightly flattened, heart-shaped, black.

DISTRIBUTION: A common weed of open areas. Johns Folly (A3934). Also on St. Croix and St. Thomas; native to the Old World, now a weed in the neotropics.

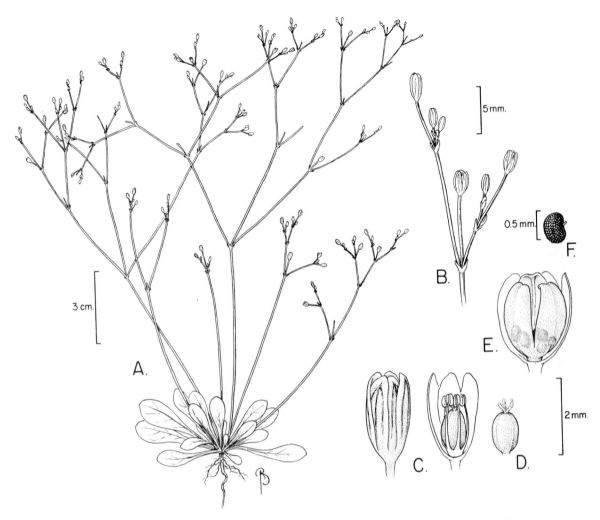

Fig. 146. *Mollugo nudicaulis*. **A.** Habit. **B.** Detail of inflorescence. **C.** Flower and l.s. flower. **D.** Pistil. **E.** Dehiscing capsule. **F.** Seed.

52. Moraceae (Mulberry Family)

Monoecious or dioecious free-standing trees, strangling (hemiepiphytic) trees, shrubs, lianas, or rarely herbs, often producing abundant milky sap. Leaves alternate or opposite, simple or rarely compound, petiolate; stipules present. Flowers minute, actinomorphic, unisexual, produced in axillary, condensed inflorescences whose axes form a common receptacle or sometimes completely enclose the flowers into a syconium; calyx of 4–5 distinct sepals, or these connate at base, petals absent; corolla wanting; stamens usually as many as the sepals, the filaments free; ovary superior, bilocular or unilocular, with 1 apical ovule per carpel, the styles 2. Fruit a dehiscent drupe, a syconium or a syncarp.

A family of 40 genera and about 1000 species, mostly in tropical and subtropical areas but present in temperate zones.

Key to the Genera of Moraceae

1. Leaves to 50 cm long, deeply lobed; fruit globose, 15–20 cm diam. .... *Artocarpus* (cultivated)
1. Leaves to 10 cm long, entire; fruit a globose or obovoid syconium <1.8 cm diam. .... 1. *Ficus*

1. FICUS L.

Trees or shrubs, free-standing, climbers, or stranglers (hemiepiphytes), producing copious milky sap. Leaves alternate, simple to lobed, usually long-petiolate; stipules deciduous, contorted into a cone-shaped hood that protects the apical meristem, often brightly colored. Flowers minute, borne in the interior of globose, axillary inflorescences formed by an enlarged receptacle (syconium); calyx reduced, membranous; stamens 2; ovary unilocular with an apical ovule. Fruit a syconium, formed by a globose receptacle containing numerous minute achenes, the apex containing an aperture or operculum.

A genus of about 800 species, with tropical and subtropical distribution.

Key to the Species of *Ficus*

1. Leaves chartaceous to subcoriaceous, elliptic, ovate, elliptic-lanceolate, or seldom oblanceolate, the apex acuminate, the base cordate to obtuse; stipules glabrous; syconia yellowish green, red-spotted, becoming red at maturity, the operculum not surrounded by a rim .... 1. *F. citrifolia*
1. Leaves coriaceous, elliptic, oblong, or obovate, the apex obtuse or acute, the base rounded to obtuse; stipules sericeous to hirsute; syconia yellowish green, not spotted, the operculum surrounded by a rim .. 2. *F. trigonata*

1. Ficus citrifolia Mill., Gard. Dict., ed. 8. 1768.

Fig. 147.

Ficus laevigata Vahl, Enum. Pl. **2**: 183. 1805.

Tree to 10 m tall, sometimes a strangler, producing abundant white latex; bark light gray. Leaf blades 4–20 × 2–12 cm, chartaceous to subcoriaceous, elliptic, ovate, elliptic-lanceolate or seldom oblanceolate, glabrous, the apex acuminate or shortly acuminate, the base obtuse, cordate or rarely truncate, the margins entire; petioles 1–7 cm long; stipules reddish, glabrous. Syconium short-peduncled, globose, depressed-globose, or obovoid, 6–12 mm diam., yellowish green, red-spotted, becoming red at maturity.

DISTRIBUTION: A common tree of moist forests to coastal scrub. Bordeaux Mountain (A1895), White Cliffs (A2042). Also on Jost van Dyke, St. Croix, St. Thomas, Tortola, and Virgin Gorda; Florida, Greater and Lesser Antilles.

COMMON NAME: white fig.

2. Ficus trigonata L., Pl. Surinam. 17. 1775.

Ficus crassinervis Willd., Sp. Pl. **4**: 1138. 1806.

Tree to 12 m tall, producing abundant white latex; bark light gray, with many lenticels; branches sparsely pubescent to hirtellous. Leaf blades 2–13 × 1.5–7 cm, coriaceous, elliptic, oblong, or obovate, glabrous, the apex rounded to obtuse, the base obtuse, rounded, to subcordate, the margins entire; petioles 0.5–2 cm long; stipules yellowish green, turning brown, sericeous to hirsute. Syconium short-peduncled, globose to depressed-globose, 8–18 mm diam., yellowish green, the operculum conspicuously rimmed.

DISTRIBUTION: A common tree of moist forests. Bordeaux Mountain (A1912), Cruz Bay Quarter along Center Line Road (A2392). Also on St. Croix, St. Thomas, and Tortola; throughout tropical America.

DOUBTFUL RECORD: *Maclura tinctoria* (L.) D. Don *ex* Steud. was reported by Britton and P. Wilson (1924) as occurring on St. John. However, no specimen has been found to confirm this record, nor has the species been collected recently on the island.

CULTIVATED SPECIES: *Artocarpus altilis* (Parkinson) Fosberg, commonly known as the breadfruit tree, is sometimes cultivated on the island.

53. Moringaceae (Horseradish Tree Family)

A monogeneric family, characterized by the following genus. Native to the Old World tropics.
REFERENCE: Verdcourt, B. 1985. A synopsis of the Moringaceae. Kew Bull. 40: 1–23.

1. MORINGA Adans.

Trees. Leaves alternate, 2–3-pinnate; leaflets opposite; stipules usually wanting. Flowers medium-sized, strongly zygomorphic, bisexual, produced in axillary panicles; calyx of 5 unequal, reflexed sepals, forming a short hypanthium; petals 5, the outermost the largest, the 2 innermost the smallest; stamens 8, some modified into staminodes, the filaments free, inserted on the hypanthium; ovary superior, of 3 connate carpels, forming a unilocular ovary, with a short gynophore at base, the placentation parietal, with 2 rows of ovules per carpel, the style terminal, truncate. Fruit an elongate explosively dehiscent capsule, with numerous winged seeds.

About 10 species, from the Old World tropics, a few species widely cultivated throughout the tropics.

1. Moringa oleifera Lam., Encycl. 1: 398. 1785.

Fig. 148.

Guilandina moringa L., Sp. Pl. 381. 1753.
Moringa pterygosperma Gaertn., Frut. Sem. Pl. **2**: 314. 1791, nom. illegit.

Small tree or shrub to 7 m tall; bark grayish, smooth; branches pubescent, fragile. Leaves 3 times pinnately compound, 20–60 cm long, nearly triangular (pinnae gradually increasing in size toward the base); pinnae opposite, often 4–5 pairs; leaflets opposite, petiolulate, 0.9–2.5 × 0.4–1.5 cm, chartaceous, elliptic, oblong, or oblanceolate, glabrous or puberulent, the apex rounded to obtuse, the base obtuse to rounded, slightly asymmetrical, the margins entire; stipules early deciduous. Flowers white, in axillary panicles. Hypanthium green; sepals white, petaloid, reflexed, oblong, pubescent, 8–12 mm long; petals white, pubescent at base, oblanceolate, the outermost erect, 12–15 mm long, the remaining ones reflexed, 6–8 mm long; stamens 5, unequal, with length of innermost decreasing, the filaments widened and pubescent at base, staminodes 3, shorter than the stamens; ovary ovoid to oblong, woolly-pubescent, the style equaling the longest stamen. Capsule explosively dehiscent by longitudinal valves, 20–40 × 1–1.5 cm, trigonous in cross section, the valves thickened and corky. Seeds numerous, globose-trigonous, 9–10 mm diam., blackish, with 3 membranous, white wings, 2.5–3 cm long.

DISTRIBUTION: Persistent after cultivation, perhaps becoming naturalized. Coral Bay (A1950), Cruz Bay Town (A4039). Also on St. Croix, St. Thomas, and Tortola; native to the Old World but widely cultivated and naturalizing throughout tropical America.

COMMON NAME: horseradish tree.

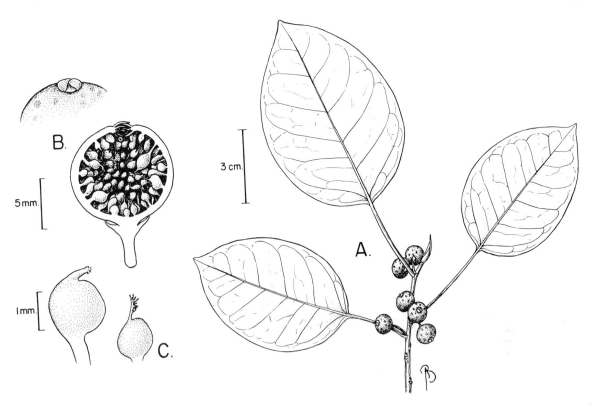

Fɪɢ. 147. *Ficus citrifolia.* **A.** Fruiting branch. **B.** Detail of syconium apex showing operculum and l.s. syconium showing developing achenes. **C.** Achene and pistil.

54. Myoporaceae (Myoporum Family)

Shrubs or small trees. Leaves alternate or seldom opposite, simple, entire or toothed; stipules wanting. Flowers large, actinomorphic to strongly zygomorphic, bisexual, in axillary cymes, or solitary; calyx bell- to cup-shaped, of 5 connate sepals; corolla tubular, with 5 unequal lobes, often bilabiate; stamens 4, the fifth modified into a staminode, the filaments free, inserted on the corolla tube, the anthers opening by longitudinal slits; ovary superior, of 2 connate carpels, the carpels with 1 to many locules, the ovules pendulous, (1–)2 per locule, the style terminal. Fruit a drupe, the stone with 2–4 locules or with many 1-seeded segments.

A family of 3–4 genera with about 125 species, mostly from Australia, Asia, and the Pacific. *Bontia* is the only genus in the Caribbean.

1. BONTIA L.

A monospecific genus characterized by the following species.

1. Bontia daphnoides L., Sp. Pl. 638. 1753.   Fig. 149.

Erect shrub, to 2(–5) m tall, many-branched from base; stems reddish brown, with persistent leaf scars. Leaves clustered at ends of branches; blades 6–9 × 1.2–2.4 cm, fleshy (chartaceous when dried), elliptic to oblong-lanceolate, glabrous, with impressed glands on lower surface, the apex acuminate, the base tapering to the petiole, the margins entire; petioles 0.5–1.2 cm long. Flowers medium-sized, zygomorphic, long-peduncled, solitary in leaf axils. Calyx nearly cup-shaped, tapering to an elongate peduncle, the sepals 5, deeply parted, ovate, ciliate, 2–2.5 mm long; corolla yellow, purple-blotched, tubular, to 2 cm long, bilabiate, the lobes pubescent within, the lower lip reddish-pubescent within; stamens 4, didynamous, as long as the petals, the filaments pubescent at base, the anthers opening by longitudinal slits; ovary ovoid, glabrous, 2-locular, with 4 ovules per locule. Drupe pendulous, ovoid, 1–1.5 cm long, turning from green to yellow, with persistent style; stone 2-celled.

Dɪsᴛʀɪʙᴜᴛɪᴏɴ: A common coastal shrub. Waterlemon Bay (A1933). Also on St. Thomas and Tortola; throughout the West Indies, south to Venezuela.

55. Myrsinaceae (Myrsine Family)

Trees, shrubs, or seldom woody vines. Leaves alternate, simple, entire or seldom serrate, usually dark-glandular-lineate or punctate; stipules wanting. Flowers minute, actinomorphic, bisexual or rarely unisexual, produced in axillary or terminal compound inflores-

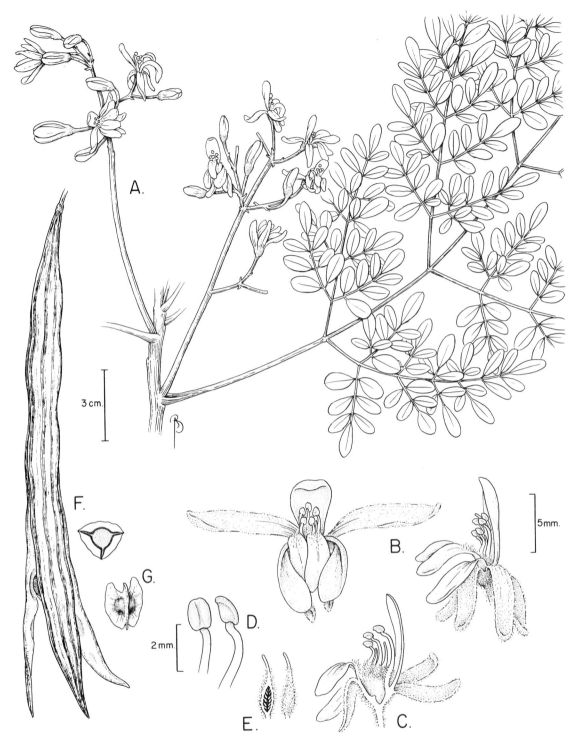

Fig. 148. *Moringa oleifera.* **A.** Flowering branch. **B.** Flower, frontal and lateral views. **C.** L.s. flower. **D.** Stamen, frontal and lateral views. **E.** L.s. pistil and pistil. **F.** Dehiscing fruit and c.s. fruit. **G.** Seed.

cences; calyx of 4–5 free or basally connate sepals; corolla of 4–5 connate or distinct petals; stamens as many as and opposite the petals, the filaments free or connate at base, the anthers opening by longitudinal slits or less often by terminal pores; ovary unilocular, of 3–5 carpels, superior or rarely half-inferior, the ovules few to numerous, the placentation free, basal, the style terminal with a capitate stigma. Fruit usually a 1-seeded berry or drupe.

A family of about 30 genera and 1000 species, with tropical and subtropical distribution.

Fɪɢ. 149. *Bontia daphnoides*. **A.** Fertile branch and detail of leaf showing impressed glands on lower surface. **B.** Flowerbud, lateral and top views. **C.** Flower. **D.** Corolla spread open to show stamens. **E.** Stamen, dorsal and frontal views. **F.** L.s. pistil. **G.** Drupe. **H.** C.s. drupe.

1. ARDISIA Sw., nom. cons.

Shrubs or small trees. Leaves alternate. Flowers medium-sized, regular, bisexual, produced in terminal panicles or axillary panicles; calyx of 5 sepals; corolla 5-parted, forming a short tube at base, the lobes reflexed at anthesis; filaments free; ovary superior; ovules numerous. Fruit a several-seeded drupe.

About 400 species, with pantropical distribution.

1. Ardisia obovata Desv. *ex* Ham., Prodr. Pl. Ind. Occid. 26. 1825.   Fig. 150.

Ardisia guadalupensis Duchass. *ex* Griseb., Abh. Königl. Ges. Wiss. Göttingen **7**: 237. 1857. *Icacorea guadalupensis* (Griseb.) Britton & P. Wilson, Bull. New York Bot. Gard. **8**: 401. 1917.

Ardisia coriacea sensu Mez *in* Urb., Symb. Ant. **2**: 398. 1901, non Sw., 1788.

Small tree to 10 m tall; bark light gray, finely fissured. Leaf blades 8–18 × 2.5–6 cm, coriaceous, obovate to elliptic-obovate, glabrous, obscurely punctate, the apex obtuse to rounded, the base cuneate, the margins slightly revolute; petioles 0.5–1.2 cm long. Flowers green with orange punctations, produced in terminal

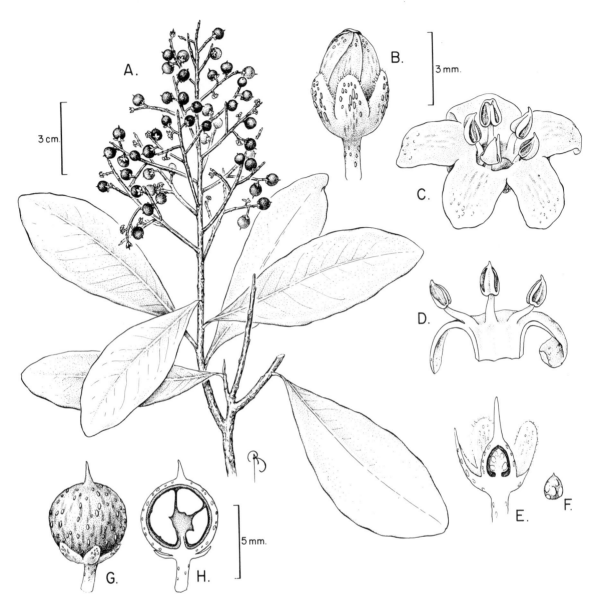

Fig. 150. *Ardisia obovata*. **A.** Fruiting branch. **B.** Flowerbud. **C.** Flower. **D.** L.s. corolla with adnate stamens. **E.** L.s. flower with developed gynoecium. **F.** Ovary. **G.** Drupe. **H.** L.s. drupe.

panicles. Calyx nearly cup–shaped, ca. 3 mm long, the sepals narrowly ovate, ciliate, 1.5–2 mm long; corolla green to greenish orange, the tube to 1.5 mm long, the petals oblong, reflexed; filaments white, the anthers yellow, mucronate, as long as the filaments; ovary ovoid, glabrous, punctate. Drupe globose to ob-late, 5–8 mm diam., black, punctate, with persistent style at apex.

DISTRIBUTION: A common shrub in secondary vegetation. Bordeaux (M17053), Gift Hill (A2464), road to Herman Farm (M17009). Also on Jost van Dyke, St. Croix, St. Thomas, Tortola, and Virgin Gorda; Puerto Rico and the Lesser Antilles.

COMMON NAME: breakbill.

56. Myrtaceae (Guavaberry Family)

Trees or shrubs. Leaves opposite or seldom alternate, simple, entire, usually with translucent oil glands; stipules wanting. Flowers small to large, actinomorphic, bisexual, produced in axillary cymes or racemes or solitary; calyx variously shaped, with 4–5 free or connate sepals, sometimes forming a calyptra and a hypanthium; corolla of 4–5 distinct petals, sometimes forming a calyptra or wanting; stamens numerous, the filaments free or connate at base, inserted on the hypanthium, the anthers opening by longitudinal slits or less often by terminal pores; ovary inferior, surrounded by a hypanthium, usually crowned by a nectary disk, the carpels 2–5, connate, with 2 to many locules, the ovules 2 or more, rarely solitary, the placentation primarily axile, rarely parietal, the style terminal with a capitate stigma. Fruit a 1- to many-seeded berry or a loculicidal capsule.

A family of 140 genera and about 3000 species, with tropical and subtropical distribution, well-represented in Australia.

REFERENCES: Landrum, L. R. 1986. *Campomanesia, Pimenta, Blepharocalyx, Legrandia, Acca, Myrrhinium,* and *Luma* (Myrtaceae). Fl. Neotrop. Monogr. 45: 1–178; McVaugh, R. 1956. Tropical American Myrtaceae. Fieldiana, Bot. 29: 145–228; McVaugh, R. 1973. Notes on West Indian Myrtaceae. J. Arnold Arbor. 54: 309–314; Wilson, K. A. 1960. The genera of Myrtaceae in the southeastern United States. J. Arnold Arbor. 41: 270–278.

NOTE: Identification of the genera of the Myrtaceae requires the presence of reproductive characters, such as inflorescence, seeds, and embryos. Since reproductive characters are rarely present, a generic key that relies on these characters is very difficult to use. In this flora, I have opted for an artificial key to the species that relies on vegetative characters. This approach has proved to be more useful than a key to the genera.

Key to the Species of *Myrtaceae*

1. Leaves sessile or nearly so, the base usually clasping the stem.
  2. Leaves >3.5 cm long, thick-coriaceous, with strongly revolute margins; flowers ca. 1.5 cm across.
    3. Flowers cauliflorous; fruit depressed-globose, purplish, 1-seeded .... 2.5. *Eugenia earhartii*
    3. Flowers solitary or paired in leaf axils; fruits ellipsoid to globose, green, many-seeded
    .... 7.1. *Psidium amplexicaule*
  2. Leaves 2–5 cm long, coriaceous, margins usually not revolute; flowers ≥5 mm wide ...... 2.4. *Eugenia cordata*
1. Leaves distinctly petiolate, the base not clasping the stem.
  4. Leaf apex obtuse or rounded.
    5. Flowers cauliflorous (at least some of them); fruits red-orange to bright red, 1.5–2 cm diam
    .... 2.11. *Eugenia sessiliflora*
    5. Flowers congested in leaf axils, in axillary cymes, or in terminal panicles, never cauliflorous; fruits green (unknown in E. xerophytica).
      6. Leaf blades with crenate margins; flowers ≥2 cm wide.
        7. Leaf blades chartaceous, involute, venation not impressed above, the base nearly cordate to truncate; young stems slightly flattened, glabrous .... 2.12. *Eugenia xerophytica*
        7. Leaf blades coriaceous, flattened, venation impressed on upper surface, the base obtuse; young branches 4-winged, pubescent .... 7.2. *Psidium guajava*
      6. Leaf blades with entire margins; flowers to 1 cm wide.
        8. Leaves rigid-coriaceous, strongly aromatic, midvein impressed on upper surface; hypanthium canescent within.
          9. Leaves 3–15 cm long, with strong bay rum smell; hypanthium glabrous without; calyx and corolla 5-merous; fruits 1–4-seeded .... 6.1. *Pimenta racemosa* var. *racemosa*
          9. Leaves 2.5–7 cm long, with strong lemony smell; hypanthium appressed-canescent without; calyx and corolla 4-merous; fruits 1–2-seeded .... 4.1. *Myrcianthes fragrans*
        8. Leaves chartaceous, slightly or not aromatic, midvein prominent on upper surface; hypanthium glabrous within... 3.1. *Myrcia citrifolia* var. *imrayana*
  4. Leaf apex acute to acuminate (at least some).
    10. Flowers in axillary panicles to 5 cm long, rusty-brown-pubescent; calyx opening by a circumscissile lid; petals wanting .... 1.1. *Calyptranthes thomasiana*
    10. Flowers congested in leaf axils or in short racemes, the main axis 0.1–3 cm long; calyx of imbricate sepals; petals present.
      11. Flowers nearly sessile... 5.1. *Myrciaria floribunda*
      11. Flowers pedicellate.
        12. Raceme axis 1–3 cm long.. 2.2. *Eugenia biflora*
        12. Raceme axis <1 cm long or flowers 1–3 in leaf axils.
          13. Reticulate venation conspicuous (prominent) at least on lower surface of leaf blade.
            14. Leaves thick-coriaceous, long-acuminate; flowers 2–6, in short axillary racemes (the axis 1–8 mm long); sepals ca. 1.5 mm long; fruits ca. 7 mm diam.
            .... 2.3. *Eugenia confusa*
            14. Leaves chartaceous, bluntly acuminate; flowers 1–3 per leaf axil (inflorescence axis ca. 1 mm long or shorter); sepals 3–3.5 mm long; fruits 1.3–2 cm diam.
            .... 2.9. *Eugenia pseudopsidium*
          13. Primary and secondary veins slightly conspicuous.
            15. Leaves with impressed midvein on upper surface; meristems covered with overlapping, elongate scales; flowers ≥1 cm across...................... 2.6. *Eugenia ligustrina*
            15. Leaves with slightly prominent midvein on upper surface; branches naked; flowers <8 mm across.

16. Pedicels 5–15 mm long; flowers appearing fasciculate, raceme axis 1–2 mm
long.
    17. Leaves 1.5–4.5 cm long, elliptic to elliptic-ovate; branchlets, petioles,
        and leaf midveins hispidulous .... 2.8. *Eugenia procera*
    17. Leaves 3–6 cm long, ovate to lanceolate; plant glabrous
        .... 2.10. *Eugenia rhombea*
16. Pedicels to 5 mm long, raceme axis to 5 mm long.
    18. Leaves 1.5–4 cm long, elliptic, elliptic-lanceolate, or ovate, with rev-
        olute base... 2.7. *Eugenia monticola*
    18. Leaves 3–8 cm long, ovate to elliptic, base not revolute.... 2.1. *Eugenia axillaris*

1. CALYPTRANTHES Sw., nom. cons.

Trees or shrubs, glabrous or pubescent, usually with simple, rusty-brown hairs; branches usually pseudodichotomous (in our area). Leaves opposite, coriaceous, with translucent oil glands. Flowers small, produced in axillary or terminal panicles; bracts and bracteoles deciduous; calyx completely closed in bud (calyptrate), with an operculum (lid) that pops up as the flower matures, then the calyx becoming truncate; petals usually wanting or inconspicuous; stamens numerous, the filaments free, inserted on the margin of the hypanthium, the anthers opening by longitudinal slits; ovary 2-locular, the locules with 2 ovules, the style filiform, the stigma puncti-form. Fruit a 1- or 2-seeded berry crowned by a hypanthium.

About 100 species, from tropical America.

1. Calyptranthes thomasiana O. Berg, Linnaea **27**: 26. 1855.   Fig. 151A–E.

Shrub 1–3 m tall, with pseudodichotomous branching (sympodial growth by substitution); bark grayish, smooth; young leaves and young stems densely covered with appressed, rusty hairs, soon becoming glabrous. Leaf blades 3.5–5.5(–8.5) × (0.7–)1.1–2.5 cm, elliptic, to elliptic-lanceolate, glabrous, the midvein slightly impressed on upper surface, the apex and base acute to obtuse, the margins revolute; petioles 3–5 mm long. Panicles axillary, the axis reddish (when fresh), 4–5 cm long, slightly flattened, densely covered with appressed rusty-brown hairs. Hypanthium cone-shaped, 3.5 mm long, rusty-brown, pubescent without, the operculum retained on one side of the hypanthium; petals wanting; filaments 5 mm long, glabrous; ovary embedded at base of hypanthium, glabrous, the style slightly longer than the stamens. Berry oblate, bilobed, 5 × 6 mm, pinkish red.

DISTRIBUTION: A rare species found only in moist forest. Bordeaux Mountain (W827, A1913, A5103). Also known from St. Thomas, Virgin Gorda; Vieques (Puerto Rico).

2. EUGENIA L.

Trees or shrubs, glabrous or pubescent, usually with simple hairs. Leaves opposite, with translucent oil glands. Flowers small to large, produced in axillary racemes, sometimes fascicle-like on shortened axes or cauliflorous; bracts and bracteoles usually persistent; hypanthium not prolonged beyond the ovary; petals 4, white, spreading at anthesis; stamens numerous, the filaments free, the anthers opening by longitudinal slits; ovary usually 2-locular, the locules with numerous ovules, the style filiform, the stigma punctiform or rarely peltate. Fruit a 1(–2)-seeded berry, crowned by the sepals.

About 550 species, of which 500 are from tropical and subtropical America; the remaining species are from the Old World tropics.

Key to the Species of *Eugenia*

1. Leaf base cordate.
    2. Leaves thick-coriaceous, brittle; flowers cauliflorous .... 5. *E. earhartii*
    2. Leaves coriaceous or chartaceous; flowers axillary.
        3. Flowers ≥2 cm wide .... 12. *E. xerophytica*
        3. Flowers ≥5 mm wide .... 4. *E. cordata*
1. Leaf base rounded, obtuse, acute, or tapering.
    4. Leaves thick-coriaceous, the apex obtuse to rounded; flowers (at least some) cauliflorous; fruits
        1.5–2 cm diam.. 11. *E. sessiliflora*
    4. Leaves coriaceous to chartaceous, the apex acute, acuminate or seldom obtuse; flowers in axillary in-
        florescences; fruits <1 cm diam.
        5. Raceme axis 1–3 cm long; the flowers in axillary racemes............................. 2. *E. biflora*
        5. Raceme axis <1 cm long; the flowers appearing fasciculate, or solitary in leaf axils.
            6. Flowers solitary or paired in leaf axil.
                7. Leaves elliptic, coriaceous, 2–5 cm long, conspicuously yellowish-punctate, the tertiary vena-
                tion inconspicuous; meristems covered with overlapping scales; flowers solitary, the pedun-
                cles 1.8–3.5 cm long, with a persistent, elongate scale at base; fruits black............... 6. *E. ligustrina*
                7. Leaves ovate or elliptic, 3.5–9 cm long, inconspicuously punctate, the tertiary venation con-

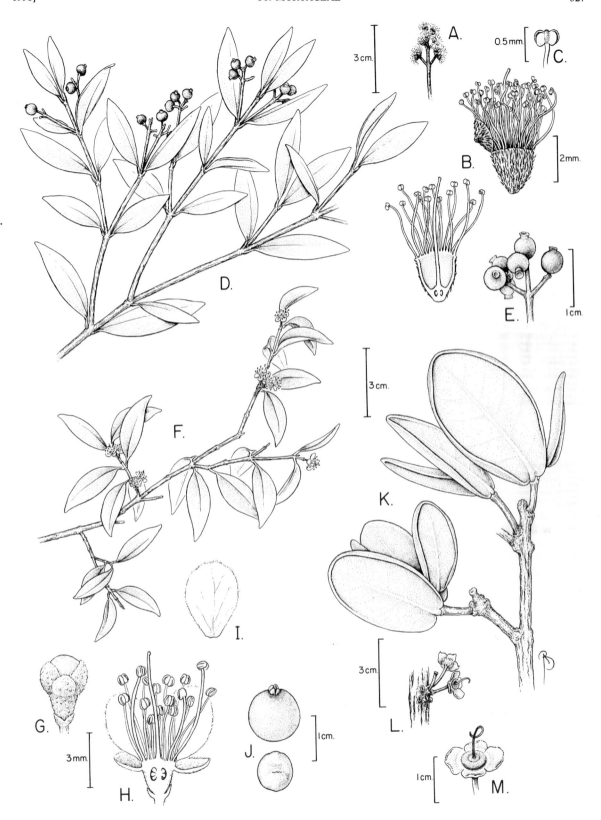

Fig. 151. A–E. *Calyptranthes thomasiana*. A. Inflorescence. B. Flower with calyptra lid and l.s. flower. C. Anther. D. Fruiting branch. E. Infructescence. F–J. *Eugenia biflora*. F. Flowering branch. G. Flowerbud. H. L.s. flower. I. Petal. J. Fruit and seed. K–M. *Eugenia earhartii*. K. Leafy branch. L. Detail of stem showing cauliflorous inflorescence. M. Flower receptacle with pistil.

spicuous; meristems naked; flowers 2–3 or less often solitary, the pedicels 1.5–2 cm long;
    fruit red-orange when mature.. 9. *E. pseudopsidium*
6. Flowers (2–)6 to numerous in leaf axils.
    8. Leaves thick-coriaceous, abruptly long-acuminate at apex...................................... 3. *E. confusa*
    8. Leaves chartaceous to coriaceous, acuminate (the acumen obtuse) or acute.
        9. Pedicels 5–15 mm long; raceme axis 1 mm long or less.
            10. Leaves 1.5–4.5 cm long, elliptic to elliptic-ovate; branchlets, petioles, lower side of
                leaf midveins, and leaf margins hispidulous .... 8. *E. procera*
            10. Leaves 3–6 cm long, ovate to lanceolate; plant glabrous........................... 10. *E. rhombea*
        9. Pedicels 1–5 mm long, raceme axis to 5 mm long.
            11. Leaves 1.5–4 cm long, elliptic, elliptic-lanceolate, or ovate, with revolute base 7. *E. monticola*
            11. Leaves 3–8 cm long, ovate to elliptic, base not revolute............................. 1. *E. axillaris*

1. Eugenia axillaris (Sw.) Willd., Sp. Pl. **2**: 960. 1799.
    Myrtus axillaris Sw., Prodr. 78. 1788.

Eugenia guadalupensis DC., Prodr. **3**: 275. 1828.

Small tree 3–7 m tall; bark light gray, smooth, slightly fissured; twigs slightly angled, rusty-pubescent when young, becoming glabrous and grayish to ash-colored. Leaves opposite or subopposite; blades 3–8 × 2–4 cm, ovate to elliptic, coriaceous, glabrous, the midvein sunken to flat on upper surface, punctations inconspicuous, the apex shortly acuminate, acute, or obtuse, the base acute to cuneate, the margins revolute; petioles 3–8 mm long. Flowers white, numerous, in short axillary racemes, the axis 2–5 mm long, slightly 4-angled; pedicels 1.5–3 mm long; bracteoles ovate to rounded, nearly clasping the pedicel. Hypanthium nearly cone-shaped, 0.5–1 mm long, puberulent, sepals rounded, gland-dotted, ciliate, in 2 dissimilar pairs, 0.5–1 mm long; petals obovate, 2.5–3 mm long, gland-dotted; disk 1–1.5 mm wide, puberulent; stamens numerous, the filaments 4–5 mm long, glabrous. Berry oblate to globose, less often ellipsoid, 0.7–1 cm diam., turning from green to nearly black.

DISTRIBUTION: An uncommon tree of moist to dry evergreen forests. Road to Bordeaux (A4053). Also on Anegada, St. Croix, Tortola, and Virgin Gorda; throughout the Caribbean and coastal Mexico.

2. Eugenia biflora (L.) DC., Prodr. **3**: 276. 1828. *Myrtus biflora* L., Syst. Nat. ed. 10, **2**: 1056. 1759.
    Fig. 151F–J.

Myrtus pallens Vahl, Symb. Bot. **2**: 57. 1791. *Eugenia pallens* (Vahl) DC., Prodr. **3**: 284. 1828.
Eugenia lancea Poir. *in* Lam., Encycl. Suppl. **3**: 123. 1813. *Eugenia biflora* var. *lancea* (Poir.) Krug & Urb., Bot. Jahrb. Syst. **19**: 631. 1895.
Myrcia thomasiana DC., Prodr. **3**: 244. 1828.
Eugenia glabrata sensu Eggers, Fl. St. Croix 51. 1879, non (Sw.) DC., 1828.

Shrub or small tree 3–6 m tall; bark grayish, smooth, becoming mottled; twigs slightly angled, whitish-pubescent, becoming glabrous and grayish to ash-colored. Leaves opposite or subopposite; blades 3–9 × 2–4 cm, ovate, elliptic, nearly coriaceous, the upper surface puberulent (appressed hairs) soon becoming glabrous, with sunken midvein, conspicuously punctate, the apex acute to acuminate, the base acute to obtuse, the margins revolute; petioles 2–5 mm long, puberulent. Flowers white, numerous, in axillary racemes, the axis 1–3 cm long, pubescent; pedicels 1.5–6(–10) mm long; bracteoles ovate clasping the pedicel. Hypanthium nearly cone-shaped, 0.5–1 mm long, pubescent, sepals rounded, pubescent, gland-dotted, ciliate, in 2 dissimilar pairs, 0.5–1 mm long;

petals obovate, 2–2.5 mm long; disk 1–1.5 mm wide; stamens numerous, the filaments 4–5 mm long, glabrous. Berry oblate to globose, 0.7–1 cm diam., glossy, turning from green to red and finally blackish.

DISTRIBUTION: A common tree of moist to dry evergreen forests. Cruz Bay Quarter along Center Line Road (A2808), trail to Petroglyphs (A2931). Also on Jost van Dyke, St. Croix, St. Thomas, Tortola, and Virgin Gorda; throughout the West Indies and Mexico to northern South America, apparently not present in Cuba.

3. Eugenia confusa DC., Prodr. **3**: 279. 1828.

Eugenia krugii Kiaersk., Bot. Tidsskr. **17**: 259. 1890.

Small tree 2.5–8 m tall; bark light brown, smooth; twigs slightly flattened, glabrous, grayish to ash-colored. Leaf blades 3–8 × 1.5–4 cm, ovate, elliptic or elliptic-ovate, thick-coriaceous, glabrous, the upper surface glossy, with sunken midvein, oil glands numerous, the apex abruptly long-acuminate, the base obtuse to rounded, the margins strongly revolute; petioles 4–9 mm long, glabrous. Flowers white, 2–6, in axillary shortened racemes, the axis 1–5 mm long, glabrous; pedicels 10–17 mm long, glabrous; bracteoles oblong. Hypanthium nearly cone-shaped, 1–1.5 mm long, glabrous, gland-dotted, sepals ovate, glabrous, gland-dotted, ca. 1.5 mm long; petals ovate, 3–3.5 mm long; disk 2–2.5 mm wide; stamens numerous, the filaments 3–5 mm long, glabrous. Berry globose, ca. 7 mm diam., turning from light green to yellowish orange.

DISTRIBUTION: A rather uncommon tree of moist or dry evergreen forests. Road to Bordeaux (A2098). Also in Florida, Greater Antilles, Bahamas, and Lesser Antilles.

4. Eugenia cordata (Sw.) DC., Prodr. **3**: 272. 1828.
    Myrtus cordata Sw., Prodr. 78. 1788.

Myrtus ramiflorus Vahl *in* H. West, Bidr. Beskr. Ste. Croix 290. 1793.
Myrciaria stirpiflora O. Berg, Linnaea **30**: 702. 1861. *Eugenia stirpiflora* (O. Berg) Krug & Urb. *ex* Urb., Bot. Jahrb. Syst. **19**: 672. 1895.

Small tree or shrub 1–5 m tall, very often many-branched from base; bark light gray, becoming rough and fissured into scaly plates; twigs puberulent, becoming glabrous, brown or grayish. Leaf blades 2–5 × 1.2–3 cm, elliptic-ovate, elliptic, or ovate, coriaceous, glabrous, the upper surface glossy, with sunken, yellowish midvein, oil glands numerous, dark, conspicuous on lower surface, the apex obtuse to rounded, the base cordate or nearly so, usually clasping the stem, the margins strongly revolute, yellowish; petioles 0–1 mm long, glabrous. Flowers sessile, along leafless stems or in axillary fascicles with overlapping bracteoles.

Hypanthium pinkish, nearly cone-shaped, 1–1.5 mm long, glabrous, gland-dotted, sepals rounded, glabrous, 1–1.5 mm long; petals white, oblong, ca. 2 mm long; disk 1.5–2 mm wide; stamens numerous, the filaments ca. 2 mm long, glabrous. Berry ellipsoid, to nearly ovoid, 8–10 mm long, bright red.

DISTRIBUTION: A common shrub of coastal scrub or seasonal deciduous forests. East side of Minna Hill (A3205), Newfound Bay (A4148), Lameshur (B636). Also on Jost van Dyke, St. Croix, St. Thomas, Tortola, and Virgin Gorda; St. Barts.

5. Eugenia earhartii Acev.-Rodr., Brittonia **45**: 133. 1993. Fig. 151K–M.

Shrub 1.5–3 m tall, many-branched from base, glabrous; bark grayish, thick, peeling off in irregular flakes; twigs reddish and slightly flattened when young, becoming grayish and cylindrical. Leaves rigid-coriaceous to brittle, ascending; blades 3.5–8.5 × 2–6.2 cm, ovate to widely elliptic, not punctate or obscurely punctate, midvein yellowish, planar on upper surface, prominent beneath, the apex rounded, the base cordate, clasping the stem, the margins strongly revolute, yellowish; petioles swollen, 1–2(–4) mm long. Flowers fasciculate in shortened, cauliflorous racemes; pedicels 4.5–9 mm long, thickened. Hypanthium nearly conical, 5–6 mm wide, glabrous, the sepals rounded, 3–4 mm long, ciliate at margins; stamens numerous, glabrous; style elongate, glabrous. Berry globose to oblate, 1.5–2 × 2–3 cm, turning from green to purple, 1-seeded.

DISTRIBUTION: A rare shrub of coastal thickets or thorny scrub. Known only from two localities on St. John: eastern side of Minna Hill (A3204), trail from Europa Bay to White Cliffs (A5233).

6. Eugenia ligustrina (Sw.) Willd., Sp. Pl. **2**: 962. 1799. *Myrtus ligustrina* Sw., Prodr. 78. 1788.

Shrub 1–2 m tall, many-branched from base; bark light gray, smooth; twigs puberulent, becoming glabrous and grayish; juvenile buds covered with overlapping, elongate, stipule-like scales. Leaf blades 2–5 × 1–2.5 cm, elliptic, coriaceous, glabrous, the upper surface glossy, with sunken midvein, punctate on lower surface, the apex obtuse to acute, often notched, the base tapering, the margins strongly revolute; petioles 2–5 mm long, glabrous. Flowers white, solitary in leaf axil, subtended by an oblong or linear scale; peduncle 1.8–3.5 cm long. Hypanthium greenish, nearly cup-shaped, ca. 1.5 mm long, glabrous, gland-dotted, the sepals oblong, glabrous, reflexed, 2.5–3 mm long; petals white, oblong, 4–5 mm long; disk 1.5–2 mm wide; stamens numerous, the filaments ca. 3 mm long, glabrous. Berry globose, 8–10 mm diam., black, with persistent elongate sepals.

DISTRIBUTION: A common shrub of coastal scrub or seasonal deciduous forests. Bethany (B347), Lameshur (A743). Also on Jost van Dyke, St. Croix, St. Thomas, Tortola, and Virgin Gorda; throughout the Caribbean and northern South America.

COMMON NAMES: birch berry, crumberry.

7. Eugenia monticola (Sw.) DC., Prodr. **3**: 275. 1828. *Myrtus monticola* Sw., Prodr. 78. 1788. Fig. 152D–F.

Eugenia maleolens Pers., Syn. Pl. **2**: 29. 1806.
Eugenia flavovirens O. Berg, Linnaea **27**: 184. 1856.

Small tree 2–12 m tall; bark light gray, fissured, with a few horizontal grooves; twigs puberulent, becoming glabrous, densely leafed. Leaves ascending; blades 1.5–4 × 0.5–2.2 cm, elliptic, elliptic-lanceolate, or ovate, chartaceous, glabrous, upper surface with puberulent, planar midvein, lower surface obscurely punctate, the apex obtuse to acute or less often acuminate, the base narrowed,

the margins revolute especially along the tapering base; petioles ca. 2 mm long, puberulent. Flowers fragrant, numerous in short axillary racemes; axis 2–5 mm long, puberulent; pedicels 3–5 mm long. Hypanthium yellowish green, cup-shaped, 1–1.3 mm long, puberulent, the sepals ovate, puberulent, concave, in unequal pairs, 1–1.5 mm long; petals white, ovate, concave, 3–4 mm long; disk ca. 1.3 mm wide; stamens numerous, the filaments ca. 4 mm long, glabrous. Berry globose, 5–7 mm diam., shiny black.

DISTRIBUTION: A common shrub of dry to moist forests. Susannaberg (A3808), trail to Brown Bay, along trail to Lameshur (A3184). Also on Jost van Dyke, St. Croix, St. Thomas, Tortola, and Virgin Gorda; throughout the West Indies and northern South America.

8. Eugenia procera (Sw.) Poir. *in* Lam., Encycl. Suppl. **3**: 129. 1813. *Myrtus procera* Sw., Prodr. 77. 1788.

Shrub 2–5 m tall; bark light gray, smooth; twigs hispidulous, becoming glabrous. Leaves opposite or subopposite; blades 1.5–4.5 × 0.5–2.5 cm, elliptic to elliptic-ovate, chartaceous, glabrous, upper surface usually with hispidulous, prominulous midvein, lower surface obscurely punctate, the apex obtuse to shortly acuminate, the acumen blunt, the base obtuse, the margins entire; petioles 2–3 mm long, hispidulous. Flowers fragrant, in short axillary racemes (appearing fasciculate); axis ca. 1 mm long, with overlapping bracts; pedicels glabrous, 5–12 mm long. Hypanthium light green, cup-shaped, 0.7–1 mm long, glabrous, the sepals ovate, ciliate, in unequal pairs, 0.5–1 mm long; petals white, obovate, concave, reflexed, 3–3.5 mm long; disk ca. 1 mm wide; stamens numerous, the filaments ca. 4 mm long, glabrous. Berry globose, ca. 5 mm diam., turning from greenish yellow to purplish black.

DISTRIBUTION: A common shrub of scrub or dry to moist forests. Bordeaux Mountain (W105), Reef Bay trail (A4123). Also on Jost van Dyke, St. Croix, St. Thomas, and Tortola; Cuba, Hispaniola, Puerto Rico, Lesser Antilles, and northern Colombia and Venezuela.

9. Eugenia pseudopsidium Jacq., Enum. Syst. Pl. 23. 1760. Fig. 152A–C.

Eugenia portoricensis DC., Prodr. **3**: 266. 1828.
Eugenia thomasiana O. Berg, Linnaea **27**: 183. 1856.

Shrub 1–5(–10) m tall, usually many-branched from base; bark light gray, smooth; twigs hispidulous, becoming glabrous. Leaf blades 3.5–9 × 2–5 cm, elliptic, ovate, to lanceolate, chartaceous, glabrous, the venation yellowish, reticulate, forming submarginal loops, prominent on lower surface, the apex obtuse to shortly acuminate, the acumen blunt, the base obtuse to nearly rounded, the margins entire, thickened, tapering onto the petiole at base; petioles 3–6 mm long, hispidulous. Flowers 1–3 in axillary racemes with axis of 1 mm or less long; bracts oblong; pedicels glabrous, 1.5–2 cm long; bracteoles triangular. Hypanthium light green, urn-shaped, 1–1.5 mm long, glabrous, the sepals ovate, in unequal pairs, 3–3.5 mm long; petals white, oblong, concave, spreading, 4.5–5 mm long; disk ca. 2 mm wide; stamens numerous, the filaments ca. 5 mm long, glabrous. Berry nearly globose to obovoid, 1.3–2 cm diam., turning from yellowish green to red-orange, with persistent green sepals.

DISTRIBUTION: A common shrub of moist to dry forests, mostly in understory, but also in disturbed, open areas. Road to Bordeaux (A2853), Bethany (B341). Also on St. Croix, St. Thomas, and Tortola; Hispaniola, Puerto Rico (including Vieques and Culebra), and Lesser Antilles.

COMMON NAMES: bastard guava, Christmas cherry, wild guava.

10. Eugenia rhombea Krug & Urb. *ex* Urb., Bot. Jahrb. Syst. **19**: 644. 1895.

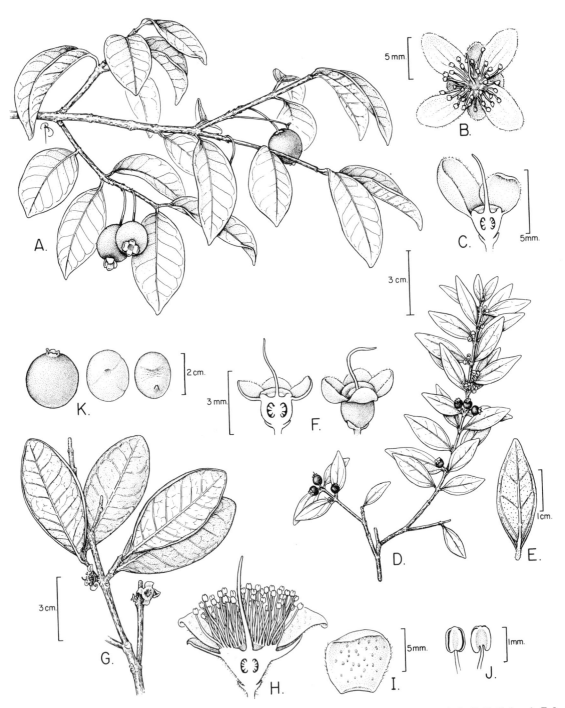

Fig. 152. A–C. *Eugenia pseudopsidium*. **A.** Fruiting branch. **B.** Flower, top view. **C.** L.s. flower. **D–F.** *Eugenia monticola*. **D.** Fertile branch. **E.** Leaf. **F.** L.s. immature fruit and immature fruit. **G–K.** *Eugenia sessiliflora*. **G.** Flowering branch. **H.** L.s. flower. **I.** Petal. **J.** Anther, frontal and dorsal views. **K.** Fruit, seed, and embryo.

Shrub or small tree 2–7 m tall; trunk slightly angled, the bark mottled whitish with grayish flakes; twigs glabrous. Leaf blades 3–6 × 1.5–3.5 cm, ovate to lanceolate, chartaceous, glabrous, midvein planar on upper surface, the apex acuminate to long-acuminate, the acumen blunt, the base obtuse to rounded, the margins entire, slightly thickened; petioles 7–10 mm long, glabrous. Flowers numerous, in short axillary racemes (appearing as fasciculate); axis ca. 1 mm long, with overlapping bracts; pedicels slender, glabrous, gland-dotted, 5–15 mm long. Hypanthium light green, cup-shaped, 1–1.5 mm long, glabrous, the sepals ovate, in unequal pairs, 2–3.5 mm long, gland-dotted; petals white, ob-ovate, concave, 4–4.5 mm long; disk 2 mm wide, pubescent; stamens numerous, the filaments ca. 5 mm long, glabrous. Berry globose or oblate, 5–6 mm diam., glandular-roughened, black.

DISTRIBUTION: An uncommon shrub of dry forests or scrub. Lameshur-Reef Bay Trail (W429). Also on St. Croix; Florida, eastern Mexico, Guatemala, Honduras, and throughout the West Indies.

COMMON NAME: crumberry.

11. Eugenia sessiliflora Vahl, Symb. Bot. 3: 64. 1794.

Fig. 152G–K.

Eugenia lateriflora Willd., Sp. Pl. **2**: 961. 1799.

Shrub or small tree 1.5–3 m tall, many-branched from base; bark light gray, smooth; twigs slightly flattened when young, glabrous. Leaf blades 4.5–8 × 3.2–5 cm, elliptic to rounded, thick-coriaceous, glabrous, midvein yellowish, sunken on upper surface, the apex obtuse or rounded, often notched, the base acute to rounded, the margins strongly revolute, thickened; petioles 3–5 mm long, glabrous. Flowers sessile, solitary or clustered, in leaf axils or cauliflorous; bracts and bracteoles deltoid. Hypanthium funnel-shaped, 4–4.2 mm long, tomentose, the sepals white, rounded, reflexed, in unequal pairs, 3.5–5 mm long, gland-dotted; petals white, rounded, reflexed, 5–5.5 mm long, gland-dotted; disk 4.5–5 mm wide, pubescent; stamens numerous, the filaments ca. 5 mm long, glabrous; style white, appressed pubescent. Berry nearly globose, 1.5–2 cm diam., smooth, orange-red to bright red.

DISTRIBUTION: An uncommon shrub of scrub, dry forests or moist forests. Concordia (A2763), vicinity of Southside Pond (A1830). Also on St. Croix, St. Thomas, Tortola, and Virgin Gorda; Puerto Rico, including Vieques and Culebra.

12. Eugenia xerophytica Britton, Bull. Torrey Bot. Club **51**: 11. 1924.

Shrub 1.5–2.5 m tall, many-branched from base; bark tan, smooth, peeling off in irregular flakes; twigs slightly flattened when young, glabrous. Leaf blades 2.8–4 × 2.1–4 cm, ovate, chartaceous, glabrous, involute, the apex obtuse to rounded, the base cordate to truncate, the margins crenate; petioles 5–9 mm long, glabrous. Flowers solitary or a few clustered in leaf axils; bracts and bracteoles ovate; peduncles 8–16 mm long. Hypanthium funnel-shaped, ca. 3 mm long, glabrous, the sepals ovate, 2–3 mm long; petals white, obovate, 1–1.2 cm long, obscurely gland-dotted; disk 4.5–5 mm wide, pubescent; stamens numerous, the filaments 7–10 mm long, glabrous. Berry not known.

DISTRIBUTION: A rare shrub of scrub or dry forests. Previously known on St. John from 3 individuals, of which only one remains. Maria Bluff (A764). Also on southern Puerto Rico and Caja de Muerto Island.

DOUBTFUL RECORDS: *Eugenia foetida* Pers. was reported by Woodbury and Weaver (1987) as occurring on St. John, but no specimen has been located to confirm this record. *Eugenia sintenisii* Kiaersk. was reported by Britton and Wilson (1925) as occurring on St. John, but no specimen has been located to confirm this record either. Neither species has been collected there recently.

3. MYRCIA DC. *ex* Guill.

Trees or shrubs, glabrous or pubescent, usually with hairs simple or less often 2-branched. Leaves opposite, with translucent oil glands. Flowers small to medium, produced in axillary panicles; bracts and bracteoles persistent; hypanthium prolonged or not beyond the ovary; sepals 5 or less often 4; petals 5 or 4, white, spreading at anthesis, deciduous; stamens numerous, the filaments free, the anthers opening by longitudinal slits; ovary 2–(3)- locular, each locule with 2 ovules, the style filiform, the stigma punctiform. Fruit a 1(–3)-seeded berry, crowned by the sepals.

About 300 species, from tropical and subtropical America.

1. Myrcia citrifolia (Aubl.) Urb. var. **imrayana** (Griseb.) Stehlé & Quentin, Fl. Guadeloupe **2**(3): 57. 1949. *Myrcia coriacea* (Vahl) DC. var. *imrayana* Griseb., Fl. Brit. W. I. 234. 1860.

Fig. 153A–F.

Shrub or small tree 2.5–7 m tall, sometimes many-branched from base; bark gray, rough, fissured, the inner bark reddish; twigs flattened and appressed-pubescent when young. Leaf blades 2.5–6 × 1.5–4 cm, elliptic to obovate, chartaceous, slightly or not aromatic, puberulent, midvein prominent on upper surface, the apex obtuse, rounded and notched, the base acute, the margins revolute; petioles 2–5 mm long, puberulent. Flowers several, in short axillary panicles, the terminal flowers sessile, the lateral pedicellate; axis 2–5 cm long, appressed pubescent; pedicels slender, 2–5 mm long, reddish; bracts and bracteoles oblong. Hypanthium funnel-shaped, 1.5–2 mm long, glabrous, the sepals 5, green, ovate, 1–1.5 mm long, obscurely gland-dotted; petals 5, white, obovate, reflexed, 4–5 mm long; disk 3 mm wide, glabrous; stamens numerous, the filaments geniculate, 3–5 mm long, glabrous; ovary 3-locular. Berry oblate, 7–9 mm diam., glandular-roughened, turning from red to black.

DISTRIBUTION: A common tree of moist forests. Bordeaux (A2598, A2861, B566). Also on St. Thomas, Tortola, and Virgin Gorda; Puerto Rico and the Lesser Antilles.

4. MYRCIANTHES O. Berg

Trees or shrubs, glabrous or pubescent, usually with simple hairs. Leaves opposite or ternate, with translucent oil glands. Flowers small to medium, produced in axillary panicles; bracts and bracteoles deciduous; hypanthium not prolonged beyond the ovary, canescent; sepals 4; petals 4, white, spreading at anthesis, deciduous; stamens numerous, the filaments free, the anthers opening by longitudinal slits; ovary 2-locular, each locule with to 20 ovules, the style filiform, the stigma punctiform. Fruit a 1- or 2-seeded berry, crowned by the sepals.

About 50 species, from tropical and subtropical America.

1. Myrcianthes fragrans (Sw.) McVaugh, Fieldiana, Bot. **29**: 485. 1963. *Myrtus fragrans* Sw., Prodr. 79. 1788. *Eugenia fragrans* (Sw.) Willd., Sp. Pl. **2**: 964. 1799. *Anamomis fragrans* (Sw.) Griseb., Fl. Brit. W. I. 240. 1860.

Fig. 153G–I.

Eugenia punctata Vahl *in* H. West, Bidr. Beskr. Ste. Croix 216. 1793.

Tree 5–20 m tall; bark reddish brown, smooth; twigs flattened and appressed-pubescent when young, becoming glabrous and

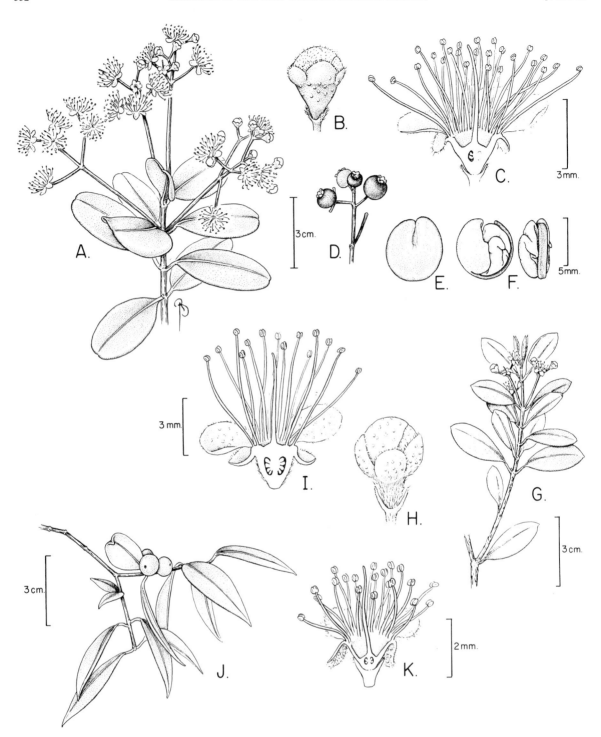

Fig. 153. A–F. *Myrcia citrifolia*. **A.** Flowering branch. **B.** Flowerbud. **C.** L.s. flower. **D.** Infructescence. **E.** Seed. **F.** Embryo, frontal and lateral views. **G–I.** *Myrcianthes fragrans*. **G.** Flowering branch. **H.** Flowerbud. **I.** L.s. flower. **J, K.** *Myrciaria floribunda*. **J.** Fruiting branch. **K.** L.s. flower.

cylindric. Leaves opposite; blades 2.5–7 × 1.4–3.2 cm, obovate, spatulate, or elliptic, coriaceous, with a strong lemony odor, puberulent, gland-dotted, midvein impressed on upper surface, the apex obtuse or rounded, seldom notched or mucronate, the base acute, the margins strongly revolute; petioles 2–10 mm long, puberulent. Flowers many, in short axillary panicles, the terminal flowers sessile, the lateral pedicellate; axis 2–5 cm long, appressed-canescent; pedicels slender, 2–7 mm long; bracts and bracteoles subulate. Hypanthium funnel-shaped, 2.5–3 mm long, appressed-canescent, the sepals ovate, 2–2.2 mm long, gland-dotted, ciliate, canescent within; petals, white, rounded, concave, canescent within, reflexed, ca. 3 mm long; disk ca. 3 mm wide, canescent; stamens numerous, the filaments ca. 3 mm long, glabrous. Berry globose to ellipsoid, 8–10 mm diam., glandular-roughened, green.

DISTRIBUTION: A common tree of moist forests. Camelberg Peak (A4249). Also on St. Croix and Tortola; throughout the West Indies and Mexico to northern South America.

5. MYRCIARIA O. Berg

Trees or shrubs, glabrous or pubescent, usually with simple hairs. Leaves opposite, with translucent oil glands. Flowers small to medium, sessile, in axillary clusters; bracts and bracteoles persistent; hypanthium prolonged beyond the ovary, circumscissile at base; sepals 4; petals 4, white; stamens numerous, the filaments free, the anthers opening by longitudinal slits; ovary 2-locular, the locules with 1 or 2 ovules, the style filiform, the stigma punctiform. Fruit a 1- or 2-seeded berry, crowned by a rounded scar.

About 40 species, from tropical and subtropical America.

1. Myrciaria floribunda (H. West *ex* Willd.) O. Berg, Linnaea **27**: 330. 1856. *Eugenia floribunda* H. West *ex* Willd., Sp. Pl. **2**: 960. 1799. Fig. 153J, K.

Small tree 5–7 m tall; bark grayish brown, smooth, thin, peeling off in long irregular flakes; twigs flattened, puberulent, becoming glabrous and cylindric. Leaf blades 3–6 × 1–2.5 cm, elliptic to lanceolate, chartaceous, glabrous, obscurely gland-dotted, midvein planar on upper surface, the apex long-acuminate, the base acute to obtuse, the margins entire; petioles 2–4 mm long, puberulent. Flowers few, nearly sessile, clustered in leaf axils; bracteoles rounded, connate around base of hypanthium. Hypanthium funnel-shaped, ca. 1.5 mm long, glabrous, the sepals rounded, 1–1.5 mm long, gland-dotted, villous; petals, white, obovate, ca. 1.3 mm long; disk ca. 1 mm wide, glabrous; stamens numerous, the filaments ca. 2.5 mm long, glabrous. Berry globose to oblate, 8–10 mm diam., smooth, turning from green to orange when mature.

DISTRIBUTION: An uncommon tree of moist forests. Bordeaux (A2626, A3823). Also on Jost van Dyke, St. Croix, St. Thomas, Tortola, and Virgin Gorda; Puerto Rico (including Vieques), Cuba, Hispaniola, Lesser Antilles, Mexico, and northern South America.

COMMON NAME: guavaberry.

6. PIMENTA Lindl.

Trees, glabrous or pubescent, the hairs simple or branched. Leaves opposite, strongly aromatic when crushed, with translucent oil glands and prominent lateral veins. Flowers small to medium, in axillary dichasial cymes or panicles; bracts and bracteoles early deciduous; hypanthium little prolonged beyond the ovary; sepals 4 or 5; petals 4 or 5, white; stamens numerous, the filaments free, the anthers opening by longitudinal slits; ovary 2-locular (rarely 1- or 3-locular), the locules with 1(–2–7) ovules, the style filiform, the stigma nearly capitate. Fruit a 1–4-seeded berry, crowned by the persistent sepals.

About 15 species, primarily in the West Indies and Central America.

1. Pimenta racemosa (Mill.) J.W. Moore var. **racemosa,** Bernice P. Bishop Mus. Bull. **102**: 33. 1933. *Caryophyllus racemosus* Mill., Gard. Dict., ed. 8. 1768.

Tree 7–12 m tall; bark tan, smooth, thin, peeling off in long irregular flakes; twigs usually 4-angled, strigose or pubescent, becoming glabrous and cylindric. Leaf blades with a strong bay rum odor, 3–15 × 1–6.5 cm, elliptic, oblong, or obovate, coriaceous, glabrous, gland-dotted, midvein impressed to plane on upper surface, the apex and base obtuse to rounded, the margins revolute; petioles 5–12 mm long, glabrous. Flowers many, in short axillary panicles, the terminal flowers sessile, the lateral pedicellate; axis 4.5–12 cm long, glabrous to sparsely pubescent; pedicels slender, 2–7 mm long; bracts and bracteoles linear. Hypanthium funnel-shaped, 2–3 mm long, glabrous, the sepals widely ovate, 0.5–1 mm long, gland-dotted, canescent within; petals white, obovate, glabrous, reflexed, 3–3.5 mm long; disk 3–4 mm wide, canescent; stamens numerous, the filaments 3–4 mm long, glabrous. Berry nearly globose, 6–10 mm diam., glandular-roughened, turning from green to black.

DISTRIBUTION: A common tree of moist secondary forests, previously widely cultivated; not known for sure whether native to the Virgin Islands or not. Bordeaux (A829), Gift Hill (A2852). This variety is also found on St. Croix, Tortola, and Virgin Gorda; Puerto Rico, Cuba, and the Lesser Antilles.

COMMON NAMES: bay leaf, bay rum tree, cinnamon, cinnamon bush.

7. PSIDIUM L.

Trees or shrubs, commonly pubescent; twigs often acutely 4-angled. Leaves opposite, with translucent oil glands. Flowers medium to large, axillary, solitary, fascicled or in dichasia; bracts and bracteoles early deciduous; hypanthium prolonged beyond the ovary; calyx lobes 4 or 5, opening by irregular fissures; petals 4 or 5, white; stamens numerous, the filaments free, the anthers opening by longitudinal slits; ovary 3–4-locular, the locules with numerous ovules, the style subulate, the stigma capitate. Fruit a many-seeded berry, crowned by the persistent sepals.

About 100 species, native to tropical and subtropical America, some in cultivation and escaped in warm regions.

Key to the Species of *Psidium*

1. Leaves nearly sessile, with base usually clasping the stem; venation inconspicuous .... 1. *P. amplexicaule*
1. Leaves distinctly petiolate, base not clasping stem; venation prominent on lower surface .... 2. *P. guajava*

1. Psidium amplexicaule Pers., Syn. Pl. **2**: 27. 1806. Fig. 154.

Shrub or small tree 2–4 m tall; bark tan, smooth, thin, peeling off in long irregular flakes; twigs cylindric slightly flattened,

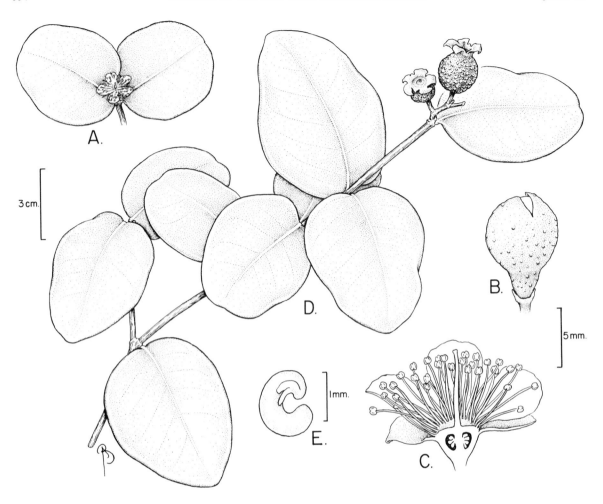

FIG. 154. *Psidium amplexicaule*. **A.** Flowering branch. **B.** Flowerbud. **C.** L.s. flower. **D.** Fruiting branch. **E.** Embryo.

glabrous, becoming cylindric at maturity. Leaf blades 3–7 × 3–6.5 cm, ovate, nearly rounded or obovate, coriaceous, glabrous, obscurely gland-dotted, midvein impressed to plane on upper surface, the apex rounded, the base cordate, clasping the stem, or less often rounded, the margins revolute; petioles 1–2.5 mm long. Flowers 1 or 2 in short axillary racemes; axis 2–3 mm long; pedicels slender, 7–10 mm long; bracts fleshy, 1.5 mm long. Hypanthium funnel-shaped, 4–5 mm long, glabrous, glandular-roughened, the sepals widely ovate, 4–4.5 mm long, puberulent within; petals white, obovate, glabrous, 6–7 mm long; disk 5–6 mm wide, glabrous; stamens numerous, the filaments 5–6 mm long, glabrous; style stout. Berry ellipsoid to globose, 1.5–2 cm diam., smooth, green.

DISTRIBUTION: A common shrub of moist forests. Bordeaux (A2862, A5216). Also on St. Thomas, Tortola, and Virgin Gorda; Puerto Rico and Nevis. According to Britton and P. Wilson (1925), this species was cultivated on St. Croix and Guadeloupe.

COMMON NAME: mountain guava.

2. Psidium guajava L., Sp. Pl. 470. 1753.

Shrub or small tree 2–8 m tall; bark tan, smooth, thin, peeling off in long irregular flakes; young stems sharply 4-angled, pubes-cent. Leaf blades 7–14 × 3–6 cm, elliptic, coriaceous, puberulent, obscurely gland-dotted, midvein impressed on upper surface, venation prominent on lower surface, the apex obtuse to rounded, shortly apiculate, the base obtuse, the margins crenate; petioles 4–7 mm long. Flowers solitary in leaf axil; peduncles stout, 2–2.5 cm long. Hypanthium ellipsoid, 5–6 mm long, pubescent, glandular-roughened, the sepals widely ovate, 6–7 mm long; petals white, oblong, pubescent, reflexed, 12–15 mm long; disk 3.5–4 mm wide; stamens numerous, the filaments 6–7 mm long, glabrous. Berry globose to ovoid, 2–4.5 cm diam., smooth, turning from green to yellow, pinkish inside.

DISTRIBUTION: Apparently introduced, a few individuals known to be growing spontaneously. Road to Bordeaux (A2896), Cinnamon Bay (A5149). Native to tropical America, now cultivated and an escape throughout the tropics.

COMMON NAME: guava.

DOUBTFUL RECORD: Syzygium jambos (L.) Alston was reported by Woodbury and Weaver (1987) to occur on St. John, but no specimen has been located to confirm this record.

57. Nyctaginaceae (Four O'Clock Family)

Herbs, shrubs, trees, vines, or lianas; stems usually showing in cross section, wood with included islands of phloem. Leaves opposite or rarely alternate, simple, usually entire, sometimes with axillary spines; stipules wanting. Flowers minute to large, actinomorphic, bisexual or rarely unisexual, produced in axillary cymes, usually with large and showy bracts; calyx of (3–)5(–8) sepals connate into a corolla-like tube; corolla wanting; stamens as many as the calyx lobes or less often fewer or more numerous, the filaments of equal or unequal length, distinct or connate at base into a short tube, the anthers opening by longitudinal slits; ovary superior, surrounded at base by an annular nectary disk, 1-carpellate, with a single basal ovule and a terminal, elongate style. Fruit an anthocarp, achene, or nut, enclosed by a hardened calyx tube.

A tropical and subtropical family with 30 genera and about 300 species.

Key to the Genera of Nyctaginaceae

1. Plants herbaceous; flowers bisexual .... 1. *Boerhavia*
1. Plants woody (trees, shrubs, or lianas); flowers unisexual.
   2. Shrub 1–3 m tall; leaves 0.5–2.5 cm long, oblanceolate to spatulate .... 3. *Neea*
   2. Trees or lianas, to over 5 m tall; leaves >2.5 cm long, variously shaped but not as above.
      3. Fruits not fleshy, covered with stipitate glands .... 4. *Pisonia*
      3. Fruit fleshy, not glandular .... 2. *Guapira*

1. BOERHAVIA L.

Spreading herbs, many-branched from a taproot; stems prostrate to erect. Leaves opposite or subopposite, with prominent linear raphides. Flowers small, in axillary or terminal panicles or pedunculate heads; bracts and bracteoles hyaline; calyx constricted above the ovary, the limb or upper portion (corolla-like) 5-lobed, variously colored; stamens 1–3, the filaments free; stigma capitate. Anthocarp dry, club-shaped, cuneiform or fusiform, 5-ribbed.

About 30 species, with pantropical distribution.

Key to the Species of *Boerhavia*

1. Anthocarps glabrous, pedicellate; flowers white or pinkish-tinged.
   2. Plant erect to decumbent; anthocarps 3.2–3.5 mm long, 5-ribbed .... 3. *B. erecta*
   2. Plant with elongate, climbing fragile stems; anthocarps 12–13 mm long, 10-striate when mature, glandular toward the apex .... 4. *B. scandens*
1. Anthocarps glandular-pubescent, sessile; flowers red to violet.
   3. Inflorescences <5 cm long, with puberulent axes; flowers 5–12 in dense subcapitate clusters ...... 1. *B. coccinea*
   3. Inflorescence >10 cm long, with glabrous axes; flowers 2–4(–7) on long secondary axes .... 2. *B. diffusa*

1. Boerhavia coccinea Mill., Gard. Dict., ed. 8. 1768.

Decumbent herb, to 50 cm long, many-branched from a thickened base; twigs cylindrical, puberulent to pilose. Leaves in unequal pairs; blades 1.2–3.5 × 0.9–2.5 cm, ovate, lanceolate to nearly rounded, chartaceous, sparsely pilose, especially on veins, lower side glaucous, ciliate, the apex obtuse to acute, the base rounded to truncate, the margins wavy; petioles pilose, 2–15 mm long. Flowers nearly sessile, 5–12 in terminal, subcapitate clusters on short, axillary racemes; axis 1–5 cm long, puberulent; bracts subulate, longer than the calyx base. Calyx base 0.5–1.5 mm long, puberulent, the limb funnel-shaped, red or violet-red, 0.6–1 mm long; stamens usually 3, slightly exserted. Anthocarp sessile, green, glandular-pubescent, sticky, short club-shaped, 2.5–3 mm long, 5-ribbed.

DISTRIBUTION: A common roadside weed. Emmaus (A2004, A2821), Reef Bay Quarter at Lameshur (B643). Also on St. Croix and St. Thomas; a pantropical weed.

2. Boerhavia diffusa Mill., Gard. Dict., ed. 8. 1768.

Boerhavia paniculata Rich., Actes Soc. Hist. Nat. Paris **1**: 105. 1792.

Prostrate or ascending herb, to 50 cm long, many-branched from a taproot; twigs cylindrical, glabrous. Leaves in unequal pairs; blades 1.2–5.5 × 1.3–4 cm, ovate to wide ovate, chartaceous, sparsely pilose, especially on veins, lower side glaucous, the apex rounded to acute, shortly apiculate, the base rounded, truncate to nearly cordate, the margins wavy, ciliate; petioles pilose, 0.5–3 cm long. Flowers nearly sessile, 2–4(–7) in terminal, subcapitate clusters on axillary racemes or terminal panicles, 10–30 cm long; the axes glabrous; bracts and bracteoles lanceolate. Calyx base 0.5–1.5 mm, puberulent, the limb funnel-shaped, red or violet, 0.6–1 mm long; stamens usually 2, slightly exserted. Anthocarp sessile, green, glandular-pubescent, sticky, short club-shaped, 2–2.5 mm long, 5-ribbed.

DISTRIBUTION: A common weed of open areas, common along sandy beaches. Reef Bay (A2727). Also on St. Croix, St. Thomas, Tortola, and Virgin Gorda; a pantropical weed.

COMMON NAMES: batta batta, kallaloo bush.

3. Boerhavia erecta L., Sp. Pl. 3. 1753. Fig. 155A–F.

Erect to decumbent herb, to 50 cm tall, many-branched from a thickened base; twigs cylindrical, puberulent, reddish-tinged, collapsing when dried. Leaves in unequal pairs; blades 2.2–6 × 1.2–5 cm, ovate, wide-ovate, or triangular, chartaceous, puberu-

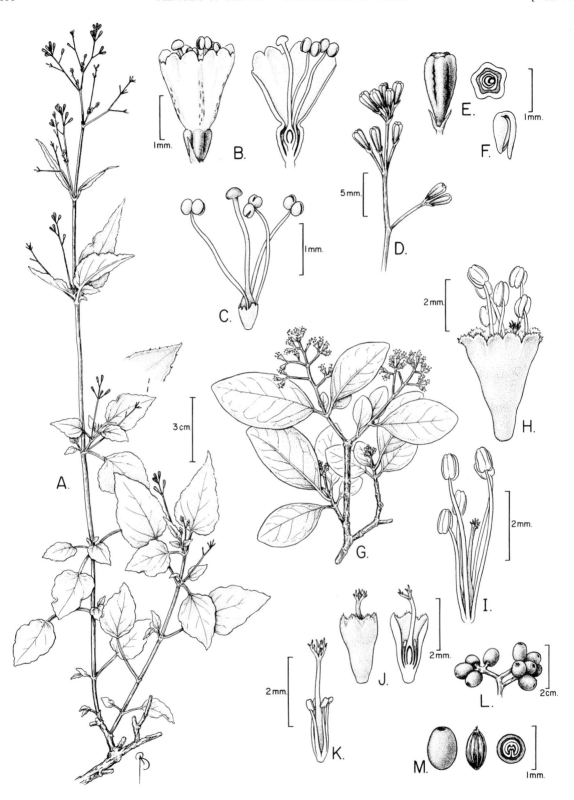

FIG. 155. A–F. *Boerhavia erecta*. **A.** Habit and detail of leaf margins. **B.** Flower and l.s. flower. **C.** Flower with perianth removed showing stamens and pistil. **D.** Infructescence branch. **E.** Anthocarp and c.s. anthocarp. **F.** Embryo. **G–M.** *Guapira fragrans*. **G.** Flowering branch. **H.** Staminate flower. **I.** Staminate flower with perianth removed showing stamens and pistillode. **J.** Pistillate flower and l.s. pistillate flower showing staminodes and pistil. **K.** Pistillate flower with perianth removed showing staminodes and pistil. **L.** Infructescence. **M.** Anthocarp, fruit, and c.s. anthocarp.

lent, lower side glaucous, the apex obtuse to acuminate, the base rounded, truncate to nearly cordate, the margins wavy, ciliate; petioles puberulent, 0.5–4 cm long. Flowers pedicellate, in terminal clusters on axillary racemes or terminal panicles to 30 cm long; the axes glabrous; bracts and bracteoles lanceolate. Calyx base 0.6–1.2 mm, glabrous, green, the limb funnel-shaped, white, pinkish-tinged, 0.6–1.2 mm long; stamens usually 3, slightly exserted, the filaments pink. Anthocarp pedicellate, turning from green to straw-colored, short club-shaped, 3.2–3.5 mm long, 5-ribbed, glabrous.

DISTRIBUTION: A common roadside weed. Great Cruz Bay (A2356), Emmaus (A2822), Fish Bay (A2812). Also on St. Croix, St. Thomas, Tortola, and Virgin Gorda; tropical America, West Africa, and Malaysia.

4. Boerhavia scandens L., Sp. Pl. 3. 1753. *Commicarpus scandens* (L.) Standl., Contr. U.S. Natl. Herb. **12:** 373. 1909.

Climbing herb, 1–2 m long, pseudodichotomously branching; twigs fragile, cylindrical, glabrous, greenish. Leaves in equal pairs; blades 2–5 × 0.8–4 cm, deltoid, ovate or wide-ovate, chartaceous, puberulent on lower side, the apex obtuse to acute, the base cordate to truncate, the margins wavy, ciliolate; petioles furrowed, 0.5–2.7 cm long. Flowers pedicellate, in axillary umbels; the axes glabrous, 5–12 cm long; bracts and bracteoles lanceolate. Calyx base 1–1.5 mm long, glabrous, green, the limb funnel-shaped, white, pinkish-tinged within, 2–2.5 mm long; stamens 2. Anthocarp pedicellate, dry, short club-shaped, 12–13 mm long, 10-striate, glabrous, turning from green to straw-colored, sticky because of apical glands.

DISTRIBUTION: A common roadside weed. Waterlemon Bay (A2520). Also on St. Croix, St. Thomas, Tortola, and Virgin Gorda; tropical to subtemperate America.

2. GUAPIRA Aubl.

Dioecious trees or shrubs. Leaves opposite or subopposite. Flowers unisexual, small, light-colored, in terminal corymbs; bracts and bracteoles minute, persistent; Staminate perianth bell-shaped; stamens 6–8, exserted, the filaments connate at base into a tube; pistillode reduced. Pistillate perianth tubular to bell-shaped; staminodes smaller than the ovary. Anthocarp fleshy without glands, the endocarp hard, striate.

About 40–50 species in tropical America.

1. Guapira fragrans (Dum. Cours.) Little, Phytologia **17:** 368. 1968. *Pisonia fragrans* Dum. Cours., Bot. Cult., ed. 2, **7:** 114. 1814. *Torrubia fragrans* (Dum. Cours.) Standl., Contr. U.S. Natl. Herb. **18:** 100. 1916.   Fig. 155G–M.

Tree or shrub 3–8(–17) m tall; bark tannish brown or grayish with numerous rounded, dark lenticels; twigs cylindrical, glabrescent, whitish gray. Leaf blades 3–14.5 × 1.7–7 cm, elliptic to nearly rounded, coriaceous, glabrous, involute, usually with insect galls, the apex obtuse, acute, or rounded, the base attenuate, the margins wavy; petioles 0.5–1.2 cm long. Flowers sessile, in

terminal corymbs; the axes puberulent, reddish-tinged; bracts and bracteoles minute, triangular. Staminate perianth 4–4.8 mm long, yellowish green, the stamens exserted; pistillate perianth 4–4.5 mm long, tubular, constricted just below the apex, the stigma many-branched, exserted. Anthocarp fleshy, ellipsoid, 10–12 mm long, smooth, glabrous, turning from green to purplish black, shiny.

DISTRIBUTION: A common tree of dry to moist forests. Bordeaux (P29306), Lind Point (A2695). Also on Jost van Dyke, St. Croix, St. Thomas, Tortola, and Virgin Gorda; the Greater Antilles, the Lesser Antilles, and northern South America.

COMMON NAMES: black mampoo, wild mampoo.

3. NEEA Ruíz & Pav.

Dioecious shrubs or trees. Leaves opposite, verticillate or rarely alternate. Flowers unisexual, usually with abortive organs of other sex, small, in axillary or terminal cymes; bracts and bracteoles minute; staminate perianth urceolate, globose, or elongate, 4–5-toothed; stamens 5–10, included, the filaments unequal, inserted at the base of the perianth; pistillode reduced; pistillate perianth urceolate or tubular, 4–5 toothed, sterile stamens inserted at base of ovary; style included, the stigma many-branched. Anthocarp fleshy without glands, the endocarp hard, striate.

About 70 species, from tropical America.

1. Neea buxifolia (Hook.f.) Heimerl, Bot. Jahrb. Syst. **21:** 633. 1896. *Eggersia buxifolia* Hook.f., Icon. Pl. **15:** 1. 1883.   Fig. 156H–K.

Shrub 1–3 m tall, many-branched from base, the branches arched; twigs cylindrical, glabrescent, whitish gray, in cross section showing xylem with internal strands of phloem. Shoots ferruginous-tomentulose. Leaves congested on short, lateral shoots; blades 0.5–2.5 × 0.3–0.9 cm, oblanceolate to spatulate, chartaceous to coriaceous, puberulent, the apex rounded, the base cune-

ate, the margins entire, slightly revolute; petioles slender, 0.5–1 mm long. Flowers sessile, 1 or 2 at end of short cymes; the axes 5–6 mm long, ferruginous-puberulent. Staminate perianth 3.5–4 mm long, yellowish, the stamens 6–8; pistillate perianth ca. 3 mm long, constricted below the apex. Anthocarp fleshy, ellipsoid-globose, ca. 5 mm long, smooth, glabrescent, shiny red, crowned by the perianth teeth.

DISTRIBUTION: An uncommon shrub of dry to moist forests. Cruz Bay Quarter along trail to Margaret Hill (A2305) and Maria Bluff area (A4706), Reef Bay Quarter at Europa Bay (A4143). Also on St. Thomas and Guana Island; Puerto Rico, including Culebra.

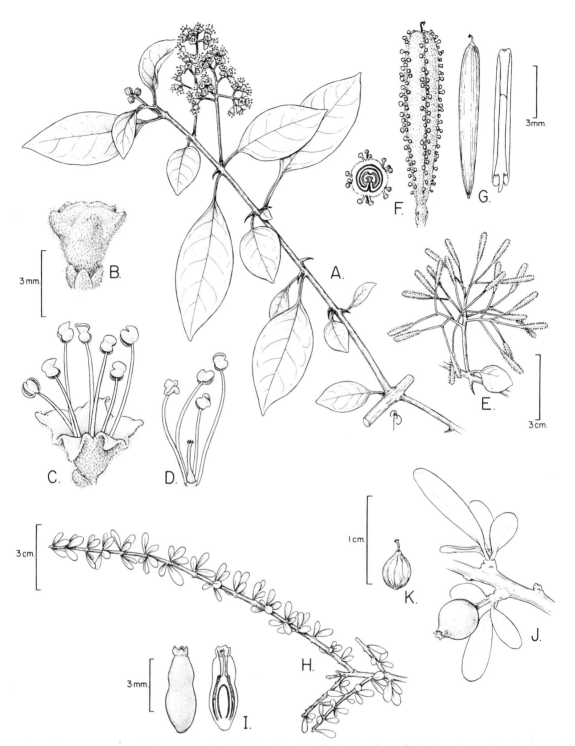

FIG. 156. A–G. *Pisonia aculeata*. **A.** Flowering branch. **B.** Staminate flowerbud. **C.** Staminate flower. **D.** Staminate flower with perianth removed showing stamens and pistillode. **E.** Infructescence. **F.** C.s. anthocarp and anthocarp. **G.** Fruit and embryo. **H–K.** *Neea buxifolia*. **H.** Leafy branch. **I.** Immature anthocarp and l.s. immature anthocarp. **J.** Fruiting branch. **K.** Fruit.

4. PISONIA L.

Dioecious or monoecious trees, shrubs, or lianas. Leaves opposite or subopposite, sometimes with axillary spines. Flowers unisexual, small, light-colored, in axillary or terminal panicles, corymbs, or cymes; bracts and bracteoles minute, persistent. Staminate perianth bell-shaped to obconical; stamens 6–8, long-exserted, the filaments unequal, connate at base. Pistillate perianth tubular, 5-

toothed; ovary ovoid, sessile, the style exserted, the stigma many-branched. Anthocarp dry, club-shaped to ellipsoid, 5-ribbed, with stipitate, viscid glands.

About 10–15 species from tropical areas. The number of species is considered to be between 35 and 75 by workers who lump the genus *Guapira* into *Pisonia*.

Key to the Species of *Pisonia*

1. Liana, with recurved spines on trunk and lateral branches; branches 4-angled, divaricate; leaves usually elliptic .... 1. *P. aculeata*
1. Tree, without spines; branches nearly cylindrical; leaves wide-ovate to rounded .... 2. *P. subcordata*

1. Pisonia aculeata L., Sp. Pl. 1026. 1753.

Fig. 156A–G.

Pisonia helleri Standl., N. Amer. Fl. **21:** 187. 1918.

Dioecious liana or climbing shrub to 20 m long, many-branched from base; trunk to 7 cm diam., with persistent recurved spines; bark grayish, smooth; branches divaricate, puberulent, and nearly cylindric, becoming glabrous, 4-angled and sulcate, developing recurved axillary spines. Leaves opposite on long, arching branches or congested on short, lateral shoots; blades 6–14 × 3–6 cm, elliptic, ovate, oblong, or rounded, chartaceous, glabrous, the apex obtuse, acute, rounded or shortly acuminate, the base cuneate, obtuse to rounded, the margins slightly wavy; petioles 1–3 cm long. Flowers sessile, clustered in axillary cymes; the axes pubescent; bracts and bracteoles minute, triangular. Staminate perianth nearly bell-shaped, 2.5–3 mm, yellowish, pubescent without, the stamens exserted, with filaments twice as long as the perianth; pistillate perianth ca. 2.5 mm long, nearly tubular, tomentose without, constricted just below the apex, 5-toothed, the stigma many-branched, exserted. Anthocarp ellipsoid, obtusely 5-angled, 10–12 mm long, tomentose, with a row or two of stipitate viscid glands along each angle.

DISTRIBUTION: An occasional liana of thickets or dry to humid forest. Lameshur floodplain (W760). Also on St. Croix and St. Thomas; widespread throughout the neotropics.

COMMON NAME: prickly mampoo.

2. Pisonia subcordata Sw., Prodr. 60. 1788.

Pisonia nigricans H. West, Bidr. Beskr. Ste. Croix 312. 1793.

Dioecious small tree or shrub, 2–8 m tall; bark grayish, smooth; twigs rusty-tomentose and flattened, becoming glabrous, cylindric and grayish, without spines. Leaves opposite; blades 2.5–12 × 2–7.5 cm, ovate to nearly rounded, chartaceous, glabrous, the apex obtuse to rounded, the base truncate, rounded or nearly cordate, the margins wavy; petioles (0.5–)1–3 cm long. Flowers sessile, clustered in terminal corymbs; the axes puberulent, to 8 cm long; bracts and bracteoles minute, subulate. Staminate perianth narrowly bell-shaped, 2.5–3.5 mm, pubescent without, the stamens exserted; pistillate perianth ca. 2.5 mm long, nearly tubular, puberulent without, pinkish-tinged, constricted just below the apex, 5-toothed, the stigma many-branched, exserted. Anthocarp cylindrical, 8–10 mm long, puberulent, with 5 longitudinal rows of stipitate viscid glands.

DISTRIBUTION: A common tree of dry forest and thickets. Denis Bay (M16586), East End Point (A657). Also on Anegada, St. Croix, St. Thomas, Tortola, and Virgin Gorda; Jamaica, Puerto Rico, including Vieques and Culebrita, and the Lesser Antilles.

COMMON NAMES: loblolly, mampoo, water mampoo.

CULTIVATED SPECIES: *Bouganvillea glabra* Choisy and *Bouganvillea spectabilis* Willd. are occasionally cultivated on the island.

58. Ochnaceae (Ochna Family)

Trees, shrubs, or rarely herbs. Leaves alternate, simple; stipules present. Flowers small to large, actinomorphic, bisexual, mostly produced in terminal panicles; calyx of 5 distinct, equal or unequal sepals; corolla of 5 distinct petals; stamens 5–10, the filaments usually borne on an androgynophore, the anthers opening by terminal pores or less often by longitudinal slits; ovary superior of 2–5 (–15) carpels, these united at least by their common gynobasic styles, the placentation axile, ovule 1 to many per carpel, the styles connate, gynobasic. Fruit usually of separate 1-seeded drupes on an enlarged, fleshy receptacle, or seldom a septicidal capsule.

A tropical and subtropical family with 30 genera and about 400 species.

1. OURATEA Aubl., nom. cons.

Glabrous shrubs or small trees. Leaves alternate, coriaceous, with secondary veins percurrent along margins; stipules minute, entire. Flowers medium-sized, in axillary or terminal racemes or panicles; bracts and bracteoles minute, persistent; calyx of 5 unequal sepals; corolla of 5 distinct petals; stamens 10, the anthers nearly sessile, elongate, opening by 2 apical pores; ovary of 5 carpels, deeply lobed, the style single, gynobasic, the locules with a single ovule. Fruit an aggregate of 1-seeded drupes (monocarps) borne on a fleshy enlarged receptacle.

About 200 species from tropical America and Africa.

1. Ouratea litoralis Urb., Symb. Ant. **1:** 363. 1899.

Fig. 157.

Shrub 2.5–7 m tall, few-branched from base; bark grayish, flaky. Leaf blades 4–10.5 × 2.1–4.2 cm, elliptic, coriaceous, glabrous, the midvein stout, the apex obtuse or acute, mucronulate, the base obtuse or acute, the margins serrulate or entire; petioles

4–6 mm long, slightly swollen at base. Flowers in axillary or terminal racemes or panicles; the axes angled, 5–10 cm long; bracts and bracteoles ovate; pedicels 3–5 mm long. Sepals green, elliptic, 5–6 mm long, concave; petals bright yellow, obovate, 7–8 mm, shortly keeled; anthers 3–4 mm long, nearly lanceolate; ovary 5-lobed, green, the style yellowish, 5–6 mm long, recurved, the stigma punctiform. Monocarp 5–8 mm long, ellipsoid to

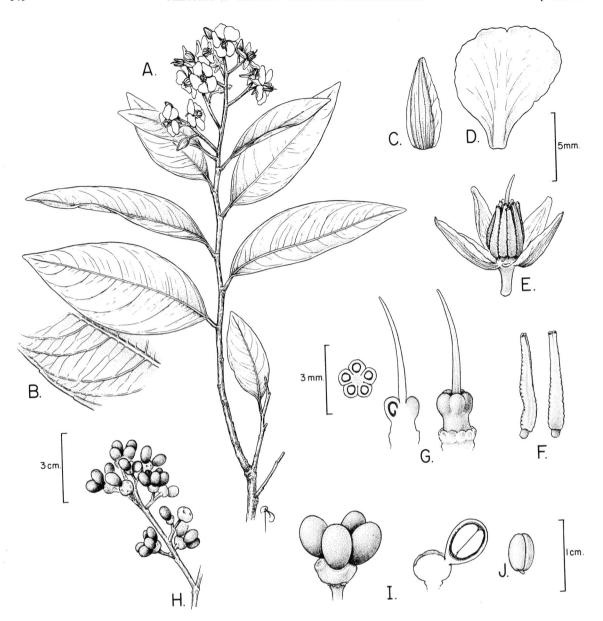

Fig. 157. *Ouratea litoralis.* **A.** Flowering branch. **B.** Detail of leaf margin. **C.** Sepal. **D.** Petal. **E.** Flower with petals removed showing anthers surrounding pistil. **F.** Stamen, frontal and lateral views. **G.** C.s. ovary, l.s. pistil, and pistil. **H.** Infructescence. **I.** Fruit monocarps on enlarged receptacle and l.s. fruit and receptacle. **J.** Embryo.

obovate, fleshy, turning from green to burgundy, finally purple; receptacle bright red.

DISTRIBUTION: An occasional shrub of moist forests. Bordeaux (A4675, A5215). Also on St. Thomas and Virgin Gorda; Puerto Rico.

59. **Olacaceae** (Olax Family)

Trees, shrubs, or woody vines, often hemiparasitic on roots of host plant. Leaves alternate, simple, entire; stipules wanting. Flowers minute, actinomorphic, bisexual or rarely unisexual, produced in axillary racemes, panicles, or fascicles; calyx minute, cup-shaped, 3–6-toothed, usually accrescent in fruit; corolla of 3–6 connate or distinct petals; stamens as many as and opposite the petals or 2–5 times as many as the petals, the filaments free or connate at base, sometimes connate into a sheath around the style, the anthers opening by longitudinal slits or less often by terminal valves; ovary superior, 3–5-locular at base, unilocular at apex, the ovules 1 per locule, pendulous, the style terminal with a capitate stigma. Fruit usually a 1-seeded berry or drupe, covered with the accrescent calyx.

A tropical and subtropical family of about 30 genera and 250 species.
REFERENCE: Sleumer, H. O. 1984. Olacaceae. Fl. Neotrop. Monogr. 38: 1–160.

Key to the Genera of of Olacaceae

1. Unarmed shrubs or trees; leaves obovate to elliptic; bracteoles connate into an epicalyx; petals connate into a bell-corolla, with a tuft of hairs at the point of attachment of anthers; drupes ovoid to ellipsoid .... 1. *Schoepfia*
1. Spiny shrubs or trees; leaves elliptic to ovate; bracteoles not forming an epicalyx; petals free to base, densely pubescent throughout most of their inner surface; drupes nearly globose .... 2. *Ximenia*

1. SCHOEPFIA Schreb.

Shrubs or trees. Leaves entire, usually drying blackish and fragile. Flowers bisexual, usually heterostylous, in axillary racemes or fascicles; bracteoles connate into an epicalyx; calyx cup-shaped, 4-lobed, adnate to the base of ovary; corolla tubular, bell-shaped to urn-shaped, 4–6-lobed; stamens as many as the corolla lobes, the filaments short, adnate to the corolla tube; ovary partially immersed in the disk, 3-locular, the locules usually with 1 ovule, the style slender, short, the stigma capitate to 3-lobed. Fruit a 1-seeded drupe, subtended by a cup-shaped epicalyx and crowned by an annular calyx scar.

A genus of 23 species, 19 of which are native to tropical America; the others are native to the Old World tropics.

Key to the Species of *Schoepfia*

1. Shrub or small tree to 5 m tall; leaves obovate to elliptic, 2–4 cm long, with obtuse apex .... 1. *S. obovata*
1. Tree to 7 m tall; leaves ovate to lanceolate, 4–8 cm long, with acute to acuminate apex .... 2. *S. schreberi*

1. Schoepfia obovata C. Wright in Sauv., Anales Acad. Ci. Méd. Habana **5**: 289. 1868.

Shrub or small tree 2–7 m tall; bark grayish, rough, furrowed; twigs angled, minutely papillose, becoming grayish at age. Leaf blades 2–4 × 0.8–2.5 cm, obovate to rarely elliptic, coriaceous, glabrous, the apex obtuse, the base cuneate, the margins revolute; petioles ca. 2 mm long. Flowers sessile, 2–3 in axillary racemes; the axes 2–8 mm long. Calyx green, cup-shaped, minute; corolla reddish or greenish yellow, bell-shaped, 4–5 mm long, with 4 ovate, spreading lobes, the tube with a tuft of hairs at point of attachment of anthers within; stamens 4, included or emergent, the filament adnate to corolla tube; ovary 3–5-lobed, the style included or exserted, the stigma nearly capitate. Drupe ovoid to ellipsoid, 6–8 mm long, yellow to red when mature.

DISTRIBUTION: A rare shrub in coastal thickets. Lameshur (W513). Also on Anegada; Bahamas, Puerto Rico, Cuba, Jamaica, and Hispaniola.

2. Schoepfia schreberi J.F. Gmel., Syst. Nat. **2**: 376. 1791. Fig. 158A–G.

Codonium arborescens Vahl, Skr. Naturhist.-Selsk. **2**: 207. 1792. *Schoepfia arborescens* (Vahl) Schult. *in* Roem. & Schult., Syst. Veg. **5**: 160. 1819.

Shrub or small tree 2–7 m tall; bark grayish, deeply furrowed; twigs glabrous, furrowed, minutely papillose, becoming grayish at age. Leaf blades 4–8 × 1.5–3.5 cm, ovate to lanceolate, folded-falcate, coriaceous, glabrous, the apex acute to acuminate, the base rounded, obtuse, or cuneate, the margins entire; petioles 3–5 mm long. Flowers sessile, 1–3 in short, axillary racemes. Calyx green, cup-shaped, minute; corolla bell-shaped, 4–4.5 mm long, the tube yellow, with a tuft of hairs at point of attachment of anthers within, the lobes 4–5, triangular-ovate, reddish; stamens 4–5, included; ovary rugose, the style included, the stigma bifid. Drupe ellipsoid, 10–13 mm long, orange, red, or purple when mature.

DISTRIBUTION: An uncommon shrub of coastal thickets and dry forests. Bordeaux (W617), Christ of the Caribbean (A2918), trail to Reef Bay (A2670). Also on St. Croix and St. Thomas; Puerto Rico (including Vieques), throughout the Caribbean region and Mexico to northern South America.

2. XIMENIA L.

Shrubs or trees, with axillary thorns. Leaves usually coriaceous, drying blackish, entire. Flowers bisexual, heterostylous, in axillary racemes, panicles, or fascicles; bracteoles not forming a cup-shaped epicalyx; calyx cup-shaped, 3–5-toothed; corolla of 4–5 free petals, pubescent within; stamens 8–10, in 2 whorls; ovary superior, 4-locular, the ovules solitary in each locule, pendulous, the style elongate, the stigma capitate. Fruit a 1-seeded, nearly round drupe.

A tropical genus of 8 species.

1. Ximenia americana L., Sp. Pl. 1193. 1753. Fig. 158H–M.

Shrub or small tree 2–7 m tall; bark reddish brown; twigs glabrous, furrowed, minutely papillose, becoming grayish at age. Leaf blades 2–6 × 1.2–3 cm, elliptic to ovate, chartaceous to coriaceous, glabrous, the apex retuse and mucronulate, the base obtuse to cuneate, the margins slightly revolute; petioles 5–7 mm long. Flowers in short, axillary, umbellate or fasciculate racemes. Calyx green, cup-shaped, minute, 3–4-toothed; petals white, linear, 7–11 mm long, pubescent within; stamens 8; ovary ovoid. Drupe ellipsoid to nearly globose, 2–3 cm long, yellow when mature.

DISTRIBUTION: An uncommon shrub of coastal thickets. Drunk Bay (W541). Also throughout tropical America and Africa.

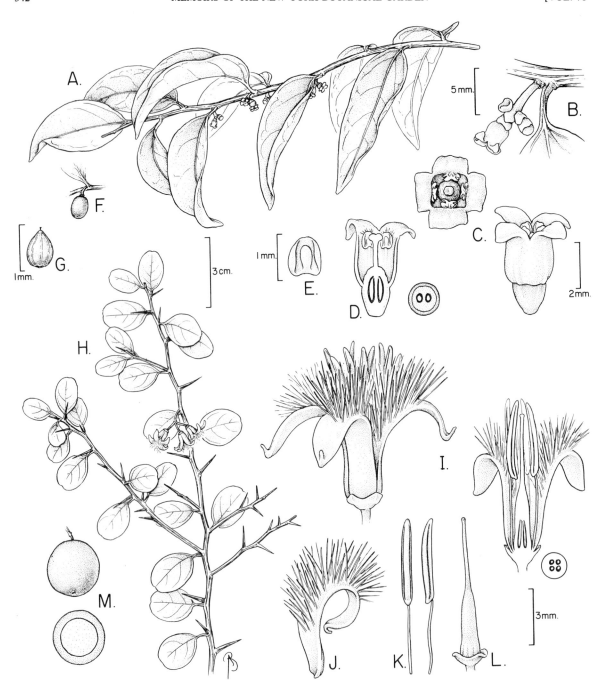

FIG. 158. A–G. *Schoepfia schreberi*. **A.** Flowering branch. **B.** Inflorescence. **C.** Flower, top and lateral views. **D.** L.s. flower and c.s. pistil. **E.** Anther. **F.** Detail of stem node showing fruit. **G.** Stone. **H–M.** *Ximenia americana*. **H.** Flowering branch. **I.** Flower, l.s. flower, and c.s. ovary. **J.** Petal. **K.** Stamen, dorsal and lateral views. **L.** Pistil. **M.** Drupe and c.s. drupe.

60. Oleaceae (Olive Family)

By S. LaGreca

Trees, shrubs (sometimes scandent), or lianas. Leaves opposite or rarely alternate, simple or pinnately compound; stipules wanting. Flowers small to medium, actinomorphic, bisexual or seldom unisexual, in axillary or terminal racemes, panicles, or cymes; calyx gamosepalous, usually 4-lobed, rarely wanting; corolla of 4 petals, these free or connate into a short to elongate tube, rarely wanting;

stamens 2(–4), the anthers opening via longitudinal slits; nectary disks usually absent; ovary superior, 2-locular, the locules with 2 to several ovules, the placentation axile, the style terminal or wanting, the stigma 2-lobed. Fruit a capsule, samara, berry, or drupe, usually 1-seeded.

A cosmopolitan family of some 24 genera and 900 species, best developed in Asia and Malaysia.

REFERENCE: Wilson, K. A. & C. E. Wood Jr. 1959. The genera of Oleaceae in the southeastern United States. J. Arnold Arbor. 40: 369–384.

Key to the Genera of Oleaceae

1. Corolla salverform.. 3. *Jasminum*
1. Corolla of 4 nearly distinct petals, or wanting.
   2. Corolla of 4 white, narrowly lanceolate petals; flowers borne in loose, drooping, axillary panicles
   .... 1. *Chionanthus*
   2. Corolla wanting; flowers borne in short (5–7 mm long) axillary racemes .... 2. *Forestiera*

1. CHIONANTHUS L.

Shrubs or trees. Leaves opposite, simple, entire. Flowers bisexual, in axillary or rarely terminal panicles, racemes, or fascicles; calyx small, 4(–5)-parted or -toothed; corolla of 4 free or nearly free, narrow, elongate petals; stamens 2(–4), the filaments short, free or adnate to the base of the corolla, the anthers lineate or ovate; ovary 2-locular, the locules with 2 ovules each, the style short, the stigma 2-lobed. Fruit an ovoid, oblong, or subglobose drupe.

A genus of about 120 species with pantropical distribution, best developed in Southeast Asia and southeastern North America, with a few species extending into the warm-temperate zone.

1. **Chionanthus compacta** Sw., Prodr. 13. 1788. *Linociera compacta* (Sw.) Roem. & Schult., Syst. Veg. **1**: 266. 1817. Fig. 159A–F.

Chionanthus caribaea Jacq., Collectanea **2**: 110, t. 6, fig. 1. 1788 [1789]. *Mayepea caribaea* (Jacq.) Kuntze, Revis. Gen. Pl. **2**: 411. 1891. *Linociera caribaea* (Jacq.) Knobl., Bot. Centralbl. **61**: 87. 1895.

Shrub to small tree, 5–10 m tall; bark light gray, smooth; young twigs appressed, yellowish-pubescent. Leaf blades 8–16.5 × 3.5–6 cm, oblong-elliptic to lanceolate, glabrous, the apex acuminate, the base acute, the margins slightly revolute; petioles appressed-pubescent, 10–15 mm long. Flowers sessile, fragrant, in axillary panicles; the axes appressed-pubescent, as long as the subtending leaf; bracteoles densely appressed-pubescent; pedicels to 1.2 mm long, densely appressed-pubescent. Calyx green, appressed pubescent, cup-shaped, 4(–5)-parted, the lobes bluntly ovate; corolla white, 8–15 mm long, of 2 pairs of narrowly lanceolate, basally connate petals with strongly involute margins; stamens 2, each one basally adnate to one of the two pairs of petals, the filaments very short, with a pronounced, apiculate connective extending beyond the anther, the anthers elongate; ovary green. Fruit ovoid, blue-black, 1.2–2.5 cm long, with white punctations.

DISTRIBUTION: Frequent in moist forests. Bordeaux Road (A1896), Susannaberg (A2800), Maho Bay Quarter along North Shore Road toward Annaberg Ruins (A1931). Also on Jost van Dyke, St. Croix, and St. Thomas; Puerto Rico (including Culebra and Vieques), Hispaniola, the Lesser Antilles, Trinidad, Tobago, Venezuela (including Margarita Island), and Colombia.

2. FORESTIERA Poir.

Dioecious or polygamodioecious shrubs or small trees. Leaves opposite, simple, short-petiolate, with entire or serrate margins. Flowers apetalous, evanescent, in short, bracteate, axillary racemes or fascicle-like clusters; calyx cup-shaped and parted into 4–8 (often irregular) lobes; corolla wanting; stamens 2 or 4, the anthers shorter than the filaments; abortive stamens 0–4; nectary disk absent; ovary with a short style and a 2-lobed or simple stigma. Fruit a small drupe.

A genus of 15 species ranging from the southern United States through Central America and the West Indies to (reportedly) South America.

1. **Forestiera eggersiana** Krug & Urb., Bot. Jahrb. Syst. **15**: 339. 1893. Fig. 159L–O.

Dioecious shrub or small tree 1.5–5 m tall; bark gray to reddish brown; twigs puberulent. Leaf blades 1.7–4.0 × 0.8–1.7 cm, elliptic, ovate, or lanceolate, chartaceous, glabrous, the upper surface bright green, the lower surface pale, the apex bluntly acute to obtuse, the base attenuate, the margins slightly revolute, crenulate toward the distal portion; petioles 2–3 mm long, glabrous. Flowers in clusters of 5–9, in axillary racemes; axes 5–7 mm long; bracteoles numerous, green, imbricate; pedicels with a fimbriolate bractlet each. Calyx green, cup-shaped, parted into 4–8 usually lanceolate (although often irregular and sometimes bifid or trifid) lobes with sharp apices; stamens 4, green, usually alternate with the calyx lobes, the anthers 4-lobed; sometimes abortive stamens present; ovary green, the stigma 2-lobed and papillose. Drupe, 1-seeded, green, narrowly ellipsoid, 8–13 mm long.

DISTRIBUTION: A common shrub of coastal thickets and dry forests. Mary Point (A2343), trail from Lameshur to Europa Bay (A3194). Also on St. Thomas and Virgin Gorda; Puerto Rico (including Culebra and Vieques) and St. Barthelemy.

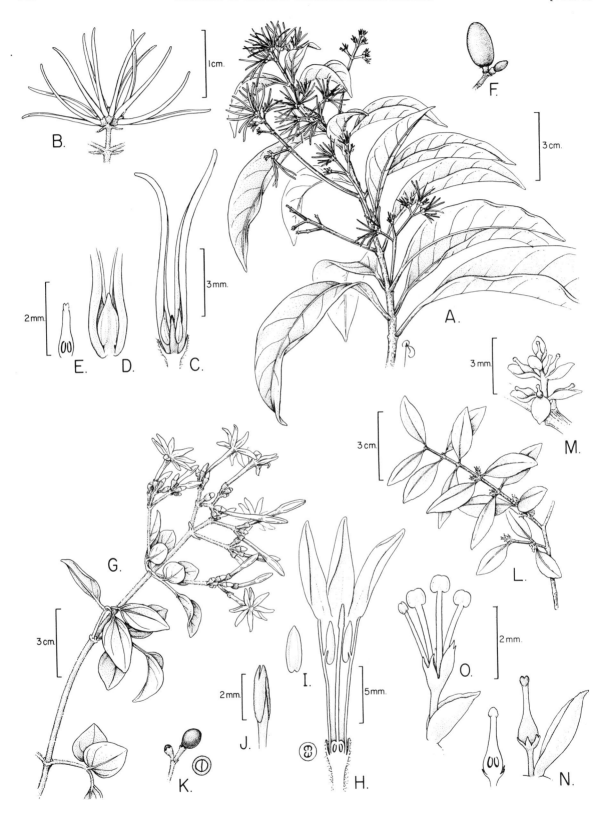

Fig. 159. A–F. *Chionanthus compacta*. **A.** Flowering branch. **B.** Portion of inflorescence. **C.** L.s. flower. **D.** Detail of base of petals showing one adnate anther. **E.** L.s. pistil. **F.** Fruit. G–K. *Jasminum fluminense*. **G.** Flowering branch. **H.** L.s. flowers and c.s. ovary. **I.** Anther. **J.** Detail of stigma. **K.** Fruit and c.s. fruit. L–O. *Forestiera eggersiana*. **L.** Flowering branch. **M.** Pistillate inflorescence. **N.** Pistillate flower and l.s. pistillate flower. **O.** Staminate flower with subtending bracteole.

3. JASMINUM L.

Climbing or erect shrubs or twining lianas. Leaves opposite or less often alternate, simple or pinnately compound. Flowers bisexual, often fragrant and showy, solitary, clustered, or in terminal or axillary cymes or panicles; calyx bell-shaped or funnelform, 4–9-lobed or toothed; corolla salverform, 4–9-lobed, the lobes slightly imbricate, the tube cylindrical; stamens 2, included in the corolla tube, the filaments short and adnate to it, the anthers 2-lobed; ovary with a long and slender style, the stigma usually 2-lobed or rarely simple. Fruit a small berry.

A genus of about 450 species native to eastern and southern Asia, Malaysia, Africa, and Australia, many of which are now naturalized in the neotropics.

REFERENCE: Green, P. S. 1965. Studies in the genus *Jasminum* L.: IV—the so-called New World species. Kew Bull. 23: 273–275.

Key to the Species of *Jasminum*

1. Leaves trifoliate; calyx lobes sharply acuminate; anther lobe apices narrowed............................ 1. *J. fluminense*
1. Leaves simple; calyx lobes linear, narrow, elongate, and erect; anther lobe apices blunt.............. 2. *J. multiflorum*

1. Jasminum fluminense Vell., Fl. Flumin. 10. 1825 [1829].   Fig. 159G–K.

Jasminum azoricum sensu authors, non L., 1753.

Twining woody vine to 6 m long, many-branched along main stem; twigs cylindric, glabrous to tomentose. Leaves opposite, trifoliolate, 5–10 cm long; leaflets 2–5 × 2–3.5 cm (terminal leaflets larger than the basal ones), broadly ovate, the upper surface puberulent, the lower surface with tufts of hairs in vein angles, the apex acute to acuminate, the base subtruncate or obtuse, the margins entire; petiolules and petioles tomentose, the petioles 0.5–2 cm long. Flowers in terminal or axillary cymes; axes densely tomentose; bracteoles elliptic-ovate, involute, sharp-tipped, densely tomentose; pedicels 3–4 mm long. Calyx green, funnelform, sparsely tomentose, 4–9-lobed, the lobes sharply acuminate; corolla white, salverform, 1.5–2.5 cm long, 4–9-lobed, the lobes narrowly oblong with bluntly acute tips; stamens 2, the filaments with an apiculate connective extending beyond the anther, the anthers elongate, the anther lobe apices narrowed; ovary 4-lobed, the style borne on raised swollen tissue, the stigma elongate, 2-lobed. Berry purple- or blue-black, shiny, globose, 5–8 mm diam.

DISTRIBUTION: A naturalized species, occasional in open, disturbed sites. Road to Bordeaux (A2892, A3839). A native of tropical Africa, but this plant was first described from Brazil and is now naturalized throughout the tropics.

2. Jasminum multiflorum (Burm.f.) Andrews, Bot. Repos. **8**: pt. 102, t. 496. 1807. *Nyctanthes multiflora* Burm.f., Fl. Indica 5, t. 3, fig. 1. 1768.

Nyctanthes pubescens Retz., Observ. Bot. **5**: 9. 1788. *Jasminum pubescens* (Retz.) Willd., Sp. Pl. **1**: 37. 1797.

Climbing shrub 2–5(–10) m long; stems slender, divaricate, tomentose. Leaves opposite, simple; blades 3–7 × 2–3.5 cm, ovate, abaxially tomentose, especially on veins, the apex acute, the base subtruncate to cordate, the margins entire; petioles densely tomentose, 6–12 mm long. Flowers fragrant and showy, clustered in short, axillary cymes; the axes densely tomentose; peduncle ca. 4 mm long, subtended by 2 small, green, ovate, densely tomentose bracts; pedicels 0–3 cm long, each one subtended by an ovate, densely tomentose bracteole. Calyx green, funnelform, about 1 cm long, densely tomentose, dissected into numerous linear, narrow, elongate, erect lobes; corolla white, salverform, about 2 cm long, 4–9-lobed, the lobes narrowly oblong with sharply acute or acuminate tips; stamens 2, the filaments with an apiculate connective extending beyond the anthers, the anthers elongate, the anther lobe apices blunt; ovary 4-lobed, the style borne on raised swollen tissue, the stigma elongate and 2-lobed. Fruit not observed on St. John population, but ovoid, 5–7 mm long, in specimens collected in Asia.

DISTRIBUTION: A naturalized species, occasional in open, disturbed sites. Road to Bordeaux (A4672), Maho Bay Quarter along Center Line Road (A3854). Native to Asia but widely cultivated throughout the tropics.

COMMON NAMES: hairy jasmine, star jessamine.

61. Onagraceae (Evening Primrose Family)

Herbs or seldom shrubs or trees. Leaves alternate, opposite, or whorled, simple, entire, serrate, or lobed; stipules minute to large, deciduous or wanting. Flowers large, actinomorphic or zygomorphic, bisexual, solitary in leaf axils or in axillary spikes, racemes, or panicles; calyx forming a hypanthium with 4 sepals at apex; corolla of 4 distinct, clawed petals; stamens twice as many as the petals, or seldom 4 or 2, the filaments borne on the hypanthium or surrounding the nectary disk that encircles the apical portion of the ovary, the anthers opening by longitudinal slits; ovary inferior, of (2–)4, united carpels, the placentation axile or parietal, the ovules 1 to many per carpel, the style terminal with a capitate to 4-lobed stigma. Fruit a loculicidal capsule, a berry, or a nut.

A family with 17 genera and about 675 species, mainly of temperate and subtropical regions.

1. LUDWIGIA L.

Erect or prostrate herbs or subshrubs. Leaves alternate or opposite, entire; stipules wanting or minute. Flowers bisexual, regular, solitary or in axillary clusters, or rarely the inflorescence appearing raceme- or spikelike because leaves reduced upward to bracts; bracteoles wanting or when present leafy or reduced; hypanthium not prolonged beyond the ovary, crowned by 4–6 sepals; petals free, 4–6, or rarely wanting; stamens as many or twice as many as the petals, in 1 or 2 whorls; ovary partially inferior, 4–6-locular, the locules with numerous ovules in 1 or numerous rows, the style elongate, the stigma capitate. Fruit a capsule, with irregular dehiscence, or by a terminal pore; seeds numerous, small.

A genus of 75 species, with nearly cosmopolitan distribution, but principally in the New World.

REFERENCE: Raven, P. H. 1963. The Old World species of *Ludwigia* (including *Jussiaea*), with a synopsis of the genus (Onagraceae). Reinwardtia 6: 327–427.

1. Ludwigia octovalvis (Jacq.) Raven, Kew Bull. **15**: 476. 1962. *Oenothera octovalvis* Jacq., Enum. Syst. Pl. 19. 1760. *Jussiaea octovalvis* (Jacq.) Sw., Observ. Bot. 142. 1791. Fig. 160.

Erect herb or subshrub, 0.3–1 m tall; twigs puberulent, slightly angled, becoming glabrous and cylindrical with age. Leaves alternate; blades 2.5–11 × 0.7–2.2 cm, narrow-elliptic to linear, chartaceous, puberulent, usually with cystoliths, the apex obtuse or acute, the base attenuate, the margins entire, ciliate; petioles 3–5 mm long. Flowers solitary in leaf axils; pedicel 1–1.5 mm long, with a pair of opposite bracteoles at base of hypanthium.

Hypanthium club-shaped, 4-angled, 3–3.5 mm long, crowned by 4 green, ovate to lanceolate sepals, 5.5–10 mm long; petals 4, yellow, 7–12 mm long, with a few cystoliths, spreading, obovate with a shallowly emarginate apex; stamens 8, the epipetalous ones shorter, the filaments 1–4 mm long; disk pubescent; ovary 4-locular with numerous ovules, the style 1.5–3.5 mm long, the stigma nearly globose, 4-lobed. Capsule cylindrical, 2.7–4.5 cm long, turning from green to brown, with 8 longitudinal ribs, irregularly dehiscent, crowned by the persistent spreading sepals. Seeds numerous, rounded, brown, ca. 0.6 mm long.

DISTRIBUTION: A common weed of wet ditches along roadsides. Road to Bordeaux (A1890). Also on St. Croix, St. Thomas, and Tortola; widespread throughout the tropics.

62. Oxalidaceae (Wood Sorrel Family)

Herbs with tubers or bulbs, or rarely subshrubs or shrubs. Leaves alternate, trifoliolate, palmately or pinnately compound; stipules minute or wanting. Flowers medium-sized, actinomorphic, bisexual, produced in axillary or terminal many- to 1-flowered cymes; calyx of 5 distinct sepals; corolla of 5 distinct or basally connate petals; stamens usually 10(–15), the filaments connate at base into a tube, the anthers opening by longitudinal slits; ovary superior, of 3–5 carpels, the placentation axile, ovules (1–)2 to many per locule, the styles distinct, terminal with a capitate stigma. Fruit usually a loculicidal capsule.

A family with 7–8 genera and about 900 species, mainly tropical and subtropical but also in temperate zones.

1. OXALIS L.

Characters of the family. By far the largest genus of the family, with about 800 species, in tropical to temperate areas.

1. Oxalis corniculata L., Sp. Pl. 435. 1753. Fig. 161.

Decumbent or prostrate herb, to 50 cm long; main stem trailing, rooting at internodes, upright branches to 15 cm tall. Leaves trifoliolate; leaflets obcordate, 0.8–1.2 cm long, membranous, sparsely pilose; petioles 3–4.5 cm long; stipules minute, ciliate. Flowers 2–3 in axillary cymes; pedicels 1–2 cm long. Sepals elliptic, 3–4 mm long, pilose, yellowish green; petals bright yellow, spreading, oblanceolate, 7–8 mm long; stamens with

filaments of unequal length, 1.5–2.5 mm long; ovary nearly cylindrical, elongate, 5-locular, the styles 5, free to base, pilose, the stigmas capitate. Capsule ascending, 9–12 mm long, cylindrical, beaked at apex, longitudinally ridged, pilose, green, with persistent sepals, the peduncle reflexed. Seeds nearly elliptic, warty on inner side, ca. 1 mm long.

DISTRIBUTION: An occasional weed of humid, shaded areas. Road to Bordeaux (A3848), Herman Farm (A870). Also on Tortola; widespread throughout the world.

63. Papaveraceae (Poppy Family)

Herbs or shrubs. Leaves alternate, subopposite, or whorled, simple, entire or lobed; stipules wanting. Flowers large, actinomorphic, bisexual, terminal, solitary or less often in cymose or umbelliform inflorescences; calyx of 2(–4) sepals, distinct or basally connate, usually deciduous; petals twice as many as the sepals, seldom more numerous or wanting; stamens 4 to many, the filaments distinct, the anthers opening by longitudinal slits; ovary superior, unilocular, of 2 carpels, the placentation parietal with numerous ovules, the stigma sessile. Fruit a valvate or poricidal capsule.

A family with 25 genera and about 200 species, from temperate to tropical areas of the Northern Hemisphere.

1. ARGEMONE L.

Erect herbs or subshrubs, stems and leaf margins spiny, producing abundant cream-colored to yellowish sap. Leaves alternate, sessile. Flowers bisexual, actinomorphic, solitary, axillary, sessile; bracts and bracteoles present, foliaceous; calyx of 2–3 deciduous sepals; petals 4–6; stamens numerous, the anther longitudinally dehiscent; ovary 1-locular, with numerous ovules, the stigma nearly sessile, capitate. Fruit a capsule, with valvate dehiscence; seeds numerous, small.

A genus of 11 species, native to southern North America.

REFERENCE: Ownbey, G. B. 1958. Monograph of the genus *Argemone* for North America and the West Indies. Mem. Torrey Bot. Club 21(1): 1–159.

1. Argemone mexicana L., Sp. Pl. 508. 1753. Fig. 162.

Erect herb, 30–60 cm tall, with a woody taproot; latex yellowish; stems glaucous, sparsely spiny. Leaf blades 7–25 × 2–10 cm, deeply sinuate to dissected, chartaceous, glaucous underneath,

sparsely spiny on both surfaces, the venation whitish, the apex obtuse to acute, the base cuneate or attenuate, the margins sparsely spiny; petioles 8–10 mm long on basal leaves, the upper leaves sessile. Flowers solitary in leaf axils, sessile or nearly so, sub-

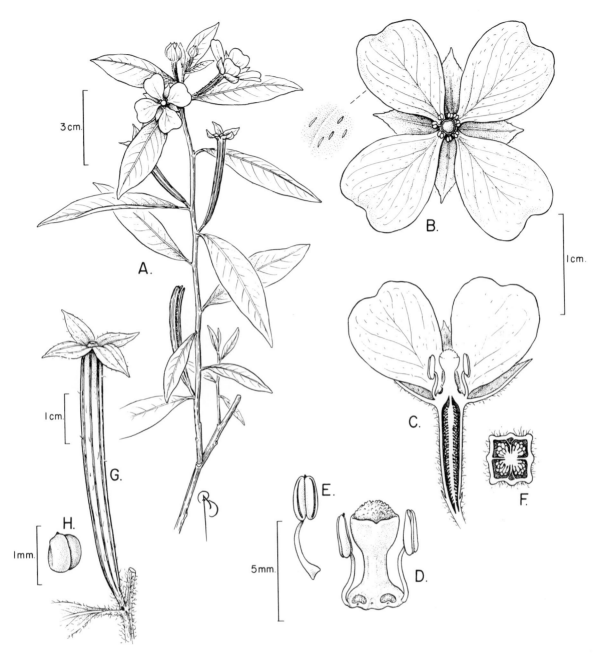

Fig. 160. *Ludwigia octovalvis*. **A.** Fertile branch. **B.** Flower from above and detail of petal surface. **C.** L.s. flower. **D.** Detail of stamens and stigma. **E.** Stamen. **F.** C.s. ovary. **G.** Capsule. **H.** Seed.

tended by a pair of foliaceous, spiny bracts; peduncle to 3.5 cm long. Sepals free to base, ovate, green, 1.8–2.2 cm long, horned at apex, early deciduous; petals yellow, broadly obovate, 2–3 cm long; stamens numerous, the filaments yellow, 7–8 mm long, the anthers lanceolate, recurved; ovary ovoid, with soft, ascending spines, green, the stigma capitate, 5-lobed, reddish. Capsule nearly cylindrical, slightly lobed, 2.7–4 cm long, turning from green to straw-colored, dehiscing from the apex. Seeds numerous, nearly elliptic, dark brown to blackish, ornamented, ca. 2 mm long.

DISTRIBUTION: A common weed of open disturbed areas. Long Bay (A672). Also on St. Croix, St. Thomas, and Tortola; widespread weed throughout the tropics.

COMMON NAMES: Mexican poppy, thistleroot, yellow-thistle.

64. **Passifloraceae** (Passion Fruit Family)

Herbaceous or woody vines climbing by axillary tendrils. Leaves alternate, simple, entire or palmately lobed; petioles often with extrafloral nectaries; stipules usually small and deciduous. Flowers large, bisexual or rarely unisexual, actinomorphic, with a saucer-shaped to tubular hypanthium, produced in axillary cymes or solitary in leaf axils, sometimes covered by an involucre of green

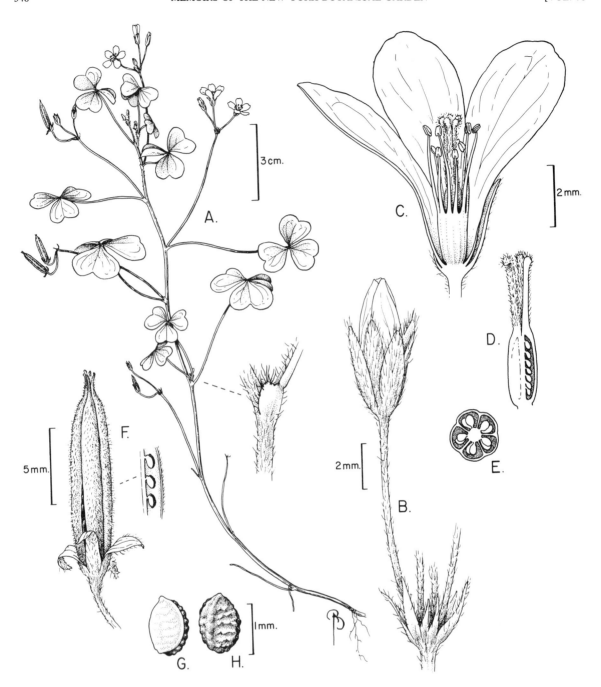

FIG. 161. *Oxalis corniculata*. **A.** Habit and detail of stipules. **B.** Portion of inflorescence with flowerbud. **C.** Flower with three sepals and two petals removed showing staminal tube and stigma. **D.** L.s. pistil. **E.** C.s. ovary. **F.** Capsule and detail of locule showing seeds. **G.** Frontal view of seed showing elastic testa. **H.** Dorsal view of seed.

foliaceous bracts; calyx of (3–)5(–8) distinct or basally connate sepals; petals as many as the sepals and alternating with them, seldom wanting; corona of 1 to several series of distinct or more or less united appendages; stamens (4–)5, the filaments distinct or raised on a stipe (an androgynophore), the anthers opening by longitudinal slits; ovary superior, usually on a gynophore or rarely sessile, unilocular, of (2–)3(–5) carpels, the placentation parietal with numerous ovules, the styles 3, distinct or connate at base, the stigma capitate or club-shaped. Fruit a berry or rarely a capsule.

A family with 16 genera and about 650 species, from tropical to warm temperate areas of the world, but mainly in tropical America and Africa.

REFERENCE: Killip, E. P. 1938. The American species of Passifloraceae. Publ. Field. Mus. Nat. Hist., Bot. Ser. 19: 1–613.

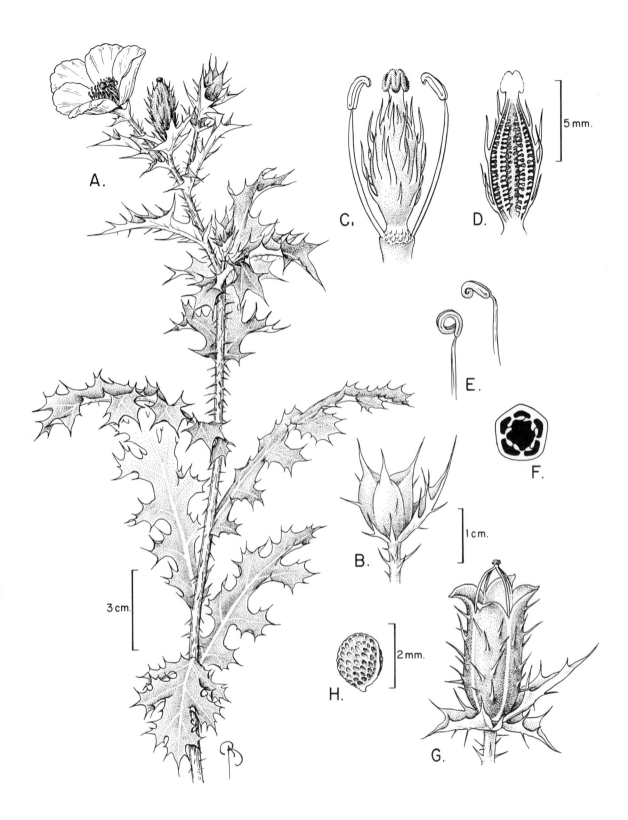

Fɪɢ. 162. *Argemone mexicana*. **A.** Fertile branch. **B.** Immature capsule. **C.** Flower with perianth removed showing stamens and pistil. **D.** L.s. pistil. **E.** Stamens. **F.** C.s. ovary. **G.** Dehisced capsule. **H.** Seed.

1. PASSIFLORA L.

Characters of the family. By far the largest genus in the family, with about 400 species, most of which (ca. 350) are native to tropical America, the others in the Old World tropics.

Key to the Species of *Passiflora*

1. Leaves oblong to elliptic-lanceolate, entire, not lobed.
    2. Plant glabrous; flowers solitary, 6–8 cm wide, subtended by an involucre of 3 large, foliaceous bracts .... 3. *P. laurifolia*
    2. Plant tomentose; flowers clustered in leaf axils, 0.8 cm wide, subtended by a minute, subulate bract .... 4. *P. multiflora*
1. Leaves variously shaped, shallowly to deeply lobed.
    3. Leaves deeply 3-lobed, with central lobe longer than the lateral ones (the leaf longer than wider); fruit yellow, orange, or purple when mature.
        4. Leaf margins serrate or deeply crenate; fruits yellow to orange when mature.
            5. Plant glabrous, odorless; petioles with 2 glands; bracts leaflike; fruit 5–7 cm long............... 1. *P. edulis*
            5. Plant pilose, with strong fetid odor; petioles glandless; bracts deeply dissected, covered with numerous sticky, glandular hairs; fruits 1.5–2.5 cm long .... 2. *P. foetida*
        4. Leaf margins entire; fruits bluish purple when mature... 6. *P. suberosa*
    3. Leaves 2-lobed, the lobes projecting beyond the apex of the blade; fruits red or reddish pink when mature .... 5. *P. rubra*

1. Passiflora edulis Sims, Bot. Mag. 45: t. 1989. 1818.
Fig. 163A–G.

Glabrous woody vine, 5–10 m long. Leaf blades 7–12 × 9–15 cm, deeply 3-lobed, coriaceous, glabrous, the lobes elliptic or oblong, with acute or acuminate apex, the base nearly cordate, the margins serrate; petioles 3–6 cm long, reddish-tinged, with a pair of globular glands near the blade; stipules filiform, ca. 5 mm long. Flowers pendulous, solitary on long peduncle in leaf axils, subtended by an involucre of 3 persistent, ovate bracts to 2 cm long; peduncle stout, 4–6 cm long. Sepals oblong, green, whitish within, 3–3.5 cm long; petals oblong, whitish within, 2.5–3 cm long; corona of biseriate appendages, violet-striped; gynophore green, ovoid-lobed, 5–7 mm long; stamens 5, spreading from base of ovary; ovary ellipsoid, green, the styles slightly reflexed, the stigmas nearly capitate. Fruit a leathery berry, ovoid, 5–7 cm long, turning from green to yellow. Seeds numerous, nearly elliptic, pitted, ca. 5 mm long, blackish, covered with a juicy orange aril.

DISTRIBUTION: Introduced for its juicy fruits, found occasionally as an escape in open disturbed areas. Road to Susannaberg (A834). Also on St. Thomas; widely cultivated throughout the tropics for its edible fruits.

COMMON NAME: passion fruit.

2. Passiflora foetida L., Sp. Pl. 959. 1753.
Fig. 163J, K.

Scarcely woody, pilose vine, 3–5 m long. Leaf blades with a strong fetid smell, 6–12 × 6–11.5 cm, deeply 3-lobed, chartaceous, softly pubescent on both surfaces, the margins crenate to serrulate, the lobes elliptic to nearly ovate, with acute or acuminate apex, the base nearly cordate; petioles 2–5.5 cm long, pilose, glandless; stipules ovate, deeply dissected, with numerous marginal glands, ca. 3 mm long. Flowers ascending, solitary on long peduncles in leaf axils, subtended by an involucre of 3 persistent, deeply dissected bracts to 4 cm long, with numerous marginal glands; peduncle slender, 1.5–5 cm long, pilose, with a few glandular hairs. Sepals oblong, greenish, whitish within, 2–2.5 cm long, with a green dorsal keel that projects beyond the obtuse apex; petals oblong, white, 2–2.5 cm long; corona multiseriate,

the inner series of segments violet at lower half; gynophore cylindrical, greenish, reddish-mottled, ca. 7 mm long; stamens 5, ascending at nearly 45°; ovary ellipsoid, green, sparsely pilose, the styles pilose, curved–reflexed, the stigmas capitate. Fruit a papery to leathery berry, widely ovoid, 1.5–2.5 cm long, turning from green to bright orange, the bracts persistent but falling when fruit fully mature. Seeds numerous, oblong to wedge-shaped, truncate at both ends, pitted, ca. 5 mm long, covered with a juicy yellow aril.

DISTRIBUTION: A common weedy species from open disturbed, dry areas. Chocolate Hole (A772), Cruz Bay (A1948). Populations on St. John seem to belong in *Passiflora foetida* var. *gossypifolia* (Ham.) Mast., which is a common variety throughout tropical America. The species was reported for St. Croix, St. Thomas, and Tortola by Britton and Wilson (1924) without reference to variety. *Passiflora foetida* var. *hispida* (DC.) Killip *ex* Gleason has been collected on St. Thomas (A5186).

COMMON NAMES: love in the mist, pap bush.

3. Passiflora laurifolia L., Sp. Pl. 956. 1753.
Fig. 163H, I.

Glabrous woody vine, 2–5(–10) m long, young parts reddish-tinged; stems cylindrical, flexible. Leaf blades 6.6–12 × 3.7–6.6 cm, oblong to elliptic-lanceolate, unlobed, coriaceous, glabrous, the apex rounded or obtuse, sometimes mucronate or shortly apiculate, the base nearly cordate, truncate to nearly rounded, the margins entire; petioles 0.7–2.5 cm long, reddish-tinged, with a pair of globular glands below the blade; stipules filiform, glandular-tipped, 8–10 mm long. Flowers hanging, solitary in leaf axils, covered by a persistent involucre of green, foliaceous bracts; the peduncle 2–3.2 cm long; bracts 3, ovate, membranous, 2.5–4 cm long, glandular-serrate along margins. Sepals oblong-elliptic, mauve within, 3–4 cm long, with a green dorsal keel projecting beyond the obtuse apex; petals oblong, whitish, with numerous mauve punctations within, 3.5–4 cm long; corona multiseriate, the segments filiform, with numerous alternate, violet and white bands, shorter than the petals; gynophore cylindrical, yellowish, lobed at base, ca. 1 cm long; stamens 5, ascending at nearly 45°; ovary ellipsoid, yellow, the styles club-shaped, curved, the stigmas capitate-bilobed. Fruit a thick-walled berry, ellipsoid, 4–6 cm

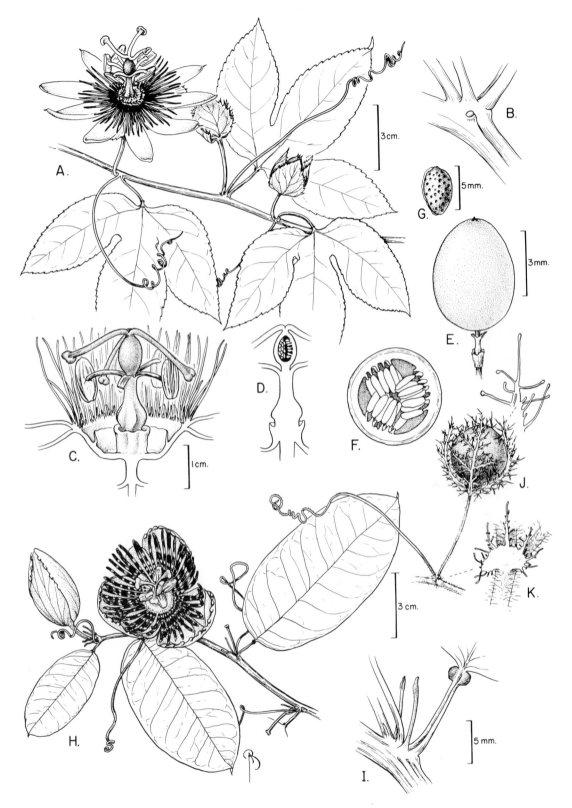

FIG. 163. **A–G.** *Passiflora edulis.* **A.** Flowering branch. **B.** Detail of node. **C.** L.s. flower. **D.** L.s. flower axis. **E.** Fruit. **F.** C.s. fruit. **G.** Seed. **H, I.** *Passiflora laurifolia.* **H.** Flowering branch. **I.** Detail of stem node showing elongate stipules at base of leaf petiole and glands at base of leaf blade. **J, K.** *Passiflora foetida.* **J.** Fruit and detail of involucral bract. **K.** Detail of stipule.

long, dull green, with numerous yellowish punctations, subtended by a persistent involucre. Seeds numerous, elliptic-triangular, pitted, ca. 4–5 mm long, beige, covered with a juicy yellowish aril.

DISTRIBUTION: A common vine of moist disturbed areas. Road to Bordeaux (A1924, A2856). Reported for St. Croix, St. Thomas (Britton and Wilson, 1924), and Tortola (D'Arcy, 1967); throughout the West Indies and northern South America.

COMMON NAMES: bell apple, water lemon.

4. Passiflora multiflora L., Sp. Pl. 956. 1753.

Scarcely woody vine, 2–5(–10) m long, all parts tomentose, many-branched along main stem; stems cylindrical, green, tomentose. Leaf blades 5–12 × 2–3 cm, oblong to lanceolate, chartaceous, unlobed, tomentose, the apex rounded or obtuse, mucronate, the base obtuse or rounded, the margins entire; petioles 4–10 mm long, with a pair of discoid glands near the blade; stipules filiform, 3–5 mm long. Flowers hanging, 2–6 clustered in leaf axils, subtended by a single subulate bract; the peduncles 5–15 mm long, tomentose; bracts and bracteoles subulate, minute. Sepals elliptic, green, ca. 3.5 × 1.5 mm; petals linear, white, 3.5 × 0.8 mm; corona of numerous yellowish, filiform segments, ca. 2.5 mm long; gynophore cylindrical, enlarged at base, 0.7–0.8 mm long; stamens 5, yellowish, spreading; ovary ellipsoid, green, the styles club-shaped, greenish, ascending, ca. 1 mm long, the stigmas capitate-bilobed, yellow. Berry fleshy, oblate, 6–8 mm diam., dull green, glabrous. Seeds numerous, nearly lenticular, wrinkled, ca. 2 mm long, light brown.

DISTRIBUTION: An occasional vine of dry to moist forests. Bordeaux (A3857). Also on Tortola; Puerto Rico, Florida, Bahamas, Cuba, and Hispaniola.

5. Passiflora rubra L., Sp. Pl. 956. 1753.

Herbaceous vine, 2–5 m long, many-branched along main stem, all parts pubescent-pilose; stems, petioles, veins, and peduncles reddish-tinged; stems slender, trigonous and striate. Leaf blades 4–12 × 4–12 cm, 2-lobed (less often with a third short, central lobe), chartaceous, softly pubescent, with 3 main veins from base, the margins entire, the lobes divergent, acute and mucronate, the base cordate; petioles 2–4 cm long, glandless; stipules subulate, ca. 5 mm long. Flowers ascending, solitary or paired in leaf axils; the peduncles 0.6–3 cm long, with a subulate bract at base. Sepals lanceolate, greenish to whitish within, pilose without, 1.3–2.5 cm long; petals linear, whitish, 0.9–1.5 cm long;

corona biseriate, the outer segments filiform, 1–1.3 cm long, pinkish to purple, the inner ones minute; gynophore cylindrical, 0.6–1 cm long; stamens 5, greenish, spreading; ovary ellipsoid-globose, greenish, white-hirsute, the styles club-shaped, reflexed, the stigmas capitate. Fruit a fleshy, tardily dehiscent capsule, to 2 cm diam., widely ellipsoid to ovoid, red or pinkish red, pilose, with 8 longitudinal ribs. Seeds numerous, ovoid, transversely ridged, ca. 3 mm long, covered with a whitish aril.

DISTRIBUTION: A common roadside weed, also in moist to dry forests. Bordeaux (A4052), Francis Bay (A905). Also on St. Croix, St. Thomas, and Tortola; throughout tropical America.

COMMON NAME: bat wing.

6. Passiflora suberosa L., Sp. Pl. 958. 1753.

Passiflora pallida L., Sp. Pl. 955. 1753. *Passiflora suberosa* var. *pallida* (L.) Mast., Trans. Linn. Soc. London **27**: 630. 1871.

Passiflora hirsuta L., Sp. Pl. 958. 1753.

Herbaceous vine, 1–3 m long; stems slender, cylindrical, puberulent to glabrous, turning whitish gray and corky at age. Leaf blades 4–10 × 4–15 cm, extremely variable, but at least some of the leaves 3-lobed (sometimes leaves ovate), nearly coriaceous, glabrous, with 3–5 main diverging veins from base, the margins entire, revolute, the lobes acute, obtuse, rounded, or shortly acuminate, the central lobe usually longer than the lateral ones, the base nearly cordate to obtuse; petioles 1.5–2 cm long, with 2 stipitate glands below the blade; stipules filiform, ca. 5 mm long. Flowers ascending, solitary or paired in leaf axils; the peduncles articulate, 1.5–2 cm long, without bracts. Sepals oblong-lanceolate, greenish to whitish within, 7.5–10 mm long; petals wanting, corona multiseriate, purple, the outer segments filiform, ca. 3 mm long, greenish toward the apex, the inner ones minute, purplish; gynophore cylindrical, ca. 5 mm long; stamens 5, greenish, ascending; ovary ellipsoid, green, the styles club-shaped, reflexed, the stigmas capitate. Fruit a fleshy, ovoid or ellipsoid berry, 10–12 mm long, turning from green to bluish purple, on a short greenish gynophore. Seeds numerous, ellipsoid, ca. 3 mm long, covered with a greenish pulp.

DISTRIBUTION: A common weed, from moist to dry forests or open disturbed areas. Bordeaux (A3133), trail to L'Esperance (A2063). Also on Anegada, St. Croix, St. Thomas, Tortola, and Virgin Gorda; common throughout the West Indies and tropical America, introduced into the Old World tropics.

COMMON NAMES: indigo berry, ink berry.

65. Phytolaccaceae (Pokeweed Family)

Herbs, shrubs, twining lianas, or rarely small trees. Leaves alternate, simple, entire; stipules minute or wanting. Flowers small, bisexual or rarely unisexual, actinomorphic, produced in axillary or leaf-opposed spikes or racemes, seldom solitary in leaf axils; calyx of 4–5 distinct or basally connate sepals; petals wanting or minute, alternating with the sepals; stamens 4 to many in 1 or 2 cycles, the filaments distinct or connate at base, the anthers opening by longitudinal slits; ovary superior, usually with an annular nectary disk at base, with 1 to many distinct to connate carpels, the ovules 1 per carpel, the placentation basal, the styles distinct. Fruit fleshy or dry with carpels separating at maturity.

A family with 18 genera and about 125 species, from tropical to subtropical areas.

Key to the Genera of Phytolaccaceae

1. Erect herbs or subshrubs; stamens 4–8.
    2. Whole plant strongly fetid; inflorescence spicate; fruits oblong-trigonous, green, with 4–6 retrorse hooks .... 1. *Petiveria*
    2. Leaves slightly fetid; inflorescence racemose; fruits globose, fleshy, red or red-orange, smooth .... 2. *Rivina*
1. Climbing shrubs or lianas; stamens 8–16 .... 3. *Trichostigma*

1. PETIVERIA L.

A monospecific genus, characterized by the following species.

1. Petiveria alliacea L., Sp. Pl. 342. 1753.
Fig. 164A–D.

Erect subshrub or herb, 50–60 cm tall, usually many-branched, from a taproot, all parts with strong fetid smell (especially the leaves); stems slender, angled to cylindrical, puberulent to glabrous, striate. Leaf blades 5–12 × 2.3–3.5 cm, elliptic to obovate, chartaceous, sparsely puberulent, the venation whitish, especially underneath, the apex obtuse, acute, or shortly acuminate, the base obtuse or attenuate, the margins entire to wavy; petioles 6–10 mm long; stipules subulate, filiform, 3–4 mm long. Flowers sessile, alternate in a terminal or axillary spike; bracts lanceolate, persistent. Tepals 4, oblong-lanceolate, nearly free to base, whitish or greenish, 2.5–3 mm long; stamens 6–8, the filaments free, unequal, the anther locules separating; ovary superior, 1-locular, tomentose, the locule with a single basal ovule, the stigma sessile, finely branched. Fruit a wedge-shaped achene, bilobed at apex, each lobe with 2 or 3 retrorse hooks, 7–8 mm long.

DISTRIBUTION: A common weed of disturbed, open areas. Road to Susannaberg (A2807), Reef Bay Quarter at Fish Bay (A3886). Also on St. Croix, St. Thomas, and Tortola; a common weed of tropical America.

COMMON NAMES: Congo root, gully root, strong man bush.

2. RIVINA L.

A monospecific genus, characterized by the following species.

1. Rivina humilis L., Sp. Pl. 121. 1753.　Fig. 164E–H.

Erect herb, to 1 m tall, branched along main stem; stems slender, angled to cylindrical, puberulent. Leaf blades with somewhat fetid smell, 3–10 × 2–4.5 cm, ovate to oblong-lanceolate, membranous to chartaceous, sparsely puberulent, the venation whitish, especially underneath, the apex acute or acuminate, the base unequal, obtuse-truncate, the margins entire; petioles 1.5–4.5 mm long, puberulent; stipules wanting. Flowers white, in axillary racemes, with a pair of minute bracteoles at base; pedicels ca. 3 mm long. Tepals 4, elliptic to oblong-elliptic, nearly free to base, 2.2–2.5 mm long; stamens 4, alternating with the tepals, the filaments free, nearly equal; ovary superior, 1-locular, the locule with a single basal ovule, the style distal, slightly curved, the stigma capitate. Fruit a globose red-orange or red fleshy drupe, 2–3.5 mm long, with persistent reflexed tepals at base.

DISTRIBUTION: A common herb in the understory of coastal forests and adjacent areas. Chocolate Hole (A784), Great Cruz Bay (A2358). Also on St. Croix, St. Thomas, and Tortola; a common weed of tropical America.

COMMON NAMES: cat's blood, jumbie pepper bush, pimba pepper, snake bush, stock ma hark.

3. TRICHOSTIGMA A. Rich.

Climbing shrubs or lianas; cross section of stem showing xylem with discrete strands of phloem. Leaves alternate or subopposite, long-petiolate; stipules minute. Flowers bisexual, actinomorphic, in axillary racemes; bracts deciduous, the bracteoles 2, persistent; perianth of 4 free tepals; stamens 8–16 in 2 whorls, the anthers opening by longitudinal slits; ovary 1-locular, with 1 ovule, the stigma sessile, many-branched. Fruit a fleshy, 1-seeded drupe, with persistent tepals at base.

A genus of 4 species, with neotropical distribution.

1. Trichostigma octandrum (L.) H. Walter, Pflanzenr. 4(83): 109. 1909. *Rivina octandra* L., Cent. Pl. II. 9. 1756.　Fig. 164I–K.

Climbing shrub or liana, 5–10(–30) m long, sometimes many-branched from base, and with numerous lateral, hanging branches along main stems; trunk 6–15 cm diam., with bark light brown, furrowed; twigs glabrous, lenticellate. Leaf blades 4–15 cm long, elliptic or oblong, coriaceous, glabrous, the apex acute or acuminate, the base obtuse or acute, sometimes unequal, the margins wavy; petioles 2–3 cm long, swollen at base; stipules wanting. Flowers in axillary racemes, with a pair of minute bracteoles at base; axes to 12 cm long; pedicels ca. 2 mm long, with a subulate bract near the base. Tepals 4, white, 3–4.5 mm long, oblong-elliptic, nearly free to base, reflexed, concave; stamens 8–16, white, the filaments free, spreading, nearly equal; ovary superior, bottle-shaped, 1-locular, yellowish, the locule with a single basal ovule, the stigma sessile, capitate, with numerous threadlike branches. Fruit a globose or ellipsoid, purple, fleshy drupe, 5–7 mm long, with persistent, spreading, reddish tepals at base.

DISTRIBUTION: A common vine from moist to dry forests or in open disturbed areas. Maria Bluff (A628). Also on St. Croix, St. Thomas, Tortola, and Virgin Gorda; common throughout tropical America.

COMMON NAMES: basket wiss, black hoopwood, hoop, hoop vine.

66. Piperaceae (Pepper Family)

By R. Quiros

Succulent erect or climbing herbs, shrubs, or rarely small trees. Leaves alternate, opposite, or whorled, petiolate or sessile, simple and entire; stipules wanting. Flowers minute, bisexual, actinomorphic, without perianth, in terminal, opposite to leaves, or axillary spikes and less often racemes; stamens 2–6; ovary superior, 1-celled, sessile or rarely short-stalked, with 1 basal ovule, the style 1, the stigmas 1–5. Fruit indehiscent, drupaceous or sticky nutlets.

About 5–8 genera and perhaps more than 3000 species of pantropical distribution, the majority in the Western Hemisphere.

REFERENCES: Boufford, D. E. 1982. Notes on *Peperomia* (Piperaceae) in the southeastern United States. J. Arnold Arbor. 63: 317–325; Howard, R. A. 1973. Notes on the Piperaceae of the Lesser Antilles. J. Arnold Arbor. 54: 377–411.

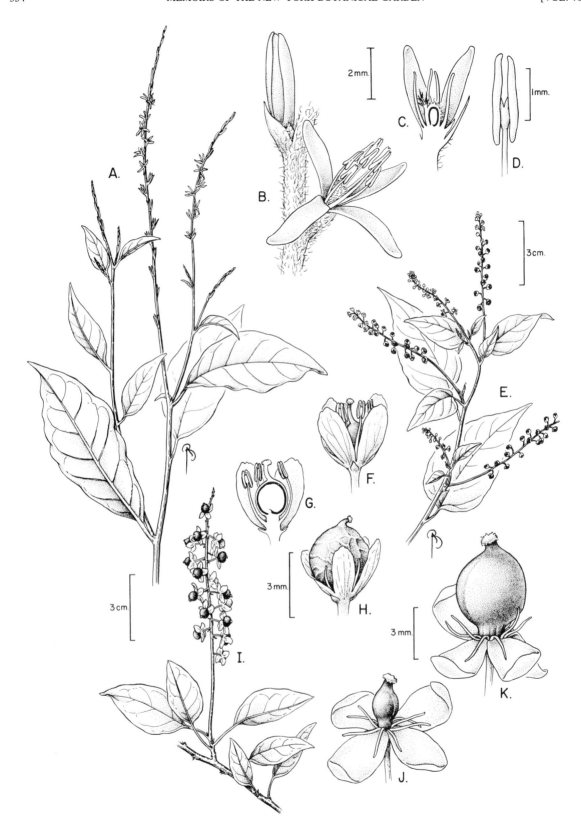

FIG. 164. A–D. *Petiveria alliacea*. **A.** Fertile branch. **B.** Detail of inflorescence. **C.** L.s. flower. **D.** Anther. **E–H.** *Rivina humilis*. **E.** Fertile branch. **F.** Flower. **G.** L.s. flower. **H.** Drupe with persistent calyx parts. **I–K.** *Trichostigma octandrum*. **I.** Fruiting branch. **J.** Fertilized flower. **K.** Drupe with persistent sepals and stamens.

Key to the Genera of Piperaceae

1. Plants herbaceous, fleshy or succulent; leaves alternate, opposite, or whorled; stigma 1 .... 1. *Peperomia*
1. Plants shrubs, more or less woody, not succulent; leaves alternate; stigmas 2–6 .... 2. *Piper*

1. PEPEROMIA Ruíz & Pav.

Terrestrial or epiphytic, erect, prostrate, or pendulous herbs; stems succulent, glabrous or variously pubescent. Leaves alternate, opposite, or whorled; the blades palmate- or pinnate-veined; petioles elongate to short or wanting. Flowers numerous, subtended by circular or elliptic, often peltate bracts, in leaf-opposed or terminal spikes; stamens 2; ovary sessile or stalked, sometimes beaked, stigma 1, terminal or lateral. Fruit a small, ellipsoid to globose drupe, sessile or stalked, smooth or verrucose, commonly viscid.

A genus of about 1000 species, in the tropics and subtropics.

Key to the Species of *Peperomia*

1. Leaf blades with black or glandular dots.
  2. Leaves opposite, or whorled in threes at the base of the inflorescences, glandular-dotted .... 2. *P. humilis*
  2. Leaves alternate, with black dots.
    3. Epiphytic herbs; leaves 3-nerved, the petioles with 2 ciliate ridges .... 1. *P. glabella*
    3. Terrestrial herbs, stem branching from the base, with a pink or reddish hue; leaves strongly 3- or 5-nerved from the base, the petioles glabrous .... 4. *P. myrtifolia*
1. Leaf blades without black or glandular dots.
  4. Terrestrial, erect or decumbent herbs; stems brownish, rooting at lower nodes; leaves 4–15 cm long, obovate or obovate-elliptic, the base attenuate into a nearly winged petiole; spikes densely flowered
    .... 3. *P. magnoliifolia*
  4. Terrestrial, erect herbs; stems translucent; leaves usually cordate at base (less often subtruncate or rounded), petioles clasping-decurrent on stem at base; spikes loosely flowered .... 5. *P. pellucida*

1. Peperomia glabella (Sw.) A. Dietr., Sp. Pl. **1**: 156. 1831. *Piper glabellum* Sw., Prodr. 16. 1788.

Epiphytic herb, to 25 cm long, climbing, creeping, or sometimes hanging, rooting at nodes. Leaves alternate; blades 2–6.5 × 0.7–4 cm, ovate-lanceolate to ovate, glabrous, 3-nerved and occasionally with 2 accessory nerves, black-dotted, the apex acute or acuminate and occasionally ciliate, the base cuneate, the margins entire and ciliate; petioles 0.2–1.3 cm long, with 2 ciliate ridges decurrent on stem. Flowers in terminal filiform spikes, often clustered at stem apex or solitary in the leaf axils, densely flowered, 5–16 cm long; anthers 0.1–0.2 mm long, often broader than long; pistil borne in a depression on the rachis. Drupe ovoid, 0.3–1.0 mm long, slightly curved upward and attached laterally to the base, the apex oblique.

DISTRIBUTION: An occasional herb of moist shaded areas, very often in forest understory. Bordeaux (A4266, B576, W762). Also on St. Thomas and Tortola; throughout tropical America.

2. Peperomia humilis A. Dietr., Sp. Pl. **1**: 168. 1831 [as a new name]. *Piper humile* Vahl, Enum. Pl. **1**: 349. 1804, nom. illegit., non Mill., 1768.

Peperomia questeliana Stehlé & Trel. *in* Stehlé, Candollea **8**: 77. 1940.

Terrestrial erect or prostrate herb, 10–20 cm long; stems simple or branched from base, purplish or pinkish, pilose, rooting at nodes. Leaves opposite or ternate at the base of inflorescence; blades 1–5 × 0.6–2.3 cm, obovate to obovate-elliptic, membranous, glandular-dotted, densely pilose, 3-nerved, the apex rounded, obtuse, or acute, the base cuneate, the margins entire and ciliate; petioles slender, 4–20 mm long, pilose. Flowers with pilose peduncle, in terminal, slender, loosely flowered spikes, 4–

14 cm long. Drupe ellipsoid-ovoid, black, glandular-papillose, ca. 0.5 mm long.

DISTRIBUTION: In shaded areas and disturbed open roadsides. Road to Ajax Peak (A2657), Center Line Road by km 5.6 (A2418), White Cliffs (A2709). Also on St. Croix, St. Thomas, and Tortola; in Florida, Central America, the Greater and Lesser Antilles.

3. Peperomia magnoliifolia (Jacq.) A. Dietr., Sp. Pl. **1**: 153. 1831. *Piper magnoliaefolium* Jacq., Collectanea **3**: 210. 1791. *Peperomia amplexicaulis* (Sw.) A. Dietr. var. *magnoliaefolia* (Jacq.) Griseb., Fl. Brit. W. I. 167. 1859.   Fig. 165A–D.

Terrestrial or rarely epiphytic herb, erect or decumbent, to 50 cm long; stems succulent, glabrous, brownish, rooting at lower nodes. Leaves alternate; blades 4–15 × 2–8 cm, obovate or obovate-elliptic, succulent, glabrous, the apex rounded or obtuse, often slightly emarginate, the base cuneate or abruptly attenuate onto the petiole, the margins slightly wavy; petioles 1–5 cm long, nearly winged or ridged. Flowers in simple or branched terminal spikes, densely flowered; pistil borne in a depression on the rachis. Drupe ellipsoid, 0.7–1.2 mm long, the beak straight, one-fourth to one-half the length of the fruit.

DISTRIBUTION: Common herb in forest understory, found on humus layer. Bordeaux Mountain (A618). Also on St. Croix, St. Thomas, and Tortola; Mexico, Central America, the West Indies, and South America.

4. Peperomia myrtifolia (Vahl) A. Dietr., Sp. Pl. **1**: 147. 1831. *Piper myrtifolia* Vahl, Enum. Pl. **1**: 341. 1804.

Piper tenuiflorum Vahl *in* H. West, Bidr. Beskr. Ste. Croix 195. 1793.

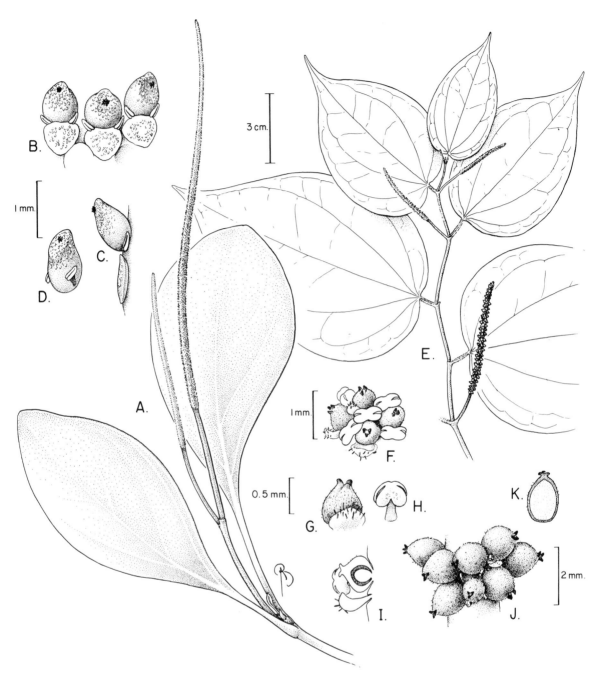

FIG. 165. A–D. *Peperomia magnoliifolia.* **A.** Habit. **B.** Detail of flowers and subtending bracteoles. **C.** Flower and bracteole, lateral view. **D.** Flower. **E–K.** *Piper amalago.* **E.** Fertile branch. **F.** Detail of inflorescence spike showing flowers and subtending bracteoles. **G.** Pistil and subtending bracteole. **H.** Stamen. **I.** L.s. pistil and bracteole. **J.** Detail of infructescence. **K.** L.s. drupe.

Erect, terrestrial, glabrous herb to 30 cm tall, many-branched from the base; stems fleshy, pink or reddish. Leaves alternate; blades 3.5–9.0 × 1.5–5.0 cm, rhombic-oblong, oval, or ovate, membranous, black-dotted and strongly 3- or 5-nerved from base, the apex obtuse or acute, the base acute, the margins entire; petiole 0.5–1.5 cm long. Flowers densely clustered in terminal spikes to 12 cm long. Drupe globose, glandular-papillose, dark brown, laterally attached.

DISTRIBUTION: Mostly on organic matter in understory of secondary forest. Hawksnest Gut (A2669); Annaberg to North Shore Road (A2869). Also on St. Croix, Virgin Gorda; Vieques and the Lesser Antilles.

5. Peperomia pellucida (L.) Kunth *in* Humb., Bonpl. & Kunth, Nov. Gen. Sp. **1:** 64. 1816. *Piper pellucidum* L., Sp. Pl. 30. 1753.

Erect, terrestrial herb, 10–35 cm tall, glabrous, many-branched at the base; stems translucent. Leaves alternate; blades 1.0–4.0 × 0.8–4.0 cm, deltoid-ovate, fleshy to membranous, the apex acuminate, the base subtruncate, rounded, or cordate, the margins slightly wavy; petioles 0.6–2.0 cm long, clasping stem at the base. Flowers in slender, terminal spikes, 1.5–5.0 cm long. Drupe subglobose, black, 0.5–0.7 mm diam.

DISTRIBUTION. Along shaded stream banks. Jeep trail to Seiben (A2054). Also on St. Croix, St. Thomas, and Tortola; pantropical.

COMMON NAME: shiny bush.

2. PIPER L.

Shrubs, rarely small trees or woody vines; stems commonly nodose. Leaves alternate, petiolate, often inequilateral at the base, sometimes with pellucid dots, the venation pinnate or palmate. Flowers usually bisexual, numerous, in typically solitary, simple leaf-opposed spikes, bracts peltate; stamens 2–6; ovary sessile; stigma 2–5, sessile or with a short or thick style. Fruit a drupe, ellipsoid to subglobose.

A tropical genus with more than 2000 species.

1. Piper amalago L., Sp. Pl. 29. 1753.   Fig. 165E–K.

Piper medium Jacq., Icon. Pl. Rar. **1:** 2. 1781.
Piper plantagineum Lam., Tabl. Encycl. **1:** 80. 1791.
Piper sieberi C. DC. *in* A. DC., Prodr. **16(1):** 248. 1869.

Erect shrub, 2–3 m tall; stems swollen at nodes, dark, lenticellate. Leaf blades 6–12 × 3.5–7 cm, ovate to oblong-ovate, membranous, glabrous, the apex long-acuminate, the base acute and asymmetrical, the margins entire; petioles slender, pubescent, 0.7–1.7 cm long. Flowers densely arranged in leaf-opposed spikes; bracts ovate; pistil with a broad base and tapering to a narrow apex. Drupe ovoid-conical, tapering to apex, ca. 1.5 mm long.

DISTRIBUTION: Mostly found in secondary forest understory. Adrian Ruins (A725), Maho Bay (A2106), Salt Pond (GTP29326). Also on St. Croix, St. Thomas, and Tortola; Mexico, Central America, Greater Antilles, Lesser Antilles, and South America.

COMMON NAME: black wattle.

67. Plantaginaceae (Plantain Family)

Herbs. Leaves commonly basal and alternate, rarely cauline, simple, entire to dentate, mostly with parallel veins; stipules wanting. Flowers small, bisexual, actinomorphic, 4-merous, produced in axillary spikes or heads; calyx lobed or cleft, 2 of the lobes usually connate; corolla gamopetalous; stamens as many as and alternate with the corolla lobes, the filaments adnate to the corolla tube, the anthers opening by longitudinal slits; ovary superior, of 2 united carpels, the placentation axile, with 1 to many ovules per carpel, the style terminal with a 2-lobed stigma. Fruit a membranous circumscissile capsule, an achene, or a nut covered by the persistent calyx.

A family of 3 genera and about 250 species, most of which belong to the genus *Plantago,* nearly cosmopolitan.

1. PLANTAGO L.

Perennial herbs, with very short or contracted stems, borne from a stout rootstock. Leaves in a basal rosette; blades usually decurrent into a sheathing base or petiolate. Inflorescence spicate. Calyx with equal lobes or 2 of them longer; corolla trumpet-shaped, with spreading or reflexed lobes; stamens 4 or 2, exserted. Fruit a membranous circumscissile capsule, with numerous tiny seeds.

A cosmopolitan genus of about 200 species, some of which are weedy.

1. Plantago major L., Sp. Pl. 112. 1753.   Fig. 166.

Acaulescent herb. Leaf blades 15–25 × 6–10 cm, ovate to widely elliptic, chartaceous, glabrous, the apex obtuse, the base narrowed, the margins coarsely dentate; petioles 2–20 cm long. Flowers white, congested in numerous scapose spikes; axes 5–25 cm long; bracts green, ovate, ca. 1 mm long. Calyx of 4 nearly equal, 1.2–2 mm long, oblong sepals; corolla 2 mm long, with spreading lobes. Capsule ovoid, 2.5–4 mm long, straw-colored, with 15–20 wedge-shaped seeds, 0.8–0.9 mm long.

DISTRIBUTION: Very uncommon on St. John, found in wet moist areas. Road to Bordeaux Mountain (W471). Also on St. Croix and Tortola; a cosmopolitan weed.

68. Plumbaginaceae (Leadwort Family)

Shrubs, woody or herbaceous vines. Leaves alternate, simple, entire; stipules present or wanting. Flowers medium to large, 5-merous, bisexual, actinomorphic, in terminal panicles, cymes, or racemes; calyx tubular, 5–10-ribbed; corolla tubular or with distinct, clawed petals; stamens as many as and alternating with the petals, the filaments distinct or adnate to corolla tube, the anthers opening by 1 or 2 longitudinal slits; ovary superior, unilocular, of 5 carpels, the ovules basal, 1 per carpel, the styles distinct or connate. Fruit an achene or a circumscissile or valvate capsule, usually enclosed by the persistent calyx.

A family with 12 genera and about 400 species, widely distributed around the world, principally in the Mediterranean and Asia.
REFERENCE: Luteyn, J. L. 1990. Plumbaginaceae. In G. Harling & L. Andersson (eds.), Fl. Ecuador 39(151): 39–46.

1. PLUMBAGO L.

Erect, climbing, or trailing shrubs or herbs. Leaves alternate, petiolate to sessile, the venation pinnate. Flowers shortly pedicellate, bibracteate, in terminal racemes or panicles; calyx 5-ribbed, these covered with stipitate glands, the lobes triangular; corolla salverform,

FIG. 166. *Plantago major*. **A.** Habit. **B.** Flower. **C.** Pistil. **D.** Capsule and seed.

the tube exceeding the calyx, the lobes obovate, rounded, or truncate; stamens 5, the filaments free, included or slightly exserted, the anthers opening by 2 longitudinal slits; style 1, included or exserted, with 5 elongate stigmas. Fruit a 1-seeded, membranous, valvate capsule.

A tropical genus of 12–20 species.

1. Plumbago scandens L., Sp. Pl., ed. 2, 1: 215. 1762. Fig. 167.

Scrambling shrub, 2–3 m long, often many-branched from base and with numerous lateral branches along main stem; stems cylindrical, striate, glabrous. Leaf blades 3–13 × 1–6 cm, ovate, oblong-lanceolate, spatulate to oblanceolate, chartaceous, gla-

brous, lepidote-punctate underneath, the apex acute, acuminate, or obtuse, the base obtuse to rounded, or cuneate to attenuate, sometimes unequal, the margins entire; petioles 5–10 mm long. Flowers heterostylous, in long terminal panicles; the axes to 30 cm long, glabrous, striate, sparsely covered with sessile glands; bracteoles ca. 2 mm long, elliptic; pedicels to 1 mm long. Calyx green, 7–11 mm long, the ribs with long stipitate, sticky glands;

Fig. 167. *Plumbago scandens*. **A.** Flowering branch. **B.** Detail of node. **C.** Flower and subtending bract and bracteoles. **D.** Stamen. (From Luteyn, J. L. 1990. Plumbaginaceae. In G. Harling & L. Andersson (eds.), Fl. Ecuador (151): 39–46.)

corolla white, the tube 1.2–2 cm long, the lobes 0.5–1 cm long, spreading, obovate and mucronate; stamens exserted, the anthers light violet. Fruit oblong, concealed by the persistent sticky, glandular calyx.

DISTRIBUTION: A common roadside weed. Road to Bordeaux (A1893), Maria Bluff (A627). Also on St. Croix, St. Thomas, and Tortola; common throughout tropical America.

COMMON NAMES: blister leaf, guinea leaf.

69. Polygonaceae (Seagrape Family)

Herbs, shrubs, twining or tendriled vines, lianas, or trees. Leaves alternate or seldom opposite or whorled, simple, entire or rarely lobed; stipules connate around the stem to form a tubular structure (an ocrea) or rarely wanting. Flowers small to medium, bisexual or unisexual, actinomorphic, not differentiated into calyx and corolla, produced in terminal or axillary spikes, panicles, or racemes; perianth of 3–6 basally connate tepals; stamens (2–)6–9, usually in 2 unequal series, the filaments free or connate at base, the anthers opening by longitudinal slits; ovary superior, unilocular, of (2–)3(–4) carpels, with basal placentation, the styles distinct or distally connate. Fruit an achene or nut, usually enclosed by the accrescent tepals.

A family of 30 genera and about 1000 species, best represented in northern temperate zones, with many woody members in the tropics.

Key to the Genera of Polygonaceae

1. Tendriled vines; flowers bisexual, the tepals showy, petal-like, pink . . . . 1. *Antigonon*
1. Trees or shrubs; flowers functionally unisexual, the tepals minute, not brightly colored . . . . 2. *Coccoloba*

1. ANTIGONON Endl.

Tendriled herbaceous to woody vines. Leaves long-petiolate; ocrea small. Flowers bisexual, actinomorphic, in ocreolate fascicles, axillary or terminal racemes, or in panicles; perianth of 5 free, unequal, petal-like tepals; stamens 8, the filaments connate at base, the anthers opening by longitudinal slits; ovary 1-locular, with a single ovule, the styles 3, free to base, the stigma peltate. Fruit a 1-seeded achene, concealed by the accrescent tepals.

A neotropical genus of 8 species.

1. Antigonon leptopus Hook. & Arn., Bot. Beechey Voy. 308, t. 69. 1838.   Fig. 168G–M.

Tendriled, subwoody vine 5–13 m long, with numerous lateral branches; stems angled, puberulent. Leaf blades 5–14.5 × 2–7 cm, ovate, triangular-ovate to nearly lanceolate, chartaceous, puberulent, especially on veins, the apex acute or acuminate, the base cordate to nearly truncate, the margins wavy, ciliate; petioles 1–5 cm long, cylindrical to narrowly winged. Flowers bisexual, clustered in axillary racemes or terminal panicles; axes 10–20 cm long, puberulent, ending in a pair of coiling tendrils; pedicels 3–4(–10) mm long. Perianth 4–7 mm long, of 5 spreading, ovate to elliptic, bright pink (or white) tepals; staminal column 2–3 mm long, with same color as the tepals. Achene ovoid, 5–8 mm long.

DISTRIBUTION: This aggressive ornamental has naturalized and is common in a few localities, where it has become a pest. Hansen Bay (A1811). Native to Mexico and Central America but now widespread throughout the Caribbean region.

COMMON NAMES: coralita, coral vine, Mexican creeper.

2. COCCOLOBA L.

Trees or shrubs (seldom scandent or liana-like). Leaves short-petiolate; ocrea small to large, usually membranous. Flowers functionally unisexual, actinomorphic, in ocreolate fascicles (staminate ones) and solitary (pistillate ones) along axillary or terminal spikes or racemes; hypanthium green or whitish, of 5 basally connate tepals; stamens 8, the filaments connate at base, the anthers longitudinally dehiscent; ovary 1-locular, uniovular, the styles 3, free to base, the stigma capitate. Fruit usually a variously shaped, 1-seeded achene, covered by a fleshy hypanthium or imbricate tepals lobes.

A neotropical genus of 400 species.

REFERENCE: Howard, R. A. 1957. Studies in the genus *Coccoloba,* IV. The species from Puerto Rico and the Virgin Islands and from the Bahama Islands. J. Arnold Arbor. 38: 211–242.

Key to the Species of *Coccoloba*

1. Leaves reniform, rounded, or obovate, almost as broad as or broader than long; pedicels 2–2.5 mm; fruit inversely pear-shaped to obovoid, 1.4–2 cm long . . . . 4. *C. uvifera*
1. Leaves variously shaped, distinctly longer than broad; pedicels to 2 mm long or wanting; fruit ellipsoid, ovoid or ovoid-trigonous, 4–10 mm long.
   2. Leaves 2–5.5 cm long.
      3. Leaves ovate (sometimes nearly elliptic), cordate at base, with entire margins; tertiary veins inconspicuous, slightly prominent, finely reticulate, the loops nearly rounded to square, 0.2–0.4 mm wide . . . . 1. *C. krugii*
      3. Leaves elliptic, oblong, or seldom ovate-lanceolate, obtuse, rounded, or rarely cordate at base, with crenate, revolute margins; tertiary veins conspicuous, prominent, coarsely reticulate, the loops nearly rectangular, 1.3–2 mm long . . . . 2. *C. microstachya*
   2. Leaves >6 cm long.
      4. Leaves elliptic to oblanceolate, 8.5–21 cm long, the upper surface with sunken secondary veins,

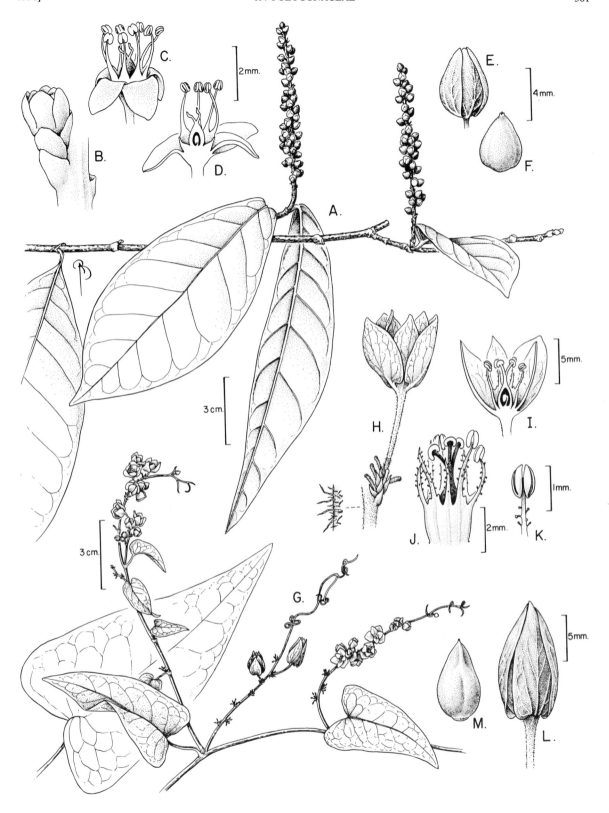

Fig. 168. A–F. *Coccoloba venosa*. A. Fruiting branch. B. Detail of spike showing flower. C. Flower. D. L.s. flower. E. Fruit enclosed by persistent tepals. F. Achene. G–M. *Antigonon leptopus*. G. Flowering branch and leaf. H. Detail of inflorescence and indument. I. L.s. flower. J. Detail of staminal tube and stigma. K. Stamen. L. Fruit with persistent tepals. M. Achene.

the base tapering to obtuse or slightly cordate lobes; fruit ca. 4.5 mm long, ovoid-trigonous, surrounded by accrescent perianth lobes .... 5. *C. venosa*

    4. Leaves elliptic, ovate, 6–10(–14.5) cm long, the upper surface flattened over the secondary veins, the base obtuse to rounded, slightly unequal; fruit 8–10 mm long, ovoid, surrounded completely by the hypanthium, the tepals minute, forming a crown at apex .... 3. *C. swartzii*

1. Coccoloba krugii Lindau, Bot. Jahrb. Syst. 13: 145. 1890.

Shrub or small tree 1–3 m tall; stems cylindric, glabrous, slightly swollen at nodes. Leaf blades 2–5.5 × 1.5–4.8 cm, ovate to widely ovate, coriaceous, glabrous, the tertiary venation inconspicuous, slightly prominent, very finely reticulate, with rounded to square loops, 0.2–0.4 mm wide, the apex obtuse, the base cordate, the margins entire; petioles 3–5 mm long, nearly cylindrical, reddish-tinged; ocrea 3.5–5 mm long, glabrous, membranous, early deciduous. Flowers greenish, solitary or paired at ocreate, widely spaced nodes along terminal racemes; axes 1.5–5(–7) cm long, glabrous; pedicels ca. 1 mm long or shorter, covered by the ocreate base and 2 bracteoles. Perianth ca. 2 mm long, the hypanthium cup-shaped, ca. 1 mm long, crowned by 5 spreading lobes; stamens yellowish. Achene ovoid to ellipsoid, 4–5 mm long, surrounded by a reddish hypanthium with imbricate, short tepal lobes at apex.

DISTRIBUTION: A common shrub of coastal scrub and dry forests. Maria Bluff (A1984, A2329). Also on Anegada; the Bahamas and the Greater and Lesser Antilles.

2. Coccoloba microstachya Willd., Sp. Pl. 2: 459. 1799.

Shrub or small tree 1.5–7 m tall; trunk to 15 cm diam.; bark grayish and rough; stems cylindric, puberulent, becoming glabrous. Leaf blades 2–5.5(–8) × 1.4–3.5(–4.5) cm, elliptic, oblong or seldom ovate-lanceolate, coriaceous, glabrous or puberulent, the tertiary venation conspicuous, prominent, coarsely reticulate, loops nearly rectangular, 1.3–2 mm long, the apex obtuse, acute, or rounded, the base obtuse, rounded, or rarely nearly cordate, the margins crenate, often revolute; petioles 4–6 mm long, nearly cylindrical, puberulent; ocrea 3–4 mm long, early deciduous. Flowers greenish, solitary or paired at ocreate, widely spaced nodes along terminal racemes; axes 4–8 cm long, puberulent; pedicels ca 0.5 mm long, covered by the ocreate base and 2 bracteoles. Perianth ca. 1.7–2 mm long, the hypanthium cup-shaped, ca. 0.5 mm long, crowned by 5 spreading, ovate lobes 1.2–1.5 mm long; stamens light yellow to whitish. Achene ovoid, 5–6 mm long, surrounded by a reddish hypanthium with short, imbricate lobes near the apex.

DISTRIBUTION: A common shrub or small tree of coastal scrub and dry forests. Monte (A2377), Reef Bay (M17072). Also on Jost van Dyke, St. Croix, St. Thomas, and Virgin Gorda; Puerto Rico (including Mona, Culebra, and Vieques) and Hispaniola.

3. Coccoloba swartzii Meisn. *in* A. DC., Prodr. 14: 159. 1856.

Coccoloba barbadensis sensu Lindau, Bot. Jahrb. Syst. 13: 148. 1890, non Jacq., 1760.

Coccoloba diversifolia sensu Lindau *in* Urb., Symb. Ant. 1: 223. 1899, non Jacq., 1760.

Shrub or small tree 1.5–7 m tall; bark grayish and smooth; stems cylindric, puberulent or glabrous. Leaf blades 6–14.5 × 4–9.5 cm, elliptic to ovate, coriaceous, glabrous, the tertiary venation reticulate, with nearly rectangular loops, 2–4 mm long, the apex obtuse to rounded, the base obtuse, rounded to nearly

cordate, usually inequilateral, the margins entire to slightly wavy; petioles 5–8 mm long, nearly cylindrical, puberulent, reddish-tinged; ocrea 7–10 mm long, membranous, puberulent to glabrous, reddish-tinged. Flowers greenish, paired at ocreate widely spaced nodes along terminal racemes; axes 5.5–15 cm long, puberulent; pedicels ca. 0.5 mm long, covered by the ocreate base and 2 bracteoles. Perianth ca. 2.4–2.7 mm long, the hypanthium cup-shaped, ca. 0.5 mm long, crowned by 5 spreading, ovate lobes 2–2.2 mm long. Achene ovoid, 8–10 mm long, surrounded by a purple hypanthium with nearly spreading lobes.

DISTRIBUTION: Occasional tree of coastal and seasonal dry forests. Caneel Bay Trail (A4118), Hawksnest Beach (A3187), Ram Head Trail (W208). Also on St. Croix, St. Thomas, and Virgin Gorda; the Greater and Lesser Antilles, Central America, and northern South America.

NOTE: A collection of this species was reported by Woodbury and Weaver (1987) in error as a hybrid between *Coccoloba uvifera* and *C. krugii* that has not yet been seen on St. John.

4. Coccoloba uvifera (L.) L., Syst. Nat., ed. 10, 2: 1007. 1759. *Polygonum uvifera* L., Sp. Pl. 365. 1753.

Shrub or small tree 2–15 m tall; bark grayish, smooth, peeling off in irregular plates, leaving a reddish brown bark underneath; stems cylindric, puberulent, pilose, or papillose. Leaf blades 7–21 × 6.5–26 cm, reniform, rounded, or obovate, thick-coriaceous, glabrous except for numerous hairs along the vein sides, the tertiary venation finely reticulate, inconspicuous, the apex rounded or notched, the base cordate to rounded, usually with one or both lobes over the petiole, the margins entire to slightly crenate, revolute; petioles 7–10 mm long, stout, pilose to papillose, reddish-tinged; ocrea 7–10 mm long, puberulent, coriaceous, persistent, reddish-tinged. Flowers greenish yellow to whitish, 3–4 at ocreate, congested nodes of terminal racemes; axes 10–25 cm long, puberulent to papillose; pedicels 2–2.5 mm long. Perianth ca. 3 mm long, the hypanthium tapering, ca. 1 mm long, crowned by 5 spreading, oblong lobes 2–2.2 mm long. Achene inversely pear-shaped to obovoid, 1.4–2 cm long, covered by a fleshy, juicy (edible) hypanthium, that turns purple at maturity.

DISTRIBUTION: A common tree of coastal fronts and sandy beaches, also inland in dry forests. Chocolate Hole (A786), Susannaberg (A4156). Also on all other Virgin Islands; a pan-Caribbean species.

COMMON NAME: seagrape.

5. Coccoloba venosa L., Syst. Nat., ed. 10, 2: 1007. 1759.   .Fig. 168A–F.

Coccoloba nivea Jacq., Enum. Syst. Pl. 19. 1760.

Shrub or small tree 3–10 m tall, usually many-branched from base; bark grayish, lenticellate; twigs angled, glabrous, with numerous lenticels. Leaf blades 8.5–21 × 3.2–10 cm, elliptic to oblanceolate, chartaceous, glabrous or sometimes with tufts of hairs at midvein axils, midvein very prominent, especially on lower surface, secondary veins sunken on upper surface, prominent on lower, the apex shortly acuminate, the base tapering into obtuse or slightly cordate lobes, unequal, the margins entire, revolute; petioles 5–12 mm long, slender, glabrous, brownish; ocrea 1–1.5 cm long, membranous, early deciduous. Flowers greenish, congested along terminal, erect racemes; axes 10–19 cm long, puberulent to papillose; pedicels 1–2 mm long. Achene ca. 4.5 mm long, ovoid-triangular, dark brown, covered from base by the accrescent, fleshy, imbricate perianth lobes, which turn reddish pink at maturity.

DISTRIBUTION: A common tree of dry to moist forests. Road to Bordeaux (A1918), Center Line Road (A2075). Also on Jost van Dyke, St. Croix,

St. Thomas, and Tortola; Hispaniola, Puerto Rico (including Vieques), the Lesser Antilles, Trinidad, Tobago, and Venezuela.

COMMON NAMES: cherry grape, chiggery grape, trible grape.

DOUBTFUL RECORD: *Coccoloba diversifolia* Jacq., was reported by Britton and Wilson (1924) as occurring on St. John; however, no specimen has been found to confirm this record, nor has the species been recently collected on the island.

70. Portulacaceae (Purslane Family)

Herbs or less often shrubs or subshrubs, often succulent. Leaves opposite or alternate, simple, entire; stipules deeply dissected, modified into tufts of hairs, or wanting. Flowers bisexual or rarely unisexual, actinomorphic or less often zygomorphic, solitary or in cymose, racemose, or paniculate inflorescences; calyx of 2(–9) sepals; corolla of 4–6(–18) free, imbricate petals; stamens as many as and opposite the petals, sometimes more or less numerous, the filaments free or basally adnate to the petals, the anthers opening by longitudinal slits; ovary superior, of 2–3 united carpels, with distinct styles, the placentation free, central, with 1 to many ovules per carpel. Fruit a membranous circumscissile or loculicidal capsule or less often indehiscent; seeds lenticular.

A cosmopolitan family of about 20 genera and 500 species.

REFERENCE: Bogle, A. L. 1969. The genera of Portulacaceae and Basellaceae in the southeastern United States. J. Arnold Arbor. 51: 431–462.

Key to the Genera of Portulacaceae

1. Prostrate, decumbent, or erect herbs; flowers solitary or fascicled; capsule circumscissile .... 1. *Portulaca*
1. Erect herb or subshrub; inflorescence racemose or paniculate; capsule loculicidal .... 2. *Talinum*

1. PORTULACA L.

Erect, prostrate, or decumbent, fleshy herbs, with tuberous or fibrous roots. Leaves alternate, opposite or whorled, succulent, short-petiolate; stipules deeply dissected or hairlike. Flowers bisexual, actinomorphic, axillary, solitary or fascicled; calyx of 2 sepals; petals 5 or seldom numerous, brightly colored, early deciduous; stamens as many as the petals, adnate to corolla at base, the anthers opening by longitudinal slits; ovary superior, 2–3-locular, the locules with many ovules, the styles 3–9, free to base, the stigma linear. Fruit a circumscissile capsule with many reniform, smooth or tuberculate, minute seeds.

A genus of about 200 species, widely distributed in the tropical and temperate zones.

REFERENCE: Legrand, C. D. 1952. Las especies americanas de *Portulaca*. Anales Mus. Nac. Montevideo, ser. 2, 70(3): 3–147 + 29 plates.

Key to the Species of Portulaca

1. Stems slender, filiform; leaves opposite .... 2. *P. quadrifida*
1. Stems succulent; leaves alternate.
    2. Leaves obovate; stipules glabrous or with few hairs .... 1. *P. oleracea*
    2. Leaves linear or oblong-linear, cylindrical or flattened; stipules densely hairy .... 3. *P. rubricaulis*

1. Portulaca oleracea L., Sp. Pl. 445. 1753.

Fig. 169A–D.

Prostrate or decumbent herb, to 30 cm long, many-branched from a taprooted base; stems succulent, glabrous or pilose at nodes, usually reddish-tinged. Leaves alternate; blades 0.7–2.5 × 0.4–1.1 cm, spatulate or obovate, thick, fleshy, glabrous, the apex rounded to truncate, the base tapering to a short petiole, the margins sometimes pinkish-tinged. Flowers solitary or a few at ends of branches. Sepals green, keeled; petals 5, 3–4.5 mm long, yellow, elliptic, emarginate. Capsule 2.5–3.7 mm wide, ovoid, opening at or below the middle. Seeds ca. 0.7 mm long, dark brown to black, tuberculate.

DISTRIBUTION: A common weed of open areas, such as roadsides and sandy coasts. Road to Lameshur (A2791), Waterlemon Bay (A1937). Also on all of the Virgin Islands; a cosmopolitan weed.

COMMON NAMES: jump up an' kiss me, purslane.

2. Portulaca quadrifida L., Mant. Pl. 1: 73. 1767.

Fig. 169E–G.

Prostrate, mat-forming herb, with numerous lateral branches to 10 cm long; stems slender (<1 mm wide), rooting and pilose

at nodes. Leaves opposite; blades 4–7 × 1–3 mm long, lanceolate to elliptic, thick, fleshy, glabrous, usually reddish-tinged, the apex acute or obtuse, the base rounded or obtuse, the margins entire; petioles minute, slender. Flowers 1 or 2 at ends of short lateral branches. Sepals green, ca. 2 mm long; petals 4, 3–4 mm long, yellow, elliptic, apiculate. Capsule ca. 6 mm long, ovoid, opening above the base. Seeds ca. 1 mm long, dark gray, tuberculate.

DISTRIBUTION: An occasional herb of open coastal and marshy areas. Trail to Fortsberg (A4089), Johns Folly (A3938). Also on Anegada, St. Croix, St. Thomas, and Tortola; throughout the Greater and Lesser Antilles and South America.

3. Portulaca rubricaulis Kunth in Humb., Bonpl. & Kunth, Nov. Gen. Sp. 6: 73. 1823.

Prostrate, decumbent, or erect herb, to 25 cm long, many-branched from base; lateral roots fleshy, tuberous; stems succulent, glabrous, grayish to light copper, with long tufts of hairs at nodes. Leaves alternate; blades 1–2.2 × 0.1–0.3 cm, oblong to narrowly elliptic, thick, fleshy, glabrous, usually reddish-tinged, the apex obtuse, the base narrowed to a minute, slender petiole, the margins

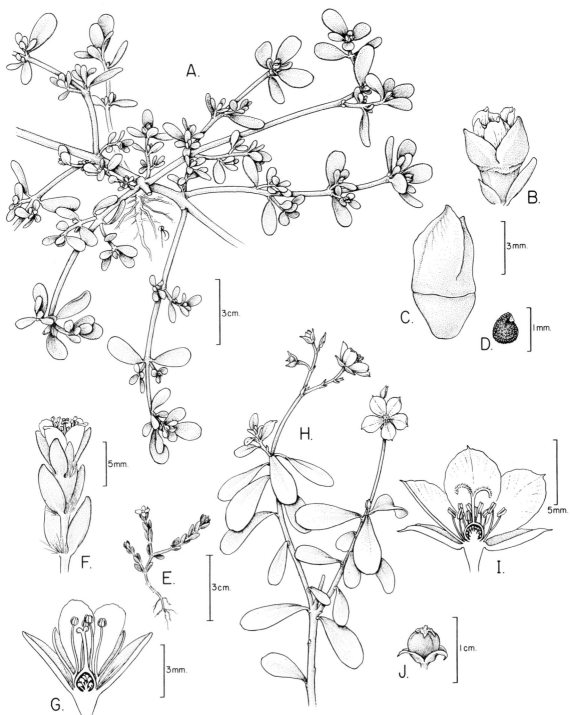

FIG. 169. A–D. *Portulaca oleracea*. **A.** Habit. **B.** Flower. **C.** Capsule. **D.** Seed. **E–G.** *Portulaca quadrifida*. **E.** Habit. **F.** Flowering branch. **G.** L.s. flower. **H–J.** *Talinum fruticosum*. **H.** Flowering branch. **I.** L.s. flower. **J.** Capsule.

entire. Flowers solitary or paired at ends of short lateral branches. Sepals green, 3–4 mm long; petals ca. 4 mm long, yellow or rarely pink, obovate. Capsule 3–5 mm wide, depressed-globose, opening above the middle, the lid saucer-shaped. Seeds ca. 0.6 mm long, reddish brown, tuberculate.

DISTRIBUTION: An occasional herb of open coastal and waterlogged areas. Trail to Fortsberg (A4078); Hermitage Ruins (A2427), Lind Pt.

(A2694). Also on St. Croix; throughout the Greater and Lesser Antilles and Mexico to northern South America.

DOUBTFUL RECORDS: *Portulaca teretifolia* Kunth was reported by Woodbury and Weaver (1987) as occurring on St. John; this record was based on a misidentification of a *P. rubricaulis* specimen. *Portulaca halimoides* L. was reported by them as well but has not been confirmed since no specimen of it was made.

2. TALINUM Adans., nom. cons.

Erect, fleshy herbs, sometimes woody at base, often with tuberous roots. Leaves alternate, slightly succulent; stipules wanting. Flowers bisexual, actinomorphic, in terminal racemes or panicles; calyx of 2 sepals; petals 5, brightly colored, early deciduous; stamens 5 to numerous, the filaments free, the anthers opening by longitudinal slits; the ovary superior, 1-locular, the locules with numerous basal ovules, the styles 3, more or less united, the stigmas linear. Fruit a unilocular loculicidal capsule, opening along 3 valves with numerous reniform to rounded, minute seeds.

A pantropical genus of about 50 species.

Key to the Species of *Talinum*

1. Inflorescence a panicle; petals 3.5–5 mm long .... 2. *T. paniculatum*
1. Inflorescence racemose; petals 6–10 mm long .... 1 *T. fruticosum*

1. Talinum fruticosum (L.) Juss., Gen. Pl. 312. 1789. *Portulaca fruticosa* L., Syst. Nat., ed. 10, **2:** 1045. 1759.   Fig. 169H–J.

Portulaca triangularis Jacq., Enum. Syst. Pl. 22. 1760. *Talinum triangulare* (Jacq.) Willd., Sp. Pl. 2: 862. 1799.

Erect subshrub to 50 cm tall, many-branched from a woody base, with a long taproot; stems succulent, glabrous, reddish-tinged to burgundy-colored. Leaf blades 2.5–4.5 × 1.1–2.2 cm, obovate to oblanceolate, fleshy, glabrous, lighter underneath, the apex rounded, obtuse, retuse and usually mucronulate, the base narrowed to a slender petiole, 3–7 mm long, the margins entire. Flowers in terminal racemose inflorescences, 2–10 cm long. Sepals 5–6 mm long, green, lanceolate, persistent; petals 5, 6–10 mm long, light yellow or magenta, imbricate, elliptic, apiculate. Capsule 5–6 mm wide, globose, irregularly opening. Seeds ca. 1 mm long, reddish brown to black, tuberculate.

DISTRIBUTION: A common plant of open disturbed areas, very often along roadsides. Lameshur (A2749), Mary Point (A2527). Also on St. Croix and St. Thomas; throughout the Greater and Lesser Antilles, Mexico, Central and South America.

2. Talinum paniculatum (Jacq.) Gaertn., Fruct. Sem. Pl. **2:** 219. 1791. *Portulaca paniculata* Jacq., Enum. Syst. Pl. 22. 1760.

Portulaca patens L., Mant. Pl. **2:** 242. 1771, nom. illegit. *Talinum patens* (L.) Willd., Sp. Pl. **2:** 863. 1799.

Erect shrub to 35(–100) cm tall, sparsely branched; stems woody at base, succulent above, glabrous. Leaf blades 2–7.5 × 1–4.5 cm, obovate to elliptic, slightly fleshy, glabrous, the apex obtuse, usually retuse and mucronulate, the base narrowed to a slender petiole 3–5 mm long, the margins entire. Flowers in terminal panicles, 20–40 cm long. Sepals orbicular, 3–4 mm long, early deciduous; petals 5, 3.5–5 mm long, light yellow or reddish, widely elliptic to rounded. Capsule 3.5–5 mm wide, nearly globose, irregularly opening. Seeds ca. 0.6 mm long, reddish brown, tuberculate, shiny.

DISTRIBUTION: An occasional shrub of rocky, open areas. East End (B1204), Lameshur (W851). Also on St. Croix and St. Thomas; throughout the Greater and Lesser Antilles, the United States, Mexico, Central and South America.

71. Rhamnaceae (Buckthorn Family)

Herbs, shrubs, twining or tendriled lianas or trees. Leaves alternate to seldom opposite, simple, pinnately veined, usually with toothed margins; stipules minute, or wanting, sometimes modified into spines. Flowers small to medium, bisexual or rarely unisexual, actinomorphic, 4–5-merous except for the 2–3-merous gynoecium, produced in terminal and or axillary cymes, panicles, or racemes or sometimes solitary; calyx of basally connate sepals forming a hypanthium; petals more or less concave, with stamens fitting into the concavity, clawed or rarely wanting; stamens opposite the petals, the filaments free, the anthers opening by longitudinal slits; nectary disk intrastaminal, sitting on top of the hypanthium, sometimes adnate to the ovary; ovary surrounded by a hypanthium, partly superior or seemingly inferior, plurilocular of 2–3(–5) united carpels, with basal placentation, the ovules solitary in each locule, the style terminal, 2–3(–5)-lobed. Fruit a capsule or a drupe, with separate stones or a plurilocular stone, sometimes separating into mericarps.

A family with 55 genera and about 900 species, cosmopolitan, but best represented in tropical and subtropical regions.

REFERENCE: Brizicky, G. K. 1964. The genera of Rhamnaceae in the southeastern United States. J. Arnold Arbor. 64: 439–463.

Key to the Genera of Rhamnaceae

1. Fruits dehiscent (splitting into 3 mericarps); leaves alternate.
   2. Woody vines, climbing by tendrils; fruit a 3-winged schizocarp, splitting into 3 indehiscent mericarps
     .... 2. *Gouania*
   2. Shrubs or small trees; fruit a depressed-globose septicidal capsule, splitting into 3, dehiscent pyrenelike mericarps .... 1. *Colubrina*
1. Fruits indehiscent, with a slightly fleshy exocarp when fresh, but drying crustaceous; leaves opposite or nearly so.
   3. Fruit 1.5–1.7 cm long, turning blackish at maturity; petals present .... 4. *Reynosia*
   3. Fruit 5–7 mm long, turning purplish red at maturity; petals wanting .... 3. *Krugiodendron*

1. COLUBRINA Brongn., nom. cons.

Trees or shrubs, commonly scandent. Leaves alternate, the blades entire or serrate, coriaceous, usually bearing glandular spots; stipules lateral and basal, or rarely intrapetiolar, deciduous. Flowers bisexual, actinomorphic, several in axillary cymes or umbel-like thyrses or rarely solitary; hypanthium crowned by 5 spreading, triangular calyx lobes; petals yellowish green to whitish, hood-shaped, sessile or shortly clawed at base, shorter than the calyx lobes; stamens as long as the petals, at some stages usually hooded by the petals, the filaments free; nectary disk annular, fleshy, adnate to the base of ovary; ovary partly superior, 3-locular, each locule with a single ovule, the style slender, trilobed. Fruit a trilocular septicidal capsule, splitting into 3 ventrally dehiscent pyrenelike mericarps; seeds obovate in outline, lustrous, brown to black.

A pantropical genus of 31 species, most of which are centered in tropical America.

REFERENCE: Johnson, M. C. 1971. Revision of *Colubrina* (Rhamnaceae). Brittonia 23: 2–53.

Key to the Species of *Colubrina*

1. Leaves mostly 6.5–15 cm long, the blades with scattered submarginal glandular spots; twigs densely to-mentose when young.. 1. *C. arborescens*
1. Leaves mostly 3.5–7.5 cm long, the blades with 1 or 2 pairs of marginal glands at base; twigs ferrugi-nous-pubescent to sparsely tomentose when young .... 2. *C. elliptica*

1. Colubrina arborescens (Mill.) Sarg., Trees & Shrubs 2: 167. 1911. *Ceanothus arborescens* Mill., Gard. Dict., ed. 8. 1768.   Fig. 170A–G.

Rhamnus colubrina Jacq., Enum. Syst. Pl. 16. 1760. *Colubrina colubrina* (Jacq.) Millsp., Publ. Field Columbian Mus., Bot. Ser. **2**: 69. 1900. *Colubrina ferruginosa* Brongn., Mém. Fam. Rham. 62. 1826.

Shrub or small tree 1–10 m tall, single-stemmed or many-branched from base; bark gray, smooth; twigs densely ferruginous-tomentose when young. Leaf blades (4.5–)6.5–15 × (2.3–)3–9.2 cm, ovate or elliptic, subcoriaceous, the lower surface sparsely to densely ferruginous-tomentose, with dispersed submarginal (less often close to the midvein) glandular dots, the apex obtuse, acute, or shortly acuminate, the base obtuse to rounded, the margins entire; petioles 0.5–2 cm long. Flowers clustered in short axillary cymes, with densely ferruginous-tomentose axes. Calyx densely ferruginous–tomentose without, the sepals 2–2.3 mm long, triangular, glabrous and greenish within, deciduous in fruit; petals yellowish, ca. 1.7 mm long, narrowed at base; stamens ca. 2 mm long, the filaments broader at base; disk bright yellow, fleshy, lobed to crenate; ovary green. Capsule 5–7 mm wide, globose-trigonous, with a persistent cuplike hypanthium at base, turning from green to black-brown. Seeds 3–4 mm long, shiny black.

DISTRIBUTION: A common coastal tree, found mostly in sandy soils. Haulover Point (A4225). Also on Anegada, Buck Island, Jost van Dyke, St. Croix, St. Thomas, and Virgin Gorda; throughout the West Indies, southern Florida, southern Mexico, and Central America.

2. Colubrina elliptica (Sw.) Brizicky & W.L. Stern, Trop. Woods **109**: 95. 1958. *Rhamnus ellipticus* Sw., Prodr. 50. 1788; Fl. Ind. Occid. **1**: 497. 1797.

Ceanothus reclinatus L'Hér., Sert. Angl. 6. 1789. *Colubrina reclinata* (L'Hér.) Brongn., Mém. Fam. Rham. 62. 1826; Ann. Sci. Nat. (Paris) **10**: 369. 1827.

Shrub or small tree 3–5 m tall, many-branched from base or less often with a single unbranched trunk; bark light brown, smooth; twigs ferruginous-pubescent to sparsely tomentose when young. Leaf blades (2.5–)3.5–7.5(–11.5) × (1.2–)2–3.5(–5) cm, elliptic, oblong-ovate, chartaceous, glabrous to densely ferrugi-nous-tomentose beneath, the apex obtuse or shortly to long-acumi-nate, the base obtuse to rounded, the margins entire, with 2–4 minute glands near the base; petioles 1–1.5 cm long, glabrous or puberulent. Flowers clustered in short axillary tomentulose cymes. Calyx densely ferruginous-tomentose without, the sepals ca. 1.2–1.5 mm, triangular, glabrous and greenish within, deciduous in fruit; petals yellowish, ca. 1 mm long, elliptic; stamens slightly longer than the petals; disk fleshy, lobed to crenate. Capsule ca. 5 mm diam., globose-trigonous, with a persistent cuplike hypanthium at base, turning from green to red at maturity. Seeds 3–4 mm long, smooth, reddish brown, flattened-convex.

DISTRIBUTION: A common shrub of moist to dry forest or coastal scrub. Road to Bordeaux (A3863), Lameshur (B518). Also on Anegada, Jost van Dyke, St. Croix, St. Thomas, Tortola, and Virgin Gorda; throughout the West Indies, southern Florida, and central Mexico to Venezuela.

COMMON NAMES: maubi, snake root.

2. GOUANIA Jacq.

Woody vines, climbing by simple, axillary tendrils, usually basal to the inflorescences. Leaves alternate, chartaceous to coriaceous, usually serrate on margins, the teeth glandular; stipules small, persistent. Flowers bisexual or less commonly unisexual too, actinomor-phic, in axillary or terminal spikes, racemes, or panicles; hypanthium obconical to bell-shaped, crowned by 5 persistent calyx lobes; petals yellowish green to whitish, hood-shaped, shortly clawed at base; stamens as long as the petals, usually hooded by the petals, the filaments adnate to the margin of nectary disk, alternating with staminodial flanges; nectary disk cup-shaped, with staminodial flanges; ovary seemingly inferior, 3-locular, the locules with a single ovule, the style slender, with 3 reflexed stigmas. Fruit a trilocular, 3-winged, septicidal schizocarp, splitting into 3 indehiscent mericarps; seeds obovate in outline, shiny.

A pantropical genus of about 50 species.

1. Gouania lupuloides (L.) Urb., Symb. Ant. **4**: 378. 1910. *Banisteria lupuloides* L., Sp. Pl. 427. 1753.   Fig. 170H–P.

Rhamnus domingensis Jacq., Enum. Syst. Pl. 17. 1760. *Gouania domingensis* (Jacq.) L., Sp. Pl., ed. 2, **2**: 1663. 1763.

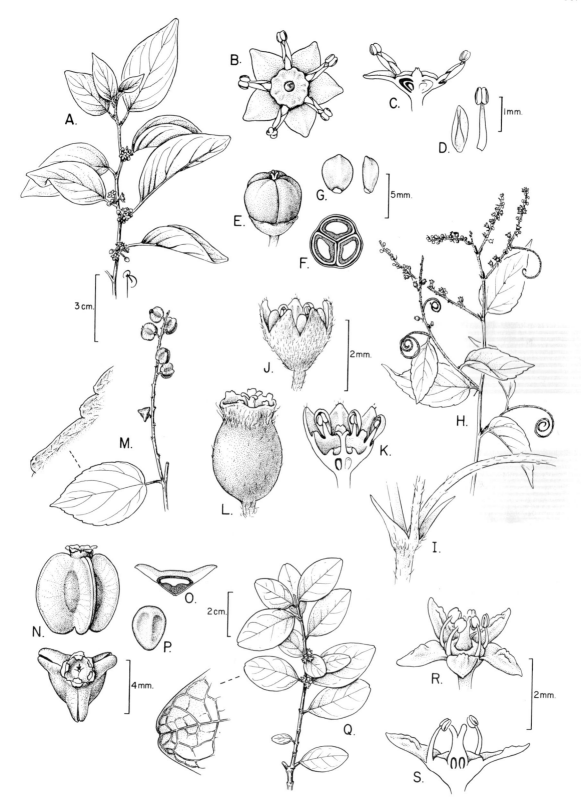

FIG. 170. A–G. *Colubrina arborescens*. **A.** Flowering branch. **B.** Flower. **C.** L.s. flower. **D.** Petal and stamen. **E.** Capsule. **F.** C.s. capsule. **G.** Seed, frontal and lateral views. **H–P.** *Gouania lupuloides*. **H.** Flowering branch. **I.** Detail of stipules. **J.** Flower. **K.** L.s. flower. **L.** Immature fruit. **M.** Fruiting branch and detail of leaf margin showing trichomes. **N.** Mature fruit, lateral and top views. **O.** C.s. mericarp. **P.** Seed. **Q–S.** *Krugiodendron ferreum*. **Q.** Flowering branch and detail of leaf apex. **R.** Flower. **S.** L.s. flower.

Liana 5–12 m long, profusely branching along main stem; twigs nearly cylindrical, glabrous to sparsely pubescent. Leaf blades 4.5–7.5 × 2–4 cm, ovate to elliptic, chartaceous, glabrous to sparsely pubescent, the apex acute to shortly acuminate, the base rounded to nearly cordate, the margins remotely serrate; petioles 0.5–1.5 cm long; stipules 2–3 mm long. Flowers in terminal racemes on short axillary branches, the axes sparsely pubescent, 5–10 cm long. Calyx pubescent to densely pubescent without, the sepals ca. 1 mm long, triangular, glabrous and greenish within, persistent in fruit; petals yellowish, ca. 1 mm long, narrowed at base; stamens slightly shorter than petals; disk ca. 1 mm high. Schizocarp 5–7 mm long, 3-winged, splitting into 3 trigonous, indehiscent mericarps. Seeds 3–4 mm long.

DISTRIBUTION: A common vine of open disturbed areas, such as roadsides and secondary forests. Bethany (B327), Susannaberg (A3812). Also on St. Croix, St. Thomas, and Tortola; throughout the West Indies, southern Florida, southern Mexico to northern South America.

3. KRUGIODENDRON Urb.

A monospecific New World genus characterized by the following species.

1. Krugiodendron ferreum (Vahl) Urb., Symb. Ant. **3:** 314. 1902. *Rhamnus ferreus* Vahl *in* H. West, Bidr. Beskr. Ste. Croix 203. 1793. *Condalia ferrea* (Vahl) Griseb., Fl. Brit. W. I. 100. 1859. Fig. 170 Q–S.

Shrub or small tree, 5–10 m tall; bark light brown, fissured; twigs nearly cylindric, puberulent. Leaves opposite or nearly so; blades 3–5.8 × 2–3.5 cm, ovate to elliptic, chartaceous, glabrous, the apex acute to obtuse, emarginate, the base obtuse or rounded, the margins entire to slightly crenate; petioles 2–5 mm long, puberulent; stipules subulate, ca. 1 mm long. Flowers greenish, 5-merous, in short axillary cymes, the axes puberulent, to 1 cm long; pedicels 2.5–3 mm long. Calyx greenish, puberulent or glabrous without, the sepals 1.2–1.4 mm long, triangular, glabrous within, deciduous in fruit; petals wanting; stamens ca. 1.2 mm long, yellowish; disk annular, bright yellow, fleshy, adnate to base of ovary; ovary superior, greenish, 2-locular, each locule with a solitary ovule. Fruit a 1-seeded, nearly globose drupe, 5–7 mm long, turning from green to purplish red at maturity. Seeds ellipsoid, 4 mm long.

DISTRIBUTION: A common tree of coastal dry forests and middle-elevation moist forests. Little Lameshur Bay (A2024), Maria Bluff (A2332). Also on Jost van Dyke, St. Croix, St. Thomas, Tortola, and Virgin Gorda; throughout the West Indies, southern Florida, southern Mexico to Honduras.

COMMON NAMES: ebony, guatafer, ironwood.

4. REYNOSIA Griseb.

Shrubs or small trees. Leaves opposite, coriaceous, with entire margins; stipules connate, persistent. Flowers bisexual, 5-merous, actinomorphic, fascicled in sessile axillary cymes or solitary; hypanthium short, bell-shaped, crowned by 5 deciduous calyx lobes; petals greenish, slightly concave, shortly clawed at base or wanting; stamens slightly longer than petals, the filaments subulate; nectary annular or lobed, fleshy; ovary superior, 2-locular, the locules with a single ovule, the style short, slender, with 2 small stigmas. Fruit a unilocular drupe; seeds 1 per fruit, ovoid.

A predominantly Greater Antillean genus with about 16 species.

1. Reynosia guama Urb., Symb. Ant. **1:** 356. 1899. Fig. 171.

Reynosia latifolia sensu Eggers, Vidensk. Meddel. Dansk Naturhist. Foren. Kjøbenhavn **9:** 173. 1878, non Griseb., 1866.

Shrub or small tree, 3–7 m tall; bark light gray, smooth; twigs nearly cylindric, puberulent. Leaves opposite to nearly alternate; blades 3–10.2 × 2–4.5 cm, elliptic, oblong, or seldom ovate, rigidly coriaceous, glabrous, midvein impressed on upper surface, the apex obtuse, deeply emarginate, the base obtuse to rounded, the margins strongly revolute; petioles 5–6 mm long, glabrous; stipules subulate, ca. 1.8 mm long. Flowers greenish, fasciculate in sessile axillary cymes; pedicels 4–6 mm long, glabrous. Calyx greenish, glabrous, the sepals 1.2–1.5 mm long, triangular, keeled within, deciduous in fruit; petals wedge-shaped, notched at apex, ca. 1.4 mm long; stamens slightly longer than the petals, the filaments adnate to the margin of disk; disk cup-shaped; ovary superior, greenish, 2-locular, each locule with a solitary ovule. Fruit a 1-seeded, ellipsoid to globose drupe, 1.5–1.7 cm long, apiculate at apex, turning from green to purplish black at maturity. Seeds nearly globose, 1 cm long.

DISTRIBUTION: An occasional tree of coastal dry forests at middle-elevation and moist forests. Bordeaux Mountain (A5121), Cruz Bay (A2937), White Cliffs (A2031). Also on Guana Island, Jost van Dyke, St. Thomas, and Virgin Gorda; Puerto Rico.

CULTIVATED SPECIES: *Ziziphus mauritiana* Lam. is known from a few cultivated trees growing in front yards.

DOUBTFUL RECORD: *Ziziphus rignonii* Delpino was reported for St. John by Woodbury and Weaver (1987), but no specimen was collected.

72. Rhizophoraceae (Red Mangrove Family)

Trees or shrubs, often with mangrove habit. Leaves opposite, simple, pinnately veined, entire; stipules forming a sheath around the terminal buds, glandular (with colleters) within at base, early deciduous, leaving a ring scar. Flowers bisexual, actinomorphic, in short axillary cymes or solitary; calyx of 4–5 valvate sepals (or more numerous), subtended by 2 connate bracts; corolla of 4–5 distinct, fleshy petals (sometimes more numerous), usually shorter than the sepals; stamens 2–3 times as many as the petals, the filaments free or connate at base, the anthers opening by longitudinal slits; nectary disk intrastaminal, perigynous; ovary partly inferior, 2–5-carpellate, 2–5-locular (seldom unilocular), with apical-axile placentation, the ovules 2(–4) in each locule, pendulous, the style usually 1, terminal, the stigmas 2 or as many as the locules. Fruit a leathery 1-seeded berry, drupe, or capsule.

A tropical family with 14 genera and about 100 species.

REFERENCE: Graham, S. A. 1964. The genera of Rhizophoraceae and Combretaceae in southeastern United States. J. Arnold Arbor. 65: 285–301.

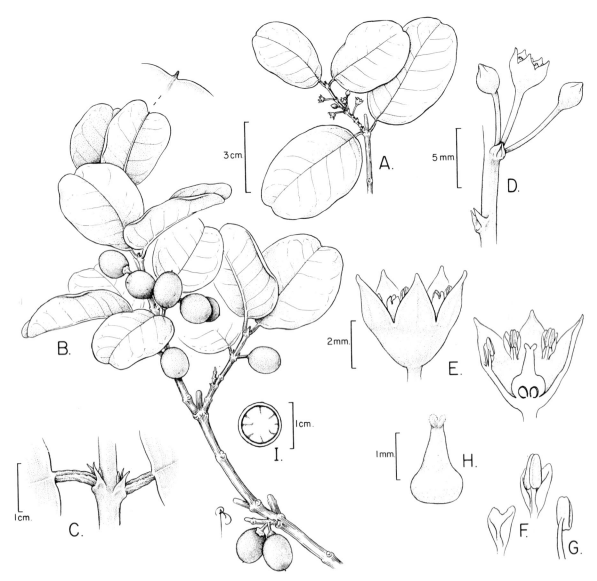

Fig. 171. *Reynosia guama*. **A.** Flowering branch. **B.** Fruiting branch with detail of leaf apex. **C.** Detail of paired stipules. **D.** Detail of inflorescence. **E.** Flower and l.s. flower showing stamens and pistil. **F.** Petal and petal surrounding stamen. **G.** Stamen, dorsal view. **H.** Pistil. **I.** C.s. fruit.

1. RHIZOPHORA L.

Shrubs or small trees; stems supported by adventitious prop roots. Leaves opposite, coriaceous, glabrous, with entire margins; stipules foliaceous, enclosing the terminal buds, early deciduous. Flowers bisexual, 4-merous, actinomorphic, in pedunculate axillary cymes; sepals leathery, keeled within; petals as long as or shorter than the sepals, densely villous along margins; stamens 8, the filaments free, short or wanting, inserted on a cup-shaped nectary disk, the anthers basifixed, opening by a longitudinal slit; ovary partly inferior, 2-locular, each locule with a pair of pendulous ovules, the style slender, the stigma bilobed. Fruit an indehiscent berry; seeds 1 per fruit, the embryo germinating while still attached to the tree, producing a long thick hypocotyl.

A pantropical genus of about 9 species, growing along coastal tidal swamps.

1. Rhizophora mangle L., Sp. Pl. 443: 1753.

Shrub or small tree 1–12(–25) m tall, with mangrove habit; bark light gray, smooth. Leaf blades 3–10 × 2–5.7 cm, elliptic to obovate, coriaceous, glabrous, the lower surface with prominent stout midvein, punctate, the apex obtuse, the base attenuate to cuneate, the margins entire, revolute; petioles 0.5–2.5 cm long, glabrous; stipules lanceolate, to 3.5 cm long, with glandular hairs at base within. Cymes 2(–5)-flowered; peduncles 2.5–5 cm long, glabrous; pedicels stout with a pair of bracteoles subtending the flower. Calyx greenish, bell-shaped, the sepals 1–1.5 cm long, lanceolate, thick, keeled within, persistent in fruit; petals 1 cm long, lanceolate, reflexed, turning from cream-colored to yellowish with age; stamens sessile, the anthers ca. 5 mm long; ovary ovoid.

Berry conical, 2–2.5 cm long, the hypocotyl 15–30 cm long, stout, fleshy.

DISTRIBUTION: A common species of mangrove swamps, along pro- tected shallow coasts. Probably on all Virgin Islands; common throughout tropical and subtropical America and West Africa.

COMMON NAMES: mangelboom, red mangrove.

73. Rosaceae (Rose Family)

Trees, shrubs, herbs, or scandent shrubs. Leaves alternate or rarely opposite, simple or variously compound or dissected; stipules present or seldom wanting. Flowers medium to large, bisexual, actinomorphic, solitary or in various sorts of cymose inflorescences; calyx usually forming a definite hypanthium, the lobes (3–)5(–10), valvate; corolla of free petals, these as many as the calyx lobes, often showy or seldom wanting; stamens often numerous, the filaments free or connate at base and attached to the hypanthium, the anthers opening by longitudinal slits or rarely by terminal pores; gynoecium of 1 to many distinct carpels, or the carpels united into a compound, inferior to superior ovary, ovules 2 or more in each carpel, the placentation axile, with 2–5 styles, usually free to base. Fruit of various types, commonly follicles, achenes, drupes, or capsules.

A family with about 100 genera and 3000 species, nearly cosmopolitan, but most diverse in temperate zones.

1. PRUNUS L.

Shrubs or trees. Leaves alternate, simple, coriaceous, glabrous, with entire or toothed margins, the base usually with a pair of glandular dots; stipules minute, lanceolate, early deciduous. Flowers bisexual, 5-merous, actinomorphic, in axillary racemes or solitary; hypanthium cone-shaped, sepals triangular, deciduous; petals longer than the sepals, deciduous; stamens 15–30, the filaments connate into a short tube; ovary superior, sessile, of a single carpel, with 2 ovules, the style slender, distal, the stigma cone-shaped to truncate. Fruit a 1-seeded, fleshy or dry drupe, ellipsoidal or bilobed.

A primarily temperate genus with about 450 species, about a dozen species naturally occurring in tropical America.

1. Prunus pleuradenia Griseb., Fl. Brit. W. I. 231. 1860.

Cerasus sphaerocarpa Hook., Bot. Mag. **59**: t. 3141. 1832, non Loisel., 1760.

Prunus dussii Krug & Urb. *in* Duss, Fl. Phan. Antill. Franç. 259. 1897.

Prunus acutissima Urb., Symb. Ant. **5**: 349. 1907.

Tree to 15 m tall, glabrous; bark reddish brown, rough; twigs angular when young, turning cylindrical and lenticellate. Leaf blades 5–9 × 2.5–4.7 cm, elliptic to elliptic-ovate, coriaceous, glabrous, V-shaped or folded, the apex shortly acuminate, the base obtuse, the margins entire, revolute; petioles 0.5–1 cm long, glabrous; stipules early deciduous. Inflorescence axillary, buds with overlapping scales.

DISTRIBUTION: Known only from a single, sterile collection (A4703) from Bordeaux Mountain, in humid remnant forest. The vegetative charac- ters of this specimen agree with those of *Prunus pleuradenia*. See Howard (1988) for a discussion of the correct name of this species, which he considered as endemic to the Lesser Antilles.

74. Rubiaceae (Coffee Family)
Reviewed by C. M. Taylor

Trees, shrubs, herbs, vines, or lianas. Leaves opposite (decussate), whorled, or rarely alternate, simple; stipules intrapetiolar or interpetiolar. Flowers bisexual or rarely unisexual, in cymose inflorescences or rarely solitary; calyx forming a hypanthium, the lobes 4–5; corolla tubular, (3–)4–5-lobed, actinomorphic or rarely zygomorphic; stamens as many as and alternate with the corolla lobes, the filaments attached to the corolla tube, the anthers opening by longitudinal slits; nectary disk intrastaminal, sitting on top of ovary; ovary inferior, of 2(–3–5) carpels, with axile placentation, each locule with 1 to many ovules, the style terminal, often 2-branched, with a capitate or bilobed stigma. Fruit of various types, commonly a berry, drupe, or capsule.

A family with about 640 genera and 10,000 species, nearly cosmopolitan, but mainly in tropical and subtropical areas.

REFERENCES: Dwyer, J. D. 1980. Flora of Panama. Rubiaceae. Ann. Missouri Bot. Gard. 67: 1–522; Robbrecht, E. 1988. Tropical woody Rubiaceae. Opera Bot. Belg. 1: 1–271; Steyermark, J. A. 1974. Rubiaceae. Fl. Venezuela, vol. 9 (3 parts). Instituto Botanico, Caracas, Venezuela.

Key to the Genera of Rubiaceae

1. Herbs, if woody only so at base.
    2. Creeping, spreading herb, rooting at nodes; fruit a fleshy drupe, bright red .... 9. *Geophila*
    2. Erect or decumbent herbs, many-branched from a nearly woody base; fruit a dry capsule separating into 2 mericarps.
        3. Mericarp (one or both) splitting full length .... 20. *Spermacoce*
        3. Mericarps indehiscent or dehiscent only at base .... 4. *Diodia*
1. Shrubs, trees, or lianas.
    4. Twining liana or scandent shrub .... 1. *Chiococca*
    4. Erect or arching shrubs or trees.
        5. Fruit a dry capsule or schizocarp.

6. Flowers 5-merous, solitary in leaf axils; corolla 6–9 cm long, the lobes linear or oblong ..... 6. *Exostema*
6. Flowers 4-merous, in axillary or terminal cymes; corolla <1.3 cm long, the lobes rounded.
    7. Plants bearing numerous, short, lateral branches and pairs of decussate, ascending spines at
        nodes; leaves <1.2 cm long; calyx lobes oblong, 1.8–2.6 mm long; corolla ca. 3 mm long,
        cream or white, villous within .... 13. *Machaonia*
    7. Plants unarmed; leaves 4–8 cm long; calyx lobes filiform, 7–8 mm long; corolla 10.5–13
        mm long, pink, with an annular thickening at the throat...................................... 18. *Rondeletia*
5. Fruit a fleshy or leathery drupe or a berry.
    8. Plants armed with 2 or more spines on short axillary branches.
        9. Leaves 4–7 mm long, coriaceous; flower 4-merous; corolla reddish brown, with yellowish
           lobes; fruit fleshy, ellipsoid, ca. 7 mm long, white... 19. *Scolosanthus*
        9. Leaves 1–3 cm long, chartaceous; flowers 5-merous; corolla white; fruit leathery to crusta-
           ceous, globose to ovoid, 1–1.2 cm diam., turning from green to white (black when dried)
           .... 17. *Randia*
    8. Plants unarmed.
      10. Fruits >4 cm long.
          11. Fruits compound, fleshy, white, ovoid, potato-shaped, with numerous eyes (scars where
             the corollas were attached); inflorescence a congested head; corolla to 1 cm long.... 14. *Morinda*
          11. Fruit simple, coriaceous, brown, globose to nearly ellipsoid, smooth; inflorescence a
             few-flowered cyme or the flowers solitary; corolla 2.2–2.6 cm long .... 8. *Genipa*
      10. Fruits <2.5 cm long.
        12. Flowers 4-merous.
          13. Shrubs; inflorescence long-racemose, arched toward the apex; calyx crowned by
             lanceolate sepals; fruits white, bluish, or lavender............................. 10. *Gonzalagunia*
          13. Shrubs or small trees; inflorescence cymose; calyx truncate at apex; fruits black.
            14. Cymes sessile, to 2 cm long, the axes reddish; corolla tube reddish, <5 mm
               long, the lobes oblong, rounded at apex .... 12. *Ixora*
            14. Cymes 5–10 cm long, the axes green; corolla tube white, ca. 7–12 mm long,
               the lobes lanceolate, acuminate at apex .... 7. *Faramea*
        12. Flowers 5(–6)-merous.
          15. Hypanthium, calyx, and corolla yellow; inflorescence axes bright orange to red
             .... 15. *Palicourea*
          15. Hypanthium and/or calyx green; corolla white, sometimes greenish- or pinkish-
            tinged; inflorescence axes green or sometimes pinkish-tinged.
            16. Corolla pinkish-tinged; bark smooth, peeling off in large irregular plates
               .... 11. *Guettarda*
            16. Corolla white or sometimes greenish-tinged; bark smooth or rough, but not
             peeling off in plates.
              17. Flowers sessile or nearly so and axillary .... 3. *Coffea*
              17. Flowers in long-branched racemes, corymbs, or panicles.
                18. Trees 8–10 m tall; fruit narrowly ellipsoid, green....................... 2. *Chione*
                18. Shrubs 1–3 m tall; fruits globose, oblate, or obovoid, red, black, or
                  white.
                  19. Corolla bell- or trumpet–shaped, 3–4 mm long, the lobes 2–3
                     mm long; fruit globose or bilobed, bright red, 2-seeded...... 16. *Psychotria*
                  19. Corolla funnel-shaped, 7–10 mm long, the tube 1–2 mm long;
                     fruit depressed-globose, globose, or ovoid, white or black,
                     8–10-seeded.. 5. *Erithalis*

1. CHIOCOCCA P. Browne

Twining lianas or shrubs. Leaves coriaceous or nearly so, entire; stipules interpetiolar, deciduous. Flowers bisexual or pistillate (the plant polygamous-dioecious), 5-merous, actinomorphic, in axillary racemes; calyx 5-lobed; corolla funnel-shaped to bell-shaped, the lobes spreading or reflexed; stamens 5, the filaments attached to the base of the corolla tube; ovary inferior, 2-carpellate, each locule with a pendulous ovule, the style filiform. Fruit a flattened drupe of 2 pyrenes.

A neotropical genus with about 20 species.

1. Chiococca alba (L.) Hitchc., Annual Rep. Missouri Bot. Gard. **4:** 94. 1893. *Lonicera alba* L., Sp. Pl. 175. 1753. Fig. 172A–H.

Chiococca racemosa L., Syst. Nat., ed. 10, **2:** 917. 1759, nom. illegit.
Chiococca racemosa var. *longifolia* DC., Prodr. **4:** 482. 1830.

Scandent shrub or liana to 10 m long, with numerous lateral decussate branches; main stem twining and furrowed; cross section of older stems with numerous cortical bundles; twigs nearly cylindrical, puberulent. Leaf blades 3–8 × 1–3.5 cm, elliptic to nearly lanceolate, coriaceous, glabrous, the apex obtuse, acute, or acuminate, the base obtuse or acute, abruptly narrowed onto the petiole, the margins entire, revolute; petioles 4–6 mm long, glabrous or puberulent; stipules ca. 2 mm long, acicular, widened at base. Flowers bisexual or pistillate (the plant polygamous-dioecious), few in axillary racemes, 2.5–5 cm long; pedicels 3.5–5 mm long, few-flowered. Hypanthium 2–2.2 mm long, elliptic, green, crowned by cup-shaped calyx 1–1.2 mm long; sepals triangular, ca. 0.5 mm long; corolla funnel-shaped (5-angled in cross section), yellow, often with reddish lines along angles without, the tube 4.5–6 mm long, the lobes reflexed, triangular, 11.5–3 mm long; stamens slightly protruding beyond the corolla tube, the filaments 1.6–2.5 mm long, unequal, pilose, connate at base into a short tube, the anthers oblong, 2–3 mm long, indehiscent in pistillate flowers; style club-shaped and long-exserted in pistillate flowers, bilobed and slightly exserted in bisexual flowers. Drupe 5–7 mm long, fleshy, nearly circular, laterally compressed, turning from green to white.

DISTRIBUTION: A common climber of open disturbed, moist environments and in mature secondary forests. Bordeaux (A1914), Rosenberg (B322). Also on St. Croix, St. Thomas, and Virgin Gorda; throughout the West Indies, Florida, and from Mexico to northern South America.

2. CHIONE DC.

Trees or shrubs. Leaves coriaceous, entire; stipules small, interpetiolar, deciduous. Flowers bisexual, 5-merous, actinomorphic, in terminal cymose inflorescences; calyx cup-shaped, with minute triangular lobes; corolla funnel-shaped, the lobes reflexed; stamens 5, the filaments attached to the base of the corolla tube, the anthers exserted; ovary inferior, 2-carpellate, each locule with a pendulous ovule, the style stout, club-shaped, the stigma slightly bilobed. Fruit a cylindric, ellipsoid, or nearly ovoid drupe of 2 pyrenes.

A genus of 10 species, from the West Indies and Central America.

1. Chione venosa (Sw.) Urb., Symb. Ant. **4:** 594. 1911. *Jacquinia venosa* Sw., Prodr. 47. 1788. Fig. 172I–M.

Tree 8–10 m tall; bark light brown; twigs slightly flattened when young, puberulent, turning cylindrical. Leaf blades 7–15.5 × 1.7–4.5 cm, elliptic to oblong, coriaceous, glabrous, the apex acute or shortly acuminate, the base acute, obtuse, or attenuate, the margins revolute; petioles 1–2 cm long; stipules early deciduous.

Flowers many in terminal, branched cymes; pedicels ca. 1 cm long. Hypanthium conical; calyx light green, cup-shaped, 1–2 mm long, with 5 triangular lobes; corolla funnel-shaped, white, the tube 5 mm long, the lobes reflexed, oblong, rounded, ca. 3 mm long; stamens long-exserted; style club-shaped, long-exserted, the stigma bifid or bilobed. Drupe 12–15 mm long, leathery, ellipsoid, with a pyrene.

DISTRIBUTION: A rare tree of moist forests. Bordeaux Mountain (A4700). Also on Tortola; Hispaniola, Puerto Rico, and the Lesser Antilles.

3. COFFEA L.

Shrubs or small trees. Leaves coriaceous to chartaceous, entire; stipules small, interpetiolar, persistent. Flowers bisexual, 4–5(–6)-merous, actinomorphic, in axillary fascicles, sessile or short-pedicellate; calyx nearly cylindrical; corolla funnel-shaped or trumpet-shaped, the lobes 4–5(–8), oblong, contorted in bud; stamens 4–5(–8), the filaments inserted on the corolla throat, the anthers exserted; ovary inferior, 2-carpellate, each locule with a single ovule, the style stout, the stigma bifid. Fruit an ellipsoid to nearly globose drupe of 2 pyrenes; seeds with a longitudinal fissure.

A genus of 40 species, native to the Old World, a few species widely cultivated throughout the tropics.

1. Coffea arabica L., Sp. Pl. 172. 1753. Fig. 173A–F.

Shrub 1.5–3 m tall, usually with numerous decussate branches; twigs slightly flattened, glabrous and green when young. Leaf blades 10–20 × 3–7 cm, elliptic, ovate-elliptic to oblong-elliptic, chartaceous, glabrous, the apex acuminate, the base acute or obtuse, the margins entire; petioles 5–15 mm long; stipules deltoid, 2–3 mm long. Flowers in axillary fascicles; pedicels 2–3 cm long. Calyx light green, nearly tubular, 1.5 mm long, with 5 minute lobes; corolla funnel-shaped, white, fragrant, the tube 7–12 mm long, the lobes 5–6, spreading, oblong, 7–10 mm long; stamens exserted; style long-exserted, the stigma bifid. Drupe 1.5–1.8 cm long, ellipsoid to globose, bright red when mature.

DISTRIBUTION: Persistent after cultivation in moist secondary forest. Gut from Bordeaux to Reef Bay (A5277). Formerly cultivated on most of the Virgin Islands; widely cultivated throughout the tropics.

COMMON NAME: coffee.

4. DIODIA L.

Annual or perennial herbs, often woody at base. Leaves opposite, chartaceous to rigid, short-petiolate or sessile; stipules interpetiolar, connate into sheath, with filiform appendages, persistent. Flowers bisexual, 4-merous, actinomorphic, sessile, in axillary glomerules; hypanthium small, obovoid, crowned by a small, 4-lobed calyx; corolla funnel-shaped to trumpet-shaped, the tube longer than the lobes, the lobes 4, valvate; stamens 4, exserted, the filaments inserted on throat of tube; ovary inferior, 2-carpellate, each carpel with an ascending ovule, the style filiform, bilobed. Fruit a dry schizocarp separating at maturity into 2 indehiscent or partly dehiscent mericarps; seeds oblong, convex.

A tropical and subtropical genus of about 50 species.

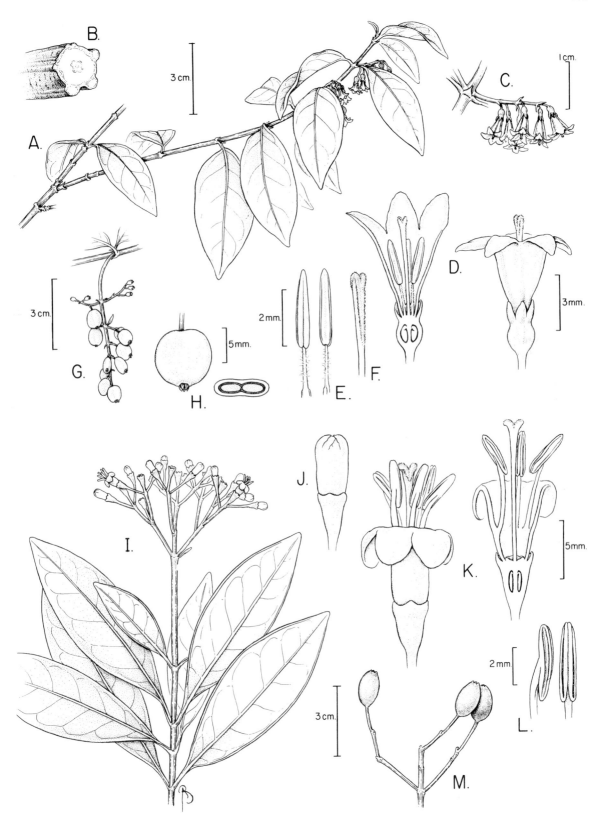

Fɪɢ. **172. A–H.** *Chiococca alba*. **A.** Flowering branch. **B.** C.s. older stem. **C.** Detail of inflorescence. **D.** L.s. flower and flower. **E.** Stamen, dorsal and lateral views. **F.** Detail of stigma. **G.** Infructescence. **H.** Drupe and c.s. drupe. **I–M.** *Chione venosa*. **I.** Flowering branch. **J.** Flowerbud. **K.** Flower and l.s. flower. **L.** Anther, lateral and frontal views. **M.** Infructescence.

1. Diodia ocymifolia (Willd.) Bremek., Recueil Trav. Bot. Neerl. **31:** 305. 1934. *Spermacoce ocymifolia* Willd. *ex* Roemer & Schult., Syst. Veg. **3:** 350. 1818. *Hemidiodia ocymifolia* (Willd.) Schum. *in* Mart., Fl. Bras. **6(6):** 9. 1888.

Erect or decumbent subshrub 30–50 cm tall, woody only at base; twigs 4-angular when young, becoming cylindrical later, ciliate on margins, pubescent to glabrescent. Leaf blades 2.5–7 × 1–3 cm, elliptic to oblong-lanceolate, chartaceous, slightly scabrous above, the apex acute to acuminate, the base cuneate, narrowed into the petiole, the margins strigillose; petioles 2–4 mm long; stipule sheath to 8 mm long, with numerous pubescent bristles. Flowers in axillary glomerules; bracts linear. Hypanthium 1.2–1.5 mm long, nearly ellipsoid, pubescent at base, the sepals ovate-lanceolate; corolla white, nearly funnel-shaped, the tube ca. 1.7 mm long, the lobes oblong, 1–1.2 mm long; stamens 4, exserted; style exserted, the stigma 2-lobed. Schizocarp ellipsoid-obovoid, pubescent, ca. 3 mm long, opening from the apex; cocci with a membranous partitioning wall that sometimes disintegrates at base. Seeds nearly ellipsoid, ca. 2.5 mm long, light brown, with a longitudinal hilum scar.

DISTRIBUTION: An uncommon weed of open disturbed areas, especially along roadsides. Road to Bordeaux (A3182, A3826). Throughout tropical America, introduced into the Old World tropics.

5. ERITHALIS P. Browne

Shrubs or small trees. Leaves coriaceous or nearly so, entire; stipules interpetiolar, connate into cup-shaped sheath, persistent. Flowers bisexual, 5–6-merous, actinomorphic, in axillary raceme, corymbs, cymes, or panicles; calyx cup-shaped; corolla funnel-shaped or trumpet-shaped, the lobes oblong, elongate; stamens 5–6, the filaments inserted at base of corolla tube, the anthers exserted; ovary inferior, 5–10-celled, each locule with a single ovule, the style stout, club-shaped. Fruit a depressed-globose or globose drupe, ridged when dry, of numerous pyrenes; seeds minute, oblong, compressed.

A genus of about 10 species in tropical and subtropical America.

1. Erithalis fruticosa L., Syst. Nat., ed. 10, **2:** 930. 1759. Fig. 173G–L.

Shrub 1–2 m tall, usually with numerous upright decussate branches; twigs slightly flattened, glabrous and green when young, older branches with leaf scars and persistent stipules. Leaf blades 4–10 × 2–6 cm, obovate, elliptic, to widely elliptic, coriaceous, glabrous, the apex obtuse or seldom acute, the base acute to attenuate, the margins entire, revolute; petioles 5–20 mm long; stipules deltoid, 2–3 mm long. Flowers many in axillary long-peduncled cymes, 5–10 cm long; pedicels 4–10 mm long. Hypanthium 1–1.5 mm long, green, ovoid, longitudinally ribbed; calyx cup-shaped, 1–1.5 mm long, truncate or obscurely toothed; corolla nearly funnel-shaped, white, the tube 1–2 mm long, the lobes 5, reflexed, oblong, 3.5–4.5 mm long; stamens long-exserted, the filaments white, the anthers lineate; style long-exserted, club-shaped. Drupe 5–7 mm long, oblate or depressed-globose and black, or ovoid to globose and white.

DISTRIBUTION: A common shrub of sandy coastal thickets and forests. Hawksnest Beach (A2904), Reef Bay (A3988), Southside Pond (A1812). Also on Anegada, St. Croix, St. Thomas, Tortola, and Virgin Gorda; Florida, Mexico (Yucatán), Belize, widespread throughout the West Indies.

NOTE: A specimen with white ovoid to globose fruits (A2903) is included under this name because of the almost identical vegetative and floral characters. Further studies may show this collection to be a different species.

6. EXOSTEMA Rich. *ex* Humb. & Bonpl.

Shrubs or small trees, glabrous or pubescent. Leaves membranous to coriaceous, entire; stipules interpetiolar, entire or bifid, persistent or deciduous. Flowers bisexual, 5(–4)-merous, actinomorphic, solitary and axillary or in terminal panicles or corymbs; hypanthium tubular or obovoid; calyx cup-shaped, 5(–4)-lobed; corolla trumpet-shaped, the tube nearly cylindrical, long or short, glabrous or pubescent at throat, the lobes oblong, elongate; stamens 5, the filaments inserted at base of corolla tube, the anthers linear, usually exserted; disk annular; ovary inferior, 2-carpellate, each locule with numerous ovules, the style filiform, the stigma capitate or bilobed. Fruit a septicidal capsule; seeds numerous, oblong, with a membranous wing.

A genus of 26–50 species, distributed throughout the West Indies and from Mexico to northern South America, south to the Peruvian Amazon.

1. Exostema caribaeum (Jacq.) Schult. *in* Roem. & Schult., Syst. Veg. **5:** 18. 1819. *Cinchona caribaea* Jacq., Enum Syst. Pl. 17. 1760. Fig. 174A–H.

Shrub 2–4 m tall; bark smooth, dark brown; twigs slightly flattened, glabrous, brown. Leaf blades 3.5–7.5 × 1.7–3 cm, elliptic to elliptic-ovate, chartaceous, glabrous or very often with a tuft of hairs in axils of midvein on lower surface, the apex acuminate, the base acute or obtuse, the margins entire or wavy; petioles 3–12 mm long; stipules ca. 4 mm long, broad-deltoid at base, with a long, subulate, coiling tip, deciduous. Flowers fragrant, solitary in leaf axils; peduncle 7–10 mm long. Hypanthium 5–6 mm long, green, bell-shaped, crowned by 5 deltoid sepals ca. 1 mm long; corolla nearly trumpet-shaped, white but later turning cream-colored, the tube 3–4 cm long, cylindrical, the lobes 5, spreading to reflexed, linear to oblong, 3–4 cm long; stamens long-exserted, the filaments white, the anthers lineate; style long-exserted, the stigma capitate. Capsule 1.3–1.7 cm long, ellipsoid, dark brown, septicidal, with each locule splitting longitudinally. Seeds ca. 4.5 mm long, lenticular, membranous, the margins winged.

DISTRIBUTION: A common shrub of coastal thickets and dry forests. Trail to Brown Bay (A1878), Fish Bay (A3911), Margaret Hill (A2308). Also on St. Croix and Tortola; widespread throughout the West Indies, the Florida Keys, and Mexico to Costa Rica.

COMMON NAMES: black torch, yellow torch.

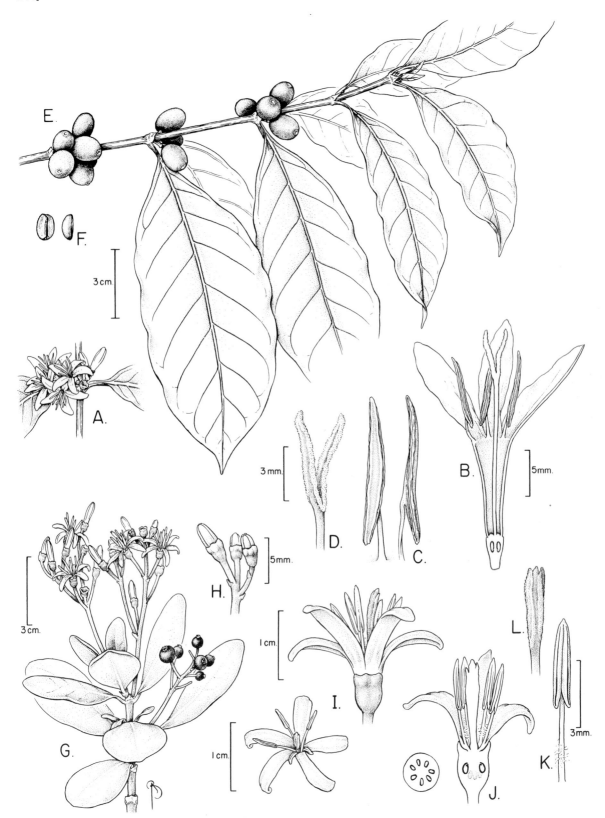

FIG. 173. A–F. *Coffea arabica.* A. Inflorescence. B. L.s. flower. C. Stamen, frontal and lateral views. D. Detail of stigma. E. Fruiting branch. F. Pyrene, frontal and lateral views. G–L. *Erithalis fruticosa.* G. Fertile branch. H. Detail of immature inflorescence. I. Flower, lateral and top views. J. L.s. flower and c.s. ovary. K. Stamen. L. Detail of stigma.

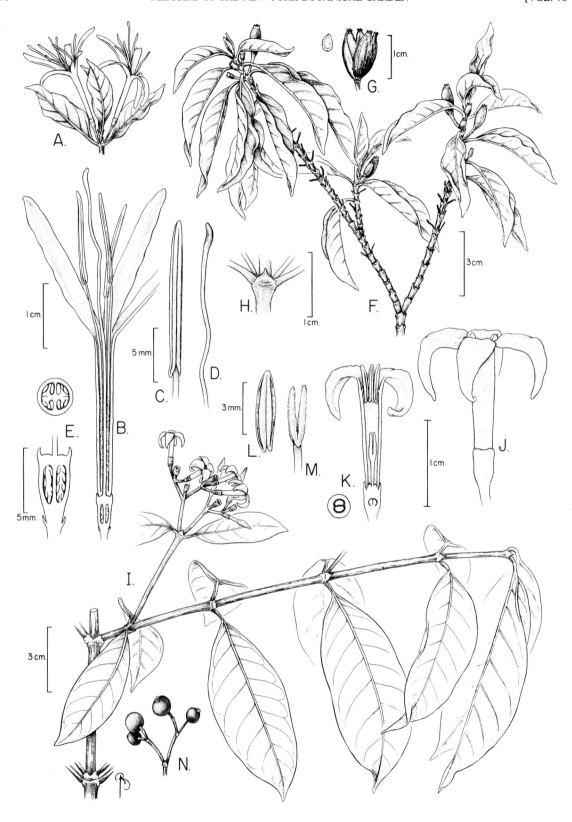

FIG. 174. A–H. *Exostema caribaeum*. **A.** Flowering branch. **B.** L.s. flower. **C.** Anther. **D.** Detail of upper portion of style showing stigma. **E.** C.s. and l.s. hypanthium. **F.** Fruiting branch. **G.** Dehiscing capsule and seed. **H.** Detail of internode showing stipule. **I–N.** *Faramea occidentalis*. **I.** Flowering branch. **J.** Flower. **K.** L.s. flower and c.s. ovary. **L.** Detail of anther. **M.** Detail of stigma. **N.** Infructescence.

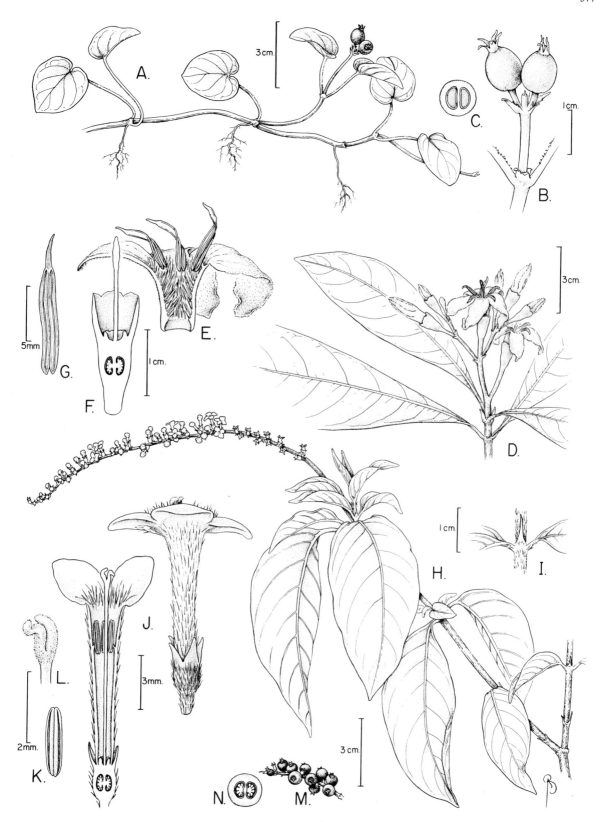

Fig. 175. A–C. *Geophila repens*. A. Habit. B. Infructescence. C. C.s. drupe. D–G. *Genipa americana*. D. Flowering branch. E. L.s. corolla showing adnate stamens. F. L.s. hypanthium showing ovary and style. G. Anther. H–N. *Gonzalagunia hirsuta*. H. Flowering branch. I. Detail of stipule. J. Flower and l.s. flower. K. Anther. L. Stigmas. M. Portion of infructescence. N. C.s. drupe.

7. FARAMEA Aubl.

Shrubs or small trees, usually glabrous. Leaves membranous to rigidly papery, entire; stipules interpetiolar, entire, connate at base, cuspidate at apex, persistent. Flowers bisexual, 4-merous, actinomorphic, in terminal or less often axillary panicles, corymbs, umbels, or heads; hypanthium ovoid to cup-shaped, crowned by a cup-shaped or tubular truncate or toothed calyx; corolla trumpet-shaped or funnel-shaped, with a glabrous throat, the lobes 4–5, oblong, elongate, valvate in bud; stamens 4–6, the filaments inserted at the throat or on the upper half of corolla tube, the anthers linear; ovary inferior, 2-carpellate, unilocular or seemingly bilocular, with 2 basal ovules, the style filiform, the stigma lineate. Fruit a 1-seeded berry.

A genus of about 130 species, of tropical America.

1. Faramea occidentalis (L.) A. Rich., Mém. Rubiac. 96. 1830. *Ixora occidentalis* L., Syst. Nat., ed. 10, **2:** 893. 1759. *Coffea occidentalis* (L.) Jacq., Enum. Syst. Pl. 16. 1760.   Fig. 174I–N.

Faramea odoratissima DC., Prodr. **4:** 496. 1830, nom. illegit.

Shrub 1.5–4 m tall, with lateral, opposite, ascending branches; bark light brown; twigs slightly flattened, glabrous, drying blackish. Leaf blades 7–14.2 × 2–6.5 cm, elliptic, rigidly chartaceous, glabrous, the apex abruptly acuminate-caudate, the base acute or obtuse, the margins entire; petioles 6–12 mm long; stipules 6–7 mm long, broadly deltoid at base, with a long caudate tip, tardily deciduous. Flowers fragrant, in terminal or axillary compound cymes, 5–10 cm long, the axes green, flattened. Hypanthium 1–1.5 mm long, green, cup-shaped, crowned by a tubular to cup-shaped, truncate calyx, 1.5–2 mm long; corolla nearly trumpet-shaped, white, the tube 7–12 mm long, cylindrical, the lobes 4, reflexed, lanceolate, acuminate, 4–8 mm long; stamens slightly exserted or included, the anthers lineate; style long-exserted or included, the stigma bilobed. Berry 8–10 mm wide, purplish black, oblate to depressed-ovoid, crowned by the persistent tubular calyx. Seeds obovoid to transversely ellipsoid, 3.5–5 × 5–7 mm.

DISTRIBUTION: A common shrub of secondary moist forests. Bordeaux (A3840), Cinnamon Bay (M17015). Also on St. Thomas and Tortola; widespread throughout the West Indies and Mexico to northern South America.

COMMON NAME: wild coffee.

8. GENIPA L.

Small to large trees, glabrous to variously pubescent. Leaves coriaceous, usually large; stipules interpetiolar, entire, often triangular, connate at base, persistent. Flowers bisexual, 5–6-merous, actinomorphic, in terminal or axillary cymes or less often solitary; hypanthium bell-shaped or cone-shaped, truncate at apex or 5-lobed; corolla shortly trumpet-shaped, pubescent or glabrous, the tube shorter than the oblong lobes; stamens 5–6, exserted, the filaments inserted in the throat, the anthers sessile or nearly so; ovary inferior, 2-carpellate, each carpel with numerous horizontal ovules, the style stout, exserted. Fruit a large coriaceous, firm, fleshy berry; seeds numerous, obtusely trigonous.

A genus of about 5 species, of tropical America.

NOTE: *Genipa* may have unisexual flowers as well. Steyermark (1974) described the genus as having bisexual flowers but then presented a figure of *G. americana* with supposedly staminate flowers.

1. Genipa americana L., Syst. Nat., ed. 10, **2:** 931. 1759.   Fig. 175D–G.

Tree 10–15 m tall; bark dark gray, smooth; twigs nearly cylindrical, rough, glabrous. Leaf blades 11–30 × 3–9.2 cm, narrowly elliptic to oblanceolate, chartaceous to coriaceous, puberulent along lower surface, the apex obtuse or acute, the base long-tapering, the margins entire; petioles stout, 1–1.5 cm long; stipules 8–13 mm long, broadly deltoid, concave, early deciduous. Flowers few in axillary compound cymes; the axes 3–7 cm long. Hypanthium 1–1.2 cm long, green, bell-shaped, 5-toothed; corolla cream-colored, appressed pubescent without, the tube 10–12 mm long, cylindrical, the lobes 5, reflexed, oblong-ovate, 12–14 mm long; stamens exserted, the anthers lineate; style long-exserted. Fruit 7–10 × 6–7 cm, brown, globose to nearly ellipsoid. Seeds numerous, nearly circular, flattened, 4–5 mm long.

DISTRIBUTION: An occasional tree of secondary moist forests. Battery Gut (A4164), Cinnamon Bay Trail (A4232). Also on St. Thomas; widespread throughout the West Indies, Central and South America.

9. GEOPHILA D. Don

Creeping herbs, with slender stems, rooting at nodes, glabrous or pubescent. Leaves chartaceous, long-petiolate, cordate at base; stipules interpetiolar, entire, orbicular to ovate, persistent. Flowers bisexual, sessile, 4–7-merous, actinomorphic, subtended by an involucre of bracts, in terminal or axillary few-flowered cymes; hypanthium short, urn-shaped or cup-shaped, crowned by 5–7 lineate or subulate sepals; corolla trumpet-shaped or funnel-shaped, white, the tube longer than the lobes; stamens 4–7, included, the filaments inserted in corolla tube; ovary inferior, of 2 carpels, each carpel with a basal ovule, the style club-shaped, included. Fruit a fleshy drupe, with 2 pyrenes; seeds 1 per pyrene.

A tropical genus of about 30 species, most of which occur in the Old World.

1. Geophila repens (L.) I.M. Johnst., Sargentia **8:** 281. 1949. *Rondeletia repens* L., Syst. Nat., ed. 10, **2:** 928. 1759.   Fig. 175A–C.

Geophila herbacea (Jacq.) Schum. *in* Engl. & Prantl, Nat. Pflanzenfam. **4(4):** 119. 1891. *Psychotria herbacea* Jacq., Enum. Syst. Pl. 16. 1760.

Creeping, mat-forming herb, sometimes covering an area of a few square meters. Leaf blades 2–4.5 × 1.8–4 cm, ovate, membranous, glabrous, the apex obtuse to rounded, the base cordate, the margins entire; petioles slender, 2–5.5 cm long, pilose; stipules ca. 2 mm long, oblong, deciduous. Flowers 2–4 in terminal involucrate cymes; the axes 1.2–4 cm long. Hypanthium 1–1.5 mm long, green, bell-shaped, the sepals 1–2.2 mm long, lanceolate; corolla white, funnel-shaped, glabrous without, the throat pilose, the tube 7–9 mm long, cylindrical, the lobes 5, oblong-ovate, to 5 mm long; stamens included, the anthers lineate; style included. Drupe ca. 1 cm long, ovoid or nearly globose, bright red, crowned by the persistent, lanceolate, green sepals. Seeds 2 per fruit, 3.5–5 mm long, ellipsoid, plano-convex, grooved.

DISTRIBUTION: A rare herb of shaded, moist forest. Trail to Reef Bay (A2672, A2927). Also on St. Thomas; widespread throughout the West Indies and Mexico to South America.

10. GONZALAGUNIA Ruíz & Pav.

Shrubs or small trees, usually with slender arching stems. Leaves chartaceous, petiolate; stipules interpetiolar, connate at base into a sheath, persistent. Flowers bisexual, 4–5-merous, actinomorphic, heterostylous, grouped into short cymes or solitary along terminal, arching racemes or spikes; hypanthium globose to bell–shaped, crowned by lanceolate or subulate sepals; corolla trumpet-shaped or narrowly funnel-shaped, white, the tube much longer than the lobes, pubescent within, the lobes imbricate or valvate in bud; stamens 4–5, included or slightly exserted, the filaments inserted on corolla tube; ovary inferior, 2–4-celled, each locule with numerous ovules, the style filiform, included or exserted, the stigma bilobed. Fruit a fleshy drupe, with 2 or 4 pyrenes; seeds minute, several per pyrene.

A neotropical genus of about 40 species.

1. **Gonzalagunia hirsuta** (Jacq.) Schum. *in* Mart., Fl. Bras. **6(6):** 291. 1889. *Justicia hirsuta* Jacq., Enum. Syst. Pl. 11. 1760. *Duggena hirsuta* (Jacq.) Britton & P. Wilson, Bot. Porto Rico **6:** 229. 1925.

Fig. 175H–N.

Gonzalagunia spicata (Lam.) M. Gómez, Anales Hist. Nat. Madrid **23:** 289. 1894. *Lygustrum spicatum* Lam., Tabl. Encycl. **1:** 286. 1792. *Duggena spicata* (Lam.) Standl., Contr. U.S. Natl. Herb. **18:** 126. 1916.

Duggena richardii Vahl *in* H. West, Bidr. Beskr. Ste. Croix 269. 1793, nom. illegit.

Shrub 1–2 m tall, many-branched from base; twigs nearly cylindric, appressed-pubescent when young. Leaf blades 5–13 × 1.2–5.3 cm, narrowly elliptic, lanceolate to ovate, chartaceous, sparsely pubescent, the apex acuminate, the base acute or obtuse, the margins entire; petioles slender, 0.3–1.5 cm long, pilose; stipules 4–7 mm long, subulate. Flowers 4–5-merous, heterostylous, appressed-pubescent without, grouped into short cymes (3-flowered) or solitary along a terminal raceme, the axis 12–25 cm long, appressed-pubescent; bracts and bracteoles subulate, 2–3 mm long; pedicels ca. 2 mm long. Hypanthium 1–1.2 mm long, green, bell-shaped; calyx tubular, 2–2.5 mm long, the sepals 1.5–2 mm long, lanceolate; corolla white, trumpet-shaped, pilose without, the throat densely pilose, the tube 6–9 mm long, narrowed at base, the lobes rounded, 2–3 mm long, spreading; stamens included or slightly exserted, the anthers oblong, ca. 2 mm long; style included or exserted, the stigma bifid. Drupe 0.6–1 cm long, depressed-globose, fleshy, white, lavender, or light blue, crowned by the persistent, green calyx; pyrenes 2, 3–5 mm long, nearly ellipsoid, with numerous dark brown, angular seeds ca. 0.5 mm long.

DISTRIBUTION: A common shrub of open, disturbed, moist areas. Road to Bordeaux (A2950), Maria Bluff (A610). Also on St. Thomas and Tortola; Hispaniola, Puerto Rico, the Lesser Antilles, Trinidad, Venezuela, the Guianas, and Brazil.

11. GUETTARDA L.

Shrubs or small trees, unarmed or less often with spiny branches. Leaves opposite or rarely 3 per node, coriaceous to rigidly coriaceous, petiolate; stipules interpetiolar, simple, persistent or deciduous. Flowers bisexual, or bisexual and unisexual on different individuals, 5–6(–9)-merous, actinomorphic, pedicellate, in axillary cymes, solitary or sessile in groups of 2 or 3; bract present or wanting; hypanthium cup-shaped or sometimes not distinguishable from the calyx; calyx tubular to ovoid, truncate or 2–9-toothed at apex; corolla trumpet-shaped, white, pink, or bright red, the tube much longer than the lobes, glabrous within, the lobes imbricate in bud; stamens 4–9, included, the filaments inserted in corolla tube; ovary inferior, 2–9-carpellate, each carpel with a solitary pendulous ovule, the style filiform, included, the stigma capitate. Fruit a leathery to fleshy drupe, with an inner stony layer containing 2–9 seeds.

A genus of about 100 species, predominantly in the West Indies, but also in Central America, South America, and a few in New Caledonia.

Key to the Species of *Guettarda*

1. Leaves 1.5–5 cm long, membranous to chartaceous, puberulent along veins in lower surface, smooth on upper surface; flowers in groups of 2 or 3 on pedunculate inflorescences .... 1. *G. odorata*
1. Leaves 3.5–15 cm long, rigidly coriaceous, densely pubescent on lower surface, scabrous (rough) on upper surface; flowers in long-peduncled cymose inflorescences .... 2. *G. scabra*

1. **Guettarda odorata** (Jacq.) Lam., Tabl. Encycl. **2:** 219. 1819. *Laugieria odorata* Jacq., Enum. Syst. Pl. 16. 1760.

Fig. 176A–D.

Guettarda parviflora Vahl, Ecogl. Amer. **2:** 26. 1798. *Guettarda parvifolia* Sw., Fl. Ind. Occid. **3:** 1958. 1806.

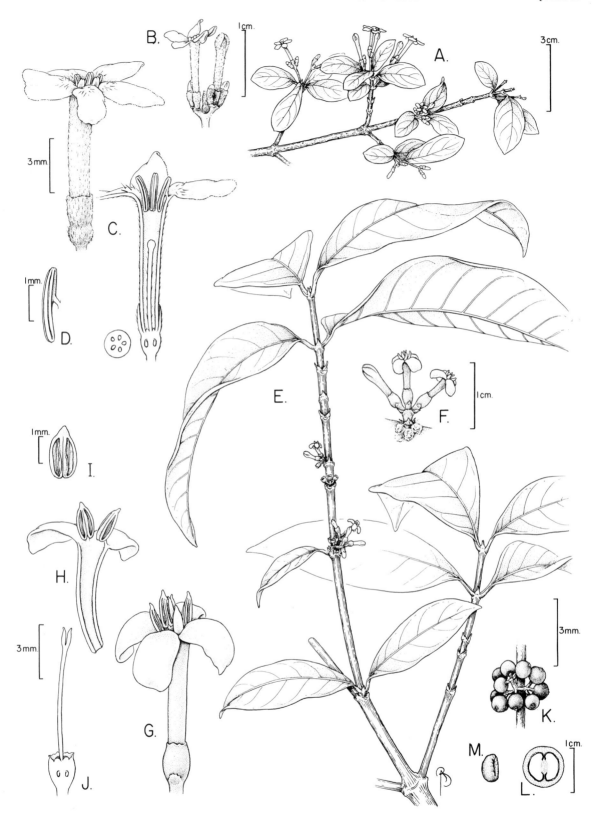

Fig. 176. A–D. *Guettarda odorata*. A. Flowering branch. B. Detail of inflorescence. C. Flower, l.s. flower, and c.s. ovary. D. Anther. E–M. *Ixora ferrea*. E. Flowering branch. F. Inflorescence. G. Flower. H. L.s. corolla showing anthers. I. Anther. J. Pistil. K. Infructescence. L. C.s. drupe. M. Seed.

Shrub or small tree 1.5–4 m tall, usually single-stemmed, with many lateral, opposite, decussate branches; bark dark brown, lenticellate; twigs cylindric, puberulent, becoming glabrous and lenticellate. Leaf blades 1.5–5 × 1.2–2.4 cm, elliptic, narrow-elliptic to oblanceolate, membranous to chartaceous, puberulent along veins on lower surface, the apex obtuse, rounded, or acute, usually mucronulate, the base acute, obtuse, or cuneate, the margins entire; petioles slender, 0.2–1.2 cm long, pilose; stipules ca. 1 mm long, subulate. Flowers not fragrant, 5–6-merous, 2–3 in shortly peduncled, axillary cymes; the axis slender, 5–20 mm long; bracteoles minute, at base of flowers; pedicels 1–1.5 mm long. Hypanthium 1–1.2 mm long, green, cup-shaped, almost indistinguishable from the calyx, glabrous; calyx vase-shaped, ca. 2 mm long, glabrous, truncate at apex; corolla trumpet-shaped, appressed-pubescent without, the tube 6–11 mm long, pinkish-tinged, the lobes rounded, ca. 2 mm long, oblong-rounded, light yellow, spreading; stamens 5–6, slightly exserted, the anthers sessile, oblong, ca. 3 mm long; style included, appressed-pubescent, the stigma capitate. Drupe 5–6 mm long, obovoid to nearly globose, leathery, green, crowned by an annular scar from the deciduous tubular calyx; stone with 5–6 cavities, each containing a single seed.

DISTRIBUTION: A common shrub of dry forest and woodlands. Cruz Bay Quarter at Lind Point (A2698), Maho Bay Quarter along road to Annaberg (A1927), Reef Bay Quarter on western slopes of Fish Bay (A3906). Also on Jost van Dyke, St. Croix, St. Thomas, Tortola, and Virgin Gorda; Puerto Rico, the Lesser Antilles, Central America, and Venezuela.

COMMON NAME: blackberry.

2. Guettarda scabra (L.) Vent., Choix Pl. t. 1. 1803.
Matthiola scabra L., Sp. Pl. 1192. 1753.

Guettarda rugosa Sw., Prodr. 59. 1788.

Shrub or small tree 1.5–5 m tall; bark grayish, smooth, peeling off in irregular flakes; twigs cylindric, reddish to ferruginous-tomentose, becoming glabrous and lenticellate. Leaf blades 3.5–15 × 1.6–9.5 cm, oblong, elliptic, obovate or ovate, rigidly coriaceous, densely pubescent underneath, very scabrous above, the apex obtuse to shortly acuminate, mucronate, the base rounded, truncate to nearly cordate, the margins entire, revolute; petioles stout, 0.5–1.6 cm long, ferruginous-tomentulose; stipules 5–10 mm long, subulate, densely pilose. Flowers 6–8-merous, sessile, in long-peduncled, axillary, compound cymes; the axis slender, 3.5–7.5 cm long, tomentulose; bracts subulate. Hypanthium green, not distinguishable from the calyx, 3–3.5 mm long, bell-shaped, tomentulose, crowned by irregular lobes; corolla trumpet-shaped, the tube 1.5–1.8 cm long, usually pinkish-tinged, the lobes obovate to oblanceolate, sometimes notched at apex, 5–7 mm long, white, spreading; stamens 6–8, included, the filaments ca. 1 mm long, the anthers oblong, 2.8–3 mm long; style included, the stigma capitate. Drupe 7–9 mm wide, oblate or depressed-globose, leathery, papillose, bright red when mature, crowned by an annular scar from the deciduous calyx; stone apparently breaking into 6–8 pyrenes.

DISTRIBUTION: A common shrub of dry to moist forests and coastal woodlands. Bordeaux (A4676), Cinnamon Bay (M17025), Susannaberg (A698). Also on St. Croix, St. Thomas, Tortola, and Virgin Gorda; throughout the West Indies, Florida, Central America, Venezuela, and Trinidad.

COMMON NAME: green heart.

12. IXORA L.

Trees or shrubs, usually glabrous. Leaves opposite or 3 per node, coriaceous, petiolate or sessile; stipules interpetiolar, simple, connate at base, persistent. Flowers bisexual, 4–5-merous, actinomorphic, pedicellate, in axillary or terminal cymes, corymbs, or panicles; bracts and bracteoles present; hypanthium not distinguishable from the calyx, 4–5-lobed; corolla trumpet-shaped, white, pink, red, or yellow, the tube much longer than the lobes, glabrous or pilose within, the lobes contorted in bud; stamens 4–5, exserted, the filaments inserted in the corolla throat; ovary inferior, 2-carpellate, each carpel with a single ovule. Fruit a fleshy, globose drupe, with 2 crustaceous or chartaceous pyrenes; seeds nearly globose, furrowed along the ventral side.

A pantropical genus of about 400 species, most of which occur in Africa and Asia.

1. Ixora ferrea (Jacq.) Benth., Linnaea **23**: 447. 1850.
Sideroxyloides ferreum Jacq., Select. Stirp. Amer. Hist. 131, t. 175, fig. 9. 1763. Fig. 176E–M.

Shrub or small tree 1.5–5 m tall, usually single-stemmed, with many lateral, opposite, ascending branches; bark dark brown; twigs glabrous, cylindric. Leaf blades 6.5–18 × 2.2–7.1 cm, elliptic or seldom oblong, chartaceous, glabrous, the apex acute or acuminate, the base acute, obtuse, or rounded, the margins entire; petioles 5–10 mm long, winged; stipules 4–9 mm long, subulate. Flowers 4-merous, sessile or very short-pedicellate, in congested, short, axillary cymes to 2 cm long; the axes <5 mm long, reddish. Hypanthium 1.8–2 mm long, bell-shaped, glabrous, green or red, lobes 4, irregular, rounded at apex; corolla trumpet-shaped, the tube 4–5 mm long, pinkish- or reddish-tinged, the lobes oblong, reflexed, ca. 2 mm long, white within but reddish underneath; stamens 4, exserted, the anthers oblong, bifid at apex; style exserted, the anthers oblong; style exserted, bifid at apex. Drupe 7–9 mm wide, fleshy, nearly globose or bilobed, black, crowned by the persistent sepals. Pyrenes 2 per fruit, ca. 5 mm long, each with a single seed.

DISTRIBUTION: A common shrub of moist secondary forests. Coral Bay Quarter at Bordeaux (A4056, A4708). Also on St. Thomas, Tortola, and Virgin Gorda; Cuba, Hispaniola, Puerto Rico, the Lesser Antilles, and northern Venezuela.

13. MACHAONIA Humb. & Bonpl.

Shrubs or small trees, glabrous or pubescent. Leaves opposite, 3 per node or fasciculate, petiolate or sessile; stipules interpetiolar, simple, triangular. Flowers bisexual, 4–5-merous, actinomorphic, pedicellate or sessile, in terminal compound cymes; bracts and bracteoles present; hypanthium not distinguishable from the calyx, crowned by 4–5 equal or unequal sepals; corolla short, funnel-shaped, white, the tube short, villous within, the lobes rounded, imbricate in bud; stamens 4–5, slightly exserted or included, the filaments inserted in the corolla throat; ovary inferior, 2-carpellate, each carpel with a single pendulous ovule, the style filiform, bifid at apex. Fruit a dry, obconical, flattened schizocarp, splitting in 2 longitudinal mericarps; seeds 1 per carpel, cylindric to ellipsoid.

A neotropical genus of about 40 species, most of which occur in the Greater Antilles.

REFERENCES: Fernández, M. & A. Borhidi. 1984. Revisión del género *Machaonia* en Cuba. Acta Bot. Acad. Sci. Hung. 30: 27–46; Standley, P. C. 1934. *Machaonia*. N. Amer. Fl. 32: 215–222.

1. Machaonia woodburyana Acev.-Rodr., Brittonia 45: 136. 1993. Fig. 177A–F.

Shrub 1.5–3 m tall, many-branched from base. Main branches twisted, bearing numerous, short, lateral branches and pairs of decussate, ascending spines at nodes; lateral branches with determinate growth and early deciduous leaves. Leaf blades 4–12 × 4–8.5 mm, chartaceous, glabrous, the margins entire; leaves of young branches opposite, ovate-rhombic, rarely nearly rounded, with obtuse apex and obtuse to rounded base; leaves of short lateral branches appearing whorled, obovate to oblanceolate, with rounded apex and tapering base; petioles 0.5–0.7 mm long; stipules minute, triangular. Flowers in terminal racemiform cymes; bracts oblong, 2.5 mm long; pedicels wanting on flowers central in the dichasia, present on lateral flowers. Hypanthium glabrous or nearly so, greenish, 1.5 × 2 mm, flattened-turbinate, each face with a depression along both sides of the medial plane; sepals 4, oblong, glabrous, rounded at apex, equal to slightly unequal, 1.8–2.6 mm long; corolla cream-colored to white, glabrous without, the tube to 1.5 mm long, the lobes 4, ovate, 1.5 mm long, papillate without; disk of 2 semiannular portions; stamens included, the filaments slightly unequal 0.8–1 mm long. Fruit flattened-turbinate, 4–5 × 2–2.1 mm, turning from green to brown, the mericarps apparently indehiscent. Seeds 2 per fruit, wedge-shaped, 3.2–3.5 mm long.

DISTRIBUTION: A rare shrub of woodlands and coastal thickets. Minna Hill (A4235), Cob Gut (A3207), trail to Southside Pond (A2661). Endemic to St. John.

14. MORINDA L.

Shrubs or small trees. Leaves opposite or 3 per node, petiolate, chartaceous to nearly coriaceous; stipules interpetiolar, simple or bifid, triangular or nearly rounded, persistent. Flowers bisexual, 4–7-merous, actinomorphic, sessile, in axillary or terminal dense heads; hypanthium not distinguishable from the calyx, truncate at apex; corolla funnel-shaped or trumpet-shaped, white, the tube longer than the lobes, villous or glabrous within, the lobes oblong, valvate in bud; stamens 4–7, exserted or included, the filaments inserted in the corolla throat; ovary inferior, 2–4-carpellate, each carpel with a single ascendant or lateral ovule, the style filiform, bifid or capitate at apex. Fruit fleshy, irregularly shaped, compound of connate drupes, each with 2–4 pyrenes; seeds oblong, obovoid, or kidney-shaped.

A pantropical genus of about 90 species, most of which are native to the Old World.

1. Morinda citrifolia L., Sp. Pl. 176. 1753.
Fig. 177G–K.

Shrub or small tree 2–4 m tall, usually with a few stems from base; bark light to dark brown, smooth; twigs glabrous, 4–angled. Leaves opposite; blades 9.5–33 × 5–20.3 cm, elliptic, widely elliptic to ovate, chartaceous, glabrous except for tufts of hairs in axils of the midveins, the apex obtuse to short-acuminate, the base obtuse to nearly rounded, the margins entire; petioles 5–25 mm long, stout; stipules 1.5–2 cm long, foliaceous, spreading. Flowers 5–6-merous, fragrant, sessile, in axillary heads, sunken into a portion of an expanded, fleshy peduncle. Hypanthium 1.8–2 mm long, cup-shaped, glabrous, green, truncate at apex; corolla trumpet-shaped, the tube 10–12 mm long, white to greenish without, pilose within, the lobes nearly lanceolate, spreading, 4–5 mm long, white; stamens 5–6, included, the filaments ca. 1 mm long, the anthers oblong, ca. 3 mm long; style stout, exserted, with 2 recurved stigmas. Compound fruit ovoid to potato-shaped, fleshy, 4–15 cm long, whitish to cream-colored, with a strong unpleasant smell especially when mature or decomposing, the surface with numerous pentagonal units (each representing a single ovary) with a central eye (scar where the corolla was attached); pyrenes 5–6 per single ovary, each containing a single seed.

DISTRIBUTION: A native of Asia and Australia. Becoming increasingly common on St. John, particularly along sandy beaches and in disturbed coastal areas. Chocolate Hole (A1843), Cinnamon Bay (A622). Also on St. Croix and St. Thomas; widely distributed throughout the Caribbean.

COMMON NAMES: headache tree, monkey apple, pain in the back bush, painkiller, starvation fruit.

15. PALICOUREA Aubl.

Shrubs or small trees. Leaves opposite, 3 or 4 per node, petiolate or sessile; stipules interpetiolar, connate at base, bifid, persistent. Flowers bisexual, distylous, 4–5-merous, actinomorphic, sessile or pedicellate, in terminal or rarely axillary, panicles or corymbs; the axes generally reddish, orange, yellow, or bluish purple; bracteoles wanting; hypanthium short, cup-shaped to tubular; calyx lobed, toothed, or truncate at apex; corolla tubular, variously colored, but usually yellow, orange, or red, the tube straight or constricted, villous at base within, the lobes very short, valvate in bud; stamens 4–5, included or exserted, the filaments inserted at middle of corolla tube; ovary inferior, 2(–6)-carpellate, each carpel with a single ascendant or lateral ovule, the style filiform, the stigma bilobed. Fruit a fleshy drupe, with 2(–6) papery or leathery pyrenes, each containing a single seed.

A neotropical genus of about 200 species.

REFERENCES: Taylor, C. M. 1989. A revision of *Palicourea* (Rubiaceae) in Mexico and Central America. Syst. Bot. Monogr. 26: 1–102; Taylor, C. M. 1993. Revision of *Palicourea* (Rubiaceae: Psychotrieae) in the West Indies. Moscosoa 7: 201–241.

1. Palicourea croceoides Ham., Prodr. Pl. Ind. Occid. 29. 1825. Fig. 178A–G.

Palicourea riparia Benth., J. Bot. (Hooker) 3: 224. 1841.

Palicourea crocea DC. var. *riparia* (Benth.) Griseb., Fl. Brit. W. I. 345. 1861.

Shrub 1–3 m tall, glabrous or nearly so; twigs nearly cylindrical, flattened and purple-tinged when young. Leaves opposite; blades 7.5–18 × 1.9–8.3 cm, elliptic, narrowly elliptic to oblong-elliptic, chartaceous, glabrous except for a few hairs on lower surface along the prominent veins and in vein axils, the apex acuminate, the base acute, sometimes unequal, the margins entire; petioles 0.5–1 cm long, slender; stipules subulate, 5–6 mm long,

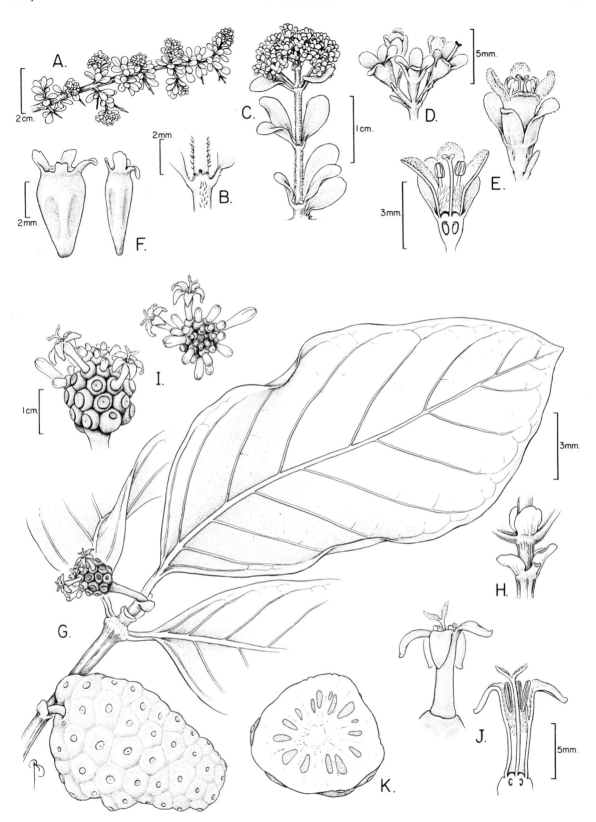

Fig. 177. **A–F.** *Machaonia woodburyana*. **A.** Flowering branch. **B.** Detail of stipules. **C.** Detail of branch with terminal inflorescence. **D.** Detail of inflorescence. **E.** Flower and l.s. flower. **F.** Schizocarp, frontal and lateral views. **G–K.** *Morinda citrifolia*. **G.** Fertile branch. **H.** Detail of stipules. **I.** Inflorescence, lateral and top views. **J.** Flower and l.s. flower. **K.** C.s. compound fruit.

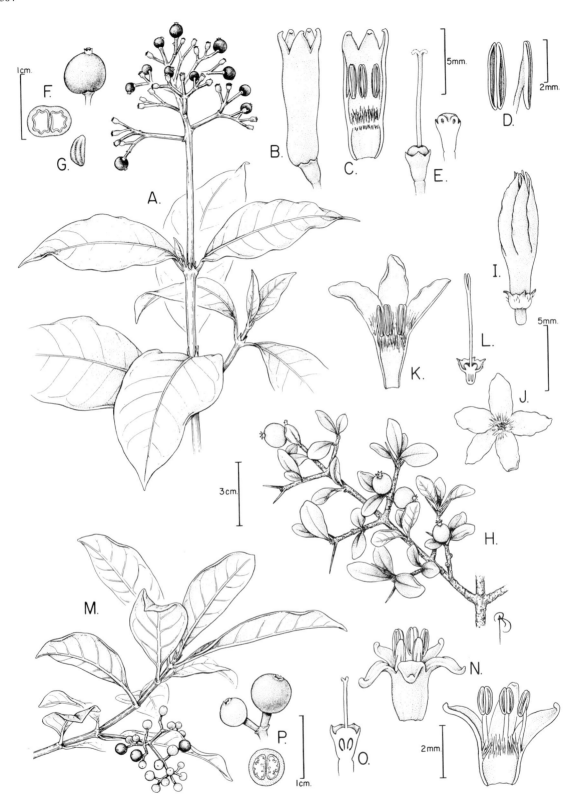

FIG. 178. A–G. *Palicourea croceoides*. **A.** Fruiting branch. **B.** Flower. **C.** L.s. corolla showing stamens. **D.** Anther, frontal and lateral views. **E.** Pistil and l.s. hypanthium. **F.** Drupe and c.s. drupe. **G.** Pyrene. **H–L.** *Randia aculeata*. **H.** Fruiting branch. **I.** Flowerbud. **J.** Flower. **K.** L.s. corolla showing stamens. **L.** L.s. hypanthium showing pistil. **M–P.** *Psychotria brownei*. **M.** Fruiting branch. **N.** Corolla with adnate stamens and l.s. corolla showing stamens. **O.** L.s. hypanthium showing pistil. **P.** Drupes and c.s. drupes.

persistent. Flowers 5-merous, in terminal corymbs; axes bright orange to red, the main axis 12–15 cm long, lateral axes 1.5–3.5 cm long; pedicels 5–8 mm long, orange, stout. Hypanthium 2.2–2.5 mm long, bell-shaped, glabrous, bright yellow, the lobes 0.8–1 mm long, ovate, apiculate; corolla yellow, tubular, the tube constricted at the middle, 10–11.5 mm long, the lobes unequal, nearly lanceolate, 2.5–2.7 mm long; stamens 5, included or slightly exserted, the anthers oblong, ca. 3.5 mm long; style filiform. Drupe ovate in outline, bilobed, laterally compressed, 5–6 mm long, shiny black or purple; pyrenes 2, each containing a single seed.

DISTRIBUTION: An occasional shrub of open disturbed moist areas or early secondary moist forests. Bordeaux Mountain (A2609, A2947). Also on St. Croix and Tortola; throughout tropical America.

COMMON NAME: yellow cedar.

16. PSYCHOTRIA L., nom. cons.

Erect or less often climbing shrubs, small trees, or creeping, erect, or epiphytic herbs. Leaves opposite, 3 or 4 per node, petiolate; stipules interpetiolar or intrapetiolar and connate at base or free, persistent or deciduous. Flowers bisexual, distylous, 4–6-merous, actinomorphic, heterostylous, sessile or pedicellate, in terminal or axillary, peduncled or sessile, panicles, corymbs, cymes, racemes, or heads; the axes greenish; bracts and bracteoles usually present; hypanthium cup-shaped, toothed or truncate at apex; corolla trumpet-shaped, funnel-shaped, or bell-shaped, variously colored, but usually white, the tube usually straight, usually pubescent within, the lobes valvate in bud; stamens 4–6, included or exserted, the filaments inserted at the corolla throat or below; ovary inferior, 2(–5)-carpellate, each carpel with a single ascendant ovule, the style included or exserted. Fruit a fleshy drupe, with 2 or rarely 5 pyrenes, each containing a single seed.

A pantropical genus of 1000–1500 species, occurring in diverse habitats.

Key to the Species of *Psychotria*

1. Scandent shrub .... 3. *P. microdon*
1. Erect shrub.
  2. Fruits black or purple when mature; leaves without pockets (domatia) in axils of midveins .... 2. *P. domingensis*
  2. Fruits orange to red when mature; leaves with pockets (domatia) in axils of midveins.
    3. Leaves long-acuminate at apex; calyx truncate or wavy at apex .... 4. *P. nervosa*
    3. Leaves acute or shortly acuminate at apex; calyx 5-lobed (lobes 0.3–0.4 mm long) .... 1. *P. brownei*

1. Psychotria brownei Spreng., Syst. Veg. **1**: 742. 1825. Fig. 178M–P.

Shrub 1.5–4 m tall, glabrous or nearly so, usually sympodially branched; twigs nearly cylindrical. Leaves opposite; blades 6.5–11.5 × 2.3–5 cm, elliptic, oblong-elliptic, to obovate, chartaceous, glabrous, with small pockets (domatia) in axils of the midveins, the apex acute or shortly acuminate, mucronulate, the base acute to obtuse, the margins entire; petioles 0.5–1.5 cm long, slender; stipules ca. 4 mm long, ovate, basally connate, pubescent at base within, early deciduous. Flowers 5-merous, in axillary panicles. Hypanthium 1.8–2 mm long (not distinguishable from the pedicel or from the calyx), bell-shaped, glabrous, green, the lobes 0.3–0.4 mm long; corolla white to greenish, tubular, the tube 2–2.5 mm long, the lobes ovate, ca. 1.5 mm long, reflexed; stamens 5, exserted, the anthers oblong, ca. 1 mm long, white; style included. Drupe globose, 5–6 mm long, smooth, turning from green to yellow and finally red; pyrenes 2, furrowed.

DISTRIBUTION: An understory shrub of moist to dry forests. Road to Bordeaux (A2886), Susannaberg (A4697). Also on Jost van Dyke, St. Thomas, and Tortola; Puerto Rico and Cuba.

2. Psychotria domingensis Jacq., Enum. Syst. Pl. 16. 1760. *Palicourea domingensis* (Jacq.) DC., Prodr. **4**: 529. 1830.

Psychotria westii DC., Prodr. **4**: 516. 1830.
Psychotria pavetta Sw., Prodr. 45. 1788. *Palicourea pavetta* (Sw.) DC., Prodr. **4**: 525. 1830.

Shrub 1–2 m tall, glabrous; twigs nearly cylindrical, sometimes constricted below the nodes. Leaves opposite; blades 6.5–17 × 3.4–7 cm, elliptic to oblong-elliptic, chartaceous, the apex acute or acuminate, the base acute to obtuse, the margins entire; petioles 1–2 cm long, slender; stipules early deciduous. Flowers 5-merous, nearly sessile, usually in groups of 3 on corymboid terminal inflorescence; axes sometimes reddish-tinged. Hypanthium 1–1.5 mm long, cup-shaped, glabrous, green; calyx 1.5–2 mm long, the sepals lanceolate, 0.5–0.6 mm long; corolla white, trumpet-shaped, the tube 9–10 mm long, the lobes unequal, nearly lanceolate, 4–5 mm long; stamens 5, included, the anthers oblong, ca. 3 mm long; style stout. Drupe compressed-subglobose, 5–6 mm long, black; pyrenes 2, furrowed.

DISTRIBUTION: An occasional shrub of open, disturbed, moist areas or in moist secondary forests. Adrian Ruins (A2461), Bordeaux Mountain (A2103). Also on St. Croix and St. Thomas; Jamaica, Hispaniola, Puerto Rico, and the Lesser Antilles.

3. Psychotria microdon (DC.) Urb., Symb. Ant. **9**: 539. 1928. *Rondeletia microdon* DC., Prodr. **4**: 408. 1830.

Psychotria pinnularis Sessé & Moç., Fl. Mexic., ed. 2, 57. 1894.

Scandent shrub 2–2.5 m long, lateral branches decussate, short to elongate; twigs 4-angled, grayish. Leaves opposite; blades 5–10 × 2–4.2 cm, obovate to oblanceolate, chartaceous, puberulent along the veins on lower surface, the apex acute to obtuse, the base acute to obtuse, the margins entire; petioles 0.5–2 cm long, slender; stipules ovate, 1.5 mm long, early deciduous. Flowers 5-merous, nearly sessile, in terminal corymbs, borne at ends of the lateral branches. Hypanthium 1.5–2 mm long, cup-shaped, glabrous, green, finely toothed; corolla white, bell-shaped, the tube ca. 5 mm long, the lobes 3–3.5 mm long, spreading; stamens 5, included, the anthers lanceolate, ca. 2 mm long; style filiform.

Drupe compressed-obovoid, 5–6 mm long, red at maturity; pyrenes 2, furrowed.

DISTRIBUTION: An occasional shrub of open, disturbed, moist areas. Adrian Ruins (A2395), road to Bordeaux (A2876). Also on St. Croix, St. Thomas, Tortola, and Virgin Gorda; Cuba, Hispaniola, Puerto Rico, the Lesser Antilles, northern South America, and along the Pacific coast to Peru.

4. Psychotria nervosa Sw., Prodr. 43. 1788.

Psychotria undata Jacq., Pl. Hort. Schoenbr. **3**: 5, t. 260. 1798.

Shrub 1.5–2 m tall, branching along main stem; twigs cylindrical, grayish. Leaves opposite; blades 5–11 × 2–4.6 cm, elliptic to obovate, chartaceous, primary and secondary veins prominent on lower surface, the apex long-acuminate, the base long-tapering, the margins wavy, revolute; petioles 4–5 mm long; stipules oblong, 9–14 mm long, sheathing terminal bud, early deciduous. Flowers 5-merous, sessile, heterostylous, in terminal corymbs. Hypanthium (including the calyx) 1–1.2 mm long, bell-shaped, glabrous, green, wavy or irregularly lobed at apex; corolla white, funnel-shaped, the tube 3–4 mm long, hirsute within, the lobes 1.5–2 mm long, spreading; stamens 5, exserted, the anthers lanceolate, ca. 0.6 mm long; style filiform. Drupe ellipsoid, 5–6 mm long, red at maturity; pyrenes 2, furrowed.

DISTRIBUTION: A common understory shrub of secondary forests. Trail to Margaret Hill (A2307), Susannaberg (A695). Also on St. Croix, St. Thomas, Tortola, and Virgin Gorda; Greater Antilles, Lesser Antilles, Central America to northern South America.

17. RANDIA L.

Shrubs or small trees, often with axillary spines. Leaves opposite or congested on short axillary branches, petiolate or sessile; stipules interpetiolar, connate at base, persistent. Flowers bisexual or unisexual (the species then dioecious), 5-merous, actinomorphic, in axillary or terminal fascicles or corymbs, if dioecious then the pistillate flower solitary; hypanthium obovoid or obconical, crowned by a tubular, toothed or truncate calyx; corolla trumpet-shaped, funnel-shaped, or rarely bell-shaped, the tube short or long, pubescent or glabrous within, the lobes contorted in bud; stamens 5, included or exserted, the filaments inserted in the corolla throat; ovary inferior, 2-carpellate, each carpel with 2 or 3 rows of horizontal ovules, the placentation parietal, the style club-shaped, many-lobed. Fruit a leathery, subglobose or ovoid berry with numerous horizontal seeds.

A neotropical genus of about 70 species.

1. Randia aculeata L., Sp. Pl. 1192. 1753.

Fig. 178H–L.

Randia mitis L., Sp. Pl. 1192. 1753.

Shrub 1.5–2 m tall, main stem with numerous lateral, decussate branches, ascending at 45° angles, the lateral secondary branches with numerous short, axillary, opposite branchlets, bearing axillary spines; bark grayish; twigs cylindrical, brownish, scabrous-puberulent. Leaves opposite on main branches but congested on short axillary ones; blades 1–3(–4.5) × 0.6–2.3 cm, obovate to oblanceolate, chartaceous, the apex rounded or obtuse, the base cuneate, the margins entire; petioles ca. 1 mm long; stipules triangular, ca. 2.5 mm long. Flowers 5-merous, sessile or nearly so, clustered at apex of short shoots. Hypanthium 1–1.2 mm long, nearly cylindrical; calyx ca. 2 mm long, cup-shaped, puberulent, the lobes 1–1.2 mm long, lanceolate or oblong; corolla white, tubular, the tube 3–4 mm long, hirsute within, the lobes 5–7 mm long, oblong, spreading; stamens 5, exserted, the anthers oblong, ca. 2.5 mm long; style stout, bifid at apex. Berry globose to ovoid, 1–1.2 cm long, leathery, turning from green to white (or black when dried), crowned by a persistent calyx. Seeds lenticular, ca. 4 mm long.

DISTRIBUTION: A common shrub of scrub and dry forests. Hansen Bay (A1803), Solomon Bay (A2282), White Cliff (A2712). Also on Jost van Dyke, St. Croix, St. Thomas, and Virgin Gorda; Greater Antilles, Lesser Antilles, Central America to northern South America.

COMMON NAMES: Christmas tree, ink berry.

18. RONDELETIA L.

Shrubs or small trees, unarmed. Leaves opposite or 3 per node, petiolate to nearly sessile; stipules interpetiolar, free or connate, forming a sheath around the stem, persistent. Flowers bisexual, actinomorphic, heterostylous, in axillary or terminal cymes, corymbs, or panicles; hypanthium rotund or oblong, crowned by a tubular, (4–)5–7-toothed calyx; corolla trumpet-shaped, (4–)5-lobed, the tube longer than the lobes, pubescent or glabrous within, the throat usually crowned by an annular thickening, the lobes imbricate in bud; stamens 5, the filaments inserted on the throat; ovary inferior, 2-carpellate, each carpel with numerous axial ovules, the style filiform, bilobed. Fruit a leathery or papery, loculicidal capsule, opening along 2 longitudinal valves; seeds numerous, minute, sometimes winged.

A neotropical genus of about 120 species.

1. Rondeletia pilosa Sw., Prodr. 41. 1788.

Fig. 179A–D.

Rondeletia triflora Vahl, Symb. Bot. **3**: 34. 1794.

Shrub 2–4 m tall, many-branched from base, the stems sometimes branched; bark dark gray, smooth; twigs cylindrical, densely pilose when young. Leaves opposite; blades 4–8 × 1–2.5(–3.5) cm, narrowly elliptic, oblong-elliptic, or rarely ovate, chartaceous, the upper surface pubescent, the lower surface densely pilose, with prominent vein network, the apex rounded, obtuse, acute, or acuminate, mucronate, the base cuneate to obtuse or rarely nearly cordate, the margins strongly revolute; petioles 2–5 mm long; stipules aristate, ca. 5 mm long, the tip fleshy, the base densely pilose. Flowers 4-merous, heterostylous, shortly pedicellate, in axillary cymes; the axis slender, 2.5–4 cm long, pilose; pedicels subtended by a pair of filiform bracteoles, ca. 5 mm long. Calyx 9–10.2 mm long, densely pilose, the hypanthium cup-shaped, 2–2.2 mm long, the lobes filiform, 7–8 mm long; corolla pink, trumpet-shaped, densely pilose, the tube 8–10(–13) mm long, the lobes 2.5–3 mm long, rounded, spreading; stamens 4, included (in short-styled flowers) or slightly exserted (in long-styled flowers), the anthers oblong, ca. 4 mm long; style filiform, the stigma bifid, slightly fleshy. Capsule ellipsoid, ca. 5 mm long, woody,

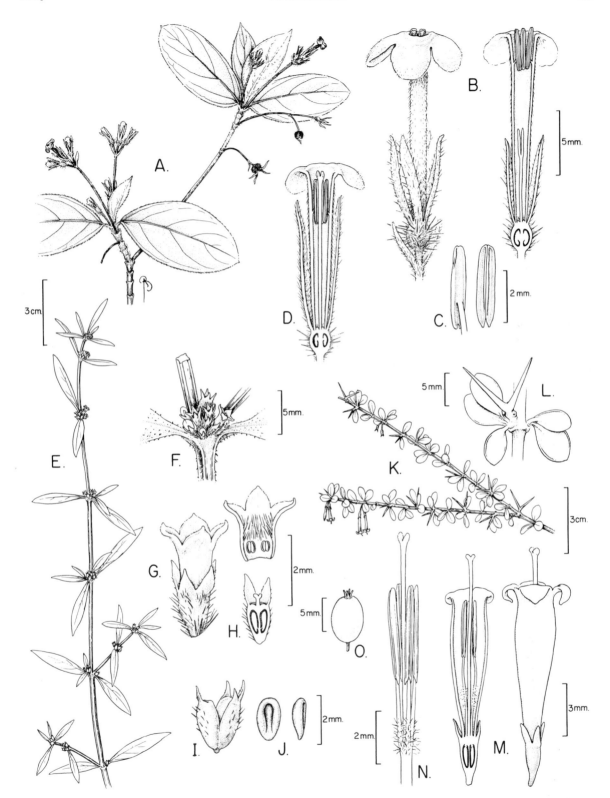

FIG. 179. **A–D.** *Rondeletia pilosa.* **A.** Fertile branch. **B.** Short-styled flower and l.s. flower. **C.** Anther, dorsal and frontal views. **D.** L.s. long-styled flower. **E–J.** *Spermacoce confusa.* **E.** Flowering branch. **F.** Detail of node showing inflorescence. **G.** Flower. **H.** L.s. corolla showing attachment of stamens and l.s. hypanthium showing style and ovary. **I.** Dehiscing capsule. **J.** seed, frontal and lateral views. **K–O.** *Scolosanthus versicolor.* **K.** Flowering branch. **L.** Detail of node showing leaves and paired axillary spines. **M.** Flower and l.s. flower. **N.** Stamens with basally connate filaments surrounding the style. **O.** Drupe.

densely pilose, crowned by the persistent, elongate sepals. Seeds cordate, ca. 3 mm long, brown, wrinkled.

DISTRIBUTION: A common shrub of scrub and dry forests. Along Cob

Gut (A3208), Maria Bluff (A592), Hawksnest (A2271). Also on St. Croix, St. Thomas, Tortola, and Virgin Gorda; Puerto Rico.

19. SCOLOSANTHUS Vahl

Shrubs, armed with axillary spines. Leaves opposite, coriaceous, short-petiolate or sessile; stipules minute, short, interpetiolar, connate, forming a sheath around the stem, persistent. Flowers bisexual, 4-merous, actinomorphic, in axillary fascicles or solitary; hypanthium obovoid or obconical, crowned by a small, 4-lobed calyx; corolla funnel-shaped, the tube longer than the lobes, the lobes imbricate in bud; stamens 4, included, the filaments connate at base; ovary inferior, 2-carpellate, each carpel with a pendulous ovule, the style filiform, bilobed. Fruit fleshy, compressed-globose drupe with 1 or 2 pyrenes, each one containing a single seed.

A genus of about 6 species, in the Greater Antilles.

1. Scolosanthus versicolor Vahl, Eclog. Amer. 1: 11. 1796 [1797]. *Chomelia versicolor* (Vahl) Spreng., Syst. Veg. 1: 410. 1824.   Fig. 179K–O.

Erect or decumbent shrub 1–3 m tall; twigs nearly cylindrical, scabrous-puberulent, armed with paired axillary spines, usually arched. Leaves opposite, but appearing as congested because of short axillary shoots; blades 4–7 × 4–5 mm, obovate to nearly spatulate, rigidly coriaceous, glabrous, the apex obtuse or rounded, the base cuneate, the margins revolute; petioles ca. 0.5 mm long; stipules 0.5–0.7 mm long, deciduous. Flowers 4-merous, shortly pedicellate, solitary or paired in leaf axils. Hypanthium ca. 2 mm long, flattened ellipsoid, glabrous, green, the sepals 1.5–1.6 mm long, lanceolate, reddish brown; corolla reddish brown, base and lobes yellow, nearly funnel-shaped, the tube 8–10 mm long, the lobes ovate, 1.5 mm long; stamens 4, included, the anthers oblong, 4 mm long, the filaments pubescent; style club-shaped, the stigma capitate. Drupe ellipsoid, 7 mm long, fleshy, white, with 1 pyrene.

DISTRIBUTION: An occasional shrub of coastal scrub. Cruz Bay Quarter along Center Line Road (A3982), Reef Bay Quarter on southern slopes of Fish Bay (A3115). Also on Guana Island, St. Croix, St. Thomas, and Virgin Gorda (fide Britton & Wilson, 1925); Puerto Rico, including Vieques.

20. SPERMACOCE L.

Annual or perennial herbs, often woody at base. Leaves opposite, appearing verticillate because of short axillary branches, chartaceous, short-petiolate or sessile; stipules interpetiolar, connate into sheath, with filiform appendages, persistent. Flowers bisexual, actinomorphic, sessile, in axillary or terminal glomerules; hypanthium small, obovoid, crowned by a small, 4-lobed calyx; corolla funnel-shaped, the tube longer than the lobes, the lobes 4, valvate; stamens 4, included or exserted, the filaments inserted on the throat; ovary inferior, 2-carpellate, each carpel with a pendulous ovule, the style filiform, bilobed. Fruit a cylindrical, septicidal capsule, separating into 2 dehiscent (sometimes one of them indehiscent) mericarps, these opening from the top; seeds ellipsoid or oblong.

A genus of perhaps 400 species (when combined with *Borreria*), with worldwide distribution.

Key to the Species of *Spermacoce*

1. Stems ridged, nearly 4-angled; seeds foveolate, not transversely furrowed.
    2. Leaves to 2 cm long, elliptic, oblong-elliptic to obovate; capsules 0.7–1 mm long; both mericarps dehiscent .... 3. *S. prostrata*
    2. Leaves 1.5–7 cm long, elliptic to lanceolate; capsules 2–2.5 mm long; only 1 mericarp dehiscent .... 2. *S. confusa*
1. Stems not ridged, nearly cylindrical; seeds transversely furrowed .... 1. *S. assurgens*

1. Spermacoce assurgens Ruíz & Pav., Fl. Peruv. 1: 60, t. 92b. 1798.

Spermacoce suffrutescens Jacq., Pl. Hort. Schoenbr. 3: 40, t. 322. 1801.

Borreria laevis sensu authors, non (Lam.) Griseb., 1861.

Erect or decumbent subshrub 30–50 cm tall, many-branched from base; twigs nearly cylindrical, sparsely pubescent to glabrate. Leaf blades 2.5–5.5 × 0.5–1.7 cm, elliptic, chartaceous, glabrous, the apex acute, the base long-tapering, narrowed to the petiole, the margins strigillose; petioles ca. 0.5 mm long; stipule sheath 4–6 mm long, with numerous bristles. Flowers in axillary and terminal heads; bracts linear. Hypanthium 2–2.5 mm long, flattened, club-shaped, pubescent along margins, the lobes ca. 0.8 mm long; corolla white, nearly funnel-shaped, the tube ca. 1.5 mm long, the lobes oblong, 0.5 mm long; stamens 4, exserted; style exserted, the stigma capitate. Capsule ellipsoid, 2–3 mm long, the cocci opening lengthwise from the apex. Seeds ellipsoid, 1.5–2 mm long, light brown, transversely furrowed.

DISTRIBUTION: A common weed of open disturbed areas, especially along roadsides. Coral Bay Quarter along road to Bordeaux (A3827), Cruz Bay Quarter along trail to Sieben (A2074). Also on St. Croix and Tortola; throughout tropical America, introduced and naturalized in tropical Africa and Asia.

2. Spermacoce confusa Rendle, J. Bot. 74: 12. 1936.
Fig. 179E–J.

Spermacoce tenuior sensu authors, non L., 1753.

Erect subshrub 30–50 cm tall, branching along main stem or many-branched from base; twigs nearly 4-angled, ridged, sparsely pubescent to glabrate. Leaf blades 1.5–7 × 0.5–2 cm, elliptic to lanceolate, chartaceous, strigose above, the apex acute, the base long-tapering, narrowed to the petiole, the margins strigillose; petioles 3–5 mm long; stipule sheath 0.5–0.8 mm long, with numerous bristles. Flowers in axillary heads, usually more developed on one side of the node. Hypanthium ca. 1 mm long, flattened-obconical, pubescent, the lobes lanceolate, ca. 1 mm long; corolla white, tubular, the tube 1–1.5 mm long, the lobes triangular ca. 1 mm long; stamens 4, included; style included. Capsule ellipsoid, 2–2.5 mm long, the cocci separating from the apex. Seeds oblong, ca. 1.5 mm long, light brown, foveolate.

DISTRIBUTION: A common weed of open disturbed areas, especially along roadsides and sandy coasts. Johnson Bay (A3155), along road to Bordeaux (A3847), Fish Bay Gut (A3876). Also on Anegada, St. Croix, St. Thomas, and Tortola; throughout tropical America.

3. Spermacoce prostrata Aubl., Hist. Pl. Guiane **1**: 58, t. 20, fig. 3. 1775.

Borreria parviflora G. Mey., Prim. Fl. Esseq. 83, t.1, figs. 1–3. 1818.

Borreria ocymoides sensu authors in part, non *Spermacoce ocymoides* Burm., 1768.

Erect or decumbent herb, to 40 cm long; twigs nearly 4-angled, ridged, sparsely pilose, especially along the ridges. Leaf blades 1–2 × 0.2–0.9 cm, elliptic, oblong-elliptic, to obovate, chartaceous, strigillose above, the apex obtuse or acute, the base acute or obtuse, the margins strigillose, revolute; petioles 0–2 mm long; stipule sheath whitish, 0.4–0.5 mm long, with numerous bristles to 2.5 mm long. Flowers in axillary or terminal heads. Hypanthium ca. 1 mm long, nearly cup-shaped, glabrous, the lobes lanceolate, unequal, ca. 1 mm long; corolla white, tubular, ca. 0.6 mm long, the lobes triangular ca. 0.2 mm long; stamens included. Capsule ellipsoid, 0.7–1 mm long, opening from the apex. Seeds nearly ellipsoid, ca. 0.5 mm long, light brown, with 4–5 rows of pits.

DISTRIBUTION: A common weed of open disturbed, moist areas. Coral Bay Quarter along road to Bordeaux (A3180), Cruz Bay Quarter along trail to Sieben (A2062). Also on Tortola; throughout tropical America and Asia.

DOUBTFUL RECORD: *Ernodea littoralis* Sw. was reported by Britton and Wilson (1925) and by Woodbury and Weaver (1987) as occurring on St. John; however, no specimen has been found to confirm these records, nor has the species been recently collected on the island.

75. Rutaceae (Orange Family)

Trees, shrubs, or herbs, usually with spines. Leaves alternate, or seldom opposite, pinnately compound, trifoliolate, or rarely simple, the blades aromatic because of numerous oil glands; stipules wanting. Flowers actinomorphic, bisexual or rarely unisexual, in cymose or racemose inflorescences or rarely solitary; calyx of 5 (less often 4 or 2–3) distinct sepals, or these connate at base; petals as many as and alternate with the sepals, distinct or basally connate; stamens 2–4 times as many as the petals, sometimes more numerous or only 2 or 3 fertile and the others staminodial, the filaments free or connate at base, the anthers opening by longitudinal slits, often gland-tipped; nectary disk intrastaminal, usually annular or rarely obsolete; ovary superior, very often plurilocular, of 5 to numerous carpels, with axile placentation, the locules with 1 to numerous ovules, the styles terminal, united, with a capitate stigma, or the styles rarely distinct. Fruit various, commonly a berry (hesperidium), drupe, capsule, or aggregate of follicles.

A family with about 150 genera and 1500 species, nearly cosmopolitan, but best represented in tropical and subtropical areas.

REFERENCES: Stone, B. 1985. Rutaceae. In M. D. Dassanayake & R. E. Fosberg (eds.), Fl. Ceylon 5: 455–462; Swingle, W. T. 1943. The botany of *Citrus* and its wild relatives of the orange subfamily. University of California Press, Berkeley and Los Angeles; Wilson, P. 1911. Rutaceae. N. Amer. Fl. 25: 173–224.

Key to the Genera of Rutaceae

1. Fruit fleshy (hesperidium), nearly globose, 2.5–4.5 cm long, turning yellow at maturity; leaves unifoliolate; petiole winged .... 2. *Citrus*
1. Fruit capsular, fleshy, or an aggregate of follicles, <1 cm long, turning brown, purplish-black, black, or red at maturity; leaves unifoliolate or with 3 to many leaflets; petioles not winged.
  2. Leaves trifoliolate (at least some of the leaves).
    3. Plant usually <1–2(–3) m tall, armed with stipular spines; fruits juicy berries, ellipsoid, bright red when mature.. 5. *Triphasia*
    3. Plants 2–8 m tall, unarmed; fruits drupes or capsules.
      4. Fruits capsular; leaflets elliptic, rounded and usually retuse at apex .... 4. *Pilocarpus*
      4. Fruits fleshy drupes, nearly globose, purplish to black when mature; leaflets rhombic, ovate to lanceolate .... 1. *Amyris*
  2. Leaves unifoliolate or with 5–21 leaflets.
    5. Plant spiny, especially the stems; fruit of aggregate dehiscent follicles .... 6. *Zanthoxylum* (in part)
    5. Plant unarmed.
      6. Leaflets with entire margins; fruit a berry with mucilaginous pulp................................... 3. *Murraya*
      6. Leaflets with crenate margins; fruit of aggregate dehiscent follicles .... 6. *Zanthoxylum* (in part)

1. AMYRIS P. Browne

Shrubs or small trees, unarmed. Leaves alternate or opposite, trifoliolate or less often unifoliolate or pinnately compound; leaflets glabrous, with numerous translucent oil glands; stipules minute, early deciduous. Flowers bisexual, 3–5-merous, actinomorphic, in terminal panicles; calyx nearly cup-shaped, the sepals free or connate at base; corolla of free, imbricate petals, early deciduous; stamens twice as many as the petals, early deciduous, the filaments free; nectary disk slightly fleshy, angular, adnate to the base of ovary; ovary unilocular with 2 pendulous ovules, the stigma nearly capitate, sessile. Fruit a 1-seeded drupe.

A neotropical genus of about 30 species.

Key to the Species of *Amyris*

1. Twigs, petioles, and inflorescence axes puberulent; leaflets rhombic to nearly ovate, with conspicuous punctations, the apex obtuse or rarely shortly acuminate, the venation conspicuously lighter color on upper surface, the margins revolute .... 1. *A. diatrypa*
1. Twigs, petioles, inflorescence, and leaves glabrous; leaflets ovate-lanceolate, the punctations very often inconspicuous, the apex long-acuminate, the venation usually of same color as the blade, the margins not revolute .... 2. *A. elemifera*

1. Amyris diatrypa Spreng., Neue Entd. **3**: 48. 1822.

Shrub 1–3 m tall, with multiple branches from base; twigs nearly cylindrical, dark gray, puberulent when young. Leaves opposite or subopposite, trifoliolate, slightly fragrant; leaflets 2–3(–5.5) × 1.2–2(–3.4) cm, rhombic to nearly ovate, coriaceous, glabrous, with numerous punctations, glossy above, the venation usually conspicuously lighter color, the apex obtuse to shortly acuminate, sometimes retuse, the base cuneate, the margins revolute, entire or crenate; petioles 5–18 mm long, puberulent. Flowers 4-merous, in terminal panicles 2–3 cm long with puberulent axes. Calyx ca. 3 mm long, glabrous or ciliate along margins; petals 1.5–1.8 mm long, obovate, punctate; stamens 8, the filaments slightly unequal, 1–1.2 mm long, the anthers ca. 0.6 mm long; ovary glabrous, the stigma papillate. Berry ellipsoid to globose, ca. 5 mm long, turning from green to purplish black.

DISTRIBUTION: A common shrub of dry forests and scrub. Cob Gut (A4234), trail from Bordeaux to Lameshur (A3864). Also on Anegada, Jost van Dyke, and St. Thomas; Hispaniola and probably Puerto Rico.

2. Amyris elemifera L., Syst. Nat., ed. 10, **2**: 1000. 1759. Fig. 180A–E.

Amyris maritima Jacq., Enum. Syst. Pl. 19. 1760.

Shrub 1–5 m tall, with multiple branches from base; bark gray, finely fissured; twigs nearly cylindrical, dark gray, glabrous. Leaves opposite or subopposite, trifoliolate or less often 5-pinnately foliolate, slightly fragrant; leaflets 2.8–7 × 1.2–4 cm, ovate-lanceolate, coriaceous, glabrous, dull, punctations very often inconspicuous, the midvein sometimes lighter, the apex long-acuminate, the base cuneate to obtuse, the margins entire or crenate; petioles 1.5–4 cm long, glabrous. Flowers 4-merous, in axillary or terminal panicles, to 5 cm long with glabrous axes. Calyx ca. 5 mm long, glabrous; petals ca. 3 mm long, obovate, punctate; stamens 8, the filaments slightly unequal, 0.5–0.7 mm long, the anthers ca. 0.6 mm long; nectary disk orange in fresh material; ovary glabrous, the stigma papillate. Berry globose, ca. 5 mm long, turning from green to black.

DISTRIBUTION: A common shrub of dry forests and scrub. Cob Gut (A4233), Maria Bluff (4707), New Found Bay (A4256). Also on Jost van Dyke, St. Croix, St. Thomas, and Virgin Gorda; throughout the West Indies and Trinidad.

COMMON NAMES: candle wood, flamboyant, torch wood.

2. CITRUS L.

Shrubs or small trees, armed with solitary, axillary spines. Leaves alternate, unifoliolate, glabrous, with numerous translucent oil glands; petioles usually winged; stipules minute, early deciduous. Flowers fragrant, bisexual or functionally staminate, 4–5-merous, actinomorphic, solitary or in short axillary racemes; calyx cup-shaped, 4–5-lobed; corolla of free, imbricate, white, thickened, punctate petals; stamens numerous, the filaments free; nectary disk annular; ovary 10–14-locular, usually with 4 or more ovules per locule, the style stout, the stigma nearly globose. Fruit a many-chambered berry (hesperidium) with leathery skin and juicy sacs inside the locules; seeds several per carpel, angular to ellipsoid.

An Old World genus with about 12 species, most of which are widely cultivated throughout tropical and subtropical regions.

1. Citrus aurantifolia (Christm.) Swingle, J. Wash. Acad. Sci. **3**: 465. 1913. *Limonia aurantifolia* Christm., Vollst. Pflanzensyst. **1**: 618. 1777. Fig. 180F–K.

Shrub to small tree 3–5 m tall; bark dark gray, with persistent spines on trunk; twigs angled when young, glabrous, green, with a single spine in axils of leaves. Leaf blades 5–11 × 3–5 cm, elliptic, chartaceous, glabrous, glandular-punctate, the apex obtuse to rounded, the base obtuse to cuneate, the margins crenate; petioles 1–1.7 cm long, glabrous, the wing wider at apex. Flowers 5-merous, few in axillary racemes; the axes puberulent, 5–7 mm long. Calyx 2–3 mm long, glabrous, punctate, the sepals ovate; petals 10–12 mm long, oblanceolate, punctate, reflexed at anthesis; stamens 20, the filaments white, thickened, slightly unequal, 6–8 mm long, the anthers linear, 2 mm long, bright yellow; nectary disk white; ovary glabrous, green, the style white, the stigma bright yellow. Fruit ellipsoid to globose, 3–4.5 cm long, turning from green to light yellow. Seeds ca. 5 mm long.

DISTRIBUTION: Persistent after cultivation. Emmaus (A2007), Reef Bay (A5251). Previously cultivated on other Virgin Islands and throughout the neotropics.

COMMON NAMES: lime, lime bush.

3. MURRAYA L., nom. cons.

Shrubs or small trees, unarmed. Leaves alternate, pinnately compound; leaflets with numerous translucent oil glands; stipules wanting. Flowers bisexual, actinomorphic, in axillary or terminal, short panicles; calyx deeply 5-lobed; corolla of 4–5 free, imbricate or valvate, white petals; stamens 8 or 10, the filaments free; nectary disk annular; ovary 2–5-locular, each locule with 1 or 2 pendulous ovules, the style slender, the stigma capitate. Fruit a few-seeded berry with leathery skin and mucilaginous pulp.

A genus of about 12 species native to eastern Asia and the Pacific islands, a few species cultivated as ornamentals.

FIG. 180. A–E. *Amyris elemifera*. A. Flowering branch. B. Flower. C. L.s. flower. D. Stamen. E. Fruiting branch and c.s. drupe. F–K. *Citrus aurantifolia*. F. Flowering branch. G. Flower. H. L.s. flower with petals removed. I. Petal. J. Stamens. K. Fruit. L–O. *Murraya exotica*. L. Flowering branch. M. L.s. flower. N. Stamen. O. Fruiting branch.

1. Murraya exotica L., Mant. Pl. **2**: 563. 1771.

Fig. 180L–O.

Murraya paniculata sensu authors, non (L.) Jack, 1820.

Shrub 2–3 m tall, many-branched from near the base; twigs nearly cylindrical, puberulent. Leaves with slightly fetid smell; leaflets 5–9, alternate along leaf rachis, 1.6–3 × 0.7–1.5 cm, oblanceolate or obovate, chartaceous, glabrous, the apex obtuse to rounded, usually apiculate, the base cuneate, the margins entire to crenate, revolute; petioles 2–3 mm long, pubescent. Flowers strongly aromatic, in terminal and axillary racemes, the axes puberulent, 1–3 cm long. Sepals 5–7 mm long; petals 1.3–1.8 cm long, white, oblong-spatulate, reflexed at anthesis; stamens 10, longer ones alternating with shorter ones, the filaments white, subulate; nectary disk adnate to the base of ovary; ovary 2–3-locular, glabrous, light green, the style cylindrical. Berry ovoid to ellipsoid, ca. 1 cm long, turning from green to red, obscurely punctate. Seeds ellipsoid, 6–7 mm long.

DISTRIBUTION: Persistent after cultivation. Adrian ruins (A4009). Probably native to China, cultivated on St. Croix, St. Thomas, and throughout the tropics, sometimes naturalized.

COMMON NAME: St. Patrick bush.

NOTE: *Murraya exotica* has long been considered a synonym of *Murraya paniculata;* however, B. C. Stone (1985) considered them to be two distinctive species.

4. PILOCARPUS Vahl

Shrubs or small trees, unarmed. Leaves alternate to subopposite, trifoliolate, pinnately compound or unifoliolate; leaflets with numerous translucent oil glands, usually notched at apex; stipules wanting. Flowers bisexual, (4–)5-merous, actinomorphic, in axillary or terminal racemes; calyx toothed or lobed; corolla of free, valvate, deciduous petals, with hooded apex; stamens (4–)5, deciduous, the filaments free; nectary disk annular to cup-shaped, adnate to the base of ovary; ovary of (4–)5 basally connate carpels, each carpel with 1 or 2 ovules, the stigma 5-lobed, with a gland in each lobe. Fruit a schizocarpic capsule, dividing into mericarps, 1–4, usually rudimentary; seeds usually reniform.

A genus of 13 species, native to tropical America.

REFERENCE: Kaastra, R. C. 1982. Pilocarpinae (Rutaceae). Fl. Neotrop. Monogr. 33: 1–198.

1. Pilocarpus racemosus Vahl, Eclog. Amer. **1**: 29. 1797.

Shrub or small tree 2–8 m tall; bark dark gray, smooth; twigs nearly cylindrical, glabrous, becoming lenticellate with age. Leaves alternate to subopposite, 1–3-foliolate; leaflets 7–11 × 4.3–6.2 cm, elliptic, chartaceous, glabrous, the apex rounded and retuse, the base obtuse, unequal in lateral leaflets, the margins entire, slightly revolute; petioles 5–6.5 cm long; rachis 2.5–4.5 cm long, furrowed; petiolules ca. 5 mm long. Flowers few to numerous along solitary axillary racemes, the axes 10–25 cm long. Calyx 5-lobed, 0.6–0.7 mm long; petals 3.2–4 mm long, inflexed at apex; stamens 5, the filaments adnate to the nectary disk; carpels adnate to the disk, glabrous, the style subapical. Capsule brownish, the mericarps with a nearly obovoid outline. Seeds black, kidney-shaped, 6–9 mm long.

DISTRIBUTION: Uncommon, known on St. John from only two collections. Eggers (*s.n.*); along Battery Gut (W357). Also throughout the Greater Antilles, (except Jamaica), Lesser Antilles and in Venezuela.

5. TRIPHASIA Lour.

Armed shrubs. Leaves alternate, trifoliolate or less often only 1 or 2, with numerous translucent oil glands; stipules modified into long spines. Flowers bisexual, 3–4-merous, solitary, or a few in axillary racemes; calyx lobed; corolla of free, imbricate petals; stamens 6 or 8, the filaments free; nectary disk annular, modified into a gynophore; ovary 3–4-locular, each locule with a single pendulous ovule, the style slender or stout, terminal, cylindrical, the stigma nearly capitate. Fruit a 1–3-seeded fleshy berry, with leathery, punctate surface.

A genus of 3 species, native to Asia and the Philippines.

1. Triphasia trifolia (Burm.f.) P. Wilson, Torreya **9**: 53. 1909. *Limonia trifolia* Burm.f., Fl. Indica 103, t. 35, fig. 1. 1768.   Fig. 181H–N.

Shrub 1–2(–3) m tall; twigs nearly cylindrical, puberulent, with (1–)2 stipular spines per node, usually arched or leaning. Leaflets 3 or less often only 1 or 2, coriaceous, glabrous or puberulent, the apex obtuse and deeply notched, the margins crenate, the terminal leaflets 2.5–5 × 1.2–3.5 cm, lanceolate, with a cuneate base, the basal leaflets 1–2.5 × 0.6–2.2 cm, elliptic, with an obtuse base; petioles 2–4 mm long, puberulent. Flowers fragrant, 3-merous, a few in axillary or terminal racemes, the axes 4–5 mm long. Calyx funnel-shaped, lobed, ca. 4 mm long; petals white, 8–9 mm long, obovate, with cuneate base; stamens 6, the filaments subulate, light green; nectary disk ca. 0.5 mm high; ovary glabrous, ellipsoid, the style terminal, stout, the stigma white, fleshy. Berry ellipsoid to globose, with mucilaginous pulp and a lemon-scented red or orange skin. Seeds 1–3 per fruit, ellipsoid, 6–7 mm long.

DISTRIBUTION: Naturalized after cultivation, a common understory shrub of coastal dry forests and open coastal areas. Sieben (A2060), Reef Bay (A2723). Native of the East Indies, cultivated and naturalized on St. Croix, St. Thomas, Tortola, and throughout the Caribbean.

COMMON NAME: sweet lime.

6. ZANTHOXYLUM L.

Dioecious shrubs or trees, very often armed with conical prickles; bark usually aromatic. Leaves alternate, pinnately compound, trifoliolate or unifoliolate; leaflets with numerous translucent oil glands, with entire or crenate margins; stipules wanting. Flowers unisexual, actinomorphic, 3–5-merous, minute, in axillary or terminal panicles or in congested axillary cymes; calyx deeply lobed, the sepals persistent or deciduous; corolla of free, valvate petals or wanting; stamens 3–5, the filaments free; nectary disk subtending the carpels; gynoecium of 1–5 free or basally connate carpels, these immersed in receptacle or stipitate. Fruit of 1–5 follicles, each carrying a single seed.

A genus of 250–300 species, of worldwide distribution.

Key to the Species of *Zanthoxylum*

1. Leaves unifoliolate .... 3. *Z. monophyllum*
1. Leaves pinnately compound.
    2. Plant not armed with prickles or spines; plant stellate-pubescent .... 1. *Z. flavum*
    2. Plants armed with prickles and/or spines; pubescence of simple hairs, papillate, or glabrous.
        3. Leaves to 8 cm long; leaflets 5–9, elliptic, ovate or obovate, rigidly coriaceous, 1.4–3.2 cm long, the margins entire or crenate, sometimes with spines on lower surface of midvein .... 4. *Z. thomasianum*
        3. Leaves to 35 cm long; leaflets 15–21, oblong to oblong-elliptic, chartaceous, 5–10 cm long, margins crenate to entire, the blades without spines .... 2. *Z. martinicense*

1. Zanthoxylum flavum Vahl, Eclog. Amer. **3:** 48. 1807. *Fagara flava* (Vahl) Krug & Urb. *in* Urb., Bot. Jahrb. Syst. **21:** 571. 1896.

Unarmed tree, 6–10 m tall; bark gray, smooth; twigs nearly cylindrical, densely stellate-pubescent. Leaves 15–30(–40) × 10–25(–30) cm; leaflets 5–7(–13), coriaceous, 3.5–11(–15) × 1.5–5(–8) cm, oblong-elliptic to nearly lanceolate, sparsely stellate-pubescent, the apex obtuse to rounded, slightly notched, the base obtuse to rounded, unequal on lateral leaflets, the margins crenate; petioles 5–6 cm long. Flowers 4-merous, in axillary panicles. Calyx of triangular lobes ca. 5 mm long; petals white, 2.5–3.2 mm long, oblong; stamens 4, early deciduous; nectary disk ca. 0.5 mm high, glabrous; ovary glabrous, 2-lobed, 2-carpellate, the carpels united but separating at maturity, the style terminal, the stigma nearly bilobed. Follicles 2, stipitate, nearly obovoid, punctate. Seeds ovoid, shiny dark brown, ca. 4.5 mm long.

DISTRIBUTION: A rare tree, known only from a single individual on Ajax Peak (A5284). Probably once common on St. John but extirpated because of timber exploitation. Also found in Florida, throughout the West Indies and northern South America.

COMMON NAMES: satin wood, yellow sandalwood.

2. Zanthoxylum martinicense (Lam.) DC., Prodr. **1:** 726. 1824. *Fagara martinicensis* Lam., Encycl. **1:** 334. 1783. Fig. 181A–E.

Zanthoxylum juglandifolium Willd., Sp. Pl. **4:** 756. 1806.
Zanthoxylum lanceolatum Poir. *in* Lam., Encycl. Suppl. **2:** 293. 1811.

Armed tree to 20 m tall; bark dark brown, smooth, with vertical stripes and conical prickles that grow with age; twigs nearly cylindrical, ferruginous-tomentose when young. Leaflets (9–)15–21, chartaceous, 5–10 × 1.4–1.8 cm, oblong to oblong-elliptic, sparsely pubescent, especially along veins, the apex acute to long-acuminate, the base unequal, obtuse-rounded, the margins crenulate to entire; petioles 5–6 cm long. Flowers 5-merous, in axillary or terminal panicles or racemes. Calyx of ovate-triangular lobes, 0.5–0.7 mm long; petals white, 2–3 mm long, oblong; stamens 5, the filaments longer than the petals; gynophore tubular, ca. 0.5 mm high, glabrous; ovary glabrous, 5-carpellate, the carpels united but separating at maturity. Follicles 5, sessile, laterally compressed, 5 mm long, punctate. Seeds nearly ovoid, shiny black, ca. 4.5 mm long.

DISTRIBUTION: A common tree of moist forests. Road to Bordeaux (A1920), Adrian Ruins (A2510). Also on St. Croix, St. Thomas, and Tortola; the Greater Antilles, Lesser Antilles, and northern South America.

3. Zanthoxylum monophyllum (Lam.) P. Wilson, Bull. Torrey Bot. Club **37:** 86. 1910. *Fagara monophylla* Lam., Encycl. **1:** 334. 1783.

Armed shrub or small tree to 7 m tall; bark dark brown or gray, with numerous spines; twigs papillate, becoming cylindrical and lenticellate at age, with numerous, slightly recurved prickles. Leaves unifoliolate; blade 3.8–11 × 1.6–5 cm, chartaceous, elliptic, glabrous, the apex obtuse to shortly acuminate, the base obtuse, the margins entire to crenulate, revolute; petioles 0.5–1.8 cm long. Flowers 5-merous, in axillary or terminal panicles. Calyx of triangular lobes, 0.5–0.6 mm long; petals white, ca. 2 mm long, elliptic; stamens 5, the filaments much shorter than the petals; gynophore tubular, ca. 0.5 mm high, glabrous; ovary 3-carpellate, covered with minute rounded scales, the carpels united. Follicles 1 or less often 2, stipitate, nearly globose, 6–7 mm long. Seeds nearly ovoid, shiny black, 5 mm long.

DISTRIBUTION: A common tree of dry forests. Carolina (A4673), Lameshur (A2744), Pen Point (A2012). Also on Jost van Dyke and St. Thomas; Puerto Rico, Hispaniola, the Lesser Antilles, Central America, and northern South America.

COMMON NAME: yellow prickle.

4. Zanthoxylum thomasianum (Krug & Urb.) P. Wilson, N. Amer. Fl. **25:** 182. 1911. *Fagara thomasiana* Krug & Urb., Bot. Jahrb. Syst. **21:** 583. 1896. Fig. 181F, G.

Armed shrub 3–4 m tall; bark dark gray, fissured, with numerous spines; twigs puberulent, with numerous straight spines with a swollen base. Leaves pinnately compound; leaflets 5–9, opposite along rachis, 1.4–3.2 × 0.8–2.4 cm, rigidly coriaceous, elliptic, ovate, or obovate, glabrous, lower surface sometimes with 1 or 2 spines along midvein, the apex apiculate, the base obtuse to rounded, the margins entire or crenate, revolute; rachis furrowed, usually with a spine at the base of lateral leaflets. Flowers 3-merous, in congested axillary racemes. Calyx of ovate lobes, ca. 1 mm long; petals white, 2.5–2.7 mm long, elliptic, concave; stamens 3, the filaments much shorter than the petals; gynophore tubular, ca. 0.5 mm high, glabrous; ovary 3-carpellate, the carpels basally united. Follicles 2, or less often 1, nearly sessile, globose, ca. 5 mm long. Seeds ovoid, shiny black, 3.5 mm long.

DISTRIBUTION: A rare, endangered shrub of dry forests and scrub. Cob Gut (A3206), Lameshur (B520), Gift Hill (A2940). Also on St. Thomas and Puerto Rico.

COMMON NAME: St. Thomas bush, St. Thomas prickly-ash.

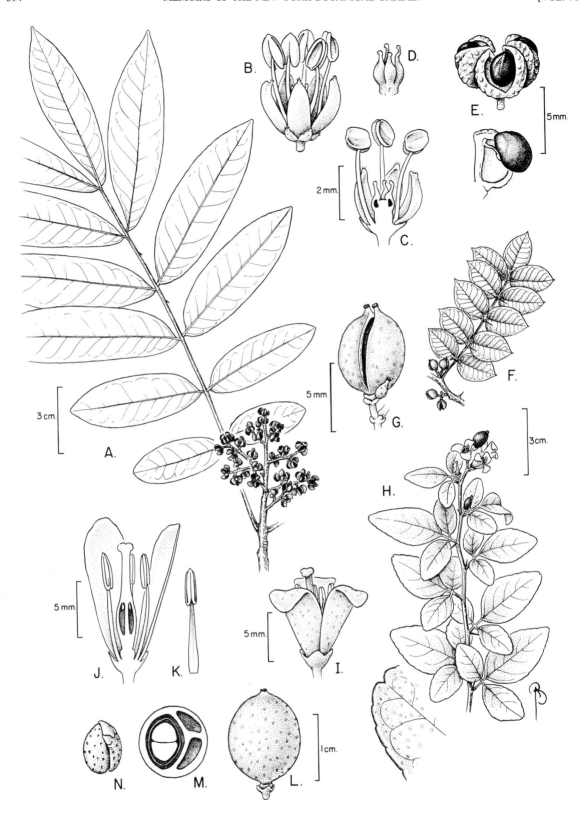

FIG. 181. A–E. *Zanthoxylum martinicense*. A. Fruiting branch. B. Flower. C. L.s. flower. D. Gynoecium. E. Dehiscing fruit and detail of fruit wall showing attached seed. F–G. *Zanthoxylum thomasianum*. F. Fruiting branch. G. Dehiscing follicle. H–N. *Triphasia trifolia*. H. Fertile branch and detail of leaf apex. I. Flower. J. L.s. flower. K. Stamen. L. Berry. M. C.s. berry. N. Embryo.

76. Sapindaceae (Soapberry Family)

Trees, shrubs, herbaceous or woody vines. Leaves alternate, pinnately compound, trifoliolate, or biternate; stipules present or wanting. Flowers 4–5-merous, small, actinomorphic or zygomorphic, bisexual or unisexual by reduction (plants monoecious or polygamous-dioecious), usually white, in terminal or axillary racemes or panicles; calyx cup–shaped, with more or less elongate sepals; corolla of distinct petals, usually bearing a petaloid appendage on inner surface; nectary disk extrastaminal, annular or unilateral; stamens (5–)8(–10), the anthers opening by longitudinal slits; ovary of (2–)3–5 united carpels, each carpel with 1 or 2 ovules, the placentation axile. Fruit a woody or membranous inflated capsule, a winged mericarp, a berry, or a drupe.

A family of about 150 genera with approximately 1500 species, mostly tropical or subtropical but extending to subtemperate zones. REFERENCE: Radlkofer, L. 1931–1934. Sapindaceae. In A. Engler (ed.), Das Pflanzenreich IV, 165(Heft 98a–h): 1–1539. Wilhelm Englemann, Leipzig.

Key to the Genera of Sapindaceae

1. Tendriled vines or lianas.
   2. Liana to 10 m long, producing milky latex; cross section of stem with 8–10 smaller, vascular cylinders, surrounding a larger one; fruit of 3 winged mericarps .... 6. *Serjania*
   2. Herbaceous vines, 1–2 m long, without milky latex; cross section of stem with a single vascular cylinder; fruit an inflated, membranous capsule .... 2. *Cardiospermum*
1. Trees or shrubs.
   3. Leaves trifoliolate .... 1. *Allophylus*
   3. Leaves pinnate, with 2–9 leaflets.
      4. Leaflets usually with serrate margins; fruit a 3-locular capsule; seeds covered with a fleshy coat .... 3. *Cupania*
      4. Leaflets with entire margins; fruit a 1–2-locular drupe or berry.
         5. Fruit subglobose or ellipsoid, >2 cm long, green; seeds with a tan, juicy, edible seed coat .... 5. *Melicoccus*
         5. Fruit spherical, 1 cm long or less, dark red, purplish, or black; seed without a fleshy coat ...... 4. *Exothea*

1. ALLOPHYLUS L.

Shrubs or trees. Leaves trifoliolate or unifoliolate, with elongate petioles; stipules wanting. Flowers zygomorphic, 4-merous, functionally pistillate or staminate (plants apparently dioecious), produced in axillary racemes, panicles, or thyrses; calyx of 2 pairs of unequal sepals; petals usually white and spatulate, with a pair of marginal appendages; nectary disk unilateral, semiannular or 4-lobed; stamens 8, the filaments unequal, connate at base into a short tube, the anthers dorsifixed; ovary 2-carpellate, each carpel with a single ovule, the style terminal, with 2 stigmatic branches. Fruit a fleshy or leathery berry, of 2 monocarps, one of which is usually abortive; seeds 1 per monocarp, usually subglobose.

A pantropical genus with approximately 230 species.

1. Allophylus racemosus Sw., Prodr. 62. 1788.

Fig. 182A–F.

Schmidelia occidentalis Sw., Fl. Ind. Occid. **2**: 665. 1798. *Allophylus occidentalis* (Sw.) Radlk., Sitzungsber. Math.-Phys. Cl. Königl. Bayer. Akad. Wiss. München **20**: 230. 1890.

Shrub or small tree 5–10 m tall, young parts covered with minute, reflexed, rusty hairs; bark light brown, smooth; stems cylindric, becoming glabrous with age. Leaves trifoliolate; leaflets 7–20 × 2.5–6 cm, chartaceous, elliptic to oblanceolate, glabrous and glossy above, pubescent below, especially along raised veins, the apex acute or acuminate, the base acute, slightly unequal on lateral leaflets, the margins serrate; petioles 2.5–6 cm long. Flowers in axillary thyrses; axes 3–7 cm long, densely pubescent. Sepals green, ovate, unequal, 1–1.5 mm long, pubescent; petals white, obovate, ca. 2 mm long; appendages simple and sericeous; nectary disk unilateral, 2-lobed; stamens 8, the filaments densely pubescent, ca. 2 mm long in staminate flowers; ovary ellipsoid, pubescent, the style slender, pubescent. Berry of 1 or 2 monocarps, these obovoid to ellipsoid, ca. 5 mm long, turning from green to red-orange. Seeds ca. 3 mm long, brown, ellipsoid.

DISTRIBUTION: An occasional tree of lowland moist forests. Road to Bordeaux (A935, A5117), Maho Bay (A1847). Also on St. Croix; Greater Antilles (except Jamaica), Lesser Antilles, and from Central America to northern South America.

2. CARDIOSPERMUM L.

Herbaceous tendriled vines, or less often erect subshrubs, with woody base, not producing milky sap; stems furrowed, cylindrical or slightly angled; cross section of stems with a single vascular cylinder. Leaves ternately or biternately compound, membranous or chartaceous; leaflets deeply serrate or lobed; petioles and rachis not winged; stipules minute, early deciduous. Flowers zygomorphic, functionally staminate or pistillate, produced in axillary thyrses; thyrses bearing tendrils at base of rachis; calyx of 4 or 5 unequal sepals; petals 4, distinct, white, with a hood-shaped appendage; nectary disk unilateral, 2-lobed, the lobes sometimes elongate; stamens 8, the filaments unequal, connate at base, the anthers dorsifixed; ovary 3-carpellate, each locule with a single ovule, the style slender,

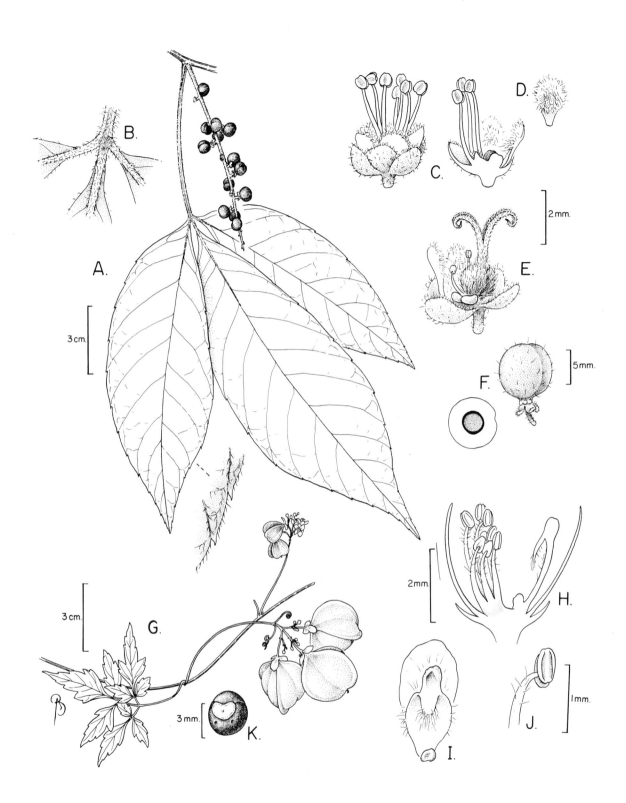

FIG. 182. A–F. *Allophylus racemosus*. **A.** Fruiting branch and detail of leaflet margin. **B.** Detail of leaflet bases. **C.** Staminate flower and l.s. staminate flower. **D.** Petal. **E.** Pistillate flower. **F.** Monocarp and c.s. monocarp. **G–K.** *Cardiospermum corindum*. **G.** Fertile branch. **H.** L.s. staminate flower. **I.** Petal with adnate appendage. **J.** Stamen. **K.** Seed showing hilum.

the stigmas 3, recurved. Fruit an inflated, membranous capsule, the septa persistent after dehiscence of valves; seeds spherical, black, with a small, white cordate hilum.

A tropical genus with about 15 species; most are native to the neotropics, but a few are pantropical. Also cultivated in North America and Europe.

Key to the Species of *Cardiospermum*

1. Plant herbaceous; leaflets deeply lobed or dissected; capsules globose to trigonous–top-shaped ... 2. *C. halicacabum*
1. Plant slightly woody; leaflets sinuate-dentate; capsules ellipsoid to nearly globose .... 1. *C. corindum*

1. Cardiospermum corindum L., Sp. Pl., ed. 2, 1: 526. 1762.   Fig. 182G–K.

Slightly woody vine, 1.5–2 m long, many-branched from base; stems 5-ribbed, pubescent. Leaves biternate, 12–16 × 9–12 cm; leaflets chartaceous, glabrous except for a few hairs on veins, sinuate-dentate at margins, the terminal leaflets rhombic, acute or acuminate at apex and attenuate at base, the lateral leaflets oblong-ovate, acute at apex and rounded at base; petioles 2–3 cm long, not winged. Thyrses axillary, shorter than the subtending leaves, bearing a pair of tendrils at base of rachis. Calyx of 5 unequal, rounded, greenish sepals, ca. 1 mm long; petals 3.5–4.5 mm long, the appendage apically yellow; stamens 8, filaments pubescent. Capsule ellipsoid to nearly globose, 2–3 cm long, turning from green to straw-colored. Seeds spherical, ca. 4 mm diam., black, with a white reniform hilum.

DISTRIBUTION: In open, disturbed areas. Coral Bay (A1844), Chocolate Hole Bay (A2345), Fish Bay (A4154). Also on St. Croix, St. Thomas, and Little St. James; weedy throughout the tropics.

COMMON NAME: balloon vine.

2. Cardiospermum halicacabum L., Sp. Pl. 366. 1753.

Herbaceous vine to 2 m long, with multiple annual branches from a semiwoody base; branches angled, 5-ribbed, glabrous or puberulous. Leaves biternate; leaflets chartaceous, ovate or oblong, with obtuse, acute, or acuminate apex and tapering base, the margins deeply lobed or dissected; petioles 2–3 cm long, not winged; stipules lanceolate, ca. 5 mm long. Thyrses axillary, shorter than the subtending leaves, bearing a pair of tendrils at base of rachis. Calyx of 5 unequal, rounded, greenish sepals; petals 3.5–4.5 mm long, the appendage apically yellow; stamens 8, the filaments pubescent. Capsule globose to trigonous–top-shaped, depressed at apex, turning from green to straw-colored, with 1 seed per locule. Seeds spherical, ca. 5 mm, black, with a white cordate hilum.

DISTRIBUTION: Disturbed areas, mostly along roadsides. Fish Bay (A2501), trail to Mennebeck Bay (A2429), Monte (A2474). Also on St. Croix and St. Thomas; elsewhere in the tropics and subtropics as a weed; cultivated as an ornamental in temperate areas.

COMMON NAME: balloon vine.

3. CUPANIA L.

Small to large trees. Leaves pinnately compound with a rudimentary distal leaflet; leaflets mostly serrate, alternate or opposite, the petiolules usually short and not enlarged; rachis cylindrical or angled; stipules wanting. Flowers 5-merous, actinomorphic, bisexual or unisexual (the plants polygamo-dioecious), in axillary or terminal panicles or thyrses; sepals short (2–4 mm long), imbricate, usually tomentose; petals as long as the sepals, with a pair of marginal tomentose appendages; nectary disk annular, usually lobed; stamens 8, the filaments of equal length, longer than the petals, tomentose and filiform, the anthers dorsifixed, ovoid or oval with retuse apex; ovary 3-carpellate, each locule with a single ovule. Fruit a woody capsule with 2 or 3 locules, each 1-seeded; seeds covered at base by a fleshy coat.

About 50 species from tropical and subtropical America.

1. Cupania triquetra A. Rich. *in* Sagra, Hist. Fis. Cuba, Bot. 119. 1845.   Fig. 183A–E.

Cupania fulva sensu Griseb., Fl. Brit. W. I. 125. 1859, non Mart., 1838.

Tree 10–15 m tall; stems covered with rusty hairs, becoming glabrous with age. Leaves with 4–8 alternate or subopposite leaflets; leaflets 5–15 × 2.5–7 cm, subcoriaceous, obovate or cuneate, lower surface covered with rusty hairs, the apex obtuse or notched, the base obtuse, the margins entire or serrate. Inflores-cence axes densely covered with rusty hairs. Calyx greenish, 2.5 mm long, covered with rusty hairs; petals white, 2.5 mm long, with 2 short marginal appendages. Capsule trigonous, turning from green to brown, stipitate, 2.5 × 2 cm, densely covered with rusty hairs outside and whitish hairs within. Seeds dark brown, 1.5 × 0.9 cm with a fleshy, orange coat at base.

DISTRIBUTION: Occasional in moist forests. Adrian Ruins (A4031), Cinnamon Bay (A2093), Rosenberg (B310). Also on St. Thomas; Hispaniola, Puerto Rico, and Lesser Antilles.

NOTE: Despite the place of publication, this species does not occur in Cuba. This species was originally collected by L. C. Richard on St. John and later described by A. Richard in his contributions to the flora of Cuba.

4. EXOTHEA Macfad.

Shrubs or small trees. Leaves pinnately compound with 1–3 pairs of opposite leaflets, the terminal leaflet rudimentary. Flowers 5-merous, actinomorphic, unisexual, with rudimentary parts of opposite sex, produced in axillary panicles; sepals free to base; petals without petaloid appendages; disk annular; stamens (7–)8(–10); ovary 2-locular, each locule with 2 or 3 ovules, the style short, slender, the stigma capitate. Fruit a globose leathery berry with persistent reflexed sepals at base; seeds usually 1 per fruit, spherical, without aril.

A genus of 3 species, in Florida, Mexico, Central America, Colombia, and the West Indies.

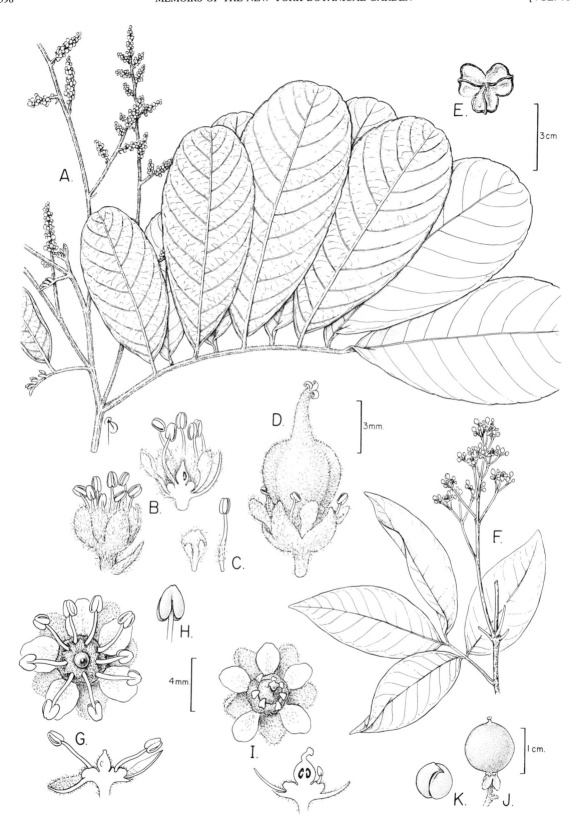

FIG. 183. A–E. *Cupania triquetra*. **A.** Flowering branch. **B.** Staminate flower, and l.s. staminate flower. **C.** Petal and stamen. **D.** Pistillate flower with developing fruit. **E.** Dehisced capsule valves, top view. **F–K.** *Exothea paniculata*. **F.** Flowering branch. **G.** Staminate flower and l.s. staminate flower. **H.** Stamen. **I.** Pistillate flower and l.s. staminate flower. **J.** Berry. **K.** Embryo.

1. Exothea paniculata (Juss.) Radlk. *in* Durand, Index Gen. Phan. 81. 1888. *Melicocca paniculata* Juss., Mém. Mus. Hist. Nat. **3**: 187, t. 5. 1817. *Hypelate paniculata* (Juss.) Cambess., Mém. Mus. Hist. Nat. **18**: 32. 1829. Fig. 183F–K.

Exothea oblongifolia Macfad., Fl. Jamaica **1**: 22. 1837.

Small tree 5–15 m tall; branches light brown, with numerous grayish lenticels. Leaves with (1–)2(–3) pairs of opposite leaflets; leaflets 5–13 × 2–4.2 cm, elliptic or oblong, chartaceous, glabrous, the apex acute or obtuse, the base obtuse, the margins entire; petioles typically 2.5 cm long or sometimes shorter; rachis not winged. Panicles to 15 cm long, pubescent. Sepals greenish, ovate, 3.5 mm long, pubescent; petals white, ovate or oval, 2.4–3 mm long, pubescent on inner surface, spreading; disk annular, pubescent; stamens (7–)8(–10), glabrous, 4 mm long in male flowers, 1.5 mm long in female flowers, the anthers ovate, basifixed; ovary ovoid, pubescent, green, with short style and capitate to bilobed stigma. Berry globose, dark red, purplish, or black, 1–1.3 cm diam., containing usually 1 seed or rarely 2. Seeds nearly globose, ca. 7 mm long.

DISTRIBUTION: Occasional in moist forest. Maho Bay Quarter along Center Line Road near intersection with Road 20 (A3992). Also in Florida, the West Indies, Guatemala, Belize, and Colombia.

5. MELICOCCUS P. Browne

Small trees. Leaves pinnately compound, distal leaflet rudimentary or wanting; leaflets 2 or 4, opposite; rachis winged; stipules wanting. Flowers 5–merous, actinomorphic, unisexual (the plants dioecious), in terminal panicles; calyx cup-shaped, of equal sepals; petals reflexed, without appendages; nectary disk prominent, annular and slightly lobed; stamens 8, filaments of equal length, longer than the petals, the anthers dorsifixed; ovary 2-carpellate, each carpel with a single ovule, the stigma lobed to subcapitate, sessile. Fruit a 1(–2)-seeded, subglobose or ellipsoid berry, with leathery pitted skin; seeds with fleshy outer coat.

A genus of 2 species native to South America.

1. Melicoccus bijugatus Jacq., Enum. Syst. Pl. 19. 1760. Fig. 184I–L.

Melicocca bijuga L., Sp. Pl., ed. 2, **1**: 495. 1762.

Dioecious tree 10–20 m tall; bark light gray, smooth, with horizontal markings. Leaves with 2 pairs of opposite or subopposite leaflets, petioles and rachis flattened and usually winged; leaflets 4–14 × 2.2–5 cm, elliptic, oblong, ovate, or obovate, chartaceous, glabrous, with stout midvein, the apex acute or obtuse, the base obtuse and usually unequal, the margins entire, slightly wavy; petioles 2.5–7 cm long. Flowers fragrant, in short panicles; calyx with 4 greenish, oblong, glabrous, sepals, 2.2 mm long; petals 5, yellowish, obovate with a narrow base, round apex and ciliate margins, lacking appendages; nectary disk annular, lobed and glabrous; stamens 8, of equal length, ca. 4 mm long in male flowers and 1 mm long in female flowers; ovary glabrous, ellipsoid or ovoid, with 2 uniovular carpels, the style minute, the stigmas bilobed. Berry subglobose or ellipsoid, green, 2–2.3 cm long. Seeds 1–(2), 1.5 cm long, covered with a tan, fleshy, edible coat.

DISTRIBUTION: Widespread throughout the island in moist to semidry areas, especially along roads. Adrian Ruins (A728, A1858). Also on Jost van Dyke, St. Croix, St. Thomas, and Virgin Gorda; native to northern South America but commonly planted and naturalized throughout Central America, and the West Indies. Apparently introduced on St. John in the eighteenth century.

COMMON NAMES: genip, kenep, keneppy tree, quenepa.

6. SERJANIA Mill.

Woody or herbaceous vines, often producing milky latex; cross section of stem with a single vascular cylinder or multiple ones. Leaves trifoliolate, biternate, triternate, or 5-foliolate; stipules small and early deciduous. Flowers zygomorphic, functionally unisexual or bisexual (plants polygamo-dioecious), in axillary or terminal thyrses, usually bearing a pair of tendrils at the base of the flowering rachis; calyx of 4–5 unequal sepals; petals 4, mostly spatulate, with basally adnate hood–shaped appendage; nectary disk unilateral, 2- or 4-lobed; receptacle sometimes enlarged into a short androgynophore; stamens 8, slightly exserted, the filaments unequal, the anthers dorsifixed; ovary of 3 uniovular carpels, the style terminal, with 3 stigmatic branches. Fruit a dry schizocarp splitting into 3 mericarps, with a proximal wing; seeds lenticular to globose.

A genus of about 230 species native to the neotropics.

1. Serjania polyphylla (L.) Radlk., Monogr. Serjania 179. 1875. *Paullinia polyphylla* L., Sp. Pl. 366. 1753. Fig. 184A–H.

Paullinia triternata Jacq., Select. Stirp. Amer. Hist. 110, t. 180. 1763. *Serjania triternata* (Jacq.) Willd., Sp. Pl. **2**: 466. 1799.

Serjania lucida Schum., Skr. Naturhist.-Selsk. **3**(2): 128. 1794.

Liana 5–10 m long, usually many-branched from base, all parts, especially stems, producing milky latex; stems cylindric or angled, furrowed, in cross section with a central vascular cylinder surrounded by 5–10 vascular cylinders. Leaves biternate, triternate, or bipinnate; leaflets 1.5–8 × 1–3 cm, subcoriaceous, glabrous, elliptic, ovate, or lanceolate, lower surface with tufts of hairs in the axils of secondary veins, the apex acute, obtuse, or acuminate, the base tapering or obtuse, the margins remotely serrate; petioles not winged, 1–3 cm long; rachis winged or not. Flowers functionally staminate or pistillate, in axillary thyrses. Sepals 5, unequal, oblong, pubescent, ca. 4 mm long; petals 4, white, obovate or spatulate, ca. 5 mm long, with a hood-shaped appendage bearing a fleshy, yellow apex; nectary disk divided into 4 ovoid glands; stamens ca. 5 mm long, the filaments pubescent, the anthers lanceolate; ovary 3-carpellate, the styles slender, with 3 recurved stigmas. Mericarps 1.5–2.2 cm long, straw-colored, the seed locule globose. Seeds globose, ca. 5 mm diam.

DISTRIBUTION: A common liana of open disturbed areas in moist to dry environments. Cruz Bay Quarter near Lind Point (A2313), East End Quarter

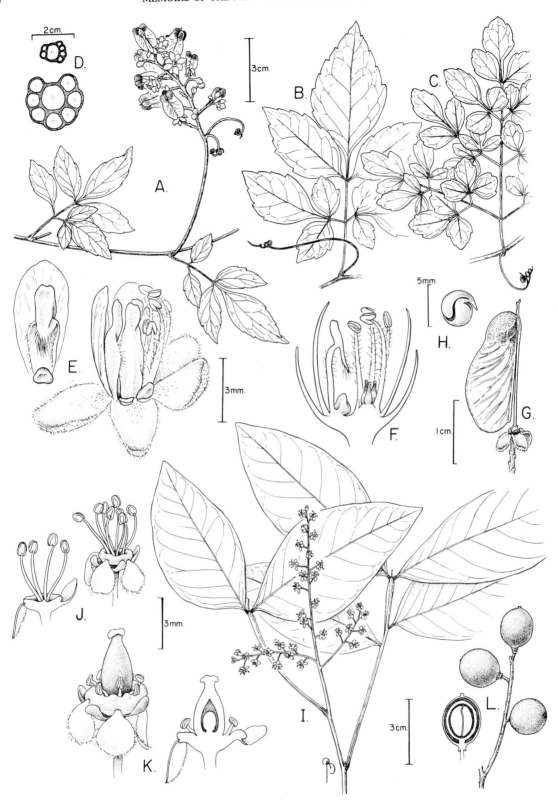

Fɪɢ. 184. A–H. *Serjania polyphylla.* **A.** Fertile branch. **B.** Biternate leaf with tendril. **C.** Bipinnate leaf with tendril. **D.** C.s. thin and thick stems. **E.** Petal with adnate appendage and staminate flower. **F.** L.s. staminate flower. **G.** Winged mericarp attached to floral axis. **H.** Embryo. **I–L.** *Melicoccus bijugatus.* **I.** Flowering branch. **J.** Staminate flower and l.s. flower. **K.** Pistillate flower and l.s. flower. **L.** Infructescence and l.s. berry showing seed and embryo.

by Southside Pond (A1828). Also on Anegada, St. Croix, St. Thomas, and Tortola; Hispaniola and Puerto Rico, including Vieques.

COMMON NAMES: basket wiss, basket wood, black wist, cabrite rotting, white root, white wist.

DOUBTFUL RECORDS: *Sapindus saponaria* L. was reported on St. John by Britton and Wilson (1924) and by Woodbury and Weaver (1987); however, no specimen has been found to confirm these records, nor has the species been recently collected on the island.

77. Sapotaceae (Sapodilla Family)

Trees or shrubs, producing abundant milky latex, unarmed or spinose. Leaves alternate, simple; stipules usually wanting. Flowers small, actinomorphic, bisexual, solitary, in axillary cymes or panicles or cauliflorous; calyx of (4–)5(–12) distinct sepals; corolla short-tubular, of 4–8 lobes, the lobes usually with a pair of lateral appendages; stamens 1–3 times as many as the lobes, sometimes some of them reduced to staminodes, the filaments adnate to the corolla tube, the anthers opening by longitudinal slits; staminodes when present, simple, lobed, divided or petaloid, alternating with the stamens or adnate to the sinuses of the corolla lobes; nectary wanting; ovary superior, the carpels 2–14, plurilocular, the placentation axile or axile-basal, the locules with 1 ovule, the styles terminal, united, with a capitate or lobed stigma. Fruit commonly a berry or less often a drupe; seeds with a large hilar scar.

A family with about 70 genera and 800 species from tropical and subtropical areas.

REFERENCES: Pennington, T. D. 1990. Sapotaceae. Fl. Neotrop. Monogr. 52: 1–770; Pennington, T. D. 1991. The genera of Sapotaceae. Royal Botanic Garden, Kew.

Key to the Genera of Sapotaceae

1. Calyx of 2 whorls of 3 sepals, the outer whorl valvate .... 2. *Manilkara*
1. Calyx of a single whorl of 4–6 imbricate sepals.
    2. Corolla lobes divided into 3 segments .... 4. *Sideroxylon*
    2. Corolla lobes simple.
        3. Fruits green or purplish; leaves elliptic, 3–10.5 cm long .... 1. *Chrysophyllum*
        3. Fruits yellow; leaves elliptic to oblanceolate, 2.5–5 cm long .... 3. *Pouteria*

1. CHRYSOPHYLLUM L.

Trees or shrubs, unarmed, producing abundant white latex. Leaves distichous or spirally arranged; stipules wanting. Flowers actinomorphic, unisexual or bisexual, in axillary fascicles, seldom solitary; calyx cup-shaped, of (4–)5(–6) imbricate, slightly unequal sepals; corolla cylindrical, bell-shaped, or funnel-shaped, the tube shorter to longer than the (4–)5(–6) lobes; stamens (4–)5(–8), included, the filaments adnate to corolla tube, the anthers dorsifixed or basifixed; staminodes absent or rarely present; ovary of (4–)5(–12) locules with axile placentation, the style terminal, with a lobed stigma. Fruit a 1- to many-seeded berry; seeds laterally compressed with a ventral scar.

A pantropical genus of about 70 species, most of which are native to the neotropics.

Key to the Species of Chrysophyllum

1. Leaves oblong-lanceolate, green underneath; fruit narrowly ovoid (1–1.5 cm long), as long as the subtending peduncle .... 2. *C. pauciflorum*
1. Leaves elliptic, rusty-brown underneath (densely sericeous); fruit ellipsoid to depressed-globose (3–7 cm long), much longer than the subtending peduncle.
    2. Fruit depressed-globose, 4–6 cm diam. .... *C. cainito* (cultivated)
    2. Fruit nearly ellipsoid, ca. 1 cm diam. .... 1. *C. bicolor*

1. Chrysophyllum bicolor Poir., Encycl. Suppl. **2**: 15. 1811.   Fig. 185 F.

Chrysophyllum eggersii Pierre *in* Urb., Symb. Ant. **5**: 155. 1904.

Tree 6–15 m tall; bark light brown to gray, fissured; branches nearly cylindrical, densely covered with rusty-brown pubescence when young. Leaf blades 5–10.7 × 2–5.5 cm, elliptic or seldom oblong-elliptic, coriaceous, densely covered with appressed rusty-brown hairs and prominent midvein underneath, the apex obtuse to shortly acuminate, the base obtuse, sometimes slightly unequal, the margins entire; petioles densely sericeous, 5–12 mm long. Flowers bisexual, 5–10, in axillary fascicles; pedicels 4–6 mm long, rusty-brown pubescent. Calyx cup-shaped, 1.5–1.7 mm long, the sepals 5, ovate-rounded, 0.5–0.9 mm long, rusty-brown

pubescent without; corolla cream-colored to white, bell-shaped, 3.5–4 mm long, the lobes 5 or 6, sparsely pubescent without, 1–1.5 mm long; stamens 5–6, the filaments ca. 0.2 mm long, the anthers ca. 0.7 mm long; ovary ovoid, densely rusty-pubescent, 6-locular, the style subulate, crowned by 6 short, rounded lobes. Berry 1-seeded, asymmetrically ellipsoid, 3–4 × 1 cm, turning from green to yellowish purple–tinged.

DISTRIBUTION: Rare, known from only a few individuals in moist secondary forests. Adrian Ruins (A2028, A2900), Sieben (A2080). Also in Puerto Rico and Hispaniola.

NOTE: This species was considered by Pennington (1990) in his monograph of Sapotaceae as a synonym of *Chrysophyllum cainito* L. However, the fruits are so strikingly different that I believe them to be two distinct species.

2. Chrysophyllum pauciflorum Lam., Tabl. Encycl. **2**: 44. 1794.   Fig. 185A–E.

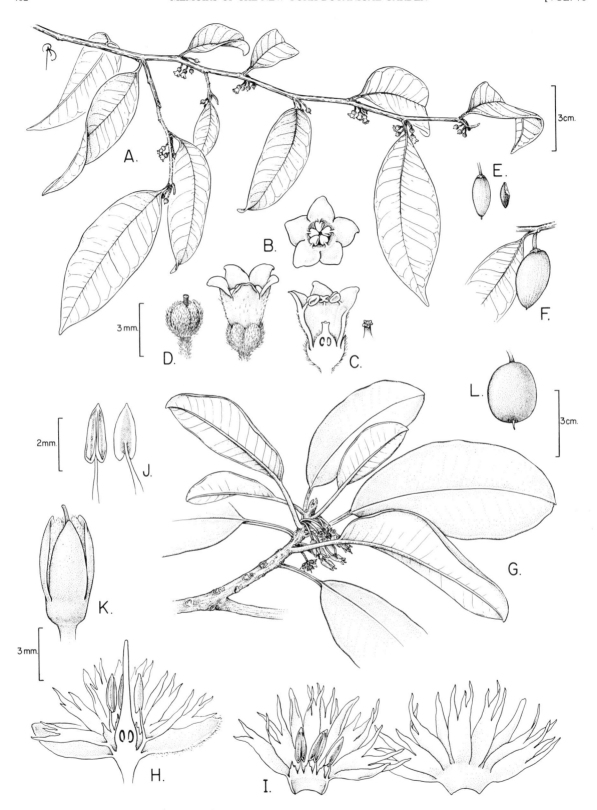

FIG. 185. A–E. *Chrysophyllum pauciflorum*. **A.** flowering branch. **B.** Flower, lateral and top views. **C.** L.s. flower and detail of stigma. **D.** Calyx and pistil after disarticulation of corolla. **E.** Berry and seed. **F.** *Chrysophyllum bicolor;* berry. **G–L.** *Manilkara bidentata*. **G.** Flowering branch. **H.** L.s. flower. **I.** L.s. corolla, dorsal and frontal views. **J.** Stamen, frontal and dorsal views. **K.** Calyx and pistil after disarticulation of corolla. **L.** Berry.

Chrysophyllum pauciflorum var. *nervosum* Pierre *in* Urb., Symb. Ant. **5**: 159. 1904.

Chrysophyllum pauciflorum var. *krugii* Pierre *in* Urb., Symb. Ant. **5**: 159. 1904.

Chrysophyllum microphyllum sensu Eggers, Fl. St. Croix 67. 1879, non Jacq., 1763.

Tree 4–9 m tall; bark light brown to gray, fissured; branches drooping, nearly cylindrical, rusty-brown pubescent when young. Leaf blades 3–9.7(–12) × 1–4(–5) cm, oblong-lanceolate, lanceolate, or oblong, coriaceous, appressed-pubescent along both surfaces when young, becoming glabrous but retaining a few hairs, especially along the prominent midvein beneath, the apex acuminate, shortly acuminate, to obtuse, the base obtuse to rounded, the margins entire; petioles densely sericeous, 5–9 mm long. Flowers bisexual, 1–6, in axillary fascicles; pedicels 5–7 mm long, rusty-brown pubescent. Calyx cup-shaped, 1.5–2.5 mm long, the sepals ovate, slightly unequal, 0.7–1.2 mm long, rusty-pubescent without; corolla cream-colored to white, bell-shaped, 3.2–4 mm long, the lobes 5 or 6, sparsely pubescent without, ca. 1 mm long; stamens 5, the filaments ca. 0.2 mm long, the anthers 0.5–0.7 mm long; ovary ovoid, 5–lobed, densely rusty-pubescent, 5-locular, the style subulate, nearly truncate at apex. Berry 1-seeded, narrowly ovoid to ellipsoid, 1.5–1.8 cm long, turning from green to reddish purple, apiculate at apex.

DISTRIBUTION: A common tree of moist and dry forests. Center Line Road (A2393), Sieben (A2067). Also on St. Croix and St. Thomas; Puerto Rico, including Vieques Island.

COMMON NAME: palmet.

CULTIVATED SPECIES: *Chrysophyllum cainito* L. is occasionally cultivated on St. John and other Virgin Islands.

2. MANILKARA Adans., nom. cons.

Trees or rarely shrubs, unarmed, producing abundant white latex. Leaves spirally arranged, rarely opposite or whorled; stipules minute and deciduous or wanting. Flowers actinomorphic, bisexual, solitary or in axillary fascicles; calyx cup-shaped, of 2 whorls of (2–)3(–4) free or basally united sepals, the outer whorl valvate to slightly imbricate; corolla glabrous, tubular, the tube shorter than to rarely equaling or exceeding the 6(–9), usually 3-parted lobes; stamens 6(–12), in a single whorl, the filaments adnate to the top of the corolla tube, the anthers basifixed; staminodes 6(–9) or rarely absent, alternating with the stamens, 2-toothed at apex; ovary of 6–12 locules with axile or axile-basal placentation, the style terminal, exserted. Fruit a 1- to several-seeded berry, variously shaped; seeds ellipsoid to ovoid, laterally compressed with a narrowly elongate ventral scar.

A pantropical genus of about 60 species, half of which are native to the neotropics.

Key to the Species of *Manilkara*

1. Flowers 2–12 in axillary fascicles; pedicels glabrous; fruits 1(–2)-seeded, ellipsoid to nearly globose, 1.5–3 cm long .... 1. *M. bidentata*
1. Flowers solitary; pedicels pubescent; fruits many-seeded, ovoid to depressed-globose, 3–4 cm long .... *M. zapota* (cultivated)

1. Manilkara bidentata (A. DC.) A. Chev., Rev. Int. Bot. Appl. Agric. Trop. **12**: 270. 1932. *Mimusops bidentata* A. DC., Prodr. **8**: 204. 1844.

Fig. 185G–L.

Tree 10–15 m tall; bark light brown, smooth; branches dark brown, glabrous, slightly angled, becoming lenticellate. Leaf blades 15–26 × 6–10.2 cm, elliptic, oblong, obovate, or oblanceolate, coriaceous, glabrous but usually with a waxy covering, midvein stout, prominent beneath, the apex obtuse, rounded, acute, shortly acuminate and usually notched, the base obtuse to attenuate, the margins entire; petioles glabrous, 1.5–3.5 cm long; stipules wanting. Flowers 2–12 in axillary fascicles; pedicels 9–25 mm long, glabrous. Calyx cup-shaped, the sepals 6, 4–6 mm long, inner ones puberulent without; corolla cream-colored to white, 3–6 mm long, glabrous, the tube 0.5–1.5 mm long, the lobes 6, divided to base into 3 segments; stamens 6, the filaments 1.5–2.5 mm long, free or basally connate to the staminodes, the anthers 1–2 mm long; staminodes 6, 1–3.5 mm long, lanceolate with variable margins; ovary broadly ovoid, glabrous, 6–12-locular, the style subulate, 3–6 mm long. Berry 1–2-seeded, ellipsoid to globose, 1.5–3 cm long, brown. Seeds 0.9–2.6 cm long, laterally compressed, smooth, brown.

DISTRIBUTION: An occasional tree of moist forests. Bordeaux (A2102, B574). Also on Tortola; the West Indies, Panama to northern South America.

COMMON NAME: bullet.

CULTIVATED SPECIES: *Manilkara zapota* (L.) P. Royen, the edible sapote, is occasionally cultivated.

3. POUTERIA Aubl.

Trees or rarely shrubs, unarmed, producing abundant white latex. Leaves spirally arranged, rarely opposite; stipules wanting. Flowers actinomorphic, often unisexual, in axillary or ramiflorous fascicles; calyx of a single whorl of 4–6 free, imbricate sepals; corolla cup-shaped to tubular, the tube shorter than, equal to, or longer than the lobes, the lobes 4–6, simple; stamens 4–6(–9), the filaments adnate to the corolla tube, the anthers basifixed; staminodes as many as corolla lobes or rarely absent, inserted on the corolla sinus; ovary of 1–6 locules, the placentation axile, the style terminal, included or exserted. Fruit a 1- to many-seeded berry; seeds broadly ellipsoid, plano-convex, with a narrowly elongate ventral scar.

A pantropical genus of about 230 species, with 188 occurring in the neotropics.

1. Pouteria multiflora (A. DC.) Eyma, Recueil Trav. Bot. Néerl. **33**: 164. 1936. *Lucuma multiflora* A. DC., Prodr. **8**: 168. 1844.   Fig. 186A–G.

Lucuma stahliana Pierre *in* Urb., Symb. Ant. **5**: 103. 1904.

Tree 20–25 m tall; bark brown, fissured and rough; twigs stout, minutely hairy to glabrous when young, becoming grayish and lenticellate. Leaves spirally arranged; blades 12–35 × 4.5–7.5 cm, elliptic to oblanceolate, subcoriaceous, glabrous, with a thickened midvein, prominent beneath, the apex obtuse to rounded, sometimes notched, the base obtuse to attenuate, the margins crenate; petioles glabrous, 1–3 cm long. Flowers 2–6 in axillary fascicles; pedicels 4–10 mm long, puberulent to glabrous. Calyx cup–shaped, the sepals 4, 4–8 mm long, ovate to rounded, puberulent without, the outer pair shorter than the inner ones; corolla white to pinkish, tubular, 7–13 mm long, the tube 4–8 mm long, the lobes simple, mostly 6, but also from 4 to 8, 3–5 mm long, oblong to ovate, papillate; stamens (4–)6(–8), the filaments 1–2 mm long, the anthers 1–2.5 mm long, lanceolate; staminodes (4–)6(–8), 1.5–3 mm long, lanceolate, papillate; ovary ovoid to globose, pubescent, 4–6-locular, the style subulate, with a capitate stigma. Berry 1- to many-seeded, ellipsoid, ovoid, or globose, 2.5–5 cm long, dark yellow at maturity. Seeds 2–3 cm long, globose to plano-convex.

DISTRIBUTION: An occasional tree of moist forests. Bordeaux (A2126). Also on Jamaica, Puerto Rico, the Lesser Antilles, Panama to northern Venezuela and Peru.

COMMON NAME: canistel.

4. SIDEROXYLON L.

Trees or shrubs, spinose or unarmed, producing abundant white latex. Leaves spirally arranged, rarely opposite; stipules wanting. Flowers actinomorphic, sessile or rarely pedicellate, bisexual or rarely unisexual, solitary or in axillary fascicles; calyx of a single whorl of 5(–8) free, imbricate sepals; corolla cup-shaped, glabrous, the tube shorter than, rarely equal to or longer than, the lobes, the lobes (4–)5(–6), divided into 3 segments; stamens (4–)5(–6), the filaments adnate to the top of corolla tube, the anthers basifixed; staminodes as many as corolla lobes, alternating with the stamens; ovary of (1–)5(–8) locules with axile-basal or basal placentation, the style terminal, included or exserted. Fruit a 1(–2)-seeded berry; seeds globose, ovoid, oblong or ellipsoid, not laterally compressed, with a small, circular, ventral hilum scar.

A pantropical genus of about 70 species, 50 of which occur in the neotropics.

Key to the Species of *Sideroxylon*

1. Often spiny shrub or small tree; leaves obovate, oblanceolate, or spatulate, 0.7–4.5 cm long .... 2. *S. obovatum*
1. Unarmed trees; leaves narrowly elliptic, elliptic, oblong-elliptic, or elliptic-lanceolate, 4.5–14.5 cm long.
   2. Leaves 1.5–4 cm wide; fruit 6–10 mm long, ellipsoid to subglobose, reddish brown .... 3. *S. salicifolium*
   2. Leaves 3.3–7.7 cm wide; fruit 1.5–2.6 cm long, ellipsoid to narrowly ovoid, yellow .... 1. *S. foetidissimum*

1. Sideroxylon foetidissimum Jacq., Enum. Syst. Pl. 15. 1760. *Mastichodendron foetidissimum* (Jacq.) H.J. Lam, Meded. Bot. Mus. Herb. Rijks. Univ. Utrecht **65**: 621. 1939.

Tree to 20 m tall; bark brown, rough, with vertical fissures; twigs minutely hairy when young, becoming glabrous. Leaves spirally arranged; blades 7.5–14.5 × 3.3–7.7 cm, elliptic, oblong-elliptic, or elliptic-lanceolate, subcoriaceous, glabrous, with a stout, yellowish midvein, prominent beneath, the apex obtuse, acute, or rounded, the base rounded, obtuse to attenuate, the margins entire; petioles glabrous, 2–4.5 cm long. Flowers bisexual, 5–20 in the axils of defoliated nodes; pedicels 4–8 mm long, puberulent to glabrous. Calyx cup-shaped, the sepals (4–)5(–6), 1.5–2 mm long, ovate to rounded, glabrous, ciliate on margins; corolla yellow, glabrous, 4–4.5 mm long, the tube 0.5–1.5 mm long, the lobes 5–6, ovate to lanceolate, with vestigial lateral segments; stamens 5–6, the filaments 1.5–2.5 mm long, the anthers 1.3–1.5 mm long, lanceolate; staminodes 5–6, 1–1.5 mm long, lanceolate, toothed; ovary ovoid, glabrous, (4–)5(–6)-locular, the style subulate, with a truncate or minutely 5-lobed stigma. Berry 1(–2)-seeded, ellipsoid to narrowly ovoid, 1.5–2.6 cm long, 2.5–5 cm long, yellow at maturity. Seeds 1.5–2 cm long, ellipsoid or ovoid.

DISTRIBUTION: An occasional tree of moist forests. Coral Bay Quarter along Center Line Road (A2114). Also on St. Croix and Tortola; throughout the West Indies, Florida, Mexico, Guatemala, and Belize.

COMMON NAMES: bully, bully-mastic.

2. Sideroxylon obovatum Lam., Tabl. Encycl. **2**: 42. 1794. *Bumelia obovata* (Lam.) A. DC., Prodr. **8**: 191. 1844.   Fig. 186H–N.

Bumelia obovata var. *portoricensis* Pierre *in* Urb., Symb. Ant. **5**: 143. 1904.
Bumelia obovata var. *thomensis* Pierre *in* Urb., Symb. Ant. **5**: 143. 1904.
Bumelia krugii Pierre *in* Urb., Symb. Ant. **5**: 146. 1904.
   Bumelia obovata var. *krugii* (Pierre) Cronquist, J. Arnold Arbor. **26**: 465. 1945.

Shrub or small tree to 7 m tall; bark brown, rough, peeling off in rectangular plates, the inner bark reddish; twigs minutely hairy when young, becoming glabrous, often with axillary, simple spines. Leaves spirally arranged; blades 0.7–4.5 × 0.2–3.8 cm, wedge-shaped, obovate, oblanceolate, to spatulate, chartaceous to coriaceous, glabrous or puberulent beneath, the apex obtuse, rounded, or truncate, sometimes notched, the base wedge-shaped or attenuate, the margins entire, usually revolute; petioles 0.5–4 mm long. Flowers bisexual, solitary or 2–5 in axillary fascicles; pedicels 0.5–5 mm long, puberulent to glabrous. Calyx cup-shaped, the sepals 5, 1.5–2.2 mm long, the outer pair smaller than the inner ones, oblong-ovate, rounded at apex, glabrous or puberulent; corolla yellow or cream-colored, glabrous, 3–4.5 mm long, the tube 1–1.5 mm long, the lobes 5, median segment ovate or elliptic, lateral segments narrowly lanceolate; stamens 5–6, the filaments 1.5–2.5 mm long, the anthers 0.7–0.9 mm long, ellipsoid; staminodes 5, 1.2–2.5 mm long, ovate or lanceolate; ovary

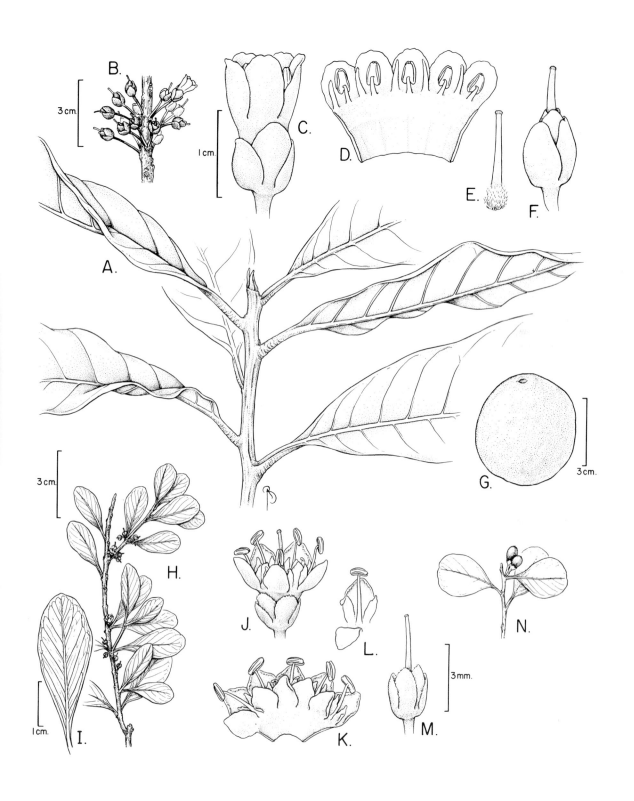

Fɪɢ. 186. A–G. *Pouteria multiflora*. **A.** Leafy branch. **B.** Inflorescence. **C.** Flower. **D.** Corolla spread open to show stamens and staminodes. **E.** Pistil. **F.** Calyx and pistil after disarticulation of corolla. **G.** Berry. **H–N.** *Sideroxylon obovatum*. **H.** Fertile branch. **I.** Leaf. **J.** Flower. **K.** Corolla spread open to show stamens and staminodes. **L.** Stamen and staminode. **M.** Calyx and pistil after disarticulation of corolla. **N.** Fruiting branch.

ovoid, pubescent, 5-locular, the style slender, elongate, with a punctiform stigma. Berry 1-seeded, ellipsoid, ovoid to globose, 0.4–1 cm long, dark red or brown at maturity. Seeds 3.7–7 mm long, subglobose to ovoid.

DISTRIBUTION: A common shrub of dry forests and scrublands. Maria Bluff (A2338), East End Point (A3792), Lameshur (B526). Also on Anegada, Jost van Dyke, St. Croix St. Thomas, Tortola, and Virgin Gorda; Hispaniola, Puerto Rico (including Vieques), and the Lesser Antilles.

3. Sideroxylon salicifolium (L.) Lam., Tabl. Encycl. **2:** 42. 1794. *Achras salicifolia* L., Sp. Pl., ed. 2, **1:** 470. 1762. *Bumelia salicifolia* (L.) Sw., Prodr. 50. 1788. *Dipholis salicifolia* (L.) A. DC., Prodr. **8:** 188. 1844.

Small tree 10–20 m tall; bark beige, smooth, the inner bark pinkish; twigs golden or whitish-pubescent, becoming glabrous, fissured and lenticellate. Leaves spirally arranged; blades 4.5–8 × 1.5–4 cm, narrowly elliptic to elliptic, chartaceous, glabrous or puberulent beneath, the apex acute to obtuse, the base attenuate, the margins entire to crenulate; petioles 0.5–1.5 cm long. Flowers bisexual, 5–12 in the axillary clusters; pedicels 1–4 mm long, tomentose. Calyx cup-shaped, the sepals 5, 2.5–3 mm long, the outer pairs smaller than the inner ones, oblong-ovate, rounded at apex, tomentose; corolla cream-colored, glabrous, 3.5–4.5 mm long, the tube 1.3–1.5 mm long, the lobes 5, median segment rounded to elliptic, clawed at base, lateral segments narrowly lanceolate; stamens 5, the anthers 1.2 mm long, ellipsoid; staminodes 5, 1.2–2 mm long; ovary ovoid, glabrous, 5-locular, the style elongate, tapering to a punctiform stigma. Berry 1–3-seeded, ellipsoid to subglobose, 0.6–1 cm long, turning from green to reddish brown to dark brown at maturity. Seeds subglobose to ellipsoid.

DISTRIBUTION: A common tree of dry to moist forests. Cinnamon Bay (A2919), Hawksnest (A2668), Susannaberg (A5264). Also on St. Croix, St. Thomas, and Tortola; Florida, West Indies, and Central America.

78. Scrophulariaceae (Figwort Family)

Herbs or shrubs. Leaves alternate or opposite or seldom whorled, simple or pinnatisect; stipules wanting. Flowers small to large, usually zygomorphic, bisexual, in thyrses, racemes, or spikes or seldom solitary; calyx (2–)4–5-lobed or cleft; corolla tubular, sometimes spurred or saccate at base, of 4–5 lobes; stamens (2–)4(–5), all fertile or one reduced to a staminode, the filaments adnate to the corolla tube, the anthers opening by longitudinal slits; nectary disk unilateral or annular; ovary superior, 2-locular, placentation axile, the locules with numerous ovules, the styles terminal, united, with a simple or a bilobed stigma. Fruit commonly a septicidal capsule; seeds usually angled and winged.

A family of 190 genera and 4000 species with cosmopolitan distribution, but most abundant in temperate regions.

Key to the Genera of Scrophulariaceae

1. Plants herbaceous, prostrate, mat-forming, with creeping stems rooting at nodes .... 1. *Bacopa*
1. Plants woody, at least at base, with erect stems.
    2. Leaves opposite; corolla 3.5–4 mm long, 4-merous, the tube with long, ascending hairs at base ..... 3. *Scoparia*
    2. Leaves alternate; corolla 6–9 mm long, 5-merous, the tube glabrous within .... 2. *Capraria*

1. BACOPA Aubl., nom. cons.

Prostrate or erect herbs. Leaves opposite or whorled, often fleshy. Flowers zygomorphic, 5-merous, bisexual, solitary in leaf axils; pedicels usually with 2 bracts; calyx of unequal sepals; corolla tubular or bell-shaped, the lobes spreading; stamens 4, didynamous, included, the anthers bilocular; ovary of 2 locules, the placentation axile, the style entire, the stigma 2-lobed or united and capitate. Fruit a many-seeded, globose, septicidal capsule; seeds minute, nearly oblong, longitudinally reticulate.

A pantropical genus of about 40 species, most of which occur in the neotropics.

1. Bacopa monnieri (L.) Pennell, Proc. Acad. Nat. Sci. Philadelphia **98:** 94. 1946. *Lysimachia monnieri* L., Cent. Pl. II. 9. 1756. *Herpestis monnieria* (L.) Kunth *in* Humb., Bonpl. & Kunth, Nov. Gen. Sp. **2:** 366. 1818.   Fig. 187A–D.

Spreading, mat-forming, succulent, glabrous herb, sometimes covering an area of a few square meters; stems prostrate, rooting at nodes. Leaves opposite; blades 4–12 × 2–5 mm, oblanceolate or obovate, fleshy, the apex rounded, the base cuneate, the margins entire; petioles 0–1 mm long. Pedicels 5–11 mm long, slender, glabrous; sepals free to base, 5–5.5 mm long, the 3 outer ones ovate, the inner ones linear-oblong; corolla white to mauve, bell-shaped, glabrous, 8–9 mm long, the tube ca. 5 mm long, greenish at base within, the lobes 5, rounded, emarginate; the 2 longer stamens nearly exserted, with filaments to 2 mm long, the shorter stamens included, with filaments ca. 0.5 mm long, the anthers 1.2 mm long, oblong; ovary flattened-ovoid, glabrous, the style cylindrical, the stigma capitate, exserted, greenish. Capsule ovoid, ca. 5 mm long, with numerous, nearly oblong seeds.

DISTRIBUTION: An occasional herb of sandy, saline, or seasonally flooded, open areas. Trail to Fortsberg (A4083), Hansen Bay (A3780), Waterlemon Bay (A1935). Also on St. Croix, St. Thomas, and Tortola; throughout tropical America.

2. CAPRARIA L.

Erect herbs or subshrubs. Leaves alternate, serrate. Flowers actinomorphic, 5-merous, bisexual, axillary; pedicels not bracteate; calyx of equal, deeply lobed sepals; corolla bell-shaped, the lobes elongate, ascending; stamens 4 or 5, the filaments of equal length, adnate to the base of corolla tube, the anthers oblong to ovoid, bilocular; ovary 2-locular with axile placentation, the style distal, elongate, exserted. Fruit a many-seeded, septicidal capsule; seeds reticulate.

A genus of 4 species in tropical America.

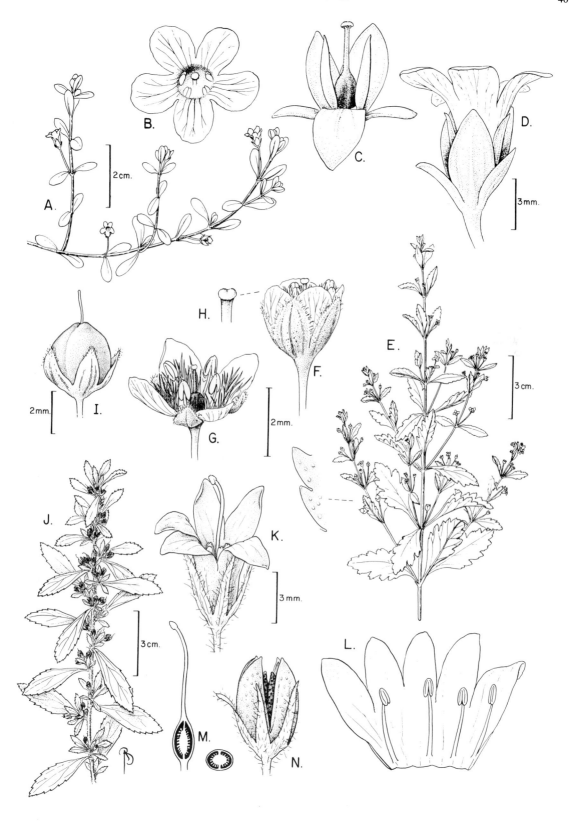

F<small>IG.</small> 187. A–D. *Bacopa monnieri*. **A.** Habit. **B.** Corolla, top view. **C.** Flower with subtending bracteoles and corolla removed to show pistil. **D.** Flower. **E–I.** *Scoparia dulcis.* **E.** Fertile branch and detail of leaf margin showing glandular-punctate surface. **F.** Closed flower. **G.** Open flower. **H.** Detail of stigma. **I.** Capsule. **J–N.** *Capraria biflora.* **J.** Fertile branch. **K.** Flower. **L.** Corolla spread open to show stamens. **M.** L.s. pistil and c.s. ovary. **N.** Dehiscing capsule. (A–I from N. Holmgren. 1984. Scrophulariaceae. In G. Harling & B. Sparre (eds.), Fl. Ecuador (177): 8–37.)

1. Capraria biflora L., Sp. Pl. 628. 1753. Fig. 187J–N.

Erect subshrub 50–70 cm tall, branching along main stem or sometimes many-branched from a subwoody base; stems cylindrical, pilose. Leaf blades 2.5–6.5 × 1–1.7 cm, oblanceolate or obovate, chartaceous, pilose, the apex acute or obtuse, the base cuneate to obtuse, the margins serrate toward distal portion; petioles 0–1 mm long. Flowers paired in leaf axils or less often solitary or 3; pedicels 2–8 mm long, slender, pilose. Sepals oblong-linear, 4–5 mm long; corolla white, glabrous, 6–9 mm long, the lobes 5, oblong, acute at apex; stamens nearly exserted, the filaments to 2 mm long, the anthers ca. 1 mm long; ovary ovoid, glabrous, the style cylindrical, elongate, the stigma capitate, exserted. Capsule ovoid, grooved, ca. 5 mm long; seeds numerous, wedge-shaped, to 0.5 mm long.

DISTRIBUTION: A common plant of open disturbed areas, especially along sandy beaches. Center Line Road (A633), Waterlemon Bay (A1936). Also on St. Croix and St. Thomas; throughout tropical America.

COMMON NAMES: goat weed, kabrita rotin.

3. SCOPARIA L.

Erect herbs or subshrubs. Leaves opposite or whorled, serrate. Flowers actinomorphic, bisexual, axillary, solitary or paired; pedicels not bracteate; calyx of 4–5 equal, deeply lobed sepals; corolla saucer-shaped, with 4 deeply parted, spreading lobes, pubescent within; stamens 4, exserted, the filaments of equal length, adnate to the base of corolla tube, the anthers oblong to ovoid, bilocular; ovary 2-locular with axile placentation, the style distal, elongate, exserted, with a capitate stigma. Fruit a many-seeded, septicidal capsule; seeds angular, obovoid.

A genus of 20 species in tropical America.

1. Scoparia dulcis L., Sp. Pl. 116. 1753.　Fig. 187E–I.

Erect subshrub 0.4–1 m tall, branching along main stem or less often many-branched from a subwoody taproot; stems cylindrical, 6-ribbed, puberulent, especially at nodes. Leaf blades 1.3–4 × 0.3–1 cm, elliptic to oblanceolate or obovate, chartaceous, glabrous, glandular-punctate, the apex acute or obtuse, the base cuneate, the margins serrate above the middle; petioles 0–3 mm long. Flowers solitary or paired in leaf axils; pedicels 5–8 mm long, slender. Sepals 4, ovate, 1.5–2 mm long, ciliate; corolla white, 3.5–4 mm long, the tube with long ascending hairs at base, the lobes 4, reflexed, oblong and rounded at apex; stamens 4, the filaments ca. 1 mm long, pinkish-tinged, the anthers ca. 1 mm long, white; ovary ovoid, glabrous, green, the style filiform, elongate, white, the stigma capitate. Capsule ovoid, 3–3.5 mm long, with a persistent style. Seeds numerous ca. 0.5 mm long, reticulate.

DISTRIBUTION: A common weed of open, disturbed areas, especially along roadsides and along trails. Bordeaux (A3824), Fortsberg (A2424). Also on St. Croix, St. Thomas, and Tortola; a pantropical weed.

COMMON NAME: teeth bush.

79. Simaroubaceae (Quassia Family)

Trees or shrubs. Leaves alternate or rarely opposite, pinnately compound or unifoliolate; stipules mostly wanting. Flowers small, actinomorphic, bisexual or unisexual, in axillary or terminal racemes, corymbs, or panicles; calyx commonly of 5 distinct or basally connate sepals; corolla commonly of 5 distinct petals or wanting; stamens commonly 5 or 10, the filaments distinct, usually with a ventral appendage, the anthers opening by longitudinal slits; nectary disk annular, sometimes modified into a gynophore or an androgynophore; ovary superior, of 2–5 weakly or firmly united carpels (or sometimes completely free), with axile placentation, the ovules solitary or paired in the carpels or locules, the styles terminal, distinct or connate. Fruit a berry, drupe, capsule, or samara.

A family of 25 genera and 150 species, with pantropical distribution.

REFERENCE: Nair, N. C. & R. K. Joshi. 1958. Floral morphology of some members of the Simaroubaceae. Bot. Gaz. (Crawfordsville) 120: 88–99.

Key to the Genera of Simaroubaceae

1. Petioles and leaf rachis winged; corolla red .... *Quassia* (cultivated)
1. Petioles and leaf rachis not winged; corolla yellowish green .... 1. *Picrasma*

1. PICRASMA Blume

Dioecious trees. Leaves alternate, odd-pinnately compound; leaflets opposite, with entire, crenate, or serrate margins; stipules wanting. Flowers actinomorphic, unisexual, 4–5-merous, in axillary panicles or corymbs; calyx of equal, free, imbricate sepals; corolla of free, valvate, reflexed petals; stamens 4–5, vestigial in pistillate flowers, the filaments of equal length, the anthers nearly ovoid; nectary disk 4–5-lobed; ovary of 2–5 free carpels united at the tip of the styles, carpellodes present or absent in staminate flowers. Fruit a drupe with crustaceous endocarp, of 1–3, basally, connate, globose monocarps; seeds 1 per monocarp.

A genus of about 6 species in tropical America and 1 in Asia and Polynesia.

1. Picrasma excelsa (Sw.) Planch., London J. Bot. **5:** 574. 1846. *Quassia excelsa* Sw., Prodr. 67. 1788. *Simarouba excelsa* (Sw.) DC., Ann. Mus. Natl. Hist. Nat. **17:** 424. 1811.　Fig. 188.

Rhus antillana Eggers, Fl. St. Croix 41. 1879. *Picrasma antillana* (Eggers) Urb., Symb. Ant. **5:** 378. 1908.

Tree 6–12 m tall; bark dark gray; twigs cylindrical, puberulent, becoming glabrous with age. Leaf rachis 15–25 cm long, puberulent; leaflets 7–13, 7–13.5 × 3–4.6 cm, elliptic, oblong, oblong-lanceolate, coriaceous, the lower surface puberulent, with prominent primary and secondary veins, the apex acute or obtusely acuminate, the base obtuse-rounded, unequal, the margins entire; petiolules 4–5 mm long. Flowers 5-merous, in axillary panicles

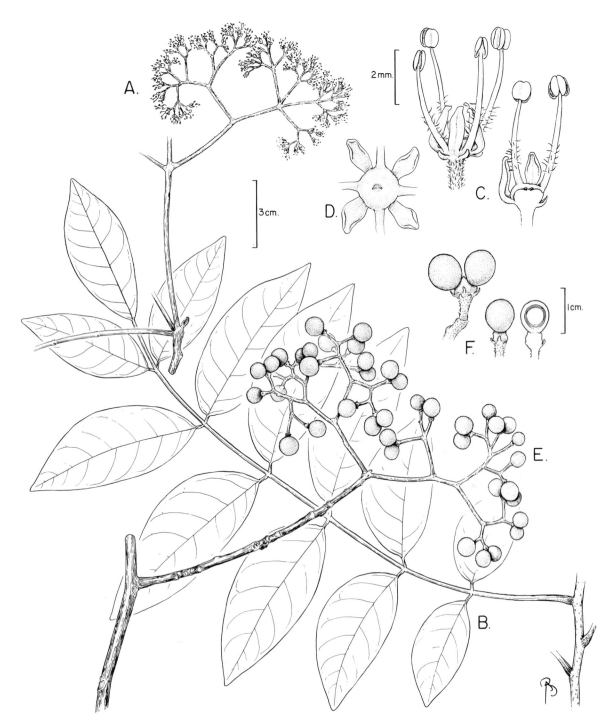

Fig. 188. *Picrasma excelsa*. **A.** Inflorescence. **B.** Branch. **C.** Staminate flower and l.s. staminate flower. **D.** Staminate flower, top view. **E.** Infructescence. **F.** Fruit of 2 and 1 monocarps and l.s. monocarp.

to 10 cm long; axes usually reddish-tinged; pedicels ca. 5 mm long, stout. Sepals triangular, 0.5–1 mm long; petals yellowish green, oblong-lanceolate, ca. 3–4 mm long. Drupe of 1 or 2 monocarps, adnate to a swollen receptacle, the monocarps globose, 7–9 mm long, turning from green to dark red and finally purplish black. Seeds 1 per monocarp, nearly globose, 4–6 mm long.

DISTRIBUTION: An occasional tree of moist forests. Road to Bordeaux (A4262), Reef Bay Trail (A4124), Maho Bay (A5266). Also on St. Croix, St. Thomas, and Tortola; Jamaica, Hispaniola, Puerto Rico, the Lesser Antilles, and northern Venezuela.

COMMON NAME: bitter ash.

CULTIVATED SPECIES: Collections of *Quassia amara* L. from St. John have been made from a few individuals in cultivation.

80. Solanaceae (Tomato Family)

Shrubs, herbs, vines, lianas, or small trees; usually with branched hairs and spines and prickles. Leaves alternate, simple or less often pinnately compound; stipules wanting. Flowers small to large, actinomorphic or slightly zygomorphic, bisexual, in extra-axillary or terminal cymes or racemes; calyx 5(–10)-lobed; corolla sympetalous, rotate to tubular, of (4–)5(–6) lobes; stamens as many as the corolla lobes or some of them reduced to staminodes, the filaments adnate to the corolla tube, the anthers opening by longitudinal slits or terminal pores or slits; nectary disk usually present; ovary superior, of 2 carpels, 2-locular or less often 4-locular, usually with axile placentation, with numerous ovules, the styles terminal, connate. Fruit a berry, a capsule, or seldom a drupe.

A cosmopolitan family of 85 genera and about 2800 species most abundant in tropical America.

Key to the Genera of Solanaceae

1. Corolla funnel- or trumpet-shaped, with a narrow elongate tube.
    2. Corolla 7–13 cm long; fruits spiny... 4. *Datura*
    2. Corolla < 6 cm long; fruits smooth.
        3. Corolla 1–1.2 cm long; fruit a fleshy, oblong-ellipsoid, black berry to 1.5 cm long .... 3. *Cestrum*
        3. Corolla 4.5–6 cm long; fruit a dry, ovoid, orange-yellow capsule to 2 cm long .... 1. *Brunfelsia*
1. Corolla bell- or saucer-shaped, or broadly funnel-shaped, not tubular.
    4. Fruit a fleshy, naked berry (not covered by an inflated calyx).
        5. Anthers opening by terminal pores.. 6. *Solanum*
        5. Anthers opening by longitudinal slits .... 2. *Capsicum*
    4. Fruit enveloped by an accrescent, membranous, inflated calyx .... 5. *Physalis*

1. BRUNFELSIA L.

Shrubs or small trees. Leaves simple, entire. Flowers zygomorphic, 5-merous, axillary or terminal, solitary or in fascicled racemes; calyx bell-shaped or tubular, lobed or toothed; corolla trumpet-shaped, with spreading lobes; stamens 4, didynamous, included, sometimes the smallest 2 sterile, the filaments of equal length, the anthers opening by longitudinal slits; ovary of 2 connate carpels, the placentation axile with numerous ovules. Fruit a partly dehiscent or indehiscent capsule; seeds numerous, nearly ovoid, reticulate or pitted.

A neotropical genus of about 40 species, 21 of which are native to the West Indies.

REFERENCE: Plowman, T. 1979. The genus *Brunfelsia*: a conspectus of the taxonomy and biogeography. In J. G. Hawkes, R. N. Lester, & A. D. Skelding (eds.), The biology and taxonomy of the Solanaceae. Linn. Soc. Symp. Ser. 7: 475–491. Academic Press, London.

1. Brunfelsia americana L., Sp. Pl. 191. 1753.

Fig. 189A–E.

Shrub 1.5–3 m tall, branching along main stem; bark gray; twigs cylindrical, pubescent, becoming glabrous and fissured with age. Leaf blades 3.7–9 × 1.5–5 cm, obovate, spatulate, or oblanceolate, chartaceous, the lower surface puberulent, with prominent midvein, the apex obtuse, rounded, truncate, or retuse, the base obtuse to cuneate, the margins entire; petioles 3–8 mm long, pubescent. Flowers in axillary fascicles or terminal on short, lateral branches; pedicels 5–6(–10) mm long, puberulent. Calyx nearly bell-shaped, 6–9 mm long, the lobes ovate-triangular; corolla tube 4.5–6 cm long, narrowed toward base, greenish and puberulent without, deep purple within, the lobes white to cream, unequal, zygomorphic, spreading, rounded at apex, 1–1.5 cm long. Capsule depressed-ovoid, shallowly bilobed, 1.5–2 cm long, turning from green to yellow-orange, the calyx persistent at base, splitting. Seeds numerous, blackish, ca. 4 mm long.

DISTRIBUTION: A common shrub of dry forests, scrub vegetation, and occasionally in open, disturbed areas. Lameshur (A2742), Threadneedle Point (A4129), White Cliff (A2715). Also on St. Croix, St. Thomas, Tortola, and Virgin Gorda; Hispaniola, Puerto Rico, and the Lesser Antilles.

COMMON NAME: rain tree.

2. CAPSICUM L.

Herbs or subshrubs. Leaves simple, entire. Flowers actinomorphic, 5-merous, axillary, solitary or in cymes; calyx bell-shaped, toothed or truncate; corolla white, saucer-shaped, with spreading lobes; stamens 5, the filaments of equal length, adnate to the base of corolla, the anthers opening by longitudinal slits; ovary of 2 or 3 connate carpels, the placentation basal-axile with numerous ovules. Fruit a fleshy to leathery berry; seeds numerous, lenticular, flattened.

A neotropical genus of about 25 species, cultivated throughout the world for its edible, often pungent peppery fruits.

1. Capsicum frutescens L., Sp. Pl. 189. 1753.

Fig. 189F–I.

Capsicum baccatum sensu many West Indian authors, non L., 1767.

Subshrub, 0.4–1.5 m tall, with drooping branches; twigs nearly cylindrical to slightly angled, puberulent, becoming glabrous with age. Leaf blades 1.5–9 × 0.9–5 cm, elliptic, ovate, to lanceolate, chartaceous, the lower surface puberulent, especially in the axils of veins, the apex short-acuminate to acute, the base acute, cuneate, or obtuse to rounded, the margins wavy; petioles 0.5–2 cm long. Flowers solitary in leaf axils or less often paired, reflexed; pedicels 4–8 mm long (much longer in fruits), puberulent or glabrous. Calyx bell-shaped, truncate to minutely toothed at apex, 1–1.5 mm long; corolla white, 2–2.5 mm long, the lobes spread-

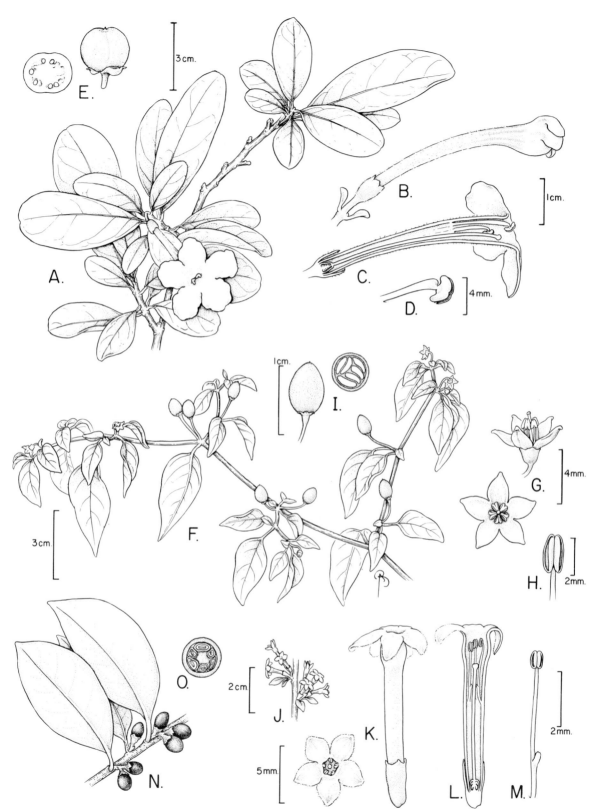

FIG. 189. A–E. *Brunfelsia americana.* **A.** Flowering branch. **B.** Flowerbud. **C.** L.s. of flower. **D.** Portion of stamen with anther. **E.** C.s. berry and berry. **F–I.** *Capsicum frutescens.* **F.** Fertile branch. **G.** Flower, top and lateral views. **H.** Stamen. **I.** Berry and c.s. berry. **J–O.** *Cestrum laurifolium.* **J.** Portion of branch with inflorescences. **K.** Flower. **L.** L.s. of flower. **M.** Stamen. **N.** Fruiting branch. **O.** C.s. berry.

ing, slightly reflexed at tips, anthers violet, connivent around the style. Berry ovoid, 6–9 mm long, turning from dark green to yellow and finally red; cultivated forms varying in shape and color. Seeds numerous, lenticular, light brown, ca. 3 mm long.

DISTRIBUTION: A common roadside weed. Center Line Road (A2122), Ajax (A2653). Also on St. Croix and St. Thomas; Puerto Rico and the Lesser Antilles.

COMMON NAME: wild pepper.

3. CESTRUM L.

Shrubs. Leaves simple, entire, petiolate. Flowers actinomorphic or zygomorphic, (4–)5-merous, in axillary or extra-axillary cymes or panicles; calyx vase-shaped or bell-shaped, shallowly lobed or toothed; corolla trumpet-shaped or funnel-shaped, with spreading lobes; stamens 5, included, the filaments of equal length, adnate to the corolla tube, the anthers opening by longitudinal slits; ovary of 2 connate carpels, the placentation axile with few ovules, the style filiform, the stigma nearly capitate. Fruit a fleshy to leathery berry; seeds few, flattened, angular, lenticular.

A neotropical genus of about 175 species, some of which are cultivated for their fragrant flowers.
REFERENCE: Francey, P. 1935–1936. Monographie du genre *Cestrum* L. Candollea 6: 46–398; 7: 1–132.

1. Cestrum laurifolium L'Hér., Stirp. Nov. 4: 69, t. 34. 1788.   Fig. 189J–O.

Shrub 1–2.5 m tall; twigs cylindrical, slightly angled when young. Leaf blades 3.5–11 × 2–5.3 cm, obovate or oblong-elliptic, coriaceous, glabrous, slightly glossy above, dull, lighter underneath, the apex short-acuminate, obtuse, acute, or seldom rounded or retuse, the base obtuse to nearly rounded, the margins entire, revolute; petioles 1–1.5 cm long. Flowers in short axillary cymes. Calyx nearly bell-shaped, ca. 4 mm long, minutely toothed; corolla yellow, 4–5-merous, trumpet-shaped, the tube 1–1.2 cm

long, narrowed at base, the lobes ovate, ca. 2 mm long, spreading; stamens included; stigma green, included. Berry ellipsoid, 1–1.2 cm long, shiny, purplish black when mature. Seeds few, angled, lenticular, light brown, ca. 4 mm long.

DISTRIBUTION: A common shrub of moist areas, especially along shaded roadsides. Road to Bordeaux (A3137), Center Line Road (A2391). Also on St. Croix, St. Thomas, Tortola, and Virgin Gorda; Greater and Lesser Antilles.

DOUBTFUL RECORDS: *Cestrum diurnum* L. and *Cestrum nocturnum* L. were reported by Woodbury and Weaver (1987) as occurring on St. John. These reports are apparently from cultivated plants, since no naturalized populations are known for either species on St. John.

4. DATURA L.

Herbs or subshrubs, with strong fetid smell. Leaves simple, entire, lobed, or incised, long-petiolate. Flowers actinomorphic, 5-merous, solitary, axillary; calyx long-tubular, usually 5-angled, deeply lobed at apex; corolla funnel-shaped, with plaited limb, 5-lobed, usually with an accessory lobe between lobes (for a total of 10); stamens 5, included, the filaments inserted on lower half of tube, the anthers opening by longitudinal slits; ovary of 2 connate carpels, the placentation axile with numerous ovules, the style filiform. Fruit a valvate capsule, subtended by the persistent, reflexed base of the calyx; seeds numerous, flattened, kidney-shaped or nearly so.

A neotropical genus of about 8 species, introduced elsewhere.

Key to the Species of *Datura*

1. Plant pilose; leaves deeply sinuate or with rounded lobes; calyx nearly tubular, not angled; capsules with reflexed peduncle .... 1. *D. inoxia*
1. Plant glabrous or puberulent; leaves lobed to deeply toothed, the teeth irregular, acuminate; calyx tubular, 5-angled, the angles narrowly keeled; capsules erect .... 2. *D. stramonium*

1. Datura inoxia Mill., Gard. Dict., ed. 8. 1768.
Fig. 190A–E.

Datura metel sensu Britton & P. Wilson, Bot. Porto Rico 6: 174. 1925, non L., 1753.

Herb to subshrub 0.5–1 m tall, with strong fetid odor, branching appearing pseudodichotomous; branches herbaceous, nearly cylindrical, soft pilose, sometimes purple-tinged. Leaf blades 6–16 × 3–11.5 cm, irregular but usually with ovate outline, chartaceous, soft-pubescent, the apex shortly acuminate, obtuse, acute, or acuminate, the base very unequal, rounded-acute, the margins deeply sinuate or lobed (the lobes rounded); petioles 3–7 cm long, pubescent. Flowers erect, sweet-scented. Calyx green, nearly tubular but widened toward the base, 5–9.5 cm long, the lobes ca. 1 cm long, ovate-triangular; corolla white, funnel-shaped, the limb pentagonous, slightly reflexed, with acute to acuminate angles, the accessory lobes acuminate, the tube 10–13 cm long, narrowed toward base; stamens included, the anthers

white or cream-colored, connivent; style white, the stigma yellow, bilobed, included. Capsule on reflexed peduncle, with ovoid outline, to 3 cm long, densely covered with sharp spines to 1 cm long, opening along 4 longitudinal valves, from top to bottom, the reflexed calyx base irregularly lobed. Seeds numerous, reddish brown, ca. 4 mm long.

DISTRIBUTION: Common in recently disturbed areas such as roadsides and waste grounds. Emmaus (A1991), Great Cruz Bay (A2362). Also on St. Croix, St. Thomas, and Tortola; southeastern United States to South America, including the West Indies, introduced throughout the tropics.
COMMON NAME: prickly bur.

2. Datura stramonium L., Sp. Pl. 179. 1753.

Erect subshrub 0.5–1 m tall, with strong fetid odor; branching appearing pseudodichotomous; branches herbaceous, nearly cylindrical, glabrous or puberulent when young. Leaf blades 7.5–17 × 4–13 cm, with ovate to lanceolate outline, chartaceous, puberulent, becoming glabrous, the apex long-acuminate, the base cuneate to rounded, sometimes unequal, the margins lobed to

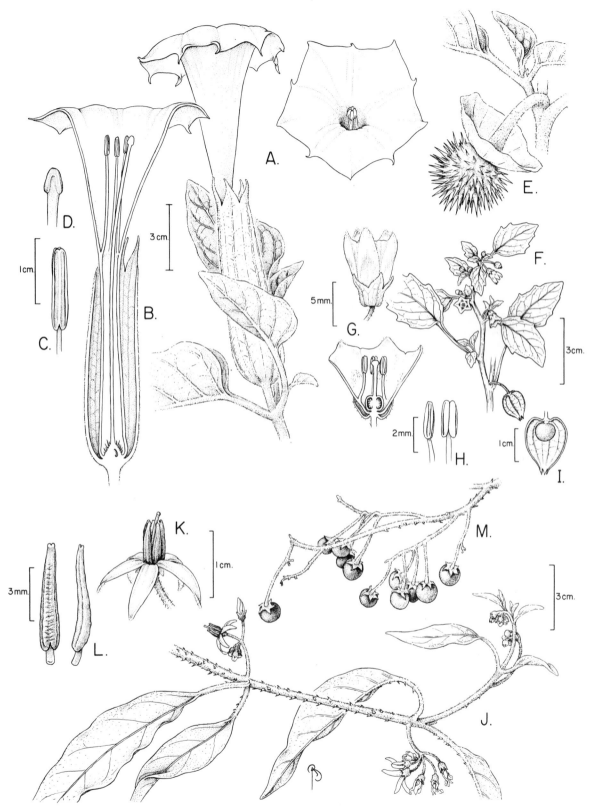

Fig. 190. A–E. *Datura inoxia*. A. Flowering branch and top view of corolla. B. L.s. of flower. C. Anther. D. Stigmatic surface. E. Capsule. F–I. *Physalis angulata*. F. Fertile branch. G. Flower and l.s. flower. H. Stamens. I. Fruit with accrescent calyx partially removed. J–M. *Solanum lancifolium*. J. Flowering branch. K. Flower. L. Stamens with poricidal anthers. M. Infructescence.

deeply toothed, the teeth irregular, acuminate; petioles 1.5–5 cm long, puberulent. Flowers erect, short-lived. Calyx green, tubular, 5-angled, keeled at angles, 3–5 cm long; corolla white, funnel-shaped, the limb obtusely 5-angled with long-acuminate lobes, the tube 7–10 cm long, narrowed toward base; stamens included. Capsule erect, with ovoid outline, 3–4 cm long, densely covered with sharp spines to 1.2 cm long, opening along 4 longitudinal valves, from top to bottom, the reflexed calyx base irregularly lobed. Seeds numerous, black, 3–4 mm long.

DISTRIBUTION: An occasional herb of open disturbed areas. Susannaberg (A4114). Also on St. Croix, St. Thomas, Tortola, and Virgin Gorda; native to Mexico but widespread as a weed throughout the world.

COMMON NAMES: deadly nightshade, jimsonweed, stinking bush, thorn apple.

DOUBTFUL RECORDS: *Datura metel* L. was reported for St. John by Woodbury and Weaver (1987) and earlier by Britton and Wilson (1925) under the name *Datura fastuosa* L. The latter record seems to be based on a cultivated plant. This species has not been seen in recent times, nor has a specimen been located to confirm these records.

5. PHYSALIS L.

Herbs or occasionally subshrubs. Leaves simple, entire, lobed, or toothed, petiolate. Flowers actinomorphic, 5-merous, solitary, axillary; calyx bell-shaped, toothed or lobed at apex; corolla bell-shaped to saucer-shaped, apically 5-angled with shallow sinuses or 5-lobed; stamens 5, included or slightly exserted, the filaments inserted near the base of corolla, the anthers opening by longitudinal slits; ovary of 2 connate carpels, the placentation axile with numerous ovules, the style filiform to club-shaped. Fruit a berry, enclosed by the accrescent, 5- or 10-angled, membranous, inflated calyx; seeds numerous, flattened, kidney-shaped.

A cosmopolitan genus of about 90 species, most of which are native to Mexico.

Key to the Species of *Physalis*

1. Fruiting calyx 10-angled; sepals triangular to triangular-ovate .... 1. *P. angulata*
1. Fruiting calyx 5-angled; sepals lanceolate to subulate.
   2. Plant glandular-pubescent; fruiting calyx with ovoid outline, 2–2.7 cm long .... 3. *P. turbinata*
   2. Plant sparsely pubescent (not glandular), with minute appressed hairs; fruiting calyx with elongate, cone-shaped outline, 4–4.5 cm long .... 2. *P. cordata*

1. Physalis angulata L., Sp. Pl. 183. 1753.

Fig. 190F–I.

Erect herb or subshrub 0.2–1 m tall, single-stemmed or many-branched from base, with long taproot; branches herbaceous to slightly fleshy, angled, sometimes reddish-tinged, puberulent when young, the pith sometimes hollow. Leaf blades 2.5–10 × 1.7–5.5 cm, with ovate to elliptic outline, chartaceous, puberulent, the apex acute, obtuse or acuminate, the base narrowed, obtuse, to rounded, unequal; the margins sinuate, lobed or serrate-dentate. Flowers reflexed; peduncle stout, 5–10 mm long. Calyx green, bell-shaped, 3–6 mm long, puberulent to glabrous, the sepals triangular to triangular-ovate; corolla broadly funnel-shaped, 6.5–9 mm long, light yellow, with 5 brown, semiannular spots within, the limb obtusely 5-angled; stamens included, the anthers bluish to purplish, connivent. Fruiting calyx cone-shaped in outline, 2–3 cm long, 10-angled, the angles usually purplish-tinged; berry globose to ovoid, 7–10 mm long, yellowish green. Seeds numerous, beige, ca. 1 mm long.

DISTRIBUTION: A common roadside weed, also found in open disturbed areas. Cruz Bay (A3076), Lameshur (A3212), Johns Folly (A3937). Also on Anegada, St. Croix, St. Thomas, Tortola, and Virgin Gorda; widespread throughout temperate and tropical areas of the world.

COMMON NAME: ground cherry.

2. Physalis cordata Mill., Gard. Dict., ed. 8. 1768.

Erect herb 40–80 cm tall, branching appearing pseudodichoto-mous; branches herbaceous to slightly fleshy, angled, sparsely pubescent. Leaf blades 2.5–7 × 2–5 cm, ovate, chartaceous, sparsely pubescent, especially along veins, the apex acute, the base rounded to truncate, slightly unequal, the margins remotely serrate, the teeth obtuse. Flowers reflexed; peduncle filiform, 5–8 mm long. Calyx green, bell-shaped, 8–9 mm long, the sepals

lanceolate to subulate, ciliate; corolla broadly funnel-shaped, 9–10 mm long, light yellow, with 5 reddish, semiannular spots within, the limb 5-lobed; stamens included. Fruiting calyx elongate, top-shaped in outline, 4–4.5 cm long, 5-angled; berry globose ca. 1.2 cm long. Seeds numerous, beige, lenticular, ca. 1.5 mm long.

DISTRIBUTION: An occasional roadside weed. Bethany (B328), Mandal (A2585). Widespread throughout tropical America.

3. Physalis turbinata Medik., Hist. & Commentat. Acad. Elect. Sci. Theod.-Palat. 4: 189. 1780.

Erect herb 30–80 cm tall, single-stemmed with pseudodichoto-mous branching or many-branched from base; branches herbaceous to fleshy, 4-angled, covered with soft, glandular hairs, the internodes slightly swollen, the pith usually hollow. Leaf blades 2.7–7 × 2–6 cm, ovate, chartaceous, sparsely pubescent, especially along veins, the apex acute to acuminate, the base rounded, truncate, or subcordate, slightly unequal, the margins remotely serrate, the teeth acute. Flowers reflexed; peduncle stout, 5–15 mm long, pubescent. Calyx green, bell-shaped, 4–5 mm long, pubescent, the sepals lanceolate; corolla broadly funnel-shaped, ca. 7 mm long, light yellow, the limb 5-angled, with round sinuses between angles; stamens slightly exserted, the anthers whitish; style club-shaped, the stigma greenish, nearly capitate. Fruiting calyx with ovoid outline, 2–2.7 cm long, 5-angled, green, pubescent; berry globose 1–1.2 cm long. Seeds numerous, orange-brown, lenticular, ca. 1.5 mm long.

DISTRIBUTION: An occasional roadside weed of moist areas. Road to Bordeaux (A2851, A3136). Widespread throughout tropical America.

DOUBTFUL RECORDS: *Physalis pubescens* L. was reported on St. John by Britton and Wilson (1925) and by Woodbury and Weaver (1987); however, no specimen has been located to confirm these records, nor has this species been seen in recent times.

6. SOLANUM L.

Herbs, shrubs, small trees, or woody to herbaceous vines, often armed with spines or prickles, glabrous or pubescent, the hairs simple or stellate. Leaves simple or compound, entire or lobed. Flowers actinomorphic, 5-merous, bisexual or rarely unisexual, produced in axillary or terminal racemes, or seldom solitary; calyx deeply lobed; corolla usually saucer-shaped, the limb apically 5-angled with shallow sinuses or with 5 deeply parted lobes; stamens 5, the filaments shorter than the anthers, the anthers yellow, connivent, opening by a terminal pore; ovary of 2 connate carpels, the placentation axile with numerous ovules, the style filiform, deciduous, the stigma bifid. Fruit a fleshy, leathery, or woody berry; seeds numerous, flattened.

A cosmopolitan genus of about 1400 species. Some species widely cultivated as food crops.

Key to the Species of *Solanum*

1. Plants unarmed.
  2. Plants herbaceous.
    3. Vine 2–5 m long; leaves simple and dissected; fruits red when mature .... *S. seaforthianum* (cultivated)
    3. Erect herb 30–80 cm tall; leaves simple and entire to sinuate; fruits purplish black when mature
    .... 1. *S. americanum*
  2. Plants shrubs.
    4. Young parts covered with appressed, multicellular hairs; leaves glabrous, 3.5–7 cm long; fruit
    ovoid-conical, 2–3 cm long .... 2. *S. conocarpum*
    4. Plant stellate pubescent or pilose; leaves stellate-pubescent underneath, 3.5–20 cm long.
      5. Upper leaf surface with simple hairs (sparsely strigose)..................................... 5. *S. mucronatum*
      5. Upper leaf surface with stellate hairs.
        6. Lower leaf surface whitish, densely stellate-tomentose....................................... 3. *S. erianthum*
        6. Lower leaf surface greenish, sparsely covered with stellate hairs .... 7. *S. racemosum*
1. Plants armed.
  7. Scandent shrubs or scrambling lianas .... 4. *S. lancifolium*
  7. Erect shrubs.
    8. Leaves deeply sinuate or lobed, 4.5–11 cm wide; fruits yellow when mature .... 8. *S. torvum*
    8. Leaves entire or nearly so, 1–5.6 cm wide; fruit orange or red when mature.
      9. Fruit depressed-globose, puberulent, orange when mature, 1.5–2 cm wide, with accrescent calyx
      at base; flowers staminate and bisexual.. 6. *S. polygamum*
      9. Fruit globose, glabrous, red when mature, 6–8 mm wide, the calyx not accrescent; flowers
      bisexual.. 7. *S. racemosum*

1. Solanum americanum Mill., Gard. Dict., ed. 8. 1768.

Solanum nodiflorum Jacq., Collectanea **1**: 100. 1788.
Solanum nigrum sensu Britton & P. Wilson, Bot. Porto
    Rico **6**: 166. 1925, non L., 1753.

Erect, unarmed herb 30–80 cm tall, single-stemmed with numerous lateral branches; stems herbaceous, slightly angled to nearly cylindrical, puberulent or glabrous, the angles usually with soft, minute conical projections. Leaf blades 5–17 × 2–10 cm, ovate to lanceolate, chartaceous, sparsely pubescent or glabrous, the apex acute to acuminate, the base attenuate, the margins entire to deeply sinuate. Flowers usually reflexed, in short, umbel-like, lateral racemes; peduncle filiform, 0.9–2 cm long. Calyx bell-shaped, ca. 1 mm long, the sepals ovate; corolla white, with a very short tube, the lobes 2.5–3.5 mm long, lanceolate, reflexed; anthers yellow, oblong, 1.5–2 mm long; style slender, the stigma nearly capitate. Berry fleshy, globose, 7–8 mm long, turning from green to purplish black when mature. Seeds numerous, beige, lenticular, ca. 1.3 mm long.

DISTRIBUTION: A common roadside weed of moist areas. Road to Bordeaux (A3140), Concordia (A4254), trail to Sieben (A2058). Also on St. Croix, St. Thomas, Tortola, and Virgin Gorda; widespread as a weed throughout the tropics.

2. Solanum conocarpum Dunal *in* Poir., Encycl. Suppl.
    3: 748. 1813.

Unarmed shrub to 3 m tall; bark ash-colored, with a few lenticels; stems nearly cylindrical, glabrous, young parts densely covered with appressed, multicellular hairs. Leaf blades 3.5–7 × 1.6–3 cm, oblong-elliptic or oblanceolate, coriaceous, glabrous, midvein yellowish, the apex obtuse or rounded, sometimes shallowly retuse, the base obtuse or narrowed, the margins entire, revolute; petioles 2–4 mm long, yellowish. Flowers usually paired, in nearly sessile, lateral or terminal cymes; pedicels filiform, 4–6 mm long. Calyx bell-shaped, greenish, ca. 3 mm long, the sepals ovate; corolla light violet, greenish at center, ca. 2 cm wide, the limb obtusely 5-angular; anthers yellow, subulate, ca. 3 mm long; style cylindrical, slender, exserted beyond the anthers, the stigma minutely capitate. Berry fleshy, ovoid-conical, 2–3 cm long, turning from green to yellow. Seeds few, reddish brown, lenticular, 4–4.5 mm long.

DISTRIBUTION: A rare, endangered species, known only from a few old collections and from two recent collections. The species is known on St. John from only a single living plant, from dry, deciduous forest. Coral Bay (the type collected by L. C. Richard); north of White Cliffs (A2710, A5437). Possibly on St. Thomas (according to Krebs, 1847) and on Virgin Gorda (sterile specimen, *Little 23836* (US), may belong this species).

COMMON NAME: marron bacora (as given by Dunal in original description of the species).

3. Solanum erianthum D. Don, Fl. Nepal. 96. 1825.

Solanum verbacifolium sensu Britton & P. Wilson, Bot. Porto Rico **6**: 168. 1925, non L., 1753.

Unarmed shrub 1.5–2.5 m tall; stems stellate-tomentulose, nearly cylindrical, becoming glabrous. Leaf blades 6–20 × 4–12 cm, ovate to nearly lanceolate, chartaceous, stellate-tomentulose, especially beneath, the apex acute, obtuse, or seldom rounded, the base obtuse to rounded, usually unequal, the margins entire; petioles 1.5–5.5 cm long, stout, stellate-tomentulose. Flowers in terminal (appearing as axillary) bifid cymes; pedicels stout, 3–5, 4–6 mm long, stellate-tomentulose. Calyx bell-shaped, ca. 5 mm long, stellate-tomentulose, the sepals triangular; corolla white, widely bell-shaped, ca. 5 mm long, deeply lobed, stellate-tomentulose without, the tube greenish within, the lobes ovate, spreading to reflexed; anthers connivent, yellow, oblong, ca. 2.7 mm long; style exserted beyond the anthers, slender, the stigma nearly capitate. Berry leathery, globose, 1–1.2 cm long, sparsely to densely covered with minute, stellate hairs, turning from green to yellowish, with brown spots. Seeds numerous, reddish brown, lenticular, ca. 1.7 mm long.

DISTRIBUTION: An occasional shrub of open disturbed areas. Ajax (A2121), Susannaberg (A706). Also on St. Croix, St. Thomas, and Tortola; native to the Old World tropics, now widespread throughout tropical America.

COMMON NAMES: turkey-berry, wild tobacco.

4. Solanum lancifolium Jacq., Collectanea **2**: 286. 1788 [1789]. Fig. 190J–M.

Solanum volubile Sw., Fl. Ind. Occid. **1**: 458. 1797.

Armed, scandent shrub or scrambling liana 2–6 m long; stems stellate-tomentulose, nearly cylindrical, densely to sparsely covered with yellowish, recurved spines. Leaf blades 10–18 × 3.7–7 cm, lanceolate, ovate, or elliptic, chartaceous, sparsely stellate-pubescent, midvein prominent and usually with recurved spines beneath, the apex acute, obtuse, or acuminate, the base obtuse to rounded, unequal, the margins entire; petioles 1–5 cm long, slender, stellate-pubescent, usually with recurved spines. Flowers in lateral racemes; the axes 2–4 cm long, stellate-pubescent; pedicels slender, 1.5–1.7 cm long, stellate-pubescent. Calyx bell-shaped, 4.5–5 mm long, stellate-pubescent, the sepals subulate; corolla white, deeply lobed, stellate pubescent without, the tube ca. 1.5 mm long, the lobes 8–11 mm long, oblong or oblong-lanceolate, ovate, spreading or slightly reflexed; anthers connivent, yellow, subulate, 6.2–6.5 mm long; style club-shaped, slightly exserted beyond the anthers, the stigma minutely capitate. Berry fleshy, globose, 0.9–1.3 cm diam., glabrous, turning from green to yellow and finally bright orange. Seeds numerous, yellowish, lenticular, 3.2–3.5 mm long.

DISTRIBUTION: A common liana of open disturbed areas such as roadsides. Bordeaux (A2848), Center Line Road (A2417). Also on St. Thomas (according to Krebs, 1847) and Tortola; Greater Antilles (except Jamaica), Lesser Antilles, and Mexico to South America.

5. Solanum mucronatum O.E. Schulz *in* Urb., Symb. Ant. **6**: 191. 1909.

Shrub to 2 m tall; stems nearly cylindrical, densely stellate-pilose when young. Leaf blades 5–20 × 2.3–7 cm, oblanceolate (young leaves nearly elliptic), chartaceous, sparsely strigose above (sometimes with a few stellate hairs), stellate-pubescent beneath, veins lighter on both surfaces, the apex short- to long-acuminate, the base attenuate-obtuse, the margins entire; petioles 5–13 mm

long, stout, stellate-pilose. Flowering and fruiting material not known.

DISTRIBUTION: Perhaps extinct. Collected once by Raunkiær (no. *1927*) in March 1906 at Dent Hill in Coral Bay. Also later collected once on St. Thomas and once on Puerto Rico.

NOTE: This species is known only from three specimens cited on the original description by O. E. Schulz, none of which had any flowering or fruiting material. The plant remains a mystery, and there are some doubts regarding its identity as a *Solanum*, since the specimens certainly look very different from any other species of *Solanum* from the region. Although this species is perhaps no longer extant on St. John (or at any of the two other original localities), it is treated in this flora in order to document its former existence. The species apparently was never very common, and most certainly habitat destruction was the cause of its disappearing before it became further known.

6. Solanum polygamum Vahl, Symb. Bot. **3**: 39. 1794.

Armed shrub 1–4 m tall; stems stellate-tomentulose, ash to rusty-brown-colored, nearly cylindrical, often with straight, slender, yellowish spines. Leaf blades 2.5–11 × 2.2–4.5 cm, lanceolate to oblong-lanceolate, coriaceous, sparsely to densely stellate-pubescent, sometimes sparsely spiny along upper surface, the apex obtuse, acute to shortly acuminate, the base unequal, rounded-obtuse or rounded-truncate to cordate, the margins entire; petioles 0.5–2.2 cm long, stout, stellate-tomentose, sometimes spiny. Flowers 4–5-merous, staminate or bisexual, with deeply lobed, white corolla. Staminate flowers in lateral cymes; the axes 2–6 cm long, stellate-pubescent; pedicels ca. 5 mm long, stellate-pubescent; calyx bell-shaped, ca. 2.5 mm long, stellate-pubescent, the sepals ovate; corolla lobes oblong, spreading, 5–7 mm long, sparsely stellate-pubescent without; anthers 4 or 5, connivent, yellow, oblong, 2.8–3 mm long. Bisexual flowers slightly larger than the staminate, solitary, subterminal; calyx 7–15 mm long, often spiny; corolla lobes 1–1.5 cm long; stamens 5–7; ovary densely tomentose. Berry fleshy, depressed-globose, 1.5–2 cm wide, stellate-pubescent, turning from green to bright orange, subtended by an accrescent calyx. Seeds numerous, beige, kidney-shaped, 4.5–5 mm long.

DISTRIBUTION: A common shrub of dry forests and open disturbed areas. Coral Bay (A2689), Susannaberg (A4155). Also on St. Croix, St. Thomas, Tortola, and Virgin Gorda; Hispaniola, Puerto Rico, including Vieques.

COMMON NAMES: kackalake berry, kakalaka, kakker lakka.

7. Solanum racemosum Jacq., Enum. Syst. Pl. 15. 1760.

Solanum ignaeum L., Sp. Pl., ed. 2, **1**: 270. 1762.
Solanum persicifolium Dunal, Hist. Nat. Solanum 185. 1813.

Armed or unarmed shrub, 1–3 m tall, many-branched from base or with numerous branches along main stem; stems sparsely to densely stellate-pubescent when young, becoming glabrous and nearly cylindrical, often with straight, stout, yellowish thorns. Leaf blades 3.5–20 × 1–5.6 cm, lanceolate to elliptic, chartaceous, sparsely stellate-pubescent, sometimes sparsely spiny along both sides of midvein, the apex acute or obtuse, the base unequal, obtuse or narrowed, the margins entire; petioles 0.2–2.5 cm long, slender, stellate-pubescent, sometimes spiny. Flowers 5-merous, in lateral to subterminal racemes; the axes 2–12 cm long, stellate-pubescent; pedicels 3–12 mm long, the lowest longer, stellate-pubescent. Calyx bell-shaped, 1.5–2 mm long, sparsely stellate-pubescent, the sepals obtuse; corolla white or light violet, yellowish at base within, deeply lobed, the lobes oblong, reflexed, 4–10 mm long, stellate-pubescent without; anthers connivent, yellow, subulate, 3–6.5 mm long, the filaments 0.2-0.3 mm long, connate into a ring; style slightly exserted. Berry fleshy, globose, 6–8 mm diam., glabrous, bright red, shiny. Seeds numerous, yellowish to beige, nearly kidney-shaped, ca. 2 mm long.

DISTRIBUTION: A common shrub of scrub, dry forests, and open disturbed areas. Emmaus (A1995), Chocolate Hole (A1976), East End (A1833). Also on Anegada, St. Croix, St. Thomas, Tortola, and Virgin Gorda; Puerto Rico including Culebra, Icacos, Vieques, and the Lesser Antilles.

COMMON NAME: cankerberry.

NOTE: This species shows a great deal of variation in leaf size and in the number of spines. Narrow-leafed, densely spiny plants have traditionally been referred to as *S. persicifolium,* whereas large-leafed, unarmed (or nearly so) plants have been called *S. racemosum.* Consistent application of either name to these plants becomes impossible because of intergradation of characters. Therefore, I am placing *S. persicifolium* as a synonym of *S. racemosum.*

8. Solanum torvum Sw., Prodr. 47. 1788.

Solanum daturifolium Dunal *in* A. DC., Prodr. **13:** 261. 1852.

Armed shrub, 1–3 m tall, branching pseudodichotomously; stems stellate-tomentulose, cylindrical, with slightly recurved, short, green prickles. Leaf blades 8–15 × 4.5–11 cm, with nearly

ovate outline, chartaceous, upper surface slightly scabrous, lower surface with soft, stellate-pubescent trichomes, sometimes with a few prickles, the apex acute or obtuse, the base slightly unequal, truncate to nearly cordate, the margins deeply sinuate or lobed; petioles 1–4(–6) cm long, stout, stellate-pubescent, sometimes spiny. Flowers 5-merous, in lateral cymes; the axes 3–5 cm long, stellate-pubescent; pedicels 5–10 mm long, glandular-pubescent. Calyx bell-shaped, 4–5 mm long, glandular-pubescent, the sepals ovate; corolla white, nearly saucer-shaped, 10–15 mm long, the lobes lanceolate, spreading, 5–8 mm long, stellate-pubescent without; anthers connivent or slightly spreading, yellow, subulate, unequal, 5–7 mm long, the filaments ca. 1 mm long, free; style exserted, whitish, the stigma minutely capitate, green. Berry leathery, globose, 10–14 mm diam., glabrous, yellow at maturity. Seeds numerous, light brown, nearly lenticular, 2.1–2.3 mm long.

DISTRIBUTION: A common shrub of open disturbed areas. Emmaus (A3215), road to Bordeaux (A5090). Also on St. Croix, St. Thomas, Tortola, and Virgin Gorda; native of tropical America but now naturalized throughout the tropics.

COMMON NAMES: plate bush, shoo shoo bush, turkey berry.

CULTIVATED SPECIES: *Solanum seaforthianum* Andr. was reported for St. John by Britton and Wilson (1925) and by Woodbury and Weaver (1987), apparently from a cultivated plant.

81. Sterculiaceae (Cacao Family)

By L. J. Dorr

Trees, shrubs, or herbs (rarely lianas), usually stellate-pubescent, but also with simple hairs or peltate scales; mucilage present. Leaves alternate, simple and entire to lobate or palmately compound, petiolate (rarely sessile); stipulate. Flowers actinomorphic or rarely zygomorphic, bisexual or unisexual, produced in axillary or terminal cymose, paniculate, umbelliform, or more-complex inflorescences or flowers solitary, sometimes cauliflorous, bracteolate, sometimes with an involucel of 3 or 4 distinct bractlets; sepals (2–)3–5(–8), usually valvate, connate (rarely distinct), often glandular at the base; petals usually 5, free, sometimes adnate at base to the staminal tube, sometimes clawed, sometimes hooded and with a terminal appendage, or absent; stamens 5–15(–45), in 2 whorls, the outer staminodial, free or more usually monadelphous in a tubular column, often on an androgynophore, sterile in female flowers; anthers 2- or 3-locular; gynoecium of 2–5(–60) syncarpous carpels, rarely apocarpous or 1-carpellate; ovary superior, usually 5-locular with axile placentation, the stigmas 1–5; styles equal in number to carpels, distinct or variously connate. Fruit various, often a capsule, follicle, or schizocarp or a leathery or woody berry, dehiscent or indehiscent; seeds 1 to many per locule.

An almost exclusively tropical family of 65–70 genera and about 1000 species.

REFERENCES: Brizicky, G. K. 1966. The genera of Sterculiaceae in the southeastern United States. J. Arnold Arbor. 47: 60–74; Robyns, A. 1964. Flora of Panama, Part VI. Family 117. Sterculiaceae. Ann. Missouri Bot. Gard. 51: 69–107.

Key to the Genera of Sterculiaceae

1. Androgynophore ≥6 cm long; fruit spirally twisted .... 3. *Helicteres*
1. Androgynophore absent or 1–1.5 mm long; fruit not spirally twisted.
    2. Staminodes 5; petals hooded, each with a terminal or subterminal appendage, deciduous.
        3. Shrubs or subshrubs; androgynophore 1–1.5 mm long; petals adnate to the apex of the staminal tube; stamens 5, each with a single 3-locular anther.. 1. *Ayenia*
        3. Trees (rarely shrubs); androgynophore absent; petals free; stamens 10 or 15, arranged in 5 groups, each group with 2 or 3 2-locular anthers.
            4. Calyx of (2)3 sepals; petal limb prolonged apically into a strap-shaped, deeply bifid appendage; flowers on axillary inflorescences; fruits indehiscent or partly dehiscent capsules 1.2–2.5 cm long, muricate... 2. *Guazuma*
            4. Calyx of 5 sepals; petal limb prolonged apically into an attenuate-spatulate, entire appendage; flowers borne on the trunk (cauliflorous) and older branches; fruits indehiscent, fleshy (drupelike) berries, >10 cm long, smooth to rugose .... 5. *Theobroma*
    2. Staminodes absent (or obsolete); petals flattened throughout, not hooded, not appendaged, withering and persisting in fruit.
        5. Corolla yellow to yellow-orange; ovary 1-carpellate, style 1, stigma plumose, brushlike; fruit a 2-valved follicle, 1-seeded .... 6. *Waltheria*
        5. Corolla pink, lavender, mauve, rose, or white with dark red veins; ovary 5-carpellate, styles 5, stigmata subulate; fruit a capsule, 5–10-seeded .... 4. *Melochia*

1. AYENIA L.

Decumbent subshrubs or herbs with simple and/or stellate hairs. Leaves simple, unlobed, dentate to crenate or serrate; stipules small, subulate, more or less persistent. Flowers bisexual, axillary, solitary or in 2–3-flowered cymes or umbels; bracts inconspicuous; sepals 5, connate at the base; petals 5, slender-clawed, the limb more or less rhomboid, hooded, inflexed and adnate to the staminal tube, often with an abaxial subterminal appendage, the entire corolla umbrella-like with a central protruding style; androgynophore present; stamens 5, alternating with 5 staminodes, connate into a staminal tube, the fertile filaments apically distinct, anthers 3-locular; ovary 5-locular, syncarpous, with 2 ovules per locule, stigma subcapitate, 5-lobed, style simple. Fruit a schizocarpic capsule separating into 5 1-seeded mericarps covered with short excrescences; seeds narrowly ovoid, slightly curved, tuberculate.

An American genus of about 70 species found in tropical and subtropical areas from the southern United States to Argentina.

REFERENCE: Cristóbal, C. L. 1960. Revisión del género *Ayenia* (Sterculiaceae). Opera Lilloana 4: 1–230.

1. Ayenia insulaecola Cristóbal, Opera Lilloana 4: 164. 1960. Fig. 191H–M.

Ayenia pusilla sensu authors, non L., 1759.

Decumbent shrub or subshrub, (10–)20–40 cm tall, much-branched from base; stems and branches puberulent, with simple, often recurved hairs. Leaf blades 0.5–3(–4) × 0.5–1.5(–2) cm, rounded or nearly so on the lower branches, ovate to ovate-elliptic toward the apex, membranous, glabrate to slightly pubescent, especially on the veins beneath, the hairs mostly stellate, the apex obtuse to acute, the base rounded to slightly cordate, the margins dentate or crenate-dentate; petioles to 1 cm long; stipules subulate, 1–2 mm long. Inflorescences axillary, 1–3-flowered umbels; pedicels 1–3 mm long; bracts inconspicuous. Sepals 2–2.5 mm long, lobes ovate-lanceolate, sparsely stellate-pubescent without; petals light yellow, ca. 1 × 1 mm, limb rhomboid with an apical sinus, claw ca. 3 mm long, appendage ca. 0.5 mm long, nipple-like; androgynophore 1–1.5 mm long, glabrous; staminal tube ca. 1 mm long; ovary subglobose, stellate-pubescent, style slightly exserted, stigmas distinct, globular. Capsule 3–4 mm long, 4–5 mm in diameter, subglobose, light green turning brown, sparsely pubescent, pendent, spines 0.5–1.5 mm long. Seeds 1.5–2 mm long, ovoid, brownish black, tuberculate.

DISTRIBUTION: Shaded banks and old cultivated grounds. South Side Pond (A4071), trail from Lameshur to Reef Bay Ruins (A5442). Also on St. Croix, St. Thomas, Tortola, and Virgin Gorda; Bahamas, Hispaniola, Puerto Rico, and the Lesser Antilles.

NOTE: *Ayenia ardua* Cristóbal and *A. insulaecola*, both restricted to the West Indies, are doubtfully distinct. The characters cited by Cristóbal (1960) as diagnostic, the shape and margins of the leaves, fail to adequately distinguish the available material. Flowers of *A. insulaecola* often are affected by insect-induced galls.

The specific epithet has been wrongly spelled as *insularis* by many authors.

2. GUAZUMA Mill.

Small to medium-sized trees, stellate-pubescent to glabrescent throughout. Leaf blades chartaceous to coriaceous, crenate-dentate, often asymmetrical at base; stipules minute, early deciduous. Flowers bisexual, in axillary, thyrsiform cymes, not borne on the trunk and older branches; calyx of (2–)3 unequal, reflexed lobes; petals 5, hooded, covering the fertile stamen groups, the apex prolonged into a deeply bifid, strap-shaped appendage; androphore or androgynophore absent; stamens (10)15, arranged in 5 antipetalous groups of (2)3, alternating with 5 staminodes, connate (stamens and staminodes) for most of their length into a staminal tube, the filaments group distinct at apex, reflexed, 1 free, 2 connate, the anthers 2-locular; staminodes distinct at apex, more or less triangular; ovary sessile, 5-carpellate, with 5 minute lobes at apex, ovules many per locule, the style simple, the stigma 1–5-lobed. Fruit a woody, warty capsule, indehiscent or dehiscent but with the 5 valves scarcely separating; seeds ovoid.

A neotropical genus of 2 species ranging from Mexico to central South America and the West Indies.

REFERENCES: Cristóbal, C. L. 1989. Comentarios acerca de *Guazuma ulmifolia* (Sterculiaceae). Bonplandia 6(3): 183–196; Freytag, G. F. 1951. A revision of the genus *Guazuma*. Ceiba 1: 193–224.

1. Guazuma ulmifolia Lam., Encycl. 3: 52. 1789. Fig. 192A–F.

Theobroma guazuma L., Sp. Pl. 782. 1753. *Guazuma guazuma* (L.) Cockerell, Bull. Torrey Bot. Club **19**: 95. 1892.
Guazuma tomentosa Kunth *in* Humb., Bonpl. & Kunth, Nov. Gen. Sp. **5**: 320. 1823.

Tree 5–10(–20) m tall; bark rough grayish brown; branches minutely stellate-pubescent but becoming glabrous. Leaf blades (3.3–)4.8–14.9 × 1.8–7.5 cm, oblong, ovate-oblong, or ovate, chartaceous to coriaceous, subglabrous to stellate-tomentose above, rarely glabrous, stellate-pubescent to tomentose beneath, less often puberulous or subglabrous, the apex acute to long-acuminate, the base rounded to distinctly cordate, asymmetrical, the margins crenate-dentate; petioles 0–3.2 cm long, densely stellate-pubescent to tomentose; stipules 2–3 mm long, triangular to lanceolate-acuminate. Flowers in axillary cymes 15–25 mm long, with densely stellate-pubescent axes; pedicels 2.5–8 mm long. Sepals 3–3.5 mm long, ovate to elliptic ovate, densely stellate-pubescent without, more or less glabrous within, reflexed, the apex obtuse to acute; petals yellow, 3–4 mm long, obovate, the appendages purple, 5–5.5 mm long, deeply bifid; stamens fused in groups of 2 or 3, alternate, with triangular staminodes at the apex of a short staminal tube; ovary ovoid, puberulous with simple hairs, especially toward apex, style and stigma simple, ca. 1–1.5 mm long. Capsule 1.2–2.5 cm long, 1.5–2.5 cm diam., oblong-ovoid to ellipsoid to subglobose-depressed ovoid, black at maturity, indehiscent or dehiscent but with valves scarcely separating, muricate, stellate-pubescent when immature, glabrescent. Seeds numerous, obovoid, 2.5–3.5 × 1.8–2 mm, obtusely angled, maculate.

DISTRIBUTION: A common tree of disturbed sites such as roadsides, pastures, and secondary vegetation. Center Line Road (A4050), Coral Bay (P29325). Also on St. Croix, St. Thomas, and Tortola; throughout tropical and subtropical America, including the West Indies (except the Bahamas); introduced in the Old World (Asia, Africa, and Hawaii).

COMMON NAMES: bastard cedar, jackass calalu, jackocalalu, jacocalalu, West Indian elm.

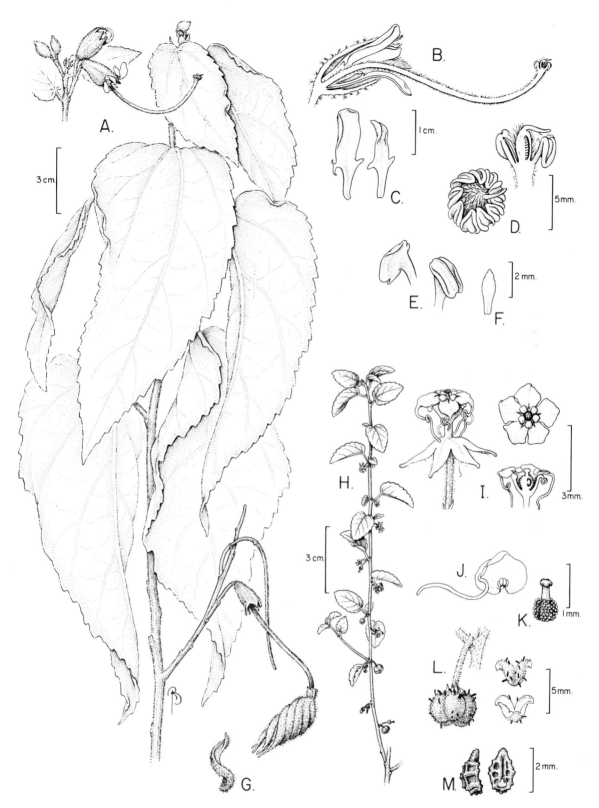

FIG. 191. A–G. *Helicteres jamaicensis*. A. Fertile branch. B. L.s. flower. C. Petals. D. Stamens and pistil from above and l.s upper portion of androgynophore with stamens and pistil. E. stamens. F. Staminode. G. Fruit valve. H–M. *Ayenia insulaecola*. H. Fertile branch. I. Flower lateral and top views, and l.s. flower. J. Hooded petal with apical tail. K. Pistil. L. Capsule and opened mericarps. M. Seeds.

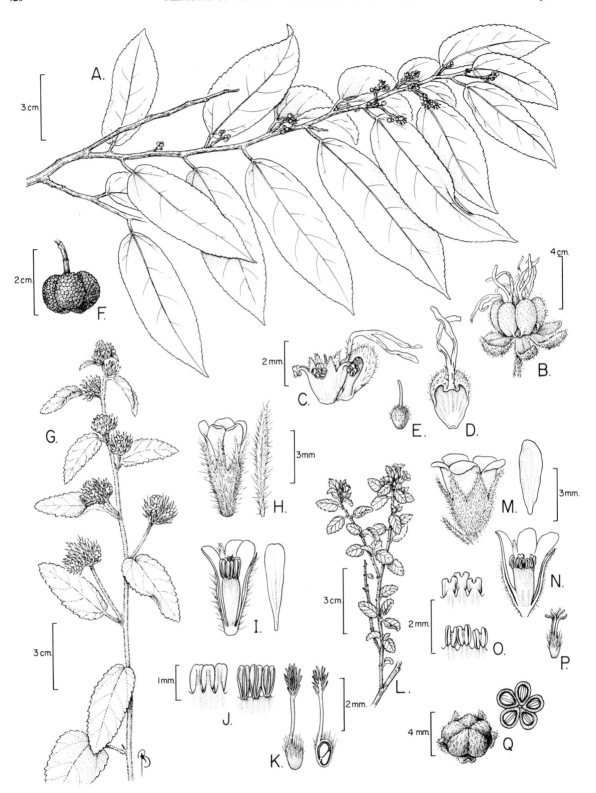

FIG. 192. **A–F.** *Guazuma ulmifolia.* **A.** Flowering branch. **B.** Flower. **C.** Staminal tube with hooded petal. **D.** Hooded petal with bifid apical tail. **E.** Pistil. **F.** Capsule. **G–K.** *Waltheria indica.* **G.** Flowering branch. **H.** Flower and bracteole. **I.** L.s. flower and petal. **J.** Upper portion of staminal tube with anthers, dorsal and frontal views. **K.** Pistil and l.s. pistil. **L–Q.** *Melochia tomentosa.* **L.** Flowering branch. **M.** Flower and petal. **N.** L.s. flower showing staminal column. **O.** Upper portion of staminal column with stamens, dorsal and frontal views. **P.** Pistil. **Q.** Capsule and c.s. capsule.

NOTE: Freytag (1951) distinguished *G. ulmifolia* from *G. tomentosa* on the basis of the dehiscence or indehiscence of the capsules, the degree of development of the claw of the petals, and the shape and pubescence of the leaves. However, Cristóbal (1989) noted that the only character that can reliably distinguish the two taxa is that of fruit dehiscence. Even then, it seems to be an unreliable character (Bornstein *in* Howard, 1989).

3. HELICTERES L.

Shrubs or small trees, with pubescence of stellate or branched hairs. Leaf blades serrate, 3- to 7-nerved at the base; stipules early deciduous. Flowers bisexual, actinomorphic or zygomorphic, pedicellate, in axillary or leaf-opposed, fasciculate inflorescences or solitary; bracteoles small and distant from the calyx; calyx tubular, glandular at base within, either erect, actinomorphic and (4–)5-lobate, or horizontal, bilabiate and (2–4–)5-lobate; petals 5, contorted, long-unguiculate, the claws often appendiculate above the base, equal or unequal; androgynophore well developed, longer than the calyx; stamens (5–)10 or numerous, the filaments in pairs, the anthers 2-locular, longitudinally dehiscent; staminodes alternating with the pairs of stamens; ovary sessile, straight or spirally twisted, 5-carpellate, each carpel with many ovules; styles 5, free or coherent, filiform, the stigmas acute or slightly capitate. Fruit borne on the elongate androgynophore, capsular, woody, straight or spirally twisted, the carpels separating at maturity by the rupture of the walls in the vicinity of the placenta; seeds numerous, small, and angular, warty or smooth.

A genus of ca. 40 species, native to the tropics of both hemispheres, exclusive of Africa.

1. Helicteres jamaicensis Jacq., Enum. Syst. Pl. 30. 1760. Fig. 191A–G.

Shrub 1.5–2.5 m tall, branching pseudodichotomous, often much-branched from near the base; twigs slender, often drooping, densely stellate-tomentose. Leaf blades 8–25 × 6.5–15 cm, broadly ovate to ovate-lanceolate or ovate-oblong, chartaceous to subcoriaceous, densely stellate-tomentose, paler green below, the apex shortly acute to acuminate, the base oblique, cordate to sagittate, the margins doubly serrate; petioles 1–2 cm long, densely stellate-tomentose; stipules to 1.5 cm long, linear, densely stellate-tomentose. Flowers solitary or few in leaf-opposed cymes; bracteoles filiform; peduncles 2–3 cm long, stellate-tomentose; pedicels shortly stellate-tomentose. Calyx 1.5–2.5 cm long, unequally 5-toothed, somewhat 2-lipped, densely stellate-tomentose; petals white to yellowish, 2.5–3 cm long, unequal; androgynophore 6–10 cm long, slightly curved; stamens 10, the filaments ca. 2 mm long; staminodes 5, spatulate, as long as the filaments; ovary ca. 10-carpellate, densely tomentose. Capsule (2.5–)3.5–5.5 × 1.5–2 cm, conelike, oblong, and spirally twisted, turning from yellowish green to brown in age, dehiscent, densely stellate-tomentose, borne on an androgynophore 6–10 cm long. Seed numerous, 2.2–2.8 × 1.2–1.5 mm, ovoid, brown, warty.

DISTRIBUTION: A common shrub of coastal thickets and woodlands. Nanny Point (A2457). Also on St. Croix, St. Thomas, Tortola, and Virgin Gorda; endemic to the West Indies from the Bahamas, Greater Antilles, and Grand Cayman Island east to St. Martin.

COMMON NAMES: cats balls, cow bush.

4. MELOCHIA L.

Trees, shrubs, or herbs, with simple and/or stellate hairs. Leaves simple, unlobed, serrate; stipules small, persistent or early deciduous. Flowers bisexual, heterostylous, borne in axillary or terminal (rarely leaf-opposed), few-flowered umbel-like peduncled cymes or many-flowered dense headlike thyrses composed of 2- or 3-flowered dichasia or monochasia, or thyrses sometimes secondarily arranged into interrupted spikelike panicles; bracts 3 or 4, distinct, forming an involucel; calyx 5-toothed or -lobed, more or less bell-shaped, persistent, nectariferous at the base; petals 5, pink to purple, usually yellow or white at base, or white, spatulate to oblong, flattened throughout (not hooded), clawed, adnate at the base or higher up to staminal tube, longer than the sepals; androphore or androgynophore absent or rarely present but very short; stamens 5, the filaments connate into a cylindrical staminal tube; staminodes 5, obsolete or toothlike, sometimes present in short-styled flowers; ovary 5-carpellate, with 2 ovules per locule, the stigmas 5, subulate or filiform to more or less clavate or shortly and racemosely branched, the styles 5, distinct, or connate to half their length. Capsule 5-locular, loculicidal and/or septicidal (rarely schizocarp), with 1 or 2 seeds per locule; seeds obovoid, nearly round in cross section or more or less angular on ventral surface, smooth to striate.

A pantropical, predominantly American genus of about 60 species, with several species extending into subtropical regions.

REFERENCE: Goldberg, A. 1967. The genus *Melochia* L. (Sterculiaceae). Contr. U.S. Natl. Herb. 34: 191–363.

Key to the Species of *Melochia*

1. Fruit depressed-globose, not prolonged into a beak, not winged; petals pink to white with dark red veins
   .... 1. *M. nodiflora*
1. Fruit pyramidal, prolonged into a beak, 5-winged; petals pink, lavender, mauve, rose, rose-purple to violet, rarely white.
   2. Inflorescences mostly axillary; leaves densely stellate-tomentose to canescent at least beneath; capsules shortly beaked .... 3. *M. tomentosa*
   2. Inflorescences mostly leaf-opposed; leaves sparsely pubescent with simple and bifurcated hairs at least beneath; capsules long beaked .... 2. *M. pyramidata*

1. Melochia nodiflora Sw., Prodr. 97. 1788.

Shrub or herb, 0.5–3(–4) m tall; branches distichous, twigs pilose with simple, bifurcate, and stellate hairs, soon glabrate.

Leaf blades (2.5–)3.6–7 × 1.2–4.2 cm, ovate to lanceolate-ovate, chartaceous, with sparse, appressed, simple hairs on both surfaces, bifurcate and stellate hairs less prevalent, the apex acute, the base rounded to cordate or subtruncate, the margins irregularly crenate-serrate; petioles 0.4–2.7 cm long; stipules 1.7–5.2 mm long, deltate to ovate, early deciduous. Flowers sessile or subsessile, in axillary, densely bracteate heads. Calyx 3–5 mm long, the lobes ovate to deltate, the apex acute to more or less caudate-acuminate, stellate-puberulous and pilose without, glabrous within; petals pink to white with dark red veins, 4–5 mm long, oblong-obovate, glabrous; staminal tube ca. 2–2.5 mm long; ovary ovoid-ellipsoid or orbicular, hirsute. Capsule 2.5–3.5 mm long, 3–4.5 mm diam., schizocarpic, ovoid to subglobose, deeply 5-lobed, not beaked, stellate-puberulous and pilose, septicidal, separating into 5 cocci, each dehiscent along a ventral suture. Seeds 2–2.2 × 1.1–1.4 mm, trigonous, brown, smooth.

DISTRIBUTION: A common weed of roadsides, pastures, and waste places. Trail to Caneel Hill (A2301), Reef Bay (A2643). Also on St. Croix, St. Thomas, and Tortola; the West Indies, Mexico, and Central America south to northern and eastern Brazil.

2. Melochia pyramidata L., Sp. Pl. 674. 1753. *Moluchia pyramidata* (L.) Britton, Brooklyn Bot. Gard. Mem. 1: 69. 1918.

Erect or spreading subshrub or herb, 0.3–1.2(–2) m, with straggling branches; stems puberulous, generally on one side, with simple and forked hairs. Leaf blades 0.5–3.1 × 1.3–7.2 cm, ovate to lanceolate, chartaceous, more or less glabrous above, glabrous to very sparsely pubescent beneath, with simple or bifurcate hairs, the apex acute to acuminate, the base rounded to broadly cuneate, the margins coarsely serrate; petioles 0.5–2 cm long, densely pilose on adaxial surface; stipules 1.5–5 mm long, deltate to lanceolate. Flowers in leaf-opposed umbelliform cymes; peduncles 0.2–1.3 cm long, pilose; pedicels 1–5 mm long, puberulous. Calyx 3–5 mm long, the lobes lanceolate to triangular, apex caudate-acuminate, ciliate and puberulous on both surfaces, the hairs primarily simple and bifurcate; petals rose-purple to violet, rarely white, 7–9 mm long, obovate-spatulate, glabrous; staminal tube 2–2.2 mm long, the free portion 1.5–2 mm long, the anthers ca. 1 mm long; ovary narrowly ellipsoid, ca. 1 mm long, borne on a gynophore <0.5 mm long, stellate-puberulous; styles 5–5.5 mm long, basally connate for ca. 2 mm. Capsule 5–9 mm long, 6–10 mm diam., pyramidal, 5-angled, shortly beaked, valves winged, the wings with acute points toward base, often purple-splotched, loculicidal, borne on a short stipe, stellate-puberulous.

Seeds 1.5–2 mm long, dull brown to black, smooth, sometimes longitudinally striate.

DISTRIBUTION: A weed along roads, in pastures, and in open places. Cruz Bay (A2323), Christ of the Caribbean (A2849). Also on St. Croix, St. Thomas (Britton & Wilson, 1924: "according to West and to Krebs"), and Tortola; Texas, Florida, the West Indies, Mexico, Central America, to northern Argentina, naturalized elsewhere in tropical and subtropical areas.

NOTE: Goldberg (1967) recognized two varieties, which he distinguished on the basis of leaf pubescence. The material from the Virgin Islands could be referred to *M. pyramidata* var. *pyramidata*.

3. Melochia tomentosa L., Syst. Nat., ed. 10, 2: 1140. 1759. *Moluchia tomentosa* (L.) Britton, Brooklyn Bot. Gard. Mem. 1: 69. 1918. Fig. 192L–Q.

Erect or spreading shrub or subshrub, 0.5–1 m tall; young stems densely stellate-tomentose, canescent, eventually glabrescent. Leaf blades generally 2–7 × (0.5–)1–4 cm (smaller on lateral branches), ovate, ovate-lanceolate, or oblong-lanceolate, coriaceous, densely stellate-tomentose to canescent above and below, the apex rounded to acute, the base rounded to subcordate or truncate, the margins crenate-serrate to serrate; petioles 0.15–2.1 cm long, densely stellate-tomentose, canescent; stipules 2–9 mm long, deltate to linear. Flowers in loose, axillary, umbelliform cymes; peduncles 1–8 mm long; pedicels 2–6 mm long. Calyx 5–8 mm long, the lobes lanceolate to linear-lanceolate, densely stellate-tomentose without, glabrous within, acuminate to caudate-acuminate at apex; petals pink, lavender, mauve, or rose, 8–9.5 mm long, oblanceolate to obovate-cuneate, glabrous; staminal tube 3–3.5 mm long, free portion of filaments 1–1.5 mm long (5–6 mm in short-styled flowers), the anthers ca. 1 mm long; ovary ellipsoid to ovoid, 2–2.5 mm long, borne on a gynophore 0–1 mm long, the styles 5–6.5 mm long (3–3.5 mm in short-styled flowers), basally connate for ca. 2 mm. Capsule pyramidal, 5-angled, 7–10 mm in diam., long-beaked. Seeds 2–3 × 1.1–2 mm, obovate, light to dark brown or reddish brown, smooth.

DISTRIBUTION: Common weed of open disturbed areas such as roadsides. Lameshur (A2760), Coral Bay (A2445). Also on Anegada, St. Croix, St. Thomas, Tortola, and Virgin Gorda (Britton & Wilson, 1924); Florida, Texas, West Indies, Mexico, Central America to Brazil.

COMMON NAMES: broom weed, broom wood.

NOTE: Goldberg (1967) recognized four varieties of this species based on leaf size and degree of pubescence. Three of the varieties are purportedly present in the West Indies, but since they intergrade, there seems to be little value in attempting to distinguish them.

5. THEOBROMA L.

Small to medium-sized trees; twigs stellate-puberulous, becoming glabrate. Leaves entire or irregularly sinuate; stipules subulate, deciduous. Flowers bisexual, solitary to many in axillary, fasciculate cymes borne on the trunk and older branches; calyx of 5 sepals connate only at base, reflexed; petals 5, contorted, hooded, the apex prolonged into a strap-shaped appendage; androphore or androgynophore absent; stamens 10, arranged in 5 groups of 2, alternating with 5 staminodes, connate (stamens and staminodes) into a short staminal tube, the free portion of filaments short, minutely bifurcate, each branch with 1 2-locular anther hidden inside the petal hood; staminodes distinct, linear-subulate, elongate; ovary sessile, 5-carpellate, with many ovules per locule, arranged in 2 rows, the styles 5, free or partially connate, the stigmas apical, short, acute. Fruit a large, indehiscent, smooth to rugose, fleshy berry, with many seeds per locule; seeds ovoid to ellipsoid, surrounded by a thick, pulpy tissue filling the cavity.

A neotropical genus of ca. 25 species, ranging from Mexico to central South America, and introduced in cultivation throughout the tropics, including the West Indies.

REFERENCE: Cuatrecasas, J. 1964. Cacao and its allies: a taxonomic revision of the genus *Theobroma*. Contr. U.S. Natl. Herb. 35: 379–614.

1. Theobroma cacao L., Sp. Pl. 782. 1753.

Tree 4–10 m tall; often with a low-spreading crown; branches whorled, densely or sparsely pubescent, but becoming glabrate. Leaf blades (12–)15–50 × 4–15 cm, obovate-elliptic to oblong-elliptic, coriaceous or chartaceous, subglabrous, but with simple, bifurcate, and stellate hairs on major veins, the apex acute to

caudate-acuminate, the base rounded to obtuse, the margins entire; petioles 1.5–2 cm long, swollen at both ends, puberulous to pubescent with simple hairs; stipules subulate, deciduous. Flowers borne on the trunk and older branches; peduncles 1–3 cm long, pubescent primarily with stellate hairs. Calyx 5–8 mm long, the sepals lanceolate to oblong-lanceolate, stellate-puberulous to glabrate without, glabrous within; petals white, greenish white, or

pale violet, 3-nerved, 2–4 mm long, strongly hooded, the appendage entire, attenuate-spatulate, ca. 4 mm long, yellowish, glabrous;
staminal tube 1–1.5 mm long, the filaments 1.5–2 mm long, the
anthers ca. 0.5 mm long; staminodes 4–6 mm long with conspicuous purple median nerve; ovary stipitate, ovoid-ellipsoid, ca. 1.5
mm long, papillose-glandular, the styles 1.5–2 mm long, adherent.
Berry fleshy, drupelike, variable in size and shape, 10–20 cm
long, to 10 cm diam., generally oblong to fusiform, 5-angled,
indehiscent. Seeds 1–2.5 mm long, brown, the surrounding pulp
white and sweet.

DISTRIBUTION: Uncommon, naturalized in Cinnamon Bay, area along
trail south of ruins (A 4153). Also on St. Croix (Eggers, 1879; Little &
Wadsworth, 1964) and Tortola (D'Arcy, 1967); native of Central and South
America, widely cultivated in the tropics.

COMMON NAMES: cacao, chocolate tree, cocoa tree.

NOTE: This species is the source of cocoa, chocolate, and cocoa butter.
Cuatrecasas (1964) recognized several infraspecific forms based on fruit
shape and texture. In addition, agriculturists recognize forms such as Criollo
and Forastero on the same basis. The material from St. John has not been
collected in fruit and therefore cannot be referred to any of these taxonomic
or agronomic forms.

6. WALTHERIA L.

Shrubs, subshrubs, or herbs, with simple and/or stellate hairs. Leaves simple, crenate-serrate or serrate; stipules small, lanceolate-
subulate, deciduous. Flowers bisexual, homostylous or heterostylous, borne in axillary or terminal compound cymes or headlike glomerules; bracts small, deciduous; calyx 5-lobed, bell-shaped–turbinate, persistent; petals 5, flattened throughout (not hooded), without
appendages, clawed, spatulate, adnate to the base of staminal tube, orange to yellow, marcescent; androphore or androgynophore
absent; stamens 5, the filaments connate into a staminal tube, anthers 2-locular; staminodes absent; ovary sessile, 1-carpellate, with 2
ovules per cavity, the style solitary, the stigma plumose, brushlike. Fruit a 2-valved, 1-seeded follicle, surrounded by the persistent
calyx and marcescent corolla; seeds small, obliquely obovoid, smooth.

A pantropical genus of about 70 species, most occurring in the New World but several in Africa and Asia; at least 1 species
extending into subtropical areas.

1. **Waltheria indica** L., Sp. Pl. 673. 1753. *Waltheria
americana* var. *indica* (L.) Schum. *in* Engl., Monogr. Afrik. Pflanzen-Fam. **5**: 47. 1900.

Fig. 192G–K.

Waltheria americana L., Sp. Pl. 673. 1753. *Waltheria
indica* var. *americana* (L.) R. Br. *ex* Hosok., Occas.
Pap. Bernice Pauahi Bishop Mus. **13**: 224. 1937.

Shrub or subshrub, 0.5–1.5 m tall; stems rigid, densely stellate-
pubescent to tomentose throughout, more or less glabrate. Leaf
blades (1–)3–10 × 1–4(–5) cm, ovate to oblong-ovate or ovate-
lanceolate, chartaceous, velvety stellate-pubescent to tomentose,
occasionally the hairs stiff, the texture hirsute to subscabrous,
grayish discolorous, paler beneath, the apex obtuse to subacute,
the base rounded to truncate to subcordate, the margins crenate-
serrate to serrate or irregularly serrate; petioles 0.4–3(–5) cm
long; stipules filiform. Flowers in axillary or terminal, generally
glomerulate but occasionally more elongate and paniculiform in-

florescences; peduncles to 5.3 cm long; bracts linear-lanceolate,
4–6 mm long. Calyx 4–4.75 mm long, stellate-sericeous without,
glabrous within, the lobes triangular-subulate, with caudate-acuminate apex; petals yellow to yellow-orange, 4–6 mm long, oblong-
spatulate to oblong-obovate; staminal tube 2–3 mm long, the
anthers subsessile, 0.75 mm long; ovary oblong-ellipsoid, ca. 1
mm long, sericeous at apex, the style ca. 1.5 mm long, the stigmas
plumose-papillate. Fruit 2.5–3 mm long, obliquely obovoid, sericeous. Seeds 1.8–2.1 × 1–1.3 mm, obovoid, reddish brown,
smooth.

DISTRIBUTION: A common weed of open places and waste areas. Fish
Bay (A2479), East End (A2782). Also on Anegada (Britton & Wilson,
1924; D'Arcy, 1971), St. Croix, St. Thomas, Tortola (D'Arcy, 1967), and
Virgin Gorda (Britton & Wilson, 1924); a pantropical weed, native to the
New World, where it occurs from Florida and the West Indies south through
continental tropical America; introduced and naturalized in the Old World.

COMMON NAME: marsh-mallow.

NOTE: Waltheria indica and *W. americana* were described simultane-
ously by Linnaeus. R. Brown (*in* Tuckey, Narr. Exped. Zaire. 484. 1818)
appears to have been the first to unite these taxa with equal priority, and
accordingly his choice of *W. indica* must be followed.

82. **Surianaceae** (Suriana Family)

Shrubs or small trees. Leaves alternate and simple; stipules small and deciduous or wanting. Flowers small, actinomorphic, bisexual
or less often unisexual, in axillary or terminal panicles or solitary; calyx of 5 distinct sepals; corolla of 5 distinct petals; stamens twice
as many as the petals or some of them reduced to staminodes or obsolete, the filaments free, the anthers opening by longitudinal slits;
nectary disk wanting; ovary superior, of 1–5 distinct carpels, the placentation basal-marginal, with 2–5 ovules per carpel, the styles
basal on ventral side, distinct. Fruit a berry or a drupe.

A family of 4 genera and 6 species, most of which are from Australia; 1 species has tropical coastal distribution.

1. SURIANA L.

A monospecific genus characterized by the following species.

1. **Suriana maritima** L., Sp. Pl. 284. 1753.   Fig. 193.

Erect shrub to 2 m tall, many-branched from base; bark dark
gray, rough; branches ascending, woolly-pubescent, cylindrical,
with numerous leaf scars. Leaves congested at ends of branches,
sessile, 1.7–3.1 × 0.3–0.5 cm, oblanceolate, slightly fleshy,

puberulent, especially along margins, the apex obtuse or acute,
the base tapered, the margins entire. Flowers 5-merous, a few in
short axillary cymes; the peduncle 4–6 mm long, pubescent;
pedicels 5–8 mm long, pubescent. Calyx green, deeply parted,
the sepals ovate, 8–10 mm long, pubescent; petals yellow, early
deciduous, obovate, 5.5–6.5 mm long; stamens 5, alternating with

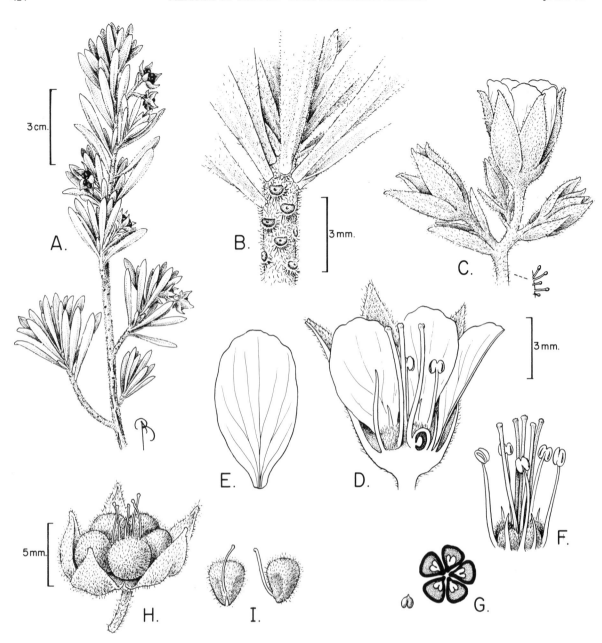

FIG. 193. *Suriana maritima*. **A.** Flowering branch. **B.** Detail of branch showing leaf scars. **C.** Portion of inflorescence and detail of glandular trichomes. **D.** L.s. flower. **E.** Petal. **F.** Stamens, staminodes, and pistils. **G.** Seed and c.s. drupelets. **H.** Drupelets on common receptacle. **I.** Drupelet, frontal and lateral views.

staminodes, shorter than the petals; carpels 5, distinct, obovoid pubescent, with a single basal-axial ovule, the styles free, arising from the ventral base of the carpels, the stigma minutely capitate. Fruit a stony, indehiscent, obovoid drupelet, ca. 4 mm long, turning from green to brown.

DISTRIBUTION: A common shrub of sandy coasts. Europa Bay (A746), Hart Bay (A809). Also on Anegada, St. Croix, St. Thomas, and Virgin Gorda; throughout coastal tropical and subtropical America, Madagascar, Polynesia, New Guinea, and Australia.

COMMON NAME: bay cedar.

83. Symplocaceae (Symplocos Family)

Shrubs or small trees. Leaves alternate and simple, entire or serrate, petiolate; stipules wanting. Flowers 5-merous, actinomorphic, bisexual, in axillary racemes or fascicles; calyx bell-shaped, 5-lobed; corolla tubular, bell-shaped, the lobes as long as the tube; stamens numerous, in 1–3 series, the filaments inserted on corolla, free or connate into a tube, the anthers short, subglobose, opening by longitudinal slits; nectary disk wanting; ovary inferior or semi-inferior, of 2–5 connate carpels, each with 2–5 pendulous ovules with axile placentation, the styles distal, slender, the stigma nearly capitate. Fruit a drupe, or rarely a berry, crowned by the persistent calyx lobes; seeds 1 per locule.

A tropical and subtropical family of a single genus with 300–400 species.

1. SYMPLOCOS Jacq.

Characters of the family.

1. Symplocos martinicensis Jacq., Enum. Syst. Pl. 24. 1760.   Fig. 194.

Tree 7–10 m tall, sometimes with numerous branches from base; bark light gray, slightly fissured; twigs glabrous or puberulent, angled. Leaf blades involute, 5–16 × 2.5–5.8 cm, obovate, elliptic, or oblong, coriaceous, glabrous, with prominent midvein along lower surface, the apex long-acuminate, the base acute to obtuse, sometimes unequal, the margins crenate to serrate; petioles 1–2 mm long, puberulent. Flowers nearly sessile, a few in axillary short racemes; peduncles <1 cm long. Calyx green, bell-shaped, 2.5 mm long, 5-lobed, the lobes ovate, ca. 1 mm long, ciliate; corolla white, the tube 5–7 mm long, the lobes as long as the tube, oblong, reflexed; stamens exserted, in 2 series, the inner series shorter, the filaments adnate to corolla tube; ovary semi-inferior, 3–5-locular, pubescent, the stigma cup-shaped. Drupe ellipsoid, 10–13 mm long, bluish black at maturity.

DISTRIBUTION: An occasional tree of moist forests. Bordeaux Mountain (A4693, A5107, W457). Also on St. Thomas and Tortola; Puerto Rico, the Lesser Antilles, and French Guiana.

84. Theaceae (Tea Family)

Trees, erect or rarely climbing shrubs. Leaves alternate or rarely opposite, simple, entire or toothed; stipules wanting. Flowers medium to large, actinomorphic, bisexual or less often unisexual, axillary, solitary or rarely in terminal panicles or racemes; calyx subtended by 2 to several bracteoles, of (4–)5(–7) distinct or basally connate sepals; corolla of (4–)5(-many) distinct or basally connate petals; stamens numerous, developing centrifugally, the filaments distinct or basally connate into a ring or into 5 bundles that are opposite to the petals, the anthers opening by longitudinal slits, rarely by terminal porelike slits; nectary disk wanting; ovary superior, of (2–)3–5(–10) united carpels, the placentation axile, commonly with numerous ovules per carpel, the styles terminal, distinct or connate into a single distally lobed style. Fruit a septicidal or loculicidal capsule or indehiscent and fleshy or dry; seeds 1 to many in each locule.

A tropical and subtropical family of about 40 genera and 600 species

1. TERNSTROEMIA L.f.

Glabrous trees or shrubs. Leaves simple, entire, spirally arranged. Flowers 5-merous, bisexual, solitary, axillary; bracteoles 2(–4), opposite; calyx denticulate, the outer 2 sepals smaller; corolla of free or basally connate petals; stamens numerous, in 2–4 series, the filaments connate, shorter than the anthers, the exterior series adnate to corolla, the anthers opening by longitudinal slits; ovary of 2–3 connate carpels, sometimes appearing 4–6-locular owing to false partitioning, the ovules pendulous, 2–20 per locule, the styles connate or 2–3-branched, the stigma lobate or minute. Fruit an indehiscent or dehiscent leathery berry, with persistent style, subtended by the persistent sepals; seeds few.

A tropical genus of about 100 species.

REFERENCE: Kobuski, C. E. 1943. Studies in the Theaceae, XIV. Notes on the West Indian species of *Ternstroemia*. J. Arnold Arbor. 24: 60–76.

1. Ternstroemia peduncularis DC., Mém. Soc. Phys. Genève **1**: 409. 1822. *Taonabo peduncularis* (DC.) Britton, Brooklyn Bot. Gard. Mem. 1: 70. 1918.

                                    Fig. 195.

Small tree or shrub 4–15 m tall, sometimes with numerous branches from base; bark light gray to light brown, usually with numerous thornlike, warty projections; twigs cylindrical, glabrous. Leaf blades 4–9.5 × 1.8–4.3 cm, obovate, coriaceous, glabrous, revolute, the apex obtuse to rounded, sometimes notched, the base tapering, the margins entire; petioles 2–6 mm long. Flowers leathery; peduncles 2–4 cm long, curved at apex; bracteoles 2, oblong-lanceolate, 2–4 mm long. Sepals spreading, pinkish-tinged at center within, ovate to nearly rounded, 7–10 mm long, apiculate, the outer ones denticulate on margins; petals oblong to obovate, 6–10 mm long, white to pinkish within, early deciduous; stamens numerous, in 2–4 series, the anthers linear, 3–3.5 mm long, early deciduous; ovary nearly discoid, 2-locular, each locule with 5–20 ovules, the style subulate, persistent, the stigma punctiform. Berry ovoid, ca. 1.5 cm long, irregularly dehiscent on distal portion, exposing 1 or a few pendulous seeds. Seeds nearly ovoid, ca. 6 mm long, covered with a bright red, fleshy aril.

DISTRIBUTION: An occasional tree of moist forests. Bordeaux Mountain (A1917, A5106). Also on St. Croix; Cuba, Hispaniola, Puerto Rico, and the Lesser Antilles.

85. Theophrastaceae (Theophrasta Family)

Trees or shrubs, often unbranched, with palmlike habit. Leaves alternate, simple, entire or spiny-toothed; stipules wanting. Flowers small to large, actinomorphic, bisexual or less often unisexual, glandular-dotted or lineate, in terminal or axillary racemes, corymbs, or panicles, or less often solitary; calyx of (4–)5 distinct or basally connate sepals; corolla fleshy, short-tubular, (4–)5-lobed; stamens

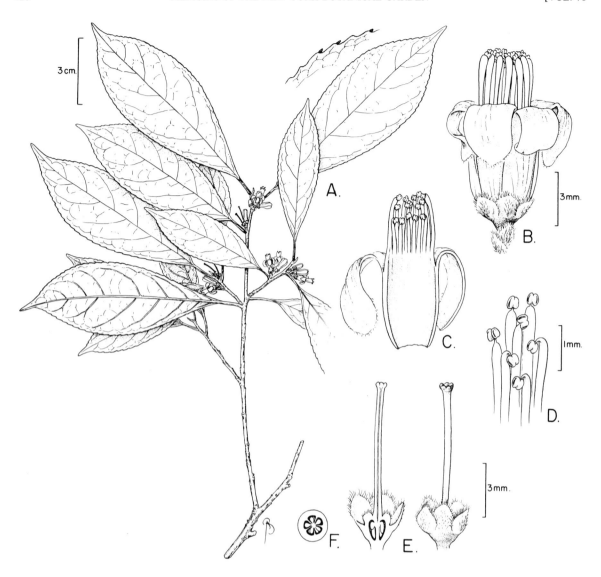

Fɪɢ. **194.** *Symplocos martinicensis*. **A.** flowering branch and detail of leaf margin. **B.** Flower. **C.** L.s. corolla and staminal tube. **D.** Upper portion of staminal tube showing anthers. **E.** L.s. calyx and pistil, and calyx and pistil. **F.** C.s. ovary. (From Mori et al., in press, Guide to the Vascular Plants of Central French Guiana, Mem. New York Bot. Gard. 76.)

twice as many as the corolla lobes, half of them reduced to staminodes, the filaments distinct or basally connate into a short tube, adnate to the base of the corolla tube, the anthers opening by longitudinal slits; nectary disk wanting; ovary superior, unilocular, of 5 carpels, the placentation free, central, with numerous ovules per carpel, the style single, the stigma discoid. Fruit a berry or a drupe; seeds 1 to several, rather large.

A family of 5 genera and about 100 species, primarily neotropical.

1. JACQUINIA L.

Small trees or shrubs. Leaves simple, entire, alternate, opposite, pseudoverticillate or aggregate near ends of branches. Flowers actinomorphic, 5-merous, bisexual, in terminal or axillary racemes or solitary; calyx of distinct or basally connate sepals; corolla short, bell-shaped; staminodes 5, petal-like, inserted on corolla tube; stamens 5, spreading, the filaments basally connate, the anthers opening by longitudinal slits; ovary ovoid to globose, of connate carpels, sometimes appearing 4–6-locular owing to false partitioning, the ovules pendulous, 2–20 per locule, the styles connate or 2–3-branched, the stigmas lobate or minute. Fruit an indehiscent, leathery berry, with persistent style, subtended by the persistent sepals; seeds few to several.

A neotropical genus of about 50 species.

Rᴇꜰᴇʀᴇɴᴄᴇ: Ståhl, B. 1992. On the identity of *Jacquinia armillaris* (Theophrastaceae) and related species. Brittonia 44: 54–60.

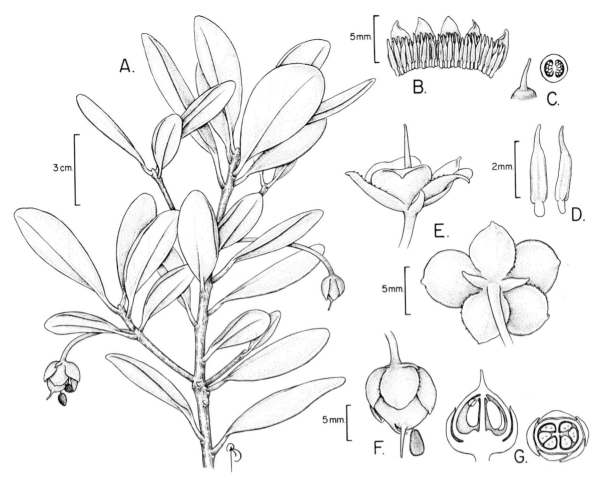

FIG. 195. *Ternstroemia peduncularis*. **A.** Fruiting branch. **B.** Corolla spread open showing stamens. **C.** Pistil and c.s. ovary. **D.** Stamen, frontal and lateral views. **E.** Flower with petals and stamens removed, lateral and bottom views. **F.** Berry with persistent calyx and pendulous seed. **G.** L.s. and c.s. berry.

Key to the Species of *Jacquinia*

1. Racemes much shorter than the subtending leaves; pedicels reflexed; leaves 2–4.5 cm long, slightly revolute .... 2. *J. berterii*
1. Racemes much longer than the subtending leaves; pedicels ascending; leaves 4–10 cm long, strongly revolute .... 1. *J. arborea*

1. Jacquinia arborea Vahl, Eclog. Amer. 1: 26. 1796 [1797]. *Jacquinia armillaris* Jacq. var. *arborea* (Vahl) Griseb., Fl. Brit. W. I. 397. 1861.

Jacquinia barbasco Mez *in* Engl., Pflanzenr., ser. 4, **236a(15)**: 32. 1903, nom. illegit.

Shrub 2–3 m tall; bark dark gray with blackish spots, rough, finely fissured; twigs angled, yellowish green to ash-colored, covered by minute, scurfy scales. Leaves spirally arranged at ends of branches or loosely whorled; blades 4–10 × 1.4–4.2 cm, obovate to spatulate, rigidly coriaceous, glabrous, minutely pitted along lower surface, the apex rounded, retuse and sometimes mucronulate, the base cuneate, the margins yellowish, entire, strongly revolute; petioles 2–6 mm long, yellowish. Flowers leathery, in terminal racemes; the axes 6–8 cm long; pedicels stout, 1–1.5 cm long, ascending. Calyx bell-shaped, green, 3–3.7 mm long, glabrous, 5-lobed, the lobes imbricate, rounded, minutely pitted, yellowish toward the margins; corolla white, 5–8 mm long, the lobes oblong, spreading, retuse at apex; staminodes white, petal-like, shorter than the corolla lobes; stamens 5, slightly exserted, the anthers white, connivent, but spreading at maturity, the filaments subulate-flattened, stout; ovary nearly ovoid, glabrous, the style short, the stigma capitate, red. Berry numerous, on long, hanging racemes, nearly globose, apiculate, 8–10 mm diam., turning from green to yellow and finally bright orange. Seeds nearly ovoid, ca. 5 mm long.

DISTRIBUTION: A common shrub of coastal scrub and dry forests, usually along sandy beaches. Hawksnest Beach (A3186), Reef Bay (A3987). Also on Anegada, St. Croix, St. Thomas, Tortola, and Virgin Gorda; Puerto Rico, Jamaica, Hispaniola, and the Lesser Antilles.

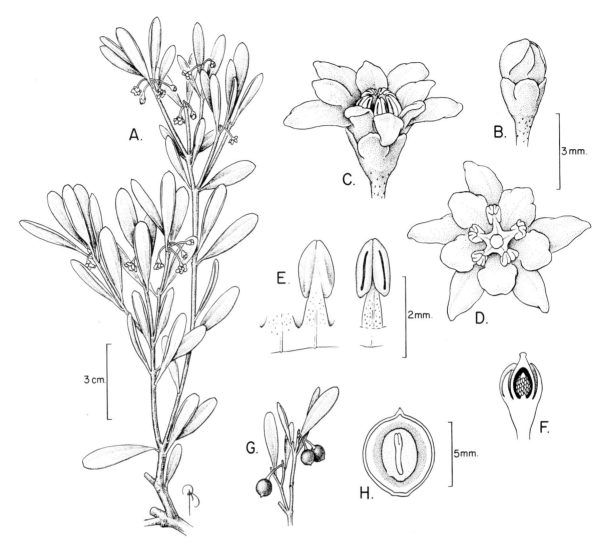

FIG. 196. *Jacquinia berterii*. **A.** flowering branch. **B.** Flowerbud. **C.** Flower. **D.** Flower, top view. **E.** Stamen, dorsal and frontal views. **F.** L.s. gynoecium with persistent calyx. **G.** Fruiting branch. **H.** L.s. berry showing seed and embryo.

NOTE: *Jacquinia arborea* was treated by Howard in his *Flora of the Lesser Antilles* (1989) as conspecific with *Jacquinia armillaris* Jacq. A recent paper by B. Ståhl (1992) recognized the two species as distinctive. *Jacquinia arborea* is a West Indian species, whereas *J. armillaris* is a species of northern South America that extends into the Lesser Antilles. *Jacquinia arborea* may be distinguished from *J. armillaris* by the glabrous sepals (vs. ciliate at margins), the inconspicuous veins on the lower surface of leaves (vs. conspicuous, secondary veins that ascend from the base), and by the spirally arranged to loosely whorled leaves (vs. whorled).

2. Jacquinia berterii Spreng., Syst. Veg. 1: 668. 1825. Fig. 196.

Shrub 2–4 m tall; bark dark gray, smooth to rough; twigs angled, rusty-brown when young, covered by minute, scurfy scales. Leaves spirally arranged at ends of branches or sometimes whorled; blades 2–4.5 × 0.6–1.6(–1.8) cm, spatulate or oblanceolate, rigidly coriaceous, glabrous, minutely pitted and gland-dotted along lower surface, the venation obscurely parallel, the apex rounded, retuse, or sometimes truncate, the base cuneate, the margins yellowish, entire, slightly revolute; petioles 2–7 mm long, yellowish. Flowers leathery, few, in short, terminal racemes; the axes 2–5 mm long, glandular-dotted; the pedicels reflexed, 5–7 mm long, gland-dotted. Calyx bell-shaped, green, 1.3–1.5 mm long, glabrous, 5-lobed, the lobes imbricate, rounded, hyaline toward the margins; corolla white to cream-colored, ca. 4.5 mm long, the lobes oblong, spreading, rounded at apex; staminodes white, petal-like, shorter than the corolla lobes; stamens 5, slightly exserted, the anthers yellow, connivent, spreading at maturity, the filaments subulate, stout; ovary nearly ovoid, glabrous, the style short, with a cup-shaped stigma. Berry with reflexed pedicels at ends of ascending branches, ovoid to nearly globose, apiculate, 5–7 mm diam., turning from green to light yellow. Seeds nearly ovoid, ca. 5 mm long.

DISTRIBUTION: A common shrub of coastal scrub, dry forests, and sandy beaches. Europa Bay (A748), Maria Bluff (A2335), White Cliffs (A2717). Also on Jost van Dyke and St. Thomas; Puerto Rico, Bahamas, Cuba, Hispaniola, and the Lesser Antilles.

86. Thymelaeaceae (Thymelaea Family)

Shrubs, trees, herbs, or lianas. Leaves alternate or opposite, simple and entire; stipules wanting or vestigial. Flowers small to medium, (3–)4–5(–6)-merous, actinomorphic or less often zygomorphic, bisexual or less often unisexual, in terminal, axillary, or extra-axillary racemes, spikes, heads, or umbels or less often solitary; calyx usually forming a hypanthium crowned by petaloid sepals or minute lobes, less often the calyx with distinct sepals; corolla of minute distinct petals, often scale-like, adnate to the throat of the hypanthium, or wanting; stamens usually as many as the sepals or twice as many, with the upper series opposite the sepals, or seldom numerous, the filaments very often short or obsolete, distinct or basally connate into a short tube, adnate to the hypanthium, the anthers opening by longitudinal slits; nectary disk cup-shaped, lobed, or wanting; ovary superior, of 2–5 united carpels, 2–5-locular, with solitary and pendulous ovules, the style single, slender, elongate to obsolete, often eccentric, the stigma capitate. Fruit a berry, a drupe, or less often a capsule.

A cosmopolitan family of about 50 genera and 500 species.

1. DAPHNOPSIS Mart.

Dioecious shrubs or small trees. Leaves simple, entire, alternate or nearly whorled; stipules wanting. Flowers actinomorphic, 4-merous, unisexual, in umbelliform, racemose, or capitate clusters on branched or unbranched extra-axillary inflorescences; hypanthium tubular, obconic to bell-shaped, the lobes unequal, reflexed or ascending; corolla of minute petals, adnate to throat of hypanthium, or wanting; stamens 8, in 2 whorls, inserted at 2 levels, the upper opposite the lobes, the lower alternating with them, the anthers sessile or on short filaments, opening by longitudinal slits; ovary on a long gynophore, ellipsoid, unilocular, the ovule solitary, pendulous, the style slender, the stigma capitate. Fruit an indehiscent, leathery, 1-seeded drupe.

A neotropical genus of about 50 species.

REFERENCE: Nevling, L. I. 1959. A revision of the genus *Daphnopsis*. Ann. Missouri Bot. Gard. 46: 257–363.

1. Daphnopsis americana (Mill.) J. Johnst. subsp. **caribaea** (Griseb.) Nevling, Ann. Missouri Bot. Gard. **46**: 315. 1959. *Daphnopsis caribaea* Griseb., Fl. Brit. W. I. 278. 1860. Fig. 197.

Shrub or small tree 3–6 m tall; bark dark gray, smooth, strongly bitter; twigs angled, puberulent. Leaves alternate; blades 5–16(–21) × 2.2–5(–7) cm, elliptic to nearly lanceolate, chartaceous, glabrous, the apex acute, acuminate, or obtuse, the base attenuate, the margins entire; petioles 4–14 mm long. Flowers apetalous, in unisexual, umbelliform, bifurcated inflorescences; axes 4–24 mm long in staminate inflorescences and 3–10 mm long in pistillate inflorescences. Staminate flowers 10–75 per inflorescence; pedicels ca. 0.5 mm long; calyx tube obconical, 3–4.5 mm long, the lobes 1–2.5 mm long, reflexed; stamens 8, the lower 4 adnate to the hypanthium throat, the upper series with each stamen adnate to a hypanthium lobe, the anthers nearly sessile; nectary disk and pistillode present. Pistillate flower 10–25 per inflorescence; pedicels 1–1.5(–3) mm long; staminodes 8, papilliform; ovary ellipsoid, green, the stigma nearly sessile, minutely pubescent. Drupe ellipsoid to nearly ovoid, 6–8 mm long, turning from green to white, crowned by the persistent stigma. Seeds ovoid, 4–6 mm long.

DISTRIBUTION: A common shrub of dry to moist forests. Center Line Road (A2487), Cinnamon Bay (M17006). Also on Jost van Dyke, St. Croix, and St. Thomas; Puerto Rico, Central America to northern South America, and the Lesser Antilles.

87. Tiliaceae (Linden Family)

By L. J. Dorr

Trees, shrubs, or rarely herbs, usually stellate-pubescent, but also with simple hairs or rarely peltate scales; mucilage present. Leaves alternate, rarely opposite, simple or lobed, petiolate, usually stipulate. Flowers actinomorphic, bisexual, rarely unisexual, produced in axillary or terminal cymose or paniculate inflorescences, or flowers solitary, bracteolate; sepals (2–)4–5(–8), usually valvate, free or sometimes connate at the base; petals 4–5, free, or absent, often glandular at the base; androgynophore present or absent; stamens 5 to more or less numerous, free or shortly connate into 5 or 10 bundles, anthers 2-locular; ovary superior with (1–)2–5(–10) syncarpous carpels, rarely apocarpous, usually with a capitate or lobed stigma and a solitary style, placentation axile (or intruded parietal). Fruit nutlike or a drupe (rarely a berry), capsular or schizocarpic, dehiscent or indehiscent, sometimes with spines or wings.

A chiefly tropical family of 50–55 genera and about 725 species but with a few genera principally temperate.

REFERENCE: Brizicky, G. K. 1965. The genera of Tiliaceae and Elaeocarpaceae in the southeastern United States. J. Arnold Arbor. 46: 286–307.

Key to the Genera of Tiliaceae

1. Fruit loculicidally dehiscent, unarmed, ellipsoid to linear-cylindric; leaves not lobed .... 1. *Corchorus*
1. Fruit indehiscent (or tardily splitting), armed with hooked spines, subglobose; leaves often palmately 3- to 5-lobed .... 2. *Triumfetta*

1. CORCHORUS L.

Annual or perennial herbs or subshrubs, with simple and/or stellate hairs. Leaves alternate, simple, usually unlobed, crenate to serrate; stipules small, linear-filiform to filiform, usually deciduous. Flowers bisexual, solitary or in few-flowered umbel-like, leaf-

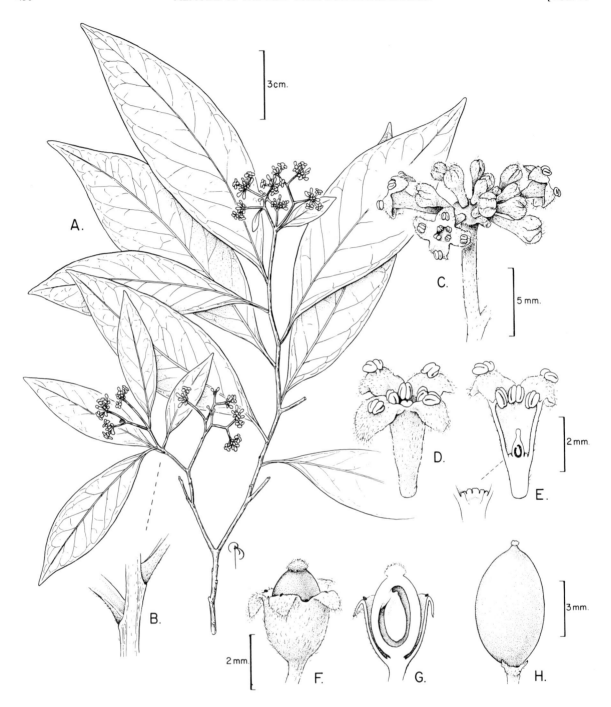

Fɪɢ. 197. *Daphnopsis americana.* **A.** Flowering branch. **B.** Detail of branch. **C.** Detail of staminate inflorescence. **D.** Staminate flower. **E.** L.s. staminate flower and detail of annular disk. **F.** Pistillate flower with developing fruit. **G.** L.s. pistillate flower with developing fruit. **H.** Drupe.

opposed cymes; bracts small, more or less persistent; sepals (4–)5, free, sometimes hooded, deciduous; petals (4–)5, free, spatulate to obovate, shorter (or longer) than or equaling the sepals in length; disk extrastaminal, ringlike or cupular or absent; stamens (10–) numerous, the filaments free or short-connate; ovary sessile (or on a short eglandular androgynophore), 2- or 3(–5)-locular, with (2–) numerous pendulous ovules in each locule, the style simple, filiform, the stigma large, discoid, irregularly dentate, crenulate, or lobulate. Fruit a loculicidal capsule, linear-oblong or subglobose, terminating in a beak, short horns, or teeth, separating into 2 or 3(–5) valves; seeds usually numerous, small, dark brown to black, irregularly 3- or 4-angled or disklike, usually truncate at both ends.

A pantropical genus of 30–40 species, with centers of diversity in Africa and Australia.

Rᴇғᴇʀᴇɴᴄᴇ: Martínez, M. D. 1981. The neotropical species of the genus *Corchorus,* Tiliaceae. Thesis. University of Kentucky, Lexington, Kentucky. 113 pp.

Key to the Species of *Corchorus*

1. Leaf blades with the 2 basal teeth prolonged into setaceous hairs; stipules conspicuous, linear-lanceolate, 5 mm or more long .... 1. *C. aestuans*
1. Leaf blades without hairlike prolongations at their base; stipules inconspicuous, linear to subulate, <5 mm long.
   2. Capsule 4-locular, 1–1.5 cm long, oblong-ellipsoid, densely tomentose, with a short, erect beak; stems densely stellate-pubescent .... 2. *C. hirsutus*
   2. Capsule 2-locular, (3.5–)5–8 cm long, linear, pubescent or glabrate, with 2 short bifid teeth; stems with lines of short hairs .... 3. *C. siliquosus*

1. Corchorus aestuans L., Syst. Nat., ed. 10, **2:** 1079. 1759.

Corchorus acutangulus Lam., Encycl. **2:** 104. 1786.

Erect or decumbent herb or subshrub, 0.3–1(–2) m tall, more or less spreading, usually much-branched from the base; stems and branches puberulent or short-pilose. Leaf blades 1.5–6.5(–7) × 2–3.5 cm, lanceolate to ovate, chartaceous, glabrous except along the veins, the apex subacute or obtuse, the base oblique or obtuse, the margins crenate to serrate with the basal pair of teeth often prolonged into setaceous hairs; petioles 0.5–3(–4) cm long; stipules linear-lanceolate, 5–7 mm long. Inflorescences 1–2(–3)-flowered; bracts stipule-like. Sepals ca. 4 mm long, linear-oblong, hooded at the apex; petals yellow, 3.5–5 mm long, obovate; androgynophore short; disk cupular; stamens (15–)30; ovary oblong-cylindric, 3-carpellate, truncate, the style trifid, the stigmas 2-lobed. Capsule solitary or paired, (1–)1.5–2.5 cm long, ca. 0.5 cm diam., narrowly oblong, angled, glabrate, green to dark brown, each valve with 2 dorsal wings that merge apically to form a single, bifid, recurved beak, 3–7 mm long. Seeds ca. 1 mm long, angular, obliquely truncate at both ends, blackish brown, smooth.

DISTRIBUTION: Occasional weed of open disturbed areas. Lind Point Research Station (A2297), Maho Bay (A1942), Great Lameshur Bay (A2564). Also on St. Croix, St. Thomas, and Tortola; tropics and subtropics from the Gulf Coast to the Amazon and in the Old World.

2. Corchorus hirsutus L., Sp. Pl. 530. 1753.

Fig. 198A–F.

Erect or less often prostrate shrub, 0.5–1.5(–2) m tall, densely branched; stems, branches, leaves, and inflorescences densely covered with brownish green stellate hairs. Leaf blades (1–)2–6 × 0.5–2 cm, ovate to oblong-lanceolate, chartaceous to subcoriaceous, the apex obtuse, the base rounded to obtuse, the margins crenate-dentate; petioles 0.5–1 mm long; stipules linear-filiform to subulate, to 5 mm long, early deciduous. Inflorescences 2–8-flowered umbelliform cymes; peduncles about as long as the pedicels. Sepals light yellow, 5–6 mm long, ovate; petals bright yellow, obovate-oblong, about as long as the sepals; androgynophore short; disk annular; stamens numerous; ovary ovoid to oblong-ovoid, 4(–5)-locular. Capsule 1–1.5 cm long, 0.5–0.6 cm diam., oblong-ellipsoid, densely tomentose, green, as long as the curved pedicels or longer, with a short, erect beak. Seeds 1–2 mm long, angular, obliquely truncate at both ends, blackish brown, smooth.

DISTRIBUTION: Common on beaches and in open areas in well-drained sandy or rocky soils. East End (A673), Coral Bay (A2580), Dittlif Point (A3962). Also on Anegada, St. Croix, St. Thomas, and Tortola; Mexico to northern South America including the West Indies, and Africa.

COMMON NAME: jack switch.

3. Corchorus siliquosus L., Sp. Pl. 529. 1753.

Erect subshrub or perennial herb, 0.3–1(–2) m tall, densely branched; young branches and leaves puberulent or pubescent; stems often with 1 or 2 lines of short hairs or glabrous. Leaf blades 1–5 × 0.5–2.5 cm, ovate to oblong-ovate, chartaceous, the apex acute or acuminate, the base mostly rounded or obtuse, the margins serrate; petioles 5–25 mm long; stipules linear-lanceolate, 1.5–5 mm long. Inflorescences 2(–3)-flowered, or flowers solitary, short-peduncled; pedicels 4–8 mm long, about twice the length of the peduncles. Sepals 6–8 mm long, oblong-lanceolate, acute; petals yellow, spatulate, 5–6 mm long; androgynophore and disk absent; stamens numerous; ovary 2-locular, narrowly oblong. Capsule (3.5–)5–8 cm long, ca. 3 mm diam., linear, silique-like, glabrous, or when young somewhat pubescent, brown, blunt, terminating in 2 short, bifid teeth at the more or less truncate apex. Seeds ca. 1.5 mm long, truncate at both ends, black, smooth.

DISTRIBUTION: A weed of cultivated and other disturbed areas. Fish Bay near mouth of Fish Bay Gut (A3877), trail to Sieben (A2641), hillside above Cinnamon Bay (M17007). Also on St. Croix, St. Thomas, and Tortola; occurring almost throughout tropical America.

COMMON NAME: broom weed.

2. TRIUMFETTA L.

Shrubs, subshrubs, or small trees, typically stellate-pubescent throughout, but also occasionally with simple hairs or peltate scales. Leaves alternate, simple or palmately 3(–5)-lobed, irregularly toothed; stipules linear-lanceolate, subpersistent. Inflorescences solitary or fascicled, few-flowered umbel-like cymes, lateral to and/or opposite the leaves or leafy bracts, sometimes terminal. Flowers bisexual (sometimes pistillate, and then the plants polygamodioecious); sepals 5, valvate, free, each with a subapical appendage; petals 5 (or absent), imbricate, free, shorter than or barely equal to the sepals in length, usually ciliate above a short claw at the base; androgynophore short (or absent), bearing 5 glands opposite the petals, usually crowned with a ciliate extrastaminal disk; stamens (5–)10 to numerous in bisexual flowers (staminodial or absent in pistillate flowers), the filaments free; ovary 2- or 3(–5)-locular, with 2 pendulous ovules per locule, the style filiform, the stigma entire or briefly 2- or 5-parted. Fruit indehiscent, nutlike (or dehiscent and capsular or schizocarpic), and (1–)2-seeded, covered by numerous rigid, glabrous to retrorsely hispidulous or featherlike-pubescent spines, each terminated by a straight or hooked hyaline hair; seeds ovoid to pyriform.

A pantropical genus of about 100 species, with centers of diversity in the Americas and Africa.

REFERENCE: Lay, K. K. 1950. The American species of *Triumfetta* L. Ann. Missouri Bot. Gard. 37: 315–395.

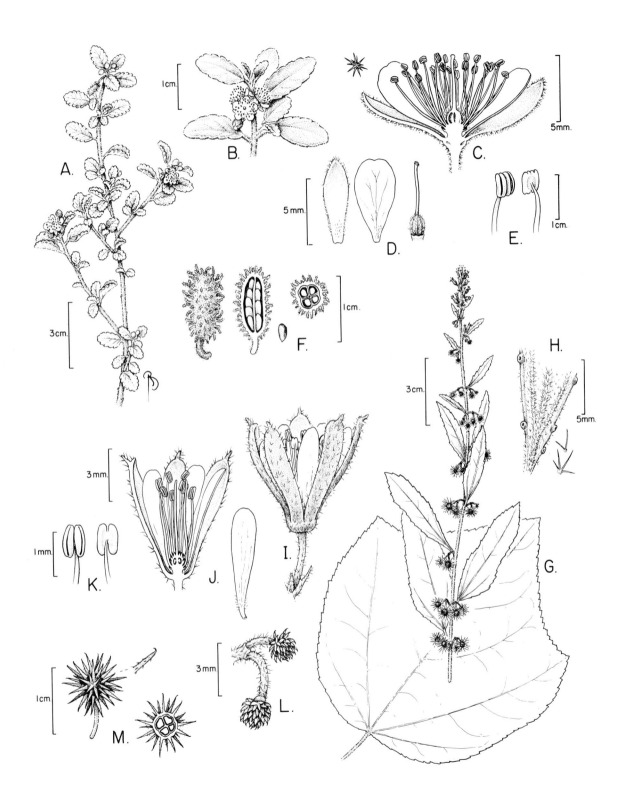

Fig. 198. A–F. *Corchorus hirsutus*. A. Fertile branch. B. Flowering branch. C. Detail of calyx trichome and l.s. flower. D. Sepal, petal, and pistil. E. Stamen, frontal and dorsal views. F. Capsule, l.s. capsule, seed, and c.s. capsule. G–M. *Triumfetta semitriloba*. G. Fruiting branch and basal leaf. H. Base of leaf with marginal glands and detail of trichomes. I. Flower. J. L.s. flower and petal. K. Stamen, frontal and dorsal views. L. Immature fruit. M. Fruit with detail of spine and c.s. fruit.

Key to the Species of *Triumfetta*

1. Petals, androgynophore, glands, and disk absent; fruit (excluding spines) conspicuously stellate-pubescent
.... 1. *T. lappula*
1. Petals, androgynophore, glands, and disk present; fruit (excluding spines) glabrous or minutely puberulent
.... 2. *T. semitriloba*

1. Triumfetta lappula L., Sp. Pl. 444. 1753.

Erect shrub or subshrub, (0.5–)1–2 m tall, much branched; stems densely stellate-pubescent to tomentose, the hairs ferruginous. Leaf blades 3.5–15.1 × 2.8–14 cm, simple or 3- to 5-lobed, the middle lobe often broadest near the apex, broadly ovate to nearly circular, chartaceous, stellate-pubescent above, densely stellate-pubescent to tomentose below, the apex acute to rounded, the middle lobe usually shallowly acuminate, the base truncate to cuneate or subcordate, the margins toothed, teeth often glandular; petioles 0.7–10.6 cm long; stipules 3.5–6 mm long, subulate. Flowers bisexual, produced in cymes of 2(–3) condensed cymules; peduncles 2–3 mm long; pedicels 1–3 mm long. Sepals yellow, 4–6 mm long, linear-oblong, the appendages ca. 0.5 mm long; petals, androgynophore, glands, and disk absent; stamens (5–)10(–15); ovary suborbicular, ca. 1–1.5 mm tall, the style ca. 3 mm long, the stigma entire or very briefly bifid. Fruit (excluding spines) ca. 2–3 mm diam., broadly ellipsoid to suborbicular, green turning brown at maturity, conspicuously stellate-pubescent, (2–)3-locular, each locule 1-seeded, the spines numerous, 1.5–3 mm long, retrorsely pilosulous. Seeds ca. 2 mm long, ovoid to pyriform, smooth.

DISTRIBUTION: A weed of waste places and thickets. Trail to Reef Bay (A2724). Also on St. Croix, St. Thomas, and Tortola; common throughout tropical America and also reported from the Cape Verde Islands and Mauritius.

COMMON NAME: maho.

2. Triumfetta semitriloba Jacq., Enum. Syst. Pl. 22. 1760.　　　　　　　　　　　　　Fig. 198G–M.

Erect shrub or subshrub, (0.3–)0.5–2 m tall, densely stellate-pubescent to tomentose, becoming glabrate with age. Leaf blades (1.3–)3.3–15.4 × 0.7–11 cm, simple or 3-lobed, usually broadly-ovate or rhombic-ovate to oblong-lanceolate or oblong-elliptic, chartaceous, stellate-pubescent to tomentose above and below, the apex acute to acuminate, the base rounded to truncate or broadly cuneate, less often subcordate, the margins irregularly toothed, teeth often glandular, especially near base; petioles 0.8–9.5 cm long; stipules 4–8 mm long, lanceolate. Flowers bisexual, produced in narrow, elongate cymes; peduncles 1–4 mm long; pedicels 2–4 mm long. Sepals green, 4–8 mm long, oblong-lanceolate, the appendages 0.5–1 mm long; petals yellow, equal to or slightly shorter than sepals, linear-elliptic; androgynophore short, with orbicular glands; disk inconspicuous; stamens 15–25; ovary ovoid to subglobose, ca. 1 mm tall, the style 6–8 mm long, the stigma entire to obscurely 3-parted. Fruit (excluding spines) 3–5 mm diam., ovoid to subglobose, glabrous or minutely puberulent, green turning reddish brown at maturity, (2–)3(–4)-locular, the locules 1- to 2-seeded, the spines numerous, 2–4 mm long, retrorsely hispidulous. Seeds 2–3 mm long, ovoid to pyriform, brown, smooth.

DISTRIBUTION: A weed in disturbed areas. Cruz Bay (A2396), L'Esperance (A2560), along road to Bordeaux (A3125). Also on St. Croix, St. Thomas, and Tortola; native to tropical America and Africa, reportedly naturalized in Africa and the Pacific.

COMMON NAME: bur bark.

88. Turneraceae (Turnera Family)

Herbs, shrubs, or less often trees. Leaves alternate, simple, entire or toothed; stipules wanting or minute. Flowers small to large, actinomorphic, bisexual, 5-merous, solitary and axillary, rarely in racemose, cymose, or paniculate inflorescences; calyx subtended by 2 bracteoles, forming a tubular hypanthium; corolla of distinct, clawed petals, inserted on distal part of hypanthium; stamens as many as and alternate with the petals, the filaments distinct, inserted at the base of the hypanthium, the anthers opening by longitudinal slits; nectary disk wanting; ovary superior or partly inferior, unilocular, of 3 united carpels, the placentation parietal, with numerous ovules per carpel, the styles terminal, distinct, with many-branched stigmas. Fruit a loculicidal capsule; seeds reticulate, pitted, with a unilateral aril.

A family of 8 genera and about 120 species, native to tropical America and Africa.

1. TURNERA Mart.

Shrubs or subshrubs. Leaves simple, entire or serrate, alternate, biglandular at base; stipules minute. Flowers actinomorphic, 5-merous, bisexual, heterostylous, solitary; bracts 2, subtending the flower; peduncle adnate to petiole; hypanthium nearly bell-shaped, the lobes equal, shorter or longer than the tube; petals free, clawed, adnate to throat of hypanthium; stamens shorter than the petals, inserted on the hypanthium, the anthers opening by longitudinal slits; ovary superior, surrounded by the hypanthium, ovoid, unilocular with 3 multiovulate parietal placentae, the styles 3, many-branched at distal portion or simple. Fruit a loculicidal, 3-valved capsule; seeds numerous, with a lateral arillode.

A neotropical genus of about 70 species ranging from the southern United States to Mexico and Argentina.

Key to the Species of *Turnera*

1. Erect or spreading shrub; leaves 0.4–1.0 cm long, not glandular; petals 4–6 mm long .... 1. *T. diffusa*
1. Erect shrub; leaves 6–10 cm long, glandular at the junction between the blade and the petiole; petals 2–3 cm long.. 2. *T. ulmifolia*

1. Turnera diffusa Willd. *in* Schult., Syst. Veg. **6:** 679. 1820.   Fig. 199I–K.

Turnera microphylla Griseb., Fl. Brit. W. I. 297. 1860.

Erect or spreading shrub, 15–70 cm tall, many-branched from a woody base; twigs puberulent, nearly cylindrical. Leaves alternate, appearing whorled because of short axillary branches; blades 4–10 × 1.5–5 mm, obovate or lanceolate, coriaceous, densely pubescent especially along lower surface, the apex obtuse or rounded, the base cuneate, not glandular, the margins serrate; petioles to 1 mm long. Flowers nearly sessile, terminal on axillary branches. Hypanthium bell-shaped, greenish, pubescent without, ca. 3.5 mm long; petals yellow, spatulate, 4–6 mm long; stamens shorter than the hypanthium, the anthers lanceolate, ca. 0.6 mm long. Capsule subglobose, the valves opening to the base.

DISTRIBUTION: A common subshrub of scrub and open dry areas. Dittlif Point (A3979), Maria Bluff (A2334). Also on Little St. James, St. Croix, St. Thomas, and Virgin Gorda; Bahamas, the Greater Antilles and continental tropical America.

COMMON NAME: old woman broom.

2. Turnera ulmifolia L., Sp. Pl. 271. 1753.   Fig. 199A–H.

Turnera angustifolia Mill., Gard. Dict., ed. 8. 1768.

Erect shrub 0.5–1.5 m tall, branching along main stem or many-branched from base; twigs appressed-pubescent, angled, becoming glabrous and cylindrical. Leaves alternate, 2.5–12 × 1.3–3 cm, lanceolate or elliptic, chartaceous, sparsely pubescent, the apex acute or acuminate, the base narrowed, with a pair of discoid glands at junction with petiole, the margins serrate; petioles 0.5–2.2 cm long. Flowers nearly sessile, terminal. Hypanthium tubular, 1.8–2 cm long, the lobes oblong, 1–1.5 cm long, greenish, pubescent without; petals yellow, spatulate, 2–3 cm long; stamens as long as the hypanthium, the anthers lanceolate, ca. 5 mm long. Capsule subglobose, 6–8 mm long, pubescent, the valves opening to the middle. Seeds numerous, oblong-ovoid, ca. 2.5 mm long, with white arillode.

DISTRIBUTION: An occasional shrub of open, disturbed areas, especially along roads. Emmaus (A2828), Fish Bay (A3900). Also on St. Croix, St. Thomas, and Virgin Gorda; throughout tropical America, naturalized throughout the tropics.

89. Ulmaceae (Elm Family)

Trees, erect or climbing shrubs, or lianas. Leaves alternate or less often opposite, simple, entire or toothed; stipules protecting the buds, deciduous. Flowers minute, actinomorphic, unisexual or less often bisexual, in axillary cymes or axillary and solitary; calyx of (2–)5(–9) distinct or connate sepals; corolla wanting; stamens as many as or rarely twice as many as the sepals, the filaments distinct, inserted on the calyx tube, the anthers opening by longitudinal slits; nectary disk wanting; ovary superior or partly inferior, 1–2-locular, of 2–3 united carpels, with pendulous, solitary ovules, the styles terminal, distinct, with decurrent stigmas. Fruit a drupe, nut, or samara.

A family of 18 genera and about 150 species, native to tropical and temperate regions.

Key to the Genera of Ulmaceae

1. Drupes 6–10 mm long; leaves puberulent, smooth on upper surface, the margins entire, crenate or coarsely serrate (3–4 teeth/cm) above the base .... 1. *Celtis*
1. Drupes to 3 mm long; leaves weakly scabrous on upper surface, the margins finely serrate (7–8 teeth/cm) from base to apex .... 2. *Trema*

1. CELTIS L.

Shrubs, small trees, or lianas, armed or unarmed. Leaves alternate, simple, usually serrate or less often crenate or entire, petiolate; stipules minute and early deciduous. Flowers 5-merous, unisexual or sometimes bisexual, the staminate flowers in axillary cymose clusters, the pistillate solitary or a few in axillary cymes; calyx of free sepals; stamens as many as the sepals, inserted in the calyx tube, the anthers opening by longitudinal slits; ovary superior, 1-locular, with a solitary, apically pendulous ovule, the styles 2. Fruit a fleshy to leathery drupe.

A genus of about 80 species, mostly from the Northern Hemisphere and southern Africa.

Key to the Species of *Celtis*

1. Scandent shrub or liana, with a pair of axillary, recurved or straight spines; leaves entire, crenate or remotely serrate .... 1. *C. iguanaea*
1. Small tree, unarmed; leaves coarsely serrate .... 2. *C. trinervia*

1. Celtis iguanaea (Jacq.) Sarg., Silva **7:** 64. 1895. *Rhamnus iguanaeus* Jacq., Enum. Syst. Pl. 16. 1760. *Momisia iguanaea* (Jacq.) Rose & Standl., Contr. U.S. Natl. Herb. **16:** 8. 1912.   Fig. 200A–G.

Celtis aculeata Sw., Prodr. 53. 1788, nom. illegit.

Polygamous monoecious scandent shrub or liana 3.5–10(–20) m long; trunk reaching 8 cm diam., the bark grayish, with light-colored lenticels; main stems with numerous, alternate, short, lateral branches; twigs puberulent, nearly cylindrical, slightly flexuous, with a pair of short, axillary, recurved or straight spines. Leaf blades 4–15 × 2.5–8.7 cm, ovate, oblong, or seldom obovate, chartaceous, puberulent, the lower surface prominently veined, with domatia in the axils of secondary veins, the apex shortly acuminate to nearly obtuse, the base cordate to truncate, the margins entire, crenate, or remotely serrate; petioles 2–15 mm long. Flowers unisexual or bisexual, the staminate flowers in

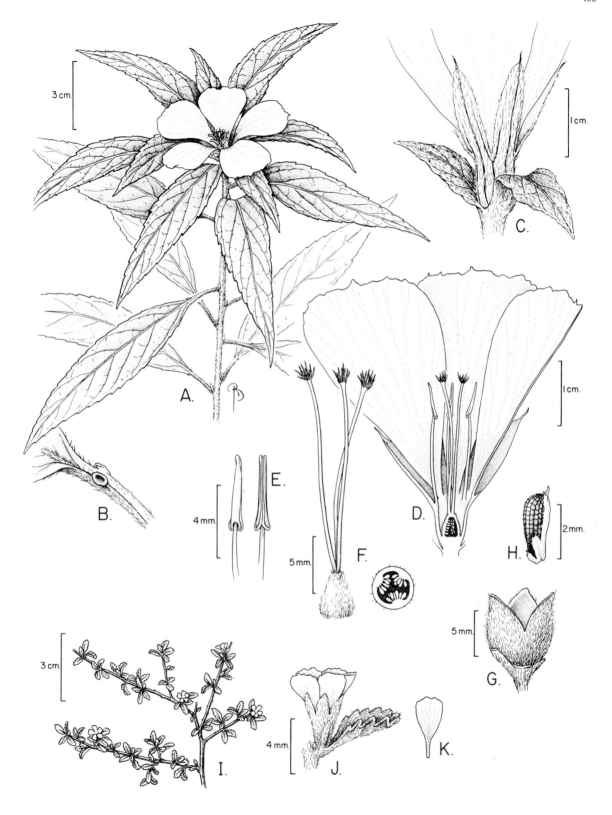

FIG. 199. A–H. *Turnera ulmifolia.* **A.** Flowering branch. **B.** Detail of leaf petiole showing glands. **C.** Detail of flower peduncle showing calyx and paired bracteoles. **D.** L.s. flower. **E.** Stamen, frontal and dorsal views. **F.** Pistil and c.s. ovary. **G.** Dehisced capsule. **H.** Seed. **I–K.** *Turnera diffusa.* **I.** flowering branches. **J.** Detail of flower in axil of leaf. **K.** Petal.

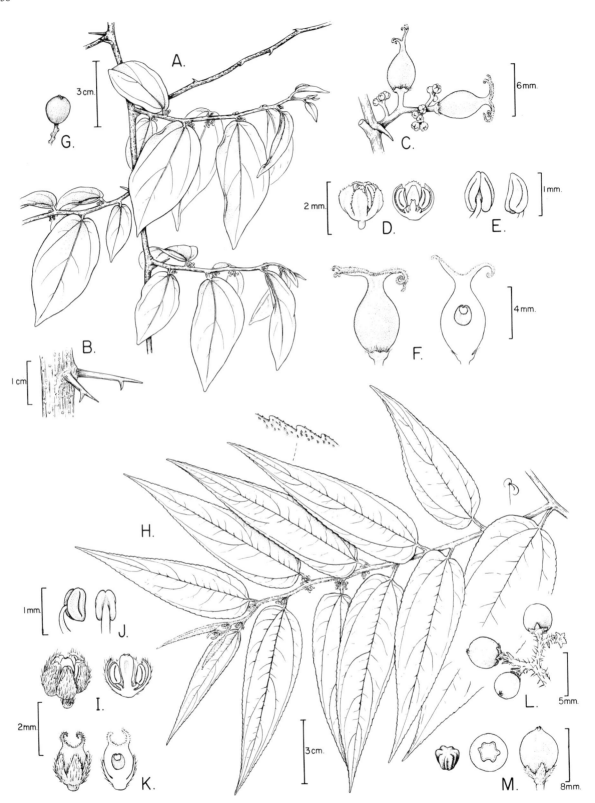

Fig. 200. A–G. *Celtis iguanaea*. A. Flowering branch. B. Detail of stem showing stipular spines. C. Inflorescence with developing fruits. D. Staminate flower and l.s. staminate flower. E. Stamen, dorsal and lateral views. F. Developing drupe and l.s. drupe. G. Drupe. H–M. *Trema micranthum*. H. Flowering branch and detail of leaf margin. I. Staminate flower and l.s. staminate flower. J. Stamen, lateral and dorsal views. K. Pistillate flower and l.s. pistillate flower. L. Infructescence. M. Stone, c.s. drupe, and drupe.

branched axillary cymes, the pistillate or bisexual flowers in few-flowered axillary cymes. Calyx deeply parted, 1–1.2 mm long, the sepals oblong, concave, with ciliate margins; stamens as long as the sepals, ascending; ovary bottle-shaped, the styles 2, pubescent, each 2-branched. Drupe nearly globose, 6–10 mm long, turning from green to orange, with persistent styles.

DISTRIBUTION: A common woody vine of scrub and dry coastal forests. Cinnamon Bay (A5150), Great Cruz Bay (A2361), Lameshur (A3237). Also on St. Croix, St. Thomas, and Tortola; the Greater and Lesser Antilles and continental tropical America.

COMMON NAMES: cat-claw, cockspur.

2. Celtis trinervia Lam., Encycl. **4:** 140. 1797. *Sponia trinervia* (Lam.) Decne., Nouv. Ann. Mus. Hist. Nat. **3:** 498. 1834.

Small tree 5–10 m tall; bark brown with light brown, rounded lenticels; twigs pubescent or puberulent, nearly cylindrical. Leaf blades 4–13 × 2.5–7.5 cm, ovate-lanceolate, chartaceous, puberulent, with 2 ascending basal secondary veins, the lower surface minutely punctate, the apex acuminate, the base rounded-cordiform, very unequal, the margins serrate; petioles 5–10 mm long; stipules early minute, deciduous. Flowers unisexual, the staminate flowers in short axillary cymes, the pistillate flowers usually solitary. Calyx deeply parted, ca. 1.5 mm long, the sepals oblong, concave, with ciliate margins; anthers nearly sessile, as long as the sepals. Drupe ellipsoid to nearly globose, 7–8 mm long, turning from green to purple-black, with 2 persistent styles.

DISTRIBUTION: A rare tree of moist to dry forests. Bordeaux (W833). Also on St. Croix, St. Thomas, and Tortola; the Greater Antilles.

2. TREMA Lour.

Monoecious or polygamous shrubs or small trees, unarmed. Leaves alternate, simple, serrate, 3-nerved at the base, petiolate; stipules minute and early deciduous. Flowers 5-merous, unisexual or bisexual, in axillary cymose clusters; calyx of free sepals; stamens as many as the sepals, inserted on the calyx tube, the anthers opening by longitudinal slits; ovary superior, sessile, 1-locular, the styles 2, undivided. Fruit a drupe; stones 1-seeded.

A genus of about 30 species, of tropical and subtropical regions.

1. Trema micranthum (L.) Blume, Ann. Mus. Bot. Lugduno-Batavum **2:** 58. 1856. *Rhamnus micranthus* L., Syst. Nat., ed. 10, **2:** 937. 1759. *Celtis micranthus* (L.) Sw., Prodr. 53. 1788. *Sponia micrantha* (L.) Decne., Nouv. Ann. Mus. Hist. Nat. **3:** 498. 1834.   Fig. 200H–M.

Monoecious shrub or small tree 1.5–8 m tall; bark light brown, smooth, with numerous lenticels; twigs pubescent or puberulent, angled to nearly cylindrical, slightly flexuous. Leaf blades 5.5–14.5 × 2–5.5 cm, lanceolate to ovate-lanceolate, chartaceous,

weakly scabrous above, the lower surface prominently veined, the apex acuminate, the base cordate to rounded, unequal, the margins finely serrate; petioles 6–10 mm long. Flowers greenish yellow, unisexual, in short, branched, axillary cymes. Calyx deeply parted, ca. 0.7 mm long, the sepals lanceolate. Drupe globose, to 3 mm long, turning from green to orange, with 2 persistent, unbranched styles.

DISTRIBUTION: A common tree of moist to dry forests. Road to Bordeaux (A2100), Maria Bluff (A619), Susannaberg (A3841). Also on Jost van Dyke, St. Croix, St. Thomas, Tortola, and Virgin Gorda; throughout tropical and subtropical America.

90. Urticaceae (Nettle Family)

Herbs, shrubs, small trees, or very rarely lianas, often with stinging hairs. Leaves alternate or opposite, simple, cystoliths usually present in epidermis, frequently with toothed margins; stipules minute or wanting. Flowers minute, actinomorphic or slightly zygomorphic, unisexual, apetalous, without nectary disk, solitary or in axillary cymes. Staminate flowers with calyx of (3–)4–5(–6) sepals; stamens as many as the sepals and opposite them, or a single stamen present, the filaments distinct, reflexed at anthesis, the anthers opening by longitudinal slits. Pistillate flowers with calyx of 4(–5) sepals or naked; staminodes scale-like; ovary superior, 1-locular, of 1(–2) united carpels, with a basal, solitary ovule, the style single, terminal, with a capitate or filiform, decurrent stigma. Fruit an achene, a nut, or a drupe, often enclosed in persistent accrescent sepals.

A family of 45 genera and about 700 species, mostly in tropical and subtropical regions.

Key to the Genera of Urticaceae

1. Leaves alternate .... 1. *Laportea*
1. Leaves opposite .... 2. *Pilea*

1. LAPORTEA Gaudich.

Monoecious, erect herbs with straight and stinging hairs. Leaves alternate, simple, serrate, long-petiolate; stipules free or connate. Flowers in axillary panicles; staminate flowers 4–5-merous, with a rudimentary ovary; pistillate flowers 4-merous, with oblique, 1-locular ovary with erect ovule, the stigma papillose; tepals imbricate. Fruit an oblique achene.

A weedy genus of about 23 species, of tropical and subtropical regions.

1. Laportea aestuans (L.) Chew, Gard. Bull. Straits Settlem. **21:** 200. 1965. *Urtica aestuans* L., Sp. Pl.,

ed. 2, **2:** 1397. 1763. *Fleurya aestuans* (L.) Gaudich., Voy. Uranie 497. 1830.

Herb to 50 cm tall, branching along main stem; branches 4-angled, fleshy, reddish-tinged. Leaf blades 5–15.5 × 3.6–14 cm, ovate or wide elliptic, membranous, with stinging hairs when young, becoming glabrous, 3-veined from base, lower surface with cystoliths, the apex acute or short-acuminate, the base cordate to truncate, the margins dentate; petioles slender, 2–13.5 cm long; stipules linear, 2.5–5 mm long. Flowers greenish, in panicles 2–20 cm long. Staminate flowers terminal on inflorescence branches; tepals oblong, ca. 1 mm long, imbricate; stamens nearly sessile. Pistillate flowers naked. Achene 1–1.5 mm long, obliquely ovoid-flattened.

DISTRIBUTION: Occasional in open disturbed areas. Bethany (B258), Susannaberg (A3807). Also on St. Croix, St. Thomas, and Tortola; throughout tropical and subtropical America.

2. PILEA Lindl.

Monoecious or dioecious, erect or prostrate, succulent herbs, lacking stinging hairs. Leaves opposite, simple, petiolate, those of a pair equal or unequal; stipules free or connate. Flowers unisexual, in the same or in separate simple or compound axillary inflorescences; staminate flowers with 4, nearly equal sepals and a rudimentary ovary; pistillate flowers with 3 unequal sepals and a nearly ovoid ovary, the stigma penicillate. Fruit a compressed–ovate achene.

A tropical genus of about 400 species.

REFERENCE: Urban, I. 1905. Flora Portoricensis. Symbo. Antill. **4**: 193–352.

Key to the Species of *Pilea*

1. Leaves <1 cm long, 1-nerved, with entire margins.
   2. Plants 15–40 cm tall; stems fleshy, drying dark-colored, showing numerous cystoliths when dried
   .... 1. *P. microphylla*
   2. Plants 5–15 cm tall; stems slender, slightly translucent, usually with a few reddish spots when dried
   .... 4. *P. tenerrima*
1. Leaves >1 cm long (at least some of them), 3-nerved from base, with toothed margins.
   3. Plants creeping, rooting at nodes; leaves 0.8–2.5 cm long, ovate to orbicular .... 2. *P. nummulariifolia*
   3. Plant erect; leaves 5–10 cm long, ovate, oblong, or elliptic .... 3. *P. sanctae-crucis*

1. Pilea microphylla (L.) Liebm., Kongel. Danske Vidensk.- Selsk. Skr. **5(2):** 296. 1851. *Parietaria microphylla* L., Syst. Nat., ed. 10, **2:** 1308. 1759.

Erect or spreading, monoecious herb, 15–40 cm tall, many- or few-branched from base; stems fleshy, usually reddish-tinged when fresh, dark on drying and showing numerous cystoliths. Leaf blades 2–9 × 2–5 mm, obovate, ovate, or oblanceolate, fleshy, glabrous, lower surface with conspicuous midvein, upper surface with numerous transverse linear cystoliths, the apex rounded or obtuse, the base cuneate, obtuse or attenuate, sometimes unequal, the margins entire, revolute; petioles slender, 0.2–2 mm long; stipules inconspicuous. Flowers greenish or whitish, reddish-tinged. Staminate flowers solitary; perianth ca. 1 mm long. Pistillate flowers in short axillary cymes; perianth ca. 0.5 mm long. Achene ca. 0.5 mm long, yellowish brown, smooth, ellipsoid.

DISTRIBUTION: A common herb of moist, shaded, open habitats. East End (A1837), Susannaberg (A2132), Waterlemon Bay (A2524). Also on Anegada, St. Croix, St. Thomas, and Tortola; throughout tropical and subtropical America.

COMMON NAME: artillery plant.

2. Pilea nummulariifolia (Sw.) Wedd., Ann. Sci. Nat. Bot., ser. 3, **18:** 225. 1852. *Urtica nummulariifolia* Sw., Kongl. Vetensk. Acad. Nya Handl. **8:** 63, t. 1, fig. 2. 1787.   Fig. 201G.

Creeping, monoecious or dioecious herb, with numerous lateral branches 10–50 cm long; rooting at nodes; stems slender, pilose. Leaf blades 0.8–2.5 × 0.8–2 cm, ovate to orbicular, chartaceous, pilose, 3-veined from base, the lower surface with numerous linear cystoliths, the apex rounded or obtuse, the base rounded, truncate, or cordate, the margins deeply crenate; petioles slender, pilose, 0.5–2.6 cm long; stipules ovate, 2–3 mm long. Flowers greenish, staminate and pistillate on different densely flowered cymose

inflorescences, shorter than the subtending leaf. Staminate flowers in sessile cymes; perianth 1.5–2.3 mm long. Pistillate flowers in pedunculed axillary cymes; perianth ca. 0.5 mm long. Achene ca. 0.5 mm long, yellowish brown, smooth, ellipsoid.

DISTRIBUTION: An occasional herb of moist, shaded habitats. Adrian Ruins (W865). Also on St. Croix and St. Thomas; throughout the West Indies.

3. Pilea sanctae-crucis Liebm., Kongel. Danske Vidensk.- Selsk. Skr. **5(2:)** 301. 1851.   Fig. 201A–F.

Erect or decumbent, monoecious or dioecious herb, 20–35 cm tall; stems fleshy, reddish-tinged, pilose. Leaf blades 5–10 × 2.7–5 cm, ovate, oblong, or elliptic, chartaceous, pilose, 3–5-veined from base, the lower surface with numerous linear cystoliths, the apex obtuse to acute, the base obtuse, attenuate, rounded, or seldom nearly cordate, the margins doubly serrate; petioles slender, pilose, 1–4 cm long; stipules oblong to obovate, ca. 6 mm long. Flowers beige or reddish-tinged, staminate and pistillate on different inflorescences. Staminate flowers congested in cyme units on long-pedunculate inflorescences; perianth 1.5–2 mm long. Pistillate flowers in branched raceme like inflorescences; perianth ca. 0.5 mm long. Achene ca. 0.5 mm long, yellowish brown, smooth, ellipsoid.

DISTRIBUTION: A common herb in moist forest understory, growing on ground or rocks. Bordeaux (A1884, A5069). Also on St. Croix, St. Thomas, and Tortola; Puerto Rico (according to Urban, 1905).

4. Pilea tenerrima Miq., Linnaea **26:** 219. 1854.

Spreading or decumbent, mat-forming, monoecious herb, 5–15 cm tall, many-branched from base; stems slender, filiform, translucent with a few reddish spots. Leaf blades 3–10 × 2–5 mm, obovate, ovate, or oblanceolate, membranous, glabrous, lower surface with conspicuous midvein, upper surface with numerous transverse linear cystoliths, the apex rounded or obtuse,

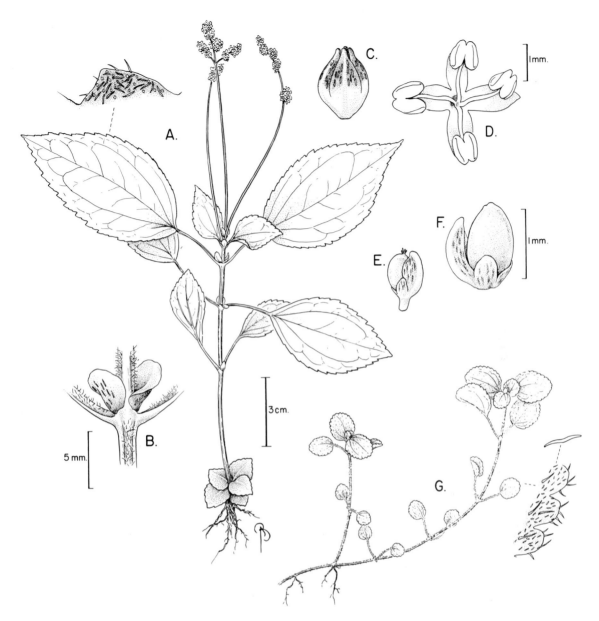

Fig. 201. A–F. *Pilea sanctae-crucis*. A. Habit and detail of leaf margin showing indument. B. Detail of stem node showing intrapetiolar stipules. C. Staminate flowerbud. D. Staminate flower, top view. E. Pistillate flower. F. Achene with persistent perianth. G. *Pilea nummulariifolia*; habit, detail of leaf margin showing indument, and detail of trichome.

the base cuneate, obtuse, or attenuate, sometimes unequal, the margins entire; petioles slender, 2–5 mm long; stipules early deciduous. Flowers greenish, in short axillary cymes. Staminate flowers usually terminal on cymes; perianth 0.5–0.7 mm long, whitish. Pistillate flowers congested in cymes; perianth ca. 0.5 mm long. Achene ca. 0.5 mm long, yellowish brown, smooth, ellipsoid.

DISTRIBUTION: An occasional herb of moist, shaded, open habitats. Coral Bay (A2583), East End (A3782). Also on St. Thomas and Tortola; Florida, Greater Antilles, the Bahamas, and South America.

91. **Verbenaceae** (Verbena Family)

Herbs, shrubs, trees, or lianas; young branches very often quadrangular. Leaves opposite or whorled, simple or pinnately or palmately compound, entire or less frequently with toothed margins; stipules wanting. Flowers small to large, actinomorphic or zygomorphic, in most genera bilabiate, bisexual or rarely unisexual, usually subtended by colored bracts, produced in terminal or axillary spikes, racemes, panicles, cymes, or sometimes heads; calyx tubular, sometimes irregular, (4–)5(–8)-lobed or toothed; corolla sympetalous, often with a narrow tube and a spreading limb, (4–)5(–8)-lobed; stamens 4 (2 smaller), or less often only 2 or 5, the filaments adnate to the corolla tube, alternate to the lobes, the anthers opening by longitudinal slits; nectary disk weakly developed or wanting; ovary superior, commonly becoming 4-locular, of 2–4(–5) united carpels, with basal placentation, usually with a single ovule per locule, the styles connate with a bilabiate stigma. Fruit a schizocarp, capsule, or drupe with 2 or 4 stones.

A family of about 100 genera and 2600 species, mostly pantropical.

REFERENCE: López Palacios, S. 1977. Verbenaceae. Fl. Venezuela 1: 1–654. Universidad de los Andes, Merida, Venezuela .

Key to the Genera of Verbenaceae

1. Plants herbaceous, sometimes woody at base, usually <60 cm tall.
  2. Flowers sessile, arranged in elongate spikes; fertile stamens 2 .... 8. *Stachytarpheta*
  2. Flowers short- to long-pedicellate, arranged in racemes; fertile stamens 4.
    3. Racemes densely flowered; fruiting calyx straw-colored, tubular to narrowly bell-shaped with setaceous lobes, puberulent, not completely enclosing the fruit .... 2. *Bouchea*
    3. Racemes loosely flowered; fruiting calyx green, ovoid, with short obtuse lobes, covered with uncinate hairs, completely enclosing the fruit .... 7. *Priva*
1. Plants woody, trees, shrubs, or vines.
  4. Woody vines or scandent shrubs.
    5. Scandent or leaning shrubs, to 3 m long, bearing 2–3 spines in the nodes; fruiting calyx not winged .... 4. *Clerodendrum*
    5. Twining, woody vine to 10 m long, unarmed; fruiting calyx with elongate, winglike, spreading lobes .... *Petrea* (cultivated)
  4. Trees or shrubs.
    6. Plant with axillary spines; fruiting calyx ovoid, narrowed toward apex, yellow-orange, fleshy, enclosing the fruit .... 5. *Duranta*
    6. Plants unarmed or if armed the spines scattered along the stem; fruiting calyx green, not enclosing the fruit.
      7. Shrubs with strongly aromatic foliage .... 6. *Lantana*
      7. Small to medium-sized, nonaromatic trees.
        8. Leaves trifoliolate (seldom unifoliolate); corolla light violet to bluish .... 9. *Vitex*
        8. Leaves simple; corolla white.
          9. Trees of brackish habitats, producing numerous erect pneumatophores; fruit elliptic, ovate, to oblong, leathery, compressed, green .... 1. *Avicennia*
          9. Trees of dry inland habitats, without pneumatophores; fruits ellipsoid to ovoid, fleshy, orange to purple at maturity .... 3. *Citharexylum*

1. AVICENNIA L.

Shrubs or small trees with a network of erect pneumatophores that protrude from the ground. Leaves opposite, thick, coriaceous, with entire margins. Flowers bisexual, in axillary or terminal cymes to nearly capitate inflorescences; calyx of 5 equal sepals; corolla 4-merous, nearly actinomorphic; stamens 4, the filaments of equal length, inserted on the corolla throat; ovary incompletely 4-locular, each locule with a single ovule, the style slender, the stigma bifid. Fruit a tardily dehiscent, fleshy to leathery capsule, usually 1-seeded, viviparous.

A pantropical genus of about 16 species, distributed in coastal tidal swamps.

1. Avicennia germinans (L.) L., Sp. Pl., ed. 3, **1**: 891. 1764. *Bontia germinans* L., Syst. Nat., ed. 10, **2**: 1122. 1759. Fig. 202A–D.

Avicennia nitida Jacq., Enum. Syst. Pl. 25. 1760.

Shrub or small tree 2–15 m tall, with mangrove habit; bark dark gray, fissured. Leaf blades 4–10.5 × 1.5–3.8 cm, ovate, elliptic, oblong, or lanceolate, coriaceous, the upper surface glabrescent and pitted, the lower surface densely covered with minute, whitish, appressed hairs and with a prominent stout midvein, the apex acute or obtuse, the base obtuse, the margins revolute; petioles 0.4–1 cm long, appressed-pubescent. Flowers sessile, congested on axillary or terminal, simple or compound dichasial cymes; peduncles 0.7–3.5 cm long, appressed-pubescent. Calyx greenish, the sepals 4–5.5 mm long, ovate, appressed-pubescent; corolla white, the tube 4–5.5 mm long, yellowish, and tomentose within, the lobes 2–2.5 mm long, oblong, reflexed; stamens and style exserted. Capsule elliptic, ovate to oblong, 1.5–2 cm long, light green, compressed, appressed-pubescent when young.

DISTRIBUTION: A common species of the inland side of brackish mangrove swamps. Lameshur (A5136). Probably on all Virgin Islands; common throughout the Caribbean, Florida, Mexico to northern South America.

COMMON NAME: black mangrove.

2. BOUCHEA Cham.

Herbs, sometimes woody at base, glabrous to pubescent. Leaves opposite, petiolate to nearly sessile, with crenate to serrate margins. Flowers bisexual, pedicellate, in axillary or terminal racemes; calyx tubular, of 5 equal setaceous teeth; corolla 5-merous, irregu-

FIG. 202. A–D. *Avicennia germinans*. **A.** Inflorescence. **B.** Detail of inflorescence. **C.** L.s. pistil and l.s. corolla with adnate stamens. **D.** Fruiting branch. **E–G.** *Bouchea prismatica*. **E.** Fertile branch. **F.** L.s. pistil and l.s. corolla with adnate stamens. **G.** Fruit with persistent calyx and mericarp. **H–M.** *Citharexylum fruticosum*. **H.** Inflorescence. **I.** Flower. **J.** L.s. corolla with adnate stamens and l.s. pistil. **K.** Stamen, lateral and frontal views. **L.** Fruiting branch. **M.** Pyrene and drupe.

lar, the tube funnel-shaped or cylindrical, the posterior lobes smaller than the anterior ones; stamens 4, the filaments in 2 pairs of unequal length, inserted on the tube; ovary 2-locular, each locule with a single ovule, the style slender, the stigma bilobed. Fruit oblong to linear, beaked, included in calyx for most of its length, longitudinally splitting into 2 mericarps of equal size, with 1 commonly sterile.

A genus of 10 species, 9 native to the neotropics.

1. Bouchea prismatica (L.) Kuntze, Revis. Gen. Pl. 2: 502. 1891. *Verbena prismatica* L., Sp. Pl. 19. 1753. Fig. 202E–G.

Bouchea ehrenbergia Cham., Linnaea 7: 253. 1882.

Erect herb 40–60 cm tall; stems somewhat woody at base, obtusely 4-angled, sulcate, glabrous to puberulent. Leaf blades 2.2–8 × 2–5 cm, ovate, membranous, with a weak fetid smell, puberulent, especially when young, the apex obtuse or acute, the base cuneate, the margins serrate to crenate-dentate; petioles 0.8–3 cm long, puberulent. Flowers nearly sessile, on axillary and terminal racemes 5–25 cm long. Calyx greenish, tubular, puberulent, 5.5–7 mm long, the setae 1–2 mm long; corolla pink to lilac, trumpet-shaped, 12–14 mm long, the tube yellowish; stamens and style included. Fruit cylindrical, 8–9.5 mm long, light brown, splitting into 2 equal mericarps; fruiting calyx covering most of fruit, straw-colored, longitudinally splitting along one side.

DISTRIBUTION: A common weed of open disturbed areas. Lameshur (A2576, A2789). Also on St. Croix; common throughout the West Indies, Mexico, Central America, and Venezuela.

3. CITHAREXYLUM L.

Shrubs or trees; branches usually 4-angled. Leaves opposite or verticillate, petiolate, with entire to serrate margins, usually with a pair of glands at base or on petiole. Flowers bisexual, pedicellate, in axillary or terminal racemes; calyx tubular, truncate or 5-lobed; corolla 5(–6)-merous, funnel- or trumpet-shaped, the lobes spreading; stamens 4(–5), the filaments in 2 pairs of unequal length, inserted on the upper half of the tube, the fifth stamen usually reduced to a staminode; ovary 4-locular (sometimes imperfectly so), each locule with a single ovule, the style slender, the stigma bifid. Fruit drupaceous, fleshy, with 2 pyrenes, each containing 2 seeds; fruiting calyx cupular and hardened.

A genus of about 70 species, from the neotropics.

1. Citharexylum fruticosum L., Syst. Nat., ed. 10, 2: 1115. 1759. Fig. 202H–M.

Shrub or small tree 1.5–10 m tall; bark dark brown to grayish, rough; young branches and petioles orange-tinged; stems obtusely 4-angled, striate, puberulent, with numerous minute, scale-like dots. Leaf blades 5–15 × 2–6 cm, elliptic, oblong, or elliptic-lanceolate, coriaceous, puberulent to pubescent, with numerous scale-like dots, the apex acute, acuminate, obtuse, or retuse, the base narrowed or obtuse, the margins revolute, entire (serrate in juvenile leaves); petioles 8–25 mm long, puberulent, with a pair of impressed glands near the blade. Flowers sessile or nearly so, on axillary and terminal spikes, 3–14 cm long. Calyx greenish, bell-shaped to tubular, puberulent, 2.5–4 mm long, the lobes ciliate; corolla white, trumpet-shaped, 4–7 mm long, the tube yellowish, the lobes spreading, ciliate; stamens and style included. Drupe ellipsoid to ovoid, 7–10 mm long, turning from green to orange to purple; fruiting calyx turning brown at maturity.

DISTRIBUTION: A common tree or shrub of dry to moist forests. Center Line Road (A3814), Emmaus (A2005), Southside Pond (A1839). Also on Anegada, Jost van Dyke, St. Croix, St. Thomas, Tortola, and Virgin Gorda; common throughout the West Indies and northern South America.

COMMON NAMES: fiddlewood, old woman bitter, susannaleche.

NOTE: A modern revision of *Citharexylum* may show *C. fruticosum* to be conspecific with *Citharexylum spinosum* L., in which case, the epithet *spinosum*, being the older name, should be adopted.

4. CLERODENDRUM L.

Trees, erect shrubs, or less often scandent shrubs or lianas, sometimes spiny at the petiole base. Leaves simple, opposite or verticillate, petiolate, with entire or dentate margins. Flowers showy, bisexual, zygomorphic, pedicellate, in axillary cymes or in terminal panicles; calyx bell-shaped or tubular, truncate, 5-lobed or 5-parted; corolla 5-merous, trumpet-shaped, with spreading limb; stamens 4, the filaments in 2 pairs of unequal length, inserted in the tube, long-exserted; ovary imperfectly 4-locular, each locule with a single ovule, the style slender with a capitate stigma. Fruit drupaceous, globose to ovoid, 4-lobed or 4-sulcate, with 4 pyrenes often coherent in pairs.

A genus of about 400 species, most of which are native to the Old World tropics.

1. Clerodendrum aculeatum (L.) Schltdtl., Linnaea 6: 750. 1831. *Volkameria aculeata* L., Sp. Pl. 637. 1753. Fig. 203A–E.

Erect or climbing shrub to 3 m long; stems cylindrical, usually finely striate, puberulent, with 2 or 3, opposite or whorled spines at the nodes. Leaves opposite or ternate; blades 1.5–7 × 0.5–2.6 cm, oblong to elliptic-obovate, chartaceous, glabrous or puberulent, pitted, the apex acute or obtuse, the base narrowed, cuneate or rounded, the margins entire; petioles 2.5–10 mm long, puberulent. Cymes axillary, 2–6 cm long. Calyx greenish, bell-shaped, puberulent, 2.5–4 mm long, the lobes spreading; corolla white, trumpet-shaped, 14–27 mm long, the tube sometimes purplish without, the lobes spreading 9–14 mm wide, ciliate; filaments twice as long as the corolla, lax, pinkish; style as long as the filaments, erect or nearly so, purplish. Drupe ovoid to depressed-ovoid, 5–7 mm long, turning from green to brown and splitting in two at maturity.

DISTRIBUTION: A common shrub of dry and disturbed habitats. East End (A5140), Emmaus (A1994), Waterlemon Bay (1934). Also on Anegada, Jost van Dyke, St. Croix, St. Thomas, and Tortola; throughout the West Indies and in Venezuela.

COMMON NAME: chuc chuc.

Fɪɢ. 203. A–E. *Clerodendrum aculeatum*. A. Flowering branch and detail of stem node showing axillary spines. B. Flower. C. Anther, frontal and dorsal views. D. L.s. corolla with adnate filaments and pistil with detail of stigma. E. Drupe, top and bottom views. F–I. *Duranta erecta*. F. Flowering branch. G. Corolla and l.s. corolla showing pistil and stamens. H. Stamen, frontal and dorsal views. I. Drupe. J–M. *Lantana involucrata*. J. Fertile branch. K. Flower, top and lateral views. L. L.s. corolla showing stamens and pistil. M. Infructescence and individual drupe.

5. DURANTA L.

Erect shrubs, sometimes with axillary spines. Leaves simple, opposite or verticillate, short-petiolate, with entire or dentate margins, usually with resinous dots beneath. Flowers small, bisexual, pedicellate, in terminal racemes or panicles or seldom axillary racemes; calyx tubular to nearly bell-shaped, truncate, 5-plicate, 5-ribbed, enclosing the fruit; corolla 5-merous, nearly trumpet-shaped, with spreading unequal lobes; stamens 4, the filaments in 2 pairs of unequal length, inserted in the middle of the tube, included; ovary of 4 bilocular carpels, each locule with a single ovule, the style slender with a capitate stigma. Fruit drupaceous, yellowish, with 4 2-seeded pyrenes.

A genus of about 17 species, from tropical and subtemperate areas of the New World.

1. Duranta erecta L., Sp. Pl. 637. 1753.   Fig. 203F–I.

Duranta repens L., Sp. Pl. 637. 1753.

Duranta plumieri Jacq., Select. Stirp. Amer. Hist. 186, t. 176, fig. 76. 1763.

Shrub 2–3 m tall; stems nearly cylindrical or obscurely 4-angled, glabrous to finely pubescent, with axillary spines, 4–18 cm long. Leaves opposite, sometimes appearing whorled on short axillary branches; blades 1–3.5 × 0.6–1.8 cm, ovate-elliptic, ovate to obovate, chartaceous, glabrous, the apex acute to rounded, the base cuneate, the margins entire or serrate toward the apex; petioles 1.5–6 mm long, puberulent. Panicles 10–30 cm long, of 5 or more recurved or spreading branches 5–10 cm long. Calyx greenish, 4.5–6 mm long, puberulent between the ribs; corolla blue, violet, or white, 10–14 mm long, the limb 8–14 mm wide; stamens and style included. Drupe ovoid, 5–7 mm long; fruiting calyx orange, ovoid, narrowed toward apex.

DISTRIBUTION: An occasional shrub of dry and disturbed habitats. Brown Bay (A1880). Also on St. Croix, St. Thomas, and Virgin Gorda; throughout tropical and subtropical America.

6. LANTANA L.

Erect herbs or shrubs, sometimes with prickles. Leaves aromatic, simple, opposite or ternate, verticillate, petiolate, with entire, serrate, dentate, or crenate margins. Flowers bisexual, variously colored, sessile, subtended by a single bracteole, produced in axillary spikes or peduncled heads; calyx short-tubular, truncate, bilabiate or 4–5-toothed; corolla 4-merous, nearly trumpet-shaped, with spreading unequal lobes; stamens 4, the filaments in 2 pairs of unequal length, included; ovary bilocular, each locule with a single ovule, the style slender with an oblique stigma. Fruit drupaceous, with 2 1-seeded pyrenes.

A genus of about 50 species, native to the tropics of the New World, introduced into the Old World.

Key to the Species of *Lantana*

1. Corollas lavender to pale pink; plant unarmed; leaves crenate .... 2. *L. involucrata*
1. Corollas orange or yellow; stems sometimes with recurved prickles; leaves serrate.
    2. Leaves sparsely pubescent beneath, the hairs coarse, and bent at base .... 1. *L. camara*
    2. Leaves densely pubescent beneath, the hairs slender, straight .... 3. *L. urticifolia*

1. Lantana camara L., Sp. Pl. 627. 1753.

Lantana aculeata L., Sp. Pl. 627. 1753. *Lantana camara* L. var. *aculeata* (L.) Moldenke, Torreya **74:** 9. 1934.

Lantana scabrida Aiton, Hort. Kew. **2:** 352. 1789.

Shrub 1–2(–3) m tall, many-branched from base; stems 4-angled, usually with numerous recurved prickles at the angles, pubescence of simple and glandular hairs. Leaves opposite; blades 3–8 × 1.7–5 cm, ovate to oblong-ovate, chartaceous, scabrous above, sparsely pubescent beneath (the hairs coarse and bent at base), the apex acute to obtuse, the base unequal, rounded, truncate, or cordate, the margins serrate; petioles 5–15 mm long, pubescent, sometimes minutely spiny. Flowers in axillary heads subtended by involucral bracts; heads 10–18 × 17–25 mm; peduncles 1.5–5.5 cm long. Calyx greenish, 1–1.5 mm long; corolla turning from yellow to orange or rarely pink-purple with a yellow throat, the tube puberulent without, yellow, 10–13 mm long, slightly curved, the limb 3–5 mm wide; stamens and style included. Drupe ovoid to obovoid, turning from green to blackish, 3–6 mm long.

DISTRIBUTION: A common shrub of dry and disturbed habitats. Emmaus (A2826), Nanny Point (A2456), Susannaberg (A3385). Also on St. Croix, St. Thomas, and Tortola; apparently native to the Greater Antilles, introduced and naturalized throughout the tropics and subtropics of the world.

COMMON NAMES: red sage, wild sage, yellow sage.

2. Lantana involucrata L., Cent. Pl. II. 22. 1756.
Fig. 203J–M.

Lantana odorata L., Syst. Nat., ed. 12, **2:** 418. 1767.

Shrub 1–2(–3) m tall, many-branched from base; stems obscurely 4-angled, glabrescent. Leaves opposite; blades 1.5–5 × 0.9–4 cm, ovate to oblong-elliptic, coriaceous, scabrous above, densely pubescent beneath, the apex obtuse to rounded, the base cuneate to rounded, the margins crenate; petioles 4–15 mm long, pubescent. Flowers in axillary heads subtended by involucral bracts; heads 8–12 × 10–15 mm; peduncles 1–5 cm long. Calyx greenish, 1–1.5 mm long; corolla lavender to pale pink, pubescent without, 6–9 mm long, the tube glabrous at base, the limb 2.2–3 mm wide; stamens and style included. Drupe globose, pinkish-violet, 2.2–3 mm long.

DISTRIBUTION: A common shrub of dry and disturbed habitats. Emmaus (A2827), Hurricane Hole (A2425), Susannaberg (A5087). Also on Anegada, St. Croix, St. Thomas, and Tortola; throughout the West Indies, Florida, and from Mexico to northern South America.

COMMON NAME: sage.

3. Lantana urticifolia Mill., Gard. Dict., ed. 8. 1768.

Lantana crocea Jacq., Pl. Hort. Schoenbr. **4:** t. 473. 1804.

Lantana arida Britton, Bull. Torrey Bot. Club. **37:** 357. 1910.

Shrub or subshrub 0.5–2(–3) m tall, many-branched from base; stems 4-angled, usually with numerous recurved prickles at the angles, pubescence of simple and glandular hairs. Leaves opposite; blades 2–7 × 0.4–2 cm, ovate to oblong-lanceolate, chartaceous, scaberulous above, densely pubescent beneath (the hairs slender and straight), the apex acute to obtuse, the base unequal, rounded, truncate, or cuneate, the margins serrate; petioles 4–15 mm long, pubescent. Flowers in axillary heads subtended by involucral bracts; heads 9–12 × 15–18 mm; peduncles 2–5.5 cm long. Calyx greenish, 1.5 mm long; corolla turning from yellow to orange, the tube puberulent without, 8–10 mm long, slightly curved, the limb ca. 4 mm wide; stamens and style included. Drupe nearly globose, 4.5–5 mm long, blackish.

DISTRIBUTION: An occasional shrub of dry and disturbed habitats. Lameshur (A2741), Rosenberg (B255). Also on St. Thomas and Tortola; throughout the West Indies and from Mexico south to Argentina.

7. PRIVA Adans.

Herbs. Leaves simple, opposite, petiolate, and membranous. Flowers bisexual, pedicellate, solitary in the axils of minute bracts, on axillary or terminal racemes; calyx tubular, with 4 unequal teeth, enclosing the fruit; corolla trumpet-shaped or funnel-shaped, with a 2-lipped limb; stamens 4, the filaments in 2 pairs of unequal length, inserted in the tube, included; ovary 4- or 2-locular, the locules with a single ovule, the style short with a bilobed stigma. Fruit a dry, often woody schizocarp of 2 cocci.

A tropical genus of about 20 species.

REFERENCE: Moldenke, H. N. 1936. A monograph of the genus *Priva*. Repert. Spec. Nov. Regni Veg. **41:** 1–76.

1. Priva lappulacea (L.) Pers., Syn. Pl. **2:** 139. 1806.

Verbena lappulacea L., Sp. Pl. 19. 1753.

Fig. 204A–F.

Priva echinata Juss., Ann. Mus. Natl. Hist. Nat. **7:** 69. 1806, nom. illegit.

Erect or spreading herb to 60 cm tall; stems 4-angled, with uncinate and straight hairs along the angles. Leaf blades 3–7 × 1.3–4.3 cm, ovate to triangular, chartaceous, pilose above, puberulent beneath, the apex acute or acuminate, the base nearly truncate or cordate, the margins serrate; petioles 6–30 mm long, puberulent. Flowers few, in terminal and axillary, erect or lax, racemes, 9–20 cm long. Calyx greenish, 2–4 mm long, densely covered with uncinate hairs; corolla light violet, 4–5 mm long; stamens and style included. Fruit stipitate, nearly ellipsoid, ca. 2 mm long; cocci with 2 rows of minute, sharp spines; fruiting calyx inflated, ovoid, 5–6 mm long, completely enclosing the fruit.

DISTRIBUTION: An occasional herb of disturbed habitats. Coral Bay along Center Line Road (A3994), Great Cruz Bay (A2359), Lameshur (B507). Also on St. Croix and St. Thomas; native to tropical and subtropical America, introduced and naturalized into the Old World.

8. STACHYTARPHETA Vahl

Herbs or subshrubs. Leaves simple, opposite or alternate, petiolate, usually toothed at margins. Flowers bisexual, solitary in the axils of minute bracts, sessile or sunken in rachis of terminal, elongate spikes; calyx tubular, with 5 equal teeth, enclosing the fruit; corolla trumpet-shaped, the limb spreading, with 5 rounded lobes; stamens 2, included, the filaments inserted on upper half of tube; staminodes 2; ovary 2–locular, each locule with a single ovule, the style elongate with a nearly capitate stigma. Fruit a dry, often woody schizocarp of 2 mericarps. A genus of about 60 species, mostly of tropical America but introduced into the Old World tropics.

Key to the Species of *Stachytarpheta*

1. Plant and inflorescence essentially glabrous or with a few scattered slender hairs, especially at nodes
.... 1. *S. jamaicensis*
1. Plant and inflorescence pilose to strigose .... 2. *S. strigosa*

1. Stachytarpheta jamaicensis (L.) Vahl, Enum. Pl. **1:** 206. 1804. *Verbena jamaicensis* L., Sp. Pl. 19. 1753. *Valerianoides jamaicense* (L.) Kuntze, Revis. Gen. Pl. **2:** 509. 1891. Fig. 204G–J.

Stachytarpheta marginata Vahl, Enum. Pl. **1:** 207. 1804.

Erect herb or subshrub 30–40 cm tall, with pseudodichotomous branching; stems obtusely 4-angled, glabrous or with a few scattered slender hairs, especially at nodes. Leaf blades 2–10 × 1.3–4.2 cm, ovate, oblong, to obovate, chartaceous, glabrous or nearly so, the apex obtuse to acute, the base long-attenuate into a petiole or less often obtuse, the margins coarsely serrate; petioles 1–2.5 cm long. Spikes 19–26 cm long, 2.5–3 mm wide, terminal, erect, the furrows much narrower than the spike; bracteoles oblong to oblong-lanceolate, 5–5.5 mm long, essentially glabrous and aristate at apex. Calyx greenish, 6.5–7 mm long, bifid, glabrous; corolla light violet to lavender, 6–9 mm long, the tube lighter, slightly curved, the lobes to 4 mm long, rounded, glandular-pubescent without; stamens and style included. Fruit nearly cylindrical, 3.5–4 mm long, with a persistent elongate style projecting beyond the fruiting calyx; cocci united at maturity.

DISTRIBUTION: A common herb of disturbed open habitats. Emmaus (A2579), Lameshur (A2739), Lind Point (A2955). Also on St. Croix, St. Thomas, and Tortola; southern United States to northern South America including the West Indies, introduced and naturalized in other tropical regions.

COMMON NAME: verbain.

2. Stachytarpheta strigosa Vahl, Enum. Pl. **1:** 207. 1804. *Valerianoides jamaicense* (L.) Kuntze f. *strigosum* (Vahl) Kuntze, Revis. Gen. Pl. **2:** 510. 1891. *Valerianoides strigosum* (Vahl) Britton *in* Britton & P. Wilson, Bot. Porto Rico **6:** 144. 1925.

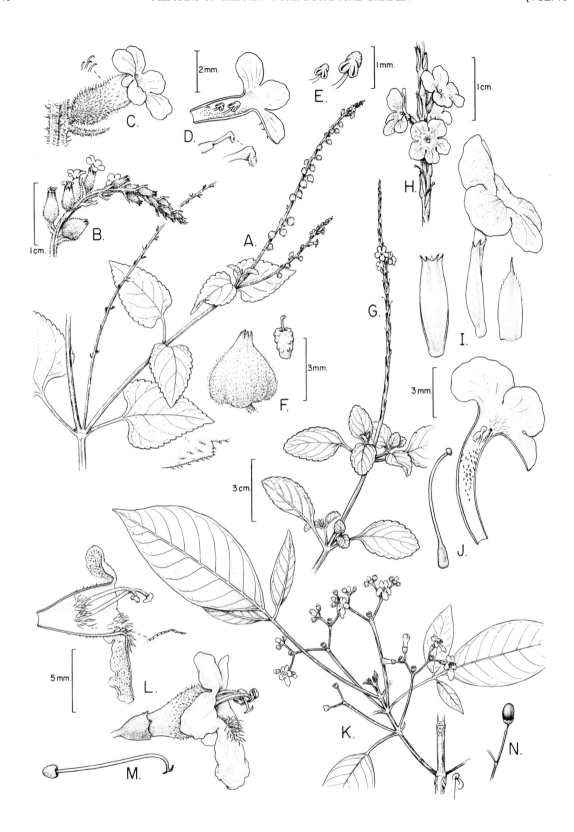

Fig. 204. A–F. *Priva lappulacea.* **A.** Fertile branch and detail of leaf margin. **B.** Inflorescence. **C.** Flower with subtending bracteole and detail of indument. **D.** L.s. corolla showing stamens, pistil, and detail of stigma. **E.** Dimorphic stamens. **F.** Accrescent calyx containing fruit and immature fruit. **G–J.** *Stachytarpheta jamaicensis.* **G.** Flowering branch. **H.** Detail of inflorescence. **I.** Calyx, flower, and bracteole. **J.** Pistil and l.s. corolla. **K–N.** *Vitex divaricata.* **K.** Flowering branch. **L.** L.s. corolla, detail of indument, and flower. **M.** Pistil. **N.** Drupe.

Erect herb or subshrub to 50 cm tall; stems obtusely 4-angled, pilose, especially when young. Leaf blades 2.5–7 × 2–4.2 cm, elliptic, ovate or oblong, chartaceous, pilose, the apex obtuse to rounded, the base long-attenuate to a petiole, the margins coarsely serrate; petioles 3–15 mm long. Spikes 10–30 cm long, 3–4 mm wide, terminal, erect, the furrows much narrower than the spike; bracteoles oblong to oblong-lanceolate, 6–6.5 mm long, pilose,

aristate at apex. Calyx greenish, ca. 6 mm long, pilose; corolla pale blue-violet, the tube ca. 7 mm long, the limb 4–6 mm wide, with rounded lobes; stamens and style included. Fruit nearly oblong, slightly compressed, 3–3.5 mm long, beaked at apex, covered by fruiting calyx; mericarps slightly separating at maturity.

DISTRIBUTION: An occasional herb of recently disturbed, open areas. Lameshur (B505). Also on St. Thomas and Tortola; Hispaniola and Puerto Rico.

9. VITEX L.

Trees, shrubs, or rarely woody vines, with obtusely 4-angled branches. Leaves palmately compound, 3–7-foliolate, seldom unifoliolate, opposite or ternate, petiolate, the leaflets entire or toothed at margins, petiolulate. Flowers bisexual, zygomorphic, in axillary cymes or terminal panicles; bracts usually smaller than the calyx; calyx bell-shaped to tubular, with 5 unequal teeth or 5-parted; corolla trumpet-shaped, the limb spreading, 2-lipped, with 5 rounded lobes; stamens 4, the filaments in 2 pairs of unequal length, inserted on tube, included or exserted; ovary 2-locular but becoming imperfectly 4-locular, each locule with a single ovule, the style elongate with a bifid stigma. Fruit a more or less fleshy drupe, subtended by an accrescent calyx.

A genus of about 250 species of tropical and temperate zones.

1. Vitex divaricata Sw., Prodr. 93. 1788. Fig. 204K–N.

Small tree to 20 m tall; bark light gray to beige, smooth; stems appressed-pubescent when young, obtusely 4-angled, grayish. Leaves opposite, trifoliolate or less often unifoliolate; leaflets 3–16 × 2–6 cm (the central one much larger than the lateral ones), elliptic or oblong-elliptic, chartaceous, puberulent on both surfaces, the lower surface with tufts of hairs on secondary vein angles, the apex obtuse to short-acuminate, the base slightly unequal, obtuse or acute, the margins entire; petioles 2–6 cm

long. Cymes axillary, 4–12 cm long, 3–4 times branched, each branch 6–31-flowered. Calyx greenish, 2–2.5 mm long, bell-shaped; corolla violet to bluish, 8–10 mm long, puberulent; stamens and style exserted. Fruit ellipsoid-ovoid, 8–10 mm long, black, subtended by a persistent cup-shaped calyx; 4-seeded.

DISTRIBUTION: An occasional tree of moist forests. Bordeaux (A2846). Also on St. Thomas and Tortola; Cuba, Hispaniola, Puerto Rico, Lesser Antilles, Trinidad, and Venezuela.

CULTIVATED SPECIES: *Petrea volubilis* L. has been collected from a plant persisting after cultivation.

92. Vitaceae (Grape Family)

Herbaceous or woody vines, with leaf-opposed tendrils, or seldom shrubs, herbs, or small trees. Leaves alternate, simple or palmately or pinnately compound, entire, lobed or with toothed margins; stipules minute, deciduous. Flowers small, actinomorphic, bisexual or less often unisexual, (3–)4–5(–7)-merous, in leaf-opposed or terminal cymes or panicles; calyx minute, cupular, or annular; corolla of distinct or basally connate, valvate petals, sometimes deciduous; stamens as many as the petals, the filaments distinct, the anthers opening by longitudinal slits; nectary disk annular, cupular, or glandular; ovary superior, sunken in the disk, 2-locular, of 2 united carpels, with basal placentation, usually with 2 ovules per locule, style single, with a discoid or capitate stigma, or less commonly the stigma sessile. Fruit a berry.

A family of about 11 genera and 700 species, mostly tropical and subtropical but with a few members extending into temperate zones.

Key to the Genera of Vitaceae

1. Inflorescences ascending and spreading, as wide or wider than long; plants fleshy, soft-stemmed; flowers 4-merous, with petals spreading at anthesis .... 1. *Cissus*
1. Inflorescences hanging, much longer than wide; stems becoming woody with age; flowers 5-merous, with petals apically coherent, falling off as a unit at anthesis .... 2. *Vitis*

1. CISSUS L.

Vines; stems often fleshy, with leaf-opposed tendrils. Leaves simple or 3-foliolate, petiolate, entire or toothed at margins; stipules deciduous. Flowers 4-merous, usually bisexual, actinomorphic, in umbelliform clusters in compound cymes opposite to a leaf; calyx bell-shaped, 4-toothed; corolla of distinct petals that spread or reflex at anthesis; stamens alternating with the petals; nectary disk cup-shaped, entire or 4-lobed; ovary 2-locular, the style elongate, stout. Berry globose, ovoid, or depressed-globose, juicy, blue to black at maturity.

A pantropical genus of about 350 species.

Key to the Species of *Cissus*

1. Leaves simple .... 3. *C. verticillata*
1. Leaves trifoliolate.
    2. Leaflets crenate to scattered-serrate above the middle; sepals and petals red .... 1. *C. obovata*
    2. Leaflets deeply serrate above the middle; sepals and petals yellowish green .... 2. *C. trifoliata*

1. Cissus obovata Vahl, Symb. Bot. **3:** 19. 1794.

Cissus caustica Tussac, Fl. Antill. **1:** 116, t. 16. 1808.

Slender liana 3–5(–15) m long, with numerous lateral branches; stems fleshy, grayish, usually zigzag-shaped, glabrous or with scattered hairs, especially at nodes. Leaves trifoliolate; leaflets 0.7–4 × 0.5–2.5 cm (the central leaflet much larger than the lateral ones), obovate, elliptic, to nearly rounded, chartaceous, glabrous or rarely pubescent, the apex rounded, mucronulate, the base cuneate to rounded, unequal on lateral leaflets, the margins crenate to scattered-serrate above the middle; petioles 0.8–5 cm long. Cymes opposite to leaves, the peduncles glabrous to puberulent, 1–3.5 cm long, many-branched; pedicels reddish-tinged. Calyx reddish, ca. 1 mm long; petals 1–2 mm long, reddish, triangular, deciduous; stamens ca. 1.5 mm long. Berry globose to globose-obovoid, black, ca. 4 mm long.

DISTRIBUTION: An occasional liana of moist to dry areas. Camelberg Peak (W293). Also on St. Croix and St. Thomas; Hispaniola, Puerto Rico, and the northern Lesser Antilles.

2. Cissus trifoliata (L.) L., Syst. Nat., ed. 10, **2:** 897. 1759. *Sicyos trifoliata* L., Sp. Pl. 1013. 1753.

Fig. 205E, F.

Cissus acida L., Sp. Pl., ed. 2, **1:** 170. 1762.

Slender liana, climbing or creeping, forming dense mats over surrounding vegetation, 1–3(–15) m long; stems slender, fleshy, cylindrical, reddish brown, puberulent or glabrous, swollen at nodes. Leaves trifoliolate; leaflets 2–3 × 1–2.6 cm (the central leaflet slightly larger than the lateral ones), ovate, obovate, or oblong, fleshy-chartaceous, glabrous, the apex obtuse or truncate, serrate, the base cuneate or shortly attenuate, the margins deeply serrate above the middle, the teeth with a sharp tip; petioles 1–3 cm long, slender; stipules lanceolate, 3.5–5 mm long, straw-

colored. Cymes opposite to leaves, the peduncles glabrous, green, 2–5 cm long, many-branched; pedicels green. Calyx green, 1.5–2 mm long; petals 2–2.5 mm long, yellowish to white, lanceolate-triangular, deciduous; disk 0.5–0.8 mm tall. Berry depressed-globose, 7–8 mm wide, turning from green to metallic blue to black, shiny.

DISTRIBUTION: A common weed of open disturbed, dry habitats. Great Lameshur Bay (A5446), Lind Point (A2693), Nanny Point (A2015). Also on St. Croix and St. Thomas; Bahamas, Cuba, Hispaniola, and Puerto Rico.

COMMON NAME: sorrel vine.

3. Cissus verticillata (L.) Nicolson & C. E. Jarvis, Taxon **33:** 727. 1984. *Viscum verticillatum* L., Sp. Pl. 1023. 1753.

Fig. 205A–D.

Cissus sicyoides L., Syst. Nat., ed. 10, **2:** 897. 1759.

Herbaceous, climbing or creeping vine 2–10 m long; stems fleshy, cylindrical to flattened on older portions, grayish, glabrous, swollen at nodes, sometimes zigzag-shaped. Leaves simple; blades 5–12 × 3.8–6.5 cm, ovate to oblong-lanceolate, fleshy-chartaceous, glabrous, the apex acute to rounded, the base cuneate, truncate, or cordate, the margins minutely toothed; petioles 2–5 cm long; stipules auriculate, 2.5–3.5 mm long, straw-colored. Cymes opposite to leaves, the peduncles glabrous, 2–4 cm long, many-branched; flowers including pedicels green to yellowish in some individuals, while others with reddish pedicels and red or pinkish flowers. Calyx green or reddish, 0.7–1.0 mm long; petals 2–2.5 mm long, yellowish or pinkish, oblong-lanceolate, deciduous; disk 0.5–0.8 mm tall, bright yellow. Berry depressed-globose, 7–10 mm wide, turning from green to black, shiny.

DISTRIBUTION: A common weed of open disturbed, dry to moist habitats. Center Line Road (A2420), Emmaus (A2820), Susannaberg (A2088). Also on St. Croix, St. Thomas, and Tortola; southern United States to northern South America, including the West Indies.

COMMON NAME: pudding vine.

2. VITIS L.

Woody vines with leaf-opposed tendrils. Leaves simple, often palmately lobed, toothed; stipules deciduous. Flowers 5-merous, bisexual or unisexual, actinomorphic, produced in umbelliform clusters in paniculate inflorescences opposite to a leaf; calyx short, 5-lobed; corolla of distinct petals apically coherent and deciduous as a unit at anthesis; stamens alternating with the petals; nectary 5-lobed; ovary 2-locular, the style short, conical. Berry globose or ellipsoid, juicy, green, blue, or purple at maturity.

A genus of about 60 species, mostly temperate in distribution, with 1 species (the European wine grape) widely cultivated for its edible fruits.

1. Vitis tiliifolia Roem. & Schult., Syst. Veg. **5:** 320. 1819.

Fig. 205G–J.

Vitis caribaea DC., Prodr. 1: 634. 1824.

Liana 10–35 m long; stems cylindrical, producing copious watery sap when cut; bark dark brown, rough, fissured. Leaves simple or 3-lobed; blades 6.5–14 × 6.5–12.5 cm, widely ovate, chartaceous, densely woolly pubescent beneath, slightly so above, the apex acute to short-acuminate, the base cordate to nearly truncate, the margins toothed; petioles 2.5–8 cm long; stipules

early deciduous. Panicles hanging, the axes woolly-tomentose to glabrous, 9–17 cm long. Calyx green, disk-shaped, the sepals ca. 0.2 mm long; petals 1.5–2.2 mm long, yellowish green, oblong-obovate; disk ca. 0.2 mm tall. Berry globose, 5–10 mm wide, turning from green to light violet. Seeds 1 or 2 per fruit, compressed, circular in outline, ca. 4 mm long.

DISTRIBUTION: An uncommon liana of wet and seepage areas. Bay Gut (A2128), Reef Bay Gut (W831). Also on St. Croix (according to Britton & Wilson, 1924); Mexico to northern South America, including the West Indies.

COMMON NAME: wild grape.

93. Zygophyllaceae (Lignum Vitae Family)

Herbs, shrubs, or small trees. Leaves opposite or less commonly alternate, pinnately compound without distal leaflets, trifoliolate or less often simple, entire or pinnately dissected; stipules well developed, sometimes modified into spines or seldom wanting. Flowers small to medium, actinomorphic to slightly zygomorphic, bisexual or less often unisexual, (4–)5(–6)-merous, in axillary cymes or seldom racemes or solitary; calyx of distinct or basally connate sepals; corolla of distinct petals or rarely wanting; stamens twice as many as the petals, or less often as many as or 3 times as many as the petals, the filaments distinct and basally glandular, the anthers

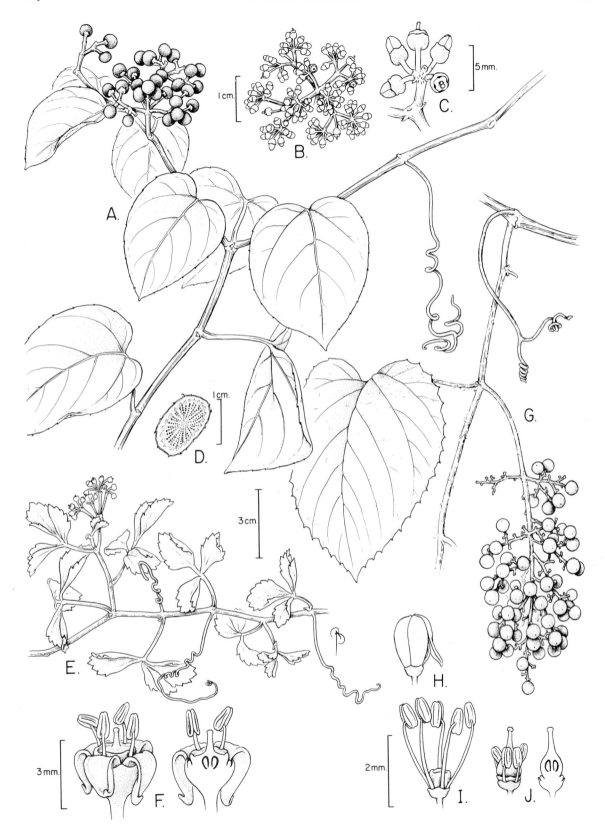

FIG. 205. A–D. *Cissus verticillata*. **A.** Fruiting branch. **B.** Inflorescence. **C.** Cymose unit of inflorescence. **D.** C.s. mature stem. **E, F.** *Cissus trifoliata*. **E.** Flowering branch. **F.** Flower and l.s. flower. **G–J.** *Vitis tiliifolia*. **G.** Fruiting branch. **H.** Flowerbud. **I.** Staminate flower. **J.** Pistillate flower and l.s. pistillate flower.

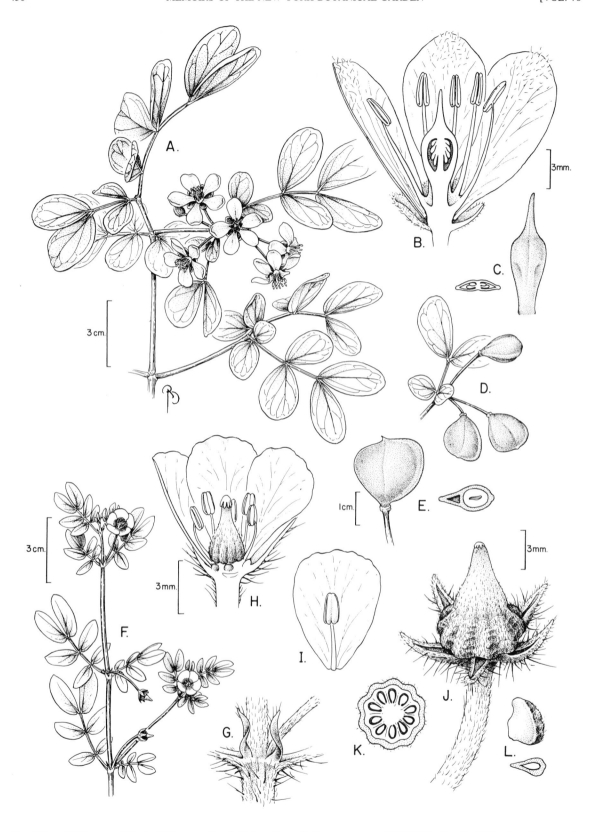

FIG. 206. A–E. *Guaiacum officinale*. **A.** Flowering branch. **B.** L.s. flower. **C.** C.s. ovary and pistil. **D.** Fruiting branch. **E.** Capsule and c.s. capsule. **F–L.** *Kallstroemia pubescens*. **F.** Fertile branch. **G.** Detail of stem node showing stipules at base of leaf petioles. **H.** L.s. flower. **I.** Petal and opposite stamen. **J.** Schizocarp with persistent calyx. **K.** C.s. schizocarp. **L.** Mericarp and c.s. mericarp.

opening by longitudinal slits; nectary disk annular, sometimes modified into an androgynophore; ovary superior, plurilocular, of (2–)5(–6) united carpels, with axile placentation, with 1 to several ovules per locule, the styles terminal, connate, with a capitate or lobed stigma. Fruit a capsule, a schizocarp, or less often a berry or a drupe.

A family of about 30 genera and 250 species, mostly of dry tropical and subtropical regions.

Key to the Genera of Zygophyllaceae

1. Trees; petals blue; fruit flattened, cordate .... 1. *Guaiacum*
1. Prostrate or decumbent herbs; petals yellow; fruit conical, of 8–12 mericarps .... 2. *Kallstroemia*

1. GUAIACUM L.

Small trees with hard wood. Leaves opposite, pinnately compound with (1–)2–3 pairs of leaflets; stipules minute, deciduous. Flowers 4–5-merous, bisexual, actinomorphic, long-pedicellate, in terminal umbelliform clusters; sepals bluish or violet; petals bluish to purple, slightly clawed; stamens twice as many as the petals, ascending; disk inconspicuous; ovary 2–5-locular, each locule with 8–10 parietal ovules, the style stout. Fruit a septicidal capsule, 2–5-winged or lobed; seeds solitary, covered by a bright, fleshy aril.

A genus of about 6 species, from tropical America.

1. Guaiacum officinale L., Sp. Pl. 381. 1753.

Fig. 206A–E.

Tree 5–10 m tall; bark dark brown, peeling off in thickened plates. Leaves pinnately compound; leaflets opposite, (1–)2–3 pairs, the distal leaflets larger than the basal ones, 1.5–4.2 × 1–2.6 cm, obovate to oblanceolate, coriaceous, glabrous, the apex obtuse to rounded, the base asymmetrical, obtuse-cuneate, the margins entire; petioles 0.7–2 cm long. Sepals purplish-tinged, oblong, ca. 3.5–4 mm long, finely pubescent on both surfaces;

petals 8–9 mm long, pale violet or bluish, oblanceolate, spreading; filaments light violet, ascending; ovary reddish purple, compressed, stipitate, the style whitish. Capsule cordate, flattened, orange, leathery, 1.5–1.7 cm long, apiculate at apex, the valves opening to expose the hanging seeds. Seeds 1 per locule, with a bright red fleshy aril.

DISTRIBUTION: An occasional tree of dry coastal forests, sometimes planted in parks and gardens. Cruz Bay (A2906). Also on Anegada, St. Croix, and St. Thomas; throughout the West Indies, Colombia, and Venezuela.

COMMON NAMES: lignum vitae, liki wiki, pockenholt.

2. KALLSTROEMIA Scop.

Prostrate, decumbent, or ascending annual herbs. Leaves opposite, evenly pinnate; leaflets oblique and opposite; stipules opposite, linear or subulate. Flowers 5–6-merous, bisexual, actinomorphic, long-peduncled, solitary, axillary; sepals distinct, greenish; petals yellow, obovate, deciduous; stamens twice as many as the petals, ascending, the ones alternating with the petals bearing a gland at base; ovary sessile, 8–12-locular, each locule with a single hanging ovule, the style short or long, stout. Fruit a schizocarp of 8–12 indehiscent mericarps.

A genus of about 23 species with disjunct distribution in tropical and subtropical areas of the New World and Australia.
REFERENCE: Porter, D. M. 1969. The genus *Kallstroemia* (Zygophyllaceae). Contr. Gray Herb. **198:** 41–153.

Key to the Species of *Kallstroemia*

1. Ovary and fruits glabrous .... 1. *K. maxima*
1. Ovary and fruits strigillose (appressed-pubescent) .... 2. *K. pubescens*

1. Kallstroemia maxima (L.) Hook. & Arn., Bot. Beechey Voy. 282. 1838. *Tribulus maximus* L., Sp. Pl. 386. 1753.

Prostrate herb, to 1 m long, many-branched from a subwoody base; stems reddish- or pinkish-tinged, cylindrical, striate. Leaflets opposite, in 3–4 pairs, the distal leaflets much larger than the basal ones, 0.7–2 × 0.4–0.9 cm, inequilateral, oblong-oblanceolate or oblong-lanceolate, chartaceous, sparsely appressed-pubescent to glabrescent, the apex obtuse, the base asymmetrical, obtuse-rounded, the margins entire, ciliate; petioles 0.5–1 cm long, appressed-pubescent; stipules subulate, 5 mm long, straw-colored. Flowers solitary, axillary; peduncles 1–4 cm long. Sepals green, lanceolate, 4–5 mm long, strigose; petals 7–8 mm long, orange-yellow, greenish at base, obovate, spreading to slightly reflexed; filaments greenish, ascending, much shorter than the petals; ovary green, turbinate-ovoid, the style conical. Schizocarp conical, 6–

6.5 mm long, 10-lobed; mericarps tuberculate, glabrous, indehiscent.

DISTRIBUTION: A common roadside weed of open, disturbed, dry areas. Enighed (A4270), Lind Point (A3162). Also on St. Croix, St. Thomas, and Tortola; Texas to northern South America including the West Indies.

COMMON NAMES: centipede root, longlo.

2. Kallstroemia pubescens (Don) Dandy *in* Keay, Kew Bull. **10:** 138. 1955. *Tribulus pubescens* Don, Gen. Hist. **1:** 769. 1831.

Fig. 206 F–L.

Kallstroemia caribaea Rydb., N. Amer. Fl. **25:** 111. 1910.

Prostrate herb to 50 cm long, many-branched from a subwoody taproot; stems reddish- or pinkish-tinged, cylindrical, striate, pubescent. Leaflets opposite, 3 pairs, the distal leaflets much larger than the basal ones, 0.5–2.2 × 0.4–1.2 cm, inequilateral, oblong-oblanceolate, chartaceous, sparsely appressed-pubescent to glabrescent, the apex obtuse, the base asymmetrical, cuneate-rounded, the margins entire, ciliate; petioles 0.5–1 cm long, appressed-pubescent; stipules subulate, 4 mm long, ciliate. Flowers solitary, axillary; peduncles 1–2.5 cm long. Sepals green, lanceolate, 4–5 mm long, strigose; petals 5–6 mm long, pale yellow, with a bright yellow base, obovate, spreading; filaments yellowish, ascending, much shorter than the petals; ovary green, turbinate-ovoid, strigillose, the style conical. Schizocarp conical, 6–6.5 mm long, 10-lobed; mericarps tuberculate, strigillose, indehiscent.

DISTRIBUTION: A common roadside weed of open, disturbed, dry areas. Fish Bay (A2814, A2381). Also on St. Croix, St. Thomas, and Tortola; Florida, Mexico to northern South America including the West Indies.

DOUBTFUL FAMILY RECORD: *Polygala hecatandra* Urb. (Polygalaceae) was reported by Woodbury and Weaver (1987) as occurring on St. John. but no specimens have been located and numerous attempts to find the plant on St. John have failed.

CULTIVATED FAMILIES: The families Bixaceae and Punicaceae are represented on St. John by a few cultivated individuals of *Bixa orellana* L. and *Punica granatum* L., respectively. As far as I am aware, these species are not naturalized on the island.

MONOCOTYLEDONS

Key to the Families of Monocotyledons

1. Plants aquatic or marine, submersed or floating.
  2. Plants floating in fresh to brackish waters.
    3. Plant with numerous leaves in a rosette at base, 3–12 cm long .... 3. *Araceae* (*Pistia*)
    3. Plants platelike, 1–2.5 mm long.................... 15. *Lemnaceae*
  2. Plants submersed in marine or brackish environment.
    4. Leaves strap-shaped, 4–12 mm wide .... 13. *Hydrocharitaceae*
    4. Leaves linear or cylindrical, to 2 mm wide.
      5. Leaves linear, with a conspicuous midvein, narrowly acuminate at apex .... 18. *Potamogetonaceae*
      5. Leaves cylindrical, without midvein, truncate to concave at the expanded apex (in ours) .... 9. *Cymodoceaceae*
1. Plants terrestrial or epiphytic.
  6. Plants woody, with a single trunk and a crown of large (>75 cm long) leaves (palmlike).................... 4. *Arecaceae*
  6. Plants herbaceous or woody; stem short and covered by a basal rosette of leaves, or the stem elongate and branched, with leaves scattered along the stem.
    7. Leaves in a rosette or rosulate at base.
      8. Leaves chartaceous to rigid-coriaceous .... 6. *Bromeliaceae*
      8. Leaves fleshy, thickened.
        9. Leaves 30–60 cm long, with a gelatinous core, not fibrous; scapes to 1 m tall .... 5. *Asphodelaceae*
        9. Leaves 90–275 cm long, with a fibrous core; scapes 5–7 m tall .... 1. *Agavaceae* (*Agave*)
    7. Leaves alternate, opposite, or spirally arranged.
      10. Plant with a definite woody stem; leaves spiny-tipped at apex .... 1. *Agavaceae* (*Yucca*)
      10. Plant herbaceous; leaves not spiny at apex.
        11. Plants climbing (vines).
          12. Plant monoecious, climbing aided by adventitious roots; leaves with a stout midvein and secondary pinnate veins; inflorescence a spadix, subtended by a spathe .... 3. *Araceae* (*Philodendron, Syngonium*)
          12. Plants dioecious, twining, or climbing by tendrils; leaves with 3–9 arcuate midveins; inflorescence spikelike, racemose, paniculate, or umbellate, not subtended by a bract.
            13. Plants with stipular tendrils; stems usually spiny; inflorescence umbellate; fruit a berry .... 19. *Smilacaceae*
            13. Plants twining, without tendrils; stems seldom spiny; inflorescence a spike, raceme, or panicle; fruit a chartaceous, 3-winged capsule .... 11. *Dioscoreaceae*
        11. Plants erect, prostrate, or decumbent, not climbing.
          14. Inflorescence a spadix subtended by a spathaceous bract.................... 3. *Araceae*
          14. Inflorescence a spike, a cyme, a raceme, or a panicle not subtended by a spathaceous bract.
            15. Flowers naked, perianth reduced to bristles or scales.
              16. Flowers spirally imbricate or less often distichous, subtended by a single scale; leaf-sheath usually closed; stems triangular or cylindrical .... 10. *Cyperaceae*
              16. Flowers distichously arranged, subtended by a pair of scales; leaf-sheath usually open; stems cylindrical .... 17. *Poaceae*

15. Flowers showy, either the sepals or the petals large, not reduced to bristles or scales.
   17. Ovary superior.
      18. Inflorescence cymose, usually enclosed by a folded, boat-shaped, spathe-like bract; flower with distinct calyx and corolla, free or basally connate .... 8. *Commelinaceae*
      18. Inflorescence racemose or paniculate, not enclosed by a boat-shaped bract; flower perianth tubular or funnel-shaped .... 12. *Dracaenaceae*
   17. Ovary inferior.
      19. Neither leaves nor stem swollen at base; plant rhizomatous .... 7. *Cannaceae*
      19. Leaves swollen at base into a bulb or corm, or the stem swollen at base into a pseudobulb.
         20. Flowers zygomorphic (in ours); anthers usually 1, sessile near the apex of a column; plants commonly epiphytic, less often epilithic or terrestrial... 16. *Orchidaceae*
         20. Flowers actinomorphic; stamens 6, with long, distinct or basally connate filaments; plants terrestrial.
            21. Perianth segments >6.5 cm long; fruit a capsule; leaves swollen at base into a bulb... 2. *Amaryllidaceae*
            21. Perianth segments to 1.5 cm long; fruit a berry; leaves not swollen .... 14. *Hypoxidaceae*

1. Agavaceae (Century Plant Family)
By M. T. Strong

Large perennial herbs, rhizomatous, stemless, short-stemmed or branching and treelike. Leaves borne in spiral rosettes at base or near branch tips; blades lanceolate, linear, or subulate, thick and fleshy, rigid or leathery, fibrous internally, the apex frequently bearing a sharp spine, the margins prickly or entire; stipules wanting. Inflorescence terminal, scapose, paniculate, racemose, or spikelike. Flowers bisexual or less often unisexual; perianth large, actinomorphic or somewhat zygomorphic, 6-parted, separate or united basally, greenish, whitish, or yellow to orange; stamens 6, the filaments distinct, adnate to the base of the tepals, the anthers introrse, longitudinally dehiscent; ovary inferior or superior, 3-locular, with 1 to many ovules per locule, the style simple, slender with a capitate or trilobate stigma. Fruit a capsule or sometimes a berry; seeds several to many, flattened and winglike, crescent-shaped or somewhat circular.

Approximately 20 genera and 670 species distributed in tropical and subtropical regions of the world with the center of distribution in the New World.

Key to the Genera of Agavaceae

1. Leaves in a rosette at base, the margins frequently prickly; ovary inferior; stamens longer than the perianth... 1. *Agave*
1. Leaves spirally arranged on an elongate stem, the margins smooth or finely toothed; ovary superior; stamens shorter than the perianth .... 2. *Yucca*

1. AGAVE L.

Robust perennial herbs, short-stemmed or stemless, frequently producing suckers at base and sometimes plantlets in the inflorescence, monocarpic or polycarpic, with hard, fibrous roots. Leaves large, succulent or fibrous, with a stiff, sharp spine at apex, the margins armed with prickles or sometimes smooth. Inflorescence paniculate, racemose, or spikelike, with a bracteate scape or scapose, the flowers borne in umbellate or cymose clusters. Flowers large, typically protandrous; perianth yellow or green, funnelform or tubular, the segments erect, rarely reflexed, imbricate in the bud; stamens long-exserted, the anthers attached at the middle; style elongate with a headlike, 3-lobed stigma; ovary inferior, the ovules in 2 rows per cell. Fruit a many-seeded, loculicidal capsule; seeds flattened, black. Approximately 300 species distributed from the southern United States south to Mexico, Central America, West Indies, and South America.

REFERENCES: Gentry, H. S. 1982. Agaves of continental North America. University of Arizona Press, Tucson; Trelease, W. 1913. *Agave* in the West Indies. Mem. Natl. Acad. Sci. 11: 1–299.

Key to the Species of *Agave*

1. Mature leaves armed with well-developed prickles; leaves 13–25 cm wide.
  2. Flowers 6–10 cm long; filaments 6–9 cm long; prickles (3–)5–10(–11) mm long, borne on raised un-

dulations of leaf margin; plants typically producing suckers at base; plantlets usually absent in inflorescence .... A. americana (cultivated)
   2. Flowers 4.5–5.5 cm long; filaments 3.5–4 cm long; prickles 3–5(–6) mm long, borne along relatively straight edge of leaf margin; plants not producing suckers at base; plantlets usually present in inflorescence .... 1. A. missionum
1. Mature leaves unarmed or with few minute prickles; leaves 7–12 cm wide..................... A. sisalana (cultivated)

1. Agave missionum Trel., Mem. Natl. Acad. Sci. **11:** 37. 1913.   Fig. 207A–E.

Agave portoricensis Trel., Mem. Natl. Acad. Sci. **11:** 38. 1913.

Plant stemless, not cespitose. Leaves lanceolate to broadly lanceolate, 90–275 × 13–23 cm, dark green or tinged with gray, somewhat glossy, occasionally pleated, acuminate to spine-tipped apex; spine awl-shaped, 1.5–2.5 cm long, straight or slightly recurved, brown, smooth, decurrent and dorsally produced into the green tissue; prickles brown or blackish, 3–5(–6) mm long, triangular, borne along relatively straight edge of margins. Inflorescence an elongate panicle, 5–7 m tall, bulbiferous; branches nearly ascending, smooth, bearing erect to ascending cymes of flowers with pedicels 15–20 mm long; bracts deltoid. Flowers yellow or greenish yellow, 4.5–5.5 cm long; perianth oblong-fusiform, fused near base, the tube ca. 7 mm deep; stamens elongate, exserted; anthers falcate, 1.4–1.7 cm long; style 4–5 cm long, with 3-lobed stigma. Capsule broadly oblong-elliptic or somewhat turbinate, 2.5–3.5 × 1.8–2.5 cm. Seeds flattened, irregularly and obtusely triangular, 6–9 × 5–6 mm, thin, black, intermixed with whitish, abortive ones.

DISTRIBUTION: Common along coastal thickets among rocky outcrops. Lameshur (B497). Also on Anegada, St. Thomas, Tortola, and Virgin Gorda; Puerto Rico (including Culebra and Vieques).

COMMON NAMES: century plant, karata.

CULTIVATED SPECIES: *Agave americana* L., the century plant, and *Agave sisalana* Perrine, both probably native to Mexico, are frequently planted in the Virgin Islands but have not become naturalized on St. John.

2. YUCCA L.

Large treelike plants, usually with a woody trunk, simple or branched. Leaves crowded at the apex of the trunk or branches, spirally arranged, linear-lanceolate, thick and rigid or sometimes flaccid, usually spine-tipped, the margins filiferous, minutely denticulate or entire. Inflorescence a large, terminal panicle. Flowers white or creamy; perianth 6-parted, bell-shaped, the segments lanceolate or ovate-lanceolate, frequently fleshy, nearly distinct to base; stamens 6, nearly free, shorter than the perianth, with thickened filaments, the anthers short, sagittate, basifixed; ovary superior, sessile or short-stipitate, 3-locular, with numerous to many ovules, the style short, stout, the stigma 3- or 6-lobed. Fruit a dry, dehiscent capsule, 6-valved above, or indehiscent and spongy or pulpy; seeds black, flattened.

Approximately 40 species with the center of distribution in the southern United States and Mexico, also occurring in Central America, West Indies, and South America.

REFERENCE: Trelease, W. 1902. The Yucceae. Annual Rep. Missouri Bot. Gard. 13: 27–133.

1. Yucca aloifolia L., Sp. Pl. 319. 1753.   Fig. 207F–H.

Yucca guatemalensis sensu Woodbury & Weaver, 1987, non Baker, 1872.

Plant unbranched (1–2 m tall) or branched (to 10 m tall). Leaves many, forming tight spirals; blades 30–75 × (1.8–)2–5 (–7) cm, linear, flat, rigid, thickened, dark green, with a sharp, cylindrical, brown or blackish spine at apex, the margins entire to minutely denticulate. Inflorescence a large, terminal panicle borne just above the leaves, 20–70 cm long. Flowers numerous; perianth creamy-white, sometimes tinged with green or purple towards base, fleshy, 3–5.8 cm long; ovary short-stipitate. Fruit an indehiscent capsule, oblong-ellipsoid, 3.8–6(–8) × 1.7–2.5(–4) cm, blackish, with dark purple pulp, the papery core wanting. Seeds round or oval, flattened, 5–6 × 6–7 mm, black and shiny.

DISTRIBUTION: Introduced, commonly planted along roadsides, becoming naturalized in open, disturbed areas. Emmaus (A2008). Also in the southern United States, Mexico, Central America, West Indies.

COMMON NAMES: Spanish bayonet, Spanish dagger.

2. Amaryllidaceae (Amaryllis Family)

By M. T. Strong

Perennial or biennial herbs with subterranean bulbs, rarely rhizomatous; stems short, herbaceous. Leaves primarily basal, distichously arranged, sheathing at base; blades flat and dorsiventral, linear to broadly elliptic, entire, usually glabrous. Inflorescence umbel-like, comprising 1 to several helicoid cymes or flowers solitary, borne on a simple, glabrous scape, subtended by 2–8 free or basally connate involucral bracts. Flowers bisexual, actinomorphic, rarely zygomorphic, often showy; perianth of 6 petaloid tepals in 2 whorls, free or connate into a tube, white, yellow, purple, or red; stamens 6(–18), the filaments inserted below the ovary or in the perianth tube, narrow and flattened, free, or expanded and connate at base into a staminal corona structure, or bearing a subulate, lateral, stipule-like appendage distally on each side of the anther, the anthers elongate, 2-locular, versatile or basifixed, longitudinally dehiscent or rarely opening by apical pores; ovary inferior, 3-carpellate, 3-locular with distinct septal nectar grooves, the ovules 2 to many in each locule, centrally inserted, the style simple, slender, the stigma punctiform, capitate or trilobate. Fruit a 3-locular capsule or indehiscent fleshy berry; seeds few to many, globose, ellipsoidal, or ovoid, or flattened.

A family of 50 genera and approximately 860 species, widely distributed throughout both hemispheres.

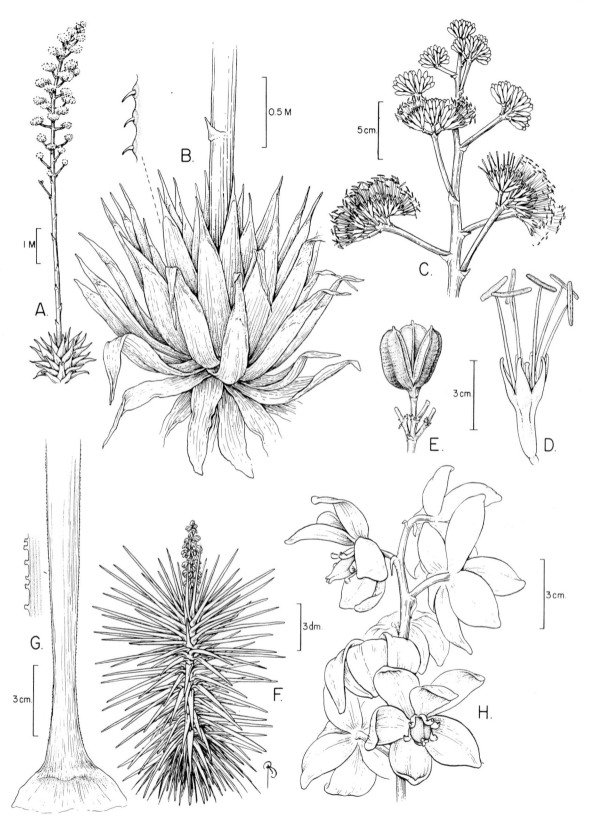

Fɪɢ. 207. A–E. *Agave missionum.* **A.** Habit. **B.** Basal rosette of leaves and detail of leaf margin. **C.** Upper portion of panicle. **D.** Flower. **E.** Dehisced capsule. **F–H.** *Yucca aloifolia.* **F.** Habit. **G.** Basal portion of leaf and detail of margin. **H.** Upper portion of panicle with open flowers.

Key to the Genera of Amaryllidaceae

1. Filaments free to base .... 1. *Crinum*
1. Filaments basally fused into a membranous cup .... 2. *Hymenocallis*

1. CRINUM L.

Plants bearing subterranean bulbs with short to long sheathing necks. Leaves succulent, linear or strap-shaped, somewhat broadened toward apex, spreading or arching, the margins entire or toothed. Inflorescence an umbellate cluster of few to many flowers subtended at base by 2 large, broad, spathelike bracts, the scape solid, erect, green or brightly colored. Flowers large, showy, short-pedicellate; perianth white, rose, or crimson, funnel-shaped or trumpet-shaped, with a long, cylindrical, straight or curved tube subtended by narrow bracts at base, the lobes linear, lanceolate, or oblong, subequal; stamens spreading or declinate, the filaments long, filiform, inserted in the throat of the perianth tube, the anthers linear, versatile; ovary globose to oblong with 2 to several ovules in each locule, the style long, filiform, the stigma minute, capitate. Fruit a subglobose, tardily dehiscent capsule; seeds large, bulbiform, fleshy, green, with very thick endosperm.

A genus of approximately 130 species, widely distributed in tropical, subtropical, and warm-temperate regions worldwide.

REFERENCE: Hannibal, L. S. 1972. Garden Crinums. Bull. Louisiana Soc. Hort. Res. 3(5): 219–322.

1. **Crinum zeylanicum** (L.) L., Syst. Nat., ed. 12, **2:** 236. 1767. *Amaryllis zeylanicum* L., Sp. Pl. 293. 1753. *Crinum latifolium* L. var. *zeylanicum* (L.) Hook.f. *in* Trimen, Handb. Fl. Ceylon **4:** 272. 1898. Fig. 208A, B.

Perennial herb, with a bulbous base; bulb ovoid to subglobose, 9–13 × 5–11 cm, the neck short; roots stout, coarse, 3–5 mm wide. Leaves many; blades 80–125 × 2.8–5.5 cm, strap-shaped, reflexed, finely veined, septate, acuminate to apex, the margins with remote, minute, blunt teeth. Scape 25–50 cm long, glabrous, reddish-tinged; spathelike bracts deltate, membranous, 6–7 × 2–3 cm. Flowers 5–10, subsessile or short-petiolate; perianth segments oblong-lanceolate, 8–12 × 2–3 cm, white with a pink or reddish median stripe, the tube curved at maturity, 8–13 cm long, 3–4 mm wide; anthers 1–2 cm long, greenish, the filaments 6.5–8 cm long, whitish or pink; style 15–20 cm long. Fruit not seen.

DISTRIBUTION: Occasional along roadsides or in disturbed secondary vegetation. Road to Ajax Peak (A5112). Also on St. Thomas; native of India and Sri Lanka, cultivated in tropical regions but frequently escaping and becoming naturalized.

COMMON NAME: spider lily.

2. HYMENOCALLIS Salisb.

Tall plants bearing subterranean bulbs. Leaves basal, linear or lanceolate, sessile or petiolate. Inflorescence an umbellate cluster of few to many flowers subtended by 2 to several spathelike, scarious bracts, the scape solid, angled or compressed. Flowers large, showy, sessile or short-pedicellate; perianth white, greenish, or red and yellow, with a long, cylindrical, straight or curved tube, subtended by 2 membranous, narrow bracts, the lobes linear, spreading or recurved, subequal; stamens inserted in the throat of the perianth tube, the filaments basally fused into a cup, the anthers linear, versatile; ovary with 2 to several ovules in each locule, the style filiform, long-exserted, the stigma minute, capitate. Fruit a fleshy capsule; seeds usually 1 or 2 per locule, large, fleshy, green.

A genus of 55 species occurring in the temperate southeastern United States, Central America, West Indies, and northern South America.

REFERENCE: Sealy, J. R. 1954. Review of the genus *Hymenocallis*. Kew Bull. 1954: 201–240.

Key to the Species of *Hymenocallis*

1. Leaf blades linear to linear-oblong, long-tapering at the base, (2.8–)4–9 cm wide .... 1. *H. caribaea*
1. Leaf blades broadly elliptic to broadly oblong-elliptic, distinctly petiolate at base, 7–15 cm wide ..... 2. *H. speciosa*

1. **Hymenocallis caribaea** (L.) Herb., Bot. Reg. **7(App.):** 44. 1821. *Pancratium caribaeum* L., Sp. Pl. 291. 1753, emend. Gawl., 1805. Fig. 208C, D.

Pancratium declinatum Jacq., Select. Stirp. Amer. Hist. 99. 1763. *Hymenocallis declinata* (Jacq.) M. Roem., Fam. Nat. Syn. Monogr. **4:** 171. 1847.

Perennial herb with bulbous base; bulb ovoid to subglobose, 4.5–7 × 4.2–9 cm, the neck short; roots stout, coarse, 2–3 mm wide. Leaf blades 23–60(–90) × (2.8–)4–9 cm, linear to linear-oblong, finely veined, septate, acute at apex, the margins smooth. Scapes 25–60 cm long, glabrous, green; spathelike bracts ovate to ovate-lanceolate, membranous, 4–6 × 1.8–3 cm. Flowers 8–20, fragrant, subsessile or short-petiolate; perianth segments linear, 7.5–12 × 0.2–0.5 cm, white, the tube 4–9 cm long; staminal cup funnel-shaped, 2–2.8 cm long; anthers 0.9–1.9 cm long, greenish, the filaments 5.7–9.5 cm long; ovary 8–20 mm long; style 14–20 cm long. Fruit ovate to oblong-ovate or subglobose, 2–2.5 cm long, 1–1.5 cm wide.

DISTRIBUTION: Occasional in disturbed areas and along roadsides. Bordeaux (A5130). Also on St. Croix and St. Thomas; Hispaniola, Puerto Rico (including Culebra), Jamaica, and the Lesser Antilles.

COMMON NAME: ladybug, white lily.

2. **Hymenocallis speciosa** (Salisb.) Salisb., Trans. Hort. Soc. London **1:** 340. 1812. *Pancratium speciosum* Salisb., Trans. Linn. Soc. London **2:** 73. 1794.

Bulbous herb; roots stout, 2–3 mm wide. Leaves 7–9, rosulate; blades 60–90 × 7–15 cm, broadly elliptic to broadly oblong-elliptic, finely veined, septate, acute at apex, tapering at base, the

Fig. 208. A, B. *Crinum zeylanicum*. A. Habit and inflorescence. B. Upper portion of stamen showing anther. C, D. *Hymenocallis caribaea*. C. Upper portion of scape with open flowers. D. Infructescence.

margins smooth; petioles 9–17 cm long. Scape 20–60 cm long, green, glaucous, compressed, 2-edged; spathelike bracts ovate to ovate-lanceolate, membranous, 4–6 × 2–3 cm. Flowers 8–14, fragrant, subsessile or short-petiolate; perianth segments linear, 6.5–12.5 × 0.4–1 cm, white, the tube 6.5–9 cm long; staminal cup funnel-shaped, 2.6–3.5 cm long; anthers 1.5–1.8 cm long, greenish, the filaments 7–8 cm long; ovary 10–15 mm long with 2 ovules in each locule; style 15–20 cm long. Fruit not seen.

DISTRIBUTION: Occasional on rocky hillsides. East End (B1207). Native to the West Indies.

NOTE: This record is based on a sterile specimen of Britton (B1207). The leaf of this specimen compares well with those of this species.

DOUBTFUL RECORDS: The following species were reported by Britton and Wilson (1924) as occurring on St. John: *Hippeastrum puniceum* (Lam.) Urb., a plant of Central America, West Indies, and northern South America; *Galatea bulbosa* (Mill.) Britton ≡ *Eleutherine bulbosa* (Mill.) Urb., a plant of Central America, West Indies, and South America; and *Zephyranthes puertoricensis* Traub, a plant from the West Indies and northern South America [the name *Atamosco tubispatha* (L'Hér.) M. Gómez ≡ *Zephyranthes tubispatha* (L'Hér.) Herb. of Argentina and Uruguay was apparently misapplied by Britton to this species]. However, no specimens have been found, nor have any of these plants been recently collected on the island.

3. Araceae (Philodendron Family)
Reviewed by D. H. Nicolson

Erect perennial herbs, often with creeping or tuberous rhizomes, or climbing robust herbs with adventitious roots. Leaves alternate, often all basal, stout, simple, compound, dissected or lobed; petioles forming a sheath at base. Flowers minute, bisexual or unisexual (monoecious), in a terminal, densely flowered spadix with a large, subtending bract (spathe); perianth of 4–6 distinct or connate tepals in 2 cycles, or these reduced or wanting; stamens (1–)4 or 6(–8), the filaments mostly short, distinct or connate, the anthers opening by terminal pores or longitudinal slits; ovary superior, with several locules, of (2–)3(–15) united carpels, with axile placentation, or unilocular with parietal placentation, the ovules 1 to many per locule, the style terminal and short or the stigma sessile. Fruit a berry or developing with the spadix maturing as a multiple fruit.

A family of about 110 genera and 1500 species, with tropical and subtropical distribution.

Key to the Genera of Araceae

1. Plants floating, aquatic; leaves subsessile, in a rosette .... 4. *Pistia*
1. Plants terrestrial or epiphytic; leaves long-petiolate, alternate.
   2. Spathe herbaceous, reflexed; flowers bisexual.. 1. *Anthurium*
   2. Spathe fleshy, erect, surrounding the spadix; flowers unisexual.
      3. Leaves pedately dissected .... 5. *Syngonium*
      3. Leaves entire, cordate or elliptic.
         4. Plants with elongate climbing stems, or the plant acaulescent or with very short stems, not growing in swamps; leaf ovate, green, cordate at base .... 3. *Philodendron*
         4. Plants with thick erect stems, of swampy or wet areas; leaves elliptic to oblong-ovate, usually variegated, rounded or cordate at base and usually asymmetric .... 2. *Dieffenbachia*

1. ANTHURIUM Schott

Erect herbs or vines climbing by adventitious roots, terrestrial or epiphytic; stems elongate, fleshy or the plant acaulescent. Leaves alternate, simple, or palmately lobed, long-petiolate, enveloped by a cataphyll in early stages. Spathe usually herbaceous, not enclosing the spadix, reflexed, usually long-lived, green, whitish to brightly colored; spadix sessile or short-stipitate, cylindrical or conical, many-flowered, flowering from base to apex. Flowers bisexual, sessile, the perianth segments 4; stamens 4; ovary 2-locular, with 1 or 2 pendulous ovules per locule, the style short or wanting, the stigma disklike to 4-lobed. Fruit a 2-locular, fleshy, bright red, white, or lavender berry; seeds oblong.

A genus of about 700 species native to the neotropics.
REFERENCE: Mayo, S. J. 1982. *Anthurium acaule* (Jacq.) Schott (Araceae) and the West Indian "bird's nest" Anthuriums. Kew Bull. 36: 691–719.

Key to the Species of *Anthurium*

1. Leaves ovate to triangular-ovate, the blades directed downward from the point of attachment at base .... 1. *A. cordatum*
1. Leaves lanceolate or elliptic, the blades ascending.
   2. Leaves lanceolate, oblong-lanceolate, or cordate-lanceolate, the base nearly cordate or truncate; petioles nearly as long as the blades.. 3. *A.* × *selloum*
   2. Leaves elliptic, oblong-elliptic, or oblanceolate, the base obtuse or rounded, tapering from the middle of the blade; petioles much shorter than the blades .... 2. *A. crenatum*

1. Anthurium cordatum (L.) Schott, Wiener Z. Kunst **3**: 828. 1829. *Pothos cordata* L., Sp. Pl., ed. 2, **2**: 1373. 1763.   Fig. 209A–G.

Anthurium cordifolium sensu Eggers, Fl. St. Croix 98. 1879, and Millsp., Publ. Field Columbian Mus., Bot. Ser. **l**: 477. 1902, non Kunth, 1841.

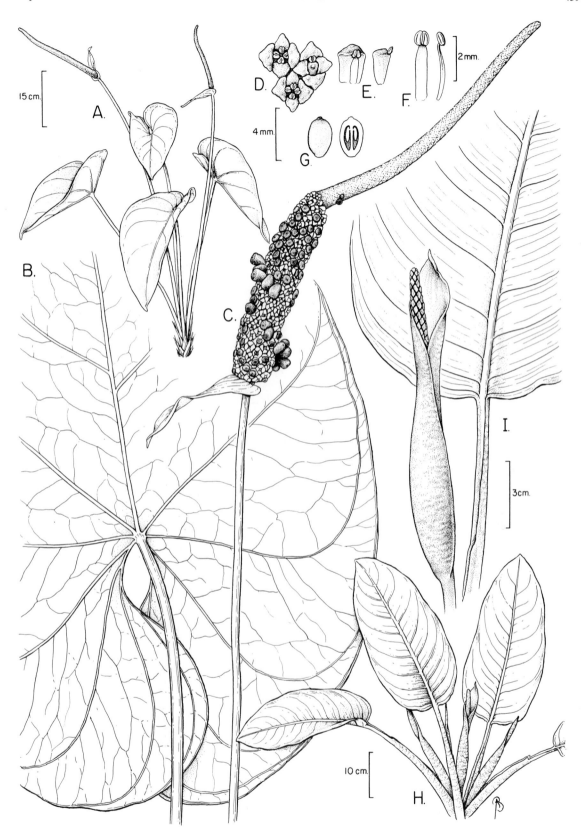

FIG. 209. A–G. *Anthurium cordatum*. A. Habit. B. Leaf. C. Infructescence. D. Flower, top view. E. Perianth segment with subtending stamen, frontal and lateral views. F. Stamen, frontal and lateral views. G. Ovary and l.s. ovary. H, I. *Dieffenbachia seguine*. H. Fertile branch. I. Inflorescence and subtending leaf.

Glabrous, erect terrestrial herb 0.5–1.5 m tall; stem erect or decumbent, cylindrical, 10–50 cm long and about 5 cm diam., with numerous, large leaf scars; cataphylls 10–20 cm long, weathering into persistent fibers. Leaves spirally arranged; blades directed downward, 30–52 × 25–30 cm, broadly ovate to triangular-ovate, coriaceous, strongly reticulate-veined, the apex obtuse, the base cordate-sagittate with overlapping sinuses, the margins entire, slightly wavy; petioles erect, cylindrical, 20–60 cm long. Inflorescence ascending; peduncles to 50 cm long; spathe reflexed, green, leafy, oblong to ligulate, to 10 cm long; spadix ascending, pendulous when fruits are mature, 10–30 cm long, dark brown, widened at base. Berry bright red, obovoid, 5–8 mm long, produced on proximal portion of spadix, embedded in the axis when immature, pendulous from a threadlike structure when ready for dispersal.

DISTRIBUTION: A common understory herb of shaded moist forests. Bordeaux (A2863). Also on St. Croix, St. Thomas, and Tortola; the Lesser Antilles.

COMMON NAME: maroon jancole.

2. Anthurium crenatum (L.) Kunth, Enum. Pl. **3**: 75. 1841. *Pothos crenata* L., Sp. Pl., ed. 2, **2**: 1373. 1763.

Anthurium acaule sensu many West Indian authors, non (Jacq.) Schott, 1832.
Anthurium huegelli sensu Eggers, Fl. St. Croix 98. 1879, and Millsp., Publ. Field Columbian Mus., Bot. Ser. **l**: 477. 1902, non Schott, 1855.

Glabrous, erect terrestrial or epiphytic herb 0.8–1.5 m tall; stem prostrate, ca. 10 cm long, with numerous adventitious roots and numerous persistent cataphylls. Leaves ascending; blades 40–80 × 15–25 cm, elliptic, oblong-elliptic, or oblanceolate, coriaceous, strongly reticulate-veined, very often weathering into a network of veins, the apex obtuse, with a short apiculum, the base obtuse or rounded, tapering from below the middle of the blade, the margins entire to slightly crenate; petioles 5–20 cm long, flattened along the ventral surface, swollen at point of attachment

with blade. Inflorescence ascending; peduncles to 80 cm long; spathe to 12 cm long, reflexed, green, leafy, lanceolate, deciduous; spadix ascending, pendulous when fruits are mature, 10–30 cm long, dark brown, widened at base. Berry bright red, obovate, 5–6 mm long, produced on proximal portion of spadix, embedded in the axis when immature, pendulous from a threadlike structure when ready for dispersal.

DISTRIBUTION: A common understory herb of shaded moist forests. Bordeaux (A2857). Also on St. Croix, St. Thomas, and Tortola; Hispaniola and Puerto Rico (including Vieques and Culebra islands).

COMMON NAME: scrub-bush.

3. Anthurium × selloum K. Koch, Index Sem. Hort. Berol. App. 8. 1855.

Anthurium macrophyllum Schott, Prodr. 516. 1860.

Glabrous, erect terrestrial or seldom epiphytic herb, 0.5–1 m tall; stem erect, short, with numerous adventitious roots; cataphylls to 18 cm long, persistent, weathering to show numerous fibers. Leaves numerous, ascending; blades 30–70 × 13–26 cm, lanceolate, oblong-lanceolate or cordate-lanceolate, coriaceous, strongly reticulate-veined, the apex obtuse to short-acuminate, the base cordate to nearly truncate, the margins entire, slightly wavy; petioles 30–50 cm long, cylindrical, canaliculate along the ventral surface, swollen at point of attachment with blade. Inflorescence ascending; peduncles 60–103 cm long, cylindrical; spathe to 15 cm long, spreading to reflexed, green, leafy, oblong-lanceolate; spadix ascending, 15–40 cm long, dark brown, cylindrical, narrowed toward the apex. Sterile, not producing berries.

DISTRIBUTION: A common understory herb of shaded moist forests. Bordeaux (A2864, A2895), Camelberg Peak (A4247). Known only from St. John and Tortola (*Proctor 41984*).

NOTE: *Anthurium × selloum* seems to be a sterile hybrid between *A. cordatum* and *A. crenatum*. It has never been described, collected, or observed in fruit; the inflorescences develop normally but later die before setting fruits. Ovules and pollen grains are produced, but the pollen is not available since the anthers do not dehisce. Although *A. × selloum* is capable of vegetative reproduction, most individuals seem to be the result of crossing from the two putative parent species. The morphology of *A. × selloum* is intermediate between that of the parent species, and the taxon is normally found where the parent species occur together.

2. DIEFFENBACHIA Schott

Erect herbs, usually of wet or swampy areas; stems thickened, elongate, usually fragile, producing a whitish, poisonous, caustic sap. Leaves simple, usually spotted or variegated, mostly borne distally on stems, long-petiolate. Spathe greenish, convolute, surrounding the spadix, constricted at the middle, adnate to the pistillate portion of the spadix, the upper portion erect, surpassing the free pistillate portion of the spadix; spadix cylindrical, many-flowered, with sterile flowers at the base. Flowers unisexual, the perianth wanting, pistillate flowers with 4 staminodes; stamens 4–5; ovary 1–3-locular, with a solitary ovule per locule, the stigma sessile, nearly globose to 3-lobed. Fruit a fleshy berry; seeds ovoid to globose.

A genus of about 30 species native to the neotropics.

1. Dieffenbachia seguine (Jacq.) Schott, Wiener Z. Kunst **3**: 803. 1829. *Arum seguine* Jacq., Enum. Syst. Pl. 31. 1760.   Fig. 209H, I.

Erect to decumbent herb to 2 m tall; stem cylindrical, ca. 3 cm diam., green with leaf scars, producing a milky or watery, caustic sap when cut. Leaves borne distally on stems; blades usually spotted or variegated, 20–50 × 12–24 cm, elliptic or oblong-ovate, chartaceous, the apex acute or acuminate, the base rounded or cordate, usually asymmetric, the margins entire; peti-

oles 30–40 cm long, sheathing. Inflorescences axillary, ascending, in groups of 2 or 3; peduncles 4–15 cm long, cylindrical; spathe fleshy, 18–30 cm long, green; spadix white, shorter than the spathe. Berry orange red.

DISTRIBUTION: An occasional herb of wet seepage areas. Bordeaux (A4705), Maho Bay Gut (W863). Also on St. Thomas and Tortola; throughout tropical America.

COMMON NAME: dumb cane.

NOTE: The sap of this plant is irritating to the skin and, if ingested, is inflammatory to the tongue and digestive tract; hence the common name, dumb cane.

3. PHILODENDRON Schott

Stout or slender vines climbing by means of adventitious roots, or less often erect herbs; stems elongate, usually producing a watery, caustic sap. Leaves simple, lobed or variously divided or pinnatifid, long-petiolate. Spathe convolute, surrounding the spadix, usually thickened, not adnate to the spadix; spadix cylindrical, erect, usually nearly sessile, with pistillate flowers on basal portion. Flowers unisexual, the perianth wanting; stamens 2–6; ovary 2- to many-locular, with 1 to many ovules per locule, the stigma sessile, entire or lobed. Fruit a fleshy, 1- to many-seeded berry; seeds ovoid to ellipsoid.

A genus of about 275 species native to the neotropics.

Key to the Species of *Philodendron*

1. Vines, with elongate stems to 10 m long and 1–2 cm diam.; petioles 10–15 cm long; cataphylls
   deciduous .... 2. *P. scandens*
1. Erect plant, with stems to 2 m long and 8–10 cm diam.; petioles to 1 m long; cataphylls weathering to a
   fibrous persistent structure .... 1. *P. giganteum*

1. Philodendron giganteum Schott, Syn. Aroid. 89. 1856.

Erect, terrestrial or epiphytic plant, to 2 m tall; stem to 20 cm long, cylindrical, 8–10 cm diam., producing a watery sap when cut; cataphylls to 60 cm long, initially entire, weathering into persistent fibers. Leaf blades horizontal or directed downward, 25–60 × 17–50 cm, lanceolate or triangular-lanceolate, nearly coriaceous, slightly paler below, the apex obtuse to acute or acuminate, the base cordate, with not-overlapping sinuses, the margins sinuate; petioles erect, to 1 m long, nearly cylindrical. Inflorescence axillary, solitary; peduncles 6–9 cm long, stout; spathe 14–21 cm long, convolute, constricted at the middle, green without, dark maroon within; spadix sessile, stout, as long as the spathe, with staminate portion whitish and pistillate portion yellowish green. Berry yellow to orange.

DISTRIBUTION: Not common, known only from a single locality in forested area. Along Reef Bay Gut (A4165). Also on St. Thomas (according to Eggers, 1879) and Tortola; Hispaniola, Puerto Rico, the Lesser Antilles, Trinidad, and Venezuela.

2. Philodendron scandens K. Koch & Sello, Index Sem. Hort. Berol. App. 4. 1853.   Fig. 210D, E.

Philodendron micans Klotzsch *ex* K. Koch, App. Gen. Sp. Nov. 7. 1854.
Philodendron oxycardium Schott, Syn. Aroid. 82. 1856.
Philodendron isertianum Schott, Syn. Aroid. 242. 1860.

Vine to 10 m long, rooting at nodes; stem cylindrical, 1–2 cm diam., producing a watery, caustic sap when cut; cataphylls to 12 cm long, deciduous. Leaf blades 14–30 × 10–20 cm, ovate, coriaceous, the apex acuminate to cuspidate, the base cordate with not-overlapping sinuses, the margins slightly sinuate; petioles curved to ascending, 10–15 cm long, nearly cylindrical, invaginate at base. Inflorescence axillary, solitary, ascending; peduncles 5–9 cm long, stout; spathe ca. 15 cm long, thickened, convolute, straight (nearly cylindrical) turning from green to yellow without at maturity, red to maroon within; spadix nearly sessile, cylindrical, stout, whitish, as long as the spathe.

DISTRIBUTION: A common vine of moist forests. Bordeaux Mountain (A2610, A2880). Also on St. Thomas and Tortola; throughout tropical America.

4. PISTIA L.

A monospecific genus, characterized by the following species.

1. Pistia stratiotes L., Sp. Pl. 963. 1753. Fig. 210A–C.

Floating acaulescent stoloniferous herb, forming large colonies. Leaves subsessile, in a rosette; blades 3–12 × 1.5–5 cm, cuneate, fleshy, with sunken parallel veins, impermeably pubescent, the apex rounded, truncate, usually notched, the base cuneate, the margins revolute. Inflorescence axillary, solitary, ascending; peduncles 5–7 mm long, slender; spathe 1.3–1.5 cm long, convolute and adnate to the spadix below, spreading above, whitish; spadix with a single pistillate flower at base, and with 2–8 staminate flowers above, shorter than the spathe. Flowers unisexual, the perianth wanting; stamens 2; ovary 1-locular, with numerous ovules, the style slender, the stigma penicillate. Fruit thin-walled, many-seeded. Seeds cylindrical, rugulose.

DISTRIBUTION: Not common, found in fresh-water cisterns or pools, sometimes cultivated. Annaberg Ruins (A2866). Also on St. Thomas; a variable species found throughout tropical and subtropical regions.

5. SYNGONIUM Schott

Epiphytes or hemiepiphytes, usually with long root-climbing stems, producing milky sap. Leaves simple or variously divided, with 5–11 leaflets; petioles sheathed toward the base. Inflorescences 1–11 per axil; peduncles erect in flower, pendent in fruit; spathe fleshy, convolute, conspicuously constricted medially, the tube ellipsoid, the blade whitish to greenish, broadly spreading at anthesis; spadix much shorter than the spathe, erect, with pistillate flowers on basal portion. Flowers unisexual, the perianth wanting; stamens 3–4, united into a synandrium; ovary (1–)2(–3)-locular, with 1(–2) ovules per locule, the stigma discoid or bilabiate. Fruit a 1-seeded berry, connate into an ovoid syncarp; seeds obovoid or ovoid.

A genus of 33 species native to the neotropics, with most species in Costa Rica and Panama.

REFERENCE: Croat, T. B. 1981. A revision of *Syngonium* (Araceae). Ann. Missouri Bot. Gard. 68: 565–651.

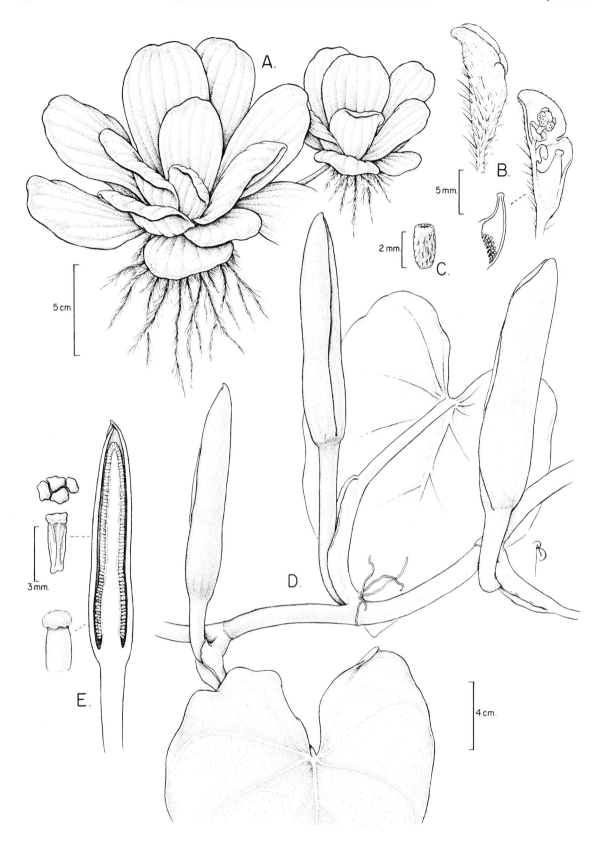

FIG. 210. A–C. *Pistia stratiotes.* **A.** Habit. **B.** Inflorescence, l.s. inflorescence showing staminate and pistillate flowers, and l.s. pistil showing ovules. **C.** Seed. **D, E.** *Philodendron scandens.* **D.** Flowering branch. **E.** L.s. inflorescence and detail of staminate and pistillate flowers.

1. Syngonium podophyllum Schott, Bot. Zeitung (Berlin) 9: 85. 1851. Fig. 211.

Vine to 10 m long, rooting at nodes; stem cylindrical, glaucous, 1–2 cm diam., producing milky sap when cut. Leaves pedately divided; leaflets 3–11, united or free to base, coriaceous, the apex acuminate, the base variously auriculate, the margins sinuate, outermost leaflets smaller, the medial leaflets 16–38 × 6–17 cm, obovate elliptic or lanceolate; petioles 15–60 cm long, nearly cylindrical, sheathed two-thirds of their length. Inflorescences 4–11 per axil, ascending; peduncles 8–9 cm long, slender; spathe ca. 10 cm long, convolute into an ellipsoid tube at base, the blade cream-colored, concave, ephemeral; spadix whitish, sessile, cylindrical, with a constriction between the pistillate and the staminate areas. Syncarp ovoid, red, reddish orange, or yellow, 3–5.5 cm long.

DISTRIBUTION: Native to Mexico and grown as an ornamental throughout the Caribbean; a few individuals have escaped cultivation on St. John. Susannaberg (A4061); naturalized throughout tropical America.

4. Arecaceae (Palm Family)

Trees, shrubs, or vines, very often with unbranched trunks. Leaves spirally arranged, usually forming a dense, terminal crown, large, simple or compound, fan-shaped or pinnatisect; petioles stout, forming a sheath around stems, the sheath sometimes prolonged as a ligule. Flowers usually small, unisexual (monoecious, dioecious, or polygamous) or less often bisexual, mostly actinomorphic, 3-merous, in axillary or less often terminal, panicles, racemes, or spikes, the inflorescence subtended by a prophyll or a spathe; perianth of 6 distinct or basally connate tepals in 2 cycles; stamens usually 6, but sometimes only 3, or more numerous, the filaments distinct or connate and adnate to the tepals, the anthers opening by longitudinal slits; gynoecium of (1–)3(–10) distinct or connate carpels forming a superior, plurilocular ovary, with axile placentation, the ovules solitary in each locule, the styles terminal, as many as the carpels or 1, the stigmas sessile. Fruit a fleshy or fibrous drupe; seeds 1–3 or rarely more per fruit.

A family of about 200 genera and 3000 species, with tropical and subtropical distribution.

REFERENCE: Uhl, N. W. & J. Dransfield. 1987. Genera Palmarum. Allen Press, Lawrence, Kansas.

Key to the Genera of Arecaceae

1. Leaves palmate.
  2. Trunk 8–12 cm diam.; leaves to 1.5 m long, silvery-white beneath; petioles ligulate at junction with blade .... 1. *Coccothrinax*
  2. Trunk 25–30 cm diam.; leaves 2.5–3 m long, green beneath; petioles not ligulate at junction with blade .... *Sabal* (cultivated)
1. Leaves pinnately compound.
  3. Trunk widest at the very base, with many obvious leaf scars; fruit 20–30 cm long, green or yellowish at maturity .... 2. *Cocos*
  3. Trunk widest just below the crown, nearly smooth; fruits 1–1.3 cm long, light brown at maturity .... 3. *Roystonea*

1. COCCOTHRINAX Sarg.

Small to medium, solitary palms; trunk slender with faint leaf scars. Leaf blades fan-shaped, with lepidote lower surface, the segments united at base; petioles flattened-convex, ligulate beyond the junction with blade, the ligule woody and entire; sheath tubular, fibrous, weathering into a network of tardily deciduous fibers. Inflorescences axillary, erect to arching, shorter than the leaves, the primary branches pendulous, paniculate; primary bracts enclosing the base of a primary branch, the lowermost one bicarinate, the others tubular, with an oblique aperture. Flowers bisexual, dispersed along the secondary branches of the inflorescence, pedicellate; perianth short, 6-merous, lobed or dentate; stamens 9–12, the filaments free to slightly connate at base, the anthers basifixed, notched at apex; ovary 1-locular, with a single ovule, the style short, the stigma laterally compressed, funnel-shaped. Fruit small, 1-seeded, depressed-globose, with thin, fleshy mesocarp; seeds globose, sulcate.

A genus of about 20 species native to the Caribbean and circum-Caribbean region.

1. Coccothrinax alta (O.F. Cook) Becc., Webbia 2: 331. 1907. *Thrincoma alta* O.F. Cook, Bull. Torrey Bot. Club 28: 540. 1901. Fig. 212A–D.

Thringis latifrons O.F. Cook, Bull. Torrey Bot. Club 28: 545. 1901.

Coccothrinax eggersiana Becc., Webbia 2: 321. 1907.

Coccothrinax sancti-thomae Becc., Webbia 2: 303. 1907.

Thrinax argentea sensu Eggers, Fl. St. Croix 100. 1879, non Lodd., 1830.

Erect solitary palm 2–6 m tall; trunk cylindrical, 8–12 cm diam. Leaf blades to 75 cm long, fan-shaped, orbicular, silvery-white beneath, cleft to beyond the middle into many narrowly lanceolate, acuminate segments; petioles as long as the blades, with the base expanded into a persistent fibrous sheath. Inflorescences paniculate, solitary in leaf axils, ascending, with pendulous primary branches, the axes light to dark brown, glabrous; pedicels 1–3 mm long. Perianth 2.5–3 mm wide; stamens usually 9. Fruit depressed-globose or globose, 5–6 mm long, turning from green to purple, apiculate at apex.

DISTRIBUTION: Common throughout St. John in moist to dry forests. Bordeaux (A5089), Susannaberg (A2131). Also on St. Croix, St. Thomas, and Virgin Gorda; Puerto Rico (including Culebra and Vieques).

COMMON NAMES: broom palm, broom teyer, fan palm, silver palm, teyer tree.

FIG. 211. *Syngonium podophyllum.* **A.** Flowering branch. **B.** Leaf. **C.** Inflorescence (spathe and spadix). (From Mori et al., in press, Guide to the Vascular Plants of Central French Guiana, Mem. New York Bot. Gard. 76.)

FIG. 212. A–D. *Coccothrinax alta*. **A.** Habit. **B.** Leaf. **C.** Infructescence. **D.** Fruit and embryo. **E–G.** *Roystonea borinquena*. **E.** Habit. **F.** Portion of leaf. **G.** Portion of infructescence.

2. COCOS L.

A monospecific genus characterized by the species given below.

1. Cocos nucifera L., Sp. Pl. 1188. 1753.

Monoecious, medium to large, solitary palm to 30 m tall; trunk stout, 25–30 cm diam., usually leaning, ringed with many obvious large leaf scars, swollen and widest toward the base. Leaves pinnately compound, arching; blades 3–6 m long, the segments numerous, 50–70 cm long (the distal ones shorter), oblong, folded, with a stout central vein; petioles stout, woody, much shorter than the blades, enlarged at base, expanding into a fibrous sheath. Inflorescences solitary in leaf axils, the main axis ascending, with spreading slender branches. Flowers unisexual, unequal in size and shape. Staminate flowers many, on distal portion of rachilla, paired or solitary, sessile or pedicellate; sepals free to base, 2 mm long; petals valvate, ca. 1 cm long; stamens 6, the anthers sagittate; pistillode prominent, trifid. Pistillate flowers 1 to few, on proximal portion of rachilla; sepals and petals similar, 2–3 cm long; ovary conical, 3-locular, the stigmas 3, sessile. Fruit ovoid to ellipsoid, 3–angled, 20–30 cm long, with a fibrous, thick mesocarp green or yellowish green. Seeds nearly globose, hard-shelled, 10–15 cm long.

DISTRIBUTION: Native to the tropical coast of the Pacific Ocean, introduced into the New World by European settlers for its edible seeds. Self-regenerating on sandy shores around the island, where they were originally planted. Distributed throughout tropical and subtropical shores or inland.

COMMON NAME: coconut palm.

3. ROYSTONEA O.F. Cook

Monoecious, medium to large, solitary palms; trunk columnar, or swollen, smooth or ringed by large leaf scars. Leaves pinnately compound; blades very large; petioles much shorter than the blade; sheaths very large, tubular, woody, smooth. Inflorescences produced at the base of the crown shaft, erect to arching, much shorter than the leaves, many-branched panicle; primary bracts deciduous, the outer ones bicarinate, the inner ones tubular, the axes whitish when young owing to farinaceous hairs. Flowers unisexual, sessile, 3-merous, usually in triads of a single pistillate and 2 lateral staminate flowers; sepals 3, imbricate; petals 3, valvate; stamens 6, the anthers dorsifixed; ovary 3-locular, usually only 1 locule functional, with a single ovule, the stigmas 3. Fruit small, 1-seeded, oblong to globose, mesocarp thin, fleshy; seeds depressed-globose.

A genus of about 12 species native to the Caribbean and circum-Caribbean region.

1. Roystonea borinquena O.F. Cook, Bull. Torrey Bot. Club 28: 552. 1901.   Fig. 212E–G.

Erect, solitary palm 10–18 m tall; trunk widest below the crown shaft portion, 50–60 cm diam., nearly smooth. Leaves 2–3 m long, ascending, the leaflets numerous, 50–80 cm long (the distal ones shorter), linear, folded, with a stout central vein; petioles stout, very short, enlarged at base, expanding into a nearly woody tubular sheath. Inflorescences paniculate, twice-branched, much shorter than the leaves. Flowers unisexual; sepals 0.8–1.5 mm long; petals 5–6 mm long, white. Fruit widely ellipsoid, ca. 13 mm long, light brown. Seeds oblong-ellipsoid to globose, ca. 9 mm long, 7 mm wide.

DISTRIBUTION: Rare, a few individuals found along gallery forests or stream banks. Maho Bay Gut (field observations). Also on St. Croix; Puerto Rico (including Culebra and Vieques).

COMMON NAMES: cabbage palm, mountain cabbage, royal palm.

CULTIVATED SPECIES: A few individuals of *Sabal casuarium* (O.F. Cook) Becc. (endemic to Puerto Rico) are cultivated on the grounds of Caneel Bay Hotel. Also, several immature individuals (persistent after cultivation) in the area of the old Reef Bay sugar plantation seem to be this species. Woodbury and Weaver (1987) reported *Phoenix dactylifera* L. from a cultivated palm at Cruz Bay, but the palm has since died.

5. Asphodelaceae (Aloe Family)

By M. T. Strong

Perennial herbs with rhizomes, or sometimes large branching trees. Leaves primarily basal, spirally arranged or sometimes 2-ranked, sheathing at base; blades thick and succulent or fleshy, gelatinous internally, flattened or cylindrical, subulate to linear-lanceolate or elliptic, with smooth, spiny or serrate margins, the apex usually spine-tipped. Inflorescences terminal, simple or compound racemes or spikes; scapes leafless or beset with small bracteal leaves. Flowers bisexual; perianth segments 6; stamens 6, with free filaments, the anthers dorsifixed, introrse, longitudinally dehiscent; ovary 3-locular, with 2 to many ovules per locule, with axile placentation, the style simple, the stigma small. Fruit a loculicidal capsule; seeds ovoid and elongate, usually arillate.

Approximately 17 genera and 814 species in subtropical and tropical regions of the Old World, with the center of distribution in southern Africa.

1. ALOE L.

Herbs or large branching trees. Leaves spirally arranged, rosulate, rarely 2-ranked; blades thick and succulent with bitter sap, margins spiny or serrate, the apex spine-tipped. Inflorescence an elongate raceme or panicle with many nodding flowers, the scape naked toward base. Perianth segments fused into a tube with spreading or recurved lobes, red, red and green, orange, or yellow; ovary ovoid to oblong-ovoid, sessile, 3-angled with numerous ovules in each locule. Capsule coriaceous; seeds many, flattened, black.

Approximately 330 species in Africa and Madagascar, with the center of distribution in South Africa. Cultivated and becoming widespread in subtropical and tropical regions of the New World.

1. Aloe vera (L.) Burm.f., Fl. Indica. 83. 1768 [6 April].
Aloe perfoliata L. var. *vera* L., Sp. Pl. 320. 1753.
Fig. 213.

Aloe barbadensis Mill. Gard. Dict., ed. 8. 1768 [16 April].
Aloe vulgaris Lam., Encycl **1**: 86. 1783, nom. illegit.

Stemless or short-stemmed herb with horizontally creeping rhizomes frequently proliferating at the nodes. Leaves narrowly lanceolate, 30–60 × 3.5–8 cm, acuminate at apex, turgid with clear, watery sap, succulent, glaucous-green, the margins with reddish-tipped, spinelike teeth. Scapes to 1 m tall, stout, with a few remote, acute scales; inflorescence a raceme 10–30 cm long; bracts ovate to lanceolate, exceeding the short pedicels. Perianth nearly cylindric, 2.5–3 cm long, 4–7 mm wide, turning from green to yellow, the tips spreading; stamens equaling or slightly exceeding the perianth; style exserted.

DISTRIBUTION: Locally common along roadsides and disturbed areas. East End (A4223). Also on St. Croix, St. Thomas, and Virgin Gorda; Vieques; probably native to the Mediterranean region, now naturalized and widespread in Florida, Central America, and the West Indies.

COMMON NAMES: aloes, bitter-aloes, sempervive.

6. Bromeliaceae (Pineapple Family)

Perennial epiphytic or terrestrial herbs or less often terrestrial subshrubs. Leaves spirally arranged, rosulate or fasciculate, simple, narrow, parallel-veined, forming a sheath around stems where water usually accumulates, the margins entire or spinose. Flowers small to medium, bisexual or functionally unisexual, mostly actinomorphic, 3-merous, in simple or compound spikes, racemes, panicles, or heads on terminal or axillary scapes, often subtended by brightly colored bracts; calyx often green, of distinct or basally connate sepals; corolla often brightly colored, of distinct or basally connate petals, often with a pair of basal appendages; stamens 6, the filaments distinct or connate and adnate to the petals, the anthers opening by longitudinal slits; gynoecium of 3 connate carpels forming a superior or partly inferior, trilocular ovary, with axile placentation, the ovules few to numerous in each locule, the style terminal, connate into a single style with 3 stigmatic branches. Fruit a berry, a dry or a fleshy septicidal capsule, or seldom the inflorescence ripening as a multiple, fleshy fruit.

A family of about 45 genera and 2000 species, with neotropical distribution.

REFERENCES: Smith, L. B. & R. J. Downs. 1974. Pitcairnioideae (Bromeliaceae). Fl. Neotrop. Monogr. 14(1): 1–658; Smith, L. B. & R. J. Downs. 1977. Tillandsioideae (Bromeliaceae). Fl. Neotrop. Monogr. 14(2): 663–1492; Smith, L. B. & R. J. Downs. 1979. Bromelioideae (Bromeliaceae). Fl. Neotrop. Monogr. 14(3): 1493–2142.

Key to the Genera of Bromeliaceae

1. Leaf margins spiny.
    2. Scapes stout, short, with congested inflorescences to 3–15 cm long, the axes flattened, densely white-farinose pubescent; hypanthium densely white-farinose pubescent; fruit a leathery yellowish berry, 2.7–3.5 cm long, with whitish or creamy pulp inside .... 2. *Bromelia*
    2. Scapes slender, elongate, with scattered inflorescences to 15–20 cm long, the axes angled-cylindrical, glabrous; hypanthium glabrous; fruit reddish, ovoid or ovoid-trigonous, <2 cm long.
        3. Leaves long-acuminate; fruit capsular; floral bracts ovate-lanceolate, acute; corolla red-orange, to 4 cm long, twice as long as the calyx.. 4. *Pitcairnia*
        3. Leaves short-acuminate to cuspidate; fruit a berry; floral bracts spiny-tipped; corolla 8–9 mm long, longer than the calyx.. 1. *Aechmea*
1. Leaf margins entire, not spiny.
    4. Capsule ovoid; mature seeds with a tuft of folded, comose hairs at apex; flowers distant to approximate in the inflorescence (not overlapping).. 3. *Catopsis*
    4. Capsule elongate, oblong, linear, or fusiform; mature seeds with a tuft of straight hairs at base; flowers close or crowded, with overlapping floral bracts, or distant to approximate in the inflorescence ...... 5. *Tillandsia*

1. AECHMEA Ruíz & Pav., nom. cons.

Epiphytic or terrestrial stemless or shortly stemmed herbs, frequently propagating by basal rhizomes. Leaves rosulate, linear, chartaceous to coriaceous, with spiny margins. Flowers bisexual, congested or scattered in simple or compound spikes or racemes on elongate scapes; sepals free or connate, usually strongly asymmetric, continuous into a hypanthium; petals free, usually bearing 2 basal appendages; stamens shorter than the petals, the filaments free or the second series adnate to the petals, the anthers dorsifixed; ovary inferior, with several ovules. Fruit a berry; seeds naked, several, small, dark.

A genus of about 170 species occurring in Central America, the West Indies, and South America.

1. Aechmea lingulata (L.) Baker, J. Bot. **17**: 164. 1879.
Bromelia lingulata L., Sp. Pl. 285. 1753. *Chevalliera lingulata* (L.) Griseb., Fl. Brit. W. I. 591. 1864.
Fig. 214A–F.

Terrestrial or epiphytic herb; stems short (to 10 cm long), erect or decumbent, clothed by the leaf bases. Leaves linear, 30–75 × 2.2–6 cm, coriaceous, short-acuminate, acuminate and spine-tipped at apex, the margins with brownish, curved, spinelike teeth, 1–2 mm long. Scapes to 1 m tall, slender; scape bracts oblong, lanceolate or subulate, spine-tipped, 2.5–6 cm long, decreasing in size toward the apex of the scape; inflorescences several ascending-curved spikes, distichous, 10–15 cm long; floral bracts subulate, 4–7 mm long, spine-tipped. Flowers sessile, distant to approximate; hypanthium green, 8–9 mm long; sepals 5–6 mm long, connate at base, the apex asymmetrical with a spine tip in subapical

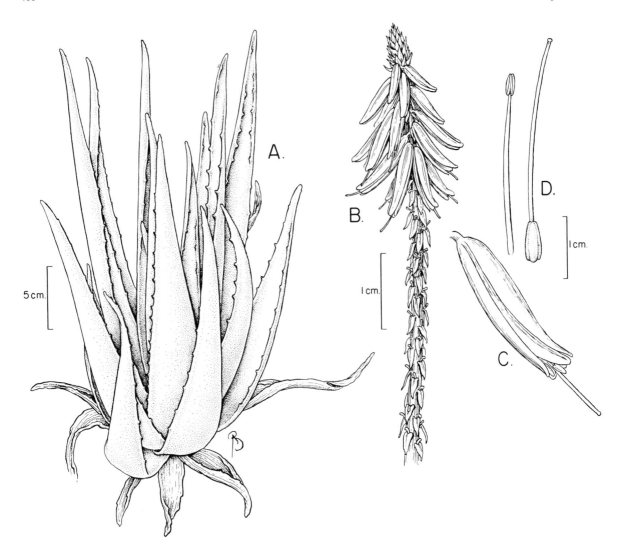

FIG. 213. *Aloe vera.* **A.** Habit. **B.** Scapose inflorescence. **C.** Flower. **D.** Stamen and pistil.

position; petals whitish, 8–9 mm long, acute; stamens of the second series, with shorter filament, adnate to the petals. Berry asymmetrically ovoid, purplish red, 8–10 mm long.

DISTRIBUTION: Occasional in moist forests. Bordeaux (A615, A5096), Center Line Road (A4035). Also on St. Thomas; Central America, West Indies, and South America.

2. BROMELIA L.

Large terrestrial, stoloniferous herbs. Leaves rosulate, narrowly linear, with spiny margins and a spine-tipped apex. Scapes very short to elongate; inflorescences racemose, capitate or paniculate; scape bracts membranous, large, overlapping. Flowers bisexual, sessile to shortly pedicellate. Sepals free or partly connate, continuous into a hypanthium; petals connate at base, convolute; stamens 6, shorter than the petals, the filaments connate at base into a short tube adnate to the petals, the anthers basifixed; ovary inferior, with several ovules. Fruit a leathery berry; seeds few to many, small, dark brown.

A genus of 47 species occurring in Central America, the West Indies, and South America.

1. Bromelia pinguin L., Sp. Pl. 285. 1753.

Fig. 214G–J.

Terrestrial large, acaulescent herb. Leaves linear, 1–2 m long, 3–5 cm wide, rigidly coriaceous, involute, long-acuminate to

caudate, and spine-tipped at apex, the margins with brownish, curved, ascending spines 0.5–1 cm long; young leaves usually bright red above. Scapes to 40 cm tall, stout, whitish farinose-pubescent; scape bracts oblong-lanceolate, 4–8 cm long, papery,

6. BROMELIACEAE

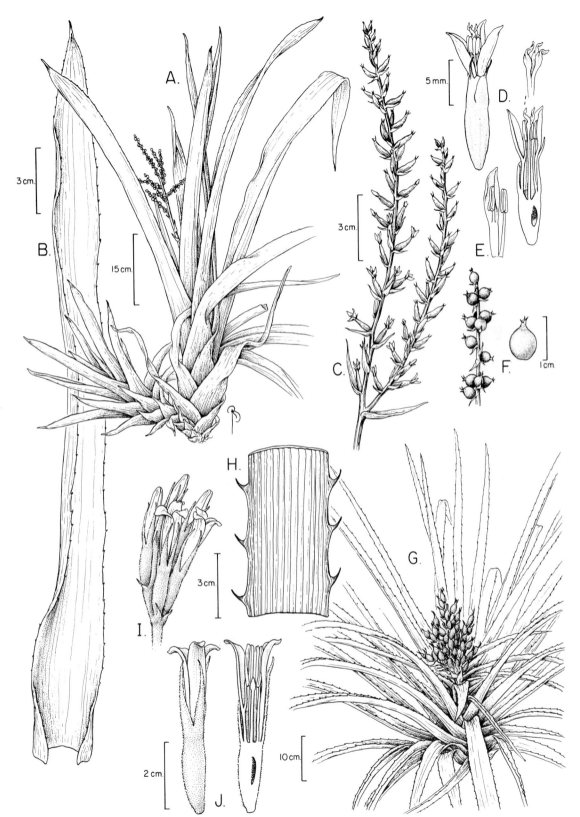

Fig. 214. A–F. *Aechmea lingulata.* **A.** Habit. **B.** Leaf. **C.** Inflorescence branch. **D.** Flower, l.s. flower, and detail of stigma. **E.** Stamen adnate to petal and free stamen. **F.** Portion of infructescence and berry. **G–J.** *Bromelia pinguin.* **G.** Habit. **H.** Section of leaf blade. **I.** Inflorescence branch. **J.** Flower and l.s. flower.

decreasing in size toward the apex of the scape; inflorescence paniculate, 3–15 cm long, whitish farinose-pubescent, 8–12-flowered, the axes flattened; floral bracts subulate, 5–9 mm long. Flowers sessile or shortly pedicellate; hypanthium fusiform, densely whitish farinose-pubescent, 2–2.5 cm long; sepals subulate, 2–2.5 cm long, white farinose-pubescent; petals pink, whitish toward the margins, 2.5–3 cm long, acute, reflexed and woolly at apex; anthers lanceolate, 8.5–9 mm long. Berry fusiform to ovoid, 2.7–3.5 cm long, densely woolly-tomentose when young, turning yellowish and glabrescent at maturity. Seeds numerous, nearly lenticular, reddish brown, ca. 4 mm long, surrounded by a white to light yellow, acid pulp.

DISTRIBUTION: Apparently introduced as a hedge plant, now spontaneous on waste grounds, roadsides, and trail sides. Lind Point (A2798), Chocolate Hole (A1979). Also on St. Croix and St. Thomas; throughout tropical America.

COMMON NAME: ping wing.

3. CATOPSIS Griseb.

Small epiphytic herbs. Leaves utriculate-rosulate, minutely appressed-lepidote, with entire, nonspiny margins. Scapes exceeding the leaves; inflorescences compound spikes or paniculate; bracts membranous. Flowers bisexual or functionally unisexual, sessile to rarely pedicellate; sepals free, strongly asymmetric; petals free, short, white or yellow; stamens 6, included, of 2 unequal series, the anthers ovoid to ellipsoid; ovary superior, ovoid or ellipsoid, with few to several ovules, the style short or wanting. Fruit a septicidal capsule; seeds several, small, with a tuft of comose, folded hairs at apex.

A genus of 19 species occurring from Mexico to South America and the West Indies.

1. Catopsis floribunda L.B. Sm., Contr. Gray Herb. **11:** 5. 1937. Fig. 215D–F.

Catopsis nutans sensu Britton & P. Wilson, Bot. Porto Rico **5:** 137. 1923, non (Sw.) Griseb., 1864.

Epiphytic, acaulescent herb. Leaves 10–30 cm long, tapering from a 3–5 cm wide base to a long-acuminate apex, chartaceous to nearly coriaceous, the margins entire. Scapes to 60 cm tall, slender; scape bracts lanceolate, membranous, 1.5–4 cm long; inflorescences paniculate, of 5–8 simple or double spikes, 5–15 cm long, 8–36-flowered; floral bracts ovate, 4–5 mm long. Flowers sessile or nearly so; sepals broadly elliptic, 4–6 mm long; petals white, ca. 7 mm long, obtuse; filaments of second series adnate to the petals at base. Capsule ovoid, 9–11 mm long, glabrous. Seeds several, oblong, ca. 3 mm long, the hairs to 2 cm long.

DISTRIBUTION: Occasional in moist forests. Bordeaux (B572), Camelberg Peak (A4248). Also on St. Thomas and Tortola; Florida, the West Indies, Trinidad, and Venezuela.

4. PITCAIRNIA L'Hér.

Small to large herbs, terrestrial or sometimes growing on rocks; stems short to elongate. Leaves fasciculate or in a dense spiral along the stem, with spiny or entire margins. Scapes usually, but not always, developed; inflorescences spicate, racemose, or paniculate; bracts conspicuous to minute. Flowers bisexual, brightly colored, red, red-orange to yellow, sessile to pedicellate; sepals free, convolute; petals free, elongate, convergent over the stamens, naked or with a single appendage; stamens 6, exserted, the anthers linear; ovary superior to inferior, with numerous ovules, the style long, slender. Fruit a capsule; seeds several, small, winged or margined.

A genus of approximately 260 species occurring from Mexico to South America, including the West Indies.

1. Pitcairnia angustifolia Aiton, Hort. Kew. **1:** 401. 1789. Fig. 215A–C.

Pitcairnia latifolia Aiton, Hort. Kew. **1:** 401. 1789.

Large terrestrial herb, usually in large colonies on rocky substrates; stems to 10 cm long, covered by the sheathing bases of the leaves; rhizome stout, prostrate, clothed by numerous adventitious roots. Leaves densely fascicled, 50–120 × 1.2–2 cm, linear, coriaceous, scaly-pubescent beneath, the apex long-acuminate, the base contracted, the margins with brown, ascending spines 2–4 mm long. Scapes 1–1.6 m tall, slender, with woolly-pubescent axes when young; scape bracts linear-lanceolate, 4–12 mm long; inflorescences several simple or branched racemes, with scattered flowers; floral bracts 5–10 mm long. Flowers red-orange, ascending; pedicels 4–12 mm long; hypanthium conical; sepals oblong, 1.1–1.7 cm long; petals linear, ascending, 3–5 cm long, without appendages; stamens slightly shorter than the petals; ovary partly inferior, the style exserted, the stigma ellipsoid. Capsule ovoid-trigonous, 1.2–2 cm long, with persistent sepals. Seeds numerous, wedge-shaped, ca. 1.5 mm long, light brown, reticulate, with whitish membranous prolongation along one edge.

DISTRIBUTION: Occasional on rocky outcrops. Waterlemon Bay (A1940), by Petroglyphs (A2934). Also on St. Croix, St. Thomas, Tortola, and Water Island; Puerto Rico (including Vieques) and the Lesser Antilles.

5. TILLANDSIA L.

Small to large epiphytic or seldom terrestrial herbs; stems short to elongate. Leaves fascicled in a rosette, or in a spiral along the stem, with entire margins, pubescence of peltate scales. Scapes usually well developed; inflorescence usually a simple distichous-flowered spike, or less often reduced to a single spike or to a single flower; bracts conspicuous to minute. Flowers bisexual, short-pedicellate; sepals free or connate, convolute; petals free, without appendages; stamens 6, included to exserted; ovary superior, with numerous caudate ovules. Fruit a septicidal capsule; seeds many, erect, cylindric to fusiform, with a basal tuft of white, straight hairs.

A genus of 400–500 species occurring from southern United States to South America, including the West Indies.

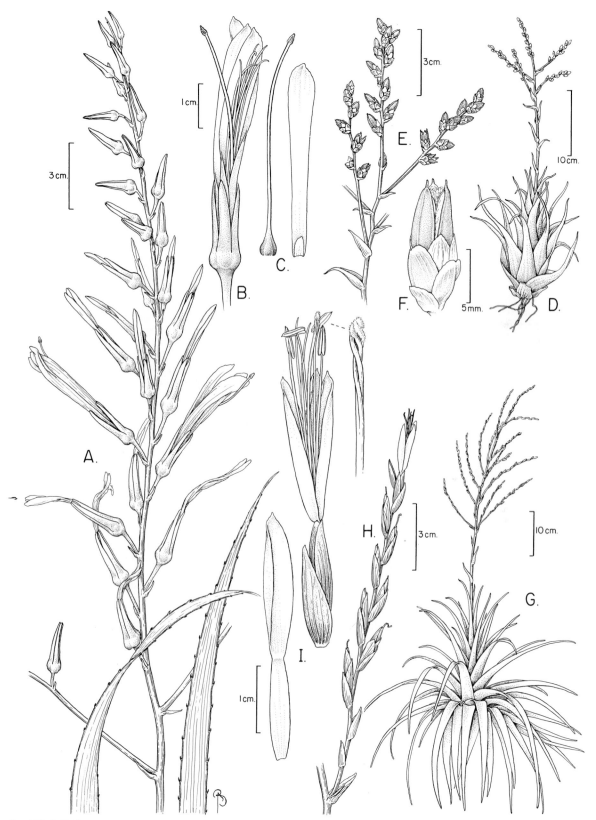

Fig. 215. A–C. *Pitcairnia angustifolia.* **A.** Inflorescence and leaf tips. **B.** Flower. **C.** Pistil and petal. **D–F.** *Catopsis floribunda.* **D.** Habit. **E.** Infructescence. **F.** Capsule. **G–I.** *Tillandsia utriculata.* **G.** Habit. **H.** Inflorescence branch. **I.** Tepal, flower, and detail of stigma.

Key to the Species of *Tillandsia*

1. Plants forming dense clusters; leaves 3–12 cm long and 0.5–2 mm wide, strongly recurvate; scapes 1–2(–5)-flowered .... 3. *T. recurvata*
1. Plants solitary or forming small clusters; leaves >30 cm long and >1 cm wide, ascending or recurved at apex; scapes bearing numerous many-flowered inflorescences.
   2. Inflorescence of close or crowded flowers with distichously overlapping floral bracts.
      3. Inflorescence 5–15 cm long, 2–3 cm wide; floral bracts 3–4 cm long, green to straw-colored .... 1. *T. fasciculata*
      3. Inflorescence 17–30 cm long, 0.8–1.0 cm wide; floral bracts 2.2–2.6 cm long, purplish-red .... 2. *T. lineatispica*
   2. Inflorescence of distant to approximate flowers, the floral bracts not overlapping .... 4. *T. utriculata*

1. Tillandsia fasciculata Sw., Prodr. 56. 1788.

Epiphytic or terrestrial acaulescent herb, forming small clusters; rhizomes short, prostrate, and stout. Leaves fasciculate, recurved, 30–45 cm long, rigidly coriaceous, tapering from a 3–4.5 cm wide base to a long-acuminate apex; upper leaves ascending, smaller. Scape erect, stout, shorter than the leaves, with densely imbricate bracts; inflorescences 4–6, simple or compound, distichous compressed spikes, 5–15 × 2–3 cm, many-flowered; floral bracts rigid-coriaceous, 3–4 cm long, lanceolate-ovate, overlapping, turning from green to straw-colored. Flowers sessile; sepals lanceolate, keeled, much shorter than the subtending bract; petals white to purple, linear, tubular, erect, 4–6 cm long; stamens and pistil exserted. Capsule fusiform, 3–4 cm long. Seeds numerous, 2–3 mm long, light brown, with hairs ca. 2 cm long.

DISTRIBUTION: Occasional on rocky outcrops or epiphytic on trees along intermittent creeks. Battery Gut (A4169), Fish Bay Gut (A2495). Also on St. Thomas (according to Britton & Wilson, 1923); throughout tropical America. A species very variable throughout its range; ours is referrable to *T. fasciculata* var. *venosispica* Mez.

2. Tillandsia lineatispica Mez *in* DC., Monogr. Phan. **9:** 699. 1896.

Terrestrial, acaulescent herb, usually in small clusters; rhizomes short, prostrate, and stout. Leaves fasciculate, ascending, 60–95 cm long, rigidly coriaceous, tapering from a 5–6.5 cm wide base to a long-acuminate to filiform apex. Scape erect, stout, longer than the leaves, with lanceolate, long-acuminate, overlapping bracts; inflorescences 6–10, simple or double, spirally arranged compressed spikes, (8–)17–30 × 0.8–1 cm, many-flowered; floral bracts membranous-coriaceous, 2.2–2.6 cm long, ovate-elliptic, overlapping, purplish red. Flowers sessile, projecting beyond the floral bract; sepals white, lanceolate, ca. 3 cm long; petals white, linear, tubular, erect, ca. 3 cm long; stamens included; pistil exserted. Capsule not known.

DISTRIBUTION: Rare, known from a few populations, on rocky outcrops. Fish Bay (A815, M17074). Known only from St. John; and from Cerro Ventana on Vieques Island (Puerto Rico).

NOTE: *Tillandsia lineatispica* is a poorly known species originally described from St. John. Its morphology and distribution are intermediate between those of *T. utriculata* and *T. fasciculata; T. lineatispica* may be a sterile hybrid between the two species.

3. Tillandsia recurvata (L.) L., Sp. Pl., ed. 2, **1:** 410. 1762. *Renealmia recurvata* L., Sp. Pl. 287. 1753.

Epiphytic herb, forming large, dense clusters; stems typically shorter than the leaves, clothed by the leaf bases, usually branched at the basal portion; rhizomes wanting; roots slender, few. Leaves distichous; blades strongly recurved, 3–12 cm long, linear-cylindric, 0.5–2 mm diam., rigidly coriaceous, densely covered with peltate grayish scales. Scape erect to recurved, slender, to 13 cm long; scape bracts linear, 1 or 2, usually immediately below the inflorescence; inflorescence a single 1–2(–5)-flowered raceme; floral bracts like the scape bracts but smaller. Flowers nearly sessile, projecting beyond the floral bract; sepals lanceolate, 4–9 mm long; petals bluish, 2–2.2 cm long, linear. Capsule linear 2–2.5 cm long. Seeds with comose hairs 2–2.5 cm long.

DISTRIBUTION: Occasional on trees and shrubs of open or dry areas; commonly found on telephone wires on other islands such as Puerto Rico. Trail to Fortsberg (A4096), Cruz Bay (A2515). Also on St. Croix and St. Thomas (according to Britton & Wilson, 1923); southernmost United States to Argentina, including the West Indies.

COMMON NAME: old man hand.

4. Tillandsia utriculata L., Sp. Pl. 286. 1753.

Fig. 215G–I.

Tillandsia sintenisii Baker, J. Bot. **26:** 12. 1888.

Large, solitary, epiphytic, herb; stems very short, clothed by the leaf bases. Leaves in a rosette; blades recurved, 35–95 cm long, coriaceous, densely covered with grayish peltate scales below, tapering from a 3–9 cm wide base to a long-acuminate apex. Scape erect, to 1.5 m long, pyramidal; scape bracts lanceolate-caudate, decreasing toward the apex; inflorescences numerous, alternate, of simple or compound spikes, with numerous distichous flowers; floral bracts lanceolate, rigid, ca. 1.5 cm long, not overlapping. Flowers nearly sessile, to 5 mm long, projecting beyond the floral bract; sepals elliptic, 1.5–1.8 cm long; petals cream-colored to light yellow, 3–4 cm long, linear. Capsule cylindrical, 3.5–5 cm long. Seeds with comose hairs 2.5–3.5 cm long.

DISTRIBUTION: A common epiphyte of moist to dry forests. Lind Point (A1971), Center Line Road (A4034); Fish Bay (A3914). Also on Anegada, St. Croix, and St. Thomas; southeastern United States, Mexico, Central America, Venezuela, and the West Indies.

COMMON NAME: wild pin.

7. Cannaceae (Arrowroot Family)

Glabrous, erect perennial herbs, often with tuberous, starchy rhizomes. Leaves spirally arranged, the blade simple, large, with a prominent midrib and numerous lateral pinnate parallel veins, the petiole forming an open sheath around stems, lacking a ligule or pulvinus. Flowers showy, bisexual, zygomorphic, 3–merous, terminal racemose or paniculate thyrses, with numerous, short, 2-flowered cymules, axillary to the bracts; calyx green or purplish, of distinct sepals; corolla brightly colored, of basally connate petals, one smaller than the other 2; stamen 1, petaloid, the anthers reduced to a single pollen sac, opening by longitudinal slits; staminode 1, petaloid, sometimes 1–4 additional staminodes present; gynoecium of 3 connate carpels, ovary inferior, trilocular, verrucose or tuberculate, with axile placentation, the ovules numerous in each locule, the styles petaloid, with papillate stigmas. Fruit a dehiscent or indehiscent capsule.

A family of a single genus and about 9 species, native to the neotropics, with some species naturalized in Africa and Asia.

1. CANNA L.

Characters as given for the family.

1. Canna indica L., Sp. Pl. 1753. Fig. 216.

Canna coccinea Mill., Gard. Dict., ed. 8. 1768.

Herb 0.5–2 m tall; rhizomes stout, often tuberous. Leaves ascending; blades 20–50 × 10–20 cm, ovate to elliptic, chartaceous, the apex acute to shortly acuminate, the base rounded, obtuse, or attenuate, the margins entire. Flowers large, showy, solitary or in 2-flowered cincinni along simple or branched inflorescences; bracts ovate to obovate, 1–3 cm long, purplish. Sepals narrowly ovate to lanceolate, 1.5–2 cm long; corolla red, 4.5–6.5 cm long, the tube 1–1.5 cm long, the lobes linear; outer staminodes narrowly elliptic, 5–7 cm long; inner staminodes recurved, narrowly oblong-ovate, 4.5–6.5 cm long; stamen 4–6 cm long; ovary tuberculate, the style as long as the stamen. Capsule ellipsoid to globose, purplish, 1.5–3 cm long, tuberculate. Seeds black, globose, numerous.

DISTRIBUTION: An occasional escape along roadsides in humid areas. Bordeaux Road (A3869). Common throughout tropical America, naturalized in Asia and tropical Africa.

COMMON NAMES: Indian-shoot, tolama, toolima.

8. Commelinaceae (Wandering Jew Family)
Reviewed by R. B. Faden

Perennial or annual herbs. Leaves alternate, less often rosulate, simple, parallel-veined, the blade or the petiole forming a closed sheath around stems. Flowers small to medium, commonly bisexual, actinomorphic or zygomorphic, 3-merous, in terminal, axillary, or leaf-opposed cymes, the inflorescence often subtended by 1 or 2 folded, spathelike bracts; calyx usually green, of distinct or basally connate sepals; corolla usually blue, pink, or white, often ephemeral, of equal or unequal, clawed petals, free or basally connate into a short tube; stamens 6, in 2 cycles, sometimes one cycle reduced to staminodes, or these on one side of the flower, the filaments distinct, rarely connate or epipetalous, the anthers opening by longitudinal slits or seldom by terminal pores; gynoecium of 3 connate carpels, ovary superior, (1–2–)3-locular, the placentation axile, the ovules 1 to few in each locule, the style simple, terminal, with a capitate or trilobed stigma. Fruit often a loculicidal capsule or a berry.

A family of about 40 genera and 650 species, mostly tropical and subtropical.

REFERENCES: Faden, R. B. & D. R. Hunt. 1991. The classification of the Commelinaceae. Taxon 40: 19–31; Hunt, D. R. 1983. Commelinaceae. Fl. Trinidad and Tobago 3(3): 255–275; Hunt, D. R. 1983. New names in Commelinaceae. Kew Bull. 38: 131–133; Tucker, G. C. 1989. The genera of Commelinaceae in the southeastern United States. J. Arnold Arbor. 70: 97–130.

Key to the Genera of Commelinaceae

1. Inflorescence subtended and enclosed by 1 or 2 spathaceous bracts.
    2. Bracts paired; flowers actinomorphic .... 3. *Tradescantia*
    2. Bract solitary; flower zygomorphic .... 2. *Commelina*
1. Inflorescence not subtended by spathaceous bracts .... 1. *Callisia*

1. CALLISIA Loefl.

Decumbent or creeping perennial herbs. Flowers bisexual or pistillate, actinomorphic, sessile in paired cymes in the axils of leaves or bracts, along terminal, racemiform and paniculiform inflorescences; sepals 2–3, free; petals free, white or pink to blue, often inconspicuous; stamens 6 or fewer, sometimes 3 or 1, the filaments glabrous or bearded, the anthers with a wide, foliaceous connective; ovary sessile, oblong, 2–3-locular, nearly trigonous to compressed, with 1 or 2 ovules per locule, the style elongate, the stigma enlarged, sometimes penicillate. Fruit a membranous 2–3-valvate capsule.

A genus of about 20 species, native to the southern United States and the neotropics.

REFERENCE: Hunt, D. R. 1986. Amplification of *Callisia* Loefl. Commelinaceae. XV. Kew Bull. 41: 407–412.

Key to the Species of *Callisia*

1. Decumbent herb; stems 5–10 mm diam.; leaves 9–28 cm long; inflorescence elongate racemelike or panicle-like; ovary and capsule trilocular .... 1. *C. fragrans*
1. Prostrate herb; stems 1–2 mm diam.; leaves 1–4 cm long; flowers in axillary, paired cymes; ovary and capsule discoid bilocular .... 2. *C. repens*

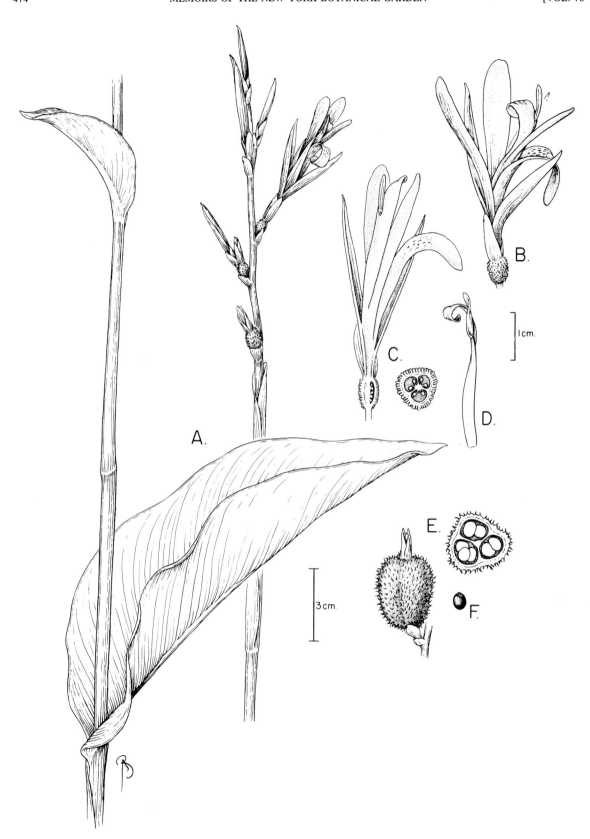

FIG. 216. *Canna indica.* **A.** Shoot and inflorescence. **B.** Flower. **C.** L.s. flower and c.s. ovary. **D.** Stamen. **E.** Capsule and c.s. capsule. **F.** Seed. (From Mori et al., in press, Guide to the Vascular Plants of Central French Guiana, Mem. New York Bot. Gard. 76.)

1. Callisia fragrans (Lindl.) Woodson, Ann. Missouri Bot. Gard. **29:** 154. 1942. *Spironema fragrans* Lindl., Bot. Reg. **26:** pl. 47. 1840. *Rectanthera fragrans* (Lindl.) O. Deg., Fl. Hawaii **1:** 62. 1932. Fig. 217A–E.

Decumbent herb 0.5–1.2 m tall, forming clumps, with basal portion rooting at nodes, stoloniferous. Leaves spreading; blades 9–28 × 3–4 cm, oblong-lanceolate to lanceolate, convolute, fleshy but drying chartaceous, the apex acute, the base slightly narrowed into the sheath, the margins entire; sheath tubular, ca. 1 cm long. Flowers small, in racemiform or paniculiform inflorescences; bracts ovate or 3-lobed, to 1 cm long, whitish; bracteoles lanceolate, 4–5 mm long. Sepals oblong, boat-shaped, hyaline, 4.5–5 mm long; petals oblong-lanceolate, white to pinkish-tinged, 4.5–5.5 mm long, papillose; filaments filiform, erect, white, the anthers elliptic, basal on a butterfly-shaped, white connective ca. 2 mm long; ovary trilocular, sharply trigonous, the style to 4 mm long, white.

DISTRIBUTION: Native to Mexico, cultivated and naturalized along roadsides and open disturbed areas. Fish Bay (A3870), Maria Bluff (A3805), trail to Solomon Beach (A2294). Also naturalized on St. Croix and Tortola; Bonaire (Dutch West Indies), Hawaii, India.

COMMON NAME: ground orchid.

2. Callisia repens (Jacq.) L., Sp. Pl., ed. 2, **1:** 62. 1762. *Hapalanthus repens* Jacq., Enum Syst. Pl. 12. 1760.

Prostrate, mat-forming herb; stems 1–2 mm diam., rooting at nodes, 10–25 cm long. Leaves spreading; blades 1–4 × 0.5–1.5 cm, cordate to lanceolate, fleshy but drying chartaceous, the apex acute, the base cordate, the margins ciliate; sheath tubular, 3–3.5 mm long, with long hairs at apex. Flowers small, in dense axillary paired cymes, enclosed by the sheath and leaf base; bracteoles filiform, ciliate, 6–7 mm long. Sepals oblong-elliptic, greenish, boat-shaped, 4–5 mm long; petals oblong, white-hyaline, as long as the sepals; stamens 0–6, the filaments ribbonlike, coiled, to 1 cm long, the anthers elliptic, basal on a kidney-shaped, white connective, ca. 0.5 mm long; ovary discoid bilocular, the style to 4.5 mm long, white, the stigma white. Capsule lenticular, ca. 1.7 mm long, opening from apex to the bottom.

DISTRIBUTION: Native to tropical and subtropical America; apparently introduced (and naturalized) on St. John; found along humid roadsides. Road to Ajax Peak (A2651), Waterlemon Bay (A1939), Lameshur (A3228). Also on St. Croix and Tortola; Texas, Mexico, Central America, West Indies, South America.

2. COMMELINA L.

Perennial or annual herbs, sometimes with tuberous roots. Leaves sessile or short-petiolate, usually with oblique bases. Flowers bisexual and/or staminate, 3-merous, zygomorphic, in single or paired cymes enclosed by a spathaceous bract, a few species with subterranean cleistogamous flowers; sepals unequal, free, the outermost one hood-shaped; petals blue or variously colored, unequal, the upper 2 clawed; stamens 6, 3 of which are fertile with ellipsoid to saddle-shaped anthers, the others modified into staminodes with X-shaped anthers, the filaments glabrous; ovary sessile, 2–3-locular, ovules 1 or 2 per locule. Fruit a dry dehiscent capsule; seeds smooth or variously patterned, with a linear hilum.

A cosmopolitan genus of about 170 species.

1. Commelina erecta L., Sp. Pl. 41. 1753. Fig. 217F–I.

Commelina elegans Kunth *in* Humb., Bonpl. & Kunth, Nov. Gen. Sp. **1:** 259. 1816.

Commelina virginica sensu many authors, non L., 1763.

Decumbent herb to 50 cm long; stems puberulent or glabrescent, rooting at nodes; ascending branches 20–25 cm long. Leaves ascending to spreading; blades 2.6–9.5 × 0.5–2 cm, lanceolate or oblong-lanceolate, chartaceous, puberulent, especially above, the apex acute to acuminate, the base acute or obtuse, sometimes unequal, the margins entire; sheath tubular, 0.8–1.6 cm long, ciliate and often auriculate at apex. Inflorescences solitary or 2–3 at end of branches; peduncle 0.7–2.2 cm long; spathaceous bract green, pilose externally, 1.7–2.5 cm long, broadly ovate, longitudinally folded inward into a boat-shaped structure, with margins connate only at base, containing 1 or 2 cymes; cymes 1–3-flowered. Flowers barely exserted from the spathaceous bract; pedicels ca. 3.5 mm long. Sepals 4–5 mm long, whitish-hyaline, the lower 2 connate; paired petals blue, 1–1.5 cm long, the claw sometimes whitish, the lower petal minute, white, linear; filaments of fertile stamens white, curved; ovary ovoid-globose, green, the style elongate, white, curved toward the minutely capitate stigma. Capsule trilocular, 5–6 mm wide, beige, 1 locule warty, indehiscent, 2 smooth, dehiscent. Seeds 1 per locule, light brown, smooth, ca. 2.7 mm long.

DISTRIBUTION: A common weedy herb of open, disturbed humid areas. Great Cruz Bay (A2357), Little Cruz Bay (A859). Also on St. Croix, St. Thomas, and Tortola; throughout temperate and tropical America and tropical Africa.

COMMON NAME: French grass.

DOUBTFUL RECORD: *Commelina diffusa* Burm.f., a common pantropical weed, was reported by Britton and Wilson (1923) and by Woodbury and Weaver (1987) as occurring on St. John. The Britton specimen has not been located, and the Woodbury record was found to be based on a misidentification of *Commelina erecta* L. *Commelina diffusa* is distinguished from *C. erecta* by the closed boat-shaped spathe with margins connate throughout their length (vs. open boat-shaped spathe with margins connate only at base).

3. TRADESCANTIA L.

Perennial herbs, sometimes with tuberous roots; stems usually not rhizomatous. Leaves alternate or rosulate, sessile or short-petiolate, with entire margins. Flowers bisexual, 3-merous, actinomorphic, in paired, fused cymes subtended and sometimes largely enclosed by a pair of spathaceous boat-shaped bracts; sepals free or less often connate at base, sometimes accrescent and fleshy in fruit; petals rose-purple to blue or white, free, sometimes clawed, or connate at base into a short tube; stamens 6, all similar and fertile, the filaments glabrous or bearded, the anthers with a broad connective; ovary 3-locular, with 1 or 2 ovules per locule, the stigma minutely capitate. Fruit a dry loculicidal capsule, rarely enclosed in fleshy sepals; seeds rugulose or rugulose-reticulate, with a linear or punctiform hilum.

A genus of about 70 species from temperate to tropical areas of the New World.

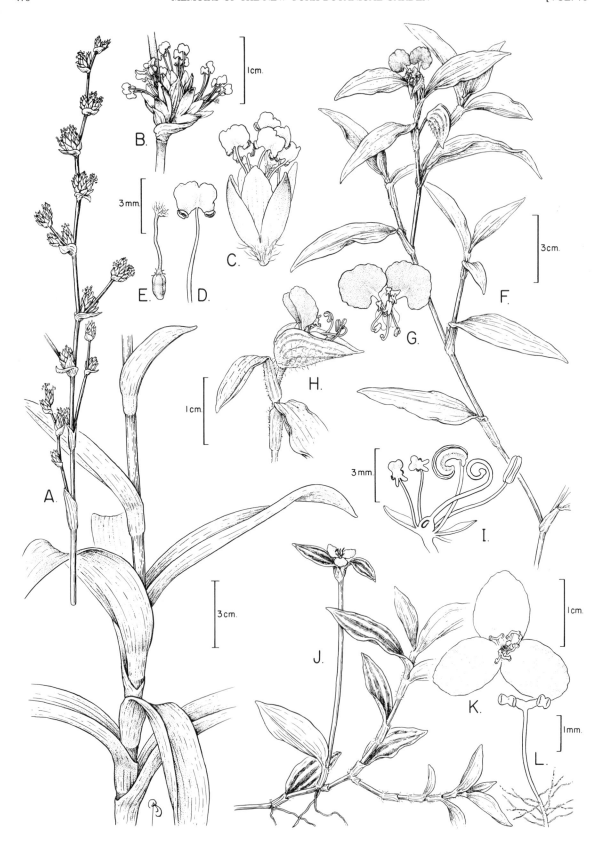

FIG. 217. A–E. *Callisia fragrans*. **A.** Habit. **B.** Inflorescence unit. **C.** Flower. **D.** Stamen. **E.** Pistil. **F–I.** *Commelina erecta*. **F.** Flowering branch. **G.** Flower. **H.** Inflorescence with subtending bract. **I.** L.s. flower. **J–L.** *Tradescantia zebrina*. **J.** Habit. **K.** Flower, top view. **L.** Stamen.

Key to the Species of *Tradescantia*

1. Decumbent herb; leaves strap-shaped, spirally arranged, 15–40 cm long, not variegated, sessile; corolla white .... 1. *T. spathacea*
1. Prostrate herb; leaves ovate-lanceolate, 3–10 cm long, distichous, variegated (an elongate whitish marking along each side of the midvein), with maroon margins, shortly petiolate; corolla rose-purple .... 2. *T. zebrina*

1. Tradescantia spathacea Sw., Prodr. 57. 1788. *Rhoeo spathacea* (Sw.) Stearn, Baileya **5**: 198. 1957.

Tradescantia discolor L'Hér., Sert. Angl. 8, t.12. 1788–1789. *Rhoeo discolor* (L'Hér.) Hance *in* Walp., Ann. Bot. Syst. **3**: 660. 1853.

Erect to decumbent herb to 35 cm tall, usually forming large colonies; stems to 1 cm diam., succulent, producing a watery, caustic sap. Leaves ascending, rosulate to spirally arranged; blades 15–41 × 2.8–6.4 cm, oblong, fleshy when fresh (chartaceous when dried), violet below, the apex acuminate, the base forming a tubular sheath, the margins entire. Inflorescences axillary; peduncles 2–4 cm long; spathaceous bracts overlapping, violet, 3–4.5 cm wide, longitudinally folded inward to conceal 2 many-flowered cymes. Flowers white; pedicels curved, 4–8 mm long. Sepals 5–6 mm long, lanceolate; petals 5–8 mm long; stamens shorter than the petals, the anther connective widely obdeltate; ovary ovoid-globose, with 1 ovule per locule, the style elongate, exserted, the stigma minutely capitate. Capsule trilobed, 4–4.5 mm long, grayish. Seeds 1 per locule, oblong-ellipsoid.

DISTRIBUTION: Cultivated but becoming naturalized. Trail to Margaret Hill (A2310). Also in Mexico, Guatemala, and Belize, naturalized in the West Indies.

2. Tradescantia zebrina Bosse, Vollst. Handb. Blumeng. **4**: 655. 1846.   Fig. 217J–L.

Zebrina pendula Schnizl., Bot. Zeitung (Berlin) **7**: 870. 1849. *Tradescantia pendula* (Schnizl.) Hunt, Kew Bull. **36**: 197. 1981.

Prostrate or trailing herb; branches few to numerous, to 35 cm long, rooting at nodes; stems slender, glabrous to pilose. Leaves spreading, alternate (distichous); blades 3–10(–12) × 1.5–3.7 cm, ovate-lanceolate, fleshy when fresh, chartaceous when dried, green, white-striped along both sides of midvein above, reddish purple below, the margins purplish, the apex acute, the base narrowed into a short petiole forming a tubular sheath 6–8 mm long. Inflorescences terminal, shortened, subtended by 2 leaflike bracts. Sepals 5–7 mm long, lanceolate; corolla rose-violet, the tube 5–6 mm long, the lobes 4–6 mm long, ovate, spreading; stamens exserted, the filaments the same color as corolla, pilose, the anthers on elongate connectives, both white; ovary 3-locular, ovules 2 per locule, the style exserted, white. Capsule trilocular. Seeds 1 or 2 per locule, ellipsoid, 1.3–1.5 × 1 mm, rugulose.

DISTRIBUTION: Cultivated but becoming naturalized. Along trail to L'Esperance (A4250). Native to Mexico but naturalized throughout the neotropics.

COMMON NAME: wandering Jew.

9. Cymodoceaceae (Manatee-Grass Family)

By M. T. Strong

Perennial, dioecious marine herbs, with elongate, creeping, monopodial or somewhat sympodial rhizomes, rooted in marine sands, the branches erect. Leaves distichous, ribbonlike or subulate; blades linear and 3- to several-nerved, or cylindrical and lacking a midvein (in ours), the apex truncate or emarginate; sheaths open, mostly ligulate. Flowers unisexual, minute, solitary and axillary or in simple or dichotomous cymes, sessile or stalked; perianth lacking; stamens with 2 partially fused filaments, the anthers 2- or 4-locular, usually dorsally fused, extrorsely dehiscent; pistils with 2 free carpels, ovule 1, apical, pendulous, the styles long, slender, and simple or 2- or 3-branched with slender stigmas. Fruit indehiscent with stony endocarp, viviparous; seeds solitary.

A family of 5 genera and 16 species, primarily in shallow subtropical and tropical marine waters, but a few in warm-temperate regions of the Old World.

1. SYRINGODIUM Kütz.

Submersed plants, with creeping, monopodial rhizomes, rooting at the nodes. Leaves 2–3, cylindrical, on short, erect branches; blades subulate, the apex truncate or concave; sheaths membranous, auricled at summit. Inflorescence cymose, the lower branches dichasial. Flowers enclosed in bracteal leaves; staminate flowers stalked; stamens 2 with fused filaments, the anthers 2-locular, attached at same height; pistillate flowers sessile or very shortly stalked, with 2 carpels, the styles short, 2-parted with filiform stigmas. Fruit drupaceous, indehiscent.

A genus of 2 species, 1 occurring in the Caribbean, the other in the Indo-Pacific region.

1. Syringodium filiforme Kütz. *in* Hohen., Algae Marinae Exs. **9**: no. 426. 1860. *Cymodocea filiforme* (Kütz.) Correll, Wrightia **4**: 74. 1968.   Fig. 218.

Cymodocea manatorum Asch., Sitzungsber Ges. Naturf. Freunde Berlin **1868**: 19, 24. 1868.

Submersed marine herb 5–25 cm tall; rhizomes reddish brown, 1–2 mm wide with erect branches; roots fibrous. Leaves 2–3; blades 2.5–30 cm long, 0.7–1.2 mm wide, green to dark green, red-spotted, the apex abruptly expanded, truncate to slightly concave, with blunt teeth; sheaths 2–5 cm long, the upper ones 1–1.6 cm long, reddish lineolate. Bracteal leaves 5–9 mm long, the

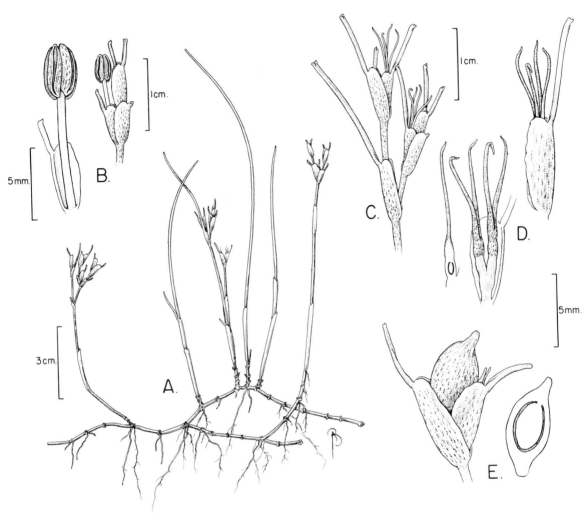

FIG. 218. *Syringodium filiforme.* **A.** Habit. **B.** L.s. bracteal leaf showing stamen, and portion of inflorescence with staminate flower. **C.** Portion of inflorescence showing pistillate flowers. **D.** L.s. carpel, l.s. bracteal leaf showing pistillate flower. **E.** Fruit with subtending bracteal leaves and l.s. fruit.

the anthers oblong-elliptic, 3–4 mm long. Pedicel of pistillate flower 1 mm long or less; style 2–3 mm long, the stigmas 4–7 mm long. Fruit ellipsoid to ellipsoid-obovoid, slightly flattened, 5–8 mm long, 2.5–3 mm wide, narrowing to a rounded base, the beak 2–5 mm long. Seeds 2.5–3 mm long.

DISTRIBUTION: Shallow saline water of lagoons and reefs. Fish Bay (*Acevedo-Rdgz. s.n.*). Probably around all islands of the Virgin Islands; throughout the Caribbean basin.

COMMON NAME: manatee-grass.

10. **Cyperaceae** (Sedge Family)

By M. T. Strong

Perennial or annual herbs, grasslike or rushlike, many with rhizomes or stolons, of aquatic or terrestrial habitats; roots fibrous; culms (stems) solitary or cespitose, 3-angled, obscurely 3(–5)-angled, or cylindrical, sometimes flattened, solid or hollow, ribbed or entire, smooth or scabrous, glabrous or sometimes pubescent. Leaves well developed or reduced to bladeless sheaths, borne at base or both basal and cauline; blades flat, pleated, folded, inrolled, or cylindrical, linear to lanceolate, glabrous or sometimes pubescent, scabrous on margins and veins or occasionally smooth; sheaths closed, the inner band membranous with a concave to convex orifice, or herbaceous, with the summit frequently prolonged as a triangular or tonguelike contraligule, the ligule short, sometimes a narrow band of hairs, or absent. Inflorescences diverse, typically umbelliform, corymbose, spicate, paniculate, racemose, or headlike, sessile or branching, bearing 1 to many spikelets; involucral bracts leaflike, reduced, or the lowest appearing as a continuation of the culm, rarely absent; spikelets 1- to many-flowered, typically with 1 to several empty (sterile) scales at base. Flowers (florets) spirally imbricate or 2-ranked, bisexual or unisexual (the plant then monoecious or rarely dioecious), borne singly, each from the axil of a scale; perianth, when present, of 3 to many smooth or barbed, hypogynous bristles or scaly segments, usually persistent at base of mature achene; stamens 1–3, the anthers elliptic to linear, basifixed, bilocular, theca parallel, longitudinally dehiscent, apiculate or appendaged at apex; ovary 2- or 3-carpellate, unilocular, style 2- or 3-branched, the unbranched portion capillary or flattened and straplike, sometimes 3-angled. Fruit an achene, 3-angled (sometimes dorsally compressed), 2-sided, globose or cylindrical, ellipsoid, oblong, or obovoid, usually apiculate at apex, smooth, reticulate, puncticulate, papillose, verrucose, or transversely rugulose, the style base deciduous or persistent at summit.

A cosmopolitan family of 110 genera and approximately 3500 species.

REFERENCES: Adams, C. D. 1992. Cyperaceae. Fl. Trinidad and Tobago 3(5): 383–561; Clarke, C. B. 1900. Cyperaceae. In I. Urban, Symb. Ant. 2: 8–169; Kral, R. 1971. A treatment of *Abildgaardia, Bulbostylis* and *Fimbristylis* (Cyperaceae) for North America. Sida 4: 57–227; Kükenthal, G. 1935. Cyperaceae-Scirpoideae-Cypereae. In A. Engler, Pflanzenr. ser. 4, 20(101): 1–671; Tucker, G. C. 1987. The genera of Cyperaceae in the southeastern United States. J. Arnold Arbor. 68: 361–445.

Key to the Genera of Cyperaceae

1. Fertile spikelet scales subtending unisexual flowers; achene (fruit) with crustaceous pericarp (with or sometimes without hypogynium at base; bristles none) .... 8. *Scleria*
1. Fertile spikelet scales subtending bisexual flowers (at least some present); achene lacking crustaceous pericarp (with or without bristles at base; hypogynium none).
   2. Spikelets bearing many flowers with only 1(–2) of the basal scales reduced and sterile.
      3. Achenes with bristles at base .... 4. *Eleocharis*
      3. Achenes without bristles at base.
         4. Scales spirally imbricate.
            5. Style base disarticulating from the summit of the achene; apex of leaf sheath entire ...... 5. *Fimbristylis*
            5. Style base persistent as a bulbous tubercle at summit of achene; apex of leaf sheath long-ciliate .... 2. *Bulbostylis*
         4. Scales 2-ranked.
            6. Unbranched portion of style fimbriate toward apex, expanded at base .... *Abildgaardia*
            6. Unbranched portion of style smooth, uniform at base .... 3. *Cyperus*
   2. Spikelets bearing 1 to several flowers with 2 to many basal scales reduced and sterile.
      7. Style base uniform, disarticulating from summit of the achene; scales 2-ranked .... 6. *Kyllinga*
      7. Style base expanded, persistent at the summit of the achene; scales spirally imbricate .... 7. *Rhynchospora*

1. ABILDGAARDIA Vahl

Plants perennial, tufted; culms obscurely angled, ribbed, ascending to deflexed, leafy toward base or leaves reduced to bladeless sheaths. Leaves basal; blades narrowly linear to filiform, scabrous on margins; blade-bearing sheaths closed, eligulate, the inner band membranous, splitting with age. Inflorescence a single spikelet or a cluster of several spikelets at the summit of the culm, or a simple, umbellate cyme, subtended by a single involucral bract; spikelets ovate, somewhat flattened, many-flowered; scales 2-ranked, at least toward the base, keeled, smooth. Flowers bisexual; bristles absent; stamens 2 or 3, the anthers narrowly oblong; style 3-branched, the branches angled, the unbranched portion narrowly wing-angled, smooth or remotely fimbriate on margins below, fimbriate toward apex, disarticulating below the expanded base. Achenes subglobose to obovoid, obscurely 3-angled, stipitate at base, verrucose; bristles absent.

Approximately 15 species in the New and Old World tropics.

1. Abildgaardia ovata (Burm.f.) Kral, Sida **4**: 72. 1971.
   Carex ovata Burm.f., Fl. Indica 194. 1768. *Fimbristylis ovata* (Burm.f.) J. Kern, Blumea **15**: 126. 1967.
                                              Fig. 219A–D.

   Cyperus monostachyus L., Mant. Pl. **2**: 180. 1771. *Abildgaardia monostachya* (L.) Vahl, Enum. Pl. **2**: 296. 1805. *Fimbristylis monostachya* (L.) Hassk., Pl. Jav. Rar. 61. 1848. *Iria monostachya* (L.) Kuntze, Revis. Gen. Pl. **2**: 751. 1891.
   Cyperus caribaeus Pers., Syn. Pl. **1**: 65. 1805.

Tufted perennial, 10–50 cm tall; rhizomes short, knotty; roots filiform; culms obtusely angled, 0.6–0.8 mm wide, 6–10-ribbed, glabrous, light green to straw-colored, smooth or upwardly scabrous near subflattened apex. Leaves several; blades 3–30 cm long, 0.8–1.1 mm wide, flat to somewhat inrolled, prominently veined on upper surface, light green to straw-colored, glabrous, scabrous on margins, abruptly acuminate at tip; sheaths glabrous, prominently veined on upper surface, ventral band open at summit, prolonged on either side into an auricle. Inflorescence of 1(–2) spikelets at summit of culm, ascending, subtended by a single involucral bract; rays when present, short, ribbed; spikelets ovate, 9–14 × 4–6 mm; scales ovate, 4–6 mm long, about as broad as long, glabrous, light green to straw-colored, keeled, with prominent midrib, mucronate at apex. Stamens 3, the anthers 1.5–2.5 mm long; style branches minutely fimbriate-scaly, one-fourth to one-third the length of the unbranched portion. Achene obscurely 3-angled, obovoid, 2.0–3.0 × 1.5–2.0 mm, stipitate at base, rounded to truncate at apex, apiculate, straw-colored to bone-white, verrucose-tuberculate.

DISTRIBUTION: Occasional in wet seepage areas. Bordeaux (A2614a), Dittlif Point (W648). Also on Anegada, St. Croix, and Virgin Gorda; pantropical.

2. BULBOSTYLIS Kunth, nom. cons.

Annuals or perennials, tufted, rarely solitary; culms ascending to deflexed, narrowly linear or filiform, coarsely ribbed. Leaves basal; blades capillary, setaceous, or narrowly linear, scabrous on margins, strongly nerved dorsally, pubescent, or glabrous; sheaths closed at summit, splitting with age, coarsely nerved, usually scaberulous, pubescent, or glabrous, the apex long-ciliate or fimbriate along margins, rarely glabrous, eligulate or ligulate. Inflorescence a simple or compound umbelliform cyme with elongate rays, a cluster of few to several spikelets at culm tips, or a solitary spikelet; involucral bracts 1–3(–5), leaflike, or wanting; rays filiform, ribbed, prophyllate, frequently with intraprophyllar buds; spikelets ovate to oblong-lanceolate, cylindrical or angled, sometimes compressed, many-flowered; scales spirally arranged, rarely somewhat 2-ranked, curvate-keeled or boat-shaped, pubescent or glabrous. Flowers bisexual; bristles absent; stamens 1– 3, the anthers oblong; styles 3-branched, with minutely papillate or scaly branches, the unbranched portion cylindrical or 3-angled, smooth or rarely ciliate distally, typically disarticulating above the swollen, bulbous base. Achenes 3-angled to obscurely 3-angled, rarely plano-convex, obovoid, with rounded angles, smooth, papillate, reticulate, or transversely rugulose; bristles absent.

Approximately 80 species distributed in warm-temperate and tropical regions of the world.

1. **Bulbostylis pauciflora** (Liebm.) C.B. Clarke, Kew Bull. **8:** 26. 1908. *Oncostylis pauciflora* Liebm., Kongel. Danske Vidensk. Selsk. Naturvidensk. Math. Afh. **5(2):** 241. 1851. Fig. 219E–H.

Stenophyllus portoricensis Britton, Torreya **13:** 216. 1913. *Bulbostylis portoricensis* (Britton) Fernald, Rhodora **40:** 392. 1938. *Fimbristylis portoricensis* (Britton) Alain, Bull. Torrey Bot. Club **92:** 290. 1965. *Bulbostylis ekmanii* Kük., Repert. Spec. Nov. Regni Veg. **23:** 197. 1926.

Tufted perennial, 10–30 cm tall; rhizomes short; roots filiform; culms obtusely angled, 0.2–0.3 mm wide, 4– 5-ribbed, wiry, reddish or chestnut-colored, glabrous, scabrous on ribs near apex. Leaves several; blades wiry, 2– 15 cm long, 0.2–0.3 mm wide, somewhat inrolled, prominently 3–5-veined on upper surface, reddish or chestnut-colored, glabrous, scabrous on margins toward apex, the apex acuminate; sheaths glabrous, prominently 3-veined on backs, long-fimbriate at apex. Inflorescence a single brown to chestnut-colored spikelet at summit of culm; involucral bracts 3, the lowest with sheathing base and apical fimbriae, elongate, overtopping the spikelet, the upper 2 typically scale-like, shorter than the spikelet; spikelet ovate-lanceolate, 6–10 × 1.5–2.5 mm; scales narrowly ovate, 3–3.5 × 1–2 mm, glabrous, light brown to chestnut-colored, curvate-keeled, the keel dark reddish brown, the apex acute, short-mucronate on basal scales. Stamens 3, the anthers 1–1.5 mm long; style 3-branched, one-half to two-thirds the length of the glabrous unbranched portion. Achene obovoid, 1.5 × 1.0 mm, 3-angled, transversely rugulose, brown, the angles lighter, yellowish, style base 3-angled, 0.3 mm long.

DISTRIBUTION: Rare, on rocky, coastal areas. Dittlif Point (W758). Also on St. Croix; Cuba, Hispaniola, Puerto Rico, and the Lesser Antilles.

3. CYPERUS L.

Perennials or annuals; rhizomes when present short or stoloniferous, rarely elongate and creeping; culms 3-angled or sometimes cylindrical, smooth or scabrous. Leaves primarily basal; blades herbaceous or stiff, flat, pleated, inrolled, cylindrical, or crescent-shaped, sometimes septate, the margins, ventral midvein, and dorsal lateral veins when present usually scabrous; sheaths rarely blade-less, eligulate or shortly ligulate, finely veined, sometimes septate, usually glabrous, the ventral band membranous. Inflorescence a simple or compound, terminal, umbel-like corymb, rarely congested or pseudolateral and headlike; involucral bracts leaflike, spreading, or the lowest erect and appearing as a continuation of the culm; rays unequal in length, glabrous, rarely scabrous, prophyllate at base, the lowermost each borne from the axil of a single involucral bract, the uppermost borne just below the base of the central spike, these frequently flexuous, becoming divergent to reflexed at maturity; spikes solitary or in umbels at ray tips, the central spike usually sessile or subsessile; spikelets ovate, ovate-lanceolate, or linear, many-flowered, flattened, subcylindrical or 4-angled, palmately, pinnately, or imbricately arranged on the rachis; scales 2-ranked, ovate, oblong-ovate, or elliptic, keeled or boat-shaped, sometimes 2-keeled, usually glabrous, the apex acute, obtuse, mucronulate, or cuspidate, the lowermost scale and subtending bract empty; rachilla winged from the persistent, decurrent base of the scale, or wingless, sometimes corky or thickened, disarticulating at base from the rachis, the whole spikelet falling entire, or disarticulating at the nodes, the internodes, scales, and achenes falling as 1-fruited segments, or rachilla persistent on the rachis, the scales and achenes deciduous. Flowers bisexual; bristles absent; stamens 1–3, the anthers oblong-elliptic, lanceolate, or linear, sometimes minutely appendaged at apex; styles 2- or 3-branched, slender, uniform, the branches minutely scaly, shorter than to exceeding the smooth unbranched portion. Achenes 3-angled or 2-sided, ovoid, obovoid, oblong-obovoid, ellipsoid, or narrowly ellipsoid, usually short-apiculate at apex, the surface puncticulate, reticulate, or smooth, sometimes transversely wrinkled; bristles absent.

A cosmopolitan genus with approximately 650 species, in temperate and tropical regions.

REFERENCE: McLaughlin, A. D. 1944. The genus *Cyperus* in the West Indies. Catholic Univ. Amer. Biol. Ser. 5: i–vii, 1–108.

Key to the Species of *Cyperus*

1. Inflorescence 5–8 mm diam.; culms 0.3–0.4 mm wide; leaves 0.3–1.6 mm wide .... 6. *C. nanus*
1. Inflorescence >2 cm diam.; culms >0.5 mm wide; well-developed leaves >2 mm wide.
   2. Stamen 1; achenes 0.8–0.9 mm long; scales 1.2–1.5 mm long .... 9. *C. surinamensis*
   2. Stamens 3; achenes >1 mm long; scales >1.5 mm long.

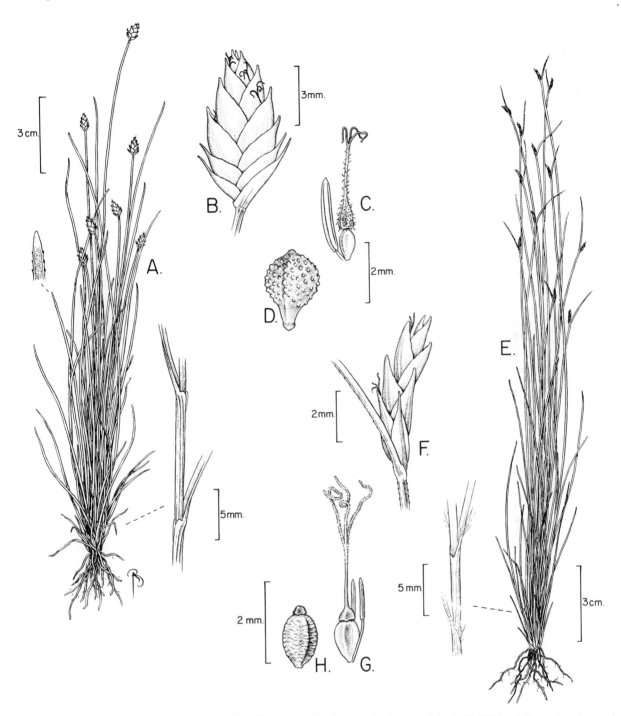

Fig. 219. A–D. *Abildgaardia ovata.* **A.** Habit, detail of leaf blade apex, and detail of culm showing apex of sheaths. **B.** Spikelet. **C.** Flower. **D.** Achene. **E–H.** *Bulbostylis pauciflora.* **E.** Habit and detail of culm showing apex of sheaths. **F.** Spikelet. **G.** Flower. **H.** Achene.

3. Inflorescence a single, sessile or subsessile glomerule of spikelets at the summit of the culm; rays wanting .... 1. *C. compressus* (part)
3. Inflorescence an umbel-like corymb; primary rays well developed.
    4. Spikelets palmately disposed in dense clusters or glomerules; leaves and inflorescence sticky; culms remotely scabrous; apex of scale with a slightly recurved cusp .... 3. *C. elegans*
    4. Spikelets spicately disposed in loose to dense spikes; plant not sticky; culms unarmed; apex of scale straight.

5. Spikelets 2.5–3 mm wide; rachilla wingless .... 1. *C. compressus*
5. Spikelets <2.5 mm wide; rachilla winged.
   6. Culms minutely papillose; leaf blades septate; spikes oblong-cylindrical to subglobose,
      congested at ray tips .... 5. *C. ligularis*
   6. Culms smooth; leaf blades not septate or obscurely so; spikes loosely disposed at ray tips.
      7. Rachilla disarticulating at the nodes (the rachilla node and achene falling together), the
         wings becoming thick and corky at maturity, enveloping the achene .... 4. *C. flexuosus*
      7. Rachilla persistent or disarticulating at base of spikelet (the spikelet falling entire), the
         wings not becoming thick and corky, thereby not enveloping achene.
         8. Spikelets 0.5–0.6 mm wide; scales 1.6–2.0 mm long, spreading at maturity .... 2. *C. distans*
         8. Spikelets 1–2 mm wide; scales 2.8–3.5 mm long, not spreading at maturity.
            9. Rhizomatous, rhizome short, thick, not bearing tubers; leaves stiff, 50–90 cm
               long, (5–)6–10(–13) mm wide, glaucous and obscurely septate on upper sur-
               face, pale and reddish brown lineolate on lower surface .... 7. *C. planifolius*
            9. Stoloniferous, stolons elongate, bearing tubers; leaves herbaceous, 5–40 cm
               long, 2–6 (–8) mm wide, green on both sides, not septate .... 8. *C. rotundus*

1. Cyperus compressus L., Sp. Pl. 46. 1753.

Densely tufted annual, 5–30(–50) cm tall; roots many, fine, capillary; culms 1–1.7(–2.0) mm wide, 3-angled, ribbed, green, smooth, glabrous, sheathing base of culm 2–5 mm wide. Leaves 2–7 from the base; blades folded to subflattened, frequently curved, 3–30 cm long, 1.5–3.0 mm wide, finely veined, glabrous, smooth to remotely scabrous on margins, acuminate at apex; sheaths finely veined, glabrous, red-tinged to purple-brown, eligulate, the ventral band red-dotted. Inflorescence an umbel-like corymb, 5–15 cm diam., or a sessile or subsessile, solitary spike; involucral bracts 3–5, 2.5–18 cm long, 1.2–3.0 mm wide, leaflike; rays 0–5, unequal, 3-angled to subcompressed, spreading, to 9 cm long, 0.5–0.9 mm wide, ribbed, smooth, glabrous; spikes hemispherical to rhomboid-obovoid, 1.1–3.5 × 1.2–5.0 cm; spikelets 3–18, condensed, oblong-elliptic to lanceolate-oblong, subcompressed, spreading, 8–25 × 2.5–3.0 mm, acute at apex, with (9–)12–38(–44) flowers; rachilla flat, wingless; scales decidu-ous, ovate to broadly ovate, folded, acute to acuminate, mucronate, 3.1–3.5 × 2.0–2.5(–3.0) mm, firm with whitish-scarious mar-gins, 11–15-nerved with broad green keel, the sides green or light brown, sometimes tinged with yellow. Stamens 3, the anthers 0.7–0.9 mm long, with a prolonged connective; style 3-branched, the branches two-thirds as long as to equaling the unbranched portion. Achene 3-angled, 1.1–1.5 × 1.0–1.1 mm, obovoid to broadly obovoid, obtuse to 3-lobed at apex, minutely apiculate, brown, shiny, minutely reticulate, the sides concave, obscurely ribbed on the angles.

DISTRIBUTION: Occasional in sandy soils of roadsides and beaches. Lameshur (W775). Also on St. Thomas and Virgin Gorda; a cosmopolitan weed.

2. Cyperus distans L.f., Suppl. Pl. 103. 1781.

Perennial, 30–100 cm tall; rhizome very short, cormlike; culms solitary or sometimes 2–3, 2–6 mm wide near base, 3-angled, finely to coarsely ribbed and channeled, green, smooth, glabrous, sheathing base of culm, (4–)5–10(–15) mm wide. Leaves 5–7; blades flat or sometimes folded, 20–60 cm long, 5–8 mm wide, finely veined, glabrous, subabruptly acuminate at apex, the mar-gins, midrib below, and the 2 lateral veins above scabrous; sheaths finely veined, septate, glabrous, pale brown to reddish brown, eligulate. Inflorescence a compound umbel-like corymb, (6–)11–21(–30) cm diam.; involucral bracts leaflike, (5–)6–7(–10), the lower to 70 cm long, 3.5–13.5 mm wide, the uppermost linear to subulate, shorter; rays 6–9, unequal, to 17 cm long, 0.5–1.2(–2)

mm wide, with tubular prophylls prolonged dorsally into 2, long, linear-subulate teeth, the primary rays elongate, secondary rays very short or the spikes sessile; spikes oblong-ovate to broadly ovate, 2.5–3.5(–4) × 1.5–3.0(–4) cm, the spikelets 9–32, loose, with (5–)7–35(–75) flowers, linear, cylindrical, divergent, 7–30(–70) × 0.5–0.6 mm (to 1.8 mm wide after spreading of mature flowers), rachilla flattened, zigzagged, winged, persistent; scales elliptic to oblong-elliptic, boat-shaped to obtusely keeled, 1.6–2 × 0.8–1.1 mm, remotely 2-ranked, spreading at maturity, membranous, with broad, scarious margins, 3–5-nerved, reddish brown on sides, with greenish keel, the apex obtuse. Stamens 3, the anthers 0.5–0.6 mm long; style 3-branched from just above the base. Achene 3-angled, oblong to oblong-ellipsoid, 1.3–1.7 × 0.4–0.5 mm, slightly curved on side facing rachilla, apiculate at apex, dark brown, minutely reticulate.

DISTRIBUTION: Occasional in seepage areas along roadsides. Bordeaux Mountain (W59, A5066), Gift Hill (A5282). Also on St. Croix, St. Thomas, and Tortola; a pantropical weed.

3. Cyperus elegans L., Sp. Pl., ed. 2, 1: 68. 1762.
Fig. 220I–K.

Cyperus viscosus Sw., Prodr. 20. 1788. *Scirpus viscosus* (Sw.) Lam., Tabl. Encycl. **1**: 142. 1791.
Cyperus confertus sensu Griseb., Fl. Brit. W. I. 563.
1864, for most part, non Swartz, 1788.

Cespitose perennial, 20–90 cm tall; rhizome short; culms 1.5–3.2 mm wide near base, obtusely 3-angled to subcylindrical, coarsely ribbed, remotely scabrous, sheathing base of culm 5–10 mm wide. Leaves several; blades inrolled along margins, laterally compressed, channeled toward apex, 7–75 cm long, 2–4 mm wide (unfolded), sticky, smooth, soft and spongy below, finely veined and septate above, scabrous on the margins, the narrowly acumi-nate bristly apex somewhat 3-angled; sheaths coarsely veined, septate, glabrous, eligulate. Inflorescence an umbel-like corymb, 9–20 cm diam.; involucral bracts leaflike, 5–6, septate, to 75 cm long, 2–4 mm wide (unfolded); rays obscurely 3-angled, remotely scabrous, the primary rays 5–9, unequal, ascending at anthesis, the lateral rays just below the central sessile spike divergent, reflexed at maturity, to 13 cm long, 0.5–1 mm wide, secondary rays 1–5, to 5 cm long, ascending at anthesis, becoming divergent (reflexed at maturity); spikes 8–30 × 10–25 mm; spikelets pal-mately disposed in dense clusters or glomerules of 5–40 at the ends of rays, ovate, oblong-ovate, or ovate-elliptic, subflattened,

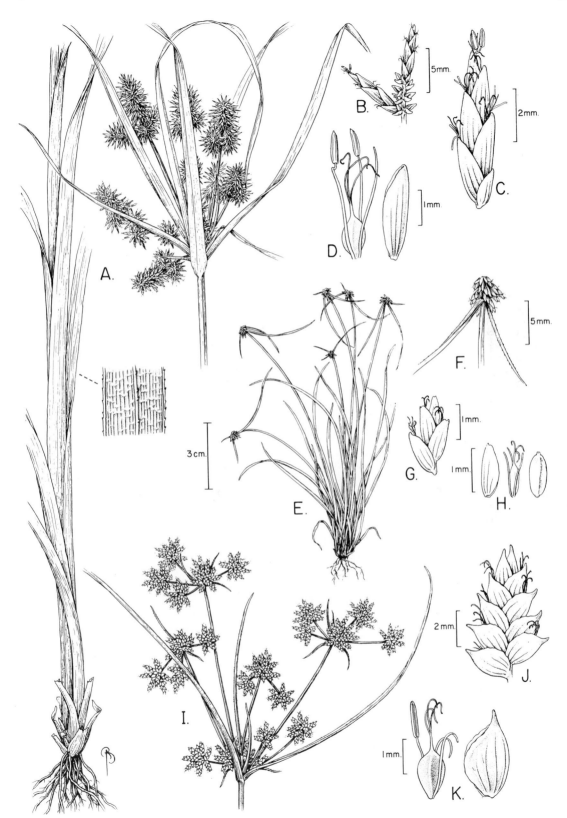

Fig. 220. A–D. *Cyperus ligularis*. **A.** Habit and detail of abaxial surface of leaf blade. **B.** Portion of inflorescence spike rachis with spikelets. **C.** Spikelet. **D.** Flower and subtending scale. **E–H.** *Cyperus nanus*, **E.** Habit. **F.** Inflorescence. **G.** Spikelet. **H.** Scale, flower, and achene. **I–K.** *Cyperus elegans*. **I.** Inflorescence. **J.** Spikelet. **K.** Flower and scale.

5–12 × 2–4 mm, with 6–20 flowers; rachilla wingless; scales broadly ovate, obtusely curvate-keeled, 2.2–2.9 × 2–2.5 mm, 7–9-nerved medially, with 2–3 coarse nerves on each side above the scarious margins, pale green, tinged with reddish brown on sides, sticky near base at maturity, scabrous on keel distally, the apex slightly recurved-cuspidate. Stamens 3, the anthers linear, ca. 1 mm long, apiculate; styles 3-branched from just above the base. Achene 3-angled, obovoid, 1.4–1.9 × 0.7–1 mm, short-beaked, maturing silvery-gray.

DISTRIBUTION: Common in open, disturbed areas. Trail to Fortsberg (A4084), Emmaus (A1999). Also on Anegada, St. Croix, St. Thomas, and Virgin Gorda; southern United States, Mexico, Central America, Peru, Argentina, and throughout the West Indies.

4. Cyperus flexuosus Vahl, Enum. Pl. 2: 359. 1806. *Torulinium flexuosum* (Vahl) T. Koyama, Phytologia 29: 74. 1974.

Cyperus insignis Kunth, Enum. Pl. 2: 92. 1837. *Cyperus flexuosus* var. *insignis* (Kunth) Kük. *in* Engl., Pflanzenr., ser. 4, 20(101): 622. 1936.
Cyperus ehrenbergii Kunth, Enum. Pl. 2: 89. 1837.
Diclidium vahlii Schrad. *ex* Nees *in* Mart., Fl. Bras. 2: 53. 1842. *Cyperus vahlii* (Schrad. *ex* Nees) Steud., Syn. Pl. Glumac. 2: 48. 1855. *Torulinium vahlii* (Schrad. *ex* Nees) C.B. Clarke *in* Urb., Symb. Ant. 2: 56. 1900.
Cyperus michauxianus Boeck., Flora 64: 77. 1881.

Perennial, 20–90 cm tall; rhizome short, woody; culms 1 to several, 3-angled, smooth, finely ribbed, 2.5–4.5 mm wide near base, the sheathing base 0.6–1 cm wide. Leaves 4–8, all basal; blades flat, 14–45 cm long, 4.5–11 mm wide, finely veined, glabrous, remotely scabrous on margins, smoothish on midvein beneath, long-acuminate at apex; sheaths finely veined, septate, glabrous, light brown, purplish near base, the inner band membranous with a convex to truncate apex. Inflorescence a simple or compound umbel-like corymb, 5–25 cm diam.; involucral bracts leaflike, 7–8, 3–51 × 0.5–1.2 cm, remotely scabrous on margins, long-acuminate at apex; rays 8–10, 0.1–14 cm long, 0.5–1.5 mm wide, obscurely 3-angled to compressed, finely ribbed, smooth; spikes widely ovate to rounded, 1.6–3.5 × 1.3–3 cm; spikelets 30–80, linear, subcylindrical, 0.5–2 cm long, 0.8–1.1 mm wide, spicate or sometimes radiate, somewhat congested, with the internodes of the rachis between spikelets <1 mm, the lower ones reflexed at maturity, with 3–12 flowers, disarticulating at base and rachilla nodes, the rachilla with broad scarious wings, these at maturity becoming thick and corky, yellowish to brown, enveloping the achene, the rachilla node and achene falling together; scales ovate to elliptic-ovate, somewhat dorsally compressed, inrolled along margins, 2.3–3.5 × 2.0–2.5 mm, 7–9-nerved, with thickly membranous, subcoriaceous, chestnut-colored, shiny, reddish brown to dark brown sides, these obscurely 1- or 2-nerved, with a narrow, scarious margin, the green median band extending as a short, prickly mucro below the obtuse apex. Stamens 3, the anthers narrowly ellipsoid, 0.5–1 mm long; styles subulate, 3-branched from below the middle. Achene 3-angled, oblong-obovoid, dorsiventrally compressed, 1.2–1.9 × 0.5–0.6 mm, slightly falcate at maturity, apiculate at apex, minutely puncticulate.

DISTRIBUTION: A common herb of moist seepage areas. Lameshur (B623), road from Bordeaux to Coral Bay (A3799). Also on St. Thomas; Cuba, Jamaica, Hispaniola, Lesser Antilles, Mexico, and Brazil.

5. Cyperus ligularis L., Pl. Jamaic. Pug. 3. 1759. *Mariscus ligularis* (L.) Urb., Symb. Ant. 2: 165. 1900.
Fig. 220A–D.

Mariscus rufus Kunth *in* Humb., Bonpl. & Kunth, Nov. Gen. Sp. 1: 216. 1816.
Cyperus sintenisii Boeck., Beitr. Cyper. 1: 12. 1888.
Cyperus trigonus Boeck., Beitr. Cyper. 1: 11. 1888.

Perennial, sometimes cespitose, 0.5–1.2 m tall; rhizome short, stout; roots thick; culms solitary, 3-angled, 3–6 mm wide near base, light green to glaucous, glabrous, minutely papillose, sheathing base of culm stout, to 3 cm wide. Leaves crowded at base; blades flattened, rarely the margins somewhat inrolled, 0.3–1.2 m long, 0.8–1.5 cm wide, finely veined, septate, glabrous, long-acuminate at apex, the margins and midvein beneath harshly scabrous; sheaths closely overlapping at base, finely veined, septate, glabrous, reddish to purplish brown at base, eligulate. Inflorescence a compound umbel-like corymb, 5–25 cm diam.; involucral bracts leaflike, 6–8, the lower elongate, to 90 cm long, 0.7–1.5 cm wide, long-acuminate at apex, the upper ones short, linear to subulate; rays 5–12, to 20 cm long, 1–2 mm wide, obscurely 3-angled, minutely papillose; spikes oblong-cylindrical to subglobose, to 3 cm long, 0.8–1.5 cm wide, reddish, congested at ray tips; spikelets oblong-elliptic, subcylindrical to subcompressed, 4–8 × 1–2 mm, disarticulating at base, falling entire, with 4–8 flowers. Scales boat-shaped, 2–3 × 1.5–2 mm, finely 9–11-nerved, glabrous, reddish to reddish brown, acute at apex, the rachilla with narrow, scarious wings. Stamens 3, the anthers ca. 1 mm long, apiculate at apex; style 3-branched from just above the base. Achene 3-angled, obovoid to broadly ellipsoid, with concave sides, 1.4–1.8 × 0.7–0.8 mm, red to reddish brown at maturity, puncticulate.

DISTRIBUTION: Common in shallow pools and wet areas along roadsides and in disturbed habitats. Bordeaux Mountain (A2094), Cruz Bay (A856), Reef Bay Trail by Petroglyphs (A2932). Also on St. Croix, St. Thomas, Tortola, and Virgin Gorda; tropical and subtropical America, tropical West Africa, Chagos Archipelago (Indian Ocean), Seychelles, and western Pacific islands and atolls.

6. Cyperus nanus Willd., Sp. Pl. 1: 272. 1797.
Fig. 220E–H.

Schoenus capillaris Sw., Prodr. 20. 1788, non *Cyperus capillaris* J. König *ex* Roxb., 1820. *Mariscus capillaris* (Sw.) Vahl, Enum. Pl. 2: 372. 1806. *Kyllinga capillaris* (Sw.) Griseb., Syst. Veg. Karaiben. 120. 1857. *Cyperus tenuis* Griseb. var. *capillaris* (Sw.) Kük., Repert. Spec. Nov. Regni Veg. 23: 188. 1926.

Cespitose perennial to 30 cm tall; rhizome short, knotty; roots filiform; culms 3-angled, wiry, finely ribbed, 0.3–0.4 mm wide, sparsely scabrous on angles near apex. Leaves crowded at base; blades narrowly linear, flattened, to 15 cm long, 0.3–1.6 mm wide, finely veined, glabrous, scabrous on margins; sheaths submembranous, finely veined, reddish brown, dark brown near base, eligulate. Inflorescence a single, globose to widely ellipsoid spike, 5–8 × 4–7 mm, with 10–30 spikelets, rarely with an additional lateral spike; involucral bracts 2–3, leaflike, reflexed, 1–8 cm long, 0.3–1 mm wide, green, scabrous on margins; spikelets oblong-ovate to oblong-lanceolate, subcompressed, 2–4 × 1 mm, reflexed at maturity, sometimes the terminal spikelet remaining erect, disarticulating at base, falling entire, the rachilla with narrow, scarious wings, bearing 3–5 flowers, the terminal flower staminate or wanting; scales ovate, 1.2–1.6 × 0.8–1 mm, 7–9-nerved, green to light reddish brown, with scarious margins and a dark green to brown median band that extends beyond the acute apex as a short mucro. Stamens 1(–2), the anthers oblong-elliptic, 0.4–0.5 mm long; style 3-branched from just above the base. Achene 3-angled, oblong-ellipsoid, 1–1.2 × 0.5–0.6 mm, with flat to slightly concave sides, apiculate at apex, light brown to purplish brown at maturity, puncticulate.

DISTRIBUTION: A rather common herb of wet, disturbed areas. Trail to Seiben (A2055), Reef Bay Trail by Petroglyphs (A2929). Also on St. Croix and Virgin Gorda; throughout the West Indies and Mexico.

7. Cyperus planifolius Rich., Actes Soc. Hist. Nat. Paris 1: 106. 1792. *Mariscus planifolius* (Rich.) Urb., Symb. Ant. 2(1): 165. 1900.

Cyperus purpurascens Vahl, Enum. Pl. 2: 359. 1806. *Mariscus purpurascens* (Vahl) C.B. Clarke *in* Urb., Symb. Ant. 2(1): 51. 1900.
Cyperus brunneus sensu Griseb., Fl. Brit. W. I. 565. 1864, in part, non Sw., 1797.

Robust perennial, 0.4–1 m tall; rhizome short and thick; culms 3-angled, (1.5–)2–4(–5) mm wide at base, smooth, the sheathing base 1–2.5 cm wide. Leaves many; blades stiff, flattish, crowded at base, 50–90 cm long, (5–)6–10(–13) mm wide, glaucous, scabrous on margins and midvein beneath, septate, sometimes obscurely so, frequently pale beneath, the surface reddish brown, lineolate, long-acuminate to obtuse tip; sheaths distinctly fine-veined, reddish to purple-brown, glabrous, eligulate, the inner band finely veined. Inflorescence an open, compound umbel-like corymb or rarely congested in 1 or 2 headlike clusters, (5–)10–20 cm diam.; involucral bracts 6–9, leaflike, the lowest 18–50 cm long, 4–10 mm wide, the upper linear, subulate; rays (6–)8–9(–12), unequal, obscurely 3-angled to slightly compressed, to 13 cm long, 0.6–1.2 mm wide, bearing clusters of spikes at apex, the primary thicker rays elongate, ascending, the secondary rays when present short, the lateral rays (just below the base of the central, subsessile spike) divergent to reflexed at maturity; spikes broadly ovate to broadly oblong-ovate, 1.3–2.5 × 1.5–2.8 cm; spikelets 6–30, narrowly linear, 6–15(–20) × 1.4–1.8 mm, spreading, falling entire, the lower ones divergent to reflexed at maturity; rachilla broadly winged, with 7–29 flowers; scales ovate-lanceolate, acutely keeled, 2.8–3.5 × 1.4–2.0 mm, 9–11-nerved, membranous, pale brown to reddish brown, with scarious margins, the narrow, 3-nerved keel ending in a short mucro below the acute apex. Stamens 3, the anthers ca. 1 mm long, with a minute, triangular, black appendage at apex; styles 3-branched to below the middle. Achene 3-angled, obovoid to obovoid-ellipsoid, 1.4–1.6 × 0.8–0.9 mm, with flat to slightly convex sides, blackish at maturity, minutely puncticulate.

DISTRIBUTION: Occasional in scrubs and dry coastal forests. Southside Pond (A1831). Also on St. Croix, St. Thomas, Tortola, and Virgin Gorda; Cuba, Hispaniola, Jamaica, Puerto Rico, and the Lesser Antilles.

8. Cyperus rotundus L., Sp. Pl. 45. 1753.

Stoloniferous perennial, 10–80 cm tall; stolons elongate, bearing tubers; culms solitary, tuberous at base, 3-angled to subcompressed, 1–3 mm wide, finely ribbed, smooth, glabrous, sheathing base of culm 3–10 mm wide. Leaves 7–17, clustered at base; blades flattish, 5–40 cm long, 2–6(–8) mm wide, finely veined, glabrous, scabrous along margins and midvein beneath toward

apex, acuminate at apex; sheaths short, finely veined, septate, glabrous, pale brown, faintly short-ligulate on lower sheaths, eligulate on upper sheaths. Inflorescence a simple to compound umbel-like corymb, 3–17 × 2.5–21 cm; involucral bracts 2–5, leaflike, to 20 cm long, 1.3–6 mm wide, scabrous along margins and midvein beneath toward the acuminate apex; rays 3–7, unequal, 4–10 cm long, 0.5–1.1 mm wide, finely ribbed, obscurely 3-angled to subcompressed; spikes broadly ovate, 1.5–4 × 1.5–7 cm; spikelets 2–12, linear-lanceolate, compressed, 10–45 × 2 mm, acute at apex; rachilla flexuous, broadly winged, bearing 11–51 flowers, persistent; scales deciduous, ovate to ovate-elliptic, 3–3.5 × 2–2.4 mm, subacutely keeled, 9-nerved (marginal ones inconspicuous) membranous, reddish brown to purplish brown, with scarious margins, the apex obtuse. Stamens 3, the anthers linear, 1.7–2.2 mm long, apiculate; styles elongate, subulate, 3-branched to below the middle, the unbranched portion subflattened. Achene 3-angled, ellipsoid to oblong-ellipsoid, 1.4–1.8 × 0.7–0.9 mm, brown to blackish, minutely puncticulate.

DISTRIBUTION: Occasional in sandy soils, along beaches, roadsides, and disturbed habitats. Coral Bay (*Raunkiær s.n.*). Also on St. Croix, St. Thomas, and Virgin Gorda; a cosmopolitan weed.

COMMON NAME: nut-grass.

9. Cyperus surinamensis Rottb., Descr. Icon. Rar. Pl. 35. 1773.

Cyperus formosus Vahl, Enum. Pl. 2: 327. 1806. *Cyperus surinamensis* var. *formosus* (Vahl) Kük., Repert. Spec. Nov. Regni Veg. 32: 74. 1933.
Cyperus surinamensis var. *lutescens* Boeck., Linnaea 35: 555. 1868.

Perennial, 0.5–1.2 m tall; rhizome short; culms cespitose or arising at short intervals along the rhizome, 2–6, 3-angled, (1–)2–4(–4.5) mm wide, retrorsely scabrous to nearly smooth, glabrous, sheathing base of culm (3–)4–7(–12) mm wide. Leaves 3–9; blades flat or sometimes laterally compressed, (12–)25–45(–65) cm long, 2–4(–6.5) mm wide, with scabrous or scaberulous margins, acuminate at apex; sheaths finely veined, septate, pale green to brownish, glabrous, eligulate, the inner band finely nerved. Inflorescence a compound, umbel-like corymb, (3–)5–18 cm diam.; involucral bracts 4–7, leaflike, flat, the lower ones to 40 cm long, 2–5 mm wide, remotely scabrous to smooth on margins and midvein beneath, acuminate at apex; rays (9–)14–16(–20), very unequal, obscurely 3-angled, cylindrical or compressed, smooth, finely ribbed, the primary ones to 15 cm long, 0.5–1 mm wide, stiff, ascending, the secondary ones 3–6, 1–2 cm long, divergent to reflexed at maturity, the lateral rays flexuous (just below the base of the central, sessile spike), divergent to reflexed at maturity; spikes broadly ovate to hemispherical, or subrounded, 0.7–1.5 × 0.7–1.8 cm; spikelets (10–)25–50(–90) per umbel, ovate, oblong-ovate to oblong-lanceolate, distinctly flattened, (3–)4–8(–12) × 1.8–2 mm; rachilla wingless, to 2 cm long, with (9–)11–31(–45) flowers; scales oblong-ovate to lanceolate, curvate-keeled, 1.2–1.5 × 0.7–0.9 mm, membranous, minutely reticulate, pale yellow, pale brown or reddish brown, the keel greenish, 3-nerved, 2-keeled near base, short-mucronate at apex, sometimes scabrous above. Stamen 1, the anther 0.7–0.8 mm long; style 3-branched to below the middle. Achene 3-angled, narrowly ovate-ellipsoid to narrowly oblong-ellipsoid, 0.8–0.9 × 0.3 mm, acute and apiculate at apex, brown to reddish brown, shiny, papillate, indistinctly rugulose, with flat to slightly convex faces.

DISTRIBUTION: Occasional in permanent shallow pools. Herman Farm

(W457), Catherinesberg (A5070). Also on St. Thomas; southern United States to South America and the West Indies.

CULTIVATED SPECIES: *Cyperus involucratus* Rottb., the umbrella-plant, has been cultivated on St. John but has not become naturalized.

DOUBTFUL RECORD: *Cyperus brunneus* Sw. was reported by Britton and Wilson (1923) as occurring on St. John; however, no specimens have been found to confirm this record, nor has the species been recently collected on the island.

4. ELEOCHARIS R. Br.

Perennials or occasionally annuals, rhizomatous, stoloniferous, or tufted; culms usually cylindrical, sometimes 3-angled, rarely quadrangular, solid or hollow, sometimes transversely septate, smooth, glabrous. Leaves reduced to bladeless sheaths; sheaths closed, tubular, the orifice truncate to obliquely truncate, apiculate at apex on the back, sometimes with a puckered, scarious appendage above the apex. Inflorescence a single, terminal spikelet at the summit of the culm; spikelets ovate to lanceolate, rarely obovate, obtuse to acute, many-flowered, with spirally imbricate scales, the lowermost 1(–2) empty; scales oblong, oblong-ovate to lanceolate, or obovate, sometimes subrounded, firm or membranous, usually with scarious margins, nerved or nerveless. Flowers bisexual; bristles usually present; stamens 3, the anthers elliptic to linear; styles capillary, 2- or 3-branched, with expanded base. Achenes 3-angled or 2-sided, obovate or elliptic, smooth or reticulate, the style base expanded, triangular or conical, sometimes spongy-thickened, persistent at the apex; bristles when present (3–)6(–12), retrorsely barbed, rarely smooth.

A cosmopolitan genus with approximately 250 species.

REFERENCES: Svenson, H. K. 1929. Monographic studies in the genus *Eleocharis*. Rhodora 31: 224–242; Svenson, H. K. 1939. Monographic studies in the genus *Eleocharis*. Rhodora 41: 1–77, 93–110.

1. Eleocharis geniculata (L.) Roem. & Schult., Syst. Veg. **2**: 150. 1817. *Scirpus geniculatus* L., Sp. Pl. 48. 1753.   Fig. 221A–E.

Scirpus caribaeus Rottb., Descr. Icon. Rar. Pl. 24. 1773. *Eleocharis caribaea* (Rottb.) S.F. Blake, Rhodora **20**: 24. 1918.

Scirpus geniculatus var. *minor* Vahl, Enum. Pl. Syst. **2**: 251. 1806. *Eleocharis geniculata* var. *minor* (Vahl) Roem. & Schult., Syst. Veg. **2**: 150. 1817.

Eleocharis capitata R. Br., Prodr. 255. 1810.

Densely tufted annual, 10–60 cm tall; roots filiform; culms obscurely angled, 0.4–0.9 mm wide, prominently ribbed, smooth, green, glabrous. Sheaths finely many-veined, smooth, green, glabrous, reddish to purplish at base, eligulate, closed at summit, with a concave to truncate orifice, the back prolonged into a short, acuminate, cuspidate tip. Inflorescence a single, erect spikelet at the summit of the culm; spikelets ovate, 2.5–4 × 2–3 mm, with obtuse apex, many-flowered, the flowers spirally imbricate; scales boat-shaped, obtuse at apex, glabrous, the basal scales elliptic to widely elliptic, the lowest one usually empty, 1.2–1.5 × 1.0–1.5 mm, 1-nerved, light green to straw-colored, smooth and shiny, with pale green midrib thickening toward the apex, tipped by an expanded callosity, the upper scales elliptic, 1.5–2 × 1–1.3 mm, dark green to brown with pale margins. Stamens 3, the anthers 0.5–0.6 mm long; styles 2-branched, the branches fimbriate-scaly, the short, unbranched portion smooth. Achene 2-sided, obovate, 1–1.1 (including style base) × 0.6–0.8 mm, smooth, lustrous, dark purplish black to black; bristles 6–8, rarely absent, whitish to tawny, retrorsely barbed, equaling or exceeding the whitish, depressed-conic, saucer-shaped tubercle.

DISTRIBUTION: Common in sandy, waterlogged areas. Concordia (A4253), Battery Gut (A4171), Fish Bay Gut (A2497). Also on St. Croix, St. Thomas, and Tortola; a pantropical weed.

DOUBTFUL RECORD: *Eleocharis mutata* (L.) Roem. & Schult. (synonym: *Scirpus mutatus* L.) was reported by Eggers (1879) and Britton and Wilson (1923) as occurring on St. John; however, no specimens have been found to confirm these records, nor has the species been recently collected on the island.

5. FIMBRISTYLIS Vahl

Perennials or annuals, rhizomatous or tufted with fibrous roots; culms cylindrical, or subcompressed to flattened, finely to coarsely ribbed, smooth, glabrous. Leaves basal; blades flat or inrolled, sometimes folded, narrowly linear to filiform, glabrous or pubescent, the margins smooth or scabrous; sheaths eligulate or ligulate, glabrous or pubescent, closed at summit, splitting with age, the apex entire. Inflorescence an open, simple or compound umbel-like cyme or sometimes a headlike cluster with congested spikelets; involucral bracts leaflike, usually smaller than the leaves, rarely overtopping the inflorescence; rays (and raylets when present) unequal, slender, finely ribbed, cylindrical to subflattened, smooth or scabrous; spikelets ovate, oblong-ovate, subglobose, or lanceolate, borne singly at ray tips or sometimes clustered, many-flowered; scales spirally imbricate, occasionally somewhat 2-ranked, ovate to oblong-ovate, obtuse to acute, 1–5-nerved, glabrous or pubescent, fertile except for the lowermost one. Flowers bisexual; bristles absent; stamens 1–3, the anthers oblong, sometimes apiculate; styles 2- or 3-branched, the unbranched portion flattened and fimbriate on margins (2-branched style) or slender, 3-angled basally, and glabrous (3-branched style), disarticulating from the summit of the achene. Achenes 2-sided or 3-angled, obovate, elliptic-obovate, or oblong, smooth, reticulate, verrucose, or warty; bristles absent.

A genus of approximately 200 species in warm-temperate and tropical regions worldwide, with the center of diversity in southeastern Asia.

Key to the Species of *Fimbristylis*

1. Unbranched portion of style slender, 3-angled at expanded base, entire; leaves eligulate; achenes verrucose-reticulate .... 1. *F. cymosa*

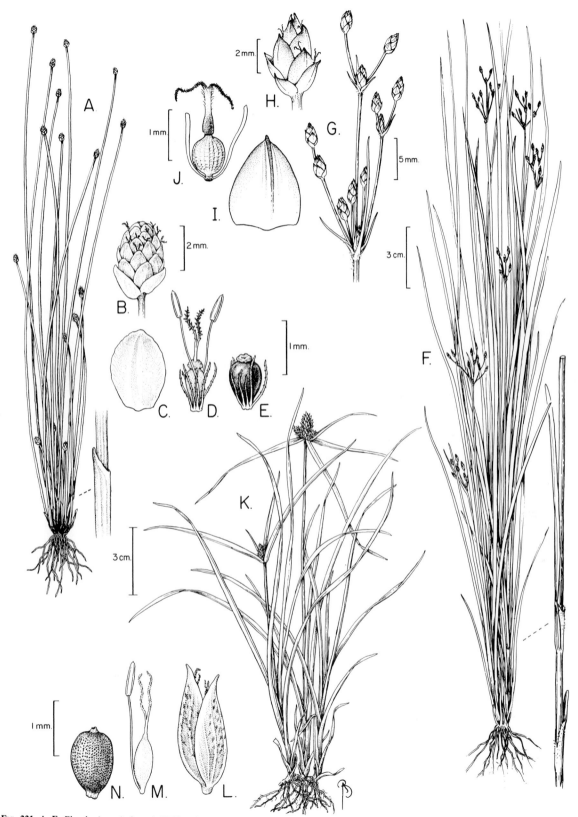

Fig. 221. A–E. *Eleocharis geniculata*. **A.** Habit and detail of culm showing sheath apex. **B.** Spikelet. **C.** Scale. **D.** Flower. **E.** Achene. **F–J.** *Fimbristylis dichotoma*. **F.** Habit and detail of culm showing apex of sheaths. **G.** Inflorescence. **H.** Spikelet. **I.** Scale. **J.** Achene with style still attached. **K–N.** *Kyllinga odorata*. **K.** Habit. **L.** Spikelet. **M.** Flower. **N.** Achene.

1. Unbranched portion of the style distinctly flattened, fimbriate along margin; leaves with a ligule of short hairs; achenes finely to coarsely reticulate.
   2. Scales glabrous; achenes coarsely reticulate with usually (5–)7–9(–12) rows of horizontally oriented, rectangular cells... 2. *F. dichotoma*
   2. Scales with dense, silvery, short-appressed hairs distally; achenes finely reticulate with usually 25–35 rows of isodiametric cells... 3. *F. ferruginea*

1. Fimbristylis cymosa R. Br., Prodr. 1: 228. 1810.

Fimbristylis spathacea Roth, Nov. Pl. Sp. 24. 1821.
   Fimbristylis cymosa ssp. *spathacea* (Roth) T. Koyama, Micronesica 1: 83. 1964.
Scirpus obtusifolius sensu Griseb., Fl. Brit. W. I. 571. 1864, non Lam., 1791.
Fimbristylis obtusifolia sensu C.B. Clarke *in* Urb., Symb. Ant. 2: 82. 1900, non Kunth, 1837.
Fimbristylis sintenisii Boeck., Bot. Jahrb. Syst. 7: 276. 1886.

Perennial, 5–60 cm tall; rhizome short, thick; culms erect, rigid, obscurely 3-angled or cylindrical, 0.5–2 mm wide at base, coarsely ribbed, green to straw-colored, the stout, sheathing bases hard and stiff, with remnants of old leaf bases. Leaves many; blades stiff, flat to somewhat inrolled, spreading, frequently falcate, 3–25 cm long, 0.8–4 mm wide, obscurely veined, green to straw-colored, glabrous, scabrous on margins, the apex with a curved tip; sheaths short, rigid, prominently veined, glabrous, pale green to dark brown, eligulate, the membranous inner band abruptly narrowing into the blade, with a concave orifice at summit. Inflorescence an open to congested, simple or compound cyme, 4–8 cm diam.; involucral bracts 1–3, leaflike, linear to subulate, shorter than the inflorescence or the lowest one sometimes exceeding the inflorescence; rays compressed, to 3 cm long, or wanting; spikelets ovate, 3–4 × 1.8–2.5 mm; scales ovate, slightly keeled, 1.2–1.6 × 0.8–1.2 mm, 3-nerved on keel, green to brown, with broad, scarious margins, the narrow, light green midrib extending as a short mucro below the obtuse to emarginate apex. Stamens 2, the anthers linear to narrowly elliptic, 0.5–0.7 mm long, apiculate at apex; styles 2-branched to below the middle, the unbranched portion glabrous, entire, 3-angled at expanded base. Achene unequally 2-sided, turgid on both faces, obovoid, 0.8–0.9 × 0.5–0.7 mm, broadly rounded to truncate at apex, narrowed to base, verrucose-reticulate, light brown, becoming dark brown to essentially black at maturity.
   DISTRIBUTION: A common herb of coastal areas, along beaches, roadsides, and disturbed habitats. Coral Bay (A2660), Great Cruz Bay (A796), Center Line Road (A3783). Also on St. Croix; a pantropical weed.

2. Fimbristylis dichotoma (L.) Vahl, Enum. Pl. 2: 287. 1805. *Scirpus dichotomus* L., Sp. Pl. 50. 1753.

Fig. 221F–J.

Scirpus diphyllus Retz., Observ. Bot. 5: 15. 1789.
   Fimbristylis diphylla (Retz.) Vahl, Enum. Pl. 2: 289. 1805.
Fimbristylis laxum Vahl, Enum. Pl. 2: 292. 1805.
Fimbristylis annua (All.) Roem. & Schult. var. *diphylla* (Retz.) Kük., Repert. Spec. Nov. Regni Veg. 23: 196. 1926.

Tufted perennial to 80 cm tall; rhizome short, nodose; culms obscurely angled, nearly cylindrical near base, compressed above, 0.8–2 mm wide near base, many-ribbed, green, glabrous. Leaves 5–7(–10); blades flat to somewhat inrolled, to 40 cm long, 1.5–3 mm wide, many-veined, green, glaucous, glabrous, rarely remotely pilose, with ciliate-scabrous margins, the apex with a curved tip; sheaths finely veined, glabrous or sometimes sparsely pilose, light reddish brown, ligulate, with a horizontal band of appressed hairs at ventral base of blade, ca. 0.5 mm wide, the inner band pilose above, ciliate along margins. Inflorescence a dense, simple to compound cyme, 2–10 cm diam.; involucral bracts 5–6, scabrous on margins, with sheathing base usually ciliate and pubescent, the lower ones leaflike, to 10 cm long, 1–2 mm wide, usually equaling to overtopping the inflorescence, the upper ones linear to filiform; rays flattened to subcompressed, remotely scabrous on margins, glabrous, the primary rays to 6 cm long, 0.2–1 mm wide, secondary rays shorter. Spikelets 10–60, ovate to ovate-lanceolate or oblong-lanceolate, acute, 4–8(–15) × 2–3.5 mm, solitary at ray tips, sometimes fasciculate. Scales ovate to broadly ovate or oblong-ovate, 2.1–3 × 1.8–2.2 mm, 3-nerved, glabrous, shiny, pale brown to reddish brown, usually dark brown toward apex, midrib pale green, the apex acute to obtuse, mucronate. Stamens 1 or 2, the anthers linear, 0.9–1 mm long, apiculate at apex; styles 2-branched, the branches one-third to one-half the length of the style, the unbranched portion flattened, ca. 0.5 mm wide, fimbriate along margins. Achene 2-sided, obovate, 1–1.3 × 0.8–1 mm, rounded and apiculate at apex, short-stipitate at base, whitish to brownish, coarsely striate-reticulate with (5–)7–9(–12) rows of horizontally arranged, rectangular cells on each side, the longitudinal ribs between the cells prominent.
   DISTRIBUTION: Occasional in wet areas along roadsides. Bordeaux (B537, A3831), Bethany (B277). Also on St. Croix, St. Thomas, Tortola, and Virgin Gorda; cosmopolitan, primarily in warm-temperate and tropical regions.

3. Fimbristylis ferruginea (L.) Vahl, Enum. Pl. 2: 291. 1805. *Scirpus ferrugineus* L., Sp. Pl. 50. 1753.

Scirpus debilis Lam., Tabl. Encycl. 1: 141. 1791, fide Vahl, Enum. Pl. 2: 292. 1805, non Pursh, 1814.
   Scirpus ferrugineus var. *debilis* (Lam.) Poir. *in* Lam., Encycl. 6: 780. 1804.
Fimbristylis ferruginea var. *compacta* Kük., Repert. Spec. Nov. Regni Veg. 23: 196. 1926.

Perennial to 1 m tall; rhizomes short, stout; culms obscurely angled, subcylindrical toward base, compressed near apex, 1.5–3 mm wide at base, many-ribbed, glabrous. Leaves 3–6; blades flat to inrolled along margins, to 40 cm long, 1–2 mm wide, light green to brown, glabrous, scabrous on margins, upper surface distinctly veined, the apex acute; sheaths prominently veined on back, pale green, ligulate at ventral base of blade, with a horizontal band 0.3–0.4 mm wide, of pale, appressed hairs. Inflorescence a simple or rarely compound, dense or sometimes somewhat headlike cyme, 2–4(–7) cm diam.; involucral bracts (2–)3–6, flattened to somewhat inrolled, to 7 cm long, 0.8–1.4 mm wide, scabrous on margins, shorter than to exceeding the inflorescence; rays compressed, short, finely ribbed; spikelets (1–)3–12, broadly to narrowly ovate, 6–12 × 3–5 mm; scales ovate to oblong-ovate, boat-shaped, 3.5–4 × 3 mm, 1-nerved, pale reddish brown except for the green or grayish green median band, with dense, silvery, short-appressed hairs and ciliate margins toward apex, the apex obtuse to rounded, shortly cuspidate. Stamens 3, the anthers linear, ca. 1 mm long, bluntly apiculate at apex; styles 2-branched,

the branches one-fourth to one-third the length of the style, the unbranched portion flattened, 0.4–0.5 mm wide, fimbriate along margins. Achene 2-sided, obovate, 1.5–1.7 × 1.1–1.2 mm, obtuse to truncate at apex, bluntly apiculate, straw-colored to pale brown, finely reticulate with 25–35 rows of isodiametric cells.

DISTRIBUTION: Occasional in wet, brackish habitats. Lameshur (B604). Also on St. Croix, St. Thomas, Tortola, and Virgin Gorda; a pantropical weed.

6. KYLLINGA Rottb.

Small, slender perennials or rarely annuals, with creeping, elongate or short knotty rhizomes, or tufted; culms 3-angled or obscurely so, smooth. Leaves basal; blades when present flat or somewhat pleated, the margins and ventral midvein scabrous; sheaths short, eligulate. Inflorescence a subglobose head of 1–4 crowded, sessile spikes; spikes cylindrical, ovoid, or globose, densely flowered with slender rachis; spikelets ovate to lanceolate, flattened, 1-flowered, disarticulating at base when mature, falling entire; scales 4, 2-ranked, the fertile scale ovate, folded, hyaline or membranous, with a smooth or scabrous keel, sometimes winged, mucronate or mucronulate at apex, the sterile scales 2, minute, basal, the apical scale similar to the fertile one, rarely staminate. Flowers bisexual and staminate; bristles absent; stamens 1–3, the anthers oblong-elliptic to linear; style 2-branched, the unbranched portion uniform at base. Achenes laterally flattened with one angle facing the rachilla, narrowly ovate to oblong or elliptic, cuneate to rounded at base, substipitate to distinctly stipitate, the apex rounded to obtuse, apiculate, the surface puncticulate; bristles absent.

A primarily pantropical genus of approximately 40 species, 8 of which occur in temperate, subtropical, and tropical regions of the New World.

1. Kyllinga odorata Vahl, Enum. Pl. 2: 382. 1806.
Fig. 221K–N.

Kyllinga sesquiflora Torr., Ann. Lyceum Nat. Hist. New York **3**: 287. 1836. *Cyperus sesquiflorus* (Torr.) Mattf. & Kük. *in* Engl., Pflanzenr., ser.4, **20(101)**: 591. 1936.

Tufted perennial, 7–30 cm tall; rhizome short, nodose, with filiform roots; culms crowded along nodes of the rhizome, obscurely 3-angled, somewhat compressed, 0.5–1.2 mm wide at base, distinctly ribbed, green, glabrous. Leaves 5–20; blades 3–18(–30) cm × 2–3 mm, many-veined, glabrous, scabrous along margins and midvein beneath, the apex long-acuminate; sheaths finely veined, green, glabrous, closed at summit, with a concave to truncate orifice, the membranous inner band red-dotted. Inflorescence a head of 1–3 short, whitish spikes, 1–1.5 cm diam., the central spike largest; involucral bracts (1–)3(–4), leaflike, 1–8(–12) cm × 1–3 mm, glabrous, exceeding the inflorescence;

spikes ovoid to cylindrical; spikelets densely crowded on rachis, ovate to ovate-lanceolate, 1-flowered, disarticulating from base at maturity, falling entire; scales broadly ovate, 2–2.5 × 1.2–2 mm, curvate-keeled, (5–)7(–9)-nerved, glabrous, whitish, minutely red-lineolate, the keel recurved at apex, short-mucronate, smooth or remotely scabrous toward base. Stamens 2, the anthers linear, 0.6–0.8 mm long, with connective prolonged at apex as a small appendage; styles 2-branched to about the middle. Achene oblong-ovate, 1.2–1.5 × 0.7–1 mm, broadly rounded to truncate at apex, apiculate, short-stipitate at base, reddish brown to dark brown, papillose.

DISTRIBUTION: Occasional along wet stream banks and roadsides. Rosenberg (B304). Also on St. Thomas; southeastern United States to South America, including the West Indies, also tropical Africa, Madagascar, Southeast Asia, Malaysia, and northern Australia.

DOUBTFUL RECORD: *Kyllinga brevifolia* Rottb. was reported by Britton and Wilson (1923), as occurring on St. John; however, no specimens have been found to confirm this record, nor has the species been recently collected on the island.

7. RHYNCHOSPORA Vahl

Perennials or sometimes annuals, vegetatively diverse; roots fibrous; culms cespitose or borne singly, 3-angled or obscurely so, sometimes channeled distally, occasionally cylindrical, smooth or sometimes scabrous distally, glabrous or sometimes hirsute. Leaves basal or basal and cauline; blades flattened or folded, sometimes inrolled, linear or filiform, herbaceous or occasionally stiff, glabrous, hirsute, or occasionally scabrous distally, rarely papillose, the margins and ventral midvein usually scabrous, ciliate, or with setose hairs; sheaths eligulate, rarely with a narrow band of trichomes, sometimes whitened at base, the inner band usually membranous, splitting with age. Inflorescence terminal or both terminal and lateral, paniculate, corymbose, or congested and headlike; involucral bracts leaflike, sometimes whitened at base; branches cylindrical, 3-angled, or subcompressed, ribbed, scabrous, ciliate, or smooth on margins; spikelets ovate, elliptic, or lanceolate, sometimes cylindrical or subcompressed, primarily 1- to several-flowered; scales spirally imbricate, ovate to lanceolate, finely nervcd, with a single, distinct midrib, light to dark brown, sometimes whitish, the lower 2 to many, sterile. Flowers bisexual above the empty basal scales, the uppermost often staminate with a rudimentary ovary; bristles 1–6(–20), rudimentary, or absent; stamens 1–3(–12), the anthers elliptic to oblong; styles subulate, 2-branched or undivided. Achenes 2-sided to subcylindrical, sometimes inrolled with winged or wavy margins, obovate, oblong-obovate, or oblong-elliptic, deeply pitted, transversely rugulose, or smooth, the expanded, usually triangular style base persistent at the summit; bristles when present antrorsely or retrorsely barbed, sometimes smooth or feathery.

A genus of approximately 250 species, with its greatest diversity in the Western Hemisphere, particularly warm-temperate North America and the neotropics.

REFERENCE: Thomas, W. W. 1984. The systematics of *Rhynchospora* section *Dichromena*. Mem. New York Bot. Gard. **37**: 1–116.

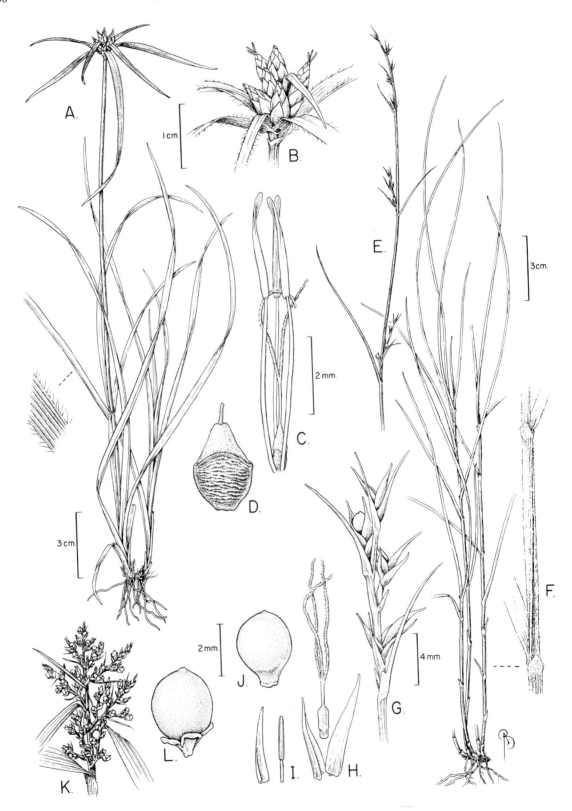

FIG. 222. **A–D.** *Rhynchospora nervosa* ssp. *ciliata*. **A.** Habit and detail of abaxial side of leaf blade. **B.** Inflorescence. **C.** Flower. **D.** Achene. **E–J.** *Scleria lithosperma*. **E.** Habit. **F.** Detail of culm showing apex of leaf sheaths with contraligule. **G.** Inflorescence. **H.** Flower and pistillate scales. **I.** Staminate scale and stamen. **J.** Achene. **K, L.** *Scleria scindens*. **K.** Inflorescence. **L.** Achene.

1. Rhynchospora nervosa (Vahl) Boeck. ssp. **ciliata** T.
    Koyama, Madroño **20**: 254. 1970 [as a new name].
    Dichromena ciliata Vahl, Enum. Pl. Syst. **2**: 240.
    1806 "*Dichroma*", nom. illegit., non Pers., 1805.
    Fig. 222A–D.

Dichromena pura Nees *in* Mart., Fl. Bras. **2**: 112.
    1842. *Rhynchospora pura* (Nees) Griseb., Fl. Brit.
    W. I. 577. 1864.
Rhynchospora vahliana Griseb., Fl. Brit. W. I. 577.
    1864.
Rhynchospora ciliata Kük., Bot. Jahrb. Syst. **56(Beibl.**
    125): 16. 1921.

Cespitose perennial, 10–70 cm tall; rhizomes short; culms
obtusely 3–angled, 1–2 mm wide at base, finely ribbed, green,
glabrous, or sometimes hirsute below, sheathing base of culms
whitened. Leaves 5–15; blades flat or canaliculate, 8–65 cm long,
1.5–4.5 mm wide, finely veined, hirsute beneath, the margins
scabrous, basally long-ciliate, the midvein scabrous beneath, the
apex narrowly acuminate; sheaths loose, 2–4 mm wide at base,
finely veined, light green to straw-colored, whitened at base,
hirsute to glabrous. Inflorescence a glomerate head of 5–16 white
spikelets, 1–1.5 cm diam.; involucral bracts 7–8, leaflike, longer
than the inflorescence, narrowly acuminate at apex, white on upper
surface at base, green and glabrous to hirsute beneath at base, the
margins scabrous, ciliate at base; spikelets ovate to widely ovate,
5–10 × 2–4 mm, several-flowered; scales 10–30, ovate, boat-
shaped, keeled above, 3.5–5 × 2–3 mm, glabrous, whitish,
frequently red-lineolate, with a narrow pale midrib, extending
beyond the acute, obtuse, rounded, or emarginate apex as a short,
straight or recurved mucro. Stamens 3, the anthers narrowly linear,
2–3 mm long, bluntly apiculate at apex; styles 2-branched to
about one-third length of style, the unbranched portion abruptly
expanding below into a shallowly, triangular base, persistent at
the summit of the achene. Achene 2-sided, turgid, depressed-
obovate, 1.8–2.0 × 1.1–1.6 mm, straw-colored to dark yellowish
brown, transversely rugulose, the style base about as wide as the
achene, convex at summit; bristles absent.

DISTRIBUTION: Occasional on grassy hillsides and roadside banks. Bordeaux (B562, A3830), Lameshur (W226). Also on St. Thomas; Mexico to
South America, and throughout the West Indies.

NOTE: This species was erroneously reported by Woodbury and Weaver
(1987) as *Dichromena radicans* Schltdl. & Cham.

8. SCLERIA Bergius

Perennials or sometimes annuals; rhizome when present elongate, or short and hardened, knotty or sometimes tuberous; culms
erect, elongating and sprawling, or climbing, 3-angled, smooth to harshly scabrous, glabrous or pubescent. Leaves well developed at
middle and upper nodes, the basal ones essentially bladeless; blades linear-elongate or sometimes lanceolate, flat to somewhat inrolled,
3-veined, herbaceous, weakly to harshly scabrous on margins and veins, glabrous or sometimes pubescent; sheaths 3-angled, closed at
summit, distinctly veined, the apex of the inner band with a rounded, obtuse, or triangular contraligule with distinct, straight or
anastomosing veins, the margins thickened or cartilaginous, sometimes with a short to elongate scarious appendage. Inflorescence
paniculate or spikelike, terminal, or terminal and axillary; involucral bracts when present, leaflike, usually shorter than the cauline
leaves; branches 3-angled, sometimes narrowly winged, scabrous or smooth; spikelets sessile or on pedicels to 1 cm long; staminate
spikelets lanceolate or narrowly oblong, cylindrical or subflattened, many-flowered, the scales spirally imbricate; pistillate spikelets
ovate to elliptic, cylindrical to subflattened, with 1 terminal flower usually subtended by a reduced scale, the scales 2-ranked, spreading
with the developing achene. Flowers unisexual, staminate and pistillate flowers in the same, or more commonly in separate spikelets;
hypogynium present or absent; stamens 1–3, the anthers appendaged at apex; styles capillary, 3-branched, the unbranched portion
glabrous. Achenes globose, rarely 3-angled, rounded, ovoid to ellipsoid, apiculate, sometimes subconic, with straight to recurved apex,
bony or crustaceous, white or sometimes variegated with purple, glabrous or pilose, the surface smooth, rugose, reticulate, trabeculate,
papillate, verrucose, or warty; hypogynium (when present) borne at base of achene, sessile or stipitate, smooth or crustaceous, entire
or 3-lobed, sometimes 3–9-tuberculate near base, the lobes entire or dissected, glabrous or ciliate along margins.

A genus of primarily warm-temperate and tropical regions worldwide, with approximately 200–225 species.

REFERENCE: Core, E. L. 1936. The American species of *Scleria*. Brittonia 2: 1–105.

Key to the Species of *Scleria*

1. Culms 1.2–1.3 mm wide near base; leaves 1–3 mm wide; achenes without a hypogynium .... 1. *S. lithosperma*
1. Culms 3–10 mm wide near base; leaves 5–17 mm wide; achenes with a 3-lobed hypogynium.
    2. Culms 3–5 mm wide near base; achenes 2–2.9 mm long (including hypogynium), 2–2.5 mm wide; ter-
        minal panicle 1–2.5 cm wide .... 2. *S. melaleuca*
    2. Culms 5–10 mm wide near base; achenes 3–3.9 mm long (including hypogynium), 2.5–2.9 mm wide;
        terminal panicle 3–8 cm wide.. 3. *S. scindens*

1. Scleria lithosperma (L.) Sw., Prodr. 18. 1788. *Scirpus*
    lithospermus L., Sp. Pl. 51. 1753, in part.
    Fig. 222E–J.

Scleria filiformis Sw., Prodr. 19. 1788. *Scleria lith-*
    osperma var. *filiformis* (Sw.) Britton, Ann. New
    York Acad. Sci. **3**: 231. 1885.
Scleria purpurea Poir. *in* Lam., Encycl. **7**: 4. 1806.
Scleria krugiana Boeck., Beitr. Cyper. **1**: 35. 1888.

Perennial, 50–90 cm tall; rhizomes short, nodulose; culms stiff
and erect, crowded together along nodes of the rhizome, 3-angled,
filiform, 1.2–1.3 mm wide near base, green, lightly scabrous on
the angles, pilose above. Leaves 6–10, shorter than the inflorescence; blades inrolled along margins, 10–20 cm long, 1–3 mm
wide, many-veined, 5-ribbed, scabrous on margins, long-acuminate at apex, the lower surface green, pilose to glabrescent, the
upper surface bluish green, shiny and glabrous except for the
pilose midrib; sheaths many-veined, pilose, red-tinged, purplish at
base, eligulate, closed at summit with a triangular or subrounded

contraligule, the apex hispid. Inflorescence of a terminal and 2–3 remote, axillary, narrow panicles, 2–4 × 1 cm; bractlets subulate, to 3 cm long; panicle branches scabrous on margins; spikelets sessile or short-stalked, bisexual or unisexual; pistillate spikelets narrowly ovate, 5–7 × 2 mm, spreading with developing achene, the 4–5 scales ovate-lanceolate, 3–5 × 1.2–1.4 mm, thin, glabrous to sparsely pilose, dark brown above, ciliolate along margins, short-mucronate at apex, the fertile scale typically bearing a reduced staminate spikelet just below attachment (cupula) of maturing fruit, usually with 1(–2) stamens, the scales membranous, about as long as the subtending pistillate scale; staminate spikelets narrowly oblong, cylindrical, 4–5 × 1 mm, the scales numerous, lanceolate-acuminate, 4–5 × 1 mm, membranous, brown. Stamen 1, the anthers 1–2 mm long, apiculate at apex; styles 3-branched from near the base. Achene ellipsoid to ellipsoid-obovoid, umbonate, 2–2.5 × 1.5–1.8 mm, shiny, smooth, bone-white when mature, with 3 basal depressions, not subtended by a hypogynium, the disk reduced to a brown, glandular ring.

DISTRIBUTION: Common herb in understory of moist to dry forest and along gallery forests and roadsides. Bordeaux (A3828), Battery Gut (A4167), Fish Bay (A1966). Also on St. Croix, St. Thomas, and Virgin Gorda; a pantropical weed.

2. Scleria melaleuca Rchb. *ex* Schltdl. & Cham., Linnaea **6**: 29. 1831. *Scleria pterota* var. *melaleuca* (Rchb. *ex* Schltdl. & Cham.) Uittien *in* Pulle, Fl. Suriname. **1(1)**: 140. 1934.

Scleria pterota C. Presl, Isis (Oken) **21**: 268. 1828, nomen nudum.

Scleria communis Kunth, Enum. Pl. **2**: 340. 1837.

Scleria pratensis Lindl. *ex* Nees *in* Mart., Fl. Bras. **2(1)**: 179. 1842, nom. illegit.

Scleria communis Liebm., Kongel Danske Vidensk. Selsk. Naturvidensk. Math. Afh. **5(2)**: 71. 1850, nom. illegit.

Scleria ottonis Boeck., Linnaea **38**: 490. 1874.

Perennial, 0.6–1 m tall; rhizome thick, nodose, purplish brown, with thickened roots; culms erect, borne singly along the rhizome, 3-angled, winged below inflorescence panicles, 3–5 mm wide at base, stiff, hard, green, glabrous to sparsely pilose, scabrous on angles above. Leaves 4–8, the upper ones equaling to exceeding the inflorescence; blades flat, 20–60 cm long, 0.6–1.7 cm wide, many-veined, 5-ribbed, obscurely septate near base, thin, glabrous or sparsely pilose near base, scabrous on margins and midvein beneath, the apex short to abruptly acuminate with blunt tip; sheaths inflated toward apex, to 8 mm wide, finely veined, indistinctly septate, scabrous on margins, sparsely pilose to glabrous, thin, friable, green, purple-tinged at base, eligulate, closed at summit with a triangular contraligule, the thickened margins hispid to glabrescent. Inflorescence a terminal and 2–3 axillary, green to purplish panicles, to 7 cm long, 1–2.5 cm wide; bractlets flattened, falcate, to 2 cm long; panicle branches wing-angled, scabrous on angles; pistillate spikelets elliptic to elliptic-obovate, subflattened, 3–4 × 3–4 mm, spreading with developing achene, the scales 3–4, ovate to widely ovate, 3 × 2–3 mm,

glabrous, shiny, abruptly acuminate at the recurved tip, short-mucronate, the margins ciliate, dark purple above; staminate spikelets narrowly oblong, cylindrical, 4–5 × 1.5 mm; the scales numerous, the basal 2 or 3 sterile, like the pistillate, narrowly ovate, 3–4 × 1–2 mm, the upper fertile, oblong-ovate to lanceolate, membranous. Stamen 1, the anthers 1–1.2 mm long, with a dark brown, triangular, bristly appendage at apex; style 3-branched, the branches two-thirds as long as to equaling the length of the unbranched portion. Achene depressed-globose to globose, subumbonate, 2–2.9 × 2.0–2.5 mm, shorter than the scales, smooth or sometimes hairy near base, white or variegated with purple; hypogynium depressed, 3-lobed, the lobes broadly rounded with glabrous margins.

DISTRIBUTION: Common along wooded hillsides, roadsides, and seepage areas. Bordeaux Mountain (W58, A2096), between Bethany and Rosenberg (B275). Also on St. Croix, St. Thomas, and Tortola; Mexico to South America, and the West Indies.

COMMON NAME: cutting grass.

3. Scleria scindens Nees *ex* Kunth, Enum. Pl. **2**: 343. 1837. Fig. 222K, L.

Scleria chlorantha Boeck., Linnaea **38**: 506. 1874.
Scleria scaberrima Boeck., Beitr. Cyper. **2**: 41. 1890.

Robust perennial, 0.5–3.5 m tall, growing in large colonies; rhizome stout, nodose; culms erect or ascending, vinelike and clambering, solitary from nodes of the rhizome, sharply 3-angled, 0.5–1 cm wide near base, scabrous on angles, sparsely strigose toward the apex, glabrous at base. Leaves numerous, overtopping the inflorescence; blades flattened, to 90 cm long, 0.5–1.5 cm wide, many-veined, obscurely septate, stiff, somewhat coriaceous, 5-ribbed, blunt to sharply scabrous on margins and midvein beneath, the apex acuminate to blunt tip; sheaths loose, inflated near summit, many-veined, glabrescent, eligulate, closed at summit with a convex to truncate contraligule, the margins cartilaginous. Inflorescence a dense, pyramidal panicle 3–8 cm diam., or sometimes with 1 or 2 additional smaller panicles in the upper axils; involucral bracts subtending the terminal panicle, typically 3, 9–50 × 0.6–1.4 cm wide, overtopping the inflorescence; bractlets setaceous, scabrous; panicle branches minutely pubescent to glabrescent; pistillate spikelets ovate-elliptic, 4–5 × 2 mm, spreading with developing achene, the scales 4, ovate-orbicular, 3.5–4 mm in diam., boat-shaped, glabrescent, green, flecked with brown, short-acuminate, mucronate, ciliate on margin; staminate spikelets narrowly oblong-ovate, cylindrical to slightly compressed, about the same size as the pistillate, the scales many, the 2–3 at base sterile, like the pistillate, 3–4 × 2–3 mm, the upper fertile, narrowly ovate to lanceolate, membranous. Stamens 3, the anthers 2–3 mm long, apiculate at apex; styles 3-branched, the branches one-third to one-half the length of the style. Achene globose, 3–3.9 × 2.5–2.9 mm, minutely apiculate, shiny white to purple-tinged; hypogynium 3-lobed with broad, rounded lobes, the margins of the lobes entire, reflexed.

DISTRIBUTION: Common in secondary vegetation along trails and roadsides. Bordeaux Mountain (A2606, A2884, A5102). Also on St. Thomas, Tortola, and Virgin Gorda; Cuba, Hispaniola, Puerto Rico, and the Lesser Antilles.

COMMON NAME: razor-grass.

11. Dioscoreaceae (Yam Family)

Twining vines or less often erect herbs, with thickened starchy tubers. Leaves alternate or opposite, simple or palmately compound; petioles usually jointed at base and forming a stipule-like flange. Flowers small, commonly unisexual (the plant dioecious) or less often bisexual, actinomorphic, 3-merous, in axillary racemes, spikes, or panicles; perianth of 6 nearly equal tepals, usually connate at base into a short tube; stamens 6, in 2 cycles, sometimes the inner cycle reduced to staminodes or vestigial, the filaments distinct, sometimes adnate to the perianth tube, the anthers opening by longitudinal slits; gynoecium 3-carpellate, ovary inferior, trilocular, with axile placentation, the ovules 2 to many per locule, the styles 3, terminal, distinct, or basally connate. Fruit commonly a capsule or less often a berry or a samara.

A family of 5 or 6 genera and about 650 species, mostly tropical and subtropical.

1. DIOSCOREA L.

Dioecious, twining, herbaceous to woody vines, with large, single or clustered tubers; stems cylindrical, angled or winged, armed or unarmed. Leaves alternate or opposite, simple or palmately lobed, with arching parallel veins, long-petiolate, usually with aerial bulbils in axils. Flowers unisexual, 3-merous, actinomorphic, in axillary spikes, racemes, or panicles. Perianth minute; staminate flowers with 6 stamens, all fertile or the inner cycle modified into staminodes; pistillate flowers bearing staminodes and an inferior ovary. Fruit a dry, chartaceous to leathery, 3-winged capsule; seeds numerous, flattened, winged.

A tropical and subtropical genus of about 600 species.

Key to the Species of *Dioscorea*

1. Stems 4-winged; leaves opposite .... 1. *D. alata*
1. Stems cylindrical, not winged; leaves alternate .... 2. *D. pilosiuscula*

1. Dioscorea alata L., Sp. Pl. 1033. 1753.

Herbaceous twining vine, 10–15 m long; stems slender, glabrous, 4-winged. Leaves mostly opposite; blades 6–26(–30) × 4–17 cm, ovate, coriaceous, glabrous, with 5–7 parallel veins, the apex acute or acuminate, the base cordate, the margins entire; petioles 4–12 cm long, 4-winged, forming an auriculate sheath around the stem. Inflorescences axillary; staminate inflorescences a panicle 5–15 cm long, with numerous lateral, flexuous, densely flowered spikes; pistillate inflorescences simple, 9–25 cm long, in distant to approximate flowered spikes. Perianth of staminate flowers 1–1.5 mm long; perianth of pistillate flowers 2–2.8 mm long. Capsule 3-winged, 2–3 cm diam.; each locule 2-seeded.

DISTRIBUTION: Collected only once on St. John and perhaps not persistent on the island. Bordeaux (W461). Native of Southeast Asia but widely cultivated throughout the tropics.

COMMON NAMES: red yam, yam.

2. Dioscorea pilosiuscula Bertero *ex* Spreng., Syst. Veg., ed. 16, **2**: 152. 1825.   Fig. 223.

Slender, herbaceous twining vine, 2.5–5 m long; stems slender, wiry, glabrous or puberulent, cylindrical to slightly angled. Leaves alternate; blades 5.5–12 × 2.5–6.5 cm, oblong-ovate to ovate, chartaceous, minutely pilose along veins below, with 5–7 parallel veins, the apex acuminate, the base cordate, the margins entire; petioles 1.5–4 cm long, slender, furrowed and swollen at both extremities; bulbils paired in leaf axils, ovoid, verrucose, 2.5–3 cm long. Inflorescences axillary; staminate inflorescences of 1–4 clustered spikes, 15–30 cm long; pistillate inflorescence a solitary spike, 9.5–30 cm long. Perianth of pistillate flowers 1.5–2 mm long. Capsule oblong, 3-winged, 1–2.5 cm long. Seeds ca. 8 mm long, with a basal wing.

DISTRIBUTION: Occasional in open disturbed areas. Maho Bay Quarter along Center Line Road (A2419), trail to Sieben (A2645). Also on St. Thomas and Tortola; throughout the Greater and Lesser Antilles and South America.

12. Dracaenaceae (She-Dragon Family)

By M. T. Strong

Trees or shrubs with woody stems or large stemless or short-stemmed plants with creeping rhizomes, the stems with secondary thickening growth. Leaves lanceolate or linear-lanceolate to ovate, flat, channeled, cylindrical, or laterally compressed, sessile, glabrous, fleshy or leathery, rigid or flexible, often concentrated in rosettes either at base or at branch tips. Inflorescence an axillary raceme or panicle. Flowers bisexual, hypogynous, often fragrant; pedicels articulate; perianth tubular or funnel-shaped, the lobes 6, narrow, subequal, spreading or rolled backward when fully expanded; stamens 6, inserted at base of the lobes; anthers dorsifixed, versatile, turned inward; ovary superior, 3-locular, the ovules 1 in each locule; style slender, the stigma trilobate or capitate. Fruit a fleshy berry or sometimes hard and woody, red or orange; seeds 1–3, bony, globose or elongate.

A family of 2 genera and 130–200 species, in subtropical and tropical regions of the Old World, becoming naturalized in the New World.

1. SANSEVIERIA Thunb.

Perennial herbs with elongate, creeping rhizomes, without aerial stems. Leaves often in rosettes, linear to widely linear, narrowed at both ends, rarely ovate or triangular, flat, terete, or crescent-shaped in cross section, entire, stiff, fleshy, coriaceous, or leathery, often variegated, fibrous. Flowers solitary on each pedicel or in clusters of 2 or more, subtended by scarious bracts; perianth tubular at base, cream-white or greenish white; stamens exserted, the filaments slender, filiform; style long and slender, filiform, about equaling the stamens; stigma capitate. Fruit a berry; seeds 1–3, globose to oblong-ellipsoid.

A genus of about 50–60 species indigenous to tropical Africa, Madagascar, and Arabia with a few species occurring in Asia.
REFERENCE: Brown, N. E. 1915. *Sansevieria*. A monograph of all the known species. Bull. Misc. Inform. 1915(5): 185–261.

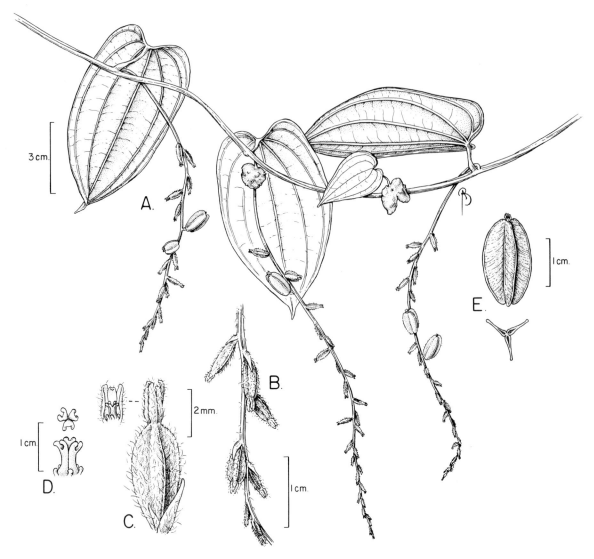

Fig. 223. *Dioscorea pilosiuscula.* **A.** Fertile branch. **B.** Detail of pistillate inflorescence. **C.** Pistillate flower and perianth of pistillate flower with two sepals removed showing style. **D.** Style, top and lateral views. **E.** Capsule and c.s. capsule.

1. Sansevieria trifasciata Prain, Bengal Pl. **2**: 1054. 1903. Fig. 224A–C.

Acaulescent, rhizomatous herb. Leaves 1 or 2, broadly linear, 30–100 cm long, 3–7 cm wide, with transverse, whitish green and dark green bands, slightly glaucous, the apex with a subulate tip, the margins green, entire. Inflorescence a raceme, 30–75 cm long, the flowers solitary or in clusters of 2–3 along main axis. Flowers pale greenish or greenish white; corolla tube 0.7–1.5 cm long, the lobes to 2 cm long, linear, rolled backward when fully expanded. Fruit a bright orange or reddish, globose berry, ca. 8 mm broad. Seeds 1, oblong-ellipsoid.

DISTRIBUTION: Along roadsides. Bordeaux Mountain (A3859). Native of Africa, cultivated and becoming naturalized. Also on St. Croix; Hispaniola, Puerto Rico, Jamaica, and Lesser Antilles.

COMMON NAMES: guana tail, lizard tail, mother-in-law tongue, rhamni, snake plant.

13. Hydrocharitaceae (Frog's-Bit Family)

By M. T. Strong

Perennial or annual, monoecious or dioecious aquatic or marine herbs of fresh- or salt water habitats, generally partially or wholly submersed, rarely free-floating, usually with elongate rhizomes. Leaves alternate, opposite, or whorled, sometimes distichous or rosulate at base; blades linear and ribbonlike, lanceolate, ovate, cordate-ovate, elliptic, or rounded, sessile and sheathing or petiolate. Flowers generally unisexual, rarely bisexual, actinomorphic, solitary, paired, or in cymose inflorescences, enclosed in distinct or united spathelike bracts or long-pedicellate from the axils of leaves; perianth free, of 3 outer sepaloid and 3 inner petaloid or 3(–2) petaloid

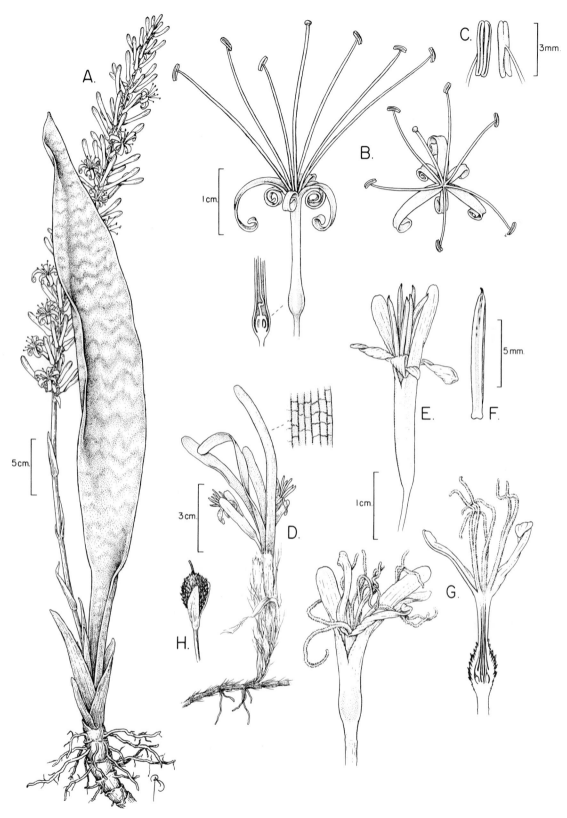

Fɪɢ. 224. A–C. *Sansevieria trifasciata*. **A.** Habit. **B.** Flower, lateral view, l.s. basal portion showing pistil, and top view. **C.** Anther, frontal and dorsal views. **D–H.** *Thalassia testudinum*. **D.** Habit and detail of reticulate venation. **E.** Staminate flower. **F.** Stamen. **G.** Pistillate flower and l.s. pistillate flower. **H.** Fruit.

segments; stamens 3-numerous, the anthers extrorse, linear, bilocular; ovary inferior, (2–)3–6(–20)-carpellate with few to many ovules; styles as many as the carpels, frequently bifurcate to the base. Fruit generally dehiscent, fleshy and berrylike, rupturing irregularly or stellately; seeds few to many, fusiform, ellipsoid or globose; embryo straight or slightly curved, lacking endosperm.

A family of 16 genera and approximately 100 species primarily of tropical, subtropical, and warm-temperate regions.

1. THALASSIA Banks & Sol. *ex* J. König

Submersed marine perennials; rhizomes elongate, scaly, with short stems produced from the nodes. Leaves tufted; blades linear, ribbonlike, falcate, eligulate, green; sheaths pale brown, the lower persistent, becoming fibrillose with age. Inflorescence of 1(–2) staminate flowers and 1 pistillate flower on peduncles arising singly from the lower leaves. Flowers unisexual with 3 tepals, enclosed in a tubular, 2-cleft spathe; staminate flowers on short pedicels with 3–12 stamens, the filaments short; pistillate flowers subsessile; ovary inferior, 6–8-carpellate, muricate; styles 6–8, bifurcate nearly to base, the stigmas pilose, longer than the styles. Fruit a globose, echinate, fleshy capsule, irregularly stellately dehiscing, the apex beaked; seeds few.

A genus of 2 species distributed along coasts; 1 occurs primarily in the Caribbean region, the other in eastern Africa, southern Asia, Indonesia, and Australia.

1. Thalassia testudinum Banks & Solander *ex* J. König, Ann. Bot. (London) **2**: 96. 1805.   Fig. 224D–H.

Submersed clonal, marine herb. Leaves 2–5 per tuft, 4–50 cm long, 4–12 mm wide, minutely serrulate at the obtuse to rounded apex; sheaths 5–10 cm long. Flowers white; staminate spathe 2–4 cm long; tepals 7–10 mm long; stamens 9, the anthers oblong to linear-oblong, 7–10 mm long; pistillate spathe 1.5–2.5 cm long; ovary 1 cm long; styles 7 or 8. Fruit ellipsoid to widely ellipsoid or globose, 1.4–2 cm long, 1.2–2 cm wide, mammillate, the beak 3–6 mm long. Seeds 2–3, pyriform, to 10 mm long.

DISTRIBUTION: Forming large colonies or prairies in shallow water of bays, beaches, and near reefs around the island. Fish Bay (*Acevedo-Rdgz. s.n.*). Also on Anegada, St. Croix, St. Thomas, Tortola, and Virgin Gorda; southern United States, Central America, West Indies, and northern South America.

COMMON NAME: turtle-grass.

14. Hypoxidaceae (Stargrass Family)

By M. T. Strong

Perennial herbs with tuberous rhizomes or with cormlike, underground, fleshy, vertically growing stems. Leaves radical, linear to lanceolate, flattened or pleated, often with uni- or multicellular hairs. Inflorescence a single flower or composed of spikes, racemes, or umbel-like clusters on a leafless, typically hairy scape. Flowers sessile or on short pedicels, bisexual, actinomorphic, the perianth 6-parted, free or sometimes fused, yellow or white, rarely red, stellately spreading, hairy on backs; stamens 6 or 3, inserted at base of perianth segments, the anthers versatile or basifixed, longitudinally dehiscent; ovary inferior, 3-locular, the placentation axile, with numerous ovules in each locule, the style short, 3-branched. Fruit a dehiscent capsule or an indehiscent berry; seeds small, globose, usually black, with a prominent raphe.

A family with 7 genera and about 120 species, widely distributed in temperate and tropical regions worldwide.

1. HYPOXIS R. Br.

Plants with underground, fleshy, cormlike, vertically growing stems. Leaves linear-lanceolate, grasslike, usually pilose. Inflorescences a solitary scape from the leaf axils, usually shorter than the leaves, with a 1- to several-flowered raceme borne distally, the flowers sessile or on short pedicels, each subtended by a setaceous bract. Perianth of similar sepals and petals, free to base, yellow or whitish within and green without, the sepals 3, typically pilose without, the petals 3, glabrous on both surfaces; stamens exserted; filaments short, the anthers versatile, rarely basifixed; style short. Fruit a dehiscent, subcylindrical, ellipsoid capsule, circumscissile below the apex, bearing the withering and persistent perianth segments at apex; seeds small, subglobose, dark brown or black, muricate, papillose, or pebbled, with a beak and rostrate hilum.

A genus of 80–100 species occurring in the Americas, subtropical Asia, and Australia, with the center of diversity in southern Africa.

REFERENCES: Brackett, A. H. 1923. Revision of the American species of *Hypoxis*. Rhodora 25: 120–147; Herndon, A. 1992. The genus *Hypoxis* (Hypoxidaceae) in Florida. Florida Sci. 55: 45–55.

1. Hypoxis decumbens L., Syst. Nat., ed. 10, **2**: 986. 1759.   Fig. 225C–G.

Plants 10–30 cm tall; underground stem vertical, cylindrical, 0.7–2 cm wide, the base above ground, 4–10 mm wide. Leaves linear-lanceolate, 10–45 × 0.3–1 cm, flattened, sparsely pilose to glabrate, overtopping the inflorescence. Scapes slender, loosely ascending or recurving, 4–20 cm long, pilose near apex, 1–4-flowered. Perianth yellow, 9–14(–15) × 9–12 mm; bracts linear-subulate; sepals and petals lanceolate, acute, 4–6(–7) mm long, the sepals pilose without, with long, tawny hairs. Capsule cylindrical, ellipsoid or club-shaped, 7–12 × 2–3 mm (not including the persistent, withered perianth), pilose, with long, tawny hairs. Seeds 0.8–1.2 mm diam., black, minutely pebbled.

DISTRIBUTION: In wet open areas. Bordeaux Mountain (W69A), between Bethany and Rosenberg (B279). Also on Tortola and Virgin Gorda; Greater Antilles, Lesser Antilles, Mexico, Central America, and South America.

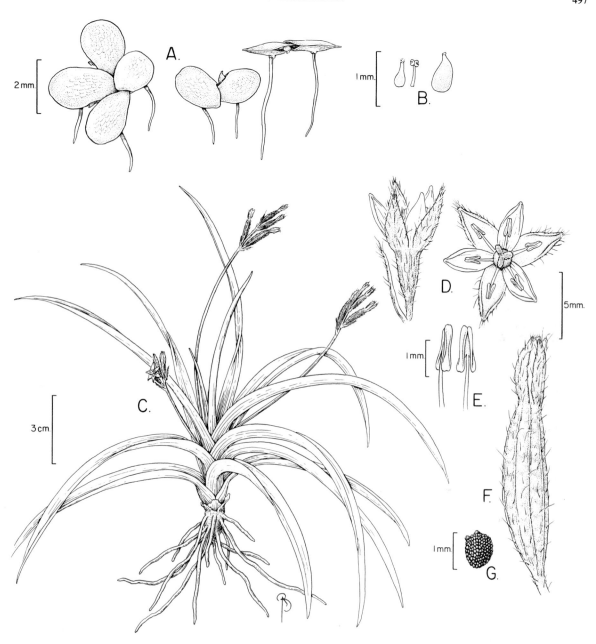

FIG. 225. **A, B.** *Lemna aequinoctialis.* **A.** Habit (colony), top and lateral views. **B.** Pistillate and staminate flower and utricle. **C–G.** *Hypoxis decumbens.* **C.** Habit. **D.** Flower, lateral and top views. **E.** Stamen, lateral and frontal views. **F.** Capsule. **G.** Seed.

15. **Lemnaceae** (Duckweed Family)

Monoecious, minute or small, floating or submersed, aquatic herbs, solitary or clustered; whole plant consisting of a platelike thallus; roots 1 to several or absent. Flowers unisexual and naked, in a short inflorescence consisting of 1 pistillate and 1 or 2 staminate flowers, usually subtended by a cuplike spathe. Staminate flowers consisting of a single stamen, the anther short, opening along longitudinal slits; pistillate flowers a sessile unilocular ovary, with 1–4 basal ovules, the style short, the stigma concave. Fruit an utricle; seeds 1–6, ellipsoid and usually ribbed.

A family of 6 genera and about 30 species, with cosmopolitan distribution.

REFERENCE: Landolt, E. 1986. The family Lemnaceae—a monographic study. Veröff. Geobot. Inst. Rübel Zürich 71: 7–566.

1. LEMNA L.

Submersed or floating aquatic herbs, forming large clusters of individuals cohering by their leaves; roots 1 or several. Plants reduced to a flat, oblong or ovate, thin or spongy thallus. Inflorescence of 1 pistillate and 2 staminate flowers. Utricle more or less ribbed.

A genus of about 15 species, with cosmopolitan distribution.

1. Lemna aequinoctialis Welw., Apont. **55**: 578. 1859.

Fig. 225A, B.

Lemna minor sensu Griseb., Fl. Br. W. I. 512. 1864, non L., 1753.

Lemna paucicostata Engelm. *in* Gray, Manual, ed. 5, 681. 1867.

Lemna perpusilla sensu authors, non Torr., 1843.

Floating herb, covering the surface of the water. Plants solitary or 3–5 in a cluster; root solitary, winged at the connection with the thallus. Thallus 1–2.5 × 0.7–2 mm, membranous, oblong, obovate, to rounded-obovate, asymmetrical, papillate along upper surface, obscurely 3-nerved.

DISTRIBUTION: Occasional in temporary bodies of fresh water or abandoned wells. In abandoned well at Emmaus (A4022), pool north of Europa Bay (A5260). Cosmopolitan, in temperate and tropical areas.

COMMON NAME: duckweed.

16. Orchidaceae (Orchid Family)

By J. D. Ackerman

Perennial herbs, rarely vines, autotrophic, mycotrophic at least at the seedling stage, rarely without chlorophyll, terrestrial, epilithic, epiphytic, rarely semiaquatic; roots adventitious, fibrous, fleshy or rarely tuberous, mostly velamentous, fasciculate or scattered along rhizome or stem; shoots monopodial or sympodial; stems variously modified, slender to pseudobulbous, rarely cormlike. Leaves alternate, rarely subopposite or whorled, convolute or conduplicate, parallel-veined, rarely reticulate. Flowers usually zygomorphic, solitary, in spikes, racemes, panicles, rarely corymbs or pseudoumbels, chasmogamous, rarely cleistogamous, hermaphroditic, rarely unisexual, usually resupinate. Perianth of 2 alternating whorls, sometimes spurred; sepals generally alike, often petaloid; lateral petals similar, median petal variously modified as a lip (labellum); androecium of 1(–2–3) anther, at or near apex of column seated in a clinandrium, pollen usually aggregated into pollinia; column (gynostemium) bilaterally symmetric, composed of androecium, style, and stigmas; ovary inferior, 1- or 3-locular, often twisted 180°, with numerous, minute ovules, the stigma lobes 3, often confluent, the median lobe usually partially sterile and modified as a rostellum. Fruit capsular, rarely a berry; seeds numerous, minute, without endosperm.

A family of about 1000 genera and between 15,000 and 20,000 species, of cosmopolitan distribution but most diverse in the tropics.

REFERENCES: Ackerman, J. D. & M. del Castillo Mayda. 1992. The orchids of Puerto Rico and the Virgin Islands, Las orquideas de Puerto Rico y las Islas Virgenes, University of Puerto Rico Press, San Juan; Ackerman, J. D. 1995. An orchid flora of Puerto Rico and the Virgin Islands. Mem. New York Bot. Gard. 73: 1–203; Cogniaux, A. 1909–1910. Orchidaceae. In I. Urban (ed.), Symb. Ant. 6: 293–696; Dressler, R. L. 1981. The orchids: natural history and classification, Harvard University Press, Cambridge, Massachusetts; Dressler, R. L. 1993. Phylogeny and classification of the Orchid family. Dioscorides Press, Portland, Oregon; Garay, L. A. & H. R. Sweet. 1974. Orchidaceae. In R. A. Howard, Fl. Lesser Antilles 1: 1–235.

Key to the Genera of Orchidaceae

1. Vines .... 9. *Vanilla*
1. Rhizomatous and often cespitose herbs.
   2. Leaf blades continuous (nonarticulate) with sheaths; pollinia soft, mealy; plants terrestrial.
      3. Flowers resupinate, lip in the lowermost position .... 1. *Cyclopogon*
      3. Flowers nonresupinate, lip in the uppermost position.
         4. Lip very fleshy, hoodlike, covering the column; sepals and petals similar, reflexed-curled; lateral sepals basally united .... 5. *Prescottia*
         4. Lip suborbicular, with margins flanking the column but not deeply hooded; sepals and petals dissimilar, not reflexed-curled; lateral sepals free .... 4. *Ponthieva*
   2. Leaf blades articulate with sheaths or stems; pollinia hard; plants terrestrial, epiphytic, or epilithic.
      5. Pseudobulbs present; leaves dorsal-ventrally flattened (conduplicate).
         6. Leaves mottled light and dark green; plants terrestrial; flowers spurred .... 3. *Oeceoclades*
         6. Leaves uniform in color; plants epiphytic or epilithic; flowers not spurred.
            7. Peduncle <50 cm long; lip white, green, or brown, adnate to the length of the column, lateral lobes of lip free .... 2. *Epidendrum*
            7. Peduncle 50–150 cm long; lip red to lavender, attached to the basal half of the column, lateral lobes of lip adnate to the sides of the column .... 6. *Psychilis*
      5. Pseudobulbs absent or inconspicuous; leaves laterally flattened or subcylindrical.
         8. Leaves laterally flattened; pollinia 2 .... 8. *Tolumnia*
         8. Leaves subcylindrical; pollinia 8 .... 7. *Tetramicra*

1. CYCLOPOGON C. Presl

Sympodial, cespitose, terrestrial herbs; roots fasciculate from a short rhizome, thick, fleshy, villous. Leaves nonarticulate, basal, petiolate. Flowers resupinate in terminal spikes or racemes; sepals subparallel, free or fused at base and forming an obscure mentum with base of column or a conspicuous sepaline nectar tube; petals connivent with dorsal sepal; lip unguiculate, sagittate to cordate, auricles often present, lateral margins appressed to sides of column; column erect; pollinia 2, clavate-oblong with an apical constriction, mealy with a relatively large disk-shaped viscidium; stylar canal entrance central, stigma lobes 2, free to approximate; rostellum soft, longer than wide; ovary sessile or subsessile. Fruit capsular.

A neotropical genus of about 70 species.

Key to the Species of *Cyclopogon*

1. Midvein and margin of petals darkened at the apex; lip broadest below apical constriction; leaf blades
   dark, velvety green above, purplish below; petioles 1–3.5 cm long .... 1. *C. cranichoides*
1. Petals unmarked at apex; lip equal or broadest above apical constriction; leaf blades green; petioles 1.5–
   10 cm long.. 2. *C. elatus*

1. Cyclopogon cranichoides (Griseb.) Schltr. , Beih. Bot. Centralbl. **37**(2): 387. 1920. *Pelexia cranichoides* Griseb., Cat. Pl. Cub. 269. 1866. *Sauroglossum cranichoides* (Griseb.) Ames, Orchidaceae **1**: 43. 1905. *Spiranthes cranichoides* (Griseb.) Cogn. *in* Urb., Symb. Ant. **4**: 338. 1909. *Beadlea cranichoides* (Griseb.) Small, Fl. S.E. U.S., ed. 2, 1375. 1913.

Spiranthes storeri Chapm., Fl. South. U.S., ed. 3, 488. 1897. *Beadlea storeri* (Chapm.) Small, Fl. S.E. U.S. 319. 1903.

Terrestrial herb to 40 cm tall; roots several, fleshy, fusiform, to 4.5 cm long, 4–8 mm thick near base. Leaves several, spreading, forming a loose, basal rosette; blades broadly elliptic to ovate, 2.5–8 × 1.3–3.8 cm, dark green above, purplish below, sometimes with whitish markings; petioles 11–35(–55) mm long. Flowers on mottled, purplish, slender, pubescent racemes, to 15 cm long, continuously elongating as fruits mature, many-flowered; floral bracts mottled, lanceolate, usually shorter than the flowers but longer than the ovary. Sepals pubescent, greenish brown, linear-elliptic to linear-pandurate, acute at apex; dorsal sepal midvein and margins dark near apex, 3.5–5 × 1.2–2 mm; lateral sepals often reflexed, 4–6 × 1–1.5 mm; petals brownish, loosely adnate to the dorsal sepal, linear, spatulate-oblanceolate, to 4.5 × 0.5–1 mm, the midvein and apical margins dark, acute at apex; lip white, basally gibbose, oblong, constricted above the middle, slightly flared at the apex, provided with a pair of basal tubercles, to 5 × 2.5 mm; column slender, clavate, ca. 3.5 mm long, pollinia yellow; ovary 4–7 mm long, pedicellate. Fruit erect, ellipsoidal, 6–8 × 3.5–4.5 mm.

DISTRIBUTION: Rare, collected once in deep humus of moist forest understory. Bordeaux Mountain (*Mrazek 4*). Also known from Florida, Bahamas, Cuba, Jamaica, Dominican Republic, Puerto Rico, Dominica, Guadeloupe, Central and South America.

2. Cyclopogon elatus (Sw.) Schltr. , Beih. Bot. Centralbl. **37**(2): 372, 387. 1920. *Satyrium elatum* Sw., Prodr. 119. 1788. *Spiranthes elata* (Sw.) Rich., De Orch. Eur. 37. 1817. *Beadlea elata* (Sw.) Small *in* Britton, Brooklyn Bot. Gard. Mem. **1**: 38. 1918.

Fig. 226A–E.

Terrestrial herbs to 70 cm tall; rhizome fleshy, erect to ascending; roots numerous, fleshy, 3–8 mm thick. Leaves 2–6, green, basal; blades 7–13 × 2.1–6.4 cm, narrowly to broadly elliptic, obtuse to acute-acuminate; petioles erect, slender, 5–9 cm long. Flowers green to coppery brown on pubescent racemes, 8–20 cm long, 7–50-flowered; bracts immaculate or minutely speckled, erect, lanceolate, often exceeding the length of the flowers, 9–15 mm long. Sepals entire, dorsally pubescent, basally connate for ca. 1 mm; dorsal sepal elliptic-lanceolate, acuminate, 5–7 × 1.5–2 mm; lateral sepals slightly sinuate-falcate, linear-oblong, 5.5–7.5 × 1–2 mm; petals glabrous, entire, linear-spatulate, adnate to the dorsal sepal, 4.5–6.5 × 0.5–0.75 mm; lip canaliculate, laterally thickened at base, with or without 2 erect callus horns, oblong-ovate then constricted above the middle, broadest at fan-shaped to kidney-shaped apex, 5.5–7 × 3–4 mm; column slender, pointed, 4–4.5 mm long; ovary subsessile, pubescent, 4–6 mm long. Capsule erect, ellipsoidal, 6–12 mm long, 4–6 mm thick.

DISTRIBUTION: A rare terrestrial herb in the shady understory of moist forests. Rosenberg (*B2899*). Also known from St. Croix, St. Thomas, Tortola; the Bahamas, Jamaica, Hispaniola, Puerto Rico, Lesser Antilles, Florida, Mexico, Central America and Venezuela.

2. EPIDENDRUM L.

Sympodial herbs, epiphytic, epilithic, or terrestrial; roots velamentous; stems slender and sometimes branched, or pseudobulbous from a creeping rhizome. Leaves articulate, distichous, conduplicate. Flowers variously colored, frequently green, from axillary or terminal racemes, panicles, or subumbels; lip usually adnate to the length of the column; anther terminal, operculate, incumbent, the pollinia 4 or rarely 2, laterally compressed, hard, waxy; stigma entire, the rostellum slitlike and oriented to the axis of the column. Fruit capsular.

A neotropical genus of several hundred species.

Key to the Species of *Epidendrum*

1. Stems slender, canelike; leaves produced along length of stem; sepals and petals 5–8 mm long; lip brown
   or green with entire lateral lobes .... 1. *E. anceps*

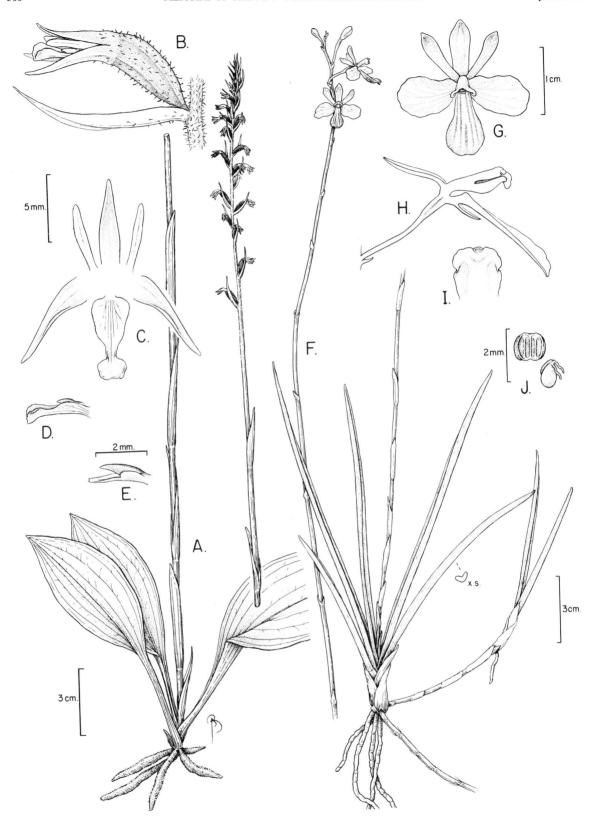

FIG. 226. A–E. *Cyclopogon elatus*. A. Habit. B. Flower with subtending bract. C. Detached perianth segments (sepals and petals) showing arrangement. D. Lip, lateral view. E. Column, lateral view. F–J. *Tetramica canaliculata*. F. Habit, with detail of c.s. leaf. G. Flower. H. L.s. flower. I. Column, dorsal view. J. Anther and pollinia.

1. Stems pseudobulbous; leaves produced at apex of stem; sepals and petals 40–50 mm long; lip white with
   dissected lateral lobes.. 2. *E. ciliare*

1. Epidendrum anceps Jacq., Select. Stirp. Amer. Hist. 224, t. 138. 1763. *Amphiglottis anceps* (Jacq.) Britton *in* Britton & P. Wilson, Bot. Porto Rico **5**: 200. 1924.

Epiphytic or epilithic herb to 80 cm tall; roots white, numerous from base of stems and short rhizomes, fleshy, 1.5–2.5 mm thick; stems slender, generally unbranched, compressed, covered by flattened leaf sheaths, to 8 dm long but generally much smaller. Leaves several to many, distichous; blades thin, flexible, deciduous on lower portions of stem, 4–12 cm long, 8–25 mm wide. Flowers olive-brown, several to many, resupinate on terminal or lateral, pendent subumbels; peduncles spreading to pendent, to 50 cm long; floral bracts inconspicuous, ca. 3 mm long. Dorsal sepal entire, elliptic-obovate, obtuse, 5–8 × 3–4 mm; lateral sepals entire, oblique, slightly broader; petals narrowly linear-oblanceolate, 5–7 × 1 mm; lip trilobed, cordate, attached to the entire length of the column, lateral lobes broad, rounded, middle lobe subquadrate, retuse with a fleshy midrib, free portion of lip 3–5 × 3–5 mm; column straight dorsally, sigmoid-arched ventrally, clavate, 4–5.5 mm long; pollinia 4, yellow; pedicellate ovary slender, 9–11 mm long. Fruit green, ellipsoidal, ca. 1.5 cm long.

DISTRIBUTION: Rare on trees in moist montane forests. Bordeaux Mountain (*Ray & Gibney s.n.*). Also on St. Thomas and Virgin Gorda; Florida, Mexico to South America including the West Indies.

2. Epidendrum ciliare L., Syst. Nat., ed. 10, **2**: 1246. 1759. *Auliza ciliaris* (L.) Salisb., Trans. Hort. Soc. London **1**: 294. 1812.   Fig. 227A–E.

Epiphytic or epilithic herb to 45 cm tall; roots velamentous, numerous, from the short, stout rhizome and base of stem, fleshy, 2–3 mm thick; stems pseudobulbous, slender, fusiform, slightly compressed, composed of several internodes, to 25 cm long. Leaves 1–3, stiff, coriaceous, to 17 × 4 cm. Flowers 1–10, resupinate, on a terminal raceme to 15 cm long; peduncles erect or arching, covered by distichous, conduplicate bracts, to 17 cm long; floral bracts lanceolate, canaliculate, spreading or clasping the ovary, 2–4.5 cm long. Sepals and petals similar, green to yellowish brown, revolute, linear, spreading, 40–50 × 3–6 mm; lip white, attached to the full length of the column, trilobed, to 3 cm long, the outer margin of the lateral lobes deeply and irregularly dissected, the middle lobe entire, linear, 2.2–3 cm long, basal callus small, bilobed, sometimes yellow, flanking the entrance to the nectar tube; column white, apically dilated, ca. 1.5 cm long; pollinia 4, yellow; pedicellate ovary slender, ca. 5 cm long. Fruit stipitate, green, ellipsoidal, 2.5–3 cm long, beak prominent, ca. 2 cm long.

DISTRIBUTION: Locally common on boulders, particularly near the coast. Enighed (A5100), Solomon Bay (A2279). Also on St. Croix, St. Thomas, Tortola, and Virgin Gorda; Mexico to South America, including the West Indies.

COMMON NAME: Christmas orchid.

3. OECEOCLADES Lindl.

Sympodial, terrestrial, cespitose herbs; roots velamentous, very thick; stems pseudobulbous. Leaves 1–3, articulate, conduplicate, coriaceous, often petiolate. Flowers resupinate in axillary, erect racemes or panicles; lip trilobed with a basal spur, the disk with a pair of callosities or with variously thickened, parallel ridges; column semicylindrical; anther terminal, operculate, incumbent, pollinia 2, hard, waxy, on a short stipe attached to a large viscidium; stigmas confluent. Fruit capsular.

A tropical African genus of about 31 species, with 1 naturalized in the neotropics.

1. Oeceoclades maculata (Lindl.) Lindl., Gen. Sp. Orchid. Pl. 237. 1833. *Angraecum maculatum* Lindl., Coll. Bot. t. 15. 1821. *Eulophidium maculatum* (Lindl.) Pfitzer, Entwurf Anordn. Orch. 88. 1887.
                                                      Fig. 227F–K.

Terrestrial, cespitose herb to 40 cm tall; roots white, numerous, fleshy, 3–6 mm diam; stems pseudobulbous, ovoid, slightly compressed, unifoliate, to 4 cm tall, often covered by scarious bracts. Leaves green with darker mottling, solitary from apex of pseudobulb, subpetiolate; blades 8–25 × 1.5–5 cm, conduplicate, coriaceous, fleshy, oblong to elliptic. Flowers resupinate, 5–15 in racemes to 15 cm long on sparsely bracteate scapes to 25 cm long, secondary racemes sometimes forming in axils of scape bracts;

floral bracts inconspicuous, linear-lanceolate, acuminate, shorter than ovary. Sepals 8–14 × 2–3 mm; dorsal sepal lanceolate to linear-elliptic, acute to acuminate; lateral sepals falcate-sigmoid, lanceolate-elliptic, acute to acuminate; petals erect, connivent with the dorsal sepal, elliptic-oblanceolate, acute, about as long as and slightly broader than the sepals; lip concave, pandurate, basal lobes erect, apical lobe subquadrate, emarginate, ca. 8 mm long; spur ca. 4 mm long; column 3–4 mm long; pollinia yellow, attached to a short stipe with a broad viscidium; pedicellate ovary slender, 7–13 mm long. Capsule green, angled, 3-sided, 2–3 cm long.

DISTRIBUTION: A rapidly invading species currently known on St. John from one site but expected to spread. Lameshur (A3236). Native to tropical Africa but naturalized in Florida, Panama, South America, and the West Indies

4. PONTHIEVA R. Br.

Sympodial, terrestrial, rarely epiphytic, cespitose herbs; roots fleshy, fasciculate, villous. Leaves nonarticulate with the sheaths, basal, subsessile to petiolate. Flowers nonresupinate in terminal, pubescent racemes; sepals spreading; dorsal sepal free or somewhat adnate to petals at apex; lateral sepals sometimes connate; petals asymmetric, often adnate to the basal flanks of the column, conspicuous, often forming a "pseudolip"; lip fleshy, uppermost in flower, adnate to the base of the column, clawed; column short, semicylindrical, dilated and slightly winged above, pointed; anther dorsal, erect behind the rostellum, pollinia 4, clavate, joined in pairs, firm but somewhat mealy, attached to caudicles and a terminal viscidium; stigma entire. Fruit capsular, nearly erect, ellipsoidal to obovoid.

A New World genus of about 25 species in tropical and subtropical areas.

FIG. 227. A–E. *Epidendrum ciliare*. **A.** Habit. **B.** Flower, top view. **C.** L.s. flower. **D.** Anthers (above) and pollinia (below). **E.** Capsule. **F–K.** *Oeceoclades maculata*. **F.** Habit. **G.** Flower. **H.** Detached perianth segments (sepals and petals) showing arrangement. **I.** Lip and column, frontal and lateral views. **J.** Pollinia. **K.** Portion of infructescence.

1. Ponthieva racemosa (Walter) Mohr, Contr. U.S. Natl. Herb. **6**: 460. 1901. *Arethusa racemosa* Walter, Fl. Carol. 222. 1788.   Fig. 228.

Neottia glandulosa Sims, Bot. Mag. **22**: t. 842. 1804. *Ponthieva glandulosa* (Sims) R. Br., Hortus Kew. **5**: 200. 1813.

Terrestrial herb, 30–60 cm tall; roots fasciculate, fleshy, villous, 2–3 mm thick. Leaves mostly basal, 2–6; blades 5–11 × 2–5 cm, delicate, elliptic to oblanceolate, rounded to acute; petioles to 6 cm long. Flowers greenish to greenish white, nonresupinate, loosely arranged in glandular pubescent racemes 5–25 cm long on scapes to 25 cm long; floral bracts pubescent, linear-lanceolate, clasping base of ovary, to 10 mm long. Perianth spreading; sepals adaxially pubescent; dorsal sepal free, ovate-elliptic to elliptic-lanceolate, obtuse, 4–7 × 2–3 mm; lateral sepals broadly and obliquely ovate, acute to obtuse, 4.3–8 × 2.5–4 mm; petals clawed, attached to base of column, obliquely triangular, glabrous, margins minutely ciliate or entire, 4–8 × 1.5–5 mm; lip fleshy, claw attached to the base of the column, lamina suborbicular, margins erect, apex short-caudate, 5–7 × 5–7 mm; column 4–5 mm long; filament erect, ca. 0.5 mm long, anther ca. 1.5 mm long, pollinia bright yellow, pedicellate ovary pubescent, 10–16 mm long. Capsule ellipsoidal, 8–13 mm long.

DISTRIBUTION: Rare; known from moist forests and thickets. Rosenberg (B289). Also reported from St. Thomas and Tortola; the West Indies, southern United States, and Mexico to tropical South America.

5. PRESCOTTIA Lindl.

Sympodial, terrestrial, cespitose herbs; roots fasciculate, fibrous or thick and fleshy, villous. Leaves nonarticulate, basal, petiolate or sessile, membranous. Flowers sessile, nonresupinate in terminal, erect spikes; sepals thin, basally connate, forming a short cup, rarely free, erect, spreading, or reflexed; petals thin, narrow, adnate to the sepaline cup; lip uppermost in flower, attached to the column foot, very fleshy, clawed, often basally auriculate above claw, deeply concave, galeate or cochleate, often enclosing the column; column and foot minute, both adnate to the sepaline cup when present; anther dorsal, erect, pollinia 4, soft, mealy, slightly flattened, caudicles absent; stigma entire. Fruit capsular.

A neotropical genus of about 25 species, particularly species-rich in Brazil.

Key to the Species of *Prescottia*

1. Petioles ≤3 cm long, erect or prostrate; leaf blades green, usually <8 cm long; lip white, 1–2 mm long
.... 1. *P. oligantha*
1. Petioles 2.5–15 cm long, erect; leaf blades green to purplish green, usually >8 cm long; lip green, pink, or straw-colored, 2.5–3.5 mm long.. 2. *P. stachyodes*

1. Prescottia oligantha (Sw.) Lindl., Gen. Sp. Orch. Pl. 454. 1840. *Cranichis oligantha* Sw., Prodr. 120. 1788.

Terrestrial herb to 40 cm tall; roots several, fingerlike, 3–8 mm thick. Leaves 2–4; blades 1.5–8 × 0.8–3 cm, entire, cuneate, elliptic to suborbicular, acute to rounded, membranous; petioles 1–3 cm long, erect or prostrate. Flowers glabrous, many in dense spikes to 17 cm long; scape 12–31 cm tall, partially covered by sheathing bracts; floral bracts ovate to lanceolate, acuminate, clasping the base of the ovary, 1.5–4 mm long. Sepals pinkish, 1-nerved, basally connate, reflexed-spreading, 1–2.2 × ca. 1 mm; dorsal sepal ovate, reflexed, curled; lateral sepals triangular to deltoid, connate with lip, forming a mentum, slightly larger than the dorsal sepal; petals white or pink, basally adnate to column and sepals, reflexed, narrowly obovate to oblong, 1–1.5 × 0.5 mm; lip uppermost, white, erect, forming a short mentum with the lateral sepals, fleshy, basally auriculate, hood-shaped, enclosing the column, shallowly saccate, apiculate, 1–2 mm long; column dorsally adnate to sepaline tube, minute, winged at apex; ovary subsessile, 1.5–3.5 mm long. Capsule erect, ellipsoidal, ca. 4 mm long.

DISTRIBUTION: Uncommon, known from shady sites in moist forests and thickets. Bordeaux (B535). Also on St. Thomas and Tortola; the West Indies, Florida, and Mexico to South America.

2. Prescottia stachyodes (Sw.) Lindl., Bot. Reg. **22**: 1916. 1836. *Cranichis stachyodes* Sw., Prodr. 120. 1788.   Fig. 229E–I.

Terrestrial, glabrous herb of variable size, often >50 cm tall; roots numerous, fasciculate, fleshy, 3–7 mm thick. Leaves green to dark purplish green, basal, 1–4, slightly fleshy, erect-spreading; blades 8–15 × 4–7 cm, elliptic, acute to acuminate, entire to denticulate; petioles erect, 2.5–15 cm long. Flowers greenish white to greenish brown, in dense, erect spikes 4–16 cm long on scapes 25–50 cm tall; floral bracts lanceolate, acuminate, to 12 mm long. Sepals basally connate, reflexed-curled; dorsal sepal linear-lanceolate, 2.5 × 0.5 mm; lateral sepals ovate-lanceolate, acute, 3 × 1.7 mm; petals adnate to the sepaline tube, filiform, reflexed-curled, ca. 2 × 0.3 mm; lip green, pink, or straw-colored, basally auriculate, adnate to sepaline tube, 2.5–3.5 mm long, very thick and fleshy, ovate and deeply concave, forming a cavity around the short column; column basally adnate to sepaline tube, minute, 1.7 mm long; ovary subsessile, 4–6 mm long. Capsule erect, ellipsoidal, 6–8 mm long.

DISTRIBUTION: Terrestrial in moist forests and thickets. Bordeaux (B559, A2604). Also reported from the Greater Antilles and Mexico to South America.

6. PSYCHILIS Raf.

Sympodial, epiphytic or epilithic herbs; roots velamentous; stems erect or ascending, pseudobulbous, composed of 3–5 subequal internodes, cylindrical, fusiform or pyriform from a short, stout rhizome. Leaves articulate, conduplicate, coriaceous to rigid, margin entire, crenate to erose-denticulate. Flowers resupinate, in short compact racemes on long, terminal, erect peduncles enclosed by

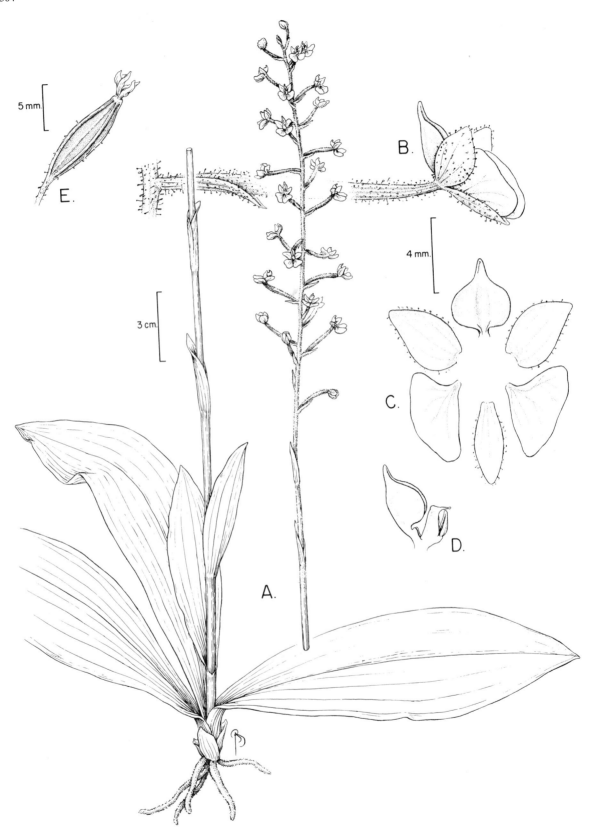

FIG. 228. *Ponthieva racemosa*. A. Habit. B. Flower. C. Detached perianth segments (sepals and petals) showing arrangement. D. Lip (labellum) and clinandrium, lateral view. E. Capsule.

FIG. 229. A–D. *Psychilis macconnelliae.* **A.** Habit. **B.** L.s. upper portion of flower. **C.** Anther and pollinia. **D.** Capsules and c.s. capsule. **E–I.** *Prescottia stachyodes.* **E.** Habit. **F.** Flower and l.s. flower. **G.** Lip, lateral and frontal views. **H.** Column, bottom and lateral views, and anther. **I.** Dehisced capsule.

scarious imbricate sheaths; sepals and petals free, spreading or reflexed; labellum clawed, basally adnate to column, trilobed, lateral lobes basally adnate to column, middle lobe variously callose, basal disk with a canaliculate callus; column cylindrical; anther terminal, operculate, incumbent, 2-celled; pollinia 4, equal, laterally compressed, hard, waxy, connected by caudicles; stigmas confluent, entire, rostellum transverse. Fruit capsular, ovoid to ellipsoidal.

A Caribbean genus of about 15 species.

1. Psychilis macconnelliae Sauleda, Phytologia 65: 18. 1988.   Fig. 229A–D.

Cespitose, epiphytic or occasionally terrestrial herb to 1.8 m tall; roots numerous from short stout rhizomes, 3–4 mm thick; pseudobulbs cylindrical to pyriform, to 18 cm long. Leaves 2–4, stiff, 10–28 × 0.8–3 cm, coriaceous, linear-oblong, margins minutely toothed. Flowers showy, of various shades of red to lavender, sometimes blotched, in short racemes, produced near apex of scapes 0.5–1.5 m long on older shoots; floral bracts scarious, ovate, acuminate, 1–3 mm long. Sepals oblong-oblanceolate, obtuse, 15–20 × 4–7 mm; petals linear-oblanceolate, obtuse, 1.5–2 cm long, 3–4 mm wide; lip with dark lines, trilobed, attached to basal half of the column, lateral lobes entire, obovate, erect, flanking the column, 12 × 8–10 mm, middle lobe obcordate, deeply notched; callus basal, bifid, elliptic, situated beneath the column; free portion of the lip with 3 parallel ribs, 2–2.8 × 1.5–2.6 cm; column straight, 5–9 mm long; pollinia yellow; pedicellate ovary slender, 2–3 cm long. Capsule ellipsoid, 2–3.2 cm long.

DISTRIBUTION: Common on small trees and shrubs in moist or dry regions, usually very near the ocean. Salt Pond (A759), East End (A1819). Also reported from St. Croix, St. Thomas, Anegada, Tortola, and Virgin Gorda; St. Kitts and Puerto Rico.

NOTE: This species has been incorrectly referred to as *Epidendrum bifidum* Aubl. or as *Epidendrum papilionaceum* Vahl by many authors in the West Indies. *Epidendrum bifidum*, with its homotypic synonym *E. papilionaceum*, is a species restricted to Haiti.

7. TETRAMICRA Lindl.

Sympodial, terrestrial or epiphytic herbs; roots velamentous; rhizomes stout, creeping; stems short, remote, rarely pseudobulbous. Leaves articulate, distichous, conduplicate, sometimes semicylindrical, fleshy, coriaceous. Flowers resupinate, in terminal, erect, long-pedunculate racemes; sepals and petals similar, free, spreading; lip trilobed, larger than the other floral segments, base forming a gibbose nectary; column semicylindrical, winged; clinandrium trilobed; anther terminal, operculate; pollinia 8, hard, waxy, connected by 4 caudicles; stigma lobes confluent, entire, transverse. Fruit a capsule.

A Caribbean genus of about 10 species.

1. Tetramicra canaliculata (Aubl.) Urb., Repert. Spec. Nov. Regni Veg. 15: 306. 1918. *Limodorum canaliculatum* Aubl., Hist. Pl. Guiane 2: 821. 1775.   Fig. 226F–J.

Cymbidium rigidum Willd., Sp. Pl. 4: 106. 1805. *Tetramicra rigida* (Willd.) Lindl., Gen. Sp. Orch. Pl. 119. 1831.
Cyrtopodium elegans Ham., Prodr. Pl. Ind. Occid. 53. 1825. *Tetramicra elegans* (Ham.) Cogn. *in* Urb., Symb. Ant. 6: 548. 1910.
Epidendrum subaequale Eggers, Fl. St. Croix 113. 1879.

Epiphytic or terrestrial, glabrous herb to 53 cm tall; roots produced from base of erect stems, gray and 1–2 mm thick when aerial, white and 2–4 mm thick when subterranean; stems slightly thickened, 4–25 mm long, produced from elongate rhizomes commonly above the soil surface on stilts of aerial roots. Leaves several, 1–21 cm long, thick, stiff, narrowly subcylindrical. Flowers few, in racemes or rarely panicles from slender scapes to 50 cm long; floral bracts thin, ovate, 2–3 mm long; sepals and petals greenish brown, sometimes spotted, spreading. Sepals oblong-lanceolate to elliptic-obovate, acute, 7–11 × 2–5 mm, the lateral sepals slightly larger than the dorsal; petals elliptic to oblanceolate, acute to obtuse, 6–9 mm long, 1–3 mm wide; lip rose-colored, prominently trilobed, 11–15 mm long; lateral lobes sometimes with dark veins, broadly elliptic, rounded, 6–8 × 4–7 mm; mid-lobe with dark veins and yellowish at the middle, obovate, 8–12 × 4–9 mm; column stout, apically expanded and conspicuously winged, ca. 6 mm long; pollinia of 4 very unequal-sized pairs; pedicellate ovary filiform, 12–18 mm long. Capsule ellipsoidal, ca. 15 mm long.

DISTRIBUTION: Locally common in exposed dry or moist sites. Maria Bluff (A4060). Also reported from St. Croix, St. Thomas, Little St. James, Anegada, Virgin Gorda, and George Dog; Puerto Rico, Hispaniola, the Lesser Antilles, and Florida.

8. TOLUMNIA Raf.

Sympodial, epiphytic, rarely epilithic or terrestrial herbs; roots velamentous; stems minute, usually hidden by leaf bases, sometimes pseudobulbous, produced from short or long and wiry rhizomes. Leaves articulate, arranged fanlike, fleshy-coriaceous, conduplicate, laterally compressed or subcylindrical, sigmoid (incorrectly referred to as "equitant"). Flowers colorful, showy, resupinate, in axillary racemes or panicles; dorsal sepal free, erect, lateral sepals free or connate; petals free, spreading, usually larger than the dorsal sepal; lip attached to the base of the column, nectary absent, trilobed, middle lobe usually emarginate, the disk callus tuberculate, glabrous to pubescent; column erect, semicylindrical, stigma flanked by lateral wings, footless; tabula infrastigmatica present; anther terminal, operculate, incumbent; pollinia 2, hard, waxy, stipe elongate, attached to a small viscidium. Fruit capsular.

A genus of about 30 species, some not well defined.

Key to the Species of *Tolumnia*

1. Flowers yellow; lip conspicuously dentate-crenate; inflorescences often forming plantlets after flowering
.... 1. *T. prionochila*

1. Flowers white to pinkish purple with brown markings; lip more or less entire; old inflorescences not forming plantlets... 2. *T. variegata*

1. Tolumnia prionochila (Kraenzl.) Braem, Orchidee (Hamburg) **37**: 58. 1986. *Oncidium prionochilum* Kraenzl. *in* Engl., Pflanzenr. IV, **50**: 233. 1922.

Fig. 230A–E.

Epiphytic, epilithic, or terrestrial, glabrous herb to 90 cm tall; roots white, mostly aerial, numerous, 0.5–1 mm thick; stems pseudobulbous, inconspicuous, generally hidden by leaf bases, fusiform, ca. 1 cm long, produced from slender, wiry rhizomes to 11 cm long or more and from the nodes of old inflorescences. Leaves several, imbricate, laterally compressed, arranged leaflike; blades 1.8–13.5 cm long or more, deeply carinate abaxially, coriaceous, short and stiff or long and lax. Flowers yellow, few, in short racemes on long, slender scapes, combined length 30–90 cm; floral bracts scarious, oblong-lanceolate, 2–5.5 mm long. Sepals 5–6 mm long, dorsal sepal oblanceolate to spatulate, lateral sepals fully connate, resulting synsepal shallowly bifid; petals spatulate, slightly larger than the dorsal sepal, margins wavy; lip trilobed, flat, 1.3–1.7 × 1.5–2 cm, lateral lobes auriculate, middle lobe fan-shaped, bifid, conspicuously dentate-crenate, the disk callus fleshy, multipronged; column erect, 4–6 mm long, wings produced opposite the stigma, conspicuous, petaloid, about half the length of the column; pollinia globose; pedicellate ovary 1.5–2.5 cm long. Capsule pendent, ellipsoidal.

DISTRIBUTION: In dry to moist scrub forests, often near the sea. Salt Pond (A758), White Cliffs (A2721). Also known from Guana Island, St. Thomas, Anegada, Tortola, and Virgin Gorda; Culebra and Cayo Luis Peña.

COMMON NAME: dancing lady.

2. Tolumnia variegata (Sw.) Braem, Orchidee (Hamburg) **37**: 59. 1986. *Epidendrum variegatum* Sw., Prodr. 122. 1788. *Oncidium variegatum* (Sw.) Sw., Kongl. Vetensk. Acad. Nya Handl. **21**: 240. 1800.

Epiphytic, glabrous herb to 65 cm tall; roots white, numerous, 0.5–1 mm thick; stems very small, inconspicuous, covered by leaf bases produced along slender, tough rhizomes 1–69 mm long. Leaves several; blades 1.5–17.2 × 3–11 mm, strongly carinate abaxially, laterally compressed, short and stiff to long and lax, falcate. Flowers mostly white, rarely pinkish purple with brown markings, in racemes or rarely panicles 2–65 cm long; floral bracts thin, spreading, lanceolate, 2–4 mm long. Dorsal sepal oblanceolate to spatulate, erect, 2–7 × 1–5 mm, lateral sepals connate, slightly shorter; petals spreading, pandurate, 4.5–10 × 2–6.5 mm; lip trilobed, the callus composed of several horns, the isthmus 1.5–5 mm wide, margin entire to minutely fimbriate, lateral lobes auriculate, ascending, more or less entire, 2–7 × 1–5 mm, middle lobe kidney-shaped, broader than long, 2.5–14 × 7–17.5 mm; column erect, white, green, or purple, 2.5–4.5 mm long, wings conspicuous, spreading, opposite the exposed concave stigma; pollinia globose; pedicellate ovary 8–26 mm long. Capsule green, pendent, ellipsoidal, ca. 1.5 cm long.

DISTRIBUTION: On twigs of shrubs and small trees, mostly in undisturbed vegetation of dry and moist areas. Trail to Sieben (A2071), Fish Bay (A3910). Also known from St. Croix, St. Thomas, Tortola, and Virgin Gorda; the Greater Antilles, except Jamaica.

9. VANILLA Plum. *ex* Mill.

Monopodial, terrestrial and epiphytic vines; roots gray-green, produced at each stem node, slender and glabrous when free, thick and villous on contact with a substrate; stems scandent, branching, naked, cylindrical, thick, fleshy, glabrous. Leaves articulate, distichous, sheathless, fleshy or coriaceous, large and persistent or scale-like and deciduous. Flowers large, showy but ephemeral, produced in succession, resupinate, in racemes or spiked, on short lateral branches; sepals free, spreading; petals free, keeled; lip clawed, adnate to base of column, basally involute, simple or lobed, disk variously ornamented; column elongate, semicylindric, often pubescent below, footless; anther terminal, incumbent, versatile, pollinia 4, soft, mealy, composed of monads, lacking accessory structures, when removed as a unit appearing triangular; stigma lobes confluent, rostellum undeveloped; pedicellate ovary articulate at perianth. Fruit elongate, cylindric to fusiform, indehiscent or tardily partly dehiscent fleshy or leathery; seeds small with a hard seed coat.

A pantropical genus consisting of about 100 species.

Key to the Species of *Vanilla*

1. Leaves shorter than the internodes, early deciduous .... 1. *V. barbellata*
1. Leaves longer than the internodes, persistent.. 2. *V. planifolia*

1. Vanilla barbellata Rchb.f., Flora **48**: 274. 1865.

Fig. 230F, G.

Terrestrial and epiphytic vine, 3–5 m long; roots gray, usually 1 or 2 per node, glabrous, 1–3 mm thick when aerial, thicker and villous when in contact with substrate; stems scandent, smooth, occasionally branched, 3–9 mm thick. Leaves early deciduous, to 40 × 8 mm, relatively thin, broad basally, otherwise linear-lanceolate, involute and usually reflexed at apex. Flowers in racemes; floral bracts fleshy, broadly ovate, 4–12 mm long. Sepals and petals green, free, somewhat spreading; sepals oblong-oblanceolate, 30–40 × 9–12 mm; petals oblong-oblanceolate, acute to obtuse, slightly falcate, dorsally keeled, 30–40 × 10–13 mm; lip adnate to the lower half of the column, apex trilobed, overall triangular-obovate, medially thickened, the lateral lobes orbicular,

involute and arching over the column, sinuses 4–5 mm deep, the middle lobe fleshy and reflexed, the disk with a tuft of rigid, retrorse bristles; column straight, semicylindrical, 2.3–3.3 cm long; pedicellate ovary 3–4.5 cm long. Fruit indehiscent, pendent, fusiform-cylindric, slightly curved, 7–9 cm long, 9–13 mm thick.

DISTRIBUTION: Climbing vine in dry to moist scrub forests. Europa Bay (A4142). Also reported from St. Thomas and Virgin Gorda; Puerto Rico, the Bahamas, Cuba, and Florida.

2. Vanilla planifolia Jacks. *in* Andrews, Bot. Repos. **8**: t. 538. 1808.

Myrobroma fragrans Salisb., Parad. Lond. t. 82. 1807, nom. illegit. *Vanilla fragrans* (Salisb.) Ames, Sched. Orch. **7**: 36. 1924.

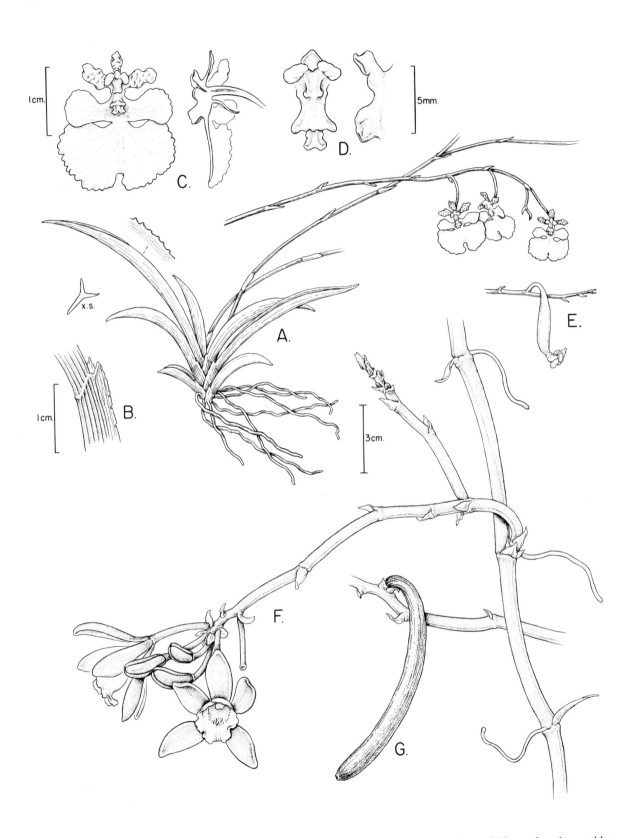

FIG. 230. A–E. *Tolumnia prionochila*. **A.** C.s. leaf, habit, and detail of leaf margin. **B.** Detail of leaf sheath apex. **C.** Flower, from above, and l.s. flower. **D.** Lip and column, top and lateral views. **E.** Developing capsule. **F, G.** *Vanilla barbellata*. **F.** Flowering branch. **G.** Fruit.

Terrestrial and epiphytic vine 5–7 m long; roots usually 1 per node, aerial portions 2–3 mm thick; stems scandent, smooth, glabrous, 5–10 mm thick, the internodes to 12 cm long. Leaves persistent, to 25 × 8 cm (longer than the internodes), rigid, fleshy, flat, oblong, elliptic to ovate, acute to acuminate; petioles ca. 1 cm long. Flowers in short-pedunculate, dense racemes to 5 cm long; floral bracts broadly ovate, to 10 mm long. Sepals and petals yellow-green, fleshy, free, spreading; sepals elliptic-oblanceolate, 3.5–5.5 cm × ca. 13 mm; petals slightly shorter and narrower than the sepals, dorsally keeled; lip clawed, cuneate, adnate to the column, dilated above, arched over column and reflexed at the apex, when spread, triangular in outline with an apical retuse lobule, 4–5 × ca. 3 cm, the disk with a tuft of long, stiff hairs and several lines of short, fleshy hairs extending to the apex; column arched, bearded abaxially beneath the stigma, 3–3.5 cm long; pollinia yellow; pedicellate ovary 3–5 cm long. Fruit indehiscent, black when mature, pendent, slender, cylindrical, fragrant, to 25 cm long, 8 mm thick.

DISTRIBUTION: Native to Mexico, introduced as a crop plant, persistent in moist habitats. Bordeaux (A4058). Also reported from St. Croix and St. Thomas; the West Indies and Mexico to South America.

COMMON NAME: vanilla.

17. Poaceae (Grass Family)
By R. D. Webster and P. M. Peterson

Herbaceous or less often woody, annuals or perennials, solitary or clustered (cespitose), often with underground stems (rhizomes) or running stems (stolons), erect or less often climbing, or aquatic; culms (stems) hollow or less often solid, terete, branching or not. Leaves alternate, usually distichous and cauline or basal; without stipules, petioles modified into a leaf sheath surrounding the stem, sheaths keeled or rounded, usually with same pubescence character as the blade, usually with a ligule at the junction with the blade or summit of the sheath; inner ligule present on adaxial surface; outer ligule present on abaxial surface in a few groups; junction of sheath and blade sometimes with an additional flange or auricle on both sides of the ligule, rarely with coarse, bristle-like appendages (oral setae); blades parallel-nerved, in some groups narrowed near base into a pseudopetiole. Inflorescences compound, terminal or axillary, open to contracted, of 1 to several spikes, racemes, or panicles. Spikelets consisting of an axis (rachilla) bearing at its base 2 sterile bracts (glumes), dorsally, ventrally, or dorsiventrally compressed, rarely plano-convex; rachilla above the glumes bearing florets; florets consisting of a lower bract (lemma) with or without a hardened or hairy callus at base and an upper appendage (palea), these together enclosing a reduced flower; lemma oriented toward the rachis or away from the rachis. Flower bisexual or unisexual, the perianth reduced to 2 or 3 minute, fleshy, scale-like lodicules; ovary unilocular, uniovulate, the styles 2, the stigmas 1–3; stamens (1)2 or 3(–6). Fruit usually a caryopsis with the pericarp and seed coat fused.

About 750 genera and 10,000 species distributed throughout the world.

REFERENCES: Clayton, W. D. & S. A. Renvoize. 1986. Genera Graminum, Grasses of the world. Kew Bull. 13: 1–389; Hitchcock, A. S. 1936. Manual of the grasses of the West Indies. U.S.D.A. Misc. Publ. 243: 1–439; Judziewicz, E. J. 1991. Poaceae. In A. R. A. Görts-van Rijn (ed.), Flora of the Guianas. Koeltz Scientific Books, Koenigstein, Germany; Watson, L. & M. J. Dallwitz. 1992. The Grass genera of the world. CAB International, Wallingford, England; Webster, R. D. 1988. Genera of the North American Paniceae (Poaceae: Panicoideae). Syst. Bot 13: 576–609; Webster, R. D., J. H. Kirkbride & J. Valdés R. 1989. New World genera of the Paniceae (Poaceae: Panicoideae). Sida 13: 393–417; Webster, R. D. & J. Valdés R. 1988. Genera of the Mesoamerican Paniceae (Poaceae: Panicoideae). Sida 13: 187–221.

Key to the Genera of Poaceae

1. Culms (1–)2–30 m tall or long, lignified and perennial; leaves dimorphic, those of the culm different from those of the foliage compliment; summit of leaf sheath bearing oral setae.
   2. Plant erect; culms 5–12 cm diam.; foliage leaves 12–30 × 1.5–4 cm, lanceolate, flat, alternately arranged on the branch; culm leaves 30–45 cm long and about as wide .... 6. *Bambusa*
   2. Plant scandent; culms <1.5 cm wide; foliage leaves 2–12 × 0.02–0.1 cm wide, filiform and tightly involute, arranged in capillary whorls; culm leaves <5 cm long, 1–3 mm wide.................. 4. *Arthrostylidium*
1. Culms usually much less than 2 m tall, herbaceous, annual or perennial; leaves the same on the culm as those borne along the branches; summit of leaf sheath never bearing oral setae.
   3. Spikelets usually paired, one sessile or subsessile and the other pedicellate, often dimorphic and with the pedicellate spikelet much reduced, if not paired, then spikelets in readily disarticulating racemes or the upper floret hyaline, geniculately awned.
      4. Pedicels and rachis internodes with a thin central longitudinal groove appearing as a translucent stripe; lower glume of the sessile spikelet with a conspicuous circular pit........................ 7. *Bothriochloa*
      4. Pedicels and rachis internodes without a thin central longitudinal groove; lower glume of the sessile spikelet without a circular pit.
         5. Inflorescence composed of a solitary raceme.
            6. Sessile spikelet with a lemmal awn 6–12 cm long, glumes dark brown; callus heavily bearded, the hairs coppery-brown .... 18. *Heteropogon*
            6. Sessile spikelet with a lemmal awn <1.3 cm long; glumes straw-colored; callus bearded or glabrous, the hairs whitish .... 29. *Schizachyrium*
         5. Inflorescences composed of 2–9 racemes.
            7. Foliage not aromatic or lemon-scented; inflorescence a large, broomlike, compound mass, the rachis straight; racemes erect to slightly divergent, borne upon unequal, terete bases .... 1. *Andropogon*

7. Foliage aromatic, lemon-scented; inflorescence an open, loosely arranged group of racemes along a zigzagging rachis; racemes usually drooping and sinuous, deflexed at maturity and borne upon subequal, flattened bases .... *Cymbopogon* (cultivated)

3. Spikelets not paired or dimorphic, or if paired, then neither in readily disarticulating racemes, nor the upper floret geniculately awned.

  8. Spikelets unisexual, dimorphic (plant monoecious).

    9. Pseudopetioles 8–60 mm long, twisted at the summit to invert the blade; blades with oblique secondary venation; pistillate spikelets with florets covered with hooked hairs .... 28. *Pharus*

    9. Pseudopetioles 4–7 mm long, not twisted at the summit; blades with parallel venation; pistillate spikelets with florets never covered with hooked hairs .... 22. *Olyra*

  8. Spikelets bisexual, monomorphic.

    10. Lemmas prominently 3-awned.. 3. *Aristida*

    10. Lemmas awnless or awned but never 3-awned.

      11. Spikelets with the lower floret usually bisexual, not reduced, the upper floret usually sterile and reduced; floret(s) falling from the persistent glumes.

        12. Inflorescence a digitate or subdigitate terminal cluster of 2 to many spikelike racemes.

          13. Spikelets with only a single bisexual floret (sterile or staminate florets also present); fruit with an adnate pericarp or outer wall, not separating upon hydration.

            14. Inflorescence with 7–15 racemes; lemmas awned; spikelets 2–several-flowered... 10. *Chloris*

            14. Inflorescence with 3–5 racemes; lemmas awnless; spikelets 1-flowered and with a rachilla internode prolonged as a sterile rudiment...................... 11. *Cynodon*

          13. Spikelets with 2 to many bisexual florets (sterile or staminate florets may also be present); fruit with a free pericarp or outer wall, this separating upon hydration.

            15. Upper glume 1-nerved and awned; primary inflorescence branches terminating in a bare point or bristle; plants often stoloniferous .... 12. *Dactyloctenium*

            15. Upper glume 3–5-nerved and unawned; primary inflorescence branches terminating in a spikelet; plants lacking stolons .... 15. *Eleusine*

        12. Inflorescence a narrow to open panicle or group of racemosely arranged racemes, the branches arranged along a central rachis, and sometimes very short.

          16. Inflorescence forming glomerules of (1–)2–5 spikelets; upper glumes with rows of hooked spines along the nerves... 33. *Tragus*

          16. Inflorescence not forming small glomerules, usually containing many more spikelets; upper glumes without hooked spines located along the nerves.

            17. Inflorescence of 1 to many, more or less spicate racemes, the spikelets arranged in 2 rows along the rachis.

              18. Spikelets not falling entire at maturity, usually disarticulating between the florets and above the glumes; plants usually <1 m tall, annuals or perennials of short duration, without rhizomes.

                19. Inflorescence 10–15 cm wide with >15 primary branches; spikelets 2.2–4.3 mm long, 3–7-flowered; lemmas 1.8–2.5 mm long, excluding the awns; blades 10–32 cm long, 7–12 mm wide, clasping near base with yellowish dewlaps; caryopsis 1–1.3 mm long; stamens 2 .... 20. *Leptochloa*

                19. Inflorescence 2–5 cm wide with only 3–7 primary branches; spikelets 6–9 mm long, 2-flowered; lemma 4.5–7 mm long, excluding the awn; blades 3–6 cm long, 2–4 mm wide, not clasping near base and without dewlaps; caryopsis 3–3.5 mm long; stamens 3 .... 8. *Bouteloua*

              18. Spikelets falling or disarticulating as a unit at maturity; plants usually 1–2 tall, stout perennials with well-developed rhizomes.

              20. Spikelets 1-flowered, 7–12 mm long; sheaths glabrous to scabrous, never with a tuft of soft hairs just below and to one side of the ligule; glumes 3–12 mm long, the upper at least as long as or longer than the spikelet; lemma 5.5–8.5 mm long; anthers 3–5 mm long .... 31. *Spartina*

              20. Spikelets 2–6-flowered, 2.5–6 mm long; sheaths ciliate with long soft hairs in tufts just below and to one side of the ligule, the hairs 1–2.5 mm long; glumes 1–2.5 mm long, both shorter than the spikelet; lemmas 1.8–2.8 mm long; anthers 1–2 mm long................ 34. *Uniola*

17. Inflorescence an open to contracted panicle, the spikelets not arranged in 2
rows along the rachis.
    21. Spikelets 1-flowered.. 32. *Sporobolus*
    21. Spikelets 2- to many-flowered.. 16. *Eragrostis*
11. Spikelets with the lower floret reduced, either staminate or sterile (rarely pistillate or bisexual), the upper floret bisexual; spikelets usually falling entire from the summit of the
pedicel.
  22. Rachis terminating in a spikelet.
    23. Lower glume fused with the callus to form a cup .... 17. *Eriochloa*
    23. Lower glume not fused with the callus or absent.
      24. Lemma of upper floret with flat margins.
        25. Disarticulation at the base of the primary branches; pedicels absent; lodi-
cules absent.. 2. *Anthephora*
        25. Disarticulation at the spikelet base; pedicels present; lodicules present.
          26. Upper floret membranous to chartaceous; spikelets laterally com-
pressed; spikelets not distinctly 1-sided or distichous .... 21. *Melinis*
          26. Upper floret cartilaginous to indurate; spikelets plano-convex; pri-
mary branches with 1-sided or distichous spikelets.
            27. Lemma of the lower floret toward the rachis; culm internodes
usually solid .... 5. *Axonopus* (part)
            27. Lemma of the lower floret away from the rachis; culm in-
ternodes usually hollow .... 12. *Digitaria*
      24. Lemma of upper floret with involute or convolute margins.
        28. Lemma of upper floret rugose... 35. *Urochloa*
        28. Lemma of upper floret not rugose.
          29. Lemma of upper floret differentiated at the apex.
            30. Spikelets terete; ligule a membrane or a ciliate membrane; pri-
mary branches with spreading secondary branches .... 19. *Lasiacis*
            30. Spikelets dorsiventrally compressed or plano-convex; ligule a
fringe of hairs or absent; primary branches with appressed secondary branches.. 4. *Echinochloa*
          29. Lemma of upper floret not differentiated at the apex.
            31. Lower glume present.
              32. Lower glume encircling the spikelet base .... 24. *Panicum*
              32. Lower glume not encircling the spikelet base.
                33. Lower glume muticous to apiculate; spikelets abaxial;
lemma of upper floret dull .... 26. *Paspalum* (part)
                33. Lower glume awned; spikelets adaxial; lemma of upper
floret shiny.. 23. *Oplismenus*
            31. Lower glume absent.
              34. Lemma of the lower floret toward the rachis; pedicels
concave .... 5. *Axonopus*
              34. Lemma of the lower floret away from the rachis; pedicels
convex .... 26. *Paspalum*
  22. Rachis terminating in a bristle.
    35. Disarticulation at the spikelet base; lemma of upper floret rugose.
      36. Main axis with distichous or 1-sided primary branches.................. 25. *Paspalidium*
      36. Main axis with primary branches neither distichous or 1-sided................ 30. *Setaria*
    35. Disarticulation at the base of the primary branches; lemma of upper floret not
rugose.
      37. Callus flared at apex.. 9. *Cenchrus*
      37. Callus prolonged but not flared at apex...................................... 27. *Pennisetum*

1. ANDROPOGON L.

Annual or perennial herbs; culms erect or decumbent. Leaves usually long and narrow, ligules membranous, fringed or not; blades flat to involute. Inflorescence often of 1 to numerous palmately arranged racemes, arising from the middle and upper nodes of the culm, raceme bases unequal and terete; rachis internodes slender to club-shaped, often ciliate, disarticulating along the rachis with spikelets falling as pairs (sessile and pedicellate) or just the pedicellate spikelet falling first. Sessile spikelet bisexual or pistillate, 2-

flowered; glumes subequal, about as long as the spikelet, membranous; lower glume keeled, 1–3-nerved, usually awnless; florets hyaline, shorter than the spikelet; lower floret sterile, lacking a palea; upper floret with an awned or unawned lemma, the palea often well developed. Pedicellate spikelet absent to well developed, usually similar to sessile spikelets when present; lower glume convex with many nerves. Caryopsis with a large embryo.

Around 100 species in tropical or temperate zones worldwide.

1. Andropogon bicornis L., Sp. Pl. 1046. 1753.

Fig. 231A–C.

Cespitose perennial herb; culms 1–2.5 m tall, erect, glabrous. Leaves coarse, shiny; sheath slightly inflated; ligules 1–2 mm long, erose-ciliolate near apex; blades 15–35(–50) × 0.2–0.7 cm, mostly glabrous or with a few hairs near base. Inflorescence 15–50 × 8–15 cm, a broom-like mass of racemes subtended by 1–4 reddish, bractlike sheaths, to 1 cm long; peduncles included or exserted, to 2.5 cm long, filiform, bearing 2–3 racemes at the summit; racemes 1.2–3.5 cm long, erect to slightly divergent; rachis internodes 2–3.5 mm long, filiform with densely spreading, ciliate hairs. Sessile spikelet 2.5–3.3 mm long, linear-lanceolate, straw-colored, firmly membranous, awnless; callus bearded, otherwise glabrous; lower glume flat to sulcate without; upper glume 1-nerved; florets 1.8–2.2 mm long; lower floret sterile; upper florets bisexual; anthers 1–1.3 mm long. Pedicellate spikelet terminal, 3.5–4.3 mm long, lanceolate, sterile or staminate; lower glume 5–7-nerved, rounded on back; upper glume 3-nerved.

DISTRIBUTION: Occasional herb of disturbed sites, clearings, old pastures, and open grasslands. Rosenberg (B305). Also on St. Croix, St. Thomas, and Tortola; Mexico to Argentina, including the West Indies.

2. ANTHEPHORA Schreb.

Cespitose, annual or weakly perennial herbs; culms erect to prostrate, sparingly branched. Leaf sheaths smooth; ligules membranous; blades linear, flat, spreading. Inflorescence reduced to a fascicle of spikelets disarticulating at base; primary branches appressed to the main axis; pedicels absent; rachis terminating in a spikelet; callus differentiated. Spikelets adaxial, dorsiventrally compressed, generally bisexual, occasionally staminate or sterile; lower glume not encircling the spikelet base, hardened; upper glume three-quarters to equal spikelet length; lower floret lemma membranous; upper floret lemma about as long as the lower lemma, chartaceous, smooth, dull, muticous, with flat, thin margins; lodicules absent. Caryopsis with white to grayish embryo.

About 12 species in the savannas of subtropical Africa, with 1 or 2 species in the American tropics.

1. Anthephora hermaphrodita (L.) Kuntze, Revis. Gen. Pl. 2: 759. 1891. *Tripsacum hermaphroditum* L., Syst. Nat., ed. 10, 2: 1261. 1759. Fig. 231D–G.

Anthephora elegans Schreb., Beschr. Gräs. 2: 105, pl. 44. 1810.

Annual or perennial herb, lacking stolons; flowering culms 2–11 dm tall, decumbent at base, sparingly branched. Leaf sheaths glabrous or hairy; ligule 1–2.5 mm long; blades linear, flat, 3–40 × 0.2–1.2 cm, smooth and glabrous on upper surface. Inflorescence linear; main axis 3–12 cm long; primary branches 3.5–6 mm long. Spikelets clustered, lanceolate to ovate, 3.5–10 × 1–2 mm; lower glume 2–9 mm long, 5–9-nerved; upper glume 1.5–8 mm long, 1-nerved; in lower floret, lemma oblong to ovate, 5–7-nerved, muticous, palea absent; in upper floret, lemma ovate to elliptic, chartaceous, smooth, yellow, 2.5–10 mm long, muticous.

DISTRIBUTION: Occasional grass of dry to moist disturbed areas, waste places, and pastures. Yawzi Point Trail (W557). Also on St. Croix, St. Thomas, Tortola, and Virgin Gorda; throughout tropical South America and the West Indies.

3. ARISTIDA L.

Cespitose, annual or perennial herbs; culms slender, solid or hollow. Leaves mostly basal; sheaths mostly glabrous; ligules ciliate membranes; blades flat or involute. Inflorescence a terminal panicle. Spikelets 1-flowered, disarticulating above the glumes; glumes lanceolate, acuminate to awn-tipped, usually 1-nerved; lemma narrowly lanceolate, 3-nerved, terete, with inrolled margins, tightly convolute around the palea, the base with a hard, sharp-pointed, bearded callus, the apex awned, sometimes beaked, usually bearing 3 stiff awns; palea minute or absent; stamens 3.

Worldwide genus of about 300 species, most of which are subtropical to tropical.

REFERENCE: McKenzie, P. M., R. E. Noble, L. E. Urbatsch & G. R. Proctor. 1989. Status of *Aristida* (Poaceae) in Puerto Rico and the Virgin Islands. Sida 13: 423–447.

1. Aristida cognata Trin. & Rupr., Mém. Acad. Imp. Sci. St.-Pétersbourg, sér. 6, Sci. Math. 5: 127. 1842. Fig. 231H, I.

Perennial herb; culms 20–50 cm tall, ascending below from a knotty base. Leaves narrow; ligules 0.1–0.3 mm long, minutely ciliate; blades 50–150 × 0.5–1.5 mm, involute, with scattered, villous hairs, 2–4 mm long near base or sheath apex. Panicles 4–15 × 1–4 cm, narrowly spreading. Spikelets 6–9 mm long, excluding the awns; glumes 4–8 mm long, the upper glume longer; lemmas 5–7 mm long, terete, the callus hairs about 0.5 mm long, the apex with elongate column or beak about 1 mm long, the awns 4–18 mm long.

DISTRIBUTION: Occasional in rocky soils and on dry hillsides. Rosenberg (E3003), Bass Gut (*Eggers s.n.*). Also on St. Thomas; Cuba, Hispaniola, Lesser Antilles, and Venezuela.

EXCLUDED SPECIES: *Aristida adscensionis* L., reported in error for St. John by Woodbury and Weaver (1987), was based on misidentifications of *A. cognata* specimens.

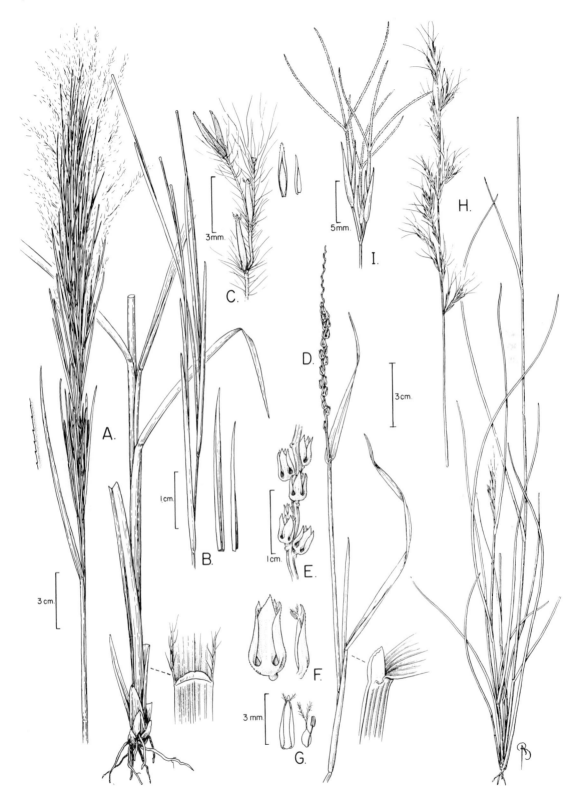

FIG. 231. A–C. *Andropogon bicornis*. **A.** Habit, detail of summit of leaf sheath showing ligule, and detail of leaf margin. **B.** Portion of inflorescence showing elongate peduncles and bractlike sheaths. **C.** Portion of the inflorescence raceme and glumes. **D–G.** *Anthephora hermaphrodita*. **D.** Habit and detail of summit of leaf sheath showing ligule. **E.** Portion of spicate inflorescence. **F.** Fascicle of spikelets and spikelet. **G.** Floret, pistil, and stamen. **H, I.** *Aristida cognata*. **H.** Habit. **I.** Portion of inflorescence panicle showing spikelets.

4. ARTHROSTYLIDIUM Rupr.

Cespitose perennial herbs with sympodial, pachymorph rhizomes; culms cylindrical, lignified, hollow, scandent or clambering; nodes with a single primary branch bud, this later producing 3 to many branches at the summit of more or less a promontory. Foliage leaves with inner and outer ligules; ligules adaxial and abaxial, membranous, or wanting, the outer ligules short and indurate; oral setae inconspicuous to prominent. Inflorescence a spicate raceme lacking bracts, with a straight or zigzagging rachis. Spikelets subsessile, 2- to many-flowered, bisexual; glumes 1 or 2, about half as long as the spikelet, persistent, chartaceous; lowermost floret usually sterile; fertile (either bisexual, pistillate or staminate) florets 1 to several, disarticulating at maturity and falling attached to the rachilla internode; uppermost florets 1 to several, successively reduced and sterile; lemma and palea subequal, the margins of the lemma clasping the rachilla internode; stamens 3.

About 30 species from Mexico to Bolivia.

REFERENCE: Judziewicz, E. J. & L. G. Clark. 1993. The South American species of *Arthrostylidium* (Poaceae: Bambusoideae: Bambuseae). Syst. Bot. 18: 80–99.

1. Arthrostylidium farctum (Aubl.) Soderstr. & Lourteig, Phytologia **64**: 163. 1987. *Arundo farcta* Aubl., Hist. Pl. Guiane **1**: 52. 1775.   Fig. 232A–C.

Arthrostylidium capillifolium Griseb., Mem. Amer. Acad. Arts **8**: 531. 1862. *Arundinaria capillifolia* (Griseb.) Hack., Osterr. Bot. Z. **53**: 69. 1903.

Scandent perennial herb; culms 3–5(–15) m long, 0.2–1 cm diam., repeatedly branching; culm leaf sheaths bearing prominent ciliate hairs; adaxial ligules 0.3–0.8 mm long, chartaceous; oral setae in 2 whorls, 5–12 mm long, white, twisted and curled above; blades mostly 1.5–4 cm long, 1–3 mm wide, flat, attenuate toward apex, reflexed. Foliage leaves crowded on short sterile branchlets arranged in capillary whorls; sheath with ciliolate margins, the

hairs 0.1–0.2 mm long; inner ligules 0.1–0.2 mm long, membranous, ciliolate, the outer ligule absent; oral setae in 2 rows, 1–8 mm long, white, mostly straight; blades mostly 2–12 cm long, 0.2–1 mm wide, filiform, tightly involute. Inflorescences 2–10 cm long, spicate, bearing 3–7 spikelets; pedicels 0.5–1.3 mm long. Spikelets 8–11 mm long, linear, appressed to the rachis; lower glume 4–5.5 mm long, 1-nerved, narrowly attenuate; upper glume 5–6.5 mm long, linear-lanceolate, 3-nerved; lemmas 3.8–5.5 mm long, 7–9-nerved with ciliolate margins near apex; paleas nearly as long as or slightly longer than the lemma. Caryopsis not seen.

DISTRIBUTION: Occasional in wooded slopes, forest margins, and dry thickets. Along road to Ajax Peak (A2652). Also on St. Thomas; Bahamas, Cuba, Hispaniola, and Puerto Rico.

NOTE: This species flowers only sporadically; it has never been collected in flower on St. John.

5. AXONOPUS P. Beauv.

Annual to robust perennial herbs; culms erect or decumbent, unbranched or branched, internodes solid. Leaf sheaths smooth or scabrous; ligule a ciliate membrane or a fringe of hairs; blades filiform to lanceolate, flat, conduplicate or involute, the margins smooth or scabrous. Inflorescence with spreading primary branches with 1-sided or distichous spikelets; secondary branches appressed; rachis terminating in a spikelet; callus not differentiated. Spikelets disarticulating at base, dorsiventrally compressed or plano-convex; lower glume absent; upper glume as long as the spikelet; lower floret lemma membranous, toward the rachis; upper floret lemma 0.5–1 times the length of the lower floret lemma, cartilaginous, smooth or muricate, shiny or dull, mostly with involute margins, muticous. Caryopsis with white or grayish embryo.

A tropical and subtropical genus with about 110 species.

1. Axonopus compressus (Sw.) P. Beauv., Ess. Agrostogr. 154. 1812. *Milium compressum* Sw., Prodr. 24. 1788.   Fig. 232D–F.

Paspalum platycaulon Poir. *in* Lam., Encycl. Suppl. **5**: 34. 1804.

Perennial, stoloniferous herb; flowering culms 1–6 dm tall, erect from the base, unbranched. Leaf sheaths glabrous; ligule 0.4–0.6 mm long; blades linear to lanceolate, flat or conduplicate,

(1–)3–8(–25) × 0.4–1 cm, glabrous, smooth on the upper surface. Inflorescence ovate; main axis present or absent, to 2 cm long; primary branches 3–8 cm long. Spikelets solitary, ovate to elliptic, 2.1–2.8 × 0.6–0.9 mm; upper glume 2.1–2.8 mm long, 4–5-nerved; lower floret, lemma lanceolate to ovate, 4-nerved, muticous; palea absent; in upper floret, lemma oblong to ovate, membranous, muricate, white, 1.7–2.4 mm long, muticous.

DISTRIBUTION: An occasional roadside weed. Bethany to Rosenberg (B282), Bordeaux (A2615), Lameshur (W311). Also on St. Croix and St. Thomas; a tropical and subtropical weed, commonly used as a lawn grass.

6. BAMBUSA Schreb.

Densely cespitose herbs, with sympodial, pachymorph rhizomes; culms erect, lignified, hollow, armed or unarmed; branches 1 to several per node with 1 branch dominant, often producing 2 subdominant branches. Culm leaves deciduous; sheath auriculate and large (in our species); ligules short; blades small, triangular, erect and persistent (in our species); foliage leaves usually in complements; sheaths with ligules adaxial and abaxial, usually with oral setae; blades petiolate. Inflorescence a series of sessile clusters of pseudospikelets borne along an elongate rachis on a leafless branch. Pseudospikelets branching into several-flowered spikelets; spikelets with sterile, glumelike bracts, disarticulating between the florets; lemmas 15–20-veined within; paleas with ciliate or glabrous keels; stamens 3 or 6.

About 100 species native to tropical Asia, some of which have been introduced in the New World.

REFERENCES: Soderstrom, T. R. and R. P. Ellis. 1988. The woody bamboos (Poaceae: Bambuseae) of Sri Lanka: A morphological-anatomical study. Smithsonian Contr. Bot. 72: 1–75.

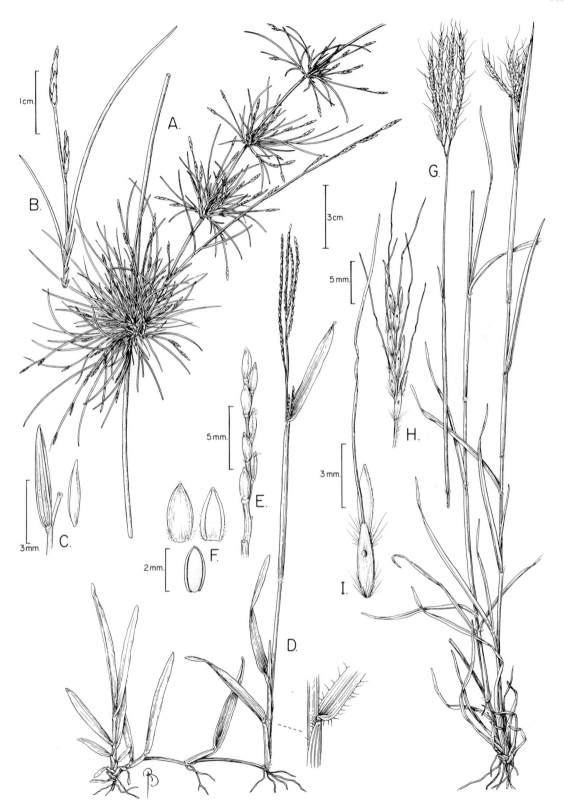

Fig. 232. A–C. *Arthrostylidium farctum*. A. Habit. B. Inflorescence raceme. C. Floret attached to rachilla, and palea. D–F. *Axonopus compressus*. D. Habit and detail of culm showing junction of leaf sheath and blade. E. Portion of inflorescence branch. F. Glume and lower lemma (above) and upper lemma (below). G–I. *Bothriochloa pertusa*. G. Habit. H. Inflorescence raceme. I. Portion of rachis showing sessile (lower) and pediceled (upper) spikelet.

1. Bambusa vulgaris Schrad. *ex* J. C. Wendl., Coll. Pl.
    2: 26, pl. 47. 1808.

Perennial, giant herb; culms 10–20 m tall, 5–12 cm diam.,
erect but arching above, green, yellowish, or yellow with green
stripes, glabrous, with lignified wall and hollow internodes. Culm
leaves alternate, 30–45 cm long and about as wide, covered with
patches of dense, blackish brown hairs, truncate at apex, bearing
prominent, ciliate, falcate auricles; inner ligules short, the outer
ligules absent; blades 6–13 cm long, triangular, erect, persistent.
Foliage leaves in complements of 6–9; sheaths glabrous to brown
hairy; inner ligules 0.5 mm long, membranous, the outer ligules
0.3–1 mm long, indurate; pseudopetioles 2–4(–6) mm long, glabrous; blades 12–30 × 1.5–4 cm, lanceolate, flat, glabrous,
truncate to rounded at base with acuminate apex. Inflorescences
loosely fasciculate, 2–3 cm long clusters borne on leafless
branches, the branches usually >50 cm long. Pseudospikelets 14–
20 mm long, sessile, ovate to lanceolate, straw-colored, 5–7-
flowered; upper floret sterile; fertile florets 8–12 mm long, ovate,
the lemmas 10–15-nerved, the paleas narrowly elliptical, shorter
than or as long as the lemmas, with pectinate-ciliate keels, the
hairs 0.4–0.6 mm long, brown; anthers 2–5 mm long. Caryopsis
not seen.

DISTRIBUTION: Uncommon, persistent after cultivation. Visually recorded at L'Esperance. Cultivated and commonly escaped in the New World
tropics; a native of Southeast Asia.

COMMON NAME: bamboo.

7. BOTHRIOCHLOA Kuntze

Cespitose, sometimes rhizomatous or stoloniferous, annual or perennial herbs; culms erect. Leaf ligules membranous, fringed or
not; blades flat. Inflorescence of several to many long-pedunculate racemes, borne along a short to elongate rachis. Spikelets paired,
falling attached to the rachis internodes; rachis internode filiform, hairy, brittle, with a thin central longitudinal groove appearing as a
translucent stripe; sessile spikelets awned, 2-flowered; lower glume dorsally compressed, about as long as the spikelet, many-nerved,
the nerves evident near apex; upper glume about as long as the spikelet, delicately membranous, 1–3-nerved, keeled; lower floret
sterile, without a palea; upper floret fertile, bisexual, the lemma reduced to a geniculate awn, the palea lacking; pedicellate spikelets
sterile or staminate, similar to sessile spikelets or reduced.

A pantropical genus of 30–40 species of which 10–12 species are native to the New World.

1. Bothriochloa pertusa (L.) Camus, Ann. Soc. Linn.
    Lyon, ser. 2, **76**: 164. 1931. *Holcus pertusus* L.,
    Mant. Pl. **2**: 301. 1771. *Andropogon pertusus* (L.)
    Willd., Sp. Pl. **4**: 922. 1806.   Fig. 232G–I.

Perennial herb; culms 25–100 cm tall, erect to decumbent or
sprawling, nodes bearded or glabrous. Leaf ligules 0.5–1.3 mm
long, a fringed membrane; blades 2–25 × 0.2–0.4 cm, usually
with papillose hairs. Inflorescence 2–10 × 1–10 cm, greenish to
purplish, with 3 to many subpalmately inserted racemes in a
fanlike pattern, usually exserted on a peduncle to 12 cm long or
partially included in the upper sheath. Sessile spikelets 3–4 mm
long; lower glume 7–11-nerved usually with a prominent circular
pit near the middle; upper glume tapering to acuminate apex; lower
lemma 2.3–2.8 mm long; palea reduced; anthers 1–1.8 mm long,
yellow. Caryopsis not seen.

DISTRIBUTION: Common weed of open disturbed areas. Dittlif Point
(A3980), East End (A3785), western slopes of Fish Bay (A3913). Also on
St. Croix and St. Thomas; native to the Old World, now naturalized in the
West Indies, Central America, Mexico, and the southern United States.

COMMON NAME: hurricane grass.

DOUBTFUL RECORD: *Bothriochloa ischaemum* (L.) Keng was listed as
occurring in the area by Woodbury and Weaver (1987). However, all
specimens of this genus seen from the West Indies had some sessile spikelets
with pitted lower glumes. The only characteristic to separate *B. ischaemum*
from *B. pertusa* is the lack of a pitted lower glume on the sessile spikelet.
Since these pits are somewhat hard to see with the naked eye, they could
easily be missed if the entire specimen is not adequately viewed.

8. BOUTELOUA Lag.

Cespitose annual or perennial herbs; culms erect to decumbent. Leaves mostly basal; ligules ciliate; blades flat or folded. Inflorescence with 1 to numerous spicate branches borne singly at the nodes, sessile to short-pedicellate. Spikelets borne in 2 rows, several-
flowered; disarticulation above the glumes; glumes 1-nerved, shorter or longer than the bisexual floret, awnless or short-awned; lower
floret bisexual, the lemma narrow, 3-nerved, usually awned at apex, sometimes the lateral nerves awned, the palea membranous with
the 2 nerves, often extending into awns; upper floret staminate or sterile; stamens 3. Caryopsis ellipsoid.

A New World genus with about 40 species, from Canada to southern South America, typically in short-grass prairies, desert
grasslands, and savannas.

REFERENCE: Gould, F. W. 1979 [1980]. The genus *Bouteloua* (Poaceae). Ann. Missouri Bot. Gard. 66: 348–416.

1. Bouteloua americana (L.) Scribn., Proc. Acad. Nat.
    Sci. Philadelphia **50**: 306. 1891. *Aristida americana*
    L., Syst. Nat., ed. 10, **2**: 879. 1759. *Aristida ad-
    scensionis* L. var. *americana* (L.) Kuntze, Revis.
    Gen. Pl. **3**: 340. 1898.   Fig. 233A–C.

Perennial herb; culms 25–68 cm tall, decumbent and rooting
at the lower nodes. Leaf ligules 0.6–0.8 mm long; blades 3–6 ×
0.2–0.4 cm, flat below and inrolled above, the lower margins
often with stiff hairs. Inflorescences 7–15 × 2–5 cm, with 3–7
loosely ascending racemelike, triquetrous deciduous branches 2–
4 cm long. Spikelets 6–9 mm long (excluding the awns), 2-
flowered, subsessile; pedicels 0.5–1 mm long; glumes 3.5–6 mm
long, arcuate, lanceolate, subaristate, subequal; lemma of bisexual
floret 4.5–7 mm long (excluding the awns), lanceolate, hyaline
with 3 green nerves, awned at apex, the awn to 2 mm long, the
callus with a tuft of hairs; lemma of upper floret rudimentary, the
nerves prolonged into awns 8–10 mm long; anthers 0.7–0.8 mm
long. Caryopsis 3–3.5 mm long.

DISTRIBUTION: Occasional weed of open disturbed areas. Bethany to
Rosenberg (B280), Coral Bay (A4030), Lameshur (B633). Also on St.
Croix, St. Thomas, and Virgin Gorda; Mexico to Brazil and the West Indies.

Fig. 233. A–C. *Bouteloua americana*. A. Habit. B. Spikelet. C. Lower glume, lemma, palea, caryopsis, sterile upper floret, and upper glume. D–F. *Cenchrus brownii*. D. Habit and detail of culm showing junction of leaf sheath and blade. E. Spiny involucre and involucre with portion cut away showing spikelets. F. Spikelet with glumes removed showing lemma and palea of lower floret, and caryopsis. G–J. *Chloris barbata*. G. Habit. H. Portion of the inflorescence raceme showing spikelets. I. Spikelet. J. Lower fertile floret lemma, palea of fertile floret, upper sterile floret, and pistil with stamens.

9. CENCHRUS L.

Annual or perennial herbs; culms erect or decumbent, unbranched or sparingly branched. Leaf sheaths smooth or scabrous; ligule a ciliate membrane or a fringe of hairs; blade filiform to linear, flat to involute, straight or recurved, the margins smooth or scabrous. Main axis of inflorescence with appressed or spreading primary branches, these reduced to a fascicle of spikelets, subtended by an involucre of spiny bracts, the fascicle and involucre falling as a unit; rachis terminating in a bristle; disarticulation at the base of the primary branches; callus differentiated, flared at apex. Spikelets dorsiventrally compressed; lower glume present or absent, not encircling the spikelet base; upper glume 0.3–1 times spikelet length; lemma of lower floret membranous; upper floret lemma about as long as the lower floret lemma, chartaceous to cartilaginous, smooth to muricate, dull, with flat margins, muticous. Caryopsis dispersed along with the spiny involucre that sticks to clothing and fur.

About 22 species of the tropics and warm-temperate regions.

Key to the Species of *Cenchrus*

1. Involucre of interlocking spines in 1 whorl, usually <4 mm wide, densely crowded on the main axis
.... 1. *C. brownii*
1. Involucre with spines emerging at irregular intervals, usually >4 mm wide, not densely crowded on the
main axis.. 2. *C. echinatus*

1. Cenchrus brownii Roem. & Schult., Syst. Veg. **2:** 258. 1817.   Fig. 233D–F.

Cenchrus inflexus R. Br., Prodr. **1:** 195. 1810, non Poir., 1804.
Cenchrus viridis Spreng., Syst. Veg. **1:** 301. 1825.

Annual herb, lacking stolons; flowering culms 3–11 dm tall, erect from the base or decumbent; sparingly branched. Leaf sheaths glabrous; ligule 0.6–1.1 mm long; blades linear, flat to conduplicate, 10–40 × 0.5–1.3 cm, smooth, glabrous. Inflorescence linear; main axis 2–6 cm long; primary branches 4–6 mm long; involucre of interlocking spines in 1 whorl, usually <4 mm wide, densely crowded on the main axis. Spikelets solitary, lanceolate to elliptic, 4.5–5.2 × 1.3–1.8 mm; lower glume 3.7–4.7 mm long, 3-nerved; upper glume 3–4 mm long, 3-nerved; in lower floret, lemma lanceolate, 3–5-nerved, muticous or mucronate, palea fully developed; in upper floret, lemma lanceolate, membranous, muricate, white, 3–5 mm long, muticous.

DISTRIBUTION: An occasional roadside weed. Chocolate Hole (A4068), Herman Farm (A822). Also on St. Croix and St. Thomas; southern United States to Brazil, the West Indies, and introduced into Australia.

COMMON NAMES: sandbur, sand-spurs.

2. Cenchrus echinatus L., Sp. Pl. 1050. 1753.

Annual herb, lacking stolons; flowering culms 1–5 dm tall, erect or decumbent at the base, sparingly branched. Leaf sheaths mostly glabrous; ligule 0.8–1.4 mm long; blades linear, flat to conduplicate, 6–20 × 0.3–1.1 cm, smooth, glabrous or hairy on the upper surface. Inflorescence linear to oblong; main axis 35–80 mm long; involucre with spines emerging at irregular intervals, usually >4 mm wide, not densely crowded on the main axis. Spikelets solitary, lanceolate to ovate, 4.9–6.5 × 1.3–2.1 mm; lower glume 1.2–3 mm long, 1-nerved; upper glume 3.8–5.7 mm long, 5-nerved, lower floret, lemma lanceolate to ovate, 5-nerved, muticous; palea fully developed; in upper floret, lemma ovate to lanceolate, hyaline, minutely papillate, yellow, 4.5–7 mm long, muticous.

DISTRIBUTION: A common herb of coastal sandy beaches. Great Cruz Bay (A788), Little Lameshur (A2020). Also on St. Croix, St. Thomas, Tortola, and Virgin Gorda; a common weed of disturbed soil sites in tropical and warm-temperate America.

DOUBTFUL RECORD: *Cenchrus tribuloides* L. was reported by Woodbury and Weaver (1987) as occurring on St. John; however, no specimens have been found to confirm this record.

10. CHLORIS Sw.

Rhizomatous, stoloniferous, or cespitose annual or perennial herbs; culms erect to decumbent. Leaf ligules membranous, ciliate, or occasionally absent; blades usually linear, flat or folded. Inflorescence a digitate or subdigitate cluster of several to many, or occasionally 1, spikelike racemes. Spikelets borne in 2 staggered rows on one side of the rachis, sessile to subsessile, several-flowered with only a single bisexual floret; glumes usually 1-nerved; upper glumes 1–4-nerved, shorter than the florets; lower floret usually bisexual; upper florets usually sterile and reduced, rarely staminate or bisexual; lemma of bisexual florets lanceolate, 3-nerved, with hairy margins, the base bearded, the apex usually awned and minutely bifid; palea shorter than the lemma, glabrous; stamens 3. Caryopsis mostly ovoid to obovoid or ellipsoid.

A tropical and warm-temperate genus with about 55 species, worldwide in grasslands and disturbed habitats.

REFERENCE: Anderson, D. 1974. Taxonomy of the genus *Chloris* (Gramineae). Brigham Young Univ. Sci. Bull., Biol. Ser. 19: 1–132.

1. Chloris barbata Sw., Fl. Ind. Occ. **1:** 200. 1797 [as a new name]. *Andropogon barbatum* L., Mant. Pl. **2:** 302. 1771, nom. illegit., non L., 1759.
          Fig. 233G–J.

Chloris inflata Link, Enum. Hort. Berol. Alt. **1:** 105. 1821.

Chloris paraguaiensis Steud., Syn. Pl. Glumac. 204. 1854.

Cespitose annual herb; culms erect, 20–90 cm tall. Leaf ligules 0.4–0.7 mm long; blades 9–25 × 0.2–0.5 cm, glabrous, sparingly pilose on the upper surface near base. Inflorescences subdigitate with 7–15 racemes, each 3–8 cm long. Spikelets 3(–4)-flowered, densely imbricate, often purplish; glumes 1 or 2, 7 mm long, linear to oblanceolate, the apex mucronate; lower sterile floret lemma 0.9–1.3 mm long, broadest at the apex, narrowly turbinate, the apex truncate, the awn 5–7 mm long; bisexual floret lemma 2–2.8 mm long, elliptical, bearing a callus with a tuft of hairs, the margins with ascending hairs, the awn 4–8 mm long, straight; anthers 0.5–0.7 mm long; uppermost sterile floret inflated, spherical; Caryopsis 1.1–1.4 mm long, ellipsoid to obovoid.

DISTRIBUTION: A common weed of open disturbed habitats, often near the coast. Center Line Road (A1856), Cruz Bay (A858), Fish Bay (A3882). Also on St. Croix, St. Thomas, Tortola, and Virgin Gorda (according to Britton and Wilson, 1924); worldwide in tropical and warm-temperate regions, native to the New World.

DOUBTFUL RECORDS: *Chloris radiata* (L.) Sw. was reported by Britton and Wilson (1924) and by Woodbury and Weaver (1987) as occurring on St. John; however, no specimens have been found to confirm these records.

11. CYNODON Rich.

Stoloniferous and rhizomatous perennial herbs; culms decumbent, creeping. Leaves often clustered near base; ligules membranous, ciliate; blades linear, soft. Inflorescence a digitate or subdigitate cluster of 2 to many spikelike racemes. Spikelets laterally compressed, borne in 2 alternating rows on one side of the rachis; sessile to subsessile; 1-flowered, with a minute prolongation of the rachilla above the floret; disarticulation above the glumes; glumes 1-nerved, subequal, shorter than the florets; lemma broadly ovate, 3-nerved; firmly membranous; palea subequal to the lemma; stamens 3. Caryopsis ellipsoid.

A genus with 8 species, of the Old World tropics.

1. Cynodon dactylon (L.) Pers., Syn. Pl. **1**: 85. 1805. *Panicum dactylon* L., Sp. Pl. 58. 1753. *Capriola dactylon* (L.) Kuntze, Revis. Gen. Pl. **2**: 764. 1891. Fig. 234A–C.

Rhizomatous and stoloniferous perennial herb; culms erect, 10–40 cm tall, freely branching near base, the plant often long-trailing. Leaf ligules 0.2–0.4 mm long; blades 2–10 × 0.1–0.3(–0.4) cm, flat to somewhat folded, scabrous above. Inflorescences digitate with 3–6 often arcuate racemes, each 1.5–6 cm long, with scabrous margins. Spikelets with pedicels 0.2–0.3 mm long; glumes 1.5–2.0 mm long; lower glumes arcuate and appressed to the rachis; upper glume straight and divergent from the rachis; lemma 2–2.5 mm long; ciliate along the upper keel; palea about as long as the lemma, narrowly elliptic; rachilla prolonged beyond the fertile floret, to 0.6 mm long; anthers 1–1.4 mm long, yellowish. Caryopsis 1.1–1.4 mm long, ellipsoid.

DISTRIBUTION: Occasional in pastures and disturbed habitats. Cinnamon Bay (M17088), Emmaus (A2003), along road from Fish Bay to Coccoloba Key (A2813). Now cosmopolitan in tropical and warm-temperate regions, but apparently native to Africa.

COMMON NAME: Bermuda grass.

12. DACTYLOCTENIUM Willd.

Cespitose to short-stoloniferous annual or perennial herbs; culms cespitose or decumbent. Leaf ligules membranous to ciliate; blades linear. Inflorescence a terminal, digitate cluster of 2–11 densely flowered racemes, each terminating in a naked point beyond the last spikelet. Spikelets sessile, 2–9-flowered, laterally compressed, borne in 2 staggered, densely imbricate rows on one side of the 3-angled rachis; glumes 1-nerved, glabrous, shorter than the spikelets, turned lateral to the rachis; the upper glume usually awned or mucronate; disarticulation above the lower glume; uppermost florets rudimentary; lowest florets fertile; lemmas ovate, 3-nerved, the apex with a short, flexuous awn or mucro; paleas shorter than the lemmas; stamens 3. Caryopsis globose to ellipsoid, transversely rugose with a thin deciduous pericarp.

A genus of the Old World tropics and subtropics, with 13 species.

1. Dactyloctenium aegyptium (L.) Willd., Enum. Pl. 1029. 1809. *Cynosurus aegyptius* L., Sp. Pl. 72. 1753.

Cespitose annual herb with short stolons; culms decumbent, 5–40 cm tall, rooting at the lower nodes. Leaf ligules 0.5–0.8 mm long, ciliate; blades 1.5–10 × 0.1–0.5 cm, papillose-pilose. Inflorescences digitate, with 2–4 racemes 0.7–6 cm long, the terminal point 1–3 mm long. Spikelets 3–4.2 mm long, 3–5-flowered, gray to purplish; glumes 2–3.4 mm long, broadly lanceolate to ovate; lower glumes persistent; upper glumes deciduous, the apex with stout divergent awn 0.8–1.5 mm long; upper 1 or 2 florets reduced and sterile; lemmas 1.9–3 mm long, ovate, the apex acuminate, scabrous along keel, the awn to 1 mm long; anthers 0.3–0.5 mm long, yellowish. Caryopsis 0.8–1 mm long, light brown.

DISTRIBUTION: Common in cultivated fields and disturbed areas. Western slope of Fish Bay (A3916); In vicinity of Bordeaux (A2629), East End (A3790). Also on Anegada, St. Croix, St. Thomas, and Tortola; native to the Old World, now widespread throughout the tropics and subtropics.

13. DIGITARIA Haller

Annual or perennial, cespitose herbs; culms erect or decumbent, unbranched or sparingly branched; internodes usually hollow. Leaf sheaths conspicuously inflated or not, smooth or scabrous; ligule a membrane or a ciliate membrane; blades filiform to lanceolate, flattened to cylindrical, the margins smooth or scabrous. Main axis of inflorescence more or less digitate, with appressed or divaricate primary branches; secondary branches appressed or spreading; rachis terminating in a spikelet; disarticulation at the spikelet base; callus not differentiated. Spikelets abaxial, plano-convex; lower glume present or absent, not encircling the spikelet base; upper glume 0.3–1 times spikelet length; lemma of lower floret membranous; lemma of upper floret 0.5–1.2 times the length of the lower floret, cartilaginous, smooth, striate, or muricate, dull, with flat, thin margins, muticous or mucronate.

About 250 species from the tropics and warm-temperate regions.

REFERENCE: Webster, R. D. 1987. Taxonomy of *Digitaria* section *Digitaria* in North America (Poaceae: Paniceae). Sida 12: 209–222.

Fig. 234. A–C. *Cynodon dactylon*. **A.** Habit and detail of adaxial junction of leaf sheath and blade showing ligule. **B.** Portion of inflorescence branch showing spikelets. **C.** Spikelet, palea, and pistil with stamen. **D–F.** *Digitaria ciliaris*. **D.** Habit and detail of junction of leaf sheath and blade showing ligule. **E.** Portion of inflorescence branch showing spikelets. **F.** Spikelet, dorsal and lateral views. **G–I.** *Echinochloa colona*. **G.** Habit. **H.** Spikelet pair. **I.** Upper glume and lower lemma (above) and floret, lateral and ventral views (below).

Key to the Species of *Digitaria*

1. Spikelets covered with long silky hairs; spikelets >3.9 mm long .... 4. *D. insularis*
1. Spikelets without long silky hairs; spikelets <3.8 mm long.
   2. Upper floret brown, muricate; primary branches not winged .... 2. *D. hitchcockii*
   2. Upper floret yellow, striate; primary branches distinctly winged.
      3. Spikelets >2.8 mm long .... 1. *D. ciliaris*
      3. Spikelets <2.8 mm long .... 3. *D. horizontalis*

1. Digitaria ciliaris (Retz.) Koeler, Descr. Gram. 27. 1802. *Panicum ciliare* Retz., Observ. Bot. **4:** 16. 1786. Fig. 234D–F.

Panicum adscendens Kunth *in* Humb., Bonpl. & Kunth, Nov. Gen. Sp. Pl. **1:** 97. 1816. *Digitaria adscendens* (Kunth) Henrard, Blumea **1:** 92. 1934.

Annual herb, lacking stolons; flowering culms 1–10 dm tall, decumbent at the base, sparingly branched. Leaf sheaths glabrous or hairy; ligule 1–4 mm long; blades linear, flat, 2–20 cm long, 3–10 mm wide, scabrous, glabrous on the upper surface (rarely with a few scattered hairs). Inflorescence obovate; main axis 14–40 cm long; primary branches 4–24 cm long, distinctly winged. Spikelets paired, lanceolate, 2.7–3.7 × 0.7–1.1 mm; lower glume 0.35–0.7 mm long, nerveless; upper glume 1.2–2.7 mm long, 3–5-nerved; lemma of lower floret lanceolate, 5–7-nerved, muticous, the palea absent; lemma of upper floret lanceolate, cartilaginous, striate, yellow (becoming purple), 2.6–3.8 mm long, muticous.
DISTRIBUTION: Occasional weed of open disturbed areas. Herman Farm (M17087), Lameshur (W334). Found throughout the tropics and subtropics worldwide.

2. Digitaria hitchcockii (Chase) Stuck., Annuaire Conserv. Jard. Bot. Genève **17:** 287. 1914. *Valota hitchcockii* Chase, Proc. Biol. Soc. Wash. **24:** 110. 1911. *Trichachne hitchcockii* (Chase) Chase, J. Wash. Acad. Sci. **23:** 454. 1933.

Perennial herb, lacking stolons; flowering culms 3–6 dm tall, erect from base, sparingly branched. Leaf sheaths glabrous; ligule 0.2–0.5 mm long; blades linear, flat, 2–6 × 0.2–0.3 cm, the upper surface smooth, glabrous or pubescent. Inflorescence linear to oblong; main axis 5–10 cm long; primary branches 2–8 cm long, not winged. Spikelets paired, ellipsoid to lanceolate, 2.5–3 × 0.7–0.9 mm; lower glume 0.2–0.3 mm long, nerveless; upper glume 2.5–3 mm long, 3-nerved; lemma of lower floret lanceolate, 5–7-nerved, muticous, the palea vestigial or absent; lemma of upper floret lanceolate to elliptic, cartilaginous, muricate, brown, 2.4–2.7 mm long, muticous.
DISTRIBUTION: Occasional in open disturbed habitats. Ram Head (W189). Known otherwise from Texas and Mexico.

3. Digitaria horizontalis Willd., Enum. Pl. 92. 1809. *Panicum horizontale* (Willd.) G. Mey., Prim. Fl. Esseq. 54. 1818.

Milium digitatum Sw., Prodr. 24. 1788, non *Digitaria digitata* Büse, 1854. *Syntherisma digitata* (Sw.) Hitchc., Contr. U.S. Natl. Herb. **12:** 142. 1908. *Digitaria setosa* Desv. *ex* Ham., Prodr. Pl. Ind. Occ. 6. 1825.

Annual herb, lacking stolons; flowering culms 2–9 dm tall, erect or decumbent at the base, sparingly branched. Leaf sheaths hairy; ligule 0.5–1.5 mm long; blades linear, flat, 2.5–12 × 0.3–0.9 cm, smooth, hairy on the upper surface. Inflorescence ovate; main axis 0.5–5.7 cm long; primary branches 6–15 cm long, distinctly winged. Spikelets paired, lanceolate, 2–2.6 × 0.47–0.6 mm; lower glume 0.1–0.25 mm long, 1-nerved; upper glume 0.8–2 mm long, 1–3-nerved; lemma of lower floret lanceolate to ovate, 5–7-nerved, muticous, the palea absent; lemma of upper floret lanceolate, cartilaginous, striate, yellow, 1.8–2.5 mm long, muticous.
DISTRIBUTION: Occasional weed of open disturbed places. Bordeaux (W189), Lameshur (A3211). Also on St. Thomas; throughout the tropics and subtropics worldwide.

4. Digitaria insularis (L.) Mez *ex* Ekman, Ark. Bot. **11:** 17. 1912. *Andropogon insularis* L., Syst. Nat., ed. 10, **2:** 1304. 1759. *Trichachne insularis* (L.) Nees, Agrost. Bras. 86. 1829. *Valota insularis* (L.) Chase, Proc. Biol. Soc. Wash. **19:** 188. 1906.

Perennial herb, lacking stolons; flowering culms 4–13 dm tall, erect, sparingly branched. Leaf sheaths usually hairy; ligule 2–6 mm long; blades linear, flat, 20–60 × 0.8–1.7 cm, scabrous, glabrous on the upper surface. Inflorescence oblong; main axis 20–35 mm long; primary branches 4–15 cm long. Spikelets paired, ovate, 4–5 × 1–1.3 mm, covered with long silky hairs; lower glume 0.4–0.7 mm long, nerveless; upper glume 3.5–4.5 mm long, 3–5-nerved; lemma of lower floret lanceolate, 7-nerved, muticous, the palea absent; lemma of upper floret lanceolate, chartaceous, muricate, brown, 3.2–3.6 mm long, muticous.
DISTRIBUTION: Common roadside weed. Bordeaux (A3800); Center Line Road (A2619), Maho Bay (A1943). Also on St. Thomas; throughout tropical and subtropical America.

DOUBTFUL RECORDS: *Digitaria sanguinalis* (L.) Scop. was reported by Britton and Wilson (1924) and by Woodbury and Weaver (1987) as occurring on St. John; however, no specimens have been found to confirm these records.

14. ECHINOCHLOA P. Beauv.

Annual to robust perennial herbs; culm erect or decumbent, unbranched or sparingly branched. Leaf sheaths conspicuously inflated, smooth or scabrous; ligule a fringe of hairs or absent; blades filiform to linear, flattened. Main axis of inflorescence with 1-sided or bilateral, appressed, or spreading primary branches; secondary branches appressed; rachis terminating in a spikelet; disarticulation at the spikelet base; callus differentiated or not differentiated. Spikelets adaxial, dorsiventrally compressed to plano-convex; lower glume encircling the spikelet base; upper glume as long as the spikelet; lower floret lemma membranous to chartaceous; upper floret lemma indurate, smooth, shiny to dull, with involute margins, the apex muticous, with different texture. Caryopsis with a whitish or grayish embryo.

A taxonomically difficult genus of about 40 species in the tropics and warm-temperate zones.

1. Echinochloa colona (L.) Link, Hort. Berol. **2:** 209. 1833. *Panicum colonum* L., Syst. Nat., ed. 10, **2:** 870. 1759.   Fig. 234G–I.

Annual herb, lacking stolons; flowering culms 1–9 dm tall, erect or decumbent at the base, unbranched or sparingly branched. Leaf sheaths glabrous; blades linear, flat, 3–22 × 0.3–0.7 cm, smooth, glabrous on the upper surface. Inflorescence oblong to lanceolate; main axis 3–13 cm long; primary branches 1–3 cm long. Spikelets paired, ovate to elliptic, 2.1–2.9 × 1.1–1.5 mm; lower glume 1.1–1.5 mm long, 3–5-nerved; upper glume 2–2.8 mm long, 5-nerved; lemma of lower floret ovate to elliptic, 5-nerved, mucronate, the palea fully developed; lemma of upper floret elliptic, indurate, smooth, yellow, 1.9–2.9 mm long, apiculate.

DISTRIBUTION: Common in moist disturbed places. Enighed (A5099); Fish Bay (A2811). Also on St. Croix, St. Thomas, and Tortola; throughout the tropics and subtropics of the world, originally native to the Old World.

15. ELEUSINE Gaertn.

Cespitose, sometimes rhizomatous annual or perennial herbs; culms erect or decumbent. Leaf sheaths glabrous or hairy; ligule a ciliate membrane; blades flat to folded. Inflorescence a digitate or subdigitate terminal cluster of spikelike racemes terminating in a spikelet. Spikelets sessile, several-flowered, strongly laterally compressed, borne in 2 staggered rows on one side of a 3-angled rachis; disarticulation above the glumes; rachilla pronounced between the florets; glumes lanceolate, shorter than the lowest floret; lower glume 1-nerved; upper glume 3–5-nerved; lemmas lanceolate, 3–5-nerved; paleas shorter than the lemmas; stamens 3. Caryopsis globose and transversely rugose with a thin, free pericarp.

A genus native to the Old World tropics, with 10 species.

REFERENCE: Phillips, S. M. 1972. A survey of the genus *Eleusine* in Africa. Kew Bull. 27: 251–270.

1. Eleusine indica (L.) Gaertn., Fruct. Sem. Pl. **1:** 8. 1788. *Cynosurus indicus* L., Sp. Pl. 72. 1753.   Fig. 235A–C.

Annual herb; culms decumbent, 15–65 cm tall, thick-walled, glabrous. Leaf ligules 0.4–1 mm long; blades 5–30 × 0.2–0.6 cm, scabrous above. Inflorescences with 2–10 digitate or subdigitate racemes 3–15 cm long. Spikelets 4–6.5 mm long, 4–7-flowered, ascending at a 45° angle from the rachis; glumes 1.5–3 mm long; lower glume 1.5–2.5 mm long; upper glume 2.2–3 mm long; uppermost florets reduced; lemmas 2–3.2 mm long, glabrous, the midvein scabrous, especially on upper portion; anthers 0.2–0.5 mm long, purplish. Caryopsis 0.7–1 mm long, dark purple.

DISTRIBUTION: Occasional in disturbed pastures and along roadsides, introduced from the Old World. Bordeaux Mountain (A2613, A3803, A3851). Also on St. Croix, St. Thomas, and Tortola; worldwide in warm climates.

COMMON NAME: Dutch grass.

16. ERAGROSTIS Wolf

Annual or perennial herbs, cespitose, rhizomatous or rarely stoloniferous; culms erect or decumbent. Leaf ligules usually ciliate, truncate; blades flat, folded or involute. Inflorescence an open to contracted panicle. Spikelets 2–15(–30)-flowered, usually laterally compressed, pedicellate; disarticulation usually above the glumes; glumes unequal to subequal, shorter than the florets, persistent or deciduous, 1(–3)-nerved; lemmas lanceolate to ovate, membranous (1–)3-nerved, the apex obtuse to acute; palea commonly hyaline with ciliate keels, longitudinally bowed out by the caryopsis; stamens 3 or 2. Caryopsis globose to ellipsoid.

A large, variable genus with about 350 species occurring in the tropics and subtropics throughout the world; about 120 species in the New World and 225 species from southern Africa.

REFERENCE: Peterson, P. M. & L. H. Harvey. In press. *Eragrostis*. In M. E. Barkworth (ed.), A manual of grasses of the United States and Canada.

Key to the Species of *Eragrostis*

1. Panicles open, cylindrical to narrowly ovate or ovate lanceolate, 1–7 cm wide; branches ascending and spreading, usually 0.5–4 cm long; base of panicle branches at junction of culm axis with a few scattered hairs; spikelets with pedicels of main culm axis bearing a few hairs; spikelets with pedicels 0.8–7 mm long .... 1. *E. amabilis*
1. Panicles contracted, narrow and glomerate, usually <1.4 cm wide; branches ascending, usually <2.5 cm long; base of panicle branches at junction of culm axis glabrous; spikelets commonly sessile or with pedicels <1 mm long .... 2. *E. ciliaris*

1. Eragrostis amabilis (L.) Wight & Arn. *in* Hook. & Arn., Bot. Beechey Voy. 251. 1838. *Poa amabilis* L., Sp. Pl. 68. 1753.

Poa tenella L., Sp. Pl. 69. 1753. *Eragrostis tenella* (L.) P. Beauv. *ex* Roem. & Schult., Syst. Veg. **2:** 576. 1817.

Eragrostis plumosa Retz., Observ. Bot. **4:** 20. 1786.

Cespitose annual herb; culms 5–40 cm tall, erect to spreading. Leaf ligules 0.2–0.3 mm long, ciliate; blades 2–8 × 0.2–0.4 cm, flat. Panicle 4–15 cm long, 1–12 cm wide, open, cylindrical to narrowly ovate; branches ascending (0.5–)2–8.5 cm long, spreading, with a few scattered hairs and sometimes with irregular

Fig. 235. A–C. *Eleusine indica*. **A.** Habit and detail of junction of leaf sheath and blade showing ciliate ligule. **B.** Portion of inflorescence branch showing glumes after disarticulation of spikelets. **C.** Spikelet and caryopsis. **D, E.** *Eragrostis ciliaris*. **D.** Habit. **E.** Detail of inflorescence branch showing spikelets and caryopsis. **F–H.** *Eriochloa punctata*. **F.** Habit. **G.** Detail of inflorescence branch showing spikelets. **H.** Upper glume, lower sterile lemma, fertile floret, and ventral view.

glandular areas at base (at junction of culm axis). Spikelets (1–) 1.5–2.2 × 0.9–1.2 mm, 4–8-flowered, ovate to oblong, reddish purple to greenish; pedicels 0.8–4(–7) mm long, with few hairs; glumes 0.4–1.1 mm long, ovate, unequal; lemmas 0.7–1.1 mm long, ovate to broadly oblong, scaberulous along keel; apex truncate to obtuse; paleas 0.6–1.1 mm long, pectinate-ciliate along the keels, the hairs 0.3–0.5 mm long; anthers about 0.2 mm long, purplish. Caryopsis 0.3–0.5 mm long, ellipsoid, translucent, light brown.

DISTRIBUTION: Occasional in open disturbed areas. Emmaus (A2000), Lameshur (A3210, B501). Also on St. Croix; native to the Old World, naturalized in the New World.

2. Eragrostis ciliaris (L.) R. Br. *in* Tuckey, Narr. Exped. Zaire 478. 1818. *Poa ciliaris* L., Syst. Nat., ed. 10, **2**: 875. 1759. Fig. 235D, E.

Cespitose annual herb; culms (3–)9–75 cm tall, erect, sometimes slightly geniculate below. Leaf ligules 0.2–0.5 mm long,

ciliate; blades 1.8–12(–15) × 0.2–0.5 cm, flat. Panicle cylindrical, spikelike, 1.7–15 × 0.2–1.4 cm, contracted and condensed into glomerate lobes, often interrupted near base; branches ascending or appressed, 0.4–2.5 cm long, densely flowered. Spikelets 1.8–3.2 × 1–2 mm, 6–11-flowered, elliptical-ovate to ovate-lanceolate, yellowish brown, sometimes purple-tinged; pedicels 0.1–0.9 mm long or spikelets sessile; glumes 0.7–1.6 mm long, ovate to lanceolate; lemmas 0.8–1.3 mm long, elliptic-ovate to lanceolate, the apex obtuse to acute; paleas 0.8–1.3 mm long, prominently ciliate-pectinate on the keels, the apex obtuse to acute; stamens 2, anthers 0.1–0.3 mm long, purplish. Caryopsis 0.4–0.5 mm long, ovoid, reddish brown, without a dorsal groove.

DISTRIBUTION: Occasional weed of grassy, open areas. Cinnamon Bay (W368), Coral Bay (W795-a), Cruz Bay Gut (W792). Native to the paleotropics, now naturalized in the New World.

DOUBTFUL RECORDS: *Eragrostis tephrosanthos* Schult. [= *Eragrostis pectinacea* (Michx.) Nees] and *Eragrostis urbaniana* Hitchc. were reported by Woodbury and Weaver (1987) as occurring on St. John; however, no specimens have been found to confirm these records.

17. ERIOCHLOA Kunth

Annual or perennial, cespitose to stoloniferous herbs; culms erect or decumbent, unbranched or sparingly branched. Leaf sheaths conspicuously inflated or not, smooth; ligules usually a fringe of hairs with a minute membranous rim at the base; blades linear to lanceolate, flattened to involute. Inflorescence with distichous or nondistichous, appressed or spreading primary branches; secondary branches appressed; rachis terminating in a spikelet; disarticulation at the spikelet base; callus differentiated. Spikelets adaxial, dorsiventrally compressed; lower glume fused with the callus and forming a cuplike structure; upper glume about as long as the spikelet; lower floret lemma membranous, sterile; upper floret lemma 0.5–0.95 times the length of the lower floret, indurate, rugose, dull, with involute margins, mucronate or awned.

About 38 species from the tropics and warm-temperate zones.
REFERENCE: Shaw, R. B. & R. D. Webster. 1987. The genus *Eriochloa* (Poaceae-Paniceae) in North America and Central America. Sida 12: 165–207.

1. Eriochloa punctata (L.) Desv. *ex* Ham., Prodr. Pl. Ind. Occ. 5. 1825. *Milium punctatum* L., Syst. Nat., ed. 10, **2**: 872. 1759. *Paspalum punctatum* (L.) Flüggé, Gram. Monogr. Paspalum 127. 1810. *Eriochloa polystachya* Kunth var. *punctata* (L.) Maiden & Betche, Cens. N.S.W. Pl. 16. 1916.

Fig. 235F–H.

Eriochloa kunthii G. Mey., Prim. Fl. Esseq. 46. 1818.

Perennial herb, lacking stolons; flowering culms 3–15 dm tall, erect or decumbent at the base, unbranched or sparingly branched.

Leaf sheaths glabrous (rarely puberulent); ligule 0.4–1 mm long; blades linear, flat, 10–50 × 0.4–1 cm, smooth, glabrous or hairy on the upper surface. Inflorescence linear; main axis 9–22 cm long; primary branches 1–6 cm long. Spikelets paired, becoming solitary toward the apex, lanceolate, 4.5–5.7 × 0.9–1.4 mm; upper glume 4.3–5.5 mm long, 5–7-nerved; lemma of lower floret lanceolate, 5–7-nerved, muticous to mucronate, the palea absent; lemma of upper floret elliptic, indurate, rugose, yellow, 2–3.5 mm long (excluding the awn), awned.

DISTRIBUTION: Occasional herb of wet to dry, open areas. Bordeaux (W66), Lameshur (W523-c). Also on St. Croix, St. Thomas, and Tortola; southern United States to South America and Greater Antilles.

18. HETEROPOGON Pers.

Annual or perennial herbs; culms branched or unbranched above. Leaf sheaths glabrous or hairy; ligules membranous with a dense white fringe near apex; blades flat. Inflorescence a solitary, terminal spikelike raceme; each raceme included within a subtending spathelike sheath below. Spikelets 1-flowered, in pairs at the nodes of inflorescence, 1 sessile and 1 pedicellate; lower few pairs of spikelets alike, staminate, awnless, the remaining sessile spikelets bisexual, awned, the pedicellate spikelets staminate or sterile, awnless; disarticulation of entire spikelet pairs along each rachis internode, the internode forming a sharp, barbed callus below the fertile (sessile) spikelet; callus hairs coppery-brown (in ours). Sessile spikelet (fertile) linear to elliptical; glumes about as long as the spikelet, indurate, dark brown to black; floret hyaline, nearly as long as the spikelet; lemma with a stout geniculate antrorsely plumose awn. Pedicellate spikelet with lower glumes lanceolate, broad to winged, finely striate, greenish, twisted near apex; upper glume 3-nerved, slightly shorter than the lower glume; florets similar to those of sessile spikelets but unawned. Caryopsis with a large embryo.

A genus with 6 species worldwide, commonly found in tropical to subtropical regions.

1. Heteropogon contortus (L.) P. Beauv. *ex* Roem. & Schult., Syst. Veg. **2**: 836. 1817. *Andropogon contortus* L., Sp. Pl. 1045. 1753. Fig. 236A, B.

Cespitose perennial herb; culms 30–75 cm tall. Leaf sheaths sharply keeled; ligules 1–1.5 mm long, light brownish on proximal

half; blades 5–30 × 0.3–0.7 cm, scabrous, lemon-scented. Inflorescence 2–7 cm long excluding the awns, 3–6 mm wide, strongly 1-sided, often long-pedunculate; peduncles included or exserted to 4 cm; rachis internodes 2–3.2 mm long. Sessile, fertile spikelet 5–8 mm long (excluding the awns); callus pungent, heavily bearded, copper-brown; lower glume narrowly elliptical, 5–7-nerved, dark

FIG. 236. **A, B.** *Heteropogon contortus.* **A.** Habit and detail of junction of leaf sheath and blade. **B.** Spikelet pair. **C–E.** *Lasiacis divaricata.* **C.** Habit and detail of culm showing ligule at junction of leaf sheath and blade. **D.** Portion of inflorescence branch showing spikelets. **E.** Lower glume, lower floret, upper floret, and upper glume. **F–H.** *Leptochloa virgata.* **F.** Habit and detail of adaxial junction of leaf sheath and blade showing ligule. **G.** Portion of inflorescence branch showing spikelets. **H.** Portion of rachis with spikelet disarticulating above glumes.

brown, truncate near apex; upper glume linear, 3-nerved, acute near apex; lemma with an awn 6–12 cm long, bent and flexuous 1–2 cm long beyond its emergence. Pedicellate spikelet 7–10 mm long; lower glume convex on back, green, glabrous below and sparingly papillose-pilose near apex; upper glume slightly smaller than the lower one; lemma awnless. Caryopsis longitudinally grooved.

DISTRIBUTION: Occasional grass of disturbed coastal habitats. East End Haulover (W596). Also on St. Thomas (Water Island) and Virgin Gorda; pantropical.

19. LASIACIS (Griseb.) Hitchc.

Perennial herbs; culms commonly lignified, climbing, clambering, decumbent or erect, branched. Leaf sheaths smooth; ligule membranous; blades linear to ovate, flattened. Main axis of inflorescence with spreading or divaricate primary branches; secondary branches spreading; rachis terminating in a spikelet. Spikelets adaxial, terete to globose; lower glume encircling the spikelet base; disarticulating at base, seldom above the lower glume; callus not differentiated; upper glume about as long as the spikelet; lemma of lower sterile floret chartaceous; lemma of upper floret as long as the lower floret, indurate, smooth, dull, with involute margins, differentiated at the apex, muticous or apiculate. Caryopsis with whitish or grayish embryo.

A neotropical genus with about 16 species.

REFERENCE: Davidse, G. 1978. A systematic study of the genus *Lasiacis* (Gramineae: Paniceae). Ann. Missouri Bot. Gard. 65: 1133–1254.

Key to the Species of *Lasiacis*

1. Ligule of the upper leaves >1.6 mm long
  2. Primary branches reflexed to spreading; sheaths glabrous or hairy .... 2. *L. ligulata*
  2. Primary branches appressed to spreading; sheaths papillose hispid or pubescent .... 3. *L. sorghoidea*
1. Ligule of the upper leaves <1.6 mm long .... 1. *L. divaricata*

1. Lasiacis divaricata (L.) Hitchc., Contr. U.S. Natl. Herb. **15**: 16. 1910. *Panicum divaricatum* L., Syst. Nat., ed. 10, **2**: 871. 1759.   Fig. 236C–E.

Panicum bambusoides Desv. *ex* Ham., Prodr. Pl. Ind. Occ. 10. 1825.

Lasiacis harrisii Nash, Torreya **13**: 274. 1913.

Perennial herb; flowering culms (5–)10–50(–70) dm tall, erect or decumbent at the base, moderately branched. Leaf sheaths glabrous; ligule 0.2–0.6 mm long; blades lanceolate to linear, flat, (3–)5–12(–16) × (0.3–)0.6–1.4(–2.0) cm, smooth, glabrous on the upper surface (except for some scabridity or puberulence along the lower part of the midrib). Inflorescence ovate; main axis 2–12(–20) cm long; primary branches 2–8(–12) cm long. Spikelets solitary, obovoid, (3.5–)3.7–4.3(–4.5) × 2.1–2.6 mm; lower glume (1.2–)1.4–2(–2.5) mm long, 7–11-nerved; upper glume 3.5–4.5 mm long, 9–11-nerved; lemma of lower floret obovate, 9–13-nerved, muticous, the palea fully developed; lemma of upper floret obovate, indurate, smooth, usually brown, 3.4–4 mm long, muticous.

DISTRIBUTION: Common in secondary vegetation. Bordeaux (A3795), Carolina (R1306), Kingshill (E3121). Also on St. Croix, St. Thomas, and Tortola; southern Florida, Mexico to South America and the West Indies.

2. Lasiacis ligulata Hitchc. & Chase, Contr. U.S. Natl. Herb. **18**: 337. 1917.

Panicum divaricatum L. var. *puberulum* Griseb., Fl. Brit. W. I. 551. 1864.

Perennial herb; flowering culms 10–50(–100) dm tall, erect, moderately branched. Leaf sheaths glabrous or hairy; ligule (1.6–) 2–3(–3.7) mm long; blades broadly to narrowly lanceolate, flat, (5–)7–14(–17) × (0.6–)1.0–2.2(–3.4) cm, scabrous, glabrous, or hairy on the upper surface. Inflorescence ovate; main axis 2–17(–21) cm long; primary branches 1–8 cm long, reflexed to spreading. Spikelets solitary, obovoid, (3–)3.2–3.8(–4.1) × 1.7–2.4 mm; lower glume 0.7–2(–2.3) mm long, 7–11-nerved; upper glume 2.7–3.5 mm long, 9–11-nerved; lemma of lower

floret obovate, 9–11-nerved, muticous, the palea fully developed; lemma of upper floret obovate, indurate, smooth, brown, 2.8–3.1(–3.5) mm long, muticous.

DISTRIBUTION: Occasional in secondary vegetation, along trails or roadsides. Trail to Sieben (A2688). Also on St. Thomas; throughout the West Indies and Central America to Bolivia and Brazil.

3. Lasiacis sorghoidea (Desv.) Hitchc. & Chase, Contr. U.S. Natl. Herb. **18**: 338. 1917. *Panicum sorghoideum* Desv., *in* Ham., Prodr. Pl. Ind. Occ. 10. 1825. *Panicum lanatum* Sw. var. *sorghoideum* (Desv.) Griseb., Fl. Brit. W. I. 551. 1864.

Panicum divaricatum L. var. *lanatum* Schltdl. & Cham., Linnaea **6**: 33. 1831.

Panicum fuscum Sieber *ex* Griseb., Fl. Brit. W. I. 552. 1864.

Perennial herb; flowering culms 10–100 dm tall, erect to decumbent at the base, moderately branched. Leaf sheaths hairy, papillose, hispid or pubescent; ligule (0.3–)0.5–1.5(–2.6) mm long; blades lanceolate, ovate to linear, flat, (6–)9–19(–23) × (0.6–)1.2–3.4(–4.6) cm, smooth, hairy on the upper surface. Inflorescence ovate; main axis (5–)9–25(–35) cm long; primary branches 4–9 cm long, appressed to spreading. Spikelets solitary, obovate to elliptic, (3–)3.4–4.1(–4.3) × 2.5–3.2 mm; lower glume (1.2–)1.5–2.1(–2.7) mm long, 7–11-nerved; upper glume 3–4.3 mm long, 9–13-nerved; lemma of lower floret elliptic to obovate, 9–11-nerved, muticous, the palea fully developed; lemma of upper floret elliptic to obovate, indurate, smooth, brown, 2.9–3.8 mm long, muticous.

DISTRIBUTION: Occasional in disturbed areas. Bordeaux (A3853), Susannaberg (A3844). Also on St. Croix and St. Thomas; throughout the West Indies and Mexico to Argentina.

EXCLUDED SPECIES: *Lasiacis maculata* (Aubl.) Urb. and *Lasiacis ruscifolia* (Kunth) Hitchc. were reported in error for St. John by Woodbury and Weaver (1987). The *L. maculata* record was based on a misidentification of a specimen of *L. sorghoidea*, and the *L. ruscifolia* record was based on a specimen of *L. divaricata*.

20. LEPTOCHLOA P. Beauv.

Cespitose annual or perennial herbs; culms erect or decumbent. Leaf sheaths glabrous or hairy; ligules membranous to ciliate; blades mostly flat, occasionally involute. Inflorescence of many racemose primary branches along an elongate, scabrous, triquetrous central axis. Spikelets mostly 3–10-flowered, laterally compressed or terete, subsessile or pedicellate; glumes shorter than the spikelets, unequal, glabrous; lower glume 1-nerved; upper glume 1–3-nerved; spikelets in 2 rows along the rachis, disarticulating above the glumes, with the lemma and palea falling as a unit; lemma 3-nerved, unawned or awned, hairy or glabrous; palea shorter than the lemma, glabrous; stamens 2 or 3. Caryopsis dorsiventrally compressed or laterally compressed.

About 45 species worldwide in tropical and warm temperate regions, 15 of these species occur in the New World.

1. Leptochloa virgata (L.) P. Beauv., Ess. Agrostogr. 71. 166, t. 5, fig. 1. 1812. *Cynosurus virgatus* L., Syst. Nat, ed. 10, **2**: 876. 1759.   Fig. 236F–H.

Perennial herb; culms erect, 50–95 cm tall. Leaf sheaths glabrous; ligules 0.4–0.8 mm long, ciliolate near the apex and indurate below; blades 10–32 × 0.7–1.2 cm, clasping near base with yellowish dewlaps. Inflorescences 10–30 × 10–15 cm wide, open; racemes loosely ascending. Spikelets (2.2–)2.5–4.3 mm long, 3–7-flowered, purplish or grayish, florets becoming smaller toward apex; glumes 1.2–2.5 mm long, 1-nerved, narrowly lanceolate; lemmas 1.8–2.5 mm long, lanceolate, ciliate on the upper half, the apex often awned to 2.2 mm long; anthers 2, 0.2–0.3 mm long, yellow or purple. Caryopsis 1–1.3 mm long, shallowly grooved on the ventral surface, brownish.

DISTRIBUTION: Occasional weed of open disturbed, moist to dry areas. Bordeaux (W436), Mamey Peak (W552). Also on St. Croix and St. Thomas; southern United States, Mexico to Argentina and the West Indies.

21. MELINIS P. Beauv.

Mostly slender perennial herbs; culms erect or decumbent, sparingly branched. Leaves without auricles; sheaths not overlapping, smooth; ligule a fringe of hairs; blades filiform to linear, flattened to involute. Main axis of inflorescence with 1-sided, spreading primary branches; secondary branches spreading; rachis terminating in a spikelet; disarticulation at the spikelet base; callus differentiated or not. Spikelets laterally compressed; lower glume not encircling the spikelet base; upper glume about as long as the spikelet; lemma of lower floret chartaceous, sterile; lemma of upper floret 0.6–0.7 times the length of the lower floret, membranous to chartaceous, fertile smooth, dull, with flat margins, muticous.

Primarily an African genus of about 20 species, 1 of which is now widely distributed as a weed.

1. Melinis repens (Willd.) Zizka, Biblioth. Bot. **138**: 55. 1988. *Saccharum repens* Willd., Sp. Pl. **1**: 322. 1797. *Rhynchelytrum repens* (Willd.) C. E. Hubb., Kew Bull. **1934**: 110. 1934. *Tricholaena repens* (Willd.) Hitchc., Man. Grasses W. Ind. 331. 1936.   Fig. 239G, H.

Tricholaena rosea Nees, Cat. Sem. Hort. Vratisl. 1836 [1835].

Annual or perennial herb, lacking stolons; flowering culms 3–10 dm tall, erect or decumbent at the base, sparingly branched. Leaf sheaths glabrous; ligule 0.5–1.4 mm long; blades linear, flat, 3–20 × 0.3–0.6 cm, smooth, glabrous on the upper surface. Inflorescence lanceolate to ovate; main axis 8–17 cm long; primary branches 2.5–6 cm long. Spikelets solitary, ovate to elliptic, 3.2–5.5 × 1.2–1.9 mm, covered with reddish to less commonly gray silky hairs; lower glumes 0.3–1.5 mm long, 0–1-nerved; upper glumes 3–5.3 mm long, 5-nerved; lemma of lower floret lanceolate to ovate, 5-nerved, muticous or awned, the palea fully developed; lemma of upper floret elliptic, chartaceous, smooth, white, 2–2.4 mm long, muticous.

DISTRIBUTION: A common roadside weed. Annaberg (W853), Peter Peak (A4106). Also on St. Thomas; widely introduced throughout the West Indies and warm to temperate North America; native to Africa.

22. OLYRA L.

Monoecious, cespitose, perennial, monocarpic herbs; culms erect or often trailing. Leaves with well-developed sheath auricles; ligules membranous; pseudopetioles present; blades usually broad. Inflorescences 1 to many panicles from the uppermost nodes; open or occasionally spicate. Spikelets 1-flowered. Pistillate spikelets large, usually borne on a clavate pedicel, at the tips of branches or upper branches, disarticulating as a unit or the floret falling from the glumes, rarely the glumes falling first; glumes acuminate to awned, equal or subequal, many-nerved, membranous, longer than the floret; floret lanceolate to ovate, indurate, white, darkening at maturity, lemma margins covering the edges of the palea. Staminate spikelets small, linear to lanceolate, hyaline, early deciduous, borne on filiform pedicels, on basal portion of branches, or on the lower branches; glumes usually absent; lemma 3–9-nerved, the apex acuminate to aristate; palea as long as lemma, 2-nerved; stamens 3.

A genus of 23 species, centered in southeastern Brazil and western Amazonia but ranging to Mexico and the West Indies south to Argentina.

REFERENCE: Soderstrom, T. R. & F. O. Zuloaga. 1989. A revision of the genus *Olyra* and the New World segregate genus *Parodiolyra* (Poaceae: Bambusoideae: Olyreae). Smithsonian Contr. Bot. 69: 1–79.

1. Olyra latifolia L., Syst. Nat., ed. 10, **2**: 1261. 1759.   Fig. 237F–I.

Perennial herb; culms 1–6 m tall, 3–12 mm diam., erect or decumbent, climbing and clambering, usually branching from the middle and upper nodes; nodes constricted, dark, pilose to glabrous; internodes glabrous, smooth. Leaf sheaths covered by papillose–pilose hairs or glabrous, auricles to 5 mm long, the lower ones deciduous, the upper persistent; ligules 0.7–5 mm long; pseudopetioles 4–7 mm long; blades 10–32 × 3–11 cm, lanceolate to ovate, obtuse to subcordate at base, acuminate at the apex. Inflorescences borne on peduncles to 15 cm long; panicles 7–20 × 4–14 cm, pyramidal; branches divergent to spreading, the lower ones whorled and bearing staminate spikelets, the upper ones alternate, with staminate spikelets below and 1 to many pistillate spikelets above. Pistillate spikelets 14–22 × 3–4 mm, borne on purplish, clavate pedicels, broadly lanceolate, aristate; glumes 5–9-nerved, sparingly pubescent; lower glume bearing an awn 5–22 mm long; upper glume 10–15 mm long, short-aristate; floret 5–6 mm long, ovate, smooth, shining, apex obtuse. Staminate spikelets 6–8 mm long, greenish to purplish; glumes absent; lemma 3-nerved, scabrous, apex aristate; palea 4–6 mm long. Caryopsis 4–4.5 mm long, ellipsoid.

DISTRIBUTION: Occasional in secondary vegetation. Bordeaux (W731), L'Esperance (M17027). Also on St. Thomas and Tortola; throughout the West Indies and Mexico to Argentina; naturalized in Africa.

23. OPLISMENUS P. Beauv.

Annual or weak decumbent perennial herbs; culms decumbent, sparingly to moderately branched. Leaf sheaths smooth; ligule a ciliate membrane; blades linear, lanceolate, or ovate, flattened. Main axis of inflorescence with 1-sided, appressed or spreading primary branches; secondary branches appressed or reduced to a fascicle of spikelets; rachis terminating in a spikelet; disarticulation at the spikelet base; callus not differentiated. Spikelets adaxial, laterally or dorsiventrally compressed; lower glume awned, not encircling the spikelet base; upper glume 0.5–0.8 times spikelet length; sterile lemma of lower floret membranous; lemma of upper floret 0.7–0.95 times the length of the lower floret, cartilaginous, smooth, shiny, with involute margins, with margins of the same texture as the body, mucronate. Caryopsis with a whitish or grayish embryo.

A tropical genus with about 5–7 species, from shaded, moist areas.

REFERENCE: Scholz, U. 1981. Monographie der Gattung *Oplismenus* (Gramineae). Phan. Monogr. 13: 1–213.

1. Oplismenus hirtellus (L.) P. Beauv., Ess. Agrostogr. 54. 1812. *Panicum hirtellum* L., Syst. Nat., ed. 10, 2: 870. 1759.   Fig. 237A–E.

Annual or perennial, stoloniferous herb; flowering culms 5–50 cm tall, decumbent, sparingly branched. Leaf sheaths glabrous or hairy; ligule 0.4–1.1 mm long; blades linear, lanceolate, or ovate, flat, 1–10 × 0.4–1.7 cm, scabrous, glabrous, or hairy on the upper surface. Inflorescence oblong; main axis 4–11 cm long; primary branches 5–15 mm long. Spikelets paired, lanceolate to elliptic, 2.1–3 × 0.8–1.2 mm; lower glume 1.4–2.1 mm long, 3–5-nerved, the awns 5–10 mm long; upper glume 1.6–2.1 mm long, 5–7-nerved; lemma of lower floret lanceolate to elliptic, 5–7-nerved, mucronate, the palea vestigial; lemma of upper floret elliptic, cartilaginous, smooth, yellow, 2–2.4 mm long, mucronate.

DISTRIBUTION: Occasional in forest understory and shaded roadbanks. Cinnamon Bay (W635), Solomon Beach (A2289), Susannaberg (A3843). Also on St. Thomas; throughout the tropics.

24. PANICUM L.

Annual or perennial herbs of varied habits; culms erect or decumbent, unbranched to moderately branched. Leaves not distinctly distichous; sheath smooth or scabrous; ligule a membrane or a ciliate membrane; blades linear to ovate, flattened to involute. Main axis of inflorescence with primary branches arising at different points; secondary branches appressed or spreading; rachis terminating in a spikelet; disarticulation at the spikelet base; callus not differentiated. Spikelets adaxial, dorsiventrally compressed; lower glume encircling the spikelet base; upper glume as long as the spikelet; lemma of lower floret membranous sterile; lemma of upper floret fertile, cartilaginous, smooth to muricate, dull, with involute margins, the margins of the same texture as the body, muticous. Caryopsis with a whitish or grayish embryo.

About 600 species worldwide, 250 of which are present in the New World.

1. Panicum diffusum Sw., Prodr. 23. 1788.
                              Fig. 238A–C.

Weak perennial herb; flowering culms 25–50 cm tall, erect or decumbent at the base, sparingly branched. Leaf sheaths glabrous or hairy; ligule 0.3–0.5 mm long; blades filiform to linear, flat, 5–20 × 0.1–0.3 cm, smooth, hairy on the upper surface. Inflorescence ovate; main axis 5–10 cm long, open; primary branches 1.5–6 cm long. Spikelets solitary, ovate, 1.5–2.3 × 0.9–1.1 mm; lower glumes 1–1.3 mm long, 5-nerved; upper glumes 1.5–2.3 mm long, 7-nerved; lemma of lower floret elliptic, 7-nerved, muticous, the palea fully developed to vestigial; lemma of upper floret elliptic, indurate, smooth, yellow, 1.4–2 mm long, muticous.

DISTRIBUTION: Known on St. John from a single collection of Eggers (*3068*) without specific locality (Hitchcock, 1936). Also on Buck Island, St. Croix, and St. Thomas; throughout the West Indies.

25. PASPALIDIUM Stapf

Annual or perennial herbs; culms erect or decumbent, unbranched to sparingly branched. Leaf sheaths smooth or scabrous; ligule a ciliate membrane; blades filiform to linear, flattened to involute. Main axis of inflorescence with distichous or 1-sided, appressed, spreading, or reflexed primary branches; secondary branches appressed; rachis terminating in a bristle; disarticulation at the spikelet base; callus not differentiated. Spikelets abaxial, laterally compressed, dorsiventrally compressed, plano-convex, or terete; lower glume encircling the spikelet base; lemma of lower floret membranous (rarely chartaceous) sterile; upper floret 0.5–1 times the length of the lower floret, cartilaginous to indurate, rugose, dull, fertile the margins involute, with same texture as the body, muticous or mucronate. Caryopsis with a grayish or whitish embryo.

About 27 species, mainly in the Old World tropics and subtropics.

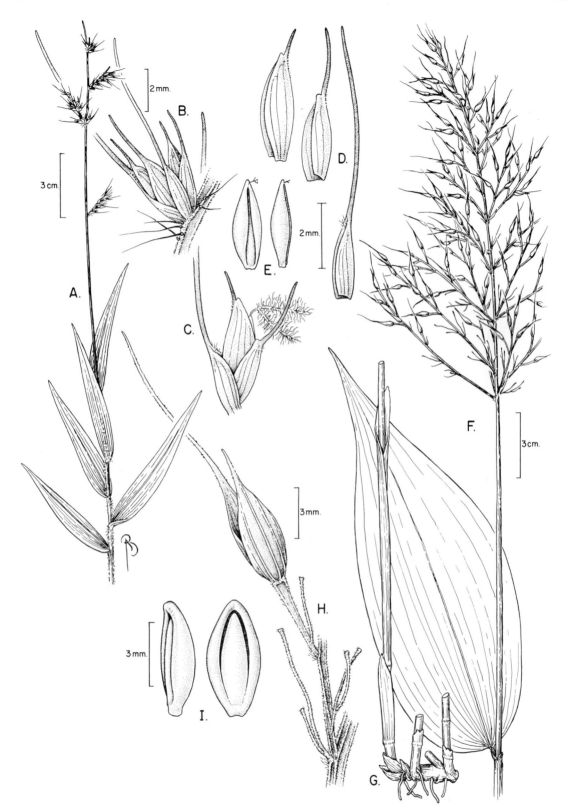

Fig. 237. **A–E.** *Oplismenus hirtellus.* **A.** Flowering culm. **B.** Inflorescence branch. **C.** Spikelet. **D.** Sterile lemma, upper glume, and lower glume. **E.** Fertile floret, ventral and lateral views. **F–I.** *Olyra latifolia.* **F.** Flowering culm. **G.** Rhizomatous base of culm. **H.** Detail of inflorescence showing pistillate spikelet. **I.** Pistillate floret, lateral and ventral views. (From Mori et al., in press, Guide to the Vascular Plants of Central French Guiana, Mem. New York Bot. Gard. 76.)

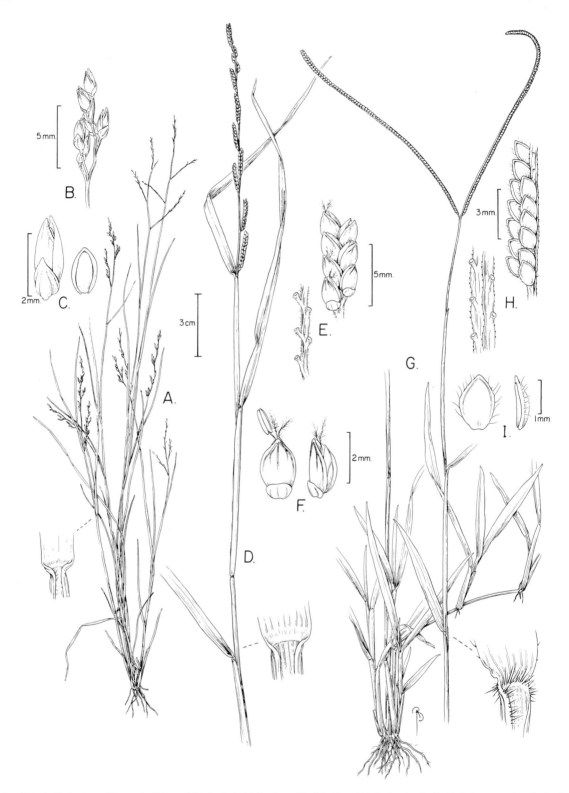

Fig. 238. A–C. *Panicum diffusum*. **A.** Habit and detail of adaxial junction of leaf sheath and blade showing ligule. **B.** Inflorescence branch showing spikelets. **C.** Spikelet and floret. **D–F.** *Paspalidium geminatum*. **D.** Habit and detail of adaxial junction of leaf sheath and blade showing ligule. **E.** Portion of inflorescence rachis before and after disarticulation of spikelets. **F.** Spikelet, dorsal and lateral views. **G–I.** *Paspalum conjugatum*. **G.** Habit and detail of adaxial junction of leaf sheath and blade showing ligule. **H.** Portion of inflorescence rachis before and after disarticulation of spikelets. **I.** Spikelet, dorsal and lateral views.

1. Paspalidium geminatum (Forssk.) Stapf, Fl. Trop. Afr. **9**: 583. 1920. *Panicum geminatum* Forssk., Fl. Aegypt.-Arab. 18. 1775.   Fig. 238D–F.

Paspalum appressum Lam., Tab. Encycl. **1**: 176. 1791.

Perennial herb; flowering culms 4–14 dm tall, decumbent, unbranched or sparingly branched. Leaf sheaths glabrous; ligule 0.7–2.7 mm long; blades linear, mostly flat, 6–20 × 0.5–0.8 cm, smooth, glabrous on the upper surface. Inflorescence linear; main axis 10–30 cm long; primary branches 1–5 cm long. Spikelets solitary, elliptic, 2–2.5 × 1.3–1.8 mm; lower glumes 0.7–0.9 mm long, faintly 3–5-nerved; upper glumes 1.7–2.3 mm long, 5–7-nerved; lemma of lower floret elliptic, 5–nerved, muticous, the palea fully developed; lemma of upper floret elliptic, indurate, rugose, yellow, 2.2–2.3 mm long, muticous or mucronate.

DISTRIBUTION: Uncommon, found in open moist areas. Lameshur (B611), John's Folly (W528). Also on St. Croix and St. Thomas; throughout the tropics and warm-temperate regions of the world.

26. PASPALUM L.

Annual or perennial herbs of diverse size and habit; culms erect or decumbent, unbranched or sparingly branched. Leaves with auricles minute or lacking; sheaths smooth or scabrous; ligule a membrane or a fringe of hairs; blades filiform to linear, flattened to conduplicate. Main axis of inflorescence with many appressed to divaricate primary branches; secondary branches appressed; rachis terminating in a spikelet; disarticulation at the spikelet base; callus not differentiated. Spikelets abaxial, plano-convex; lower glume typically absent, or when present not encircling the spikelet base; upper glumes nearly as long as the spikelet; lemma of lower floret membranous, sterile; lemma of upper floret as long as the lower one, indurate, fertile, smooth or striate, dull, with involute margins, muticous. Caryopsis with whitish to grayish embryo.

About 250 species, mainly in the New World.

Key to the Species of *Paspalum*

1. Upper glume and lower lemma broadly winged to produce a circular spikelet .... 2. *P. fimbriatum*
1. Upper glume and lower lemma not broadly winged.
   2. Racemes 2, conjugate.
      3. Spikelets <2 mm long.. 1. *P. conjugatum*
      3. Spikelets >2 mm long.
         4. Lower lemma membranous; leaves distinctly distichous; spikelets typically <1.6 mm wide .... 6. *P. vaginatum*
         4. Lower lemma chartaceous; leaves not distinctly distichous; spikelets typically 1.1–2.4 mm wide .... 5. *P. notatum*
   2. Racemes not conjugate, usually more than 2.
      5. Pedicels as long as the spikelets; spikelets loosely imbricate.............................. 4. *P. molle*
      5. Pedicels much shorter than the spikelets; spikelets closely imbricate .... 3. *P. laxum*

1. Paspalum conjugatum Berg., Acta Helv. Phys.-Math. **7**: 129. 1762.   Fig. 238G–I.

Perennial, stoloniferous herb; flowering culms 2–10 dm tall, erect from the base, sparingly branched. Leaf sheaths glabrous; ligule 0.3–1.5 mm long; blades linear to lanceolate, flat, 6–20 × 0.5–1.5 cm, smooth, glabrous or hairy on the upper surface. Inflorescence ovate; primary branches 4–16 cm long; racemes 2, conjugate. Spikelets solitary, elliptic, 1.2–1.8 × 0.8–1.2 mm; upper glume 1.2–1.8 mm long, 2-nerved; lemma of lower floret elliptic, 2-nerved, mucronate, the palea absent; lemma of upper floret elliptic, cartilaginous, smooth to striate, yellow, 1.2–1.8 mm long, muticous.

DISTRIBUTION: Occasional grass of open disturbed areas. Cinnamon Bay (W369). Also on St. Croix and St. Thomas; throughout the tropics and subtropics.

2. Paspalum fimbriatum Kunth *in* Humb., Bonpl. & Kunth, Nov. Gen. Sp. **1**: 93. 1816.

Annual herb, lacking stolons; flowering culms 2–6 dm tall, erect from the base, unbranched or sparingly branched. Leaf sheaths glabrous; ligule 2 mm long; blades linear to lanceolate, flat, 10–24 × 0.6–1.3 cm, scabrous and glabrous on the upper surface. Inflorescence ovate; main axis 6–10 cm long; primary branches 1–1.5 mm long. Spikelets paired, ovate, 3–3.5 × 3–3.5 mm, with winged glumes and lemma to produce a circular spikelet; upper glumes 3–3.5 mm long, 3-nerved, broadly winged; lemma of lower floret ovate, 3-nerved, broadly winged, muticous, the palea vestigial; lemma of upper floret ovate to elliptic, coriaceous, smooth, yellow, 2–2.2 mm long, muticous.

DISTRIBUTION: Common roadside weed. Bordeaux (A2630); Center Line Road (A2704), Susannaberg (M16577). Also on St. Croix; Central America to Brazil and the West Indies.

3. Paspalum laxum Lam., Tabl. Encycl. **1**: 176. 1791.

Perennial herb lacking stolons; flowering culms 3–8 dm tall, erect from the base, unbranched. Leaf sheaths glabrous; ligule 0.4–1 mm long; blades linear, flat, 20–30 × 0.4–0.7 cm, smooth, glabrous on the upper surface. Inflorescence ovate; main axis 2–5 cm long; primary branches 4–8 cm long, with more than 2 racemes, not conjugate. Spikelets paired, closely imbricate, elliptic, 1.6–1.8 × 0.9–1.1 mm; upper glumes 1.6–1.8 mm long, 3–5-nerved; lemma of lower floret elliptic, 3–5-nerved, muticous, the palea absent; lemma of upper floret elliptic, cartilaginous, smooth, yellow, 1.5–1.7 mm long, muticous.

DISTRIBUTION: A common weed of disturbed open areas. Reef Bay Trail (A2933), Rosenberg (B292), trail to Sieben (A2642). Also on Anegada, St. Croix, St. Thomas, Tortola, and Virgin Gorda; throughout the West Indies and southern United States to Central America.

4. Paspalum molle Poir. *in* Lam., Encycl. **5**: 34. 1804.

Paspalum caespitosum Flüggé, Gram. Monogr. Paspalum 161. 1810.

Perennial herb lacking stolons; flowering culms 2–5(–9) dm tall, erect, unbranched. Leaf sheaths glabrous; ligule 0.5–1 mm long; blades linear, flat, 5–40 × 0.4–1.5 cm, smooth, glabrous on the upper surface. Inflorescence ovate; main axis 2–4 cm long; primary branches 1.5–6 cm long; with more than 2 racemes, not conjugate. Spikelets paired, elliptic, 1.5–2.2 × 0.7–0.9 mm; upper glumes 1.5–2.2 mm long, 3–5-nerved; lemma of lower floret elliptic, 3–5-nerved, muticous, the palea absent; lemma of upper floret elliptic, cartilaginous, smooth, yellow, 1.4–2.1 mm long, muticous.

DISTRIBUTION: Uncommon herb of moist open areas. Bordeaux (W62). Also on St. Thomas; southern Florida, Mesoamerica to northern South America and the West Indies.

5. Paspalum notatum Flüggé, Gram. Monogr. Paspalum 106. 1810.

Perennial herb, lacking stolons; flowering culms 2–10 dm tall, erect, unbranched. Leaf sheaths glabrous; ligule 0.3–1.5 mm long; blades linear, not distinctly distichous, flat or conduplicate, 6–25 × 0.4–1 cm, smooth, glabrous on the upper surface. Inflorescence obovate; main axis absent; primary branches 2.5–13 cm long;

racemes 2, conjugate. Spikelets solitary, ovate or elliptic; 2.8–3.7 × 1.1–2.4 mm; upper glumes 2.8–3.7 mm long, 5-nerved; lemma of lower floret ovate to elliptic, chartaceous, 5-nerved, muticous, the palea absent; lemma of upper floret ovate or elliptic, cartilaginous, striate, yellow, 2.5–3.5 mm long, muticous.

DISTRIBUTION: Uncommon herb of disturbed sites. Lameshur (W847), Maria Bluff (W801). Also on St. Thomas; Central and South America.

6. Paspalum vaginatum Sw., Prodr. 21. 1788.

Perennial, stoloniferous herb; flowering culms 0.6–6 dm tall, erect or decumbent at the base, unbranched or sparingly branched. Leaf sheaths glabrous; ligule 0.5–1.2 mm long; blades linear, flat, conduplicate or convolute, 5–22 × 0.1–0.4 cm, smooth, glabrous on the upper surface. Inflorescence ovate; main axis 0–12 mm long; primary branches 2–5 cm long; racemes 2, conjugate. Spikelets solitary, elliptic, 2.5–3.7 × 0.9–1.5 mm; upper glumes 2.5–3.7 mm long, 4–6-nerved; lemma of lower floret lanceolate, membranous, 5-nerved, muticous, the palea absent; lemma of upper floret elliptic, indurate, smooth to striate, yellow, 2.3–3.5 mm long, muticous.

DISTRIBUTION: Uncommon coastal herb. Fish Bay (W808). Also on St. Croix and Tortola; throughout the tropics and subtropics.

DOUBTFUL RECORDS: *Paspalum virgatum* L. was reported by Britton and Wilson (1924) and by Woodbury and Weaver (1987) as occurring on St. John; however, no specimens have been found to confirm these records, nor has the species been recently collected on the island.

27. PENNISETUM Rich.

Hermaphroditic, annual or perennial herbs, sometimes with stolons and/or rhizomes; flowering culms cespitose or not, erect or decumbent from base, sparingly branched, usually rooting at the lower nodes. Leaves without auricles; sheaths overlapping or not, smooth or scabrous, glabrous or variously pubescent; ligule a ciliate membrane or a fringe of hairs; blades filiform to linear, flat, conduplicate, or involute, appressed to the culm or spreading, smooth or scabrous, glabrous or pubescent, the margins planar or undulating. Inflorescence a solitary panicle or raceme; peduncle glabrous or pubescent, fully exserted at maturity to fully enclosed in the leaf sheath; main axis straight, stout or slender; primary branches originating at any point, appressed to the main axis or spreading, and reduced to a fascicle of spikelets, disarticulating at base of primary branches; rachis terminating in a bristle; spikelets 1–5, subtended by 1–10 bristles; pedicels present or absent; callus prolonged into a pronounced stipe, not flared and forming a discoid receptacle; lower glume when present ovate, membranous and awned, not encircling the spikelet base; upper glume 0.5–1.1 times spikelet length; lemma of lower floret sterile or staminate, lanceolate, ovate, or elliptic, not keeled, muticous, or awned; palea of lower floret fully developed, or absent, 0.4–0.9 times lemma length, ovate, lanceolate, or elliptic; stamens 3, the anthers purple; stigmas purple. Caryopsis gray, with punctiform hilum.

A cosmopolitan genus of 80 species.

1. Pennisetum clandestinum Chiov., Ann. Inst. Bot. Roma **8**: 41. 1903.   Fig. 239D–F.

Perennial herb; stoloniferous and/or rhizomatous; flowering culms 5–45 cm long, decumbent at base, with 6–15 nodes. Leaves distinctly distichous; sheaths smooth, glabrous or pubescent, laterally compressed, not keeled; ligule a fringe of hairs, 1.3–2.2 mm long; blades linear, flat or conduplicate, 1.5–3 × 0.2–0.6 cm, smooth, glabrous or pubescent, the margins ciliate or glabrous. Inflorescence partially enclosed in the leaf sheath; main axis straight, 4–5 mm long, smooth; primary branches reduced to a fascicle of spikelets; bristles 5–10, 10–15 mm long, antrorsely scabrous or glabrous, not fused; callus differentiated, not flared to

form a discoid receptacle; spikelets dorsiventrally compressed, usually 1 per primary branch. Lower glume not encircling the spikelet base, muticous; rachilla not pronounced below the upper glume; upper glume 0.1–0.2 times the length of the lower one, ovate, hyaline, not nerved, with glabrous margins; lemma of lower floret lanceolate, membranous, 10–20 × 0.8–1.1 mm, 9–13-nerved, the palea wanting; stamens wanting; lemma of upper floret lanceolate, smooth, 10–20 mm long, glabrous, acuminate, muticous; stamens 3, the anthers 4–7 mm long.

DISTRIBUTION: An occasional herb of open disturbed areas. Cinnamon Bay (W855). Native of tropical Africa, introduced into the New World as a forage grass.

28. PHARUS P. Br.

Monoecious, cespitose to stoloniferous and/or rhizomatous perennial herbs (apparently some monocarpic); culms solid. Leaf sheaths glabrous; ligules membranous; pseudopetioles present, twisted 180° at summit and inverting the blades; blades with secondary nerves diverging obliquely from the midvein. Inflorescence an open, terminal panicle; branches deciduous, covered with minute hooked hairs; rachis prolonged into a naked bristle or tipped with a staminate spikelet. Spikelets 1-flowered, appressed on short branchlets, spikelets of both sexes paired or with pistillate spikelet solitary. Pistillate spikelets larger than the staminate spikelets, subsessile; glumes shorter than the floret, unequal, lanceolate, persistent, several-nerved, purple to green; floret cylindrical, linear to sigmoid-curved, indurate, with some minute, hooked hairs, deciduous; lemma 7-nerved with inrolled, usually free margins concealing the palea. Staminate spikelets borne on long pedicels, appressed to the pistillate spikelet, membranous, elliptical, persistent; glumes unequal, sometimes the lower absent; lemma ovate, 3–5-nerved; stamens 6. Pistillate florets with hooked hairs that readily adhere to fur and clothing.

A genus of 7 species, from Mexico and southeastern United States to Argentina and Uruguay in moist to wet, lowland forests.

1. Pharus lappulaceus Aubl., Hist. Pl. Guiane 2: 859. 1775.   Fig. 239A–C.

Pharus glaber Kunth *in* Humb., Bonpl. & Kunth, Nov. Gen. Sp. **1**: 196. 1816.

Cespitose, perennial herb; culms 25–50 cm tall, erect or occasionally decumbent and rooting at the nodes. Leaf ligules 1–2 mm long; pseudopetioles 8–60 mm long; blades 8–25 × 2–4 cm, narrowly lanceolate to elliptic. Inflorescence 10–20 cm long, sparsely flowered; branches borne singly at the nodes, the terminal bristly, 1–3 cm long, usually tipped with a staminate spikelet.

Pistillate spikelets 7.5–11 mm long; glumes about half as long as the spikelet; lower glumes 4–6 mm long, 5-nerved; upper glumes 5–7 mm long, 3-nerved; lemma linear-oblong, straight, with an abrupt beak about 1 mm long, covered with minute hooked hairs. Staminate spikelet 2–3 mm long, purple; pedicels 4–11 mm long; lower glume absent or to 2 mm long; upper glume 1.5–3 mm long, 3-nerved. Caryopsis with persistent lemma bearing hooked hairs that disperse as a unit by adhering to fur and clothing.

DISTRIBUTION: A common herb of forest understory. Bethany (B223), Bordeaux (A3852). Found throughout the West Indies and Mexico to Argentina and Paraguay.

29. SCHIZACHYRIUM Nees

Annual or perennial herbs; culms erect or decumbent. Leaf sheaths hairy; ligules membranous, with a fringe near apex; blades flat. Inflorescence a single raceme enclosed in a bractlike sheath, usually several per culm. Spikelets several to many pairs, falling attached to the more or less thickened adjacent rachis internode; rachis internode narrow at base and thickened near the apex, the margins with irregular lobes; rachis ending in a triad of 1 sessile and 2 pedicellate spikelets. Sessile spikelets bisexual; glumes subequal, concealing the florets; lower glume membranous, dorsally compressed, flat to strongly convex, entire to bidentate, 2- to many-nerved, marginal nerves more conspicuous than the midvein, the margins enfolding the upper glume and florets; upper glume membranous to hyaline, 1- or 3-nerved, keeled; lower floret sterile, represented by a hyaline lemma, lacking a palea; upper floret fertile, bisexual, the lemma hyaline, bilobed, usually awned, the awn arising from between the lobes, geniculate below, sharply bent after emerging from the spikelet, the palea absent. Pedicellate spikelets sterile, similar to sessile spikelets or reduced. Caryopsis with a large embryo.

About 60 species of tropical and temperate grasslands with 15 or so in the New World.

REFERENCES: Hatch, S. 1975. A biosystematic study of the *Schizachyrium cirratum-S. sanguineum* complex. Dissertation. Texas A&M University, College Station; Turpe, A. M. 1984. Revision of South American species of *Schizachyrium* (Gramineae). Kew Bull. 39: 169–178.

1. Schizachyrium sanguineum (Retz.) Alston, Suppl. Fl. Ceylon 334. 1931. *Rottboellia sanguinea* Retz., Observ. Bot. **3**: 25. 1783.   Fig. 240A–C.

Schizachyrium semiberbe Nees, Agrost. Bras. 336. 1829. *Andropogon semiberbis* (Nees) Kunth, Révis. Gramin. **1 (Suppl.)**: 39. 1830.

Cespitose perennial herb; culms 40–180 cm tall, unbranched below, branched above, glabrous. Leaf ligules 0.6–1.7 mm long; blades 11–25 × 0.2–0.6 cm, glabrous to sparingly pilose. Inflorescence axillary and terminal, reddish to grayish, often on filiform peduncles to 10 cm long; bractlike sheaths subtending the racemes

to 5 cm long; racemes 4.5–13 × 0.4–1.3 cm, wider at the apex, glabrous to densely pubescent with spreading hairs. Sessile spikelets 4.5–8 mm long (excluding the awn), linear-lanceolate, densely pubescent to glabrous; callus bearded or glabrous, the hairs whitish; lower floret 3–5 mm long; upper floret 3.8–6.5 mm long, awn brown, sharply bent 1–2 mm after leaving the spikelet, 8–10 mm long after the bend. Pedicellate spikelet 3–4.5 mm long, lanceolate, pubescent to scabrous; pedicel 4–6 mm long, pubescent or occasionally glabrous; upper floret awnless or with an awn to 3 mm long; glumes straw-colored; anthers 1.4–2 mm long.

DISTRIBUTION: Occasional grass of coastal areas. Ditlif Point (W650). Also on St. Croix; pantropical.

30. SETARIA P. Beauv.

Annual or perennial herbs of diverse size and habitats; culms erect or decumbent, unbranched or sparingly branched. Leaves without auricles; sheath smooth; ligule a ciliate membrane or a fringe of hairs; blades linear, flattened, conduplicate or plicate. Main axis of inflorescence with appressed or spreading primary branches; secondary branches appressed or spreading or reduced to a fascicle of spikelets; rachis terminating in a bristle; disarticulation at the spikelet base; callus not differentiated. Spikelets abaxial, dorsiventrally compressed; lower glume encircling the spikelet base; upper glume 0.4–1 times spikelet length; lemma of lower floret membranous; lemma of upper floret as long as the lower floret, cartilaginous to indurate, rugose, dull, with involute margins, muticous. Caryopsis with a whitish or grayish embryo.

A tropical and subtropical genus with about 105 species.

Key to the Species of *Setaria*

1. Bristles present only at the termination of the primary branch .... 2. *S. utowanaea*
1. Bristles present beneath each spikelet .... 1. *S. setosa*

FIG. 239. A–C. *Pharus lappulaceus.* **A.** Habit and flowering culm. **B.** Detail of inflorescence branch showing staminate and pistillate spikelets. **C.** Pistillate spikelet lemma and palea with gynoecium. **D–F.** *Pennisetum clandestinum.* **D.** Habit. **E.** Detail of inflorescence enclosed in the leaf sheath. **F.** Pistillate spikelet. **G, H.** *Melinis repens.* **G.** Habit and flowering culm. **H.** Fertile floret and spikelet.

FIG. 240. A–C. *Schizachyrium sanguineum*. A. Habit. B. Detail of inflorescence branch showing spikelets. C. Detail of raceme showing sessile and pediceled spikelets. D–F. *Spartina patens*. D. Habit and detail of culm showing junction of leaf sheath and blade. E. Portion of raceme inflorescence with spikelets. F. Spikelet, lateral view. G–J. *Setaria setosa*. G. Habit. H. Detail of inflorescence branch. I. Detail of inflorescence branch after disarticulation of spikelets showing the rachis terminating in a bristle. J. Spikelet, lateral view.

1. Setaria setosa (Sw.) P. Beauv., Ess. Agrostogr. 51, 171. 1812. *Panicum setosum* Sw., Prodr. 22. 1788. Fig. 240G–J.

Panicum caudatum Lam., Tabl. Encycl. **1:** 171. 1791. *Setaria caudata* (Lam.) Roem. & Schult., Syst. Veg. **2:** 49. 1817.

Perennial herb, lacking stolons; flowering culms 4–10 dm tall, erect or decumbent at the base, sparingly branched. Leaf sheaths glabrous or hairy; ligule 0.7–1 mm long; blades linear, flat, 10–25 × 0.1–2.5 cm, smooth, glabrous or hairy on the upper surface. Inflorescence linear; main axis 7–22 cm long; primary branches 4–10 mm long. Spikelets solitary, ovate, 2–2.4 × 1–1.2 mm, subtended by more than one bristle, beneath each spikelet; lower glumes 0.9–1.1 mm long, 3-nerved; upper glumes 1.6–1.8 mm long, 3–5-nerved; lemma of lower floret ovate, 5-nerved, muticous, the palea fully developed; lemma of upper floret ovate, indurate, rugose, yellow, 1.7–2.2 mm long, apiculate.

DISTRIBUTION: A common herb of open, dry to moist areas. Fish Bay (W470), Lameshur (W430), Margaret Hill (W530). Also on St. Croix, St. Thomas, and Virgin Gorda; West Indies to northern South America.

2. Setaria utowanaea (Scribn. *ex* Millsp.) Pilg., Nat. Pflanzenfam., ed. 2, **14e:** 72. 1940. *Panicum utowanaeum* Scribn. *ex* Millsp., Publ. Field Columbian Mus., Bot. Ser. **2:** 25. 1900.

Panicum sintenisii Nash, Bull. Torrey Bot. Club **30:** 382. 1903.

Perennial, stoloniferous herb; flowering culms 2.5–5 dm tall, erect, unbranched. Leaf sheaths glabrous; ligule 0.5–0.8 mm long; blades filiform, conduplicate, 10–20 × 0.1–0.3 cm, scabrous, glabrous on the upper surface. Inflorescence linear; main axis 1–3 cm long; primary branches 1–2.5 cm long. Spikelets solitary, elliptic, 1.9–2.2 × 0.7–0.8 mm, subtended by more than one bristle at the termination of the primary branch; lower glumes 1 mm long, 3-nerved; upper glumes 1.4–1.6 mm long, 5-nerved; lemma of lower floret elliptic, 5-nerved, apiculate, the palea fully developed; lemma of upper floret elliptic, indurate, rugose, yellow, 1.8–2 mm long, muticous.

DISTRIBUTION: Uncommon grass of dry forest understory. Francis Bay (W537). Also on Anegada; Cuba, Puerto Rico, and Lesser Antilles.

DOUBTFUL RECORDS: *Setaria geniculata* (Lam.) P. Beauv. was reported by Woodbury and Weaver (1987) and by Britton and Wilson [1924; as *Chaetochloa geniculata* (Lam.) Millsp. & Chase] as occurring on St. John; however, no specimens have been found to confirm these records. In addition, *Setaria pradana* (Léon) Léon was listed by Woodbury and Weaver (1987; as *S. padana*) as occurring on St. John, but no specimen has been seen to confirm this record.

EXCLUDED SPECIES: *Setaria leiophylla* (Nees) Kunth [= *Setaria setosa* var. *leiophylla* (Nees) Arechav.], reported in error for St. John by Woodbury and Weaver (1987), was based on a specimen of *S. setosa*.

31. SPARTINA Schreb.

Cespitose or rhizomatous perennial herbs; culms often over 1 m tall. Leaves coarse; sheaths mostly scabrous; ligules a line of hairs; blades elongate, flat to involute, often sharp-pointed. Inflorescence of a few to many, 1-sided, ascending, densely flowered spikes. Spikelets borne in 2 rows along 2 sides of the triquetrous rachis, 1-flowered, strongly laterally compressed, membranous to firm; disarticulation below the glumes, spikelet falling as an entire unit; glumes lanceolate, unequal; lower glumes 1-nerved, shorter than the lemma; upper glumes 3–5-nerved, as long as or longer than the lemma; lemma 1–3-nerved, firm; palea membranous, about as long as or longer than the lemma; stamens 3. Caryopsis rarely seen.

A genus of 16 species, surrounding the Atlantic Ocean and found inland in North America to the West Coast south to Mexico and Chile.

REFERENCE: Mobberley, D. G. 1956. Taxonomy and distribution of the genus *Spartina*. Iowa State Coll. J. Sci. 30: 471–574.

1. Spartina patens (Aiton) Muhl., Descr. Gram. 55. 1817. *Dactylis patens* Aiton, Hort. Kew. **1:** 104. 1789. Fig. 240D–F.

Perennial herb with slender, wiry rhizomes; culms 0.7–2 m tall. Leaf ligules 0.5–1 mm long; blades 15–60 × 0.1–0.5 cm, glabrous below and scabrous above, the apex acuminate. Inflorescences 3–15 × 0.4–8 cm; spikes 2–15, 1–7 cm long, ascending and appressed to loosely spreading, sessile or occasionally pedunculate. Spikelets 10–30 per spike, 7–12 mm long,

tightly appressed, closely imbricate; glumes glabrous to sparingly hispidulous, the keels hispid; lower glumes 3–8 mm long, linear; upper glumes 7–12 mm long, linear-lanceolate; lemma 5.5–8.5 mm long, lanceolate, glabrous to sparingly hispidulous, the apex obtuse to irregularly lobed; palea longer than the lemma, thin and papery, glabrous; anthers 3–5 mm long.

DISTRIBUTION: Occasional on sandy beaches. Caneel Bay (W177), Solomon Bay (A2285). Also on St. Croix; eastern North America to the Caribbean, introduced to Europe.

NOTE: Not known to flower on St. John.

32. SPOROBOLUS R. Br.

Cespitose to stoloniferous or rhizomatous annual or perennial herbs; culm erect. Leaf sheaths glabrous or hairy; ligules usually short, a line of hairs; blades flat, involute, or terete. Inflorescence an open or contracted panicle. Spikelets 1-flowered; disarticulation usually above the glumes; glumes usually 1-nerved, the lower one often nerveless, usually shorter than the spikelets; lemma 1-nerved, membranous, the apex unawned; palea subequal to the lemma; stamens 2 or 3. Fruit with a free pericarp.

Approximately 45 species in the New World and over 100 in the Old World.

REFERENCES: Baaijens, G. J. & J. F. Veldkamp. 1991. *Sporobolus* (Gramineae) in Malesia. Blumea 35: 393–458; Clayton, W. D. 1965. Studies in the Gramineae: VI. The *Sporobolus indicus* complex. Kew Bull. 19: 287–296.

Key to the Species of *Sporobolus*

1. Plants strongly rhizomatous and/or stoloniferous; culms widely creeping in rows with virgate branching; leaves strongly distichous or appearing subopposite; plants predominantly coastal, in saline habitats .... 3. *S. virginicus*

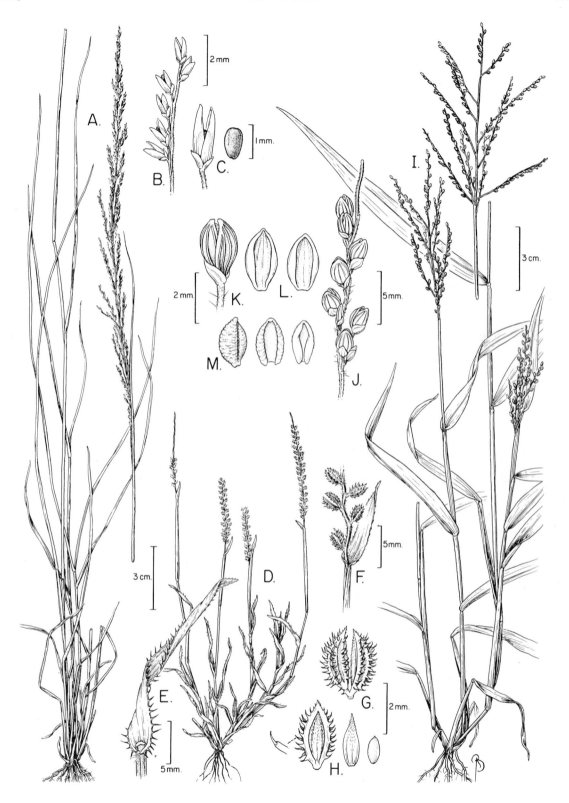

FIG. 241. A–C. *Sporobolus indicus*. A. Habit. B. Detail of inflorescence branch showing spikelets. C. Spikelet and caryopsis, lateral view. D–H. *Tragus berteronianus*. D. Habit. E. Detail of summit of leaf sheath and blade. F. Detail of proximal portion of inflorescence spike showing spikelets. G. Spikelet. H. Glume with detail of surface, lemma, and caryopsis, dorsal view. I–M. *Urochloa fasciculatum*. I. Habit. J. Detail of inflorescence branch showing spikelets. K. Spikelet. L. Upper glume and sterile lemma of lower floret. M. Lemma and palea of fertile floret, ventral view.

1. Plants cespitose, without widely creeping rhizomes or stolons, not branching as above; leaves not distichous; plants of various habitats.
   2. Cespitose, long-lived annuals or perennials; spikelets 1.3–2.7 mm long; panicles narrow, contracted, with ascending and appressed branches, floriferous near base, the pedicels appressed; caryopses 0.8–1.1 mm long, oblong .... 1. *S. indicus*
   2. Delicate, short-lived annuals; spikelets 0.7–1.1 mm long; panicles diffuse, with spreading to reflexed branches, naked near base, the pedicels widely spreading; caryopses 0.5–0.7 mm long, broadly pyriform .... 2. *S. tenuissimus*

1. Sporobolus indicus (L.) R. Br., Prodr. **1**: 170. 1810.
   Agrostis indica L., Sp. Pl. 63. 1753.

Fig. 241A–C.

Cespitose, long-lived annual or perennial herb; culms 30–100 cm tall, glabrous. Leaves borne all along the culm; ligules 0.2–0.5 mm long; blades 6–15 × 0.2–0.7 cm, flat or folded. Panicles 7–50 × 0.3–2 cm, narrow, contracted; branches ascending and appressed, floriferous near base; pedicels appressed. Spikelets 1.3–2.7 mm long; glumes 0.3–1.7 mm long, ovate to elliptic; lower glume nerveless; upper glume longer than the lower one, the apex acute to obtuse often irregularly toothed; lemma 1.3–2.7 mm long, ovate to oblong; anthers 0.5–1.1 mm long. Caryopsis 0.8–1.1 mm long, oblong.

DISTRIBUTION: Occasional in open disturbed habitats. East End (A2744). Also on St. Croix and St. Thomas; pantropical, a very polymorphic species as currently recognized.

2. Sporobolus tenuissimus (Schrank) Kuntze, Revis. Gen. Pl. **3**: 369. 1898. *Panicum tenuissimum* Schrank, Denkschr. Königl. Bayer. Bot. Ges. Regensburg **3**: 26. 1822.

Delicate, cespitose, short-lived annual herb; culms 40–100 cm tall, glabrous, often shiny. Leaves mostly basal, reduced upward; ligules 0.2–0.3 mm long; blades 5–23 × 0.2–0.4 cm, flat or folded. Panicles 15–30 × 3.5–8 cm, diffuse; branches spreading to reflexed, naked near base; pedicels widely spreading. Spikelets 0.7–1.1 mm long; glumes 0.2–0.5 mm long, ovate, the apex acute to obtuse; lower glumes veinless; upper glumes longer than the lower ones; lemma 0.7–1 mm long, elliptic; anthers 0.2–0.3 mm long. Caryopsis 0.5–0.7 mm long, broadly pyriform.

DISTRIBUTION: Uncommon. Known on St. John from only a single collection by Raunkiær (*1383*). Also on Tortola and St. Croix; now pantropical, but perhaps originally native to the New World.

3. Sporobolus virginicus (L.) Kunth, Révis. Gramin. **1**: 67. 1829. *Agrostis virginica* L., Sp. Pl. 63. 1753.

Strongly rhizomatous and stoloniferous perennial herb; culms widely creeping in rows and branching virgately, glabrous, often yellowish, the upright portions 20–60 cm tall. Leaves distichous or appearing subopposite, overlapping; ligules 0.1–0.3 mm long; blades 4–16 × 0.2–0.5 cm, flat, widest at base. Panicles 3–10 × 0.6–1.5 cm, linear, contracted; branches ascending and appressed; pedicels appressed. Spikelets 2–3 mm long; glumes 1.5–2.8 mm long, ovate-oblong, scaberulous along keel, the apex acute; lemma 2.1–2.9 mm long, elliptic; anthers 1–1.7 mm long. Caryopsis usually absent.

DISTRIBUTION: A common grass of sandy beaches, coastal dunes, and saline habitats. Dittlif Point (A3963), Little Lameshur (A2019), Reef Bay (A2729). Also on Anegada, St. Croix, and St. Thomas; pantropical.

33. TRAGUS Haller

Annual or perennial herbs, usually stoloniferous; culms usually creeping. Leaf sheaths with scabrous margins; ligules a ciliate membrane or a fringe of hairs; blades flat. Inflorescence a narrow, terminal false spike with clusters or glomerules of (1–)2–5 spikelets on spicate racemes; racemes disarticulating as a unit. Spikelets 1-flowered, sessile to subsessile on a short, zigzag rachis. Glumes very unequal and dissimilar; 5–7-nerved; lower glumes often lacking; upper glume large and hard, with rows of hooked spines along the nerves; lemma flat, membranous to hyaline; 3-nerved; palea convex, membranous to hyaline; stamens 3. Caryopsis ellipsoid.

A genus of 7 species, distributed primarily in tropical to subtropical Africa.

REFERENCE: Gibbs-Russell, G. E. 1991. *Tragus*. In Gibbs-Russell et al. (eds.), Grasses of southern Africa. Mem. Bot. Surv. S. Afr. 58: 337–339.

1. Tragus berteronianus Schult. *in* Roem. & Schult., Syst. Veg. **2**: 205. 1824. Fig. 241D–H.

Cespitose annual herb, spreading near base; culms 5–60 cm tall. Leaf ligules 0.5–1.1 mm long, a fringe of hairs; blades 1–6 × 0.2–0.5 cm, the margins white, cartilaginous, stiffly ciliate. Inflorescences 2–10 × 0.3–0.7 cm, narrowly spikelike. Spikelets 2.3–3.7 mm long, sessile or pedicellate, the pedicel to 0.5 mm long; glumes lanceolate, 5-nerved with 5 rows of hooked spines along the nerves, the spines 0.3–0.8 mm long; anthers 0.4–0.6 mm long. Caryopsis 0.9–1.1 mm long, light yellowish.

DISTRIBUTION: An occasional weed of open, dry disturbed habitats. Drunk Bay (A5064), Nanny Point (A2452). Also on St. Croix and St. Thomas; native to the Old World, now a pantropical weed.

34. UNIOLA L.

Perennial herbs with long and thick rhizomes or stolons; culms erect. Leaf ligules a line of hairs; blades flat to involute, tough in texture, tapering to a point. Inflorescence of 1 to many spicate racemes; branches terminating in a spikelet; disarticulation below the glumes. Spikelets 2–20-flowered, falling as a unit, solitary, strongly laterally compressed, pedicellate, in 2 rows along one side of the rachis; rachilla pronounced between florets; glumes 2–5-nerved, shorter than the spikelets, glabrous, unawned. Sterile florets above and below the fertile ones, the lower 2–6 empty and without paleas; lemma entire, coriaceous, strongly keeled, glabrous, 3–9-nerved, unawned; palea chartaceous, glabrous, unawned, the margins ciliate; stamens 3. Caryopsis terete.

A genus of 4 species, occurring in the New World from the southeastern United States to Central America, Cuba, and south to Colombia and Ecuador.

REFERENCE: Yates, H. O. 1966. Revision of grasses traditionally referred to *Uniola*, 1. *Uniola* and *Leptochloopsis*. Southw. Naturalist 11: 372–394.

1. Uniola virgata (Poir.) Griseb., Fl. Brit. W. I. 531. 1864. *Poa virgata* Poir. *in* Lam., Encycl. **5**: 78. 1804. *Leptochloöpsis virgata* (Poir.) Yates, Southw. Naturalist **11**: 384. 1966.

Cespitose rhizomatous perennial herb; culms 0.6–2 m tall, 4–7 mm thick. Leaf sheaths ciliate with long soft hairs in tufts just below and to one side of the ligule, the hairs 1–2.5 mm long; ligule a dense circle of hairs to 1.5 mm long; blades 15–70 cm long, involute, the tip curled, the margins scabrous. Inflorescence 20–70 × 1–5 cm wide; branches ascending 1.5–4 cm long, floriferous to base. Spikelets 2–6-flowered, 2.5–6 mm long; glumes 1–2.5 mm long, the apex truncate, sometimes mucronate; lemmas of fertile florets 1.8–2.8 mm long, weakly 3–5-nerved, scabrous along the keel especially near the mucronate apex; paleas about as long as the lemmas; anthers 1–2 mm long. Caryopsis 0.9–1.1 mm long, ellipsoid.

DISTRIBUTION: Locally common, in open exposed areas especially along the coast. Maria Bluff (A1980), Reef Bay (W326). Also on Little St. James and St. Thomas; throughout the West Indies.

35. UROCHLOA P. Beauv.

Annual or perennial herbs; culms erect or decumbent, unbranched or sparingly branched. Leaf sheaths overlapping, smooth; ligule a fringe of hairs; blades linear to lanceolate, flattened or conduplicate. Main axis of inflorescence with appressed or divaricate primary branches; secondary branches appressed or spreading; rachis terminating in a spikelet; disarticulation at the spikelet base; callus differentiated or not. Spikelets adaxial or abaxial, dorsiventrally compressed or plano-convex; lower glume not fused with the callus, encircling the spikelet base; upper glumes about as long as the spikelet; lemma of lower floret membranous to chartaceous, sterile; lemma of upper floret shorter than or about as long as the lower floret, cartilaginous to indurate, fertile, rugose, dull, with involute margins, the apex with same or different texture as the body, muticous, mucronate, or awned.

A tropical and subtropical genus with about 40 species.

REFERENCE: Morrone, O. & F. O. Zuloaga. 1991. Revisión de las especies sudamericanas nativas e introducidas de los géneros *Brachiaria* y *Urochloa* (Poaceae: Panicoideae: Paniceae). Darwiniana 31: 43–109.

Key to the Species of *Urochloa*

1. Primary branches paniculate, with spreading secondary branches, whorled at the lowermost inflorescence node... 3. *U. maxima*
1. Primary branches racemose, with appressed secondary branches, not whorled at the lowermost inflorescence node.
   2. Upper glume and lower lemma reticulate-nerved, glabrous; spikelets typically <3 mm long... 2. *U. fasciculatum*
   2. Upper glume and lower lemma not reticulate-nerved, pubescent; spikelets typically >3 mm long 1. *U. adspersa*

1. Urochloa adspersa (Trin.) R.D. Webster, Syst. Bot. **13**: 607. 1988. *Panicum adspersum* Trin., Gram. Panic. 146. 1826. *Brachiaria adspersa* (Trin.) Parodi, Darwiniana **15**: 96. 1969.

Annual herb, lacking stolons; flowering culms 3–12 dm tall, erect or decumbent, sparingly branched. Leaf sheaths glabrous; ligule 0.5–1 mm long; blades linear to lanceolate, flat, 5–20 × 0.8–1.5 cm, smooth, glabrous on the upper surface. Inflorescence ovate; main axis 5–18 cm long; primary branches 1.5–8 cm long, with appressed secondary branches. Spikelets paired, elliptic, 2.9–3.8 × 1.2–1.4 mm; lower glumes 0.9–1.3 mm long, 3–5-nerved; upper glumes 2.7–3.6 mm long, pubescent, 5–7-nerved; lemma of lower floret elliptic, 5-nerved, muticous, the palea fully developed; lemma of upper floret 2.1–2.9 mm long, elliptic, indurate, rugose, yellow, muticous.

DISTRIBUTION: Occasional in open disturbed areas. Bethany (B334), Fish Bay (W491). Also on St. Croix, St. Thomas, Tortola, and Virgin Gorda; warm-temperate North America to tropical South America and the West Indies.

2. Urochloa fasciculatum (Sw.) R.D. Webster, Australian Paniceae 235. 1987 [1988]. *Panicum fasiculatum* Sw., Prodr. 22. 1788. *Brachiaria fasiculata* (Sw.) Parodi, Darwiniana **15**: 96. 1969.   Fig. 241I–M.

Panicum fuscum Sw., Prodr. 23. 1788.

Annual herb, lacking stolons; flowering culms 2–6 dm tall, erect or decumbent, sparingly branched. Leaf sheaths glabrous or hairy; ligule 0.5–1.5 mm long; blades linear, flat, 3–20 × 0.5–1.5 cm, smooth, glabrous or hairy on the upper surface. Inflorescence ovate; main axis 6–15 cm long; primary branches 2–8 cm long, with appressed secondary branches. Spikelets paired, elliptic, 2.5–3.1 × 1.4–1.6 mm; lower glumes 1–1.2 mm long, 1–3-nerved; upper glumes 2.5–3.1 mm long, 7-nerved; lemma of lower floret elliptic, 7-nerved, mucronate, the palea fully developed; lemma of upper floret elliptic, indurate, rugose, yellow, 2.3–2.9 mm long, mucronate.

DISTRIBUTION: A common roadside weed. Bordeaux (A3798), Great Cruz Bay (A2354), Fish Bay (A3884). Also on St. Croix, St. Thomas, and Tortola; southern United States to South America and the West Indies, introduced in Australia.

3. Urochloa maxima (Jacq.) R.D. Webster, Australian Paniceae 241. 1987 [1988]. *Panicum maximum* Jacq., Icon. Pl. Rar. **1**: 2. 1781.

Perennial herb, stoloniferous or lacking stolons; flowering culms 6–20 dm tall, erect or decumbent at the base, sparingly branched. Leaf sheaths glabrous or hairy; ligule 1–3 mm long; blades linear, flat, 15–60 × 1–3 cm, smooth, glabrous, or hairy on the upper surface. Inflorescence obovate; main axis 20–60 cm long; primary branches 12–35 cm long, with spreading secondary branches. Spikelets solitary or paired, oblong, 2.7–3.5 × 0.9–1.1 mm; lower glumes 0.8–1.1 mm long, 1–3-nerved; upper glumes 2.1–3.5 mm long, 5-nerved; lemma of lower floret oblong, 5-nerved, muticous or mucronate, the palea fully developed; lemma of upper floret elliptic, indurate, rugose, yellow, 1.9–2.4 mm long, mucronate.

DISTRIBUTION: An occasional weed. Fish Bay (A3887), Lameshur (W845). Also on St. Croix and St. Thomas; originally native to Africa, widely introduced and naturalized in the tropics and subtropics for pasture improvement.

COMMON NAME: guinea grass.

CULTIVATED SPECIES: *Cymbopogon citratus* (DC.) Stapf, commonly known as lemon grass, was reported by Woodbury and Weaver (1987) as occurring on St. John; the species, however, is known only in cultivation.

DOUBTFUL RECORD: *Lithachne pauciflora* (Sw.) P. Beauv. was listed by Woodbury and Weaver (1987) as occurring on St. John, but no voucher specimen has been seen to confirm this record.

EXCLUDED SPECIES: *Stenotaphrum secundatum* (Walter) Kuntze, reported for St. John by Woodbury and Weaver (1987), was based on a misidentification of a specimen of *Pennisetum clandestinum*. *Stenotaphrum secundatum* is otherwise known in the Virgin Islands from St. Croix and St. Thomas.

18. Potamogetonaceae (Pondweed Family)

By M. T. Strong

Aquatic fresh to saline water herbs, perennials or sometimes annuals, with weak, slender, submersed or floating stems. Leaves opposite or alternate (2-ranked), flat, narrowly linear to broadly ovate, often polymorphic depending on whether submersed or floating, sessile or sometimes petiolate in floating leaves, green, glabrous; sheaths tubular, frequently stipular at apex, free or partially adnate to petiole. Inflorescences in axillary or terminal pedunculate spikes, the peduncles sheathing at base. Flowers bisexual, small, 4-parted, rarely 2- or 3-parted; perianth of 4 short-clawed segments or wanting; stamens inserted on the claws of the segments, the anthers sessile, extrorse, bilocular, longitudinally dehiscent; carpels generally 4, separate, 1-celled; ovules solitary; styles short, the stigma rounded or peltate. Fruit a drupelet, achene, or rarely a berry, sessile or long-stipitate, 1-seeded, indehiscent; seeds lacking endosperm.

A somewhat cosmopolitan family of 3 genera and approximately 100 species.

1. RUPPIA L.

Submersed perennials of brackish or saline ponds, pools, lagoons, and marshes, with long, slender, branching stems. Leaves alternate (distichous) or subopposite, linear with a single midvein; sheaths expanded at base, hyaline along margin, the apex estipular. Peduncles long and spirally twisted or short; spikes generally 2-flowered; stipes elongating after flowering. Perianth segments lacking; stamens 2, the filaments short, broad; stigma peltate; ovule campylotropous, apically inserted. Fruit an oblique drupelet, the short or elongate style persistent at apex.

A genus of approximately 1–7 species, distributed throughout temperate and subtropical regions.

1. Ruppia maritima L., Sp. Pl. 127. 1753.

Fig. 242A–F.

Submersed or floating plants, forming dense masses, the light green to whitish stems to 1 m long. Leaves linear to filiform, 3–11 cm long, 0.3–0.9 mm wide, dark-green, narrowly acuminate to apex; sheaths 6–10(–14) mm long, membranous. Peduncle short, 7–30(–70) mm long, not elongating and spirally twisting after anthesis. Drupelet ovoid, 2–3 mm long, 1–1.5 mm wide with persistent style base, the stipe to 30 mm long.

DISTRIBUTION: In shallow brackish to saline waters of lagoons, ponds, and pools. Coral Bay Quarter in intermittent brackish pool by entrance to Fortsberg (A2421). Also on St. Croix and St. Thomas; cosmopolitan.

19. Smilacaceae (Sarsaparilla Family)

Herbaceous to slender woody vines aided by tendrils, or less often erect herbs or shrubs, arising from creeping, starchy rhizomes. Leaves alternate or sometimes opposite; petioles usually with a pair of tendrils at junction with short stipule–like flanges. Flowers small, commonly unisexual (the plant dioecious) or less often bisexual, actinomorphic, 3–merous, in axillary umbels, racemes, or spikes bearing lateral umbels; staminate flowers lacking pistillodes; pistillate flowers bearing staminodes. Perianth of 6 equal, petaloid tepals in 2 series, free or connate at base into a short tube; nectaries commonly at the inner base of the tepals or at the base of the stamens or staminodes; stamens commonly 6 or less often more numerous or only 3, the filaments distinct, sometimes adnate to the perianth tube, or more or less connate into a column, the anthers opening by longitudinal slits; gynoecium 3-carpellate, ovary superior, trilocular or unilocular, the placentation accordingly axile or parietal, the ovules 1 to many per locule, the styles terminal, distinct, or basally connate. Fruit a 1–6-seeded berry.

A family of about 12 genera and about 375 species, mostly with tropical and subtropical distribution, but occurring in subtemperate zones as well.

REFERENCE: Howard, R. A. 1979. The genus *Smilax* L. in the Lesser Antilles. Taxon 28: 55–58. 1979.

1. SMILAX L.

Dioecious, herbaceous to woody vines, with starchy rhizomes; stems cylindrical, wiry, armed or unarmed. Leaves alternate, simple, with arching parallel veins; petioles with a pair of filiform tendrils at junction with the open sheaths. Flowers unisexual, 3-merous, actinomorphic, with 6 perianth segments, produced in axillary umbels. Staminate flowers with 6 stamens, the anthers basifixed; pistillate flowers usually bearing staminodes; ovary 3-carpellate with 1 or 2 ovules per carpel. Fruit a fleshy berry; seeds 1–6.

A genus of about 350 species, from tropical and subtropical areas of the world.

1. Smilax coriacea Spreng., Syst. Veg. **2:** 103. 1825.

Fig. 242G–L.

Slightly woody vine, to 5 m long; stems slender, cylindrical, wiry, puberulent or slightly scabrous, with scattered curved spines.

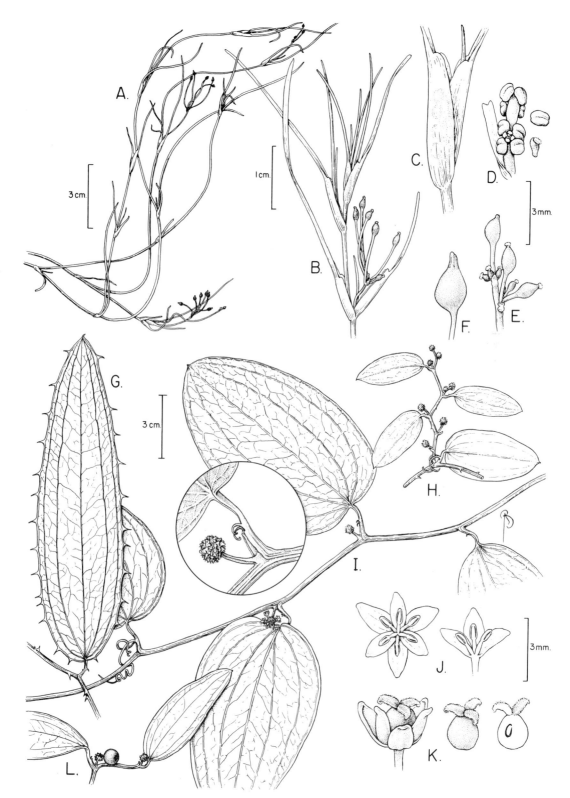

FIG. 242. A–F. *Ruppia maritima*. A. Habit. B. Portion of flowering stem. C. Detail of stem with inflorescence hidden by leaf sheaths. D. Inflorescence, stamen, and carpel. E. Infructescence. F. Drupelet. G–L. *Smilax coriacea*. G. Leaf with spiny margins. H. Fertile lateral branch. I. Fertile main branch and detail of stem node showing expanded receptacle of inflorescence after fruits are shed. J. Staminate flower, from above, and l.s. staminate flower. K. Pistillate flower, pistil, and l.s. pistil. L. Fruiting branch.

Leaf blades 2–17 × 1–12 cm, ovate, elliptic, oblong, lanceolate or linear, thick–coriaceous, glabrous, with 3–7 parallel veins, the margins spiny or entire, the apex acute to rounded, usually mucronate, the base rounded, obtuse, or cordate; petioles 0.5–2 cm long, slightly swollen, with a pair of filiform tendrils near the base. Inflorescence a flexuous spike bearing lateral umbels. Perianth yellowish, in staminate flowers 2–3 mm long, in pistillate flowers 1.5–1.8 mm long. Berry globose or depressed-globose, 5–7 mm diam., turning from green to black at maturity.

DISTRIBUTION: Occasional in open disturbed areas. Coral Bay Quarter along road to Bordeaux (A3818), Reef Bay Quarter at Lameshur (B522). Also on St. Croix, St. Thomas, Tortola, and Virgin Gorda; Hispaniola, Puerto Rico (including Vieques), and the Lesser Antilles.

NOTE: In a previous publication (Acevedo-Rodriguez & Woodbury, 1985), *Smilax coriacea* Spreng. was treated as a synonym of *Smilax havanensis* Jacq., following Liogier's (Liogier & Martorell, 1982) *Synoptical Flora of Puerto Rico and Adjacent Islands*. However, they are distinct species, easily recognized by the venation pattern. The secondary veins of *S. coriacea* are borne at an angle of 45° to 90° from the primary veins, whereas those of *S. havanensis* are ascending, being borne at 25° to 35°. *Smilax havanensis* occurs in Cuba and Hispaniola but does not reach Puerto Rico or the Virgin Islands.

DOUBTFUL RECORD: *Smilax domingensis* Willd. was reported by Woodbury and Weaver (1987) as occurring on St. John, but no specimen has been located to confirm this record.

DOUBTFUL FAMILY RECORD: *Typha domingensis* Pers. (Typhaceae) was reported by Woodbury and Weaver (1987) as occurring on St. John; however, no specimens have been located, and numerous attempts to find these plants on St. John have failed.

Appendix 1
Illustration Vouchers

Fig. 1.
Psilotum nudum: (A–C) *Little 13540.*
Lycopodium cernuum: (D–F) *Woodbury 322.*

Fig. 2.
Adiantum fragile var. *fragile:* (A, B) *Acevedo-Rdgz. 696.*
Blechnum occidentale: (C) *Acevedo-Rdgz. 2596.*
Asplenium pumilum: (D, E) *Acevedo-Rdgz. 2526.*

Fig. 3.
Doryopteris pedata: (A, C) *Acevedo-Rdgz. 714 & 2416.*
Hemionitis palmata: (D, E) *Acevedo-Rdgz. 3518.*
Nephrolepis multiflora: (F, G) *Acevedo-Rdgz. 2085.*

Fig. 4.
Odontosoria aculeata: (A, B) *Acevedo-Rdgz. 2095.* (C) *Mori 17023.*
Pityrogramma calomelanos: (D) *Acevedo-Rdgz. 2070.* (E) *Acevedo-Rdgz. 704.*
Polypodium aureum: (F–H) *Acevedo-Rdgz. 2070.*

Fig. 5.
Polypodium phyllitidis: (A, B) *Acevedo-Rdgz. 694.*
Pteris biaurita: (C–E) *Shafer 764.*

Fig. 6.
Thelypteris dentata: (A–C) *Correll 47630.*
Thelypteris poiteana: (D, E) *Proctor 39693.*

Fig. 7.
Asystasia gangetica: (A–E) *Acevedo-Rdgz. 3083.*
Blechum pyramidatum: (F–J) *Acevedo-Rdgz. 2912 & 2656.*
Barleria lupulina: (K–P) *Acevedo-Rdgz. 1869.*

Fig. 8.
Dicliptera sexangularis: (A–E) *Acevedo-Rdgz. 3150 & 3203.*
Justicia periplocifolia: (F–J) *Acevedo-Rdgz. 2747.*
Justicia mirabiloides: (K–O) *Acevedo-Rdgz. 2597.*

Fig. 9.
Ruellia tuberosa: (A–C) *Acevedo-Rdgz. 2756.* (D) *Acevedo-Rdgz. 2476.*
Oplonia microphylla: (E–J) *Acevedo-Rdgz. 1813.*
Oplonia spinosa: (K) *Acevedo-Rdgz. 760.*
Ruellia coccinea: (L–N) *Acevedo-Rdgz. 3174.*

Fig. 10.
Thunbergia fragrans: (A–F) *Acevedo-Rdgz. 1928.*
Stenandrium tuberosum: (G–J) *Britton 352, Woodbury 761.*
Siphonoglossa sessilis: (K–P) *Acevedo-Rdgz. 2746 & 2648.*

Fig. 11.
Sesuvium portulacastrum: (A) Field sketch. (B–D) *Acevedo-Rdgz. 2915.*
Trianthema portulacastrum: (E) *Acevedo-Rdgz. 2790.* (F–J) *Acevedo-Rdgz. 3160.*
Cypselea humifusa: (K–N) *Acevedo-Rdgz. 2832.*

Fig. 12.
Amaranthus viridis: (A–D) *Acevedo-Rdgz. 2905.*
Achyranthes aspera: (E–J) *Acevedo-Rdgz. 2926.*
Alternanthera tenella: (K) *Acevedo-Rdgz. 2728.* (L–O) *Acevedo-Rdgz. 3075.*

Fig. 13.
Gomphrena serrata: (A–I) *Acevedo-Rdgz. 2935.*
Blutaparon vermiculare: (J–M) *Acevedo-Rdgz. 2921.*

Fig. 14.
Iresine angustifolia: (A–F) *Acevedo-Rdgz. 779.*
Celosia nitida: (G–K) *Woodbury 172.*

Fig. 15.
Anacardium occidentale: (A,E) Photo. (B–D) *Acevedo-Rdgz. 605.*
Comocladia dodonaea: (F, I) *Mori 16601.* (G, H) *Acevedo-Rdgz. 745.*

Fig. 16.
Mangifera indica: (A) *Mori & Mitchell 18778.* (B–D) *De La Cruz s.n.* (E) Photo.

Fig. 17.
Schinus terebinthifolius: (A–C) Field sketch + *Correll 46698.* (D, E) *Howard 9992.* (F, G) *Zanoni 25786.*

Fig. 18.
Spondias mombin: (A) *Mori & Boom 15324.* (B, C) *De La Cruz 3835.* (D) *Lanjouw 3152.*

Fig. 19.
Annona glabra: (A) Field sketch. (B–F) *Acevedo-Rdgz. 2796.*
Annona squamosa: (G, H) *Acevedo-Rdgz. 2692.*

Fig. 20.
Cyclospermum leptophyllum: (A–E) *Acevedo-Rdgz. 3034.*

Fig. 21.
Nerium oleander: (A) Field sketch. (B–D) Specimen not cited.
Catharanthus roseus: (E) Photos. (F–H) *Acevedo-Rdgz. 677.*
Plumeria alba: (I, J) *Prance 29299* + photo.

Fig. 22.
Pentalinon luteum: (A–C) *Acevedo-Rdgz. 685.* (D) *Acevedo-Rdgz. 1835.*
Prestonia agglutinata: (E–J) *Acevedo-Rdgz. 902.* (K) *Zanoni 35972.*

Fig. 23.
Rauvolfia nitida: (A–F) *Acevedo-Rdgz. 2878.*

Fig. 24.
Ilex urbaniana: (A) *Mori 17078.* (B, C) *Acevedo-Rdgz. 2683.* (D–G) *Fishlock 80.*

Fig. 25.
Schefflera morototoni: (A) *Acevedo-Rdgz. 2887.* (B–G) *Bro. Hioram s.n.* (Jan 1914). (H, I) *Mejia 31358.*

Fig. 26.
Aristolochia elegans: (A–F) *Acevedo-Rdgz. 4136.*
Aristolochia trilobata: (G) *Duss 2285.* (H) *Marcano 8217.*

Fig. 27.
Asclepias curassavica: (A, B) Field sketch.
Calotropis procera: (C–G) Field sketch, pickled *s.n.*

Fig. 28.
Cryptostegia grandiflora: (A–D) Field sketch.
Matelea maritima: (E–H) *Acevedo-Rdgz. 2737.*
Metastelma grisebachianum: (I–K) *Acevedo-Rdgz. 1815.*

Fig. 29.
Acanthospermum hispidum: (A–F) *Liogier 24335, Zanoni 14874.*

Fig. 30.
Ageratum conyzoides: (A–C, F) *Eggers s.n.* (Feb 1887). (D, E) *Fishlock 281.* (F) *Correll 49537.*

Fig. 31.
Bidens cynapiifolia: (A–E) *Britton 1922.*

Fig. 32.
Chaptalia nutans: (A–E) *Eggers s.n.* (Feb 1887).

Fig. 33.
Chromolaena odorata: (A–D) *Acevedo-Rdgz. 2360.*

Fig. 34.
Conyza bonariensis: (A) *Bro. León 72700.* (B) *Dalton 13.206.* (C–G) *Pere Duss 586.*

Fig. 35.
Eclipta prostrata: (A–E) *Shafer 32.*

Fig. 36.
Elephantopus mollis: (A, F) *Britton 727.* (B–E, G) *Boom 6858.*

Fig. 37.
Emilia fosbergii: (A, B) Field sketch. (C, D) *Britton 589.* (E) *Stevenson 2341.*
Erigeron cuneifolius: (F) *Crosby 73.* (G–J) *Harris 12088.*

Fig. 38.
Lepidaploa glabra: (A, D) *Acevedo-Rdgz. 3148.* (B, C) *Acevedo-Rdgz. 2910.*
Launaea intybacea: (E–G) *Correll 48865.* (H) *Shafer 327.*

Fig. 39.
Mikania cordifolia: (A–D) *Piere Duss 2814.*

Fig. 40.
Neurolaena lobata: (A–D) *Shafer 1338.* (E) *Shafer 1155.*

Fig. 41.
Parthenium hysterophorus: (A–E) *W. Kings G.C. 302.*

Fig. 42.
Pectis humifusa: (A, B) *Howard 19276.* (C–F) *Acevedo-Rdgz. 3968.*
Piptocoma antillana: (G) Field sketch. (H–K) *Acevedo-Rdgz. 2220.*

Fig. 43.
Pluchea odorata: (A) *Leonard 14086.* (B–E) *Liogier35436.*

Fig. 44.
Pseudelephantopus spicatus: (A–E) *Liogier 15745.* (F) *Shafer 274.*

Fig. 45.
Pterocaulon virgatum: (A) *Urban 1868.* (B) *Britton 844.* (C–F) *Britton 3.*

Fig. 46.
Sonchus oleraceus: (A, C, D) *Brown 1914.* (B) *Collins 315.* (E) *Leonard 11378.*

Fig. 47.
Sphagneticola trilobata: (A–E) *Acevedo-Rdgz. 2794.* (F) *Zanoni 10417.*

Fig. 48.
Synedrella nodiflora: (A) *Mejia 10293.* (B) *Acevedo-Rdgz. 4041.* (C–G) *Burch 2522.*

Fig. 49.
Tridax procumbens: (A–F) *Acevedo-Rdgz. 4025.*

Fig. 50.
Anredera vesicaria: (A, B) *Zanoni 37125.* (C-G) *Acevedo-Rdgz. 1870.*

Fig. 51.
Batis maritima: (A–H) From Cronquist, A. 1981. An integrated system of classification of flowering plants. Columbia University Press, New York.

Fig. 52.
Cydista aequinoctialis: (A–C) *Acevedo-Rdgz. 2810* + photo. (D) *Questel 575.*
Amphitecna latifolia: (E, F) *Fosberg 55388* + photo. (G) *Wright s.n.*

Fig. 53.
Arrabidaea chica: (A–C) *Proctor 27391.* (D, E) *Saunders 604.*
Crescentia linearifolia: (F) *Acevedo-Rdgz. 2342.* (G) *Jiménez 2323.*

Fig. 54.
Macfadyena unguis-cati: (A, B) *Holdridge 1074.* (C) *Acevedo-Rdgz. 2803.*
Spathodea campanulata: (D) B. Angell photo. (E, F, H) *Acevedo-Rdgz. 2511.* (G) *Jack 4358.*

Fig. 55.
Tecoma stans: (A–E) Field sketch.
Tabebuia heterophylla: (F–H) Field sketch.

Fig. 56.
Ceiba pentandra: (A) Photo. (B) *Little 22001.* (C) Photo. (D, E) *Acevedo-Rdgz. 4231.* (F) *Morrow s.n.* (Nov 1921).
Quararibea turbinata: (G) *Box 1009.* (H–J) *Fairchild 2805.* (K) *Little 13111.*

Fig. 57.
Bourreria succulenta: (A–D) *Acevedo-Rdgz. 2350.*
Argusia gnaphalodes: (E) Field sketch + *Acevedo-Rdgz. 4062.* (F–H) *Britton 652.* (I) *Acevedo-Rdgz. 4062.*

Fig. 58.
Cordia sebestena: (A, B) Field sketch. (C) Photo. (D) Field sketch. (E–G) Pickled, *s.n.*

Cordia rickseckeri: (H) *Fosberg 54062.* (I) Photo.
Cordia collococca: (J) *Prance 29264.* (K–M) *Ricksecker 275A.*

Fig. 59.
Heliotropium curassavicum: (A–F) *Acevedo-Rdgz. 2781.* (G) *Acevedo-Rdgz. 795.*
Rochefortia acanthophora: (H, K) *Little 23824.* (I) Photo + *Acevedo-Rdgz. 3909.* (J) *Little 22034.*

Fig. 60.
Tournefortia hirsutissima: (A–D) *Acevedo-Rdgz. 2865.* (E) *Acevedo-Rdgz. 5448.*
Tournefortia microphylla: (F–I) *Acevedo-Rdgz. 777.* (J) *Acevedo-Rdgz. 2326.*

Fig. 61.
Cakile lanceolata: (A, E, F) *Acevedo-Rdgz. 780.* (B–D) *Acevedo-Rdgz. 3969.*
Lepidium virginicum: (G, K) *Acevedo-Rdgz. 866.* (H–J) *Liogier 3131.*

Fig. 62.
Bursera simaruba: (A–C) *Acevedo-Rdgz. 840.* (D–G) *Mori 17020.*

Fig. 63.
Hylocereus trigonus: (A–C) *Acevedo-Rdgz. 2874.*
Selenicereus grandiflorus: (D) *Acevedo-Rdgz. 2835.* (E) *Rose 6,210* (photo).

Fig. 64.
Melocactus intortus: (A–H) Field sketch + photos.
Mammillaria nivosa: (I) Photo. (J, K) Redrawn from Priscilla Fawcett, *Fl. Bahamas.* (L) *Acevedo-Rdgz. 4364.*

Fig. 65.
Opuntia repens: (A) Field sketch. (B) *Acevedo-Rdgz. 2907.*
Opuntia rubescens: (C, E–G) *Acevedo-Rdgz. 4132.* (D) Field sketch.

Fig. 66.
Pereskia aculeata: (A–C, E) *Acevedo-Rdgz. 4267.* (D) Redrawn from Mem. New York Bot. Gard. 41: 1–141.
Pilosocereus royenii: (F–H) *Acevedo-Rdgz. 4668.* (I) Pickled *s.n.*

Fig. 67.
Hippobroma longiflora: (A) Field sketch. (B–F) *Acevedo-Rdgz. 3130.*

Fig. 68.
Canella winterana: (A, F, G) *Sauleda 2669.* (B–E) *Acevedo-Rdgz. 2936.*

Fig. 69.
Capparis cynophallophora: (A–D) *Acevedo-Rdgz. 1798 & 1808.* (E) *Acevedo–Rdgz. 1821.*
Capparis flexuosa: (F) *Prance 29290.* (G) *Luteyn 5145.* (H) *Acevedo-Rdgz. 2521.*
Morisonia americana: (I, J) *Prance 29339.* (K) *Mori 17058.*

Fig. 70.
Cleome spinosa: (A–E) Field sketch + *Acevedo-Rdgz. 3777.*
Cleome viscosa: (F, G) *Woodbury s.n.* (Aug 22, 1979).

Fig. 71.
Carica papaya: (A–H) Field sketch, Saül, French Guiana.

Fig. 72.
Cecropia schreberiana: (A–C) *Acevedo-Rdgz. 2845.* (D) *Acevedo-Rdgz. 2956.* (E–G) Pickled, *s.n.*

Fig. 73.
Crossopetalum rhacoma: (A–D) *Acevedo-Rdgz. 699 & 1804.*
Cassine xylocarpa: (E–J) *Acevedo-Rdgz. 662 & 1824.*

Fig. 74.
Maytenus laevigata: (A–D) *Prance 29265.* (E) *Acevedo-Rdgz. 2069.*
Schaefferia frutescens: (F, G, K, L) *Mori 17065.* (H) *Correll 45438.* (I, J) *Woodbury 1977.*

Fig. 75.
Chenopodium ambrosioides: (A–D) *Shafer 2745.*
Atriplex cristata: (E–G) *Von Reis 235.*

Fig. 76.
Chrysobalanus icaco: (A–E) *Acevedo-Rdgz. 2674, Taylor 7633.*

Fig. 77.
Clusia rosea: (A–C) Field sketch, *Acevedo-Rdgz. 4016.* (D, E) Photo.
Mammea americana: (F) *Acevedo-Rdgz. 2042.* (G–I) *Acevedo-Rdgz. 2842.* (J, K) *Acevedo-Rdgz. 2078.*

Fig. 78.
Bucida buceras: (A, B) *Acevedo-Rdgz. 2566.* (C) *Zanoni 29955.* (D, E) *Liogier 21455.* (F) *Zanoni 10806.* (G) *Liogier 31490.*
Buchenavia tetraphylla: (H–J) *Britton 8093.* (K–M) *Liogier 30331.*

Fig. 79.
Conocarpus erectus: (A–D) *Britton 976.* (E) *Acevedo-Rdgz. 4082.* (F, G) *Fishlock 127.*
Laguncularia racemosa: (H–J) *Acevedo-Rdgz. 791.* (K) *Ricksecker 325.*

Fig. 80.
Quisqualis indica: (A–C) Field sketch. (D, E) *Ricksecker 337.*
Terminalia catappa: (F) Field sketch. (G, H) *Yuncker 18412.*

Fig. 81.
Convolvulus nodiflorus: (A) *Acevedo-Rdgz. 3960.* (B–E) *Acevedo-Rdgz. 3960 & 2378.*
Evolvulus nummularius: (F–I) *Acevedo-Rdgz. 3151 & 2322.*

Fig. 82.
Ipomoea repanda: (A–D) Field sketch.
Ipomoea eggersii: (E, F) *Acevedo-Rdgz. 4043.*
Ipomoea triloba: (G, H) *Acevedo-Rdgz. 3099.* (I) *Acevedo-Rdgz. 3079.*

Fig. 83.
Ipomoea nil: (A) *Acevedo-Rdgz. 2315 & 3100.* (B–D) *Acevedo-Rdgz. 3080* (pickled). (E) *Eggers s.n.* (Feb 1887).
Ipomoea indica var. *acuminata:* (F) *Acevedo-Rdgz. 4051.* (G–I) *Acevedo-Rdgz. 4127.*
Ipomoea hederifolia: (J–L) *Acevedo-Rdgz. 3031.* (J, M) *Acevedo-Rdgz. 3082.*

Fig. 84.
Jacquemontia pentanthos: (A, B) *Acevedo-Rdgz. 4037.* (C, D) *Acevedo-Rdgz. 687.*
Jacquemontia cumanensis: (E) *Acevedo-Rdgz. 3971.*
Jacquemontia havanensis: (F, G, I) *Acevedo-Rdgz. 4077.* (H) *Proctor 42607.*

Fig. 85.
Merremia aegyptia: (A–E) *Acevedo-Rdgz. 3080.*
Merremia quinquefolia: (F, J, K) *Acevedo-Rdgz. 3081.* (G–I) *Acevedo-Rdgz. 2296.*

Fig. 86.
Merremia umbellata: (A–C) *Acevedo-Rdgz. 4019.* (D, E) *Acevedo-Rdgz. 2472.*
Stictocardia tiliifolia: (F–H) *Acevedo-Rdgz. 4008. & 3120 (photo).* (I) *Proctor 43448.*

Fig. 87.
Bryophyllum pinnatum: (A–D) *Acevedo-Rdgz. 3959.*

Fig. 88.
Doyerea emetocathartica: (A–C) *Acevedo-Rdgz. 4044.* (D) *Acevedo-Rdgz. 1944.*
Cayaponia americana: (E, F) *Acevedo-Rdgz. 3856.*
Cucumis anguria: (G–J) *Acevedo-Rdgz. 4023.*

Fig. 89.
Luffa aegyptiaca: (A, C) *Acevedo-Rdgz. 4066.* (B) *Bro. León 502.*
Melothria pendula: (D) *Fosberg 55376.* (E, F) *Sauleda 2894.*

Fig. 90.
Momordica charantia: (A–F) *Mori et al. 21055.*

Fig. 91.
Cuscuta americana: (A–H) *Acevedo-Rdgz. 2569.*

Fig. 92.
Erythroxylum brevipes: (A–H) *Acevedo-Rdgz. 742.*

Fig. 93.
Argythamnia candicans: (A–G) *Mori 16597.*
Acalypha poiretii: (H–O) *Ricksecker 147.*

Adelia ricinella: (P,Q) *Garber 29.* (R, S) *Lavestre 2025.* (T, U) *Zanoni 19742.*

Fig. 94.
Chamaesyce hyssopifolia: (A) *Acevedo-Rdgz. 2499.* (B–D) *Acevedo-Rdgz. 3078* (pickled).
Chamaesyce mesembrianthemifolia: (E–J) *Acevedo-Rdgz. 1871.*
Croton fishlockii: (K–N) *Fishlock 311.* (O–R) *Acevedo-Rdgz. 752.*

Fig. 95.
Dalechampia scandens: (A, D) *Acevedo-Rdgz. 1882.* (B) *Thompson 1089.* (C) *Bro. León 630.* (E) *Acevedo-Rdgz. 660.*
Drypetes alba: (F, I) *Mejia 5075.* (G, H) *Acevedo-Rdgz. 571.*
Euphorbia petiolaris: (J–N) *Britton 234.*

Fig. 96.
Gymnanthes lucida: (A–E) *Acevedo-Rdgz. 2232.* (F, G) *VanderWerff 8693.*
Flueggea acidoton: (H–K) *Thompson 346.*

Fig. 97.
Hippomane mancinella: (A, B, F) *Arnoldo-Broeders 3768.* (C–E) *Acevedo-Rdgz. 781.* (G) *Liogier 34346.*
Hura crepitans: (H–L) *Acevedo-Rdgz. 4010.* (M) *Zanoni 30602.* (N) Photo.

Fig. 98.
Jatropha gossypifolia: (A–C) *Acevedo-Rdgz. 1951.* (D, E) Pickled *s.n.*

Fig. 99.
Margaritaria nobilis: (A–E) *Harris 3183.* (F, G) *Zanoni 16285.*
Pedilanthus tithymaloides subsp. *angustifolius:* (H) *León 5025.* (I–K) *Correll 47246.*
Phyllanthus niruri: (L–O) *Woodbury 113.* (P) *Duss 2921.*

Fig. 100.
Ricinus communis: (A–E) *Boom 7039.* (F, G) *Boom 553.*
Sapium caribaeum: (H–K) *Beard 504.* (L, M) *Acevedo-Rdgz. 2127.*

Fig. 101.
Savia sessilflora: (A–C) *Proctor 40537.* (D) *Liogier 18109.*
Tragia volubilis: (E–I) *Acevedo-Rdgz. 703.*

Fig. 102.
Caesalpinia bonduc: (A) *Acevedo-Rdgz. 2570.* (B, C) *Acevedo-Rdgz. 4021.*
Caesalpinia ciliata: (D–F) *Acevedo-Rdgz. 4227.*
Chamaecrista glandulosa var. *swartzii:* (G–I) *Acevedo-Rdgz. 904.*

Fig. 103.
Delonix regia: (A–D) Field sketch.

Fig. 104.
Hymenaea courbaril: (A–D) *Acevedo-Rdgz. 5073.* (E) Photo.
Parkinsonia aculeata: (F–I) *Acevedo-Rdgz. 4040.*

Fig. 105.
Erythrina eggersii: (A) Field sketch. (B–D) *Britton 1950.* (E) *Otero 604* (pod), *Crawford 9799* (seed).
Peltophorum pterocarpum: (F, G) *Manuel 109.* (H) *Mejia 8670.*

Fig. 106.
Senna bicapsularis: (A–C) Field sketch. (D–F) *Acevedo-Rdgz. 3917.*
Tamarindus indica: (G–I) *Acevedo-Rdgz. 1810.* (J) Field sketch + photo.

Fig. 107.
Abrus precatorius: (A) *Acevedo-Rdgz. 710.* (B, C) *Yuncker 17303.* (D, E) Pickled *s.n.*
Alysicarpus vaginalis: (F–I) *Shafer 34.*
Aeschynomene americana: (J–N) *Britton 778.*

Fig. 108.
Andira inermis: (A–F) *Mori 17086.*

Fig. 109.
Cajanus cajan: (A–E) *Acevedo-Rdgz. 1952.*
Centrosema virginianum: (F) *Mori 17093.* (G–K) *Acevedo-Rdgz. 1929.*
Canavalia rosea: (L–O) *Acevedo-Rdgz. 776.* (P) *Zanoni 17093.*

Fig. 110.
Clitoria ternatea: (A–C) *Acevedo-Rdgz. 2623.* (D) *Britton 920.*
Crotalaria pallida var. *ovata:* (E–J) *Acevedo-Rdgz. 689.*
Coursetia caribaea: (K) *Eggers 387.* (L, M) *Acevedo-Rdgz. 2341.*

Fig. 111.
Dalbergia ecastaphyllum: (A, E) *Acevedo-Rdgz. 2039.* (B–D) *Acevedo-Rdgz. 2041.*
Desmodium incanum: (F–I) *Acevedo-Rdgz. 4159.*
Galactia dubia: (J–L) *Andrews 575.*

Fig. 112.
Indigofera suffruticosa: (A-E) *Acevedo-Rdgz. 3183.*
Gliricidia sepium: (F–I) *Yuncker 8814.*

Fig. 113.
Lablab purpureus: (A–F) *Acevedo-Rdgz. 2123.*
Machaerium lunatum: (G, K, L) *Heller 823, Woodbury 327.* (H–J) *Acevedo-Rdgz. 2829.*

Fig. 114.
Macroptilium lathyroides: (A–E) *Acevedo-Rdgz. 2117.*
Piscidia carthagenensis: (F–H) *Acevedo-Rdgz. 763.* (I) *Acevedo-Rdgz. 1885 & 2581* (photo).
Pictetia aculeata: (J, K) *Acevedo-Rdgz. 2046.* (L, M) *Prance 29303.* (N) *Fishlock 123.*

Fig. 115.
Phaseolus peduncularis: (A) *Acevedo-Rdgz. 3796.* (B, E) *Acevedo-Rdgz. 3868.* (C, D) *Hess 5115.*
Sesbania sericea: (F–I) *Acevedo-Rdgz. 3925.* (J) *Correll 46738.*

Fig. 116.
Sophora tomentosa: (A–C) *Eggers s.n.* (D) *Maxon 1583.*
Stylosanthes hamata: (E–J) *Britton 662.*
Poitea florida: (K–M) *Prance 29286.* (N) *Acevedo-Rdgz. 851.*

Fig. 117.
Pueraria phaseoloides: (A–C) *Lavestre 1826.* (D, E) *Liogier 28042.*
Rhynchosia reticulata: (F–J) *Mori 17094.*

Fig. 118.
Vigna luteola: (A–D) *Stimson 3158.* (E, F) *Fosberg 48305.*
Tephrosia cinerea: (G–K) *Woodbury 28.*
Teramnus labialis: (L–O) *Acevedo-Rdgz. 1956.* (P) *Acevedo-Rdgz. 716.*

Fig. 119.
Acacia retusa: (A–C) *Acevedo-Rdgz. 5050.* (D) *Acevedo-Rdgz. 4055.*
Adenanthera pavonina: (E–H) *Ekman 10335.* (I,J) *Acevedo-Rdgz. 2930.*

Fig. 120.
Albizia lebbeck: (A–E) Field sketch + *Acevedo-Rdgz. 3981.*
Zapoteca portoricensis: (F–H) *Acevedo-Rdgz. 5062.*

Fig. 121.
Desmanthus virgatus: (A, G) *U.S. Forest Dept. 6793.* (B–F) *Acevedo-Rdgz. 2784.*

Fig. 122.
Leucaena leucocephala: (A–D) *Acevedo-Rdgz. 654.* (E, F) *Acevedo-Rdgz. 1801.*
Inga laurina: (G, H) *Cowell 535.* (I, J) *Acevedo-Rdgz. 932.* (K) *Thompson 429.*

Fig. 123.
Mimosa ceratonia: (A–E) *Mori 17024.* (F) *Zanoni 29326.*
Pithecellobium unguis-cati: (G, H, L, M) *Acevedo-Rdgz. 2022.* (I–K) *Prance 29336.*

Fig. 124.
Casearia decandra: (A) *Acevedo-Rdgz. 5075.* (B, C) *Acevedo-Rdgz. 735.* (D) *Acevedo-Rdgz. 2802.*
Prockia crucis: (E–I) *Alain 4314.* (J) *Woodbury s.n.* (22 Jun 1966).

Fig. 125.
Samyda dodecandra: (A–D) *Allard 15743.*
Xylosma buxifolia: (E, H) *Ekman 3983.* (F) *Goll 1026.* (G) *Ekman 8415.*

Fig. 126.
Scaevola plumieri: (A, E) Field sketch. (B–D) *Cowan 1669.*

Fig. 127.
Leonotis nepetifolia: (A) Field sketch. (B, C) *Acevedo-Rdgz. 3878.*
Hyptis pectinata: (D) *Acevedo-Rdgz. 2587.* (E, F) *Acevedo-Rdgz. 4018.*

Fig. 128.
Leonurus sibiricus: (A) *Acevedo-Rdgz. 5086.* (B, C) *Acevedo-Rdgz. 3811.*
Ocimum campechianum: (D–F) *Mori s.n.* (G) *Acevedo-Rdgz. 3771.*
Salvia serotina: (H–J) *Acevedo-Rdgz. 671.*

Fig. 129.
Cassytha filiformis: (A) *Acevedo-Rdgz. 4372.* (B–F) *Acevedo-Rdgz. 3952.*
Cinnamomum elongatum: (G–I) *Acevedo-Rdgz. 4207.* (J) *Gregory 39.*

Fig. 130.
Ocotea coriacea: (A, G) *Acevedo-Rdgz. 5109.* (B–F) *Acevedo-Rdgz. 636.*
Licaria triandra: (H) *Acevedo-Rdgz. 4122, Pimentel 957.* (I–K) *Gonzáles 555.*

Fig. 131.
Spigelia anthelmia: (A–F) *Acevedo-Rdgz. 2407.*

Fig. 132.
Dendropemon caribaeus: (A–F) *Acevedo-Rdgz. 744.*

Fig. 133.
Ginoria rohrii: (A, B, F) *Acevedo-Rdgz. 5132.* (C–E) Pickled *s.n.*
Ammannia coccinea: (G, K) *Acevedo-Rdgz. 4080.* (H–J) *Acevedo-Rdgz. 3199.*

Fig. 134.
Byrsonima spicata: (A–E) *Acevedo-Rdgz. 2939.* (F) *Taylor 408.*
Heteropteris purpurea: (G–K) *Acevedo-Rdgz. 664 & 2658.*
Bunchosia glandulosa: (L, P, Q) *Acevedo-Rdgz. 2574.* (M–O) *Howard 14540.*

Fig. 135.
Malpighia linearis: (A–F) *Acevedo-Rdgz. 3229.*
Stigmaphyllon emarginatum: (G–N) Field sketch + *Acevedo-Rdgz. 4042.*

Fig. 136.
Abutilon umbellatum: (A, D) *Acevedo-Rdgz. 2775.* (B, C) *Acevedo-Rdgz. 3933.*
Bastardia viscosa var. *viscosa:* (E–H) *Acevedo-Rdgz. 2901.*
Gossypium barbadense: (I–K) *Acevedo-Rdgz. 4013.*

Fig. 137.
Herissantia crispa: (A–D) *Eggers 339.*
Malachra alceifolia: (E) *Liogier & Solaro 31437.* (F-I) *Acevedo-Rdgz. 3098.*
Malvastrum americanum: (J) *Acevedo-Rdgz. 2446.* (K–N) Pickled *s.n.*

Fig. 138.
Pavonia spinifex: (A) *Mori et al. 17033.* (B, C) *Acevedo-Rdgz. 3124.* (D, E) *Acevedo-Rdgz. 3836.*
Sida glomerata: (F–J) *Acevedo-Rdgz. 3778.*
Sida repens: (K–M) *Acevedo-Rdgz. 2290.*

Fig. 139.
Thespesia populnea: (A–D) *Acevedo-Rdgz. 4028.*
Sidastrum multiflorum: (E–G) *Acevedo-Rdgz. 2646.* (H) *Boom 6955.*

Fig. 140.
Urena lobata: (A) *Acevedo-Rdgz. 2680.* (B–F) *Acevedo-Rdgz. 3177.*
Wissadula periplocifolia: (G) Field sketch. (H–J) *Acevedo-Rdgz. 3127.*

Fig. 141.
Tetrazygia eleagnoides: (A–G) *Acevedo-Rdgz. 855.* (H) *Britton 390.*
Miconia laevigata: (I–N) *Acevedo-Rdgz. 845.*

Fig. 142.
Melia azedarach: (A–E) *Acevedo-Rdgz. 4033.* (F, G) *Acevedo-Rdgz. 4137.*
Cedrela odorata: (H–M) *Acevedo-Rdgz. 2850.* (N) *Acevedo-Rdgz. 2608.*

Fig. 143.
Swietenia mahagoni: (A–D) *Acevedo-Rdgz. 2514.* (E) *Acevedo-Rdgz. 5143.* (F) *Boom 9993.*

Fig. 144.
Cissampelos pareira: (A) *Mori 17021.* (B–D) *Acevedo-Rdgz. 826.* (E–J) *Acevedo-Rdgz. 2439.*

Fig. 145.
Hyperbaena domingensis: (A, B, F) *Daly 5301.* (C–E) *Díaz 333.* (G) *Howard 19774.* (H) *Duss 3682.*

Fig. 146.
Mollugo nudicaulis: (A–F) *Acevedo-Rdgz. 3934.*

Fig. 147.
Ficus citrifolia: (A–C) *Acevedo-Rdgz. 1967.*

Fig. 148.
Moringa oleifera: (A) *Melo 52.* (B–G) *Acevedo-Rdgz. 4039.*

Fig. 149.
Bontia daphnoides: (A, G, H) *Prance 29324.* (B–F) *Morton s.n.*

Fig. 150.
Ardisia obovata: (A, G, H) *Prance 29340.* (B-F) *Mori 17009.*

Fig. 151.
Calyptranthes thomasiana: (A–C) *Woodbury 827.* (D, E) *Acevedo-Rdgz. 5103.*
Eugenia biflora: (F–I) *Acevedo-Rdgz. 2931.* (J) *Acevedo-Rdgz. 2808.*
Eugenia earhartii: (K–M) *Acevedo-Rdgz. 2030* + photo.

Fig. 152.
Eugenia pseudopsidium: (A) Field sketch. (B, C) *Acevedo-Rdgz. 2853.*
Eugenia monticola: (D, E) *Acevedo-Rdgz. 1897.* (F) *Acevedo-Rdgz. 1877.*
Eugenia sessiliflora: (G) *Acevedo-Rdgz. 1830.* (H–J) *Acevedo-Rdgz. 2763.*

Fig. 153.
Myrcia citrifolia: (A–C) *Acevedo-Rdgz. 2861.* (D–F) *Acevedo-Rdgz. 1846.*
Myrcianthes fragrans: (G–I) *Ekman 13041.*
Myrciaria floribunda: (J, K) *Acevedo-Rdgz. 2823.*

Fig. 154.
Psidium amplexicaule: (A) *Acevedo-Rdgz. 2673.* (B, C) *Acevedo-Rdgz. 2862.* (D) *Acevedo-Rdgz. 5098.* (E) *Acevedo-Rdgz. 1915.*

Fig. 155.
Boerhavia erecta: (A–F) *Acevedo-Rdgz. 2822.*
Guapira fragrans: (G–I) *Acevedo-Rdgz. 5041.* (J, K) *Proctor 43468.* (L, M) *Acevedo-Rdgz. 2695.*

Fig. 156.
Pisonia aculeata: (A–D) *Acevedo-Rdgz. 4209.* (E–G) *Rose 3548.*
Neea buxifolia: (H) *Acevedo-Rdgz. 4143.* (I) *Little 21592.* (J) *Acevedo-Rdgz. 4706.* (K) *Acevedo-Rdgz. 4706.*

Fig. 157.
Ouratea litoralis: (A–F) *Acevedo-Rdgz. 2847.* (G–J) *Acevedo-Rdgz. 4675.*

Fig. 158.
Schoepfia schreberi: (A, B) Field sketch. (C–E) *Britton 1271.* (F, G) *Proctor 43680.*
Ximenia americana: (H, M) *Winters 2248.* (I–L) *Liogier 33621.*

Fig. 159.
Chionanthus compacta: (A–E) *Acevedo-Rdgz. 5128.* (F) *Acevedo-Rdgz. 2800.*
Jasminum fluminense: (G) *Acevedo-Rdgz. 2829.* (H–K) *Acevedo-Rdgz. 3839.*
Forestiera eggersiana: (L–N) *Acevedo-Rdgz. 3865.* (O) *Acevedo-Rdgz. 3194.*

Fig. 160.
Ludwigia octovalvis: (A–H) *Fosberg 54099.*

Fig. 161.
Oxalis corniculata: (A-H) *Zanoni 18607 & 24981.*

Fig. 162.
Argemone mexicana: (A–H) *Acevedo-Rdgz. 672.*

Fig. 163.
Passiflora edulis: (A–D) *Acevedo-Rdgz. 834.* (E–G) *Maxon 10244.*
Passiflora laurifolia: (H, I) *Mori 17054.*
Passiflora foetida: (J, K) *Acevedo-Rdgz. 1948.*

Fig. 164.
Petiveria alliacea: (A) *Fishlock 133.* (B–D) *Acevedo-Rdgz. 1863.*
Rivina humilis: (E–H) *Woodbury M-124.*
Trichostigma octandrum: (I–K) *Zanoni 25547.*

Fig. 165.
Peperomia magnoliifolia: (A–D) *Acevedo-Rdgz. 618.*
Piper amalago: (E–K) *Acevedo-Rdgz. 725.*

Fig. 166.
Plantago major: (A–D) *Stimson 1472.*

Fig. 167.
Plumbago scandens: (A-D) *Hudson 754.*

Fig. 168.
Coccoloba venosa: (A, E, F) *Acevedo-Rdgz. 2075.* (B–D) *Liogier 35195.*
Antigonon leptopus: (G–M) *Acevedo-Rdgz. 1811.*

Fig. 169.
Portulaca oleracea: (A) Field sketch. (B–D) *Acevedo-Rdgz. 4015.*
Portulaca quadrifida: (E) *Acevedo-Rdgz. 5276.* (F, G) *Acevedo-Rdgz. 3938.*
Talinum fruticosum: (H) Field sketch. (I, J) *Acevedo-Rdgz.4014.*

Fig. 170.
Colubrina arborescens: (A–D) *Acevedo-Rdgz. 2442.* (E–G) *Zanoni 26350.*
Gouania lupuloides: (H–L) *Woodbury 17003.* (M–P) *Acevedo-Rdgz. 2295.*
Krugiodendron ferreum: (Q) *Acevedo-Rdgz. 2129.* (R, S) *Britton 9473.*

Fig. 171.
Reynosia guama: (A–D, I) *Acevedo-Rdgz. 5121.* (E–H) *Morrow 172.*

Fig. 172
Chiococca alba: (A, C) *Acevedo-Rdgz. 5077.* (B) *Acevedo-Rdgz. 1914.* (D–F) *Acevedo-Rdgz. 5077.* (G, H) *Acevedo-Rdgz. 3821.*
Chione venosa: (I–L) *Smith 635, Acevedo-Rdgz. 4700.* (M) *Boom 6945.*

Fig. 173.
Coffea arabica: (A–D) *Liogier 33979.* (E, F) *Acevedo-Rdgz. 5277.*
Erithalis fruticosa: (G, H) Field sketch. (I–L) *Acevedo-Rdgz. 2731, 3988 + 3988, 3017.*

Fig. 174.
Exostema caribaeum: (A–E) *Acevedo-Rdgz. 3111.* (F–H) *Acevedo-Rdgz. 4049.*
Faramea occidentalis: (I) Field sketch + photo. (J–N) Pickled *s.n.*

Fig. 175.
Geophila repens: (A) Field sketch. (B, C) *Acevedo-Rdgz. 4126.*
Genipa americana: (D–G) *Little 13653.*
Gonzalagunia hirsuta: (H, I) Field sketch. (J–N) Pickled *s.n.*

Fig. 176.
Guettarda odorata: (A, B) *Mori 17026.* (C, D) *Acevedo-Rdgz. 2507* + pickled *s.n.*
Ixora ferrea: (E, F) *Acevedo-Rdgz. 4050.* (G–J) *Acevedo-Rdgz. 4056.* (K–M) *Acevedo-Rdgz. 4708.*

Fig. 177.
Machaonia woodburyana: (A–E) *Acevedo-Rdgz. 2838.* (F) *Acevedo-Rdgz. 4235.*
Morinda citrifolia: (G, H) Field sketch. (I–K) *Acevedo-Rdgz. 2778 & 2694.*

Fig. 178.
Palicourea croceoides: (A) Field sketch. (B–G) *Acevedo-Rdgz. 2609.*
Randia aculeata: (H) *Mori 17090.* (I–L) *Prance 29280.*
Psychotria brownei: (M) *Acevedo-Rdgz. 3811.* (N, O) *Acevedo-Rdgz. 2886.* (P) *Acevedo-Rdgz. 3811.*

Fig. 179.
Rondeletia pilosa: (A, D) *Acevedo-Rdgz. 2836.* (B, C) *Proctor 42603.*
Spermacoce confusa: (E, I, J) *Acevedo-Rdgz. 2565.* (F–H) *Acevedo-Rdgz. 3847.*
Scolosanthus versicolor: (K, M, N) *Acevedo-Rdgz. 2327.* (L, O) *Acevedo-Rdgz. 3115.*

Fig. 180.
Amyris elemifera: (A) *Acevedo-Rdgz. 4707.* (B–D) *Eggers s.n.* (Nov 1880). (E) *Acevedo-Rdgz. 3864.*
Citrus aurantifolia: (F) *Acevedo-Rdgz. 642.* (G–J) *Acevedo-Rdgz. 642 & 2007,* (K) *Acevedo-Rdgz. 2007.*
Murraya exotica: (L–N) *Correll 47258.* (O) *Acevedo-Rdgz. 4009.*

Fig. 181.
Zanthoxylum martinicense: (A–E) *Acevedo-Rdgz. 2510.*
Zanthoxylum thomasianum: (F, G) *Woodbury s.n.*
Triphasia trifolia: (H–K) Field sketch. (L–N) *Leucht 1520.*

Fig. 182.
Allophylus racemosus: (A, B) *Acevedo-Rdgz. 1847.* (C–F) *Acevedo-Rdgz. 935.*
Cardiospermum corindum: (G–K) *Correll 43324, Acevedo-Rdgz. 2373.*

Fig. 183.
Cupania triquetra: (A) *Acevedo-Rdgz. 2093, Britton 310.* (B–D) *Britton 310.* (E) *Acevedo-Rdgz. 721.*
Exothea paniculata: (F, J, K) *Proctor 21819.* (G–I) *Zanoni 27880.*

Fig. 184.
Serjania polyphylla: (A, H) *Acevedo-Rdgz. 709.* (B) *Woodbury 314.* (C) *Acevedo-Rdgz. 2200.* (D–G) *Acevedo-Rdgz. 629.*
Melicoccus bijugatus: (I, J) *Prance 29276.* (K) *Acevedo-Rdgz. 728.* (L) *Acevedo-Rdgz. 1858.*

Fig. 185.
Chrysophyllum pauciflorum: (A) *Acevedo-Rdgz. 5088.* (B–D) *Acevedo-Rdgz. 2079.* (E) *Acevedo-Rdgz. 4166.*
Chrysophyllum bicolor: (F) *Acevedo-Rdgz. 2900.*
Manilkara bidentata: (G) *Acevedo-Rdgz. 2102.* (H–K) *Ekman 15087.* (L) *Duke 7513.*

Fig. 186.
Pouteria multiflora: (A) Field sketch. (B–G) *Acevedo-Rdgz. 7018.*
Sideroxylon obovatum: (H, I) *Acevedo-Rdgz. 2338.* (J–M) *Acevedo-Rdgz. 3792.* (N) *Acevedo-Rdgz. 4329.*

Fig. 187.
Bacopa monnieri: (A, B) *Asplund 8624.* (C, D) *Holmgren & Holmgren 10087.*
Scoparia dulcis: (E–I) *Asplund 5250.*
Capraria biflora: (J–N) Specimens not cited.

Fig. 188.
Picrasma excelsa: (A, C, D) *Proctor 32917.* (B, E) *Howard 18797.* (F) *Acevedo-Rdgz. 4152.*

Fig. 189.
Brunfelsia americana: (A, B) *Acevedo-Rdgz. 4011.* (C, D) *Acevedo-Rdgz. 2742.* (E) *Acevedo-Rdgz. 4129.*
Capsicum frutescens: (F, G, I) *Acevedo-Rdgz. 5092.* (H) *Acevedo-Rdgz. 2433.*
Cestrum laurifolium: (J–M) *Acevedo-Rdgz. 3137.* (N, O) *Acevedo-Rdgz. 3835 & 2699.*

Fig. 190.
Datura inoxia: (A–D) *Acevedo-Rdgz. 4064.* (E) *Acevedo-Rdgz. 2362.*
Physalis angulata: (F–I) *Acevedo-Rdgz. 3937.*
Solanum lancifolium: (J–M) *Acevedo-Rdgz. 5108.*

Fig. 191.
Helicteres jamaicensis: (A, G) Field sketch. (B–F) *Acevedo-Rdgz. 2834.*
Ayenia insulaecola: (H) *Proctor 43428.* (I–M) *Acevedo-Rdgz. 3164.*

Fig. 192.
Guazuma ulmifolia: (A) *Acevedo-Rdgz. 5043.* (B–E) *Acevedo-Rdgz. 5043.* (F) *Acevedo-Rdgz. 4050.*
Waltheria indica: (G) *Proctor 42577.* (H–K) *Acevedo-Rdgz. 2780.*
Melochia tomentosa: (L) *Acevedo-Rdgz. 1842.* (M–Q) *Acevedo-Rdgz. 3845.*

Fig. 193.
Suriana maritima: (A, B, H, I) *Acevedo-Rdgz. 746.* (C–G) *Liogier 32663.*

Fig. 194.
Symplocos martinicensis: (A–F) *Mori et al. 15037.*

Fig. 195.
Ternstroemia peduncularis: (A, F, G) *Acevedo-Rdgz. 5106.* (B–E) *Acevedo-Rdgz. 941.*

Fig. 196.
Jacquinia berterii: (A–F) *Acevedo-Rdgz. 1814.* (G, H) *Acevedo-Rdgz. 1978.*

Fig. 197.
Daphnopsis americana: (A–E) *Acevedo-Rdgz. 2487.* (F, G) *Mori 17006.* (H) *Mori 17034.*

Fig. 198.
Corchorus hirsutus: (A) *Acevedo-Rdgz. 2050.* (B) *Acevedo-Rdgz. 4293.* (C–F) *Acevedo-Rdgz. 3962.*
Triumfetta semitriloba: (G–M) *Acevedo-Rdgz. 3817.*

Fig. 199.
Turnera ulmifolia: (A–C) *Acevedo-Rdgz. 2480.* (D–H) *Fosberg 54053.*
Turnera diffusa: (I–K) *Acevedo-Rdgz. 1834.*

Fig. 200.
Celtis iguanaea: (A–C) *Acevedo-Rdgz. 2691.* (D–F) *Acevedo-Rdgz. 2691.* (G) *Acevedo-Rdgz. 2021.*
Trema micranthum: (H, K) *Acevedo-Rdgz. 619.* (I, J) *Prance 29314.* (L, M) *Acevedo-Rdgz. 3841.*

Fig. 201.
Pilea sanctae-crucis: (A–F) *Mori 16997.*
Pilea nummulariifolia: (G) *Woodbury 865.*

Fig. 202.
Avicennia germinans: (A–C) *Acevedo-Rdgz. 626.* (D) *Acevedo-Rdgz. 5136.*
Bouchea prismatica: (E–G) *Acevedo-Rdgz. 2576.*
Citharexylum fruticosum: (H) Photo. (I–K) *Acevedo-Rdgz. 3814.* (L, M) *Acevedo-Rdgz. 5097.*

Fig. 203.
Clerodendrum aculeatum: (A, E) *Acevedo-Rdgz. 5140.* (B–D) *Acevedo-Rdgz. 2818.*
Duranta erecta: (F–I) *Ricksecker 57.*
Lantana involucrata: (J) *Acevedo-Rdgz. 5087.* (K, L) *Acevedo-Rdgz. 2741.* (M) *Acevedo-Rdgz. 2828.*

Fig. 204.
Priva lappulacea: (A–F) *Acevedo-Rdgz. 2825.*
Stachytarpheta jamaicensis: (G–J) *Acevedo-Rdgz. 2739.*
Vitex divaricata: (K–M) *Proctor 17797.* (N) *Little 13275.*

Fig. 205.
Cissus verticillata: (A, D) *Acevedo-Rdgz. 5114.* (B, C) *Acevedo-Rdgz. 4685.*
Cissus trifoliata: (E) *Acevedo-Rdgz. 5127.* (F) *Acevedo-Rdgz. 2693.*
Vitis tiliifolia: (G, J) *Augusto 1574.* (H, I) *Taylor 8086.*

Fig. 206.
Guaiacum officinale: (A) Field sketch. (B, C) *Croat 61041.* (D, E) Field sketch.
Kallstroemia pubescens: (F–L) *Ricksecker 138.*

Fig. 207.
Agave missionum: (A–D) Photo. (E) *Dewey s.n.*
Yucca aloifolia: (F–H) *Acevedo-Rdgz. 2008.*

Fig. 208.
Crinum zeylanicum: (A, B) *Acevedo-Rdgz. 5712.*
Hymenocallis caribaea: (C, D) *Acevedo-Rdgz. 5730.*

Fig. 209.
Anthurium cordatum: (A, C–G) *Acevedo-Rdgz. 2863.* (B) *Mori 16998.*
Dieffenbachia seguine: (H, I) *Liogier 31015.*

Fig. 210.
Pistia stratiotes: (A–C) *Acevedo-Rdgz. 2866.*
Philodendron scandens: (D, E) *Acevedo-Rdgz. 2680.*

Fig. 211.
Syngonium podophyllum: (A, B) *Mori 15131.* (C) *Acevedo-Rdgz. 4061.*

Fig. 212.
Coccothrinax alta: (A) Field sketch. (B–D) *Acevedo-Rdgz. 1911.*
Roystonea borinquena: (E) Photo. (F) *Britton 348.* (G) *Britton 1813.*

Fig. 213.
Aloe vera: (A) Photo. (B–D) *Abbott 915a.*

Fig. 214.
Aechmea lingulata: (A, B, F) *Acevedo-Rdgz. 5096.* (C–E) *Acevedo-Rdgz. 615.*
Bromelia pinguin: (G) *Acevedo-Rdgz. 2798.* (H) *Acevedo-Rdgz. 1979.* (I, J) *Acevedo-Rdgz. 2798.*

Fig. 215.
Pitcairnea angustifolia: (A–C) Field sketch.
Catopsis floribunda: (D, E) *Britton 572.* (F) *Acevedo-Rdgz. 4248.*
Tillandsia utriculata: (G) *Acevedo-Rdgz. 1971.* (H) *Acevedo-Rdgz. 2496.* (I) Pickled *s.n.*

Fig. 216.
Canna indica: (A-F) *Mori & Gracie 18429.*

Fig. 217.
Callisia fragrans: (A–E) *Acevedo-Rdgz. 3870.*
Commelina erecta: (F, I) *Acevedo-Rdgz. 859.* (G, H) Photo.
Tradescantia zebrina: (J–L) *Allard 13731.*

Fig. 218.
Syringodium filiforme: (A, B) *Fosberg 56809.* (C, D) *Liogier 14410.* (E) *Wright 3719.*

Fig. 219.
Abildgaardia ovata: (A–C) *Stevenson 5655.* (D) *Jiménez 2969.*
Bulbostylis pauciflora: (E–H) *Ekman 9069.*

Fig. 220.
Cyperus ligularis: (A–D) *Acevedo-Rdgz. 856.*
Cyperus nanus: (E) *Acevedo-Rdgz. 2055.* (F–H) *Proctor 46512.*
Cyperus elegans: (I-K) *Sargent 678.*

Fig. 221.
Eleocharis geniculata: (A–E) *Acevedo-Rdgz. 794.*
Fimbristylis dichotoma: (F–J) *Mori & Gracie 18405.*
Kyllinga odorata: (K–N) *Blake 7438.*

Fig. 222.
Rhynchospora nervosa ssp. *ciliata:* (A–D) *Heller 11.*
Scleria lithosperma: (E–J) *Acevedo-Rdgz. 2500.*
Scleria scindens: (K, L) *Ernst 1577.*

Fig. 223.
Dioscorea pilosiuscula: (A–E) *Acevedo-Rdgz. 3991.*

Fig. 224.
Sansevieria trifasciata: (A) Field sketch. (B, C) *Acevedo-Rdgz. 3859.*
Thalassia testudinum: (D–H) *Acevedo-Rdgz. 2732* (pickled).

Fig. 225.
Lemna aequinoctialis: (A, B) *Acevedo-Rdgz. 5260.*
Hypoxis decumbens: (C) *Heller 182.* (D, E) *Underwood 977.* (F, G) *Stevenson s.n.*

Fig. 226.
Cyclopogon elatus: (A–E) *Britton 4459.*
Tetramicra caniculata: (F–J) *Acevedo-Rdgz. 4060.*

Fig. 227.
Epidendrum ciliare: (A) Life plant from New York Botanical Garden propagation house. (B, C) *Acevedo-Rdgz. 2279.* (D, E) *Acevedo-Rdgz. 5100.*
Oeceoclades maculata: (F–J) *Axelrod 594.* (K) *Acevedo-Rdgz. 3236.*

Fig. 228.
Ponthieva racemosa: (A) *Britton 7524.* (B–D) *Proctor 41051.* (E) *Urban 3994.*

Fig. 229.
Psychilis macconnelliae: (A–D) Field sketch.
Prescottia stachyodes: (E–I) *Acevedo-Rdgz. 2604, Britton 2131.*

Fig. 230.
Tolumnia prionochila: (A, C–E) Field sketch + pickled *s.n.* (B) *Acevedo-Rdgz. 758.*
Vanilla barbellata: (F) *Liogier 15640.* (G) *Liogier 15047.*

Fig. 231.
Andropogon bicornis: (A–C) *Mori & Gracie 18388.*
Anthephora hermaphrodita: (D–G) *Ricksecker 253.*
Aristida cognata: (H, I) *Eggers s.n.*

Fig. 232.
Arthrostylidium farctum: (A–C) *Ekman 9286.*
Axonopus compressus: (D–F) *Hitchcock 16308.*
Bothriochloa pertusa: (G–I) *Acevedo-Rdgz. 2302.*

Fig. 233.
Bouteloua americana: (A–C) *Eggers s.n.*
Cenchrus brownei: (D–F) *Lewis s.n.*
Chloris barbata: (G–J) *Acevedo-Rdgz. 2618.*

Fig. 234.
Cynodon dactylon: (A–C) *Kemp 46.*
Digitaria ciliaris: (D–F) *Mori 17087.*
Echinochloa colona: (G–I) *Ricksecker 31.*

Fig. 235.
Eleusine indica: (A–C) *Acevedo-Rdgz. 3803.*
Eragrostis ciliaris: (D, E) *Ricksecker 33.*
Eriochloa punctata: (F–H) *Ricksecker 467.*

Fig. 236.
Heteropogon contortus: (A, B) *Hitchcock 9839.*
Lasiacis divaricata: (C–E) *Acevedo-Rdgz. 2681.*
Leptochloa virgata: (F–H) *Chase 6701.*

Fig. 237.
Oplismenus hirtellus: (A–E) *Mori et al. 18630.*
Olyra latifolia: (F–I) *Mori & Gracie 18848.*

Fig. 238.
Panicum diffusum: (A–C) *Eggers s.n.*
Paspalidium geminatum: (D–F) *Thomson s.n.*
Paspalum conjugatum: (G–I) *Ricksecker 223.*

Fig. 239.
Pharus lappulaceus: (A) *Acevedo-Rdgz. 2064.* (B, C) *Prance 29338.*
Pennisetum clandestinum: (D–F) *Asplund 10712.*
Melinis repens: (G, H) *Acevedo-Rdgz. 4106.*

Fig. 240.
Schizachyrium sanguineum: (A–C) *Hitckcock 10194.*
Spartina patens: (D–F) *Fishlock 57.*
Setaria setosa: (G–J) *Brown 313.*

Fig. 241.
Sporobolus indicus: (A) *Hitchcock 16314.* (B, C) *Acevedo-Rdgz. 2774.*
Tragus berteronianus: (D–H) *Acevedo-Rdgz. 2452.*
Urochloa fasciculatum: (I–M) *Mori 17074.*

Fig. 242.
Ruppia maritima: (A–E) *Shafer 2849.* (F) *Acevedo-Rdgz. 2421.*
Smilax coriacea: (G) *Martorell s.n.* (H, K) *Fuertes 444.* (I) *Britton 1376.* (J) *Liogier 35155.* (L) *Acevedo-Rdgz. 3818.*

Note: Field sketches were made by B. Angell on St. John. Pickled collections and photos were made by P. Acevedo-Rodríguez on St. John.

Glossary

abaxial Situated facing away from the axis of the plant, as the undersurface of a leaf. *Syn.* dorsal. *Ant.* ventral or adaxial.

acaulescent Stemless, or apparently so; sometimes the short stem is underground or protrudes only slightly.

accrescent Increasing in size with age, especially any increase in calyx size after pollination.

achene A simple, dry, one-celled, one-seeded, indehiscent fruit; seed coat is not attached to the pericarp.

acicular Needle-shaped.

acroscopic Facing toward the apex.

actinomorphic Having flowers that are radially symmetrical, e.g., a star pattern, capable of bisection into identical halves forming mirror images. *Syn.* regular symmetry, radial symmetry. *Ant.* zygomorphic.

acumen A tapering point.

acuminate Having a long, slender, sharp point with a terminal angle less than 45°.

acute Sharp-pointed; margins straight to convex, forming a terminal angle between 45° and 90°.

adaxial The side toward the axis of the plant; the surface of the leaf that faces the stem during development; upper side of the leaf. *Syn.* ventral. *Ant.* dorsal or abaxial.

adnate Having united unlike parts; organically united or fused with another dissimilar part, e.g., ovary to calyx tube or stamens to petals. *Contrast* connate.

adventitious As applied to roots, arising from some part of a plant other than the base of the stem or other roots.

allomorphic Existing in more than one form. *Ant.* heteromorphic.

ament A deciduous spike of apetalous pistillate or staminate flowers, usually bracteate, as in *Cecropia*.

amentiferous Amentlike, bearing an ament.

anadromous Refers to a fern frond in which the first secondary division (or vein) of all the pinnae is on the acroscopic side; also the type of venation pattern in which the first set of nerves in each segment of the frond is given off on the upper side of the midrib toward the apex.

anastomosing Pertaining to leaf veins that unite repeatedly to form a network; reticulate.

androecium A collective term referring to the stamens of a flower; the stamens as a unit of the flower; the male portion of a flower.

androgynophore A stalk bearing both stamens and pistil above the point of perianth attachment.

andromonoecious A plant with staminate and perfect flowers, but lacking pistillate flowers.

androphore A stalk supporting the androecium or stamens.

annual A plant that completes its life cycle within one year or one growing season; i.e., it grows vegetatively, produces flowers, and sets seed in one season.

annular In a ring or ring-shaped; forming a ring or circle.

annulus In some ferns, a specialized ring of thick-walled cells extending from the stalk over the sporangium.

anther The part of a stamen that forms the pollen.

anthesis The time at which a flower comes into full bloom; when a flower is fully expanded.

anthocarp A dry, indehiscent fruit formed by the union of the perianth, or part of it, with the fruit itself, as in Nyctaginaceae.

antipetalous Opposite the petals, as in stamens that are situated in front of petals.

antrorse Bent or directed upward or forward. *Ant.* retrorse.

apical Pertaining to the tip or apex of any structure.

apiculate With a short, abrupt, or acute point.

apiculum A short, sharp, flexible point.

apocarpous The condition in which carpels are separate and not connate.

appendage An attached subsidiary or secondary part, as a projecting or hanging organ.

appendiculate Having a small appendage, as a stamen.

appressed Lying flat against another structure.

arborescent Of treelike habit; resembling a tree in growth or appearance.

arcuate Moderately curved or arched, like a bow.

areolate Divided into many angular or squarish spaces.

areole A small spine-bearing area in most members of Cactaceae.

aril A pulpy or fleshy outer covering of some seeds.

arillate Having an aril. *Ant.* exarillate.

arillode *See* aril.

aristate Bearing a stiff bristle-like awn or seta; tapered to a narrow elongated apex.

assurgent Ascending, rising; growing upward at an angle, but not erect.

attenuate Narrowly tapering; drawn out to a long, slender point, more gradual than acuminate.

auricle 1. An ear-shaped appendage. 2. Found on one or both sides of the summit of the leaf sheath in some grasses (Poaceae). 3. In orchids (Orchidaceae), a small lateral outgrowth on the anther.

auriculate Having a basal lobe; bearing auricles.

autogamous Self-fertilizing.

autotrophic Refers to organisms that are able to manufacture all of their own food from organic compounds.

awn 1. A terminal or dorsal, slender, often rigid or stiff bristle or bristle-like appendage. 2. In grasses (Poaceae), a bristle-like appendage borne on a spikelet bract; most commonly a prolongation of the midvein of the lemma; glumes may also be awned.

awned With awns.

axil The angle formed between the axis and any organ or other appendage attached to it, e.g., a stem and a leaf petiole.

axile Of, belonging to, or located in the axis; central in position.

axis The main or central line of development of any plant or organ. In those inflorescences consisting of a group of racemes, refers to the central stem on which the racemes are arranged (*see* rachis).

barbate Bearded with long stiff hairs, usually in a tuft, line, or zone.

barbellate With short stiff hairs.

basal At or near the base of an organ or part.

basifixed Referring to a structure attached or fixed at its base to a support.

basionym The name, replaced by another, making use of the same epithet, as a result of a change in position and/or rank of the taxon to which it refers.

basiscopic Toward the basal or proximal (as opposed to distal) end; facing basally.

bi- A prefix meaning two or twice, as in bipinnate (twice-pinnate).

bibracteate Having two bracts; two-bracteate.

bicarinate Having two ridges or keels.

bicarpellate Composed of two carpels.

bicolor, bicolorous Two-colored.

bifid Forked; two-cleft; divided into two parts or lobes.

bifurcate Divided into two forks or branches; Y-shaped.

bilabiate Having two lips, usually applied to the corolla or calyx.

bilateral Having two equal sides; two-sided; arranged on two sides.

bilobed   Two-lobed.

bilocular   Two-celled, or with two locules.

biovular   Containing two ovules.

bipinnate   Doubly or twice-pinnate; the condition of a leaf in which both primary and secondary divisions are pinnate.

bipinnatifid   More or less deeply cut in a bipinnate fashion.

biseriate   In two whorls or cycles; in two rows or series, as a perianth composed of a calyx and a corolla.

bisexual   Having both stamens and pistils, possessing perfect, i.e., hermaphroditic, flowers. *Compare* unisexual.

biternate   Arranged in three groups of three; ternate, with each division itself again ternate.

blade   1. The expanded portion of the leaf and often referred to as the lamina. 2. In monocots, e.g., sedges (Cyperaceae) and grasses (Poaceae), the portion of the leaf located above the sheath, ligule(s), and pseudopetiole (if present); often flattened, and usually broader than the sheath.

bract   1. Small, modified, usually bladeless leaves. The modified leaves intermediate in characteristics between the calyx and the normal leaves. 2. In grasses (Poaceae), used with respect to a spikelet, may refer generically to the glumes, lemmas, and paleas.

bracteal   Pertaining to a bract.

bracteole   A secondary bract, often very small.

bulb   A short, usually globose underground stem, covered by fleshy overlapping leaf bases or scales that function as storage organs.

bulbiferous   Producing bulbs.

bulbil   A small vegetative propagule that develops in a leaf axil, inflorescence, or other unusual location.

bullate   Having a blistered, swollen, or puckered appearance.

callose   Forming callosities, hardened spots, or protuberances.

callosity   A leathery or hard thickening of part of an organ.

callus   1. A tissue composed of large thin-walled parenchyma cells that develop on or below a wounded surface, often resulting in a firm thickening or protuberance. 2. In orchids (Orchidaceae), a crest or fleshy outgrowth of the lip. 3. In grasses (Poaceae), the hardened, often pointed base of a floret (e.g., *Aristida*) or a spikelet (*Heteropogon*) with any associated structures (*Paratheria*).

calyptra   A hood- or cap-shaped calyx or corolla The calyptrate calyx splits open into irregular segments; the calyptrate corolla usually falls as a unit.

calyptrate   Having a calyptra.

campanulate   Bell-shaped; with a flaring tube about as broad as long and with a flaring limb or lobes.

canaliculate   Channeled or grooved longitudinally.

canescent   Covered with dense, fine, grayish white hairs; becoming hoary, usually with a gray pubescence.

capitate   Formed like a head; aggregated into very dense clusters or heads.

capsule   A dry, dehiscent fruit derived from a compound ovary of two or more carpels.

carinate   Having a ridge or keel.

carpel   In angiosperms, the ovule-bearing structure of a flower; usually consists of an ovary, a style, and a stigma. Often several may be fused into a single ovary.

carpellodes   The reduced carpels occurring in some staminate flowers.

cartilaginous   Like cartilage in texture; firm and elastic.

caruncle   A wart or protuberance near the hilum of a seed.

carunculate   Having a caruncle.

caryopsis   A dry, indehiscent, one-seeded fruit, typical of the grasses (Poaceae), in which the pericarp is fused to the seed coat.

castaneous   Chestnut-colored; dark brown.

cataphyll   1. Small scale leaves of the rhizomes of monocots and dicots. 2. Protective bud-scales of shrubs and trees.

caudate   Bearing a slender terminal or basal tail or tail-like appendage.

caudex   A thickened, often woody, vertical, perennial stem, such as in palms or tree ferns.

caudicle   The strap-shaped stalk that connects a mass of pollen (pollinium) to the stigma in many orchids.

cauliflorous   Having flowers arising from axillary buds on leafless regions of the main stem or older branches. *Contrast* ramiflorous.

cauline   Belonging to the stem or arising from it.

centrifugal   In inflorescences, having the youngest flowers toward the outside.

cephalium   A woody enlargement at the apex of the stem in some Cactaceae, from which the flowers appear.

cespitose   Matted, growing in tufts or small dense clumps.

chartaceous   Papery in texture, opaque and thin.

chasmogamous   Referring to pollination that takes place in open flowers. *Ant.* cleistogamous.

chlorophyll   The green pigments found in the thylakoids of chloroplasts, which are essential for the utilization of light energy in photosynthesis.

ciliate   Fringed with conspicuous hairs along the margin.

ciliolate   Diminutive of ciliate.

cincinnus (pl., cincinni)   A tight, unilateral scorpioid cyme.

circumscissile   Opening or dehiscing by a line around a fruit or anther, the valve coming off as a lid.

clathrate   Lattice-like.

clavate   Club-shaped; like a baseball bat.

claviform   See clavate.

claw   The long narrow petiole-like base or limb of the petals or sepals in some flowers.

cleistogamous   Having flowers that remain unopened that undergo self-fertilization. *Ant.* chasmogamous.

clinandrium   The anther bed; that portion of the column under or surrounding the anther.

clonal   Propagated from a bud; *see* clone.

clone   A group of genetically identical individuals resulting from asexual, vegetative multiplication, i.e., by mitosis; any plant that has propagated vegetatively and therefore is considered a genetic duplicate of its parent.

coccus (pl., cocci)   A berry; in particular, one of the parts of a lobed, sometimes leathery or even dry fruit with one-seeded cells.

cochleate   Coiled like a snail shell.

collar   The outer (abaxial) surface of the junction of the leaf sheath and blade.

colleter   A multicellular glandular appendage that secretes a sticky substance.

columella   The central axis or stalk of a fruit (schizocarpic or capsular) that remains after dehiscence of the fruit.

column   1. In the orchids (Orchidaceae), the central structure of the flower made up of the style and the filaments of one or more anthers. *Syn.* gynostemium. 2. In the grasses (Poaceae), the basal, twisted portion of an awn.

commissure   The place of joining or meeting.

comose   Hairy; having a tuft of hairs.

concolorous   Similarly colored on both sides.

conduplicate   Folded together lengthwise.

conjugate Coupled; as a pinnate leaf of two leaflets.

conical Cone-shaped.

connate United or fused with another like part or organ. *Contrast* adnate.

connivent Coming into contact or converging, often clinging together by hairs.

conspecific Within or belonging to the same species.

contraligule A short, prolonged appendage opposite the leaf blade at the apex of the inner band of the leaf sheath in some sedges (Cyperaceae).

convolute Rolled up longitudinally and usually twisted apically.

cordate, cordiform Heart-shaped.

coriaceous Thick, tough, and leathery.

corm A bulblike fleshy stem or base of stem.

corona 1. A crown. 2. Any appendage between the corolla and stamens; may be petaloid or staminal in origin, as in the staminal corona in Asclepiadaceae.

corymb A flat-topped or convex, indeterminate, racemose inflorescence. The lower or outer pedicels longer and their flowers opening first.

corymbiform, corymboid, corymbose Arranged in corymbs.

costa (pl., costae) A midrib. In ferns, sometimes applied to the vascular axis of the blade in simple fronds or more often to the axis of a primary pinna. *See* costule.

costal, costular Referring to or having a costa.

costate Conspicuously and coarsely ribbed.

costule The axis or midvein of a pinnule.

crenate Referring to a margin with scalloped or rounded teeth.

crenulate Minutely or finely crenate.

cristate Crested or ridged.

crustaceous Hard, thin, and brittle.

culm The main stem of sedges (Cyperaceae), grasses (Poaceae), and other monocots.

culm leaves In the Bambusoideae and some bamboolike genera such as *Lasiacis* and *Olyra* (Poaceae), the leaves found along the main culm. At maturity they are usually nonphotosynthetic, with broad sheaths and small triangular blades.

cuneate, cuneiform Wedge-shaped, triangular.

cupula A little cup; the cup-shaped receptacle on which the hypogynium and ovary of *Scleria* (Cyperaceae) are attached.

cupule A cuplike structure at the base of some fruits.

cupuliform Nearly hemispherical like a cup.

cusp An abrupt, sharp, often rigid point.

cuspidate With an apex somewhat abruptly and sharply constricted into an elongated sharp-pointed tip or cusp.

cyathium (pl., cyathia) A cuplike structure (involucre) enclosing minute flowers in the Euphorbiaceae; the cyathium usually looks like a flower.

cyme A type of inflorescence consisting of a broad, more or less flat-topped, determinate flower cluster, with the central flowers opening first.

cymose, cymosely Having a cyme; as a cyme.

cymule Diminutive of cyme, usually few-flowered.

cystoliths Intercellular concretions that develop within the epidermal cells of certain plants; usually composed of calcium carbonate, appearing as whitish, minute lines on the surface of leaves; typical of Acanthaceae and Urticaceae.

declinate Bent or directed downward or forward.

decompound A general term for leaves that are more than once divided or compound.

decumbent Reclining, but with the summit ascending.

decurrent Extending down a stem or rachis below the point of attachment. *Compare* excurrent.

decurved Bent down, deflexed, curved downward.

decussate Opposite leaves alternating at right angles with those above and below.

deflexed Reflexed; bent or turned abruptly downward.

dehiscent Splitting open at maturity releasing contents.

deltate, deltoid Shaped like an equilateral triangle.

dendritic, dendroid Having a branched appearance; treelike.

dentate Referring to a leaf margin with sharp teeth pointing outward at right angles to the midrib.

denticulate Minutely or finely dentate.

determinate growth Growth of limited duration, characteristic of floral meristems and leaves; when the season's growth ends with a bud. *Compare* indeterminate growth.

dewlap A triangular flange of frequently yellowish or darkened tissue found on both sides of the junction of the leaf sheath and blade in grasses (Poaceae).

diadelphous With two groups of stamens; describing stamens united by their filaments into two bundles or clusters. *Compare* monadelphous.

diaphragmal pith A solid core of pith cells transversed by distinct layers (diaphragms) of firm-walled cells, forming small compartments in the pith.

dichasium (pl., dichasia) A determinate type of cymose inflorescence having a central older flower that develops first and a pair of opposite lateral branches bearing younger flowers.

dichotomous Branching by repeated, more or less equal, forking in pairs.

didynamous With four stamens in two pairs of two different (unequal) lengths, as most members of Lamiaceae.

digitate Fingerlike; shaped like an open hand; palmate.

dimorphic Occurring in two distinct forms, sizes or shapes within the same species; e.g., having two forms of leaves, juvenile and adult, or having two kinds of pollen produced by the same plant. *Compare* monomorphic.

dioecious Unisexual, the male and female floral elements on different plants. *Compare* monoecious.

disarticulate To separate at a preexisting joint; e.g., disarticulation above or below the glumes in grasses (Poaceae).

disciform Having heads with all flowers regular, but with outermost pistillate.

discolor, discolorous Unlike in color, used most often in reference to the two surfaces of leaves.

distal Opposite from the point of origin or attachment; toward the apex. *Compare* proximal.

distichous Two-ranked; referring to leaves, bracts, or scales that are alternate in the same plane.

distylous Referring to the flowers of a species which possess one of two style types: a long style ("pin" flowers) or a short style ("thrum" flowers). E.g., *Cordia sebestena*. These function as a mechanism to promote outcrossing. *Compare* homostylous.

divaricate Widely spreading.

domatia Depressions or small pockets (usually hairy) at the axil of leaf veins.

dorsal The surface turned away from the axis; the lower or undersurface of a leaf. *Syn.* abaxial. *Ant.* ventral or adaxial.

dorsifixed Attached to the back or dorsal surface.

dorsiventral, dorsiventral symmetry Flattened and having distinct dorsal and ventral surfaces, as a leaf.

drupe A fruit, such as a plum, with a fleshy or leathery exocarp or mesocarp and a stony endocarp (stone) that contains one or more seeds.

drupelet 1. One drupe of a fruit composed of aggregate drupes, as in blackberries and raspberries. 2. A diminutive of drupe.

e- A prefix indicating lack of a structure or characteristic.

echinate Covered with spines, prickly.

eglandular Having no glands.

eligulate Having no ligule.

emarginate Having a notch cut out at the apex.

epicalyx A whorl of bracts adjacent to and resembling a true calyx.

epidermis The outer cellular layer of the primary tissues of a plant, e.g., of leaves, young stems, and young roots. Usually consists of one layer.

epilithic Growing on rocks.

epipeltate Having the base of the limb on a superior face.

epipetalous Having stamens attached to or inserted on the petals or corolla.

epiphytic Relating to an organism that grows on another plant but is not parasitic on it.

equitant Leaves two-ranked, with overlapping bases, usually sharply folded lengthwise along midrib.

erose Having a margin that is irregularly or shallowly toothed and/or lobed; appearing gnawed or jagged.

evanescent Disappearing quickly, lasting only a short time.

exarillate Without an aril.

excurrent Extending beyond the apex or margin of a leaf into a mucro or awn. *Compare* decurrent.

exocarp The outermost layer of the pericarp or fruit wall.

exserted Projected beyond, stuck out, or protruding.

extra- A prefix meaning outside of or beyond.

extrafloral Situated outside the flower.

extrastaminal Outside the whorls of stamens. *Compare* interstaminal *and* intrastaminal.

extrorse Opening or facing outward; as with anthers that dehisce toward the outside of the flower. *Ant.* introrse.

exudate Fluids excreted or produced by a plant.

falcate Sickle-shaped.

farinaceous, farinose Covered with a mealy powder; mealy in texture. Containing starch or starchlike material.

fascicle A close cluster or bundle of flowers, leaves, stems, or roots.

fasciculate Clustered or bundled, originating from a common point.

ferruginous Rust-colored.

fibrillose Covered with little fibers.

-fid A suffix meaning deeply cut or cleft.

filamentiferous Bearing filaments; in a staminal column, the stamen filaments that are free toward the upper portion.

filamentous With a threadlike cellular growth form; elongate, formed of filaments or fibers.

filiferous With coarse marginal fibers or threads; thread-bearing, filamentous.

filiform Threadlike.

fimbria (pl., fimbriae) A fringe or fringelike border.

fimbriate Having a fimbria.

fimbriolate Minutely fimbriate.

flexuous Bent alternately in opposite directions; zigzag.

floret In the grasses (Poaceae), the lemma and palea together with the enclosed flower; more accurately, the lemma plus the tiny branchlet it subtends, this branchlet bearing the palea as the first appendage and terminating in a flower. Spikelets with *x* number of florets are referred to as *x*-flowered.

floriferous Bearing or producing flowers.

flower 1. The specialized reproductive structure of the angiosperms. 2. In the sedges (Cyperaceae), includes the bristles or hypogynium (if present), androecium, and gynoecium. 3. In the grasses (Poaceae), includes the lodicules, androecium, and gynoecium.

foliaceous Having the texture or shape of a leaf; leaflike.

foliage leaves In the Bambusoideae (Poaceae), the leaves found along the ultimate branches. They are photosynthetic, with narrow sheaths and usually broad blades.

foliar Pertaining to leaves or leaflike parts.

foliate Having leaves or leaflike structures.

foliolate Having leaflets.

foliole A leaflet, the secondary division of a compound leaf.

follicle A dry, dehiscent fruit derived from one carpel that splits along one suture; any small saclike cavity.

-form A suffix meaning having the form of; shaped like.

foveate Pitted.

foveolate Marked with small pits.

free veins Veins not uniting to form a network.

friable Easily crumbled or crushed into powder.

frond The foliage of ferns or palms or other similar flattened leaflike structures.

funiculus The basal stalk of an ovule arising from the placenta in the angiosperms.

furcate Forked.

furfuraceous Covered with soft scales. *Syn.* lepidote.

fusiform Spindle-shaped.

galeate Helmet-shaped.

gametangium (pl., gametangia) Any cell or organ that produces gametes.

gametophyte Inconspicuous structure that represents one of the stages in the life cycle of ferns and fern allies that bears male and female sex organs and sex cells (gametes).

gamopetalous Having petals that are partly or completely connate.

gamosepalous Having sepals that are partly or completely connate.

geniculate Abruptly bent at a joint, like a bent knee.

geophytic Producing underground buds or organs, such as bulbs or rhizomes.

gibbose Enlarged or swollen unequally, that is, more on one side than the other; inflated on one side near the middle.

glabrate Nearly or becoming glabrous at maturity.

glabrescent *See* glabrate.

glabrous Lacking hairs.

gland A multicellular secretory structure on or near the plant surface, e.g., leaf gland, glandular hair, or nectary.

glaucous Covered with a waxy coating that gives the surface a whitish or bluish cast.

globose Spherical or rounded.

glochid A minute barbed bristle or hair; often in tufts, as in certain cacti.

glochidiate Pubescent with trichomes barbed at the tip.

glomerate In dense or compact clusters or heads.

glomerulate Arranged in clusters.

glomerule A small dense indeterminate cluster of more or less sessile flowers.

glumes The sterile lowest bracts of a spikelet in the grasses (Poaceae). Most grasses have two glumes, but some lack the lower glume, and in a few genera (e.g., *Leersia*) both are absent. In some Bambusoideae, more than two are present and are sometimes referred to as "transitional glumes."

granular Finely covered with very small mealy granules.

granule Any small particles, e.g., pollen grains.

gynobasic Applied to a style that originates between the lobes of a deeply lobed ovary as in Lamiaceae and Boraginaceae.

gynoecium A collective term for all the carpels (female parts) of a flower.

gynophore A stalk or stipe bearing an ovary or fruit.

gynostegium A protective covering for the gynoecium, in the Asclepiadaceae consisting of a central column formed by filaments, anthers, and the thickened style head.

gynostemium The united pistil and stamens in orchids. *Syn.* column.

hastate Shaped like an arrow head, with basal lobes turned outward.

haustorium (pl., haustoria) A specialized root of parasitic plants capable of penetrating and absorbing food and other materials from host tissue.

head A dense inflorescence composed of a determinate or indeterminate group of sessile or subsessile flowers on a compound receptacle.

helicoid Curved or spiraled like a snail shell.

hemiepiphyte A plant whose seeds germinate on another plant but later send roots to the ground.

hemiparasitic Parasitic, but containing chlorophyll and so partly self-sustaining.

herb A nonwoody plant (annual, biennial, or perennial) whose aerial portion does not persist after the end of the growing season.

herbaceous With the texture, color, and other characteristics of an herb.

hesperidium A type of berry having a leathery pericarp, as in citrus fruits (Rutaceae; e.g., oranges and lemons).

heterogamous Bearing two or more kinds of flowers. *Compare* homogamous.

heteromorphic Morphologically different; with different forms; polymorphic. *Ant.* allomorphic.

heterostylous Having flowers with styles differing in length or shape.

hilar Relating to the hilum.

hilum The scar left on a seed where it was formerly attached to the funiculus.

hirsute Covered with long distinct hairs.

hirtellous Minutely hirsute.

hispid Covered with rough hairs or bristles; usually stiff enough to penetrate the skin.

hispidulous Somewhat or minutely hispid.

homogamous Having flowers all alike; having stamens and pistils mature at the same time. *Compare* heterogamous.

homosporous Having only one type of spore.

homostylous Having styles of the same lengths and/or shapes. *Compare* distylous.

homotypic Having the same nature or character; homologous.

hyaline Thin and translucent or transparent, like glass.

hybrid The offspring of two plants or animals of different breeds, races, forms, varieties, subspecies, species, or genera.

hydathode The enlarged tip of a vein; an epidermal structure, associated with a vein tip, supposedly for the secretion of water.

hygroscopic Readily absorbing and retaining moisture from the atmosphere; results in changes in form or position of certain cells or structures.

hypanthial Relating to the hypanthium.

hypanthium A floral tube formed by the fusion of the basal portions of the sepals, petals, and stamens and from which the rest of the floral parts emanate.

hypocarp An enlarged, fleshy growth of the peduncle beneath the fruit.

hypocotyl That portion of an embryo or seedling below the cotyledon and above the radicle; the embryonic stem in a seed.

hypogynium A perianth-like structure of bony scales subtending the ovary, as in the sedge genus *Scleria* (Cyperaceae).

hypogynous Having the flower parts attached below the base of the ovary and free from it; flowers with this arrangement have a superior ovary.

imbricate Overlapping, as the tiles on a roof.

imparipinnate Pinnate with an odd terminal leaflet. *Contrast* paripinnate.

incumbent In orchids, refers to an anther that bends downward during the development of the flower.

incurved Curved inward or upward.

indehiscent Remaining closed at maturity.

indeterminate growth Unrestricted growth, as with a vegetative apical meristem capable of producing an unlimited number of lateral organs. *Compare* determinate growth.

indument Any covering of a plant surface, such as hairs or scales.

indurate Hardened.

indusiate Possessing an indusium.

indusioid John Smith's expression for any indusium-like covering in ferns.

indusium (pl., indusia) The scale-like or membranous flap of tissue covering the sorus of many ferns.

inequilateral Asymmetrical, unequal-sided.

inferior ovary An ovary that is situated below the point of insertion of the other floral organs; one that is adnate to the hypanthium and situated below the calyx lobes.

inflexed Bent abruptly inward.

inflorescence The flowering part of a shoot; the floral structure in which the flowers are arranged.

infra- A prefix meaning below or beneath.

inframedial Below the middle.

infraspecific Referring to any taxon below the species level.

infructescence A mature inflorescence bearing fruit; a composite or confluent fruit.

inserted Growing upon or attached to.

intercostal Space between ribs or costae.

intergrade The gradual merging of one form into another by a series of intermediate forms, kinds, or types.

internode 1. The portion of the stem between two successive nodes. 2. In sedges (Cyperaceae) and grasses (Poaceae), that portion of a culm or other structure (e.g., the rachilla of the spikelet) separating two nodes.

interpetiolar Between the petioles; enclosed by the expanded base of a petiole; also applied to free or sometimes connate stipules between the petioles of two opposite leaves.

interstaminal Between stamens. *Compare* extrastaminal.

intra- A prefix meaning within.

intrapetiolar Within the petiole or between it and the stem.

introrse Turned inward toward the axis. *Ant.* extrorse.

invaginate Enclosed within a sheath; introverted; concave.

involucel A small involucre.

involucellar Of the involucre.

involucrate Having an involucre.

involucre 1. A ring of bracts surrounding several flowers or their supporting axes. 2. The bracts, usually in a whorl, subtending a flower cluster, as in heads of Asteraceae.

involute With the margins rolled or turned in over the upper, ventral, surface. *Ant.* revolute.

isodiametric With equal diameters; length and width are equal.

labellum In orchid flowers, the lower of the three petals, which is usually greatly modified and enlarged; the lip.

laciniate Slashed; cut into narrow pointed segments.

laminar Thin and flattened, as in a leaf blade.

lanate Woolly, cottony; covered with long, fine, intertwined trichomes (hairs).

lanceolate Lance-shaped; several times longer than wide, tapering at both ends, the wider part toward the base.

latex A usually milky-looking (although it may also be clear, yellow, orange, or brown), water-soluble liquid contained in laticifers.

leaf complement In the Bambusoideae (Poaceae), the collective term for a discrete group of foliage leaves.

lemma In the grass (Poaceae) floret, the lower of the two bracts that enclose the flower; usually with a well-developed midvein. Morphologically, a modified leaflike bract that subtends the tiny branchlet bearing the palea and flower.

lemmatal Of the lemma.

lenticel Spongy areas in the cork (phellum) surfaces of stems, roots, certain fruits, and other plant parts, which allow interchange of gases between internal tissues and the atmosphere.

lenticellate Bearing lenticels.

lenticular Lens-shaped.

lepidote Covered with small, scurfy scales; the scales are usually peltate. *Syn.* furfuraceous.

leptomorph A long, hard rhizome with indeterminate growth.

-let A suffix typically used in botany to express the diminutive (smaller size) of a structure or form (e.g., leaflet).

liana A large, woody, tropical climbing plant or vine.

ligneous Woody.

lignified Converted into wood.

ligulate Having a ligule.

ligulate-pseudobilabiate The condition of capitula in *Chaptalia* wherein the corollas are trimorphic, varying from outer ones obscurely ligulate to inner ones weakly bilabiate.

ligule 1. A strap- or tongue-shaped structure. 2. A membranous flange or hairy fringe on the adaxial leaf surface at the junction of sheath and blade in the grasses (Poaceae) and some sedges (Cyperaceae). In most grasses, only an **inner ligule,** found on the inner (adaxial) surface, is present; in descriptions in this treatment, *ligule* always refers to this appendage. The Bambusoideae also have a membranous external or **outer ligule** found on the outer (abaxial) surface.

liguliform Strap-shaped.

lineate Marked with parallel lines.

lineolate Marked with fine lines.

lip 1. One of the two divisions of a bilabiate corolla or calyx, i.e., one in an upper (superior) and one in a lower (inferior) portion as in Lamiaceae and Scrophulariaceae. 2. The labellum of the flower in orchids (Orchidaceae).

lobate Divided into lobes.

lobulate Divided into small lobes.

-locular A suffix meaning to have compartments, chambers, or cavities; divided into locules. A bilocular structure has two locules, a trilocular structure has three, etc.

locule A compartment, chamber, or cavity of the ovary.

loculicidal Splitting along the walls of the locules of a capsule, as contrasted with splitting along the septae. *Compare* septicidal.

lodicules In grass (Poaceae) florets, the minute tepals; bambusoid grasses often have three lodicules, while most other grasses have two.

loment, lomentum A legume that is contracted between the seeds and at maturity falls apart at the constrictions into one-seeded segments, e.g., *Desmodium.*

lunate Crescent-shaped; half-moon-shaped.

lyrate Lyre-shaped.

malpighiaceous Referring to the peculiar T-shaped hairs of many species of Malpighiaceae and other families.

mammillate Having nipples or teat-shaped processes.

marcescent 1. Withering but not falling off, as corollas and stamens in certain kinds of flowers. 2. Leaves of short duration, dying at the end of the growing season.

marginal Attached to or on the edge.

marginate Having a distinct margin.

medial Situated in the middle.

membranaceous, membranous Thin, soft, flexible, and more or less translucent.

mentum A chinlike extension at the base of the orchid flower, made up of the column foot and the lateral sepals.

mericarp 1. A portion of a fruit that seemingly matured and split away as a separate fruit. 2. One of the two seedlike carpels of fruits in Apiaceae.

meristem Nascent tissue generally localized within the plant body, capable of being transformed into special forms, as cambium, etc.

-merous A suffix indicating the number of parts.

mesocarp The middle layer of cells of the pericarp or fruit wall.

monadelphous Stamens connate by fusion of their filaments into a single group, forming a tube or column. *Compare* diadelphous.

monocarp Fruit formed from individual or seemingly individual carpels.

monocarpic Referring to a perennial or annual blooming and fruiting only once and then dying.

monocephalous Bearing a single head.

monochasia A cymose inflorescence with main axes producing one branch each.

monochasium A cyme reduced to a single flower on each axis.

monoecious Having both male and female reproductive structures on the same plant but in separate flowers. *Compare* dioecious.

monolete Pertaining to a fern spore with an unbranched fissure in its integral wall.

monomorphic 1. Producing spores of only one form or type. 2. Having the same size and shape. *Compare* dimorphic.

monopodial Branching from one primary axis. *Contrast* sympodial.

mucro or **mucron** A short, sharp or blunt, abrupt tip.

mucronate Possessing a sharp, straight tip.

mucronulate Diminutive of mucronate.

multi- A prefix meaning several to many.

multilocular Many-celled, as an ovary.

multiovular Having several to many ovules.

multiseriate Having many rows or many series. *Compare* uniseriate.

muricate Rough; covered with short hard projections.

muriculate Slightly rough.

muticous Pointless, blunt, awnless.

mycorrhiza A symbiotic, nonpathogenic association between fungi and roots of many types of plants.

mycotrophic Plants living symbiotically with fungi.

nectariferous Nectar-producing; having a nectary.

nectary Any structure that secretes nectar.

nervate Having veins.

node Of a stem or culm, the point of attachment of the branches or leaves.

nodose Having knots or being knobby.

nodulose Having small knobs or knots.

nutlet 1. A small nut; one of a group of small seeds. 2. The stone formed in drupaceous fruits.

ob- A prefix usually meaning in the reversed direction or inverted position.

obconic Inversely conical in outline, but attached at the narrower end.

obcordate Inversely heart-shaped in outline, with the notch at the apex.

obdeltate Inversely deltate in outline, broadest toward apex.

oblanceolate Inversely lanceolate in outline, broadest toward apex.

oblate Nearly spherical in outline, but compressed at the poles.

oblique Slanted; unequal-sided.

obovate Inversely ovate in outline, broadest above the middle.

obovoid Inversely ovoid three-dimensionally; roughly egg-shaped, with narrow end downward.

obtuse Blunt or rounded at the end, with an angle of greater than 90°.

ocrea A stipular sheath that encloses the base of a leaf.

ocreate Having a stipular sheath surrounding a stem above the insertion of a petiole or leaf.

ocreolate Having smaller or secondary sheaths.

operculate Covered by a lid or cap (operculum).

operculum A lid, cap, or cover.

oral setae Coarse, bristle-like appendages found at the summit of the leaf sheaths in the grass *Pariana* and most Bambusoideae (Poaceae).

orbicular, orbiculate More or less circular in outline or shape.

ovary The ovule-bearing region of a carpel in a simple pistil or of a gynoecium composed of connate carpels (compound ovary).

ovate Egg-shaped in outline.

ovulate Bearing or possessing ovules.

ovule The young seed in the ovary; the organ that, after fertilization, develops into a seed.

pachymorph A shortened thick and fleshy rhizome with determinate growth. *Ant.* leptomorph.

palea In the grass (Poaceae) floret, the upper of the two bracts that enclose the flower; usually two-veined, bicarinate, and lacking a midvein. Morphologically, the palea is the first appendage (prophyll) of the tiny branch that bears the flower; in the Asteraceae a receptacular bract.

paleate Having small membranous scales; chaffy.

palmate Having lobes, veins (nerves), or divisions radiating from a common point, as in palmately lobed, palmately veined, or palmately compound. *Contrast* pinnate.

palmately compound Referring to a compound leaf having leaflets radiating from a common point. *Syn.* digitate.

pandurate Fiddle-shaped.

panicle An indeterminate inflorescence with a branched main axis and pedicellate flowers borne on the secondary branches.

paniculate Having a panicle-like inflorescence.

paniculiform Panicle-shaped.

papilla (pl., papillae) A soft, nipple-like projection or protuberance.

papillate, papillose Bearing papillae.

papilliform Shaped like a papilla.

pappus A modified outer perianth borne on the ovary in the Asteraceae, being plumose, or of bristles, scales, awns, or otherwise.

paraphyses Sterile hairs of various shapes arising among sporangia.

parietal Borne on or belonging to a wall.

paripinnate Evenly pinnate, without a terminal leaflet. *Contrast* imparipinnate.

pectinate Comblike; having closely parallel, narrow, toothlike projections.

pedate Palmately divided or parted, with the lateral lobes cleft or divided.

pedicel 1. The stalk of an individual flower in an inflorescence or the stalk of a sedge (Cyperaceae) or grass (Poaceae) spikelet. 2. The stalk of sporangia in ferns.

pedicellate Borne on a pedicel.

peduncle The stalk of an inflorescence or the stalk of a solitary flower.

peduncled Having a peduncle.

pellucid Nearly transparent.

peltate Umbrella or shield-shaped, with a stalk attached to the lower surface near the center of the structure.

penicillate Ending in a tuft of fine hairs or branches.

pentagonal, pentagonous Having five angles.

pepo A modified berry with a leathery nonseptate rind. Derived from an inferior ovary. Characteristic fruit of the cucumber family.

percurrent Extending throughout the entire length.

perennial A plant that lives for more than 2 years; herbaceous perennials have stems and/or leaves that are produced and die back annually, with their underground stems and/or roots remaining alive; woody perennials, e.g., trees and shrubs, have aerial stems that may live for many years.

perianth A collective term for the floral envelope, usually the combined calyx and corolla, or tepals of a flower.

pericarp The mature fruit wall that develops from the ovary wall, and frequently of several layers: exocarp, mesocarp, and endocarp.

perigynous Borne around the gynoecium.

perispore A wrinkled outer covering of some fern spores.

persistent Remaining attached; not falling free.

petaloid Resembling a petal in shape, texture, and/or color.

petiolar Pertaining to a petiole.

petiole The stem of a leaf; the stalk attaching the leaf to the stem.

petiolulate Having a petiolule.

petiolule The stalk of a leaflet.

phloem The principal food-conducting tissue of a vascular plant, that usually composed of sieve elements, various kinds of parenchyma cells, fibers, and sclereids.

phyllary An involucral bract.

phylloclade A flattened branch assuming the form and function of foliage.

pilose Having long, soft hairs.

pilosulous Diminutive of pilose.

pinna (pl., pinnae) A primary subdivision, or leaflet, of a compound leaf or frond. May be further subdivided into pinnules.

pinnate Shaped like a feather; having leaflets of a compound leaf arranged on opposite sides of a common axis. *Contrast* palmate.

pinnatifid. Pinnately cleft or divided; simple leaves lobed halfway to midrib.

pinnatisect Simple leaf of laminar structure pinnately cut to the midrib into distinct segments.

pinnule Subdivision of the pinna or leaflet of a compound leaf.

pistil The female reproductive organ of a flower, composed of an ovary, style, and stigma.

pistillate Having only pistils in a flower; a female flower lacking stamens. A functionally pistillate flower has fertile pistils but vestigial or rudimentary stamens (staminode).

pistillode A rudimentary or vestigial pistil present in some staminate flowers.

placentation The arrangement and distribution of placentae and ovules within the ovary.

plano-convex Planar on one side and convex on the other.

plantlet A small plant, usually produced by asexual reproduction. *See* bulbil.

plumose Covered with a fine, featherlike pubescence.

pluri- A prefix meaning several or many.

pluricellular Many-celled.

plurilocular Having several to many locules or cells, referring to the ovary.

pneumatophores Vertical extensions of the buried roots of certain trees (e.g., mangroves) that exist in marsh or swamp habitats.

pollinium (pl., pollinia) The structure formed when individual pollen grains remain massed together and are transported as a unit during pollination; a pollen mass.

polycarpic With many separate carpels; producing seed season after season, as in perennials.

polygamo-dioecious Plants functionally dioecious but having a few perfect flowers on otherwise staminate or pistillate plants.

polygamous With hermaphroditic and unisexual flowers on the same, or on different, individuals of the same species.

polymorphic Having several to many variable forms within the same species.

poricidal Opening or dehiscing by pores.

prickle A sharp-pointed epidermal or cortical outgrowth.

procumbent Trailing or lying on the ground without rooting at the nodes. *See also* prostrate.

prominulous Pertaining to leaf or frond veins that are slightly raised above the surrounding surface, forming minute ridges.

prophyll One of the first leaves of a lateral branch. A bracteole or small scale-like appendage.

prophyllate Having a prophyll.

prostrate Trailing on the ground with rooting at the nodes. *See also* procumbent.

protandrous Said of flowers in which the anthers mature and shed their pollen before the stigma of the same flower is receptive.

protogynous Said of the flowers in which the stigma becomes receptive prior to maturation of anthers and dehiscence of pollen in same flower.

proximal Closest to the point of attachment. *Compare* distal.

pseudanthium (pl., pseudanthia) Reduced cymose inflorescences that consist of both male and female flowers, all of them reduced to one stamen or one carpel, respectively.

pseudo- A prefix meaning false.

pseudobilabiate A tubular corolla of Asteraceae that is slightly zygomorphic and weakly bilabiate or appearing to be bilabiate.

pseudobulb In orchids (Orchidaceae), a solid, aboveground, thickened or bulblike stem.

pseudodichotomous Branching by repeated forking in pairs.

pseudopetiole The constricted portion of the leaf found between the ligule(s) and the blade; not well developed in most grasses, but characteristic of the Bambusoideae and other broad-bladed groups such as *Pharus*.

pseudoraceme Racemelike inflorescence, consisting of a central axis or stalk with flowers borne from swollen nodes.

pseudospikelet In the grass family, a group of indeterminate spikeletlike structures that develop by way of buds in their glumelike, sterile lower bracts.

pseudostaminodia Hairlike appendages that alternate with the filaments in the staminal column of many members of Amaranthaceae.

pseudostipule A rudimentary or aborted prophyll (the first leaf on a lateral branch) in the axil of a petiole, as in *Aristolochia*.

puberulent, puberulous Slightly hairy.

pubescent Covered with fine, soft hairs, often used to refer generally to the presence of hairs without specifying the nature of the hairs.

pulvinate Cushion-shaped or cushionlike.

pulvinulus The enlarged base of a petiolule.

pulvinus The enlarged base of a petiole; functions in the movement of a leaf or leaflet.

punctate Dotted with minute depressions or pits.

punctation Minute depression or pit.

puncticulate Minutely punctate.

punctiform In the shape of a point or dot.

pustulate Having pustules.

pustule A pimple or a blister.

pyrene The pit or seed of a drupe that is surrounded by a bony endocarp.

pyriform Pear-shaped.

raceme An unbranched, indeterminate inflorescence in which the individual flowers are borne on pedicels along the main axis. In grasses (Poaceae), refers to a spikelike inflorescence in which all the spikelets are borne on short pedicels; an individual inflorescence may consist of one to many racemes borne along a common axis (*see* rachis).

racemiform, racemose Having the form of a raceme.

rachilla The diminutive of rachis; in particular, the axis of the spikelet of sedges (Cyperaceae) and grasses (Poaceae) that bears the flowers or florets.

rachis The major axis of an inflorescence; the axis of a compound leaf or frond. In the grasses (Poaceae), in an inflorescence consisting of a series of racemes, the term is reserved for the axes of the racemes, and the main stem is termed the axis.

ramiflorous Flowering on the branches. *Contrast* cauliflorous.

raphides Needle-shaped crystals of calcium oxalate in the cells of plants.

ray 1. One of the primary branches of an umbel. 2. In the Asteraceae, a ray flower.

receptacle The expanded region at the end of a peduncle to which the floral parts are attached; the point on a leaf or thallus where reproductive organs are borne.

receptacular bracts Bracts that are attached to the receptacle.

reclining Leaning on other vegetation, leaning over.

recurved, recurvate Bent or curved downward or backward.

reflexed Abruptly bent or curved downward or backward.

reniform Kidney-shaped.

repand With slightly wavy or uneven margins. *See* undulate.

replum A framelike placenta from which the valves fall away in dehiscence.

resupinate So twisted that parts are upside down. In orchids, the flower having the lip on the lower side.

reticulate Forming a netted pattern or network.

retrorse Bent or directed downward or backward.

retuse Having a shallow notch at a rounded apex.

revolute With the margins rolled downward or toward the lower side. Ant. involute.

rhizoid A rootlike structure in function and appearance, but lacking vascular tissue.

rhizomatous Producing or bearing rhizomes; like a rhizome.

rhizome An underground, more or less horizontal stem.

rhombic Diamond-shaped.

rootlet A small root.

rosette A crowded, circular cluster of leaves or other organs; often in reference to a growth habit in which leaves radiate from a crown, close to the ground.

rostellum In orchids, a portion of the stigma that aids in gluing the pollinia to the pollinating agent; the tissue that separates the anther from the fertile stigma; sometimes beaklike.

rostrate Having a beak.

rosulate Having the form of or being in a rosette.

rotate Shaped like a wheel or saucer.

rufous Reddish brown.

rugose Wrinkled; covered with coarse reticulate lines.

rugulose Finely wrinkled.

saccate Shaped like a sac or pouch.

sagittate Shaped like an arrowhead.

salverform Shaped like a trumpet.

samara An indehiscent winged fruit.

samaroid Resembling a samara.

sarcotesta Softer fleshy outer portion of testa or outer coat of a seed.

scaberulous Slightly rough.

scabrid Roughened.

scabridulous Slightly rough; minutely scabrous.

scabrous Having a surface that is rough to the touch because of the presence of short stiff hairs.

scale 1. Any thin, usually small and dry, scarious to coriaceous bract. In the sedges (Cyperaceae), the bract that subtends the flower. 2. A usually disk-shaped trichome attached by a stalk.

scandent With climbing stems or branches.

scape A leafless floral axis or peduncle arising from the ground.

scapose In the form of a scape.

scar A mark left by the natural separation of one organ from another, e.g., as on a stem after abscission of a leaf, bud, flower, or fruit or on a seed after its detachment from a fruit.

scarious Thin, dry, and membranous, not green.

schizo- A prefix meaning split.

schizocarp A dry, dehiscent fruit that splits into two or more separate one-seeded portions (mericarps) at maturity.

schizocarpic Having a schizocarp.

sclerophyllous Having hard and stiff leaves.

scorpioid Coiled like the tail of a scorpion.

scrub Vegetation characterized by stunted trees or shrubs.

scurfy Having a minute scaly-powdery covering; minutely scaly-roughened.

secund Arranged on one side by twisting or torsion; often used in reference to an arrangement of flowers on one side of an inflorescence axis.

segment A portion or lobe of a leaf or calyx.

semi- A prefix meaning half.

semiparasite, semiparasitic. A partial parasite, as a plant that derives only a part of its nutriment from its host.

senescent Aging.

sepal One of the outermost, sterile appendages of a flower, which normally encloses the other floral parts in the bud.

sepaline, sepaloid Sepal-like in shape, texture, and/or color.

septal Pertaining to a septum.

septate Divided by cross-walls or partitions. The cross-walls

between the veins in the sheaths and leaves of some sedges (Cyperaceae)

septicidal When a capsule dehisces through the septa or lines of junction between the locules. *Compare* loculicidal.

septiferous Having septa.

septum (pl., septa) A dividing cross-wall or partition.

seriate In series or rows.

sericeous Silky; having fine, straight, long soft hairs that are usually appressed.

serrate Having a saw-toothed margin with sharp teeth pointing forward.

serrulate Minutely or finely saw-toothed.

sessile Without a stalk; sitting directly on its base.

seta (pl., setae) A bristle-like hair.

setaceous Having bristle-like hairs; bristly.

setose Covered with bristles.

setulose Having minute bristles.

sheath In the monocots, particularly sedges (Cyperaceae) and grasses (Poaceae), the unexpanded basal portion of the leaf, usually clasping the culm; found below the blade or ligule(s).

shrub A perennial woody plant usually with several main stems arising from or near the ground; a bush.

sigmoid S-shaped, doubly curved.

silique A dry, dehiscent fruit, that is two-celled. The valves split from the bottom up and leave the placentae with the false partition (replum) between them. Fruit narrow and longer than wide; characteristic of many species in Brassicaceae (mustard family).

sinuate, sinuous Having a deep wavy margin.

soriate Bearing sori.

sorus (pl., sori) A cluster of sporangia in ferns and fern allies.

spadix A succulent spike; a spike borne on a succulent axis surrounded by a spathe.

spathaceous Spathelike.

spathe A large enveloping leaf protecting a spadix.

spatheate Having a spathe.

spatulate Spoon- or spatula-shaped.

spicate Having the form of or produced in a spike.

spiciform *See* spicate.

spikelet A small or secondary spike. The characteristic flowering spike of the sedges (Cyperaceae) and grasses (Poaceae).

spine A hard, sharp-pointed structure, usually a modified leaf, portion of a leaf, or stipule. *Compare* thorn.

spinescent Spine-tipped, having spines.

spinose Spinelike or having spines.

spinulose Having small spines.

sporangiophore A specialized structure bearing sporangia.

sporangium (pl., sporangia) A minute capsule containing spores.

spore A haploid reproductive cell, produced as a result of mitosis or meiosis, capable of developing into an adult without fusion with another cell; spores are usually unicellular, but may be multicellular, and may undergo mitosis to produce gametes.

sporophyll A modified leaf or leaflike structure that bears sporangia.

squamellae Small scales.

stamen In angiosperms, the pollen-producing structure in a flower, usually consisting of an anther and a filament. Collectively, the stamens are called the androecium.

staminal Referring to a stamen.

staminate Having only stamens in a flower; a male flower lacking a pistil. A functionally staminate flower has fertile stamens but vestigial or rudimentary pistils (pistillode).

staminode A sterile stamen that does not produce pollen.

staminodial Having a staminode.

stellate Star-shaped or radiating like points of a star.

stipe A supporting stalk, such as the stalk of a pistil or the petiole of a fern frond.

stipel A stipule-like appendage at the base of a petiolule.

stipitate Borne on a stipe or stalk.

stipule A small structure or appendage found at the base of some leaf petioles; usually present in pairs; they are morphologically variable and appear as scales, spines, glands, or leaflike structures.

stolon An aerial stem that grows horizontally along the ground, often forming adventitious roots at the nodes.

stoloniferous Producing or bearing stolons.

stomium 1. The lip cell region where dehiscence occurs in fern sporangia. 2. The fissure or pore through which pollen dehiscence occurs in an anther locule.

striate Marked with fine longitudinal parallel grooves.

strigillose Minutely strigose.

strigose Having straight, sharp, stiff appressed hairs or bristles.

strobilus (pl., strobili) A compact cluster of sporophylls; a cone.

stylar Pertaining to the style.

sub- A prefix meaning nearly or almost.

subquadrate Nearly square.

subsessile Nearly sessile; with almost no stalk.

subshrub An undershrub or small shrub that may have partially herbaceous stems.

subtending Occurring immediately below and close to.

subtruncate Terminating rather abruptly.

subulate Awl-shaped, tapering from base to apex.

succulent 1. A plant that accumulates water in fleshy, water-storing stems, leaves, or roots. 2. Juicy, fleshy; in reference to texture or appearance.

suffrutescent A perennial plant that is slightly woody only at the base.

sulcate Grooved or furrowed.

superior ovary When all the floral parts are attached below the ovary.

supramedial Above the middle.

syconium (pl., syconia) A hollow, multiple fruit in which flowers and achenes are borne on the inside of a receptacle or peduncle, e.g., the fig, *Ficus*.

sympetalous Having petals that are partly or completely connate. *Syn.* gamopetalous.

sympodial, sympodial branching Branching in which the apical bud withers at the end of the growing season and growth continues the following season at the lateral bud immediately below. *Contrast* monopodial.

synandrium An androecium in which the anthers are superficially joined but not connate.

synangium A compound structure formed by the coalescence (fusion) of sporangia.

syncarp An aggregate fruit with connate carpels.

syncarpous Having connate carpels.

synonym One of two or more names applied to a taxon (e.g., species).

synsepal A compound organ formed by the union of the two lateral sepals.

synusia A group of plants of the same life form occurring together in the same habitat; the synusia may be composed of the same or unrelated species.

tabula infrastigmatica In *Tolumnia* and related genera, a structure that is basal and ventral on the column, often convex and different in texture or color from the rest of the column; perhaps actually derived from the lip.

taxon (pl., taxa) A taxonomic group of any rank, e.g., genus, species.

tendril A long, slender, coiling, modified leaf (or rarely stem), by which a climbing plant attaches to its support.

tepal A perianth member or segment; term used for perianth parts undifferentiated into distinct sepals and petals.

terete More or less circular in cross section; cylindrical and elongate.

ternate Arranged in threes.

testa The outer seed coat.

tetrahedral Four-sided, as a pyramid.

thorn A hard, sharp-pointed, modified branch. *Compare* spine.

thyrse A panicle-like inflorescence that has one main indeterminate axis and many lateral determinate axes, as in lilac.

thyrsoid Resembling a thyrse.

tomentose Covered with dense, matted, woolly hairs.

tomentulose Covered with relatively short, fine woolly hairs.

torose Cylindrical, with contractions or swellings at more or less regular intervals.

tortuous Irregularly twisted or twining.

torulose Minutely torose.

trabeculate Having cross-bar markings.

transverse Across; perpendicular to the long axis.

trapezoid Four-sided figure, no two sides of which are parallel.

treelet A small tree.

tri- A prefix meaning three.

tricarpellate Having three carpels.

trichome Any hairlike or scale-like epidermal outgrowth.

tricoccate Referring to a three-carpellate fruit, of three cocci.

trifid Three-cleft nearly to the middle.

trifoliolate With three leaflets growing from the same point.

trigonous Three-angled.

trilete Pertaining to a fern spore with a three-branched fissure in its integral wall; having three scar lines forming a Y.

trilobate Having three lobes.

trimerous Composed of three or multiples of three, as parts of a flower.

trimorphic A characteristic that has three different forms within the same species.

tripartite Divided into three parts.

triplinerved Three-nerved from near base; as in leaves triplinerved.

triquetrous Having three edges, with the faces between them concave.

triternate Three times ternate; as leaflets in three sets, each ternately compound.

trullate Trowel-shaped.

truncate Terminating abruptly, as if cut off.

tubercle A small tuberlike swelling, nodule, or projection.

tuberculate Having knobby projections or tubercles.

tuberous Having tubers or resembling a tuber.

turbinate Shaped like a top; inversely conical.

turgid Swollen or inflated, said of a cell that is firm owing to uptake of water.

umbel An inflorescence in which a cluster of pedicels spring from the same point, as the ribs of an umbrella.

umbellate Arranged in umbels.

umbelliferous Bearing umbels.

umbelliform Resembling an umbel; umbel-shaped.

umbo A blunt or rounded projection arising from a surface.

umbonate Bearing an umbo in the center.

uncinate Hooked at the tip.

undulate A wavy margin; less pronounced than sinuate. *See* repand.

unguiculate Clawed; contracted at the base into a claw, forming a stalk.

uni- A prefix meaning one or single.

uniaperturate With a single aperture.

unifoliate Having one leaf.

unifoliolate Having only one leaflet.

unilabiate Having a corolla with only one lip.

unilocular Having a single locule or chamber.

uniovular Having a solitary ovule.

uniseriate Having a single, horizontal row or series of cells. *Compare* multiseriate.

unisexual Having flowers of one sex only, staminate or pistillate. *Compare* bisexual.

urceolate Urn-shaped, constricted just below the mouth.

urticating hairs The stinging hairs of nettles (Urticaceae) and other plants.

utricle A small bladder; a one-seeded, dry fruit, often dehiscent by a lid.

utriculate Having utricles.

valvate Opening by valves; when margins of perianth parts of a flower bud meet exactly without overlapping.

vascular Refers to any plant tissue or area that consists of or gives rise to conducting tissue, e.g., xylem, phloem, or vascular cambium.

vein A strand of vascular tissue (a vascular bundle), that is part of a network of supporting and conducting tissue of an expanded organ, as a leaf, petal, etc.

veinlets *See* free veins.

velamen, velamentum One or more layers of spongy cells on the outside of an orchid root; a specialized moisture-absorbing tissue.

venation The arrangement of veins.

ventral Pertaining to the surface nearest the axis; the upper surface of a leaf; the inner surface of an organ. *Syn.* adaxial. *Ant.* dorsal.

verrucate, verrucose Covered with small wartlike projections.

versatile An anther attached near its middle to the apex of the filament and capable of swinging more or less freely.

verticil A whorl or circular arrangement of similar parts around an axis.

verticillate Arranged in whorls.

villous Covered with long, soft fine hairs.

virgate, virgately Wandlike, slender, straight and erect.

viscid, viscous Sticky; glutinous.

viscidium A viscid part of the rostellum that is clearly defined and removed with the pollinia as a unit and serves to attach the pollinia to an insect or other agent.

viscidulous Slightly sticky.

viviparous Bearing young plants, sprouting or germinating while attached to the parent plant.

whorl A group of three or more parts at a node; the arrangement of organs in a circle around an axis.

zygomorphic Bilaterally symmetrical; divisible into equal halves in one plane only, i.e., most leaves can be divided into similar halves only by cutting along the midrib. *Syn.* irregular symmetry. *Ant.* actinomorphic.

Index to New Combinations

Index to Scientific and Vernacular Names

Accepted scientific names are in **boldface** type; vernacular names, synonyms, and names for doubtful and excluded species are in normal type; and family names are in all capitals. Page numbers in **boldface** type indicate primary reference for treated taxa.

Corrections to the Second Printing

Page 36. Add vines to the second couplet of number 25 in the key.

Page 82. Replace the name and citation of **Cryptostegia grandiflora** with **Cryptostegia madagascariensis** Bojer ex Decne. in DC., Prodr. 8: 492. 1844. (*Cryptostegia grandiflora* does not occur on St. John or in the Virgin Islands.)

Page 121. Replace the name and citation of **Wedelia fruticosa** Jacq. with **Wedelia calycina** Rich. (*Wedelia fruticosa* is a South American species not found on St. John.)

Page 176. Replace description of fruit in **Terminalia** with: Fruit characteristically 2-winged, less often 5-winged or a 1-seeded dry or fleshy drupe.

Page 192. **Cucurbitaceae**: Change tendrils description as lateral to leaves instead of opposite.

Page 222. **Fabaceae**: Add trifoliolate to leaf description.

Page 235. **Abrus precatorius**: leaflet length ranges from 0.8–1.5 cm.

Page 243. **Crotalaria pallida**: Replace the name of var. **ovata** and its basyonym to obovata as follows. **Crotalaria pallida** Dryand. var. **obovata**; *Crotalaria obovata*.

Page 251. Replace name and citation of **Phaseolus peduncularis** to **Vigna antillana** (Urb.) Fawcett & Rendle, Fl. Jam. 4(2): 69. 1920. (*Phaseolus peduncularis* does not occur on St. John.)

Page 253. Fig. 115. A-E. Replace name **Phaseolus peduncularis** with **Vigna antillana**.

Page 257. **Rhynchosia minima**: Change calyx measurement to 2.5–3 mm long.

Pages 293–294. Change spelling of **Heteropteris** to **Heteropterys**.

Page 310. Replace name and citation of **Wissadula amplissima** with **Wissadula hernandioides** (L'Hér.) Garcke, Zeitschr. Naturwiss. 63: 124. 1890. (*Wissadula amplissima* is a South American species not found in the West Indies.)

Page 316. Change size of fruits in **Hyperbaena domingensis** to 10–15 mm.

Page 335. Change author and reference of **Boerhavia diffusa** to L., Sp. Pl. 3. 1753.

Page 397. Change description of calyx in **Cardiospermum corindum** to read: Calyx of 4 unequal, rounded greenish sepals, outer sepals ca. 1 mm long, inner sepals ca. 2.5 mm long. Change description of flower in **Cardiospermum halicacabum**: Calyx of 4 . . . sepals; petals 2.5–3.5 mm long. . . .

Page 399. Use the name **Serjania lucida** Schum. instead of **S. polyphylla**. Serjania polyphylla is a species now known to be restricted to Puerto Rico and Hispaniola but not found on the Virgin Islands.

Page 456. Change petiolate to pedicelled in flower description of **Crinum zeylanicum** and **Hymenocallis caribaea**.

Page 494. Replace name and citation of **Sansevieria trifasciata** with: **Sansevieria hyacinthoides** (L.) Druce, Bot. Exch. Club Brit. Isles 3: 423. 1914; *Aloe hyacinthoides* ß *guineensis* L., Sp. Pl. 321. 1753; *Sansevieria guineensis* (L.) Willd., Sp. Pl. 2: 159. 1799; *Cordyline guineensis* (L.) Britton, Mem. Brooklyn Bot. Gard. 1: 35. 1918. (*Sansevieria trifasciata* does not occur on St. John.)

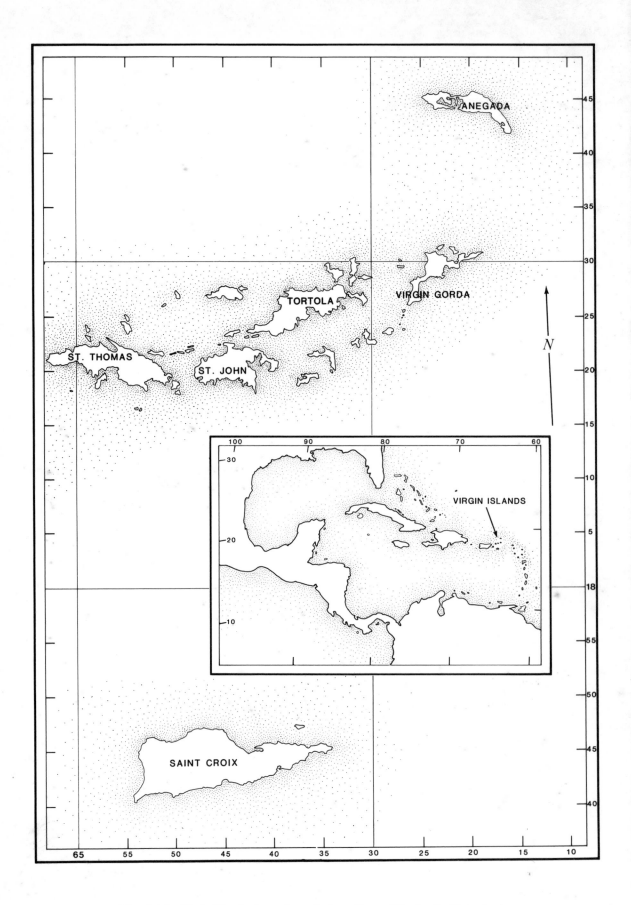

MAP 1. The Virgin Islands. Inset shows their position within the Caribbean basin.